MATHEMATICAL STATISTICS

Parimal Mukhopadhyay

CELEBRATING 60 YEARS 1960-2020

Since BOOKS AND ALLIED 1960

[*For Undergraduate (Honours/Major) & Postgraduate students of Statistics & Mathematics*]

MATHEMATICAL STATISTICS

Dr. Parimal Mukhopadhyay

Formerly, Professor, Indian Statistical Institute (ISI), Kolkata, India.

BOOKS & ALLIED (P) Ltd.

No. 1E(1), 'SHUBHAM PLAZA'

83/1, Dr. S.C. Banerjee Road, Beliaghata, Kolkata 700010

Tel : (033) 23740062 • ***e-mail*** : booksandallied1960@gmail.com

website : www.booksandallied.com

MATHEMATICAL
STATISTICS

First Published : 1996
Second Edition : 2000
Reprinted : 2002
Third Edition (thoroughly revised) : 2006
Reprinted : 2009, 2015, 2016, 2018
Reprinted : 2022

Publisher : Ruba Sen
BOOKS & ALLIED (P) LTD.
No. 1E(1), 'Shubham Plaza'
83/1, Dr. S.C. Banerjee Road, Beliaghata, Kolkata 700010

Typesetter : BOOKS & ALLIED (P) LTD.
No. 1E(1), 'Shubham Plaza'
83/1, Dr. S.C. Banerjee Road, Beliaghata, Kolkata 700010

Printer : EUREKA
91/A, Baithakkhana Road, Kolkata 700009

Project Supervisor : Salil Das

ISBN 978-8187134-93-0

***Price* : ₹935.00**

To

my family

Manju, Pabak, Jayita, Pralay *and* Shilpi

Preface to the Third Edition

In this edition three new sections have been added,—Section 8.8 (location and scale families of distributions), Section 15.9 (robust estimators) and Section 15.10 (estimators based on resampling methods). Some errors in the earlier version have also been removed.

I am thankful to the readers for their support and encouragement.

September 30, 2006
Kolkata **Parimal Mukhopadhyay**

Preface to the Second Edition

I am greatful to my readers whose encouragement has led to a new edition of the book within a very short span of time. I have used this opportunity to add some new materials to the book and rearrange some of the contents of the first version. Some more results on the topics, Strong law of large numbers (section 9.7), Multivariate central limit theorem (section 13.4), Orthogonal polynomials (section 21.9), among others, have been added. New sections on sample quantiles (section 11.8), Dirichlet distribution (section 13.8), criteria for comparison among estimators in multiparameter set up (section 15.12), some more hypotheses relating to multivariate normal distributions (sections 18.8, 18.9), Wilk's statistic involved in likelihood ratio tests (section 18.11), among others, have been added. Whereever found suitable, more examples and exercises have been included. Attempts have been made to rectify the mistakes that crept into the first edition, as far as possible.

Thanks are due to Mr. Arunabha Sen of Books and Allied (P) Ltd., Calcutta, for this untiring efforts in bringing out the treatise within a very short time.

March 2, 2000
The University of South Pacific **Parimal Mukhopadhyay**
Suva, Fiji Islands

Preface to the First Edition

This is a sequel to may book *Theory of Probability* (1991). The encouragement I received from the users of that book—students, teachers, reviewers in India and also in some universities abroad—led me to extend the work to mathematical methods of statistics.

The book is intended mainly for Undergraduate (Honours) and Postgraduate students of Statistics and Mathematics. However, students of other disciplines (e.g., Economics, Agricultural Sciences, Engineering) who have to learn this subject as a part of their curricula may also find many parts of this book useful.

The first nine chapters (1–9) have been adopted from the *Theory of Prabability* with modifications.

Chapter 0 deals with some mathematical preliminaries with emphasis on the theory of matrices. Chapter 1 deals with elements of theory of probability. Chapters 2 and 3 introduce the notion of random variables, probability distributions and some of their properties. Chapter 4 considers different generating functions, probability, moments and cumulants. Different discrete and continuous univeriate distributions have been addressed in Chapters 5 and 6. Chapters 7 and 10 consider general properties of n-dimensional probability distributions, including rank of a distribution, concentration ellipsoid, conditional distributions, multiple regression, partial correlation, etc. Multinomial distribution and bivariate normal distribution are also considered in Chapter 7. Chapter 8 deals with probability distributions of functions of random variables. Different limit theorems have been considered in Chapter 9 and also in parts of Chapter 13, Chapters 11 through 1 deal with basic distributions (c^2, t and F) are sampling distributions of statistics arising out of univeriate bivariate and multivariate normal distributions. Chapter 13 considers in details multivariate normal distribu tions, and also discusses in brief, multivariate t distribution, Spherical and Elliptic distributions (the last thre items are for advanced readers who may like to apply these results in their research, specially in study o robustness). Chapters 15 and 16 deal exhaustively with theory and methods of point estimation. Testing o hypotheses (in parametric set up) has been considered in Chapter 17 and 18. Chapter 19 addresses confi dence interval estimation. Chapter 20 considers some large sample procedures. Theory of linear estimatior is dealt with in Chapter 21. The next two chapters–22–23–consider nonparametric inference and decisior theory, respectively. Chapter 24 discusses some elements of sequiential inference. The last chapter–25- addresses some preliminaries of Stochastic Processes.

Attempt has been made at each step to give a systematic exposition of the theory augmented by a number of examples and exercises.

In preparing the book I have unhesitatingly taken the help of number of publications, a list of which has been enclosed as reference.

Thanks are due to the Indian Statistical Institute for their suppose and facilities provided during the preparation of the manuscript.

My sincere thanks are due to Shri Arunabha Sen of New Central Book Agency (P) Ltd. who took great care in bringing out this treatise within an extremely short span of time.

I acknowledge may indebtedness to may wife and sons Pabak and Pralay who bestowed invaluable encouragement in preparing the manuscript.

February 1, 1996
Indian Statistical Institute
Kolkata, INDIA

Parimal Mukhopadhyay

Contents

Chapter 0

Mathematical Preliminaries

0.1 Introduction

In this chapter we consider some preliminary mathematical results which have been used extensively in this treatise. These results are not exhaustive and the reader is advised to refer to the relevant works cited in the bibliography for further details. Section 0.2 considers elements of theory of sets, Section 0.3 elements of measure theory and Section 0.4 elements of theory of matrices and determinants. The concept of Lebesgue-Stieltjes integral has been considered in Section 2.6, Cauchy-Schwartz inequality, Holder's inequality, Minkowski's inequality in appendix to Chapter 3, convergence of power series in Section 15.7. The results cited in this chapter are not exhaustive and the interested reader may refer to relevant works on the subject e.g., Halmos (1964) [for theory of sets and measure], Bodewig (1959) [for determinants and Matrices], Piskunov (1981) [for differential and integral calculus], Apostol (1974) [for analysis of real and complex variable], Korovkin (1975), for a review of different inequalities, among others.

0.2 Elements of Set Theory

We consider a few definitions.

Set : A set is a collection of some elements which are its members. Thus we may have a set of college students, a set of some books, a set of some digits, a set of houses, etc. A set can be denoted either by enumerating all its elements given in braces or by indicating (also in braces) a rule which associates an element with a set. Thus the set of all odd numbers not exceeding 15 can be written as $\{1, 3, 5, 7, 9, 11, 13, 15\}$ or $\{2x + 1; x = 1, 2, \cdots, 7\}$. A set is often pictured as an oval shaped space with points denoting its elements [Fig. 0.1a]. The diagram representing a set is called a Venn diagram.

Depending on how many elements it has, a set may be finite or infinite. The set $M = \{1, 2, \cdots, 50\}$ is finite and contains 50 elements. The set of all natural numbers $N_1 = \{1, 2, \cdots, n, \cdots\}$ is infinite. The set of all even numbers $N_2 = \{2, 4, \cdots, 2n, \cdots\}$ is also infinite. An infinite set is said to be countable if all its elements can be enumerated Both the sets N_1, N_2 above are countable. The set C of all points within or on a circl of radius $r > 0$, $C = \{(x, y) : x^2 + y^2 \leq r^2\}$ is infinite and uncountable. Its elements ca not be enumerated.

Universal Set : It is the largest possible set of interest under a given condition All other sets are parts (subsets) of the universal set. Thus if we are interested in som students of a college, a set containing all the students of this college may constitut an universal set. The set of all first year commerce students, set of all second yea chemistry-honours students are all subsets of the above universal set. An universal se is often represented as $\mathcal{U}$ (Fig. 0.1b).

Null Set : It is a set containing no element. It is also called an empty set and is often represented as ϕ. It is a member of all other sets (Fig. 0.1b).

Disjoint Sets : Two sets A, B are disjoint if they have no common element Similarly sets $A_1, \cdots, A_m$ are mutually disjoint if no two of them have any commor element (Fig. 0.1c).

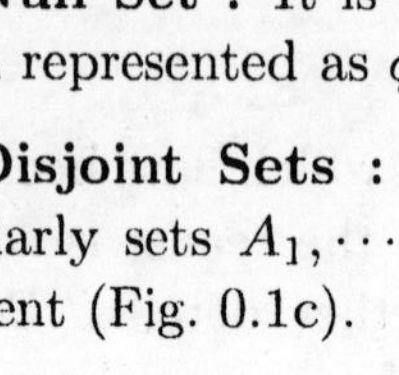

Fig. 0.1a

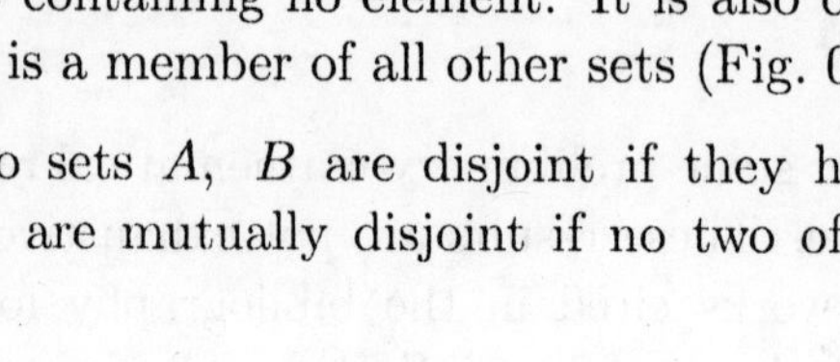

Fig. 0.1b

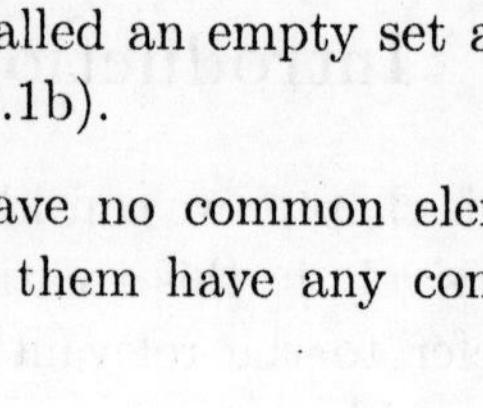

Fig. 0.1c

Union of Sets : Let A, B be two sets. The union of A and B, denoted as $A \cup B$ is the set of all elements contained in either A or B or both. In fig. 0.2a, $A \cup B$ is the region covered by either horizontal bars or dots or both.

Similarly the union of m sets $A_1, \cdots, A_m$, namely, $\bigcup_{1}^{m} A_i$ is the set of all elements contained in at least one of $A_1, \cdots, A_m$. In fig. 0.2c, $A \cup B \cup C$ is the region covered by either horizontal bars or dots or vertical bars or by any two of them or by all. When the sets $A_1, \cdots, A_m$ are mutually disjoint we shall sometime write

$$\bigcup_{1}^{m} A_i = A_1 + A_2 + \cdots + A_m.$$

Intersection : The intersection of two sets A and B, denoted as $A \cap B$ or simply AB is the set of all elements contained in both A and B. In fig. 0.2a, $A \cap B$ is the region covered by both horizontal bars and dots.

Similarly the intersection of m sets $A_1, \cdots, A_m$, denoted as $\bigcap_{i=1}^{m} A_i$ or simply $A_1 A_2 \cdots A_m$ is the set of all elements contained in all of $A_1, \cdots, A_m$. In fig. 0.2c, ABC is the region covered by all of the three markings — vertical and horizontal bars and dots. If the sets $A_1, \cdots, A_m$ are mutually disjoint, $A_i \cap A_j = \phi$ for all $i \neq j = 1, \cdots, m$.

Complementation : The complement of set A, denoted as A^c or A' or $\bar{A}$ is the set of all elements not contained in A. Hence $A^c = \mathcal{U} - A$. In fig. 0.2b, A^c is the entire shaded region.

Clearly $\mathcal{U}^c = \phi, \phi^c = \mathcal{U}, (A^c)^c = A$.

Subset : A set C is a subset of a set D if every member of C is necessarily a member of D. This is denoted as $C \subseteq D$. When D contains at least one element not contained in C, C is a proper subset of D and we write $C \subset D$.

For two sets A, B, clearly, $A \cap B \subset (A \cup B)$. For three sets $A, B, C, ABC \subset (AB)$, $ABC \subset A \cup (BC)$, etc.

Difference : The difference $A - B$ is the set of all elements contained in A but not in B. Similarly $B - A$ is the set of all elements contained in B but not in A. In fig. 0.2a, $A - B$ is the region covered by bars but not by dots; $B - A$ is the region covered by dots but not by bars. Clearly $A - B \neq B - A$. Also $(A - B) \subset A, (B - A) \subset B$. Clearly $A - B$ and $B - A$ are disjoint sets.

It follows that $A - B = A \cap B^c$, $B - A = B \cap A^c$.

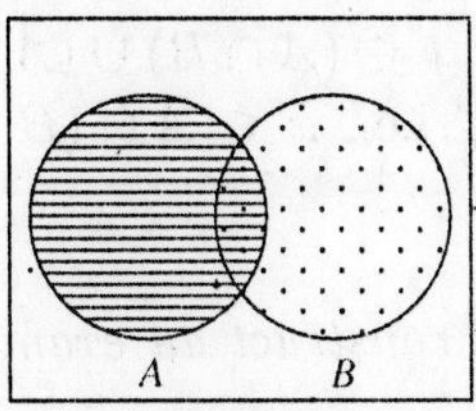

Fig. 0.2a

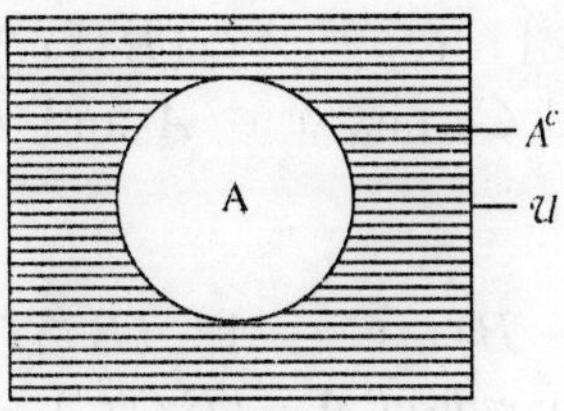

Fig. 0.2b

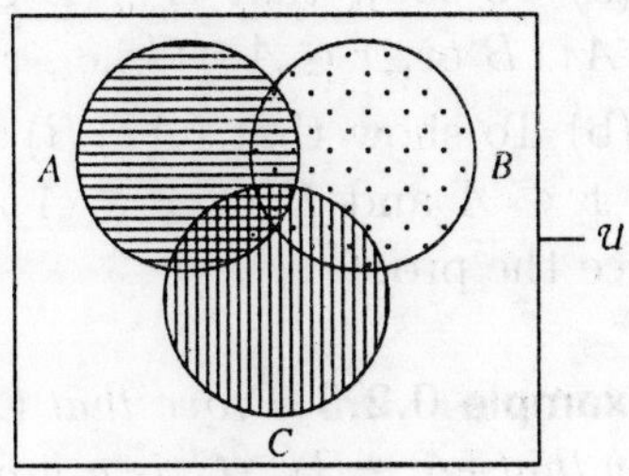

Fig. 0.2c

The operations of union and intersection satisfy the following properties:

Commutativity : $A \cup B = B \cup A$; $A \cap B = B \cap A$.

Associativity : $A \cup (B \cup C) = (A \cup B) \cup C$; $A \cap (B \cap C) = (A \cap B) \cap C$.

Distributivity : $A \cap (B \cup C) = (A \cap B) \cup (A \cap C)$; $A \cup (B \cap C) = (A \cup B) \cap (A \cup C)$.

Idempotency : $A \cup A = A$; $A \cap A = A$.

They also obey De Morgan's laws:

(a) $(A \cup B)^c = A^c \cap B^c$.

For m sets, $A_i, i = 1, \cdots, m$

$$\left(\bigcup_{i=1}^{n} A_i\right)^c = \bigcap_{i=1}^{m} A_i^c.$$

(b) $(A \cap B)^c = A^c \cup B^c$.

For m sets, $A_i, i = 1, \cdots, m$

$$\left(\bigcap_{i=1}^{m} A_i\right)^c = \bigcup_{i=1}^{m} A_i^c.$$

► **Example 0.2.1** *Prove that* $A \cap (B - C) \subset A - (B \cap C)$.

Proof: Let $x \in A \cap (B - C)$; i.e., $x \in A$ and $x \in (B - C)$ i.e., $x \in A$, $x \in B$ bu $\notin B \cap C$.

Hence $x \in A - (B \cap C)$. Thus $\{x \in A \cap (B - C)\} \Rightarrow \{x \in (A - B \cap C)\}$.

Hence $A \cap (B - C) \subset \{A - (B \cap C)\}$.

To establish that $A \cap (B - C)$ is a proper subset of $A - (B \cap C)$ consider the followin example. Let $A = \{1, 4, 5, 7\}$, $B = \{2, 4, 6, 7\}$, $C = \{3, 5, 6, 7\}$. Here $A \cap (B - C) = \{4\}$ $A - (B \cap C) = \{1, 4, 5\}$.

► **Example 0.2.2** *Prove that* $A \cap (B \cap C) = (A \cap B) \cup (A \cap C)$.

(a) To show that $A \cap (B \cup C) \subset (A \cap B) \cup (A \cap C)$: Let $x \in A \cap (B \cup C)$; i.e. $x \in A \cap B$ or, $x \in A \cap C$ i.e., $x \in (A \cap B) \cup (A \cap C)$.

(b) To show that $(A \cap B) \cup (A \cap C) \subset A \cap (B \cup C)$: Let $x \in (A \cap B) \cup (A \cap C)$ i.e., $x \in A$ and B or $x \in A$ and C, i.e., $x \in A$ and B or C i.e., $x \in A \cap (B \cup C)$ Hence the proof.

► **Example 0.2.3** *Prove that* $(A - B) \cap C \subset A - (B \cap C)$ *and construct an example t show that* $(A - B) \cap C$ *is a proper subset* $A - (B \cap C)$.

Let $x \in (A - B) \cap C$ i.e., $x \in (A - B)$ and $C \Rightarrow x \in A - B \cap C$. Hence $(A - B) \cap C \subset A - (B \cap C)$.

Let $A = \{1, 2, 3, 4\}$, $B = \{3, 4, 5, 6\}$, $C = \{1, 6, 7, 8\}$. $(A - B) \cap C = \{1\}$; $A - B \cap C = \{1, 2, 3, 4\}$.

► **Example 0.2.4** *Prove that* $(A - B) \cap (C - D) \subset (A \cap C) - (B \cap D)$.

Let $x \in (A - B) \cap (C - D)$ i.e., $x \in (A - B)$ and $(C - D)$ i.e., $x \in A, x \notin B$, $x \in C, x \notin D$ i.e., $x \in A, x \in C, x \notin B \cup D$ i.e., $x \in A \cap C, x \notin B \cup D$ i.e., $x \in A \cap C - B \cup D \Rightarrow x \in A \cap C - B \cap D$.

► **Example 0.2.5** *Prove that* $(A - B) \cap (A - C) \subset A - (B \cap C)$.

Let $x \in (A - B) \cap (A - C)$ i.e., $x \in A$, $x \notin B$, $x \notin C$ i.e., $x \in A$ and $x \notin B \cup C$ i.e., $x \in A - (B \cup C) \Rightarrow x \in A - (B \cap C)$.

).3 Some Elements of Theory of Measure

'he union and intersection of a countable number of sets $A_1, A_2, \cdots$ are written as

$$\bigcup_{i=1}^{\infty} A_i \text{ and } \bigcap_{i=1}^{\infty} A_i \text{ respectively.}$$

'iven n sets $A_1, \cdots, A_n$ we can write

$$\bigcup_{i=1}^{n} A_i = A_1 + A_1'A_2 + A_1'A_2'A_3 + \cdots + A_1' \cdots A_{n-1}'A_n \tag{0.3.1}$$

ıs the sum of n disjoint sets $A_1, A_1'A_2, A_1'A_2'A_3, \cdots$ etc. For a countably infinite number)f sets $A_1, A_2, \cdots$

$$\bigcup_{i=1}^{\infty} A_i = A_1 + A_1'A_2 + A_1'A_2'A_3 + \cdots.$$

Definition 0.3.1 *Sequence and limits of sets. To every integer* $n = 1, 2, \cdots$ *we assign a set. The ordered class of sets* $A_1, A_2, \cdots$ *is a sequence* $\{A_n\}$ *of sets.*

$\{A_n\}$ is monotonically increasing if $A_n \subseteq A_{n+1}$. In this case

$$\bigcup_{k=1}^{n} A_k = A_n; \quad \bigcup_{k=1}^{\infty} A_k = A$$

is called the limit of $\{A_n\}$. Symbolically $A_n \uparrow A$.

$\{A_n\}$ is monotonically decreasing if $A_n \supseteq A_{n+1}$. Here

$$\bigcap_{k=1}^{n} A_k = A_n; \quad \bigcap_{k=1}^{\infty} A_k = A$$

is called the limit of $\{A_n\}$. Symbolically $A_n \downarrow A$.

Now we define limit of arbitrary sequence of sets. For any sequence $\{A_n\}$ define

$$\begin{aligned} B_n &= \inf_{k\geq n} A_k = \bigcap_{k=n}^{\infty} A_k \\ &= \{w : w \text{ belongs to all } A_k \text{ except perhaps for } A_1, \cdots, A_{n-1}\}. \\ C_n &= \sup_{k\geq n} A_k = \bigcup_{k=n}^{\infty} A_k \\ &= \{w : w \text{ belongs to at least one of } A_n, A_{n+1}, \cdots\}. \end{aligned}$$

Thus B_n is a monotonically increasing sequence with limit

$$B = \bigcup_{n=1}^{\infty} \bigcap_{k=n}^{\infty} A_k = \liminf A_k = \underline{\lim} A_n.$$

B is the set of all points which belong to almost all A_n (all but any finite number of sets). C_n is a monotonically decreasing sequence with limit

$$C = \bigcap_{n=1}^{\infty} \bigcup_{k=n}^{\infty} A_k = \limsup A_k = \overline{\lim} A_n.$$

C is the set of all those points which belongs to infinitely many A_n.

Since every point which belongs to almost all A_n belongs to infinitely many A_n, $\underline{\lim} A_n \subseteq \overline{\lim} A_n$. If $\underline{\lim} A_n = \overline{\lim} A_n = A$ (say), the limit of $\{A_n\}$ is said to exist and A is called the limit of $\{A_n\}$.

Definition 0.3.2 *Real valued point function on a set. A real valued point function f defined on a set S is a rule that associates with each element x of S a real number $f(x)$ called the value of f for the point x. The set S is called the domain of f and the set of values $\{f(x); x \in S\}$ is called the range of f. We denote the set containing the range of f as R.*

If for each x, $f(x)$ is different, the function is one-to-one $(1:1)$. If two or more x's have the same value $f(x)$, the function is many-to-one. An example of a 2-1 function is $f(x) = x^2$.

Thus a function f which maps S to R (we denote it as $S \xrightarrow{f} R$) assigns to each $x \in S$, a value $f(x) \in R$. $f(x)$ is called the image of x under f and x is called the pre-image of $f(x)$.

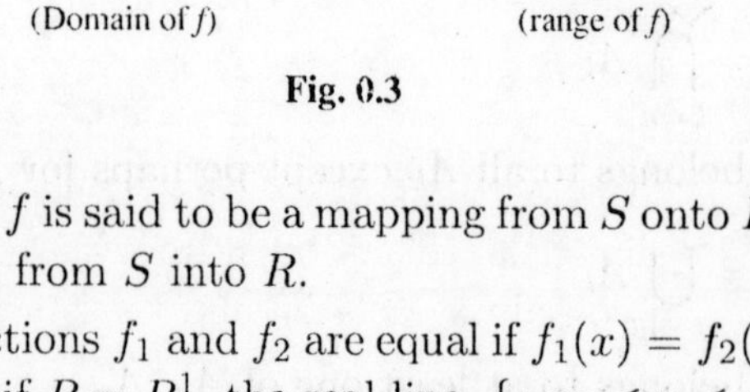

Fig. 0.3

If $R = \{f(x) : x \in S\}$, f is said to be a mapping from S onto R. If $\{f(x) : x \in S\} \in R$, f is said to be a mapping from S into R.

Real-valued point functions f_1 and f_2 are equal if $f_1(x) = f_2(x) \; \forall \; x \in S$. The function f is a numerical function if $R = R^1$, the real line, $\{-\infty < y < \infty\}$.

Definition 0.3.3 *Inverse function of a point function on a set. If the set $B \in R$, then $\{x : f(x) \in B\}$ is called the inverse image of B under f and is denoted as $f^{-1}(B)$. Thus with each point function f (with domain S and range R) there is a set function*

f^{-1} *whose domain is R and range is S. f^{-1} is called the inverse function of f. The following relations hold :*

$$f^{-1}\left(\bigcap_k B_k\right) = \bigcap_k f^{-1}(B_k),\ f^{-1}\left(\bigcup_k B_k\right) = \bigcup_k f^{-1}(B_k),\ f^{-1}(B') = (f^{-1}(B))'.$$

Definition 0.3.4 *Class of sets. A class of sets is a collection of sets. The elements of this class are sets. An example is $\mathcal{C} = \Big\{\{a\}, \{b\}, \{a, b\}\Big\}$.*

Definition 0.3.5 *Real valued set function. Let $\mathcal{C}$ be a class of sets. A rule f that associates with each element set C of $\mathcal{C}$ a real number $f(C)$ is called a real valued set function, defined on $\mathcal{C}$. The class $\mathcal{C}$ is called the domain of f, the set of values $\{f(C), C \in \mathcal{C}\}$ is called the range of f.*

Definition 0.3.6 *Finitely additive class of sets. A class $\mathcal{C}$ of sets is called finitely additive if it has the following properties :* (i) $\phi \in \mathcal{C}$ (ii) *if $C \in \mathcal{C}$ then $C' \in \mathcal{C}$* (iii) *if C_i, $i = 1, \cdots, n \in \mathcal{C}$, then $\bigcup_{i=1}^{n} C_i \in \mathcal{C}$. It follows that $\bigcap_{i=1}^{n} C_i \in \mathcal{C}$. Clearly the universal set $\in \mathcal{C}$. Such a class is also called a Boolean field of sets (* or a *Boolean algebra).*

Definition 0.3.7 *Completely additive class of sets. A class $\mathcal{C}$ of sets is completely additive if it has the following properties :* (i) $\phi \in \mathcal{C}$ (ii) *if $C \in \mathcal{C}$, then $C' \in \mathcal{C}$* (iii) *if $C_i, i = 1, 2, \cdots \in \mathcal{C}$ then $\bigcup_{i=1}^{\infty} C_i \in \mathcal{C}$. It follows that $\bigcap_{i=1}^{\infty} C_i \in \mathcal{C}$. Such a class is also called a Borel field or a σ-field (or a σ-algebra) of sets.*

Given a class $\mathcal{C}$ of sets, the smallest (minimal) (Boolean field) σ-field containing $\mathcal{C}$ is the intersection of all (Boolean fields) σ-fields containing $\mathcal{C}$. There is one and only one minimal (Boolean field) σ-field containing $\mathcal{C}$. This is the minimal (Boolean field) borel field of subsets of $\mathcal{C}$. This field can be generated out of an initial subclass of sets in $\mathcal{C}$.

If $S = R^1$, the real line, $\mathcal{C}$ is the class of all real half-open intervals $(a, b]$, minimal σ-field containing $\mathcal{C}$ is the borel field $\mathcal{B}_1$ of subsets of the real line and its sets are borel sets of R^1. The same borel field is obtained if $\mathcal{C}$ is taken as the class of all intervals of the type (a, b) or $[a, b)$ or $[a, b]$.

If $\mathcal{S} = R^n$, n-dimensional Eucledian space (the set of all n-tuples of finite real numbers), $\mathcal{C}$ is the class of all n-dimensional half-open intervals $\{(x_1, \cdots, x_n),\ a_i < x_i \le b_i,\ i = 1, \cdots, n\}$, minimal σ-field containing $\mathcal{C}$ is $\mathcal{B}_n$ and its sets are borel sets of R^n.

Definition 0.3.8 *Finitely additive real-valued set function. Let $\mathcal{C}$ be a finitely additive class of sets. A real valued set function μ defined on $\mathcal{C}$ is finitely additive if* (i) $\mu(\phi) = 0$ (ii) *if $C_i \in \mathcal{C}$, $i = 1, \cdots, n$ are disjoint for any n, then $\mu\left(\bigcup_{i=1}^{n} C_i\right) = \sum_{i=1}^{n} \mu(C_i)$.*

Definition 0.3.9 *Completely additive real-valued set function. Let $\mathcal{C}$ be a σ-field of sets. A real-valued set function μ defined on $\mathcal{C}$ is called completely additive if* (i) $\mu(\phi) = 0$ (ii) *if $C_i \in \mathcal{C}$, $i = 1, 2, \cdots$ are disjoint then* $\mu\left(\bigcup_{i=1}^{\infty} C_i\right) = \sum_{i=1}^{\infty} \mu(C_i)$.

Definition 0.3.10 *Measure and measurable sets. A non-negative, additive, real-valued set function defined on an additive class of sets is called a measure. Thus if $\mathcal{C}$ is an additive class of subsets of a set $\mathcal{U}$ (universal set), a real-valued set function defined on $\mathcal{C}$ is called a measure if*

(i) $\mu(C) \geq 0 \quad \forall\, C \in \mathcal{C}$,

(ii) $\mu(\phi) = 0$,

(iii) if $C_i \in \mathcal{C}$ $(i = 1, 2, \cdots)$ are disjoint such that $\bigcup_1^{\infty} C_i \in \mathcal{C}$ then $\mu\left(\bigcup_1^{\infty} C_i\right) = \sum_1^{\infty} \mu(C_i)$.

The measure μ is finitely additive if (iii) is valid for finite number of disjoint sets; it is completely additive if (iii) also extends to a countable number of disjoint sets. The sets C are called measurable sets w.r.t. measure μ. The measure μ is called bounded if $\mu(\mathcal{U}) < \infty$.

Definition 0.3.11 *Lebesgue measure and Lebesgue measurable sets. On the borel field $\mathcal{B}_1$ of subsets of R^1 a unique finite measure can be defined such that*

$$\mu[(a, b]] = b - a.$$

The measure μ is called the Lebesgue measure. Sometimes one may take the σ-field obtained by adjoining to $\mathcal{B}_1$ all subsets of measure zero and here μ is defined over this larger σ-field, $\mathcal{B}_1'$ (say). The members of $\mathcal{B}_1'$ are called Lebesgue measurable sets. μ defined over $\mathcal{B}_1$ or $\mathcal{B}_1'$ is called the Lebesgue measure.

Definition 0.3.12 *A property which is true everywhere, except for a set of measure zero is said to hold almost everywhere (abbreviated to a.e.).*

In the theory of probability, the universal set corresponds to the sample space $\mathcal{S}$ underlying a random experiment and the σ-field $\mathcal{A}$ corresponds to the class of all events (subsets of $\mathcal{S}$) under consideration.

0.3.1 $\bigcirc(.)$, $\circ(.)$ and $\sim$

A function $f(x)$ is at most of order $g(x)$, as $x \to L$, denoted as $f(x) = \bigcirc(g(x))$, $x \to L$ (not necessarily finite) if $|f(x)/g(x)| < c$ for some constant c as $x \to L$.

$f(x)$ is of smaller order than $g(x)$ as $x \to L$, written as $f(x) = \circ(g(x))$, $x \to L$ if

$$\lim_{x \to L} \frac{f(x)}{g(x)} = 0.$$

We write $f(x) \sim g(x)$ as $x \to L$ if

$$\lim_{x \to L} \frac{f(x)}{g(x)} = 1.$$

Example 0.3.1 *Consider the function*

$$f(x) = 1 - \left(1 - \frac{1}{x}\right)\left(1 - \frac{2}{x}\right).$$

Clearly $f(x) \to 1$ as $x \to \infty$. Check that

$$\begin{aligned} f(x) &= \frac{3}{x} + O(x^{-2}), \quad \text{as } x \to \infty \\ &= \frac{3}{x} + o(x^{-1}), \quad \text{as } x \to \infty \\ &\sim \frac{3}{x}, \quad \text{as } x \to \infty. \end{aligned}$$

0.4 Elements of Theory of Matrices and Determinants

A rectangular array of real or complex numbers is called a matrix. It will be assumed throughout that the numbers are real. Let

$$\boldsymbol{A} = ((a_{ij})) = \begin{bmatrix} a_{11} & \cdots & a_{1n} \\ \vdots & \cdots & \vdots \\ a_{p1} & \cdots & a_{pn} \end{bmatrix}$$

denote a $p \times n$ matrix. If $p = n$, $\boldsymbol{A}$ is a square matrix. If $n = 1$, $\boldsymbol{A}$ is a column vector $(p \times 1)$, if $p = 1$, $\boldsymbol{A}$ is a row vector $(1 \times n)$. If $a_{ij} = 0 \; \forall \; i, j$, $\boldsymbol{A}$ is a zero matrix. If $n = p$ and $a_{ij} = 1(0)$ if $i = j$ (otherwise), $\boldsymbol{A}$ is an identity matrix, often denote as I_n.

A square matrix whose off-diagonal elements $a_{ij} = 0 \; \forall \; i \neq j$ is called a diagonal matrix,

$$\text{Diag}\,(a_{11}, a_{22}, \cdots, a_{pp}).$$

The elements just to the right (left) of those on the diagonal of a square matrix are called the elements of a super (sub) diagonal, while those along the diagonal are called the elements of principal diagonal.

The transpose of $\boldsymbol{A}$ is the matrix obtained by interchanging the rows and columns of $\boldsymbol{A}$ and is denoted by $\boldsymbol{A}'$. A square matrix $\boldsymbol{A}$ is symmetric if $\boldsymbol{A} = \boldsymbol{A}'$.

An orthogonal matrix $\boldsymbol{A}$ is a $p \times p$ matrix in which

$$\left.\begin{aligned} &\text{(i)} \sum_{j=1}^{p} a_{ij}a_{kj} = 0 \qquad (i \neq k) \\ &\text{and (ii)} \sum_{j=1}^{p} a_{ij}^2 = 1 \qquad (i = 1, \cdots, p) \end{aligned}\right\} \text{ or } \quad \begin{aligned} &\text{(i)} \sum_{i=1}^{p} a_{ij}a_{ik} = 0 \qquad (j \neq k) \\ &\text{(ii)} \sum_{i=1}^{p} a_{ij}^2 = 1 \qquad (j = 1, \cdots, p) \end{aligned}$$

Equivalently, $\boldsymbol{A}$ is orthogonal, if

$$\boldsymbol{AA'} = \boldsymbol{A'A} = \boldsymbol{I}.$$

► **Example 0.4.1** *The following matrix called Helmert matrix is an orthogonal matrix.*

$$\boldsymbol{A} = \begin{bmatrix} \frac{1}{\sqrt{p}} & \frac{1}{\sqrt{1\cdot 2}} & \frac{1}{\sqrt{2\cdot 3}} & \cdots & \frac{1}{\sqrt{(p-1)p}} \\ \frac{1}{\sqrt{p}} & -\frac{1}{\sqrt{1\cdot 2}} & \frac{1}{\sqrt{2\cdot 3}} & \cdots & \frac{1}{\sqrt{(p-1)p}} \\ \frac{1}{\sqrt{p}} & 0 & -\frac{2}{\sqrt{2\cdot 3}} & \cdots & \frac{1}{\sqrt{(p-1)p}} \\ \cdots & \cdots & \cdots & \cdots & \cdots \\ \frac{1}{\sqrt{p}} & 0 & 0 & \cdots & -\frac{(p-1)}{\sqrt{(p-1)p}} \end{bmatrix}.$$

This matrix has also the property that all the columns except the first sum to zero.

A $p \times p$ symmetric matrix whose off-diagonal elements are common and whose diagonal elements are also common is said to be a matrix of intraclass correlation pattern.

$$\boldsymbol{A} = \begin{bmatrix} a & b & \cdots & b \\ b & a & \cdots & b \\ \cdots & \cdots & \cdots & \cdots \\ b & b & \cdots & a \end{bmatrix}, \quad b \geq -a/(p-1). \tag{0.4.1}$$

A *Jacobi* matrix is a square matrix in which the elements not on the principal, super or sub-diagonals are zeros.

A symmetric matrix in which all the elements except those on the super, principal and subdiagonals are zeros is called a *tri-diagonal* matrix. This is given by

$$\begin{bmatrix} a & b & 0 & \cdots & \cdots & 0 \\ b & a & b & \cdots & \cdots & 0 \\ 0 & b & a & b & \cdots & 0 \\ \cdots & \cdots & \cdots & \cdots & \cdots & \vdots \\ 0 & 0 & 0 & \cdots & a & b \\ 0 & 0 & 0 & \cdots & b & a \end{bmatrix} \tag{0.4.2}$$

A *circular* matrix is one which has the structure

$$\begin{bmatrix} a_1 & a_2 & \cdots & a_p \\ a_p & a_1 & \cdots & a_{p-1} \\ \cdots & \cdots & \cdots & \cdots \\ a_2 & a_3 & \cdots & a_1 \end{bmatrix}. \tag{0.4.3}$$

.4.1 Matrix Algebra

(1) Let $\boldsymbol{A} = ((a_{ij}))$, $\boldsymbol{B} = ((b_{ij}))$ both $p \times n$. Then

$$\boldsymbol{A} \pm \boldsymbol{B} = \boldsymbol{C}$$

here $\boldsymbol{C} = ((c_{ij})), c_{ij} = a_{ij} \pm b_{ij}$.

(2) Let $\boldsymbol{D} : p \times n, \boldsymbol{E} : n \times q$. Then the product $\boldsymbol{F} = \boldsymbol{DE}$ is defined and

$$f_{ij} = \sum_{\alpha=1}^{n} d_{i\alpha} e_{\alpha j}, \quad \begin{array}{l} i = 1, \cdots, p \\ j = 1, \cdots, q \end{array}$$

Note that even for square matrices it is not generally true that $\boldsymbol{AB} = \boldsymbol{BA}$.

(3) The following properties hold

(i) $c\boldsymbol{A} = ((ca_{ij}))$ where c is a scalar;
(ii) $(\boldsymbol{A} + \boldsymbol{B})' = \boldsymbol{A}' + \boldsymbol{B}'$;
(iii) $(\boldsymbol{AB})' = \boldsymbol{B}'\boldsymbol{A}'$ if $\boldsymbol{A}, \boldsymbol{B}$ are conformable for multiplication;
(iv) $(\boldsymbol{A}')' = \boldsymbol{A}$;
(v) $\boldsymbol{A}(\boldsymbol{B} + \boldsymbol{C}) = \boldsymbol{AB} + \boldsymbol{AC}$;
(vi) $\boldsymbol{AI} = \boldsymbol{A}$ for all $\boldsymbol{A}$.

(4) The length of a (column) vector $\boldsymbol{a}$ is defined as

$$||\boldsymbol{a}|| = \sqrt{\boldsymbol{a}'\boldsymbol{a}} = \sqrt{\sum_{k=1}^{n} a_k^2}. \tag{0.4.4}$$

(Unless otherwise stated we shall denote by $\boldsymbol{a}$ the column-vector $\boldsymbol{a}$). Clearly, the length of the vector $\boldsymbol{a}/\sqrt{\sum_{k=1}^{n} a_k^2}$ is unity.

Schwarz Inequality of two vectors $\boldsymbol{a}, \boldsymbol{b}$ is

$$(\boldsymbol{a}'\boldsymbol{b})^2 \leq (\boldsymbol{a}'\boldsymbol{a})(\boldsymbol{b}'\boldsymbol{b}), \tag{0.4.5}$$

the equality holds, iff either $\boldsymbol{a}$ is proportional to $\boldsymbol{b}$ or $\boldsymbol{a} = \boldsymbol{0}$.

0.4.2 Determinants

A square matrix $\boldsymbol{A} = ((a_{ij}))$ of order m, has associated with it a scalar real-valued function of its elements, called its *determinant* denoted by $|\boldsymbol{A}|$.

$$|\boldsymbol{A}| = \sum \pm a_{1i_1} a_{2i_2} \cdots a_{mi_m},$$

where the summation is taken over all the $m!$ permutations $(i_1, i_2, \cdots, i_m)$ of the integers $(1, 2, \cdots, m)$. The plus (minus) sign applies if the numbers of subscripts rearranged in permutation is even or zero (otherwise).

Elementary properties

(1) $|A| = 0$ if all the elements of a row (column) are zero; or if all the elements of a row (column) are proportional to the elements of another row (column).

(2) $|A| = |A'|$.

(3) The value of $|A|$ remains unchanged if a linear combination of any number of rows (columns) is added to any other row (column).

(4) If two rows (columns) are interchanged the determinant is multiplied by (-1).

(5) If all elements of a row (column) are multiplied by a constant factor, the determinant is multiplied by the same factor.

(6) If every element in any row (column) is the sum of two quantities, the determinant may be expanded as the sum of two determinants of the same order. Thus, if

$$B = \begin{bmatrix} a_{11}+b_{11} & \cdots & a_{1n}+b_{1n} \\ a_{21} & \cdots & a_{2n} \\ \vdots & \cdots & \cdots \\ a_{n1} & \cdots & a_{nn} \end{bmatrix}$$

then

$$|B| = \begin{vmatrix} a_{11} & \cdots & a_{1n} \\ a_{21} & \cdots & a_{2n} \\ \vdots & \cdots & \cdots \\ a_{n1} & \cdots & a_{nn} \end{vmatrix} + \begin{vmatrix} b_{11} & \cdots & b_{1n} \\ a_{21} & \cdots & a_{2n} \\ \vdots & \cdots & \cdots \\ a_{n1} & \cdots & a_{nn} \end{vmatrix}.$$

Minors and Cofactors

The minor of a_{ij} of A is the value of the determinant obtained from A by deleting the ith row and jth column.

The cofactor of a_{ij}, denoted as $A_{ij} = (-1)^{i+j}$ minor of the a_{ij}.

A *principal minor* of A is a minor obtained by deleting from A the same rows and columns.

Laplace Expansion of a determinant:

$$\begin{aligned} |A| &= \sum_{j=1}^{p} a_{ij}A_{ij} \text{ for any fixed } i = 1, \cdots, p \\ &= \sum_{i=1}^{p} a_{ij}A_{ij} \text{ for any fixed } j = 1, \cdots, p. \end{aligned} \quad (0.4.6)$$

If $|A| = 0$, A is called a singular matrix.

Some Special Properties

(1) Let $A_1, \cdots, A_k$ be all of order $p \times p$. then

$$|A_1 \cdots A_k| = |A_1| \cdots |A_k|. \tag{0.4.7}$$

(2) If $A : p \times n$, $B : n \times p$ then

$$|I_p + AB| = |I_n + BA|. \tag{0.4.8}$$

In particular if $n = 1$ so that A is a column vector a and B is a row vector b', then

$$|I_p + ab'| = 1 + b'a. \tag{0.4.9}$$

Determinants of Some Patterned Matrices

If A is orthogonal, then

$$|A| = \pm 1. \tag{0.4.10}$$

If $D = \text{Diag}\ (\lambda_1, \cdots, \lambda_p)$, then

$$|D| = \lambda_1 \cdots \lambda_p. \tag{0.4.11}$$

If A $(p \times p)$ has a intraclass correlation structure (0.4.1)

$$|A| = (a - b)^{p-1}[a + b(p-1)]. \tag{0.4.12}$$

More generally, if G $(pk \times pk)$ has the structure of the *matrix intraclass correlation matrix,*

$$G = \begin{bmatrix} A & B & \cdots & B \\ B & A & \cdots & B \\ \cdots & \cdots & \cdots & \cdots \\ B & B & \cdots & A \end{bmatrix} \tag{0.4.13}$$

where A, B are each of order $k \times k$, then

$$|G| = |A - B|^{p-1}[|A + (p-1)B|]. \tag{0.4.14}$$

(Here G is "partitioned matrix" (vide subsection 0.4.4)).

0.4.3 Matrix Inversion

For a square matrix $A = ((a_{ij}))$ of order p, the transpose of the matrix $((A_{ij}))$ is called the adjoint or adjugate of A. This will be denoted by adj.A.

If $|A| \neq 0$, A has associated with it an unique $(p \times p)$ matrix, denoted by $A^{-1} = ((a^{ij}))$ such that

$$A^{-1}A = AA^{-1} = I. \tag{0.4.15}$$

A^{-1} is called the inverse of A. Here

$$a^{ij} = A_{ji}/|A|.$$

If A is singular, A^{-1} does not exist.

Properties of Matrix Inverses

(1) $(\boldsymbol{A}')^{-1} = (\boldsymbol{A}^{-1})'$.

(2) Let $\boldsymbol{A}$, $\boldsymbol{B}$ be both non-singular and of order p. Then

$$(\boldsymbol{AB})^{-1} = \boldsymbol{B}^{-1}\boldsymbol{A}^{-1}. \tag{0.4.16}$$

(3) If $\boldsymbol{A}$ is non-singular,

$$|\boldsymbol{A}^{-1}| = |\boldsymbol{A}|^{-1}. \tag{0.4.17}$$

(4) If $\boldsymbol{D}_\lambda = \text{Diag}\ (\lambda_1, \cdots, \lambda_n)$,

$$\boldsymbol{D}_\lambda^{-1} = \text{Diag}\ (\lambda_1^{-1}, \cdots, \lambda_n).$$

(5) If $\boldsymbol{A}$ is an orthogonal matrix,

$$\boldsymbol{A}^{-1} = \boldsymbol{A}'. \tag{0.4.18}$$

(6) *Binomial Inversion Theorem :* Assume $\boldsymbol{A} : p \times p$, $\boldsymbol{U} : p \times q$, $\boldsymbol{B} : q \times q$, $\boldsymbol{V} : q \times p$
Then if $\boldsymbol{A}$ and $\boldsymbol{B}$ are non-singular,

$$(\boldsymbol{A}+\boldsymbol{UBV})^{-1} = \boldsymbol{A}^{-1} - \boldsymbol{A}^{-1}\boldsymbol{UB}(\boldsymbol{B}+\boldsymbol{BVA}^{-1}\boldsymbol{UB})^{-1}\boldsymbol{BVA}^{-1}. \tag{0.4.19}$$

In particular, let $\boldsymbol{B} = \boldsymbol{I}$ and $q = 1$. Then $\boldsymbol{U}$ is a column vector $\boldsymbol{u}$ and $\boldsymbol{V}$ is a row vector $\boldsymbol{v}'$. The theorem implies that

$$(\boldsymbol{A}+\boldsymbol{uv}')^{-1} = \boldsymbol{A}^{-1} - \frac{\boldsymbol{A}^{-1}\boldsymbol{uv}'\boldsymbol{A}^{-1}}{1+\boldsymbol{v}'\boldsymbol{A}^{-1}\boldsymbol{u}}. \tag{0.4.20}$$

If, further, $\boldsymbol{A} = \boldsymbol{I}$,

$$(\boldsymbol{I}+\boldsymbol{uv}')^{-1} = \boldsymbol{I} - \frac{\boldsymbol{uv}'}{1+\boldsymbol{v}'\boldsymbol{u}}. \tag{0.4.21}$$

(7) Consider the intraclass correlation matrix defined in (0.4.1). $\boldsymbol{A}$ may be written as

$$\boldsymbol{A} = (a-b)\boldsymbol{I} + b\boldsymbol{11}'$$

where $\boldsymbol{1}$ is a column vectors of ones. Hence, if $\boldsymbol{A}$ is of order p, applying, (0.4.20),

$$\boldsymbol{A}^{-1} = \frac{\boldsymbol{I}}{a-b} - \frac{b\boldsymbol{11}'}{(a-b)[a+(p-1)b]} \tag{0.4.22}$$

which is again a matrix with intraclass correlation structure.

Generalised Inverses

Let $\boldsymbol{A}$ be of order p and assume $|\boldsymbol{A}| = 0$. Hence, $\boldsymbol{A}^{-1}$ does not exist. However, it is possible to define a generalisation of the usual inverse, $\boldsymbol{A}^*$, as a matrix that exists evens when $\boldsymbol{A}^{-1}$ does not, that equals $\boldsymbol{A}^{-1}$ when $\boldsymbol{A}^{-1}$ exists and that satisfies the property

$$\boldsymbol{A}\boldsymbol{A}^*\boldsymbol{A} = \boldsymbol{A} \tag{0.4.23}$$

$\boldsymbol{A}^*$ is called a generalised inverse (g-inverse) of $\boldsymbol{A}$. This matrix is not necessarily unique, although it is possible, by imposing additional constraints, to define it uniquely.

Moore (1935) and Penrose (1955) defined a particular generalised inverse (often called pseudo-inverse) as a matrix $\boldsymbol{A}^+$ satisfying

$$\begin{aligned} &\text{(a) } \boldsymbol{A}\boldsymbol{A}^+\boldsymbol{A} = \boldsymbol{A}, \quad &&\text{(c) } \boldsymbol{A}^+\boldsymbol{A}\boldsymbol{A}^+ = \boldsymbol{A}^+, \\ &\text{(b) } (\boldsymbol{A}\boldsymbol{A}^+)' = \boldsymbol{A}\boldsymbol{A}^+, &&\text{(d) } (\boldsymbol{A}^+\boldsymbol{A})' = \boldsymbol{A}^+\boldsymbol{A}. \end{aligned} \tag{0.4.24}$$

Such a matrix not only always exists, but it is unique. For details on generalised inverses, the reader may refer to Rao (1965), for instance.

We give below a method of finding pseudo-inverse.

Let $\boldsymbol{A}$ be a $p \times n$ matrix ($p \leq n$). Using orthogonal factorisation technique [vide (0.4.52)] decompose $\boldsymbol{A}$ as

$$\boldsymbol{A} = \boldsymbol{B}(\boldsymbol{D}_\lambda \boldsymbol{O})\boldsymbol{C}'.$$

Replace non-zero elements of $\boldsymbol{D}_\lambda$ by their reciprocals and call the resulting matrix as $\boldsymbol{D}_{\lambda^*}$. Then

$$\boldsymbol{A}^+ = \boldsymbol{C}(\boldsymbol{D}_{\lambda^*}\boldsymbol{O})\boldsymbol{B}'.$$

▶ **Example 0.4.2** *Let* $\boldsymbol{A} = \begin{bmatrix} 4 & 2 \\ 2 & 1 \end{bmatrix}$.

Latent roots of $\boldsymbol{A}$ *are* $\lambda_1 = 5$, $\lambda_2 = 0$.

The matrix with columns as latent vector of $\boldsymbol{A}$ *is*

$$\boldsymbol{B} = \frac{1}{\sqrt{5}} \begin{bmatrix} 2 & -1 \\ 1 & 2 \end{bmatrix}.$$

Here $\boldsymbol{C} = \boldsymbol{B}$.

Hence,

$$\boldsymbol{A}^+ = \boldsymbol{B}\boldsymbol{D}_{\lambda^*}\boldsymbol{B}' = \frac{1}{25}\boldsymbol{A}.$$

0.4.4 Partitioned Matrices

A matrix $\boldsymbol{A}$ of the form

$$\boldsymbol{A} = \begin{bmatrix} \boldsymbol{A}_{11} & \boldsymbol{A}_{12} & \cdots & \boldsymbol{A}_{1k} \\ \dots & \dots & \dots & \dots \\ \boldsymbol{A}_{k1} & \boldsymbol{A}_{k2} & \cdots & \boldsymbol{A}_{kk} \end{bmatrix}$$

where $\boldsymbol{A}_{ij}$ are *submatrices* of $\boldsymbol{A}$ is called a *partitioned* matrix.

If

$$\boldsymbol{A} = \begin{bmatrix} \boldsymbol{A}_{11} & \boldsymbol{A}_{12} \\ \boldsymbol{A}_{21} & \boldsymbol{A}_{22} \end{bmatrix}, \quad \boldsymbol{B} = \begin{bmatrix} \boldsymbol{B}_{11} & \boldsymbol{B}_{12} \\ \boldsymbol{B}_{21} & \boldsymbol{B}_{22} \end{bmatrix}$$

when $\boldsymbol{A}_{11}$ is $k \times l$, $\boldsymbol{A}_{12}$ is $k \times (q-l)$, $\boldsymbol{A}_{21}$ is $(p-k) \times l$, $\boldsymbol{A}_{22}$ is $(p-k) \times (q-l)$ and $\boldsymbol{B}_{ij}$ are of similar order,

$$\boldsymbol{A}+\boldsymbol{B} = \begin{bmatrix} \boldsymbol{A}_{11}+\boldsymbol{B}_{11} & \boldsymbol{A}_{12}+\boldsymbol{B}_{12} \\ \boldsymbol{A}_{21}+\boldsymbol{B}_{21} & \boldsymbol{A}_{22}+\boldsymbol{B}_{22} \end{bmatrix}.$$

Also, if $\boldsymbol{C}$ is $q \times r$ matrix partitioned as

$$\boldsymbol{C} = \begin{bmatrix} \boldsymbol{C}_{11} & \boldsymbol{C}_{12} \\ \boldsymbol{C}_{21} & \boldsymbol{C}_{22} \end{bmatrix}$$

when $\boldsymbol{C}_{11}$ is $l \times m$, $\boldsymbol{C}_{12}$ is $l \times (r-m)$, $\boldsymbol{C}_{21}$ is $(q-l) \times m$, $\boldsymbol{C}_{22}$ is $(q-l) \times (r-m)$, the

$$\boldsymbol{AC} = \begin{bmatrix} \boldsymbol{A}_{11} & \boldsymbol{A}_{12} \\ \boldsymbol{A}_{21} & \boldsymbol{A}_{22} \end{bmatrix} \begin{bmatrix} \boldsymbol{C}_{11} & \boldsymbol{C}_{12} \\ \boldsymbol{C}_{21} & \boldsymbol{C}_{22} \end{bmatrix} = \begin{bmatrix} \boldsymbol{A}_{11}\boldsymbol{C}_{11}+\boldsymbol{A}_{12}\boldsymbol{C}_{21} & \boldsymbol{A}_{11}\boldsymbol{C}_{12}+\boldsymbol{A}_{12}\boldsymbol{C}_{22} \\ \boldsymbol{A}_{21}\boldsymbol{C}_{11}+\boldsymbol{A}_{22}\boldsymbol{C}_{21} & \boldsymbol{A}_{21}\boldsymbol{C}_{12}+\boldsymbol{A}_{22}\boldsymbol{C}_{22} \end{bmatrix}.$$

It follows that, if $\boldsymbol{D}_{11}$ is $p \times p$, $\boldsymbol{D}_{12}$ is $p \times q$, $\boldsymbol{D}_{21}$ is $q \times p$, $\boldsymbol{D}_{22}$ is $q \times q$, then

$$\begin{vmatrix} \boldsymbol{D}_{11} & \boldsymbol{D}_{12} \\ \boldsymbol{O} & \boldsymbol{D}_{22} \end{vmatrix} = \begin{vmatrix} \boldsymbol{D}_{11} & \boldsymbol{O} \\ \boldsymbol{D}_{21} & \boldsymbol{D}_{22} \end{vmatrix} = |\boldsymbol{D}_{11}| \cdot |\boldsymbol{D}_{22}|.$$

Inverse of a partitioned matrix

Assume $\boldsymbol{A}$ is an arbitrary $(p+n) \times (p+n)$ matrix with $|\boldsymbol{A}| \neq 0$. Partition $\boldsymbol{A}$ as

$$\boldsymbol{A} = \begin{bmatrix} \boldsymbol{A}_{11} & \boldsymbol{A}_{12} \\ \boldsymbol{A}_{21} & \boldsymbol{A}_{22} \end{bmatrix} \tag{0.4.25}$$

where $\boldsymbol{A}_{11}$ is $p \times p$ and $\boldsymbol{A}_{22}$ is $n \times n$. Let

$$\boldsymbol{B} = \boldsymbol{A}^{-1} = \begin{bmatrix} \boldsymbol{B}_{11} & \boldsymbol{B}_{12} \\ \boldsymbol{B}_{21} & \boldsymbol{B}_{22} \end{bmatrix}.$$

when $\boldsymbol{B}_{11}$ is $p \times p$, $\boldsymbol{B}_{22}$ is $n \times n$. It follows that

$$\boldsymbol{B}_{11} = (\boldsymbol{A}_{11} - \boldsymbol{A}_{12}\boldsymbol{A}_{22}^{-1}\boldsymbol{A}_{21})^{-1} \equiv \boldsymbol{A}_{11\cdot 2}^{-1} \tag{0.4.26}$$

$$\boldsymbol{B}_{22} = (\boldsymbol{A}_{22} - \boldsymbol{A}_{21}\boldsymbol{A}_{11}^{-1}\boldsymbol{A}_{12})^{-1} \equiv \boldsymbol{A}_{22\cdot 1}^{-1} \tag{0.4.27}$$

$$\boldsymbol{B}_{12} = -\boldsymbol{A}_{11}^{-1}\boldsymbol{A}_{12}\boldsymbol{A}_{22\cdot 1}^{-1} \tag{0.4.28}$$

$$\boldsymbol{B}_{21} = -\boldsymbol{A}_{22}^{-1}\boldsymbol{A}_{21}\boldsymbol{A}_{11\cdot 2}^{-1}. \tag{0.4.29}$$

It is simple to check that if $\boldsymbol{A}_{12} = \boldsymbol{A}_{21} = \boldsymbol{O}$, then

$$\boldsymbol{A}^{-1} = \begin{bmatrix} \boldsymbol{A}_{11}^{-1} & \boldsymbol{O} \\ \boldsymbol{O} & \boldsymbol{A}_{22}^{-1} \end{bmatrix}.$$

Determinant of a partitioned matrix

(a) $\begin{vmatrix} \boldsymbol{A} & \boldsymbol{O} \\ \boldsymbol{O} & \boldsymbol{B} \end{vmatrix} = |\boldsymbol{A}|\cdot|\boldsymbol{B}|.$

(b) $\begin{vmatrix} \boldsymbol{A} & \boldsymbol{C} \\ \boldsymbol{O} & \boldsymbol{B} \end{vmatrix} = |\boldsymbol{A}|\cdot|\boldsymbol{B}|$ if $|\boldsymbol{A}| \neq 0$.

(c) Suppose $\boldsymbol{A}$ is partitioned as in (0.4.25). Then

(i) if $|\boldsymbol{A}_{22}| \neq 0$,

$$|\boldsymbol{A}| = |\boldsymbol{A}_{22}| \cdot |\boldsymbol{A}_{11} - \boldsymbol{A}_{12}\boldsymbol{A}_{22}^{-1}\boldsymbol{A}_{21}| = |\boldsymbol{A}_{22}| \cdot |\boldsymbol{A}_{11\cdot 2}|; \tag{0.4.30}$$

(ii) if $|\boldsymbol{A}_{11}| \neq 0$,

$$|\boldsymbol{A}| = |\boldsymbol{A}_{11}| \cdot |\boldsymbol{A}_{22} - \boldsymbol{A}_{21}\boldsymbol{A}_{11}^{-1}\boldsymbol{A}_{12}| = |\boldsymbol{A}_{11}| \cdot |\boldsymbol{A}_{22\cdot 1}|. \tag{0.4.31}$$

Transpose of a partitioned matrix

Let $\boldsymbol{A}$ be as in (0.4.25). Then

$$\boldsymbol{A}' = \begin{bmatrix} \boldsymbol{A}'_{11} & \boldsymbol{A}'_{21} \\ \boldsymbol{A}'_{12} & \boldsymbol{A}'_{22} \end{bmatrix}.$$

0.4.5 Rank

The rank of a matrix $\boldsymbol{A}$ often denoted as $r(\boldsymbol{A})$ is the largest order of the submatrices with non-vanishing determinants.

Properties:

(1) $r(\boldsymbol{A}) = r(\boldsymbol{A}')$.

(2) If $\boldsymbol{D} = \text{Diag}\,(\lambda_1, \cdots, \lambda_p), r(\boldsymbol{D}) =$ number of non-zero λ's.

(3) Let $\boldsymbol{A}$ and $\boldsymbol{B}$ be two conformable matrices. Then

$$r(\boldsymbol{AB}) \leq \min[r(\boldsymbol{A}), r(\boldsymbol{B})]. \tag{0.4.32}$$

(4) If $\boldsymbol{A} : p \times n$ and $\boldsymbol{B} : p \times n$, then

$$r(\boldsymbol{A} + \boldsymbol{B}) \leq r(\boldsymbol{A}) + r(\boldsymbol{B}). \tag{0.4.33}$$

(5) If $\boldsymbol{A} : p \times n$,

$$r(\boldsymbol{A}) \leq \min[p, n]. \tag{0.4.34}$$

(6) If $\boldsymbol{P} : m \times m$, $\boldsymbol{Q} : n \times n$, $|\boldsymbol{P}| \neq 0, |\boldsymbol{Q}| \neq 0$, then for $\boldsymbol{A} : m \times n$

$$r(\boldsymbol{PAQ}) = r(\boldsymbol{A}). \tag{0.4.35}$$

(7) If $\boldsymbol{A} : p \times n$, $\boldsymbol{B} : n \times r$ such that $\boldsymbol{AB} = \boldsymbol{O}$, then

$$r(\boldsymbol{B}) \leq n - r(\boldsymbol{A}). \tag{0.4.36}$$

0.4.6 Eigenvalues and Eigenvectors

Let $\boldsymbol{A}$ be a square matrix of order n. The transformed vector $\boldsymbol{Ax}$ may be proportiona[l] to the original vector $\boldsymbol{x}$ i.e.,

$$\begin{aligned} \boldsymbol{Ax} &= \lambda \boldsymbol{x} \\ \text{or,} \quad (\boldsymbol{A} - \lambda \boldsymbol{I})\boldsymbol{x} &= \boldsymbol{0} \end{aligned} \tag{0.4.37}$$

where λ is a scalar. The linear system (0.4.37) has solution $\boldsymbol{x} \neq 0$ iff

$$\begin{aligned} & |\boldsymbol{A} - \lambda I| = 0 \\ \text{i.e.,} \quad & \lambda^n + b_{n-1}\lambda^{n-1} + \cdots + b_1\lambda + b_0 = 0 \\ \text{or,} \quad & \prod_{j=1}^{n} (\lambda - \lambda_j) = 0 \end{aligned} \tag{0.4.38}$$

(0.4.38) is called the characteristic equation of $\boldsymbol{A}$. For each of its root λ_i, the equation (0.4.37) has a solution $\boldsymbol{x}_i \neq \boldsymbol{0}$. The λ_i's are called the eigenvalues (or latent roots or characteristic values) of $\boldsymbol{A}$, the $\boldsymbol{x}$'s, the eigenvectors (or the characteristic vectors) of $\boldsymbol{A}$. Thus

$$\begin{aligned} \boldsymbol{A}[\boldsymbol{x}_1 \cdots \boldsymbol{x}_n] &= [\lambda_1 \boldsymbol{x}_1 \cdots \lambda_n \boldsymbol{x}_n] \\ \text{or,} \quad \boldsymbol{AX} &= \boldsymbol{XD} \end{aligned}$$

where $\boldsymbol{X} = [\boldsymbol{x}_1 \cdots \boldsymbol{x}_n]$, $\boldsymbol{D} = \text{Diag}\ (\lambda_1 \cdots \lambda_n)$. If $\boldsymbol{x}_i$'s are linearly independent

$$\boldsymbol{X}^{-1}\boldsymbol{AX} = \boldsymbol{D}. \tag{0.4.39}$$

Some of the λ's may be equal. An eigenvector $\boldsymbol{x}$ with real entries is called standardized if $\boldsymbol{x}'\boldsymbol{x} = 1$.

The following results hold:

(1) $|A| = \prod_{j=1}^{n} \lambda$. Therefore, $r(A) < n$ iff at least one latent root is zero. The rank of a symmetric matrix is the number of non-zero latent roots it possesses.

(2) In general, latent roots are complex numbers (occurring in complex conjugate pairs when the roots are not real). However, if $\boldsymbol{A}$ is real and symmetric, all the eigenvalues and therefore, all the eigenvectors are real.

(3) If $\boldsymbol{A}$ has the eigenvalue λ and eigenvector $\boldsymbol{x}$, then $R(\boldsymbol{A})$ where R is a rational function has eigenvalue $R(\lambda)$ and the eigenvector $\boldsymbol{x}$. In particular $\boldsymbol{A}^m$ has the eigenvalue λ^m with the same multiplicity.

(4) If $\boldsymbol{A}$ has the eigenvalue λ and eigenvector $\boldsymbol{x}$ then $\boldsymbol{TAT}^{-1}$ has the eigenvalue λ and eigenvector $\boldsymbol{Tx}$.

(5) Latent roots of $\boldsymbol{A}$ and $\boldsymbol{A}'$ are the same.

(6) If $\boldsymbol{A} > 0$, $\boldsymbol{B} > 0$ are $n \times n$ matrices then all the non-zero eigenvalues of $\boldsymbol{B}^{-1}\boldsymbol{A}$ are positive.

(7) $\boldsymbol{AB}$ and $\boldsymbol{BA}$ have the same eigenvalues while their vectors are $\boldsymbol{x}_i$ and $\boldsymbol{Bx}_i$ (or $\boldsymbol{A}^{-1}\boldsymbol{x}_i$).

(8) If $\boldsymbol{A}$ and $\boldsymbol{B}$ are $n \times n$ and $\boldsymbol{A}$ is non-singular, then the latent roots of $\boldsymbol{AB}$ and $\boldsymbol{BA}$ are the same.

(9) If $\boldsymbol{A}$ is $m \times n$, the non-zero latent roots of $\boldsymbol{AA}'$ and $\boldsymbol{A}'\boldsymbol{A}$ are the same.

(10) $tr(\boldsymbol{A}) = \sum_{i=1}^{n} \lambda_i$; $tr(\boldsymbol{A}^m) = \sum_{i=1}^{n} \lambda_i^m$. (Here $tr(\boldsymbol{A})$ means trace of $\boldsymbol{A}$, defined in subsection (0.4.7)

If $\boldsymbol{A}$ is symmetric, for $m = 2$, $tr(\boldsymbol{A}^2) = tr(\boldsymbol{AA}') = \Sigma\lambda_i^2$. If therefore, all $\lambda_i = 0$, all $a_{ik} = 0$, i.e., $\boldsymbol{A} = \boldsymbol{O}$.

(11) *Theorem of Cayley.* If $f(\lambda) = 0$ is the characteristic equation of $\boldsymbol{A}$, then also $f(\boldsymbol{A}) = \boldsymbol{O}$ i.e., every matrix satisfies its characteristic equation.

Extremum Properties of the Eigenvalues

The symmetrical form $\boldsymbol{A}$ has n real eigenvalues λ_i and real eigenvectors $\boldsymbol{x}_i$ which are characterised by the following properties:

(12) The largest eigenvalue λ_1 is the maximum of the quadratic form $\boldsymbol{x}'\boldsymbol{Ax}$ for all vectors $\boldsymbol{x}$ of length 1, i.e.,

$$\lambda_1 = \max_{\boldsymbol{x}} \frac{\boldsymbol{x}'\boldsymbol{A}\boldsymbol{x}}{\boldsymbol{x}'\boldsymbol{x}}.$$

The vector $\boldsymbol{x}_1$ which gives this maximum is the eigenvector corresponding to λ_1:

$$\lambda_1 = \frac{\boldsymbol{x}_1'\boldsymbol{A}\boldsymbol{x}_1}{\boldsymbol{x}_1'\boldsymbol{x}_1} \quad \text{and} \quad \boldsymbol{A}\boldsymbol{x}_1 = \lambda_1\boldsymbol{x}_1.$$

(13) The eigenvalue λ_i is the maximum of the form $\boldsymbol{x}'\boldsymbol{A}\boldsymbol{x}$ for all $\boldsymbol{x}$ of length 1 for which

$$\boldsymbol{x}'\boldsymbol{x}_1 = \boldsymbol{x}'\boldsymbol{x}_2 = \cdots = \boldsymbol{x}'\boldsymbol{x}_{i-1} = 0.$$

The vector $\boldsymbol{x}_i$ for which the form assumes this maximum is the eigenvector belonging to λ_i.

(14) The smallest eigenvalue λ_n is the minimum of the quadratic form $\boldsymbol{x}'\boldsymbol{A}\boldsymbol{x}$ for all vectors of length 1, i.e.,

$$\lambda_n = \min_{\boldsymbol{x}} \frac{\boldsymbol{x}'\boldsymbol{A}\boldsymbol{x}}{\boldsymbol{x}'\boldsymbol{x}}.$$

Latent vectors associated with distinct latent roots of a symmetric matrix are orthogonal. If a latent root has multiplicity m, it is always possible to find m mutually orthogonal latent vectors of unit length that are also orthogonal to the latent vectors corresponding to the distinct latent roots.

If $\boldsymbol{A}$ is orthogonal, all its latent roots have unit modulus (i.e., all its latent roots lie on the unit circle in the complex plane). Also the reciprocals of its latent roots are latent roots.

If $\boldsymbol{A}$ is the pth order intraclass correlation matrix given in (0.4.1), its latent roots are given by

$$\lambda_1 = a + (p-1)b, \quad \lambda_2 = \cdots = \lambda_p = a - b.$$

The latent vectors of the intraclass correlation matrix do not depend on the elements of $\boldsymbol{A}$. Any p mutually orthogonal vectors of unit length, the first of which have components that are all identical, will do. For example, the columns of the Helmert matrix in example 0.4.1 are a set of latent vectors for $\boldsymbol{A}$.

Kantorovich inequality: Let $\boldsymbol{A}$ be an $n \times n$ real symmetric positive definite matrix (defined in subsection 0.4.9) and $\boldsymbol{x}$ be an $n \times 1$ real vector satisfying $\boldsymbol{x}'\boldsymbol{x} = 1$.

Then

$$\boldsymbol{x}'\boldsymbol{A}\boldsymbol{x}\boldsymbol{x}'\boldsymbol{A}^{-1}\boldsymbol{x} \le \frac{(\lambda_1 + \lambda_n)^2}{4\lambda_1\lambda_n}$$

when $\lambda_1 \ge \cdots \ge \lambda_n$ are eigenvalues of $\boldsymbol{A}$. The inequality has been generalised by Bloomfield and Watson (1975), Knott (1975), Khatri and Rao (1975), among others.

0.4.7 Trace of a Matrix

The sum of the diagonal elements of a square matrix $\boldsymbol{A}$ is called its trace. Thus, if $\boldsymbol{A} = ((a_{ij}))$, $i, j = 1, \cdots p$,

$$tr(\boldsymbol{A}) = \sum_{j=1}^{p} a_{jj}. \tag{0.4.40}$$

Properties

(1) If $\boldsymbol{A} : p \times n$, $\boldsymbol{B} : n \times p$, then $tr(\boldsymbol{AB}) = tr(\boldsymbol{BA})$.

(2) The trace of a square matrix is equal to the sum of its latent roots.

(3) If $\boldsymbol{A} : p \times n$, $\boldsymbol{B} : p \times n$, then

$$tr(\boldsymbol{A}+\boldsymbol{B}) = tr(\boldsymbol{A}) + tr(\boldsymbol{B}).$$

(4) If α is a scalar and $\boldsymbol{A} : p \times p$, then

$$tr(\alpha\boldsymbol{A}) = \alpha\, tr(\boldsymbol{A}).$$

(5) Let $\boldsymbol{S}$ be a non-singular $p \times p$ matrix. Then

$$tr(\boldsymbol{S}^{-1}\boldsymbol{AS}) = tr(\boldsymbol{A}).$$

The transformation $\boldsymbol{S}^{-1}\boldsymbol{AS}$ is often called a *similarity transformation.*

(6) If $A : p \times p$ be of rank 1, then its only non-zero latent root is $tr(\boldsymbol{A})$.

(7) Let $\boldsymbol{A}$ be a $k \times k$ symmetric matrix and $\boldsymbol{x}$ be a $k \times 1$ vector. Then $\boldsymbol{x}'\boldsymbol{Ax} = tr(\boldsymbol{Axx}')$.

0.4.8 Direct Product

Let $\boldsymbol{A}$ be of order $p \times p$ and $\boldsymbol{B}$ of order $n \times n$. Then the $pn \times pn$ matrix $\boldsymbol{C} = \boldsymbol{A} \otimes \boldsymbol{B}$ is called the direct product of $\boldsymbol{A}$ and $\boldsymbol{B}$ and is given by

$$\boldsymbol{C} = \boldsymbol{A} \otimes \boldsymbol{B} = \begin{bmatrix} a_{11}\boldsymbol{B} & a_{12}\boldsymbol{B} & \cdots & a_{1p}\boldsymbol{B} \\ \vdots & \vdots & \vdots & \vdots \\ a_{p1}\boldsymbol{B} & a_{p2}\boldsymbol{B} & \cdots & a_{pp}\boldsymbol{B} \end{bmatrix} \tag{0.4.41}$$

Properties

(1) $\alpha(\boldsymbol{A} \otimes \boldsymbol{B}) = (\alpha\boldsymbol{A}) \otimes \boldsymbol{B} = \boldsymbol{A} \otimes (\alpha\boldsymbol{B})$ for all scalar α.

(2) $(\boldsymbol{A} \otimes \boldsymbol{B})' = \boldsymbol{A}' \otimes \boldsymbol{B}'$.

(3) $\boldsymbol{A} \otimes (\boldsymbol{B} \otimes \boldsymbol{D}) = (\boldsymbol{A} \otimes \boldsymbol{B}) \otimes \boldsymbol{D}$.

(4) Let $\boldsymbol{A} : p \times p$ and $\boldsymbol{B} : n \times n$. Then $(\boldsymbol{A} \otimes \boldsymbol{B})^{-1} = \boldsymbol{A}^{-1} \otimes \boldsymbol{B}^{-1}$, for any nonsingular $\boldsymbol{A}, \boldsymbol{B}$.

(5) Let $\boldsymbol{A} : p \times p$ and $\boldsymbol{B} : n \times n$. Then $|\boldsymbol{A} \otimes \boldsymbol{B}| = |\boldsymbol{A}|^n \times |\boldsymbol{B}|^p$.

(6) $(\boldsymbol{A}+\boldsymbol{B}) \otimes \boldsymbol{D} = \boldsymbol{A} \otimes \boldsymbol{D} + \boldsymbol{B} \otimes \boldsymbol{D}$.

(7) If $\boldsymbol{A}$ and $\boldsymbol{C}$ are of order p and $\boldsymbol{B}$ and $\boldsymbol{D}$ are of order n, then

$$(\boldsymbol{A}\otimes\boldsymbol{B})(\boldsymbol{C}\otimes\boldsymbol{D}) = (\boldsymbol{AC})\otimes(\boldsymbol{BD}).$$

(8) $\boldsymbol{A}\otimes(\boldsymbol{B}+\boldsymbol{D}) = \boldsymbol{A}\otimes\boldsymbol{B}+\boldsymbol{A}\otimes\boldsymbol{D}$.

(9) Suppose the latent roots of $\boldsymbol{A} : p\times p$ are $\lambda_1,\cdots,\lambda_p$ and those of $\boldsymbol{B} : n\times n$ are $\mu_1,\cdots,\mu_n$.

Then
$$|\boldsymbol{A}\otimes\boldsymbol{B}| = (\lambda_1\cdots\lambda_p)(\mu_1\cdots\mu_n).$$

(10) $tr(\boldsymbol{A}\otimes\boldsymbol{B}) = (tr\boldsymbol{A})(tr\boldsymbol{B})$.

0.4.9 Quadratic Forms and Definiteness

Let $\boldsymbol{x} = (x_1,\cdots,x_p)'$ denote a $p\times 1$ vector and $\boldsymbol{A} = ((a_{ij}))$ denote a $p\times p$ symmetric matrix. Then

$$Q = \boldsymbol{x}'\boldsymbol{A}\boldsymbol{x} = \sum_{i=1}^{p}\sum_{j=1}^{p} a_{ij}x_ix_j \tag{0.4.42}$$

is called a *quadratic form* in the variables $x_1,\cdots,x_p$.

If $Q\equiv \boldsymbol{x}'\boldsymbol{A}\boldsymbol{x} > 0(<0)$ for all vectors $\boldsymbol{x}(\neq \boldsymbol{0}), Q$ is called a *positive* (*negative*) *definite quadratic form* and the matrix $\boldsymbol{A}$ is called *positive* (*negative*) definite; it is then written as $\boldsymbol{A} > (<)0$. If $Q\geq 0\ \forall\ \boldsymbol{x}(\neq\boldsymbol{0}), Q$ is called a positive semi-definite (*p.s.d.*) quadratic form and $\boldsymbol{A}$ is called a p.s.d. matrix ($\boldsymbol{A}\geq 0$). Similarly, a negative semi-definite (*n.s.d.*) quadratic form and a *n.s.d.* matrix is defined. If none of these conditions hold, Q is called an *indefinite quadratic* form and $\boldsymbol{A}$ is called an *indefinite* matrix.

Consider all submatrices of a symmetric matrix $\boldsymbol{A}(p\times p)$ whose principal diagonals are parts of the principal diagonals of $\boldsymbol{A}$. The determinants of these *principal submatrices* are called the *principal minors* of $\boldsymbol{A}$. A necessary and sufficient condition for $\boldsymbol{A} > 0$ is that $|\boldsymbol{A}| > 0$ and all principal minors of $\boldsymbol{A}$ are positive.

Properties of Positive Definite Matrices

(1) If
$$\boldsymbol{A} > 0,\ \ \boldsymbol{A}^{-1} > 0. \tag{0.4.43}$$

(2) Let $\boldsymbol{A} : n\times n$, $\boldsymbol{B} : p\times n$, $p\leq n$; $\boldsymbol{A} > 0$ and rank $(\boldsymbol{B}) = r$. Then

$$\boldsymbol{BAB}' > 0 \quad\text{if}\quad r = p \quad\text{and}\quad \boldsymbol{BAB}'\geq 0 \quad\text{if}\quad r < p. \tag{0.4.44}$$

(3) If $\boldsymbol{A} > 0$, $\boldsymbol{B} > 0$,
$$|\boldsymbol{A}| + |\boldsymbol{B}| \leq |\boldsymbol{A}+\boldsymbol{B}| \tag{0.4.45}$$

with equality iff $\boldsymbol{A}$ is proportional to $\boldsymbol{B}$.

(4) If $A > 0$, $B > 0$ and $A - B > 0$, then

$$|A| > |B|. \qquad (0.4.46)$$

(5) If $A \geq 0$, there exists at least one matrix B such that $A = BB'$. Conversely, if for some B, $A = BB'$, then $A \geq 0$.

(6) *Hadamard's Inequality* (1893): Let $A : p \times p = (a_{ij})$ and assume $A > 0$. Then

$$|A| \leq \prod_{1}^{p} a_{ii}. \qquad (0.4.47)$$

(7) If $A > 0$, there exists a non-singular matrix M such that

$$MAM' = I. \qquad (0.4.48)$$

(8) If $A > 0$, all its latent roots are positive, and conversely.

(9) If $A > 0$, $B > 0$ and $A - B > 0$, then

$$B^{-1} - A^{-1} > 0. \qquad (0.4.49)$$

(10) Let $\lambda_i(A), \lambda_i(B)$ denote the ith latent roots of A and B, respectively, where $A = A'$, $B = B'$. Then, if $A - B \geq 0$, $\lambda_i(A) \geq \lambda_i(B) \; \forall \, i$.

Factorisation of Symmetric Matrices (Spectral Decomposition or Jordan Decomposition)

If $A = A'$, there exists an orthogonal matrix B such that

$$A = BD_\lambda B' \qquad (0.4.50)$$

where $D_\lambda = \text{Diag}\,(\lambda_1, \cdots, \lambda_p)$ and λ's are the latent roots of A (some of which may be zero and some of which may be multiple roots). The columns of B form an orthonormal set of latent vectors X_i $(i = 1, \cdots, p)$ of A.

$$B = (X_1, \cdots, X_p)$$

$$A = \sum_{i=1}^{p} \lambda_i X_i X_i'.$$

$X_i X_i'$ is called the ith spectral component of A. If A is a non-singular symmetric matrix, then

$$A^k = \sum_{i=1}^{p} \lambda_i^k X_i X_i' = B\,\text{Diag}\,(\lambda_i^k,\ i = 1, \cdots, n) B' \qquad (0.4.50')$$

for any integer k, positive or negative. If all eigenvalues of A are positive, we can define the rational powers $A^{r/s} = BD^{r/s}B'$ where

$$D^{r/s} = \text{Diag}\,(\lambda_i^{r/s},\ i = 1, \cdots, n) \qquad (0.4.50'')$$

for integer $s(> 0)$ and r. If some of the eigenvalues of A are zero, then (0.4.50′) and (0.4.50″) hold if the exponents are restricted to be non-negative.

Simultaneous Diagonalisation

Let $\boldsymbol{A} > 0$ and $\boldsymbol{B} \geq 0$. Then there exists a matrix $\boldsymbol{V}$, $|\boldsymbol{V}| \neq 0$, such that

$$\boldsymbol{VAV'} = I \quad \text{and} \quad \boldsymbol{VBV'} = \text{Diag}\ (\lambda_1, \cdots, \lambda_p), \tag{0.4.51}$$

where λ's are the roots of $|\boldsymbol{B} - \lambda \boldsymbol{A}| = 0$.

Orthogonal Factorisation (Singular Value Decomposition)

For $\boldsymbol{A} : p \times n$, $p \leq n$, there exist orthogonal matrices $\boldsymbol{B} : p \times p$, $\boldsymbol{C} : n \times n$ such that

$$\boldsymbol{A} = \boldsymbol{B}(\boldsymbol{D}_\lambda \boldsymbol{O})\boldsymbol{C}' \tag{0.4.52}$$

where $\boldsymbol{D}_\lambda = \text{Diag}\ (\lambda_1, \cdots, \lambda_p)$ and $|\boldsymbol{AA}' - \lambda^2 I| = 0$.

Note that if $n = p$ and $\boldsymbol{A} = \boldsymbol{A}'$, by taking $\boldsymbol{C} = \boldsymbol{B}$, this property reduces to (0.4.50)

Powers of Symmetric Matrices

Suppose $\boldsymbol{A} = \boldsymbol{A}'$. From (0.4.50), $\boldsymbol{A} = \boldsymbol{BD}_\lambda \boldsymbol{B}'$.
Then

$$\begin{aligned} \boldsymbol{A}^2 = \boldsymbol{A} \cdot \boldsymbol{A} &= (\boldsymbol{BD}_\lambda \boldsymbol{B}')(\boldsymbol{BD}_\lambda \boldsymbol{B}') = \boldsymbol{BD}_\lambda^2 \boldsymbol{B}' \\ &= \boldsymbol{BD}_{\lambda^2} \boldsymbol{B}'. \end{aligned} \tag{0.4.53}$$

It follows that for any integer n, positive or negative,

$$\boldsymbol{A}^n = \boldsymbol{BD}_{\lambda^n} \boldsymbol{B}'.$$

Simultaneous Orthogonal Diagonilisation

If $\boldsymbol{A} = \boldsymbol{A}', \boldsymbol{B} = \boldsymbol{B}'$ and $\boldsymbol{AB} = \boldsymbol{BA}$, there exists an orthogonal matrix $\boldsymbol{P}$, such that

$$\begin{aligned} \boldsymbol{P}'\boldsymbol{AP} &= \text{Diag}\ (\lambda_1, \cdots, \lambda_p) \\ \boldsymbol{P}'\boldsymbol{BP} &= \text{Diag}\ (\mu_1, \cdots, \mu_p) \end{aligned}$$

where λ, μ's are the latent roots of $\boldsymbol{A}$, $\boldsymbol{B}$ respectively.

0.4.10 Idempotent Matrices

A square matrix $\boldsymbol{A} : p \times p$ is said to be *idempotent* if $\boldsymbol{A}^2 = \boldsymbol{A}$. If $\boldsymbol{y} = \boldsymbol{Ax}$, and $\boldsymbol{A}$ ($p \times p$ is idempotent then $\boldsymbol{y}$ is a projection of $\boldsymbol{x}$.

Properties

(1) If $\boldsymbol{A}$ is idempotent, its latent roots are either zero or one.

(2) If $\boldsymbol{A}$ is idempotent and $\boldsymbol{A} \geq 0$,

$$r(\boldsymbol{A}) = tr(\boldsymbol{A}).$$

(3) If $\boldsymbol{A}$ is idempotent, $(I - \boldsymbol{A})$ is also so.

(4) If $\boldsymbol{A}$ is idempotent and $\boldsymbol{A} = \boldsymbol{A}'$, then $\boldsymbol{A} \geq 0$.

(5) If $\boldsymbol{A}$ is idempotent and $\boldsymbol{A} \geq 0$, there exists an orthogonal matrix $\boldsymbol{B}$ such that

$$\boldsymbol{BAB}' = \begin{pmatrix} \boldsymbol{I}_r & \boldsymbol{O} \\ \boldsymbol{O} & \boldsymbol{O} \end{pmatrix}$$

where $r(\boldsymbol{A}) = r$.

0.4.11 Differentiation

The derivative of a scalar function f of a matrix $\boldsymbol{X} = ((x_{ij}))_{p \times n}$ is defined as

$$\frac{d}{d\boldsymbol{X}} f(\boldsymbol{X}) = \left(\frac{df(\boldsymbol{X})}{dx_{ij}} \right)_{\substack{i=1,\cdots,p \\ j=1,\cdots,n}}. \tag{0.4.54}$$

Some properties are:

(1)

$$\frac{d|\boldsymbol{X}|}{d\boldsymbol{X}} = |\boldsymbol{X}| \cdot \boldsymbol{X}^{-1} \quad \text{for} \quad \boldsymbol{X} : p \times p, |\boldsymbol{X}| \neq 0. \tag{0.4.55}$$

(2)

$$\frac{d}{d\boldsymbol{X}} |\boldsymbol{X}|^a = a|\boldsymbol{X}|^{a-1} \frac{d}{d\boldsymbol{X}} |\boldsymbol{X}| \quad \text{for} \quad \boldsymbol{X} : p \times p. \tag{0.4.56}$$

(3)

$$\frac{d}{d\boldsymbol{X}} [f(\boldsymbol{X}) g(\boldsymbol{X})] = f(\boldsymbol{X}) \frac{dg(\boldsymbol{X})}{d\boldsymbol{X}} + g(\boldsymbol{X}) \frac{df(\boldsymbol{X})}{d\boldsymbol{X}} \tag{0.4.57}$$

for scalar functions $f(\cdot)$ and $g(\cdot)$.

(4) If $\boldsymbol{x} : p \times 1$ and $\boldsymbol{A} : p \times p$ then

$$\frac{d}{d\boldsymbol{x}} (\boldsymbol{x}' \boldsymbol{A} \boldsymbol{x}) = 2\boldsymbol{A}\boldsymbol{x}. \tag{0.4.58}$$

(5) If $\boldsymbol{x} : p \times 1$ and $\boldsymbol{a} : p \times 1$ then

$$\frac{d}{d\boldsymbol{x}} (\boldsymbol{a}' \boldsymbol{x}) = \boldsymbol{a}. \tag{0.4.59}$$

(6) If $\boldsymbol{X} : p \times p$, $|\boldsymbol{X}| \neq 0$, $\boldsymbol{a} : p \times 1$ then

$$\frac{d}{d\boldsymbol{X}} (\boldsymbol{a}' \boldsymbol{X}^{-1} \boldsymbol{a}) = -\boldsymbol{X}^{-1} \boldsymbol{a}\boldsymbol{a}' \boldsymbol{X}^{-1}. \tag{0.4.60}$$

(7)

$$\frac{d}{d\boldsymbol{X}} tr(\boldsymbol{A}'\boldsymbol{X}) = \boldsymbol{A}. \tag{0.4.61}$$

(8)

$$\frac{d}{d\boldsymbol{X}} tr(\boldsymbol{A}\boldsymbol{X}') = \boldsymbol{A}. \tag{0.4.62}$$

The derivative of a matrix $\boldsymbol{X} = ((x_{ij}))_{p\times n}$ with respect to an element y is defined as

$$\frac{d\boldsymbol{X}}{dy} = \left(\left(\frac{dx_{ij}}{dy}\right)\right). \tag{0.4.63}$$

If $\boldsymbol{I}_{ij}$ be a matrix with a one in the (i, j)th place and zero elsewhere, then for $\boldsymbol{X}(p\times p)$ and $|\boldsymbol{X}| \neq 0$

$$\frac{d\boldsymbol{X}^{-1}}{dx_{ij}} = -\boldsymbol{X}^{-1}\boldsymbol{I}_{ij}\boldsymbol{X}^{-1}. \tag{0.4.64}$$

Second derivative of a scalar function of a vector with respect to a vector :

Let $f(\boldsymbol{x})$ be a scalar function of a $p \times 1$ vector $\boldsymbol{x}$. The derivative of $f(\boldsymbol{x})$ w.r.t. $\boldsymbol{x}$ is a $p \times 1$ vector

$$\frac{df(\boldsymbol{x})}{d\boldsymbol{x}} = \left(\frac{df(\boldsymbol{x})}{dx_i},\ i = 1, \cdots, p\right)'.$$

The second derivative of $f(\boldsymbol{x})$ w.r.t. $\boldsymbol{x}$ is obtained by differentiating each element of $\frac{df(\boldsymbol{x})}{d\boldsymbol{x}}$ w.r.t. the row vector $\boldsymbol{x}'$.

Thus

$$\frac{d^2 f(\boldsymbol{x})}{d\boldsymbol{x}'d\boldsymbol{x}} = \frac{d}{d\boldsymbol{x}'}\left(\frac{df(\boldsymbol{x})}{d\boldsymbol{x}}\right) = \left[\frac{d}{dx_1}\left(\frac{df(\boldsymbol{x})}{d\boldsymbol{x}}\right), \cdots, \frac{d}{dx_p}\left(\frac{df(\boldsymbol{x})}{d\boldsymbol{x}}\right)\right]. \tag{0.4.65}$$

It can be easily checked that if $f(\boldsymbol{x}) = \boldsymbol{x}'\boldsymbol{x}$, then

$$\frac{d^2 f(\boldsymbol{x})}{d\boldsymbol{x}'d\boldsymbol{x}} = 2I_p.$$

If $\boldsymbol{x}$ is $p \times 1$ and $\boldsymbol{A}$ is $p \times p$ then

$$\frac{d^2(\boldsymbol{x}'\boldsymbol{A}\boldsymbol{x})}{d\boldsymbol{x}'d\boldsymbol{x}} = \boldsymbol{A}. \tag{0.4.66}$$

The symmetric matrix

$$\frac{d^2 f(\boldsymbol{x})}{d\boldsymbol{x}'d\boldsymbol{x}} = \boldsymbol{H} = ((h_{ij})) \quad \text{(say)} \tag{0.4.67}$$

is defined as the *Hessian matrix* of the scalar function $f(\boldsymbol{x})$.

If $\left.\frac{df(\boldsymbol{x})}{d\boldsymbol{x}}\right]_{\boldsymbol{x}_0} = \boldsymbol{0}$ then $\boldsymbol{x}_0$ is a stationary point of $f(\boldsymbol{x})$.

If in addition $\boldsymbol{H} > (<) 0$ for all $\boldsymbol{x}$, then $\boldsymbol{x}_0$ corresponds to a global minimum naximum) of $f(\boldsymbol{x})$. If $\boldsymbol{H} > 0(< 0)$ at least in a neighbourhood of the point $\boldsymbol{x}_0$, then $\boldsymbol{x}_0$ orresponds to local minimum (maximum).

Functions $f(\boldsymbol{x})$ for which $\boldsymbol{H} > 0(< 0)$ are called convex (concave).

In general, $f(\boldsymbol{x})$ is convex (concave) if for any two points $\boldsymbol{x}_1$ and $\boldsymbol{x}_2$ and for any λ, $0 < \lambda < 1$, $f[\lambda \boldsymbol{x}_1 + (1-\lambda)\boldsymbol{x}_2] \leq (\geq) \lambda f(\boldsymbol{x}_1) + (1-\lambda) f(\boldsymbol{x}_2)$.

.4.12 System of Equations

onsider a set of m linear non-homogeneous equations in n unknowns, $x_1, \cdots, x_n$:

$$\boldsymbol{A}\boldsymbol{x} = \boldsymbol{b} \tag{0.4.68}$$

here $\boldsymbol{A} = ((a_{ij}, i = 1, \cdots, m; j = 1, \cdots, n))$, $\boldsymbol{x} = (x_1, \cdots, x_n)'$, and $\boldsymbol{b} = (b_1, \cdots, b_m)'$.

The equations are consistent (i.e., at least one set of values of $\boldsymbol{x}$ exists that satisfy ll the equations) iff $r = r'$, where

$$r = r(\boldsymbol{A}), \quad r' = r(\boldsymbol{B}), \quad \text{and} \quad \boldsymbol{B} = (\boldsymbol{A}, \boldsymbol{b}).$$

If $r(\boldsymbol{A}) = r = r' < n$, there exists an infinite number of solutions of the equation 0.4.68). If $n = r = r'$, the equation (0.4.68) have a unique solution.

Consider the set of m homogeneous equations

$$\boldsymbol{A}\boldsymbol{x} = \boldsymbol{0}. \tag{0.4.69}$$

Iere $r = r'$ always and hence the equations are always consistent. When $r = n$, the nly solution of (0.4.69) is $\boldsymbol{x} = \boldsymbol{0}$. When $r = m < n$ we reduce the solution of (0.4.69) o solution of

$$\sum_{t=1}^{r} a_{it} x_t = - \sum_{t=r+1}^{n} a_{it} x_t \quad (i = 1, \cdots, r) \tag{0.4.70}$$

uch that the determinant of the coefficients on the left-hand side are not zero. In this ase, all solution of (0.4.69) are given by assigning arbitrary values to $x_{r+1}, \cdots, x_n$ and hen solving (0.4.70) for $x_1, \cdots, x_r$.

Let $\boldsymbol{X}_K$ be the solution obtained by putting $(x_{r+1} = 0, \cdots, x_{r+K-1} = 0, x_{r+K} = 1,$ $\cdots, x_n = 0)$, $K = 1, \cdots, n-r$.

The general solution $\boldsymbol{X}$ of (0.4.69) is given by

$$\boldsymbol{X} = \sum_{K=1}^{n-r} x_{r+K} \boldsymbol{X}_K \tag{0.4.71}$$

vhere $x_{r+1}, \cdots, x_n$ are given arbitrary values. Solutions $\boldsymbol{X}_1, \cdots, \boldsymbol{X}_{n-r}$, are all linearly ndependent (i.e., no member of it can be expressed as a sum of multiples of the others). Such a set of solutions $\boldsymbol{X}_1, \cdots, \boldsymbol{X}_{n-r}$ is called a fundamental set. The rank of the matrix vhose columns form a fundamental set of solutions is $(n-r)$.

0.4.13 Orthogonal Projection of a Vector

Let $\boldsymbol{\alpha} \notin \mathcal{M}$, where $\mathcal{M}$ is a finite-dimensional manifold and $\boldsymbol{\beta} \in \mathcal{M}$. Then there exis an $\boldsymbol{\gamma}$ which is perpendicular to $\mathcal{M}$ and is such that $\boldsymbol{\alpha} = \boldsymbol{\beta} + \boldsymbol{\gamma}$. Furthermore, th representation is unique. $\boldsymbol{\beta}$ is called the projection of $\boldsymbol{\alpha}$ on $\mathcal{M}$ and $\boldsymbol{\gamma}$ is called th orthogonal or perpendicular projection of $\boldsymbol{\alpha}$ on $\mathcal{M}$. The perpendicular projection $\boldsymbol{\gamma}$ ha the property

$$\inf_{\boldsymbol{x} \in \mathcal{M}} ||\boldsymbol{\alpha} - \boldsymbol{x}|| = ||\boldsymbol{\gamma}||$$

where $||\boldsymbol{\delta}||$ denotes the length of the vector $\boldsymbol{\delta} = (\delta_1, \delta_2, \cdots, \delta_n)$, $||\boldsymbol{\delta}|| = \sqrt{\Sigma \delta_k^2}$.

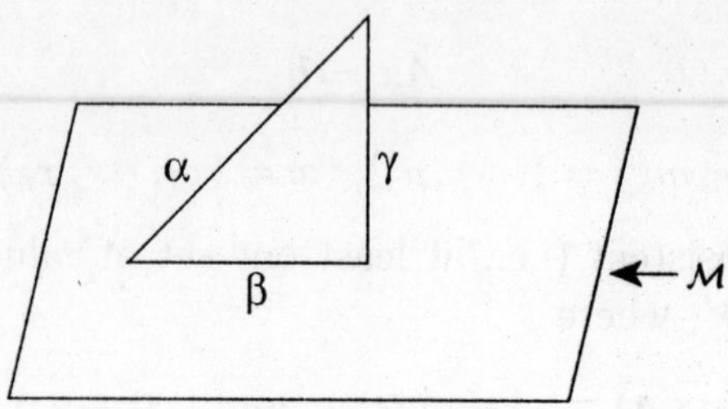

Fig. 0.4

Projective Transformation

Let $\boldsymbol{X} (\neq \boldsymbol{0})$ be a vector which does not belong to $\mathcal{M}$. Let the projective transformatio of $\boldsymbol{X}$ on $\mathcal{M}$ be $\boldsymbol{Y}$ so that

$$\boldsymbol{Y} = \boldsymbol{P}\boldsymbol{X}. \tag{0.4.72}$$

Since the projection of $\boldsymbol{Y}$ (which belongs to $\mathcal{M}$) on $\mathcal{M}$ is $\boldsymbol{Y}$ itself,

$$\boldsymbol{Y} = \boldsymbol{P}\boldsymbol{Y} = \boldsymbol{P}(\boldsymbol{P}\boldsymbol{X}) = \boldsymbol{P}^2\boldsymbol{X}. \tag{0.4.73}$$

Comparing (0.4.72) and (0.4.73),

$$\boldsymbol{0} = (\boldsymbol{P} - \boldsymbol{P}^2)\boldsymbol{X} \quad \Rightarrow \quad \boldsymbol{0} = \boldsymbol{P} - \boldsymbol{P}^2 \quad \Rightarrow \quad \boldsymbol{P}^2 = \boldsymbol{P}.$$

Hence $\boldsymbol{P}$ must be an idempotant matrix.

Let us write

$$\boldsymbol{X} = \boldsymbol{Y} + \text{ a vector perpendicular to } \mathcal{M}$$
$$\Rightarrow \boldsymbol{X} - \boldsymbol{Y} = \text{a vector perpendicular to } \mathcal{M}$$
$$\Rightarrow (\boldsymbol{X} - \boldsymbol{Y}) \text{ is perpendicular to } \boldsymbol{Y} (\because \ \boldsymbol{Y} \in \mathcal{M})$$
$$\Rightarrow (\boldsymbol{X} - \boldsymbol{Y})'\boldsymbol{Y} = 0$$
$$\Rightarrow (\boldsymbol{X} - \boldsymbol{P}\boldsymbol{X})'\boldsymbol{Y} = 0 \quad \Rightarrow \quad [(I - \boldsymbol{P})\boldsymbol{X}]'\boldsymbol{P}\boldsymbol{X} = 0$$
$$\Rightarrow \boldsymbol{X}'(I - \boldsymbol{P})'\boldsymbol{P}\boldsymbol{X} = 0 \quad \Rightarrow \quad \boldsymbol{X}'\boldsymbol{P}\boldsymbol{X} - \boldsymbol{X}'\boldsymbol{P}'\boldsymbol{P}\boldsymbol{X} = 0$$
$$\Rightarrow \boldsymbol{X}'(\boldsymbol{P} - \boldsymbol{P}'\boldsymbol{P})\boldsymbol{X} = 0 \quad \Rightarrow \quad \boldsymbol{P} = \boldsymbol{P}'\boldsymbol{P} \tag{0.4.74}$$
$$\Rightarrow \boldsymbol{P}' = \boldsymbol{P}'\boldsymbol{P} \tag{0.4.75}$$

(0.4.74) and (0.4.75) imply that $\boldsymbol{P} = \boldsymbol{P}'$.

Hence $\boldsymbol{P}$ is symmetric and idempotent.

Conversely, every symmetric and idempotent matrix is a projection matrix. This can e seen as follows. Let $\boldsymbol{P}$ be any symmetric and idempotent matrix. If

$$\boldsymbol{Y} = \boldsymbol{P}\boldsymbol{X} \tag{0.4.76}$$

ıen

$$\boldsymbol{Y} = \boldsymbol{P}^2\boldsymbol{X} = \boldsymbol{P}(\boldsymbol{P}\boldsymbol{X}) = \boldsymbol{P}\boldsymbol{Y} \tag{0.4.77}$$

).4.76), (0.4.77) $\Rightarrow$

$$\boldsymbol{0} = \boldsymbol{P}(\boldsymbol{Y} - \boldsymbol{X}) \quad \Rightarrow \quad \boldsymbol{0} = \boldsymbol{P}'(\boldsymbol{X} - \boldsymbol{Y})$$

$\Rightarrow$ $(\boldsymbol{X} - \boldsymbol{Y})$ is perpendicular to the linear manifold generated by the column vector f $\boldsymbol{P}$.

$\Rightarrow$ $\boldsymbol{X} = \boldsymbol{Y} +$ a vector perpendicular to the vector space spanned by the columns f $\boldsymbol{P}$

$\Rightarrow$ $\boldsymbol{Y}$ is a projection of $\boldsymbol{X}$ on the vector space spanned by the columns of $\boldsymbol{P}$

$\Rightarrow$ $\boldsymbol{Y} = \boldsymbol{P}\boldsymbol{X}$

$\Rightarrow$ $\boldsymbol{P}$ is a projection matrix.

Note that if $\mathcal{M} = C(\boldsymbol{X})$, column space of $\boldsymbol{X}$, then $\boldsymbol{P}_{\mathcal{M}} = \boldsymbol{X}(\boldsymbol{X}'\boldsymbol{X})^{-1}\boldsymbol{X}'$ is the rthogonal projection matrix onto $\mathcal{M}$ and $\boldsymbol{P}_{\mathcal{M}}\boldsymbol{\alpha}$ is the projection of $\boldsymbol{\alpha}$ onto $\mathcal{M}$.

Exercise

1. (a) Verify the following relations for sets A, B, C.

   (i) $(A \cup B \cup C)' = A'B'C'$;

   (ii) $A' \cup B' \cup C' = (ABC)'$;

   (iii) $(A \cup B) \cap C = AC \cup BC$;

   (iv) $(A \cap C) \cap (B \cap C) = ABC$;

   (v) $(A \cup B) - (A \cap B) = (A - B) \cup (B - A)$;

   (vi) $A \cap (B - C) = (A \cap B) - (A \cap C)$;

   (vii) $(A \cup B)' \cap C = C - C \cap (A \cup B)$;

   (viii) $(A \cup B) - C \subset A \cup (B - C)$;

   (ix) $A \cap (B \cup C) \subset (A \cap B) \cup C$.

   (b) Verify the correctness or otherwise of the statements:

   (i) $(A \cap B) - C = A \cap B \cap C$; [incorrect]

   (ii) $(A \cup B) - C = (A - C) \cup (B - C)$; [correct]

(iii) $(A \cup B)'C = (A' \cap C) \cup (B' \cap C)$; [incorrec

(iv) $(A \cap B) \cap (A' \cap C) = \phi$; [correc

(v) $(A - B) \cup B = A$. [incorrec

2. Show that the Vandermonde determinant defined by $\begin{bmatrix} 1 & 1 & \cdots & 1 \\ a_1 & a_2 & \cdots & a_n \\ \cdots & \cdots & \cdots & \cdots \\ a_1^{n-1} & a_2^{n-1} & \cdots & a_n^{n-1} \end{bmatrix}$ ha the value $\prod_{i<j}(a_i - a_j)$.

3. If $\boldsymbol{A}, \boldsymbol{B}, \boldsymbol{C}$ are any three matrices such that the $\boldsymbol{AB}, \boldsymbol{BC}$ and $\boldsymbol{ABC}$ are defined Show that

$$\text{rank } (\boldsymbol{AB}) + \text{rank } (\boldsymbol{BC}) \leq \text{rank } (\boldsymbol{B}) + \text{rank } (\boldsymbol{ABC}).$$

4. Define $\boldsymbol{A} = 8\begin{bmatrix} 1 & \frac{1}{4} & \frac{1}{2} \\ \frac{1}{4} & 1 & \frac{1}{3} \\ \frac{1}{2} & \frac{1}{3} & 1 \end{bmatrix}$. Find latent roots and all sets of normalised laten vectors of $\boldsymbol{A}$. Is $\boldsymbol{A}$ singular? What are the latent roots of $\boldsymbol{A}^2$?

5. (a) Find three nonsingular 3×3 matrices such that their sum is singular.

(b) Show that if $\boldsymbol{A}^2 + 3\boldsymbol{A} + I = \boldsymbol{0}$ then $\boldsymbol{A}^{-1} = -\boldsymbol{A} - 3I$.

(c) Let $\boldsymbol{A}$ and $\boldsymbol{B}$ be $n \times n$ matrices such that $\boldsymbol{AB} = \boldsymbol{0}$. Show that if $\boldsymbol{B} \neq 0$ ther $\boldsymbol{A}$ is singular.

6. Let $\boldsymbol{Ax} = \boldsymbol{b}$ be a system of linear equations, where $\boldsymbol{A}$ is a $n \times n$ matrix.

(a) Suppose $\boldsymbol{x} = \boldsymbol{0}$ is the only solution to the system when $\boldsymbol{b} = \boldsymbol{0}$. How many solutions, if any, does the system have when $\boldsymbol{b}' = (1, 2, \cdots, n)$? Justify.

(b) Suppose $\boldsymbol{x} = \boldsymbol{0}$ is not the only solution of the system when $\boldsymbol{b} = \boldsymbol{0}$. Does the system have precisely one solution when $\boldsymbol{b}' = (1, 2, \cdots, n)$? Justify.

7. (a) Determine the nullspace of the following matrix.

$$\begin{bmatrix} 1 & 1 & -1 & 2 \\ 2 & 2 & -3 & 1 \\ -1 & -1 & 0 & -5 \end{bmatrix}.$$

(b) Let S be the set of all ordered pairs of real numbers. Define scalar multiplication and addition on S by

$$\alpha(x_1, x_2) = (\alpha x_1, \alpha x_2)$$
$$(x_1, x_2) + (y_1, y_2) = (x_1 + y_1, 0).$$

Is S a vector space? Justify.

8. Let $\boldsymbol{A}$ and $\boldsymbol{B}$ be symmetric real positive definite matrices of order n and let $\boldsymbol{A} - \boldsymbol{B}$ be positive. Show that $\boldsymbol{B}^{-1} - \boldsymbol{A}^{-1}$ is positive definite.

9. Show that the following matrix identities are true:

   (a) $(\boldsymbol{D} + \boldsymbol{AB})^{-1} = \boldsymbol{D}^{-1} - \boldsymbol{D}^{-1}\boldsymbol{A}(\boldsymbol{I} + \boldsymbol{BD}^{-1}\boldsymbol{A})^{-1}\boldsymbol{BD}^{-1}$;
   (b) $(\boldsymbol{D} + \boldsymbol{EFE}')^{-1} = \boldsymbol{D}^{-1} - \boldsymbol{D}^{-1}\boldsymbol{E}(\boldsymbol{E}'\boldsymbol{D}^{-1}\boldsymbol{E} + \boldsymbol{F}^{-1})^{-1}\boldsymbol{E}'\boldsymbol{D}^{-1}$;
   (c) $(\boldsymbol{D} + \boldsymbol{B})^{-1} = \boldsymbol{D}^{-1} - \boldsymbol{D}^{-1}(\boldsymbol{D}^{-1} + \boldsymbol{B}^{-1})^{-1}\boldsymbol{D}^{-1}$;
   (d) $(\boldsymbol{D} + \boldsymbol{B})^{-1}\boldsymbol{B} = \boldsymbol{I} - (\boldsymbol{D} + \boldsymbol{B})^{-1}\boldsymbol{D}$;

   where it is assumed that all matrices are of appropriate (conformable) dimensions and that all the stated inverses exist.

10. Let $\boldsymbol{x}, \boldsymbol{a}, \boldsymbol{b}$ be $k \times 1$ vectors, and $\boldsymbol{A}, \boldsymbol{B}$ be $k \times k$ symmetric matrices such that $(\boldsymbol{A} + \boldsymbol{B})^{-1}$ exists. Show that

$$(\boldsymbol{x}-\boldsymbol{a})'\boldsymbol{A}(\boldsymbol{x}-\boldsymbol{a})+(\boldsymbol{x}-\boldsymbol{b})'\boldsymbol{B}(\boldsymbol{x}-\boldsymbol{b}) = (\boldsymbol{x}-\boldsymbol{c})'(\boldsymbol{A}+\boldsymbol{B})(\boldsymbol{x}-\boldsymbol{c})+(\boldsymbol{a}-\boldsymbol{b})'\boldsymbol{A}(\boldsymbol{A}+\boldsymbol{B})^{-1}\boldsymbol{B}(\boldsymbol{a}-\boldsymbol{b})$$

   where $\mathbf{c} = (\boldsymbol{A} + \boldsymbol{B})^{-1}(\boldsymbol{Aa} + \boldsymbol{Bb})$.

11. Let $\boldsymbol{V}$ be a $n \times n$ positive definite symmetric matrix with distinct characteristic roots $\lambda_1, \cdots, \lambda_r$ having multiplicities $m_1, \cdots, m_r$ respectively,

$$\sum_{i=1}^{r} m_i = n.$$

   Then show that $\boldsymbol{V}$ can be expressed as

$$\sum_{i=1}^{r} \lambda_i \boldsymbol{A}_i$$

   where $\boldsymbol{A}_i$ are mutually orthogonal symmetric idempotent matrices satisfying $\boldsymbol{VA}_i = \lambda_i \boldsymbol{A}_i$, $i = 1, \cdots, r$. Show also that the matrix $\boldsymbol{V}^{-1/2}$ defined by

$$\boldsymbol{V}^{-1/2} = \sum_{i=1}^{r} \lambda_i^{-1/2} \boldsymbol{A}_i.$$

   satisfies

$$\boldsymbol{V}^{-1/2}\boldsymbol{V}\boldsymbol{V}^{-1/2} = \boldsymbol{I}_n.$$

   [Fuller and Battese, 1973]

12. Let $\lambda_1(\boldsymbol{C}) \geq \lambda_2(\boldsymbol{C}) \geq \cdots \geq \lambda_n(\boldsymbol{C})$ be the eigenvalues of a $n \times n$ symmetric matrix $\boldsymbol{C}$. Show that for $n \times n$ symmetric matrices $\boldsymbol{A}$ and $\boldsymbol{B}$

$$\lambda_i(\boldsymbol{A}) + \lambda_n(\boldsymbol{B}) \leq \lambda_i(\boldsymbol{A} + \boldsymbol{B}) \leq \lambda_i(\boldsymbol{A}) + \lambda_i(\boldsymbol{B}),$$
$$\lambda_n(\boldsymbol{B})\lambda_i(\boldsymbol{A}^2) \leq \lambda_i(\boldsymbol{ABA}) \leq \lambda_1(\boldsymbol{B})\lambda_i(\boldsymbol{A}^2), \quad i = 1, \cdots, n.$$

   [Wang et al, 1990]

13. Let $\lambda_1, \lambda_2, \cdots, \lambda_n$ be the eigenvalues of a symmetric matrix $\boldsymbol{A}$ such that $|\lambda_1| > |\lambda_2| > \cdots > |\lambda_n|$. Show that for large k,

$$\lim_{k\to\infty}\left(\frac{\boldsymbol{A}^k}{\lambda_1^k}\right) = \boldsymbol{X}_1\boldsymbol{X}_1'; \quad \lim_{k\to\infty}\left(\frac{\boldsymbol{A}-\lambda_1\boldsymbol{X}_1\boldsymbol{X}_1'}{\lambda_2}\right)^k = \boldsymbol{X}_2\boldsymbol{X}_2'.$$

$$\cdots\cdots\cdots\cdots\cdots\cdots\cdots\cdots$$

$$\lim_{k\to\infty}\left(\frac{\boldsymbol{A}-\lambda_1\boldsymbol{X}_1\boldsymbol{X}_1'-\lambda_2\boldsymbol{X}_2\boldsymbol{X}_2'-\cdots-\lambda_{n-1}\boldsymbol{X}_{n-1}\boldsymbol{X}_{n-1}'}{\lambda_n}\right)^k = \boldsymbol{X}_n\boldsymbol{X}_n'.$$

$$\lambda_1 \simeq \left(\frac{\boldsymbol{A}^k}{\boldsymbol{X}_1\boldsymbol{X}_1'}\right)^{1/k}$$

$$\lambda_2 \simeq \left[\frac{(\boldsymbol{A}-\lambda_1\boldsymbol{X}_1\boldsymbol{X}_1')^k}{\boldsymbol{X}_2\boldsymbol{X}_2'}\right]^{1/k}$$

$$\cdots\cdots\cdots\cdots\cdots\cdots\cdots\cdots$$

$$\lambda_n \simeq \left[\frac{(\boldsymbol{A}-\lambda_1\boldsymbol{X}_1\boldsymbol{X}_1'-\lambda_2\boldsymbol{X}_2\boldsymbol{X}_2'-\cdots-\lambda_{n-1}\boldsymbol{X}_{n-1}\boldsymbol{X}_{n-1}')^k}{\boldsymbol{X}_n\boldsymbol{X}_n'}\right]^{1/k}$$

where $\boldsymbol{X}_i$ is the characteristic vector corresponding to λ_i $(i = 1, \cdots, n)$.

14. If rank $(\boldsymbol{A}) = r < n$ for a real symmetric matrix $\boldsymbol{A}(n \times n)$ with distinct non-zero characteristic roots λ_i $(i = 1, \cdots, r)$ and $\boldsymbol{X}_i$ is the chracteristic vector corresponding to λ_i, then show that

$$\boldsymbol{A}^* = \sum_{i=1}^{r} \lambda_i^{-1}\boldsymbol{X}_i\boldsymbol{X}_i'$$

is a g-inverse of $\boldsymbol{A}$.

15. Suppose that $\boldsymbol{\Sigma}$ is positive definite so that $\boldsymbol{\Sigma}^{-1}$ exists. Show that if $(\lambda, \boldsymbol{e})$ is an eigenvalue-eigenvector pair of $\boldsymbol{\Sigma}$, $\left(\frac{1}{\lambda}, \boldsymbol{e}\right)$ is the corresponding eigenvalue-eigenvector pair of $\boldsymbol{\Sigma}^{-1}$. Also prove that $\boldsymbol{\Sigma}^{-1}$ is positive definite.

16. Consider an arbitrary $n \times p$ matrix $\boldsymbol{A}$. Then $\boldsymbol{A}'\boldsymbol{A}$ is a symmetric $p \times p$ matrix. Show that $\boldsymbol{A}'\boldsymbol{A}$ is necessarily nonnegative definite.
[*Hints:* Put $\boldsymbol{y} = \boldsymbol{A}\boldsymbol{x}$.]

17. Let $(\lambda_i, \boldsymbol{e}_i)$ $(i = 1, \cdots, m)$ be the eigenvalue-eigenvector pairs of a matrix $\boldsymbol{A}(m\times m)$. Show that

$$\sum_{i=1}^{m} \boldsymbol{e}_i\boldsymbol{e}_i' = \boldsymbol{I}_m.$$

Chapter 1

Elements of Theory of Probability

1.1 Introduction

We shall first consider the classical concept of theory of probability. For this we examine a few definitions.

Event : An event is the result of an experiment. In throwing a die, the results—face 5 turns up, an even number turns up are both events. Events may be of two types:

(a) **Elementary Event** (or **Case**) **:** It is an event which cannot be decomposed further into smaller events. It is an 'atom' of an event. Thus, in throwing a die, each of the events $1, \cdots, 6$ is an elementary event. Symbolically, an elementary event will often be represented by lower case letters a, b, etc.

(b) **Compound (contingent) Event :** A compound (contingent) event is obtained through the combination of several elementary events. Thus, in throwing a die, the event that an odd number turns up is a combination of three elementary events: 1 or 3 or 5 turns up. This event occurs if any of these three cases takes place. Symbolically, a compound event will often be represented by upper case letters A, B, etc.

We shall often use simply the word event, whether it is elementary or compound would be clear from the context.

Mutually Exclusive Events : Events are said to be mutually exclusive (or incompatible) with respect to an experiment if the occurrence of one of them precludes the occurrence of all the others every time the experiment is performed; in other words, the two events cannot materialize simultaneously. Thus in throwing a die, the events, '2 occurs' and '5 occurs' are mutually exclusive.

Exhaustive Events : A set of events in relation to a random experiment is said to be exhaustive if one of them must necessarily materialize everytime the experiment is performed. Thus in throwing a die cases $1, \cdots, 6$ form an exhaustive set of cases.

Equally likely Cases : Cases are said to be equally likely when we have no reason to believe that one is more likely to occur than the other. Thus in drawing a card at random from a full pack of well-shuffled cards, one may believe that each of the cards is equally likely to appear and in this case, each of the 52 cases is equally likely.

Favourable Cases: A case 'a' is said to be favourable to an event A if whenever 'a' occurs, 'A' occurs. Thus in throwing a die, each of the cases 1, 3, 5 is favourable to the event, 'an odd face turns up'.

▸ **Example 1.1.1** *A head and a tail being denoted by H and T respectively, the cases HH, HT, TH, TT are mutually exclusive, exhaustive and equally likely cases for the random experiment of throwing a fair coin twice.*

▸ **Example 1.1.2** *In throwing two balanced dice, mutually exclusive, exhaustive and equally likely cases are 36 in number, namely,* $(1,1)$, $(1,2)$, $\cdots$, (1.6), $(2,1)$, $\cdots$, $(2,6)$, $\cdots$, $(6,1)$, $\cdots$, $(6,6)$, *where in* (x,y), x,y *denote the result of first and second throw respectively.*

In the probability theory an elementary event or a case resulting from a random experiment is represented as a point in a set. A compound event A is represented as a set containing all the points which stand for the elementary events of which A is made up (i.e., which are favourable to A). Thus in throwing a die, each of the cases $1, \cdots, 6$ is a point in the set. The compound event, an odd face turns up is represented by the set A (Fig. 1.1). Two events A and B are mutually exclusive if the sets representing them are disjoint.

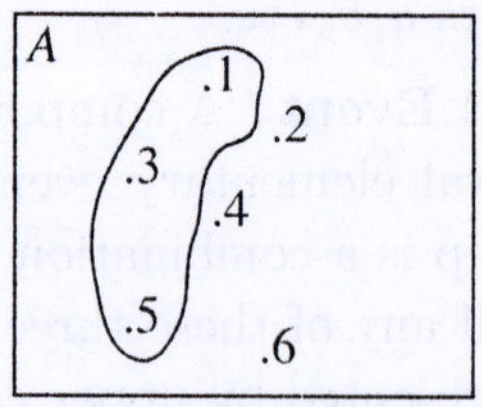

Fig. 1.1

The set containing all the possible points representing the elementary events of a random experiment is called the 'sample space'. It is often represented as S. Thus, in tossing a coin once, $S = \{H, T\}$. In tossing a coin twice,

$$S = \{HH, HT, TH, TT\}.$$

In throwing two dice,

$$S = \{(1,1), (1,2), \cdots, (6,6)\}.$$

1.2 The Classical Definition of Probability

We assume that the total number of elementary events N in the sample space of the experiment is finite. We also assume that all the elementary events are equally likely to occur.

If an experiment can result in N mutually exclusive, exhaustive and equally likely cases of which $n(A)$ are favourable to an event A then the probability of occurrence of A is

$$P(A) = \frac{n(A)}{N}. \tag{1.2.1}$$

▶ **Example 1.2.1** *Two unbiased coins are tossed. What is the probability of getting at least one head?*

The possible cases are HH, HT, TH, TT. All these cases are mutually exclusive and equally likely. Thus $N = 4$. The event A (at least one head) occurs if either of HH, HT, TH occurs.

Thus

$$n(A) = 3.$$

Hence

$$P(A) = \frac{3}{4}.$$

The following properties immediately follow from equation (1.2.1).

(a) Since $0 \leq n(A) \leq N$, probability $P(A)$ of an event A satisfies $0 \leq P(A) \leq 1$. If $P(A) = 0$ then $n(A) = 0$, i.e., there is no favourable case for A and hence A is impossible under the given circumstances. If $P(A) = 1$, $n(A) = N$, i.e., all the cases are favourable to A and hence A is sure to occur.

(b) The number of cases which are not favourable for A is $N - n(A)$. Hence probability of non-occurrence of A is $1 - P(A)$.

(c) If occurrence of an event A implies occurrence of an event B, then the cases favourable for A form a subset of the cases favourable for B; hence $n(A) \leq n(B)$. Thus

$$P(A) \leq P(B).$$

The classical definition of probability, however, suffers from some serious disadvantages. The statistical or empirical definition of probability takes care of these drawbacks to some extent (vide Section 1.7).

We now consider union and intersection of events and complement of an event.

Union: Let A, B be two events. The union, $A \cup B$ represents the event in which at least one of A and B occurs. Similarly the union of m events $A_1, \cdots, A_m$, namely, $\bigcup_1^m A_i$ represents the event that at least one of $A_1, \cdots, A_m$ occurs. $\bigcup_{i=1}^m A_i$ is also called a total event.

Intersection : The intersection of two events A and B, denoted as $A \cap B$ or simply AB denotes the event that both A and B occur. Similarly the intersection of m events $A_1, \cdots, A_m$, denoted as $\bigcap_{i=1}^{m} A_i$ or simply $A_1 A_2 \cdots A_m$ represents the event that all of $A_1, \cdots, A_m$ occur. If the sets $A_1, \cdots, A_m$ are mutually disjoint, $A_i \cap A_j = \phi$ for all $i \neq j = 1, \cdots, m$.

Complementation : The complement of an event A, denoted as A^c or A' or $\bar{A}$ is the event that A does not occur.

1.3 Theorems on Probability of Union of Events

We now consider some results which give probability of union of certain events $A_1, \cdots, A_n$ in terms of probabilities of events $A_1, \cdots, A_n$, probabilities of intersection of events, $A_1A_2, A_1A_3, \cdots$, etc.

Theorem 1.3.1 *If $A_1, \cdots, A_n$ are all mutually exclusive then probability of union of events $A_1, \cdots, A_n$ is*

$$P\left(\bigcup_{i=1}^{n}\right) = P(A_1) + P(A_2) + \cdots + P(A_n). \tag{1.3.1}$$

Proof: Let M be the total number of mutually exclusive and equally likely cases in the sample space of which m_i are favourable for the event A_i, $i = 1, \cdots, n$. Since $A_1, \cdots, A_n$ are mutually exclusive events, the sets representing them are mutually disjoint. Hence the total number of cases favourable for the event

$$\bigcup_{i=1}^{n} A_i \left(= \sum_{i=1}^{n} A_i \right) \text{ is } \sum_{i=1}^{n} m_i.$$

Thus

$$P\left(\bigcup_{i=1}^{n} A_i\right) = \frac{m_1 + \cdots + m_n}{M} = \frac{m_1}{M} + \cdots + \frac{m_n}{M} = P(A_1) + \cdots + P(A_n).$$

Corollary 1.3.1 *If an event A consists of n mutually exclusive forms $A_1, \cdots, A_n$ so that A happens whenever any of these occurs and vice-versa, then*

$$A = A_1 + \cdots + A_n \quad \textit{and} \quad P(A) = P(A_1) + \cdots + P(A_n).$$

Corollary 1.3.2 *Let the event B imply A. Then the set B is a subset of A. Now A can be decomposed into two disjoint subsets B and $A - B$.*

Hence

$$P(A) = P(B) + P(A - B).$$

Thus

$$P(A - B) = P(A) - P(B), \; B \subset A. \tag{1.3.2}$$

▶ **Example 1.3.1** 4 *cards are drawn randomly from a well-shuffled pack of* 52 *cards. What is the probability all are black or red.*

Let A_1, A_2 be the events, all the cards are black, red respectively. The two events are mutually exclusive. Here

$$P(A_1) = P(A_2) = \binom{26}{13} \Big/ \binom{52}{13} = a \text{ (say)}.$$

Hence

$$P(A_1 + A_2) = 2a.$$

We now consider the case when $A_1, \cdots, A_n$ are not necessarily mutually exclusive events. For simplicity we introduce the following notations. Let

$$\sum_{i=1}^{n} P(A_i) = S_1, \quad \sum_{i<j=1}^{n}\sum P(A_iA_j) = S_2, \quad \sum_{i<j<k=1}^{n}\sum\sum P(A_iA_jA_k) = S_3$$

....

$$P(A_1A_2 \cdots A_n) = S_n.$$

Thus S_1 is the sum of n terms each of probability of an event A_i; S_2 is the sum of $\binom{n}{2}$ terms each of probability of joint occurrence of a pair A_i, A_j; S_3 is the sum of $\binom{n}{3}$ terms each of probability of joint occurrence of a triplet of events A_i, A_j, A_k etc. We have

Theorem 1.3.2 *Let* $A_1, \cdots, A_n$ *be events not necessarily mutually exclusive. The probability that at least one of* $A_1, \cdots, A_n$ *occurs is*

$$P\left(\bigcup_{i=1}^{n} A_i\right) = S_1 - S_2 + S_3 - \cdots + (-1)^{n-1}S_n. \tag{1.3.3}$$

Proof: (By method of induction). First consider the events A_1, A_2.

Now

$$A_1 \cup A_2 = A_1 + (A_2 - A_1 \cap A_2).$$

Hence by theorem 1.3.1,

$$P(A_1 \cup A_2) = P(A_1) + P(A_2 - A_1 \cap A_2).$$

Since $A_1 \cap A_2 \subset A_2$, by corollary 1.3.2

$$P(A_1 \cup A_2) = P(A_1) + P(A_2) - P(A_1 \cap A_2). \tag{1.3.4}$$

Thus the theorem is true for $n = 2$. Assume (1.3.3) holds for $n = t (\geq 2)$.
Thus

$$P\left(\bigcup_{i=1}^{t} A_i\right) = \sum_{i=1}^{t} P(A_i) - \sum\sum_{i<j=1}^{t} P(A_iA_j) + \sum\sum\sum_{i<j<k=1}^{t} P(A_iA_jA_k) - \cdots$$
$$+(-1)^{t-1}P(A_1 \cdots A_t). \tag{1.3.5}$$

Now

$$\bigcup_{i=1}^{t+1} A_i = \bigcup_{i=1}^{t} A_i + \left(A_{t+1} - \left(\bigcup_{1}^{t} A_i\right) \cap A_{t+1}\right) = \bigcup_{i=1}^{t} A_i + \left(A_{t+1} - \bigcup_{1}^{t} (A_i \cap A_{t+1})\right).$$

Hence

$$P\left(\bigcup_{i=1}^{t+1} A_i\right) = P\left(\bigcup_{i=1}^{t} A_i\right) + P\left(A_{t+1} - \bigcup_{1}^{t} (A_i \cap A_{t+1})\right)$$
$$= P\left(\bigcup_{i=1}^{t} A_i\right) + P(A_{t+1}) - P\left(\bigcup_{1}^{t} (A_i \cap A_{t+1})\right). \tag{1.3.6}$$

Since, by assumption, the theorem holds for t sets,

$$P\left(\bigcup_{1}^{t} (A_i \cap A_{t+1})\right) = \sum_{1}^{t} P(A_i \cap A_{t+1}) - \sum\sum_{i<j=1}^{t} P\{(A_i \cap A_{t+1})(A_j \cap A_{t+1})\}$$
$$+ \sum\sum\sum_{i<j<k=1}^{t} P\{(A_i \cap A_{t+1})(A_j \cap A_{t+1})(A_k \cap A_{t+1})\} - \cdots$$
$$+(-1)^{t-1}P\{(A_1 \cap A_{t+1}) \cdots (A_t \cap A_{t+1})\} \tag{1.3.7}$$

Now

$$(A_i \cap A_{t+1})(A_j \cap A_{t+1}) = A_iA_jA_{t+1}$$
$$(A_i \cap A_{t+1})(A_j \cap A_{t+1})(A_k \cap A_{t+1}) = A_iA_jA_kA_{t+1} \text{ etc.}$$

Using in (1.3.6) the expressions in (1.3.5) and (1.3.7), after rearrangement of terms, we get

$$P\left(\bigcup_{i=1}^{t+1} A_i\right) = \sum_{i=1}^{t+1} P(A_i) - \sum\sum_{i<j=1}^{t+1} P(A_iA_j) + \sum\sum\sum_{i<j<k=1}^{t+1} P(A_iA_jA_k) - \cdots$$
$$+(-1)^{t}P(A_1 \cdots A_{t+1}).$$

Hence, if the theorem holds for $n = t$, it holds for $n = t + 1$. The theorem has been shown to hold for $n = 2$. Hence it holds for all integral values of n.

Example 1.3.2 *Find the probability that in a bridge game at least one of the players will get a complete suit of cards.*

Let A_i be the event that player i gets a complete suit of cards.

Thus

$$P\left(\bigcup_{1}^{4} A_i\right) = \sum_{1}^{4} P(A_i) - \sum\sum_{i<j=1}^{4} P(A_iA_j) + \sum\sum\sum_{i<j<k=1}^{4} P(A_iA_jA_k) - P(A_1A_2A_3A_4).$$

Let

$$P(A_i) = \frac{4}{\binom{52}{13}} = p_1, \qquad P(A_iA_j) = \frac{4.3}{\binom{52}{13}\binom{39}{13}} = p_2,$$

$$P(A_iA_jA_k) = \frac{4.3.2}{\binom{52}{13}\binom{39}{13}\binom{26}{13}} = p_3, \; P(A_1A_2A_3A_4) = \frac{4!}{\binom{52}{13}\binom{39}{13}\binom{26}{13}} = p_4.$$

Hence required probability $= 4p_1 - 6p_2 + 4p_3 - p_4$.

Theorem 1.3.3 (***Boole's inequality***). *Given $n(>1)$ events $A_1, \cdots, A_n$*

$$P\left(\bigcup_{i=1}^{n} A_i\right) \le \sum_{i=1}^{n} P(A_i). \tag{1.3.8}$$

Proof: Consider first A_1 and A_2. Now $A_1 \cup A_2 = A_1 + (A_2 - A_1)$, the events A_1, $A_2 - A_1$ being disjoint. Hence

$$P(A_1 \cup A_2) = P(A_1) + P(A_2 - A_1) \le P(A_1) + P(A_2)$$

since $(A_2 - A_1) \subset A_2$. Thus the inequality is proved for $n = 2$. Now

$$\bigcup_{i=1}^{n} A_i = \left(\bigcup_{i=1}^{n-1} A_i\right) \bigcup A_n$$

$$\Rightarrow P\left(\bigcup_{i=1}^{n} A_i\right) \le P\left(\bigcup_{i=1}^{n-1} A_i\right) + P(A_n) \le P\left(\bigcup_{i=1}^{n-2} A_i\right) + P(A_{n-1}) + P(A_n)$$

....

$$\le \sum_{i=1}^{n} P(A_i).$$

Corollary 1.3.3 *Given n events $A_1, \cdots, A_n$*

$$P\left(\bigcap_{i=1}^{n} A_i\right) \ge 1 - \sum_{i=1}^{n} P(A_i^c) = \sum_{i=1}^{n} P(A_i) - (n-1).$$

Proof: Since

$$\bigcap_{i=1}^{n} A_i = \left(\bigcup_{i=1}^{n} A_i^c\right)^c \text{ (by De Morgan's law of sets)}$$

$$P\left(\bigcap_{i=1}^{n} A_i\right) = 1 - P\left(\bigcup_{i=1}^{n} A_i^c\right) \geq 1 - \sum_{i=1}^{n} P(A_i^c)$$

by (1.3.8). The second expression follows by putting $P(A_i^c) = 1 - P(A_i)$.

Theorem 1.3.4 (***Bonferroni's inequality***). *Given* $n(>1)$ *events* $A_1, \cdots, A_n$,

$$\sum_{i=1}^{n} P(A_i) - \sum_{i<j=1}^{n}\sum P(A_iA_j) \leq P\left(\bigcup_{i=1}^{n} A_i\right) \leq \sum_{i=1}^{n} P(A_i). \tag{1.3.9}$$

We only prove the left side. The right side follows by (1.3.8). For $n = 2$.

$$P(A_1) + P(A_2) - P(A_1A_2) = P(A_1 \cup A_2)$$

and thus the inequality is true for $n = 2$ (by equality). For $n = 3$

$$P\left(\bigcup_{i=1}^{3} A_i\right) = \sum_{i=1}^{3} P(A_i) - \sum_{i<j=1}^{3}\sum P(A_iA_j) + P(A_1A_2A_3)$$

and the result holds. Assuming that it holds for m events $(3 < m \leq n-1)$, we show that it holds for $m + 1$

$$\begin{aligned}
P\left(\bigcup_{i=1}^{m+1} A_i\right) &= P\left(\bigcup_{i=1}^{m} A_i \bigcup A_{m+1}\right) \\
&= P\left(\bigcup_{i=1}^{m} A_i\right) + P(A_{m+1}) - P\left\{A_{m+1} \bigcap \left(\bigcup_{i=1}^{m} A_i\right)\right\} \\
&\geq \left(\sum_{i=1}^{m} P(A_i)\right) - \sum_{i<j=1}^{m}\sum P(A_iA_j) + P(A_{m+1}) - P\left(\bigcup_{i=1}^{m}(A_i \cap A_{m+1})\right) \\
&\geq \sum_{i=1}^{m+1} P(A_i) - \sum_{i<j=1}^{m}\sum P(A_iA_j) - \sum_{i=1}^{m} P(A_iA_{m+1}) \\
&\qquad \text{(by Boole's inequalities).} \\
&= \sum_{i=1}^{m+1} P(A_i) - \sum_{i<j=1}^{m+1}\sum P(A_iA_j).
\end{aligned}$$

Theorem 1.3.5 *The probability that exactly* r *of the events* $A_1, \cdots, A_n$ *occur is,* $(1 \leq r \leq n)$,

$$P_{[r]} = S_r - \binom{r+1}{r} S_{r+1} + \binom{r+2}{r} S_{r+2} - \cdots \pm \binom{n}{r} S_n. \tag{1.3.10}$$

Proof: To calculate $P_{[r]}$ we should add the probabilities of all points which occur in exactly r of the sets $A_1, \cdots, A_n$, each point taken only once. Let E be an arbitrary point contained in exactly m of the events $A_1, \cdots, A_n$. We calculate its contribution to the expression in the right-hand side of (1.3.10).

If $m < r, P(E)$ does not contribute. If $m = r$, $P(E)$ occurs only in S_m and its coefficient in the right-hand side of (1.3.10) is 1. For $m > r$, coefficient of $P(E)$ in the right-hand side of (1.3.10) is

$$\binom{m}{r} - \binom{r+1}{r}\binom{m}{r+1} + \binom{r+2}{r}\binom{m}{r+2} \cdots \pm \binom{m}{r}\binom{m}{m} \tag{1.3.11}$$

since $P(E)$ occurs $\binom{m}{r}$ times in S_r, $\binom{m}{r+1}$ times in S_{r+1}, etc.

Now (1.3.11)

$$= \binom{m}{r}\left\{1 - \binom{m-r}{1} + \binom{m-r}{2} - \cdots \pm \binom{m-r}{m-r}\right\} = 0.$$

Hence the proof.

Corollary 1.3.4 *The probability that at least r of the n events $A_1, \cdots, A_n$ occur is*

$$P_{[r]} + P_{[r+1]} + \cdots + P_{[n]}.$$

1.4 Conditional Probability: Theorem of Compound Probability

Suppose in the sample space of a random experiment there are N mutually exclusive, exhaustive and equally likely cases. Of these, N_A are favourable to event A. Out of these N_A cases N_{AB} cases are favourable to B. Thus N_{AB} cases are favourable to both A and B. (Fig. 1.2). The proportion N_{AB}/N_A is the proportion of cases among the cases favourable to A which are favourable to B and is defined as the conditional probability of occurrence of B given that the event A has occurred. This is denoted as $P(B/A)$. The proportion N_{AB}/N is the probability of joint occurrence of the events A and B and is $P(AB)$.

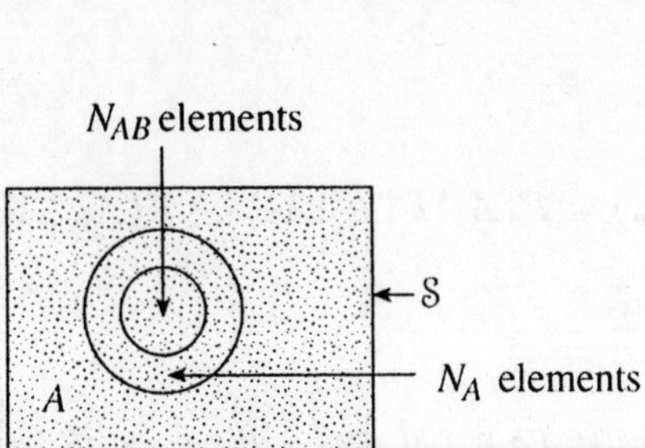

Fig. 1.2

Now

$$\frac{N_{AB}}{N} = \left(\frac{N_A}{N}\right)\left(\frac{N_{AB}}{N_A}\right), \quad \text{i.e., } P(AB) = P(A)P(B/A). \tag{1.4.1}$$

Hence we have

Theorem 1.4.1 *The probability of the simultaneous occurrence of two events A and B is equal to the probability of A multiplied by the conditional probability of B given that A has occurred (it is also equal to the probability of B multiplied by the conditional probability of A given that B has occurred).*

Clearly (1.4.1) can also be written as

$$P(AB) = P(B)P(B/A). \tag{1.4.2}$$

From (1.4.1) we have the conditional probability

$$P(B/A) = \frac{P(AB)}{P(A)} \tag{1.4.3}$$

provided $P(A) > 0$. The conditional probability $P(B/A)$ is not defined if $P(A) = 0$.

Corollary 1.4.1 *For three events* A_1, A_2, A_3,

$$\begin{aligned} P(A_1A_2A_3) &= P(A_1A_2)P(A_3|A_1A_2) \\ &= P(A_1)P(A_2|A_1)P(A_3|A_1A_2) \end{aligned}$$

provided $P(A_1A_2) > 0$.

Similarly for four events A_1, A_2, A_3, A_4

$$\begin{aligned} P(A_1A_2A_3A_4) &= P(A_1A_2A_3)P(A_4|A_1A_2A_3) \\ &= P(A_1)P(A_2|A_1)P(A_3|A_1A_2)P(A_4|A_1A_2A_3) \end{aligned}$$

provided $P(A_1A_2A_3) > 0$.
In general for m events $A_1, \cdots, A_m$,

$$P(A_1 \cdots A_m) = P(A_1)P(A_2|A_1) \cdots P(A_m|A_1 \cdots A_{m-1}) \tag{1.4.4}$$

provided $P(A_1 \cdots A_{m-1}) > 0$.

▶ **Example 1.4.1** *An urn contains 6 red and 4 black balls. Two balls are drawn without replacement. What is the probability that the second ball is red if it is known that the first is red?*

Let A and B be the events that the first and the second ball are red respectively. ence,

$$P(AB) = \binom{6}{2} \Big/ \binom{10}{2} = \frac{1}{3}; \ P(A) = \frac{6}{10}.$$

hus

$$P(B|A) = \frac{5}{9}.$$

heorem 1.4.2 *Suppose the event A can occur only along with the event B. Suppose lso B can occur only in m mutually exclusive ways $B_1, \cdots, B_m$. Then*

$$P(A) = \sum_{i=1}^{m} P(B_i)P(A|B_i) \tag{1.4.5}$$

rovided $P(B_i) > 0, i = 1, \cdots, m$.

Proof: The events $AB_1, \cdots, AB_m$ are mutually exclusive. Thus

$$P(A) = P(AB) = \sum_{i=1}^{m} P(AB_i) = \sum_{i=1}^{m} P(B_i)P(A|B_i).$$

orollary 1.4.2 *For any event B which can occur along (not necessarily always) with A,*

$$P(A) = P(AB) + P(AB^c).$$

▸ **Example 1.4.2** *An urn contains a white balls and b black balls; another contains c hite balls and d black balls. One ball is transferred from the first into the second and hen one ball is drawn from the later. What is the probability that it is a white ball?*

The ball transferred from the first urn may be a black (B) one or a white (W) one. hus the possible events are BW, WW. Now

$$P(BW) = \frac{bc}{(a+b)(c+d+1)}, \quad P(WW) = \frac{a(c+1)}{(a+b)(c+d+1)}.$$

Ience the required probability $= P(BW) + P(WW) = \dfrac{ac+bc+a}{(a+b)(c+d+1)}.$

1.5 Independence of Events

We have seen in the previous section that $P(B|A)$ does not generally equal $P(B)$. Thus he information whether A has happened changes the probability of occurrence of B. f, however, $P(B|A) = P(B)$, probabilities of B does not depend on whether A has or has not happened. In this case events A, B are said to be stochastically independen Hence we have the following definition.

Independence of a pair of events: Two events A and B are said to b independent if the conditional probability of occurrence of B given the occurrence o A is equal to the unconditional probability of B, i.e., if $P(B|A) = P(B)$, provide $P(B|A)$ is defined.

Thus for independent events A and B,

$$P(AB) = P(A)P(B|A) = P(A)P(B).$$

It follows that

$$P(B|A) = P(B) \Leftrightarrow P(A|B) = P(A),$$

because

$$P(A)P(B|A) = P(AB) = P(B)P(A|B).$$

Thus the independence of A and B implies that the unconditional probability o either event is equal to its conditional probability given the other event.

Events A and B are stochastically independent if and only if

$$P(AB) = P(A)P(B). \tag{1.5.1}$$

The definition of independence of events can easily be extended to any finite numbe of events. Given n events

$$A_1, \cdots, A_n, P(A_i|A_1, \cdots, A_{i-1}, A_{i+1}, \cdots, A_n)$$

denotes the conditional probability of A_i given that the events $A_1, \cdots, A_{i-1}, A_{i+1}, \cdots, A$ have occurred.

$$P(A_i|A_1 \cdots A_{i-1}A_{i+1} \cdots A_n) = \frac{P(A_1 \cdots A_n)}{P(A_1 \cdots A_{i-1}A_{i+1} \cdots A_n)}$$

if $P(A_1 \cdots A_{i-1}A_{i+1} \cdots A_n) > 0$.

Independence of a number of events: n events $A_1, \cdots, A_n$ are said to b independent if the conditional probabilities of any one of these events, say A_i, given on or more of the remaining events, is equal to the unconditional probability of A_i.

Theorem 1.5.1 (***Compound probability for independent events***). *If $A_1, \cdots, A_n$ ar independent events,*

$$P(A_1 \cdots A_n) = P(A_1)P(A_2) \cdots P(A_n).$$

Proof: We have, for any set of events $A_1, \cdots, A_n$

$$P(A_1 \cdots A_n) = P(A_1)P(A_2|A_1) \cdots P(A_n|A_1 \cdots A_{n-1}).$$

Since $A_1, \cdots, A_n$ are independent we have

$$P(A_2|A_1) = P(A_2), \cdots, P(A_n|A_1 \cdots A_{n-1}) = P(A_n).$$

'he theorem follows.

'hree events A, B, C are independent, by definition, if

$$\begin{aligned} P(A|BC) &= P(A|B) = P(A|C) = P(A) \\ P(B|AC) &= P(B|A) = P(B|C) = P(B) \\ P(C|AB) &= P(C|A) = P(C|B) = P(C). \end{aligned}$$

It is easy to verify that three events, A, B, C are stochastically independent if and only if the following conditions hold.

$$\left.\begin{aligned} &P(AB) = P(A)P(B),\ P(AC) = P(A)P(C),\ P(BC) = P(B)P(C) \\ &P(ABC) = P(A)P(B)P(C). \end{aligned}\right\}. \tag{1.5.2}$$

Similarly m events, $A_1, \cdots, A_m$ are stochastically independent if the following conditions hold.

$$\begin{array}{lcll} P(A_iA_j) & = & P(A_i)P(A_j), & i < j = 1, \cdots, m \\ P(A_iA_jA_k) & = & P(A_i)P(A_j)P(A_k), & i < j < k = 1, \cdots, m \\ \cdots\cdots\cdots & & \cdots\cdots\cdots\cdots\cdots\cdots & \cdots\cdots\cdots\cdots\cdots\cdots \\ P(A_1 \cdots A_m) & = & P(A_1)P(A_2)\cdots P(A_m). & \end{array} \tag{1.5.3}$$

▶ **Example 1.5.1** *Consider families with 3 children. Sample space, here, consists of 8 points*

$$\{(bbb), (bbg), (bgb), (bgg), (gbb), (gbg), (ggb), (ggg)\},$$

where 'b' stands for a boy and 'g' for a girl. Assume the cases are all equally likely. Let A be the event that the family has children of both sexes and B the event there is at most one girl. Examine if A and B are stochastically independent.

Here

$$P(A) = \frac{3}{4},\quad P(B) = \frac{1}{2}.$$

The event AB means the family has exactly one girl.

$$P(AB) = \frac{3}{8} = P(A)P(B).$$

Thus A, B are independent.

▶ **Example 1.5.2** *This is an example of pairwise independent events that are not independent. An urn contains four tickets bearing numbers* 1234, 2341, 3412 *and* 4123 *and one ticket is drawn. Let the events A, B, C be as follows:*

A = 1st digit of the ticket drawn is 1 or 4.
B = 2nd digit of the ticket drawn is 2 or 4.
C = 3rd digit of the ticket drawn is 3 or 4.

Thus

$$P(A) = P(B) = P(C) = \frac{1}{2}, \;\; P(AB) = P(AC) = P(BC) = \frac{1}{4}, \;\; P(ABC) = \frac{1}{4}.$$

Obviously A, B, C are pairwise independent but $P(ABC) \neq P(A)\,P(B)\,P(C)$. Hen A, B, C are not independent.

1.6 Bayes Theorem

Suppose an event A can occur if and only if one of the hypotheses $B_1, \cdots, B_k$ is true. Th probability $P(B_i)$ of occurrence of B_i is known for each $i, i = 1, \cdots, k$. Also known is th conditional probability $P(A|B_i)$ of occurrence of A given that B_i has already occurre $i = 1, \cdots, k$. We want to find the conditional probability $P(B_i|A)$ of occurrence of B given that A has already occurred. This is given in Bayes theorem.

Theorem 1.6.1 (*Bayes theorem*). *Let an event A occur only if one of the hypothese $B_1, \cdots, B_k$ is true. Known are the prior probabilities $P(B_1), \cdots, P(B_k)$ of occurren of $B_1, \cdots, B_k$ respectively. The conditional probabilities $P(A|B_i)$, $i = 1, \cdots, k$, are als known. The posterior probability $P(B_i|A)$ is given by*

$$P(B_i|A) = \frac{P(B_i)P(A|B_i)}{\sum_{j=1}^{k} P(B_j)P(A|B_j)}$$

provided at least one $P(B_t) > 0$, $t = 1, \cdots, k$.

Proof: We have

$$P(AB_i) = P(B_i)P(A|B_i) = P(A)P(B_i|A).$$

Hence

$$P(B_i|A) = \frac{P(B_i)P(A|B_i)}{P(A)}.$$

Now

$$P(A) = P(AB_1) + P(AB_2) + \cdots + P(AB_k) = \sum_{j=1}^{k} P(B_j)P(A|B_j) \text{ [by (1.4.5)].}$$

Hence the theorem.

▶ **Example 1.6.1** *The first of three urns contains 7 white and 10 black balls, the secon contains 5 white and 12 black balls and the third contains 17 white balls (and no blac ball). A person chooses an urn at random and draws a ball from it. The ball is white Find the probabilities that the ball came from (i) the first; (ii) the second and (iii) th third urn.*

Let H_i be the hypothesis that the ith urn was chosen and E be the event a white ball is drawn.

$$P(H_i) = \frac{1}{3}, i = 1, 2, 3, \quad P(E|H_1) = \frac{7}{17}, \quad P(E|H_2) = \frac{5}{17}, \quad P(E|H_3) = 1$$

$$P(H_1|E) = \frac{\frac{7}{51}}{\frac{7}{51} + \frac{5}{51} + \frac{1}{3}} = \frac{7}{29}, \quad P(H_2|E) = \frac{5}{29}, \quad P(H_3|E) = \frac{17}{29}.$$

1.7 Statistical or Empirical Definition of Probability

The classical definition of probability suffers from the drawbacks that this definition is not applicable to situations in which the sample space contains an infinite number of points or even if the sample space is finite, all the outcomes are not equally likely. The following is the empirical or statistical definition of probability.

Definition 1.7.1 *Let $f_n(A)$ be the number of times in which an event A, the outcome of an experiment, occurs in a series of n repetitions of the trial conducted under identical conditions. The relative frequency of A is $r_n(A) = f_n(A)/n$. The probability of the event A is defined as*

$$P(A) = \lim_{n \to \infty} \frac{f_n(A)}{n}$$

provided the limit exists and is unique.

1.8 Geometric Probability

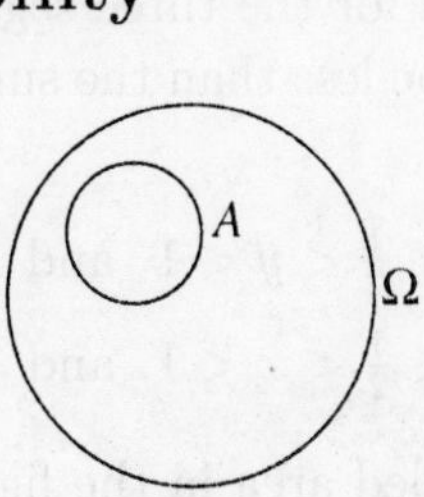

Fig. 1.3

Geometrical approach to the calculation of probabilities is employed when the sample space Ω includes an uncountable set of elementary events ω and none of them is more likely to occur than the other. Suppose as in Fig. 1.3 the sample space Ω is a domain in a plane and the elementary events ω are points within Ω. If an event A is represented by the event A within Ω, so that all ω belonging to the region A are favourable to the event A, then probability of A is

$$P(A) = \frac{\text{area of subdomain } A}{\text{area of domain } \Omega} \tag{1.8.1}$$

In general,

$$\text{Probability of an event } A = \frac{S_A}{S_\Omega}, \tag{1.8.2}$$

where S_A and S_Ω are measures of the specified parts of the region representing the even A and the sample space Ω respectively. The measure may be length, area, volume etc according as the region is in one, two, three, ..., dimensions.

Formula (1.8.1) is a generalisation of classical formula (1.2.1) to an uncountable se of elementary events. The symmetry of the experimental conditions with respect to th elementary outcomes ω is usually formulated by the assumption of randomness.

► **Example 1.8.1** *A straight line of unit length, say, (interval) (0, 1) is divided into thre intervals by choosing two points at random. What is the probability that the three lin segments form a triangle?*

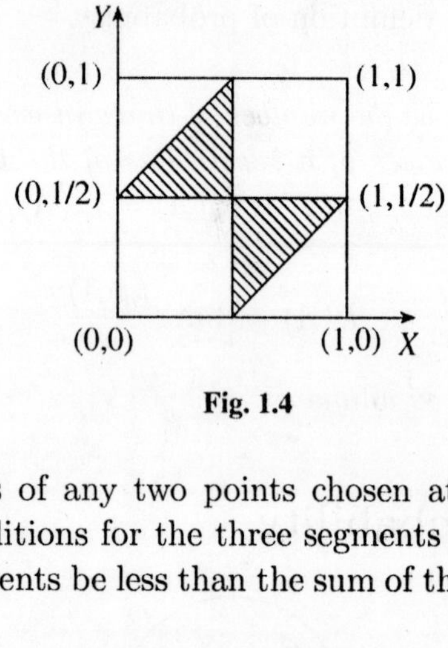

Fig. 1.4

Let x, y be the abscissas of any two points chosen at random on (0,1). A set o necessary and sufficient conditions for the three segments to form a triangle is that th length of any one of the segments be less than the sum of the other two. Hence we shoul have either

$$0 < x < \tfrac{1}{2} < y < 1 \text{ and } \quad y - x < \tfrac{1}{2}$$
$$\text{or } 0 < y < \tfrac{1}{2} < x < 1 \text{ and } \quad x - y < \tfrac{1}{2}.$$

This is represented by the shaded area in the figure 1.4. Hence the probability is $\frac{1}{4}$.

The short-comings of the classical definition of probability were overcome by axiomatic approach to the theory of probability, first formulated by S. N. Bernstei and developed by A. Khintchine, A. Kolmogoroff, among others.

1.9 Axiomatic Definition of Probability

Let (i) $\mathcal{S}$ be any abstract set, the sample space underlying a random experiment; (ii) $\mathcal{A}$ be a σ-field (σ-algebra) of subsets of $\mathcal{S}$ (events resulting from the experiment) (iii) P be a measure defined on $\mathcal{A}$ such that $P(\mathcal{S}) = 1$. Then P is called a probability measure defined on $\mathcal{A}$. The triplet $(\mathcal{S}, \mathcal{A}, P)$ is called a probability space.

Example 1.9.1 *Suppose $\mathcal{S}$ is finite containing N elementary events $w_i, i = 1, \cdots, N$. We assign to each w_i an arbitrary non-negative number p_i such that $\sum_{i=1}^{N} p_i = 1$. As the σ-field we take the class of all elements in $\mathcal{S}$, the total number of such events being 2^N. If A is an event*

$$P(A) = \sum_{w_i \in A} P(w_i).$$

It follows from the definition that we have postulated the following axioms:

(a) With each event $A \in \mathcal{A}$ there is associated a non-negative real quantity $P(A)$, called its probability.

(b) $P(\mathcal{S}) = 1$.

(c) If events $A_1, A_2, \cdots$ are pairwise mutually exclusive,

$$P\left(\bigcup_{i=1}^{\infty} A_i\right) = \sum_{i=1}^{\infty} P(A_i).$$

It is obvious that if axiom (c) is accepted, the result in axiom (c) also holds for a finite number of disjoint events, $A_1, \cdots, A_n$, i.e.,

(c)′ $P\left(\bigcup_{i=1}^{n} A_i\right) = \sum_{i=1}^{n} P(A_i)$ if $A_iA_j = \phi$, $i \neq j = 1, \cdots, n$ (a proof has been given below).

Theorem 1.9.1 (***Continuity theorem***). *If $\{A_n\}$ is a monotone sequence of events belonging to $\mathcal{A}$, then*

$$P\left\{\lim_{n\to\infty} A_n\right\} = \lim\{P(A_n)\}.$$

Proof: Suppose that $\{A_n\}$ is monotonically non-decreasing. Here

$$\lim A_n = \bigcup_{n=1}^{\infty} A_n.$$

Now

$$A_n = A_1 + (A_2 - A_1) + \cdots + (A_n - A_{n-1}) = \sum_{k=1}^{n} (A_k - A_{k-1}),$$

sum of n disjoint subsets. Hence

$$\lim A_n = \sum_{k=1}^{\infty} (A_k - A_{k-1}).$$

$$\therefore P(\lim A_n) = P\left(\sum_{k=1}^{\infty} (A_k - A_{k-1})\right) = \sum_{k=1}^{\infty} P(A_k - A_{k-1}) \text{ [by axiom (c)]}$$

$$= \lim_{n\to\infty} \sum_{k=1}^{n} P(A_k - A_{k-1}) = \lim P(A_n).$$

Suppose now $\{A_n\}$ is monotonically non-increasing. The sequence $\{A_n^c\}$ is the monotonically non-decreasing and each A_n^c belongs to $\mathcal{A}$.
Hence

$$P(\lim A_n^c) = \lim P(A_n^c),$$
$$\text{i.e., } P\left[(\lim A_n)^c\right] = \lim[1 - P(A_n)]$$
$$\left[\text{Since } \lim A_n^c = \bigcup_{n=1}^{\infty} A_n^c = \left(\bigcap_{n=1}^{\infty} A_n\right)^c = (\lim A_n)^c\right],$$
$$\text{i.e., } 1 - P(\lim A_n) = 1 - P \lim P(A_n)$$
$$\text{and hence } P(\lim A_n) = \lim P(A_n).$$

1.10 Conditional Probability (Axiomatic Definition)

Consider the probability space $(\mathcal{S}, \mathcal{A}, P)$. Let A_1, A_2 be two events in $\mathcal{A}$ such tha $P(A_i) > 0$, $i = 1, 2$. The ratio

$$P(A_2|A_1) = \frac{P(A_1 \cap A_2)}{P(A_1)} \tag{1.10.1}$$

is called the conditional probability of the event A_2 given that A_1 occurs. Whe $P(A_1) = 0$, $P(A_1 \cap A_2) = 0$ and $P(A_2|A_1)$ remains undefined.

For fixed A_1 in $\mathcal{A}$ such that $P(A_1) > 0$, $(\mathcal{S}, \mathcal{A}, P(\cdot|A_1))$ is a probability space wher $P(\cdot|A_1)$ is a probability measure which takes the value $P(A_2|A_1)$ on the event $A_2 \in \mathcal{A}$ To prove this it is enough to prove

Theorem 1.10.1 $P(A_2|A_1)$ *given in* (1.10.1) *satisfies axioms* (a), (b), (c) *provide* $P(A_1) > 0$.

Proof:

(a) $P(A_2|A_1) \geq 0$ since $P(A_1 \cap A_2) \geq 0$

(b) $P(\mathcal{S}|A_1) = \frac{P(A_1)}{P(A_1)} = 1$

(c) Let $\{B_i\}$ be a sequence of disjoint events in $\mathcal{A}$. We have

$$\left\{\bigcup_{i=1}^{\infty} B_i\right\} \cap A_1 = \bigcup_{i=1}^{\infty} (B_i \cap A_1).$$

Since $\{B_i\}$ is a sequence of disjoint events in $\mathcal{A}$, $\{B_i \cap A_1\}$ is a sequence of disjoin events in $\mathcal{A}$. Hence

$$P\left[\left\{\bigcup_{i=1}^{\infty} B_i\right\} \cap A_1\right] = \sum_{i=1}^{\infty} P(B_i \cap A_1).$$

Dividing both sides by $P(A_1)$

$$P\left(\left(\bigcup_{i=1}^{\infty} B_i\right) \middle| A_1\right) = \sum_{i=1}^{\infty} P(B_i|A_1).$$

Hence the proof. □

(1.10.1) can also be written as

$$P(A_1 \cap A_2) = P(A_1)P(A_2|A_1). \tag{1.10.2}$$

If $P(A_2|A_1) = P(A_2)$, then (1.10.2) reduces to the case of independent events A_1, A_2 and we have

$$P(A_1 \cap A_2) = P(A_1)P(A_2).$$

More generally, the following result can be shown.

If $A_1, A_2, \cdots$ is a finite or countably infinite sequence of events in $\mathcal{A}$ having probability measure P such that

$$P(A_1) > 0, \;\; P(A_1 \cap A_2) > 0, \;\; P(A_1 \cap A_2 \cap A_3) > 0, \;\; \cdots,$$

then

$$\begin{aligned} P(A_1 \cap A_2 \cap \cdots) = {} & P(A_1)P(A_2|A_1)P(A_3|A_1 \cap A_2)\cdots \\ & P(A_n|A_1 \cap A_2 \cap \cdots \cap A_{n-1})\cdots. \end{aligned} \tag{1.10.3}$$

In the case of mutual independence of $A_1, A_2, \cdots$, (1.10.3) reduces to

$$P(A_1 \cap A_2 \cap \cdots) = P(A_1)P(A_2)P(A_3)\cdots.$$

Theorem 1.10.2 *If A_1, A_2, $\cdots$ is a finite or countably infinite sequence of disjoint events in $\mathcal{A}$ having non-zero probabilities, where $\bigcup_{i=1}^{\infty} A_i = \mathcal{S}$ (sample sapce), and if A is any event in $\mathcal{A}$ then*

$$P(A) = P(A_1)P(A|A_1) + P(A_2)P(A|A_2) + \cdots.$$

Proof: Since $A_1, A_2, \cdots$ is a sequence of disjoint events in $\mathcal{A}$ and

$$\cup A_i = \mathcal{S}, \;\; A \cap A_1, \;\; A \cap A_2, \;\; \cdots,$$

is a sequence of disjoint events in $\mathcal{A}$ and

$$\bigcup_{i=1}^{\infty}(A \cap A_i) = A.$$

Hence

$$P(A) = \sum_{i=1}^{\infty} P(A \cap A_i) = \sum_{i=1}^{\infty} P(A_i)P(A|A_i).$$ □

We recall another important theorem.

Theorem 1.10.3 *If* $\{A_i\}$ *is a sequence of disjoint events in* $\mathcal{A}$,

$$P\left(\bigcup_{i=1}^{\infty} A_i\right) = \sum_{i=1}^{\infty} P(A_i) - \sum_{i<j=1}^{\infty}\sum P(A_i \cap A_j) + \sum_{i<j<k=1}^{\infty}\sum\sum P(A_i \cap A; \cap A_k) - \cdots.$$

□

This chapter is only a introductory chapter on the theory of probability, covering the major concepts and the theorems in the area. For detailed exposition on the theory of probability, the reader is asked to refer to Mukhopadhyay (1991).

Chapter 2

Probability Distributions on $\mathbb{R}^1$

2.1 Introduction

In this chapter we examine the concept of a random variable. In section 2.3 we study the notion of probability distribution function of a random variable; section 2.4 deals with some special types of random variables. Section 2.5 considers independent random variables. In section 2.6 and 2.7 we consider Lebesgue-Stieltjes integration.

2.2 Random Variables

In chapter 0 we considered the notion of a point function defined on an arbitrary set, inverse function of a point function, set function, measure and measurable sets with respect to (w.r.t) an arbitrary σ-field of sets. We consider now the concept of a measurable function.

Definition 2.2.1 *Consider $\mathbb{R}^1$, the real line and $\mathcal{B}_1$, the borel field of subsets of $\mathbb{R}^1$. If a function X which maps $S \to \mathbb{R}^1$ is such that $X^{-1}(B) \subset \mathcal{C}\ \forall B \subset \mathcal{B}_1$, where $\mathcal{C}$ is the σ-field over S, then X is called a $\mathcal{C}$-measurable function (measurable w.r.t. $\mathcal{C}$).*

If $S = \mathbb{R}^1$ (or a subset of $\mathbb{R}^1$) then $\mathcal{C} = \mathcal{B}_1$; if X is measurable w.r.t. $\mathcal{B}_1$ then X is a borel measurable function. Here $X^{-1}(B) \subset \mathcal{B}_1\ \forall B \in \mathcal{B}_1, X^{-1}$ is defined on $\mathcal{B}_1$, having its range in $\mathcal{B}_1$.

Definition 2.2.2 (***Random Variable***). *Consider $(\mathcal{S}, \mathcal{A})$, $\mathcal{S}$ being the sample space of a random experiment. A finite single-valued function X which maps $\mathcal{S}$ into $\mathbb{R}^1$ is called a random variable if the inverse image under X of all borel sets in $\mathcal{B}_1$ are events, i.e., if*

$$X^{-1}(B) = \{w : X(w) \in B\} \in \mathcal{A}, \text{ for all } B \in \mathcal{B}_1,$$

i.e., if X is a $\mathcal{A}$-measurable function. Any $\mathcal{A}$-measurable function is a random variable on $(\mathcal{S}, \mathcal{A})$.

The following theorem gives a necessary and sufficient condition for a function X to be a random variable.

Theorem 2.2.1 *X is a random variable if and only if for each $x \in \mathbb{R}^1$, $\{w : X(w) \leq x\} \in \mathcal{A}$.*

► **Example 2.2.1** (a) *A coin is tossed three times. Here $S = \{w_1 = (HHH), w_2 = (HHT), w_3 = (HTH), w_4 = (HTT), w_5 = (THH), w_6 = (THT), w_7 = (TTH), w_8 = (TTT)\}$. Let X denote the number of heads. Here*

$$X^{-1}(-\infty, x] = \begin{cases} \phi & x < 0 \\ \{w_8\}, & 0 \leq x < 1 \\ \{w_8, w_4, w_6, w_7\}, & 1 \leq x < 2 \\ \{w_8, w_4, w_6, w_7, w_2, w_3, w_5\}, & 2 \leq x < 3 \\ S & 3 \leq x. \end{cases}$$

Thus, X is a random variable.

(b) *A coin is tossed until a head appears. Let X be the number of tosses (including the toss in which H occurs) required. Here*

$$S = \{w_1 = (H), w_2 = (TH), w_3 = (TTH), \cdots\},$$

$$X^{-1}(-\infty, x] = \begin{cases} \phi & x < 1 \\ \{(H)\}, & 1 \leq x < 2 \\ \{(H), (TH)\}, & 2 \leq x < 3 \\ \{(H), (TH), (TTH)\}, & 3 \leq x < 4 \\ \cdots \end{cases}$$

Thus X is a random variable.

Theorem 2.2.2 *A constant function $X(w)$ defined on $(\mathcal{S}, \mathcal{A})$ is a random variable.*

Proof: Let $X(w) = c$ (a real quantity), $w \in \mathcal{S}$. Let $B \subset \mathcal{B}_1$ be any borel set. Either $c \in B$ or $c \notin B$. If $c \in B$, then $X^{-1}(B) = \mathcal{S}$. If $c \notin B, X^{-1}(B) = \phi$. Since $\phi, \mathcal{S} \in \mathcal{A}, X$ is a $\mathcal{A}$-measurable function and hence is a random variable.

Theorem 2.2.3 *If $X(w)$ is a random variable on $(\mathcal{S}, \mathcal{A}), cX(w)$ is also a random variable on $(\mathcal{S}, \mathcal{A})$.*

Proof: Let x be any arbitrary but fixed real number. Then $(-\infty, x] \subset \mathcal{B}_1$. For $c = 0$, the theorem is trivial (by theorem 2.2.2). Assume $c > 0$.

$$(cX)^{-1}(-\infty, x] = \{w : cX(w) \leq x\} = \left\{w : X(w) \leq \frac{x}{c}\right\} = X^{-1}\left(-\infty, \frac{x}{c}\right] \subset \mathcal{A}.$$

$(\because X \text{ is a r.v.})$

Hence $cX(w)$ is $\mathcal{A}$-measurable and hence is a random variable on $(\mathcal{S}, \mathcal{A})$. The case $c < 0$ can similarly be proved.

The following theorems hold:

Theorem 2.2.4 *If $\{c_i\}$, $i = 1, \cdots, n$ are real numbers and $X_1(w), \cdots, X_n(w)$ are r.v.'s on $(\mathcal{S}, \mathcal{A})$ then $\sum_1^n c_i X_i$ is a r.v. on $(\mathcal{S}, \mathcal{A})$.*

Theorem 2.2.5 *Any polynomial of a r.v. is also a r.v.*

Theorem 2.2.6 *If $X(w)$ and $Y(w)$ are r.v.'s on $(\mathcal{S}, \mathcal{A})$ and $\{w : Y(w) = 0\} = \phi$, then $\frac{X(w)}{Y(w)}$ is a r.v. on $(\mathcal{S}, \mathcal{A})$.*

Theorem 2.2.7 *If $X(w), Y(w)$ are r.v.'s on $(\mathcal{S}, \mathcal{A})$ then max $(X(w), Y(w))$ and min $(X(w), Y(w))$ are r.v.'s on $(\mathcal{S}, \mathcal{A})$.*

Definition 2.2.3 (***Indicator random variable***). *For an event A associated with a random experiment we define a random variable $I_A(w)$ on each point w in the sample space $\mathcal{S}$ such that*

$$I_A(w) = 1(0) \text{ if } w \in A (w \notin A).$$

$I_A(w)$ is called the indicator r.v. (function) of the set A.

Theorem 2.2.8 *The indicator function $I_A(w)$ is a r.v. on $(\mathcal{S}, \mathcal{A})$ if and only if $A \in \mathcal{A}$ (if A is not measurable, I_A need not be a r.v.).*

Proof: Let B be any borel set in $\mathcal{B}_1$. We shall have one of the four cases: (i) $1 \in B$, $0 \in B$, (ii) $1 \in B, 0 \notin B$, (iii) $1 \notin B, 0 \in B$, (iv) $1 \notin B, 0 \notin B$. If A is a measurable set, $I_A^{-1}(B)$ is $\mathcal{S}, A, A', \phi$ respectively in cases (i), (ii), (iii) and (iv) and each of these is a $\mathcal{A}$-measurable set.

Theorem 2.2.9 *Let $X(w)$ be a r.v. on $(\mathcal{S}, \mathcal{A})$ and $g(X(w))$ be a borel measurable function,*
$g : \mathbb{R}^1 \to \mathbb{R}^1$. Then $g(X(w))$ is a r.v.

Proof: Let B be any borel set.

$$g(X)^{-1}(B) = \{w : g(X(w)) \in B\} = \{w : X(w) \in g^{-1}(B)\} = X^{-1}[g^{-1}(B)].$$

Since g is borel-measurable, $g^{-1}(B)$ is a borel set. Thus $X^{-1}[g^{-1}(B)] \subset \mathcal{A}$.

Corollary 2.2.1 *If $X(w)$ is a r.v. then $g(X(w))$ is also a r.v. when g is a continuous function from $\mathbb{R}^1$ to $\mathbb{R}^1$.*

Proof: Continuity of g implies g is borel-measurable.

Corollary 2.2.2 *If $X(w)$ is a r.v. then $g(X(w))$ is also a r.v. when g is a monotonic function from $\mathbb{R}^1$ to $\mathbb{R}^1$.*

Proof: Monotonicity of g implies g is borel-measurable.

We note that the concept of probability is not required in the definition of a r.v. We now consider the probability distribution of a r.v. X.

Theorem 2.2.10 *The r.v. X defined on $(\mathcal{S}, \mathcal{A}, P)$ induces a probability space $(\mathbb{R}^1, \mathcal{B}_1, P_X)$ for the r.v. X. The probability measure P_X is given by*

$$P_X(B) = P\{X^{-1}(B)\} = P\{w : X(w) \in B\}$$

for all $B \in \mathcal{B}_1$. P_X is called the probability distribution of X.

Proof: Clearly $P_X(B) \geq 0$ for all $B \in \mathcal{B}_1$. $P_X(\mathbb{R}^1) = P\{X \in \mathbb{R}^1\} = P(\mathcal{S}) = 1$.

Let $B_i \in \mathcal{B}_1$, $i = 1, 2, \cdots$ with $B_i \cap B_j = \phi$ $(i \neq j)$. Now $X^{-1}\left(\bigcup_{i=1}^{\infty} B_i\right) = \bigcup_{i=1}^{\infty} X^{-1}(B_i)$ and $X^{-1}(B_i)$ and $X^{-1}(B_j)$ are disjoint sets in $\mathcal{A}$.

$$P_X\left(\bigcup_1^{\infty} B_i\right) = P\left\{X^{-1}\left(\bigcup_1^{\infty} B_i\right)\right\} = P\left\{\left(\bigcup_{i=1}^{\infty} X^{-1}(B_i)\right)\right\}$$

$$= \sum_{i=1}^{\infty} P\{X^{-1}(B_i)\} = \sum_{i=1}^{\infty} P_X(B_i).$$

Thus P_X satisfies all the three axioms and hence is a probability measure. $(\mathbb{R}^1, \mathcal{B}_1, P_X)$ is, therefore, a probability space.

► **Example 2.2.2** *A die is tossed. Let X be a r.v. which takes the value 1 if the result is an odd number, 2 if the result is an even number.*

Here $\mathcal{S} = \{1, 2, \cdots, 6\}$, i denoting the event, i comes up, $\mathcal{A}$ the σ-field over $\mathcal{S}$. Range space of X is $\{1, 2\}$. Let $B_1 = \{1\}$, $B_2 = \{2\}$, $\mathcal{B} = \{\{1\}, \{2\}, \{1, 2\}\}$. $X^{-1}(B_1) = \{1, 3, 5\} \in \mathcal{A}$, $X^{-1}(B_2) = \{2, 4, 6\} \in \mathcal{A}$. Thus X is a r.v.

$$P_X(B_1) = P(X^{-1}(1)) = \frac{1}{2}, \ \ P_X(B_2) = P(X^{-1}(2)) = \frac{1}{2}.$$

The sets B_1, B_2 are disjoint and form the range space of X.

$$P_X(B_1 \cup B_2) = P_X(B_1) + P_X(B_2) = 1$$

► **Example 2.2.3** *Let $A_1, \cdots A_n$ be disjoint subsets with $\bigcup_1^n A_i = \mathcal{S}$; X is a function defined on $(\mathcal{S}, \mathcal{A})$ such that $X(w) = a_i$ for each $w \in A_i$ $(i = 1, \cdots, n)$. Thus one can write*

$$X(w) = \sum_{i=1}^{n} a_i I_{A_i}(w) \tag{2.2.1}$$

where I_{A_i} is the indicator function of set A_i. A function of the type (2.2.1) is called a simple function.

Consider any borel set $B \in \mathcal{B}_1$. B includes either some of a_i's, say, $a_{i_1}, \cdots, a_{i_r}$ $(r \leq n)$ or none at all. In the former case $X^{-1}(B)$ is $\bigcup_{j=1}^{r} A_{i_j}$ and in the later ϕ. If each of $A_i (i = 1, \cdots, n) \in \mathcal{A}$, X is a r.v.

▶ **Example 2.2.4** *Consider $\{A_n, n = 1, 2, \cdots\}$, a sequence of disjoint subsets of $\mathcal{S}$, $\bigcup_1^\infty A_n = \mathcal{S}$, $A_n \in \mathcal{A}$ $(n = 1, 2, \cdots)$. X is a function on $(\mathcal{S}, \mathcal{A})$ such that $X(w) = a_n$ whenever $w \in A_n$ $(n = 1, 2 \cdots)$. Such a function can be written as*

$$X(w) = \sum_1^\infty a_n I_{A_n}(w), \tag{2.2.2}$$

where I_{A_n} is the indicator function of A_n and is called an elementary function. As in example 2.2.3, *it follows that $X(w)$ is $\mathcal{A}$-measurable and hence a r.v.*

2.3 Probability Distribution Function of a r.v.

We have seen that the r.v. X induces a set function $P_X(x)$ on $(\mathbb{R}^1, \mathcal{B}_1)$. We now consider a point function

$$P(X \leq x) = P_X((-\infty, x]) = F_X(x) \text{ (say)}$$

on $(\mathbb{R}^1, \mathcal{B}_1)$. $F_X(x)$ is called the distribution function (d.f.) or cumulative distribution function (c.d.f.) of the r.v. X.

Theorem 2.3.1 (i) *F is monotonically non-decreasing;* (ii) $F(-\infty) = 0$; (iii) $F(\infty) = 1$; (iv) *F is continuous to the right, i.e., $F(x + 0) = F(x)$.*

Proof:

(i) Let $x < y$ be two real numbers. Denote the events $[X \leq x]$, $[X \leq y]$ by B_x and B_y respectively. Now $B_x \subseteq B_y$. Hence $P(B_x) \leq P(B_y)$, i.e., $F_X(x) \leq F_X(y)$.

(ii) Let for integral $n (> 0)$, B_n be the event $[X \leq -n]$. We have $B_1 \supset B_2 \supset B_3 \cdots$ and $\lim_{n\to\infty} B_n = \phi$. Also each of $B_1, B_2 \cdots$ is a measurable event under X. Hence by the continuity theorem (Theorem 1.9.1) $\lim_{n\to\infty} P(B_n) = P(\phi) = 0$, i.e., $F(-\infty) = 0$.

(iii) Let for integral $n (> 0)$, B_n be the event $[X \leq n]$. Then $B_1 \subset B_2 \subset B_3 \subset \cdots$ and $\lim_{n\to\infty} B_n = \mathcal{S}$. Also each of $B_1, B_2, \cdots$ is a measurable event under X. By the continuity theorem, therefore, $\lim_{n\to\infty} P(B_n) = P(\mathcal{S}) = 1$, i.e., $F(\infty) = 1$.

(iv) Let for integral $n(>0)$, $C_n = \left[X \le x + \frac{1}{n}\right]$. Then $C_1 \supset C_2 \supset \cdots$ and $\lim_{n\to\infty} C_n = [X \le x]$. Hence by the continuity theorem,

$$\lim_{n\to\infty} P[C_n] = P(X \le x), \quad \text{i.e., } \lim_{n\to\infty} P\left[X \le x + \frac{1}{n}\right] = F_X(x),$$
$$\text{i.e., } \lim_{\epsilon\downarrow 0} F(x+\varepsilon) = F_X(x), \qquad \text{i.e., } F_X(x+0) = F_X(x).$$

We now prove that $F(x-0)$ is not necessarily equal to $F_X(x)$. Consider $D_n = \left[X \le x - \frac{1}{n}\right]$. Then $D_1 \subset D_2 \subset \cdots$ and $\lim_{n\to\infty} D_n = [X < x]$. Thus,

$$\begin{aligned} \lim_{n\to\infty} P\left[X \le x - \frac{1}{n}\right] &= P(X < x), \\ \text{i.e., } \lim_{\epsilon\downarrow 0} F_X(x-\varepsilon) &= P(X < x), \\ \text{i.e., } F_X(x-0) &= P(X < x). \end{aligned}$$

Thus $F_X(x-0)$ is not necessarily equal to $F_X(x)$ and hence $F_X(x)$ is not continuous to the left.

The quantity $F_X(x) - F_X(x-0) = P[X \le x] - P[X < x] = P[X = x]$ is called the saltus (or jump) of the distribution function $F_X(x)$. If $P[X = x] > 0$, $F_X(x)$ has a discontinuity at x with the saltus $P(X = x)$. If $P(X = x) = 0$, $F_X(x)$ is continuous at x (see Fig. 2.1).

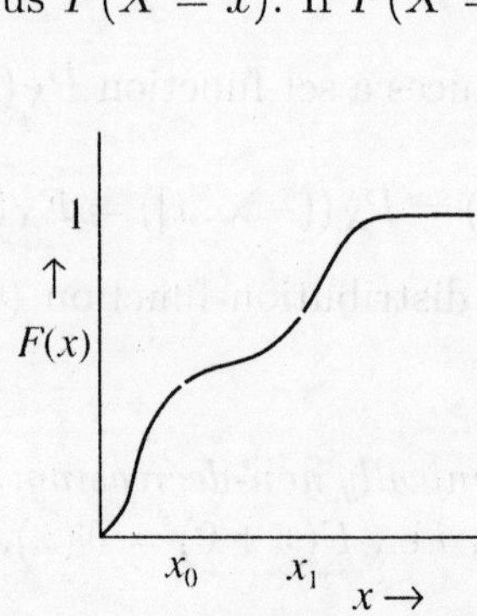

Fig. 2.1 F(x) has saltus at x = x_0, x_1 but is continuous at other points of x

■ NOTE 2.3.1 *In fact, any real valued function F defined on* $\mathbb{R}^1$ *that satisfies the conditions* (i)–(iv) *of theorem* 2.3.1 *is called a distribution function.*

Theorem 2.3.2 *The points of discontinuity of F form at most a countable set.*

Proof: Let $(a, b]$ be a finite interval with at least n discontinuity points

$$a < x_1 < x_2 < \cdots < x_n \le b.$$

Then

$$F(a) < F(x_1-) < F(x_1) \le \cdots \le F(x_n-) < F(x_n) \le F(b).$$

Let $p_k = F(x_k) - F(x_k-)$ (= Saltus of $F_X(x)$ at x_k), $k = 1, \cdots, n$. Then

$$\sum_{k=1}^{n} p_k \leq F(b) - F(a). \tag{2.3.1}$$

Let ν_ε be the number of discontinuity points with jump $> \varepsilon$. Then

$$\sum_{k=1}^{n} p_k \geq \varepsilon\nu_\varepsilon + \sum{}' p_k \tag{2.3.2}$$

$\sum'$ being summation over other points. Hence from (2.3.1), (2.3.2),

$$\varepsilon\nu_\varepsilon \leq F(b) - F(a), \quad \text{i.e., } \nu_\varepsilon \leq \frac{F(b) - F(a)}{\varepsilon} \text{ (a finite quantity).}$$

Thus for every integer n, number of discontinuity points with jump> $\frac{1}{n}$ is finite. Therefore, there are no more than a countable number of discontinuity points in any finite interval $(a, b]$. Again $\mathbb{R}^1$ is union of countable number of such intervals. Hence the number of discontinuity points is at most countable.

Theorem 2.3.3 (***Converse of theorem* 2.3.1**). *If F is a function on $\mathbb{R}^1$ such that it has all the properties* (i)–(iv) *mentioned in theorem* 2.3.1, *then there exists a r.v. X such that $F(x)$ is the d.f. of X.*

Theorem 2.3.4 *Two r.v.s. have the same distribution if and only if they have the same distribution functions.*

□ **Remark 2.3.1** *If X is a r.v. on $(\mathcal{S}, \mathcal{A}, P)$, theorem* 2.3.1 *states that $F(x) = P\{X \leq x\}$ is a d.f. associated with X. Theorem* 2.3.3 *states that given a d.f. $F(x)$ there exists a r.v. X whose d.f. is $F(x)$. Thus given a r.v. there exists a d.f. and conversely.*

2.4 Discrete and Continuous Random Variables

We shall consider two types of r.v.'s: (i) discrete r.v. (ii) continuous r.v.

Definition 2.4.1 *Let X be a r.v. on $(\mathcal{S}, A) \to (\mathbb{R}^1, \mathcal{B}_1)$. If the number of elements in the range of X in $\mathbb{R}^1$ is finite or countably infinite then $X(w)$ is called a* (*one dimensional*) *discrete r.v.*

Let $x_1, x_2, \cdots$ be the elements in the range sapce of X.

Let $p(x_i) = P[X(w) = x_i]$, $x_i \in \mathbb{R}^1$. $p(x_i)$ is called the probability mass function (p.m.f.) of X at the mass point x_i if (i) $p(x_i) \geq 0$ for all $x_i \in \mathbb{R}^1$ (ii) $\sum_{x_i \in \mathbb{R}^1} p(x_i) = 1$.

The distribution function is given by

$$F_X(x) = P[X \leq x]$$
$$= \sum_{x_i \leq x} p(x_i).$$

Thus F is a step function which remains constant over every interval not containing any mass point, but has at each mass point x_i a step of height p_i (Fig. 2.3).

Fig. 2.2 shows the column diagram of p.m.f. of a discrete r.v. X.

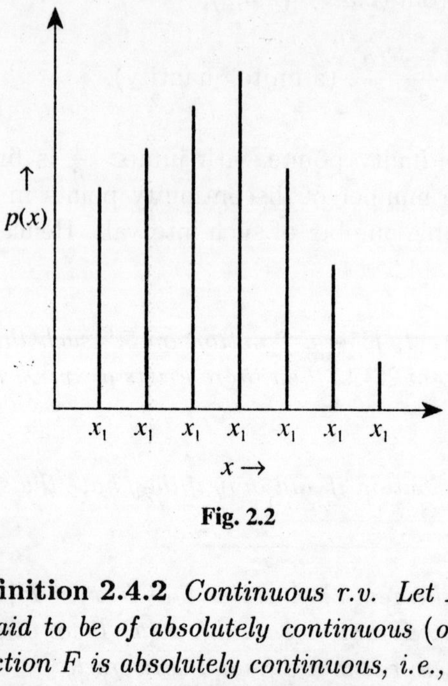

Fig. 2.2

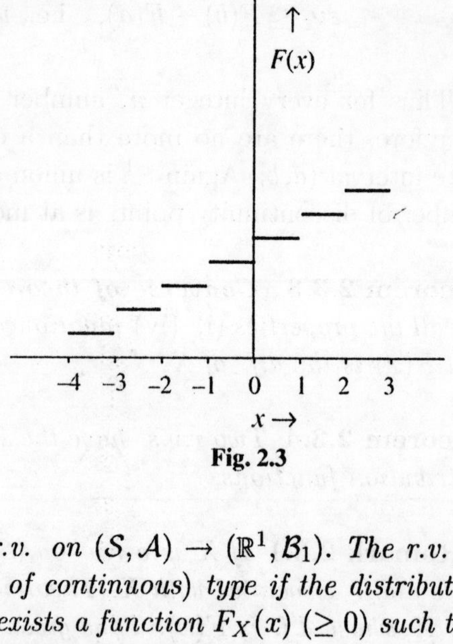

Fig. 2.3

Definition 2.4.2 *Continuous r.v. Let X be a r.v. on $(\mathcal{S}, \mathcal{A}) \to (\mathbb{R}^1, \mathcal{B}_1)$. The r.v. X is said to be of absolutely continuous (or simply of continuous) type if the distribution function F is absolutely continuous, i.e., if there exists a function $F_X(x)$ (≥ 0) such that for every real number x we have*

$$F_X(x) = \int_{-\infty}^{x} f_X(y)\, dy.$$

In this case, $\frac{dF}{dx}$ exists and $\frac{dF}{dx} = f_X(x)$ except for a set of values of X of probability 0.

The function $f_X(x)$ is called the probability density function (p.d.f.) of r.v. X. Since $F_X(x)$ is continuous everywhere, it has no point of jump (Fig. 2.4). Thus the probability X takes the value x, $P[X = x] = 0$.

For any real number α, β

$$P[\alpha < X < \beta) = \int_{\alpha}^{\beta} f(x)\, dx = F(\beta) - F(\alpha).$$

$P[\alpha < X < \beta]$ represents the area under the curve of $f(x)$ between the lines $x = \alpha$ and $x = \beta$ (Fig. 2.5).

It follows (from the mean value theorem of calculus) that

$$P(x \le X \le x+dx) = \int_x^{x+dx} f(t)\,dt = dx\,f(\xi), \quad x \le \xi \le x+dx. \tag{2.4.1}$$

If dx is small, (2.4.1) is approximately equal to

$$f(x)\,dx. \tag{2.4.2}$$

The properties of $f(x)$ are: (a) $f(x) \ge 0 \;\forall\; x$; (b) f is integrable (in the Riemann sense) over the domain of x; (c) $\int_{-\infty}^{\infty} f(x)\,dx = 1$

In fact any function $f(x)$ satisfying (a), (b), (c) above is the p.d.f. of an absolutely continuous distribution.

Clearly the range space of a continuous r.v. X in $\mathbb{R}^1$ is an interval or the union of two or more disjoint intervals in $\mathbb{R}^1$.

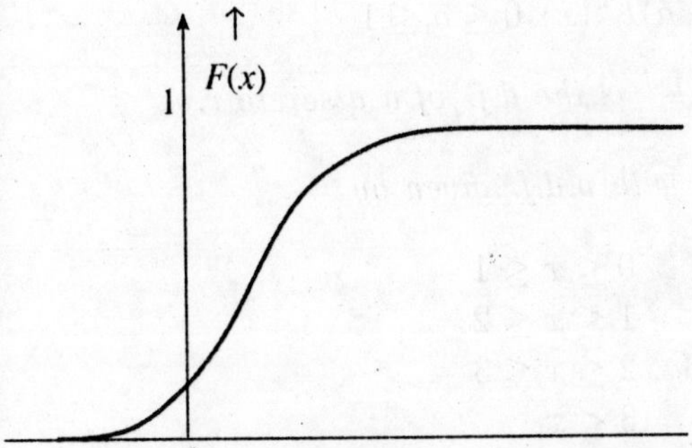

Fig. 2.4

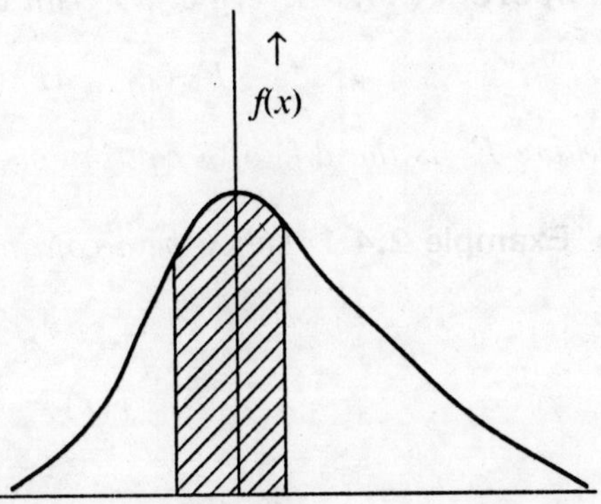

Fig. 2.5

Mixed distributions: Suppose the r.v. X assumes certain discrete values $x_1, \cdots, x_n$ with positive probabilities and assumes all values in an interval say, $a \le X \le b$. Thus X is discrete for certain range of values of X and continuous over an interval or some intervals not containing $x_1, \cdots, x_n$. The distribution of X here is of mixed type.

To each value x_i assign a non-negative number $p(x_i)$ (probability, $X = x_i$) such that

$$p(x_i) \ge 0, \;\; i = 1, \cdots, n \;\; \text{and} \;\; \sum_{i=1}^{n} p(x_i) = p(<1).$$

Let f be a function defined over (a, b) such that $f(x) \ge 0$ for all $x \in (a, b)$, f is integrable and

$$\int_a^b f(x)\,dx = 1 - p.$$

The probability (mass or density) function defined in this way satisfies $P(-\infty < X < \infty) = 1$.

Definition 2.4.3 *Let $0 < p < 1$. The function F defined by*

$$F(t) = pG(t) + (1 - p)H(t), \tag{2.4.3}$$

where

$$G(t) = \sum_{x_i \le t} p(x_i), \quad H(t) = \int_{-\infty}^{t} h(x)\,dx, \quad -\infty < t < \infty$$

is also a distribution function of a r.v. X and is a mixture of d.f. G of a discrete r.v and d.f. H of a continuous r.v. The distribution of X here is of mixed type.

Such a distribution may arise if one, after making a random experiment with probability of success p and probability of failure $(1 - p)$, chooses d.f. $G(t)$ if there is a success and $H(t)$ if there is a failure.

Fig. 2.1 shows the d.f. of a mixed distribution.

Theorem 2.4.1 *Every d.f. F can be decomposed into two parts according to*

$$F_X(x) = aF^c(x) + (1 - a)F^d(x), 0 \le a \le 1$$

where F^c is the d.f. of a continuous r.v. and F^d is the d.f. of a discrete r.v.

► **Example 2.4.1** *Let X be a continuous r.v. with p.d.f. given by*

$$f(x) = \begin{cases} ax, & 0 \le x \le 1 \\ a, & 1 \le x \le 2 \\ -ax + 3a, & 2 \le x \le 3 \\ 0. & 3 < x \end{cases}$$

Determine a and $F(x)$.

We have

$$\int_{-\infty}^{\infty} f(x)\,dx = a\int_0^1 x\,dx + a\int_1^2 dx + \int_2^3 (3a - ax)\,dx = 1$$

when $a = 2$.

$$F_X(x) = \begin{cases} 0, & x < 0 \\ 2\int_0^x t\,dt = x^2, & 0 \le x \le 1 \\ 2\int_0^1 x\,dx + 2\int_1^x dt = 2x - 1, & 1 \le x \le 2 \\ 2\int_0^1 x\,dx + 2\int_1^2 dx + \int_2^x (6 - 2t)\,dt = 6x - x^2 - 5, & 2 \le x \le 3 \\ 1. & x > 3 \end{cases}$$

▶ **Example 2.4.2** *A number is selected at random from the interval* $[0, 1]$. *Suppose* X *is the square of the number selected. The c.d.f. of* X *is*

$$F_X(x) = \begin{cases} 0, & x < 0 \\ \sqrt{x}, & 0 \le x \le 1 \\ 1, & 1 < x \end{cases}$$

F is continuous and is differentiable everywhere except at $t = 0$ and $t = 1$.

$$f(x) = \begin{cases} \frac{1}{2\sqrt{x}}, & 0 \le x \le 1 \\ 0 & \text{otherwise.} \end{cases}$$

▶ **Example 2.4.3** *Let* X *have the triangular p.d.f.* [Fig. 2.6(a)]

$$f(x) = \begin{cases} x, & 0 < x \le 1 \\ 2 - x, & 1 \le x \le 2 \\ 0, & \text{elsewhere.} \end{cases}$$

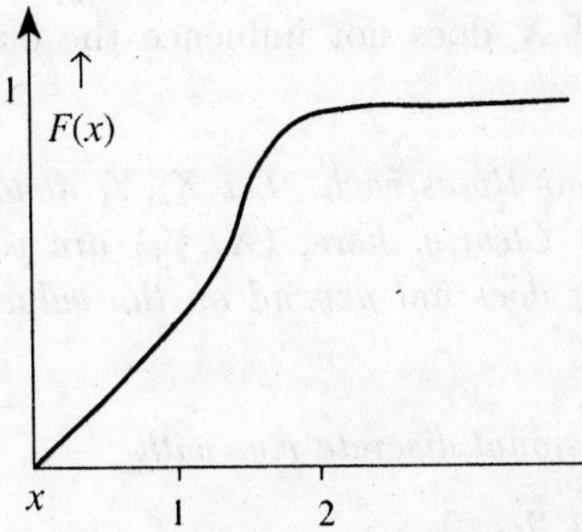

Fig. 2.6 (a) Graph of f

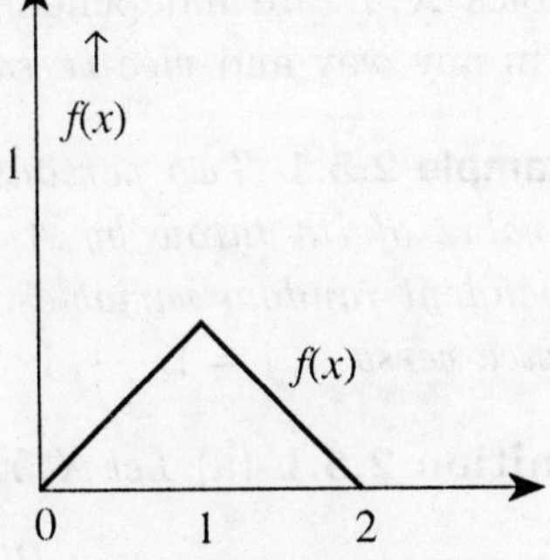

Fig. 2.6 (b) Graph of F

It is seen that f is a p.d.f. The d.f. [Fig. 2.6(b)] is

$$F_X(x) = \begin{cases} 0, & x \le 0 \\ \int_0^x t\,dt = \frac{x^2}{2}, & 0 < x \le 1 \\ \int_0^1 t\,dt + \int_1^x (2-t)\,dt = 2x - \frac{x^2}{2} - 1, & 1 < x \le 2 \\ 1, & x \ge 2 \end{cases}$$

► **Example 2.4.4** *Let for a r.v. X,*

$$F_X(x) = \begin{cases} 0, & x < 0 \\ \frac{x}{2}, & 0 \le x < 1 \\ \frac{3}{4}, & 1 \le x < 2 \\ 1, & 2 \le x \end{cases}$$

Here $F_X(x) = \frac{1}{2}(G(x) + H(x))$ where

$$G(x) = \begin{cases} 0, & x < 1 \\ \frac{1}{2}, & 1 \le x < 2 \\ 1, & 2 \le x \end{cases}, \quad H_X(x) = \begin{cases} 0, & x < 0 \\ x, & 0 \le x \le 1 \\ 1, & 1 < x \end{cases}$$

Thus X has a mixture distribution.

2.5 Independent Random Variables

Just as we defined the concept of independence between two events A and B, we may define the concept of independence between random variables. Intuitively, random variables X, Y are independent if the outcome of X does not influence the outcome of Y in any way and *vice-versa.*

► **Example 2.5.1** *Two persons A, B toss a die four times each. Let X_i, Y_i denote the face value of ith throw by A and B respectively. Clearly, here, (X_i, Y_j) are pairs of independent random variables, because value of X_i does not depend on the value of Y_j and vice versa, $i, j = 1, \cdots, 4$.*

Definition 2.5.1 (a) *Let X, Y be each one-dimensional discrete r.v. with*

$$P(X = x_i) = p_i \ \ (i = 1, 2, \cdots),$$
$$P(Y_j = y_j) = q_j \ \ (j = 1, 2, \cdots).$$

We say X, Y are independent r.v.'s if and only if the joint probability $P(X = x_i, Y = y_j) = p_i q_j$ for all i, j.

(b) *Let X, Y be each one-dimensional continuous r.v. with p.d.f.'s $g(x)$, $h(y)$ respectively. We say, X, Y are independent r.v.'s if and only if*

$$\begin{aligned} & P\left\{x - \frac{1}{2}\,dx \le X \le x + \frac{1}{2}\,dx, \ \ y - \frac{1}{2}\,dy \le Y \le y + \frac{1}{2}\,dy\right\} \\ &= P\left\{x - \frac{1}{2}\,dx \le X \le x + \frac{1}{2}\,dx\right\} \ P\left\{y - \frac{1}{2}\,dy \le Y \le y + \frac{1}{2}\,dy\right\} \\ &= g(x)h(y)\,dx\,dy \end{aligned}$$

for all x, y in the domain of X, Y respectively.

The definition can similarly be extended to n variables $X_1, \cdots, X_n$. $X_1, \cdots, X_n$, when they are discrete, are said to be independent if and only if

$$P(X_1 = x_1, \cdots, X_n = x_n) = P(X_1 = x_1) \cdots P(X_n = x_n)$$

for all $x_1, \cdots, x_n$ in the domain of $X_1, \cdots, X_n$ respectively.

In case $X_1, \cdots, X_n$ are continuous, they are said to be independent if and only if

$$P\left\{x_1 - \frac{1}{2}\,dx_1 \le X_1 \le x_1 + \frac{1}{2}\,dx_1, \cdots, x_n - \frac{1}{2}\,dx_n \le X_n \le x_n + \frac{1}{2}\,dx_n\right\}$$
$$= P\left\{x_1 - \frac{1}{2}\,dx_1 \le X_1 \le x_1 + \frac{1}{2}\,dx_1\right\} \cdots P\left\{x_n - \frac{1}{2}\,dx_n \le X_n \le x_n + \frac{1}{2}\,dx_n\right\}$$
$$= g_1(x_1) \cdots g_n(x_n)\,dx_1 \cdots dx_n$$

for all $x_1, \cdots, x_n$ in the domain of $X_1, \cdots, X_n$ respectively, where $g_i(x_i)$ is the p.d.f. of X_i, $i = 1, \cdots, n$.

The concept will be discussed further in chapter 7.

2.6 Lebesgue-Stieltjes Integral

In the case of a continuous variable the d.f. is given by an integral and in the case of a discrete variable by a sum. In stating results equally true for a continuous and a discrete variable it is advantageous to have a single representation valid for both the cases. For this, we first examine a few types of integrations.

(a) **Riemann-Stieltjes Integral:** Let $F_X(x)$ be a never decreasing function over $x \in [a, b]$. For any particular $x_0 \in (a, b)$ both the limits $\lim\limits_{\epsilon \downarrow 0} F(x_0 + \varepsilon) = F(x_0 + 0)$ and $\lim\limits_{\epsilon \downarrow 0} F(x_0 - \varepsilon) = F(x_0 - 0)$ exist, since $F(x_0 - \varepsilon) \le F(x_0) \le F(x_0 + \varepsilon)$. If x_0 is a point of continuity $F(x_0 - 0) = F(x_0 + 0)$. If $F(x_0 - 0) < F(x_0 + 0)$, x_0 is a point of discontinuity (increase) of $F_X(x)$ and $F(x_0 + 0) - F(x_0 - 0) = h_0$ (say) is a measure of discontinuity at x_0 [$F(x_0 + 0) - F(x_0)$ = right hand discontinuity; $F(x_0) - F(x_0 - 0)$ = left hand discontinuity]. For any number of points of discontinuity, $y_0, y_1, \cdots, y_n$ the sum of discontinuities $h_0 + \cdots + h_n \le F(b) - F(a)$ and the points of discontinuities form a measurable set.

Let $g(x)$ be a continuous function in $[a, b]$. Divide the range $[a, b]$ into n subintervals by inserting points $x_1, x_2, \cdots, x_{n-1}$. In order of these arbitrarily choose points $\xi_1, \xi_2, \cdots, \xi_n$ in these intervals and form the sum

$$S = \sum_1^n g(\xi_i)\,[F(x_i) - F(x_{i-1})], \quad x_0 = a_1, x_n = b.$$

If M_i, m_i respectively are the upper and lower bound of $g(x)$ in (x_{i-1}, x_i), define upper and lower sum, respectively, as

$$S_U = \sum_1^n M_i\,[F(x_i) - F(x_{i-1})], \quad S_L = \sum_1^n m_i\,[F(x_i) - F(x_{i-1})].$$

In the same way as for the ordinary integral (Riemann (R) integral), it is proved when max $(x_i - x_{i-1}) \to 0$, S tends to a definite limit and is called the Riemann-Stieltje (*RS*) integral of $g(x)$ w.r.t. $F(x)$ over $[a, b]$ and is given by

$$\lim_{n\to\infty} S = \lim_{n\to\infty} S_U = \lim_{n\to\infty} S_L = \int_a^b g(x)\, dF(x) \qquad (2.6.1$$

whatever ξ_i in (x_{i-1}, x_i), provided max $(x_i - x_{i-1}) \to 0$.

The *RS* integral (2.6.1) exists in the more general case when $g(x)$ is bounded in $[a, b$ and has at most a finite number of discontinuities r_i, provided $F(x)$ is continuous a every r_i. When $F(x) = x$, the RS integral (2.6.1) becomes the ordinary *R*-integral.

■ NOTE 2.6.1 *In case $F(x)$ is continuous everywhere in $[a, b]$ and has continuou derivative $F'(x)$ except at most in a finite number of points, it can be shown that* (2.6.1 *becomes the ordinary R-integral*

$$\int_a^b g(x)F'(x)\, dx. \qquad (2.6.2$$

■ NOTE 2.6.2 *If $f(x)$ is a step function with discontinuity $p_1, p_2, \cdots$ at points x_1, x_2 $\cdots$, RS integral becomes the sum $\sum p_i g(x_i)$, which is a finite sum or absolutely convergen infinite series according to the set of points $\{x_i\}$ is finite or infinite.*

■ NOTE 2.6.3 *The RS-integral extends to the case of a complex-valued function $g(x$ and also to infinite interval (a, b) provided $g(x)$ is integrable over (a, b) w.r.t. $F(x)$. Her*

$$\lim_{n\to\infty} \sum_1^n g(\xi_i)\,[F(x_i) - F(x_{i-1})] = \int_{-\infty}^{\infty} g(x)\, dF(x) \qquad (2.6.3$$

when max $(x_i - x_{i-1}) \to 0$ as $n \to \infty$ and $a \to -\infty$, $b \to \infty$ as $n \to \infty$.

(b) **Lebesgue-Stieltjes Integral :** Lebesgue-Stieltjes (LS) integral of $g(x)$ over a set S w.r.t. a given (non-negative and additive set function) measure μ is similarly defined as

$$\lim_{n\to\infty} S_L = \lim_{n\to\infty} S_U = \int_S g(x)\, d\mu(S) \qquad (2.6.4)$$

provided $g(x)$ is bounded for all $x \in S$ and $\mu(S) < \infty$, where S is divided into n disjoint subsets S_i, M_i, m_i being respectively the upper and lower limits of $g(x_i)$ in S_i

$$\left[S_U = \sum_{i=1}^n M_i \mu(Si), \;\; S_L = \sum_{i=1}^n m_i \mu(S_i) \right].$$

Since to any measure $\mu(B)$ defined for all borel sets $B \in \mathcal{B}_1$ there exists a non-decreasing point function $f(x)$ such that $F(b) - F(a) = \mu\{(a, b]\}$, (2.6.4) can also be written as

$$\int_S g(x)\, dF(x). \qquad (2.6.5)$$

In particular when μ is a Lebesgue measure the integral (2.6.4) or (2.6.5) becomes Lebesgue-integral.

2.7 Integration of a Random Variable

Consider the probability space $(\mathcal{S}, \mathcal{A}, P)$ and $X(w)$, a r.v. relative to $\mathcal{A}$. When $X(w)$ is a simple function (r.v.) [for an explanation, vide example 2.2.3], the LS-integral of $X(w)$ w.r.t. $(\mathcal{S}, \mathcal{A}, P)$ over $\mathcal{S}$ is defined by

$$\int_{\mathcal{S}} X(w)\, dP(w) = \sum_{i=1}^{n} x_i P(A_i).$$

In general for any subset $E \in \mathcal{A}$,

$$\int_E X(w)\, dP(w) = \sum_{i=1}^{n} x_i P(E \cap A_i). \tag{2.7.1}$$

When $X(w)$ is bounded and can take infinitely many values, there exists a sequence of simple r.v.'s $\{X_n(w)\}$ such that

$$\lim_{n\to\infty} X_n(w) = X(w) \tag{2.7.2}$$

[i.e., $|X_n(w) - X(w)| < \varepsilon$ for every $\varepsilon > 0$ for sufficiently large n] uniformly for all $w \in \mathcal{S}$. In this case LS-integral of $X(w)$ w.r.t. $(\mathcal{S}, \mathcal{A}, P)$ over any $E \subset \mathcal{A}$ is defined by

$$\int_{\mathcal{S}} X(w)\, dP(w) = \lim_{n\to\infty} \int_{\mathcal{S}} X_n(w)\, dP(w) \tag{2.7.3}$$

and this limit is independent of the choice of the sequence $\{X_n(w)\}$ satisfying (2.7.2).

When $X(w)$ is not necessarily bounded, but non-negative, it is integrable if there exists a non-decreasing sequence of bounded r.v.'s $\{X_n(w)\}$ such that (2.7.2) holds for all $w \in \mathcal{S}$ and if

$$\lim_{n\to\infty} \int_E X_n(w)\, dP(w) < \infty. \tag{2.7.4}$$

This limit is the LS-integral of $X(w)$ w.r.t. $(S, \mathcal{A}, P)$ over $E \in \mathcal{A}$. This limit when it exists does not depend on the choice of $\{X_n(w)\}$ satisfying (2.7.2) and (2.7.4).

If $X(w)$ is arbitrary, it is integrable if it can be written as $X(w)=X_1(w) - X_2(w)$, where $X_1(w), X_2(w)$ are both non-negative integrable r.v.'s and in this case

$$\int_E X(w)\, dP(w) = \int_E X_1(w)\, dP(w) - \int_E X_2(w)\, dP(w) \tag{2.7.5}$$

and its value is independent of choice of $X_1(w)$ and $X_2(w)$.

The LS-integral of a r.v. $X(w)$ w.r.t. $(\mathcal{S}, \mathcal{A}, P)$ over $E \subset \mathcal{A}$ has the following important properties.

(a) If $X(w) = k$, a constant for all $w \in \mathcal{S}$ except possibly for a set of probability measure zero (i.e., *a.e*)

$$\int_E X(w)\, dP(w) = kP(E).$$

(b) If $\alpha \leq X(w) \leq \beta$ for all $w \in \mathcal{S}$ (a.e.) where α, β are constants, then

$$\alpha P(E) \leq \int_E X(w)\, dP(w) \leq \beta P(E).$$

(c) If $X_1(w), X_2(w)$ are integrable and if α, β are constants, $\alpha X_1(w) + \beta X_2(w)$ is integrable and

$$\int_E (\alpha X_1(w) + \beta X_2(w))\, dP(w) = \alpha \int_E X_1(w)\, dP(w) + \beta \int_E X_2(w)\, dP(w).$$

(d) If $X_1(w) \leq X(w) \leq X_2(w)$ for all $w \in \mathcal{S}$ (a.e.) where $X_1(w), X(w), X_2(w)$ are integrable, then

$$\int_E X_1(w)\, dP(w) \leq \int_E X(w)\, dP(E) \leq \int_E X_2(w)\, dP(w).$$

(e) $X(w)$ is integrable if and only if $|X(w)|$ is integrable and further,

$$\left| \int_E X(w)\, dP(w) \right| \leq \int |X(w)|\, dP(w).$$

(f) If $\{X_n(w)\}$ is a sequence of r.v.s. such that

$$\lim_{n\to\infty} X_n(w) = X(w) \quad \text{and} \quad |X_n(w)| \leq Y(w),$$

where $Y(w)$ is integrable, then $X(w)$ is integrable and

$$\lim_{n\to\infty} \int_E |X_n(w) - X(w)|\, dP(w) = 0.$$

Also

$$\lim_{n\to\infty} \int_E X_n(w)\, dP(w) = \int_E X(w)\, dP(w)$$

uniformly for all $E \in \mathcal{A}$.

The readers interested in the theory of measure and integration can refer to books by Cramer (1950), Halmos (1950), Loève (1963), Munroe (1953), McShane (1944), Pitt (1963), among others.

Exercise

1. A r.v. X has an uniform distribution

$$f(x) = \begin{cases} 0, & x < a \\ \frac{1}{b-a}, & a \leq x \leq b \\ 0, & x > b. \end{cases}$$

Find its d.f.

2. A r.v. X has the p.d.f. $f(x) = \lambda e^{-\lambda x}, x > 0$. Find its d.f.

3. A r.v. X has p.d.f.

$$f(x) = \begin{cases} Kx^2 e^{-\frac{x}{2}}, & x > 0 \\ 0 & \text{otherwise} \end{cases}$$

where K is a suitable constant. Find its d.f.

4. If X has a continuous distribution with p.d.f.

$$f(x) = \begin{cases} \frac{2}{a}\left(1 - \frac{x}{a}\right), & 0 < x < a \\ 0, & \text{otherwise} \end{cases}$$

Find its d. f. Find $P\left(\frac{a}{2} < X < a\right)$. (This is called 'right triangular' distribution).
[Prob. $= \frac{1}{4}$]

5. A r.v. X has the p.d.f. $f(x) = Axe^{-h^2x^2}, x > 0$. Find the factor A. Find its d.f. (This is called Rayleigh distribution). [$A = 2h^2$]

6. Is the function $f(x)$ defined as

$$f(x) = \begin{cases} 0, & x < 2 \\ \frac{1}{18}(3 + 2x), & 2 \le x \le 4 \\ 0, & x > 4 \end{cases}$$

a density function? Find the probability that a r.v. having this density will fall in the interval [2, 3]. [Yes. $\frac{4}{9}$]

7. If $f(x)$ has the density function $f(x) = \frac{x}{2}, 0 \le x \le 2$, find the probability that (a) both of two sample values exceed 1, (b) exactly two out of four sample values will exceed 1. [(a) $\frac{9}{16}$, (b) $\frac{27}{128}$]

8. Does the function $f(x) = \frac{x+1}{\theta(\theta+1)} e^{-x/\theta} (x > 0)$, where $\theta > 0$, represent a p.d.f. ? Find the corresponding d.f.

9. Do the following functions define d.f.'s ? In case they are d.f.'s, determine the p.d.f., $f(x)$. Draw the graphs of $F_X(x)$ and $f(x)$.

(a) $F_X(x) = 0, x \le 1; 1 - \frac{1}{x}, 1 < x$.
(b) $F_X(x) = 1 - e^{-x}, x \ge 0; 0, x < 0$.
(c) $F_X(x) = \frac{2}{\pi}\sin^{-1}(\sqrt{x}), 0 \le x \le 1$.
(d) $F_X(x) = e^{3x}, -\infty < x \le 0$.
(e) $F_X(x) = \frac{x^3}{2} + \frac{1}{2}, -1 \le x \le 1$.
(f) $F_X(x) = \frac{1}{1+\exp\{-\frac{x-\alpha}{\beta}\}}, \beta > 0$.

(This is called Logistic distribution.)

10. A unbiased die is thrown. If the die shows 1 or 2, a number is selected at random from the interval [0, 1]. If the die shows 3, 4, 5 or 6, a number is selected at random from the interval [1, 2]. Show that the distribution function of the number selected is

$$F_X(x) = \begin{cases} 0, & x < 0 \\ \frac{x}{3}, & 0 \leq x < 1 \\ \frac{2x}{3} - \frac{1}{3}, & 1 \leq x < 2 \\ 1, & 2 < x. \end{cases}$$

11. Two unbiased coins are tossed. The score is 1 for each head and 0 for each tail. Find the distribution function of the total score.

12. Two unbiased dice are tossed. Find the distribution function of the total score.

13. Let X be a r.v. with d.f.

$$F_X(x) = \begin{cases} 0, & x < 0 \\ \frac{1}{2}, & x = 0 \\ \frac{1}{2} + \frac{x}{2}, & 0 < x < 1 \\ 1, & 1 \leq x. \end{cases}$$

Show that $F_X(x)$ can be written as a mixture of two d.f.'s $F_1(x)$ and $F_2(x)$ when

$$F_1(x) = \begin{cases} 0, \ x < 0 \\ 1, \ x \geq 0 \end{cases} \quad \text{and} \quad F_2(x) = \begin{cases} 0, \ x \leq 0 \\ x, \ 0 < x < 1 \\ 1, \ 1 \leq x \end{cases}$$

$[F(x) = \frac{1}{2}(F_1(x) + F_2(x))]$

14. The percentage of a component $(100X)$ in a certain medical compound may be considered as a r.v., where X has the p.d.f.

$$F_X(x) = 20x^3(1-x), \ \ 0 < x < 1.$$

(a) Find the d.f. and sketch its graph; (b) Find $P(X \leq \frac{2}{3})$; (c) When $x \leq \frac{1}{3}$ the compound sells for Rs. 10.00 per litre. When $x > \frac{1}{3}$ it sells for Rs. 12.00 per litre. If the production cost is Rs. 9.50 per litre find the probability distribution of net profit per litre. [(a) $5x^4 - 4x^5$, (c) Hints: P_r (Profit $= 0.50$ Rs./litre) $= \frac{8}{243}$]

15. A r.v. X has a Laplace distribution $f(x) = ae^{-\lambda|x|}$, $\lambda > 0$, $-\infty < x < \infty$. Find the factor a. Find $F_X(x)$. [$a = \frac{1}{2}$, $F_X(x) = \frac{1}{2}e^{\lambda x}$, $x \leq 0$; $1 - \frac{1}{2}\ e^{\lambda x}$, $x > 0$]

16. The r.v. X has a Cauchy distribution $f(x) = \frac{a}{\pi(1+x^2)}$, $-\infty < x < \infty$. Find the factor a. Find the d.f. Find $P(-1 < X < 1)$. [$\frac{1}{\pi}$, $F_X(x) = \frac{1}{\pi}$ arc tan $x + \frac{1}{2}$, Prob. $= \frac{1}{2}$]

17. A variable X has p.d.f.

$$f(x) = \begin{cases} ae^{-a(x-x_0)}, & x > 10 \\ 0, & x < 10 \end{cases}$$

Find the factor a. Find its d.f.

18. An angle θ is chosen at random from $[-\frac{\pi}{2}, \frac{\pi}{2}]$ and a line is drawn through the point $(0, a)$ making an angle θ with the y-axis. Find the density function for the point x at which this line crosses the x-axis. $[f(x) = \frac{a}{\pi(a^2+x^2)}]$

19. A r.v. X has a density $f_1(x)$ with probability p_1 and a density $f_2(x)$ with probability $p_2, p_1 + p_2 = 1$. Find the p.d.f. and distribution function of X.

20. A r.v. X has possible values $1, 2, \cdots, j \cdots$ with probability $P(X = j) = \frac{1}{2^j}$, $j = 1, 2, \cdots$. Calculate (a) $P(X$ is odd); (b) $P(X \geq 6)$; (c) $P(X$ is divisible by 5). [(a) $\frac{2}{3}$]

21. Suppose f and g are p.d.f. on the same interval, $\alpha \leq x \leq \beta$.

(a) Show that $f + g$ is not a p.d.f. on that interval,

(b) Show that for every number $a, 0 < a < 1$, $af(x)+ (1 - a)\ g(x)$ is a p.d.f. on that interval.

22. If X_1 and X_2 are independently and identically distributed random variables, prove that

(a) $P\{|X_1 - X_2| > t\} \leq 2P\{|X_1| > \frac{t}{2}\}$;

(b) if $a \geq 0$ such that $P\{X_1 \geq a\} \leq 1 - p$ and $P\{X_1 \leq -a\} \leq 1 - p$, then $P\{|X_1 - X_2| \geq \varepsilon\} \geq pP\{|X_1| > a + \varepsilon\}$.

[*Hints:* (a) $\{|X_1| > \frac{t}{2}\} \cup \{|X_2| > \frac{t}{2}\} \supset \{|X_1 - X_2| > t\}$

$\therefore P\{|X_1 - X_2| > t\} \leq P\{|X_1| > \frac{t}{2}\} + P\{|X_2| > \frac{t}{2}\} = 2P\{|X_1| > \frac{t}{2}\}$

(b) $\{X_1 \geq a + \varepsilon, X_2 \leq a\} \cup \{X_1 \leq -a - \varepsilon, X_2 \geq -a\} \subset \{|X_1 - X_2| \geq \varepsilon\}$

$\therefore P\{|X_1 - X_2| \geq \varepsilon\} \geq P\{X_1 > a + \varepsilon,\ X_2 \leq a\} + P\{X_1 < -a - \varepsilon,\ X_2 \geq -a\}$

$\geq pP\{X_1 \geq a + \varepsilon\} + pP\{X_1 \leq -a - \varepsilon\} = pP\{|X_1| > a + \varepsilon\}$.]

23. For any real r.v. X with finite expectation, show that

$$\sum_{n=1}^{\infty} P(|X| \geq n) \leq E(|X|) < 1 + \sum_{n=1}^{\infty} P(|X| \geq n).$$

[*Hints:* Let $Y = |X|$ and suppose G is the d.f. of Y.

$$E(Y) = \int_0^{\infty} ydG(y) = \sum_{i=0}^{\infty} \int_i^{(i+1)} y\,dG(y).$$

Now

$$i \int_i^{(i+1)-} dG(y) \le \int_i^{(i+1)-} y\, dG(y) \le (i+1) \int_i^{(i+1)-} dG(y),$$

$$\text{i.e., } iP[i \le Y < i+1] \le \int_i^{(i+1)-} y\, dG(y) \le (i+1)P[i \le Y < i+1].$$

Summing over all i

$$\sum_{i=0}^{\infty} iP[i \le Y < i+1] \le E(Y) \le \sum_{i=0}^{\infty} (i+1)P[i \le Y < i+1].$$

Now

$$\sum_{i=0}^{\infty} iP[i \le Y < i+1] = \sum_{n=1}^{\infty} P(|X| \ge n).$$

$$\sum_{i=0}^{\infty} (i+1)P[i \le Y < i+1] = 1 + \sum_{n=1}^{\infty} P(|X| \ge n).]$$

Chapter 3

Some Characteristics of a Probability Distribution

3.1 Introduction

This chapter deals with some characteristics of probability distribution of a r.v., namely, its expectation, moments, measures of central tendency and dispersion and some probability inequalities. We study expectation in section 3.2, moments in sections 3.3–3.6, measures of central tendency in section 3.7, measures of dispersion in section 3.8, measures of Skewness and Kurtosis in section 3.9. Lastly some probability inequalities are discussed in section 3.10.

3.2 Expectation

Let X be a r.v. defined on a probability space $(\mathcal{S}, \mathcal{A}, P)$ and let F be its distribution function. The expectation of X is defined by

$$E(X) = \int_{-\infty}^{\infty} x\,dF(x) \tag{3.2.1}$$

$$= \int_{\mathcal{S}} X dP \tag{3.2.2}$$

$$= \int X dP_X, \tag{3.2.3}$$

where the integral (3.2.1) is the Reimann-Stieltjes (RS)-integral of x w.r.t. $F(x)$ over R^1, the integral (3.2.2) is the Lebesgue-Stieltjes (LS)-integral of the r.v. X w.r.t. $(\mathcal{S}, \mathcal{A}, P)$ over $\mathcal{S}$ and the integral (3.2.3) is the LS-integral of X w.r.t. the induced probability space $(R^1, \mathcal{B}_1, P_X)$ over R^1. $E(X)$ is said to exist if and only if

$$\int_{-\infty}^{\infty} |x|\,dF(x) < \infty. \tag{3.2.4}$$

Thus if X be a r.v. of discrete type with p.m.f. $P(X = x_k) = p(x_k)$, $k = 1, 2, \cdots$ the series

$$E(X) = \sum_{k=1}^{\infty} x_k p(x_k) \tag{3.2.5}$$

is called the expectation of X provided the condition

$$\sum_{k=1}^{\infty} |x_k| p(x_k) < \infty \tag{3.2.6}$$

is satisfied. If X is a continuous r.v. with p.d.f. $f(x)$, the integral

$$E(X) = \int_{-\infty}^{\infty} x f(x)\, dx \tag{3.2.7}$$

is called the expectation of X provided the condition

$$\int_{-\infty}^{\infty} |x| f(x)\, dx < \infty \tag{3.2.8}$$

is satisfied.

□ **Remark 3.2.1** *The expectation of a r.v. X may not always exist. Before concluding that $E(X)$ exists one must check if* (3.2.6) *or* (3.2.8) *is satisfied.*

The notion of mathematical expectation is due to Huygens (1629–95), a Dutch mathematician.

► **Example 3.2.1** *A variate X has p.d.f.*

$$\begin{aligned} f(x) &= \frac{1}{b-a} \quad \textit{if } a \le x \le b \\ &= 0 \qquad \textit{otherwise} \\ E(X) &= \int_a^b \frac{x\, dx}{b-a} = \frac{a+b}{2}. \end{aligned}$$

► **Example 3.2.2** *A variate X has p.d.f.* $f(x) = \frac{1}{x\sqrt{2\pi}} e^{-\frac{1}{2}(\log x)^2}$, $x > 0$

$$\begin{aligned} E(X) &= \int_0^{\infty} \frac{1}{\sqrt{2\pi}} e^{-\frac{1}{2}(\log x)^2}\, dx = \frac{1}{\sqrt{2\pi}} \int_{-\infty}^{\infty} e^{-\frac{t^2}{2}} e^t\, dt \quad (\textit{putting } t = \log x) \\ &= \frac{\sqrt{e}}{\sqrt{2\pi}} \int_{-\infty}^{\infty} e^{-\frac{1}{2}(t-1)^2}\, dt = \sqrt{e}. \quad \left[\because \frac{1}{\sqrt{2\pi}} \int_{-\infty}^{\infty} e^{-y^2/2}\, dy = 1\right] \end{aligned}$$

► **Example 3.2.3** *The following example shows that* (3.2.5) *is finite though the condition* (3.2.6) *is not satisfied.*

Let X be a discrete r.v. with p.m.f. $P(X = x) = \frac{6}{\pi^2 x^2}$ (0) for $x = 1, -2, +3, -4, \cdots$ otherwise).

$$E(X) = \frac{6}{\pi^2} \sum_{x=1}^{\infty} \frac{(-1)^{x-1}}{x} < \infty.$$

However,

$$E|X| = \frac{6}{\pi^2} \sum_{x=1}^{\infty} \frac{1}{x} = \infty.$$

Hence $E(X)$ does not exist.

▶ **Example 3.2.4** *The distribution* $f(x) = \frac{1}{\pi(1+x^2)}$, $-\infty \leq x \leq \infty$ *is known as the Cauchy distribution. Here* $E(X)$ *exists if*

$$E\{|X|\} = \lim_{\substack{a \to -\infty \\ b \to \infty}} \int_a^b \frac{|x|}{\pi(1+x^2)}\, dx < \infty.$$

However,

$$\int_0^b \frac{|x|}{\pi(1+x^2)}\, dx = \frac{1}{2\pi}\left(\log(b^2+1)\right) \to \infty$$

as $b \to \infty$. *Hence* $E(X)$ *does not exist for this distribution.*

We have seen in theorem 2.2.9, if X is a r.v. on $(\mathcal{S}, \mathcal{A}, P)$, so is any borel-measurable function $g(X)$. $E[g(X)]$, if it exists, is given by

$$E[g(X)] = \int_{-\infty}^{\infty} g(x)\, dF(x) \tag{3.2.9}$$

Thus if X is discrete with p.m.f. $p(x_i)$,

$$E[g(X)] = \sum_{i=1}^{\infty} g(x_i) p(x_i); \tag{3.2.10}$$

if X is continuous with p.d.f. $f(x)$,

$$E[g(X)] = \int_{-\infty}^{\infty} g(x) f(x)\, dx \tag{3.2.11}$$

provided the expectation exists.

3.2.1 Some Results on Expectation

1. If the r.v. X is a constant c, i.e., if $X(w) = c$ a.e., i.e., if $P\{X = c\} = 1$, $E(X) = c$.

   Proof:

$$E(X) = \int_{-\infty}^{\infty} x\, dF(x) = \int_{-\infty}^{\infty} c\, dF(x) = c\int_{-\infty}^{\infty} dF(x) = c.$$

2. If $E(X)$ exists $E(aX+b)$ exists and is given by

$$E(aX+b) = aE(X)+b$$

where a, b are arbitrary constants.

Proof:

$$\int_{-\infty}^{\infty} |ax+b|\, dF(x) < \int_{-\infty}^{\infty} |ax|\, dF(x) + |b| \int_{-\infty}^{\infty} dF(x) < |a| \int_{-\infty}^{\infty} |x|\, dF(x) + |b| < \infty$$

since $E|X| < \infty$. Also

$$E(aX+b) = \int_{-\infty}^{\infty} (ax+b)\, dF(x) = a \int_{-\infty}^{\infty} x\, dF(x) + b \int_{-\infty}^{\infty} dF(x) = aE(X)+b.$$

3. If $g(X) = g_1(X) + g_2(X)$ where $g_1(X)$, $g_2(X)$ are two borel measurable functions o r.v. X, then

$$E[g(X)] = E[g_1(X)] + E[g_2(X)]$$

provided the expectations exist.

Proof:

$$\begin{aligned} E[g(X)] &= \int_{-\infty}^{\infty} [g_1(x) + g_2(x)]\, dF(x) \\ &= \int_{\infty}^{\infty} g_1(x)\, dF(x) + \int_{-\infty}^{\infty} g_2(x)\, dF(x) = Eg_1(X) + Eg_2(X). \end{aligned}$$

4. If $g_1(X) \leq g_2(X)$ for all possible values of X, then
(i) $Eg_1(X) \leq Eg_2(X)$; (ii) $|Eg_1(X)| \leq E|g_1(X)|$.

Proof:

$$\text{(i)} \quad Eg_1(X) = \int_{-\infty}^{\infty} g_1(x)\, dF(x) \leq \int_{-\infty}^{\infty} g_2(x)\, dF(x) = Eg_2(X)$$

$$\text{(ii)} \quad |Eg_1(X)| = \left| \int_{-\infty}^{\infty} g_1(x)\, dF(x) \right| \leq \int_{-\infty}^{\infty} |g_1(x)|\, dF(x) = E|g_1(X)|$$

5. ***Conditional expectation of X given A.*** Let A be a measurable subset of the sampl sapce S. Consider the r.v. XI_A where I_A is the indicator function so that for a point w

$$\begin{aligned} XI_A(w) &= X(w) \text{ if } w \in A \\ &= 0 \qquad \text{otherwise.} \end{aligned}$$

The conditional expected value of X given A is defined as

$$E(X|A) = \frac{E(XI_A)}{P(A)} \tag{3.2.12}$$

provided $E(XI_A)$ exists and $P(A) > 0$. Since $|XI_A| \leq |X|$, $E(XI_A)$ exists if $E(X)$ exists (Clearly $E(XI_A) = \Sigma_{w \in A} X(w)p(w)$).

Theorem 3.2.1 *Let $\{A_i, i = 1, \cdots, k\}$ be a measurable partition of the sample sapce S i.e., $\bigcup_{i=1}^k A_i = S$, $A_iA_j = \phi$, $i \neq j = 1, \cdots, k$) such that $P(A_i) > 0$ for all i and let $E(X)$ exist. Then*

$$E(X) = \sum_{i=1}^{k} E(X|A_i)P(A_i) \; (= E_A[E(X|A)]) \tag{3.2.13}$$

$E_A(\cdot)$ denotes expectation with respect to A i.e., for different subsets $A_i, i = 1, \cdots, k$).

Proof:

$$\begin{aligned} E(X) &= \int XdP = \int_{\Sigma A_i} XdP = \sum_i \int XI_{A_i}dP \\ &= \sum_i E(XI_{A_i}) = \sum_i E(X|A_i)P(A_i) \end{aligned} \tag{3.2.14}$$

by (3.2.12).

Theorem 3.2.2 *If $X_1, \cdots, X_n$ are each one-dimensional r.v.s, then*

$$E(X_1 + \cdots, X_n) = E(X_1) + \cdots + E(X_n).$$

Theorem 3.2.3 *If $X_1, \cdots, X_n$ are independent r.v.s, $E(X_1 \cdots X_n) = E(X) \cdots E(X_n)$. Proof of these theorems will be given in chapter 7.*

6. ***An inequality for convex functions.*** A continuous function $f(x)$ is said to be convex in an interval I, if for any two values x_1, x_2 in I,

$$\lambda f(x_1) + (1-\lambda)f(x_2) \geq f(\lambda x_1 + (1-\lambda)x_2)$$

for any $\lambda, 0 < \lambda < 1$, i.e., any point lying on the straight line joining any two points of $f(x)$, $x \in I$ lies above the point with the same value of abscissa on the curve (Fig. 3.1). For such a function for every point x_0 in I there exists a straight line which passes through $(x_0, f(x_0))$ and lies on or below the graph of $f(x)$. The following inequality holds in the case of a convex function.

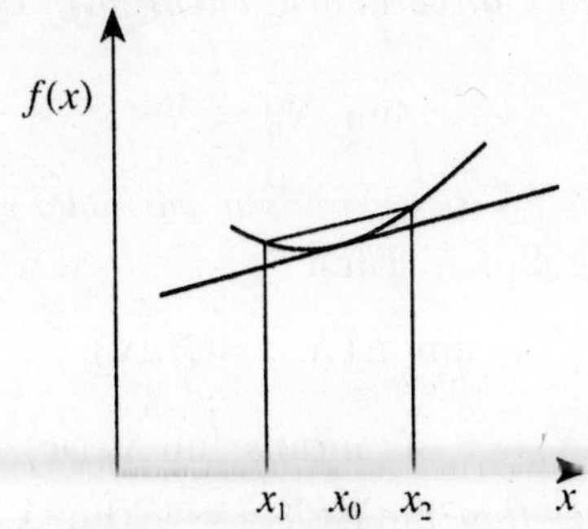

Fig. 3.1

Jensen's Inequality. Let X be a r.v. and $f(X)$ a convex function of X. Then

$$E[f(X)] \geq f[E(X)]$$

provided the expectations exist. The equality holds if and only if X has a degenerat distribution [i.e., $X(w) = k$, a constant, a.e.].

Proof: Since $f(X)$ is continuous and convex there exists a line $l(X) = a + bX$ whic passes through the point $[E(X), f(E(X))]$ and lies below it, i.e., $l(X) \leq f(X)$ for a values of X and $l[E(X)] = f[E(X)]$. Now

$$l[E(X)] = a + bE(X) = E[l(X)].$$

Hence

$$f[E(X)] = E[l(X)] \leq E[f(X)], \text{ i.e., } E[f(X)] \geq f[E(X)].$$

► **Example 3.2.5** $f(x) = x^2$ *is a convex function of* x. *Hence*

$$E(X^2) \geq (E(X))^2, \text{ i.e., } V(X) \geq 0.$$

► **Example 3.2.6** *Since* $|X|^r$ *is a convex function of* X *for* $r > 1$, *we have*

$$E|X|^r \geq [E|X|]^r.$$

We consider two limiting properties of expectation.

Let $X_1, X_2, \cdots$ be a sequence of random variables all defined on the same sampl space $\mathcal{X}$. Also let $\lim_{n\to\infty} X_n = X$ mean $\lim_{n\to\infty} X_n(w) = X(w)$ for every $w \in \mathcal{X}$.

Theorem 3.2.4 (***Monotone Convergence Theorem***). *If* $X_1, X_2, \cdots$ *is a sequence o random variables such that*

$$0 \leq X_1 \leq X_2 \leq \cdots$$

and

$$\lim_{n\to\infty} X_n = X$$

then

$$\lim_{n\to\infty} E(X_n) = E(X).$$

Theorem 3.2.5 (***Dominated Convergence Theorem***). *Let* $X_1, X_2, \cdots$ *be a sequence o random variables such that*

$$\lim_{n\to\infty} X_n = X.$$

Moreover, assume that there exists a random variable Y *with* $E(Y) < \infty$ *such tha* $|X| \leq Y$, $|X_n| \leq Y$, *for* $n = 1, 2, \cdots$. *Then*

$$\lim_{n\to\infty} E(X_n) = E(X).$$

In both the theorems, $\lim_{n\to\infty} X_n = X$ *means* $\lim X_n(w) = X(w)$ *for every* $w \in \mathcal{X}$. *Also, both the theorems assert that* $\lim E(X_n) = E(\lim X_n)$, *which means that we ma interchange the order of integration and passage to the limit.*

3.3 Moments

Let X be a r.v. with p.m.f. $p(x_k)$ or p.d.f. $f(x)$. The rth raw (non-central) moment of the distribution of X, if it exists, is defined as

$$\mu'_r = E(X^r) = \sum_{i=1}^{\infty} x_i^r p(x_i) \tag{3.3.1}$$

$$\text{or} \int_{-\infty}^{\infty} x^r f(x)\, dx \tag{3.3.2}$$

according as X is discrete or continuous. The series (3.3.1) or the integral (3.3.2) must be absolutely convergent for μ'_r to exist. The first moment μ'_1, if it exists, is the expectation of X. Clearly $\mu'_0 = 1$.

The rth moment of X about a point 'a', if it exists, is given by,

$$\mu'_r(a) = E(X-a)^r = \sum_{i=1}^{\infty} (x_i - a)^r p(x_i) \tag{3.3.3}$$

$$\text{or} \int_{-\infty}^{\infty} (x-a)^r f(x)\, dx \tag{3.3.4}$$

according as X is discrete or continuous.

In particular if $E(X) = \mu'_1 = \mu$ (say) exists, then one may take $a = \mu$. The expectation of $(X-\mu)^r$, when it exists, is called the rth central moment of X and is generally denoted as μ_r. Thus

$$\mu_r = \mu'_r(\mu) = E(X-\mu)^r = \sum_{i=1}^{\infty} (x_i - \mu)^r p(x_i) \tag{3.3.5}$$

$$\text{or} \int_{-\infty}^{\infty} (x-\mu)^r f(x)\, dx \tag{3.3.6}$$

according as X is discrete or continuous. Expanding $(X-\mu)^r$ in the integral (or summation), we get

$$\begin{aligned}\mu_r &= \int_{-\infty}^{\infty} \left[x^r - \binom{r}{1} x^{r-1}\mu + \binom{r}{2} x^{r-2}\mu^2 - \cdots + (-1)^r \mu^r\right] f(x)\, dx \\ &= \mu'_r - \binom{r}{1}\mu'_{r-1}\mu + \binom{r}{2}\mu'_{r-2}\mu^2 - \cdots + (-1)^r \mu^r. \end{aligned} \tag{3.3.7}$$

The relation (3.3.7) gives the expressions for the central moments in terms of raw moments. Thus,

$$\mu_1 = \mu'_1 - \mu'_0\mu = 0; \quad \mu_3 = \mu'_3 - 3\mu\mu'_2 + 2\mu^3;$$

$$\mu_2 \text{ (or } \sigma^2) = \mu'_2 - \mu^2; \ \mu_4 = \mu'_4 - 4\mu\mu'_3 + 6\mu^2\mu'_2 - 3\mu^4$$

Similarly one may express μ'_r in terms of central moments by considering

$$\int_{-\infty}^{\infty} x^r \, dF(x) = \int_{-\infty}^{\infty} \{(x-\mu)+\mu\}^r \, dF(x)$$

when

$$\mu'_r = \mu_r + \binom{r}{1}\mu_{r-1}\mu + \binom{r}{2}\mu_{r-2}\mu^2 + \cdots + \mu^r. \tag{3.3.8}$$

In particular,

$$\mu'_1 = \mu; \qquad \mu'_3 = \mu_3 + 3\mu_2\mu + \mu^3;$$
$$\mu'_2 = \mu_2 + \mu^2; \; \mu'_4 = \mu_4 + 4\mu_3\mu + 6\mu_2\mu^2 + \mu^4.$$

In theorems 3.3.1–3.3.5 below we consider the problem of existence of moments.

Theorem 3.3.1 *If μ'_r exists for a r.v. X, then μ'_s exists, $0 < s < r$.*

Proof: We have for $|X| \geq 1, |X|^s \leq |X|^r$; for $|X| < 1$, $1 > |X|^s > |X|^r$. Now

$$E|X|^s = \int_{|x|\leq 1} |x|^s \, dF(x) + \int_{|x|>1} |x|^s \, dF(x)$$
$$\leq \int_{|x|\leq 1} dF(x) + \int_{|x|>1} |x|^r \, dF(x) < 1 + E|X|^r < \infty.$$

The following theorems can be proved similarly.

Theorem 3.3.2 *If $\mu'_r(a)$ exists for a r.v. X, then $\mu'_s(a)$ exists where a is an arbitrary constant, $0 < s < r$.*

Theorem 3.3.3 *If μ_r exists for a r.v. X, then μ_s exists, $0 < s < r$.*

Theorem 3.3.4 *If μ'_r exists, $\mu'_r(a)$ exists and is given by*

$$\mu'_r(a) = \mu'_r - \binom{r}{1}\mu'_{r-1}a + \binom{r}{2}\mu'_{r-2}a^2 - \cdots + (-1)^r a^r. \tag{3.3.9}$$

Proof: (3.3.9) follows as in (3.3.7) by expanding $(x-a)^r$ in

$$\int_{-\infty}^{\infty} (x-a)^r \, dF(x).$$

Existence follows from the fact

$$\int_{-\infty}^{\infty} |x-a|^r \, dF(x) \leq \int_{-\infty}^{\infty} |x|^r \, dF(x) + \binom{r}{1}|a| \int_{-\infty}^{\infty} |x|^{r-1} \, dF(x)$$
$$+ \binom{r}{2}|a|^2 \int_{-\infty}^{\infty} |x|^{r-2} \, dF(x) + \cdots + \cdots |a|^r$$

and since μ'_r exists, $\mu'_{r-1}, \mu'_{r-2} \cdots$ exist.

Theorem 3.3.5 *If $\mu_r'(a)$ exists, μ_r' exists and is given by*

$$\mu_r' = \mu_r'(a) + \binom{r}{1} a\mu_{r-1}'(a) + \binom{r}{2} a^2 \mu_{r-2}'(a) + \cdots + a^r. \tag{3.3.10}$$

Proof: Follows as in (3.3.8) and theorem 3.3.4.

The quantity $E|X|^r = \nu_r$ (say), when it exists, is called the rth absolute moment of r.v. X.

▶ **Example 3.3.1** *Find the rth non-central moment of the rectangular distribution*

$$f(x) = \frac{1}{b-a}, a < x < b; = 0, \text{ otherwise}$$

$$\mu_r' = \frac{1}{b-a}\int_a^b x^r\,dx = \frac{b^{r+1} - a^{r+1}}{(b-a)(r+1)}.$$

Thus

$$\mu = \frac{b+a}{2}, \quad \sigma^2 = \frac{(b-a)^2}{12}.$$

▶ **Example 3.3.2** *A Continuous r.v. X has the distribution function $F(x)$ proportional to*

$$\alpha x^\beta - \beta x^\alpha, \quad \alpha > \beta \geq 1, \quad 0 \leq X < 1.$$

Show that the rth moment of X about the origin is $\alpha\beta/(\alpha + r)(\beta + r)$.

Let $F(x) = k(\alpha x^\beta - \beta x^\alpha)$ when k is a suitable constant.

$$f(x) = \frac{dF(x)}{dx} = k\alpha\beta\left(x^{\beta-1} - x^{\alpha-1}\right).$$

Since

$$\int_0^1 f(x)\,dx = 1, \quad k = \frac{1}{\alpha - \beta}.$$

$$\mu_r' = \frac{(\alpha\beta)}{\alpha - \beta}\int_0^1 x^r\left(x^{\beta-1} - x^{\alpha-1}\right)dx = \frac{k\alpha\beta}{(\alpha + r)(\beta + r)}.$$

▶ **Example 3.3.3** *Find the rth non-central moment of the distribution*

$$f(x) = \frac{1}{\Gamma(n)}\alpha^n e^{-\alpha x} x^{n-1}, \quad x \geq 0, n > 0$$

(This is known as Gamma distribution. $\Gamma(n) = (n-1)!$ for integral n).

$$\mu_r' = \frac{\alpha^n}{\Gamma(n)}\int_0^\infty e^{-\alpha x} x^{n+r-1}\,dx = \frac{\Gamma(n+r)}{\alpha^r\,\Gamma(n)}.$$

Hence

$$\mu_1' = \mu = \frac{n}{\alpha}, \quad \mu_2 = \mu_2' - \mu^2 = \frac{n}{\alpha^2}.$$

For $n = 1$, the distribution is known as exponential distribution when $\mu = \frac{1}{\alpha}$, $\sigma^2 = \frac{1}{\alpha^2}$.

► **Example 3.3.4** *If $f(x)$ is an odd function of x of period $\frac{1}{2}$, show that*

$$\int_0^{\infty} x^r x^{-\log x} f(\log x)\, dx = 0$$

for all integral values of r. Hence show that

$$dF = x^{-\log x}\left[1 - \lambda \sin(4\pi \log x)\right] dx, \quad 0 < x < \infty, \quad 0 \le \lambda \le 1$$

have the same moments whatever be the value of λ.

Let

$$I = \int_0^{\infty} x^{r - \log x} f(\log x)\, dx.$$

Putting $\log_e x = y + \frac{1}{2}$,

$$I = \int_{-\infty}^{\infty} e^{(y+\frac{1}{2})(r-y-\frac{1}{2})} f\left(y + \frac{1}{2}\right) e^{y+\frac{1}{2}}\, dy$$

$$= e^{\frac{1}{4}(r+1)^2} \int_{-\infty}^{\infty} e^{-\left(y - \frac{1}{2}r\right)^2} f(y)\, dy, \quad \left[\text{since } f\left(y + \tfrac{1}{2}\right) = f(y)\right]. \qquad \text{(i)}$$

Again putting $\log_e x = -y + \frac{1}{2}$

$$I = -e^{\frac{1}{4}(r+1)^2} \int_{-\infty}^{\infty} e^{-\left(y + \frac{r}{2}\right)^2} f(y)\, dy, \quad [\text{since } f(-y) = -f(y)]. \qquad \text{(ii)}$$

Adding (i) and (ii),

$$2I = e^{\frac{1}{4}(r+1)^2} \left[\int_{-\infty}^{\infty} e^{-\left(y - \frac{r}{2}\right)^2} f(y)\, dy - \int_{-\infty}^{\infty} e^{-\left(y + \frac{r}{2}\right)^2} f(y)\, dy\right]. \qquad \text{(iii)}$$

Putting $y - \frac{r}{2} = u$ in the first integral and $y + \frac{r}{2} = v$ in the second integral in (iii) and observing that

$$f\left(u + \frac{r}{2}\right) = f(u), \quad f\left(v - \frac{r}{2}\right) = f(v),$$

$$2I = e^{\frac{1}{4}(r+1)^2} \left[\int_{-\infty}^{\infty} e^{-u^2} f(u) du - \int_{-\infty}^{\infty} e^{-v^2} f(v) dv\right] = 0.$$

Hence $I = 0$. Now $\sin\,(4\pi x)$ is an odd function of x of period $\frac{1}{2}$.

Thus

$$\int_0^{\infty} x^{r - \log x} \sin(4\pi \log x) = 0.$$

Hence the moments of the given distribution do not depend on the value of λ.

3.4 Some Inequalities Concerning Moments

First we consider *Cauchy-Schwartz inequality*. If X, Y are two random variables both defined on $(\mathcal{S}, \mathcal{A}, P)$ and with finite variances then

$$E(X^2)E(Y^2) \geq (E(XY))^2. \tag{3.4.1}$$

Proof: Since

$$|XY| \leq \frac{1}{2}(X^2+Y^2), \quad E|XY| < \infty,$$

as $E(X^2), E(Y^2)$ exist and hence $E(XY)$ exists.

For any real number t, let $h(t) = E(tX - Y)^2$. Clearly $h(t)$ is real and ≥ 0. Hence the equation

$$h(t) = t^2E(X^2) - 2tE(XY) + E(Y^2) = 0 \tag{3.4.2}$$

has either no (real) solution or only one solution. (3.4.2) has no solution if and only if

$$(E(XY))^2 < E(X^2)E(Y^2).$$

(3.4.2) has exactly one solution if and only if

$$\left(E(XY)^2\right) = E(X^2)E(Y^2).$$

Hence the inequality (3.4.1). Equality in (3.4.1) holds if and only if there exists some (real) t for which $E(tX - Y)^2 = 0$, i.e., $P[tX - Y = 0 \text{ for some } t] = 1$, i.e., X is proportional to Y a.e.

■ NOTE 3.4.1 *If in* (3.4.1) *we take* $X - E(X), Y - E(Y)$ *in place of* X, Y *respectively the inequality reduces to*

$$V(X)V(Y) \geq [Cov\ (X,Y)]^2. \tag{3.4.3}$$

■ NOTE 3.4.2 *The relation* (3.4.1) *holds if* X, Y *are replaced respectively by some borel measurable functions* $g_1(X), g_2(X)$ *of r.v.* X *defined on* $(\mathcal{S}, \mathcal{A}, P)$ *provided* $E[g_1(X)]^2 < \infty$, $E[g_2(X)^2] < \infty$.

[Algebraic versions of $C - S$ inequality have been given in the appendix to this chapter.]

Theorem 3.4.1 *If both* μ'_{2r}, μ'_{2s} *exist then*

$$\text{(i) } \mu'_{2r}\mu'_{2s} \geq (\mu'_{r+s})^2; \tag{3.4.4}$$

$$\text{(ii) } \mu_{2r}\mu_{2s} \geq (\mu_{r+s})^2. \tag{3.4.5}$$

In particular,

$$\text{(iii) } \beta_2 = \frac{\mu_4}{\mu_2^2} \geq 1, \text{ provided } \mu_2 > 0. \tag{3.4.6}$$

Proof: If μ'_{2r}, μ'_{2s} exist μ_{2r}, μ_{2s} exist. Also μ_{r+s} exists (theorems 3.3.2 and 3.3.4).

Putting in (3.4.1) X^r, X^s in place of X and Y respectively the inequality (3.4.4 follows. Putting in (3.4.1), $(X - E(X))^r$, $(X - E(X))^s$ in place of X, Y respectivel (3.4.5) follows. (3.4.5) gives $\mu_4 \geq \mu_2^2$ which is (3.4.6), when $r = 2, s = 0$.

■ NOTE 3.4.3 *The inequality* (3.4.6) *can also be proved as follows: Since*

$$V(X^2) = E\left[X^2 - E(X^2)\right]^2 \geq 0 \tag{a}$$

$$E(X^4) = E\left(E(X^2)\right)^2 \geq 0, \text{ i.e., } \mu'_4 \geq (\mu'_2)^2.$$

Putting $X - E(X)$ in place of X in (a), (3.4.6) follows—equality holds if and only if

$$(X - E(X))^2 = k \text{ (a constant)}$$

for all values of X. This occurs if X takes only two values, x_1 and x_2 with equa probabilities when

$$(X - E(X))^2 = \frac{1}{4}(x_1 - x_2)^2.$$

Theorem 3.4.2 *For any r.v. X,*

$$\begin{vmatrix} \mu_{2a} & \mu_{a+b} & \mu_{a+c} \\ \mu_{a+b} & \mu_{2b} & \mu_{b+c} \\ \mu_{a+c} & \mu_{b+c} & \mu_{2c} \end{vmatrix} = \delta(a, b, c) \text{ (say) } \geq 0, \tag{3.4.7}$$

where a, b, c are non-negative integers. Hence, provided μ_4 exists,

$$\beta_2 - \beta_1 - 1 \geq 0 \tag{3.4.8}$$

(where $\beta_2 = \frac{\mu_4}{\mu_2^2}, \beta_1 = \frac{\mu_3^2}{\mu_2^3}$).

Proof: We have

$$E\left[AX^a + BX^b + CX^c\right]^2 \text{ is always } \geq 0,$$

i.e.,

$$E\left[A^2X^{2a} + B^2X^{2b} + C^2X^{2c} + 2ABX^{a+b} + 2ACX^{a+c} + 2BCX^{b+c}\right] \geq 0 \tag{3.4.9}$$

for all values of A, B and C. If X is measured from $E(X)$, (3.4.9) reduces to

$$A^2\mu_{2a} + B^2\mu_{2b} + C^2\mu_{2c} + 2AB\mu_{a+b} + 2AC\mu_{a+c} + 2BC\mu_{b+c} \geq 0 \tag{3.4.10}$$

for all values of A, B and C, i.e., the left hand side of (3.4.10) is a positive semi-definite quadratic form in A, B, C.

A set of necessary and sufficient conditions for this quadratic form to be positive semi-definite is that all the principal minors of δ are non-negative, i.e.,

$$\mu_{2a} \geq 0, \quad \begin{vmatrix} \mu_{2a} & \mu_{a+b} \\ \mu_{a+b} & \mu_{2b} \end{vmatrix} \geq 0, \quad \delta \geq 0$$

For $a = 0$, $b = 1$, $c = 2$, (3.4.7) reduces to

$$\begin{vmatrix} \mu_0 & \mu_1 & \mu_2 \\ \mu_1 & \mu_2 & \mu_3 \\ \mu_2 & \mu_3 & \mu_4 \end{vmatrix} \geq 0 \quad \text{or } \mu_2\mu_4 - \mu_3^2 - \mu_2^3 \geq 0 \quad \text{or } \beta_2 - \beta_1 - 1 \geq 0$$

Since $\beta_1 \geq 0$, $\beta_2 \geq 1$.

Corollary 3.4.1 *For any r.v. X,*

$$\begin{vmatrix} \mu_0 & \mu_1 & \mu_2 & \mu_3 \\ \mu_1 & \mu_2 & \mu_3 & \mu_4 \\ \mu_2 & \mu_3 & \mu_4 & \mu_5 \\ \mu_3 & \mu_4 & \mu_5 & \mu_6 \end{vmatrix} \geq 0.$$

Theorem 3.4.3 *For any r.v. X, $\nu_r^{\frac{1}{r}}$ is a nondecreasing function of r, i.e., $\nu_{r+1}^{\frac{1}{(r+1)}} \geq \nu_r^{\frac{1}{r}}$, $r = 1, 2, \cdots$.*

Proof: We have

$$E\left[|X|^{\frac{r+1}{2}} u + |X|^{\frac{r-1}{2}} v\right]^2 \geq 0, \tag{3.4.11}$$

where u, v are arbitrary constants. Thus (3.4.9) means

$$u^2\nu_{r+1} + v^2\nu_{r-1} + 2uv\nu_r \geq 0$$

is a positive semi-definite quadratic form in u, v. A set of necessary and sufficient conditions for this is

$$\begin{vmatrix} \nu_{r+1} & \nu_r \\ \nu_r & \nu_{r-1} \end{vmatrix} \geq 0, \text{ i.e., } \nu_{r-1}\nu_{r+1} \geq \nu_r^2 \text{ or } \nu_{r-1}^r \nu_{r+1}^r \geq \nu_r^{2r}. \tag{3.4.12}$$

Putting in (3.4.12), $r = 1, 2, \cdots, r$ we have

$$\nu_0\nu_2 \geq \nu_1^2 \tag{i}$$
$$\nu_1^2\nu_3^2 \geq \nu_2^4 \tag{ii}$$
$$\nu_2^3\nu_4^3 \geq \nu_3^6 \tag{iii}$$
$$\cdots\cdots\cdots\cdots \qquad \cdots$$
$$\nu_{r-1}^r \nu_{r+1}^r \geq \nu_r^{2r}. \tag{r}$$

Multiplying (i), ..., (r),

$$\nu_{r+1}^r \geq \nu_r^{r+1}, \quad \text{or, } \nu_{r+1}^{\frac{1}{r+1}} \geq \nu_r^{\frac{1}{r}}.$$

Thus $\frac{1}{r+1}\log\nu_{r+1} \geq \frac{1}{r}\log\nu_r$, or, $\frac{1}{r}\log\nu_r$ is an increasing function of r for positive r.

Theorem 3.4.4 *For any r.v. X,*

$$\nu_{\alpha_1}\nu_{\alpha_2}\cdots\nu_{\alpha_k} \geq \left[\left(\nu_{\frac{\alpha_1+\cdots+\alpha_k}{k}}\right)\right]^k, \tag{3.4.13}$$

where $\alpha_1, \cdots, \alpha_k$ are positive integers.

Proof: Consider generalised Cauchy-Schwartz inequality: for positive r.v.s $Y_1, \cdots, Y_k$

$$E\left(Y_1^k\right) E\left(Y_2^k\right) \cdots E\left(Y_k^k\right) \geq \{E(Y_1 \cdots Y_k)\}^k \tag{3.4.14}$$

(also see Holder's inequality in section 3.12).

Putting in (3.4.14), $Y_1 = |X-\mu|^{\alpha_1/k}, \cdots, Y_k = |X-\mu|^{\alpha_k/k}$, the inequality (3.4.13) follows.

Theorem 3.4.5 (***Liapounoff's inequality***). *For any r.v. X,*

$$\nu_b^{a-c} \leq \nu_c^{a-b}\nu_a^{b-c} \tag{3.4.15}$$

for $a \geq b \geq c \geq 0$.

Proof: Put in (3.4.13),

$$k = a-c; \;\; \alpha_1 = \alpha_2 = \cdots = \alpha_{a-b} = c; \;\; \alpha_{a-b+1} = \alpha_{a-b+2} = \cdots = \alpha_{a-c} = a.$$

Thus $\frac{\alpha_1+\cdots+\alpha_k}{k} = b$ and (3.4.15) readily follows.

3.5 Moments of a Symmetric Probability Distribution

A r.v. X is said to be symmetric about a point a if

$$P(X \geq a+x) = P(X \leq a-x)$$

for all x. In terms of the distribution function of X this means, if

$$F_X(a-x) = 1 - F_x(a+x) + P\{X = a+x\}$$

holds for all $x \in \mathcal{R}^1$, we say that the d.f. of r.v. X is symmetric with a as the centre of symmetry. If $a = 0$, then for every x, this means

$$F_X(-x) = 1 - F_X(x) + P\{X = x\}.$$

If X is a r.v. of continuous type, X is symmetric with centre a if and only if the p.d.f. $f(x)$ of X satisfies

$$f(a-x) = f(a+x) \text{ for all } x \in R^1.$$

If $a = 0$, we say simply that X is symmetric (or F is symmetric).
It follows immediately that if X is symmetric about a, $E(X)$ exists and is a.

Theorem 3.5.1 *If X is symmetric about $a, E(X) = a$ and μ_{2r+1} (if it exists) $= 0$,* $r = 0, 1, 2, \cdots$.

Suppose X is a continuous r.v. with p.d.f. $f(x)$.

$$E(X) = \int_{-\infty}^{\infty} xf(x)\,dx = \int_{-\infty}^{a} xf(x)\,dx + \int_{a}^{\infty} xf(x)\,dx = I_1 + I_2 \text{ (say)}.$$

$$\text{In } I_1 \text{ put } y = a - x;\ I_1 = \int_0^{\infty} (a - y) f(a - y)\,dy.$$

$$\text{In } I_2 \text{ put } y = x - a;\ \ I_2 = \int_0^{\infty} (a + y) f(a + y)\,dy.$$

Hence $E(X) = a$.

Again

$$\mu_{2r+1} = \int_{-\infty}^{\infty} (x-a)^{2r+1} f(x)\,dx$$

$$= \int_{-\infty}^{a} (x-a)^{2r+1} f(x)\,dx + \int_{a}^{\infty} (x-a)^{2r+1} f(x)\,dx = I_3 + I_4 \text{ (say)}.$$

Now

$$I_3 = -\int_0^{\infty} z^{2r+1} f(a - z)\,dz \text{ (putting } z = a - x);$$

$$I_4 = \int_0^{\infty} z^{2r+1} f(a + z)\,dz \text{ (putting } z = x - a).$$

Hence $\mu_{2r+1} = 0$. In case X is discrete, the result can similarly be proved.

3.6 Factorial Moments

Let X be a discrete r.v. with p.m.f. $P(X = k) = p_k$, $k = 0, 1, 2, \cdots$. Then $E\left[(X)_r\right]$, if it exists, is called the rth factorial moment of X about origin and will be denoted as $\mu_{(r)}$.

$$\mu_{(r)} = E\left[X(X-1)\cdots(X-r+1)\right] = \sum_{x=r}^{\infty} x(x-1)\cdots(x-r+1)p_x.$$

It readily follows that $\mu_{(r)}$ exists if and only if μ'_r exists. It is easy to check

$$\mu_{(1)} = \mu'_1; \qquad \mu_{(3)} = \mu'_3 - 3\mu'_2 + 2\mu'_1;$$
$$\mu_{(2)} = \mu'_2 - \mu'_1; \ \mu_{(4)} = \mu'_4 - 6\mu'_3 + 11\mu'_2 - 6\mu'_1.$$

Again since $x^2 = (x)_2 + x$, $x^3 = (x)_3 + 3(x)_2 + x$, $x^4 = (x)_4 + 6(x)_3 + 7(x)_2 + x, \cdots$ it follows that

$$\mu'_1 = \mu_{(1)}; \qquad \mu'_3 = \mu_{(3)} + 3\mu_{(2)} + \mu_{(1)};$$
$$\mu'_2 = \mu_{(2)} + \mu_{(1)}; \ \mu'_4 = \mu_{(4)} + 6\mu_{(3)} + 7\mu_{(2)} + \mu_{(1)}.$$

▸ **Example 3.6.1** *Consider the Poisson distribution*

$$P(X=x)=e^{-\lambda}\frac{\lambda^x}{x!},\quad x=0,1,2,\cdots,\lambda>0.$$

Hence

$$\mu_{(r)}=\sum_{0}^{\infty}x_{(r)}e^{-\lambda}\frac{\lambda^x}{x!}=\sum_{x=r}^{\infty}\frac{e^{-\lambda}\lambda^x}{(x-r)!}=\sum_{y=0}^{\infty}e^{-\lambda}\frac{\lambda^{y+r}}{y!}=\lambda^r.$$

Thus $\mu=\lambda$, $\sigma^2=\lambda^2+\lambda-\lambda^2=\lambda$.

▸ **Example 3.6.2** *Consider the discrete uniform distribution* $P(X=x)=\frac{1}{N}$, $x=1,2,\cdots,N$.

Hence

$$\mu_{(r)}=\frac{1}{N}\sum_{x=1}^{N}(x)_r.$$

Now

$$\sum_{j=a}^{b}(j)_n=\frac{(b+1)_{(n+1)}-(a)_{(n+1)}}{n+1}.$$

Hence

$$\mu_{(r)}=\frac{(N+1)_{(r+1)}}{N(r+1)},\quad r=1,\cdots,\quad N-1;\quad \mu_{(N)}=(N-1)!.$$

Thus

$$\mu=\frac{N+1}{2},\quad \mu_{(2)}=\frac{N^2-1}{3},\quad \sigma^2=\frac{(N+1)(N-1)}{12}.$$

3.7 Different Measures of Central Tendency

The r.v. X takes values in the borel sets in $\mathcal{B}_1$ according to some probability law which is called the probability distribution of X. However, the different values of X will have a tendency to cluster around some central value. This value may be taken as a representative value of the distribution of X and measures the central tendency (location parameter) of the distribution. Different measures of central tendency are given below.

Mean: This is the most commonly used measure of central tendency of the probability distribution of a r.v. X. This is also called the expectation of X and if it exists, is given by

$$E(X)=\int_{-\infty}^{\infty}x\,dF(x).$$

Median: If μ_e is a point which divides the whole probability mass of the distribution into two equal parts, each containing probability mass equal to $\frac{1}{2}$, μ_e is called the median of the distribution.

Consider the curve of probability distribution function $F(x)$ of a continuous r.v. X. The abscissa of the point of intersection of this curve with the straight line $y = \frac{1}{2}$ is the unique median μ_e of this distribution. Thus μ_e satisfies

$$F(\mu_e - 0) \leq \frac{1}{2} \leq F(\mu_e). \tag{3.7.1}$$

For a discrete distribution it may so happen that $F(x) = \frac{1}{2}$ for all x in the interval $a \leq x \leq b$. Then median is taken to be $\frac{1}{2}(a+b)$.

The mean does not exist always, whereas a median always exists (an example is Cauchy distribution discussed in section 3.2). Even in case where mean does exist, the median is sometimes preferred as a measure of central tendency, since the value of the mean may sometimes be largely affected by the presence of some stray probability masses at either end.

Mode: Mode is that item or value in the population which occurs most frequently. It corresponds to the maximum ordinate of the frequency curve $y = f(x)$. If the p.m.f. or p.d.f. of X has a single relative maximum (unique mode) m_0, then m_0 may be used as a measure of the central tendency of the distribution of X. Clearly for a discrete r.v., m_0, if it exists, will be the value of X which has the highest probability. For a continuous r.v. with p.d.f. $f(x)$, if $f(x)$ is twice differentiable, m_0 can be found out from the conditions $f'(m_0) = 0$, $f''(m_0) < 0$. In case $f(x)$ is not differentiable twice, m_0 has to be found out from the condition $f(m_0) > f(x)$ for all $x \neq m_0$. For a multimodal distribution mode cannot be used as a measure of central tendency.

For a non-negative r.v. X one may sometimes use the geometric mean (G.M.), G as a measure of central tendency. This is given by

$$\log G = \int_{-\infty}^{\infty} (\log x)\, dF(x).$$

Sometimes harmonic mean (H.M.), H is used as a masure. This is given by

$$H = \frac{1}{\int_{-\infty}^{\infty} \left(\frac{1}{x}\right) dF(x)}.$$

In a perfectly symmetrical distribution the mean, median and the mode coincide. For a moderately asymmetric distribution occuring in practice, it is found empirically

$$\text{Mean} - \text{Mode} = 3(\text{Mean} - \text{Median}).$$

▸ **Example 3.7.1** *Find the arithmetic mean* (A.M.), *geometric mean* (G.M.) *and the harmonic mean* (H.M.) *of the distribution*

$$dF(x) = \frac{1}{B(p,q)}(1-x)^{p-1}x^{q-1}\, dx, \quad 0 \leq x \leq 1; \;\; p,q > 0,$$

where

$$B(p,q)=\int_0^1(1-x)^{p-1}x^{q-1}\,dx=\frac{\Gamma(p)\,\Gamma(q)}{\Gamma(p+q)}$$

and $\Gamma(r)=(r-1)!$ *for integral* r.

The A.M. is given by $E(X)=\frac{p}{p+q}$.
The G.M., G is given by

$$\log G=E(\log X)=\frac{1}{B(p,q)}\int_0^1\log x(1-x)^{p-1}x^{q-1}\,dx.$$

Since

$$\int_0^1(1-x)^{p-1}x^{q-1}\,dx=B(p,q)$$

differentiating both sides w.r.t. q (the differentiation under the sign of integration being permissible as the integral converges uniformly)

$$\int_0^1(1-x)^{p-1}x^{q-1}\log x\,dx=\frac{\partial}{\partial q}B(p,q).$$

Hence

$$\log G=\frac{1}{B(p,q)}\frac{\partial}{\partial q}B(p,q)=\frac{\partial\log B(p,q)}{\partial q}$$

$$=\frac{\partial}{\partial q}\log\left[\frac{\Gamma(p)\,\Gamma(q)}{\Gamma(p+q)}\right]=\frac{\partial}{\partial q}[\log\Gamma(q)-\log\Gamma(p+q)].$$

The H.M. is given by

$$\frac{1}{H}=E\left(\frac{1}{X}\right)=\frac{1}{B(p,q)}\int_0^1(1-x)^{p-1}x^{q-2}\,dx$$

$$=\frac{B(p,q-1)}{B(p,q)}=\frac{\Gamma(q-1)\,\Gamma(p+q)}{\Gamma(p+q-1)\,\Gamma(q)}=\frac{p+q-1}{q-1}.$$

Hence

$$H=\frac{q-1}{p+q-1}.$$

▶ **Example 3.7.2** *Find the position of the mode and median of the curve*

$$f(x)=\frac{\alpha\beta x^{\alpha-1}}{(1+\beta x^{\alpha})^2},\quad\beta>0,\quad\alpha>1,\quad x\geq 0.$$

Let x_m,μ_e be the modal value and the median of the distribution respectively.

$$\frac{df(x)}{dx}=\frac{\alpha\beta x^{\alpha-2}(1+\beta x^{\alpha})\{\alpha-1-\beta x^{\alpha}(1+\alpha)\}}{(1+\beta x^{\alpha})^4}.$$

Hence

$$\frac{df(x)}{dx} = 0 \Rightarrow x_m = \left[\frac{\alpha - 1}{\beta(1+\alpha)}\right]^{1/\alpha}.$$

It can be checked

$$\left[\frac{d^2 f(x)}{dx^2}\right]_{x_m} < 0.$$

Again,

$$\int_0^{\mu_e} \frac{\alpha\beta x^{\alpha-1}}{(1+\beta x^\alpha)^2}\, dx = \frac{1}{2}.$$

Putting

$$1 + \beta x^\alpha = t, \quad \int_1^{1+\beta\mu_e^\alpha} \frac{dt}{t^2} = \frac{1}{2} \text{ when } \mu_e = \left(\frac{1}{\beta}\right)^{1/\alpha}.$$

Example 3.7.3 *For the distribution in example* 3.3.2, *find the median when* $\alpha = 2\beta$.

When $\alpha = 2\beta$, $f(x) = 2\beta\left(x^{\frac{\alpha}{2}-1} - x^{\alpha-1}\right)$. If $\hat{x}$ is the median of the distribution,

$$\int_0^{\hat{x}} f(x)\, dx = \frac{1}{2}, \text{ i.e., } 2\hat{x}^\alpha - 4\hat{x}^{\alpha/2} + 1 = 0 \text{ when } \hat{x} = \left(1 - \frac{1}{\sqrt{2}}\right)^{2/\alpha}.$$

3.8 Measures of Dispersion

Measures of spread or dispersion (sometimes called measures of concentration) are parameters which give an idea of the scatter (or cluster) of the probability mass of the distribution about the location parameter. These are given below.

Range: If X has a distribution such that

$$\begin{aligned} F(x) &= 0 && \text{if } x < a \\ 0 < F(x) &< 1 && \text{if } a \le x \le b \\ F(x) &= 1 && \text{if } x \ge b \end{aligned}$$

then the difference $b - a$, called the range of X, may be taken as a measure of dispersion. This measure does not take into account how the probability mass is distributed over $[a, b]$ and is thus very unsuitable as a good measure of the dispersion of distribution of X.

Mean deviation: It is the expected value of the absolute deviation of X about A, $E|X - A| = MD_A$ (say), where A is an arbitrary number. The absolute value of $(X - A)$ is taken because dispersion should be indicated by the magnitude of deviation and not by its sign. Its value is large when there is a great scattering and small when there is a great concentration about A. Mean deviation is generally taken about the mean.

Theorem 3.8.1 *Mean deviation is least when taken about the median, i.e.,*

$$E|X - A| \ge E|X - \mu_e|,$$

where A is any number.

Proof: Suppose: $A < \mu_e$. Let us write $\mu_e = m$ (say) for simplicity.

$$\begin{aligned}
E|X-A| &= \int_{-\infty}^{\infty} |x-A|\,dF(x) = \int_{-\infty}^{A}(A-x)\,dF(x) + \int_{A}^{\infty}(x-A)\,dF(x)\\
&= \int_{-\infty}^{A}(A-m+m-x)\,dF(x) + \int_{A}^{\infty}(x-m+m-A)\,dF(x)\\
&= (A-m)F(A) + (m-A)(1-F(A)) + \int_{-\infty}^{A}(m-x)\,dF(x)\\
&\quad + \int_{A}^{\infty}(x-m)\,dF(x)\\
&= (m-A)(1-2F(A)) + \int_{-\infty}^{m}(m-x)\,dF(x) - \int_{A}^{m}(m-x)\,dF(x)\\
&\quad + \int_{A}^{m}(x-m)\,dF(x) + \int_{m}^{\infty}(x-m)\,dF(x)\\
&= (m-A)(1-2F(A)) + E|X-m| + \int_{A}^{m}[(x-m)-(m-x)]\,dF(x)\\
&= E|X-m| + (m-A)(1-2F(A)) + 2\int_{A}^{m}(x-m)\,dF(X)\\
&= E|X-m| + (m-A)(1-2F(A)) + 2\int_{A}^{m}[(x-A)+(A-m)]\,dF(x)\\
&= E|X-m| + (m-A)(1-2F(A)) - 2(m-A)\left[\int_{-\infty}^{m}dF(x) - \int_{-\infty}^{A}dF(x)\right]\\
&\quad +2\int_{A}^{m}(x-A)\,dF(x)\\
&= E|X-m| + (m-A)(1-2F(m)) + 2\int_{A}^{m}(x-A)\,dF(x).
\end{aligned}$$

Since $F(m) = \frac{1}{2}$, the middle term vanishes. Since $A < m$ the integral is also positive. Hence the result.

The result for the case $\mu_e < A$ can be proved similarly.

Variance, Mean Square Error: For measuring the deviation of X around A one may also consider $E(X-A)^2$ which like $E|X-A|$ is free from sign and measures the deviations of X around A. This is called the mean square error of X around $A\,[MSE_A(X)]$. When A is $E(X) = \mu$, this is called the variance of X, often denoted as $V(X)$.

$$V(X) = E(X-\mu)^2 = \mu_2.$$

Clearly

$$E(X-A)^2 = E(X-\mu)^2 + (\mu-A)^2$$

so that

$$MSE_A(X) = V(X) + (\mu-A)^2.$$

Another measure is the positive square root of $V(X)$, called the standard deviation (s.d.) of X, often denoted as

$$\sigma_X \left(\sigma_X = +\sqrt{V(X)}\right).$$

The positive square root of $MSE_A(X)$ is called the root mean square deviation of X about A.

Theorem 3.8.2 *$V(X)$ exists if and only if $E(X)$ and $E(X^2)$ exists.*

Proof is easy.

Theorem 3.8.3 *For any r.v. X, $V(X) \leq E(X-A)^2$, $A \neq \mu$.*

Theorem 3.8.4 *For any r.v. X, $E|X-\mu| \leq \sqrt{E(X-\mu)^2}$ where $\mu = E(X)$.*

Proof: In Cauchy-Schwartz inequality put $|X-\mu|$ in place of X and 1 in place of Y. Taking positive square root of both sides of the inequality, the result follows.

Conditional variance of X given A: Let A be a measurable subset of $\mathcal{S}$. The conditional variance of X given A is

$$V(X|A) = E\left[\{X - E(X|A)\}^2 |A\right] = \frac{E\left\{(X - E(X|A))^2 I_A\right\}}{P(A)}.$$

Theorem 3.8.5 *Let $\{A_i\}$ be a measurable partition of the sample space $\mathcal{S}$ such that $P(A_i) > 0$ for each $i = 1, \cdots, k$ and let $V(X)$ exist. Then*

$$\begin{aligned} V(X) &= \sum_{i=1}^{k} P(A_i)V(X|A_i) + \sum_{i=1}^{k} P(A_i)\left[E(X|A_i) - E(X)\right]^2 \qquad (3.8.1) \\ &= E_A\left[V(X|A)\right] + V_A\left[E(X|A)\right], \end{aligned}$$

where $E(X|A_i)$, $V(X|A_i)$ are respectively the conditional expected value of X given A_i (defined in section 3.2) and conditional variance of X given A_i.

Proof:

$$V(X) = E(X - E(X))^2 = \sum_i P(A_i)E\left\{(X - E(X))^2\middle| A_i\right\} \text{ (by theorem 3.2.1).}$$

Now

$$\begin{aligned} P(A_i)E\left\{(X - E(X))^2\middle| A_i\right\} &= E\left[\left\{(X - E(X))^2\right\} I_{A_i}\right] \text{ (by 3.2.12)} \\ &= \int_{A_i} (X - E(X))^2 \, dP \end{aligned}$$

$$= \int_{A_i} \{X - E(X|A_i) + E(X|A_i) - E(X)\}^2 \, dP$$

$$= \int_{A_i} (X - E(X|A_i))^2 \, dP + \int_{A_i} (E(X|A_i) - E(X))^2 \, dP$$

$$= P(A_i)V(X|A_i) + (E(X|A_i) - E(X))^2 \, P(A_i).$$

Hence the theorem.

Note that we can also write the right hand side of (3.8.1) as $E_A\left[V(X|A)\right]+$ $V_A\left[E(X|A)\right]$ where $V_A(\cdot)$ denotes variance of $(\cdot)$ w.r.t. A (i.e., for different subset of A), because, by theorem 3.2.1, $E(X) = E_A E(X|A)$.

Quartile deviation : Consider a quantity $0 < p < 1$. The value $x = \zeta_p$ (say) of r.v X which satisfies

$$F(\zeta_p - 0) \leq p \leq F(\zeta_p) \tag{3.8.2}$$

is said to be a quantile or fractile of X of order p. Comparing with (3.7.1), median is quantile of order $\frac{1}{2}$.

If X is a discrete r.v., ζ_p may be defined as the smallest value of X (with non-zero probability) satisfying $F(x) \geq p$.

The quantiles $\zeta_{1/4}$, $\zeta_{2/4}$, $\zeta_{3/4}$ are called the first quartile (Q_1) second quartile (Q_2 also median), third quartile (Q_3) respectively. The quantiles $\zeta_{.1}$, $\zeta_{.2}$, $\cdots$, $\zeta_{.9}$ are called the first, second, $\cdots$, nineth decile respectively. The quantiles $\zeta_{.01}$, $\zeta_{.02}$, $\cdots$ are called the first, second, $\cdots$ percentiles respectively.

The quantity

$$Q = \frac{Q_3 - Q_1}{2}$$

called the quartile deviation (or semi-interquartile range) is also taken as a measure of dispersion. The higher the dispersion in values of X the higher is the value of Q.

Gini's Mean difference : The Italian Statistician Corrado Gini suggested the quantity Δ_1 based on the expected absolute difference between pair of units as a measure of dispersion.

$$\Delta_1 = \int_{-\infty}^{\infty} \int_{-\infty}^{\infty} |x - y| f(x) f(y) \, dx \, dy \tag{3.8.3}$$

if X is a continuous r.v. with p.d.f. $f(x)$, where x, y are two arbitrary values of X.

$$\Delta_1 = \sum_{i=1}^{\infty} \sum_{j=1}^{\infty} |x_i - x_j| p(x_i) p(x_j) \tag{3.8.4}$$

if X is discrete with p.m.f. $P(X = x_k) = p(x_k)$. Clearly Δ_1 does not depend on any measure of central tendency.

Another measure Δ_2 based on the square of difference of pairs of observations is obtained by replacing $|x - y|$ by $(x - y)^2$ and $|x_i - x_j|$ by $(x_i - x_j)^2$ in (3.8.3) and (3.8.4) respectively.

Lemma 3.8.1 $\Delta_2 = 2\sigma_X^2$.

Proof: Let X be continuous with $E(X) = \mu$.

$$\begin{aligned}
\Delta_2 &= \int_{-\infty}^{\infty}\int_{-\infty}^{\infty} [(x-\mu)-(y-\mu)]^2 f(x)f(y)\,dx\,dy \\
&= \int_{-\infty}^{\infty} (x-\mu)^2 f(x)\,dx + \int_{-\infty}^{\infty} (y-\mu)^2 f(y)\,dy \\
&\quad +2\int_{\infty}^{\infty} (x-\mu)f(x)\,dx \int_{-\infty}^{\infty} (x-\mu)f(y)\,dy \\
&= 2\sigma_X^2.
\end{aligned}$$

Proof follows similarly when X is discrete.

Corollary 3.8.1

$$\frac{1}{N}\sum_{i=1}^{N}(x_i-\bar{x})^2 = \frac{1}{2N^2}\sum_{i=1}^{N}\sum_{j=1}^{N}(x_i-x_j)^2, \quad \bar{x} = \frac{1}{N}\sum_{i=1}^{N} x_i. \tag{3.8.5}$$

Proof: If X takes N values x_i $(i = 1, \cdots, N)$ each with probability $\frac{1}{N}$ the left hand side of (3.8.5) is σ_x^2 and the right hand side is $\Delta_2/2$.

Lemma 3.8.2 *If X is a continuous r.v. with distribution function $F(x)$,*

$$\Delta_1 = 2\int_{-\infty}^{\infty} F(x)\{1-F(x)\}\,dx. \tag{3.8.6}$$

Proof:

$$\begin{aligned}
\Delta_1 &= \int_{-\infty}^{\infty}\int_{-\infty}^{\infty} |x-y|\,dF(x)\,dF(y) = 2\int_{-\infty}^{\infty}\int_{y=-\infty}^{x} (x-y)\,dF(x)\,dF(y) \\
&= 2\int_{-\infty}^{\infty} dF(x)\int_{-\infty}^{x} (x-y)\,dF(y).
\end{aligned}$$

Now

$$\int_{-\infty}^{x} (x-y)\,dF(y) = \int_{-\infty}^{x} F(y)\,dy$$

(on integration by parts taking $(x-y)$ as the first function).

Hence

$$\begin{aligned}
\Delta_1 &= 2\int_{-\infty}^{\infty} dF(x)\int_{-\infty}^{x} F(y)\,dy = 2\int_{-\infty}^{\infty} dF(x)\phi(x), \quad \text{where } \phi(x) = \int_{\infty}^{x} F(y)\,dy \\
&= 2F(x)\phi(x)]_{-\infty}^{\infty} - 2\int_{-\infty}^{\infty} F(x)F(x)\,dx \text{ (integration by parts)} \\
&= 2\int_{-\infty}^{\infty} F(x)\,[1-F(x)]\,dx
\end{aligned}$$

since

$$F(x)\phi(x)\Big]_{-\infty}^{\infty} = \int_{-\infty}^{\infty} F(x)\,dx.$$

Coefficient of Variation: If $E(X)$ and σ_X exists coefficient of variation of a r.v X is defined as $\frac{\sigma_X}{E(X)} = C_X$ (say) and is taken as a measure of dispersion. Unlike the previous measures it has no unit and is often expressed in percentages ($100C_X\%$).

Gini's coefficient of concentration defined as $G = \frac{\Delta_1}{2E(X)}$ is another measure often used by the scientists in measuring social inequality in terms of concentration of money resources, etc.

► **Example 3.8.1** *X is a continuous r.v. with p.d.f. $f(x) = Ce^{-x/\sigma}$, $x \geq 0$, $\sigma > 0$. Find its mean, median, variance and quartile deviation.*

Since

$$\int_0^{\infty} f(x)\,dx = 1, \;\; C = \frac{1}{\sigma}$$

$$\mu = \frac{1}{\sigma}\int_0^{\infty} xe^{-x/\sigma}\,dx = -xe^{-x/\sigma}\Big]_0^{\infty} + \int_0^{\infty} e^{-x/\sigma}\,dx = \sigma.$$

If m be the median,

$$\int_0^m f(x)\,dx = \frac{1}{2}, \quad \text{i.e.,} \; \frac{1}{\sigma}\int_0^m e^{-x/\sigma}\,dx = \frac{1}{2}.$$

This gives $e^{m/\sigma} = 2$ or $m = \sigma \log_e 2$.

Again

$$\int_0^{\zeta_{1/4}} \frac{1}{\sigma}e^{-x/\sigma}\,dx = \frac{1}{4}.$$

This gives $\zeta_{1/4} = \sigma(\log 4 - \log 3)$.

Similarly

$$\int_0^{\zeta_{3/4}} \frac{1}{\sigma}e^{-x/\sigma}\,dx = \frac{3}{4}, \quad \text{implies } \zeta_{3/4} = \sigma \log 4.$$

Hence quartile deviation is

$$\frac{1}{2}\left(\zeta_{3/4} - \zeta_{1/4}\right) = \frac{\sigma \log 3}{2}, \quad \mu_2' = \frac{1}{\sigma}\int_0^{\infty} x^2 e^{-x/\sigma}\,dx = 2\sigma^2.$$

Thus $\mu_2 = \sigma^2$.

3.9 Measures of Skewness and Kurtosis

By skewness we mean the asymmetry of the distribution. A probability density curve (or pm-histogram) is positively skewed if it has more probability mass to the right of the mode (Fig. 3.2); is negatively skewed if more mass is distributed to the left of the mode (Fig. 3.3). A measure of skewness should be unit-free and should take positive (negative) values for a positively (negatively) skewed distribution. For a symmetrical distribution $\mu_3 = 0$; μ_3 is positive (negative) for a positively (negatively) skewed distribution. To make it dimensionless we divide it by σ^3. Thus

$$\sqrt{\beta_1} = \frac{\mu_3}{\sigma^3} = \frac{\mu_3}{\mu_2^{3/2}} \tag{3.9.1}$$

or β_1 is often taken as a measure of skewness (if μ_3 exists). For a symmetrical distribution $\beta_1 = 0$. The value of β_1 gives a measure of departure from symmetry, i.e., of skewness. It can take any value between $-\infty$ and ∞.

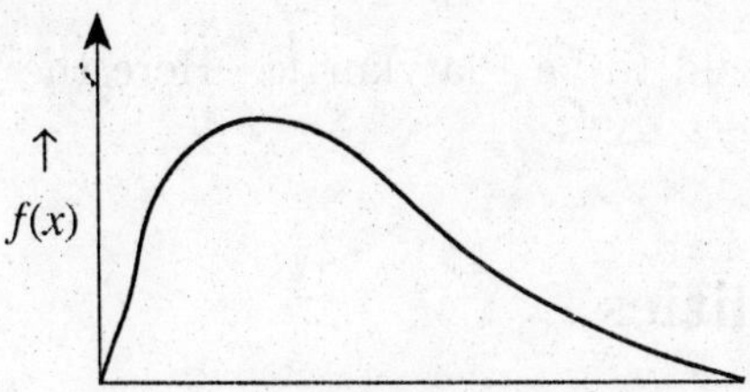

Fig. 3.2

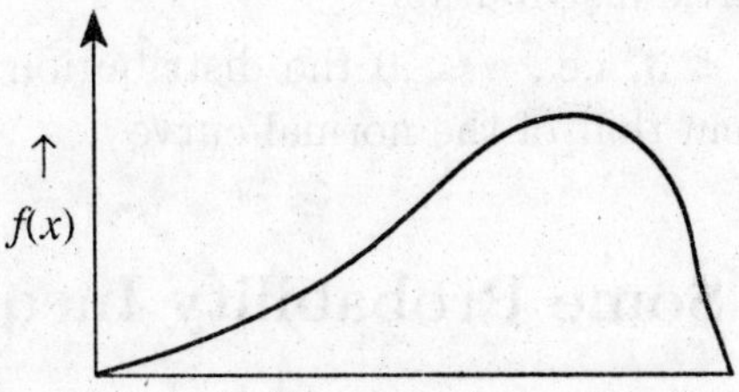

Fig. 3.3

Other measures are:

(i)

$$\frac{E(X) - m_0}{\sigma} \tag{3.9.2}$$

(ii)

$$\frac{3\,(E(X) - \mu_e)}{\sigma} \quad \text{[since mean } - \text{ mode} \simeq 3(\text{ mean } - \text{ median }) \text{ empirically]} \tag{3.9.3}$$

(iii) For a symmetrical distribution, $Q_3 - \mu_e = \mu_e - Q_1$. Hence $(Q_3 - \mu_e) - (\mu_e - Q_1)$ can be taken as a measure of skewness. To make it dimensionless we divide it by $Q_3 - Q_1$ to obtain

$$\frac{Q_3 - 2Q_2 + Q_1}{Q_3 - Q_1} \tag{3.9.4}$$

as another measure.

For a symmetrical distribution each measure has value zero. For a positively skewed distribution, $\mu_3 > 0$, $E(X) > \mu_e > m_0$, $Q_3 - Q_2 > Q_2 - Q_1$ and each measure takes positive value. For a negatively skewed distribution the inequalities are reversed and each measure takes a negative value.

Since $|E(X) - \mu_e| \leq E|X - \mu_e| \leq E|X - \mu| \leq \sigma$, (3.9.3) varies between -3 and 3. The measures (3.9.2) and (3.9.4) being approximately equal have the same limits. For a extremely positively (negatively) skewed distribution, $Q_2 - Q_1 \to 0$ ($Q_3 - Q_2 \to 0$) and hence (3.9.4) $\to 1(-1)$.

3.9.1 Measures of Kurtosis

By Kurtosis we mean the flatness or peakedness of the curve of probability distribution. The quantity

$$\beta_2 - 3 = \frac{\mu_4}{\mu_2^2} - 3 = \gamma \text{ (say)}$$

is taken as a measure of Kurtosis.

For a normal distribution (section 6.7), $\beta_2 = 3$ and $\gamma = 0$. If for a distribution $\beta_2 > 3$ the peak of the distribution is sharper than that of a normal curve and it is called a Lepto-kurtic distribution.

If $\beta_2 < 3$, i.e., $\gamma < 0$ the distribution is said to be Platykurtic. Here the peak is flatter than that of the normal curve.

3.10 Some Probability Inequalities

Theorem 3.10.1 *Let $g(X)$ be a non-negative borel-measureable function of a r.v. X. If $Eg(X)$ exists, then, for every $\varepsilon > 0$*

$$P\{g(X) \geq \varepsilon\} \leq \frac{E(g(X))}{\varepsilon}.$$

Proof:

$$Eg(X) = \int_A g(x)\,dF(x) + \int_{A^c} g(x)\,dF(x)$$

where $A = \{x : g(x) \geq \varepsilon\}$.

$$Eg(X) \geq \int_A g(x)\,dF(x) \geq \varepsilon \int_A dF(x) = \varepsilon P\{g(X) \geq \varepsilon\}.$$

Corollary 3.10.1 *If $g(X) = |X|^r$ and $\varepsilon = t^r$, $t > 0$, $r > 0$ and ν_r exists,*

$$P\{|X| \geq t\} \leq \frac{E|X|^r}{t^r}.$$

For $r = 1$, we have

$$P\{|X| \geq t\} \leq \frac{E|X|}{t} \tag{3.10.1}$$

which is known as Markov's inequality.

Taking $g(X) = (X - \mu)^2$, $t = \varepsilon^2\sigma^2$, we have

$$P\{|X - \mu| \geq \varepsilon\sigma\} \leq \frac{1}{\varepsilon^2}$$

which is known as Chebychev's inequality, where $E(X) = \mu$, $V(X) = \sigma^2$.

Theorem 3.10.2 (*Chebychev's inequality*). *If X be a r.v. with finite variance $\sigma^2 (> 0)$ and $E(X) = \mu$, then*

$$P\{|X - \mu| \geq \varepsilon\} \leq \frac{\sigma^2}{\varepsilon^2} \tag{3.10.2}$$

for any $\varepsilon > 0$.

Chebyshev's inequality gives an upper bound for the probability that X deviates from its mean by more than ε units. No assumption on the distribution of X is required except that it has a finite variance. Many probability inequalities have been developed starting from Chebychev's inequality and many sharper bounds for $P(|X - \mu| \geq \varepsilon)$ than the one given by this inequality exist. However, in general, the bound given by the Chebychev's inequality can hardly be improved upon.

▶ **Example 3.10.1** *Let $X_1, \cdots, X_n$ be independent and identically distributed r.v.s with mean zero and unit variance. Then their sum $X = X_1 + \cdots + X_n$ is also a r.v. with $E(X) = 0$, $V(X) = n$. Hence from (3.10.2),*

$$P\{|X| \geq \varepsilon\} \leq \frac{n}{\varepsilon^2}.$$

When $\varepsilon = \sqrt{n}$,

$$P\{|X| \geq \sqrt{n}\} \leq 1.$$

▶ **Example 3.10.2** *Suppose $X_1, \cdots, X_{n+1}$ are the results of $(n+1)$ Bernoulli trials with $P(X_i = 1) = p$, $P(X_i = 0) = 1 - p$, $i = 1, \cdots, n+1$. Suppose also that Y_n is equal to be number of $i(i = 1, \cdots, n)$ such that $X_i = X_{i+1} = 1$. Using Chebychev's inequality find an upper bound to*

$$P\left\{\left|\frac{Y_n}{n} - p^2\right| > \varepsilon\right\}, \quad \varepsilon > 0.$$

Let $Z_i = 1$ if $X_i = 1$, $X_{i+1} = 1$ and 0 otherwise. Then $Y_n = Z_1 + \cdots + Z_n$. $E(Z_i) = p^2$. $V(Z_i) = p^2(1 - p^2)$. $\text{Cov}(Z_i, Z_{i+1}) = E(Z_iZ_{i+1}) - p^4 = p^3(1 - p)$, because $E(Z_iZ_{i+1}) = P(Z_iZ_{i+1} = 1) = P(Z_i = 1)P(Z_{i+1} = 1|Z_i = 1) = p^2\left(\frac{p^3}{p^2}\right) = p^3$. Hence $E(Y_n) = np^2$, $V(Y_n) = p^2[n(1 + 3p)(1 - p) + 2p(p - 1)]$. By (3.10.2),

$$P\left(\left|\frac{Y_n}{n} - p^2\right| > \varepsilon\right) < \frac{V(Y_n)}{n^2\varepsilon^2} = \frac{p^2(1 + 3p)(1 - p)}{n\varepsilon^2} + \frac{2p^3(p - 1)}{n^2\varepsilon^2} < \frac{p^2(1 + 3p)(1 - p)}{n\varepsilon^2}.$$

Theorem 3.10.3 (***One-sided Chebychev inequality***). *If X be a random variable with finite variance $\sigma^2(>0)$ and $E(X)=\mu$ then*

$$P\left[X \geq \mu + t\sigma\right] \leq E\left(\frac{X-\mu-a\sigma}{t\sigma-a\sigma}\right)^2 \tag{3.10.3}$$

for any $t>0$ and any a.

Proof: Define a r.v. Y such that

$$Y = \begin{cases} 1 \text{ if } X \geq \mu + t\sigma \\ 0 \text{ otherwise.} \end{cases}$$

Hence

$$E(Y) = P(X \geq \mu + t\sigma).$$

Now $\left(\frac{X-\mu-a\sigma}{t\sigma-a\sigma}\right)^2 \geq Y$ for all values of X a.e. Taking expectation of both sides (3.10.3) follows.

It can be seen that equality in (3.10.3) holds when X takes only two values $\mu+t\sigma$ and $\mu+a\sigma$ with probabilities p and $1-p$ respectively. The conditions $E(X)=\mu$ and $V(X)=\sigma^2$ give, in this case,

$$a = -\frac{1}{t}, \quad p = \frac{1}{1+t^2}$$

in which case

$$E\left(\frac{X-\mu-a\sigma}{t\sigma-a\sigma}\right)^2 = p.$$

The inequality (3.10.3) can thus be put in the alternative form:

$$P[X \geq \mu + t\sigma] \leq \frac{1}{1+t^2} \tag{3.10.4}$$

Corollary 3.10.2

$$P[X \leq \mu - t\sigma] = P[-X \geq -\mu + t\sigma] \leq \frac{1}{1+t^2}. \tag{3.10.5}$$

Corollary 3.10.3 *The distribution function $F(x)$ of a r.v. X with mean μ and variance σ^2 satisfies the inequalities:*

(i) $F(x) \leq \frac{\sigma^2}{\sigma^2+(x-\mu)^2}$ for $x \leq \mu$

(ii) $F(x) \geq \frac{(x-\mu)^2}{\sigma^2+(x-\mu)^2}$ for $x \geq \mu$.

Proof: (i) Put $x=\mu-t\sigma$ in (3.10.5); (ii) Put $x=\mu+t\sigma$ in (3.10.4).

Exercise

1. Show that for the distribution $P(X = k) = \frac{1}{2^k}$, $k = 1, 2, \cdots$, $E(X)$ exists. Find its value.

2. For the continuous distribution $dF = \frac{3}{4}x(2-x)\,dx$, $0 \le x \le 2$, find the first four moments, mean deviation about mean, skewness and Kurtosis.

$$\left[\mu = 1, \mu_2 = \tfrac{1}{5}, \mu_3 = 0, \mu_4 = \tfrac{3}{35}, \text{ m.d. } = \tfrac{3}{8}, \beta_1 = 0, \beta_2 = -\tfrac{6}{7}\right]$$

3. Find the mean, variance, mean deviation about mean of the r.v. X with p.d.f.

(a) $f(x) = \frac{1}{2}e^{-|x|}(-\infty < x < \infty)$

(b) $f(x) = \begin{cases} \frac{1}{2b} \text{ if } |x-a| \le b \\ 0 \text{ otherwise.} \end{cases}$ $\quad$ [(a) $\mu = 0, \mu_2 = 2; 1$ (b) $\mu = a, \mu_2 = \frac{b^2}{3}; \frac{b}{2}$]

4. Show that for the triangular distribution

$$dF = \frac{1}{\alpha}\left[1 - \frac{|x-\beta|}{\alpha}\right]dx, \ |x-\beta| > \alpha$$

the mean is β and variance is $\frac{\alpha^2}{6}$.

5. X has a continuous distribution with $dF = k(x - x^2)\,dx$, $0 \le x \le 1$, k being a constant. Find the arithmetic mean, harmonic mean, mode and median.

[$k = 6$, a.m.$= \frac{1}{2}$, h.m.$= \frac{1}{3}$, mode $= \frac{1}{2}$, median $= \frac{1}{2}$]

6. Show that for the rectangular distribution $dF = dx$, $0 \le x \le 1$ $\mu_1' = \frac{1}{2}$, $\mu_2 = \frac{1}{12}$, mean deviation $= \frac{1}{4}$, $\Delta_1 = \frac{1}{3}$. Show also that $E\left(\frac{1}{X}\right)$ does not exist.

7. X has p.d.f. $f(x) = k(1-x)$, $0 \le x \le 1$. Find k and determine the arithmetic mean, median and variance. [$k = 2$, mean $= \frac{1}{3}$, median $= 1 - \frac{1}{\sqrt{2}}$, variance $= \frac{1}{18}$]

8. Obtain (quartile deviation)/(standard deviation) for the distribution with p.d.f.

$$f(x) = \begin{cases} 2(b+x)/b(a+b), & -b \le x \le 0; \\ 2(a-x)/a(a+b), & 0 \le x \le a. \end{cases}$$

9. Show that the geometric mean G of the distribution $dF = 6(2-x)(x-1)\,dx$, $1 \le x \le 2$ is given by $6\log(16G) = 19$.

10. A continuous r.v. X has the p.d.f. proportional to $(\alpha + |x|)e^{-\beta|x|}$, $-\infty < x < \infty$, where α, β are positive constants. Find β_2, the coefficient of kurtosis of X and show that $\frac{1}{3} < \beta_2 < 3$, irrespective of the values of α and β.

11. A continuous r.v. X has the distribution function $F(x) = 1 - e^{\alpha \tan x}, 0 \le x \le \frac{\pi}{2}$, $0 < \alpha < 1$. Find the p.d.f. of X and show that the mode of the distribution is $m_0 = \frac{1}{2}(\pi - \sin^{-1}\alpha)$.

12. For a r.v. X with p.d.f.

$$f(x) = \begin{cases} \frac{x}{2}, & 0 \le x \le 1 \\ \frac{1}{2}, & 1 < x \le 2 \\ \frac{1}{2}(3-x), & 2 < x \le 3 \end{cases}$$

show that moments of all orders exist. Find the mean and variance of X.

13. For a continuous r.v. X defined in the range $a \le x \le b$ and having p.d.f proportional to $g(x)$, prove that $\tilde{m}$, the median of distribution satisfies the relation

$$2\int_a^{\tilde{m}} g(x)\,dx = \int_a^b g(x)\,dx.$$

14. For the Poisson r.v. with p.m.f. $P(X = x) = e^{-\lambda}\frac{\lambda^x}{x!}$, $x = 0, 1, 2\cdots$, show that

$$\begin{aligned} &E(X) = \lambda, && E(X^2) = \lambda + \lambda^2, \\ &E(X^3) = \lambda + 3\lambda^2 + \lambda^3, && E(X^4) = \lambda + 7\lambda^2 + 6\lambda^3 + \lambda^4, \\ &\mu_2 = \mu_3 = \lambda, && \mu_4 = \lambda + 3\lambda^2. \end{aligned}$$

15. For a continuous r.v. X the distribution function is $F(x) = 1 - e^{-\beta x^2}, \beta > 0$, $0 \le x \le \infty$. Find the median of the distribution. Also if m, m_0, σ denote respectively the mean, mode and the s.d. of the distribution, show that $2m_0^2 - m^2 = \sigma^2$ and $m_0 = m\sqrt{\frac{2}{\pi}}$. $\left[\text{median} = \sqrt{\frac{\log_e 2}{\beta}}, m = \frac{1}{2}\sqrt{\frac{\pi}{\beta}}, \sigma^2 = \frac{4-\pi}{4\beta}, m_0 = \frac{1}{\sqrt{2\beta}}\right]$

16. If X is a continuous r.v. with p.d.f. $f(x)$ and if m is the median of the distribution of X and c is a real constant, show that $E|X - c| = E|X - m| + 2\int_m^c (c - x) f(x)\,dx$.

17. Let X be a r.v. with p.d.f. $f(x) = \frac{ak^a}{x^{a+1}}$, $a > 0$, $x \ge k > 0$. (This is called Pareto distribution). Show that the d.f. is $F(x) = 1 - \left(\frac{k}{x}\right)^a$; $E|X|^r < \infty$ for $r < a$ and $\mu_{r'} = \frac{ak^r}{a-r}$; $\mu = \frac{ak}{a-1}$ (if $a > 1$) and $V(X) = \frac{ak^2}{(a-1)(a-2)}$ (if $a > 2$), the mean deviation is $\frac{2k}{a-1}(1 - a^{-1})^{a-1}$; the geometric mean is $G = ke^{1/a}$; the harmonic mean is $H = k\left(1 + \frac{1}{a}\right)$; median is $m = k2^{1/a}$.

18. If X has p.d.f. proportional to $x/(1+x)^3 (0 \le x \le \infty)$, show that the median of the distribution is $(1 + \sqrt{2})$; prove that mode is $\frac{1}{2}$ and hence deduce the skewness of the distribution.

19. Show that if a distribution $dF = \frac{1}{\sqrt{2\pi}} e^{-x^2/2}\,dx,\ -\infty \le x \le \infty$ is transformed by $x = a\log(y-b)+c$, the distribution of y has $\dfrac{\text{mean} - \text{mode}}{\text{mean} - \text{median}} = -\exp\left(\frac{1}{2a^2}\right)$.

20. Show that for the Cauchy p.d.f. $f(x) = \frac{1}{\pi(1+x^2)},\ -\infty \le x \le \infty$, $E(X)$ does not exist and the median of X is zero.

21. For the Poisson variate at exercise 14, find $E\left(kXe^{-kX}\right)$, where k is a constant.

$$\left[E\left(kXe^{-kX}\right) = \sum_0^\infty (kxe^{-kx})\left(e^{-\lambda}\frac{\lambda^x}{x!}\right) = \lambda k \exp\left\{\lambda(e^{-k}-1)-k\right\}\right]$$

22. For the distribution $dF = k\left(1-\frac{x^2}{a^2}\right)^{-h} dx,\quad 0 < h < 1,\quad |x| < a,\ k, a$ constant, show that

$$\mu'_{r+1} = \frac{ra^2}{r+2-2h}\mu'_{r-1}.$$

[Integrate by parts]

23. For the distribution $dF = k\left(1+\frac{x}{\alpha}\right)^{m-1} e^{-\frac{mx}{\alpha}}\,dx, -\alpha \le x \le \infty$ find the constant k and the first four moments. Show that $8\beta_2 = 4 + 9\beta_1$.

$$\left[k = m^m e^{-m}/a\overline{|(m)}\,, \mu_1' = \alpha, \mu_2 = \frac{\alpha^2}{2m}, \mu_3 = \frac{\alpha^3}{3m^2}, \mu_4 = \frac{\alpha^4(m+2)}{8m^2}\right]$$

24. A continuous r.v. X has p.d.f. $f(x) = \frac{ak(x-x_0)^{a-1}}{\{1+k(x-x_0)^a\}^2}, k > 0, a > 1, x_0 \le x \le \infty$. Find the median and modal value.

$$\left[\text{median} = x_0 + \left(\frac{1}{k}\right)^{1/a},\ \text{mode} = x_0 + \left\{\frac{a-1}{k(a+1)}\right\}^{1/2}\right]$$

25. For the curve $y = y_0 x^r(1-x)^s,\ 0 \le x \le 1$ find skewness $S_k = $ (mean−median)/s.d. and $\sqrt{\beta_1}$.

$$\left[S_k = \frac{s-r}{s+r}\sqrt{\frac{r+s+3}{(r+1)(s+1)}}, \sqrt{\beta_1} = \frac{2(s-r)}{r+s+4}\sqrt{\frac{r+s+3}{(r+1)(s+1)}}\right]$$

26. For the distribution $dF = \frac{x}{\sigma^2}e^{-\frac{x^2}{2\sigma^2}}\,dx,\ x > 0$ find $Q_3 - Q_1$ and $(Q_3 - Q_1)/\text{s.d.}$

$$\left[\sigma\sqrt{2}\left[\sqrt{\log 4} - \sqrt{\log \tfrac{4}{3}}\right]; \left(\sqrt{\log 4} - \sqrt{\log \tfrac{4}{3}}\right)\Big/\sqrt{1-\tfrac{\pi}{2}}\right]$$

27. The distribution function of a r.v. X is

$$F(x) = \begin{cases} 0 & x < -1 \\ a + b \text{ arc } \sin x & -1 \le x \le 1 \\ 1 & x > 1 \end{cases}$$

Find $E(X), V(X)$. $\left[E(X) = 0, V(X) = \frac{1}{2}\right]$

28. A r.v. X has p.d.f. $f(x) = k \cdot e^{-b(x-a)},\ a \le x \le \infty\ k, a, b$ being constants. Show that $k = b = \frac{1}{\sigma},\ a = m - \sigma$ where m, σ respectively are the mean and standard deviation of the distribution. Find β_1, β_2 of the distribution. $[\beta_1 = 4, \beta_2 = 9]$

29. Prove that if $y = g(x)$ be a monotone function and μ_e be the median of X, the $g(\mu_e)$ is median of $Y = g(X)$.

30. A target is made of 3 concentric circles of radii $\frac{1}{\sqrt{3}}$, 1 and $\sqrt{3}$ feet respectivel Shots within the inner circle count 4 points, shots within the next ring outsid this circle 3 points, shots within the next ring 2 points and shots outside coun zero. Supposing the probability of a hit at a distance r from the centre is given b $p(r) = \frac{2}{\pi(1+r)^2}$, find the expectation of the score after 5 shots and the probabilit that the score exceeds 15. $\left[\text{Probability} = \frac{2}{\pi}\tan^{-1} r,\ \text{Expected score} = \frac{65}{6}\right.$

31. If for a r.v. X, μ_{2r+2} exists then show that $(\mu_{2r+1})^2 \geq \mu_{2r}\mu_{2r+2}$.
[*Hints:* In Cauchy-Schwartz inequality $E(Y^2)E(Z^2) \geq \{E(YZ)\}^2$, put $Y = (X-\mu)^r$, $Z = (X-\mu)^{r+1}$ where $\mu = E(X)$].

32. Suppose X is a non-negative r.v. whose mean (μ) exists. Prove that

    (a) $\lim_{t\to\infty} t[1 - F(t)] = 0$ and hence in case X is absolutely continuous;

    (b) $\mu = \int_0^\infty [1 - F(x)]\, dx$.

[*Hints:* We prove the results when X is absolutely continuous with p.d.f. f.

$$\mu = \int_0^\infty x f(x)\, dx = \lim_{t\to\infty} \int_0^t x f(x)\, dx. \quad \text{(i}$$

On integration by parts,

$$\int_0^t x f(x)\, dx = tF(t) - \int_0^t F(x)\, dx = -t[1 - F(t)] + \int_0^t [1 - F(x)]\, dx \quad \text{(ii}$$

but

$$t[1 - F(t)] = t\int_t^\infty f(x)\, dx < \int_t^\infty x f(x)\, dx \quad \text{(iii}$$

Since $E|X| < \infty$, taking limit of both sides of (iii), $\lim_{t\to\infty} t[1 - F(t)] = 0$ which proves (a). Henc from (i) and (ii)

$$\mu = \lim_{t\to\infty} \int_0^t [1 - F(x)]\, dx = \int_0^\infty [1 - F(x)]\, dx \quad \text{(iv}$$

which proves (b)]

33. Show that if the mean exists, $\Delta_1 = 4\int_{-\infty}^{\infty} x\left[F(x) - \frac{1}{2}\right] dF$.

[*Hints:*

$$\Delta_1 = 2\int_{-\infty}^{\infty}\int_{-\infty}^{\infty} (x - y)\, dF(x)\, dF(y) = 2\int_{-\infty}^{\infty} dF(x) \int_{-\infty}^{x} (x - y)\, dF(y)$$

$$= 2\int_{-\infty}^{\infty} dF(x) \int_{-\infty}^{x} F(y)\, dy$$

[on integration by parts taking $(x - y)$ as the first factor].

$$= 2\int_{-\infty}^{\infty} dF(x)\left\{[yF(y)]_{-\infty}^{x} - \int_{-\infty}^{x} y\,dF(y)\right\}$$

$$= 2\int_{-\infty}^{\infty} xF(x)[dF(x)] - 2\int_{-\infty}^{\infty} dF(x)\int_{-\infty}^{x} y\,dF(y) \qquad \text{(i)}$$

The least term in

$$\text{(i)} = 2\int_{-\infty}^{\infty} y\,dF(y)\int_{y}^{\infty} dF(x) = 2\int_{-\infty}^{\infty} y\,dF(y)(1-F(y))$$

Hence

$$\text{(i)} = 2\int_{-\infty}^{\infty} xF(x)\,dF(x) - 2\int_{-\infty}^{\infty} x(1-F(x))\,dF(x) = 4\int_{-\infty}^{\infty} x\left[F(x)-\frac{1}{2}\right] dF(x).]$$

34. Show that if m_0 be the mode of X, then $a+bm_0 (b\neq 0)$ is the mode of $Y=a+bX$.

35. Prove that for any positive random variable Z,

$$\sum_{n=1}^{\infty} P(Z\geq n) \leq E(Z) \leq 1+\sum_{n=1}^{\infty} P(Z\geq n). \qquad \text{(Chung, 1974)}$$

36. Let S_n be the total score obtained in n tosses in a symmetric die. Using Chebychev's inequality, make an upper estimation of $P\left\{\left|\frac{S_n}{n}-3.5\right|>\varepsilon\right\}, \varepsilon>0.$ $\left[\frac{35}{12n\varepsilon^2}\right]$

37. Let X have a Poisson distribution with parameter λ. Prove that

a) $P\left(X\geq \frac{\lambda}{2}\right)\leq \frac{4}{\lambda}$; b) $P(X\geq 2\lambda)\leq \frac{1}{\lambda}$.

38. For the geometric distribution $p_n = 2^{-n}$, $n=1,2,\cdots$, show that Chebychev's inequality gives $P[|X-2|>2]<\frac{1}{2}$ while the actual probability is $\frac{1}{16}$.

39. If $X_1,\cdots,X_k$ are independently and identically distributed r.v.s having zero means and unit variance then prove that $P\left(\sum_1^k X_i^2 \geq k\lambda\right)\leq \frac{1}{\lambda}$ for $\lambda>0$.

[Apply Markov inequality].

40. Show that for any distribution

$$\int_{-\infty}^{\infty}\left(1-\frac{x^2}{t^2}\right) dF(x) \leq \int_{-t}^{t} dF(x)$$

and hence deduce that $P\,[|X-E(X)|>k\sigma]\leq \frac{1}{k^2}$ when $k>0$ and $\sigma^2=V(X)$.

[*Hints:*

$$\int_{-\infty}^{\infty}\left(1-\frac{x^2}{t^2}\right) dF(x) = \int_{-\infty}^{-t}\left(1-\frac{x^2}{t^2}\right) dF(x) + \int_{-t}^{t}\left(1-\frac{x^2}{t^2}\right) dF(x)$$

$$+\int_{t}^{\infty}\left(1-\frac{x^2}{t^2}\right) dF(x)$$

$$\leq \int_{-t}^{t}\left(1-\frac{x^2}{t^2}\right) dF(x) \leq \int_{-t}^{t} dF(x)$$

Now

$$P(|X| > k\sigma) = 1 - \int_{-k\sigma}^{k\sigma} dF(x) \le 1 - \int_{-\infty}^{\infty}\left(1 - \frac{x^2}{k^2\sigma^2}\right) dF(x)$$
$$= \int_{-\infty}^{\infty} \frac{x^2}{k^2\sigma^2}\, dF(x) = \frac{1}{k^2} \text{ when the origin is at the mean.]}$$

41. (a) Let X be a non-negative r.v. Considering the discontinuous Heaveside function
$H(X-a)$ where $H(Y) = 1(Y > 0), = 0(Y \le 0)$ 'a' being a positive constant prove Markov inequality $P(X > a) < \frac{E(X)}{a}$.
(b) Let X be a r.v. with finite $E(X^2)$. Considering the inequality $H(|X-b|-a) \le \left(\frac{X-b}{a}\right)^2$ where $a(> 0), b$ are arbitrary constants, show that $P(|X-b| > a) \le \frac{E(X-b)^2}{a^2}$.

[*Hints:* (a) $H(X-a) = 1(0)$ if $\frac{X}{a} > 1$ (otherwise). Hence $H(X-a) \le \frac{X}{a}(X \ge 0)$. Thus $E[H(X-a)] = P(X > a) \le \frac{E(X)}{a}$.
(b) $H(|X-b|-a) = 1(0)$ if $\frac{|X-b|}{a} > 1$ (otherwise). Hence $H(|X-b|-a) \le \frac{(X-b)^2}{a^2}$. The result follows by taking expectation.]

42. Show that if $M(X)$ is an increasing positive function of $|X|$, then $P[|X| > a] \le \frac{E[M(X)]}{M(a)}$, provided $E[M(X)]$ exists.

[*Hints:* Define a r.v. Y such that $Y = 1(0)$ if $|X| > a$ (otherwise). Hence $E(Y) = P(|X| > a)$.
Now $M(X) \ge YM(a)$. The result follows by taking expectation.]

43. Let X be such that $E(e^{aX})$ exists, $a > 0$. Show that for any t, $P(X > t) < \frac{E(e^{aX})}{e^{at}}$

[*Hints:* Proceed as in 41; e^{ax} is a non-negative function of x.]

44. Let g be a non-negative non-decreasing function. Prove that if $E[(g(|X-\mu|)$ where $\mu = E(X)$ exits, then $P(|X-\mu| > t) < \frac{E[g(|X-\mu|)]}{g(t)}$.

[*Hints:* Proceed as in 41.]

3.11 Appendix (Some Inequalities)

3.11.1 Cauchy-Schwartz Inequality

Let $(x_1, \cdots, x_n)$, $(y_1, \cdots, y_n)$ be two sets of real numbers. Then

$$\left(\sum_1^n x_i^2\right)\left(\sum_1^n y_i^2\right) \ge \left(\sum_1^n x_i y_i\right)^2 \tag{3.11.1}$$

equality holds when and only when $a_i \propto b_i$, $i = 1, \cdots, n$.

Proof: Consider the quadratic form in two real variables a, b

$$\sum_{i=1}^{n}(ax_i + by_i)^2 = a^2\sum_{i=1}^{n} x_i^2 + 2ab\sum_{i=1}^{n} x_iy_i + b^2\sum_{i=1}^{n} y_i^2$$

which is always ≥ 0 for all values of a, b and hence has a non-negative determinant Δ. The equality is attained when $\Delta = 0$, which means there exists real numbers a, b such that $ax_i + by_i = 0$ for all $i = 1, \cdots, n$.

(3.12.1) also applies to infinite sums, i.e., $\Sigma x_i^2 \Sigma y_i^2 \geq (\Sigma x_i y_i)^2$ and also to complex numbers, i.e., $\Sigma|x_i|^2\Sigma|y_i|^2 \geq \Sigma|x_iy_i|^2$ where x_i, y_i are arbitrary complex numbers and $|t|$ denotes modulus of complex number t.

Inequality (3.12.1) also applies to integrals. Let $f(x)$ and $g(x)$ be two arbitrary real functions of x. Then

$$\int f^2\,dx \int g^2\,dx \geq \left\{\int fg\,dx\right\}^2. \tag{3.11.2}$$

3.11.2 Holder's Inequality

Let $(x_1, \cdots, x_n)$, $(y_1, \cdots, y_n)$ be two sets of non-negative real numbers. Then

$$\left(\sum_{i=1}^{n} x_i^p\right)^{\frac{1}{p}} \left(\sum_{i=1}^{n} y_i^q\right)^{\frac{1}{q}} \geq \sum_{i=1}^{n} x_iy_i \tag{3.11.3}$$

where $\frac{1}{p} + \frac{1}{q} = 1$, $p > 1$. The equality holds if and only if $y_i \propto x_i^{p-1}$.

(3.12.3) also applies to integrals. Let $f(x), g(x)$ be two arbitrary real non-negative functions of x. Then

$$\left(\int f^p\,dx\right)^{\frac{1}{p}} \left(\int g^q\,dx\right)^{\frac{1}{q}} \geq \int fg\,dx$$

provided $\frac{1}{p} + \frac{1}{q} = 1$, $p > 1$.

More generally, we have, for sets of non-negative real numbers $(x_1, \cdots, x_n)$, $(y_1, \cdots, y_n)$, $(z_1, \cdots, z_n)$, $\cdots$

$$\left(\sum x_i^p\right)^{\frac{1}{p}} \left(\sum y_i^q\right)^{\frac{1}{q}} \left(\sum z_i^q\right)^{\frac{1}{r}} \cdots \geq \sum x_iy_iz_i \cdots$$

if $\frac{1}{p} + \frac{1}{q} + \frac{1}{r} + \cdots = 1$.

3.11.3 Minkowski's Inequality

For two sets of non-negative real numbers $(x_1, \cdots, x_n)$, $(y_1, \cdots, y_n)$ and $p \geq 1$,

$$\left(\sum x_i^p\right)^{\frac{1}{p}} + \left(\sum y_i^p\right)^{\frac{1}{p}} \geq \left[\sum (x_i + y_i)^p\right]^{\frac{1}{p}}.$$

Chapter 4

Generating Functions

.1 Introduction

/e consider different generating functions in this chapter. Section 4.2 deals with robability generating function of a random variable. We consider moment generating nction in section 4.3, factorial moment generating function in section 4.4, cumulants nd cumulant generating function in section 4.5 and characteristic function in section 3.

.2 Probability Generating Function

et X be an integer-valued r.v. with $P(X = k) = p_k$, $k = 0, 1, 2, \cdots$, and $\sum_{k=0}^{\infty} p_k = 1$.

Definition 4.2.1 *The function defined by*

$$P_X(t) = E(t^X) = \sum_{k=0}^{\infty} p_k t^k, \quad |t| < 1 \tag{4.2.1}$$

called the probability generating function (p.g.f.) *of X. Since $P_X(1) = 1$, series* (4.2.1) *s uniformly and absolutely convergent in $|t| \leq 1$ and $P_X(t)$ is a continuous function of t.*

Every p.g.f. determines a unique probability distribution, i.e., a unique set of robabilities $\{p_k\}$. Coefficient of t^k in the expansion of $P_X(t)$ gives $P(X = k)$. The .g.f. is used for discrete variables only.

Example 4.2.1 *X is uniformly distributed over $0, 1, \cdots, n$ with*

$$P(X = k) = \frac{1}{n+1}, \quad k = 0, 1, \cdots, n$$

$$P_X(t) = \frac{1}{n+1} \sum_{k=0}^{n} t^k = \frac{1}{(1-t)(n+1)} \left[1 - t^{n+1}\right], \quad |t| < 1.$$

► **Example 4.2.2** *X has a geometric distribution,*

$$P(X = k) = pq^k, \ k = 0, 1, \cdots, \ 0 < p < 1, \ q = 1 - p,$$
$$P_X(t) = \sum_0^{\infty} t^k pq^k = \frac{p}{1 - tq}, \ |t| < 1.$$

► **Example 4.2.3** *X has a Poisson distribution,*

$$P(X = k) = e^{-\lambda}\frac{\lambda^k}{k!}, \ k = 0, 1, \cdots$$
$$P_X(t) = \sum_0^{\infty} (\lambda t)^k \frac{e^{-\lambda}}{k!} = e^{-\lambda(1-t)}, \ |t| < 1.$$

► **Example 4.2.4** *X has a Binomial distribution,*

$$P(X = k) = \binom{n}{k} p^k (1 - p)^{n-k}, \ 0 < p < 1.$$
$$P_X(t) = \sum_0^{n} t^k \binom{n}{k} p^k (1 - p)^{n-k} = [1 + p(t - 1)]^n.$$

► **Example 4.2.5** *Negative Binomial distribution with parameters n, p :*

$$P(X = x) = \binom{-n}{x} p^n (p - 1)^x, x = 0, 1, 2, \cdots$$

$$P_X(t) = \sum_0^{\infty} t^x \binom{-n}{x} p^n (p - 1)^x = p^n \sum_0^{\infty} \binom{-n}{x} [t(p - 1)]^x$$
$$= p^n [1 + t(p - 1)]^{-n}, \ \textit{provided} \ |t(p - 1)| < 1.$$

Note that when $n = 1$, the p.g.f. becomes identical with the p.g.f. of a geometric distribution with parameter p (example 4.2.2). Hence if $X_1, \cdots, X_n$ are independently and identically geometrically distributed r.v.'s with parameter p, their sum $X_1 + \cdots + X_n$ follows a negative binomial distribution with parameters n, p (follows by theorem 4.2.2).

Theorem 4.2.1 *Let X be an integer-valued r.v. Consider $Y = aX + b, a, b$ being arbitrary constants. The p.g.f. of Y is*

$$P_Y(t) = E\left(t^{aX+b}\right) = t^b P_{aX}(t). \tag{4.2.2}$$

Theorem 4.2.2 *Let $X_1, \cdots, X_n$ be n independent r.v.s with p.g.f.s $P_1(t), \cdots, P_n(t)$ respectively. Consider the sum $X_1 + \cdots + X_n = S_n$. The p.g.f. of S_n is*

$$P_{Sn}(t) = P_1(t) \cdots P_n(t). \tag{4.2.3}$$

Proof:

$$P_{S_n}(t) = E\left(t^{S_n}\right) = E\left(t^{X_1}\right)\cdots E\left(t^{X_n}\right) = P_1(t)\cdots P_n(t), \quad |t|<1,$$

nce each of $t^{X_1}, \cdots, t^{X_n}$ are independent r.v.s.

Example 4.2.6 *Independent Bernoullian trials with probability p of a success are erformed, $0 < p < 1$. Let X_k denote the number of failures following the $(k-1)$th uccess and preceding the kth success. Then $S_r = X_1 + \cdots + X_r$ is the total number of ailures preceding the rth success. Here $X_1, \cdots, X_r$ are independently and indentically istributed r.v.s., each X_i having the geometric distribution mentioned in example 4.2.2. Ience*

$$P_{S_r}(t) = \{P_X(t)\}^r = \left(\frac{p}{1-tq}\right)^r, \quad |t|<1.$$

Coefficient of t^k is

$$\binom{-r}{k} p^r(-q)^k, \quad k = 0, 1, \cdots$$

which is the probability of k failures preceding the rth success. (*vide section* 5.8)

We now consider the p.g.f. of a sum of indicator r.v.s.

Theorem 4.2.3 *Suppose $I = I_1 + \cdots + I_n$ where each I_i is an indicator r.v. Let A_i be he event for which I_i is an indicator r.v., i.e., $P(I_i = 1) = P(A_i)$. The p.g.f. of I is*

$$P_I(t) = 1 + (t-1)B_1 + (t-1)^2B_2 + \cdots + (t-1)^nB_n, \quad |t|<1 \tag{4.2.4}$$

where

$$B_k = \sum\cdots\sum_{i_1<\cdots<i_k=1}^{n} P\left(A_{i_1}\cdots A_{i_k}\right)$$

Proof: Consider the random function

$$Y = [1 + (t-1)I_1]\cdots[1 + (t-1)I_n]. \tag{4.2.5}$$

Suppose for a sample point w, $I(w) = k$. For this w, $Y(w) = t^k$. Hence

$$P[Y = t^k] = P(I = k), \quad k = 0, 1, \cdots, n.$$

Thus

$$\begin{aligned} E(Y) &= t^0P(Y = t^0) + tP(Y = t) + \cdots + t^nP(Y = t^n) \\ &= t^0P(I = 0) + tP(I = 1) + \cdots t^nP(I = n) = P_I(t). \end{aligned} \tag{4.2.6}$$

Again, from (4.2.5),

$$Y = 1 + (t-1)\sum_i I_i + (t-1)^2 \sum\sum_{i<j} I_i I_j + \cdots + (t-1)^n I_1 \cdots I_n$$

$$E(Y) = 1 + (t-1)B_1 + (t-1)^2 B_2 + \cdots + (t-1)^n B_n. \tag{4.2.7}$$

Combining (4.2.6) and (4.2.7) the result follows.

► **Example 4.2.7** *A coin is tossed n times. Find the p.g.f. of number of heads. Let $X_i = 1(0)$ if the ith toss results in a head (tail); then $X = X_1 + \cdots + X_n$ = total number of heads.*

Here

$$B_k = \text{Probability of getting } k \text{ heads} = \binom{n}{k}\frac{1}{2^k}$$

$$P_X(t) = \sum_{k=0}^{n} \binom{n}{k}\frac{(t-1)^k}{2^k} = \left(\frac{1+t}{2}\right)^n, \quad |t| < 1.$$

► **Example 4.2.8** *Let X be a r.v. with $P(X = k) = p_k$, $k = 1, 2, \cdots$. Let*

$$P(X > k) = q_k, k = 0, 1, \cdots \text{ and } Q(t) = \sum_{j=0}^{\infty} q_j t^j.$$

Show that the p.g.f. of tail probabilities q_k is

$$Q(t) = \frac{1 - P(t)}{1 - t}, \quad |t| < 1 \tag{4.2.8}$$

where $P(t)$ is the p.g.f. of X. Show also that

$$\lim_{t \uparrow 1} \frac{1 - P(t)}{1 - t} = E(X) \tag{4.2.9}$$

when $t \uparrow 1$ means t approaches 1 from left.

Here

$$\begin{aligned} Q(t) &= 1 + q_1 t + q_2 t^2 + \cdots = 1t^0 + (1 - p_1)t + (1 - p_1 - p_2)t^2 + \cdots \\ &= (1 + t + t^2 + \cdots) - p_1 t(1 + t + t^2 + \cdots) - p_2 t^2(1 + t + t^2 + \cdots) - \cdots \\ &= \frac{1 - P(t)}{1 - t}. \end{aligned}$$

Hence

$$\sum_{j=0}^{\infty} t^j P(X > j) = \frac{1 - P(t)}{1 - t}.$$

Taking limit of both sides as $t \uparrow 1$, (4.2.9) follows.

If a distribution is specified by its p.g.f. its factorial moments (about origin) can be obtained as follows:

$$\frac{\partial^r P_X(t)}{\partial t^r} = \sum_x x(x-1)\cdots(x-r+1)t^{x-r}p_x.$$

Hence

$$\left.\frac{\partial^r P_x(t)}{\partial t^r}\right]_{t=1} = \mu_{(r)}.$$

Example 4.2.9 *In example* 4.2.2,

$$P'(t) = \frac{pq}{(1-tq)^2},\ \ P''(t) = \frac{2pq^2}{(1-tq)^3}.$$

Hence

$$E(X) = \frac{q}{p},\ \ E(X^2) = \frac{q}{p} + \frac{2pq^2}{p^3}$$

when $V(X) = \frac{q}{p^2}$.

In example 4.2.3,

$$P'(t) = \lambda e^{-\lambda(1-t)},\ \ P''(t) = \lambda^2 e^{-\lambda(1-t)}.$$

Hence

$$E(X) = \lambda,\ \ E\left(X(X-1)\right) = \lambda^2,$$

when $V(X) = \lambda$.

Theorem 4.2.4 *Suppose X is non-negative and integer-valued and that moments of all orders exist. Then*

$$P_X(t) = \sum_{k=0}^{\infty}\{E(X)_k\}\frac{(t-1)^k}{k!}, |t| \le 1. \tag{4.2.10}$$

Proof: We have

$$P_X(t) = p_0 + p_1 t + p_2 t^2 + \cdots.$$

Now

$$t^i = (t-1+1)^i = \sum_{k=0}^{i}\binom{i}{k}(t-1)^k.$$

Hence

$$P_X(t) = \sum_{i=0}^{\infty} p_i \sum_{k=0}^{i}\binom{i}{k}(t-1)^k = \sum_{k=0}^{\infty}(t-1)^k \sum_{i=k}^{\infty}\binom{i}{k}p_i.$$

The rearrangement in the order of summation is valid because (4.2.10) is absolutely convergent for $|t| \le 1$. Since

$$\sum_{i=k}^{\infty}\binom{i}{k}p_i = \frac{1}{k!}E\{(X)_k\},$$

the result follows.

4.3 Moment Generating Function

Definition 4.3.1 *Let X be a r.v. defined on $(\mathcal{S}, \mathcal{A}, P)$. The function*

$$M_X(t) = E(e^{tX})$$

defined for all real values of t is known as the moment generating function (m.g.f.) of X if the expectation of e^{tX} exists in some neighbourhood of the origin. For a discrete r. with p.m.f.

$$P(X = x_k) = p_k, \quad k = 0, 1, 2, \cdots$$

$$M_X(t) = \sum_0^\infty e^{tx_k} p_k.$$

In the case of a continuous probability distribution, specified by a p.d.f. f, the m.g. is

$$M_X(t) = \int_{-\infty}^{\infty} e^{tx} f(x)\, dx.$$

Since for every fixed t, e^{tx} is a positive function of x, it follows that $M_X(t)$ is eithe finite or infinite. We say that the probability law possesses a m.g.f. if there exists positive number T such that $M_X(t)$ is finite for $|t| \leq T$. In this case all moments distribution of X exist and are given by the successive derivatives of $M_X(t)$ at $t = 0$.

If a m.g.f. exists for some t in $[-T, T]$, then one may form its successive derivative by successively differentiating under the integral or summation sign. Thus

$$M'(t) = \frac{d}{dt} M(t) = E\left(\frac{\partial}{\partial t} e^{tX}\right) = E\left(Xe^{tX}\right)$$

$$M''(t) = \frac{d^2}{dt^2} M(t) = E\left(\frac{\partial}{\partial t} Xe^{tX}\right) = E\left(X^2 e^{tX}\right)$$

$$\cdots\cdots\cdots\cdots\cdots\cdots\cdots\cdots\cdots\cdots\cdots\cdots\cdots\cdots\cdots.$$

Letting $t = 0$ we have

$$M^{(k)}(0) = E(X^k), \text{ for all positive integral } k.$$

Thus

$$M'(0) = E(X), \quad M''(0) = E(X^2).$$

If the m.g.f. $M(t)$ is finite for some $|t| \leq T$ (for some $T > 0$) one can express $M(t)$ (uniquely) in a Maclaurin series expansion:

$$M(t) = \sum_{j=0}^{\infty} \frac{M^{(j)}(0)}{j!} t^j = \sum_{j=0}^{\infty} \frac{\mu'_j}{j!} t^j \tag{4.3.1}$$

so that μ'_j is the coefficient of $\frac{t^j}{j!}$ in the expresion (4.3.1).

We have seen there exists probability laws without finite moments (e.g., Cauch distribution). Consequently there also exist probability laws which do not possess m.g.f. The following results hold for a m.g.f.:

1. **Theorem 4.3.1** *A m.g.f., if it exists, uniquely determines the distribution function of a r.v. X and conversely if the m.g.f. exists it is unique.*

   Thus if X, Y be two r.v.s with m.g.f. $M_X(t)$, $M_Y(t)$ respectively, such that $M_X(t) = M_Y(t)$ for all values of t, then X, Y have the same probability distribution.

2. Even if the moments of all orders exist for a r.v. X it does not follow that m.g.f. exists.

3. If not all the moments of a r.v. X exist one can not find the distribution function of X from the moments, even from the moments that do exist.

Since there exist probability laws for which m.g.f. does not exist, utility of m.g.f. in determining the distribution function is somewhat limited. It is more convenient to use the characteristic function (discussed in section 4.6), $\varphi_X(t) = E(e^{itX})$, where $i = \sqrt{-1}$, the imaginary quantity and t is any real number, which always exists and uniquely determines the distribution of X. $\varphi_X(t)$ can also be used, like m.g.f., to determine those moments which do exist.

▶ **Example 4.3.1** *Consider again Binomial distribution,*

$$P(X = k) = \binom{n}{k} p^k q^{n-k}, \quad 0 < p < 1, \; q = 1 - p.$$

$$M(t) = \sum_{k=0}^{n} e^{tk} \binom{n}{k} p^k q^{n-k} = \left(q + pe^t\right)^n \text{ exists for } -\infty < t < \infty$$

$$M'(t) = npe^t \left(pe^t + q\right)^{n-1}$$

$$M''(t) = npe^t \left(pe^t + q\right)^{n-1} + n(n-1)p^2 e^{2t} \left(pe^t + q\right)^{n-2}.$$

Thus

$$M'(0) = E(X) = np$$

$$M''(0) = E(X^2) = npq + n^2p^2$$

$$V(X) = npq.$$

▶ **Example 4.3.2** *Poisson distribution with parameter λ:*

$$P(X = k) = e^{-\lambda} \frac{\lambda^k}{k!}, \quad k = 0, 1, 2, \cdots; \; M(t) = \sum_0^{\infty} e^{-\lambda} \frac{\lambda^k}{k!} e^{tk} = e^{\lambda(e^t - 1)}$$

$$M'(t) = e^{\lambda(e^t-1)} \lambda e^t, \qquad M''(t) = \lambda^2 e^{\lambda(e^t-1)} e^{2t} + \lambda e^t e^{\lambda(e^t-1)}.$$

Thus

$$E(X) = M'(0) = \lambda, \; E(X^2) = M''(0) = \lambda^2 + \lambda, \; V(X) = \lambda.$$

For truncated Poisson distribution, truncated at $X = 1$,

$$P(X = k) = \frac{e^{-\lambda}\lambda^k}{k!(1 - e^{-\lambda})}, \quad k = 1, 2, \cdots,$$
$$M_X(t) = \left(e^{\lambda e^t} - 1\right) \Big/ \left(e^{\lambda} - 1\right).$$

▶ **Example 4.3.3** *Geometric distribution with parameter* $p, 0 < p < 1$:

$$P(X = k) = pq^k, \quad q = 1 - p, \ k = 0, 1, 2, \cdots$$
$$M(t) = \frac{p}{1 - qe^t}$$

exists for t such that $qe^t < 1$. *It follows*

$$E(X) = \frac{q}{p}, \quad V(X) = \frac{q}{p^2}.$$

▶ **Example 4.3.4**

$$f(x) = \frac{6}{\pi^2 x^2}, \quad x = 1, 2, \cdots, = 0 \text{ otherwise}$$
$$M_X(t) = \sum_{1}^{\infty} \frac{6e^{tx}}{\pi^2 x^2} \text{ is infinite for every } t > 0.$$

The m.g.f., therefore, does not exist.
The m.g.f. of $Z = \frac{X-a}{h}$ (a, h are arbitrary constants) is

$$M_z(t) = Ee^{tZ} = e^{-\frac{at}{h}} M_X\left(\frac{t}{h}\right).$$

The moment generating function of X about its mean μ, $M_{X-\mu}(t)$ gives the central moments of X, μ_r.

$$M_{X-\mu}(t) = \sum_{j=0}^{\infty} \frac{\mu_j}{j!} t^j,$$

provided the m.g.f. exists for some $|t| \leq T$.

Theorem 4.3.2 *The m.g.f. of the sum* $S_n = X_1 + \cdots + X_n$ *of n independent r.v.s* $X_1, \cdots, X_n$ *is*

$$M_{S_n}(t) = M_{X_1}(t) \cdots M_{X_n}(t)$$

where $M_{X_i}(t)$ *is the m.g.f. of* X_i, *provided all the m.g.f.s exist.*

Proof:

$$MS_n(t) = E\left(e^{t\Sigma_1^n X_i}\right) = E\left(\prod_{i=1}^{n} e^{tX_i}\right) = \prod_{i=1}^{n} E\left(e^{tX_i}\right)$$

because of independence.

▶ **Example 4.3.5** *X is normally distributed with mean μ and s.d. σ.*

$$f(x) = \frac{1}{\sigma\sqrt{2\pi}} e^{-\frac{(x-\mu)^2}{2\sigma^2}}, \quad -\infty \le x \le \infty.$$

The m.g.f. of X about μ is

$$M_{X-\mu}(t) = E\left[e^{t(X-\mu)}\right] = \frac{1}{\sigma\sqrt{2\pi}} \int_{-\infty}^{\infty} e^{t(x-\mu)} e^{-\frac{(x-\mu)^2}{2\sigma^2}}\, dx$$

$$= \frac{e^{t^2\sigma^2/2}}{\sigma\sqrt{2\pi}} \int_{-\infty}^{\infty} e^{-\frac{1}{2\sigma^2}\{(x-\mu)-\sigma^2 t\}^2}\, dx = e^{t^2\sigma^2/2} = \sum_{j=0}^{\infty} \frac{(t^2\sigma^2)^j}{2^j j!}.$$

Hence

$$E(X-\mu) = 0 \text{ or } E(X) = \mu$$

$$E(X-\mu)^{2r+1} = 0, \quad E(X-\mu)^{2r} = \mu_{2r} = \frac{\sigma^{2r}(2r)!}{2^r \cdot r!}$$

$$= (2r-1)(2r-3)\cdots 3.1\cdots\sigma^{2r} \text{ for } r = 1, 2, \cdots.$$

The m.g.f. of X is $e^{\mu t + \frac{t^2\sigma^2}{2}}$.

▶ **Example 4.3.6** *Consider example 4.3.5. Let $Y = \alpha X + \beta$ when α, β are constants. The m.g.f. of Y is*

$$M_Y(t) = e^{\beta t} M_X(\alpha t) = e^{(\beta+\alpha\mu)t + (\alpha\sigma)^2\frac{t^2}{2}}.$$

Thus Y follows a normal distribution with mean $\beta + \alpha\mu$ and variance $\sigma^2\alpha^2$ (by theorem 4.3.1).

▶ **Example 4.3.7** *Consider n independent Poisson variables $X_1, \cdots, X_n$ (vide example 4.3.2). By theorem 4.3.2, the m.g.f. of $S_n = X_1 + \cdots + X_n$ is*

$$\exp\left\{\sum_i^n \lambda_i(e^t - 1)\right\}.$$

Thus by theorem 4.3.1, S_n has a Poisson distribution with parameter $\sum_{i=1}^n \lambda_i$ (This is called the reproductive property of Poisson distribution. The Poisson distribution is discussed in details in section 5.5).

4.4 Factorial Moment Generating Function

Definition 4.4.1 *The factorial moment generating function (f.m.g.f.) of a discrete probability distribution of X about the origin is defined as*

$$\phi_X(t) = E\left[(1+t)^X\right] = E\left[e^{X\log_e(1+t)}\right] = M_X\left[\log_e(1+t)\right]. \tag{4.4.1}$$

Since

$$(1+t)^x = \sum_{r=0}^{\infty} \frac{(x)_r}{r!} t^r,$$

we get $\mu_{(r)}$ = coefficient of $\frac{t^r}{r!}$ in the expansion of $\phi_X(t)$ in ascending powers of t. Also

$$\frac{d^r}{dt^r}\phi_X(t)\bigg]_{t=0} = \mu_{(r)}.$$

▶ **Example 4.4.1** *In the familar matching problem with n cards, the probability distribution of number of matches X is*

$$P[X=m] = \frac{1}{m!}\sum_{k=0}^{n-m}(-1)^k\frac{1}{k!}, \quad m = 0, 1, \cdots, n.$$

The m.g.f. of X is

$$M_X(t) = E(e^{tX}) = \frac{1}{m!}\sum_{k=0}^{n-m}(-1)^k\frac{e^{tX}}{k!} = \sum_{r=0}^{n}\frac{1}{r!}(e^t-1)^r.$$

Hence the f.m.g.f. is

$$\phi_X(t) = \sum_{r=0}^{n}\frac{t^r}{r!}.$$

4.5 Cumulant Generating Function

Definition 4.5.1 *If the logarithm of the m.g.f. of the distribution of a r.v. X can be expanded in a convergent series in powers of t viz.,*

$$K_X(t) = \log_e M_X(t) = \sum_{r=1}^{\infty} k_r(X)\frac{t^r}{r!},$$

then the coefficients $k_1(X)$, $k_2(X)$, $\cdots$, are called the first cumulant, second cumulant,$\cdots$, etc. of the distribution of X and $K_X(t)$ is called the cumulant generating function (c.g.f.) *of the distribution of X.*

We have

$$k_r(X) = \frac{d^r}{dt^r}\log_e M_X(t)\bigg]_{t=0}.$$

If the m.g.f. exists, so does the c.g.f. Evidently $k_r(X)$ is a function of the moments of X. The first few cumulants are as follows:

$$\begin{aligned} k_1 &= \mu_1' = \mu \\ k_2 &= \mu_2' - (\mu_1')^2 = \mu_2 \\ k_3 &= \mu_3' - 3\mu_2'\mu_1' + 2(\mu_1')^3 = \mu_3 \\ &\cdots\cdots\cdots\cdots\cdots\cdots \end{aligned}$$

nd conversely,

$$\begin{aligned}
\mu_1' &= k_1 \\
\mu_2' &= k_2 + k_1^2 \\
\mu_3' &= k_3 + 3k_1k_2 + k_1^3 \\
&\cdots\cdots\cdots\cdots\cdots\cdots
\end{aligned}$$

Now, for any constants a and h, the m.g.f. of $Z = \frac{X-a}{h}$ is

$$M_Z(t) = E\left[e^{t\left(\frac{(X-a)}{h}\right)}\right] = e^{-\frac{at}{h}} M_X\left(\frac{t}{h}\right).$$

Hence

$$\begin{aligned}
K_Z(t) &= \log_e M_{\frac{X-a}{h}}(t) = -\frac{at}{h} + \log_e M_X\left(\frac{t}{h}\right) \\
&= -\frac{at}{h} + \frac{t}{h}k_1(X) + \frac{t^2}{2!h^2}k_2(X) + \frac{t^3}{3!h^3}k_3(X) + \cdots \\
&= (-a + k_1(X))\frac{t}{h} + \frac{t^2}{2!h^2}k_2(X) + \frac{t^3}{3!h^3}k_3(X) + \cdots. \qquad (4.5.1)
\end{aligned}$$

Thus

$$k_1(Z) = \frac{1}{h}[-a + k_1(X)], \ k_r(Z) = \frac{k_r(X)}{h^r}, \quad r \geq 2.$$

Thus only the first cumulant of Z is affected by the change of origin. All the cumulants of Z are affected by the change of scale. For this reason, the cumulants are sometimes called semi-invariants. The cumulants were originally defined and studied by Thiele (1903).

Putting $a = \mu$ and $h = 1$ in (4.5.1) we get the central cumulants (cumulants about the mean μ) which are the same as the cumulants about zero except for the first central cumulant which becomes zero.

Theorem 4.5.1 *If $X_1, \cdots, X_n$ are independent r.v.s and if the m.g.f. $M_{X_i}(t)$ exists $(i = 1, \cdots, n)$, the c.g.f. of $S_n = X_1 + \cdots + X_n$ is*

$$K_{Sn}(t) = \sum_{i=1}^{n} K_{X_i}(t).$$

Proof:

$$K_{Sn}(t) = \log_e\{M_{Sn}(t)\} = \log_e\left\{\prod_{i=1}^{n} M_{X_i}(t)\right\}.$$

Hence the proof.

It follows that

$$k_r\left(\sum_{i=1}^{n} X_i\right) = \sum_{i=1}^{n} k_r(X_i) \text{ for all } r.$$

Thus the cumulant of the sum is the sum of the cumulants. This makes the name 'cumulant' appropriate.

► **Example 4.5.1** (***Poisson distribution***). $P(X = k) = e^{-\lambda}\frac{\lambda^k}{k!}$. *The c.g.f. is* $\lambda(e^t - 1)$. *Hence*

$$k_r = \lambda, \quad r = 1, 2, \cdots.$$

For normal distribution (vide example 4.3.5) the c.g.f. about mean μ is $\frac{t^2\sigma^2}{2}$. Hence

$$k_1 = \mu, \quad k_2 = \sigma^2, \quad k_r = 0, \quad r \geq 2.$$

4.6 Characteristic Function

Definition 4.6.1 *The characteristic function (c.f.) of a r.v. X is defined as*

$$\psi_X(t) = E(e^{itX}) \tag{4.6.1}$$

where t is any real number and $i = \sqrt{-1}$. *Hence for a discrete r.v. with p.m.f.* p_k $(k = 0, 1, 2, \cdots,)$

$$\psi_X(t) = \sum_{k=0}^{\infty} e^{itx_k} p_k \tag{4.6.2}$$

and for a continuous r.v. with p.d.f. $f(x)$

$$\psi_X(t) = \int_{-\infty}^{\infty} e^{itx} f(x)\, dx. \tag{4.6.3}$$

Since $e^{itx} = \cos tx + i \sin tx$ and $\cos tx$ and $\sin tx$ are both integrable over R^1 for any real t, $\psi(t)$ is a complex number whose real and imaginery parts are finite for every value of t.

The r th non-central moment μ_r' of X, when it exists, can be determined from the c.f. by the formula

$$\left.\frac{d^r}{i^r\, dt^r}\right]_{t=0} = \mu_r'.$$

The properties of c.f. are:

1. $\psi(0) = 1$.

2. $|\psi(t)| \leq 1$ for all values of t, because $|\psi(t)| = \left|E(e^{itX})\right| \leq E\left|e^{itX}\right| = E(1) = 1$.

3. If $\psi(t)$ is the c.f. of X,$\overline{\psi(t)}$, complex conjugate of $\psi(t)$, is the c.f. of $(-X)$.

   Proof: $E(e^{-itX}) = E(\cos tX - i \sin tX) = E(\cos tX) - iE \sin(tX) = \overline{\psi(t)}$.

4. $\psi(t)$ is uniformly continuous on R^1.

Proof:

$$\psi(t) = \int_{-\infty}^{\infty} e^{itx} dF(x).$$

$$|\psi(t+h) - \psi(t)| = \int_{-\infty}^{\infty} e^{itx}(e^{ihx} - 1) dF(x)$$

$$\leq \int_{-\infty}^{\infty} \left| e^{ihx} - 1 \right| dF(x) = 2\int_{-\infty}^{\infty} \left| \sin \frac{hx}{2} \right| dF(x).$$

Hence

$$|\psi(t+h) - \psi(t)| \leq 2\int_{-\infty}^{-A} \left| \sin \frac{hx}{2} \right| dF(x) + 2\int_{-A}^{B} \left| \sin \frac{hx}{2} \right| dF(x)$$

$$+2\int_{B}^{\infty} \left| \sin \frac{hx}{2} \right| dF(x) \text{ for } A > 0, B > 0. \quad (4.6.4)$$

The right hand side of (4.6.4) is independent of t. By choosing A and B sufficiently large and h sufficiently small the first and the third term on the right hand side of (4.6.4) can be made sufficiently small when the proof follows.

We state below the following properties without proof.

5. If μ_r' exists for the distribution of X, then $\psi^{(j)}(t)$ exists and is continuous for $1 \leq j \leq r$ and

$$\psi(t) = \sum_{j=0}^{r} \frac{(it)^j}{j!} \mu_j' + o(t^r)$$

where $\lim_{t \to 0} \frac{o(t^r)}{t^r} = 0$.

6. Suppose the c.f. $\psi(t)$ has a derivative of order $2r$ at zero, $\psi^{(2r)}(0)$ (2r is an even integer). Then the $(2r)$th moment μ_{2r}' exists and there is a $2r$th derivative for all real t given by

$$\psi^{(2r)}(t) = (-1)^r \int_{-\infty}^{\infty} e^{itx} x^{2r} dF(x).$$

7. The necessary and sufficient conditions for a bounded and continuous function to be the characteristic function of a distribution are that $\psi(0) = 1$ and the function

$$\Psi(x, A) = \int_0^A \int_0^A \varphi(t-u) e^{i(t-u)x} \, dt \, du$$

is real and non-negative for all x and $A > 0$.

8. **Theorem 4.6.1** (***The Uniqueness theorem***). *Two distribution functions $F_1(x)$ and $F_2(x)$ are identical if and only if their characteristic functions $\psi_1(t)$ and $\psi_2(t)$ are identical.*

The uniqueness theorem states that, given a function $\psi(t)$, if $\psi(t)$ is the c.f. of a r.v. X, we can determine its d.f. $F(x)$ from it. The process of determining the distribution function from the c.f. is given by the following inversion theorem.

9. **Theorem 4.6.2** (*Inversion theorem*). *Let $\psi(t)$ be the characteristic function of the distribution $F(x)$. Then for any two continuity points a and $a+h$ $(h>0)$ of $F(x)$,*

$$F(a+h)-F(a)=\lim_{T\to\infty}\frac{1}{2\pi}\int_{-T}^{T}\frac{1-e^{-ith}}{it}e^{-ita}\psi(t)\,dt. \tag{4.6.5}$$

Chosing a $= x-\delta$ and $h=2\delta$, (4.6.5) can be written as:

$$\frac{F(x+\delta)-F(x-\delta)}{2\delta}=\lim_{T\to\infty}\frac{1}{2\pi}\int_{-T}^{T}\frac{\sin t\delta}{t\delta}e^{-itx}\psi(t)\,dt \tag{4.6.6}$$

provided $x+\delta$, $x-\delta$ are continuity points of $F(x)$.

10. If c.f. $\psi(t)$ is absolutely integrable over $\mathcal{R}^1$ (i.e., $\int_{-\infty}^{\infty}|\psi(t)|\,dt<\infty$) then the derivative of $F(x)$ exists, which is bounded and continuous on $\mathcal{R}^1$ and is given by

$$f(x)=F'(x)=\frac{1}{2\pi}\int_{-\infty}^{\infty}e^{-itx}\psi(t)\,dt \tag{4.6.7}$$

for every x in $\mathcal{R}^1$.

$f(x)$ is continuous for all values of x on $\mathcal{R}^1$. (4.6.7) enables one to determine the density function $f(x)$ by means of the c.f. when the latter is absolutely integrable.

11. $\psi(t)$ is real if and only if $F(x)$ is symmetric.

Proof:

$$\psi_X(t)=\int_{-\infty}^{\infty}(\cos tx+i\sin tx)dF(x).$$

If X is symmetric, the contribution of sine term vanishes, because

$$\int_{-\infty}^{0-}\sin tx\,dF_X(x)+\int_{0+}^{\infty}\sin tx\,dF_X(x)+(\sin 0)(P(X=0))$$
$$=\int_{0+}^{\infty}-\sin tx\,dF_X(x)+\int_{0+}^{\infty}\sin tx\,dF_X(x)=0.$$

Conversely, if $\psi_X(t)$ is real, consider the r.v. $Y=-X$.

$$\psi_Y(t)=\int_{-\infty}^{\infty}e^{-itx}dF_X(x)=\psi_X(-t)=\overline{\psi_X(t)}.$$

Since $\psi_X(t)$ is real, $\psi_X(t)=\overline{\psi_X(t)}$. Thus X and Y have the same c.f.
Again,

$$P(X\le -x)=P(-X\ge x)=P(Y\ge x),$$

i.e., $F_X(-x)=1-P(Y<x)=1-F_Y(x-0)=1-F_X(x-0)$ (by theorem 4.6.1).
Hence F_X is symmetric about zero.
For proof of these different theorems, the reader may refer to Lukacs (1960).

Example 4.6.1 *For the Normal distribution*

$$f(x) = \frac{1}{\sqrt{2\pi}\sigma}e^{-\frac{1}{2}\frac{(x-\mu)^2}{\sigma^2}}, -\infty < x < \infty$$

$$\psi(t) = \int_{-\infty}^{\infty} \frac{e^{itx}}{\sqrt{2\pi}\sigma}e^{-\frac{1}{2}\frac{(x-\mu)^2}{\sigma^2}}\, dx. \quad Let\ \frac{x-\mu}{\sigma} = y$$

$$\psi(t) = \int_{-\infty}^{\infty} \frac{1}{\sqrt{2\pi}}e^{it(\mu+\sigma y)}e^{-\frac{y^2}{2}}\, dy = \frac{e^{it\mu}}{\sqrt{2\pi}}\int_{-\infty}^{\infty} \cdot e^{-\frac{1}{2}(y^2-2it\sigma y)}\, dy$$

$$= e^{it\mu-\frac{t^2\sigma^2}{2}}\int_{-\infty}^{\infty} \frac{1}{\sqrt{2\pi}} \cdot e^{-\frac{1}{2}(y-it\sigma)^2}\, dy = e^{it\mu-\frac{t^2\sigma^2}{2}},$$

he integral being unity, since it is the integral of a normal distribution.

Ience,

$$\psi'(t) = e^{it\mu-\frac{t^2\sigma^2}{2}}(i\mu - t\sigma^2)$$

$$\mu''(t) = e^{it\mu-\frac{t^2\sigma^2}{2}}\left[(i\mu - t\sigma^2)^2 - \sigma^2\right].$$

Thus

$$\psi'(0) = i\mu, \quad \psi''(0) = i^2(\mu^2 + \sigma^2).$$

▶ Example 4.6.2 *For the Gamma distribution*

$$f(x) = \frac{\lambda^n}{\Gamma(n)}e^{-\lambda x}x^{n-1}, 0 < x < \infty, \lambda > 0$$

$$\psi(t) = \int_0^{\infty} \frac{\lambda^n}{\Gamma(n)}e^{-(\lambda-it)x}x^{n-1}\, dx = \frac{\lambda^n}{(\lambda-it)^n}\int_0^{\infty}\frac{(\lambda-it)^n}{\Gamma(n)}e^{-(\lambda-it)x}x^{n-1}\, dx$$

$$= \left(\frac{\lambda}{\lambda - it}\right)^n = \left(1 - \frac{it}{\lambda}\right)^{-n}$$

For the exponential distribution, which is a gamma distribution with parameter $n = 1$,

$$\psi(t) = \left(1 - \frac{it}{\lambda}\right)^{-1}, \quad \psi'(t) = (-n)\left(1 - \frac{it}{\lambda}\right)^{-n-1}\left(-\frac{i}{\lambda}\right)$$

$$\psi''(t) = (-n)(-n-1)\left(1 - \frac{it}{\lambda}\right)^{-n-2}\left(\frac{i^2}{\lambda^2}\right)$$

$$\psi'(0) = \frac{in}{\lambda}, \quad \psi''(0) = \frac{n(n+1)i^2}{\lambda^2}.$$

► **Example** 4.6.3 *Laplace distribution*

$$f(x) = \frac{1}{2}e^{-|x|}, \quad -\infty \leq x \leq \infty$$
$$\psi(t) = \frac{1}{2}\int_{-\infty}^{\infty} e^{itx}e^{-|x|}\,dx = \frac{1}{2}\int_{-\infty}^{\infty}(\cos tx + i\sin tx)e^{-|x|}\,dx = \int_{0}^{\infty} \cos tx\, e^{-x}\,dx.$$

Now

$$\int_0^\infty e^{-x}\cos tx\,dx = 1 - t\int_0^\infty e^{-x}\sin tx\,dx$$

(on integration by parts, taking $\cos tx$ as first function).
Similarly

$$\int_0^\infty e^{-x}\sin tx\,dx = t\int_0^\infty e^{-x}\cos tx\,dx.$$

Hence

$$\psi(t) = \frac{1}{1+t^2}.$$

Note that $\psi(t)$ is absolutely integrable and by (4.6.6)

$$\int_{-\infty}^{\infty}\frac{e^{-itx}}{1+t^2}\,dt = \pi e^{-|x|}. \qquad \text{(a}$$

► **Example** 4.6.4 *For the Cauchy distribution*

$$f(x) = \frac{1}{\pi(1+x^2)}, \quad -\infty < x < \infty, \quad \psi(t) = \int_{-\infty}^{\infty}\frac{e^{itx}dx}{\pi(1+x^2)}.$$

To evaluate this consider the result (a) in example 4.6.1. Replacing t by $-t$ an
interchanging t and x

$$\int_{-\infty}^{\infty}\frac{e^{itx}}{1+x^2}\,dx = \pi e^{-|t|}.$$

Hence

$$\psi(t) = e^{-|t|}.$$

$\psi(t)$ is not differentiable at the point $t = 0$. This shows that for the cauchy distribution no moment exists.

Theorem 4.6.3 *If a and b are real constants, then c.f. of $Y = aX + b$ is*

$$\psi_Y(t) = e^{itb}\psi_X(at).$$

In particular

$$\Psi_{\frac{X-\mu}{\sigma}} = e^{-it\mu/\sigma}\psi_X\left(\frac{t}{\sigma}\right).$$

Example 4.6.5 *Consider the normal distribution in example* 4.6.1. *The c.f. of*

$$Y = \frac{X-\mu}{\sigma} \text{ is } e^{-\frac{t^2}{2}} = \sum_{j=0}^{\infty} \frac{1}{j!}\left(-\frac{t^2}{2}\right)^j.$$

Since the coefficient of $\frac{(it)^j}{j!}$ gives the jth moment, μ_j of distribution of Y is

$$\frac{j!}{2^{j/2}\left(\frac{j}{2}\right)!} \text{ if } j \text{ is even,}$$
$$0 \quad \text{if } j \text{ is odd.}$$

Exercise

1. Find the p.g.f. of the r.v. X when

   (a) $P(X=k) = [e^{-\mu}/(1-e^{-\mu})]\frac{\mu^k}{k!}$, $k = 1, 2, \cdots, \lambda > 0$.

   (b) $P(X=k) = pq^k/(1-q^{N+1})$, $k = 0, 1, \cdots, N; 0 < p < 1, q = 1-p$.

2. Let X be the number of tosses required to obtain a head in tossing a unbiased coin. Find p.g.f. of X. $\left[\frac{1}{2-t}, |t| < 1\right]$

3. In the familiar matching problem with 2 decks of n cards each let X be the number of matches. Show that p.g.f. of X is $\sum_{k=0}^{n} \frac{(t-1)^k}{k!}$.

4. (a) If $P(t)$ is a p.g.f. show that $\frac{1}{2-P(t)}$ is a p.g.f.

   (b) Show that $P(t) = \frac{2}{1+t}$ is not a p.g.f. even though $P(1) = 1$.

   [*Hints:* (a) $Q(t) = \frac{1}{2-P(t)} = \frac{1}{2}\left(1 - \frac{P(t)}{2}\right)^{-1} = \frac{1}{2}\left[1 + \frac{1}{2}P(t) + \left(\frac{P(t)}{2}\right)^2 + \left(\frac{P(t)}{2}\right)^3 + \cdots\right]$

   $Q(1) = 1$, since $P(1) = 1$; also coefficient of $(P(t))^i$ is non-negative, $i = 1, 2, \cdots$.

   (b) Coefficient of t^i, $i = 1, 2, \cdots$, are not all non-negative.]

5. Balls are distributed among m cells. Let Z be the number of balls that must be distributed to occupy all m cells. Find the p.g.f. of Z.

   [*Hints:* $Z = m + X_1 + X_2 + \cdots + X_{m-1}$ where X_i is the number of balls following the occupancy of i cells for the first time and preceding the occupancy of $(i+1)$ cells for the first time.

   X_1 follows a geometric distribution with probability of success $p = \frac{m-1}{m}$. The p.g.f. of X_1 is $\frac{(m-1)/m}{1-\frac{t}{m}}$ (by example 4.2.2).

   Similarly X_r follows a geometric distribution with $p = \frac{m-r}{m}$, $r = 2, 3, \cdots, m-1$. Also $X_1, \cdots, X_{m-1}$ are independent. Hence p.g.f. of Z is $t^m \prod_{r=1}^{m-1} \frac{(m-r)}{(m-tr)}$.]

6. n balls are randomly distributed among m cells. Find the p.g.f. of X, the numbe of unoccupied cells.

[*Hints:* Let $X_i = 1(0)$ if the ith cell is empty (occupied); $X = \sum_1^m X_i$ is the number of unoccupie cells. By the notation of theorem 4.2.3, $B_i = \binom{m}{i}\left(\frac{m-i}{m}\right)^n$. Hence by (4.2.4)

$$P_X(t) = \sum_{i=0}^{m} \binom{m}{i}\left(\frac{m-i}{m}\right)^n (t-1)^i.$$

Hence

$$P(X = k) = \sum_{i=0}^{m-k} (-1)^i \binom{k+i}{k}\binom{m}{k+i}\left(\frac{m-k-i}{m}\right)^n.]$$

7. For any integer-valued r.v. X, show that $\sum_{n=0}^{\infty} t^n P(X \le n) = (1-t)^{-1}P(t)$, wher $P(t)$ is the p.g.f. of X.

8. Let X, Y be independent r.v.s each assuming the values $1, 2, \cdots, a$ with probabilit $\frac{1}{a}$. Find the p.g.f. of $X - Y$ and hence determine the probability that X exceeds Y

$\Big[$*Hints:* $P_{X-Y}(t) = E\left(t^{X-Y}\right) = E(t^X)E(t^{-Y})$, $(X, -Y)$ are independent), $E(t^X) = \frac{t(1-t^a)}{a(1-t)}$ $E(t^{-Y}) = \frac{1-t^a}{at^a(1-t)}$. Hence $P_{X-Y}(t) = \frac{1}{a^2t^{a-1}}\left(\frac{1-t^a}{1-t}\right)^2$. Also

$$P(X > Y) = \sum_{r=1}^{a-1} \text{coefficient of } t^r \text{ in } P_{X-Y}(t)$$

$$= \sum_{r=1}^{a-1} \text{coefficient of } t^{r+a-1} \text{ in } \frac{1}{a^2}\left(\frac{1-t^a}{1-t}\right)^2 = \frac{1}{a^2}\sum_{r=1}^{a-1}(a-r) = \frac{1}{2} - \frac{1}{2a}.\Big]$$

9. Find the p.g.f. of the total score obtained when n balanced dice are rolled.

[*Hints:* Let X_i denote the score on the ith die $(i = 1, \cdots, n)$, X_i has a uniform distribution ove $1, \cdots, 6$ and its p.g.f. is $P_i(t) = \frac{t(1-t^6)}{6(1-t)}$. Hence p.g.f. of $S_n = \sum_{i=1}^n X_i$ is

$$P(t) = \left[\frac{t(1-t^6)}{6(1-t)}\right]^n = \left(\frac{t}{6}\right)^n (1-t^6)^n(1-t)^{-n}$$

$$= \left(\frac{t}{6}\right)^n \left\{\sum_{i=0}^{n}\binom{n}{i}(-1)^i t^{6i}\right\}\left\{\sum_{0}^{\infty}\binom{-n}{j}(-1)^j t^j\right\}$$

$$= \left(\frac{t}{6}\right)^n \sum_{i=0}^{n}\sum_{j=0}^{\infty}\binom{n}{i}\binom{-n}{j}(-1)^{i+j}t^{6i+j}$$

$$= \left(\frac{t}{6}\right)^n \sum_{k=0}^{\infty} t^k \sum_{i}\binom{n}{i}\binom{-n}{k-6i}(-1)^{i+k} \text{ where } k = j + 6i.$$

Prob $(S_n = n + k)$ = coefficient of t^{n+k} in $P(t)$

$$= \frac{1}{6^n}\sum_{i}\binom{n}{i}\binom{-n}{k-6i}(-1)^{i+k} \qquad \text{(a}$$

where the ith term in the summed in (a) is zero whenever $i > n$ or $6i > k$.]

10. Let X, Y be independent variables with p.g.f. $U(t), V(t)$, respectively. Show that $P(X-Y)=j$ is the coefficient of t^j in $U(t)V\left(\frac{1}{t}\right)$, when $j=0, \pm 1, \pm 2, \cdots$.

11. A r.v. X has p.m.f.

$$P(X=x)=\frac{1}{x!}\left[1-\frac{1}{1!}+\frac{1}{2!}\cdots \pm \frac{1}{(n-x)!}\right], \quad x=0,1,\cdots,n.$$

Show that p.g.f. is $P(t)=\sum_{i=0}^{n}\frac{(t-1)^i}{i!}$. Hence find the mean, variance, factorial moments of the distribution of X.

12. Let $X_1, \cdots, X_n$ be i.i.d. r.v. with $P(X_i=k)=\frac{1}{a}, k=0,1,\cdots,a-1; i=1,\cdots,n$. Let $X=X_1+\cdots+X_n$. Show that p.g.f. of X is $P_X(t)=\left\{\frac{1-t^a}{a(1-t)}\right\}^n$ and hence $P\{X=j\}$

$$=\frac{1}{a^n}\sum_{u=0}^{\infty}(-1)^{u+j+au}\binom{n}{u}\binom{-n}{j-au}.$$

The probability $P\{X\leq j\}$ has the p.g.f. $\frac{P(t)}{1-t}$ and hence

$$P\{X\leq j\}=\frac{1}{a^n}\sum_{u}(-1)^u\binom{n}{u}\binom{j-au}{n}.$$

If, further, $a\to\infty$, $j\to\infty$, such that $\frac{j}{a}\to x$, then

$$P\{X\leq j\}\to\frac{1}{n!}\sum_{u}(-1)^u\binom{n}{u}(x-u)^n.$$

13. Suppose X has p.g.f. $P(t)$ and that $E(X^3)$ exists. Show that (i) $E(X^3)=P^{(3)}(1)+3P^{(2)}(1)+P^{(1)}(1)$ (ii) $E(X^2)=P^{(2)}(1)+P^{(1)}(1)$, when $P^{(1)}(1)$ denotes the ith derivative of $P(t)$ at $t=1$.

14. Show that the function $E\left(\frac{1}{1-tX}\right)$ can be used to generate the moments of the r.v. X. State the condition on t.

15. Find the m.g.f., expectation and variance of X having p.d.f. $f(x)=\frac{1}{2}e^{-\frac{x}{2}}, x>0$.

$$\left[m.g.f.=\frac{1}{1-2t}\right]$$

16. Show that

$$e^{itx}=1+(e^{it}-1)x_{(1)}+(e^{it}-1)^2\frac{x_{(2)}}{2!}+\cdots+(e^{it}-1)^r\frac{x_{(r)}}{r!}+\cdots$$

where $x_{(r)}=x(x-1)\cdots(x-r+1)$. Hence show that $\mu_{(r)}=\Delta^r\phi(t)\Big]_{t=0}$ where $\Delta=\frac{d}{d(e^t)}$, $\phi(t)=E(e^{itX})$.

[*Hints:* Let $e^{itx} = A_0 + A_1 x_{(1)} + A_2 \frac{x_{(2)}}{2!} + \cdots + A_r \frac{x_{(r)}}{r!} + \cdots$ an identity in $\frac{x_{(i)}}{i!}$.
Putting $x = 0, 1, 2, \cdots$ and solving for $A_0, A_1, A_2, \cdots$, $A_0 = 1$, $A_1 = e^{it} - 1$, $A_2 = (e^{it} - 1)^2, \cdots$.
Hence

$$\begin{aligned} E(e^{itX}) &= A_0 + A_1 E(X_{(1)}) + \frac{A_2}{2!} E(X_{(2)}) + \cdots + \frac{A_r}{r!} E(X_{(r)}) + \cdots \\ &= A_0 + A_1 \mu_{(1)} + \frac{A_2}{2!} \mu_{(2)} + \cdots + \frac{A_r}{r!} \mu_{(r)} + \cdots \end{aligned}$$

Hence $\Delta^r \phi(t)]_{t=0} = \mu_{(r)}$.]

17. Define a factorial m.g.f. and show that if the factorial m.g.f. of a function $f(x)$ is $G(t)$, then the factorial m.g.f. of $(-\nabla)^r f(x)$ is $t^r G(t)$ where $\nabla f(x) = f(x) - f(x-1)$.

[*Hints:* Factorial m.g.f., $G(t) = E\left[(1+t)^X\right] = \sum_x (1+t)^x f(x)$.

$$\begin{aligned} f.m.g.f. \text{ of } f(X-1) - f(X) &= \sum_x (1+t)^x \{f(x-1) - f(x)\} \\ &= (1+t) \sum_x (1+t)^{x-1} f(x-1) - \sum_x (1+t)^x f(x) = tG(t). \end{aligned}$$

Now $\nabla^2 f(x) = \nabla(\nabla f(x)) = f(x) - 2f(x-1) + f(x-2)$. f.m.g.f. of $\nabla^2 f(x) = t^2 G(t)$.

Thus the result follows by induction.]

18. With usual notations, prove that

$$\frac{\partial \mu'_r}{\partial k_j} = \binom{r}{j} \mu'_{r-j}.$$

[*Hints:* We have,

$$\exp\left\{k_1 t + k_2 \frac{t^2}{2!} + \cdots + k_r \frac{t^r}{r!} + \cdots\right\} = 1 + \mu'_1 t + \mu'_2 \frac{t^2}{2!} + \cdots.$$

Differentiating w.r.t. k_j,

$$\begin{aligned} &\frac{t^j}{j!} \exp\left\{k_1 t + k_2 \frac{t^2}{2!} + \cdots + k_r \frac{t^r}{r!} + \cdots\right\} = \sum_{r=0}^{\infty} \frac{\partial \mu'_r}{\partial k_j} \frac{t^r}{r!} \\ \text{or} \quad &\frac{t^j}{j!} \left\{1 + \mu'_1 t + \mu'_2 \frac{t^2}{2!} + \cdots\right\} = \sum_{r=0}^{\infty} \frac{\partial \mu'_r}{\partial k_j} \frac{t^r}{r!} \end{aligned}$$

The result follows equating the coefficient of t^r from both sides.]

19. For the logistic distribution in exercise 9(f) of section 2.7, find the m.g.f. Hence show that $E(X) = \alpha$, $V(X) = \beta^2 \pi^2 / 3$.

[*Hints:* Substituting $Y = \frac{X-\alpha}{\beta}$, the m.g.f. of Y is $M_Y(t) = B(1-t, 1+t)$ where $B(p, q)$ is the Beta function with parameters p, q. Beta function has been discussed in section 6.5.]

20. For the power series distribution

$$P\{X=j\}=a_j\frac{\theta^j}{f(\theta)},\quad j=0,1,2,\cdots,$$

$\theta>0$, where $a_j\geq 0$, $f(\theta)=\sum\limits_{j=0}^{\infty}a_j\theta^j$, find the p.g.f. and m.g.f.

Show that, for this distribution

$$\mu'_{r+1}=\theta\frac{d\mu'_r}{d\theta}+\mu'_r\mu'_1$$

$$\theta\frac{d\mu_r}{d\theta}=\mu_{r+1}-r\theta\frac{d\mu'_1}{d\theta}\mu_{r-1}$$

$$\mu'_1=\theta\frac{d\log f(\theta)}{d\theta},\quad \mu_2=\theta^2\frac{d^2\log f(\theta)}{d\theta^2}+\mu'_1=\theta^2\frac{f''(\theta)}{f(\theta)}-\theta^2\left(\frac{f'(\theta)}{f(\theta)}\right)^2+\theta\frac{f'(\theta)}{f(\theta)}.$$

21. Show that for a symmetric distribution all the cumulants of odd order except the first vanish.

[*Hints:* If the distribution of X is symmetric about a, we transform from X to $X-a$. The m.g.f. about a is

$$M_{X-a}(t)=E\left(e^{t(X-a)}\right)=\left[1+\mu_2\frac{t^2}{2!}+\mu_4\frac{t^4}{4!}+\cdots+\mu_{2r}\frac{t^{2r}}{2r!}+\cdots\right],$$

since all the existent moments of odd order about a vanish.

$$K_{X-a}(t)=\log\left[1+\mu_2\frac{t^2}{2!}+\cdots\right].$$

Hence $k_1(X)=a+k_1(X-a)=a$, $k_{2r-1}(X)=k_{2r-1}(X-a)=$ coefficient of $\frac{t^{2r-1}}{(2r-1)!}=0$.]

22. A variate X has moments $\mu'_r=r!$. Find its characteristic function and deduce its distribution. $\left[\phi(t)=\frac{1}{1-it};\ f(x)=e^{-x},\ 0\leq x\leq\infty.\right]$

23. Show that the distribution for which the characteristic function is $e^{-|t|}$ has p.d.f.

$$f(x)=\frac{1}{\pi(1+x^2)},\quad -\infty\leq x\leq\infty.$$

Chapter 5

Some Discrete Distributions on $\mathbb{R}^1$

5.1 Introduction

We consider in this chapter some special univariate discrete distributions and study their properties. The Discrete Uniform distribution is considered in section 5.2, Bernoulli distribution in 5.3, Binomial distribution in 5.4 and Hypergeometric distribution in 5.5. Sections 5.6 through 5.8 consider respectively Poisson distribution, Geometric distribution and Negative Binomial distribution. Along with each distribution its truncated form has been considered. We conclude this chapter with a brief discussion on the Power-Series distribution of which all the above distributions come as special cases.

An inportant discrete distribution not considered here is multinomial distribution which being a multivariate distribution, would be taken up in chapter 7.

5.2 The Discrete Uniform Distribution

A r.v. X is said to have a discrete uniform distribution over the range $[1, N]$ if its p.m.f. is

$$P(X = x) = \begin{cases} \frac{1}{N} & \text{for } x = 1, 2, \cdots, N \\ 0 & \text{otherwise.} \end{cases} \tag{5.2.1}$$

N is the parameter of the distribution and lies in the set of all positive integers. (5.2.1) is also called a discrete rectangular distribution.

$$E(X) = \frac{1}{N}\sum_1^N i = \frac{N+1}{2}, E(X^2) = \frac{1}{N}\sum_1^N i^2 = \frac{(N+1)(2N+1)}{6}$$

$$V(X) = E(X^2) - (E(X))^2 = \frac{(N+1)(N-1)}{12}.$$

The p.g.f. is

$$P_X(t) = E(t^X) = \frac{1}{N}\sum_1^N t^x = \frac{t(1-t^N)}{N(1-t)}, \quad t \neq 1 (= 1 \text{ if } t = 1).$$

The m.g.f. is

$$M_X(t) = E(e^{tX}) = \frac{1}{N}\sum_{1}^{N} e^{ti} = \frac{e^t(1-e^{Nt})}{N(1-e^t)}.$$

When X is distributed uniformly over the range $[0, N]$, its p.m.f. is

$$P[X = x] = \begin{cases} \frac{1}{N+1}, & x = 0, 1, \cdots, N \\ 0 & \text{otherwise.} \end{cases}$$

Here the p.g.f. is

$$P_X(t) = \frac{1}{N+1} \cdot \frac{1-t^{N+1}}{1-t}, \quad t \neq 1 (= 1, \text{ if } t = 1).$$

The m.g.f. is

$$M_X(t) = \frac{1}{N+1} \cdot \frac{e^{(N+1)t} - 1}{e^t - 1}.$$

► **Example 5.2.1** *Let $X_i (i = 1, \cdots, n)$ be mutually independent r.v.'s each havin[g] rectangular distribution $P(X_i = x) = \frac{1}{N}$, $x = 1, \cdots, N$, $i = 1, \cdots, n$. Find the distributio[n] of M_n and N_n, where $M_n = \min_{i=1,\cdots,n} X_i$, and $N_n = \max_{i=1,\cdots,n} X_i$.*

$$\begin{aligned} P(M_n = r) &= P(X_i \geq r, i = 1, \cdots, n) - P(X_i \geq r+1, i = 1, \cdots, n) \\ &= \left(\frac{N-r+1}{N}\right)^n - \left(\frac{N-r}{N}\right)^n = \frac{(N-r+1)^n - (N-r)^n}{N^n}. \end{aligned}$$

Similarly,

$$\begin{aligned} P(N_n = r) &= P(X_i \leq r, i = 1, \cdots, n) - P(X_i \leq r-1, i = 1, \cdots, n) \\ &= \left(\frac{r}{N}\right)^n - \left(\frac{r-1}{N}\right)^n = \frac{r^n - (r-1)^n}{N^n}. \end{aligned}$$

5.3 The Bernoulli Distribution

A r.v. X is said to have a Bernoulli distribution with parameter p if its p.m.f. is give[n] by

$$P(X = x) = \begin{cases} p^x(1-p)^{1-x} & \text{for } x = 0, 1 \\ 0 & \text{otherwise.} \end{cases} \tag{5.3.1}$$

The parameter p satisfies $0 \leq p \leq 1$. Often $(1-p)$ is denoted as q.

$$E(X) = 1 \cdot p + 0 \cdot q = p, \quad E(X^2) = p, V(X) = pq.$$

The p.g.f. is

$$P_X(t) = pt + q.$$

The m.g.f. is

$$M_X(t) = q + pe^t.$$

A random experiment whose outcomes are of two types, success S and failure F, occurring with probabilities p and q respectively is called a Bernoulli trial. If for this experiment, a r.v. X is defined such that it takes value 1 when S occurs and 0 if F occurs, then X follows a Bernoulli distribution.

Example 5.3.1 *For an arbitrary probability space $(\mathcal{S}, \mathcal{A}, P)$ and for an event $A \in \mathcal{A}$ consider the indicator r.v. $I_A(w) = 1(0)$ if $w \in A(\notin \mathcal{A})$. Then $P(I_A = 1) = P(w \in A)$ $= P(A)$. I_A follows a Bernoulli distribution with parameter $p = P(A)$.*

5.4 The Binomial Distribution

A r.v. X is said to have a Binomial distribution with parameters n and p if its p.m.f. is given by

$$P(X = x) = b(x; n, p) = \begin{cases} \binom{n}{x} p^x q^{n-x} & \text{for } x = 0, 1, \cdots, n \\ 0 & \text{otherwise,} \end{cases} \quad (5.4.1)$$

n ranges over the set of positive integers and p satisfies $0 \le p \le 1$. The probabilities in (5.4.1) are the terms of binomial expansion of $(q + p)^n$. For this reason the distribution is called a binomial distribution.

For $n = 1$, (5.4.1) reduces to the Bernoulli distribution. For this reason, sometimes Bernoulli distribution is called a point binomial distribution.

A binomial distribution arises in the following circumstances. Consider n independent Bernoulli trials each having probability of success p. Let X be the r.v. denoting the number of successes in these n trials. The sample space for such an experiment is given by

$$\mathcal{S} = \{(t_1, \cdots, t_n);\ t_i = S \text{ or } F,\ (i = 1, \cdots, n)\},$$

t_i denoting the result of the ith trial. Since the trials are independent, the probability of any specified outcome, say,

$$(S, S, F, F, \cdots, S) \text{ is } p.p.q.q.\cdots p.$$

Hence, $P(X = x) =$ Prob [there are x successes and $(n - x)$ failures] $= \binom{n}{x} p^x q^{n-x}$, since any outcome with x successes and $(n-x)$ failures has probability $p^x q^{n-x}$ and there are $\binom{n}{x}$ such mutually exclusive outcomes. Hence, X follows a binomial distribution.

▶ **Example 5.4.1** *An urn contains a white balls and b black balls. From this n balls are sampled with replacement. Find the probability that the sample contains x white balls.*

Since sampling is with replacement probability that a white ball is selected at any particular trial is $\frac{a}{a+b}$. We have, therefore, a series of n independent bernoulli trials with probability of success $\frac{a}{a+b}$ at each trial. Hence, X, the number of white balls in the sample follows a binomial distribution with parameters n and $\left(\frac{a}{a+b}\right)$, when

$$P(X=x) = \binom{n}{x}\left(\frac{a}{a+b}\right)^x\left(\frac{b}{a+b}\right)^{n-x}, \quad x = 0, 1, \cdots, n.$$

Considering (5.4.1),

$$\sum_{x=0}^{n} P(X=x) = \sum_{x=0}^{n}\binom{n}{x}p^x q^{n-x} = (p+q)^n = 1.$$

The rth factorial moment of X is

$$\mu_{(r)} = E\left[(X)_r\right] = \sum_{x=0}^{n}(x)_r\binom{n}{x}p^x q^{n-x} = (n)_r p^r \sum_{x'=0}^{n-r}\binom{n-r}{x'}p^{x'}q^{n-r-x'}$$

where $x' = x - r$

$$= (n)_r p^r (p+q)^{n-r} = (n)_r p^r.$$

Thus

$$\begin{aligned}
\mu_1' &= np \\
\mu_2' &= \mu_{(2)} + \mu_{(1)} = np + n(n-1)p^2 \\
\mu_3' &= \mu_{(3)} + 3\mu_{(2)} + \mu_{(1)} = np + 3n(n-1)p^2 + n(n-1)(n-2)p^3 \\
\mu_4' &= \mu_{(4)} + 6\mu_{(3)} + 7\mu_{(2)} + \mu_{(1)} \\
&= np + 7n(n-1)p^2 + 6n(n-1)(n-2)p^3 + n(n-1)(n-2)(n-3)p^4.
\end{aligned}$$

The central moments are

$$\mu_2 = npq(<\mu_1'), \quad \mu_3 = npq(q-p), \quad \mu_4 = 3(npq)^2 + npq(1-6pq),$$

$$\sqrt{\beta_1} = \frac{q-p}{\sqrt{npq}}, \quad \beta_2 = 3 + \frac{1-6pq}{npq}, \quad \beta_2 - \beta_1 - 3 = -\frac{2}{n}.$$

Thus, the skewness of the distribution is positive (negative) if $p < (>)\frac{1}{2}$. The distribution is symmetrical if and only if $p = \frac{1}{2}$. For a fixed value of p the point (β_1, β_2) falls on the straight line

$$\frac{\beta_2 - 3}{\beta_1} = 1 - \frac{2pq}{(q-p)^2}.$$

As $n \to \infty$, p remaining finite, the point (β_1, β_2) approaches the point (0, 3) as a limit, i.e., the distribution tends to be symmetrical and mesokurtic ($\beta_2 = 3$). Figures 5.1(a) and 5.1(b) represent binomial distributions with parameters (10, 0.35) and (10, 0.55) respectively.

The following recurrence relation holds for the moments.

Lemma 5.4.1 *For the Binomial distribution,*

$$\mu_{r+1} = pq\left(nr\mu_{r-1} + \frac{d\mu_r}{dp}\right). \tag{5.4.2}$$

Proof:

$$\begin{aligned}
\mu_r &= \sum_{j=0}^{n} \binom{n}{j} p^j q^{n-j}(j-np)^r \\
\frac{d\mu_r}{dp} &= \sum \binom{n}{j} jp^{j-1}q^{n-j}(j-np)^r - \sum (n-j)\binom{n}{j} p^j q^{n-j-1}(j-np)^r \\
&\quad -nr\sum \binom{n}{j} p^j q^{n-j}(j-np)^{r-1} \\
&= \sum \binom{n}{j} p^{j-1}q^{n-j-1}(j-np)^r(jq-np+jp) - nr\mu_{r-1}.
\end{aligned}$$

Hence,

$$pq\frac{d\mu_r}{dp} = \sum \binom{n}{j} p^j q^{n-j}(j-np)^{r+1} - npqr\mu_{r-1}$$

when $\mu_{r+1} = pq\left(\frac{d\mu_r}{dp} + nr\mu_{r-1}\right)$.

A similar relation holds for the non-central moments:

$$\mu'_{r+1} = pq\left(\frac{n}{q}\mu'_r + \frac{d\mu'_r}{dp}\right). \tag{5.4.3}$$

The mean deviation about mean is

$$E\left[|X-np|\right] = n\binom{n-1}{[np]} p^{[np]+1} q^{n-[np]} \tag{5.4.4}$$

where $[\theta]$ denotes the highest integer contained in θ.

The ratio

$$\frac{b(k+1;n,p)}{b(k;n,p)} = \frac{n-k}{k+1}\cdot\frac{p}{q}, \quad k = 0, 1, \cdots, n-1$$

shows that

$$b(k+1;n,p) > (<) b(k;n,p) \text{ if } k < (>)(n+1)p - 1.$$

Hence, as k increases the quantities $P(X=k)$ increase to a maximum value at the integer k^* satisfying

$$(n+1)p - 1 < k^* < (n+1)p,$$

provided $(n+1)p$ is not an integer, and thereafter they decrease. k^* is the mode of the distribution.

If $(n+1)p$ is an integer,

$$P[X=(n+1)p-1]=P[X=(n+1)p]$$

and this common value is the maximum among the values of $P[X=k]$. Here mode is at

$$X=(n+1)p-1 \quad \text{and} \quad X=(n+1)p.$$

The p.g.f. is

$$P_X(t)=E(t^X)=(q+pt)^n, \quad |t|\le 1 \text{ (vide example 4.2.4).}$$

The m.g.f. about origin is

$$M_X(t)=E(e^{tX})=(q+pe^t)^n, \quad |t|<\infty \text{ (vide example 4.3.1).}$$

The m.g.f. about mean is

$$M_{X-\mu_1'}(t)=(qe^{-pt}+pe^{tq})^n.$$

The c.g.f. about mean is

$$\begin{aligned} K_{X-\mu_1'}(t) &= n\log(q+pe^t)^n \\ &= n\log\left[1+pq\frac{t^2}{2!}+pq(q-p)\frac{t^3}{3!}+pq(q^2-pq+p^2)\frac{t^4}{4!}+\cdots\right]. \end{aligned}$$

Equating the coefficients of $\frac{t^2}{2!}, \frac{t^3}{3!}, \cdots$, on either side we obtain, after some simplifications,

$$k_2=npq, \quad k_3=npq(q-p), \quad k_4=npq(1-6pq).$$

The factorial m.g.f.

$$=E\left[(1+t)^X\right]=\sum_0^n (1+t)^x\binom{n}{x}p^x q^{n-x}=(1+pt)^n.$$

The rth factorial moment is

$$\mu_{(r)}=(n)_r p^r.$$

Theorem 5.4.1 *Let X_i $(i=1,\cdots,k)$ be independent Binomial variables with parameters (n_i,p). Then $S_k=\sum_{i=1}^k X_i$ has a Binomial distribution with parameters (n,p), where $n=\sum_1^k n_i$.*

Proof: The m.g.f. of S_k

$$= P_{S_k}(t) = \prod_{i=1}^{k} P_{X_i}(t) = \prod_{i=1}^{k} (q + pe^t)^{n_i} = (q + pe^t)^n$$

which is the m.g.f. of a binomial r.v. with parameters (n, p). The result follows from the uniqueness theorem of m.g.f. (Theorem 4.3.1).

For large n, the calculation of values of the cumulative probabilities $F_X(k) = P(X \leq k)$ is facilitated by the use of K. Pearson's Tables of the Incomplete Beta Function.

$$\sum_{x=0}^{k} b(x, ; n, p) = \sum_{x=0}^{k} \binom{n}{x} p^x q^{n-x}$$

$$= \frac{n!}{k!(n-k-1)!} \sum_{x=0}^{k} \binom{k}{x} \frac{(n-k-1)!(k-x)!}{(n-x)!} p^x q^{n-x}$$

$$= \frac{1}{B(n-k, k+1)} \sum_{x=0}^{k} \binom{k}{x} B(n-k, k-x+1) p^x q^{n-x},$$

since $B(a, b) = \frac{(a-1)!(b-1)!}{(a+b-1)!}$ for integral values of a and b [The Beta function $B(a, b)$ has been discussed in section 6.5].

$$= \frac{1}{B(n-k, k+1)} \sum_{x=0}^{k} \binom{k}{x} \left[\int_0^1 y^{n-k-1} (1-y)^{k-x} dy \right] p^x q^{n-x}$$

$$= \frac{1}{B(n-k, k+1)} \int_0^1 y^{n-k-1} q^{n-k} \left[\sum_{x=0}^{k} \binom{k}{x} p^x \{q(1-y)\}^{k-x} \right] dy$$

(by changing the order of summation and integration)

$$= \frac{1}{B(n-k, k+1)} \int_0^1 (qy)^{n-k-1} (1-qy)^k d(qy)$$

$$= \frac{\int_0^q z^{n-k-1} (1-z)^k dz}{B(n-k, k+1)}, \text{ where } z = qy$$

$$= I_q(n-k, k+1) = 1 - I_p(k+1, n-k). \tag{5.4.5}$$

The function $I_x(p, q)$ has been tabulated by K. Pearson in his Tables of Incomplete Beta function.

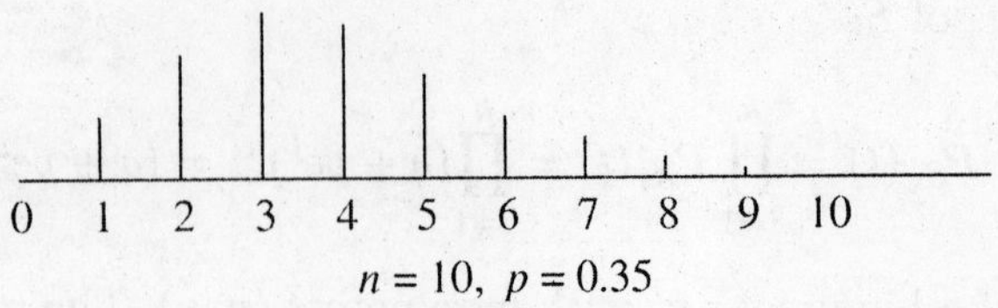

$n = 10,\ p = 0.35$

(a)

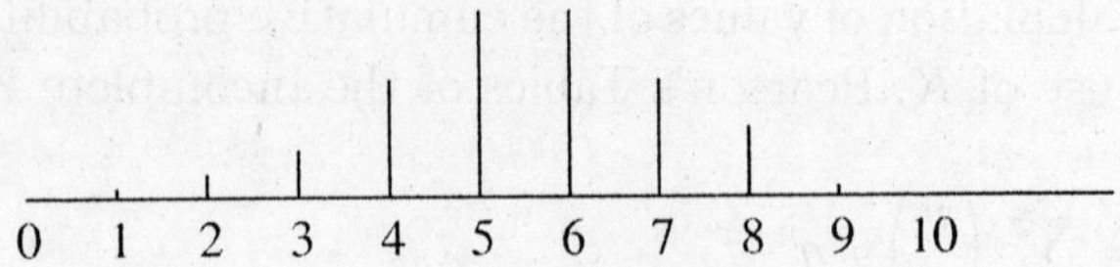

$n = 10,\ p = 0.55$

(b)

Fig. 5.1 Binomial Probabilities

Consider two independent variables X_1, X_2 distributed as $b(x; n_1, p)$ and $b(x; n_2, p)$ respectively. The conditional probability $X_1 = k$ given that $X_1 + X_2 = x$ is

$$P[X_1 = k | X_1 + X_2 = x] = \frac{P(X_1 = k, X_2 = x - k)}{P(X_1 + X_2 = x)}$$

$$= \frac{\binom{n_1}{k} p^k q^{n_1 - k} \binom{n_2}{x-k} p^{x-k} q^{n_2 - x + k}}{\binom{n_1+n_2}{x} p^x q^{n_1+n_2-x}} = \frac{\binom{n_1}{k}\binom{n_2}{x-k}}{\binom{n_1+n_2}{x}}, \qquad (5.4.6)$$

$\max\ (0, x - n_2) \leq k \leq \min\ (n_1, x)$. (5.4.6) is known as a hypergeometric distribution (discussed in section 5.5).

The distribution of the difference $(X_1 - X_2)$ is

$$P(X_1 - X_2 = k) = \sum_{x_1} \binom{n_1}{x_1} \binom{n_2}{x_1 - k} p^{2x_1 - k} q^{n_1 + n_2 - (2x_1 - k)} \qquad (5.4.7)$$

where the summation is over the limits, $\max\ (0, k) \leq x_1 \leq \min\ (n_1, n_2 + k)$. When $p = q = \frac{1}{2}$, (5.4.7) reduces to

$$P(X_1 - X_2 = k) = \frac{\binom{n_1+n_2}{n_2+k}}{2^{n_1+n_2}}, \quad (-n_2 \leq k \leq n_1). \qquad (5.4.8)$$

The distribution of the standardised binomial variate

$$X' = \frac{X - np}{\sqrt{npq}}$$

tends to the unit normal distribution as $n \to \infty$. This has been discussed in section 6.

If $n \to \infty$ and $p \to 0$ in such a way that $np = \lambda$ remains constant, then

$$b(x; n, p) \to e^{-\lambda} \frac{\lambda^x}{x!}.$$

This approximation has been discussed in lemma 5.6.2.

5.4.1 The Truncated Binomial Distribution

A singly truncated binomial distribution is formed if the binomial distribution is truncated at only one end, i.e., if only the values $0, 1, \cdots, K_1 - 1 (K_1 \geq 1)$ or the values $n - K_2 + 1, \cdots, n$ $(K_2 \geq 1)$ are omitted from the values of a binomial variable. The distribution obtained by omitting the value 0 only, is

$$P(X = r) = \frac{\binom{n}{r} p^r q^{n-r}}{1 - q^n}, \quad r = 1, \cdots, n. \tag{5.4.9}$$

The jth raw moment of the truncated distribution (5.4.9) is obtained by dividing the corresponding moment of the complete distribution (5.4.1) by $1 - q^n$. In particular, its mean and variance are $\frac{np}{1-q^n}$ and $\frac{npq}{1-q^n} - \frac{n^2p^2q^n}{(1-q^n)^2}$ respectively.

A binomial distribution truncated on the left at $x = K$ is

$$\frac{\binom{n}{x} p^x q^{n-x}}{\sum_K^n \binom{n}{x} p^x q^{n-x}} = \frac{1}{H_K(z)} \binom{n}{x} z^x, \quad x = K, K+1, \cdots, n \tag{5.4.10}$$

where $z = \frac{p}{q}$, $H_k(z) = \sum_K^n \binom{n}{x} z^x$.

Mean of the distribution is

$$\mu_1' = \frac{1}{H_K(z)} \sum_K^n x \binom{n}{x} z^x = \frac{zH_K'(z)}{H_K(z)}. \tag{5.4.11}$$

For the distribution (5.4.10) the recurrence relation (5.4.2) holds, i.e.,

$$\mu_{r+1} = pq \left(nr\mu_{r-1} + \frac{d\mu_r}{dp} \right). \tag{5.4.12}$$

Proof:

$$\mu_r = \frac{1}{H_K(z)} \sum_K^n (x - \mu_1')^r \binom{n}{x} z^x,$$

$$z\frac{d\mu_r}{dz} = \frac{d}{dz} \left\{ \sum_K^n (x - \mu_1')^r \binom{n}{x} z^x \right\} \frac{z}{H_K(z)} - \left\{ \sum_K^n (x - \mu_1')^r \binom{n}{x} z^{x+1} \right\} \frac{H_K'(z)}{(H_K(z))^2}$$

$$= \frac{1}{H_K(z)} \sum_K^n \binom{n}{x} (x - \mu_1')^r xz^x - zr\frac{d\mu_1'}{dz} \frac{1}{H_K(z)} \sum_K^n \binom{n}{x} z^x (x - \mu_1')^{r-1}$$

$$-\frac{zH_K'(z)}{H_K(z)} \cdot \frac{1}{H_K(z)} \sum_K^n (x - \mu_1')^r \binom{n}{x} z^x. \tag{5.4.13}$$

The first and the last term together on the r.h.s. of (5.4.11) give

$$\frac{1}{H_K(z)}\sum_{K}^{n}\binom{n}{z}z^x(x-\mu_1')^{r+1} = \mu_r + 1, \quad \text{using (5.4.13).}$$

Hence, (5.4.13) reduces to

$$\mu_{r+1} = z\left[\frac{d\mu_r}{dz} + r\frac{d\mu_1'}{dz}\mu_{r-1}\right].$$

(5.4.12) follows, using $\frac{dz}{dp} = \frac{1}{q^2}$.

The relation (5.4.3) holds similarly for the distribution (5.4.10).

A doubly truncated binomial variable is formed from (5.4.1) by omitting the values of x such that $0 \le x < K_1$, and $x - K_2 < x < n$ (with $0 \le K_1 \le n - K_2 \le n$). For the resulting distribution

$$P(X = j) = \frac{\binom{n}{j}p^j q^{n-j}}{\sum\limits_{r=K_1}^{n-K_2}\binom{n}{r}p^r q^{n-r}}, \quad j = K_1, K_1+1, \cdots, n-K_2.$$

5.5 The Hypergeometric Distribution

A r.v. X is said to have a hypergeometric distribution with parameters N, M and n if its p.m.f. is given by

$$P(X = x) = H(x; N, M, n) \text{ (say) } = \begin{cases} \dfrac{\binom{M}{x}\binom{N-M}{n-x}}{\binom{N}{n}} & \text{for max } (0, n-N+M) \\ & \le x \text{ (integer)} \le \min(M, n) \\ 0 & \text{otherwise} \end{cases} \tag{5.5.1}$$

where N is a positive integer, M is a positive integer not exceeding N and n is a positive integer that is at most N.

The formula in (5.5.1) is valid for all integer values of x such that, for each symbol of the form $\binom{a}{b}$ in (5.5.1), $a \ge b$. This means that

$$0 \le x \le M \text{ and } 0 \le (n - x) \le N - M;$$

that is,

$$\max\,(0, n - N + M) \le x \le \min\,(M, n).$$

Note that the same distribution is obtained if n and M are interchanged.

The quantities in the r.h.s. of (5.5.1) are successive terms in the expansion of

$$\frac{(N-n)!(N-M)!}{N!(N-M-n)!}F(-n, -M, N-M-n+1, 1)$$

where

$$F(\alpha, \beta, \gamma, t) = 1 + \frac{\alpha\beta}{\gamma}\frac{t}{1!} + \frac{\alpha(\alpha+1)\beta(\beta+1)}{\gamma(\gamma+1)}\frac{t^2}{2!} + \cdots$$

is a hypergeometric function. That is why the distribution is called a hypergeometric distribution.

The distribution (5.5.1) arises in the following circumstances. Suppose an urn contains M white balls and $N - M$ black balls. From this n balls are selected without replacement. The probability that the sample contains x white balls and $(n - x)$ black balls is given by (5.5.1). The hypergeometric distribution, therefore, arises in sampling from a dichotomous population (a population consisting only of two categories of objects) when sampling is without replacement whereas a binomial distribution arises when sampling is with replacement (vide example 5.4.1).

We have

$$H(x; N, M, n) = \binom{n}{x}\frac{(M)_x(N-M)_{(n-x)}}{(N)_n} \text{ for } x = 0, 1, \cdots, n.$$

The rth factorial moment about origin is

$$\begin{aligned}
\mu_{(r)} &= \sum_0^n (x)_r \frac{\binom{n}{x}(M)_x(N-M)_{(n-x)}}{(N)_n} \\
&= \frac{1}{(N)_n}\sum_{y=0}^{n-r}(n-y)_r\binom{n}{y}(M)_{(n-y)}(N-M)_y, \text{ (writing } n - x = y) \\
&= \frac{(M)_r(n)_r}{(N)_n}\sum_{y=0}^{n-r}\binom{n-r}{y}(M-r)_{(n-r-y)}(N-M)_y \\
&= \frac{(M)_r(n)_r}{(N)_n}(N-r)_{(n-r)} \quad \left[\because \sum_{x=0}^{n} H(x; N, M, n) = 1\right] \\
&= \frac{(M)_r(n)_r}{(N)_r}.
\end{aligned}$$

Hence, $E(X) = \dfrac{Mn}{N} = np \quad \left(\text{writing } p = \dfrac{M}{N}, q = 1 - p\right)$

$$V(x) = \frac{N-n}{N-1}npq$$

$$\sqrt{\beta_1} = \frac{q-p}{\left[\left(\frac{N-n}{N-1}\right)npq\right]^{1/2}}\frac{N-2n}{N-2}$$

$$\beta_2 = \frac{3(N-1)(N+6)}{(N-2)(N-3)} + \frac{N(N^2-1)}{(N-n)(N^2-5N+6)}$$

$$\left[1 - \frac{6N}{N+1}\left(pq + \frac{n(N-n)}{N^2}\right)\right] \cdot \frac{1}{npq}.$$

The following recurrence relations hold:

$$H(x+1;N,M,n)=\frac{(M-x)(n-x)}{(x+1)(N-M-n+x+1)}H(x;N,M,n) \tag{5.5.2}$$

$$H(x;N,M+1,n)=\frac{(M+1)(N-M-n+x)}{(N-M)(M+1-x)}H(x,N,M,n) \tag{5.5.3}$$

$$H(x;N,M,n+1)=\frac{(N-M-n+x)(n+1)}{(n+1-x)(N-n)}H(x;N,M,n) \tag{5.5.4}$$

$$H(x;N+1,M,n)=\frac{(N+1-n)(N+1-M)}{(N+1-n-M+x)(N+1)}H(x;N,M,n). \tag{5.5.5}$$

From (5.5.2) it follows that $H(x+1;N,M,n) \gtrless H(x;N,M,n)$ according as

$$x \lessgtr \frac{(n+1)(M+1)}{N+2}-1.$$

Hence, the distribution has mode at k^*, where k^* is the greatest integer which does not exceed $\frac{(n+1)(M+1)}{N+2}$.

If $\frac{(n+1)(M+1)}{N+2}$ is an integer, the distribution is bimodal having modes at

$$x=\frac{(n+1)(M+1)}{N+2}-1 \quad \text{and} \quad x=\frac{(n+1)(M+1)}{N+2}.$$

The following binomial approximation to a hypergeometric distribution holds.

$$H(x;N,M,n)\simeq\binom{n}{x}p^x(1-p)^{n-x} \tag{5.5.6}$$

where $p=\frac{M}{N}$, when $N\to\infty$, provided $\frac{n}{N}$ is negligible.

If $\frac{M}{N}$ is small but n is large, the Poisson approximation

$$H(x;N,M,n)\simeq\frac{(np)^x}{x!}e^{-np} \tag{5.5.7}$$

where $p=\frac{M}{N}$, holds.

5.5.1 The Positive Hypergeometric Distribution

This is a truncated hypergeometric distribution, where the zero observed value is omitted. However, if $n>(N-M)$, there is no zero observed value. If $n\le(N-M)$, the distribution is

$$P(X=x)=\frac{\binom{M}{x}\binom{N-M}{n-x}}{\left[\binom{N}{n}-\binom{N-M}{n}\right]},\quad x=1,\cdots,\min(n,M).$$

.5.2 The Negative Hypergeometric Distribution

This distribution arises in the following circumstances. Suppose we have b black balls and w white balls in an urn. Balls are drawn one a time, without replacement. We are interested in the probability distribution of the number R of draws needed to obtain a specific number k of black balls.

$$\begin{aligned}
P(R=r) &= P\{(k-1) \text{ black balls are drawn in the first } (r-1) \text{ draws}\} \\
&\quad \times P\{\text{a black ball is drawn from an urn containing } (b-k+1) \\
&\quad \text{black balls and } (w-r+k) \text{ white balls}\} \\
&= \binom{r-1}{k-1}\frac{b(b-1)\cdots(b-k+2)w(w-1)\cdots(w-r+k+1)}{(b+w)(b+w-1)\cdots(b+w-r+2)} \\
&\quad \times\frac{b-k+1}{b+w-r+1} \\
&= \binom{r-1}{k-1}\frac{(b)_k(w)_{r-k}}{(b+w)_{(r)}}, \quad r=k,k+1,\cdots.
\end{aligned}$$

The distribution of R is a negative hypergeometric distribution with parameters b, w, k.

The jth ascending factorial moment of R is

$$E[R(R+1)\cdots(R+j-1)] = \frac{k^{(j)}(b+w+1)^{(j)}}{(b+1)^{(j)}}$$

where $n^{(t)} = n(n+1)\cdots(n+t-1)$.

In particular

$$E(R) = \frac{k(b+w+1)}{b+1}, \quad V(R) = \frac{k(b+w+1)w(b+1-k)}{(b+1)^2(b+2)}.$$

5.6 The Poisson Distribution

A r.v. X is said to have a Poisson distribution with parameter $\lambda(>0)$ if its p.m.f. can be written as

$$P(X=x) = P(x;\lambda) = \begin{cases} e^{-\lambda}\frac{\lambda^x}{x!}, & x=0,1,2,\cdots \\ 0 & \text{otherwise.}\end{cases}$$

The p.g.f. is

$$\begin{aligned}
P_X(t) &= e^{-\lambda(1-t)}, \quad |t|\le 1 \text{ (vide example 4.2.3)} \\
\mu_1' &= \sum_0^\infty xe^{-\lambda}\frac{\lambda^x}{x!} = \lambda e^{-\lambda}\left(1+\frac{\lambda}{1!}+\frac{\lambda^2}{2!}+\cdots\right) = \lambda e^{-\lambda}e^{\lambda} = \lambda \\
\mu_2' &= \mu_1' + \mu_{(2)} = \lambda + \sum_0^\infty x(x-1)e^{-\lambda}\frac{\lambda^x}{x!} = \lambda+\lambda^2.
\end{aligned}$$

Hence, $\mu_2 = \lambda$.

The m.g.f. about zero is

$$M_X(t) = e^{\lambda(e^t-1)} \text{ (vide example 4.3.2).}$$

Hence,

$$\mu_3' = \lambda + 3\lambda^2 + \lambda^3, \ \ \mu_4' = \lambda + 7\lambda^2 + 6\lambda^3 + \lambda^4, \ \ \mu_3 = \lambda, \ \ \mu_4 = \lambda + 3\lambda^2,$$
$$\sqrt{\beta_1} = \lambda^{-\frac{1}{2}}, \ \ \beta_2 = 3 + \tfrac{1}{\lambda}.$$

Thus for any Poisson distribution,

$$\mu_1' = \mu_2 \text{ and } \beta_2 - \beta_1 - 3 = 0.$$

Lemma 5.6.1 *The central moments of a Poisson distribution satisfy the followin recurrence relation*

$$\mu_{r+1} = \lambda\left(r\mu_{r-1} + \frac{d\mu_r}{d\lambda}\right). \quad (5.6.1$$

Proof:

$$\mu_r = \sum_{x=0}^{\infty}(x-\lambda)^r e^{-\lambda}\frac{\lambda^x}{x!}$$
$$\frac{d\mu_r}{d\lambda} = \sum_{x=0}^{\infty}\left[-r(x-\lambda)^{r-1} + (x-\lambda)^r\left(\frac{x}{\lambda}-1\right)\right]e^{-\lambda}\frac{\lambda^x}{x!} = -r\mu_{r-1} + \frac{1}{\lambda}\lambda_{r+1}.$$

Hence, the proof.
The ratio

$$\frac{P(x;\lambda)}{P(x-1;\lambda)} = \frac{\lambda}{x}$$

shows that $P(x;\lambda) > (<) P(x;\lambda)$ according as $x < (>) \lambda$.

Hence, if λ is not an integer, the distribution has mode at the integral value x satisfying

$$\lambda - 1 < x^* < \lambda.$$

If λ is an integer, $P[X = \lambda - 1] = P[X = \lambda]$ and the distribution has modes at $x = \lambda - 1$ and $x = \lambda$.
The mean deviation of the Poisson distribution is

$$E[|X-\lambda|] = \frac{2e^{-\lambda}\lambda^{[\lambda]+1}}{[\lambda]!}.$$

The figures 5.2(a), (b) and (c) represent Poisson distributions with $\lambda = 0.75$, 1.5 an 3 respectively.

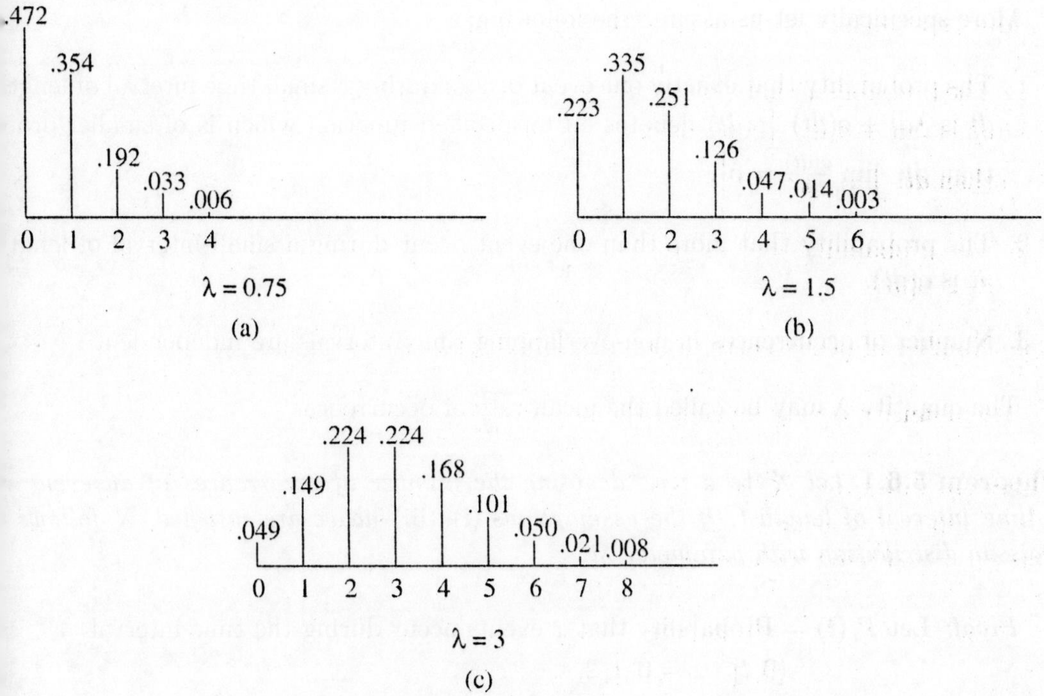

Fig. 5.2 Poisson Probabilities

We have seen in example 4.3.7, if $X_1, \cdots, X_n$ are n independent Poisson variates, X_i being Poisson with parameter λ_i, the sum $S_n = X_1 + \cdots + X_n$ has m.g.f., $M_{S_n}(t) = e^{\Sigma_1^n \lambda_i(e^t-1)}$. It, therefore, follows by the uniqueness property of m.g.f. (theorem 4.3.1) that S_n follows a Poisson distribution with parameter $\sum_{i=1}^{n} \lambda_i$. This is called the reproductive roperty of the Poisson distribution.

The Poisson distribution provides a realistic model for many random phenomena. ince the values of a Poisson variable are non-negative integers, the distribution may e used to describe experiments in which observed variable is a count. The Poisson istribution may also be used to describe events occurring randomly and independently n time (space), e.g., number of telephone calls comming into the telephone exchange in ime of length t, number of fatal traffic accidents occurring per week in a given region, umber of cars passing through a street in time t, number of α particles emitted by radio-active substance reaching a given portion of sapce during time t, number of lying-bomb hits over a region divided into several small areas. In all such cases we may ssume that the conditions of the experiment remain constant in time (space) and that he non-overlapping intervals of time (space) are stochastically independent in the sense hat the information concerning the number of events in one interval does not give any nformation about the other. These two assumptions provide the basis for a random ariable to obey Poisson law.

More specifically let us assume the following:

1. The probability that exactly one event occurs during a small time interval of lengt dt is $\lambda dt + o(dt)$ $\left[o(dt)\right.$ denotes an unspecified function which is of smaller orde than dt; $\left.\lim_{dt\to 0}\frac{o(dt)}{dt} = 0\right]$.

2. The probability that more than one event occur during a small interval of lengt dt is $o(dt)$.

3. Number of occurrences in non-overlapping time intervals are independent.

The quantity λ may be called the mean rate of occurrence.

Theorem 5.6.1 *Let X be a r.v. denoting the number of occurrences of an event i a time interval of length t. If the assumptions* (i)–(iii) *above are satisfied, X follows Poisson distribution with parameter λt.*

Proof: Let $P_x(t)$ = Probability that x events occur during the time interval $(0, t]$, $x = 0, 1, 2, \cdots$.

Hence,

$$\begin{aligned}P_0(t+dt) &= P\,(\text{no event in } (0, t+dt])\\ &= P\,(\text{no event in } (0,t] \text{ and no event in } (t, t+dt])\\ &= P\,(\text{no event in } (0,t])\,P\,(\text{no event in } (t, t+dt])\end{aligned} \tag{5.6.2}$$

Now

$$\begin{aligned}&P\,(\text{no event in } (t, t+dt])\\ &= 1 - P\,(\text{one event in } (t, t+dt]) - P\,(\text{two or more events in } (t, t+dt])\\ &= 1 - \lambda dt - o(dt)\end{aligned}$$

Hence, from (5.6.2),

$$P_0(t+dt) = P_0(t)\,(1 - \lambda dt - o(dt))$$

$$\text{or}\quad \frac{P_0(t+dt) - P_0(t)}{dt} = -\lambda P_0(t) + \frac{o(dt)}{dt}P_0(t).$$

Taking limit as $dt \to 0$

$$P_0'(t) = -\lambda P_0(t).$$

The solution of this differential equation is

$$P_0(t) = e^{-\lambda t} \text{ since } P_0(0) = 1.$$

or $x = 1, 2, \cdots$,

$$P_x(t+dt) = P\,(x \text{ events in } (0,t] \text{ and no event in } (t, t+dt]) \\ +P\,(x-1 \text{ events in } (0,t] \text{ and one event in } (t,t+dt]) \\ +\sum_{h=2}^{x} P\,(x-h \text{ events in } (0,t] \text{ and } h \text{ events in } (t,t+dt])$$

$$= P_x(t)\,(1-\lambda dt - o(dt)) + P_{x-1}(t)\,(\lambda dt + o(dt)) + \sum_{h=2}^{x} P_{x-h}(t)\,(o(dt))$$

$$\therefore P_x(t+dt) = P_x(t)(1-\lambda dt) + P_{x-1}(t)\lambda dt + o(dt)$$

$$\text{or } \frac{P_x(t+dt) - P_x(t)}{dt} = -\lambda[P_x(t) - P_{x-1}(t)] + \frac{o(dt)}{dt}.$$

Taking limit as $dt \to 0$

$$P'_x(t) = \lambda[P_{x-1}(t) - P_x(t)]. \tag{5.6.3}$$

Hence, For $x = 1$

$$P'_1(t) + \lambda P_1(t) = \lambda e^{-\lambda t}.$$

Multiplying both sides by $e^{\lambda t}$ and integrating, we obtain

$$e^{\lambda t} P_1(t) = \lambda t + K$$

where K is a constant.
The initial conditions $P_1(0) = 0$ implies

$$P_1(t) = \lambda t e^{-\lambda t}. \tag{5.6.4}$$

Similarly, $P_2(t)$ is found from (5.6.3) by using (5.6.4). Induction on n shows that

$$P_n(t) = e^{-\lambda t}\frac{(\lambda t)^n}{n!}.$$

▶ **Example 5.6.1** *A telephone switch board receives* 20 *calls on an average during an hour. Find the probability that during a period of* 5 *minutes* (a) *no call is received,* (b) *exactly* 3 *calls are received and* (c) *more that* 5 *calls are received. Assume that the time is measured in minutes.*

We assume that the number of incoming calls during any time period obeys a Poisson law (which follows on the basis of the assumptions (i)–(iii) above). 20 calles per hour is equivalent to 0.33 calls per minute, which is the mean rate of occurrence. Hence, number of calls in a 5 minute period follows a Poisson distribution with parameter $\lambda = 1.65$.

1. P (no call in a 5 minute period) $= e^{-1.65} = 0.192$;

2. P (3 calls in a 5 minute period) $= e^{-1.65}\frac{(1.65)^3}{3!} = 0.144$;

3. P (more than 5 calls in a 5 minute period) $= \sum_{x=6}^{\infty} e^{-1.65}\frac{(1.65)^x}{x!} = 0.007$.

► **Example 5.6.2** *A traffic light has a constant probability λdt of changing from a re light to a green light or from a green light to a red light in an infinitesimal interval c length dt. Prove that a car arriving at a random instant has a probability $\frac{1}{2}$ of passin through without waiting and a probability element $\frac{1}{2}\lambda e^{-\lambda w}$ of waiting for a time $w(> 0$ before passing.*

A car would pass at any moment with or without waiting. Since both the cases ar mutually exclusive and equally likely the probability of each case is $\frac{1}{2}$.

The car waits for a time period w if (i) the traffic light is red when it arrives and (ii does not change to green within the next time period w and (iii) thereafter (within tim dt) changes. The three events are mutually independent and probability of (i) and (iii are $\frac{1}{2}$ and λdt respectively.

Let $\frac{w}{dt} = n$. The chance that the traffic light does not change in time w is

$$(1 - \lambda dt)^n = (1 - \lambda dt)^{\frac{w}{dt}} = (1 - x)^{\frac{w\lambda}{x}} \text{ where } x = \lambda dt,$$

which on taking limit as $dt \to 0$ (i.e., $x \to 0$) reduce to $e^{-w\lambda}$.

The required probability is, therefore, $\frac{1}{2}e^{-w\lambda}\lambda dt$.

► **Example 5.6.3** *Let X_1, X_2 be two independent Poisson variates with parameters λ_1 and λ_2 respectively. Show that the conditional distribution of X_1 given $X_1 + X_2$ is binomial.*

For positive integers $k, n(k < n)$,

$$P(X_1 = k|X_1 + X_2 = n) = \frac{P(X_1 = k, X_2 = n - k)}{P(X_1 + X_2 = n)}$$

$$= \frac{\left(e^{-\lambda_1}\frac{\lambda_1^k}{k!}\right)\left(e^{-\lambda_2}\frac{\lambda_2^{n-k}}{(n-k)!}\right)}{e^{-(\lambda_1+\lambda_2)}\frac{(\lambda_1+\lambda_2)^n}{n!}}$$

(since X_1, X_2 are independent and $X_1 + X_2$ is a Poisson $(\lambda_1 + \lambda_2)$).

$$= \binom{n}{k}p^k(1-p)^{n-k}, \text{ where } p = \frac{\lambda_1}{\lambda_1 + \lambda_2};\ k = 0, 1, \cdots, n.$$

Hence, the conditional distribution of X_1 given $X_1 + X_2 = n$, is binomial with parameters n and $p = \frac{\lambda_1}{\lambda_1+\lambda_2}$.

► **Example 5.6.4** *Suppose that the number of eggs laid by an insect follows a Poisson distribution with parameter λ and the probability of an egg developing is p. Assuming mutual independence of the events, show that the number of surviving insects follows a Poisson distribution with parameter λp.*

Let X, Y be r.v.'s denoting respectively the number of eggs laid and number of urvivors.

$$P(X = n, Y = k) = \left(e^{-\lambda}\frac{\lambda^n}{n!}\right)\left\{\binom{n}{k}p^k q^{n-k}\right\}, \quad q = 1 - p, \quad k \leq n$$

(by independence)

$$P(Y = k) = \sum_{n=k}^{\infty} P(X = n, Y = k) = e^{-\lambda}\left(\frac{p}{q}\right)^k \sum_{n=k}^{\infty} \frac{(q\lambda)^n}{k!(n-k)!}$$

$$= e^{-\lambda}\left(\frac{p}{q}\right)^k (q\lambda)^k \sum_{n-k=0}^{\infty} \frac{(q\lambda)^{n-k}}{k!(n-k)!} = e^{-\lambda}\left(\frac{p}{q}\right)^k \frac{(q\lambda)^k}{k!} e^{q\lambda}$$

$$= e^{-\lambda p}\frac{(\lambda p)^k}{k!}, \quad k = 0, 1, \cdots.$$

▶ **Example 5.6.5** *The number of female insects in a given region follows a Poisson distribution with parameter λ. The number of eggs laid by each female insect follows a Poisson distribution with mean μ. Find the probability distribution of number of eggs n the region.*

Let X denote the number of female insects and Y the total number of eggs laid down n the region $P(X = t) = e^{-\lambda}\frac{\lambda^t}{t!}$.

Let Y_i denote the number of eggs laid down by the ith insect. Hence,

$$P(Y = s|X = t) = P\left(Y_1 = s_1, Y_2 = s_2, \cdots, Y_t = s_t; \sum_1^t s_i = s\right)$$

$$= \sum_{\substack{s_1, \cdots, s_t: \\ \Sigma_1^t s_i = s}} \left(e^{-\mu}\frac{\mu^{s_1}}{s_1!}\right)\cdots\left(e^{-\mu}\frac{\mu^{s_t}}{s_t!}\right)$$

$$= \sum{}' e^{-\mu t}\frac{\mu^s}{s_1!\cdots s_t!} \text{ (say)} = \frac{1}{s!}e^{-\mu t}(\mu t)^s.$$

Hence,

$$P(X = t; Y = s) = \left(e^{-\lambda}\frac{\lambda^t}{t!}\right)\left(e^{-\mu t}\frac{(\mu t)^s}{s!}\right)$$

$$P(Y = s) = \sum_{t=0}^{\infty} P(X = t; Y = s) = \frac{e^{-\lambda}\mu^s}{s!}\sum_{t=0}^{\infty}\frac{\lambda^t t^s}{t!}e^{-\mu t}$$

which is the distribution of Y. The m.g.f. of Y is

$$M_Y(h) = E\left(e^{hY}\right) = e^{-\lambda}\sum_{s=0}^{\infty}\frac{e^{hs}e^{-\lambda}\mu^s}{s!}\sum_{t=0}^{\infty}\frac{\lambda^t t^s}{t!}e^{-\mu t} = e^{-\lambda}\sum_{t=0}^{\infty}\sum_{s=0}^{\infty}e^{-\mu t}\frac{(\mu t e^h)^s}{s!}\frac{\lambda^t}{t!}$$

$$= e^{-\lambda}\sum_{t=0}^{\infty}\frac{\{\lambda e^{\mu(e^h-1)}\}^t}{t!} = e^{-\lambda}e^{\lambda e^{\mu(e^h-1)}} = e^{\lambda(e^{\mu(e^h-1)}-1)}.$$

$$= \exp\left[\lambda\left\{e^{\mu(h+\frac{h^2}{2!}+\cdots)} - 1\right\}\right]$$

$$= \exp\left[\lambda\left\{\mu\left(h + \frac{h^2}{2!} + \cdots\right) + \frac{\mu^2}{2!}\left(h + \frac{h^2}{2!} + \cdots\right)^2 + \cdots\right\}\right]$$

$$= 1 + \lambda\left[\mu\left(h + \frac{h^2}{2!} + \cdots\right) + \frac{\mu^2}{2!}\left(h + \frac{h^2}{2!} + \cdots\right)^2 + \cdots\right]$$

$$+ \frac{\lambda^2}{2!}\left[\mu\left(h + \frac{h^2}{2!} + \cdots\right) + \frac{\mu^2}{2!}\left(h + \frac{h^2}{2!} + \cdots\right)^2 + \cdots\right]^2 + \cdots$$

Hence,

$$\mu_1' = \text{coefficient of } h = \mu\lambda,$$
$$\mu_2' = \text{coefficient of } \frac{h^2}{2!} = \lambda(\mu^2 + \mu + \lambda\mu^2),$$
$$\mu_2 = \lambda\mu(\mu + \lambda).$$

The following lemma shows that a binomial distribution tends to a Poisso distribution under certain conditions.

Lemma 5.6.2 *Let X be a binomial r.v. with parameters n and p. If $n \to \infty$ and $p \to$ in such a way that $np = \lambda$ remains constant then*

$$b(x; n, p) \to e^{-\lambda}\frac{\lambda^x}{x!}$$

for fixed integer x.

Proof:

$$b(x; n, p) = \binom{n}{x} p^x q^{n-x} = \frac{n!}{x!(n-x)!}\left(\frac{\lambda}{n}\right)^x \left(1 - \frac{\lambda}{n}\right)^{n-x}$$

$$= \left\{\frac{\lambda^x}{x!}\left(1 - \frac{\lambda}{x}\right)^n \left(1 - \frac{\lambda}{n}\right)^{-x}\right\}\left\{\frac{n!}{n^x(n-x)!}\right\}.$$

Hence,

$$\lim_{n\to\infty} b(x; n, p) = e^{-\lambda}\frac{\lambda^x}{x!}\left[\lim_{n\to\infty}\frac{n!}{n^x(n-x)!}\right]. \tag{5.6.5}$$

Using Stirling's approximation for $n!$,

$$n! \simeq \sqrt{2\pi}e^{-n}n^{n+\frac{1}{2}}$$

the quantity in (5.6.5) is

$$\lim_{n\to\infty}\left[\frac{e^{-x}n^{n+\frac{1}{2}-x}}{(n-x)^{n-x+\frac{1}{2}}}\right] = \lim_{n\to\infty}\left[e^x\left(1 - \frac{x}{n}\right)^n \left(1 - \frac{x}{n}\right)^{-x+\frac{1}{2}}\right]^{-1} = e^x \cdot e^{-x} = 1.$$

Hence, the proof.

Corollary 5.6.1

$$\lim_{\substack{n\to\infty\\ np=\lambda}} \sum_A b(x;n,p) = \sum_A e^{-\lambda}\frac{\lambda^k}{k!}$$

where $\sum_A$ *denotes summation over any (finite or infinite) subset A of the non-negative integer* $0, 1, 2, \cdots$.

5.6.1 The Truncated Poisson Distribution

Consider a truncated Poisson variable, truncated at $x = 0$, whose p.m.f. is

$$P(X = x) = e^{-\lambda}\frac{\lambda^x}{x!}(1 - e^{-\lambda})^{-1} = \frac{\lambda^x}{x!(e^\lambda - 1)}, \quad x = 1, 2, \cdots$$

$$E(X) = \frac{\lambda}{1 - e^{-\lambda}}, \quad V(X) = (1 - e^{-\lambda})^{-1}\left[1 - \lambda e^{-\lambda}(1 - e^{-\lambda})^{-1}\right].$$

The rth factorial moment is

$$\mu_{(r)} = \frac{\lambda^r}{(1 - e^{-\lambda})}.$$

5.7 The Geometric Distribution

A r.v. X is said to have a geometric distribution with parameter p if its p.m.f. is

$$P(X = x) = g(x;p) = \begin{cases} p(1-p)^x, & x = 0, 1, \cdots \\ 0 & \text{otherwise.} \end{cases} \tag{5.7.1}$$

The parameter p satisfies $0 < p < 1$.

The geometric distribution arises in the following circumstances. Bernoulli trials with constant probability p of success are performed until the first success occurs. The probability that exactly x failures precede the first success is given by (5.7.1).

The p.g.f. is

$$P_X(t) = p(1 - tq)^{-1} \quad \text{(vide example 4.2.2).}$$

The m.g.f. about origin is

$$M_X(t) = p(1 - qe^t)^{-1} \quad \text{(vide example 4.3.3).}$$

Thus,

$$\mu_1' = \frac{q}{p}, \quad \mu_2' = \frac{q}{p} + 2\frac{q^2}{p^2}, \quad \mu_2 = \frac{q}{p^2}$$

$$\sqrt{\beta_1} = \frac{1+q}{\sqrt{q}}, \quad \beta_2 = 9 + \frac{p^2}{q}, \quad \beta_2 - \beta_1 = 5.$$

The factorial moment of order r is $\mu_{(r)} = r!\left(\frac{q}{p}\right)^r$.

A very interesting property of the geometric distribution is that the condition probability that a geometric r.v. X is greater than or equal to $j+k$ given that X is greater than j is equal to the unconditional probability that it will be greater than or equal to k. Thus, if a sequence of repeated, independent Bernoulli trials, each with probability p of success, is performed until the first success occurs, the probability that the number of failures X preceding the first success will be $\geq j+k$ given that $X \geq j$ is equal to the unconditional probability that $X \geq k$. The event that the waiting time for the first success has already been at least j does not affect the probability of the future waiting time. As a crude example if the arrival of a bus of a particular route at a specified bus stop in each minute (unit of time) is considered as a Bernoulli trial with constant probability p and if these trials are assumed independent, the probability that a person has to wait, say, at least for 5 minutes, remains independent of whether he has already waited for 3 minutes or is just coming. This is called the "lack of memory" property of the geometric distribution. This is proved in

Theorem 5.7.1 *If X is a geometric distribution then for any two positive integer j and k,*

$$P\{X \geq j+k|X \geq j\} = P\{X \geq k\}.$$

Proof:

$$P\{X \geq j+k|X \geq j\} = \frac{p\{X \geq j+k\}}{P\{X \geq j\}} = \frac{\sum_{x=j+k}^{\infty} pq^x}{\sum_{y=j}^{\infty} pq^y} = q^k.$$

Again

$$P\{X \geq k\} = p\sum_{y=k}^{\infty} q^y = q^k.$$

Some authors define a geometric random variable as the waiting time till the first success, including the trial on which the success occurs. Here $P(X=x) = pq^{x-1}, x = 1, 2, \cdots$.

Hence

$$E(X) = \frac{1}{p}, \quad V(X) = \frac{q}{p^2}, \quad M_X(t) = \frac{pe^t}{1-qe^t}.$$

► **Example 5.7.1** *Let X_1, X_2 be independent r.v.'s each having geometric distribution*

$$q^x p, \; x = 0, 1, \cdots.$$

Show that the conditional distribution of X_1 given $X_1 + X_2$ is rectangular.

$$P(X_1 = r|X_1 + X_2 = n) = \frac{P(X_1 = r, X_2 = n-r)}{P(X_1 + X_2 = n)}$$

$$= \frac{P(X_1 = r)P(X_2 = n-r)}{P(X_1 + X_2 = n)} \text{ (by independence).} \quad (5.7.2)$$

Now

$$P(X_1 + X_2 = n) = \sum_{x=0}^{n} P(X_1 = x)P(X_2 = n - x)$$
$$= \sum_{x=0}^{n} q^x p q^{n-x} p = p^2 q^n (n+1).$$

Hence, (5.7.2) $= \frac{1}{n+1}$.

▶ **Example 5.7.2** *Let X_1, X_2 be two independent r.v.'s each having geometric distribution $q^k p$, $k = 0, 1, \cdots$. Find the probability distribution of $Z = \max(X_1, X_2)$.*

$$P(Z = j) = P(Z = j, X_1 = j, X_2 < j) + P(Z = j, X_2 = j, X_1 < j)$$
$$+P(Z = j, X_1 = j, X_2 = j)$$
$$= 2q^j p \sum_{k=0}^{j-1} q^k p + (q^j p)^2 = 2q^j p^2 \left(\frac{1 - q^j}{1 - q}\right) + q^{2j} p^2$$
$$= 2q^j p(1 - q^j) + q^{2j} p^2 = 2q^j p - q^{2j} p - q^{2j+1} p.$$

5.8 The Negative Binomial Distribution

A r.v. X is said to have a negative binomial distribution with parameters k and p if its p.m.f. is given by

$$P(X = x) = b'(x; k, p) = \begin{cases} \binom{k+x-1}{x} p^k q^x = \binom{-k}{x} p^k (-q)^x, & x = 0, 1, 2, \cdots \\ 0 & \text{otherwise} \end{cases} \tag{5.8.1}$$

where the parameters satisfy $0 < p < 1, k = 1, 2, 3, \cdots$.

A negative binomial distribution arises in the following circumstances. An urn contains np white and nq black balls. Balls are drawn at random with replacement. The probability that exactly $x + k$ trials will be required to produce k white balls is

$$\binom{x+k-1}{k-1} q^x p^k = \frac{(x+k-1)(x+k-2)\cdots k}{x!} q^x p^k$$
$$= (-1)^x \frac{(-k)(-k-1)\cdots(-k-x+1)}{x!} q^x p^k = (-1)^x \binom{-k}{x} p^k q^x.$$

Clearly

$$\sum_{x=0}^{\infty} b'(x; k, p) = p^k \sum_{0}^{\infty} \binom{-k}{x} (-q)^x = p^k (1 - q)^{-k} = 1.$$

The distribution (5.8.1) remains meaningful even when k is not an integer. When k is an integer the distribution is sometimes called a Pascal distribution. In this case (5.8.1) is also called a discrete waiting time distribution as it gives the probability that the waiting time is x (in terms of failures) to attain the kth success in an unending sequence of Bernoulli trials. For $k = 1$, the distribution reduces to the geometric distribution.

It is conventional to express the distribution in terms of parameters $Q = \frac{1}{p}$, $P = \frac{1-p}{p}$ so that $Q - P = 1$. The expression (5.8.1) reduces to

$$\binom{k+x-1}{x}\left(\frac{P}{1+P}\right)^x (1+P)^{-k} = Q^{-k}\binom{k+x-1}{x}\left(\frac{P}{Q}\right)^x$$

$$= \binom{k+x-1}{x}\left(\frac{P}{Q}\right)^x\left(1-\frac{P}{Q}\right)^k,$$

$$x = 0, 1, 2, \cdots \tag{5.8.2}$$

which is the $(x+1)$th term in the expansion of $(Q-P)^{-k}$.

The p.g.f. is

$$E(t^X) = (Q - P_t)^{-k} = p^k(1-tq)^{-k}, |t| \leq 1 \text{ (vide example 4.2.5.).}$$

The m.g.f. about origin is

$$M_X(t) = E\left(e^{tX}\right) = \sum_{x=0}^{\infty}\binom{k+x-1}{x}\left(\frac{Pe^t}{Q}\right)^x\left(1-\frac{P}{Q}\right)^k$$

$$= Q^{-k}\sum_{x=0}^{\infty}\binom{-k}{x}\left(-\frac{Pe^t}{Q}\right)^x = (Q-Pe^t)^{-k}$$

$$= p^k(1-qe^t)^{-k} \text{ for } qe^t < 1.$$

Thus,

$$\mu_1' = M_X'(t)\Big]_{t=0} = \left[kPe^t(Q-Pe^t)^{-(k+1)}\right]_{t=0} = kP$$

$$\mu_2' = M_x''(t)\Big]_{t=0} = \left[kPe^t(Q-Pe^t)^{-(k+1)} + k(k+1)P^2e^{2t}(Q-Pe^t)^{-(k+2)}\right]_{t=0}$$

$$= kP + k(k+1)P^2.$$

Hence,

$$\mu_2 = kPQ = k\frac{q}{p^2} (> \mu_1').$$

Note that the variance of the negative binomial distribution is greater than its mean (for binomial distribution, variance is less than its mean).

The jth factorial moment about zero is

$$\begin{aligned}\mu_{(j)} &= \sum_{x=j}^{\infty}(x)_j\binom{k+x-1}{k-1}\left(\frac{P}{Q}\right)^x\left(1-\frac{P}{Q}\right)^k\\ &= (k+j-1)_{(j)}\left(\frac{P}{Q}\right)^j Q^{-k}\sum_{x=j}^{\infty}\binom{k+j+x-j-1}{k+j-1}\left(\frac{P}{Q}\right)^{x-j}\\ &= (k+j-1)_{(j)}P^j.\end{aligned}$$

From these it follows

$$\sqrt{\beta_1} = \frac{Q+P}{\sqrt{kPQ}},\quad \beta_2 = 3+\frac{1+6PQ}{kPQ}.$$

The points (β_1, β_2) lie on the straight line

$$\beta_2 - \beta_1 - 3 = \frac{2}{k}.$$

The cumulants satisfies the following recurrence relation

$$k_{r+1} = PQ\frac{\partial k_r}{\partial Q},\quad r \geq 1.$$

The ratio

$$\frac{b'(x+1;k,p)}{b'(x;k,p)} = \frac{(k+x)P}{(x+1)Q}$$

shows that $b'(x+1;k,p) \gtrless b'(x;k,p)$ according as $x \lessgtr kP-Q$. Hence, if $kP > Q$ there is a mode at x^* where x^* satisfies $kP-Q-1 < x^* < kP-Q$. If $kP < Q$ the mode is at $x = 0$.

Lemma 5.8.1 *As $k \to \infty$ and $P \to 0$ with kP remaining fixed at λ, $b'(x;k,p) \to e^{-\lambda}\frac{\lambda^x}{x!}$.*

Proof:

$$\begin{aligned}\lim_{\substack{k\to\infty\\ P\to 0}} b'(x;k,p) &= \lim_{\substack{k\to\infty\\ P\to 0}} \frac{(k+x-1)\cdots k}{x!}\left(\frac{P}{1+P}\right)^x(1+P)^{-k}\\ &= \lim_{\substack{k\to\infty\\ P\to 0}} \frac{1}{x!}\left(1+\frac{x-1}{k}\right)\left(1+\frac{x-2}{k}\right)\cdots 1\cdot\lambda^x\left(1+\frac{\lambda}{k}\right)^{-x}\left(1+\frac{\lambda}{k}\right)^{-k}\\ &= e^{-\lambda}\frac{\lambda^x}{x!}.\end{aligned}$$

Theorem 5.8.1 *Let $X_1, \cdots, X_n$ be independently distributed negative binomial variables, X_i having parameters $(k_i, p), i = 1, \cdots, n$. Then $S_n = X_1 + \cdots + X_n$ has a negative binomial distribution with parameters (k, p) when*

$$k = \sum_{i=1}^{n} k_i.$$

Proof: m.g.f. of S_n is

$$M_{Sn}(t) = \prod_{i=1}^{n} M_{X_i}(t) = \prod_{i=1}^{n} p^{k_i}(1-qe^t)^{-k_i} = p^k(1-qe^t)^{-k}$$

which is the m.g.f. of a negative binomial distribution with parameters k, p. Hence, the result follows by the uniqueness property of m.g.f. (theorem 4.3.1).

Corollary 5.8.1 *Let $X_1, \cdots, X_n$ be independent geometric variables each with parameter p. Then S_n has a negative binomial distribution with parameters (n, p).*

5.8.1 The Truncated Negative Binomial Distribution

Consider a negative binomial distribution truncated at $x = 0$. The p.m.f. of the truncated distribution is

$$\begin{aligned} b_t'(x; k, p) &= \binom{k+x-1}{x} \frac{p^k q^x}{1-p^k} \\ &= \binom{k+x-1}{x} (1-Q^{-k})^{-1} \left(\frac{P}{Q}\right)^x \left(1-\frac{P}{Q}\right)^k, \quad x = 1, 2, \cdots. \end{aligned}$$

The moments about zero of this distribution are obtained by multiplying the raw moments of the untruncated distribution by $(1-Q^{-k})^{-1}$. Thus for $b_t'(x; k, p)$,

$$E(X) = \frac{kP}{1-Q^{-k}}, \quad E(X^2) = \frac{kPQ + (KP)^2}{1-Q^{-k}}$$

$$V(X) = \frac{kPQ}{1-Q^{-k}} \left[1 - k\left(\frac{P}{Q}\right)\left\{\left(1-Q^{-k}\right)^{-1} - 1\right\}\right].$$

5.9 The Power Series Distribution

Consider the power series

$$f(z) = \sum_{x=0}^{\infty} a_x z^x$$

where a_x's are real non-negative constants, $0 < z < r$ (radius of convergence of $f(z)$) so that $f(z)$ is finite. Define a r.v. X with p.m.f.

$$P\{X = x\} = P_x = \frac{a_x z^x}{f(z)}, \quad x = 0, 1, 2, \cdots. \tag{5.9.1}$$

Clearly

$$P_x \geq 0, \quad \sum_{0}^{\infty} P_x = 1$$

so that X has a probability distribution P_x. The distribution of X is called the power series distribution (p.s.d.)

$$\mu_1' = E(X) = \frac{1}{f(z)}\sum_0^\infty x a_x z^x.$$

Now

$$f'(z) = \frac{1}{z}\sum_1^\infty x a_x z^x.$$

Hence,

$$\mu_1' = \frac{zf'(z)}{f(z)} = z\frac{d\log f(z)}{dz}, \quad \mu_r' = \frac{1}{f(z)}\sum_0^\infty x^r a_x z^x, \quad r = 1, 2, \cdots. \tag{5.9.2}$$

Now

$$\frac{d\mu_r'}{dz} = \frac{1}{f(z)}\sum_x a_x x^{r+1} z^{x-1} - \left(\sum_x x^r a_x z^x\right)\frac{f'(z)}{(f(z))^2}.$$

Thus

$$z\frac{d\mu_r'}{dz} = \frac{1}{f(z)}\sum_x a_x x^{r+1} z^x - \left(\sum_x x^r a_x z^x\right)\frac{zf'(z)}{(f(z))^2},$$

$$\text{i.e., } z\frac{d\mu_r'}{dz} = \mu_{r+1}' - \mu_r'\mu_1'.$$

Thus we have the recurrence relation

$$\mu_{r+1}' = z\frac{d\mu_r'}{dz} + \mu_r'\mu_1'. \tag{5.9.3}$$

The central moments are

$$\mu_r = \frac{1}{f(z)}\sum_0^\infty (x - \mu_1')^r a_x z^x.$$

Now

$$z\frac{d\mu_r}{dz} = \frac{d}{dz}\left\{\sum_0^\infty (x-\mu_1')^r a_x z^x\right\}\frac{z}{f(z)} - \left\{\sum_0^\infty (x-\mu_1')^r a_x z^{x+1}\right\}\frac{f'(z)}{(f(z))^2}.$$

Again,

$$\frac{d}{dz}\left\{\sum_x (x-\mu_1')^r a_x z^x\right\} = \sum_x a_x\left[z^x r(x-\mu_1')^{r-1}\left(-\frac{d\mu_1'}{dz}\right) + (x-\mu_1')^r x z^{x-1}\right].$$

Thus

$$z\frac{d\mu_r}{dz} = \frac{1}{f(z)}\sum_x a_x \cdot x(x-\mu_1')^r z^x - zr\frac{d\mu_1'}{dz}\frac{1}{f(z)}\sum_x a_x z^x$$

$$(x-\mu_1')^{r-1} - \frac{zf'(z)}{(f(z))^2}\sum_x (x-\mu_1')^r a_x z^x. \tag{5.9.4}$$

The first and the last term in the r.h.s. of (5.9.4) together simplify to, by virtue of (5.9.1),

$$\frac{1}{f(z)}\sum_x a_x z^x (x-\mu_1')^{r+1} = \mu_{r+1}.$$

Hence, we can write (5.9.4) as

$$z\frac{d\mu_r}{dz} = \mu_{r+1} - rz\frac{d\mu_1'}{dz}\mu_{r-1}$$

when we have the recurrence relation between the central moments,

$$\mu_{r+1} = z\left[\frac{d\mu_r}{dz} + r\frac{d\mu_1'}{dz}\mu_{r-1}\right]. \tag{5.9.5}$$

Putting $r = 1$, $\mu_0 = 1$, $\mu_1 = 0$ we get the variance of X as

$$\mu_2 = z\frac{d\mu_1'}{dz} = z\frac{d}{dz}\left\{z\frac{d}{dz}\log f(z)\right\} = z^2\frac{d^2\log f(z)}{dz^2} + \mu_1'. \tag{5.9.6}$$

Now

$$\frac{d^2\log f(z)}{dz^2} = \frac{f''(z)f(z) - (f'(z))^2}{(f(z))^2}.$$

Hence,

$$\mu_2 = z^2\frac{f''(z)}{f(z)} - z^2\left(\frac{f'(z)}{f(z)}\right)^2 + z\frac{f'(z)}{f(z)}. \tag{5.9.7}$$

From (5.9.7), (5.9.5) can be written as

$$\mu_{r+1} = z\frac{d\mu_r}{dz} + r\mu_2\mu_{r-1}. \tag{5.9.8}$$

The characteristic function of X is

$$\psi_X(t) = \sum_x e^{itx}P_x = \frac{1}{f(z)}\sum_x e^{itx}a_x z^x = \frac{f(e^{it}z)}{f(z)}. \tag{5.9.9}$$

Some Special Cases

1. ***Poisson distribution***: Let

$$f(z) = e^z = 1 + z + \frac{z^2}{z!} + \frac{z^3}{3!} + \cdots.$$

Here $a_x = \frac{1}{x!}$. Thus

$$P_x = \frac{a_x z^x}{f(z)} = e^{-z}\frac{z^x}{x!}, \quad x = 0, 1, 2, \cdots,$$

which is the Poisson distribution.

From (5.9.2) and (5.9.6) respectively, $\mu_1' = z, \mu_2 = z$.

The recurrence relation (5.9.5) assumes the form

$$\mu_{r+1} = z\left(\frac{d\mu_r}{dz} + r\mu_{r-1}\right)$$

2. ***Binomial distribution***: Let

$$f(z) = (1+z)^n = 1 + \binom{n}{1}z + \binom{n}{2}z^2 + \cdots + z^n.$$

Here

$$a_x = \binom{n}{x}, \quad x = 0, 1, \cdots, n$$

$$P_x = \binom{n}{x}\left(\frac{z}{1+z}\right)^x\left(\frac{1}{1+z}\right)^{n-x}$$

when X follows a Binomial distribution with parameter n and $p = \frac{z}{1+z}$.

From (5.9.2)

$$\mu_1' = z\frac{f'(z)}{f(z)} = np.$$

From (5.9.6)

$$\mu_2 = z\frac{d\mu_1'}{dz} = \frac{nz}{(1+z)^2} = np(1-p).$$

The recurrence relations (5.9.3) and (5.9.5), respectively, assume the form

$$\mu_{r+1}' = z\frac{d\mu_r'}{dz} + n\left(\frac{z}{1+z}\right)\mu_r' = pq\left(\frac{d\mu_r'}{dp} + \frac{n}{q}\mu_r'\right) \text{ (relation 5.4.3)}$$

and

$$\mu_{r+1}' = \frac{z}{(1+z)^2}\left[(1+z)^2\frac{d\mu_r}{dz} + nr\mu_{r-1}\right] = pq\left[\frac{d\mu_r}{dp} + nr\mu_{r-1}\right] \text{ (relation 5.4.2)}$$

3. ***Negative Binomial distribution***: Let $f(z) = (1-z)^{-k}, k > 0, 0 < z < 1$. Hence,

$$f(z) = 1 + \binom{k}{1}z + \binom{k+1}{2}z^2 + \binom{k+2}{3}z^3 + \cdots + \binom{k+x-1}{x}z^x + \cdots.$$

Here

$$a_x = \binom{k+x-1}{x}.$$

Hence,

$$P_x = \frac{a_x z^x}{f(z)} = \binom{k+x-1}{x} z^x (1-z)^k$$

which is the negative binomial distribution. By (5.9.2)

$$E(X) = z\frac{d}{dz}\log\left\{(1-z)^{-k}\right\} = \frac{zk}{1-z}.$$

The recurrence relation for the central moments is

$$\mu_{r+1} = z\left[\frac{d\mu_r}{dz} + r\mu_{r-1}\frac{k}{(1-z)^2}\right].$$

Hence,

$$\mu_2 = \frac{kz}{(1-z)^2}, \quad \mu_3 = \frac{k(1+z)z}{(1-z)^3}$$

$$\mu_4 = z\left[\frac{d\mu_3}{dz} + 3\mu_2\frac{k}{(1-z)^2}\right] = \frac{kz(1+4z+3kz+z^2)}{(1-z)^4}.$$

The characteristic function for this distribution is

$$\psi(t) = \frac{f(e^{it}z)}{f(z)} = \left(\frac{1-e^{it}z}{1-z}\right)^{-k}.$$

4. ***Geometric distribution***: Let

$$f(z) = (1-z)^{-1} = 1 + z + z^2 + \cdots, \quad |z| < 1.$$

Here $P_x = z^x(1-z)$ which is the geometric distribution with probability of success $(1-z)$.

$$\mu_1' = z\frac{f'(z)}{f(z)} = \frac{z}{1-z}, \quad \mu_2 = z\frac{d\mu_1'}{dz} = \frac{z(1-2z)}{(1-z)^2}.$$

5. ***Logarithmic series distribution*:** Let

$$f(z) = -\log(1-z) = z + \frac{z^2}{2} + \frac{z^3}{3} + \cdots, \quad 0 < z < 1.$$

Hence,

$$P_x = -\frac{z^x}{x\log(1-z)}$$

which is the logarithmic series distribution.

$$\mu_1' = z\frac{f'(z)}{f(z)} = -\frac{z}{(1-z)\log(1-z)}$$

$$\frac{d\mu_1'}{dz} = -\frac{\log(1-z)+z}{(1-z)^2\{\log(1-z)\}^2}.$$

Hence,

$$\mu_2 = -\frac{z^2 + z\log(1-z)}{(1-z)^2\{\log(1-z)\}^2}.$$

Thus

$$\mu_{r+1} = z\left[\frac{d\mu_r}{dz} - r\mu_{r-1}\frac{z+\log(1-z)}{(1-z)^2\{\log(1-z)\}^2}\right].$$

Characteristic function is

$$\psi(t) = \frac{f(e^{it}z)}{f(z)} = \frac{\log(1-e^{itz})}{\log(1-z)}.$$

We now consider an interesting theorem characterising the property of a Poisson distribution.

Theorem 5.9.1 *The equality of mean and variance is a necessary and sufficient condition for a p.s.d. to be Poisson.*

Proof: For any p.s.d. given by (5.9.1), $\mu_2 = z\dfrac{d\mu_1'}{dz}$.

When $f(z) = e^z$, the p.s.d. gives Poisson distribution for which $\mu_1' = z\dfrac{f'(z)}{f(z)} = z$ and hence, $\mu_2 = z$. Hence, the necessity.

To prove the sufficiency, let $\mu_2 = \mu_1'$ which implies $\dfrac{d\mu_1'}{dz} = \dfrac{\mu_1'}{z}$.

Integrating both sides,

$$\log\mu_1' = \log(cz)$$

where c is a positive constant. Hence $\mu_1' = cz$.

Again, by (5.9.2)

$$z\frac{d\log f(z)}{dz} = cz \quad \text{or} \quad \frac{d\log f(z)}{dz} = c.$$

Integrating both sides,

$$\log f(z) = cz + \log k, k \text{ is a positive constant.}$$

Thus $f(z) = ke^{cz}$ from which it follows

$$P_x = e^{-cz}\frac{(cz)^x}{x!}$$

so that X follows a Poisson distribution with parameter cz.

Exercise

1. A die is cast repeatedly until an ace turns up. Find the expected number of throws necessary.

2. Find the expected number of aces drawn in a bridge hand of 13 cards. Also find the expected number of face cards drawn.

3. Let X have a probability distribution defined by the function $f(x) = \frac{1}{\pi(1+x^2)}$. Let $Y = 1(0)$ if $X \geq 1(< 1)$. Find probability distribution of Y. [Bernoulli, $p = \frac{1}{4}$]

4. Let X, Y be independent geometric r.v.'s. Show that minimum (X, Y) and $X - Y$ are independent.

5. Let $X_1, \cdots, X_n$ be independent geometric r.v.'s with parameters $p_1, \cdots, p_n$ respectively. Show that $Z = \min(X_1, \cdots, X_n)$ is a geometric r.v. with parameter $p = 1 - \prod_{i=1}^{n}(1 - p_i)$.

6. Let X and Y be two independently and identically distributed r.v.'s with

$$P\{X = x\} = p_x, \quad x = 0, 1, 2, \cdots.$$

   Suppose

$$P\{X = s|X + Y = s\} = P\{X = s - 1|X + Y = s\} = \frac{1}{s+1}, \quad s \geq 0.$$

   Show that X and Y are geometric r.v.'s.

7. Let X be a non-negative integer-valued random variable satisfying

$$P\{X > j + k|X > j\} = P\{X \geq k\}.$$

   for any $j(> 0), k(> 0)$. Then X must have a geometric distribution.

8. Show that for a binomial distribution $b(x; n, p) = \binom{n}{x} p^x q^{n-x}$,

$$P\{X \geq j\} \leq b(j, n, p) \frac{q(j+1)}{j+1-(n+1)p} \text{ for } j \geq np$$

$$P\{X \leq j\} \leq b(j, n, p) \frac{(n-j+1)}{(n+1)p-j} \text{ for } j \leq np.$$

9. Let $X_1, \cdots, X_n$ be n independently and identically distributed Bernoulli variates each with parameter p. Show that their sum $S_n = \sum_{i=1}^{n} X_i$ is distributed as a binomial variate with parameter (n, p).

10. Let $X_1, \cdots, X_n$ be independent Bernoulli variates with common parameter

$$p = P\{X_i = 1\}, \ 0 < p < 1, \ i = 1, \cdots, n.$$

Set

$$S_j = X_1 + \cdots + X_j, \ 1 \leq j \leq n.$$

Show that

$$P[S_m = r | S_n = s], \ m < n$$

does not depend on p. Identify the probability distribution and obtain the first two moments of it. [Hypergeometric $H(x; n, m, s)$.]

11. Let X_1, X_2 be independent binomial wih parameters $\left(n_i, \frac{1}{2}\right)$, $i = 1, 2$. Find the p.m.f. of $X_1 - X_2 + n_2$.

[*Hints:* $(n_2 - X_2)$ is binomial with parameters $\left(n_2, \frac{1}{2}\right)$.

$$P(X_1 - X_2 + n_2 = k) = \binom{n_1+n_2}{n_2+k} \Big/ 2^{n_1+n_2}, (-n_2 \leq k \leq n_1).]$$

12. In a population of n elements n_1 are red and $n_2 (= n - n_1)$ are black. A group of r elements are chosen at random without replacement. Find out the probability

$$q_k (k = 0, 1, \cdots, \ \min(n_1, r))$$

that the group contains exactly k red elements; hence, show that if $np = n_1$, $q = 1 - p$,

$$\binom{r}{k} \left(p - \frac{k}{n}\right)^k \left(q - \frac{r-k}{n}\right)^{r-k} < q_k < \binom{r}{k} p^k q^{r-k} \left(1 - \frac{r}{n}\right)^{-r}.$$

Justify how you can approximate this q_k by a binomial mass function.

13. Show that

$$\frac{\binom{Np}{x}\binom{Nq}{n-x}}{\binom{N}{n}} \to \binom{n}{x} p^x q^{n-x}, \quad q = 1-p, \; 0 < p < 1$$

as $N \to \infty$ for any fixed x, provided $\frac{n}{N}$ is negligible.

14. An urn contains a white balls and b black balls. A sample of 2 balls is chosen wihout replacement. Under what conditions will the probability that both the chosen balls are of the same colour be greater than $\frac{1}{2}$? [$(a-b)^2 > a+b$]

15. Show that

$$\binom{k+x-1}{x} p^k q^x \to e^{-\lambda} \frac{\lambda^x}{x!}$$

as $p \to 1$ and $k \to \infty$ in such a way that $k(1-p) = \lambda$ remains constant.

16. Obtain by two methods, the probability that in a series of Bernoulli trials (with probability p of success) 'a' successes occur before at most b failures. Hence, prove the identity

$$\sum_{k=0}^{b-1} \binom{a+b-1}{a+k} p^k q^{b-k-1} = \sum_{k=0}^{b-1} \binom{a+k-1}{k} q^k.$$

[*Hints:*. Prob. that in $(a+b-1)$ trials, $(a+k)$ successes and $b-k-1$ failures occur is

$$\binom{a+b-1}{a+k} p^{a+k} q^{b-k-1}.$$

Hence, probability of getting 'a' successes before 'b' failures is

$$\sum_{k=0}^{b-1} \binom{a+b-1}{a+k} p^{a+k} q^{b-k-1}. \qquad \text{(i)}$$

Alternatively, probability that in $(a+k)$ trials the last trial is a success and the preceding $(a+k-1)$ trials contain $(a-1)$ successes is

$$\binom{a+k-1}{a-1} p^a q^k.$$

The required probability is therefore,

$$= p^a \sum_{k=0}^{b-1} \binom{a+k-1}{k} q^k \qquad \text{(ii)}$$

Hence, the identity.]

17. Let X_1, X_2 be two independent negative binomial variables, X_i having parameters $k_i, p, i = 1, 2$. Find the conditional probability that $X_1 = t$, given that $X_1 + X_2 = r$. Show also that when $k_1 = k_2 = 1$, then $P(X_1 = t | X_1 + X_2 = r) = \frac{1}{r+1}$, $t = 0, 1, \cdots, r$; $r = 0, 1, 2, \cdots$.

[*Hints:* $P(X_1 = t | X_1 + X_2 = r) = \dfrac{\binom{t+k_1-1}{t}\binom{r+k_2-t-1}{r-t}}{\binom{r+k_1+k_2-1}{r}}, r = 0, 1, 2, \cdots$.]

18. Consider a r.v. X distributed as $H(x; N, M, n)$. Show that for fixed n and fixed $p = \frac{M}{N}$, the rth factorial moment of this distribution tends to that of a binomial distribution with parameters (n, p) as $N \to \infty$.

19. An urn contains N balls numbered $1, \cdots, N$. From these balls n balls are drawn at random without replacement. Let X_i denote the number drawn on the ith ball $(i = 1, \cdots, n)$ and $S_n = \sum_{i=1}^{n} X_i$. Find $E(S_n)$, $V(S_n)$.

[*Hints:* There are $\binom{N}{n}$ possible sets of n sampled balls $s = (i_1, \cdots, i_n)$, i_t denoting the number marked on a sampled ball. Let T = sum of the numbers on the sampled balls $= \sum_{t=1}^{n} i_t$. Probability of selecting each s is M_0^{-1}, $s \in \mathcal{S}$ (sample sapce), where $M_j = \binom{N-j}{n-j}$, $j = 0, 1, 2, \cdots$.

$$E(T) = \sum_{s \in \mathcal{S}} \left[\sum_{t=1}^{n} i_t \right] \frac{1}{M_0} = \frac{1}{M_0} \sum_{i=1}^{N} i \times \text{(Number of samples containing the ball marked } i)$$

$$= \frac{M_1}{M_0} \sum_{i=1}^{N} i = \frac{n(N+1)}{2}.$$

$$E(T^2) = E\left[\sum_{t=1}^{n} i_t^2 + \sum_{t \neq t'=1}^{n} \sum i_t i_{t'} \right]$$

$$E\left(\sum_{t=1}^{n} i_t^2 \right) = \sum_{S \in \mathcal{S}} \left[\sum_{t=1}^{n} i_t^2 \right] \cdot \frac{1}{M_0} = \frac{M_1}{M_0} \sum_{i=1}^{N} i^2 = \frac{n(N+1)(2N+1)}{6}.$$

$$E\left(\sum_{t \neq t'=1}^{n} \sum i_t i_{t'} \right) = \frac{1}{M_0} \sum_{S \in \mathcal{S}} \left[\sum_{t \neq t'=1}^{n} \sum i_t i_{t'} \right]$$

$$= \frac{1}{M_0} \sum_{t \neq t'=1}^{N} \sum i i' \times \begin{array}{l}\text{(number of samples containing} \\ \text{the balls marked } i \text{ and } i')\end{array}$$

$$= \frac{M_2}{M_0} \cdot \frac{N(N+1)(3N+2)(N-1)}{12} = \frac{n(n-1)(N+1)(3N+2)}{12}.$$

Hence,

$$E(T^2) = \frac{n(N+1)}{6} \left[2N + 1 + \frac{(n-1)(3N+2)}{2} \right]$$

$$V(T) = E(T^2) - (E(T))^2 = \frac{n(N+1)(N-n)}{12}.]$$

20. Telephone calls arrive at a certain exchange at an average rate of 3 calls per minute. Assuming that the number of incoming calls follow a Poisson distribution, find the probability that 3 or more calls arrive in a 15-second time interval. [0.0406]

21. A Geiger Muller counter registrars on an average 50 counts per minute in th neighbourhood of a certain radioactive substance. Find the probability there wi be (i) 3 counts in a 6-second period; (ii) x counts in a T-second period.

[(i) 0.140; (ii) $e^{-0.833T}\frac{(0.833T)^x}{x!}$

22. Find the probability generating function of the sum S of the numbers obtaine when one ball is drawn from each of m urns, each of which contains n bal numbered $1, 2, \cdots, n$. Hence, find the probability distribution of S.

[*Hints:* Let X_i be the number on the ball drawn from the ith urn ($i = 1, \cdots, m$). Each X_i has uniform distribution $P(X_i = k) = \frac{1}{n}$, $k = 1, \cdots, n$. Hence, proceed as in exercise 9 of chapter 4.

Chapter 6

Some Continuous Distributions on $\mathbb{R}^1$

6.1 Introduction

We consider some univariate continuous distributions in this chapter. Uniform distribution is discussed in the section 6.2; sections 6.3–6.6 discuss respectively exponential, gamma (chi-square), beta and cauchy distributions. Section 6.7 deals with the celebrated normal distribution. We consider log-normal, Laplace, Pareto, Weibull, extreme-value and logistic distributions in sections 6.8 to 6.13 respectively.

6.2 Continuous Uniform Distribution

A random variable X is said to have a continuous uniform (rectangular) distribution on the interval $[a, b]$ $(-\infty < a < b < \infty)$ if its p.d.f. is given by

$$f(x; a, b) = \begin{cases} \dfrac{1}{b-a}, & \text{if } a \le x \le b \\ 0, & \text{otherwise,} \end{cases} \tag{6.2.1}$$

a, b are the parameters of the distribution. The c.d.f. is

$$F_X(x) = \begin{cases} 0, & x \le a \\ \dfrac{x-a}{b-a}, & a < x < b \\ 1, & x \ge b. \end{cases} \tag{6.2.2}$$

$$E(X) = \int_a^b \frac{x}{b-a}\,dx = \frac{a+b}{2}, \quad E(X^2) = \frac{b^2+ab+a^2}{3}, \quad V(X) = \frac{(b-a)^2}{12}.$$

The m.g.f. is

$$M_X(t) = \int_a^b \frac{e^{tx}}{b-a}\,dx = \frac{e^{bt}-e^{at}}{(b-a)t}, \quad t \neq 0.$$

The distribution can also be defined over the open interval (a, b) or semi-closed intervals $[a, b), (a, b]$. The c.d.f. remains (6.2.2) for all these distributions (Fig. 6.1).

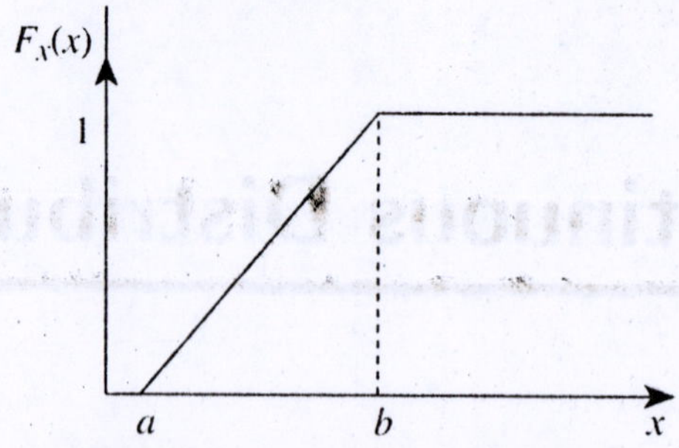

Fig. 6.1 c.d.f. of rectangular distribution

6.3 Exponential Distribution

A r.v. X is defined to have an (negative) exponential distribution with parameter $\lambda > 0$ if its p.d.f. is given by

$$f(x, \lambda) = \begin{cases} \lambda e^{-\lambda x}, & 0 < x < \infty,\ \lambda > 0 \\ 0, & \text{otherwise.} \end{cases} \tag{6.3.1}$$

The c.d.f. is

$$F_X(x) = \begin{cases} 0, & x < 0 \\ 1 - e^{-\lambda x}, & 0 \le x. \end{cases}$$

The m.g.f. of the distribution is

$$M_X(t) = \frac{1}{1 - \left(\frac{t}{\lambda}\right)}$$

$$\mu_1' = \frac{1}{\lambda},\quad \mu_r' = \frac{r!}{\lambda^r},\quad r = 2, 3, \cdots$$

$$\mu_2 = \frac{1}{\lambda^2},\quad \mu_3 = \frac{2}{\lambda^3},\quad \mu_4 = \frac{9}{\lambda^4}$$

$$\sqrt{\beta_1} = 2,\quad \beta_2 = 6.$$

The mean deviation is $\frac{2}{e\lambda}$. The ratio (mean deviation)/s.d. $= \frac{2}{e} = 0.736$. The median of the distribution is $\frac{1}{\lambda}\log_e 2$. The mode is at $x = 0$.
The cumulant generating function is

$$-\log\left(1 - \frac{t}{\lambda}\right),\quad k_1 = \frac{1}{\lambda},\quad k_r = (r-1)!\lambda^{-r},\quad r > 1.$$

Fig. 6.2 shows an exponential density.

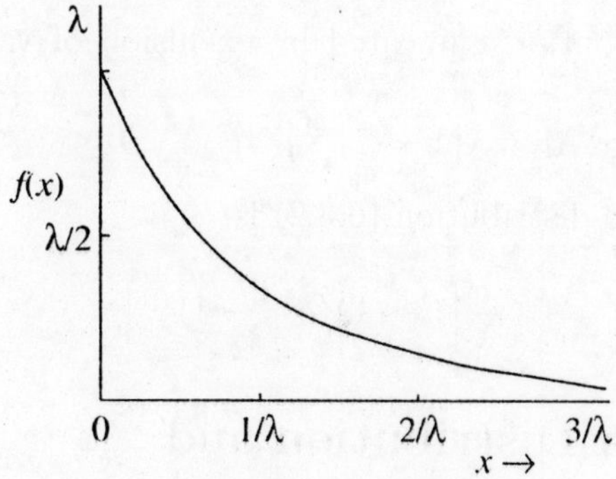

Fig. 6.2 Exponential Density

Theorem 6.3.1 *For an exponential r.v. X with parameter* λ,

$$P[X > a+b|X > a] = P[X > b] \text{ for } a > 0,\ b > 0.$$

Proof:

$$P[X > a+b|X > a] = \frac{P[X > a+b]}{P[X > a]} = \frac{e^{-\lambda(a+b)}}{e^{-\lambda a}} = e^{-\lambda b} = P[X > b].$$

If X represents the lifetime of a given component, say, an electric bulb, theorem 6.3.1 states that the probability that a bulb will be operative for at least $(a+b)$ units given that it has already run for a units is the same as its intial probability for lasting for at least b units. This means that the future lifetime of an individual has the same distribution, no matter how old it is at present. This is the "lack of memory" property of the exponential distribution. A similar property was noted for the geometric distribution (vide section 5.7).

Suppose that the events occur randomly in time in such a way that the number of happenings in a fixed time interval $[0, t]$ has a Poisson distribution with parameter λt. Suppose one happening has just occured. Probability that the next happening will occur not before time T is P [waiting time $> T$] $= P$ [no happening in time interval of length T] $= e^{-\lambda T}$. Hence $P(X \leq T) = 1 - e^{-\lambda T} (T > 0)$ so that X, the waiting time has an exponential distribution.

Thus exponential distribution and Poisson distribution are related (vide also example 6.4.1).

Because of the simplicity in mathematics associated with it, a negative exponential distribution is often used as an approximate representation of more complex models suitable for a particular application. The lifetime of an component can often be represented by an exponential random variable. The distribution is often used in the field of life-testing.

An exponential distribution truncated from below is still an exponential distribution, with the same scale parameter.

An exponential distribution truncated by exclusion of values exceeding A has the p.d.f.

$$f_t(x;\lambda) = \lambda\{1 - e^{-\lambda A}\}^{-1} e^{-\lambda x}, \quad 0 < x < A. \tag{6.3.2}$$

The expected value of the distribution (6.3.2) is

$$\frac{1}{\lambda}\left[1 - A\lambda(e^{\lambda A} - 1)^{-1}\right].$$

6.4 The Gamma Distribution and Chi-squared Distribution

1. **Gamma Function:** For $\alpha > 0$, the gamma function is defined as

$$\Gamma(\alpha) = \int_0^\infty t^{\alpha-1} e^{-t} dt (\alpha > 0). \tag{6.4.1}$$

$\Gamma(1) = 1$. For $\alpha > 1$, integration by parts yields

$$\Gamma(\alpha) = (\alpha - 1)\Gamma(\alpha - 1). \tag{6.4.2}$$

If $\alpha = n$, a positive integer

$$\Gamma(n) = (n-1)!. \tag{6.4.3}$$

$$\Gamma\left(\frac{1}{2}\right) = \sqrt{\pi}. \tag{6.4.4}$$

Gauss' multiplication formula:

$$\Gamma(n\alpha) = (2\pi)^{-\frac{n-1}{2}} n^{n\alpha - \frac{1}{2}} \prod_{j=0}^{n-1} \Gamma(\alpha + jn^{-1}). \tag{6.4.5}$$

Duplication formula ($n = 2$):

$$\Gamma(2\alpha) = (2\pi)^{-\frac{1}{2}} 2^{2\alpha - \frac{1}{2}} \Gamma(\alpha)\Gamma\left(\alpha + \tfrac{1}{2}\right). \tag{6.4.6}$$

The ***incomplete gamma function:***

$$\Gamma_y(\alpha) = \int_0^y e^{-t} t^{\alpha-1} dt. \tag{6.4.7}$$

The ***incomplete gamma function ratio:***

$$\frac{\Gamma_y(\alpha)}{\Gamma(\alpha)}. \tag{6.4.8}$$

Stirling's formula:

$$\Gamma(\alpha+1) = \sqrt{2\pi}\,\alpha^{\alpha+\frac{1}{2}} e^{-\left(\alpha + \frac{\theta_\alpha}{12\alpha}\right)} \tag{6.4.9}$$

for some θ_α, $0 < \theta_\alpha < 1$.

2. **Gamma Distribution:** A r.v. X is defined to have a gamma distribution with parameters $(\lambda, p)(\lambda > 0, p > 0)$ if its p.d.f. is given

$$f_G(x;\lambda,p) = \begin{cases} \dfrac{\lambda}{\Gamma(p)} e^{-\lambda x}(\lambda x)^{p-1}, & 0 < x < \infty \\ 0, & \text{otherwise.} \end{cases} \tag{6.4.10}$$

The m.g.f. is

$$\begin{aligned} M_X(t) = E(e^{tX}) &= \int_0^\infty \frac{\lambda^p}{\Gamma(p)} e^{tx} e^{-\lambda x} x^{p-1}\, dx \\ &= \left(\frac{\lambda}{\lambda - t}\right)^p \int_0^\infty \frac{(\lambda - t)^p}{\Gamma(p)} x^{p-1} e^{-(\lambda - t)x}\, dx = \left(\frac{\lambda}{\lambda - t}\right)^p. \end{aligned} \tag{6.4.11}$$

Hence

$$\mu_1' = \frac{p}{\lambda}, \qquad \mu_2' = \frac{p(p+1)}{\lambda^2}, \; \mu_r' = \frac{\Gamma(p+r)}{\Gamma(p)\,\lambda^r}$$

$$\mu_2 = \frac{p}{\lambda^2}, \qquad \mu_3 = \frac{2p}{\lambda^3}, \qquad \mu_4 = \frac{3p^2 + 6p}{\lambda^4}$$

$$\sqrt{\beta_1} = \frac{2}{\sqrt{p}}, \; \beta_2 = 3 + \frac{6}{p}.$$

The cumulative distribution function is given by the incomplete gamma function ratio

$$F(x) = P(X < x) = \frac{\Gamma_x(p)}{\Gamma(p)}.$$

The distribution has a mode at $x = \frac{1}{\lambda}(p-1)$. There are points of inflexion, equidistant from the mode, at $x = p - 1 \pm \sqrt{p-1}$ (provided the values are real and positive). When $p = 1$, (6.4.10) reduces to the exponential distribution (6.3.1).

The standard form of the gamma distribution is obtained by putting $\lambda = 1$, giving

$$f_G(x;p) = \frac{e^{-x} x^{p-1}}{\Gamma(p)} \quad (x \geq 0). \tag{6.4.12}$$

The standard distribution (6.4.12) has a single mode at $x = p - 1$ if $p \geq 1$. If $p < 1$, $f_G(x;p) \to \infty$ as $x \to 0$.

Some typical standard gamma distributions are shown in Fig. 6.3. It is seen as p increases the shape of the curve becomes similar to that of the normal probability curve (discussed in section 6.7).

3. **χ^2-distribution:** A r.v. X is said to have a chi-square distribution with n degrees of freedom if its p.d.f. is given by

$$f(x) = \frac{1}{2^{n/2}\Gamma\left(\frac{n}{2}\right)} e^{-\frac{x}{2}} x^{\frac{n}{2}-1}, \quad 0 < x < \infty. \tag{6.4.13}$$

4. Clearly $\frac{x}{2}$ has a standard gamma distribution (6.4.12) with parameter $p = \frac{n}{2}$. Thi distribution will be taken up in greater details in section 11.2.

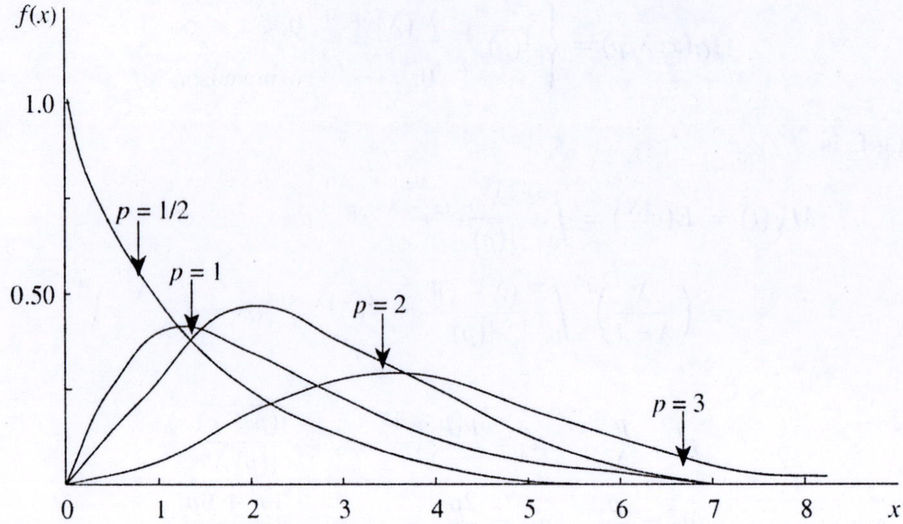

Fig. 6.3 Standard Gamma Density

Theorem 6.4.1 *If $X_1, \cdots, X_n$ are independent r.v.'s such that X_j is distributed a gamma with parameters (λ, p_j) then their sum $S_n = \sum_{j=1}^{n} X_j$ is distributed as gamm with parameters (λ, p) where $p = \sum_{j=1}^{n} p_j$.*

Proof: The m.g.f. of

$$S_n = M_{S_n}(t) = \prod_{j=1}^{n} M_{X_j}(t) = \prod_{j=1}^{n} \left(\frac{\lambda}{\lambda - t}\right)^{p_j} = \left(\frac{\lambda}{\lambda - t}\right)^{p}.$$

The result follows by the uniqueness property of the m.g.f.

Corollary 6.4.1 *If $X_1, \cdots, X_n$ are independent exponential variables each with param eter λ their sum $S_n = \sum_{j=1}^{n} X_j$ is a gamma variable with parameters (λ, n).*

Corollary 6.4.2 *If $X_1, \cdots, X_n$ are independent chi-square variables, X_j having degre of freedom n_j, their sum $S_n = \sum_{j=1}^{n} X_j$ has a chi-square distribution with degree of freedon $n = \sum_{j=1}^{n} n_j$.*

Theorem 6.4.2 *Let X be a Poisson r.v. with parameter λ. Then*

$$P\{X \le t\} = \frac{1}{\Gamma(t+1)} \int_{\lambda}^{\infty} e^{-x} x^t \, dx \tag{6.4.14}$$

(6.4.14) expresses the d.f. of X in terms of an incomplete gamma function.

Proof: Integrating by parts

$$\frac{1}{t!}\int_{\lambda}^{\infty} e^{-x} x^t \, dx = \frac{1}{t!}\lambda^t e^{-\lambda} + \int_{\lambda}^{\infty} \frac{1}{(t-1)!} x^{t-1} e^{-x} \, dx = \cdots$$

$$= e^{-\lambda}\left(1 + \lambda + \frac{\lambda^2}{2!} + \cdots + \frac{\lambda^t}{t!}\right).$$

Example 6.4.1 *The life-time T of a component has an exponential distribution*

$$f_T(t) = \lambda e^{-\lambda t} \quad (t \ge 0, \lambda > 0).$$

Assume that as soon as a component fails it is replaced by another component with exactly the same distribution of life-time. Show that the probability distribution of number of failures X in a period of fixed length τ is Poisson with parameter $\lambda\tau$.

Let T_i denote the life-time of the i-th component.

$$P(X = k) = P(T_1 + \cdots + T_{k+1} > \tau) - P(T_1 + \cdots + T_k > \tau) \tag{i}$$

For a fixed m, $\sum_{j=1}^{m} T_j$ follows a gamma distribution with parameter (λ, m). Hence

$$P(T_1 + \cdots + T_{k+1} > \tau) = \int_{\tau}^{\infty} \frac{\lambda}{\Gamma(k+1)} e^{-\lambda y} (\lambda y)^k \, dy$$

$$= \frac{1}{\Gamma(k+1)} \int_{\lambda\tau}^{\infty} e^{-z} z^k \, dz \quad \text{where } z = \lambda y$$

$$= \sum_{j=0}^{k} e^{-\lambda\tau} \frac{(\lambda\tau)^j}{j!} \quad \text{(by theorem 6.4.2).}$$

Hence from (i),

$$P(X = k) = \sum_{j=0}^{k} e^{-\lambda\tau}\frac{(\lambda\tau)^j}{j!} - \sum_{j=0}^{k-1} e^{-\lambda\tau}\frac{(\lambda\tau)^j}{j!} = e^{-\lambda\tau}\frac{(\lambda\tau)^k}{k!}.$$

6.5 Beta Distribution

1. The Beta function:

$$B(p,q) = \int_0^1 t^{p-1}(1-t)^{q-1}dt \ \ (p,q>0) \tag{6.5.1}$$

$$= 2\int_0^{\pi} 2\sin^{2p-1}\theta\cos^{2q-1}\theta\, d\theta. \tag{6.5.2}$$

In terms of gamma function

$$B(p,q) = \frac{\Gamma(p)\,\Gamma(q)}{\Gamma(p+q)} = B(q,p). \tag{6.5.3}$$

The incomplete beta function

$$B_y(p,q) = \int_0^y t^{p-1}(1-t)^{q-1}dt. \tag{6.5.4}$$

The incomplete beta function ratio

$$I_y(p,q) = \frac{B_y(p,q)}{B(p,q)}. \tag{6.5.5}$$

2. **Beta distribution:** A r.v. X is said to have a beta distribution with parameters a, b $(a>0,\ b>0)$ if its p.d.f. is given by

$$f(x;a,b) = \begin{cases} \frac{1}{B(a,b)},\ x^{a-1}(1-x)^{b-1}, & 0<x<1,\ a>0, b>0 \\ 0, & \text{otherwise.} \end{cases} \tag{6.5.6}$$

The rth raw moment

$$\mu'_r = \frac{1}{B(a,b)}\int_0^1 x^{r+a-1}(1-x)^{b-1}\,dx = \frac{B(r+a,b)}{B(a,b)} = \frac{\Gamma(r+a)\,\Gamma(a+b)}{\Gamma(a)\,\Gamma(r+a+b)}$$

In particular,

$$\mu'_1 = \frac{a}{a+b}, \qquad \mu_2 = \frac{ab}{(a+b)^2(a+b+1)},$$

$$\mu_3 = \frac{2ab(b-a)}{(a+b)^3(a+b+1)(a+b+2)}, \quad \mu_4 = \frac{3ab\{ab(a+b-6)+2(a+b)^2\}}{(a+b)^4(a+b+1)(a+b+2)(a+b+3)},$$

$$\sqrt{\beta_1} = \frac{2(b-a)\sqrt{a+b+1}}{\sqrt{ab}(a+b+2)}, \qquad \beta_2 = \frac{3(a+b+1)\{ab(a+b-6)+2(a+b)^2\}}{ab(a+b+2)(a+b+3)}.$$

The m.g.f. of the distribution is

$$M_X(t) = E(e^{tX}) = \frac{1}{B(a,b)}\int_0^1 e^{tx}x^{a-1}(1-x)^{b-1}\,dx.$$

The m.g.f. does not have a simple form. However the moments of all orders exist nd hence one can write

$$M_X(t) = \sum_{j=0}^{\infty} \mu_j' \frac{t^j}{j!} = \sum_{j=0}^{\infty} \frac{t^j}{j!} \cdot \frac{\Gamma(a+j)\,\Gamma(a+b)}{\Gamma(a)\,\Gamma(a+b+j)}.$$

he cumulative distribution function is

$$F_X(u) = P(X \le u) = \begin{cases} 0, & u \le 0 \\ \frac{1}{B(a,b)} \int_0^u x^{a-1}(1-x)^{b-1}\, dx, & 0 < u < 1 \\ \frac{B_u(a,b)}{B(a,b)} = I_u(a,b) & \\ 1, & u > 1 \end{cases}$$

The incomplete Beta function $I_u(a, b)$ has been tabulated extensively.

If $a = b = 1$, (6.5.6) reduces to the continuous uniform distribution over (0, 1).

Fig. 6.4 shows some beta density curves for different values of (a, b).

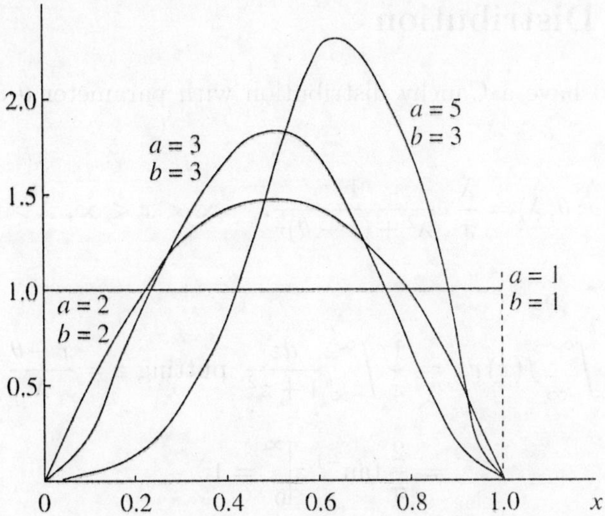

Fig. 6.4 Beta Densities

The beta density can assume a variety of shapes depending on different values of parameters and hence this distribution has an wide application to model experiments of various types.

The distribution (6.5.6) is also called beta distribution of the first type.

3. **Beta distribution of the second type:** A r.v. X is said to have a beta distribution of the second type with parameters $a, b (a > 0, b > 0)$ if its p.d.f. is given by

$$f(x; a, b) = \begin{cases} \frac{1}{B(a,b)} \frac{x^{a-1}}{(1+x)^{a+b}}, & x \ge 0, \quad a > 0, b > 0 \\ 0, & \text{otherwise} \end{cases} \qquad (6.5.7)$$

The rth raw moment is,

$$\mu'_r = \frac{1}{B(a,b)} \int_0^\infty \frac{x^{a+r-1}}{(1+x)^{a+b}}\, dx.$$

Putting

$$(1+x) = \frac{1}{y} = \frac{1}{B(a,b)} \int_0^1 y^{b-r-1}(1-y)^{a+r-1}\, dy$$

$$= \frac{B(b-r, a+r)}{B(a,b)} = \frac{\overline{|(b-r)}\,\overline{|(a+r)}}{\overline{|(a)}\,\overline{|(b)}}.$$

In particular,

$$\mu'_1 = \frac{a}{b-1}, \quad \mu_2 = \frac{a(a+b-1)}{(b-1)^2(b-2)}.$$

6.6 Cauchy Distribution

A r.v. X is said to have a Cauchy distribution with parameter θ and $\lambda(\lambda > 0)$ if it p.d.f. is given by

$$f(x;\theta,\lambda) = \frac{\lambda}{\pi} \cdot \frac{1}{\lambda^2 + (x-\theta)^2}, \quad -\infty < x < \infty,\ \lambda > 0. \qquad (6.6.1$$

We check that

$$\int_{-\infty}^{\infty} f(x)\, dx = \frac{1}{\pi} \int_{-\infty}^{\infty} \frac{dz}{1+z^2} \quad \text{putting } z = \frac{x-\theta}{\lambda}$$

$$= \frac{2}{\pi} \tan^{-1} z \Big|_0^\infty = 1.$$

The cumulative distribution of X is

$$F_X(x) = \frac{1}{2} + \frac{1}{\pi} \tan^{-1}\left(\frac{x-\theta}{\lambda}\right).$$

The parameters θ and λ are location and scale parameters respectively. The distribution is symmetrical about $x = \theta$ and hence the median is at $x = \theta$; the upper and lower quantiles are $\theta \pm \lambda$. The p.d.f. has points of inflexion at $\theta \pm \frac{\lambda}{\sqrt{3}}$. The density function is maximum at $x = \theta$, hence $x = \mu$ is mode of the distribution.

The distribution does not possess finite moments of order greater than or equal to 1 Hence the distribution does not possess finite expected value or standard deviation (see example 3.2.4).

Considering,

$$\mu_1' = \frac{\lambda}{\pi}\int_{-\infty}^{\infty} \frac{x\,dx}{\lambda^2 + (x-\theta)^2}\,dx$$

$$= \frac{1}{\pi}\int_{-\infty}^{\infty} \frac{(\theta + \lambda z)}{1+z^2}\,dz \text{ (putting } \tfrac{x-\theta}{\lambda} = z)$$

$$= \theta + \frac{\lambda}{\pi}\int_{-\infty}^{\infty} \frac{z\,dz}{1+z^2} = \theta + \frac{\lambda}{\pi}\lim_{\substack{A\to\infty\\ B\to\infty}} \int_{-A}^{B} \frac{z\,dz}{1+z^2}. \tag{6.6.2}$$

The integral in (6.6.2) does not exist if A, B approach infinity independently. Consequently the mean of the distribution does not exist. However, when $A = B$, the limit of this integral exists and the limit is zero. Hence if we take the principal value of this integral, the value of the expression in (6.6.2) is θ.

The characteristic function of the distribution is

$$\Psi_X(t) = E(E^{itX}) = e^{it\theta - |t|\lambda} \text{ (vide example 4.6.4).}$$

The standard form of the distribution, represented in Fig. 6.5 is obtained by putting $\theta = 0$, $\lambda = 1$.

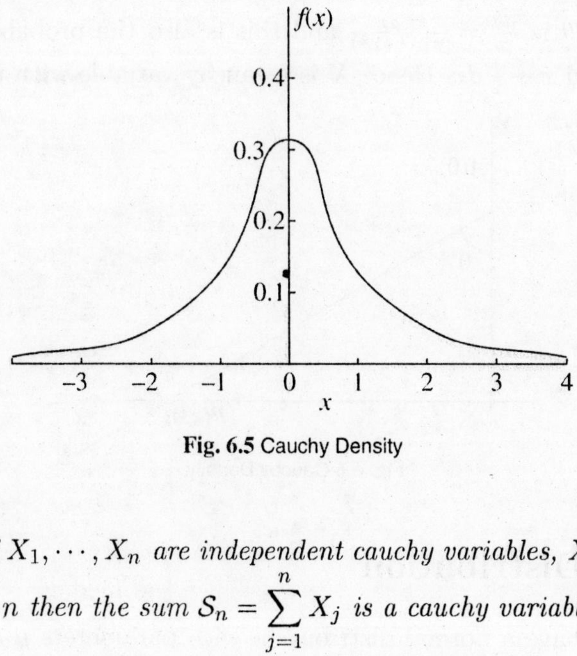

Fig. 6.5 Cauchy Density

Theorem 6.6.1 *If $X_1, \cdots, X_n$ are independent cauchy variables, X_j having parameters (λ_j, θ_j), $j = 1, \cdots, n$ then the sum $S_n = \sum_{j=1}^{n} X_j$ is a cauchy variable with parameters*

$$\lambda' = \sum_{j=1}^{n} \lambda_j, \quad \theta' = \sum_{j=1}^{n} \theta_j.$$

Proof: The characteristic function of $\mathcal{S}_n$ is

$$\Psi_{\mathcal{S}_n}(t) = \prod_{i=1}^{n} \Psi_{X_i}(t)$$

$$= \exp\left[it\sum_{j=1}^{n}\theta_j - |t|\sum_{j=1}^{n}\lambda_j\right],$$

which is the characteristic function of a cauchy variable with parameters λ', θ'. Hence the theorem follows by the uniqueness property of the characteristic function (theorem 4.6.1).

Corollary 6.6.1 *The arithmetic mean* $\bar{X} = \frac{1}{n}\sum_{i=1}^{n} X_i$ *of* n *independently and identically distributed cauchy variables with parameters* (λ, θ) *is a cauchy variable with the same set of parameters* (λ, θ).

► **Example 6.6.1** *A straight line AB is allowed to move through a fixed point A with coordinates* $(0, \lambda)$ *and the length of the intersection X it makes with the x-axis is noted (Fig. 6.6). AB makes an angle* $\theta(= \angle OAP)$ *with the y-axis. Assuming that* θ *has an uniform distribution between* $-\frac{\pi}{2}$ *(corresponding to* $OP = -\infty$*) and* $\frac{\pi}{2}$ *(corresponding to* $OP = \infty$*), find the probability density function of X.*

Relation between θ and X is $\theta = \tan^{-1}\left(\frac{X}{\lambda}\right)$. Probability that θ takes value between $\theta - \frac{1}{2}\,d\theta$ and $\theta + \frac{1}{2}\,d\theta$ is $\frac{d\theta}{\pi} = \frac{\lambda\,dx}{\pi(\lambda^2+x^2)}$ and this is also the probability that X will lie between $x - \frac{1}{2}\,dx$ and $x + \frac{1}{2}\,dx$. Hence X is a cauchy variable with parameters $(\lambda, 0)$.

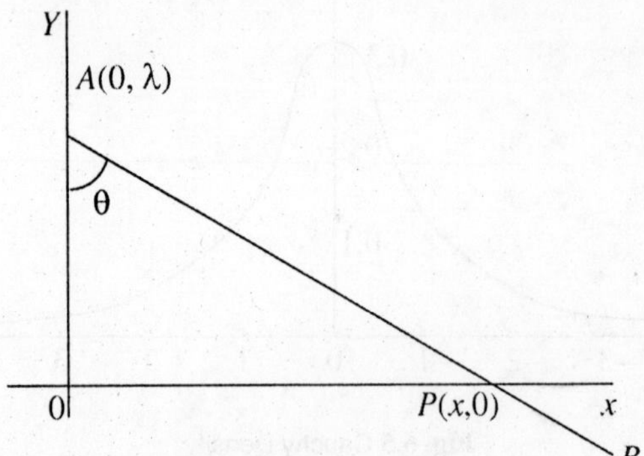

Fig. 6.6 Cauchy Density

6.7 Normal Distribution

A r.v. X is said to have a normal distribution with parameters μ and σ if its p.d.f. is given by

$$f(x;\mu,\sigma) = \frac{1}{\sqrt{2\pi}\sigma}e^{-\frac{1}{2\sigma^2}(x-\mu)^2}, \quad -\infty < x < \infty, \tag{6.7.1}$$

where the parameters satisfy $-\infty < \mu < \infty$, $\sigma > 0$.

o show that

$$\int_{-\infty}^{\infty} f(x;\mu,\sigma)\,dx = 1, \tag{6.7.2}$$

e make the transformation $\frac{x-\mu}{\sigma} = z$. The L.H.S. of (6.7.2) is

$$\frac{1}{\sqrt{2\pi}}\int_{-\infty}^{\infty} e^{-\frac{z^2}{2}}\,dz = I \text{ (say)}$$

$$I^2 = \frac{1}{2\pi}\int_{-\infty}^{\infty} e^{-\frac{z^2}{2}}\,dz \int_{-\infty}^{\infty} e^{-\frac{y^2}{2}}\,dy = \frac{1}{2\pi}\int_{-\infty}^{\infty}\int_{-\infty}^{\infty} e^{-\frac{y^2+z^2}{2}}\,dy\,dz.$$

Ve make now polar transformation $y = r\cos\theta$, $z = r\sin\theta$, $0 < r < \infty$, $0 < \theta < 2\pi$

$$I^2 = \frac{1}{2\pi}\int_0^{\infty}\int_0^{2\pi} re^{-\frac{1}{2}r^2}\,dr\,d\theta = \int_0^{\infty} re^{-\frac{1}{2}r^2}\,dr = 1 \text{ when } I = 1.$$

The probability curve (6.7.1) is symmetrical about $x = \mu$, because $f(\mu+\lambda) = f(\mu-\lambda)$ or all values of λ. The ordinate decreases rapidly as $(x-\mu)$ increases numerically. The urve extends to infinity on either side of the ordinate at $x = \mu$. The distribution is nimodal with mode at $x = \mu$. The modal value of the probability density function is $\sigma\sqrt{2\pi})^{-1} = \frac{0.3979}{\sigma}$. Since the ordinate at $x = \mu$ divides the area under the normal curve nto two equal parts, the median is also μ. The mean ($= \mu$, shown below), median and he mode coincide for this distribution. The density function has points of inflexion at $= \mu \pm \sigma$. (The points of inflexion are obtained by solving the equation $\frac{d^2 f(x)}{dx^2} = 0$ and erifying that at these points $\frac{d^3 f(x)}{dx^3} \neq 0$). These are the points at which the curve of a ormal distribution changes from concave upwords to convex downwards.

The Fig. 6.7 show some normal density curves for various values of μ and σ.

The mean deviation about mean μ is

$$\begin{aligned}
E|X-\mu| &= \frac{1}{\sigma\sqrt{2\pi}}\int_{-\infty}^{\infty} |x-\mu| e^{-\frac{1}{2\sigma^2}(x-\mu)^2}\,dx \\
&= \frac{\sigma}{\sqrt{2\pi}}\int_{-\infty}^{\infty} |z| e^{-\frac{z^2}{2}}\,dz \text{ (putting } \tfrac{x-\mu}{\sigma} = z) \\
&= \frac{\sigma}{\sqrt{2\pi}}\left[-\int_{-\infty}^{0} ze^{-\frac{z^2}{2}}\,dz + \int_0^{\infty} ze^{-\frac{z^2}{2}}\,dz\right] \\
&= \sigma\sqrt{\frac{2}{\pi}}\int_0^{\infty} ze^{-\frac{z^2}{2}}\,dz = \sigma\sqrt{\frac{2}{\pi}} = 0.7979\sigma.
\end{aligned}$$

Considering the odd order central moments

$$\begin{aligned}
\mu_{2r+1} &= \frac{1}{\sigma\sqrt{2\pi}}\int_{-\infty}^{\infty} (x-\mu)^{2r+1} e^{-\frac{1}{2\sigma^2}(x-\mu)^2}\,dx \\
&= \frac{1}{\sigma\sqrt{2\pi}}\left[\int_{-\infty}^{\mu} (x-\mu)^{2r+1} e^{-\frac{1}{2\sigma^2}(x-\mu)^2}\,dx + \int_{\mu}^{\infty} (x-\mu)^{2r+1} e^{-\frac{1}{2\sigma^2}(x-\mu)^2}\,dx\right] \\
&= \frac{1}{\sigma\sqrt{2\pi}}[I_1 + I_2] \text{ (say)}.
\end{aligned}$$

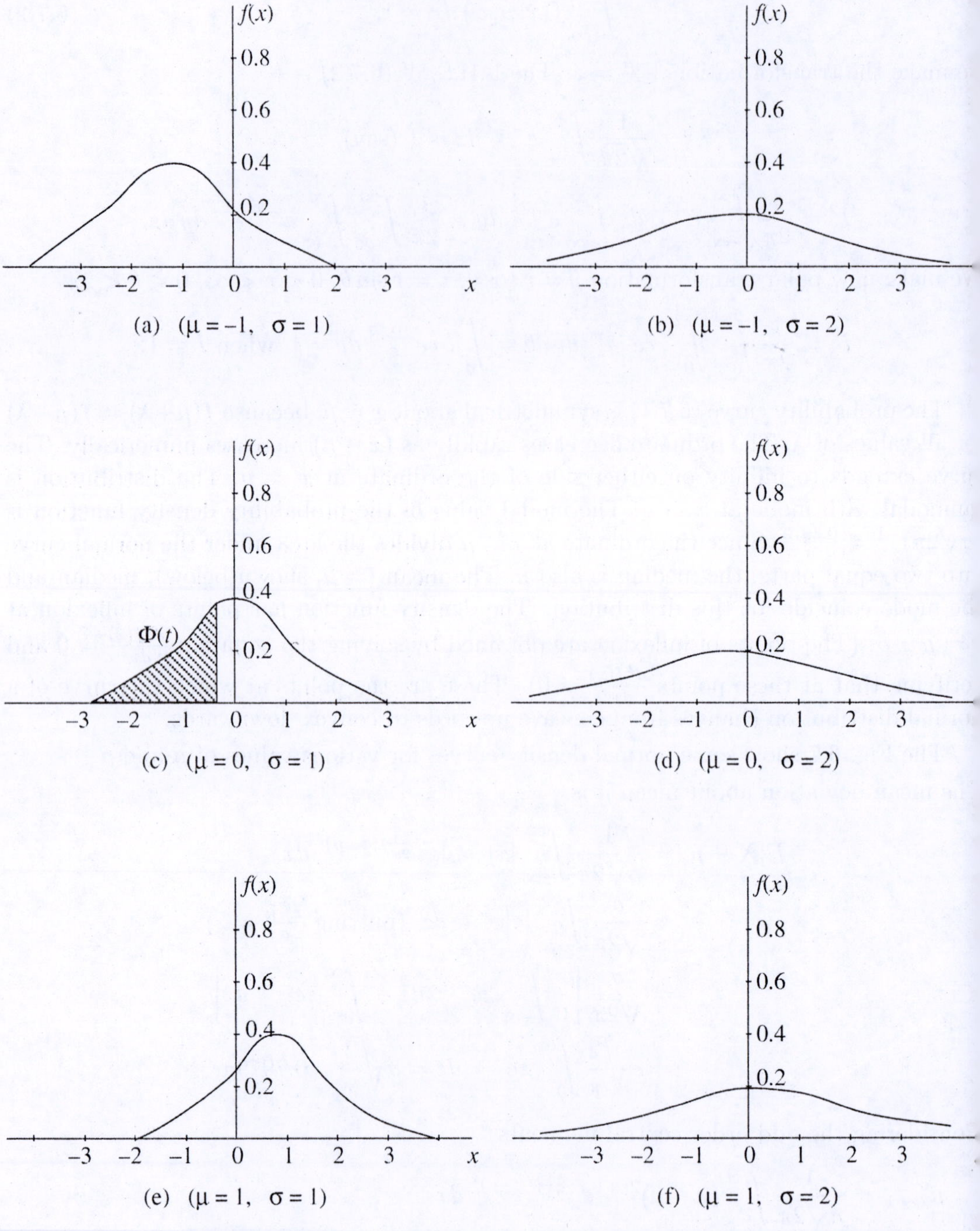

Fig. 6.7 Normal Density

In I_1 put $\mu - x = y$ and in I_2 put $x - \mu = y$, when $I_1 = -I_2$ and central moments of all odd orders,

$$\mu_{2r+1} = 0, r = 1, 2, \cdots. \tag{6.7.3}$$

Similarly,

$$\mu_{2r} = \frac{2}{\sigma\sqrt{2\pi}} \int_{-\infty}^{\infty} (x-\mu)^{2r} e^{-\frac{1}{2\sigma^2}(x-\mu)^2}\, dx. \tag{6.7.4}$$

Putting $\frac{x-\mu}{\sigma} = z$, R.H.S. of (6.7.4) reduces to

$$\begin{aligned}\sqrt{\frac{2}{\pi}}\sigma^{2r} \int_0^{\infty} z^{2r} e^{-\frac{z^2}{2}}\, dz &= \frac{2^r}{\sqrt{\pi}}\sigma^{2r}\Gamma\left(r+\frac{1}{2}\right) \quad \text{[using gamma function in (6.4.1)]}\\ &= \frac{\sigma^{2r}}{\sqrt{\pi}}2^r\left(r-\frac{1}{2}\right)\left(r-\frac{3}{2}\right)\cdots\frac{1}{2}\Gamma\frac{1}{2} \quad \text{[using (6.4.3) and (6.4.4)]}\\ &= \sigma^{2r}(2r-1)(2r-3)\cdots 3.1.\end{aligned}$$

In particular,

$$\mu_1' = \mu, \;\; \mu_2 = \sigma^2, \;\; \mu_4 = 3\sigma^4, \;\; \beta_1 = 0, \;\; \beta_2 = 3.$$

The moment generating function about $x = 0$ is

$$M_X(t) = e^{\mu t + \frac{t^2\sigma^2}{2}} \quad \text{(vide example 4.3.5).} \tag{6.7.5}$$

The m.g.f. about mean, μ is

$$M_{X-\mu}(t) = e^{\frac{1}{2}t^2\sigma^2}. \tag{6.7.6}$$

The cumulant generating function about $x = 0$ is

$$K_X(t) = \log_e M_X(t) = \mu t + \frac{1}{2}t^2\sigma^2.$$

Consequently

$$k_1 = \mu, \;\; k_2 = \sigma^2, \;\; k_r = 0, \;\; r \geq 3.$$

The characteristic function about $x = 0$ is

$$\psi_X(t) = e^{i\mu t - \frac{t^2\sigma^2}{2}}. \tag{6.7.7}$$

The c.f. about mean is

$$\psi_{X-\mu}(t) = e^{-\frac{1}{2}t^2\sigma^2}. \tag{6.7.8}$$

When X is normally distributed with mean μ and standard deviation σ it is customary to write X is distributed as $N(\mu, \sigma)$ or $X \sim N(\mu, \sigma)$.

Theorem 6.7.1 *If $X_1, \cdots, X_n$ are independent normal variables, X_i being $N(\mu_i, \sigma_i$ then the variate $Y = \sum_{i=1}^{n} a_i X_i$ is an $N(\mu', \sigma')$ r.v. with*

$$\mu' = \sum_{i=1}^{n} a_i \mu_i, \quad \sigma' = \sqrt{\sum_{i=1}^{n} a_i^2 \sigma_i^2},$$

where a_i's are constants.

Proof: The characteristic function of Y is

$$\begin{aligned}\psi_Y(t) &= E\left(e^{itY}\right) = E\left(e^{it\Sigma_{i=1}^{n} a_i X_i}\right) \\ &= E\left\{\prod_{i=1}^{n} e^{ita_i X_i}\right\} = \prod_{i=1}^{n} E\left(e^{ita_i X_i}\right) \quad \text{(by virtue of independence)} \\ &= \prod_{i=1}^{n} \psi_{a_i X_i}(t) = \prod_{i=1}^{n} \left(e^{a_i \mu_i t - \frac{1}{2} t^2 \sigma_i^2 a_i^2}\right) = e^{t\mu' - \frac{1}{2} t^2 \sigma'^2}\end{aligned}$$

which is the characteristic function of an $N(\mu', \sigma')$ r.v. Hence the result follows by th uniqueness property of characteristic function (theorem 4.6.1).

Corollary 6.7.1 *Let $X_1, \cdots, X_n$ be independent normal variables, X_i being $N(\mu_i, \sigma_i$ Then the sum $S_n = \sum_{i=1}^{n} X_i$ is an $N(\mu', \sigma')$ variable where*

$$\mu' = \sum_{i=1}^{n} \mu_i \quad \textit{and} \quad \sigma' = \sqrt{\sum_{i=1}^{n} \sigma_i^2}.$$

Corollary 6.7.2 *If $X_1, \cdots, X_n$ are independently and identically distributed $N(\mu, \sigma$ variables, the arithmetic mean $\bar{X} = \frac{S_n}{n}$ is an $N\left(\mu, \frac{\sigma}{\sqrt{n}}\right)$ r.v.*

The variable $\frac{X-\mu}{\sigma} = \tau$ (say) is called a standard or unit normal variable. Its p.d.f. is

$$\frac{1}{\sqrt{2\pi}} e^{-\frac{\tau^2}{2}}. \tag{6.7.9}$$

Clearly $E(\tau) = 0, V(\tau) = 1,$

$$M_\tau(t) = e^{\frac{t^2}{2}}, \quad \psi_\tau(t) = e^{-\frac{t^2}{2}}. \tag{6.7.10}$$

The ordinate (6.7.9) of the unit normal variate is denoted as $\phi(\tau)$. Clearly $\phi(-t) = \phi(t$ for any $t > 0$. Cumulative distribution function of τ,

$$F_\tau(u) = P[\tau \le u] = \frac{1}{\sqrt{2\pi}} \int_{-\infty}^{u} e^{-\frac{\tau^2}{2}} d\tau \tag{6.7.11}$$

is denoted as $\Phi(u)$, $\Phi(u)$ represents the area of a standard normal curve to the left of the ordinate at $\tau = u$ [Fig. 6.7(c)]. Clearly

$$\Phi(0) = 0.5,\ \ \Phi(\infty) = 1,\ \ \Phi(-t) = 1 - \Phi(t),\ \ P[|\tau| < t] = 2\Phi(t) - 1$$

for any $t > 0$. The ordinates $\phi(t)$ and the area $\Phi(t)$ under the curve (6.7.11) up to the left of the ordinate at t have been extensively tabulated and are reproduced in tables 1 and 2 respectively at the appendix. It is seen

$$P\{|\tau| > 1\} = 0.318,\ \ P\{|\tau| > 2\} = 0.046,\ \ P\{|\tau| > 3\} = 0.0027.$$

Hence for a $N(\mu, \sigma)$ variable X, $P\{|X - \mu| > 3\sigma\} = 0.002$. Thus practically a normal distribution is concentrated within three standard deviation units from the mean. It is also seen

$$\Phi(1.96) = 0.9750,\ \ \Phi(2.58) = 0.9951.$$

If X is $N(\mu, \sigma)$, for any two integers $a < b$,

$$P(a \le X \le b) = \frac{1}{\sigma\sqrt{2\pi}} \int_a^b e^{-\frac{(x-\mu)^2}{2\sigma^2}}\, dx = \frac{1}{\sqrt{2\pi}} \int_{\frac{a-\mu}{\sigma}}^{\frac{b-\mu}{\sigma}} e^{-\frac{\tau^2}{2}}\, d\tau,\ \ \tau = \frac{x-\mu}{\sigma}$$

$$= \Phi\left(\frac{b-\mu}{\sigma}\right) - \Phi\left(\frac{a-\mu}{\sigma}\right). \tag{6.7.12}$$

Normal distribution arises as a reasonable approximation to many random phenomena happening in nature. Many distributions such as those relating to measurements on plant and animal, gradation of students in test are approximately of the normal form. Many theoretical reasons for the use of the normal distribution are based on the forms of *central limit theorems* (discussed in chapter 9). These theorems state conditions under which the distribution of the standardised sum of random variables tends to a unit normal distribution as the number of variables in the sum increases. Many theoretical distributions, often of a complex nature, are approximated by a nomal distribution with the same expected value and standard deviation.

▶ **Example 6.7.1** *The results of a particular examination are as follows:*

RESULT	PERCENTAGE OF STUDENTS
Passed with distinction	15
Passed without distinction	42
Failed	43

It is known a candidate fails if he obtains less than 40% marks while he must obtain at least 75% marks in order to pass with distinction. Hence determine the mean and standard deviation of the distribution of marks assuming it to be normal.

Let μ and σ be the mean and standard deviation of the distribution of X, the marks of a student. Hence

$$F_X(40) = \Phi\left(\frac{40-\mu}{\sigma}\right) = 0.43 = \Phi(-0.18), \quad F_X(75) = \Phi\left(\frac{75-\mu}{\sigma}\right) = 0.85 = \Phi(1.04)$$

because 43% of students secured less than 40 marks, 85% of students secured less than 75 marks. Hence

$$\frac{40-\mu}{\sigma} = -0.18, \quad \frac{75-\mu}{\sigma} = 1.04 \text{ when } \mu = 45.16, \quad \sigma = 28.68.$$

Theorem 6.7.2 (DeMoivre-Laplace theorem). *Let X be a binomial variate with mean np and variance npq. If $n \to \infty$, the probability distribution of the standardised binomial variable $z = \frac{(X-np)}{\sqrt{npq}}$ tends to that of a unit normal variate.*

Proof:

$$P(X=r) = P(n,r) = \frac{n!}{r!(n-r)!}p^r q^{n-r}.$$

Applying Stirling's approximation (6.4.9) to factorials,

$$\lim_{n\to\infty} P(n,r) = \frac{e^{-n}n^{n+\frac{1}{2}}p^r q^{n-r}}{\sqrt{2\pi}e^{-r}r^{r+\frac{1}{2}}e^{-(n-r)}(n-r)^{n-r+\frac{1}{2}}} = \frac{A}{\sqrt{2\pi npq}} \text{ (say).}$$

where

$$A = \left(\frac{np}{r}\right)^{r+\frac{1}{2}}\left(\frac{nq}{n-r}\right)^{n-r+\frac{1}{2}}$$

$$= \left(1 + z_r\sqrt{\frac{q}{np}}\right)^{-(np+z_r\sqrt{npq}+\frac{1}{2})}\left(1 - z_r\sqrt{\frac{p}{nq}}\right)^{-(nq-z_r\sqrt{npq}+\frac{1}{2})},$$

where $z_r = \frac{(r-np)}{\sqrt{npq}}$.

$$\log_e A = -\left(np + z_r\sqrt{npq} + \frac{1}{2}\right)\log_e\left(1 + z_r\sqrt{\frac{q}{np}}\right)$$
$$-\left(nq - z_r\sqrt{npq} + \frac{1}{2}\right)\log_e\left(1 - z_r\sqrt{\frac{p}{nq}}\right).$$

Assuming $|z_r| < \min\left(\sqrt{\frac{np}{q}}, \sqrt{\frac{nq}{p}}\right)$,

$$\log_e A = -\left(np + z_r\sqrt{npq} + \frac{1}{2}\right)\left[z_r\sqrt{\frac{q}{np}} - \frac{z_r^2 q}{2np} + \frac{z_r^3 q^{\frac{3}{2}}}{3(np)^{\frac{3}{2}}} + 0(n^{-2})\right]$$
$$+\left(nq - z_r\sqrt{npq} + \frac{1}{2}\right)\left[z_r\sqrt{\frac{p}{nq}} + \frac{z_r^2 p}{2nq} + \frac{z_r^3 p^{\frac{3}{2}}}{3(nq)^{\frac{3}{2}}} + 0(n^{-2})\right]$$
$$= -\frac{z_r^2}{2} - \frac{z_r(q-p)}{2\sqrt{npq}} + \frac{z_r^2}{4npq}(p^2+q^2) + \frac{z_r^3}{6\sqrt{npq}}(q-p)$$
$$-\frac{z_r^3}{6(npq)^{\frac{3}{2}}}(q^3-p^3) + 0(n^{-2}). \tag{6.7.13}$$

As $n \to \infty$, (6.7.13) $\to -\frac{z_r^2}{2}$.

As r takes values $0, 1, \cdots, n$ and $n \to \infty$, z_r takes values between $-\infty$ and ∞. Also as r increases by value 1, z_r increased by the amount $\frac{1}{\sqrt{npq}}$, which for large n, would be taken as dz. Hence z_r can be taken as a continuous variable taking values between $-\infty$ and ∞ and

$$\lim_{n\to\infty} P(n, r) = \frac{1}{\sqrt{2\pi}} e^{-\frac{z_r^2}{2}} \, dz, -\infty \le z_r \le \infty. \tag{6.7.14}$$

For an alternative proof see section 9.2.

Corollary 6.7.3 *Let X be a binomial variable with parameters (n, p). For a fixed a and b,*

$$P[a \le X \le b] \to \Phi\left(z_b + \frac{1}{2}\right) - \Phi\left(z_a - \frac{1}{2}\right) \tag{6.7.15}$$

as $n \to \infty$.

Proof: By (6.7.14)

$$P[a \le X \le b] \simeq \sum_{r=a}^{b} \phi(z_r) \tag{6.7.16}$$

as $n \to \infty$.

By mean value theorem of integral calculus

$$\left[\int_a^b f(x)\, dx = (b-a) f(x_0) \text{ for an } x_0 \in [a, b]\right],$$

$$\Phi\left(z_r + \frac{1}{2}\right) - \Phi\left(z_r - \frac{1}{2}\right) = dz\phi(z_r') \text{ for some } z_r - \tfrac{1}{2} < z_r' < z_r + \tfrac{1}{2}.$$

Hence

$$\phi(z_r)\, dz = dz\phi(z_r')e^{\frac{1}{2}(z_r'^2 - z_r^2)} = \left[\Phi\left(z_r + \frac{1}{2}\right) - \Phi\left(z_r - \frac{1}{2}\right)\right] e^{\frac{1}{2}(z_r'^2 - z_0^2)}.$$

Now

$$\frac{1}{2}|z_r'^2 - z_r^2| = \frac{1}{2}|z_r' - z_r||z_r' + z_r| < dz\left[|z_r| + \frac{dz}{4}\right] < \epsilon$$

for sufficiently large n. Hence

$$\phi(z_r)\, dz \simeq \Phi\left(z_r + \frac{1}{2}\right) - \Phi\left(z_r - \frac{1}{2}\right)$$

when the result follows from (6.7.16).

Equation (6.7.15) shows that for large n, an approximate value for the probability that a binomial random variable falls in an interval can be obtained from the standard normal distribution.

The following approximations hold for $\Phi(x)$.

(a) $\frac{1}{2}\left[1+\sqrt{1-e^{\frac{-x^2}{2}}}\right] \leq \Phi(x) \leq \frac{1}{2}\left[1+\sqrt{1-e^{-x^2}}\right]$

(b) $\Phi(x) \simeq \frac{1}{2}\left[1+\left\{1-\exp\left(-\frac{2x^2}{\pi}\right)\right\}^{\frac{1}{2}}\right].$

The following property holds.

Lemma 6.7.1 *If X, Y are independent random variables whose sum $X+Y$ is normal distributed, then both X, Y follow normal distribution.*

A proof of this lemma is given in Feller (1971), vol. II, section 15.8.

6.7.1 Truncated Normal Distribution

A r.v. X has a doubly truncated normal distribution if its p.d.f. is given by

$$f_t(x;\mu,\sigma) = \frac{\frac{1}{\sigma\sqrt{2\pi}}e^{-\frac{1}{2\sigma^2}(x-\mu)^2}}{\int_A^B \frac{1}{\sigma\sqrt{2\pi}}e^{-\frac{1}{2\sigma^2}(x-\mu)^2}\,dx}, \quad (A \leq x \leq B)$$

$$= \left[\frac{1}{\sigma}\Phi\left(\frac{x-\mu}{\sigma}\right)\right] \Big/ T \qquad (6.7.1\text{'}$$

where $T = \Phi\left(\frac{B-\mu}{\sigma}\right) - \Phi\left(\frac{A-\mu}{\sigma}\right)$.

The distribution is singly truncated from below or above if B is replaced by ∞ or is replaced by $-\infty$ respectively. The expected value and variance of the distribution i (6.7.17) is

$$E(X) = \mu + \frac{\sigma}{T}[\phi_1 - \phi_2]$$

$$V(X) = \sigma^2\left[1+\frac{\left(\frac{A-\mu}{\sigma}\right)\phi_1 - \left(\frac{B-\mu}{\sigma}\right)\phi_2}{T} - \left\{\frac{\phi_1-\phi_2}{T}\right\}^2\right]$$

where $\phi_1 = \phi\left(\frac{A-\mu}{\sigma}\right)$, $\phi_2 = \phi\left(\frac{B-\mu}{\sigma}\right)$.

► **Example 6.7.2** *A normal distribution $f(x) = \frac{1}{\sigma\sqrt{2\pi}}e^{-\frac{1}{2\sigma^2}(x-\mu)^2}, -\infty < x < \infty$ truncated at $x = A$, values $x \leq A$ are discarded. Find the mean and the variance the truncated distribution.*

$$\mu_1' = \left[\int_A^\infty x f(x)\,dx\right] \Big/ \left[\int_A^\infty f(x)\,dx\right]$$

$$= \left[\int_{A_0}^\infty (\mu + z\sigma)e^{-\frac{z^2}{2}}\,dz\right] \Big/ \left[\int_{A_0}^\infty e^{-\frac{z^2}{2}}\,dz\right] \quad (\text{where } z = \tfrac{x-\mu}{\sigma}, A_0 = \tfrac{A-\mu}{\sigma})$$

$$= \mu + \frac{\sigma}{M\sqrt{2\pi}} e^{-\frac{A_0^2}{2}}, \quad \text{where } M = 1 - \Phi(A_0)$$

$$\mu_2' = \left[\int_A^\infty x^2 f(x)\,dx\right] \Big/ \left[\int_A^\infty f(x)\,dx\right] = \frac{1}{M\sqrt{2\pi}} \int_{A_0}^\infty (\mu + \sigma_z)^2 e^{-\frac{z^2}{2}}\,dz$$

$$= \mu^2 + \frac{2\mu\sigma}{M\sqrt{2\pi}} e^{-\frac{A_0^2}{2}} + \frac{\sigma^2}{M\sqrt{2\pi}} \int_{A_0}^\infty z^2 e^{-\frac{z^2}{2}}\,dz.$$

Now

$$\int_{A_0}^\infty z^2 e^{\frac{-z^2}{2}}\,dz = z \int e^{\frac{-z^2}{2}} d\left(\frac{z^2}{2}\right)\Bigg|_{A_0}^\infty - \int_{A_0}^\infty \left[\int e^{\frac{-z^2}{2}} d\left(\frac{z^2}{2}\right)\right] dz$$

$$= -ze^{\frac{-z^2}{2}}\Bigg|_{A_0}^\infty + \int_{A_0}^\infty e^{\frac{-z^2}{2}}\,dz = A_0 e^{\frac{-A_0^2}{2}} + M\sqrt{2\pi}.$$

Hence

$$\mu_2 = \mu_2' - \mu_1'^2 = \sigma^2 \left[1 + \frac{A_0 e^{\frac{-A_0^2}{2}}}{M\sqrt{2\pi}} - \frac{1}{2\pi M^2} e^{-A_0^2}\right].$$

6.8 Log-Normal Distribution

A random variable X is said to have a log-normal distribution if $\log X$ has a normal distribution ($-\infty < \log X < \infty$). If $\log X$ has mean μ and s.d. σ, the probability differential of $\log X$ is

$$f(\log x)d(\log x) = \frac{1}{\sigma\sqrt{2\pi}} e^{-\frac{1}{2\sigma^2}(\log x - \mu)^2}\, d(\log x), \quad -\infty < \log x < \infty. \tag{6.8.1}$$

and hence

$$f(x)\,dx = \frac{1}{\sigma^x\sqrt{2\pi}} e^{-\frac{1}{2\sigma^2}(\log x - \mu)^2}\, dx, 0 < x < \infty \tag{6.8.2}$$

$$\mu_r' = \int_0^\infty \frac{x^r}{\sigma^x\sqrt{2\pi}} e^{-\frac{1}{2\sigma^2}(\log x - \mu)^2}\, dx. \tag{6.8.3}$$

Substituting $\dfrac{(\log x - \mu)}{\sigma} = z$

$$\mu_r' = \frac{1}{\sqrt{2\pi}} \int_{-\infty}^{\infty} e^{r\sigma z + \mu r - \frac{z^2}{2}}\, dz$$

$$= e^{\mu r + \frac{r^2\sigma^2}{2}} \cdot \frac{1}{\sqrt{2\pi}} \int_0^{\infty} e^{-\frac{1}{2}(z - r\sigma)^2}\, dz = e^{\mu r + \frac{r^2\sigma^2}{2}}; \tag{6.8.4}$$

μ_r', therefore, increases very rapidly with r.

In particular,

$$\mu_1' = e^{\mu + \frac{\sigma^2}{2}}, \;\; \mu_2' = e^{2\mu + 2\sigma^2}, \;\; \mu_3' = e^{3\mu + \frac{9}{2}\sigma^2}, \;\; \mu_4' = e^{4\mu + 8\sigma^2}.$$

Denoting $e^{\sigma^2} = \alpha$,

$$\mu_2 = \alpha(\alpha - 1)e^{2\mu}, \;\; \mu_3 = \alpha^{\frac{3}{2}}(\alpha - 1)^2(\alpha + 2)e^{3\mu},$$

$$\mu_4 = \alpha^2(\alpha - 1)^2(\alpha^4 + 2\alpha^3 + 3\alpha^2 - 3)e^{4\mu}.$$

In general

$$\mu_r = \frac{\mu_2^{\frac{n}{2}}}{(\alpha - 1)^{\frac{n}{2}}} \sum_{j=0}^{r} (-1)^j \binom{r}{j} \alpha^{\frac{(r-j)(r-j-1)}{2}} \tag{6.8.5}$$

$$\sqrt{\beta_1} = \sqrt{(\alpha - 1)(\alpha + 2)},$$

$$\beta_2 = \alpha^4 + 2\alpha^3 + 3\alpha^2 - 3.$$

The curve is therefore positively skewed. Note that neither $\sqrt{\beta_1}$, nor β_2 depend on μ.

The coefficient of variation is $\sqrt{(\alpha - 1)}$ and does not depend on μ.

The distribution is unimodal because $\dfrac{df(x)}{dx} = 0$ gives

$$-\frac{1}{\sigma x^2 \sqrt{2\pi}} e^{-\frac{1}{2\sigma^2}(\log x - \mu)^2} \left[\frac{1}{\sigma^2}(\log x - \mu) + 1\right] = 0 \tag{6.8.6}$$

when the mode is $x_m = e^{(\mu - \sigma^2)}$.

The value X_α such that $P[X \le X_\alpha] = \alpha$ is related to the corresponding percentile z_α of the unit normal distribution

$$z = \frac{\log X - \mu}{\sigma} \tag{6.8.7}$$

as follows:

$$P[X \leq X_\alpha] = \alpha \Rightarrow P[\log X \leq \log X_\alpha] = \alpha$$

$$\Rightarrow P\left[\frac{(\log X - \mu)}{\sigma} \leq \frac{(\log X_\alpha - \mu)}{\sigma}\right] = \alpha$$

when

$$z_\alpha = \frac{\log X_\alpha - \mu}{\sigma} \quad \text{or} \quad X_\alpha = e^{(\mu + \sigma z_\alpha)}.$$

In particular for $\alpha = 0.50$, $z_\alpha = 0$, when the median of the distribution is e^μ. For this distribution, therefore,

$$\mu_1' > \text{median} > \text{mode} \quad \text{and} \quad \frac{\text{mode}}{\mu_1'} = \left(\frac{\text{median}}{\mu_1'}\right)^3.$$

The standardised log-normal distribution is

$$\xi = \frac{X - \mu_1'}{\sqrt{\mu_2}} = \frac{e^{z\sigma} - e^{\frac{\sigma^2}{2}}}{\left[e^{\sigma^2}(e^{\sigma^2} - 1)\right]^{\frac{1}{2}}}, \tag{6.8.8}$$

where z is as defined in (6.8.8). As $\sigma \to 0$, distribution of ξ tends to a unit normal distribution.

Fig. 6.8 shows some standardised log-normal distributions.

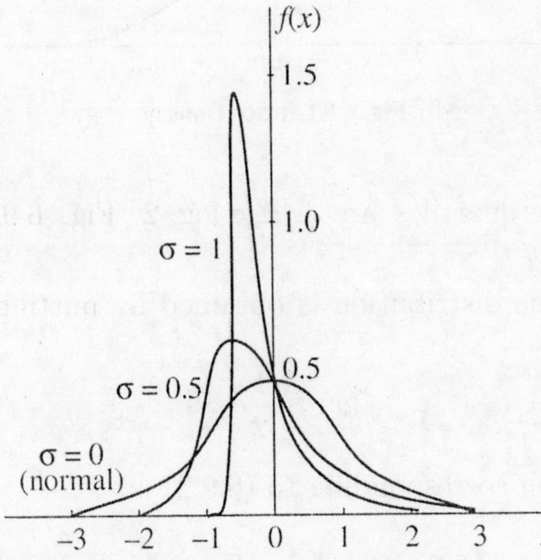

Fig. 6.8 Standardized Log-normal Distribution

A log-normal distribution often gives a good approximation to a normal distribution which has a small absolute value of coefficient of variation.

6.9 Double Exponential (Laplace) Distribution

A r.v. X is defined to have a double exponential distribution if its p.d.f. is given by

$$f(x;\mu,\sigma) = \frac{1}{2\sigma}e^{-\frac{|x-\mu|}{\sigma}} \quad (\sigma > 0) \tag{6.9.1}$$

when $-\infty < \mu < \infty$ and $\sigma > 0$. Here $E(X) = \mu, V(X) = 2\sigma^2$.

The distribution is symmetrical about $x = 0. \sqrt{\beta_1} = 0, \beta_2 = 6$. The mean deviation is

$$E(|X|) = 1.$$

The mode is at $x = \mu$. The cumulative distribution function is

$$F_X(x) = \begin{cases} \frac{1}{2}\exp\left[-\frac{(\mu - x)}{\sigma}\right] & (x \leq \mu), \\ 1 - \frac{1}{2}\exp\left[-\frac{(x-\mu)}{\sigma}\right] & (x \geq \mu). \end{cases}$$

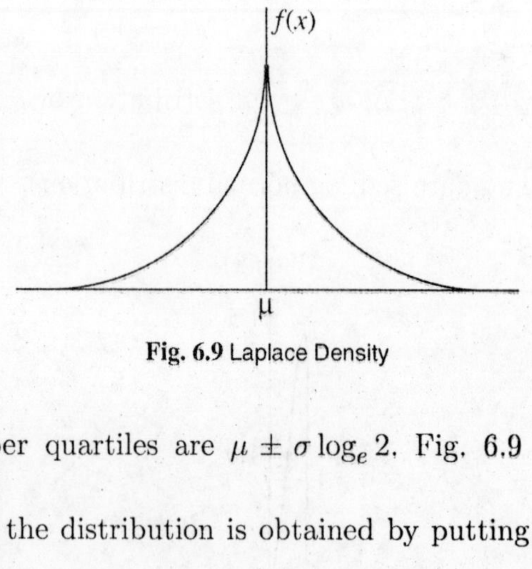

Fig. 6.9 Laplace Density

The lower and upper quartiles are $\mu \pm \sigma \log_e 2$. Fig. 6.9 shows the form of the distribution.

A standard from of the distribution is obtained by putting $\mu = 0, \sigma = 1$ when the p.d.f. is

$$f(x) = \frac{1}{2}e^{-|x|}. \tag{6.9.2}$$

The characteristic function corresponding to (6.9.2) is

$$\psi_X(t) = E(e^{itX}) = \frac{1}{1+t^2} \text{ (vide example 4.6.3)}$$

The m.g.f. is

$$M_X(t) = \frac{1}{1-t^2}.$$

he central moments are

$$\mu_r = \begin{cases} 0, & \text{if } r \text{ is odd} \\ r!, & \text{if } r \text{ is even.} \end{cases}$$

he c.g.f. is $K_X(t) = -\log(1-t^2)$.

hus

$$k_r = \begin{cases} 0, & \text{if } r \text{ is odd} \\ 2[(r-1)!], & \text{if } r \text{ is even.} \end{cases}$$

.10 Pareto Distribution

r.v. X is said to have a Pareto distribution with parameters α, β if its p.d.f. is given y

$$f(x) = \frac{\alpha\beta^\alpha}{x^{\alpha+1}} \quad \alpha > 0, \;\; x \geq \beta > 0. \tag{6.10.1}$$

he c.d.f. is

$$F_X(x) = 1 - \left(\frac{\beta}{x}\right)^\alpha.$$

rovided $r < \alpha$,

$$\mu'_r = \frac{\alpha\beta^r}{\alpha - r}.$$

n particular,

$$\mu'_1 = \frac{\alpha\beta}{\alpha-1}(\alpha > 1), \;\; \mu_2 = \frac{\alpha\beta^2}{(\alpha-1)^2(\alpha-2)} \;\; (\alpha > 2).$$

The mode of the distribution is at $x = \beta$. The median is $x_m = \beta 2^{\frac{1}{\alpha}}$. The population eometric mean is $\beta \cdot e^{\frac{1}{\alpha}}$. The population harmonic mean is $\beta\left(1 + \frac{1}{\alpha}\right)$.

In many socio-economic investigations, like distribution of population sizes, occurrence of natural resources, size of firms, distribution of income over population, pareto urve gives a good fit to the data towards the extremes of the range, while log-normal istribution fits well over the other part of the data.

6.11 Weibull Distribution

A r.v. X has an Weibull distribution with parameters $c(>0), a(>0), \mu$ if its p.d.f. is

$$f(x; c, a, \mu) = \frac{c}{a}\left(\frac{x-\mu}{a}\right)^{c-1} \exp\left[-\left(\frac{x-\mu}{a}\right)^c\right], \;\; x > \mu. \tag{6.11.1}$$

The standard form of the distribution is obtained by putting $\mu = 0, a = 1$ when th density is

$$f(x) = cx^{c-1}e^{-x^c} \quad (x > 0) \qquad (6.11.2$$

$$E(X) = \left\lceil\left(\tfrac{1}{c}+1\right), \quad V(X) = \left\lceil\left(\tfrac{2}{c}+1\right) - \left(\left\lceil\left(\tfrac{1}{c}+1\right)\right)^2.$$

For $c = 1$ (6.11.2) reduces to the exponential density.
The distribution finds frequent use in the theory of reliability.

6.12 Extreme Value Distribution

The p.d.f. for this distribution with parameters $\theta(> 0), \mu$ is

$$f(x) = \theta^{-1}e^{-\frac{(x-\mu)}{\theta}} \exp\left[-e^{-\frac{(x-\mu)}{\theta}}\right]$$

$$E(X) = \mu - 0.57721\sigma, \quad V(X) = \pi^2\frac{\theta^2}{6}. \qquad (6.12.1$$

mode $= \mu$, median $= \mu + \sigma \log\log 2$.

6.13 Logistic Distribution

The c.d.f. for this distribution is

$$F_X(x) = \frac{1}{1+e^{-\frac{(x-\mu)}{\sigma}}}, \quad \sigma > 0. \qquad (6.13.1$$

The corresponding p.d.f. is

$$f_X(x) = \frac{e^{-(x-\mu)/\sigma}}{\sigma\left[1+e^{-\frac{(x-\mu)}{\sigma}}\right]^2}, \quad -\infty < x < \infty, \ \sigma > 0$$

$$E(X) = \mu, \quad V(X) = \frac{\sigma^2\pi^2}{3}. \qquad (6.13.2$$

The function (6.13.1) is widely used to represent growth functions with x representin time.

Exercise

1. Show that for the rectangular distribution $f(x) = \frac{1}{b}$, $a \le x \le a+b$ the m.g.f. i $e^{at}\frac{[e^{bt}-1]}{bt}$, $\mu_1' = a + \frac{b}{2}$ (= median), $\mu_2 = \frac{b^2}{12}$, $\mu_r = 0$ for r odd, $= \left(\frac{b}{2}\right)^r/(r+1)$ for even.

2. X has a negative exponential distribution with parameter λ. If $P[X \leq 1] = P[X > 1]$, find var (X).

3. $\bar{X}$ is mean of n independent $N(\mu, \sigma^2)$-variables. Show that for any

$$\epsilon > 0, \lim_{n\to\infty} P(|\bar{X} - \mu| \geq \epsilon) = 0.$$

[*Hints:* $\bar{X} \sim N\left(\mu, \frac{\sigma}{\sqrt{n}}\right); P(|\bar{X} - \mu| \geq \epsilon) = 1 - \frac{1}{\sqrt{2\pi}} \int_{-\epsilon\sqrt{n}/\sigma}^{\epsilon\sqrt{n}/\sigma} e^{-\frac{t^2}{2}} dt \left(t = \frac{(\bar{x}-\mu)\sqrt{n}}{\sigma}\right)$.

$\therefore \lim_{n\to\infty} P|\bar{X} - \mu| \geq \epsilon) = 1 - \frac{1}{\sqrt{2\pi}} \int_{-\infty}^{\infty} e^{-\frac{t^2}{2}} dt = 0.$]

4. The random variable X takes the values $0, 1, \cdots, n$ with

$$P(X = x) = \binom{n}{x} \frac{B(a + x, n + b - x)}{B(a, b)}.$$

Calculate the expectation and variance of X.

5. Show that for the triangular distribution (Fig. 6.10)

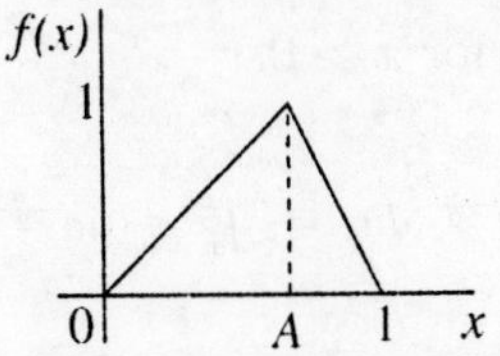

Fig. 6.10 Triangular Distribution

$$f(x) = \begin{cases} \dfrac{2x}{A}, & 0 \leq x \leq A \\ \dfrac{2(1-x)}{(1-A)}, & A \leq x \leq 1 \end{cases}$$

$$\mu_1' = \frac{A+1}{3}, \quad \mu_2 = \frac{1}{18}(1 - A + A^2).$$

6. Show that for Erlang distribution

$$f(x) = \frac{\left(\frac{x}{b}\right)^{c-1} e^{-\frac{x}{b}}}{b[(c-1)!]}, \quad 0 \leq x \leq \infty, \ b > 0, \ c(\text{integer}) > 0$$

m.g.f. is $(1 - bt)^{-c}, t > \frac{1}{b}; \mu_1' = bc, \mu_2 = b^2c, \sqrt{\beta_1} = \frac{2}{\sqrt{c}}, \beta_2 = 3 + \frac{6}{c}$, mode is $b(c - 1)$.

7. For a beta distribution with density

$$f(x) = \frac{1}{B(p,q)}(1 - x)^{p-1}x^{q-1}, \quad 0 < x < 1, \ p > 0, \ q > 0$$

verify that the harmonic mean is less than the arithmetic mean.

8. Show that for the power function distribution

$$f(x) = cx^{c-1}, \quad 0 \leq x \leq 1,$$

$$\mu_r' = \frac{c}{c+r}, \quad \mu_2 = \frac{c}{(c+2)(c+1)^2},$$

mode $= 1$ for $c > 1, 0$ for $c < 1$, median $= \left(\frac{1}{2}\right)^{\frac{1}{c}}$.

9. Let N be a non-negative random integer with geometric distribution, $P(N = n)$ $(1 - e^{-\mu})\, e^{-n\mu}$, $n = 0, 1, 2, \cdots$. Let M be a positive integer with p.m.f.

$$P(M = m) = \frac{e^{-\mu}\mu^m}{m!(1 - e^{-\mu})}, \quad m = 1, 2, \cdots.$$

Let $(X_1, \cdots, X_M)$ be M independent variables, each uniform over (0, 1). Sho that

$$Y = \mu\{N + \min(X_1, \cdots, X_N)\}$$

has the standard exponential distribution.

10. Show that for $x > 0$.

$$1-\Phi(x) = \frac{1}{\sqrt{2\pi}} e^{\frac{-x^2}{2}} \left\{ \frac{1}{x} - \frac{1}{x^3} + \frac{1.3}{x^5} - \frac{1.3.5}{x^7} + \cdots + (-1)^k \frac{1.3 \cdots (2k-1)}{x^{2k+1}} + \cdots \right\}$$

Hence,

$$1 - \Phi(x) < \frac{\phi(x)}{x}, \quad \text{for } x > 0. \quad (i$$

[*Hints:* $1 - \Phi(x) = \frac{1}{\sqrt{2\pi}} \int_x^\infty e^{\frac{-u^2}{2}} du$; also, $\int_x^\infty e^{\frac{-u^2}{2}} du = \int_x^\infty \frac{1}{u}(ue^{\frac{-u^2}{2}}) du = \frac{1}{x} e^{\frac{-x^2}{2}}$ $\int_x^\infty \frac{1}{u^3}(ue^{\frac{-u^2}{2}}) du$. (i

(Integrating by parts, taking $\frac{1}{u}$ as the first factor.)

Again, evaluating $\int_x^\infty \frac{1}{u^3}(ue^{\frac{-u^2}{2}}) du$ by integration by parts (with $\frac{1}{u^3}$ as the first factor) a proceeding similarly the first part follows.

From (iii), $\int_x^\infty e^{\frac{-u^2}{2}} du < \frac{1}{x} e^{-\frac{x^2}{2}}$ for $x > 0$. Hence the second part.]

Chapter 7

Probability Distribution on $\mathbb{R}^n$

7.1 Introduction

In chapter 2 we considered a r.v. X and its probability distribution on $\mathbb{R}^1$. In many experiments, however, observations are expressed as a number of several numerical quantities. In tossing a pair of coins, for example, observations are (x, y), $x(y)$ denoting the face value turned up by the first (second) die. In such cases we have, therefore, a vector of random variables $\boldsymbol{X} = (X_1, \cdots, X_n)$ where each X_i is a random variable. In this chapter we shall study the probability distribution of n-dimensional random vector $\boldsymbol{X}$ on $\mathbb{R}^n$. Section 7.2 deals with the concept of joint densities, marginal densities and conditional densities and the respective distribution functions. The concept of stochastic independence is also considered. Section 7.3 considers expectation of functions of random variables and moments, section 7.4 the generating functions. Conditional expectation, conditional variance, regression are considered in 7.5. The remaining two sections deal with multinomial distribution and bivariate normal distribution respectively.

7.2 Probability Distribution of a Random Vector

Definition 7.2.1 *Random Vector. Consider the probability space $(S, \mathcal{A}, P)$ arising out of a random experiment. A vector of functions $(X_1, \cdots, X_n)$ which maps S into $\mathbb{R}^n$ is called a random vector if and only if for each $a_i \in \mathbb{R}^1 (i = 1, \cdots, n)$,*

$$\{w : X_1(w) \leq a_1, \cdots, X_n(w) \leq a_n\} \subset \mathcal{A}, \tag{7.2.1}$$

i.e.,

$$\boldsymbol{X}^{-1}[I = \{(x_1, \cdots, x_n) : -\infty < x_i \leq a_i, i = 1, \cdots, n\}] \subset \mathcal{A}, \tag{7.2.2}$$

i.e., if $\boldsymbol{X}$ is a $\mathcal{A}$-measureable function. Here I is an n-dimensional interval.

► **Example 7.2.1** *A fair coin is tossed twice. Let us assign the score $1(2)$ for $H(T)$. Let X_1, X_2 denote respectively the number at the first toss, sum of scores at the two tosses. Here $\boldsymbol{X} = (X_1, X_2)$ and*

$$\boldsymbol{X}^{-1}\{(-\infty, a_1], (-\infty, a_2]\} = \begin{cases} \phi, & a_1 < 0, a_2 < 1 \\ (HH), & 0 < a_1 \leq 1, 1 < a_2 \leq 2 \\ (HT), & 0 < a_1 \leq 1, 2 < a_2 \leq 3 \\ (TH), & 1 < a_1 \leq 2, 2 < a_2 \leq 3 \\ (TT), & 1 < a_1 \leq 2, 3 < a_2 \leq 4 \\ S, & a_1 > 2, a_2 > 4. \end{cases}$$

Thus X is a random vector.

We shall first consider in details the two dimensional random vector (X, Y). The same abbreviation r.v. will be used to denote a random vector as well as a random variable.

Definition 7.2.2 *The joint cumulative distribution function of* (X, Y) *is defined as*

$$F_{X,Y}(x, y) = P[X \leq x, Y \leq y] \text{ for all } (x, y) \in \mathbb{R}^2. \tag{7.2.3}$$

As in the case of a single r.v., the following properties hold for $F_{X,Y}(.,.)$.

1. (a) $F_{X,Y}(-\infty, y) = \lim\limits_{x \to -\infty} F(x, y) = 0$ for all y;
   (b) $F_{X,Y}(x, -\infty) = \lim\limits_{y \to -\infty} F(x, y) = 0$ for all x;
2. $F_{X,Y}(\infty, \infty) = \lim\limits_{\substack{x \to \infty \\ y \to \infty}} F(x, y) = 1$;
3. If $x_1 < x_2,\ y_1 < y_2,\ P[x_1 < X \leq x_2,\ y_1 < Y \leq y_2] = F(x_2, y_2) - F(x_2, y_1) - F(x_1, y_2) + F(x_1, y_1) \geq 0$;
4. $F(x, y)$ is continuous from the right in each argument; that is, $\lim\limits_{\varepsilon \downarrow 0} F(x + \varepsilon, y) = \lim\limits_{\varepsilon \downarrow 0} F(x, y + \varepsilon) = F(x, y)$.

Any function satisfying the properties (1)–(4) above is a bivariate d.f.

Definition 7.2.3 *If* (X, Y) *takes only finite or countably infinite number of pair of values* (x, y) *in* $\mathbb{R}^2$ *then the r.v. is of discrete type.*

Let $(x_i, y_j), i = 1, 2, \cdots; j = 1, 2, \cdots$ be the values assumed by (X, Y) with probability

$$P[X = x_i, Y = y_j] = p_{ij}.$$

p_{ij} is called the joint p.m.f. of (X, Y) at the mass point (x_i, y_j) if

1. $p_{ij} \geq 0$ for all $i = 1, 2, \cdots;\ j = 1, 2, \cdots,$
2. $\sum\limits_i \sum\limits_j p_{ij} = 1.$

The d.f. is given by

$$F_{X,Y}(x,y) = P[X \le x, Y = y] = \sum_{x_i \le x} \sum_{y_j \le y} p_{ij}. \tag{7.2.4}$$

Every point (x_i, y_j) where $p_{ij} > 0$ is a jump-point of $F_{X,Y}(x,y)$ and p_{ij} is its jump at (x_i, y_j).

Definition 7.2.4 *A r.v. (X,Y) is of continuous type if $F_{X,Y}(x,y)$ is continuous, i.e., if there exists a function $f(x,y)(\ge 0)$ such that for every $(x,y) \in \mathbb{R}^2$,*

$$F_{X,Y}(x,y) = \int_{-\infty}^{x} \left[\int_{-\infty}^{y} f(u,v)\,dv\right] du. \tag{7.2.5}$$

If $F(x,y)$ has partial derivatives of order up to two at (x,y)

$$\frac{\partial^2 F(x,y)}{\partial x \partial y} = f(x,y). \tag{7.2.6}$$

The function $f(x,y)$ is called the joint p.d.f. of (X,Y).
Clearly

$$F_{X,Y}(\infty,\infty) = \lim_{\substack{x\to\infty \\ y\to\infty}} \int_{-\infty}^{x}\int_{-\infty}^{y} f(u,v)\,du\,dv = \int_{-\infty}^{\infty}\int_{-\infty}^{\infty} f(u,v)\,du\,dv = 1.$$

Again

$$\lim_{\substack{\delta x\to 0 \\ \delta y\to 0}} \frac{F(x+\delta x, y+\delta y) - F(x, y+\delta y) - F(x+\delta x, y) + F(x,y)}{\delta x \delta y} = \frac{\partial^2 F(x,y)}{\partial x \partial y}$$

i.e.,

$$\lim_{\substack{\delta x\to 0 \\ \delta y\to 0}} \frac{P[x < X \le x+\delta x, y < Y \le y+\delta y]}{\delta x \delta y} = f(x,y),$$

i.e.,

$$P[x < X \le x+dx,\ \ y < Y \le y+dy] = f(x,y)\,dx\,dy. \tag{7.2.7}$$

Also

$$P[x < X \le x+dx, y \le Y] = F(x+dx, y) - F(x,y) = \frac{\partial F}{\partial x}\,dx. \tag{7.2.8}$$

Similarly

$$P[X \le x, y < Y \le y+dy] = \frac{\partial F}{\partial y}\,dy. \tag{7.2.9}$$

For any two pair of real murnbers $(a_1,b_1),(a_2,b_2), a_i < b_i (i=1,2)$,

$$P[a_1 < X \le b_1, a_2 < Y \le b_2] = \int_{a_2}^{b_2}\left[\int_{a_1}^{b_1} f(u,v)\,du\right] dv. \tag{7.2.10}$$

► **Example 7.2.2** *A pair of fair dice are tossed. Let X denote the face value of the first die, Y the larger of two face values. Here (X, Y) is a discrete r.v. The joint p.m.f. is given in Table 7.1.*

Table 7.1

Y \ X	1	2	3	4	5	6	Total
1	1/36	0	0	0	0	0	1/36
2	1/36	2/36	0	0	0	0	3/36
3	1/36	1/36	3/36	0	0	0	5/36
4	1/36	1/36	1/36	4/36	0	0	7/36
5	1/36	1/36	1/36	1/36	5/36	0	9/36
6	1/36	1/36	1/36	1/36	1/36	1/6	11/36
Total	1/6	1/6	1/6	1/6	1/6	1/6	1

Thus $P(X = 2, Y = 4) = \frac{1}{36}$. The d.f. can be calculated from the above table. For example,

$$F_{X,Y}(3, 4) = \frac{12}{36}.$$

Definition 7.2.5 ***Marginal p.m.f.*** *If (X, Y) is a discrete r.v. with joint p.m.f.*

$$P(X = x_i, Y = y_j) = p_{ij}, \quad i = 1, 2, \cdots; \; j = 1, 2, \cdots$$

then $\{p_{io}\}$ where $p_{io} = \sum_j p_{ij}$ is called the marginal p.m.f. of X. Clearly

$$p_{io} = P(X = x_i) \geq 0 \;\; and \;\; \sum_i p_{io} = 1.$$

Similarly $\{p_{oj}\}$ where $p_{oj} = \sum_i p_{ij}$ is the marginal p.m.f. of Y. $p_{\cdot j} = P(Y = y_j)$.

For example 7.2.2, the column (row) totals give the marginal p.m.f. of $X(Y)$. Thus

$$p_{io} = \frac{1}{6}, \; i = 1, \cdots, 6; \; p_{01} = \frac{1}{36}, \; p_{02} = \frac{3}{36}, \; p_{03} = \frac{5}{36}, \; p_{04} = \frac{7}{36}, \; p_{05} = \frac{9}{36}, \; p_{06} = \frac{11}{36}.$$

Definition 7.2.6 ***Marginal p.d.f.*** *If (X, Y) is a continuous r.v. with joint p.d.f. $f(x, y)$, then*

$$f_X(x) = \int_{-\infty}^{\infty} f(x, y)\, dy \tag{7.2.11}$$

$$f_Y(y) = \int_{-\infty}^{\infty} f(x, y)\, dx \tag{7.2.12}$$

are called the marginal p.d.f. of X and the marginal p.d.f. of Y respectively.

early

$$f_X(x) \geq 0, \ f_Y(y) \geq 0, \ \int_{-\infty}^{\infty} f_X(x)\,dx = \int_{-\infty}^{\infty} f_Y(y)\,dy = 1.$$

Example 7.2.3 *Let (X, Y) have joint p.d.f.*

$$f_{X,Y}(x, y) = 2, \ 0 < x < y < 1.$$

$$f_X(x) = \int_x^1 f(x, y)\,dy = 2(1 - x), \quad 0 < x < 1$$

$$f_Y(y) = \int_0^y 2\,dx = 2y, \quad 0 < y < 1.$$

Example 7.2.4 *(X, Y) is distributed with a constant density inside a square R of side Find $f(x, y), F(x, y)$ and the marginal p.d.f.s.*

$$f(x, y) = \begin{cases} \dfrac{1}{b^2} & \text{for } (x, y) \in R \\ 0 & \text{otherwise} \end{cases}$$

$$F(x, y) = 0 \text{ for } x \leq 0, \ y \leq 0$$

$$F(x, y) = \int_0^x \int_0^y \frac{dx\,dy}{b^2} = \frac{xy}{b^2} \text{ for } 0 < x \leq b, 0 < y \leq b$$

$$F(x, y) = \frac{by}{b^2} = \frac{y}{b} \text{ for } x > b, 0 < y \leq b$$

$$F(x, y) = \frac{x}{b} \text{ for } 0 < x \leq b, y > b$$

$$F(x, y) = 1 \text{ for } x > b, y > b$$

$$f_X(x) = \int_0^b \frac{dy}{b^2} = \frac{1}{b} \text{ for } x \in (0, b).$$

imilarly, $f_Y(y) = \frac{1}{b}$ for $y \in (0, b)$.

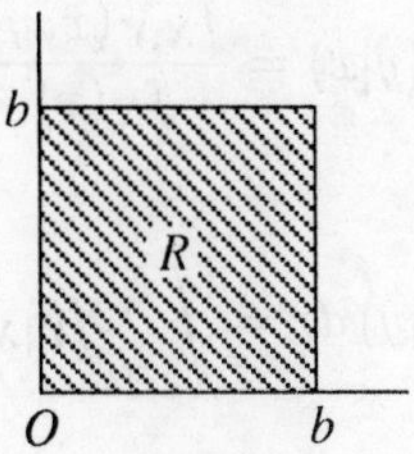

Fig. 7.1

Definition 7.2.7 ***Conditional p.m.f.*** *If (X, Y) is a discrete r.v. with $P(Y = y_j) > 0$ hen*

$$P(X = x_i | Y = y_j) = \frac{P(X = x_i, Y = y_j)}{P(Y = y_j)} = \frac{p_{ij}}{p_{.j}} = p_{X|Y}(i|j) \ (say). \qquad (7.2.13)$$

The set of values $\left\{\frac{p_{ij}}{p_{\cdot j}}\right\}$ is called the conditional p.m.f. of X given that $Y = y_j$. Clearly

$$\frac{p_{ij}}{p_{\cdot j}} \geq 0, \quad \sum_i \left(\frac{p_{ij}}{p_{\cdot j}}\right) = 1.$$

Similarly the set of values$\left\{\frac{p_{ij}}{p_{i\cdot}}\right\}$ is the conditional p.m.f. of Y given that $X = x_i$. W shall denote $\frac{p_{ij}}{p_{i\cdot}}$ as $P_{Y|X}(j|i)$.

Note that joint p.m.f.

$$p_{ij} = p_{X|Y}(i|j)p_{\cdot j} = p_{i\cdot}p_{Y|X}(j|i). \qquad (7.2.1$$

Definition 7.2.8 ***Conditional d.f.*** *(discrete case). If X, Y are discrete r.v.'s wi $P(Y = y_j) > 0$, the conditional d.f. of X given that $Y = y_i$ is*

$$F_{X|Y}(x|y_j) = P(x \leq x|Y = y_j) = \frac{1}{p_{\cdot j}} \sum_{x_i \leq x} p_{ij}. \qquad (7.2.1$$

Similarly the conditional d.f. of Y given that $X = x_i$ is

$$F_{Y|X}(y|x_i) = P(Y \leq y|X = x_i) = \frac{1}{p_{i\cdot}} \sum_{y_j \leq y} p_{ij} \qquad (7.2.1$$

provided $p_{i\cdot} > 0$.

Definition 7.2.9 ***Conditional p.d.f.*** *Let X, Y be continuous r.v.'s. At every poi (x, y) where f is continuous and $f_Y(y) > 0$, and is continuous, the conditional p.d.f. X given $Y = y$ is*

$$f_{X|Y}(x|y) = \frac{f_{X,Y}(x,y)}{f_Y(y)} (\geq 0) \qquad (7.2.1$$

(7.2.17) remains undefined at points where $f_Y(y) = 0$. Similarly the conditional p.d.f. Y given $X = x$ is

$$f_{Y|X}(y|x) = \frac{f_{X,Y}(x,y)}{f_X(x)} (\geq 0)$$

provided $f_X(x) > 0$. Clearly

$$\int_{-\infty}^{\infty} f_{X|Y}(x|y)\,dx = \int_{-\infty}^{\infty} f_{Y|X}(y|x)\,dy = 1.$$

Note that

$$f(x, y) = f_{X|Y}(x|y)f_Y(y) = f_{Y|X}(y|x)f_X(x). \qquad (7.2.1$$

Definition 7.2.10 ***Conditional d.f.*** *(continuous case). Let (X, Y) be a continuous r. The conditional d.f. of X given Y is defined as*

$$F_{X|Y}(x|y) = \int_{\infty}^{x} f_{X|Y}(u|y)\,du \qquad (7.2.1$$

for all y for which $f_Y(y) > 0$. Similarly

$$F_{Y|X}(y|x) = \int_{-\infty}^{y} f_{Y|X}(v|x)\, dv$$

for all x for which $f_X(x) > 0$.

■ NOTE 7.2.1 *If one variable, say, X is discrete and Y is continuous, then also the joint density, marginal densities, conditional densities are similarly defined.*

▶ **Example 7.2.5** *The random variables X, Y have joint p.d.f. $f(x,y) = xe^{-x(y+1)}$ $(x \geq 0,\ y \geq 0)$. Find the marginal and conditional p.d.f's.*

$$f_X(x) = xe^{-x}\int_0^{\infty} e^{-xy}\, dy = e^{-x}$$

$$f_Y(y) = \int_0^{\infty} xe^{-x(y+1)}\, dx = \frac{1}{(y+1)^2}\int_0^{\infty} ze^{-z}\, dz \ \textit{(putting } x(y+1) = z\textit{)}$$

$$= \frac{1}{(y+1)^2}$$

$$f_X(x|y) = (y+1)^2 xe^{-x(y+1)}$$

$$f_Y(y|x) = xe^{-xy}.$$

▶ **Example 7.2.6** *For a given λ, X follows Poisson with parameter λ. λ follows a gamma distribution*

$$f_\lambda(\lambda) = \frac{r^p}{\Gamma(p)} e^{-r\lambda}\lambda^{p-1}, \quad \lambda > 0, r > 0, p \textit{ (integer) } > 0.$$

Find the overall distribution of X, its mean and variance.

Here (X, λ) is a r.v. with

$$f_X(x|\lambda) = e^{-\lambda}\frac{\lambda^x}{x!}.$$

The joint p.d.f.

$$f_{X,\lambda}(x,\lambda) = \frac{r^p}{\Gamma(p)\, x!}\lambda^{x+p-1}e^{-\lambda(r+1)}$$

$$f_X(x) = \int_0^{\infty} f_{X,\lambda}(x,\lambda)d\lambda = \frac{r^p\,\Gamma(x+p)}{x!\,\Gamma(p)\,(r+1)^{x+p}}$$

$$= \left(\frac{r}{r+1}\right)\frac{1}{(r+1)^x}\,\frac{(x+p-1)!}{(p-1)!x!} = \binom{x+p-1}{x}\left(\frac{1}{r+1}\right)^x\left(\frac{r}{r+1}\right)^p,$$

a negative binomial distribution with parameters p and $\left(\frac{r}{r+1}\right)$

$$E(X) = \frac{p}{r},\ \ V(X) = \frac{p(r+1)}{r^2}.$$

► **Example 7.2.7** *Suppose the conditional density (p.m.f.)*

$$f_{X|Y}(x|y) = e^{-y}\frac{y^x}{x!}, \quad x = 0, 1, 2, \cdots,$$

a poisson with parameter y. The r.v. Y is continuous with marginal p.d.f.

$$f_Y(y) = e^{-y}, \quad y \geq 0.$$

Show that $f_X(x) = 2^{-(x+1)}, x = 0, 1, \cdots$.

Joint p.d.f. is

$$f_{X,Y}(x, y) = f_Y(y), \ \ f_{X|Y}(x|y) = e^{-2y}\frac{y^x}{x!}.$$

Hence

$$\begin{aligned} f_X(x) &= \int_0^\infty f(x,y)\,dy = \int_0^\infty e^{-2y}\frac{y^x\,dy}{x!} \\ &= \frac{\overline{|(x+1)}}{x!\,2^{x+1}} = \frac{1}{2^{x+1}} \ \text{(using gamma function).} \end{aligned}$$

Consider now the conditional probability of an event A given the value of a r.v. X, both A and X being defined on the same probability space $(\mathcal{S}, \mathcal{A}, P)$.
If X is discrete

$$P(A|X = x) = \frac{P[A \cap (X = x)]}{P(X = x)}. \tag{7.2.20}$$

Hence

$$P(A) = \sum_x P(A|X = x)P(X = x) \ \text{(by theorem 1.4.2).} \tag{7.2.21}$$

Also the joint probability

$$P[A; X \in B] = \sum_{x \varepsilon B} P(A|X = x)P(X = x).$$

If X is continuous $P(X = x) = 0$. In this case

$$P(A|X = x) = \lim_{\varepsilon \downarrow 0} P[A|x - \varepsilon < X < x + \varepsilon] \tag{7.2.22}$$

provided $P[x - \varepsilon < X < x + \varepsilon] > 0$ for every ε and the limit in (7.2.22) exists. Thus

$$P(A) = \int_{-\infty}^{\infty} P[A|X = x]f_X(x)\,dx. \tag{7.2.23}$$

Again

$$P(A; X \in B) = \int_{x \varepsilon B} P[A|X = x]f_X(x)\,dx.$$

Example 7.2.8 *What is the probability that three points selected randomly on the cumference of a circle will lie on a semi-circle.*

Without loss of generality assume that the first point P_1 falls on the positive x-is. The position of the second point P_2 is X (radian) measured from OP_1 in the ti-clockwise direction (Fig. 7.2): The p.d.f. of X is

$$f(x) = \frac{1}{2\pi}, \quad 0 < x \leq 2\pi$$

ce points are located at random. Let A be the event, P_1, P_2, P_3 are on a semi-circle. ow

$$P(A|x) = \begin{cases} \dfrac{2\pi - x}{2\pi}, & (0 < x < \pi), \text{ since } A \text{ occurs if and only if } P_3 \text{ falls between } P_1' \text{ and } P_2' \text{ i.e., } P_3 \text{ falls on an arc of length } \pi - (x - \pi) = 2\pi - x. \\ \dfrac{x}{2\pi}, & (\pi < x < 2\pi). \end{cases}$$

ence

$$P(A) = \int_0^{2\pi} P(A|x) f(x)\, dx = \frac{1}{2\pi}\left[\int_0^{\pi} \frac{2\pi - x}{2\pi}\, dx + \int_{\pi}^{2\pi} \frac{x}{2\pi}\, dx\right] = \frac{3}{4}.$$

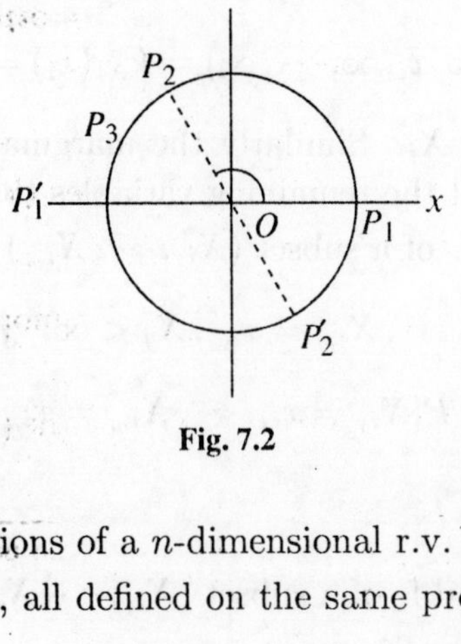

Fig. 7.2

We now consider distributions of a n-dimensional r.v. $\boldsymbol{X} = (X_1, \cdots, X_n)$.

Let $X_1, \cdots, X_n$ be n r.v.'s, all defined on the same probability space $(\mathcal{S}, \mathcal{A}, P)$. The int c.d.f. of $\boldsymbol{X}$ is

$$F_{\boldsymbol{X}}(x_1, \cdots, x_n) = P[X_1 \leq x_1, \cdots, X_n \leq x_n] \tag{7.2.24}$$

r all $(x_1, \cdots, x_n)$.

$F_{\boldsymbol{X}}(x_1, \cdots, x_n) = 0$ if at least one of x_i's is equal to $-\infty$ and $F(\infty, \cdots, \infty) = 1$. $\boldsymbol{X}$ is n-dim. discrete r.v. if $\boldsymbol{X}$ assumes only a countable number of values $\boldsymbol{X} = (x_1, \cdots, x_n)$ $\mathbb{R}^n$. In this case the joint p.m.f. of $X_1, \cdots, X_n$ is

$$p_{1,2,\cdots,n} = P[X_1 = x_1, \cdots, X_n = x_n] \geq 0 \tag{7.2.25}$$

satisfying

$$\sum_{i_1}\cdots\sum_{i_n} p_{i_1\cdots i_n} = 1.$$

The random vector $\boldsymbol{X}$ is a n-dimensional continuous r.v. if the corresponding d.f. continuous and if there exists a $f_{\boldsymbol{X}}(x_1, \cdots, x_n)(\geq 0)$ such that for all $(x_1, \cdots, x_n) \in \mathbb{R}^n$

$$F_{\boldsymbol{X}}(x_1, \cdots, x_n) = \int_{-\infty}^{x_n} \cdots \int_{-\infty}^{x_1} f(u_1, \cdots, u_n)\, du_1 \cdots du_n. \tag{7.2.2}$$

If F has partial derivatives up to order n,

$$\frac{\partial^n F_{\boldsymbol{X}}(x_1, \cdots, x_n)}{\partial x_1 \cdots \partial x_n} = f_{\boldsymbol{X}}(x_1, \cdots, x_n).$$

$f_{\boldsymbol{X}}(x_1, \cdots, x_n)$ is defined to be the joint p.d.f. of $X_1, \cdots, X_n$ and satisfies

$$\int_{-\infty}^{\infty} \cdots \int_{-\infty}^{\infty} f_X(x_1, \cdots, x_n)\, dx_1 \cdots dx_n = 1. \tag{7.2.2}$$

For any set A

$$P\{\boldsymbol{X} \in A\} = \int_A \int f_{\boldsymbol{X}}(x_1, \cdots, x_n)\, dx_1 \cdots dx_n. \tag{7.2.2}$$

The function

$$F_{\boldsymbol{X}}(\infty, \cdots, \infty, x_i, \infty, \cdots, \infty) = F_{X_i}(x_i) = P(X_i \leq x_i) \tag{7.2.2}$$

is called the marginal d.f. of X_i. Similarly the marginal d.f. of any subset $(X_{i_1}, \cdots X_{i_m})$ is obtained by setting all the remaining variables to be ∞.

For discrete $\boldsymbol{X}$ marginal p.m.f. of a subset $(X_{i_1}, \cdots, X_{i_m})$ of $\boldsymbol{X}$ is

$$\begin{aligned} p_{i_1,\cdots,i_m} &= P[X_{i_1} = x_{i_1}, \cdots, X_{i_m} = x_{i_m}, X_j < \infty,\ \ j \neq (i_1, \cdots, i_m)] \\ &= \sum_{\text{all values of } X_j}\cdots\sum P[X_{i_1} = x_{i_1}, \cdots, X_{i_m} = x_{i_m}, X_j,\ \ j \neq (i_1, \cdots, i_m)]. \end{aligned} \tag{7.2.3}$$

For continuous $\boldsymbol{X}$, marginal p.d.f. of a subset $(X_{i_1}, \cdots, X_{i_m})$ of $\boldsymbol{X}$ is

$$f_{X_{i_1},\cdots,X_{i_m}}(x_{i_1}, \cdots, x_{i_m}) = \int_{-\infty}^{\infty} \cdots \int_{-\infty}^{\infty} f_{\boldsymbol{X}}(x_1, \cdots, x_n) \prod_{\substack{j \\ (\neq i_1,\cdots,i_m)}} dx_j. \tag{7.2.3}$$

For discrete $\boldsymbol{X}$, the conditional p.m.f. of $X_{i_1}, \cdots, X_{i_k}$ given $X_{j_1} = x_{j_1}, \cdots, X_{j_{n-k}} = x_{j_{n-k}}$ is

$$\begin{aligned} &P\{X_{i_1} = x_{i_1}, \cdots, X_{i_k} = x_{i_k} | X_{j_1} = x_{j_1}, \cdots, X_{j_{n-k}} = x_{j_{n-k}}\} \\ &= \frac{p_{12\cdots n}}{p_{j_1,\cdots,j_{n-k}}} = p_{X_{i_1}\cdots,X_{i_k}|X_{j_1},\cdots X_{j_{n-k}}}(i_1, \cdots, i_k | j_1, \cdots, j_{n-k}). \end{aligned} \tag{7.2.3}$$

The quantities in (7.2.32) are non-negative and their summation over all values of $_{i_1}, \cdots, X_{i_k}$ is unity.

or continuous $\boldsymbol{X}$, the conditional p.d.f. of $X_{i_1}, \cdots, X_{i_k}$ given $X_{j_1}, \cdots, X_{j_{n-k}}$ is

$$f_{X_{i_1},\cdots,X_{i_k}|X_{j_1}\cdots,X_{j_{n-k}}}(x_{i_1}, \cdots, x_{i_k}|x_{j_1} \cdots, x_{j_{n-k}}) = \frac{f_{X_1,\cdots,X_n}(x_1, \cdots, x_n)}{f_{X_{j_1},\cdots,X_{j_{n-k}}}(x_{j_1}, \cdots, x_{j_{n-k}})}. \tag{7.2.33}$$

efinition 7.2.11 (***Stochastic Independence***). *The r.v.'s $X_1, \cdots, X_n$ are said to be tochastically independent if and only if*

$$F_{\boldsymbol{X}}(x_1, \cdots, x_n) = F_{X_1}(x_1)F_{X_2}(x_2)\cdots, F_{X_n}(x_n) \tag{7.2.34}$$

or all $(x_1, \cdots, x_n) \in \mathbb{R}^n$.

$\boldsymbol{X}$ is a n-dimensional discrete r.v., $X_1, \cdots, X_n$ are stochastically independent if

$$P(X_1 = x_1, \cdots, X_n = x_n) = P(X_1 = x_1)\cdots P(X_n = x_n) \tag{7.2.35}$$

or all $(x_1, \cdots, x_n) \in \mathbb{R}^n$.

$\boldsymbol{X}$ is a n-dimensional continuous r.v., $X_1, \cdots, X_n$ are stochastically independent if

$$f_{\boldsymbol{X}}(x_1, \cdots, x_n) = f_{X_1}(x_1)\cdots f_{X_n}(x_n) \tag{7.2.36}$$

or all $(x_1, \cdots, x_n) \in \mathbb{R}^n$.

n case of two random variables, (7.2.35) implies that X, Y are independent if

$$\left.\begin{aligned} P(X = x|Y = y) = P(X = x) \\ P(Y = y|X = x) = P(Y = y) \end{aligned}\right\} \text{ for all } (x, y).$$

when X, Y are discrete. For continuous r.v.'s, (7.2.36) implies X, Y are inpendent if

$$\left.\begin{aligned} f_{X|Y}(x|y) = f_X(x) \\ f_{Y|X}(y|x) = f_Y(y) \end{aligned}\right\} \text{ for all } (x, y)$$

7.2.35) is equivalent to the following definition: $X_1, \cdots, X_n$ are independent if

$$P[X_1 \in A_1, X_n \in A_n] = P(X_1 \in A_1)\cdots P(X_n \in A_n) \tag{7.2.37}$$

or all sets $A_1, \cdots, A_n \in \mathcal{B}_1$.

The following important theorem holds:

Theorem 7.2.1 *Let $X_1, \cdots, X_n$ be independent r.v.'s. Let $g_i(X_i)$ be a function of X_i uch that $g_i(X_i)$ is a r.v. $(i = 1, \cdots, n)$. Then $g_1(X_1), \cdots, g_n(X_n)$ are independent r.v.s.*

Proof: Let $A_1, \cdots, A_n$ be some borel sets in $\mathcal{B}_1$. For the sample space $(\mathcal{S}, \mathcal{A})$

$$\{w : g_i(X_i(w)) \in A_i\} \equiv \{w : X_i(w) \in g_i^{-1}(A_i)\}.$$

Hence

$$\begin{aligned}
& P[g_1(X_1) \in A_1, \cdots, g_n(X_n) \in A_n] \\
&= P[X_1 \in g_1^{-1}(A_1), \cdots, X_n \in g_n^{-1}(A_n)] \\
&= P[X_1 \in g_1^{-1}(A_1)] \cdots, P[X_n \in g_n^{-1}(A_n)] \\
&\qquad \text{(by independence of } X_1, \cdots, X_n) \\
&= P[g_1(X_1) \in A_1] \cdots P[g_n(X_n) \in A_n].
\end{aligned}$$

▶ **Example 7.2.9** *(X, Y) have joint d.f.*

$$F_{X,Y}(x,y) = \begin{cases} (1-e^{-ax})y^2, & (0 \le x < \infty, 0 \le y < 1) \\ 1 - e^{-ax}, & (0 \le x < \infty, 1 < y < \infty) \\ 0, & \text{(otherwise)} \end{cases}$$

where a (> 0) is a constant. Here

$$F_X(x) = \begin{cases} 1 - e^{-ax}, & (0 \le x < \infty) \\ 0, & \text{otherwise} \end{cases}$$

$$F_Y(y) = \begin{cases} y^2, & (0 \le y \le 1) \\ 1, & (1 < y < \infty) \\ 0, & \text{otherwise} \end{cases}$$

$$f(x,y) = \begin{cases} 2aye^{-ax}, & (0 \le x < \infty, 0 \le y \le 1) \\ 0, & \text{otherwise} \end{cases}$$

$$f_X(x) = \begin{cases} ae^{-ax}, & (0 \le x < \infty) \\ 0, & \text{otherwise} \end{cases}$$

$$f_Y(y) = \begin{cases} 2y, & (0 \le y \le 1) \\ 0, & \text{otherwise.} \end{cases}$$

Note

$$\left.\begin{aligned} F_{X,Y}(x,y) &= F_X(x)F_Y(y) \\ f_{X,Y}(x,y) &= f_X(x)f_Y(y) \end{aligned}\right\} \text{ for all } (x,y).$$

Hence X, Y are independent.

▶ **Example 7.2.10** *Let X_1, X_2 be independently and identically distributed r.v.'s with*

$$P(X_i = \pm 1) = \frac{1}{2}, \quad i = 1, 2.$$

Let $X_3 = X_1X_2$. Show that X_1, X_2, X_3 are pairwise independent but not independent

$$P(X_3 = 1) = P(X_1 = 1, X_2 = 1) + P(X_1 = -1, X_2 = -1)$$
$$= P(X_1 = 1)P(X_2 = 1) + P(X_1 = -1)P(X_2 = -1) = \frac{1}{4} + \frac{1}{4} = \frac{1}{2}.$$

Similarly

$$P(X_3 = -1) = \frac{1}{2}.$$

$$P(X_1 = 1, X_3 = 1) = P(X_1 = 1)P(X_3 = 1 | X_1 = 1) = P(X_1 = 1)\frac{P(X_1 = 1, X_3 = 1)}{P(X_1 = 1)}$$
$$= P(X_1 = 1)\frac{P(X_1 = 1, X_2 = 1)}{P(X_1 = 1)} = \frac{1}{4} = P(X_1 = 1)P(X_3 = 1).$$

Thus it can be proved X_1, X_3 and X_2, X_3 are independent. However,

$$P(X_1 = 1, X_2 = 1, X_3 = -1) = P(X_1 = 1, X_2 = 1)P(X_3 = -1 | X_1 = 1, X_2 = 1) = 0.$$

Hence X_1, X_2, X_3 are not independent.

▶ **Example 7.2.11** $f(x, y) = \frac{1}{2}xy, 0 < y < x, 0 < x < 2.$

$$f_X(x) = \frac{1}{2}x\int_0^x y\,dy = \frac{x^3}{4}, 0 < x < 2$$

$$f_Y(y) = \frac{1}{2}y\int_y^2 x\,dx = y - \frac{y^3}{4}, 0 < y < 2.$$

Thus $f_{X,Y}(x, y) \neq f_X(x)f_Y(y)$ and hence, X, Y are not stochastically independent.

7.3 Expectation and Moments of Random Variables

Let $g(\boldsymbol{X})$ be a (borel-measurable) function of $\boldsymbol{X}$ such that $g(\boldsymbol{X})$ is a random variable.

$$E\{g(\boldsymbol{X})\} = \int_{-\infty}^{\infty} \cdots \int_{-\infty}^{\infty} g(x_1, \cdots, x_n)dF_{\boldsymbol{X}}(x_1, \cdots, x_n) \tag{7.3.1}$$

provided $|g(x_1, \cdots, x_n)|$ is integrable (in the Riemann-Stieltjes sense) with respect to $F_{\boldsymbol{X}}(x_1, \cdots, x_n)$.
In particular,

$$E\{X_k\} = \int_{-\infty}^{\infty} \cdots \int_{-\infty}^{\infty} x_k dF_{\boldsymbol{X}}(x_1, \cdots, x_n) \tag{7.3.2}$$

provided the integral is absolutely convergent.
In case $\boldsymbol{X}$ is discrete

$$E\{g(\boldsymbol{X})\} = \sum_{x_1 \cdots x_n}\sum g(x_1 \cdots x_n)P[X_1 = x_1, \cdots, X_n = x_n] \tag{7.3.3}$$

provided the series in (7.3.3) is absolutely convergent.

In case $\boldsymbol{X}$ is continuous

$$E\{g(\boldsymbol{X})\} = \int_{-\infty}^{\infty}\int_{-\infty}^{\infty} g(x_1, \cdots, x_n) f(x_1, \cdots, x_n)\, dx_1, \cdots, dx_n \tag{7.3.4}$$

provided

$$\int_{-\infty}^{\infty}\int_{-\infty}^{\infty} |g(x_1, \cdots, x_n)| f(x_1, \cdots, x_n)\, dx_1, \cdots, dx_n < \infty.$$

When $g(\boldsymbol{X}) = X_1^{k_1} \cdots X_n^{k_n}$, $E\{g(\boldsymbol{X})\}$ in (7.3.1) defines the non-central moment

$$\mu'_{k_1,\cdots,k_n} \tag{7.3.5}$$

of the distribution $F_{\boldsymbol{X}}(x_1, \cdots, x_n)$.

In the case of two variables X, Y we have bivariate non-central moments

$$\mu'_{r,s} = E(X^r Y^s) \tag{7.3.6}$$

when

$$\left.\begin{aligned} E(X) &= \mu'_{1,0} = \mu_X \text{ (say)} \\ E(Y) &= \mu'_{0,1} = \mu_Y \text{ (say)} \end{aligned}\right\} \tag{7.3.7}$$

which are mean of X and mean of Y respectively.

The bivariate central moments are

$$\mu_{rs} = E(X - \mu_X)^r (Y - \mu_Y)^s. \tag{7.3.8}$$

We have

$$\left.\begin{aligned} \mu_{20} &= E(X - \mu_X)^2 = \sigma_X^2 \\ \mu_{02} &= E(Y - \mu_Y)^2 = \sigma_Y^2 \end{aligned}\right\} \tag{7.3.9}$$

which are variances of X and Y respectively.

The product moment

$$\mu_{11} = \mu_{X,Y} = E(X - \mu_X)(Y - \mu_Y) = E(XY) - \mu_X \mu_Y \tag{7.3.10}$$

is called the covariance between X and Y, often denoted as Cov (X, Y) or σ_{XY}.

It has been proved in section 3.4 that

$$\mu_{X,Y}^2 \leq \sigma_X^2 \sigma_Y^2.$$

The quantity

$$\rho(X, Y) = \frac{\mu_{X,Y}}{\sigma_X \sigma_Y} \tag{7.3.11}$$

is defined as the correlation coefficient between X and Y, provided $\mu_{X,Y}, \sigma_X, \sigma_Y$ exist and $\sigma_X > 0$, $\sigma_Y > 0$.

Hence

$$\rho^2 \leq 1 \text{ or } -1 \leq \rho \leq 1 \tag{7.3.12}$$

$\rho^2 = 1$ if and only if X or Y is an exact linear function of the other (vide theorem 7.5.3).

▶ **Example 7.3.1**

$$f_{X,Y}(x,y) = \frac{3x+y}{4}e^{-x-y}, \ \ x>0, \ \ y>0.$$

Find the marginal densities, conditional densities and $\rho(X,Y)$.

$$f_X(x) = e^{-x}\int_0^\infty \frac{3x+y}{4}e^{-y}\,dy = \frac{3x+1}{4}e^{-x}$$
$$f_Y(y) = e^{-y}\int_0^\infty \frac{3x+y}{4}e^{-x}\,dx = \frac{y+3}{4}e^{-y}$$
$$f_{X|Y}(x|y) = \frac{3x+y}{y+3}e^{-x}$$
$$f_{Y|X}(y|x) = \frac{3x+y}{3x+1}e^{-y}, y>0$$
$$\mu'_{1,0} = \mu_X = \frac{1}{4}\int_0^\infty x(3x+1)e^{-x}\,dx = \frac{7}{4}$$
$$\mu'_{2,0} = 5, \sigma_X^2 = \frac{31}{16}, \mu'_{0,1} = \frac{5}{4}$$
$$\mu'_{0,2} = 3, \sigma_Y^2 = \frac{23}{16}$$
$$\mu'_{1,1} = \frac{1}{4}\int_0^\infty\int_0^\infty xy(3x+y)e^{-x-y}\,dx\,dy = 2$$
$$\mu_{1,1} = -\frac{3}{16}$$
$$\rho(X,Y) = -\frac{3}{\sqrt{31\times 23}} = -0.112.$$

Theorem 7.3.1 *If $X_1,\cdots,X_n$ are any given set of r.v.'s with finite means and if*

$$Y = c_0 + c_1X_1 + \cdots + c_nX_n$$

where $c_0, c_1, \cdots, c_n$ are any given set of constants, then

$$E(Y) = c_0 + c_1E(X_1) + \cdots + c_nE(X_n). \tag{7.3.13}$$

Proof:

$$E(Y) = \int_{-\infty}^\infty \cdots \int_{-\infty}^\infty \left(c_0 + \sum_{i=1}^n c_iX_i\right) dF_{\mathbf{X}}(x_1,\cdots,x_n)$$
$$= c_0 + \sum_{i=1}^n c_i \int_{-\infty}^\infty \cdots \int_{-\infty}^\infty x_i dF_{\mathbf{X}}(x_1,\cdots,x_n) = c_0 + \sum_{i=1}^n c_iE(X_i)$$

by (7.3.2) and by the fact

$$\int_{-\infty}^\infty \cdots \int_{-\infty}^\infty dF_{\mathbf{X}}(x_1,\cdots,x_n) = 1.$$

Theorem 7.3.2 *If $X_1, \cdots, X_n$ are independent r.v.'s with finite means*

$$E(X_1 \cdots X_n) = E(X_1) \cdots E(X_n). \tag{7.3.14}$$

Proof:

$$\begin{aligned} E(X_1 \cdots X_n) &= \int_{-\infty}^{\infty} \cdots \int_{-\infty}^{\infty} x_1 \cdots x_n dF_X(x_1, \cdots, x_n) \\ &= \int_{-\infty}^{\infty} \cdots \int_{-\infty}^{\infty} x_1 \cdots x_n dF_{X_1}(x_1) \cdots dF_{X_n}(x_n) \text{ (by 7.2.34)} \\ &= \left[\int_{-\infty}^{\infty} x_1 dF_{X_1}(x_1)\right] \cdots \left[\int_{-\infty}^{\infty} x_n dF_{X_n}(x_n)\right] = E(X_1) \cdots E(X_n). \end{aligned}$$

Corollary 7.3.1 *If $X_1, \cdots, X_n$ are independent r.v.'s and if $g_i(X_i)$ is a* (Borel-measureable) *function of X_i such that $g_i(X_i)$ $(i = 1, \cdots, n)$ is a r.v. then*

$$E\{g_1(X_1) \cdots g_n(X_n)\} = E(g_1(X_1)) \cdots E(g_n(X_n)). \tag{7.3.15}$$

Proof follows from the fact that $g_1(X_1), \cdots, g_n(X_n)$ are independent r.v.'s.

► **Example 7.3.2**

$$E(e^{tX_1 + \cdots + tX_n}) = E\left(\prod_{i=1}^{n} e^{tX_i}\right) = \prod_{i=1}^{n} E(e^{tX_i})$$

i.e., the m.g.f. of sum of n independent r.v.'s is the product of m.g.f. of each r.v.

Two r.v.'s X, Y are said to be uncorrelated if

$$E(XY) = E(X)E(Y) \tag{7.3.16}$$

i.e., if Cov $(X, Y) = 0$.

It follows from theorem 7.3.2 that if X, Y are independent with finite means and if $\mu'_{X,Y}$ exists then X, Y are uncorrelated. The converse is not necessarily true (vide exercises 11, 12 of section 7.8).

If X and Y are uncorrelated then in general

$$E\{g(X)h(Y)\} \neq E\{g(X)\}E\{h(Y)\}$$

whereas, if they are independent

$$E\{g(X)h(Y)\} = E\{g(X)\}E\{h(Y)\}.$$

Definition 7.3.1 *Two random variables are orthogonal if*

$$E(XY) = 0. \tag{7.3.17}$$

f X, Y are uncorrelated,

$$E(X - E(X))(Y - E(Y)) = 0$$

e., $(X - E(X))$ and $(Y - E(Y))$ are orthogonal.

f X, Y are orthogonal,

$$E(X + Y)^2 = E(X^2) + E(Y^2).$$

If X, Y are uncorrelated and mean of at least one of them is zero then they are also rthogonal.

heorem 7.3.3 (*Approximate evaluation of expectation and variance of $g(X)$*). *Let X e a r.v. with $E(X) = \mu, V(X) = \sigma^2$.*

hen

$$E\{g(X)\} \simeq g(\mu) + \frac{g''(\mu)}{2}\sigma^2 \tag{7.3.18}$$

$$V\{g(X)\} \simeq [g'(\mu)]^2\sigma^2 \tag{7.3.19}$$

rovided $g(X)$ is atleast twice differentiable at $X = \mu$.

Proof: Expanding $g(X)$ in a Taylor series about $X = \mu$,

$$g(X) = g(\mu) + (X - \mu)g'(\mu) + \frac{(X - \mu)^2}{2!}g''(\mu) + \Delta \tag{7.3.20}$$

vhere Δ is a remainder. Discarding Δ and taking expectation of both sides (7.3.18) ollows.

Ignoring the last two terms in the right hand side of (7.3.20) and taking variance of oth sides, (7.3.19) follows.

Theorem 7.3.4 *Let X, Y be r.v.'s with finite variances σ_X^2, σ_Y^2.*

Then

$$V(X + Y) = V(X) + V(Y) + 2 \text{ Cov } (X, Y).$$

Proof:

$$\begin{aligned}
V(X + Y) &= \int_{-\infty}^{\infty}\int_{-\infty}^{\infty} \{x + y - E(X) - E(Y)\}^2 dF_{X,Y}(x, y) \\
&= \int_{-\infty}^{\infty}\int_{-\infty}^{\infty} \{(x - E(X)) + (y - E(Y)\}^2 dF_{X,Y}(x, y) \\
&= \int_{-\infty}^{\infty}\int_{-\infty}^{\infty} (x - E(X))^2 dF(x, y) + \int_{-\infty}^{\infty}\int_{-\infty}^{\infty} (y - E(Y))^2 dF(x, y) \\
&\quad + 2\int_{-\infty}^{\infty}\int_{-\infty}^{\infty} (x - E(X))(y - E(Y)) dF(x, y) \\
&= V(X) + V(Y) + 2 \text{ Cov } (X, Y)
\end{aligned}$$

ince means and covariances exist as variances exist.

Theorem 7.3.5 *Let $X_1, \cdots, X_n$ be r.v.'s with $\mu_i = E(X_i), V(X_i) = \sigma_i^2$, Cov $(X_i, X_j) = \sigma_{ij}$ $(i \neq j)$, all finite.*

Let

$$Y = c_0 + \sum_{i=1}^{n} c_i X_i$$

where $c_j (j = 0, \cdots, n)$ are arbitrary constants. Then

$$V(Y) = \sum_{i=1}^{n} c_i^2 \sigma_i^2 + \sum_{i \neq j=1}^{n}\sum c_i c_j \sigma_{ij}.$$

Proof: Follows as in the proof of theorem 7.3.4.

Theorem 7.3.6 *Let*

$$U = \sum_{i=1}^{n} c_i X_i, V = \sum_{i=1}^{n} d_i X_i$$

be two linear functions of $X_1, \cdots, X_n, c_i, d_i$ $(i = 1, \cdots, n)$ being arbitrary constants.

Then

$$\text{Cov }(U, V) = \sum_{i=1}^{n}\sum_{j=1}^{n} c_i d_j \sigma_{ij}$$

where $\sigma_{ii} = V(X_i), \sigma_{ij} = \text{Cov }(X_i, X_j), i \neq j$, provided the variances and covariance exist. Proof is straight forward.

Corollary 7.3.2 *If $U = a + cX, V = b + dY$,*

1. Cov $(U, V) = cd$ Cov (X, Y).
2. $\rho(U, V) = \{sign\ (cd)\}\rho(X, Y)$.

▶ **Example 7.3.3** *Let $X_1, \cdots, X_n$ be independent r.v.'s with*

$$E|X_i^2| < \infty \quad and \quad V(X_i) = \sigma_i^2, i = 1, \cdots, n.$$

Let $S = \sum_{1}^{n} a_i X_i, a_1, \cdots, a_n$ are constants, $\sum_{i=1}^{n} a_i = 1$. Find a_i's so that $V(S)$ is minimum

$$V(S) = \sigma \ (say) \ = \sum_{i=1}^{n} a_i^2 \sigma_i^2 = \sum_{j=1}^{n-1} a_j^2 \sigma_j^2 + (1 - a_1 - \cdots - a_{n-1})^2 \sigma_n^2.$$

For minimising σ w.r.t. a_i's we have

$$\frac{\partial \sigma}{\partial a_1} = 2a_1\sigma_1^2 - 2(1 - a_1 - \cdots - a_{n-1})\sigma_n^2 = 0$$

$$\cdots\cdots\cdots\cdots\cdots\cdots\cdots\cdots\cdots\cdots\cdots$$

$$\frac{\partial \sigma}{\partial a_{n-1}} = 2a_{n-1}\sigma_{n-1}^2 - 2(1 - a_1 - \cdots - a_{n-1})\sigma_n^2 = 0.$$

It follows that

$$a_j\sigma_j^2 = a_n\sigma_n^2, \quad j = 1, \cdots, n-1$$

i.e., $a_j = \frac{\lambda}{\sigma_j^2}$, λ the constant of proportionality.

$$\sum_1^n a_j = 1 \text{ gives } \lambda = \left(\sum_1^n \frac{1}{\sigma_j^2}\right)^{-1}.$$

The minimum value of σ is

$$\sigma_{\min} = \lambda^2 \sum_1^n \frac{1}{\sigma_i^2} = \left(\sum_1^n \frac{1}{\sigma_i^2}\right)^{-1}.$$

As in the proof of theorem 7.3.3 the following theorem can be proved.

Theorem 7.3.7

$$E\{g(X,Y)\} \simeq g(\mu_X, \mu_Y) + \frac{1}{2}\left[\mu_{20}\frac{\partial^2 g}{\partial x^2} + 2\mu_{11}\frac{\partial^2 g}{\partial x \partial y} + \mu_{02}\frac{\partial^2 g}{\partial y^2}\right]_{\mu_X, \mu_Y} \tag{7.3.21}$$

$$V\{g(X,Y)\} \simeq \left[\left(\frac{\partial g}{\partial x}\right)^2 \mu_{20} + \left(\frac{\partial g}{\partial y}\right)^2 \mu_{02} + 2\left(\frac{\partial g}{\partial x}\right)\left(\frac{\partial g}{\partial y}\right)\mu_{11}\right]_{\mu_X, \mu_Y} \tag{7.3.22}$$

when $\mu_X = E(X), \mu_Y = E(Y)$, provided the function $g(\mu_X, \mu_Y)$ is well-defined.

7.4 Generating Functions

Definition 7.4.1 *Joint moment generating function. The joint m.g.f. of the r.v. $\mathbf{X} = (X_1, \cdots, X_n)$ is defined as*

$$M_{\mathbf{X}}(t_1, \cdots, t_n) = M_{\mathbf{X}}(\mathbf{t}) = E\left[e^{\Sigma_1^n t_i X_i}\right] \text{ (where } \mathbf{t} = (t_1, \cdots, t_n)) \tag{7.4.1}$$

if the expectation exists for all values of $t_1, \cdots, t_n$ such that $|t_j| < T$ for some $T(>0), j = 1, \cdots, n$.

The raw moments are obtained as

$$\mu'_{r_1, \cdots, r_n} = \left.\frac{\partial^{r_1 + \cdots + r_n}}{\partial t_1^{r_1} \cdots \partial t_n^{r_n}} M_{\mathbf{X}}(\mathbf{t})\right|_{\mathbf{t}=\mathbf{0}} \tag{7.4.2}$$

where $\mathbf{0} = (0, \cdots, 0)$, provided the required partial derivatives exist and provided the order of differentiation and integration are interchangeable. In particular

$$\mu'_r = \left.\frac{\partial^r}{\partial t_1^r} M_{\mathbf{X}}(\mathbf{t})\right|_{\mathbf{t}=\mathbf{0}}; \quad \mu'_{r,s} = \left.\frac{\partial^{r+s}}{\partial t_1^r \partial t_2^s} M_{\mathbf{X}}(\mathbf{t})\right|_{\mathbf{t}=\mathbf{0}}.$$

Thus expanding $M_{\mathbf{X}}(\mathbf{t})$ in a Maclaurin series, one obtains

$$M_{\mathbf{X}}(\mathbf{t}) = \sum_{r_1,\cdots,r_n} \frac{t_1^{r_1}\cdots t_n^{r_n}}{r_1!\cdots r_n!}\mu'_{r_1,\cdots,r_n}. \quad (7.4.3)$$

The marginal m.g.f.'s are obtained as

$$\begin{aligned} M_{X_1}(t_1) &= M_{\mathbf{X}}(t_1, 0, \cdots, 0) = \lim_{\substack{t_i\to 0\\ i=2,\cdots,n}} M_{\mathbf{X}}(t_1, t_2, \cdots, t_n) \\ M_{X_2}(t_2) &= M_{\mathbf{X}}(0, t_2, 0, \cdots, 0) = \lim_{\substack{t_i\to 0\\ i(\neq 2)=1,\cdots,n}} M_{\mathbf{X}}(t_1, \cdots, t_n). \end{aligned}$$

If the variables $X_1, \cdots, X_n$ are mutually independent,

$$M_{\mathbf{X}}(\mathbf{t}) = \prod_{i=1}^{n} E(e^{t_i X_i}) = \prod_{i=1}^{n} M_{X_i}(t_i), \quad (7.4.4)$$

i.e., the joint m.g.f. is the product of separate m.g.f.'s.

Theorem 7.4.1 *The m.g.f. $M(t_1, t_2)$, when it exists, determines the joint distributio[n] of (X, Y) and conversely if the m.g.f. exists, it is unique.*

Proof: Omitted.

Theorem 7.4.2 *Two jointly distributed r.v.'s. X, Y are independent if and only if*

$$M_{X,Y}(t_1, t_2) = M_X(t_1) M_Y(t_2) \quad (7.4.5)$$

for all t_1, t_2 such that $|t_i| < T(> 0), i = 1, 2$.

Proof: If X, Y are independent (7.4.5) follows from (7.4.4). If (7.4.5) holds,

$$\iint e^{t_1X+t_2Y} dF(x,y) = \int e^{t_1x} dF(x) \int e^{t_2y} dF(y) = \iint e^{t_1x+t_2y} dF(x) dF(y).$$

Hence

$$dF(x, y) = dF(x) dF(y)$$

by the uniqueness property of m.g.f. (theorem 7.4.1).

The theorem can be generalised to n r.v.'s.

▶ **Example 7.4.1** *Let (X, Y) have a bivariate uniform distribution over the rectangle (R) defined by*

$$a \le x \le a + k, \quad b \le y \le b + l, \quad k > l > 0.$$

The p.d.f. is

$$f_{X,Y}(x,y) = \begin{cases} \frac{1}{kl}, & \text{for } (x,y) \in R \\ 0, & \text{otherwise.} \end{cases}$$

The m.g.f. is

$$M_{X,Y}(t_1,t_2) = \int_a^{a+k} \int_b^{b+l} e^{t_1 x + t_2 y} \cdot \frac{1}{kl}\, dx\, dy = -\frac{\exp[at_1 + bt_2]}{klt_1 t_2}(e^{kt_1} - 1)(e^{lt_2} - 1).$$

▶ **Example 7.4.2** *X_1, X_2, X_3, X_4 are independently and identically distributed with p.d.f. e^{-x}, $x > 0$. Find the m.g.f. of the joint distribution of U, V, W where*

$$U = X_1 + X_2, \quad V = X_3 + X_2, \quad W = X_4 + X_2.$$

$$\begin{aligned}
\text{m.g.f.} &= \int_0^\infty \int_0^\infty \int_0^\infty e^{ut_1+vt_2+wt_3} e^{-x_1-x_2-x_3-x_4}\, dx_1 \cdots dx_4 \\
&= \int_0^\infty \int_0^\infty \int_0^\infty e^{x_1(t_1-1)+x_2(t_1+t_2+t_3-1)+x_3(t_2-1)+x_4(t_3-1)}\, dx_1 \cdots dx_4 \\
&= \int_0^\infty e^{-x_1(1-t_1)}\, dx_1 \int_0^\infty e^{-x_2(1-t_1-t_2-t_3)}\, dx_2 \int_0^\infty e^{-x_3(1-t_2)}\, dx_3 \int_0^\infty e^{-x_4(1-t_3)}\, dx_4 \\
&= \{(1-t_1)(1-t_2)(1-t_3)(1-t_1-t_2-t_3)\}^{-1}.
\end{aligned}$$

Definition 7.4.2 *Bivariate probability generating function. For a pair of integer-valued r.v.'s. X, Y with a joint density function $P\{X = i, Y = j\} = p_{ij}, i, j = 0, 1 \cdots$. we define the p.g.f.*

$$P_{X,Y}(t_1,t_2) = \sum_i \sum_j p_{ij} t_1^i t_2^j, |t_i| < 1, i = 1, 2. \tag{7.4.6}$$

The p.g.f. of marginal distributions of X, Y are respectively,

$$P_X(t_1) = P(t_1, 1), P_Y(t_2) = P(1, t_2).$$

It follows that p.g.f. of $X + Y$ is $P(t, t)$. X, Y are independent if and only if

$$P(t_1, t_2) = P(t_1, 1)P(1, t_2) \text{ for all } t_1, t_2 \text{ with } |t_i| < 1, i = 1, 2.$$

▶ **Example 7.4.3** *Consider*

$$P(t_1, t_2) = e^{\mu_1(t_1-1)+\mu_2(t_2-1)+\lambda(t_1t_2-1)}, \mu_1, \mu_2 > 0$$

which has a power series expansion with positive coefficients adding up to unity. $P(t_1, t_2)$ is the p.g.f. of a bivariate probability distribution. $P(t_1, 1)$ (or $P(1, t_2)$) is the p.g.f. of a Poisson distribution with parameter $\mu_1 + \lambda$ (or $\mu_2 + \lambda$).

▶ **Example 7.4.4** *$\{X_n\}$ is a sequence of independently and identically distributed (iid) r.v.'s with p.m.f. $P\{X_k = j\} = p_j, j = 1, 2, \cdots$ and the p.g.f. $P_X(t) = \sum p_j t^j, |t| < 1$. Let*

$$S_N = X_1 + \cdots + X_N,$$

where N is a r.v. with p.m.f. $P\{N = n\} = g_n$ and p.g.f. $G_N(t) = \sum_0^\infty g_n t^n, |t| < 1$. Find the p.g.f. of S_N.

In particular find the p.g.f. of S_N when X_i's are Bernoulli variables with $P(X_i = 1) = p$, $P(X_i = 0) = q$ and N is a Poisson variable with mean λ.

$$P(S_N = j) = \sum_{n=0}^{\infty} P(N = n)P(S_n = j) = h_j \text{ (say)}.$$

Hence p.g.f. of S_N

$$\begin{aligned}
&= H_{S_N}(t) = \sum_0^\infty h_j t^j = \sum_{j=0}^{\infty}\left\{\sum_{n=0}^{\infty} g_n P(S_n = j)\right\} t^j \\
&= \sum_{n=0}^{\infty} g_n \left\{\sum_{j=0}^{\infty} P(S_n = j)t^j\right\} = \sum_{n=0}^{\infty} g_n P_{S_n}(t) \text{ (where } P_{S_n}(t) = \text{p.g.f. of } S_n) \\
&= \sum_{n=0}^{\infty} g_n [P(t)]^n \text{ (since } X_1, \cdots, X_n \text{ are } iid) \\
&= G_N(P(t)).
\end{aligned}$$

When X is a Bernoulli variable, its p.g.f. $P_X(t) = q + pt, |t| < 1$. N being a poisson variable, its p.g.f.

$$G_N(t) = e^{-\lambda(1-t)}, |t| < 1.$$

Hence p.g.f. of S_N is

$$G_N(P(t)) = e^{-\lambda(1-q-pt)} = e^{-\lambda p(1-t)}.$$

Therefore S_N is a poisson variable with mean λp.

The joint cumulant generating function $\log_e M_{\mathbf{X}}(\mathbf{t})$ is defined similarly as in the case a single random variable, details of which are omitted here. The joint characteristic function $\psi_{\mathbf{X}}(\mathbf{t})$ is analogously defined. Multivariate cumulants have been discussed in details in chapter 13.

7.5 Conditional Expectation, Variance, Regression

Definition 7.5.1 *Let (X, Y) be two r.v.'s. The conditional expected value of $g(X, Y)$ given $X = x$ is defined as*

$$E[g(X, Y)|X = x] = \int_{-\infty}^{\infty} g(x, y) f_{Y|X}(y|x)\, dy \tag{7.5.1}$$

f (X,Y) is a continuous r.v. For discrete (X,Y),

$$E[g(X,Y)|X=x]=\sum_y g(x,y)P[Y=y|X=x] \tag{7.5.2}$$

vhere the summation is over all possible values of Y.

In particular if $g(X,Y) = Y$, (7.5.1), (7.5.2) define respectively the conditional expectation of Y given x,

$$E(Y|x)=\mu_{Y|x}=\int_{-\infty}^{\infty} y f_{Y|x}(y|x)\,dy \text{ or } \sum_y yP[Y=y|X=x] \tag{7.5.3}$$

according as (X,Y) is continuous or discrete.

Similarly the conditional expectation of X given $Y = y, E(X|y) = \mu_{X|y}$ can be defined.

In general $\mu_{Y|x}$ is a function of x and similarly $\mu_{X|y}$ a function of y.

Definition 7.5.2 *The graph of $E(Y|x)$ as a function of x,*

$$y=E(Y|x) \tag{7.5.4}$$

is called the regression curve of Y on X. The graph of $E(X|y)$ as a function of y,

$$x=E(X|y) \tag{7.5.5}$$

is called the regression curve of X on Y.

Theorem 7.5.1 *For two r.v.'s (X,Y)*

$$E[E\{g(X)|Y\}]=E[g(X)] \tag{7.5.6}$$

provided $E[g(x)] < \infty$.

Proof: We prove the result when (X,Y) is jointly continuous. The left hand side

$$=\int_{-\infty}^{\infty}\left[\int_{-\infty}^{\infty} g(x)f_{X|Y}(x|y)\,dx\right] f_Y(y)\,dy=\int_{-\infty}^{\infty}\int_{-\infty}^{\infty} g(x)f(x,y)\,dx\,dy=E[(g(X)].$$

As a particular case,

$$E[E(X|Y)]=E(X).$$

Similarly

$$E[E(Y|X)]=E(Y).$$

Theorem 7.5.2 *If one of the regression curves is a constant, $\sigma_{X,Y}=\rho(X,Y)=0$.*

Proof: Suppose $E(Y|x) = k$ (a constant). Then by theorem 7.5.1, $E(Y) = k$. Hence assuming X, Y are continuous,

$$\begin{aligned}\sigma_{X,Y} &= E(X-\mu_X)(Y-\mu_Y) = E(XY) - k\mu_X \\ &= \int_{-\infty}^{\infty}\int_{-\infty}^{\infty} xyf(x,y)\,dx\,dy - k\mu_X \\ &= \int_{-\infty}^{\infty} x\left[\int_{-\infty}^{\infty} yf_{Y|X}(y|x)\right] f_X(x)\,dx - k\mu_X = k(\mu_X - \mu_X) = 0.\end{aligned}$$

(The case when X, Y are discrete can be proved similarly.)

Theorem 7.5.3 $\rho(X,Y) = \pm 1$ *if and only if there exist constants a, b such that*

$$P(Y = aX + b) = 1,$$

i.e., $Y = aX + b$ a.e. Further, if $a < 0$, then $\rho = -1$ and if $a > 0, \rho = 1$.

Proof:

1. Let $Y = aX + b$ a.e. Then $V(Y) = a^2V(X)$, Cov $(X,Y) = aV(x)$ and hence $\rho = \frac{a}{|a|} = \pm 1$.

2. If $\rho = \pm 1$, $[\text{Cov }(X,Y)]^2 = V(X)V(Y)$ which holds by the condition of equality in Cauchy-Schwartz inequality, if and only if $t(X - \mu_X) = Y - \mu_Y$ a.e. for some (vide section 3.4), i.e., if and only if $Y = aX + b$ a.e. for some constants a and b i.e., if X, Y are perfectly linearly related.

Definition 7.5.3 ***Conditional variance.*** *The conditional variance of Y given $X = x$ i*

$$V(Y|X=x) = E[Y^2|X=x] - [E(Y|X=x)]^2.$$

Theorem 7.5.4

$$V(Y) = E[V(Y|X)] + V[E(Y|X)]. \tag{7.5.7}$$

Proof:

$$\begin{aligned}E[V(Y|X)] &= E[E(Y^2|X)] - E[E(Y|X)]^2 \\ &= \left\{E(Y^2) - (E(Y))^2\right\} - \left\{E[E(Y|X)]^2 - (E(Y))^2\right\} \\ &= V(Y) - \left[E[E(Y|X)]^2 - \{EE(Y|X)\}^2\right] = V(Y) - V\{E(Y|X)\}.\end{aligned}$$

Theorem 7.5.5 *The following theorem is easy to prove.*

$$\text{(i)}\ E[\{g_1(X) + g_2(X)\}|Y] = E[g_1(X)|Y] + E[g_2(X)|Y] \tag{7.5.8}$$

$$\text{(ii)}\ E[g_1(X)g_2(Y)|X=x] = g_1(x)E[g_2(Y)|X=x]$$

The definitions of conditional expectation, conditional variance can be extended to more than two dimensions. For example

$$E[g(X_1,\cdots,X_n)|X_1 = x_1,\cdots,X_m = x_m], m < n$$
$$= \int_{-\infty}^{\infty}\int_{-\infty}^{\infty} g(x_1,\cdots,x_n) f_{X_{m+1},\ldots,X_n|X_1,\ldots,X_m}(x_{m+1},\ldots,x_n|x_1,\ldots,x_m)\,dx_{m+1}\cdots dx_n$$

in case $(X_1,\cdots,X_n)$ is a n-dimensional random vector.

▶ **Example 7.5.1** *For the p.d.f. in example 7.2.3,*

$$f_{X|Y}(x|y) = \frac{1}{y},\quad 0 < x < 1$$
$$f_{Y|X}(y|x) = \frac{1}{1-x},\quad 0 < y < 1$$
$$E(X|y) = \int_0^1 \frac{x}{y}\,dx = \left.\frac{x^2}{2y}\right|_0^1 = \frac{1}{2y}$$
$$E(Y|X) = \int_0^1 y\frac{dy}{1-x} = \left.\frac{y^2}{2(1-x)}\right|_0^1 = \frac{1}{2(1-x)}.$$

Definition 7.5.4 (***Linear Regression***). *Consider (X, Y) Regression of Y on X is linear if the corresponding regression curve is a straight line, i.e., if*

$$y = E(Y|x) = \eta_x \text{ (say) } = \alpha + \beta x, \tag{7.5.9}$$

where α, β are suitable constants. Similarly regression of X on Y is linear if

$$x = E(X|y) = \xi_y \text{ (say) } = \alpha' + \beta' y \tag{7.5.10}$$

α', β' being suitable constants.

Theorem 7.5.6 $E(Y - \eta_x) = E(X - \xi_y) = 0$. *Proof follows readily by theorem* 7.5.1.

Theorem 7.5.7 *If the regression of Y on X is linear and if $\sigma_X^2, \sigma_Y^2, \rho$ exist then the constants α, β in* (7.5.9) *are given by*

$$\alpha = \mu_Y - \rho\frac{\sigma_Y}{\sigma_X}\mu_X,\quad \beta = \rho\frac{\sigma_Y}{\sigma_X} \tag{7.5.11}$$

Proof: Assume X, Y are continuous.

$$\eta_x = \alpha + \beta x,\quad E(Y|X) = \alpha + \beta X. \tag{7.5.12}$$

Taking expectation of both sides with respect to X, by theorem 7.5.1,

$$E(Y) = \mu_y = \alpha + \beta E(X) = \alpha + \beta\mu_X. \tag{7.5.13}$$

Again, multiplying both sides of (7.5.12) by X and taking expectation w.r.t. X assuming without loss of generality (X, Y) are continuous,

$$\int_{-\infty}^{\infty} x \left[\int_{-\infty}^{\infty} y f_{Y|X}(y|x)\, dy \right] f_X(x)\, dx = \alpha E(X) + \beta E(X^2),$$

i.e.,

$$E(XY) = \alpha E(X) + \beta E(X^2). \tag{7.5.14}$$

Solving equations (7.5.13) and (7.5.14)

$$\beta = \frac{\mu_{X,Y}}{\sigma_X^2} = \rho \frac{\sigma_Y}{\sigma_X}, \quad \alpha = \mu_Y - \beta \mu_X = \mu_Y - \rho \frac{\sigma_Y}{\sigma_X} \mu_X.$$

Corollary 7.5.1 *Similarly, α', β' in (7.5.10) are given by*

$$\alpha' = \mu_X - \rho \frac{\sigma_X}{\sigma_Y} \mu_Y, \quad \beta' = \frac{\mu_{X,Y}}{\sigma_Y^2}.$$

Theorem 7.5.8 *If regression of Y on X is linear and if the conditional variance $V(Y|X = x)$ is independent of x (i.e., conditional distribution of Y given x is homoscedastic), then*

$$V(Y|x) = \sigma_Y^2(1 - \rho^2).$$

Proof:

$$E[Y - E(Y|X)]^2 = \int_{-\infty}^{\infty} \int_{-\infty}^{\infty} \left\{ Y - \mu_Y - \rho \frac{\sigma_Y}{\sigma_X}(X - \mu_X) \right\}^2 f_{X,Y}(x, y)\, dx\, dy$$
$$= \sigma_Y^2(1 - \rho^2)$$

Now $\quad E[Y - E(Y|X)]^2 = E[E\{Y - E(Y|X)\}^2|X] = E[V(Y|X)].$

Since $V(Y|x)$ is constant for all x, $E[V(Y|X)] = \sigma_Y^2(1 - \rho^2)$ implies $V(Y|x) = \sigma_Y^2(1 - \rho^2)$ for all x.

Coefficient of dependence between X and Y: We have noted in definition 7.5.2 that $E(Y/x) = \mu_Y(x)$ is the true regression of Y on x. Therefore, $E[Y - \mu_Y(X)]^2$ would measure the departure of $\mu_Y(x)$ from the true value of Y. Here the expectation is taken with respect to the joint distribution of X and Y.

The quantity

$$1 - \frac{E[Y - \mu_Y(X)]^2}{\sigma_Y^2} = \eta_{YX}^2 \tag{7.5.15}$$

is called the *correlation ratio* of Y on X.

Now,

$$\sigma_Y^2 = E[Y - \mu_Y]^2 = E[Y - \mu_Y(X)]^2 + E[\mu_Y(X) - \mu_Y]^2.$$

Hence,

$$0 \le \eta_{YX}^2 \le 1.$$

Also

$$\eta^2_{YX} = \frac{E[\mu_Y(X) - \mu_Y]^2}{\sigma^2_Y}. \tag{7.5.16}$$

Therefore, if $\mu_Y(X) = \mu_Y$ with probability 1, $\eta^2_{YX} = 0$. If $Y = \mu_Y(X)$ with probability 1, $\eta^2_{YX} = 1$.

We shall now show that

$$\eta^2_{YX} \geq \rho^2.$$

To prove this we are to show

$$E[Y - \alpha - \beta X]^2 \geq E[Y - \mu_Y(X)]^2.$$

Now,

$$E[Y - \alpha - \beta X]^2 = E[Y - \mu_Y(X)]^2 + E[\mu_Y(X) - \alpha - \beta X]^2 \\ +2E[Y - \mu_Y(X)][\mu_Y(X) - \alpha - \beta X].$$

Again,

$$E[Y - \mu_Y(X)][\mu_Y(X) - \alpha - \beta X] = E[\mu_Y(X) - \alpha - \beta X]\, E[Y - \mu_Y(X)|X] = 0.$$

Hence

$$E[Y - \alpha - \beta X]^2 \geq E[Y - \mu_Y(X)]^2$$

equality holds when $\mu_Y(X) = \alpha + \beta X$ with probability unity i.e., $\mu_Y(X)$ is a perfect linear function of X. In this case, there exists a true linear relationship between Y and X and we can not improve our prediction by fitting a higher degree curve.

The coefficient $\eta^2_{YX} - \rho^2$ is a measure of departure of true regression from linearity and is called *coefficient of linearity.*

Least Square Parabolic Regression: Consider a bivariate distribution of (Y, X). Suppose we consider Y as the response variable, X as the explanatory variable and wish to express the regression of Y on X as a pth degree polynomial in X. We, therefore, want to express

$$E(Y/x) = \beta_0 + \beta_1 x + \cdots + \beta_p x^p \tag{7.5.17}$$

where $\beta_0, \beta_1, \cdots, \beta_p$ are constants to be suitably determined. The constants are so determined that

$$E[Y - \beta_0 - \beta_1 X - \cdots - \beta_p X^p]^2 \tag{7.5.18}$$

is minimum. Note that in (7.5.17) the expectation is taken w.r.t. the joint distribution of X and Y. The equation (7.5.17) is known as the parabolic regression equation of pth degree of Y on X.

For determining the constants we differentiate (7.5.18) w.r.t. $\beta_0, \beta_1, \cdots, \beta_p$ and equate them to zero. Denoting $E(X^r Y^s)$ as $\mu'_{r,s}$, the normal equations are:

$$\left.\begin{aligned} \beta_0 + \beta_1\mu'_{1,0} + \cdots + \beta_p\mu'_{p,0} &= \mu'_{0,1} \\ \beta_0\mu'_{1,0} + \beta_1\mu'_{2,0} + \cdots + \beta_p\mu'_{p+1,0} &= \mu'_{1,1} \\ \cdots\ \cdots\ \cdots\ \cdots\ \cdots &= \cdots \\ \beta_0\mu'_{p,0} + \beta_1\mu'_{p+1,0} + \cdots + \beta_p\mu'_{2p,0} &= \mu'_{p,1} \end{aligned}\right\} \tag{7.5.19}$$

These equations are uniquely solvable if the matrix

$$\begin{bmatrix} \mu'_{0,0} & \mu'_{1,0} & \cdots & \mu'_{p,0} \\ \cdots & \cdots & \cdots & \cdots \\ \mu'_{p,0} & \mu'_{p+1,0} & \cdots & \mu'_{2p,0} \end{bmatrix} = \boldsymbol{\mu}$$

is non-singular. To determine the conditions under which this happens consider

$$Q = E\{u_0 + u_1X + \cdots + u_pX^p\}^2 = \sum_i \sum_j u_iu_j\mu'_{i+j,0} \tag{7.5.20}$$

where $u_0, u_1, \cdots, u_p$ are some arbitrary constants, not all zero. The expression (7.5.20) is a non-negative definite quadratic form with $\boldsymbol{\mu}$ as its matrix. Therefore, an investigation into the condition when Q is non-zero will throw light on the non-singularity of $\boldsymbol{\mu}$.

Lemma 7.5.1 *The quadratic form Q is positive definite iff the probability mass of X-distribution is not confined to p or less distinct points.*

Proof: ***Sufficiency.*** The equation

$$u_0 + u_1x + \cdots + u_px^p = 0$$

would have at most p distinct real roots. Since the total probability of X is not confined to p or less distinct points we cannot have

$$u_0 + u_1X + \cdots + u_pX^p = 0$$

with probability one. This means $Q > 0$.

Necessity. If the X-distribution is confined to $q(\leq p)$ distinct points we can always find a set $(u_0, u_1, \cdots, u_p)(\neq (0, \cdots, 0))$ such that

$$u_0 + u_1x + \cdots + u_px^p = 0$$

at the q-admissible points of this distribution. In this case

$$E\{u_0 + u_1X + \cdots + u_pX^p\}^2 = Q = 0.$$

Hence the proof.

[Suppose the bivariate distribution is an observed distribution $(x_1, y_1), \cdots, (x_n, y_n)$. In such a case the constants are solvable if the number of distinct points $x_1, \cdots, x_n$ is greater than p. In fact, if there are q distinct values $x_{(1)}, \cdots, x_{(q)}$ so that each value appears several times in $\{(x_1, y_1), \cdots, (x_n, y_n)\}$ then one can fit a pth degree equation iff $q > p$].

If $\varepsilon_p(X)$ denotes the residual $Y - \sum_{j=0}^{p} \hat{\beta}_j X^j$, then it follows from (7.5.18) and (7.5.19) hat

$$\begin{aligned}\operatorname{Var}(\varepsilon_p(X)) &= E\left[Y - \hat{\beta}_0 - \hat{\beta}_1 X - \cdots - \hat{\beta}_p X^p\right]^2 \\ &= \mu'_{0,2} - \beta_0\mu'_{0,1} - \beta_1\mu'_{1,1} - \cdots - \beta_p\mu'_{p,1} \end{aligned} \tag{7.5.21}$$

'he quantity,

$$\rho_p^2 = 1 - \frac{\operatorname{Var}(\varepsilon_p(x))}{\sigma_Y^2} \tag{7.5.22}$$

nay be used as a measure of goodness of the polynomial regression in describing the true egression $\mu_Y(x)$. Its positive square root, ρ_p is called the *correlation index* of degree p f Y on X.

It can be proved that

$$0 \le \rho^2 \le \rho_2^2 \le \cdots \le \eta_{YX}^2. \tag{7.5.23}$$

'he correlation ratio η_{YX}^2 serves two purposes. It may be used as a measure to which ' may be said to depend on X. Secondly, $\eta_{YX}^2 - \rho^2$ or $\eta_{YX}^2 - \rho_p^2$ indicates the degree to vhich the true regression of Y on X deviates from the linear regression or the polynomial egression.

Some other statistics defined on a multivariate distribution have been considered in hapter 10.

We conclude this chapter by considering two important probability distributions, ne is a discrete multinomial distribution (on $\mathbb{R}^n$), the other is continuous—a bivariate ormal distribution.

7.6 Multinomial Distribution

Suppose an experiment may result in k possible outcomes $0_1, \cdots, 0_k$ with probabilities $p_1, p_2, \cdots, p_k$ respectively, $\sum_{1}^{k} p_i = 1$. Such independent experiments, N in number, are performed. Let X_i be the number of experiments that result in outcome $O_i (i = 1, \cdots, k)$, $\sum_{i=1}^{k} X_i = N$.

$$P(X_1 = x_1, \cdots, X_k = x_k) = P(x_1, \cdots, x_k) = \frac{N!}{\prod_{i=1}^{k} x_i!} \prod_{i=1}^{k} p_i^{x_i}, \tag{7.6.1}$$

where $x_i = 0, 1, \cdots, N$, $\sum_{i=1}^{k} x_i = N$.

By virtue of independence, any specific order of N outcomes with 0_i occurring x_i times ($i = 1, \cdots, k$) has probability $\prod_{i=1}^{k} p_i^{x_i}$ and since there are $\frac{N!}{\prod_{i=1}^{k} x_i!}$ such orderings, (7.6.1) follows.

The random vector $(X_1, \cdots, X_k)$ is said to follow a multinomial distribution with parameters $N, p_1, \cdots, p_{k-1}$ when the p.m.f. is given by (7.6.1).

The expression $P(x_1, \cdots, x_k)$ shows that it can be regarded as the coefficient of $\prod_{j=1}^{k} t_j^x$ in the expansion of

$$(t_1 p_1 + \cdots + t_k p_k)^N.$$

The moment generating function is

$$M_{X_1,\cdots,X_k}(t_1, \cdots, t_k) = E(e^{t_1 X_1 + \cdots + t_k X_k})$$

$$= \sum_{x_1,\cdots,x_k} \frac{N!}{\prod_{i=1}^{k} x_i!} \prod_{i=1}^{k} (p_i e^{t_i})^{x_i} = (p_1 e^{t_1} + \cdots + p_k e^{t_k})^N. \tag{7.6.2}$$

The marginal m.g.f. of X_i, $(i = 1, \cdots, k)$ is

$$M_{X_i}(t_i) = M(0, \cdots, 0, t_i, 0, \cdots 0) = (p_i e^{t_i} + q_i)^N, \quad q_i = 1 - p_i$$

which is the m.g.f. of a binomial distribution with parameters (N, p_i). Hence marginal distribution of X_i is binomial (N, p_i).

The marginal m.g.f. of a set $(X_1, \cdots, X_m)$, $m < k$ is

$$M_{X_1,\cdots,X_m}(t_1, \cdots, t_m) = M(t_1, \cdots, t_m, 0, \cdots, 0) = (p_1 e^{t_1} + \cdots + p_m e^{t_m} + 1 - p_1 - \cdots - p_m)^N \tag{7.6.3}$$

so that $(X_1, \cdots, X_m)$ follows a $(m+1)$-nomial distribution with parameters $N, p_1, \cdots, p_m$. Calculating directly, the marginal p.m.f. of a set $(X_1, \cdots, X_m), m < k$ is

$$\begin{aligned} f_{X_1,\cdots,X_m}(x_1, \cdots, x_m) &= \sum_{x_{m+1},\cdots,x_k} \frac{N!}{x_1! \cdots x_k!} p_1^{x_1} \cdots p_k^{x_k} \\ &= \frac{N!}{x_1! \cdots x_m! \left(N - \sum_{j=1}^{m} x_j\right)!} p_1^{x_1} \cdots p_m^{x_m} Q^{N - \Sigma_{j=1}^{m} x_j} \\ &\times \left[\sum_{x_{m+1}\cdots,x_k} \frac{\left(N - \sum_{j=1}^{m} x_j\right)!}{x_{m+1}! \cdots x_k!} \left(\frac{p_{m+1}}{Q}\right)^{x_{m+1}} \cdots \left(\frac{p_m}{Q}\right)^{x_k} \right] \end{aligned} \tag{7.6.4}$$

where $Q = 1 - \sum_{j=1}^{m} p_j$. The quantity in [ ] being the sum of the probabilities of a multinomial distribution with parameter $\left(N - \sum_{j=1}^{m} x_j\right)$, $\frac{p_{m+j}}{Q}(\geq 0)$, $j = 1, \cdots, k-m$, is unity. Hence the marginal distribution of $X_1, \cdots, X_m$ is multinomial with parameters $N, p_1, \cdots, p_m$. We now consider the conditional distribution of $X_{m+1}, \cdots, X_k$ given $(X_1 = x_1, \cdots, X_m = x_m)$.

$$f_{X_{m+1},\ldots,X_k|X_1,\cdots,X_m}(x_{m+1}, \cdots, x_k|x_1, \cdots, x_m)$$

$$= \frac{f_{X_1,\cdots,X_k}(x_1, \cdots, x_k)}{f_{X_1,\cdots,X_m}(x_1, \cdots, x_m)}$$

$$= \frac{\dfrac{N!}{\prod_{i=1}^{k} x_i!} \prod_{i=1}^{k} p_i^{x_i}}{\dfrac{N!}{x_1!\cdots x_m!\left(N - \sum_{j=1}^{m} x_j\right)!} \prod_{j=1}^{m} p_j^{x_j} \left(1 - \sum_{j=1}^{m} p_j\right)^{N - \Sigma_{j=1}^{m} x_j}}$$

$$= \frac{\left(N - \sum_{j=1}^{m} x_j\right)!}{x_{m+1}! \cdots x_k!} \left(\frac{p_{m+1}}{Q}\right)^{x_{m+1}} \cdots \left(\frac{p_k}{Q}\right)^{x_k}. \tag{7.6.5}$$

This is a multinomial distribution with parameters

$$\left(N - \sum_{j=1}^{m} x_j\right), \quad \frac{p_{m+j}}{Q}, \quad j = 1, \cdots, k-m.$$

We now consider the moments of the distribution (7.6.1) from its m.g.f. (7.6.2).

$$\frac{\partial M(t_1, \cdots, t_k)}{\partial t_i} = N p_i e^{t_i} (p_1 e^{t_1} + \cdots + p_k e^{t_k})^{N-1}$$

$$\frac{\partial^2}{\partial t_i^2} M(t_1, \cdots, t_k) = N(N-1) p_i^2 e^{2t_i} (p_1 e^{t_1} + \cdots + p_k e^{t_k})^{N-2}$$
$$+ N p_i e^{t_i} (p_1 e^{t_1} + \cdots + p_k e^{t_k})^{N-1}$$

$$\frac{\partial^2}{\partial t_i \partial t_j} M(t_1, \cdots, t_k) = N(N-1) p_i p_j e^{t_i + t_j} (p_1 e^{t_1} + \cdots + p_k e^{t_k})^{N-2},$$
$$(\text{for } i \neq j).$$

Hence, evaluating the derivatives at $(0, \cdots, 0)$

$$E(X_i) = Np_i, E(X_i^2) = Np_i + N(N-1)p_i^2$$
$$E(X_iX_j) = N(N-1)p_ip_j$$
$$V(X_i) = Np_i(1-p_i)$$
$$\text{Cov}\,(X_i, X_j) = -Np_ip_j \quad i \neq j$$
$$\rho(X_i, X_j) = -\sqrt{\frac{p_ip_j}{(1-p_i)(1-p_j)}}. \tag{7.6.6}$$

The regression of X_i on X_j is linear:

$$E[X_i|X_j = x_j] = (N - x_j)p_i(1-p_j)^{-1}.$$

The modes of the distribution (7.6.1) are near the expected values, are at

$$x_{1m} \simeq Np_1, \cdots, x_{km} \simeq Np_k.$$

The joint factorial moments of this distribution is

$$\mu_{(r_1,\cdots,r_k)} = E[(X_1)_{r_1} \cdots (X_k)_{r_k}] = \left[(N) \sum_{j=1}^{k} r_j \right] p_1^{r_1} \cdots p_k^{r_k} \tag{7.6.7}$$

where $^{(a)}b$ denotes $a(a-1)\cdots(a-b+1)$.
The probability generating function is

$$(p_1t_1 + \cdots + p_kt_k)^N \tag{7.6.8}$$

The cumulant generating function is

$$\log_e M(t_1, \cdots, t_k) = N \log_e \sum_{j=1}^{k} p_j e^{t_j}. \tag{7.6.9}$$

7.7 Bivariate Normal Distribution

The two-dimensional random variable (X, Y) is said to follow a bivariate normal distribution with parameters $\mu_X, \mu_Y, \sigma_X, \sigma_Y$, and ρ if their joint p.d.f. is given by

$$f_{X,Y}(x, y) = \frac{1}{2\pi\sigma_X\sigma_Y\sqrt{1-\rho^2}} \times \exp\left[-\frac{1}{2(1-\rho^2)} \left\{ \left(\frac{x-\mu_X}{\sigma_X}\right)^2 \right.\right.$$
$$\left.\left. -2\rho\left(\frac{x-\mu_X}{\sigma_X}\right)\left(\frac{y-\mu_Y}{\sigma_Y}\right) + \left(\frac{y-\mu_Y}{\sigma_Y}\right)^2 \right\} \right] \tag{7.7.1}$$

$-\infty < x < \infty$, $-\infty < y < \infty$, when the parameters satisfy $-\infty < \mu_X < \infty$, $-\infty < \mu_Y < \infty$, $\sigma_X > 0$, $\sigma_Y > 0$, $|\rho| < 1$.

To show that

$$\int_{-\infty}^{\infty}\int_{-\infty}^{\infty} f(x,y)\,dx\,dy = 1 \tag{7.7.2}$$

substitute

$$\frac{x-\mu_X}{\sigma_X} = u, \quad \frac{y-\mu_Y}{\sigma_Y} = v. \tag{7.7.3}$$

The left-hand side of (7.7.2) becomes

$$\frac{1}{2\pi\sqrt{1-\rho^2}}\int_{-\infty}^{\infty}\int_{-\infty}^{\infty} e^{-\frac{1}{2(1-\rho^2)}[u^2-2uv+v^2]}\,du\,dv \tag{7.7.4}$$

$$= \frac{1}{2\pi\sqrt{1-\rho^2}}\int_{-\infty}^{\infty}\int_{-\infty}^{\infty} e^{-\frac{1}{2(1-\rho^2)}[(u-\rho v)^2+(1-\rho^2)v^2]}\,du\,dv. \tag{7.7.5}$$

Letting $w = \frac{u-\rho v}{\sqrt{1-\rho^2}}$, (7.7.5) becomes

$$\left[\frac{1}{\sqrt{2\pi}}\int_{-\infty}^{\infty} e^{-\frac{w^2}{2}}\,dw\right]\left[\frac{1}{\sqrt{2\pi}}\int_{-\infty}^{\infty} e^{-\frac{v^2}{2}}\,dv\right] = 1.$$

NOTE 7.7.1 *If $\rho^2 = 1$, the p.d.f. (7.7.1) is not defined. In this case it is known that there exist constants a, b such that $P(Y = aX + b) = 1$. The distribution then is said to be a bivariate degenerate (or singular) normal distribution.*

We shall here consider non-degenerate bivariate normal distribution, where $\rho^2 < 1$. The moment generating function is

$$M_{X,Y}(t_1,t_2) = E\left(e^{t_1X+t_2Y}\right) = E\left(e^{t_1(\mu_X+U\sigma_X)+t_2(\mu_Y+V\sigma_Y)}\right)$$

where U and V are given in (7.7.3).

$$= \frac{e^{t_1\mu_X+t_2\mu_Y}}{2\pi\sqrt{1-\rho^2}}\int_{-\infty}^{\infty}\int_{-\infty}^{\infty} e^{t_1u\sigma_X+t_2v\sigma_Y-\frac{1}{2(1-\rho^2)}[u^2-2\rho uv+v^2]}\,du\,dv. \tag{7.7.6}$$

The exponent in (7.7.6) is

$$-\frac{1}{2(1-\rho^2)}[u^2-2\rho uv+v^2-2(1-\rho^2)\{t_1u\sigma_X+t_2v\sigma_Y\}]$$

$$= -\frac{1}{2}S^2-\frac{1}{2}Z^2+\frac{1}{2}(t_1^2\sigma_X^2+2\rho t_1t_2\sigma_X\sigma_Y+t_2^2\sigma_Y^2)$$

where $S = \frac{\{u-\rho v-(1-\rho^2)t_1\sigma_X\}}{\sqrt{1-\rho^2}}$, $Z = v-\rho t_1\sigma_X - t_2\sigma_Y$.

Hence the integral becomes

$$e^{t_1\mu_X+t_2\mu_Y+\frac{1}{2}(t_1^2\sigma_X^2+2\rho t_1t_2\sigma_X\sigma_Y+t_2^2\sigma_Y^2)}\left[\frac{1}{\sqrt{2\pi}}\int_{-\infty}^{\infty} e^{-\frac{s^2}{2}}\,ds\right]\left[\frac{1}{\sqrt{2\pi}}\int_{-\infty}^{\infty} e^{-\frac{z^2}{2}}\,dz\right]$$

$$= \exp\left[t_1\mu_X+t_2\mu_Y+\frac{1}{2}(t_1^2\sigma_X^2+2\rho t_1t_2\sigma_X\sigma_Y+t_2^2\sigma_Y^2)\right]. \tag{7.7.7}$$

The m.g.f. of the distribution (7.7.1) about the point (μ_X, μ_Y) is

$$\begin{aligned} M_{X-\mu_X, Y-\mu_Y} &= e^{-\mu_X t_1 - \mu_Y t_2} M_{X,Y}(t_1, t_2) \\ &= \exp\left[\frac{1}{2}(t_1^2\sigma_X^2 + 2\rho t_1 t_2 \sigma_X \sigma_Y + t_2^2\sigma_Y^2)\right]. \end{aligned} \tag{7.7.8}$$

From (7.7.7), the following results follow:

$$\begin{aligned} E(X) &= \left.\frac{\partial M(t_1, t_2)}{\partial t_1}\right|_{t_1=t_2=0} &&= \mu_X \\ E(X^2) &= \left.\frac{\partial^2 M(t_1, t_2)}{\partial t_1^2}\right|_{t_1=t_2=0} &&= \mu_X^2 + \sigma_X^2 \\ E(XY) &= \left.\frac{\partial^2 M(t_1, t_2)}{\partial t_1 \partial t_2}\right|_{t_1=t_2=0} &&= \rho\sigma_X\sigma_Y + \mu_X\mu_Y. \end{aligned}$$

Hence $E(Y) = \mu_Y, V(X) = \sigma_X^2, V(Y) = \sigma_Y^2$, Cov $(X, Y) = \rho\sigma_X\sigma_Y$, Corr. Coeff $(X, Y) = \rho$.

Writing $\frac{(X-\mu_X)}{\sigma_X} = U$ and $\frac{(Y-\mu_Y)}{\sigma_Y} = V$ we have the standardised bivariate norma density with parameter ρ,

$$f(u, v) = \frac{1}{2\pi\sqrt{1-\rho^2}} e^{-\frac{1}{2(1-\rho^2)}\{u^2 - 2\rho uv + v^2\}} \quad -\infty < u < \infty, -\infty < v < \infty.$$

The m.g.f. of a standardised bivariate normal density is

$$M_{U,V}(t_1, t_2) = e^{\frac{1}{2}(t_1^2 + 2\rho t_1 t_2 + t_2^2)}$$

For this distribution

$$\begin{aligned} &\mu_{21} = \mu_{12} = 0, && \mu_{22} = 1 + 2\rho^2, \\ &\mu_{31} = \mu_{13} = 3\rho, && \mu_{32} = \mu_{23} = 0, && \mu_{33} = 3\rho(3 + 2\rho^2), \\ &\mu_{41} = \mu_{14} = 0, && \mu_{42} = \mu_{24} = 3(1 + 4\rho^2), \\ &\mu_{51} = \mu_{15} = 15\rho. \end{aligned}$$

All the cumulants of order higher than 2 vanish.
The marginal m.g.f.'s of the distribution (7.7.1) are:

$$\begin{aligned} M_X(t_1) &= M_{X,Y}(t_1, 0) = e^{\mu_X t_1 - \frac{1}{2}t_1^2\sigma_X^2} \\ M_Y(t_2) &= M_{X,Y}(0, t_2) = e^{\mu_Y t_2 - \frac{1}{2}t_2^2\sigma_Y^2} \end{aligned}$$

so that the marginal distribution of X is univariate normal $N(\mu_X, \sigma_X)$ and similarly for Y.

We have already noted that if two random variables are uncorrelated they are not necessarily independent.

Theorem 7.7.1 *If (X, Y) have a bivariate normal distribution, then X, Y are independent if and only if they are uncorrelated.*

Proof: If X, Y are uncorrelated, i.e., if $\rho = 0$, the bivariate density (7.7.1) can be written as the product of two univariate normal densities so that X, Y are independent. Of course, if X, Y are independent, they must be uncorrelated.

Normal variables, like other variables, may however be dependent with zero correlations. Thus X, Y may each have univariate normal density with zero correlation, but the joint density of X, Y may not be a bivariate normal.

Theorem 7.7.2 *If (X, Y) have a bivariate normal distribution, the marginal density of X is $N(\mu_X, \sigma_X)$ and of Y is $N(\mu_Y, \sigma_Y)$.*

Proof:

$$f_X(x) = \int_{-\infty}^{\infty} f(x, y)\, dy.$$

Substituting $v = \dfrac{y - \mu_Y}{\sigma_Y}$

$$= \frac{1}{2\pi\sigma_X\sqrt{1-\rho^2}} \int_{-\infty}^{\infty} \exp\left[-\frac{1}{2}\left(\frac{x-\mu_X}{\sigma_X}\right)^2 - \frac{1}{2(1-\rho^2)}\left(v - \rho\frac{x-\mu_X}{\sigma_X}\right)^2\right] dv.$$

Substituting $w = \left\{v - \rho\left(\dfrac{x-\mu_X}{\sigma_X}\right)\right\} \Big/ \sqrt{1-\rho^2}$,

$$f_X(x) = \frac{1}{\sigma_X\sqrt{2\pi}} \exp\left[-\frac{1}{2}\left(\frac{x-\mu_X}{\sigma_X}\right)^2\right].$$

Similarly for Y.

Theorem 7.7.3 *If (X, Y) has a bivariate normal distribution, then the conditional distribution of Y given $X = x$ is univariate normal*

$$N\left(\mu_X + \rho\frac{\sigma_Y}{\sigma_X}(x - \mu_X),\ \sigma_Y^2(1-\rho^2)\right).$$

Similarly the conditional distribution of X given $Y = y$ is univariate normal

$$N\left(\mu_X + \rho\frac{\sigma_Y}{\sigma_X}(y - \mu_Y),\ \sigma_X^2(1-\rho^2)\right).$$

Proof:

$$f_{Y|X}(y|x) = \frac{f_{X,Y}(x,y)}{f_X(x)} = \frac{1}{\sigma_X\sqrt{2\pi(1-\rho^2)}} e^{-\frac{1}{2\sigma_Y^2(1-\rho^2)}\{y-\mu_Y-\rho\frac{\sigma_Y}{\sigma_X}(x-\mu_X)\}^2} \tag{7.7.9}$$

obtained by the substitutions as in (7.7.3) and proceeding similarly as in the proof of theorem 7.7.2. Hence the conditional distribution of Y is univariate normal with parameters as stated above. Similarly for $f_{Y|X}(y|x)$.

Corollary 7.7.1 *The regression curve of Y on X is*

$$y = \mu_Y + \rho\frac{\sigma_Y}{\sigma_X}(x - \mu_X) \tag{7.7.10}$$

which is a linear function of x. More specifically, the regression curve of Y on X is a straight line with gradient $\rho\frac{\sigma_Y}{\sigma_X}$ and intercept

$$\left(\mu_Y - \rho\frac{\sigma_Y}{\sigma_X}\mu_X\right).$$

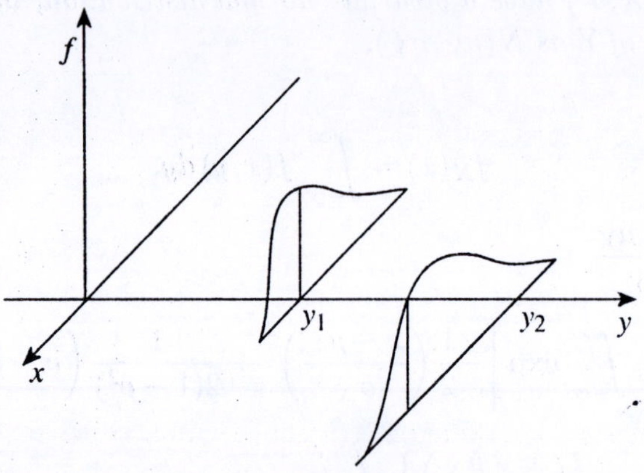

Fig. 7.3

Similarly the regression curve of X on Y is a straight line (Fig. 7.3)

$$x = \mu_X + \rho\frac{\sigma_X}{\sigma_Y}(y - \mu_Y). \tag{7.7.11}$$

■ NOTE 7.7.2 *The variance of the conditional distribution of Y given x (X given y) does not depend on x(y). The conditional distributions are homoscedastic.*

The locus of the points in the xy-plane such that $f(x, y)$ is a constant is an ellipse

$$\frac{(x-\mu_X)^2}{\sigma_X^2} - 2\rho\frac{(x-\mu_X)(y-\mu_Y)}{\sigma_X\sigma_Y} + \frac{(y-\mu_Y)^2}{\sigma_Y^2} = k^2 \tag{7.7.12}$$

with centre (μ_X, μ_Y), where k is a constant (independent of x and y). Thus if the probability density surface (7.7.1) is cut by the plane $z = z_0$ (a sufficiently small possitive quantity), parallel to the xy-plane, the resulting contour curves are similar and concentric ellipses (7.7.12) with centre (μ_X, μ_Y). The curves (7.7.12) are called equi-probability contours. k is a decreasing function of z_0. The density $f(x, y)$ is maximum at the centre and is equal to

$$f(\mu_X, \mu_Y) = \frac{1}{2\pi\sigma_X\sigma_Y(1-\rho^2)}.$$

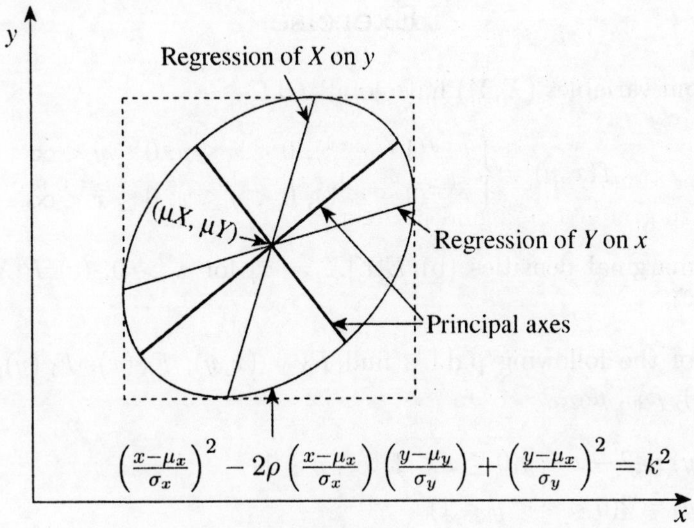

Fig. 7.4

The line of regression of y on $x(x$ on $y)$ passes through the two points where the tangent to the ellipse are vertical (horizontal) [Fig. 7.4].

As the value of $|\rho|$ tends to one, the ellipse becomes elongated and both the lines of regression approaches the same diagonal of the circumscribed rectangle.

If $\rho = 0$, the principal axes of each ellipse are parallel to the co-ordinate axes.

If $t_x = \frac{x-\mu_X}{\sigma_X}$, $t_y = \frac{y-\mu_y}{\sigma_y}$ and t_y is graphed as a function of t_x, the curves (7.7.12) will still be ellipses; however, they will be centred at the new origin and the semi-major axis will be along the 45° line though the origin if $\rho > 0$, and along the 135° line if $\rho < 0$. For $\rho > 0$, the semi-major and semi-minor axes are proportional to $\sqrt{1+\rho}$ and $\sqrt{1-\rho}$ respectives, with the reverse holding for $\rho < 0$.

Consider a bivariate normal r.v. (X, Y) with $\rho(X, Y)$ not necessary equal to zero. For the transformed variables

$$U = X\cos\theta + Y\sin\theta, \quad V = -X\sin\theta + Y\cos\theta \tag{7.7.13}$$

$$\text{Cov}\,(U, V) = \frac{1}{2}(\sigma_Y^2 - \sigma_X^2)\sin 2\theta + \rho\sigma_X\sigma_Y\cos 2\theta.$$

Hence U, V are uncorrelated if

$$2\theta = \tan^{-1}\left(\frac{2\rho\sigma_X\sigma_Y}{\sigma_X^2 - \sigma_Y^2}\right). \tag{7.7.14}$$

Therefore, by the transformation of co-ordinates from (X, Y) to (U, V) as in (7.7.13), (7.7.14), the new axes of co-ordinates become parallel to the principal axes of the ellipse (7.7.12).

Exercise

1. The random variables (X, Y) have joint p.d.f.

$$f(x,y) = \begin{cases} e^{-y}(1-e^{-x}), \ 0 < x \leq y, \ 0 \leq y < \infty \\ e^{-x}(1-e^{-y}), \ 0 < y \leq x, \ 0 \leq x < \infty. \end{cases}$$

Find (a) marginal densities (b) $E[Y|X = x]$ for $x > 0$, (c) $P[X \leq 2, Y \leq 2]$, (d) $\rho(X, Y)$.

2. For each of the following p.d.f.'s find $F_{X,Y}\ (x, y)$, $F_X(x)$, $F_Y(y)$, $f_X(x)$, $f_Y(y)$, $F_{Y|X}\ (y|x)$, μ_{20}, μ_{02}, ρ.

   (a) $f(x, y) = 2 - x - y (0 \leq x \leq 1, 0 \leq y \leq 1)$
   (b) $f(x, y) = 2 (0 < x < y < 1)$
   (c) $f(x, y) = \frac{6-x-y}{8} (0 < x < 2, 2 < y < 4)$
   (d) $f(x, y) = 2e^{-x-y} (0 < x < y < \infty)$
   (e) $f(x, y) = x + y (0 < x < 1, 0 < y < 1)$
   (f) $f(x, y) = \frac{(n-1)(n-2)}{n(1+x+y)}, x > 0, y > 0, n(\text{a fixed constant}) > 2.$
   (g) $f(x, y) = x^2 + \frac{xy}{3}, (0 < x < 1, 0 < y < 2).$

3. (X, Y) have bivariate cumulative distribution function

$$F_{X,Y}(x, y) = \begin{cases} 1 - e^{-x} - e^{-y} + e^{-(x+y)}, \ x, y \geq 0 \\ 0 \qquad \text{elsewhere.} \end{cases}$$

Find joint p.d.f. and marginal p.d.f.'s.

4. (X, Y) have joint p.d.f.

$$f(x, y) = \begin{cases} \frac{1}{2} \text{ for } |x| + |y| \leq 1 \\ 0 \quad \text{elsewhere.} \end{cases}$$

Find marginal p.d.f.'s and conditional p.d.f.'s.

$$\Big[f_X(x) = \int_{|x|+|y| \leq 1} \tfrac{1}{2}\, dy = \int_{|x|-1}^{1-|x|} \tfrac{1}{2}\, dy = 1 - |x|, |x| \leq 1.$$
$$= 0 \text{ (elsewhere).}$$

$$f_X(x|y) = \begin{cases} \frac{1}{2(1-|y|)} & \text{for } |x| \leq 1 - |y| \text{ if } |y| \leq 1 \\ 0 & \text{for } |x| > 1 - |y| \text{ if } |y| \leq 1 \\ \text{undefined for all } x \text{ if } |y| > 1. \end{cases}$$

5. (X,Y) have joint p.d.f. $f(x,y) = 3x^2y + 3y^2x, 0 \le x \le 1, 0 \le y \le 1; = 0$ elsewhere. Find the marginal p.d.f.'s. Find the conditional $P\left(\frac{1}{2} \le Y \le \frac{3}{4} | \frac{1}{3} \le X \le \frac{2}{3}\right)$

$$\left[f_X(x) = x + \tfrac{3}{2}x^2,\ f_Y(y) = y + \tfrac{3}{2}y^2,\ \text{Prob.} = \tfrac{515}{4096}\right].$$

6. X has a discrete probability distribution $P(X = x) = \frac{x}{3}, x = 1, 2; f_{Y|X}(y|x)$ is binomial with parameters $\left(x, \frac{1}{2}\right), y = 0, 1, \cdots, x$. Find $E(X), V(X), f(x,y), E(Y)$.

7. The joint p.d.f. of bivariate gamma variables (X,Y) is

$$f(x,y) = \frac{\lambda^{r+p}}{\lceil(r)\lceil(p)} e^{-\lambda x} y^{p-1}(x-y)^{r-1}, \quad 0 < y < x < \infty; \quad \lambda, p, r > 0.$$

Find the marginal p.d.f.'s and conditional p.d.f.'s.

8. Consider (X,Y) with $Y = 2X$. Show that

$$F_{X,Y}(x,y) = \begin{cases} F_X(x) & \text{for } x \le \frac{y}{2} \\ F_X\left(\frac{y}{2}\right) = F_Y(y) & \text{for } x > \frac{y}{2}. \end{cases}$$

[*Hints:* (a) Suppose $2x \le y$. The events, $\{X \le x\} \Leftrightarrow \{Y \le 2x\}$. Hence

$$\begin{aligned} P\{X \le x, Y \le y\} &= P\{X \le x, Y \le 2x\} + P\{X \le x, 2x < Y \le y\} \\ &= P\{X \le x, Y \le 2x\} = P\{X \le x\} = F_X(x). \end{aligned}$$

(b) Suppose $y < 2x$. The events, $\{Y \le y\} \Leftrightarrow \left\{X \le \frac{y}{2}\right\}$.

$$\begin{aligned} P\{X \le x, Y \le y\} &= P\left\{X \le \tfrac{y}{2}, Y \le y\right\} + P\left\{\tfrac{y}{2} < X \le \tfrac{x}{2}, Y \le y\right\} \\ &= P\left\{X \le \tfrac{y}{2}, Y \le y\right\} = P\left\{X \le \tfrac{y}{2}\right\} = F_X\left(\tfrac{y}{2}\right)]. \end{aligned}$$

9. Consider (X,Y) with $Y = g(X)$. Show that if

(a) $g(t)$ is monotonically increasing

$$F_{X,Y}(x,y) = \begin{cases} F_Y(y) & \text{if } y < g(x) \\ F_X(x) & \text{if } y \ge g(x). \end{cases}$$

(b) $g(t)$ is monotonically decreasing

$$F_{X,Y}(x,y) = \begin{cases} 0 & \text{if } y < g(x) \\ F_X(x) + F_Y(y) - 1 & \text{if } y \ge g(x). \end{cases}$$

10. (X, Y) have joint p.d.f. $f(x, y)$.

 (a) Show that for the following p.d.f.'s

  (i) X, Y are stochastically independent:

   A. $f(x,y) = 6xy^2, 0 < x, y < 1.$

   B. $f(x,y) = 12xy(1-y), 0 < x, y < 1.$

   C. $f(x,y) = 4xy, 0 < x, y < 1.$

  (ii) X, Y are stochastically dependent:

   A. $f(x,y) = \frac{1}{2}, |x| + |y| \le 1.$

   B. $f(x,y) = 8xy, 0 \le y \le x < 1.$

   C. $f(x,y) = 2e^{-(x+y)}, 0 < x < y < \infty.$

   D. $f(x,y) = \frac{1}{4\pi^2}[1 - \sin(x+y)], -\pi \le x, y \le \pi.$

   E. $f(x,y) = x + y, 0 < x, y < 1.$

 (b) Verify if X, Y are stochastically independent when

  (i) $f(x,y) = p_1^x(1-p_1)^{1-x}p_2^y(1-p_2)^{1-y}[1 + \lambda(x-p_1)(y-p_2)],\ 0 < x,\ y < 1$
  $0 < p_1,\ p_2 < 1,\ |\lambda| \le 1.$

  (ii) $f(x,y) = \frac{\sqrt{3}}{2\pi}\exp\{-[(x-7)^2 + (x-7)y + y^2]\}.$

11. Show that the r.v. (X, Y) with joint p.d.f.

$$f(x,y) = \frac{1}{2\pi\sqrt{1+y^2}}\exp\left\{-\frac{x^2}{2(1+y^2)} - \frac{y^2}{2}\right\}$$

have covariance zero and yet X, Y are not independent.

12. Show that X, Y are dependent r.v.'s but uncorrelated, when

 (a) (X, Y) is a discrete r.v. with

$$P(X = i, Y = j) = \frac{1}{8},\ i, j = 1, -1;\ P(X = 0, Y = 0) = \frac{1}{2}.$$

 (b) (X, Y) are continuous r.v.'s with $f(x,y) = \frac{1}{2}$ inside the square with corners at points $(1,0)$, $(0,1)$, $(-1,0)$, $(0,-1)$.

 (c) $X = \sin 2\pi R, Y = \cos 2\pi R$ when R has a uniform distribution over $(0,1)$.

13. Let (X, Y) have joint p.d.f.

$$f(x,y) = \begin{cases} \dfrac{1+xy}{4}, & |x| < 1, |y| < 1 \\ 0, & \text{elsewhere.} \end{cases}$$

Show that X, Y are not independent. The r.v.'s X^2, Y^2 are, however, independent.

14. X and θ are independently and uniformly distributed in the intervals $(0, a)$ and $(0, \frac{\pi}{2})$ respectively. Show that

$$P[X < b\cos\theta] = \frac{2b}{\pi a}, b < a.$$

15. Show that r.v.'s X and $g(X)$ are not independent when $g(X)$ is a monotonic function of X.

16. (X, Y) have joint p.d.f. $f(x, y) = \frac{3}{2\sqrt{x}}, 0 < y < x < 1$.

Find

$$f_{Y|X}\left(y|x = \frac{1}{2}\right), f_{X|Y}\left(x|y = \frac{1}{2}\right); \left[2, 0 < y < \frac{1}{2}; \frac{1}{(2-\sqrt{2})\sqrt{x}}, \frac{1}{2} < x < 1\right].$$

17. A r.v. X has p.d.f. $f(x) = \frac{e^{-\frac{x}{2}}}{\sqrt{2\pi x}}, x > 0$.

Conditional p.d.f. of Y given x is

$$f_{X|Y}(y|x) = \frac{\sqrt{x}}{\sqrt{2\pi}} e^{-\frac{xy^2}{2}}, -\infty < y < \infty.$$

Show that the marginal density of Y is

$$f_Y(y) = \frac{1}{\pi(1+y^2)}, -\infty < y < \infty.$$

18. (X, Y) have joint p.d.f. $f(x, y) = A\frac{x}{y}, 0 < x < 1, 1 < y < 2; = 0$ otherwise; A, a constant. Evaluate A.

Find

$$f_{Y|X}(y|x), f_{X|Y}(x|y), \left[\frac{2}{\log_e 2}; 2x, 0 < x < 1; \frac{1}{y\log_e 2}, 1 < y < 2\right].$$

19. For the bivariate distribution

$$f(x, y) = 1, |y| < x, 0 < x < 1.$$

Find the conditional densities, $E(Y|x), \rho(X, Y)$.

$$\begin{aligned}[f_{X|Y}(x|y) &= (1+y)^{-1}, -1 < y \le 0; (1-y)^{-1}, 0 < y < 1 \\ f_{Y|X}(y|x) &= \tfrac{1}{2x}, 0 < x < 1 \\ E(Y|x) &= 0, \rho = 0.]\end{aligned}$$

20. Show that for the distribution

$$f(x, y) = x + y, 0 < x < 1, 0 < y < 1$$

$$\mu'_{r,s} = \frac{1}{(r+2)(s+1)} + \frac{1}{(r+1)(s+2)}, r, s, \text{ being positive}$$

integers; hence $\rho(X, Y) = -\frac{1}{11}$.

21. (X, Y) follows a bivariate normal distribution with $(0, 0, \sigma_X^2, \sigma_Y^2, \rho)$. Show that $U = \frac{X}{\sigma_X} - \rho\frac{Y}{\sigma_Y}$ and $V = (\sqrt{1-\rho^2})Y/\sigma_Y$ are independent variates.

22. Let X_1, X_2 be independent normal with parameters $(\mu_1, \sigma_1), (\mu_2, \sigma_2)$ respectively. Show that the correlation coefficient between U and V where $U = X\cos\theta + Y\sin\theta$, $V = -X\sin\theta + Y\cos\theta$ is given by

$$\rho^2 = 1 - \frac{4\sigma_1^2\sigma_2^2}{4\sigma_1^2\sigma_2^2 + (\sigma_1^2 - \sigma_2^2)^2\sin^2 2\theta}.$$

23. X, Y are standard normal variates with correlation coefficient ρ. Show that correlation coefficient $(X^2, Y^2) = \rho^2$.

[*Hints:* $\rho(X^2, Y^2) = \frac{E(X^2Y^2) - E(X^2)E(Y^2)}{\sqrt{E(X^4)-(E(X^2))^2}\sqrt{E(Y^4)-(E(Y^2))^2}}$.

The m.g.f. $M_{X,Y}(t_1, t_2) = \exp\left[\frac{1}{2}(t_1^2 + 2\rho t_1 t_2 + t_2^2)\right]$. $E(X^2Y^2) =$ coefficient of $\frac{t_1^2}{2!}\frac{t_2^2}{2!}$ in $M_{X,Y}(t_1, t_2) = 2\rho^2 + 1$. Similarly $E(X^4) = E(Y^4) = 3$.]

24. Let (X, Y) have a bivariate normal distribution $(\mu_1, \mu_2, \sigma_1^2, \sigma_2^2, \rho)$. Show that $Z = aX + bY + c$ is a univariate normal $N(a\mu_1 + b\mu_2 + c, a^2\sigma_1^2 + 2ab\rho\sigma_1\sigma_2 + b^2\sigma_2^2)$.

[*Hints:* Use m.g.f. of (X, Y) to find m.g.f. of Z.]

25. (X, Y) follow bivariate $N(\mu_x, \mu_y, \sigma_x, \sigma_y, \rho)$-distribution. Show that $U = \frac{X-\mu_x}{\sigma_x} + \frac{Y-\mu_y}{\sigma_y}$ and $V = \frac{X-\mu_x}{\sigma_x} - \frac{Y-\mu_y}{\sigma_y}$ are independently normally distributed with zero means and variances $2(1+\rho)$ and $2(1-\rho)$ respectively.

26. Multivariate Chebychev inequality: Let $(X_1, \cdots, X_n)$ be jointly distributed with $E(X_i) = \mu_i, V(X_i) = \sigma_i^2, i = 1, \cdots, n$. Define $A_i = \{|X_i - \mu_i| \le \sqrt{n}t\sigma_i\}$. Prove that

$$P\left(\bigcap_{i=1}^n A_i\right) \ge 1 - \frac{1}{t^2} \text{ for } t > 0.$$

[*Hints:* We have by Chebychev inequality $P(A_i) \ge 1 - \frac{1}{nt^2}$; hence $P(A_i^c) \le \frac{1}{nt^2}$. Now by corollary 1.3.3

$$P\left(\bigcap_{i=1}^n A_i\right) \ge 1 - \sum_{i=1}^n P(A_i^c) \ge 1 - \frac{n}{nt^2} = 1 - \frac{1}{t^2}, \quad t > 0.]$$

27. X_1, X_2 are r.v.'s with means 0, variances 1 and correlation coefficient ρ. Show that

$$E[\max(X_1^2, X_2^2)] \le 1 + \sqrt{1-\rho^2}.$$

Hence show that for r.v.'s X, Y that are jointly distributed with means μ_X and μ_Y and variances σ_X^2 and σ_Y^2 respectively and correlation coefficient ρ and for any $t > 0$,

$$P[|X - \mu_X| \ge t\sigma_X \quad \text{or} \quad |Y - \mu_Y| \ge t\sigma_Y] \le \frac{1}{t^2}\{1 + \sqrt{1-\rho^2}\}.$$

[*Hints: First part*: 2 max $(X_1^2, X_2^2) = |X_1^2 - X_2^2| + X_1^2 + X_2^2$.

Hence

$$E\{\max(X_1^2, X_2^2)\} = \frac{1}{2}E|X_1^2 - X_2^2| + \frac{1}{2}E(X_1^2 + X_2^2). \qquad \text{(i)}$$

Now

$$(E|X_1^2 - X_2^2|)^2 = (E\{|X_1 - X_2||X_1 + X_2|\})^2 \le E\{|X_1 - X_2|\}^2 E\{|X_1 + X_2|^2\}$$
$$= \{2(1-\rho)\}\{2(1+\rho)\} \qquad \text{(ii)}$$

(the inequality in (ii) follows from Cauchy-Schwartz inequality).

Hence the result follows from (i).

Second part: Let $\frac{X-\mu_X}{\sigma_X} = Z_1, \frac{Y-\mu_Y}{\sigma_Y} = Z_2$. Hence $E(Z_1) = E(Z_2) = 0, V(Z_1) = V(Z_2) = 1, \rho(Z_1, Z_2) = \rho$.

Let max $(Z_1^2, Z_2^2) = Z$, which is a random variable. By Markov inequality

$$P(Z \ge t^2) \le \frac{E(Z)}{t^2} \le \frac{1+\sqrt{1-\rho^2}}{t^2},$$

i.e., $P[|X - \mu_X| \ge t\sigma_X \text{ or } |Y - \mu_Y| \ge t\sigma_Y] \le \frac{1+\sqrt{1-\rho^2}}{t^2}$.]

28. For the bivariate normal distribution $N(0,0,1,1,\rho)$, show that the moments obey the recurrence relation

$$\mu_{r,s} = (r+s-1)\rho\mu_{r-1,s-1} + (r-1)(s-1)(1-\rho^2)\mu_{r-2,s-2}.$$

Hence or otherwise show that $\mu_{r,s} = 0$ if $r+s$ is odd; $\mu_{3,1} = 3\rho, \mu_{22} = 1 + 2\rho^2$.

[*Hints:* Consider m.g.f. $M(t_1, t_2) = \exp\left\{\frac{1}{2}(t_1^2 + 2\rho t_1 t_2 + t_2^2)\right\}$. Find $\frac{\partial^2 M}{\partial t_1 \partial t_2}, t_1\frac{\partial M}{\partial t_1} + t_2\frac{\partial M}{\partial t_2}$ and get relationship between these quantities. Identify coefficients.]

29. (X, Y) is a discrete r.v. with p.m.f. $P(X = i, Y = j) = p_{ij}, i, j = 0, 1$. Find the m.g.f. of X, m.g.f. of Y, m.g.f. of (X, Y). Show that X, Y are independent if and only if

$$p_{00}p_{11} - p_{01}p_{10} = 0.$$

30. Find the m.g.f. of

(a) $f(x, y) = x + y, 0 < x, y < 1$,

(b) $f(x, y) = e^{-(x+y)}, 0 < x, y < \infty$.

31. Determine the m.g.f. of the r.v. with p.d.f.

$$f(x) = \begin{cases} k_1 & \text{for } a \le x \le a + c_1 \\ k_2 & \text{for } b \le x \le b + c_2 \\ 0 & \text{otherwise.} \end{cases}$$

where a, b, c_1, c_2, k_1, k_2 are constants such that

$$c_1 > 0, c_2 > 0, k_1 > 0, k_2 > 0, a + c_1 \le b, k_1c_1 + k_2c_2 = 1.$$

32. If $f_{X,Y}(x,y) = \frac{1}{8}(6 - x - y), 0 \le x \le 2, 2 \le y \le 4$ calculate

(a) $P(X + Y < 3)$,

(b) $P\left(X < \frac{3}{2}, Y < \frac{5}{2}\right)$. $\left[\frac{3}{8}, \frac{9}{32}\right]$.

33. A r.v. X has an exponential density $f(x) = \lambda e^{-\lambda x}, x > 0, \lambda > 0$. Find (a) $E(Y), V(Y)$ where $Y = e^{-X}$. (b) $E(Z), V(Z)$ when $Z = e^X$ and examine the conditions for their existence.

[*Ans.*

(a) $\frac{\lambda}{\lambda+1}, \frac{\lambda}{(\lambda+1)(\lambda+2)}$;

(b) $\frac{\lambda}{\lambda-1}$ (provided $\lambda > 1$), $\frac{\lambda}{\lambda-2} - \left(\frac{\lambda}{\lambda-1}\right)^2$ (provided $\lambda > 2$).]

34. X has a Poisson distribution with parameter λ. Conditional distribution of Y given x is binomial with parameter (x, p). Show that the marginal distribution of Y is Poisson with parameter λp.

[*Hints:*

$$f(x,y) = h(x)g(y|x) = \left\{e^{-\lambda}\frac{\lambda^x}{x!}\right\}\left\{\binom{x}{y}p^y q^{x-y}\right\}$$

$$x = 0, 1, 2, \cdots, y = 0, 1, \cdots, x, \lambda > 0, p > 0.$$

Let $x - y = z, z = 0, 1, 2, \cdots$

$$f(z,y) = e^{-\lambda}\frac{\lambda^{y+z}}{(y+z)!}\binom{y+z}{y}p^y q^z.$$

Marginal distribution of y is

$$g(y) = \sum_{z=0}^{\infty} f(z,y) = \frac{(\lambda p)^y}{y!}e^{-\lambda}\sum_{z=0}^{\infty}\frac{(\lambda q)^z}{z!} = e^{-\lambda p}\frac{(\lambda p)^y}{y!}, y = 0, 1, \cdots.]$$

35. $X_1, \cdots, X_k$ are k independent Poisson variates with parameters $\lambda_1, \cdots, \lambda_k$ respectively. Show that the conditional distribution of $X_1, \cdots, X_k$ given $\sum_{i=1}^{k} X_i = x$ is multinomial.

[*Hints:*

$$P\left(X_1 = x_1, \cdots, X_k = x_k \,\middle|\, \sum_{i=1}^{k} X_i = x\right)$$

$$= \frac{P\left(X_1 = x_1, \cdots, X_k = x_k; \sum_1^k X_i = x\right)}{P\left(\sum_1^k X_i = x\right)}$$

$$= \frac{\left(e^{-\lambda_1}\frac{\lambda_1^{x_1}}{x_1!}\right)\cdots\left(e^{-\lambda_k}\frac{\lambda_k^{x_k}}{x_k!}\right)}{e^{-\lambda}\frac{\lambda^x}{x!}}, \quad \lambda = \sum_{i=1}^{k}\lambda_i$$

(by the reproductive property of Poisson distribution)

$$= \frac{x!}{x_1!\cdots x_k!}\left(\frac{\lambda_1}{\lambda}\right)^{x_1}\cdots\left(\frac{\lambda_k}{\lambda}\right)^{x_k}$$

which is multinomial with parameters x and $\frac{\lambda_i}{\lambda}, i = 1, \cdots, k - 1$.]

36. Show that for positive values of the parameters, $p_0^a\{1-p_1t_1-p_2t_2\}^{-a}$ is the p.g.f. of (X, Y) such that marginal distributions of X, Y and $X+Y$ have negative binomial distributions.

37. (X, Y) have joint p.d.f. $f(x, y) = 2, 0 < x < y < 1$. Find $E(X|y), V(X|y)$,

$$P\left(X < \frac{1}{2}\middle|Y = \frac{3}{4}\right), P\left(X < \frac{1}{2}\right) \left[\frac{y}{2}, \frac{y^2}{12}, \frac{2}{3}, \frac{3}{4}\right].$$

38. X, Y are jointly distributed discrete r.v.'s. with

$$p_{x,y} = \frac{2}{n(n+1)}(0), x = 1, \cdots, n; y = 1, \cdots, x \text{ (otherwise)}.$$

Show that X, Y are not independent. Find regression of Y on X and of X on Y.

[*Hints:* Marginal densities:

$$f_X(x) = \tfrac{2x}{n(n+1)}, \quad f_Y(y) = \tfrac{2}{n+1}$$
$$E(X|y) = \tfrac{n+1}{2}, \quad E(Y|x) = \tfrac{x+1}{2}.$$

Hence regression of Y on x is $y = \frac{x+1}{2}$.]

39. X, Y have joint p.d.f. $f(x, y) = 1, 0 < x < 1, |y| < x$. Show that on the set of positive probability density, the graph of $E(Y|x)$ is a straight line, whereas that of $E(X|y)$ is not a straight line.

40. Show that the angle θ between the regression lines (of Y on X and of X on Y) is

$$\tan^{-1}\left(\frac{1-\rho^2}{\rho}\frac{\sigma_X\sigma_Y}{\sigma_X^2+\sigma_Y^2}\right).$$

What can be concluded from this about the relative location of the two lines of regression for ρ close to 0 and for ρ close to 1?

41. (X, Y) has a p.d.f. $f(x, y) = k \cdot e^{-(ax+by)}, 0 \le y \le x+1, a(>0), b(>0)$ are given constants. Find the normalising constant k; find the regression curve of Y on X. Sketch this regression curve for $a = 2, b = 1$.

42. X, Y, Z have distribution

$$dF = Ax^{l-1}y^{m-1}z^{n-1}, x, y, z \ge 0, x+y+z \le 1.$$

Show that $A = \dfrac{\Gamma(l+m+n+1)}{\Gamma(l)\,\Gamma(m)\,\Gamma(n)}$.

43. The r.v. (X_1, X_2, X_3) has the trivariate normal probability density

$$f_{X_1,X_2,X_3}(x_1, x_2, x_3) = k \exp\left[-\frac{1}{2}\left(2X_1^2 + X_2^2 + 4X_3^2 - X_1X_2 - 2X_1X_3\right)\right].$$

Write the matrix of coefficients of the quadratic from in the expression and compute k. Obtain (a) marginal p.d.f. of (X_1, X_3), (b) the conditional p.d.f. of (X_1, X_3) given X_2.

44. Let $f_1(x_1, y_1), f_2(x_2, y_2)$ be two bivariate normal density functions with zero means unit variances and correlation coefficients ρ_1, ρ_2 respectively ($\rho_1 \neq \rho_2$). Show tha the density function $\frac{1}{2}[f_1(x_1, y_1) + f_2(x_2, y_2)]$ is not normal but its marginal densit functions are normal.

45. Let $(X_1, X_2)'$ have $N_2(0, \boldsymbol{\Sigma})$ distribution where $\boldsymbol{\Sigma} = \begin{bmatrix} 1 & \rho \\ \rho & 1 \end{bmatrix}$. Put $X_1 = r\cos\theta$ $X_2 = r\sin\theta (r > 0, 0 < \theta < 2\pi)$.

   (a) Show that the marginal density function of θ is

$$\sqrt{1-\rho^2}/\{2\pi(1 - 2\rho\sin\theta\cos\theta)\}, 0 < \theta < 2\pi.$$

   (b) Show that

$$P(X_1 > 0, X_2 > 0) = \frac{1}{2} - \frac{1}{2\pi}\cos^{-1}\rho.$$

   (c) Show that

$$P(X_1 X_2 > 0) = \frac{1}{2} + \frac{1}{\pi}\sin^{-1}\rho.$$

   (d) Show that

$$P(X_1 X_2 < 0) = \frac{1}{\pi}\cos^{-1}\rho.$$

[Muirhead, 1982

Chapter 8

Probability Distribution of Functions of Random Variables

8.1 Introduction

In this chapter we consider the problem of finding the joint probability distribution of several functions, $Y_1 = g_1(X_1, \cdots, X_n), \cdots, Y_n = g_n(X_1, \cdots, X_n)$ of random variables $(X_1, \cdots, X_n)$ starting from the joint distribution of $(X_1, \cdots, X_n)$. Section 8.2 considers the function $Y = g(X)$ of a single r.v. X. The probability integral transformation is considered in section 8.3. Section 8.4 considers functions $Y_1 = g_1(X_1, X_2), Y_2 = g_2(X_1, X_2)$ of two random variables (X_1, X_2)—first dealing with some simple univariate functions, $X_1 + X_2, X_1 - X_2, X_1X_2, X_1/X_2$ (section 8.4.1) and then considering the technique of transformation of variables (section 8.4.2). Section 8.5 considers functions of n random variables, 8.6 the distribution of maximum and minimum of several random variables. Since the moment generating function, when it exists, uniquely determines the distribution of random variables, such functions can be used to find the joint distribution of $Y_1, \cdots, Y_n$ starting from the distribution of $(X_1, \cdots, X_n)$. The technique has been discussed in section 8.7.

8.2 Functions of One Random Variable

Let X be a r.v. and $g(X)$ a (borel measurable) function of X. $Y = g(X)$ is a r.v. (vide theorem 2.2.9). We want to find the d.f. and p.d.f. of Y.

$$F_Y(y) = P[Y \leq y] = P[X \in I_y] = \int_{I_y} dF_X(x)$$

where $I_y = \{x : g(x) \leq y\}$ and $X \in C$ denotes the event $\{w : X(w) \in C\}$. $x \in I_y$ if and only if $g(x) \leq y$.

If X is a discrete r.v.,

$$\begin{aligned} P[Y=y] &= P[g(X)=y] \\ &= \sum_x P\{X(w)=x : g(x)=y\} \end{aligned}$$

provided g has a finite number of inverses for each y.

► **Example 8.2.1** $Y = aX + b$,

(i) *If* $a > 0, ax + b \le y$ *if and only if* $x \le \frac{y-b}{a}$.

(ii) *If* $a < 0, ax + b \le y$ *if and only if* $x \ge \frac{y-b}{a}$.

Hence, in case (i)

$$F_Y(y) = F_X\left(\frac{y-b}{a}\right)$$

in case (ii)

$$F_Y(y) = 1 - F_X\left(\frac{y-b}{a} - 0\right).$$

If $a = 0$,

$$F_Y(y) = \begin{cases} 0, \; y < b \\ 1, \; y \ge b. \end{cases}$$

► **Example 8.2.2** $Y = X^2$.

$$\begin{aligned} F_Y(y) &= 0 \textit{ for } y < 0 \\ F_Y(y) &= P(X^2 \le y) = P(-\sqrt{y} \le X \le \sqrt{y}) \\ &= F_X(\sqrt{y}) - F_X(-\sqrt{y} - 0), \;\; y \ge 0. \end{aligned}$$

If X is absolutely continuous, X^2 is absolutely continuous and differentiation gives

$$f_Y(y) = \begin{cases} \frac{1}{2\sqrt{y}}[f_X(\sqrt{y}) + f_X(-\sqrt{y})], & (y \ge 0) \\ 0, & (y < 0). \end{cases}$$

If X is discrete, p.m.f. of Y is

$$f_Y(y) = f_X(-\sqrt{y}) + f_X(\sqrt{y}).$$

In particular if X is a standard normal deviate

$$F_Y(y) = \Phi(\sqrt{y}) - \Phi(-\sqrt{y}).$$

Example 8.2.3 *Let X be a $N(\mu, \sigma)$-variable and $Y = g(X) = |X - \mu|$.*

$$F_Y(y) = P(|X-\mu| \le y) = P(\mu - y \le X \le \mu + y)$$

$$= \frac{1}{\sigma\sqrt{2\pi}} \int_{\mu-y}^{\mu+y} e^{-(x-\mu)^2/2\sigma^2}\,dx = \frac{2}{\sigma\sqrt{2\pi}} \int_0^y e^{-t^2/2\sigma^2}\,dt, \quad y > 0.$$

ence

$$f_Y(y) = \begin{cases} 0, & y \le 0 \\ \frac{\sqrt{2}}{\sigma\sqrt{\pi}}\, e^{-y^2/2\sigma^2}, & y > 0. \end{cases}$$

Example 8.2.4 *Let $Y = g(X) = aX^2 + bX + c, a \neq 0$.*

(i) Let $a > 0$. $g(x) < y$ iff $ax^2 + bx + (c - y) \le 0$ i.e. iff $\alpha \le x \le \beta$ where

$$\alpha = \frac{-b - \sqrt{b^2 - 4a(c-y)}}{2a}, \quad \beta = \frac{-b + \sqrt{b^2 - 4a(c-y)}}{2a}.$$

ence

$$F_Y(y) = P(\alpha \le X \le \beta) = F_X(\alpha) - F_X(\beta - 0).$$

(ii) Similarly if $a < 0$,

$$F_Y(y) = 1 - F_X(\alpha - 0) + F_X(\beta).$$

heorem 8.2.1 *Let X be a r.v. of continuous type with p.d.f. $f_X(x)$. Let $g(x)$ be a ontinuous function of x and the equation $y = g(x)$ have a finite number of solutions $x_1, \cdots, x_n$ for every y. Assume also $F_X(x)$ is differentiable at all these points. Then*

$$f_Y(y) = \sum_{i=1}^{n} \frac{f_X(x_i)}{|g'(x_i)|} \tag{8.2.1}$$

where $g'(x_i) = \left.\frac{dg(x)}{dx}\right|_{x_i}$.

f, in particular, there is only one solution of the equation

$$y = g(x), \quad f_Y(x) = \frac{f_X(x)}{g'(x)}.$$

Proof: Since $y = g(x)$ whenever $x = x_1, x_2, \cdots, x_n$,

$$\begin{aligned} f_Y(y)\,dy &= P\{y < Y < y + dy\} \\ &= P\{x_1 < X < x_1 + dx_1\} + \cdots + P\{x_n < X < x_n + dx_n\} \\ &= f(x_1)|\,dx_1| + \cdots + f(x_n)|\,dx_n|, \end{aligned} \tag{8.2.2}$$

vhen in some of these terms in (8.2.2), where y is a decreasing function of x around x_i, dx_i will be negative (and the inequality will be $x_i + dx_i < X < x_i$).

Now

$$g'(x_i)\,dx_i = dy, \quad i = 1, \cdots, n.$$

Hence

$$f_Y(y) = \sum_{i=1}^{n} \frac{f_X(x_i)}{|g'(x_i)|}.$$

■ NOTE 8.2.1 *The theorem is valid if the equation $y = g(x)$ has only a countable numbe of solutions $x_1, x_2, \cdots$.*

► **Example 8.2.5** *If $Y = aX + b, a \neq 0, f_Y(y) = \frac{1}{|a|} f_X\left(\frac{y-b}{a}\right)$.*

In particular if X is $N(\mu, \sigma)$ and $Y = a + bX, b \neq 0$,

$$\begin{aligned} f_Y(y) &= \frac{1}{|b|\sigma\sqrt{2\pi}} \exp\left[-\frac{1}{2\sigma^2}\left\{\frac{1}{b}(y-a) - \mu\right\}^2\right] \\ &= \frac{1}{|b|\sigma\sqrt{2\pi}} \exp\left[-\frac{1}{2b^2\sigma^2}\{y - (a + b\mu)\}^2\right], \quad -\infty < y < \infty. \end{aligned}$$

Hence Y is normally distributed with mean $a + b\mu$ and variance $b^2\sigma^2$.

► **Example 8.2.6** *Let $g(x) = |x|$. Here the equation $y = g(x)$ has no solution if $y <$ and has two solutions*

$$x_1 = x, x_2 = -x \text{ for } y > 0.$$

Hence

$$\begin{aligned} f_Y(y) &= f_X(x) + f_X(-x), \quad y > 0. \\ f_Y(0) &= f_X(0). \end{aligned}$$

Again

$$F_Y(y) = \begin{cases} P\{|X| \leq y\} = F_X(y) - F_X(-y-0), & y \geq 0 \\ 0, & y < 0. \end{cases}$$

► **Example 8.2.7** $g(x) = x(0)$ *if $x > 0$ (otherwise).*

If $y > 0$,

$$\begin{aligned} P[Y \leq y] &= P[g(X) \leq y] = P[g(X) = 0] + P[0 < g(X) \leq y] \\ &= P[X \leq 0] + P[0 < X \leq y] = F_X(y). \end{aligned}$$

If $y < 0, F_Y(y) = 0$.
If $y = 0, F_Y(0) = P(Y = 0) = P(X \leq 0) = F_X(0)$. (Fig. 8.1).

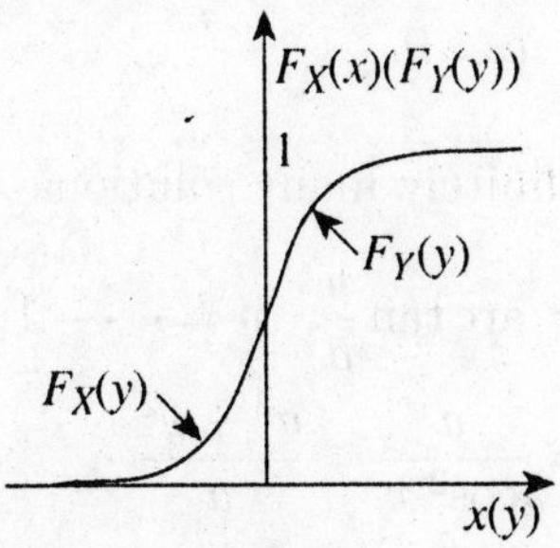

Fig. 8.1

Thus $F_Y(y)$ is discontinuous with discontinuity at $y = 0$.

$$f_Y(y) = \begin{cases} f_X(y), & \text{for } y > 0 \\ F_X(0), & \text{for } y = 0 \\ 0, & \text{for } y < 0. \end{cases}$$

Example 8.2.8 $y = a\cos(x + \theta), a > 0$.

f $|y| < a$, the equation $y = a\cos(x+\theta)$ has infinitely many solutions

$$x_n, \quad n = \cdots - 1, 0, 1, \cdots$$

$$|g'(x_n)| = a\sin(x_n + \theta) = \sqrt{a^2 - y^2}.$$

Hence

$$f_Y(y) = \begin{cases} \dfrac{1}{\sqrt{a^2 - y^2}} \sum\limits_{n=-\infty}^{\infty} f_X(x_n), & |y| < a \\ 0, & |y| > a. \end{cases} \tag{8.2.3}$$

Note that $f_Y(\pm a) = \infty$.

In particular if X is uniformly distributed over $(-\pi, \pi)$, $f_X(x) = \frac{1}{2\pi}(0)$ for $|x| \leq \pi$ (otherwise). The equation $y = a\cos(x+\theta)$, has just two solutions in $(-\pi, \pi)$, $f_X(x) = \frac{1}{2\pi}$ at both these solutions. Hence from (8.2.3) we obtain

$$f_Y(y) = \frac{2}{\sqrt{a^2 - y^2}} \cdot \frac{1}{2\pi} \text{ for } |y| < a. \tag{8.2.4}$$

The distribution function obtained by integrating (8.2.4) is

$$F_Y(y) = \begin{cases} 0, & y < -a \\ \frac{1}{\pi}\arccos\frac{y}{a} - 1, & |y| < a \\ 1, & y > a. \end{cases}$$

Note that $f_Y(\pm a) = \infty$; however $P(Y = \pm a) = 0$.

► **Example 8.2.9** $Y = a\tan X,\ \ a > 0.$

The equation $y = a\tan x$ has infinitely many solutions

$$\begin{aligned} x_n &= \arctan\frac{y}{a},\ \ n = \cdots -1, 0, 1, \cdots \\ g'(x) &= \frac{a}{\cos^2 x} = \frac{a^2+y^2}{a} \\ f_Y(y) &= \frac{a}{a^2+y^2}\sum_{n=-\infty}^{\infty} f_X(x_n). \end{aligned}$$

If in particular X is uniformly distributed over $(-\pi, \pi)$ then $y = a\tan x$, has onl two solutions in $(-\pi, \pi)$ and

$$f_Y(y) = \frac{2a}{a^2+y^2}\cdot\frac{1}{2\pi} = \frac{a}{\pi(a^2+y^2)},$$

a Cauchy distribution.

► **Example 8.2.10**

$$g(x) = \begin{cases} k\cdot e^{-\beta x},\ x \geq 0, & \beta > 0, k > 0, \\ 0, & x < 0 \text{ (Fig. 8.2)}. \end{cases}$$

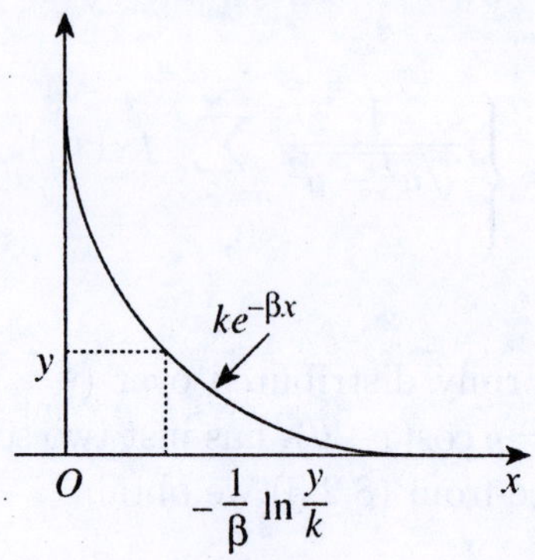

Fig. 8.2

The equation $y = g(x)$ has no solution for $y < 0$ and $y > k$. For $0 < y < k$,

$$\begin{aligned} x &= -\frac{1}{\beta}\ln\frac{y}{ak} \\ g'(x) &= -\beta y \\ f_Y(y) &= \begin{cases} \frac{1}{\beta y} f_X\left(-\frac{1}{\beta}\ln\frac{y}{k}\right), & 0 < y < k \\ 0, & \text{otherwise.} \end{cases} \end{aligned}$$

For $0 < y < k, g(x) \le y$ for $x \le 0$ and $x > -\frac{1}{\beta}\log\left(\frac{y}{k}\right)$

$$F_Y(y) = P[X \le 0] + P\left[X > -\frac{1}{\beta}\log\frac{y}{k}\right] = F_X(0) + 1 - F_X\left(-\frac{1}{\beta}\log\frac{y}{k}\right), 0 < y < k.$$

In particular if X is rectangular on (a, b), $f_X(x) = (b-a)^{-1}, a < x < b$,

$$f_Y(y) = \frac{1}{\beta y}, \quad k\,e^{-b\beta} < y < k\,e^{-a\beta}$$

and is zero elsewhere.

▶ **Example 8.2.11** *X is a $N(\mu, \sigma)$ variable and $Y = bX^2$, $b > 0$*

$$\begin{aligned}
f_Y(y) &= \frac{1}{2a\sqrt{\frac{y}{b}}}\left[f_X\left(\sqrt{\frac{y}{b}}\right) + f_X\left(-\sqrt{\frac{y}{b}}\right)\right] \\
&= \frac{1}{2\sigma\sqrt{2\pi by}}\left[\exp\left\{-\frac{1}{2\sigma^2}\left(\sqrt{\frac{y}{b}} - \mu\right)^2\right\} + \exp\left\{-\frac{1}{2\sigma^2}\left(-\sqrt{\frac{y}{b}} - \mu\right)^2\right\}\right] \\
&= \frac{1}{2\sigma\sqrt{2\pi by}}\, e^{-\frac{1}{2\sigma^2}\left(\mu^2 + \frac{y}{b}\right)}\left[e^{\frac{\mu}{\sigma^2}\sqrt{\frac{y}{b}}} + e^{-\frac{\mu}{\sigma^2}\sqrt{\frac{y}{b}}}\right] \\
&= \frac{e^{-\frac{1}{2\sigma^2}\left(\mu^2 + \frac{y}{b}\right)}}{\sigma\sqrt{2\pi b}} y^{-\frac{1}{2}} \cosh\left[\frac{\mu}{\sigma^2}\sqrt{\frac{y}{b}}\right], 0 < y < \infty.
\end{aligned}$$

In particular if $\mu = 0$,

$$f_Y(y) = \frac{y^{\frac{-1}{2}}\, e^{\frac{-y}{2b\sigma^2}}}{\Gamma\left(\frac{1}{2}\right)\sqrt{2b\sigma^2}}, \quad y > 0$$

so that Y has a gamma distribution with parameters $\left(\frac{1}{2b\sigma^2}, \frac{1}{2}\right)$. If $b = 1, \sigma^2 = 1$, Y follows a χ^2-distribution with one degree of freedom. If $\mu \ne 0, b = 1, \sigma^2 = 1$, the resulting distribution (8.2.5) is called a non-central χ^2-distribution with one d.f.

8.3 Probability Integral Transformation

Consider a r.v. X with d.f. $F_X(x), 0 \le F_X(x) \le 1$. We define an inverse function F_X^{-1} on $0 \le y \le 1$ by setting $F_X^{-1}(y)$ equal to the smallest value of x for which $F_X(x) \ge y$.

Theorem 8.3.1 *If X is a r.v. with continuous distribution function $F_X(x)$, then the r.v. $Y = F_X(x)$ has a rectangular distribution over $(0, 1)$.*

Conversely if Y be a rectangular variable over $(0, 1)$ then $X = F_X^{-1}(Y)$ has probability distribution function F_X.

Proof: The d.f. of Y is

$$F_Y(y) = P[Y \le y] = P[F_X(x) \le y] = P[X \le F_X^{-1}(y)]$$
$$= F_X(F_X^{-1}(y)) = y, \ \ 0 \le y \le 1.$$

Hence, Y follows a uniform distribution over (0, 1).

Conversely, suppose Y has a uniform distribution over (0, 1). The d.f. of

$$X, P[X \le x] = P[F_X^{-1}(Y) \le x] = P[Y \le F_X(x)] = F_X(x),$$

so that X has d.f. $F_X(x)$.

The transformation $Y = F_X(x)$ is called probability integral transformation.

The theorem 8.3.1 is of fundamental importance and states that every continuous probability distribution can be transformed into a normalised uniform distribution. Theoretically, therefore, every continuous distribution can be transformed into every other distribution by applying the above-mentioned transformations. The transformation $Y = F_X(x)$ is very useful in drawing a random sample from the distribution of X. Since $0 \le Y \le 1$, if one draws a random number y between [0, 1] and calculates $x = F_X^{-1}(y)$ then x is a random sample from the distribution of X.

▶ **Example 8.3.1** *$X_1, \cdots, X_n$ have each continuous uniform distribution*

$$dF = dx, \ \ 0 \le x \le 1.$$

Show that

$$Z = -2\sum_{i=1}^{n} \log_e X_i \text{ has a } \chi^2_{(2n)} \text{ distribution.}$$

Consider transformed variable

$$Y = -2\log_e X; \ \ 0 < Y < \infty.$$

Hence

$$x = e^{\frac{-y}{2}}, \ \ dx = -\frac{1}{2}\, e^{\frac{-y}{2}}\, dy.$$

The p.d.f. of Y is

$$f_Y(y) = \frac{1}{2}\, e^{\frac{-y}{2}}, \ \ 0 < y < \infty$$

so that Y follows a $\chi^2_{(2)}$ distribution. Hence $Z = \sum_{i=1}^{n} Y_i$ follows $\chi^2_{(2n)}$-distribution by the reproductive property of this distribution.

Example 8.3.2 *A variable X has a uniform distribution over $[0, 1]$. Find a function of X say Y, with probability element*

$$f_Y(y)\,dy = e^{-y}\,dy,\ \ 0 \le y \le \infty.$$

It is known that the d.f. of Y follows a uniform distribution over [0, 1]. Now

$$F_Y(y) = \int_0^y e^{-t}\,dt = 1 - e^{-y}.$$

Hence $1 - e^{-Y}$ is uniform over [0, 1].
Therefore, $X = 1 - e^{-Y}$ with probability one, when

$$Y = \log_e\left(\frac{1}{1-X}\right).$$

▶ **Example 8.3.3** *$X_1, \cdots, X_n$ are random samples from a uniform distribution over $(0, 1)$. Find the p.d.f. of $u = (X_1, \cdots, X_n)^{1/n}$.*

It is known that

$$Z = -2\sum_{i=1}^{n} \log_e X_i$$

has a χ^2-distribution with $2n$ degrees of freedom.

$$f_Z(z)\,dz = \frac{1}{2^n \boxed{(n)}}\, e^{-\frac{1}{2}z} z^{n-1}\,dz,\ \ 0 \le z < \infty.$$

Make the transformation

$$u = e^{\frac{-z}{2n}};\ \ Z = 2n\log_e\left(\frac{1}{u}\right)$$

$$\frac{du}{dz} = -\frac{1}{2n}\, e^{\frac{-z}{2n}} = -\frac{u}{2n};\ \ 0 < u < 1$$

$$\begin{aligned} f_U(u)du &= \frac{1}{2^n \boxed{(n)}}\, e^{-\frac{1}{2}[2n\log_e \frac{1}{u}]}\left(2n\log_e \frac{1}{u}\right)^{n-1} \frac{2n}{u} du \\ &= \frac{n^n u^{n-1}}{\boxed{(n)}}\left(\log_e \frac{1}{u}\right)^{n-1} du;\ \ 0 < u < 1. \end{aligned}$$

8.4 Functions of Two Random Variables

8.4.1 Using Distribution Functions

Let (X, Y) be two jointly continuous random variables with p.d.f. $f_{X,Y}\,(x, y)$. We first consider the problem of finding probability distributions of some simple functions $U = X + Y$, $V = X - Y$, $W = XY$, $Z = X/Y$ by using the procedure of finding the marginal distributions.

Theorem 8.4.1 *Let (X,Y) be jointly distributed continuous random variables with p.d.f $f_{X,Y}(x,y)$. The p.d.f. of $U = X+Y, V = X-Y$ are, respectively, given by*

$$\text{(i) } f_U(u) = \int_{-\infty}^{\infty} f_{X,Y}(x, u-x)\, dx = \int_{-\infty}^{\infty} f_{X,Y}(u-y, y)\, dy \qquad (8.4.1)$$

$$\text{(ii) } f_V(v) = \int_{-\infty}^{\infty} f_{X,Y}(x, x-v)\, dx = \int_{-\infty}^{\infty} f_{X,Y}(v+y, y)\, dy. \qquad (8.4.2)$$

Proof: The d.f. of $U = X + Y$ is

$$\begin{aligned} F_U(u) &= P(X+Y \le u) \\ &= \iint_{x+y\le u} f_{X,Y}(x,y)\, dx\, dy = \int_{-\infty}^{\infty} \left[\int_{-\infty}^{u-x} f_{X,Y}(x,y)\, dy \right] dx \\ &= \int_{-\infty}^{\infty} \left[\int_{-\infty}^{u} f_{X,Y}(x, z-x)\, dz \right] dx \text{ (Putting } y = z - x) \\ &= \int_{-\infty}^{u} \left[\int_{-\infty}^{\infty} f_{X,Y}(x, z-x)\, dx \right] dz = \int_{-\infty}^{u} f_U(z)\, dz. \end{aligned} \qquad (8.4.3)$$

Differentiating both sides we get the p.d.f. of U. The change from a double integra to a repeated integral, as well as the change of order of integration are justified sinc $f(x,y)$ is abolutely integrable over the entire xy-plane. The second expression in (8.4.1 can be derived similarly. The result (8.4.2) is proved in an analogous manner.

Corollary 8.4.1 *If X, Y are independent continuous random variables, the p.d.f. o $U = X + Y$ is*

$$f_U(u) = \int_{-\infty}^{\infty} f_X(x) f_Y(u-x)\, dx = \int_{-\infty}^{\infty} f_X(u-y) f_Y(y)\, dy. \qquad (8.4.4)$$

The density $f_U(.)$ is called, here, the convolution of the densities $f_X(.), f_Y(.)$.

► **Example 8.4.1** *X, Y are independent uniform variables over $(0,1)$. Find the p.d.f. of (i) $U = X + Y$, (ii) $V = X - Y$ (iii) $W = |X - Y|$.*

$$\text{(i)} \qquad f_U(u) = \int_{-\infty}^{\infty} f_X(x) f_Y(u-x)\, dx, \quad 0 \le u \le 2.$$

When $0 \le u \le 1, 0 \le x \le u$

$$f_U(u) = \int_0^u dx = u.$$

When $1 \le u \le 2, u-1 \le x \le 1$

$$f_U(u) = \int_{u-1}^1 dx = 2 - u.$$

Hence

$$f_U(u) = \begin{cases} u, & 0 \leq u \leq 1 \\ 2-u, & 1 \leq u \leq 2. \end{cases}$$

This is known as the triangular distribution.

(ii) $$f_V(v) = \int_{-\infty}^{\infty} f_X(x) f_Y(x-v)\,dx, \quad -1 \leq v \leq 1.$$

When $-1 \leq v \leq 0$, $0 \leq x \leq 1+v$

$$f_V(v) = \int_0^{1+v} dx = 1+v.$$

When $0 \leq v \leq 1$, $v \leq x \leq 1$

$$f_V(v) = \int_v^1 dx = 1-v.$$

Hence

$$f_V(v) = \begin{cases} 1+v, & -1 \leq v \leq 0 \\ 1-v, & 0 \leq v \leq 1 \end{cases} \tag{i}$$

as in Fig. 8.3.

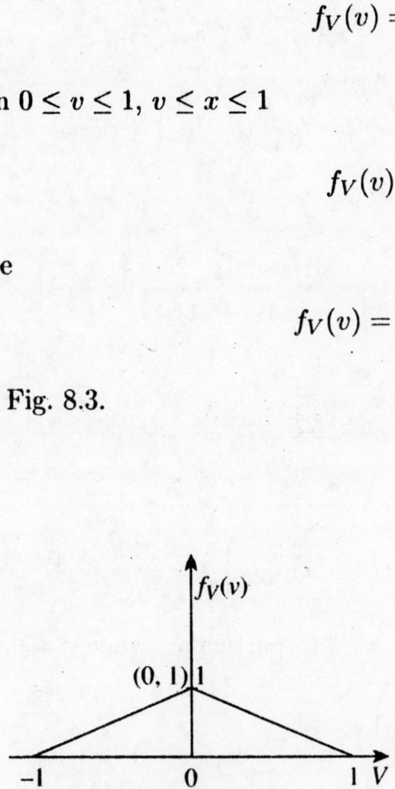

Fig. 8.3

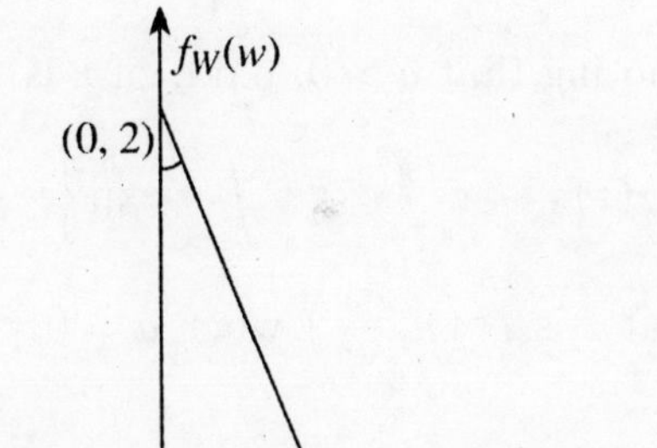

Fig. 8.4

(iii) Let $w = |x-y|; 0 \leq w \leq 1$. Solving the equations

$$w = |v|, \quad w = +v \text{ or } -v, \quad \left|\frac{dw}{dv}\right| = 1$$

at each solution.

Hence from (i),

$$f_W(w) = 2(1-w), \quad 0 \leq w \leq 1$$

as in Fig. 8.4.

► **Example 8.4.2** *(X, Y) has a bivariate normal density with parameters $(\mu_1, \mu_2, \sigma_1, \sigma_2, \rho)$. Find the p.d.f. of $Z = X + Y$.*

$$f(x,y) = k\exp\left[-\frac{1}{2(1-\rho^2)}\left\{\left(\frac{x-\mu_1}{\sigma_1}\right)^2 - 2\rho\left(\frac{x-\mu_1}{\sigma_1}\right)\left(\frac{y-\mu_2}{\sigma_2}\right)+\left(\frac{y-\mu_2}{\sigma_2}\right)^2\right\}\right]$$
$$= k\exp\left[-\frac{Q(x,y)}{2}\right] \text{ (say).}$$

Writing $x + y = Z$, $\mu_1 + \mu_2 = \mu$,

$$A_{11} = [\sigma_1^2(1-\rho^2)]^{-1},\quad A_{12} = -\rho[\sigma_1\sigma_2(1-\rho^2)]^{-1},\quad A_{22} = [\sigma_2^2(1-\rho^2)]^{-1},$$
$$Q(x,y) = A_{11}(x-\mu_1)^2 + 2A_{12}(x-\mu_1)(y-\mu_2) + A_{22}(y-\mu_2)^2$$
$$= A_{11}[(z-y)-(\mu-\mu_2)]^2 + 2A_{12}[(z-y)-(\mu-\mu_2)](y-\mu_2) + A_{22}(y-\mu_2)^2$$
$$= \left(A_{11} - \frac{(A_{12}-A_{11})^2}{A_{11}-2A_{12}+A_{22}}\right)(z-\mu)^2$$
$$+\left[(A_{11}-2A_{12}+A_{22})^{\frac{1}{2}}(y-\mu_2) + \frac{A_{12}-A_{11}}{(A_{11}-2A_{12}+A_{22})^{\frac{1}{2}}}(z-\mu)\right]^2.$$

Writing the coefficient of $(z-\mu)^2$ in (i) as a,

$$b = \sqrt{A_{11}-2A_{12}+A_{22}},\quad c = \frac{A_{12}-A_{11}}{b}$$

and noting that $a > 0$, p.d.f. of z is

$$f_Z(z) = k\,e^{-\frac{a}{2}(z-\mu)^2}\int_{-\infty}^{\infty}\exp\left\{-\frac{1}{2}[b(y-\mu_2)+c(z-\mu)]^2\,dy\right\} = \frac{k\sqrt{2\pi}}{b}\,e^{-\frac{a}{2}(z-\mu)^2}$$

so that z is $N\left(\mu, \frac{1}{\sqrt{a}}\right)$ when $a = (\sigma_1^2+\sigma_2^2+2\rho\sigma_1\sigma_2)^{-1}$. In particular, when $\rho = 0$, z is

$$N\left(\mu, \sqrt{\sigma_1^2+\sigma_2^2}\right).$$

Theorem 8.4.2 *Let X, Y be jointly distributed continuous random variables with p.d.f. $f_{X,Y}(x,y)$. The p.d.f. of $W = XY$ and $Z = \frac{X}{Y}$ are respectively,*

$$\text{(i)}\quad f_W(w) = \int_{-\infty}^{\infty}\frac{1}{|x|}f_{X,Y}\left(x,\frac{w}{x}\right)dx = \int_{-\infty}^{\infty}\frac{1}{|y|}f_{X,Y}\left(\frac{w}{y},y\right)dy \tag{8.4.5}$$

$$\text{(ii)}\quad f_Z(z) = \int_{-\infty}^{\infty}|y|f_{X,Y}(yz,y)\,dy,\ \text{provided } P(Y=0)=0. \tag{8.4.6}$$

We omit the proof.

In particular if $f_{X,Y}(x,y) = f_{X,Y}(-x,-y)$, (8.4.6) reduces to

$$f_Z(z) = 2\int_0^{\infty} y f_{X,Y}(yz,y)\,dy. \tag{8.4.7}$$

▶ **Example** 8.4.3 (X, Y) *follows bivariate normal with* $(0, 0, \sigma_1^2, \sigma_2^2, \rho)$. *Find the p.d.f. of*

$$Z = \frac{X}{Y}.$$

By (8.4.3)

$$f_Z(z) = \frac{1}{\pi\sigma_1\sigma_2\sqrt{1-\rho^2}}\int_0^\infty y\, e^{-\frac{y^2}{2(1-\rho^2)}\left[\frac{z^2}{\sigma_1^2}-\frac{2\rho z}{\sigma_1\sigma_2}+\frac{1}{\sigma_2^2}\right]}\,dy = \frac{\sqrt{1-\rho^2}}{\pi\sigma_1\sigma_2\left[\frac{z^2}{\sigma_1^2}-\frac{2\rho z}{\sigma_1\sigma_2}+\frac{1}{\sigma_2^2}\right]}$$

$$= \frac{\sigma_1\sigma_2\sqrt{1-\rho^2}}{\pi\left[\sigma_2^2\left(z-\rho\frac{\sigma_1}{\sigma_2}\right)^2+\sigma_1^2(1-\rho^2)\right]}$$

so that Z is a Cauchy variable with parameters $\lambda = \frac{\sigma_1}{\sigma_2}\sqrt{1-\rho^2}$ and $\theta = \rho\frac{\sigma_1}{\sigma_2}$.
In particular if $\sigma_1 = \sigma_2 = 1$, $\rho = 0$,

$$f_Z(z) = \frac{1}{\pi(z^2+1)}, -\infty < z < \infty.$$

▶ **Example** 8.4.4 X, Y *are independent variables, each uniform over* $(0, 1)$. *Find p.d.f. of* (i) $W = XY$, (ii) $Z = X/Y$.

(i) $f_W(w) = \int_{-\infty}^{\infty}\frac{1}{|x|}f_X(x)f_Y\left(\frac{w}{x}\right)dx = \int_w^1\frac{1}{x}\,dx \quad \left(\because x = \frac{w}{y} \text{ and hence } w < x < 1\right)$

$$= -\log_e w, \quad 0 < w < 1.$$

(ii) $f_Z(z) = \int_{-\infty}^{\infty}|y| f_X(yz) f_Y(y)\, dy.$

When $0 < z < 1, 0 < y < 1$.

$$f_Z(z) = \int_0^1 y\, dy = \frac{1}{z}.$$

When $1 \le z < \infty, 0 < y < \frac{1}{z}$

$$f_Z(z) = \int_0^{\frac{1}{z}} y\, dy = \frac{1}{2z^2}.$$

Hence

$$f_Z(z) = \begin{cases} \frac{1}{2}, & 0 < z < 1 \\ \frac{1}{2z^2}, & 1 \le z < \infty. \end{cases}$$

We now consider the problem of finding the distribution of general functions of two random variables by using the method of transformation of variables.

8.4.2 Transformation of Variables

Consider random variables (X_1, X_2) having joint density $f_{X_1,X_2}(x_1, x_2)$. Let

$$Y_1 = g_1(X_1, X_2), \quad Y_2 = g_2(X_1, X_2)$$

be two functions of (X_1, X_2). We want to find the joint distribution of (Y_1, Y_2). The joint d.f. of (Y_1, Y_2) is

$$F_{Y_1,Y_2}(y_1, y_2) = P(Y_1 \leq y_1, \quad Y_2 \leq y_2) = \iint\limits_{R_{y_1 y_2}} f_{X_1,X_2}(x_1, x_2)\, dx_1\, dx_2$$

where $R_{y_1 y_2} = \{(x_1, x_2) : Y_1 \leq y_1, \quad Y_2 \leq y_2\}$.
Define

$$J = \begin{vmatrix} \dfrac{\partial x_1}{\partial y_1}, & \dfrac{\partial x_2}{\partial y_1} \\ \\ \dfrac{\partial x_1}{\partial y_2}, & \dfrac{\partial x_2}{\partial y_2} \end{vmatrix}$$

as the Jacobian of transformation.
To find the joint p.d.f. of (Y_1, Y_2), solve the simultaneous equations

$$g_1(x_1, x_2) = y_1, \quad g_2(x_1, x_2) = y_2 \tag{8.4.8}$$

giving (x_1, x_2) in terms of (y_1, y_2). Let

$$x_{1i} = g_{1i}^{-1}(y_1, y_2), \quad x_{2i} = g_{2i}^{-1}(y_1, y_2), \quad i = 1, \cdots, m$$

be the m sets of solutions, all real. Calculate the Jacobians,

$$J_i = \begin{vmatrix} \dfrac{\partial g_{1i}^{-1}}{\partial y_1}, & \dfrac{\partial g_{2i}^{-1}}{\partial y_1} \\ \\ \dfrac{\partial g_{1i}^{-1}}{\partial y_2}, & \dfrac{\partial g_{2i}^{-1}}{\partial y_2} \end{vmatrix}.$$

The joint p.d.f. of (Y_1, Y_2) is

$$f_{Y_1,Y_2}(y_1, y_2) = \sum_{i=1}^{m} f_{X_1,X_2}(x_{1i}, x_{2i})|J_i|. \tag{8.4.9}$$

We have assumed that the first partial derivatives of g_{1i}^{-1}, g_{2i}^{-1} are continuous on the domain of (Y_1, Y_2) and $|J_i| \neq 0$ on the domain of $(Y_1, Y_2), i = 1, \cdots, m$.

If for certain (y_1, y_2), the equations (8.4.8) have no real solution, $f(y_1, y_2) = 0$. The proof is omitted.

f $(x_{1i}, x_{2i}) = (x_1, x_2)$ for all $i = 1, \cdots, m$,

$$f_{Y_1,Y_2}(y_1, y_2) = f(x_1, x_2)|J|.$$

If only distribution of Y_1 is desired, a variable $Y_2 = g_2(X_1, X_2)$ is suitably chosen, joint distribution of (Y_1, Y_2) is obtained and distribution of Y_1 is found out by integrating over Y_2.

▶ **Example 8.4.5** *X_1, X_2 are independent gamma variables with parameters (α, p_1), (α, p_2) respectively. Show that $U = X_1 + X_2, V = X_1/(X_1 + X_2)$ are independently distributed, U following a gamma distribution with parameters $(\alpha, p_1 + p_2), V$ a beta distribution with parameters (p_1, p_2). Also find the distribution of $W = \frac{X_1}{X_2}$.*

The probability element of (X_1, X_2) is

$$dF = \frac{\alpha^{p_1+p_2}}{\Gamma(p_1)\,\Gamma(p_2)}\, e^{-\alpha(x_1+x_2)} x_1^{p_1-1} x_2^{p_2-1}\, dx_1\, dx_2, \quad 0 < x_1, x_2 < \infty.$$

Use transformation

$$\left.\begin{aligned} u &= x_1 + x_2 \\ v &= \tfrac{x_1}{x_1+x_2} \end{aligned}\right\} \text{i.e.,} \quad \begin{aligned} x_1 &= uv, & 0 \le u < \infty \\ x_2 &= u(1-v), & 0 \le v < 1 \end{aligned}$$

$|J| = u.$

Joint probability element of (U, V) is

$$\begin{aligned} dF &= \frac{\alpha^{p_1+p_2}}{\Gamma(p_1)\,\Gamma(p_2)}\, e^{-\alpha u}(uv)^{p_1-1}[u(1-v)]^{p_2-1} u\,du\,dv \\ &= \left\{\frac{\alpha^{p_1+p_2}}{\Gamma(p_1+p_2)} u^{p_1+p_2-1}\, e^{-\alpha u} du\right\}\left\{\frac{1}{B(p_1,p_2)} v^{p_1-1}(1-v)^{p_2-1} dv\right\} \\ &= f_U(u) f_V(v). \end{aligned}$$

Again $w = \frac{v}{1-v}$; hence

$$v = \frac{w}{1+w}; \quad dv = \frac{dw}{(1+w)^2}.$$

The p.d.f. of W is obtained from the p.d.f. of V as

$$f_W(w) = \frac{1}{B(p_1,p_2)}\left[\frac{w}{1+w}\right]^{p_1-1} \frac{1}{(1+w)^{p_2+1}} = \frac{1}{B(p_1,p_2)} \frac{w^{p_1-1}}{(1+w)^{p_1+p_2}}, \quad 0 \le w < \infty$$

so that W follows a beta distribution of the second kind with parameters p_1, p_2.

[Alternatively, put

$$X_1 = r\cos^2\theta, \quad X_2 = r\sin^2\theta; \quad |J| = 2r\cos\theta\sin\theta, \quad X_1 + X_2 = r, \quad \frac{X_1}{(X_1+X_2)} = \cos^2\theta;$$

show that r and $\cos^2\theta$ are independently distributed.]

▶ **Example 8.4.6** *Let X, Y be independent beta variates with parameters (a, b), (c, d) respectively such that $a = c+d$. Show that XY is a beta variate with parameters $(c, b+d)$.*

$$f_{X,Y}(x, y) = kx^{a-1}(1-x)^{b-1}y^{c-1}(1-y)^{d-1} \text{ where } k = [B(a, b)B(c, d)]^{-1}.$$

Let $u = x, z = xy$; hence

$$x = u, y = \frac{z}{u} \cdot 0 < u, \quad z < 1. \ |J| = \frac{1}{|u|}.$$

The joint density of U and Z is

$$\begin{aligned} f_{U,Z}(u, z) &= ku^{a-c-d}(1-u)^{b-1}z^{c-1}(u-z)^{d-1} \\ &= k(1-u)^{b-1}z^{c-1}(u-z)^{d-1} \text{ (since } a = c+d). \end{aligned}$$

The marginal density of z is

$$\begin{aligned} f_z(z) &= kz^{c-1}\int_0^1 (1-u)^{b-1}(u-z)^{d-1}du \\ &= kz^{c-1}(1-z)^{b+d-1}\int_0^1 t^{b-1}(1-t)^{d-1}\,dt \left[\text{where } t = \frac{1-u}{1-z}\right] \\ &= \frac{B(b, d)}{B(a, b)B(c, d)}z^{c-1}(1-z)^{b+d-1} = \frac{1}{B(c, b+d)}z^{c-1}(1-z)^{b+d-1}. \end{aligned}$$

In general product of Beta variates with parameters $(a_1, b_1), \cdots, (a_k, b_k)$ such that

$$a_i = a_{i+1} + b_{i+1}, i = 1, \cdots, k-1$$

is distributed as a Beta variate with parameters

$$\left(a_k, \sum_{i=1}^{k} b_i\right).$$

▶ **Example 8.4.7** *Let X, Y be independent, each $N(0, \sigma)$-variable. Find the joint p.d.f of*

$$Z = \sqrt{X^2 + Y^2}, \ W = \frac{X}{Y}.$$

Solving the equations

$$z = \sqrt{x^2 + y^2}, \quad w = \frac{x}{y}$$

we have two solutions

$$\left(x_1 = \frac{zw}{\sqrt{1+w^2}}, \ y_1 = \frac{z}{\sqrt{1+w^2}}\right), \ (-x_1, -y_1)$$

r any w. Here

$$\begin{vmatrix} \dfrac{\partial z}{\partial x}, & \dfrac{\partial z}{\partial y} \\ \dfrac{\partial w}{\partial x}, & \dfrac{\partial w}{\partial y} \end{vmatrix} = \begin{vmatrix} \dfrac{x}{\sqrt{x^2+y^2}}, & \dfrac{y}{\sqrt{x^2+y^2}} \\ \dfrac{1}{y}, & \dfrac{-x}{y^2} \end{vmatrix} = \frac{1+w^2}{-z}; \text{ hence } |J_i| = \frac{|z|}{1+w^2},$$

or each point $(x_i, y_i), i = 1, 2$.

$$f_{X,Y}(x, y) = \frac{1}{2\pi\sigma^2} e^{-\frac{1}{2\sigma^2}(x^2+y^2)}.$$

ince $f(x_1, y_1) = f(x_2, y_2)$

$$f_{Z,W}(z, w) = \frac{2}{2\pi\sigma^2} e^{\frac{-z^2}{2\sigma^2}} \left(\frac{z}{1+w^2}\right) = \left\{\frac{z}{\sigma^2} e^{\frac{-z^2}{2\sigma^2}}\right\} \left\{\frac{1}{\pi(1+w^2)}\right\}$$

hich shows that (Z, W) are independently distributed, Z having a Rayleigh distribution nd W a Cauchy distribution.

***Alternative Solution*:** Consider polar transformation from (x, y) to

$$(z, \theta) : x = z\cos\theta, \ \ y = z\sin\theta, \ \ z > 0, \ \ 0 < \theta < 2\pi, \ \ w = \cot\theta, \ \ J = z.$$

The joint probability element of (Z, θ)

$$dF(z, \theta) = \frac{z}{2\pi\sigma^2} e^{\frac{-z^2}{2\sigma^2}} \, dz \, d\theta.$$

Hence p.d.f. of Z is

$$f_Z(z) = \frac{z}{\sigma^2} e^{-z^2/2\sigma^2}; \text{ p.d.f. of } \theta \text{ is } f(\theta) = \frac{1}{2\pi}, 0 < \theta < 2\pi. \tag{8.4.10}$$

For each value of w there are two values of θ in $(0, 2\pi)$ satisfying $\cot\theta = w$.

$$\frac{dw}{d\theta} = \operatorname{cosec}^2\theta = (1+w^2).$$

Hence from (8.4.9), p.d.f. of W is

$$f_W(w) = \frac{1}{\pi(1+w^2)}, -\infty < w < \infty.$$

8.5 Functions of n Random Variables

Let $X_1, \cdots, X_n$ be jointly continuous random variables with joint p.d.f. $f_{X_1,\cdots,X}$ $(x_1, \cdots, x_n)$ and let $\mathcal{X} = \{(x_1, \cdots, x_n) : f(x_1, \cdots, x_n) > 0\}$. Let $Y_1 = g_1(X_1, \cdots, X_n$ $\cdots, Y_n = g_n(X_1, \cdots, X_n)$. Let $\mathcal{Y}$ = domain of $(Y_1, \cdots, Y_n)$.

Suppose $\mathcal{X}$ can be decomposed into m disjoint subsets $\mathcal{X}_1, \cdots, \mathcal{X}_m$ such that $(x_1, \cdots$ $x_n) \to (y_1, \cdots, y_n)$ is a one-one transformation of $\mathcal{X}_i$ onto $\mathcal{Y}$. Let $x_1 = g_{1i}^{-1}(y_1, \cdots, y_n$ $\cdots, x_n = g_{ni}^{-1}(y_1, \cdots, y_n)$ denote the inverse transformations of $\mathcal{Y}$ onto $\mathcal{X}_i, i = 1, \cdots, m$ Define the jacobian,

$$J_i = \begin{vmatrix} \dfrac{\partial g_{1i}^{-1}}{\partial y_1} & \cdots & \dfrac{\partial g_{ni}^{-1}}{\partial y_1} \\ \\ \dfrac{\partial g_{1i}^{-1}}{\partial y_n} & \cdots & \dfrac{\partial g_{ni}^{-1}}{\partial y_n} \end{vmatrix}$$

We assume all the partial derivatives exist and are continuous over $\mathcal{Y}$ and $J_i \neq 0, i =$ $1, \cdots, m$. Then

$$f_{Y_1,\cdots,Y_n}(y_1, \cdots, y_n) = \sum_{i=1}^{m} f_{X_i,\cdots,X_n}(x_{1i}, \cdots, x_{ni})|J_i|,$$

for $(y_1, \cdots, y_n) \in \mathcal{Y}$.

If joint distribution of $Y_1, \cdots, Y_k (k < n)$ are desired, additional $(n - k)$ variables

$$Y_{k+1} = g_{k+1}(X_1, \cdots, X_n), \cdots, \; Y_n = g_n(X_1, \cdots, X_n)$$

are suitably chosen, joint p.d.f. of $(Y_1, \cdots, Y_n)$ are obtained and then distribution of $(Y_1, \cdots, Y_k)$ is found by integrating out $(Y_{k+1}, \cdots, Y_n)$.

► **Example 8.5.1** *Let X_1, X_2, X_3 be three independently and identically distribute random variables, each with p.d.f. $f(x) = e^{-x} (x > 0)$. Show that the variables*

$$Y_1 = X_1 + X_2 + X_3, \; Y_2 = \frac{X_1 + X_2}{X_1 + X_2 + X_3}, \; Y_3 = \frac{X_1}{X_1 + X_2}$$

are independently distributed.

Using the transformation

$$y_1 = x_1 + x_2 + x_3, \; y_2 = \tfrac{x_1+x_2}{x_1+x_2+x_3}, \qquad y_3 = \tfrac{x_1}{x_1+x_2},$$

$$x_1 = y_1 y_2 y_3, \qquad x_2 = y_1 y_2 (1 - y_3), \; x_3 = y_1(1 - y_2),$$

$$|J| = y_1^2 y_2.$$

$$f_{X_1,X_2,X_3}(x_1, x_2, x_3) = e^{-y_1}.$$

Hence

$$f_{Y_1,Y_2,Y_3}(y_1, y_2, y_3) = e^{-y_1} y_1^2 y_2$$

which shows that Y_1, Y_2, Y_3 are independently distributed.

► **Example 8.5.2** *Let (X, Y, Z) have joint p.d.f.*

$$f(x, y, z) = 6(1 + x + y + z)^{-4}, \quad x, y, z > 0.$$

Find the p.d.f. of $X + Y + Z$.

Use the transformation: $u = x + y + z, \quad uv = y + z, \quad uvw = z.$

$$J = u^2 v; \quad 0 < u < \infty, \quad 0 < v < 1, \quad 0 < w < 1.$$

$$f(u, v, w) = \frac{6u^2 v}{(1+u)^4}.$$

Hence integrating over v and w,

$$f_U(u) = \frac{Cu^2}{(1+u)^4}, \quad 0 < u < \infty$$

where C is a suitable constant.

► **Example 8.5.3** *Let $X_i, i = 1, \cdots, n$ be n independently distributed normal variables with parameters (μ_i, σ_i). Find the p.d.f. of $Y = a + \sum_{i=1}^{n} b_i X_i$ where $a, b_1, \cdots, b_n$ are arbitrary constants.*

The probability element of the joint distribution of $(X_1, \cdots, X_n)$ is

$$dF = \frac{1}{(2\pi)^{\frac{n}{2}} \sigma_1 \cdots \sigma_n} \exp\left\{ -\frac{1}{2} \sum_{i=1}^{n} \left(\frac{x_i - \mu_i}{\sigma_i^2} \right)^2 \right\}. \tag{8.5.1}$$

We make the following transformation from $(x_1, \cdots, x_n)$ to $(y_1, \cdots, y_n)$

$$\left.\begin{aligned}
y_1 &= \frac{b_1\sigma_1\left(\frac{x_1 - \mu_1}{\sigma_1}\right) + \cdots + b_n\sigma_n\left(\frac{x_n - \mu_n}{\sigma_n}\right)}{\sqrt{\sum_1^n b_i \sigma_i^2}} \\
y_2 &= l_{21}\left(\frac{x_1 - \mu_1}{\sigma_1}\right) + \cdots + l_{2n}\left(\frac{x_n - \mu_n}{\sigma_n}\right) \\
&\cdots\cdots\cdots\cdots\cdots\cdots\cdots\cdots \\
y_n &= l_{n1}\left(\frac{x_1 - \mu_1}{\sigma_1}\right) + \cdots + l_{nn}\left(\frac{x_n - \mu_n}{\sigma_n}\right)
\end{aligned}\right\} \tag{8.5.2}$$

where $(l_{i1}, \cdots l_{in}), i = 2, \cdots, n$, are mutually orthogonal vectors of unit length and each is also orthogonal to the vector

$$\left(\frac{b_1\sigma_1}{\sqrt{\sum b_i^2\sigma_i^2}}, \quad \cdots, \quad \frac{b_n\sigma_n}{\sqrt{\sum b_i^2\sigma_i^2}}\right).$$

The transformation (8.5.2) can be written as

$$\begin{bmatrix} y_1 \\ \vdots \\ y_n \end{bmatrix} = L \begin{bmatrix} \frac{(x_1-\mu_1)}{\sigma_1} \\ \vdots \\ \frac{(x_n-\mu_n)}{\sigma_n} \end{bmatrix} \quad \text{i.e., } \boldsymbol{Y} = L\boldsymbol{Z} \text{ (say)}$$

where L is an orthogonal matrix, $LL' = L'L = I = \text{Diag }(1, \cdots, 1)$.
Hence

$$\sum_1^n y_i^2 = \boldsymbol{Y}'\boldsymbol{Y} = \boldsymbol{Z}'L'L\boldsymbol{Z} = \boldsymbol{Z}'\boldsymbol{Z} = \sum_1^n \frac{(x_i - \mu_i)^2}{\sigma_i^2}$$

$$J\left(\frac{x_1 \cdots x_n}{y_1 \cdots y_n}\right) = \sigma_1 \cdots \sigma_n |L| = \pm\sigma_1 \cdots \sigma_n.$$

Hence p.d.f. of $y_1, \cdots, y_n$ is

$$\frac{1}{(2\pi)^n} \exp\left\{-\frac{1}{2}\sum_1^n y_i^2\right\}, -\infty < y_i < \infty, \quad i = 1, \cdots, n$$

which shows that $Y_1, \cdots, Y_n$ are independent, each $N(0, 1)$. In particular, Y_1 is $N(0, 1)$. Now

$$Y = a + \sum b_i X_i = a + \sum b_i \mu_i + Y_1 \sqrt{\sum_1^n b_i^2 \sigma_i^2}$$

$$E(Y) = a + \sum b_i \mu_i, \quad V(Y) = \sum b_i^2 \sigma_i^2.$$

Hence Y is $N(a + \sum b_i\mu_i, \sum b_i^2\sigma_i^2)$ since if Z is normal (μ, σ), any linear function

$$\alpha Z + \beta, \quad (\alpha \neq 0)$$

itself follows a normal distribution $N(\alpha\mu + \beta, \alpha^2\sigma^2)$ (vide Example 4.3.6).

[The results in this example were proved in theorem 6.7.1 by using characteristic function.]

8.6 Distributions of Maximum and Minimum

Let $X_1, \cdots, X_n$ be r.v.'s each defined on the probability space $(\mathcal{S}, \mathcal{A}, P)$. Let

$$Y_1 = \max(X_1, \cdots, X_n), \quad Y_2 = \min(X_1, \cdots, X_n).$$

Clearly Y_1, Y_2 are r.v.'s, because

$$\{w : Y_2(w) \le z\} = \{w : X_1(w) \ge z, \cdots, X_n(w) \ge z\} = \bigcap_{i=1}^{n} \{w : X_i(w) \ge z\} \in \mathcal{A}.$$

Thus Y_2 is a $\mathcal{A}$-measurable function and hence is a r.v. Similarly Y_1 is a r.v.

Theorem 8.6.1 *If $X_1, \cdots, X_n$ are independent r.v.'s, X_i having d.f. $F_{X_i}(x)$, the d.f. of Y_1, Y_2, respectively, are*

1. $F_{Y_1}(x) = \prod_{i=1}^{n} F_{X_i}(x),$
2. $F_{Y_2}(x) = 1 - \prod_{i=1}^{n} [1 - F_{X_i}(x)].$

Proof:

$$\begin{aligned} F_{Y_1}(x) = P(Y_1 \le x) &= P(X_1 \le x, \cdots, X_n \le x) \\ &= P(X_1 \le x) \cdots P(X_n \le x) \text{ [by independence]} \\ &= \prod_{i=1}^{n} F_{X_i}(x). \\ F_{Y_2}(x) = P(Y_2 \le x) &= 1 - P(Y_2 > x) = 1 - P(X_1 > x, \cdots, X_n > x) \\ &= 1 - P(X_1 > x) \cdots P(X_n > x) \text{ [by independence]} \\ &= 1 - \prod_{i=1}^{n} [1 - F_{X_i}(x)]. \end{aligned}$$

Corollary 8.6.1 *If $X_1, \cdots, X_n$ are iid variables each having d.f. $F_X(x)$,*

$$F_{Y_1}(x) = (F_X(x))^n, \quad F_{Y_2}(x) = 1 - (1 - F_X(x))^n.$$

Corollary 8.6.2 *If, further, $X_1, \cdots, X_n$ are continuous with common p.d.f. f, the p.d.f. of Y_1, Y_2 are respectively given by*

1. $f_{Y_1}(x) = nF^{n-1}(x) f(x)$
2. $f_{Y_2}(x) = n[1 - F(x)]^{n-1} f(x).$

▶ **Example 8.6.1** *Let X, Y be independent r.v.'s with negative exponential distributions having respective parameters λ_1, λ_2. Find $E[\max\,(X, Y)]$.*

Let $Z = \max(X, Y)$. The d.f. of Z is

$$
\begin{aligned}
F_Z(z) &= F_X(z)F_Y(z) = (1-e^{-\lambda_1 z})(1-e^{-\lambda_2 z}) \\
f_Z(z) &= \frac{d}{dz}(F_Z(z)) = \frac{d}{dz}[1-e^{-\lambda_1 z}-e^{-\lambda_2 z}+e^{-(\lambda_1+\lambda_2)z}] \\
&= \frac{e^{-\lambda_1 z}}{\lambda_1}+\frac{e^{-\lambda_2 z}}{\lambda_2}+\frac{e^{-(\lambda_1+\lambda_2)z}}{-(\lambda_1+\lambda_2)} = \frac{e^{-\lambda_1 z}}{\lambda_1}+\frac{e^{-\lambda_2 z}}{\lambda_2}-\frac{e^{-(\lambda_1+\lambda_2)z}}{(\lambda_1+\lambda_2)}. \\
E(Z) &= \int_0^\infty z f_Z(z)\,dz.
\end{aligned}
$$

Now

$$\int_0^\infty z\frac{e^{-\lambda_1 z}}{\lambda_1}\,dz = \int_0^\infty \frac{e^{-\lambda_1 z}(\lambda_1 z)^{2-1}}{\lambda_1^3}d(\lambda_1 z) = \frac{\boxed{(2)}}{\lambda_1^3} = \frac{1}{\lambda_1^3}.$$

Thus

$$E(Z) = \frac{1}{\lambda_1^3}+\frac{1}{\lambda_2^3}-\frac{1}{(\lambda_1+\lambda_2)^3}.$$

▶ **Example 8.6.2** *Show that a necessary and sufficient condition for independent r.v.'s $X_1, \cdots, X_n$ to be geometrically distributed with parameter p is that $Y = \min(X_1, \cdots, X_n)$ has a geometric distribution with parameter $(1-q^n), q = 1-p$.*

Suppose X has a geometric distribution with parameter p. The d.f. is

$$F_X(x) = P(X \le x) = 1 - q^{x+1}.$$

Hence d.f. of Y is

$$F_Y(x) = 1-[1-F_X(x)]^n = 1-q^{n(x+1)} \tag{8.6.1}$$

so that Y is geometric with parameter $(1-q^n)$. Conversely, suppose Y has d.f. (8.6.1). Now

$$F_Y(x) = 1-[1-F_X(x)]^n.$$

Hence

$$1-F_X(x) = q^{x+1} \text{ when } F_X(x) = 1-q^{x+1}$$

so that each X is geometric with parameter p.

Theorem 8.6.2 *If $X_1, \cdots, X_n$ are each independent r.v.'s, the joint d.f. of Y_1, Y_2 is*

$$F_{Y_1,Y_2}(y_1,y_2) = \begin{cases} \prod_{i=1}^n F_{X_i}(y_1) - \prod_{i=1}^n [F_{X_i}(y_1)-F_{X_i}(y_2)], & y_1 > y_2 \\ \prod_{i=1}^n F_{X_i}(y_1), & y_1 \le y_2. \end{cases}$$

Proof:

$$\left.\begin{aligned} F_{Y_1,Y_2}(y_1,y_2) &= P[Y_1 \le y_1, Y_2 \le y_2] = P[Y_1 \le y_1] - P[Y_1 \le y_1, Y_2 \ge y_2] \\ &= P[X_1 \le y_1, \cdots, X_n \le y_1] - P[y_2 \le X_1, \cdots, X_n \le y_1] \\ &= \prod_{i=1}^{n} F_{X_i}(y_1) - \prod_{i=1}^{n} [F_{X_i}(y_1) - F_{X_i}(y_2)] \\ &\qquad \text{(by independence).} \end{aligned}\right\} \quad (8.6.2)$$

If $y_1 \le y_2$, the last term in the right hand side of (8.6.2) vanishes. Hence the result.

Corollary 8.6.3 *If $X_1, \cdots, X_n$ are iid with common d.f. F then*

$$F_{Y_1,Y_2}(y_1,y_2) = \begin{cases} F^n(y_1) - [F(y_1) - F(y_2)]^n, & y_1 \le y_2 \\ F^n(y_1), & y_1 \le y_2. \end{cases}$$

Corollary 8.6.4 *If $X_1, \cdots, X_n$ are continuous with common density f, the joint p.d.f. of Y_1, Y_2 is*

$$f_{Y_1,Y_2}(y_1,y_2) = \begin{cases} n(n-1)[F(y_1) - F(y_2)]^{n-2} f(y_1) f(y_2), & y_1 > y_2 \\ 0, & y_1 \le y_2. \end{cases}$$

Example 8.6.3 $X_1, \cdots, X_n$ *are independent exponential variables with the same parameter λ. Find the joint distribution of Y_1, Y_2.*

$$F_{Y_1,Y_2}(y_1,y_2) = \begin{cases} (1 - e^{-\lambda y_1})^n - [e^{-\lambda y_2} - e^{-\lambda y_1}]^n, & y_1 > y_2 \\ (1 - e^{-\lambda y_1})^n, & y_1 \le y_2. \end{cases}$$

The joint p.d.f. is

$$f_{Y_1,Y_2}(y_1,y_2) = \begin{cases} n(n-1)\lambda^2 [e^{-\lambda y_2} - e^{-\lambda y_1}]^{n-2} e^{-\lambda(y_1+y_2)}, & y_1 > y_2 \\ 0, & y_1 \le y_2. \end{cases}$$

8.7 Use of Moment Generating Functions

Since moment generating function, when it exists, uniquely determines the distribution function, one may use the m.g.f. to determine the distribution. Thus to find the joint distribution of $Y_1 = g_1(X_1, \cdots, X_n), \cdots, Y_n = g_n(X_1, \cdots, X_n)$, one may find the joint m.g.f. of $(Y_1, \cdots, Y_n)$ with the help of the joint p.d.f. of $(X_1, \cdots, X_n)$. If the m.g.f. so obtained can be identified to be the m.g.f. of a known distribution ξ, then ξ would be the joint distribution of $(Y_1, \cdots, Y_n)$. However, since only m.g.f. of some one-dimension random variables and some two-dimensional random variables are readily recognisabl the use of m.g.f. to obtain the distribution will be limited in use to univariate functio or bivariate functions of random variables only.

Mostly by using the property that the m.g.f. of a sum of n independent rando variables is the product of their m.g.f.'s we have been able to prove several properti (including the reproductive property) of binomial (theorem 5.4.1), Poisson (Examp 4.3.7), geometric, negative binomial (theorem 5.8.1), exponential, gamma (theore 6.4.1), cauchy (theorem 6.6.1) and normal variables (theorem 6.7.1) and hence have bee able to identify the distributions of the relevant functions of these variables. Here, w consider the m.g.f. of arbitrary functions of random variables, restricting our discussio to the case of jointly continuous two-dimensional variables only.

Let (X_1, X_2) be two continuous variables having joint p.d.f. $f_{X_1,X_2}(x_1, x_2)$. The joi moment generating function of $Y_1 = g_1(X_1, X_2), Y_2 = g_2(X_1, X_2)$ is

$$M_{Y_1Y_2}(t_1, t_2) = E(e^{t_1Y_1+t_2Y_2}) = \int_{-\infty}^{\infty}\int_{-\infty}^{\infty} e^{t_1g_1(x_1,x_2)+t_2g_2(x_1,x_2)} f(x, y)\, dx\, dy. \quad (8.7.1)$$

The case of discrete random-variables is treated analogously.

▶ **Example 8.7.1** *Let X, Y be independent standard normal variables. Find the m.g. of XY.*

$$M_{XY}(t) = E(e^{txy}) = \frac{1}{2\pi}\int_{-\infty}^{\infty}\int_{-\infty}^{\infty} e^{txy-\frac{x^2+y^2}{2}}\, dx\, dy$$

$$= \frac{1}{2\pi}\int_{-\infty}^{\infty} e^{-\frac{y^2}{2}(1-t^2)}\left[\int_{-\infty}^{\infty} e^{-\frac{1}{2}(x-ty)^2}\, dx\right] dy = \frac{1}{\sqrt{1-t^2}}.$$

▶ **Example 8.7.2** *(X, Y) follows a bivariate normal distribution with parameters $(\mu_1, \mu_2, \sigma_1, \sigma_2, \rho)$. Show that $Z = aX + bY + c$ where a, b, c are constants, is*

$$N\left(a\mu_1 + b\mu_2 + c, \sqrt{a\sigma_1^2 + b\sigma_2^2 + 2ab\rho\sigma_1\sigma_2}\right).$$

Moment generating function of Z is

$$M_Z(t) = \int_{-\infty}^{\infty}\int_{-\infty}^{\infty} e^{t(ax+by+c)} f_{X,Y}(x, y)\, dx\, dy. \quad (8.7.2)$$

Now it is known

$$M_{X,Y}(t) = E(e^{t_1X+t_2Y}) = e^{\mu_1t_1+\mu_2t_1+\frac{1}{2}(\sigma_1^2t_1^2+\sigma_2^2t_2^2+2\rho\sigma_1\sigma_2t_1t_2)}. \quad (8.7.3)$$

Putting $t_1 = at, t_2 = bt$ in (8.7.3), we obtain

$$M_Z(t) = \exp\left[c + at\mu_1 + bt\mu_2 + \frac{t^2}{2}(a^2\sigma_1^2 + b^2\sigma_2^2 + 2\rho\sigma_1\sigma_2ab)\right]$$

which is the m.g.f. of a $N(a\mu_1 + b\mu_2 + c, a^2\sigma_1^2 + b^2\sigma_2^2 + 2\rho ab\sigma_1\sigma_2)$. Hence the result follow by the uniqueness property of the m.g.f.

8.8 Location and Scale Families of Distributions

In this section we discuss three techniques for constructing families of distributions. The three types of families are called *location* families, *scale* families and *location-scale* families.

Each of these families is constructed by specifying a single pdf, say, $f(x)$ called the standard pdf of the family. All other pdf's in the family are generated by transforming the standard pdf in a prescribed way. We first consider a simple theorem about pdf's.

Theorem 8.8.1 *Let $f(x)$ be any pdf and let μ and $\sigma(> 0)$ be any given constants. Then the function*

$$h(x \mid \mu, \sigma) = \frac{1}{\sigma} f\left(\frac{x-\mu}{\sigma}\right)$$

is a pdf.

Proof: To prove the theorem we must check that $h(x \mid \mu, \sigma)$ is nonnegative and integrales to 1. Since $f(x) \geq 0$, $h(x \mid \mu, \sigma) \geq 0$, for all values of x, μ and σ. Now

$$\int_{-\infty}^{\infty} \frac{1}{\sigma} f\left(\frac{x-\mu}{\sigma}\right) dx = \int_{-\infty}^{\infty} f(y)dy \quad \left(\text{substituting } y = \frac{x-\mu}{\sigma}\right)$$
$$= 1.$$

Hence of proof.

We first consider the definition of location parameter and scale parameter of a pdf $f(x, \theta)$. The parameter θ is called the location parameter if

$$f(x, \theta) = h(x - \theta)$$

for some probability density or mass function $h(\cdot)$.
The parameter θ is called the scale parameter if

$$f(x, \theta) = \frac{1}{\theta} h\left(\frac{x}{\theta}\right)$$

for some probability density or mass function $h(\cdot)$.

The mean of a distribution may or may not be its location parameter. For normal distribution, mean is the location parameter. For exponential distribution, mean is a scale parameter. For gamma distribution, the mean is neither a location parameter, nor a scale parameter.

We now consider the definition of location families of distributions.

Definition 8.8.1 *Let $f(x)$ be any pdf. Then the family of pdf's, $f(x - \mu)$, indexed by the parameter μ $(-\infty < \mu < \infty)$ is called the location family with standard pdf $f(x)$ and μ is called the location parameter of the family.*

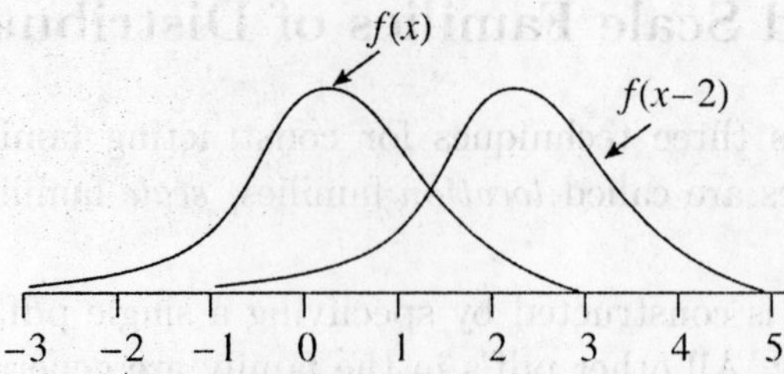

Fig. 8.5 Two members of the location family {f(x−μ)} with μ = 0, μ = 2

It is seen that the location parameter μ simply shifts the pdf $f(x)$ so that the shape of the graph remains unchanged but the point of the graph that was above $x = a$ for $f(x)$ is now above $x = \mu + a$ for $f(x - \mu)$. The figure 8.5 shows that the area under the graph of $f(x)$ between $x = -2$ and 1 is the area under the graph of $f(x - 2)$ between $x = 0$ (i.e., $\mu - 2$) and 3 (i.e., $\mu + 1$).

As an example, if $\sigma(> 0)$ is specified and we define

$$f(x) = \frac{1}{\sigma\sqrt{2\pi}} e^{-x^2/2}, \quad -\infty < x < \infty$$

then the location family generated by this standard pdf is the set of normal distributions with unknown mean μ and known variance σ^2. Similarly, the Cauchy family (with λ a specified value) and double exponential family (with σ a specified value) are examples of location families.

The definition 8.8.1 states that we can start with any pdf $f(x)$ and generate a family of pdf's by introducing a location parameter.

Definition 8.8.2 *Let $f(x)$ be any pdf. Then for any $\sigma(> 0)$. The family of pdf's $(1/\sigma)f(x/\sigma)$, indexed by the parameter σ, is called the scale family with standard pdf $f(x)$ and σ is called the scale parameter of the family.*

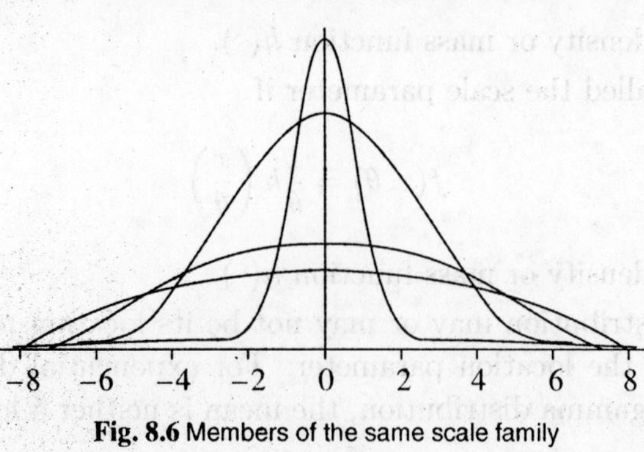

Fig. 8.6 Members of the same scale family

The effect of the scale parameter σ is either to stretch ($\sigma > 1$) or contract ($\sigma < 1$) the graph of $f(x)$ while maintaining the same basic shape of the graph.

Several of the families of the distributions introduced so far are scale families. These are the gamma family if p is a fixed value and λ is the scale parameter, the normal family

$\mu = 0$ and σ is the scale parameter, the exponential family and the double exponential mily if $\mu = 0$ and σ is the scale parameter. In each case, the standard pdf is the pdf tained by setting the scale parameter equal to 1.

efinition 8.8.3 *Let $f(x)$ be any pdf. Then, for any $\mu(-\infty < \mu < \infty)$ and any $\sigma(> 0)$, e family of pdf's $(1/\sigma)f((x - \mu)/\sigma)$, indexed by the parameter (μ, σ) is called the cation-scale family with standard pdf $f(x)$. The parameter μ is called the location rameter and σ is called the scale parameter.*

The effect of introducing both the location and the scale parameters is to stretch > 1) or contract ($\sigma < 1$) the graph of $f(x)$ and then shift the graph, so obtained so at the point that was above $x = a$ in $f(x)$ is now above $x = a + \mu$ for $f(x - \mu)$. The rmal family and double-exponential family are examples of location-scale families.

The following theorem shows the relationship between the original variable Z with lf $f(z)$ and the transformed variable X with pdf $\frac{1}{\sigma}f\left(\frac{x-\mu}{\sigma}\right)$, which is a member of the cation-scale family generated by $f(z)$.

heorem 8.8.2 *Let $f(\cdot)$ be any pdf. Let μ be any real number and $\sigma(> 0)$ be any sitive real number. Then X is a random variable with pdf $\frac{1}{\sigma}f\left(\frac{x-\mu}{\sigma}\right)$ iff there exists a ndom variable Z with pdf $f(z)$ such that $Z = \frac{X-\mu}{\sigma}$.*

Proof: If part. Suppose Z is a random variable with pdf $f(z)$. We are to find the df of X. Let $X = g(Z)$. Then $Z = g^{-1}(X)$ is a monotonic function, $\left|\frac{dg^{-1}(x)}{dx}\right| = \sigma^{-1}$. he pdf of X,

$$f_X(x) = f_Z\left\{g^{-1}(x)\right\}\left|\frac{d}{dx}g^{-1}(x)\right| = f\left(\frac{x-\mu}{\sigma}\right)\cdot\frac{1}{\sigma}.$$

nly if part. We have X a random variable with pdf $\frac{1}{\sigma}f\left(\frac{x-\mu}{\sigma}\right)$. Let $Z = h(X)$. Then $^{-1}(Z) = \sigma z + \mu$, $\left|\frac{d}{dz}h^{-1}(z)\right| = \sigma$ and the pdf of Z is

$$f_Z(z) = f_X(h^{-1}(z))\left|\frac{d}{dz}h^{-1}(z)\right| = \frac{1}{\sigma}\,f\left(\frac{\sigma z + \mu - \mu}{\sigma}\right)\sigma = f(z).$$

ence the proof.

Exercise

1. X is a r.v. with d.f. $F_X(x)$. Let

$$g(X) = \begin{cases} x, & |x| \geq c \\ 0, & |x| < c. \end{cases}$$

Show that d.f. of $Y = g(x)$ is $F_Y(y) =$

(i) $F_X(y)$ for $y \geq c$, (ii) $F_X(c)$ for $0 \leq y < c$, (iii) $F_X(-c)$ for $-c \leq y < 0$, (iv) $F_X(y)$ for $y < -c$.

2. X is uniformly distributed over $(-a, a)$. Find the distribution of $Y =$ (i) $\frac{1}{X^2}$, (ii) $\sqrt{|X|}$, (iii) $\cos X$, (iv) e^X, (v) $bX^2 + cX + d$.

3. X has a uniform distribution over $(0, 1)$. Find the probability distribution of Y (i) X^2, (ii) $\frac{1}{X}$, (iii) $\sqrt{X}$, (iv) $\sin(\pi X)$, (v) $-2\log_e X$, (vi) $\tan \pi X$, (vii) aX^2+bX+ (viii) e^X, (ix) $\frac{X}{(1+X)}$.

4. Let X be a r.v. with p.d.f.

$$f(x) = \begin{cases} \lambda\, e^{-\lambda x}, & x > 0 \\ 0, & x \leq 0 \end{cases}, \quad \lambda > 0.$$

Find the p.d.f. of Y where Y is (i) $\sin X$, (ii) $\cos X$, (iii) $\tan X$.

5. X has a uniform distribution over $(0, 1)$. Find the distribution of $\frac{1}{X}$.

6. X, Y are independent, each $N(0, \sigma)$ variable. Find the p.d.f. of $w = \tan^{-1}\left(\frac{x}{y}\right)$

$\left[f(w) = \frac{1}{\pi}, \quad |w| <\right.$

7. X has a p.d.f. $f_X(x) = C\, e^{-|x|}$. Find the constant C and then find the distributi of $Y = X^2$.

8. X had p.d.f.

$$f_X(x) = \begin{cases} 0, & \text{for } X \leq -k \\ \frac{1}{k^2}(k + X), & \text{for } -k \leq X \leq 0 \\ \frac{1}{k^2}(k - X), & \text{for } 0 \leq X \leq k \\ 0, & \text{for } k \leq X. \end{cases}$$

Find the p.d.f. of $Y = e^X$.

9. Let X be a r.v. with p.d.f. $f(x) = \frac{2x}{\pi^2}(0), 0 < x < \pi$ (otherwise). Find the d.f. a p.d.f. of $Y = \sin X$.

10. Let X have a binomial distribution with parameters (n, p). Find the p.d.f. of Y $(X - np)^2$.

11. X, Y are independent standard normal variates. Find the p.d.f. of $T = |X - Y|$

12. Let X_1, X_2 be independent $\chi^2_{(2)}$-variables. Show that $Y = \frac{1}{2}(X_1 - X_2)$ has p.d.f

$$g_Y(y) = \frac{1}{2}\, e^{-|y|}, \quad -\infty < y < \infty.$$

3. X_1, X_2 are random samples from rectangular population

$$dF = dx, \quad 0 \leq x \leq 1.$$

Find the p.d.f. of standard deviation $S = \frac{|X_1 - X_2|}{2}$. [$4(1-2s)$]

4. The speed of a molecule of gas at equilibrium is a random variable X with density

$$f(x) = \begin{cases} Kx^2 e^{-bx^2}, & x > 0 \\ 0 & , \quad x \leq 0. \end{cases}$$

where K is a normalising constant and b depends on the temperature of the gas and the mass of the molecule. Find the probability density of the kinetic energy $E = mX^2/2$ of the molecule. [$g_E(u) = m^{-3/2} K\sqrt{2u}\, e^{-2bu/m}, u > 0$]

5. Let X, Y, Z be independent random variables with the same geometric distribution $\{q^k p\}$. Find

(a) $P\{X = Y\}$, (b) $P\{X \geq 2Y\}$, (c) $P\{X + Y \leq Z\}$.

$\left[\text{(a) } \frac{p}{1+q}, \text{ (b) } \frac{1}{1+q+q^2}, \text{ (c) } \frac{1}{(1+q)^2}\right]$

16. The r.v.'s. X, Y are uniformly distributed over the circle C of radius 1 so that

$$f_{X,Y}(x, y) = \frac{1}{\pi}(0) \text{ if } (x, y) \in C \text{ (otherwise).}$$

Find the p.d.f. of $R = \pm\sqrt{X^2 + Y^2}$.

[*Hints:* Use polar transformation; $g(r) = 2r$.]

17. X, Y have joint density

$$f_{X,Y}(x, y) = \frac{1}{4}, \quad -1 \leq x, y \leq 1.$$

Find (i) $P(X + Y \leq t)$ where $0 < t \leq 1$, (ii) $P(X^2 + Y^2 \leq r^2)$, (iii) the d.f. and p.d.f. of $R^2 (= X^2 + Y^2)$.

$\left[\text{(i) } \frac{t^2}{4}, \text{ (ii) } r^2, \text{ (iii) } dF = \frac{\pi}{4} d(r^2)\right]$

18. X has a Rayleigh distribution $f(x) = \frac{x}{\sigma^2}\, e^{-\frac{x^2}{2\sigma^2}}, x > 0$. Find p.d.f. of $Y = e^{-x^2}$.

$\left[\frac{1}{2y\sigma^2} y^{\frac{1}{2\sigma^2}}, \quad 0 < y < 1\right]$

19. X has a Cauchy distribution $f(x) = \frac{1}{\pi(1+x^2)}, -\infty < x < \infty$. Find p.d.f. of $Y = \frac{1}{X}$.

$\left[\frac{1}{\pi(1+y^2)}, \quad -\infty < y < \infty\right]$

20. X had p.d.f., $f_1(x) = \sqrt{\frac{2}{\pi}}\, e^{-\frac{x^2}{2}}, 0 \leq x < \infty$, Y has p.d.f. $f_2(y) = 2y^3\, e^{-y}, 0 < y < \infty$. Find distribution of $U = \frac{X}{Y}$. $\left[\frac{6}{(u^2+2)^{5/2}}, \quad 0 < u < \infty\right]$

21. X, Y are each independent binomial variables with parameter (n, p). Find t p.m.f. of (i) $X+Y$, (ii) $X-Y$, (iii) $U = \frac{X}{Y+1}$.

22. Let X have a binomial distribution with parameters (n, p). Find the p.m.f. $Y = X^2$. $\left[\frac{n!}{(\sqrt{y})!(n-\sqrt{y})!} p^{\sqrt{y}} q^{n-\sqrt{y}}, \ p+q=\right.$

23. Two independent variables X, Y are each uniformly distributed over the ra $(-a, a)$. Show that the sum $Z = X + Y$ has a triangular density.

$$f(z) = \begin{cases} \dfrac{2a+z}{4a^2}, & (-2a \le z \le 0) \\ \dfrac{2a-z}{4a^2}, & (0 \le z \le 2a). \end{cases}$$

Show that the m.g.f. of Z is $\left(\frac{1}{at} \sinh \ at\right)^2$.

24. (X, Y) are continuous random variables with joint p.d.f.

$$f_{X,Y}(x, y) = \begin{cases} \frac{1}{4a^2}[1 + xy(x^2 - y^2)], & |x|, \ |y| \le a, \ a > 0 \\ 0, & \text{elsewhere.} \end{cases}$$

Show that the p.d.f. of $Z = X + Y$ is triangular with p.d.f. given in (i).

25. Show that the p.d.f. of $Z = X_1 + X_2$ when X_1, X_2 are independent, each rectangu over (a, b) is,

$$f_Z(z) = \begin{cases} \dfrac{z-2a}{(b-a)^2}, & 2a \le z \le a+b \\ \dfrac{2b-z}{(b-a)^2}, & a+b \le z \le 2b. \end{cases}$$

26. X is a continuous r.v. with p.d.f. $f_X(x)$. Another r.v. Y, distributed independe of X has a uniform distribution over $(-a, a)$. Show that p.d.f. of $Z = X + Y$ i

$$f_Z(z) = \frac{1}{2a}[F_X(z+a) - F_X(z-a)].$$

27. X, Y are independent normal, each $N(0, \sigma^2)$-variable. Show that

(a) $Z = +\sqrt{X^2 + Y^2}$ and $W = \arctan \frac{X}{Y}$ are independently distributed, Z hav a Rayleigh distribution (vide exercise 18) and W a uniform distribution $\left(-\frac{\pi}{2}, \frac{\pi}{2}\right)$.

(b) Z and $V = \frac{X}{Y}$ are independently distributed, Z having a Rayleigh distribu and V a Cauchy distribution (Note that if X, Y are normal, but their expec values are not zero, Z and V are not independent).

28. Let X be a non-negative r.v. of the continuous type with p.d.f. $f_X(x)$. Show that p.d.f. of $Y = X^m, m > 0$ is

$$f_Y(y) = \begin{cases} \frac{1}{m} y^{\frac{1}{m}-1} f(y^{\frac{1}{m}}), & y > 0 \\ 0, & y \le 0. \end{cases}$$

29. Let X be a standard normal variate. Show that p.d.f. of $Y = X^{2m}$ is

$$f_Y(y) = \begin{cases} \frac{1}{\sqrt{2\pi} m y^{1-\frac{1}{2}m}} \cdot \exp\left\{-\frac{y^{\frac{1}{m}}}{2}\right\}, & y > 0 \\ 0, & y \le 0. \end{cases}$$

30. Let $X_1, \cdots, X_n$ be iid rectangular variables over $(0, 1)$. Let $X_{(1)} = \min\{X_1, \cdots, X_n\}$, $X_{(n)} = \max\{X_1, \cdots, X_n\}$.

   (a) Find the d.f. and p.d.f. of (i) $X_{(1)}$, (ii) $X_{(n)}$.

   (b) Find the joint d.f. and joint p.d.f. of $X_{(1)}, X_{(n)}$.

   (c) Find the d.f. and p.d.f. of the range, $X_{(n)} - X_{(1)}$.

   (d) Find $E[X_{(1)}], \sigma^2_{X_{(1)}}, E[X_{(n)}], \sigma^2_{X_{(n)}}, \rho_{X_{(1)}, X_{(n)}}$.

   (e) Find the conditional p.d.f. of $X_{(1)}$ given $X_{(n)}$ and the conditional mean and variance of $X_{(1)}$, given $X_{(n)}$.

31. Suppose $X_1, \cdots, X_n$ are iid rectangular variables over (a, b). Extend the results of the previous problem to this probability density function.

32. Life of an electric bulb manufactured by a certain company is assumed to be exponentially distributed with mean 200 hours. Bulbs, nine in number are bought. What is the chance that no bulb will get extinct before 100 hours? Find the average life of the bulb that lasts the longest.

33. X, Y, Z have joint p.d.f.

$$g(x, y, z) = \frac{1}{\sqrt{xyz}} f(x + y + z), 0 < x, y, z < \infty.$$

Derive the distribution of (i) $U = X + Y + Z$, (ii) $S = \frac{Y}{X}$, (iii) $T = \frac{Z}{X+Y}$.

[*Hints:* (i) transformation: $u = x + y + z, uv = y + z, uvw = z$. (ii) and (iii) transformation: $u = x + y + z, uv = x + y, uvw = x$.]

34. Find the distribution of $\bar{X} = \frac{1}{n}\sum_{i=1}^{n} X_i$, where $X_1, \cdots, X_n$ are independently and identically distributed r.v.'s with common p.d.f. $f(x) = \frac{1}{\theta}\, e^{-\frac{x}{\theta}}, 0 < x < \infty, \theta > 0$.

[$f(\bar{x}) = \frac{n}{\theta} \exp\left(-\frac{n\bar{x}}{\theta}\right)$]

35. Let X, Y be independently distributed with a common p.d.f. $f(x) = \frac{1}{\lambda}\, e^{-\frac{x}{\lambda}}$, $x > 0, \lambda > 0$. Define $U = X + Y, V = \frac{X}{(X+Y)}, W = \frac{X}{Y}$. Check if (i) U is independent of V (ii) U is independent of W.

36. X_1, X_2 are independent standard normal deviates. Let $Y = \min(X_1, X_2), Z = \max(X_1, X_2)$. Find $E(Y), E(Z), V(Y), V(Z)$.

37. X is a $N(0,1)$ r.v. Find the moment generating function of $Y = X^2$.

$$\left[M_Y(t) = E(e^{tX^2}) = \frac{1}{\sqrt{1-2t}},\ \ t < \frac{1}{2}\right]$$

38. (X, Y) have joint p.d.f.

$$f(x,y) = \frac{y}{(1+x)^4} \exp\left(-\frac{y}{1+x}\right),\ \ x, y > 0.$$

Show that joint density of U and V where $U = \frac{Y}{1+X}, V = \frac{1}{1+X}$ is

$$dF = u\, e^{-u},\ \ 0 \le u \le \infty,\ \ 0 \le v \le 1.$$

39. X_1, X_2, X_3 are each independent standard normal deviates. By using polar transformations, $x_1 = r\cos\theta_1 \cos\theta_2$, $x_2 = r\cos\theta_1 \sin\theta_2$, $x_3 = r\sin\theta_1$,

$$\left(0 \le r < \infty,\ \ -\frac{\pi}{2} < \theta_1 < \frac{\pi}{2},\ \ 0 < \theta_2 < 2\pi\right)$$

show that the variates corresponding to r, θ_1, θ_2 are independently distributed and distribution of r is

$$dF = \sqrt{\frac{2}{\pi}}\, e^{-\frac{r^2}{2}} r^2 dr,\ \ r \ge 0.$$

40. X_1, X_2, X_3 are independent random samples from the distribution $f(x) = \frac{x}{\lambda^2}\, e^{-\frac{x}{\lambda}}$, $x > 0, \lambda > 0$. find the joint p.d.f. of $Y_1 = X_1 + X_2 + X_3$, $Y_2 = X_2$, $Y_3 = X_3$ and the marginal p.d.f. of Y_1. $\left[\frac{y_1^5}{120\lambda^6}\, e^{-\frac{y_1}{\lambda}}\right]$

41. Let X_1, X_2 be independent random variables, each rectangular over $(0,1)$. Show that

$$Y = \sqrt{-2\log X_1}\cos 2\pi X_2 \ \text{ and } \ Z = \sqrt{-2\log X_1}\sin 2\pi X_2$$

are independent random variables, each $N(0,1)$.

42. (X, Y) have joint p.d.f.

$$f_{X,Y}(x,y) = \frac{e^{-y}}{\Gamma(k_1)\,\Gamma(k_2)} x^{k_1-1}(y-x)^{k_2-1},\ \ 0 < x < y < \infty$$

which is zero otherwise. Show that marginal distribution of X and Y are gamma distributions $G(k_1)$ and $G(k_1 + k_2)$ respectively. Find the correlation coefficient between (X, Y). $\left[\sqrt{\frac{k_1}{k_1+k_2}}\right]$

43. X, Y are independent random variables with p.d.f.

$$f_X(x) = \alpha\, e^{-\alpha x}, \quad x \geq 0, \quad f_Y(y) = \beta\, e^{-\beta y}, \quad y \geq 0.$$

Show that p.d.f. of $Z = X + Y$ is

$$f_Z(z) = \begin{cases} \frac{\alpha\beta}{\beta-\alpha}(e^{-\alpha z} - e^{-\beta z}), & \beta \neq \alpha \\ \alpha^2 z\, e^{-\alpha z}, & \beta = \alpha. \end{cases}$$

44. X, Y are independent random variables, each with Reyleigh density

$$f_X(x) = \frac{x}{\alpha^2}\, e^{-\frac{x^2}{2\alpha^2}}, \quad x \geq 0$$

$$f_Y(y) = \frac{y}{\beta^2}\, e^{-\frac{y^2}{2\beta^2}}, \quad y \geq 0.$$

Show that the p.d.f. and d.f. of $Z = \frac{X}{Y}$ are, respectively.

$$f_Z(z) = \frac{2\alpha^2}{\beta^2} \cdot \frac{z}{\left(z^2 + \frac{\alpha^2}{\beta^2}\right)^2}, \quad z \geq 0.$$

$$F_Z(z) = \frac{z^2}{z^2 + \frac{\alpha^2}{\beta^2}}, \quad z > 0.$$

45. Let X, Y, Z be mutually independent random variables with the p.d.f.

$$f(u) = \begin{cases} \alpha\, e^{-\alpha u}, \ 0 \leq u < \infty, \alpha > 0 \\ 0, \qquad \text{otherwise.} \end{cases}$$

Find the p.d.f. of (i) $X + Y + Z$, (ii) $\frac{(X+Y)}{Z}$, (iii) $\frac{(X+Y)}{X}$.

46. Consider the one-one transformation $(x_1, \cdots, x_n) \rightarrow (r, \theta_1, \cdots, \theta_{n-1})$ such that

$$x_1 = r\cos\theta_1, \qquad 0 \leq r < \infty$$

$$x_2 = r\sin\theta_1\cos\theta_2, \qquad -\tfrac{\pi}{2} < \theta_i < \tfrac{\pi}{2}, i = 1, \cdots, n-2$$

$$x_3 = r\sin\theta_1\sin\theta_2\cos\theta_3, \ 0 < \theta_{n-1} < 2\pi$$

....

$$x_{n-1} = r\sin\theta_1\sin\theta_2\sin\theta_3\cdots\sin\theta_{n-2}\cos\theta_{n-1} x_n = r\sin\theta_1\sin\theta_2\sin\theta_3 \cdots\sin\theta_{n-2}\sin\theta_{n-1}.$$

Find the Jacobian of the transformation.

$$\left[|J| = r^{n-1}(\sin\theta_1)^{n-2}(\sin\theta_2)^{n-3}\cdots\sin\theta_{n-2}\right].$$

47. Consider the pdf $f(x) = y_0\, x^3(1-x)^4, 0 \leq x \leq 1$. Draw graph of $(1/\sigma)f((x-\mu)/$ for each of the following on the same axes.

    (a) $\mu = 0, \sigma = 1$

    (b) $\mu = 2, \sigma = 1$

    (c) $\mu = 4, \sigma = 2.$

48. Show that if $f(x)$ is a pdf symmetric about 0, then μ is the median of the locatio scale pdf $(1/\sigma)f((x-\mu)/\sigma), -\infty < x < \infty.$

Chapter 9

Convergence of a Sequence of Random Variables

9.1 Introduction

In this chapter we consider a sequence of random variables $\{X_n\}, n = 1, 2, \cdots$, all defined on the same probability space $(\mathcal{S}, \mathcal{A}, P)$ and study their different modes of convergence. For each point w we have, therefore, an infinite number of values $\{X_n(w)\}$. Section 9.2 discusses convergence of a sequence $\{X_n\}$ to a r.v. X in distribution. Sections 9.3 through 9.5 deals, respectively, with convergence in probability, almost sure convergence, mean-square convergence and gradation of different modes of convergence. Weak and strong laws of large numbers are considered in Sections 9.6 and 9.7 respectively. The section 9.8 deals with the celebrated Central Limit Theorem—which discusses the situations when a sequence of random average converges in distribution to a normal law.

There are situations when a random average does not converge to a normal distribution. Results on such topics have not been considered. Convergence of sequences of random vectors are also not considered. •

9.2 Convergence of a Sequence of Distribution Functions

Let $\{X_n\}$ be a sequence of r.v.'s all defined on the probability space $(\mathcal{S}, \mathcal{A}, P)$, with $F_n(x)$ as the d.f. of $X_n, n = 1, 2, \cdots$.

Definition 9.2.1 *Let F_n be a sequence of d.f.'s. If there exists a d.f. F such that*

$$F_n(x) \to F(x)$$

at every point x where $F(x)$ is continuous, then F_n is said to converge weakly to F and we write

$$F_n \xrightarrow{w} F.$$

If $\{X_n\}$ be a sequence of r.v.'s with $\{F_n\}$ as the corresponding sequence of d.f.'s and if there exists a r.v. X with d.f. F such that $F_n \xrightarrow{w} F$ then we say X_n converges in law to X and write

$$X_n \xrightarrow{L} X.$$

■ NOTE 9.2.1 *If the sequence $\{F_n\}$ converges to F, it is not necessary that F will be a d.f. always.*

■ NOTE 9.2.2 *As the definition 9.2.1 implies, it is not necessary that $\{F_n(x)\}$ would converge to $F(x)$ at every point of discontinuity of $F(x)$, even if $F_n \xrightarrow{w} F$.*

► **Example 9.2.1** *Consider $\{X_n\}$ where $P(X_n = n) = 1$, $n = 1, 2, \cdots$.*

Here

$$F_n(x) = \begin{cases} 0, \ x < n \\ 1, \ x \geq n \end{cases}$$

$$\lim_{n\to\infty} F_n(x) = 0 \ (-\infty < x < \infty).$$

Thus $\{F_n\}$ does not converge to a d.f.

► **Example 9.2.2** *Let $\{X_n\}$ be a sequence of independent r.v.'s each uniform over $(0, \theta)$. Let*

$$X_{(n)} = \max_{1\leq i\leq n} X_i.$$

The d.f. of $X_{(n)}$ is

$$F_{(n)}(x) = \begin{cases} 0, & x \leq 0 \\ \left(\frac{x}{\theta}\right)^n, & 0 < x < \theta \\ 1, & x \geq \theta. \end{cases}$$

Consider $Y_n = \frac{X_{(n)}}{n}$. The d.f. of Y_n is

$$F_n(x) = P(Y_n \leq x) = P(X_{(n)} \leq nx) = \begin{cases} 0, & x \leq 0 \\ \left(\frac{x}{\theta}\right)^n, & 0 < x < \frac{\theta}{n} \\ 1, & x \geq \frac{\theta}{n}. \end{cases}$$

Hence

$$\lim_{n\to\infty} F_n(x) = F(x) = \begin{cases} 0, \ x < 0 \\ 1, \ x \geq 0. \end{cases}$$

which is the d.f. of a degenerate r.v. X with $P(X = 0) = 1$. Hence $F_n \xrightarrow{w} F$ and $X_n \xrightarrow{L} X$. Note that $x = 0$ is a point of discontinuity of $F(x)$, $F_n(0) = 0$ for all n and $F(0) = 1$, so that F_n does not converge to F at the point $x = 0$.

[A r.v. X is said to be degenerate if it has a degenerate distribution $F(x)$ i.e., if there exists a real number 'c' such that

$$F(x) = \begin{cases} 0, \ x < c \\ 1, \ x \geq c. \end{cases}$$

Here $P(X = c) = 1$].

Let the sequence of d.f.'s $F_n(x) \xrightarrow{w} F$, the d.f. of a r.v. X. Then

$$\lim_{n\to\infty} P(a < X_n \leq b) = F(b) - F(a), \tag{9.2.1}$$

for any two points of continuity a, b of $F(x), a < b$. Thus

$$\lim_{n\to\infty} F_n(t) = F(t) \tag{9.2.2}$$

for any t at which $F(x)$ is continuous.

■ NOTE 9.2.3 *Whenever we say that a standardised variate*

$$\frac{\{t_n - E(t_n)\}}{\sqrt{V(t_n)}} = Y_n \to \tau \text{ as } n \to \infty,$$

where τ is a $N(0,1)$ variate, we actually say, $Y_n \xrightarrow{L} \tau$, i.e., $F_n(y) \xrightarrow{w} \Phi(y)$ for all $y, -\infty < y < \infty$ (since $\Phi(y)$ has no point of discontinuity).

The following theorems hold.

Theorem 9.2.1 (a) *If $X_n (n = 1, 2, \cdots), X$ are continuous r.v.'s, with density functions $f_n(x), f(x)$ respectively, then $X_n \xrightarrow{L} X$ iff*

$$f_n(x) \to f(x) \text{ almost everywhere as } n \to \infty.$$

(b) *If $X_n (n = 1, 2, \cdots), X$ are discrete r.v.'s with p.m.f.'s,*

$$P(X_n = j) = p_n(j), \ \ P(X = j) = p(j), \ \ j = 0, 1, 2, \cdots,$$

then $X_n \xrightarrow{L} X$ iff

$$p_n(j) \to p(j) \text{ for all } j = 0, 1, 2, \cdots.$$

Theorem 9.2.2 *Let $\{X_n\}$ be a sequence of r.v.'s such that $X_n \xrightarrow{L} X$. Then for any constant 'a',*

$$a + X_n \xrightarrow{L} a + x, \ \ aX_n \xrightarrow{L} aX (a \neq 0).$$

Theorem 9.2.3 *Let $\{X_n\}, \{Y_n\}$ be two sequences of r.v.'s with $X_n \xrightarrow{L} X$, $Y_n \xrightarrow{L} c$ where X is a r.v. and 'c' is a constant; then* (i) $X_n + Y_n \xrightarrow{L} X + c$, (ii) $X_n Y_n \xrightarrow{L} cX$, (iii) $\frac{X_n}{Y_n} \xrightarrow{L} \frac{X}{c}$ $(c \neq 0)$.

It will often be tedious to determine if $F_n \xrightarrow{w} F$ by calculating the limit of $F_n(x)$ at every continuity point of $F(x)$. The following theorem stated without proof shows that this difficulty can be overcome by part by the use of characteristic function. The proof can be seen in the treatise by Lukacs (1960).

Theorem 9.2.4 (***Levy-Cramer***). *Let $\{X_n\}(n = 1, 2, \cdots)$ be a sequence of r.v.'s and let $F_n(x)$, $\Psi_n(t)$ be respectively the d.f. and characteristic function of X_n. The sequence $\{F_n(x)\}$ tends to the d.f. $F(x)$ as $n \to \infty$ iff the sequence $\{\Psi_n(t)\}$ converges uniformly to a function $\Psi(t)$ for all t, where $\Psi(t)$ is continuous at $t = 0$. Then the limiting function $\Psi(t)$ is the c.f. of the limiting d.f. $F(x)$.*

A weaker version of the above theorem stated in terms of moment generating function is as follows (Rohatgi, 1976, p. 277).

Theorem 9.2.5 *Let $\{X_n\}$ be a sequence of r.v.'s and let $F_n(x)$ and $M_n(t)$ be respectively the d.f. and moment generating function of $F(x)$, where $M_n(t)$ exists for $|t| < T$ for every n. If there exists a d.f. F with corresponding m.g.f. $M(t)$ which exists for $|t| \le t_1 < T$ such that $M_n(t) \to M(t)$ as $n \to \infty$ for every $|t| \le t_1$, then $F_n(x)$ converges to the d.f. $F(x)$.*

► **Example 9.2.3** *Consider $\{X_n\}$ of example 9.2.1, where*

$$F_n(x) = \begin{cases} 0, & x < n \\ 1, & x \ge n. \end{cases}$$

Here

$$\Psi_n(t) = e^{itn} = \cos nt + i \sin nt.$$

For $t = \frac{\pi}{2}$, $\Psi_n(t)$ takes the values $1, i, -i, -1$ only and $\Psi_n(t)$ does not converge. Hence $F_n(x)$ does not converge.

► **Example 9.2.4** *Let $\{X_n\}$ be a sequence of independent Cauchy variates. Consider the sequence $Y_n = \frac{S_n}{n^2}$ where $S_n = \sum_{i=1}^{n} X_i$. The c.f. of $Y_n = \Psi_n(t) = e^{-|t|/n}$. Hence*

$$\lim_{n\to\infty} \Psi_n(t) = \Psi(t) \text{ (say) } = \begin{cases} 1, & t = 0 \\ 0, & t \ne 0 \end{cases}$$

$\Psi(t)$ is not continuous at $t = 0$ and cannot be a characteristic function. Therefore, $\frac{S_n}{n^2}$ does not converge in law.

► **Example 9.2.5** *Let $\{X_n\}$ be a sequence of independent r.v.'s such that*

$$P\left(X_k = \frac{k}{n}\right) = \frac{1}{n}, \quad k = 1, 2, \cdots, n.$$

he c.f. of X_n is

$$\Psi_n(t) = E(e^{itX_n}) = \frac{1}{n}\sum_{1}^{n} e^{itk/n} = \frac{e^{it/n}}{n}\left(\frac{1-e^{it}}{1-e^{it/n}}\right).$$

ow

$$\lim_{n\to\infty} \frac{n(1-e^{it/n})}{e^{it/n}} = -it.$$

ence

$$\lim_{n\to\infty} \Psi_n(t) = \frac{e^{it}-1}{it} = \Psi(t),$$

hich is the c.f. of the uniform density on [0, 1]. Hence X_n converges in law to a uniform istribution.

We now apply theorem 9.2.4 to show that for many distributions, limiting distribution s a standard normal distribution.

Example 9.2.6 *Let $\{X_n\}$ be a sequence of binomial variables,*

$$P(X_n = r) = \binom{n}{r} p^r q^{n-r}, \quad 0 < p < 1, \quad q = 1-p, \quad r = 0, 1, \cdots, n$$

$E(X_n) = np, V(X_n) = n\,pq$. Consider the sequence of standardised binomial variables

$$Y_n = \frac{(X_n - np)}{\sqrt{n\,pq}}.$$

The c.f. of Y_n is

$$\begin{aligned}\Psi_n(t) &= E(e^{it(X_n - np)/\sqrt{n\,pq}}) \\ &= e^{-it\sqrt{np/q}}(pe^{it/\sqrt{n\,pq}} + q)^n.\end{aligned}$$

$$\begin{aligned}\log_e \Psi_n(t) &= -it\sqrt{np/q} + n\log_e(pe^{it/\sqrt{n\,pq}} + q) \\ &= -it\sqrt{np/q} + n\log_e\left[1 + p\frac{it}{\sqrt{n\,pq}} - \frac{pt^2}{2n\,pq} + o\left(\frac{t^2}{n}\right)\right] \\ &\left(\because \quad \log_e(1+x) = x - \frac{x^2}{2} + \frac{x^3}{3} - \cdots, |x| < 1\right) \\ &= -it\sqrt{np/q} + n\left[\frac{ipt}{\sqrt{n\,pq}} - \frac{pt^2}{2n\,pq} + \frac{p^2t^2}{2n\,pq} + 0\left(\frac{t^3}{n^{3/2}}\right)\right] \\ &= -\frac{t^2}{2} + 0\left(\frac{t^3}{\sqrt{n}}\right).\end{aligned}$$

Thus for any finite t, $\log_e \Psi_n(t)$ converges to $(-t^2/2)$ uniformly and hence $\Psi_n(t) \to e^{-t^2/2}$ uniformly in t. The latter being the c.f. of a standard normal deviate, it follow by theorem 9.2.4 that the d.f. of Y_n converges to that of a standard normal variable.

We have thus proved De-Moivre Laplace theorem (theorem 6.7.2). The same theorem is stated below in a slightly alternative form.

Theorem 9.2.6 *Let $\{X_n\}$ be a sequence of binomial variables with parameters (n, p),*

$$Y_n = \frac{(X_n - np)}{\sqrt{n\,pq}}$$

and $F_n(Y)$ be the d.f. of Y_n. If $0 < p < 1$, the relation

$$\lim_{n\to\infty} F_n(Y) = \frac{1}{\sqrt{2\pi}} \int_{-\infty}^{y} e^{-\frac{z^2}{2}}\,dz \tag{9.2.3}$$

holds for every value of y (since the d.f. of a normal variable has no point of discontinuity convergence considered in (9.2.3) holds for every value of y).

The r.v. X_n is said to have an asymptotically normal distribution $N(np, \sqrt{n\,pq})$ (vid definition 9.8.1).

Corollary 9.2.1 *The sequence $\left\{\frac{X_n}{n}\right\}$ is asymptotically $N\left(p, \sqrt{\frac{pq}{n}}\right)$.*

For a yet-another proof see Example 9.8.2.

► **Example 9.2.7** *Let $\{X_n\}$ be a sequence of binomial variables with parameters (n, p) Let $n \to \infty$ in such a way that np remains constant at λ.*

The c.f. of X_n is

$$\Psi_n(t) = (q + pe^{it})^n = \left(1 - \frac{\lambda}{n} + \frac{\lambda}{n}e^{it}\right)^n$$

$$= \left(1 + \frac{\lambda}{n}(e^{it} - 1)\right)^n \to \exp[\lambda(e^{it} - 1)] = \Psi(t)$$

as $n \to \infty$. $\Psi(t)$ is the c.f. of a Poisson distribution with parameter λ. Hence the binomia d.f. approaches a poission d.f. (an alternative proof was given in lemma 5.6.2).

► **Example 9.2.8** *Let $\{X_n\}$ be a sequence of poisson variables with parameter λ. Consider the standardised variable $Y_n = \frac{(X_n - \lambda)}{\sqrt{\lambda}}$. The c.f. of Y_n is*

$$\Psi_n(t) = e^{-it\sqrt{\lambda}} \exp[\lambda(e^{it\sqrt{\lambda}} - 1)]$$

$$\log_e \Psi_n(t) = -it\sqrt{\lambda} + \lambda(e^{it/\sqrt{\lambda}} - 1) = -it\sqrt{\lambda} + \lambda\left(\frac{it}{\sqrt{\lambda}} - \frac{t^2}{2\lambda} - \frac{it^3}{6\lambda^{\frac{3}{2}}} + \cdots\right)$$

$$= -\frac{t^2}{2} - \frac{it^3}{6\lambda^{\frac{1}{2}}} + \cdots.$$

Hence

$$\Psi_n(t) \to \exp(-t^2/2) \text{ as } \lambda \to \infty.$$

Thus Y_n tends in distribution to a $N(0,1)$-variable as $\lambda \to \infty$.

▶ **Example 9.2.9** *Let $\{X_n\}$ be a sequence of independently and identically distributed $\chi^2_{(1)}$ r.v.'s. Let $Y_n = \frac{(S_n - n)}{\sqrt{2n}}$, where $S_n = \Sigma_{i=1}^n X_i$.*

The m.g.f. of Y_n

$$= M_n(t) = E(e^{tY_n}) = \left(1 - \sqrt{\frac{2}{n}}t\right)^{-\frac{n}{2}} \exp\left[\left(-t\sqrt{\frac{n}{2}}\right)\right], \quad |t| < \sqrt{\frac{n}{2}}$$

$$\log_e M_n(t) = -t\sqrt{\frac{n}{2}} - \frac{n}{2}\log_e\left(1 - \sqrt{\frac{2}{n}}t\right)$$

$$= -t\sqrt{\frac{n}{2}} + \frac{n}{2}\left[\sqrt{\frac{2}{n}}t + \frac{1}{2}\left(\sqrt{\frac{2}{n}}t\right)^2 + \cdots\right]$$

$$= \frac{t^2}{2} + 0\left(\frac{1}{n}\right).$$

Hence $M_n(t) \to e^{\frac{t^2}{2}}$ and $Y_n \xrightarrow{L} \tau$ where τ is a $N(0,1)$-variable.

An important question regarding the weak convergence of F_n to F is whether the pointwise convergence holds uniformly. The following result is quite useful.

Theorem 9.2.7 (***Polya***). *If $F_n \xrightarrow{w} F$ and F is continuous, then*

$$\lim_{n\to\infty} \sup_t |F_n(t) - F(t)| = 0.$$

9.3 Convergence in Probability

Definition 9.3.1 *A sequence of r.v.'s $\{X_n\}$ is said to converge in probability to a r.v. X if, for every given $\varepsilon(>0)$ and $\delta(>0)$, there exists a N such that*

$$P\{|X_n - X| > \varepsilon\} < \delta \text{ for all } n \geq N \text{ i.e., } \lim_{n\to\infty} P\{|X_n - X| > \varepsilon\} = 0 \qquad (9.3.1)$$

for every $\varepsilon(>0)$. This is written as $X_n \xrightarrow{P} X$ or $p\lim X_n = X$. $\{X_n\}$ is said to converge stochastically to X.

In particular, if the r.v. X is a constant i.e., $P(X = \mu) = 1$, $X_n \xrightarrow{P} \mu$ if $P\{|X_n - \mu| < \varepsilon\} \to 1$ for every fixed $\varepsilon(>0)$. In this case, it is said that the Weak Law of Large Numbers (WLLN) holds for $\{X_n\}$.

► **Example 9.3.1** *Consider the probability space* $(\mathcal{S}, \mathcal{A}, P)$ *where* $\mathcal{S} = [0,1]$ *and* P *is such that* $P[a \le w \le b] = (b-a)^{-1}, 0 \le a \le b \le 1$. *Define* $\{X_n\}$ *such that* $X_n(w) = 1(0)$ *if* $w \in \left[0, \frac{1}{n}\right]$ *(otherwise). Then*

$$P(X_n = 1) = \frac{1}{n}, P(X_n = 0) = 1 - \frac{1}{n}$$

$$P(|X_n| \le \varepsilon) = \begin{cases} P(X_n = 0) = 1 - \frac{1}{n}, & \varepsilon < 1 \\ 1, & \varepsilon \ge 1. \end{cases}$$

Thus $P(|X_n| \ge \varepsilon) \to 1$ as $n \to \infty$ for every $\varepsilon (> 0)$. Hence $X_n \xrightarrow{P} X$ where X is a degenerate r.v. with $P(X = 0) = 1$.

The following results hold:

Theorem 9.3.1 .

(a) $X_n \xrightarrow{P} X \quad \Rightarrow \quad X_n - X \xrightarrow{P} 0,$

(b) $X_n \xrightarrow{P} X \quad \Rightarrow \quad X_n - X_m \xrightarrow{P} 0$ as $n, m \to \infty$.

Theorem 9.3.2 *Let* $\{X_n\}, \{Y_n\}$ *be two sequences of r.v.'s with* $X_n \xrightarrow{P} X, Y_n \xrightarrow{P} Y$. *Then*

(a) $cX_n \xrightarrow{P} cX$ ('c', a real constant),

(b) $X_n + Y_n \xrightarrow{P} X + Y$,

(c) $X_n Y_n \xrightarrow{P} XY$,

(d) $\frac{X_n}{Y_n} \xrightarrow{P} \frac{X}{Y}$ if $P[Y_n = 0] = 0 \ \forall n$ and $P[Y = 0] = 0$.

Theorem 9.3.3 *If* $g(x)$ *is a continuous real valued function, and* $X_n \xrightarrow{P} X$ *then* $g(X_n) \xrightarrow{P} g(X)$.

Theorem 9.3.4 (Cramer, 1946, p. 254). *Let* $\{X_n\}, \{Y_n\}$ *be two sequences of r.v.'s with*

$$X_n \xrightarrow{L} X, \ Y_n \xrightarrow{P} c$$

where X *is a r.v. and 'c'* (> 0) *a constant. Then*

(a) $X_n + Y_n \xrightarrow{L} X + c$, (b) $X_n Y_n \xrightarrow{L} cX$, (c) $\frac{X_n}{Y_n} \xrightarrow{L} X/c$ if $c \ne 0$.

9.4 Almost Sure Convergence

Definition 9.4.1 *The sequence* $\{X_n\}$ *converges almost surely (or almost everywhere) to a r.v.* X *if*

$$P\left\{\lim_{n \to \infty} X_n = X\right\} = 1 \tag{9.4.1}$$

$$\text{i.e. if } P\left\{w : \lim_{n\to\infty} X_n(w) = X(w)\right\} = 1$$

$$\text{i.e. if } P\{|X_m - X| < \varepsilon;\ m \geq n\} \to 1.$$

s $n \to \infty$, for arbitrarily small positive ε holds.

This is denoted as a.s. lim $X_n = X$ or $X_n \xrightarrow{a.s.} X$. When $X_n \xrightarrow{a.s.} c$, a constant, we ay that the strong law of large numbers (SLLN) holds for the sequence $\{X_n\}$. To verify $X_n \xrightarrow{P} X$ one has to calculate the probability, $P\{|X_n - X| < \varepsilon\} = p_n(\varepsilon)$ and show that $p_n(\varepsilon) \to 1$ as $n \to \infty$. To verify $X_n \xrightarrow{a.s.} X$ one has to show that the set of points w for hich $\lim_{n\to\infty} X_n(w) = X(w)$ has probability one.

The following results elucidate definition 9.4.1.

Theorem 9.4.1 $X_n \xrightarrow{a.s.} X$ *iff* $\lim_{n\to\infty} P\left\{\sup_{m\geq n}|X_m - X| > \varepsilon\right\} = 0$ *for all* $\varepsilon > 0$.

Theorem 9.4.2 (***Cauchy Criterion***). $X_n \xrightarrow{a.s.} X$ *iff*

$$\lim_{n\to\infty} P\left\{\sup_m |X_{m+n} - X_n| \leq \varepsilon\right\} = 1 \ \textit{for every}\ \varepsilon > 0.$$

The following theorem holds.

Theorem 9.4.3 *If* $X_n \xrightarrow{a.s.} X$ *and* $Y_n \xrightarrow{a.s.} Y$, *then* $X_n + Y_n \xrightarrow{a.s.} X + Y$ *and* $X_nY_n \xrightarrow{a.s.} XY$.

9.4.1 Kolmogorov's inequality and its refinements

Theorem 9.4.4 (***Kolmogorov's inequality***). *Suppose that*

$$E(X_j) = E(X_j|X_1, X_2, \cdots, X_{j-1}) = 0 \ (j = 1, 2, \cdots). \tag{9.4.2}$$

Then

$$P(|S_j| \leq u; j = 1, \cdots, n) \geq 1 - \frac{E(S_n^2)}{u^2} = 1 - \frac{\sum_1^n E(X_j^2)}{u^2} \tag{9.4.3}$$

where $S_n = \sum_1^n X_j$.

Proof: Let A denote the event $[|S_j| < u, j = 1, \cdots, n]$ and B_i the event

$$[|S_1| < u, \cdots, |S_{i-1}| < u,\ \ |S_i| \geq u].$$

Then

$$\bigcup_1^n B_i = \bar{A}.$$

Hence

$$P(A) = 1 - \sum_1^n P(B_i) \tag{9.4.4}$$

since B_i's are mutually disjoint. Now

$$E(S_n^2) = \sum_1^n P(B_i)E(S_n^2|B_i) + P(A)E(S_n^2|A) \quad \text{(by theorem 3.2.1)} \tag{9.4.5}$$

Now

$$E(S_n^2|A) \geq 0 \tag{9.4.6}$$

$$\begin{aligned} E(S_n^2|B_i) &= E[(S_n - S_i + S_i)^2|B_i] \\ &= E[(S_n - S_i)^2|B_i] + 2E[(S_n - S_i)S_i|B_i] + E(S_i^2|B_i) \geq u^2 \end{aligned} \tag{9.4.7}$$

since the first term on the right-hand side of (9.4.7) is positive and the second term is zero, because

$$\begin{aligned} E\left[(S_n - S_i)S_i|B_i\right] &= E\left[E\{(S_n - S_i)S_i|X_1, \cdots, X_i, B_i, \}|B_i\right] \\ &= E\left[S_i E\{(S_n - S_i)|X_1, \cdots, X_i, B_i\}|B_i\right] \end{aligned}$$

Now, $\quad E(S_n - S_i|X_1, \cdots, X_i, B_i) = E(S_n - S_i|X_1, \cdots, X_i) = 0$

by (9.4.2). Hence by (9.4.5), (9.4.6) and (9.4.7)

$$E(S_n^2) \geq u^2 \sum_1^n P(B_i) \tag{9.4.8}$$

and (9.4.3) follows by (9.4.4).

■ NOTE 9.4.1 *If* (9.4.3) *is a bound simply for* $P(|S_n| \leq u)$ *then one gets Chebychev's inequality. The inequality* (9.4.3) *gives a lower bound for the probability that* $S_1, S_2, \cdots, S_n$ *are simultaneously (and not simply* S_n*) less than* u *in modulus.*

■ NOTE 9.4.2 *The inequality is not derived on the assumption of independence of the r.v.'s* $X_1, \cdots, X_n$*, but on the weaker assumption* (9.4.2) *that the conditional mean of* X_j *should have the same value as the unconditional mean* $E(X_j)$*, which is zero. However, if* $E(X_j) = \mu_j (\neq 0)$*, one can consider* $X_j - \mu_j$ *in place of* X_j*.*

■ NOTE 9.4.3 *We have assumed that the second moment of* X_j *exists* $(j = 1, 2, \cdots,)$.

Theorem 9.4.5 (***The Hájek-Rényi generalisation of Kolmogorov's inequality***). *Under assumption* (9.4.2)

$$P(|S_j| \leq u_j, j = 1, \cdots, n) \geq 1 - \sum_1^n \frac{E(X_j^2)}{u_j^2} \tag{9.4.9}$$

for any positive increasing sequence $\{u_j\}$.

Proof: Consider the expression

$$T = \sum_{1}^{n}(\alpha_j - \alpha_{j+1})S_j^2 \tag{9.4.10}$$

vhere the α_j constitutes a non-negative decreasing sequence, with $\alpha_{n+1} = 0$. Then

$$E(T) = \sum_{1}^{n} \alpha_j E(X_j^2) \geq \sum_{1}^{n} P(B_i)E(T|B_i) \tag{9.4.11}$$

vhere B_i is the event

$$[|S_j| \geq u_j \text{ first for } j = i]$$

and

$$A = \left(\bigcup_{i=1}^{n} B_i\right)^c = [|S_j| < u_j, j = 1, \cdots, n].$$

Now

$$E(S_j^2|B_i) \geq \begin{cases} 0 & (j < i) \\ u_i^2 & (j \geq i). \end{cases} \tag{9.4.12}$$

Hence, by (9.4.11) and (9.4.12)

$$\sum_{1}^{n} \alpha_j E(X_j^2) \geq \sum_{1}^{n} P(B_i) \sum_{j=i}^{n}(\alpha_j - \alpha_{j+1})u_i^2 = \sum_{1}^{n} \alpha_i u_i^2 P(B_i).$$

The ineqality (9.4.9) follows by choosing $\alpha_i = u_i^{-2}$. For a generalisation of Kolmogorov-Hájek-Rényi (KHR) inequality see exercise 17. The KHR-inequality will be useful in deriving the strong laws of large numbers for the random sum $S_n = \sum_{j=1}^{n} X_j$ in section 9.7.

9.5 Convergence in rth mean

Definition 9.5.1 *Let $\{X_n\}$ and X be r.v.'s having moments of rth order. Then $|X_n - X|$ has also moments of rth order. $\{X_n\}$ is said to converge in rth mean to X if*

$$E|X_n - X|^r \to 0 \text{ as } n \to \infty, \quad r > 0. \tag{9.5.1}$$

This is also called L_r-convergence, $X_n \xrightarrow{L_r} X$. For $r = 2$, this is called mean square (or L_2) convergence. $X_n \xrightarrow{ms} X$ if $E|X_n - X|^2 \to 0$.

Lemma 9.5.1 $X_n \xrightarrow{L_r} X \Rightarrow E|X_n|^r \to E|X|^r$.

► **Example 9.5.1** *Consider* $\{X_n\}$ *where*

$$P(X_n = n) = \frac{1}{n^r}, \quad P(X_n = 0) = 1 - \frac{1}{n^r}, \quad r > 0.$$

Then

$$P\{|X_n| > \varepsilon\} = \begin{cases} 0, & \varepsilon \geq n \\ P(X_n = n) = \frac{1}{n^r}, & \varepsilon < n. \end{cases}$$

Hence $X_n \xrightarrow{P} 0$. Again $E|X_n|^r = 1$ so that $X_n \xrightarrow{L_r} 0$. If $r = 3, E|X_n|^2 = \frac{1}{n}$ so tha $X_n \xrightarrow{L_2} Y$ where $P(Y = 0) = 1$.

Lemma 9.5.2 *Let* X, Y *be two r.v.'s with d.f.'s* $F(x), G(y)$ *respectively. Let* ε, δ *be tw positive constants such that*

$$P(|X - Y| \leq \varepsilon) \geq 1 - \delta. \quad (9.5.2)$$

Then $|F(x) - G(x)| \leq [F(x + \varepsilon) - F(x - \varepsilon)] + \delta$.

Proof: Let A be the event $[|X - Y| \leq \varepsilon]$, B the event $[Y \leq x]$. Then

$$P[|X - Y| \leq \varepsilon, Y \leq x] \geq P[Y \leq x] - \delta. \quad [\because P(AB) \geq P(B) + P(A) - 1]$$

Hence,

$$P[X \leq x + \varepsilon] \geq P[Y \leq x] - \delta. \quad (9.5.3)$$

Similarly considering the event A and $\bar{B}$,

$$P[X > x - \varepsilon] \geq P[Y > x] - \delta. \quad (9.5.4)$$

From (9.5.3) and (9.5.4) it follows that

$$F(x - \varepsilon) - \delta \leq G(x) \leq F(x + \varepsilon) + \delta.$$

Since $F(x - \varepsilon) \leq F(x) \leq F(x + \varepsilon)$,

$$F(x - \varepsilon) - F(x + \varepsilon) - \delta \leq F(x) - G(x) \leq F(x + \varepsilon) - F(x - \varepsilon) + \delta.$$

The following theorem shows the gradation of different types of convergence.

Theorem 9.5.1 *Let* $\{X_n\}$ *be a sequence of r.v.'s and* X *another r.v. Then*

(a) $[X_n \xrightarrow{L_r} X] \Rightarrow [X_n \xrightarrow{L_s} X; 0 \leq s \leq r] \Rightarrow [X_n \xrightarrow{P} X] \Rightarrow [X_n \xrightarrow{L_r} X]$,

(b) $[X_n \xrightarrow{a.s.} X] \Rightarrow [X_n \xrightarrow{P} X]$.

Proof: (a) (i) Lr-convergence implies Ls-convergence ($r \geq s$):

$$[X_n \xrightarrow{L_r} X] \Rightarrow E|X_n - X|^r \to 0.$$

Similarly

$$\left[X_n \xrightarrow{L_s} X\right] \Rightarrow E|X_n - X|^s \to 0.$$

Now

$$(E|X_n - X|^r)^{\frac{1}{r}} \geq (E|X_n - X|^s)^{\frac{1}{s}}, \quad r \geq s > 0 \text{ (by theorem 3.4.3).} \qquad (9.5.5)$$

Hence if $X_n \xrightarrow{L_r} X$, the left-hand side of (9.5.5) $\to 0$ and hence the right-hand side of (9.5.5) $\to 0$, which in turn implies $X_n \xrightarrow{L_s} X$.

(ii) Lr convergence implies P-convergence.

By Markov's inequality (3.10.1)

$$P\{|X_n - X| > \varepsilon\} \leq \frac{E|X_n - X|^r}{\varepsilon^r}, \quad r > 0,$$

because $|X_n - X|^r$ is a positive increasing function of $|X_n - X|$; but $E|X_n - X|^r \to 0$, hence $P\{|X_n - X| > \varepsilon\} \to 0$ for every ε.

(iii) P-convergence implies convergence in law.

Proof: Suppose $X_n \xrightarrow{P} X$ so that

$$P(|X_n - X| \leq \varepsilon) \geq 1 - \delta$$

for every $\varepsilon(> 0), \delta(> 0)$ and n sufficiently large.

By lemma 9.5.2

$$|F(x) - F_n(x)| \leq F(x + \varepsilon) - F(x - \varepsilon) + \delta. \qquad (9.5.6)$$

If x is a continuity point of F, the right-hand side of (9.5.6) can be made arbitrarily small by choosing ε and δ sufficiently small. Hence $F_n(x) \xrightarrow{w} F(x)$.

(b) $X_n \xrightarrow{a.s.} X$ means $P\{X_n \to X\} = 1$ i.e., $P\{|X_n - X| < \varepsilon$ for $n = n_0 + 1, n_0 + 2, \cdots, n_0 + r\} > 1 - \eta$ for all r. Putting $r = 1$, we get $P\{|X_n - X| < \varepsilon\} > 1 - \eta$ for $n \geq n_0$, which is convergence in probability.

■ NOTE 9.5.1 *The converses in theorem 9.5.1 are not true.*

■ NOTE 9.5.2 *Although convergence in rth mean is stronger than convergence in probability, it is neither implied by nor implies convergence a.s.*

▶ **Example 9.5.2** *Show that if $X_n \xrightarrow{P} a$, where 'a' is a constant, then $[X_n \xrightarrow{P} a] \Leftrightarrow [X_n \xrightarrow{L} a]$.*

(i) $[X_n \xrightarrow{P} a] \Rightarrow [X_n \xrightarrow{L} a]$ follows by theorem 9.5.1. This can also be seen as follows:

$$[X_n \xrightarrow{P} a] \Rightarrow P\{|X_n - a| \leq \varepsilon\} \to 1$$

for any $\varepsilon(> 0)$.

i.e., $P\{X_n \leq a - \varepsilon, X_n \geq a + \varepsilon\} \to 0$

i.e., $P\{X_n \leq a - \varepsilon\}P\{X_n \geq a + \varepsilon\} \to 0$ (the events being independent)

i.e., $F_n(a - \varepsilon) \to 0, 1 - F_n(a + \varepsilon - 0) \to 0,$ (i)

F_n being d.f. of X_n. Since ε is arbitrary, taking limit of (i) as $n \to \infty$

$$\lim F_n(x) = F(x) \text{ (say) } = \begin{cases} 0, & x < a \\ 1, & x \geq a \end{cases} \tag{ii}$$

i.e., $X_n \xrightarrow{L} X$ where $P(X = a) = 1$.

(ii) Let $X_n \xrightarrow{L} a$ i.e., limiting distribution of X_n is $F(x)$ in (ii).

$$P\{|X_n - a| \geq \varepsilon\} = P\{X_n \leq a - \varepsilon\}P\{X_n \geq a + \varepsilon\} = F_n(a - \varepsilon)\{1 - F_n(a + \varepsilon - 0)\} \tag{iii}$$

Taking limit of both sides of (iii).

$$\lim P\{|X_n - a| > \varepsilon\} = F(a - \varepsilon)(1 - F(a + \varepsilon - 0)) = 0(1 - 1) = 0.$$

► **Example 9.5.3** *$\{X_n\}$ is a sequence of r.v.'s with*

$$P(X_n = 0) = 1 - \frac{1}{n}, \quad P(X_n = n^{2/r}) = \frac{1}{n}.$$

Show that P-convergence does not imply L_r-convergence.

$$P\{|X_n| < \varepsilon\} = P\{X_n = 0\} = 1 - \frac{1}{n} > 1 - \delta \text{ for } n > \frac{1}{\delta}.$$

Hence $X_n \xrightarrow{P} 0$. Now $E|X_n|^r = n \to \infty$. Hence $X_n \not\xrightarrow{L_r} 0$.

9.6 Weak Laws of Large Numbers

We have already defined when a sequence of variables $\{X_n\}$ is said to satisfy the weak law of large numbers (WLLN) and strong law of large numbers (SLLN). These are the situations when $X_n \xrightarrow{P} c$ or $X_n \xrightarrow{a.s.} c$ ('c' a constant) respectively. We will now consider in details the situations when a sequence of random averages, $S_n/n = \sum_{i=1}^{n} X_i/n = \bar{X}_n$ satisfies the weak laws of large numbers.

We recall Chebychev's inequality: If X is a r.v. with $E(X) = \mu, V(X) = \sigma^2$,

$$P\{|X - \mu| \geq \varepsilon\sigma\} \leq \frac{1}{\varepsilon^2} \tag{9.6.1}$$

for any arbitrary real $\varepsilon(> 0)$.

(a) **Bernoulli's WLLN:** Consider n Bernoulli trials each with probability p of success. Let $X_i = 1(0)$ if the ith trial results in success (failure). $E(\bar{X}_n) = p, V(\bar{X}_n) = \frac{pq}{n}$ where

$$\bar{X}_n = \frac{1}{n}\sum_{i=1}^{n} X_i, q = 1 - p.$$

pplying (9.6.1) to the sequence of r.v.'s $\{\bar{X}_n\}$,

$$P\{|\bar{X}_n - p| \le \varepsilon\} \ge 1 - \frac{pq}{n\varepsilon^2} > 1 - \frac{1}{n\varepsilon^2}$$

$$\text{i.e.} \quad \lim_{n\to\infty} P\{|\bar{X}_n - p| \le \varepsilon\} = 1 \qquad (9.6.2)$$

or every $\varepsilon > 0$). Hence $\bar{X}_n \xrightarrow{P} p$ (i.e., $\bar{X}_n \xrightarrow{L} X$ where X is a degenerate r.v. at $x = p$). This is known as Bernoulli's WLLN.

(b) **Chebychev's WLLN :** Let $\{X_n\}$ be a sequence of uncorrelated r.v.'s for which

$$E(X_i) = \mu_i, \quad V(X_i) = \sigma_i^2.$$

Then

$$E(\bar{X}_n) = \frac{1}{n}\sum_{i=1}^{n} \mu_i = \bar{\mu}_n, \quad V(\bar{X}_n) = \frac{1}{n^2}\sum_{i=1}^{n} \sigma_i^2.$$

f the condition

$$\frac{1}{n^2}\sum_{i=1}^{n} \sigma_i^2 \to 0 \text{ as } n \to \infty \qquad (9.6.3)$$

s satisfied then $\bar{X}_n \xrightarrow{P} \bar{\mu}_n$.

Proof: Follows immediately by applying (9.6.1) to $\{\bar{X}_n\}$, when

$$P\{|\bar{X}_n - \bar{\mu}_n| > \varepsilon\} \le \sum_{1}^{n} \sigma_i^2 \Big/ n^2\varepsilon^2$$

or every $\varepsilon (> 0)$. Assumption (9.6.3) is immediately satisfied if $V(X_i) = \sigma^2 (< \infty)$ for all i.

We note that here $E|\bar{X}_n - \bar{\mu}_n|^2 \to 0$ (by assumption (9.6.3)) and hence $\bar{X}_n \xrightarrow{ms} \bar{\mu}_n$.

The condition (9.6.3) is a sufficient (not necessary) condition for $\{S_n\}$ to satisfy the WLLN.

(c) **Poisson's WLLN :** Consider Poisson's scheme of sampling: n independent Bernoulli trials are performed, the ith trial with probability of success as p_i. Let $X_i = 1(0)$ if the ith trial results in success (failure).

$$E(\bar{X}_n) = \frac{\Sigma p_i}{n} = \bar{p}, \quad V(\bar{X}_n) = \frac{1}{n^2}\sum_{i=1}^{n} p_i q_i \le \frac{1}{4n}$$

$\to 0$ as $n \to \infty$. Hence by inequality (9.6.1), $\bar{X}_n \xrightarrow{P} \bar{p}$.

(d) **Khintchine's WLLN :** Assume $\{X_n\}$ to be a sequence of independently and identically distributed r.v.'s with $E(X_i) = \mu \ \forall\, i$. Then the sequence $\bar{X}_n \xrightarrow{P} \mu$.

Proof: Let $\Psi(t)$ be the c.f. of $X_i (i = 1, \cdots, n)$. The c.f. of $\bar{X}_n$ is

$$\Psi_n(t) = \left[\Psi\left(\frac{t}{n}\right)\right]^n.$$

Expanding $\Psi(t)$ in Maclaurin series about $t = 0$,

$$\Psi(t) = 1 + i\mu t + o(t).$$

Hence

$$\Psi_n(t) = \left[1 + \frac{\mu it}{n} + o\left(\frac{t}{n}\right)\right]^n, \quad \log_e \Psi_n(t) = n\left[\frac{\mu it}{n} + o\left(\frac{t}{n}\right)\right] = \mu it + no\left(\frac{t}{n}\right).$$

Hence $\log_e \Psi_n(t) \to \mu it$ and $\Psi_n(t) \to \exp(\mu it)$, as $n \to \infty$. Now $\exp(\mu it)$ is the c.f of a degenerate variable Y with $P(Y = \mu) = 1$. Hence by theorem 9.2.4, $\bar{X}_n \xrightarrow{P} \mu$.

(e) **Markoff's WLLN:** Let $\{X_n\}$ be a sequence of variables not necessarily uncorrelated. Then

$$\bar{X}_n \xrightarrow{P} \bar{\mu}_n \text{ if } V(\bar{X}_n) \to 0 \text{ as } n \to \infty.$$

9.7 Strong Laws of Large Numbers

We will consider in details the situations when a sequence of random averages,

$$\frac{S_n}{n} = \frac{\sum_{i=1}^{n} X_i}{n} = \bar{X}_n$$

satisfies the strong laws of large numbers (SLLN).

We first consider Borel-Cantelli Lemmas. For this we first note a definition and a few relations.

Definition 9.7.1 *Let A_n be a sequence of subsets of $\mathcal{S}$, $A_n \in \mathcal{A}$. We define (vide section 0.3)*

$$\limsup_n A_n = \bigcap_{m=1}^{\infty} \bigcup_{n=m}^{\infty} A_n$$

$$\liminf_n A_n = \bigcup_{m=1}^{\infty} \bigcap_{n=m}^{\infty} A_n.$$

Note that

$$\liminf_n A_n = \left(\limsup_n A_n^c\right)^c. \tag{9.7.1}$$

Lemma 9.7.1 (i) *A point belongs to $\limsup_n A_n$ iff it belongs to infinitely many terms of the sequence $\{A_n, n \geq 1\}$.*

(ii) *A point belongs to $\liminf_n A_n$ iff it belongs to all terms of the sequence from a certain term on (all but finitely many terms).*

Proof: (i) If w belongs to infinitely many A_n's then w belongs to

$$B_m = \bigcup_{n=m}^{\infty} A_n \quad \text{for every } m;$$

hence it belongs to

$$\bigcap_{m=1}^{\infty} B_m = \limsup_n A_n.$$

Conversely, if w belongs to $\bigcap_{m=1}^{\infty} B_m$, then w belongs to B_m for every m. If w were to belong to only a finite number of the A_n's there would be an m such that $w \notin A_n$ for $n \geq m$, so that

$$w \notin \bigcup_{n=m}^{\infty} A_n = B_m.$$

This contradiction proves that w must belong to an infinite number of A_n's.

(ii) If w belongs to all but a finite number of A_n's then w belongs to

$$C_m = \bigcap_{n=m}^{\infty} A_n$$

for at least one m and hence w belongs to $\bigcup_{m=1}^{\infty} C_m$.

Conversely, if w belongs to $\bigcup_{m=1}^{\infty} C_m$, then w belongs to C_m for at least one m. Let m_0 be the smallest of all such integers m. Then w belongs to all A_n's except for $A_1, \cdots, A_{m_0-1}$.

Hence the proof.

■ NOTE 9.7.1 *We say that the event* $\limsup_n A_n$ *occurs iff the event* A_n *occurs infinitely often. Thus*

$$P(\limsup_n A_n) = P(A_n i.o.)$$

where the abbreviation i.o. stands for infinitely often.

Lemma 9.7.2 *If each* $A_n \in \mathcal{A}$, *then*

$$P\left(\limsup_n A_n\right) = \lim_{m\to\infty} P\left(\bigcup_{n=m}^{\infty} A_n\right)$$

$$P\left(\liminf_n A_n\right) = \lim_{m\to\infty} P\left(\bigcap_{n=m}^{\infty} A_n\right).$$

Proof: Note that B_m is a decreasing sequence and C_m an increasing sequence. Hence by the monotone property of probability measures (vide Continuity theorem, Theorem 1.9.1),

$$P\left(\limsup_n A_n\right) = P\left(\bigcap_{m=1}^{\infty} B_m\right) = \lim_{m\to\infty} P(B_m);$$

$$P\left(\liminf_n A_n\right) = P\left(\bigcup_{m=1}^{\infty} C_m\right) = \lim_{m\to\infty} P(C_m).$$

Theorem 9.7.1 (i) ***First Borel-Cantelli Lemma.***

Let $\{A_n, n \geq 1\}$ be a sequence of events in $(S, \mathcal{A}, P)$ such that $\sum_{n=1}^{\infty} P(A_n) < \infty$. Then

$$P(A_n i.o.) = 0.$$

(ii) ***Second Borel-Cantelli Lemma.*** *Let $\{A_n, n \geq 1\}$ be a sequence of pair-wise independent events in $(S, \mathcal{A}, P)$ such that $\sum_{n=1}^{\infty} P(A_n) = \infty$. Then*

$$P(A_n i.o.) = 1.$$

(iii) *If $\{A_n, n \geq 1\}$ is a sequence of (pairwise or mutually) independent events, then $P(A_n i.o.) = 0$ or 1 according as $\sum_{n=1}^{\infty} P(A_n) < \infty$ or $= \infty$.*

Proof: (i) By Lemma 9.7.2,

$$\begin{aligned} P\left(\limsup_n A_n\right) &= \lim_{m\to\infty} P\left(\bigcup_{n=m}^{\infty} A_n\right) \\ &\leq \lim_{m\to\infty} \sum_{n=m}^{\infty} P(A_n) \quad \text{(by Boole's inequality)} \\ &= 0. \end{aligned}$$

(ii) Let I_n denote the indicator function of the event A_n,

$$\begin{aligned} I_n(w) &= 1(0) \quad \text{if} \quad w \in A_n \quad \text{(otherwise)}; \\ E(I_n(w)) &= P(A_n) \\ E(I_n I_{n'}) &= E(I_n)E(I_{n'}), \quad n \neq n'. \end{aligned}$$

We have

$$\sum_{n=1}^{\infty} I_n(w) = \infty$$

if w occurs in an infinite number of A_n's. Therefore,

$$P(A_n i.o.) = P\left(\sum_{n=1}^{\infty} I_n(w) = \infty\right). \tag{9.7.2}$$

Consider the partial sum

$$S_m = \sum_{n=1}^{m} I_n.$$

By Chebychev's inequality, we have, with $N > 0$,

$$P\left[|S_m - E(S_m)| \leq N\sigma(S_m)\right] \geq 1 - \frac{1}{N^2}$$

$$\text{i.e.,}\quad P\left[E(S_m) - N\sigma(S_m) \leq S_m \leq E(S_m) + N\sigma(S_m)\right] \geq 1 - \frac{1}{N^2} \tag{9.7.3}$$

where $\sigma^2(S_m)$ denotes the variance of S_m.

Let $p_n = P(A_n)$. Hence

$$E(S_m) = \sum_{n=1}^{m} p_n$$

$$\sigma^2(S_m) = \sum_{n=1}^{m} p_n(1 - p_n) \leq \sum_{n=1}^{m} p_n.$$

Therefore,

$$\frac{\sigma(S_m)}{E(S_m)} \leq \frac{1}{\sqrt{E(S_m)}} \to 0 \quad \text{as } m \to \infty \tag{9.7.4}$$

since $\sum_{n=1}^{m} p_n \to \infty$ as $m \to \infty$. It follows, therefore, that for given $N(> 1)$ and $0 < \alpha < 1$, there exists a $m_0 = m_0(N, \alpha)$ such that for $m \geq m_0$,

$$\frac{\sigma(S_m)}{E(S_m)} \leq \alpha_1 = \frac{\alpha}{N}(< 1).$$

Therefore,

$$N\sigma(S_m) \leq \alpha E(S_m) \quad \text{for } m \geq m_0.$$

From (9.7.3),

$$P\left[E(S_m) - N\sigma(S_m) \leq S_m\right] \geq 1 - \frac{1}{N^2} \;\forall\, m \geq m_0$$

which implies

$$P\left[\beta E(S_m) \leq S_m\right] \geq 1 - \frac{1}{N^2} \;\forall\, m \geq m_0$$

where $\beta = 1 - \alpha(> 0)$. Therefore, by monotonicity of S_m,

$$P\left[\beta \lim_{m\to\infty} E(S_m) \leq \lim_{m\to\infty} S_m\right] \geq 1 - \frac{1}{N^2}.$$

Letting $m \to \infty$ and then $N \to \infty$, so that $\beta \to 1$. This gives

$$P\left[\lim_{m\to\infty} S_m = \infty\right] = 1.$$

This establishes the result because of (9.7.2).

(iii) By Lemma 9.7.2,

$$P\left(\liminf_n A_n^c\right) = \lim_{m\to\infty} P\left(\bigcap_{n=m}^{\infty} A_n^c\right).$$

The events A_n^c's are independent, since A_n's are so.
Hence, for $m' > m$,

$$P\left(\bigcap_{n=m}^{m'} A_n^c\right) = \prod_{n=m}^{m'} P(A_n^c) = \prod_{n=m}^{m'} (1 - P(A_n)). \tag{9.7.5}$$

Again, for any $t > 0$,

$$1 - t < e^{-t}.$$

Hence

$$(9.7.5) \leq \prod_{n=m}^{m'} e^{-P(A_n)} = e^{-\sum_{n=m}^{m'} P(A_n)} \to 0 \text{ as } m' \to \infty,$$

since $\sum_{n=1}^{\infty} P(A_n) = \infty$. Therefore,

$$P\left(\bigcap_{n=m}^{\infty} A_n^c\right) = \lim_{m'\to\infty} P\left(\bigcap_{n=m}^{m'} A_n^c\right) = 0.$$

Hence

$$P\left(\liminf_n A_n^c\right) = 0$$

and consequently

$$P(A_n i.o.) = 1$$

by (9.7.1).

As an application of Borel-Cantelli Lemma we consider the following application of SLLN.

Theorem 9.7.2 *If $X_1, X_2, \cdots,$ are iid r.v.'s with common mean μ and finite fourth order moment, then*

$$P\left\{\lim_{n\to\infty} \frac{S_n}{n} = \mu\right\} = 1,$$

where $S_n = X_1 + \cdots + X_n$.

Proof: We have

$$E\left\{\sum (X_i - \mu)\right\}^4 = nE(X_1 - \mu)^4 + 6\binom{n}{2}\sigma^4 \leq Kn^2$$

where K is a constant. By Markov's inequality

$$P\left\{\left|\sum_1^n (X_i-\mu)\right| > n\varepsilon\right\} \le \frac{E\left\{\sum_1^n (X_i-\mu)\right\}^4}{(n\varepsilon)^4} \le \frac{Kn^2}{(n\varepsilon)^4} = \frac{K'}{n^2}.$$

Therefore,

$$\sum_{n=1}^{\infty} P\{|S_n - n\mu| > n\varepsilon\} \le K' \sum_{n=1}^{\infty} \frac{1}{n^2} < \infty.$$

Therefore, it follows by the Borel-Cantelli Lemma that with probability 1 only finitely many of the events $\left\{w : \left|\frac{S_n}{n} - \mu\right| > \varepsilon\right\}$ occur, i.e., $P(A_\varepsilon) = 0$ where

$$A_\varepsilon = \limsup_{n\to\infty} \left\{\left|\frac{S_n}{n} - \mu\right| > \varepsilon\right\}.$$

The set A_ε increases, as $\varepsilon \to 0$, to the set of w's on which $\frac{S_n}{n} \to \mu$. Letting $\varepsilon \to 0$ through a countable set of values, we have

$$P\left\{\frac{S_n}{n} - \mu \nrightarrow 0\right\} = P\left\{\bigcup_K A_{1/K}\right\} = 0.$$

Theorem 9.7.3 ***Kolmogorov's SLLN*** **(*First form of SLLN*):** *Let the independent random variables $X_k (k = 1, 2, \cdots, n)$ have zero means and variances σ_k^2 such that*

$$\sum_1^\infty \sigma_k^2 \Big/ k^2 < \infty. \tag{9.7.6}$$

Then the sequence $\bar{X}_n \xrightarrow{a.s.} 0$. That is, given $\varepsilon > 0, \eta > 0$, however small, there exists a n_0 depending on ε and η such that

$$P\{|\bar{X}_j| < \varepsilon, j \ge n_0\} > 1 - \eta.$$

Proof: Define the random variables

$$Y_m = \frac{S_m - S_{\nu_i - 1}}{\nu_i} \text{ for } \nu_i = 2^{i-1} \cdot n \le m \le 2^i \cdot n.$$

Then

$$E(Y_m) = 0, \quad V(Y_m) = \frac{1}{\nu_i^2} \sum_{e=\nu_i}^{m} \sigma_e^2.$$

Let

$$Z_i = \max_m Y_m^2, \quad \nu_i \le m \le \nu_{i+1}.$$

By Kolmogorov's inequality

$$P\left(Z_i > \frac{\varepsilon^2}{4}\right) = q_i \text{ (say) } < \frac{4}{\varepsilon^2 \nu_i^2} \sum_{l=\nu_i}^{\nu_{i+1}-1} \sigma_e^2 = \frac{4}{\varepsilon^2 n^2 2^{2i-2}} \sum_{e=\nu_i}^{\nu_{i+1}-1} \sigma_e^2$$

and

$$q_1 + q_2 + q_3 + \cdots < 4\varepsilon^{-2} \sum_{i=1}^{\infty} \frac{1}{n^2 2^{2i-2}} \sum_{e=2^{i-1}n}^{2^i \cdot n-1} \sigma_e^2 < 16\varepsilon^{-2} \sum_{i=1}^{\infty} \sum_{e=2^{i-1}n}^{2^i n-1} \frac{\sigma_e^2}{e^2}$$

or

$$q_1 + q_2 + \cdots < 16\varepsilon^{-2} \sum_{k=n}^{\infty} \frac{\sigma_k^2}{k^2}$$

Therefore, by Boole's inequality

$$P\left[Z_i \leq \frac{\varepsilon^2}{4} \text{ for all } i = 1, 2, \cdots\right] > 1 - \frac{16}{\varepsilon^2} \sum_{k=n}^{\infty} \frac{\sigma_k^2}{k^2}.$$

Now the inequalities $\left|\frac{S_k}{k}\right| \leq \varepsilon,\ k = n, n+1, n+2, \cdots$ are satisfied when simultaneousl

$$Z_i < \frac{\varepsilon^2}{4}, i = 1, 2, 3, \cdots \text{ and } \left|\frac{S_{n-1}}{n}\right| \leq \frac{\varepsilon}{2}.$$

Now

$$P\left(\left|\frac{S_{n-1}}{n}\right| \leq \frac{\varepsilon}{2}\right) \geq 1 - \frac{4\sum_{k=1}^{n} \sigma_k^2}{n^2 \varepsilon^2} \quad \text{(by Chebychev's inequality).}$$

Hence

$$P\left[\left|\frac{S_k}{k}\right| \leq \varepsilon, k = n, n+1, \cdots\right] > 1 - \frac{16}{\varepsilon^2} \sum_{k=n}^{\infty} \frac{\sigma_k^2}{k^2} - \frac{4\sum_{k=1}^{n} \sigma_k^2}{n^2 \varepsilon^2} \quad \text{(by Corollary 1.3.3)}$$

$\to 1$ as $n \to \infty$, since

$$\sum_{k=1}^{\infty} \sigma_k^2 \Big/ k^2 < \infty.$$

Corollary 9.7.1 *Let the sequence of independent random variables $\{X_k\}$ be such that*

$$E(X_k) = \mu_k \text{ and } V(X_k) = \sigma_k^2.$$

Then

$$\sum_{k=1}^{\infty} (\sigma_k^2/k^2) < \infty \Longrightarrow \bar{X}_n - \bar{\mu}_n \xrightarrow{a.s.} 0$$

where $\bar{\mu}_n = \frac{1}{n} \sum_{k=1}^{n} \mu_k$. This means that under condition (9.7.1), $\{\bar{X}_n\}$ obeys the SLLN.

Corollary 9.7.2 *Every sequence $\{X_k\}$ of independent r.v.'s with uniformly bounded variances obeys the SLLN.*

Proof: If $V(X_k) \le A \ \forall\, k$, then

$$\sum_{k=1}^{\infty} \frac{\sigma_k^2}{k^2} \le A \sum_{k=1}^{\infty} \frac{1}{k^2} < \infty.$$

Hence $\bar{X}_n - \bar{\mu}_n \xrightarrow{a.s.} 0$.

The SLLN holds necessarily for a sequence of independently and identically distributed r.v.'s X_k having finite variance, since then the condition $\sum_1^{\infty} \sigma_k^2/k^2 < \infty$ is automatically satisfied.

Theorem 9.7.4 (***Borel's SLLN***). *Consider n Bernoulli trials each with probability p of success. Let $X_i = 1(0)$ if the ith trial results in success (failure). Then*

$$\frac{S_n}{n} \xrightarrow{a.s.} p.$$

Proof: Here

$$E(X_k) = p, \ \ V(X_k) = p(1-p) \le \frac{1}{4}, \quad 0 < p < 1.$$

The result follows by corollary 9.7.1.

Theorem 9.7.5 (***Main SLLN***). *Let $\{X_n, n \ge 1\}$ be independent random variables on $(\mathcal{S}, \mathcal{A}, P)$ with a common distribution and $S_n = \sum_{k=1}^{n} X_k$. Then $S_n/n \to \alpha$, a constant a.s. iff $E(|X_1|) < \infty$, in which case $\alpha = E(X_1)$. On the other hand, if $E(|X_1|) = \infty$, then $\limsup_{n\to\infty}(|S_n|/n) = \infty$ ae.*

Theorem 9.7.6 *Let $X_1, X_2, \cdots$ be a sequence of independent random variables on $(\mathcal{S}, \mathcal{A}, P)$ with means $\mu_1, \mu_2, \cdots$ and variance $\sigma_1^2, \sigma_2^2, \cdots$. Let*

$$S_n' = \sum_{k=1}^{n} (X_k - \mu_k)$$

and $\sigma^2 = \sum_{k=1}^{\infty} \sigma_k^2$. Suppose $\sigma^2 < \infty$ and $\sum_{k=1}^{\infty} \mu_k$ converges. Then $\sum_{k=1}^{\infty} X_k$ converges a.s. and in the mean of order 2 to a r.v. X. Moreover, $E(X) = \sum_{k=1}^{\infty} \mu_k$, $\text{Var}(X) = \sigma^2$ and for any $\varepsilon > 0$,

$$P\left[\sup_{n\ge 1} |S_n'| \ge \varepsilon\right] \le \frac{\sigma^2}{\varepsilon^2}.$$

Theorem 9.7.7 *Let $X_1, X_2, \cdots$ have means $\mu_1, \mu_2, \cdots$, variances $\sigma_1^2, \sigma_2^2, \cdots$ and covariance $\text{Cov}(X_i, X_j)$ satisfying*

$$\text{Cov}(X_i, X_j) \leq \rho_{j-i}\sigma_i\sigma_j \ \ (i \leq j)$$

where $0 \leq \rho_k \leq 1$ for all $k = 0, 1, 2, \cdots$. If the series $\sum_1^\infty \rho_k$ and $\sum_1^\infty \sigma_i^2 (\log i)^2/i^2$ are both convergent, then

$$\frac{1}{n}\sum_1^n X_i - \frac{1}{n}\sum_1^n \mu_i \xrightarrow{a.s.} 0.$$

For proof of these results and further details the reader may refer to Feller (1968), vol. I and Rao(1984).

9.8 Central Limit Theorems

Let $\{X_n\}$ be a sequence of independent random variables. In laws of large numbers we consider the cases when a sequence of random averages $\bar{X}_n$ converges to $E(\bar{X}_n)$ either in probability or almost certainly. Here we consider some different situations, namely, the situations when $\bar{X}_n \xrightarrow{L} \tau$ where τ is a standard normal variable. If the sequence $\bar{X}_n \xrightarrow{L} \tau, \bar{X}_n$ is said to follow the central limit law or normal convergence. In this section we consider different Central Limit Theorems (CLT).

Definition 9.8.1 *A sequence of random variables $\{Y_n\}$ is asymptotically normal with mean a_n and variance b_n^2 if $b_n > 0$ for all n sufficiently large and*

$$\frac{Y_n - a_n}{b_n} \xrightarrow{L} N(0, 1).$$

We write Y_N is $AN(a_n, b_n^2)$.

Here, $\{a_n\}$ and $\{b_n\}$ are sequences of constants. It is not necessary that a_n and b_n^2 be the mean and variance of Y_n, not even that Y_n possesses such moments. Note that if Y_n is asymptotically $N(a_n, b_n^2)$, it does not necessarily mean that $\{Y_n\}$ converges in distribution to anything. The sequences a_n and b_n may or may not depend on n.

The following lemmas hold:

Lemma 9.8.1 *If Y_n is $AN(a_n, b_n^2)$, then also Y_n is $AN(\bar{a}_n, \bar{b}_n^2)$ iff*

$$\frac{\bar{b}_n}{b_n} \to 1 \quad \text{and} \quad \frac{\bar{a}_n - a_n}{b_n} \to 0.$$

Lemma 9.8.2 *If Y_n is $AN(a_n, b_n^2)$ then also $c_nY_n + d_n$ is $AN(a_n, b_n^2)$ iff*

$$c_n \to 1, \quad \frac{a_n(c_n - 1) + d_n}{b_n} \to 0.$$

Example 9.8.1 *If Y_n is $AN(n, 2n)$ then so is $\frac{n-1}{n}Y_n$ but not $\frac{\sqrt{n-1}}{\sqrt{n}}Y_n$.*

heorem 9.8.1 (*Lindeberg-Levy Central Limit Theorem*). *Let $\{X_n\}$ be a sequence of dependently and identically distributed random variables, with*

$$E(X_i) = \mu, \quad V(X_i) = \sigma^2 < \infty, \quad i = 1, 2, \cdots.$$

hen the average $\bar{X}_n = \frac{1}{n}\sum_{i=1}^{n} X_i$ is asymptotically normal $N\left(\mu, \frac{\sigma^2}{n}\right)$.

Proof: Consider the standardised variable

$$Z_n = \frac{1}{\sqrt{n}\sigma}\sum_{i=1}^{n}(X_i - \mu).$$

he r.v.'s $Y_i = (X_i - \mu)$ are iid with $E(Y_i) = 0, V(Y_i) = \sigma^2, (i = 1, \cdots, n)$. et $\Psi_Y(t)$ be the characteristic function of Y_i. The c.f. of Z_n is

$$\Psi_n(t) = E[e^{it(S_n - n\mu)/\sigma\sqrt{n}}] = \left[\Psi_Y\left(\frac{t}{\sigma\sqrt{n}}\right)\right]^n.$$

eveloping $\Psi_Y(t)$ in Maclaurin's series about $t = 0$

$$\Psi_Y(t) = \Psi_Y(0) + t\Psi'_Y(0) + \frac{t^2}{2!}\Psi''_Y(0) + o(t^2) = 1 - \frac{t^2}{2!}\sigma^2 + o(t^2).$$

ence

$$\Psi_n(t) = \left[1 - \frac{t^2}{2n} + o\left(\frac{t^2}{n}\right)\right]^n$$

$$\log_e \Psi_n(t) = n\log_e\left[1 + \left\{\frac{-t^2}{2n} + o\left(\frac{t^2}{n}\right)\right\}\right]$$

$$= n\left[-\frac{t^2}{2n} + o\left(\frac{t^2}{n}\right)\right] = -\frac{t^2}{2} + no\left(\frac{t^2}{n}\right).$$

Hence $\log_e \Psi_n(t) = -\frac{t^2}{2}$ (since $\lim_{n\to\infty} no\left(\frac{t^2}{n}\right) = 0$ for every fixed t) and $\Psi_n(t) \to e^{-\frac{t^2}{2}}$, he c.f. of τ, a standard normal deviate. Therefore, by the continuity theorem, $Z_n \xrightarrow{L} \tau$ nd hence $\bar{X}_n$ is asymptotically $N(\mu, \sigma^2/n)$.

Corollary 9.8.1 *Under the conditions of theorem 9.8.1, the standardised sum $\frac{(S_n - n\mu)}{\sigma\sqrt{n}} = \frac{\sqrt{n}(\bar{X}-\mu)}{\sigma}$ is asymptotically*

$$N(0, 1).$$

► **Example 9.8.2** *Consider binomial r.v.'s X_n with parameters (n,p). Now $X_n = Y_1$ $\cdots + Y_n$ where Y_i denotes the result of the ith bernoulli trial, $Y_i = 1(0)$ if a succ (failure), $Y_1, \cdots, Y_n$ being independent with $E(Y_i) = p, V(Y_i) = p(1-p)$. Hence, applying theorem 9.8.1, $\bar{X}_n$ is asymptotically normal $(p, pq/n)$.*

► **Example 9.8.3** *Let $\{X_n\}$ be a sequence of iid standard Cauchy variables. Since Cauchy variable has no finite mean, theorem 9.8.1 does not apply.*

The c.f. of each X is $\Psi_X(t) = e^{-|t|}$. The c.f. of $\frac{S_n}{n} = \bar{X}_n$ is $\Psi_n(t) = e^{-|t|}$. Hen $\bar{X}_n$ has a Cauchy distribution.

In case the independent variables $X_1, \cdots, X_n$ are not identically distributed, th sum need not converge to a normal distribution even if all the variables have fin standard deviations.

The following theorem due to Liapounov gives a sufficient condition that a sum independent r.v.'s should have an asymptotically normal distribution.

Theorem 9.8.2 (***Liapounov***). *Let $\{X_n\}$ be a sequence of mutually independent rand variables such that for some $\delta(>0), E\{|X_k - \mu_k|^{2+\delta}\}$ exists for every $k = 1, 2, \cdots$. Th if the condition*

$$\lim_{n\to\infty} \frac{1}{\left(\sum_1^n \sigma_k^2\right)^{1+\frac{\delta}{2}}} \sum_{k=1}^n E\left\{|X_k - \mu_k|^{2+\delta}\right\} = 0 \qquad (9.8$$

is satisfied, then $\frac{(S_n - \mu_{(n)})}{\sigma_{(n)}}$ is asymptotically normal with mean 0 and variance 1, whe

$$\mu_{(n)} = \sum_{k=1}^n \mu_k, \quad \sigma_{(n)}^2 = \sum_{k=1}^n \sigma_k^2.$$

Proof is omitted.

The Liapounov theorem gives only a sufficient condition for $\{S_n\}$ to follow cent limit law. The Lindeberg-Feller theorem stated below (without proof) gives a necessa and sufficient condition for the sum S_n to have an asymptotically normal distributio This is the Central Limit Theorem in the most general form.

Theorem 9.8.3 (***Lindeberg-Feller***). *Let $\{X_n\}$ be a sequence of independent r.v.'s w the cumulative d.f. of X_k as $F_k(x), E(X_k) = \mu_k, V(X_k) = \sigma_k^2 < \infty$. Then*

1. S_n is asymptotically $N(\mu_{(n)}, \sigma_{(n)})$;
2. $\lim_{n\to\infty} \max_{1\le k\le n} \frac{\sigma_k}{\sigma_{(n)}} = 0$

if and only if

$$A_n(\varepsilon) = \frac{1}{\sigma_{(n)}^2} \sum_{k=1}^n \int_{|x_k - \mu_k| \ge \varepsilon\sigma_{(n)}} (x - \mu_k)^2 dF_k(x) \to 0 \qquad (9.8.$$

as $n \to \infty$, holds for every $\varepsilon(>0)$.

rollary 9.8.2 *If X_k's are iid random variables each with finite variance σ^2, condition 8.2) is always satisfied; because,*

$$A_n(\varepsilon) = \frac{1}{n\sigma^2}\sum_{k=1}^{n}\int_{|x_k-\mu_k|\geq\varepsilon\sigma\sqrt{n}}(x-\mu)^2 dF(x) = \frac{1}{\sigma^2}\int_{y^2\geq n\varepsilon^2\sigma^2} y^2 dF(y) \to 0$$

$n\varepsilon^2\sigma^2 \to \infty$ as $n \to \infty$.

rollary 9.8.3 *If Liapounov's condition (9.8.1) is satisfied, then Lindeberg-Feller ndition (9.8.2) is also satisfied. For*

$$\begin{aligned} A_n(\varepsilon) &= \frac{1}{\sigma^2_{(n)}}\sum_{k=1}^{n}\int_{|x-\mu_k|\geq\varepsilon\sigma_{(n)}}(x-\mu_k)^2 dF_k(x) \\ &\leq \frac{1}{\varepsilon^\delta\sigma^{2+\delta}_{(n)}}\sum_{k=1}^{n}\int_{|x-\mu_k|\geq\varepsilon\sigma_{(n)}}|x-\mu_k|^{2+\delta} dF_k(x) \\ &\leq \frac{1}{\varepsilon^\delta\sigma^{2+\delta}_{(n)}}\sum_{k=1}^{n} E|X_k-\mu_k|^{2+\delta} \to 0 \text{ as } n\to\infty \end{aligned}$$

(9.8.1) and hence (9.8.2) holds.

NOTE 9.8.1 *The condition (9.8.2) can also be stated as follows. Define truncated ndom variables,*

$$U_k = \begin{cases} X_k - \mu_k, & |X_k-\mu_k| \geq \varepsilon\sigma_{(n)} \\ 0, & |X_k-\mu_k| < \varepsilon\sigma_{(n)}. \end{cases} \tag{9.8.3}$$

hen (9.8.2) reduces to the condition

$$B_n(\varepsilon) = \frac{1}{\sigma^2_{(n)}}\sum_{k=1}^{n} E(U_k^2) \to 0, \quad \text{as } n\to\infty. \tag{9.8.4}$$

The CLT has also been studied for the case when X_i's are not all mutually dependent random variables. The CLT for multivariate distributions has been onsidered in Section 13.4.

It is of interest to assess the error of approximation in CLT.

enote by

$$G_n(t) = P(S_n^* \leq t)$$

here

$$S_n^* = \frac{\sum_1^n X_i - E\left(\sum_1^n X_i\right)}{\sqrt{\text{Var}\sum_1^n X_i}}.$$

For the iid case, an exact bound on the error of approximation is provided by the followi theorem due to Berry (1941) and Esséen (1945).

Theorem 9.8.4 *Let $\{X_i\}$ be iid with mean μ and variance $\sigma^2 > 0$. Then*

$$\sup_t |G_n(t) - \Phi(t)| \leq \frac{33}{4} \frac{E|X_i - \mu|^3}{\sigma^3 \sqrt{n}}.$$

► **Example 9.8.4** *Examine if the law of large numbers and CLT hold for the sequen of mutually independent random variables $\{X_k\}$, where*

$$P[X_k = k] = P[X_k = -k] = \frac{1}{2\sqrt{k}}, \quad P[X_k = 0] = 1 - \frac{1}{\sqrt{k}}.$$

Taking $\delta = 1$ in (9.8.1), CLT holds if $\lim_{n\to\infty} \frac{\rho}{\sigma} = 0$ where

$$\rho^3 = \sum_{k=1}^{n} E|X_k - \mu|^3 \text{ and } \sigma = \sqrt{\sum_{k=1}^{n} \sigma_k^2}.$$

Here $E(X_k) = 0, E|X_k^3| = k^{\frac{5}{2}}$,

$$\lim_{n\to\infty} \frac{\rho}{\sigma} = \lim_{n\to\infty} \frac{\left(\sum_{k=1}^{n} k^{\frac{5}{2}}\right)^{\frac{1}{3}}}{\left(\sum_{k=1}^{n} k^{\frac{3}{2}}\right)^{\frac{1}{2}}} = \lim_{n\to\infty} \frac{\left(\int_0^n x^{\frac{5}{2}}\, dx\right)^{\frac{1}{3}}}{\left(\int_0^n x^{\frac{3}{2}}\, dx\right)^{\frac{1}{2}}} = \lim_{n\to\infty} \frac{A}{\sqrt{n}}$$

(where $A = (\sqrt{5})/(\sqrt[3]{7})(\sqrt[6]{2})) = 0$.

Hence CLT holds. Here

$$\lim_{n\to\infty} \frac{V(S_n)}{n^2} = \frac{2}{5}\sqrt{n}$$

does not tend to zero as $n \to \infty$. Hence condition (9.6.3) does not hold.

► **Example 9.8.5** *For the sequence of r.v.'s $\{X_n\}$ with*

$$P\{X_k = k^\lambda\} = P\{X_k = -k^\lambda\} = \frac{1}{2}, \quad \lambda > 0,$$

determine if the WLLN holds, central limit law holds.

$$E(X_k) = 0, \quad \sigma_{(n)}^2 = \sum_{k=1}^{n} k^{2\lambda} \sim \int_0^n x^{2\lambda}\, dx = \frac{n^{2\lambda+1}}{2\lambda+1}.$$

Hence $\sigma_{(n)}^2/n^2 = n^{2\lambda-1}/(2\lambda+1) \to 0$ as $n \to \infty$ if $\lambda < \frac{1}{2}$. Hence WLLN holds f $\lambda < \frac{1}{2}$.

To verify if CLT holds consider the truncated variables U_k defined in (9.8.3)

$$|X_k| = k^\lambda \le n^\lambda.$$

Now $n^\lambda \le \varepsilon\sigma_{(n)} \Rightarrow n \ge \frac{2\lambda+1}{\varepsilon^2} = n_0(\varepsilon)$ (say).

Hence

$$U_k = \begin{cases} X_k, & n \le n_0(\varepsilon) \\ 0, & n > n_0(\varepsilon) \end{cases}.$$

$B_n(\varepsilon) = \sum_1^{n_0(\varepsilon)} k^{2\lambda} / \sum_1^n k^{2\lambda} \to 0$ as $n \to \infty$ for every $\varepsilon(>0)$ and every $\lambda(>0)$. Hence CLT holds for every $\lambda > 0$.

S_n is likely to be of the order of $n^{\lambda+\frac{1}{2}}$ ($\because S_n$ lies within $E(S_n) \pm 3\sigma_{(n)}$ with high probability). Hence for $\lambda \ge \frac{1}{2}$, S_n does not converge in probability to zero.

▶ **Example 9.8.6** *Show that if the variables X_k are uniformly bounded then* (i) $S_n = \sum_{k=1}^n X_k$ *satisfies WLLN* (ii) S_n *obeys the central limit property, provided* $\sigma_{(n)} \to \infty$.

1. $V(S_n) = \sum_1^n \sigma_k^2 \le 2nA^2$ where $|X_k| < A \ \forall\, k$. Hence

$$\frac{V(S_n)}{n^2} \le \frac{2A^2}{n} \to 0 \text{ as } n \to \infty.$$

Therefore WLLN holds.

2. Suppose $\sigma_{(n)}^2 \to \infty$. Here $|X_k| < A \ \forall\, k$. Hence, the truncated variables in (9.8.3) are:

$$\begin{aligned} U_k &= 0 \text{ if } n \text{ is so large that } \sigma_{(n)} > \frac{2A}{\varepsilon} \text{ i.e., for } n \ge n_0(\varepsilon) \text{ (say)} \\ &= X_k - \mu_k \text{ otherwise,} \end{aligned}$$

because $|X_k - \mu_k| \le |X_k| + |\mu_k| \le 2A$ and hence $|X_k - \mu_k| \le \varepsilon\sigma_{(n)}$ holds if $2A \le \varepsilon\sigma_{(n)}$. Therefore, for

$$n \ge n_0(\varepsilon), \quad P\{(U_k = 0), k = 1, \cdots, n\} = 1.$$

Hence

$$\lim_{n\to\infty} B_n(\varepsilon) = 0$$

for every ε and the central limit law holds.

The condition $\sigma_{(n)}^2 \to \infty$ is a necessary condition for CLT to hold. If $\sigma_{(n)}^2 \not\to \infty$, there exists a constant B such that $\sigma_{(n)}^2 \uparrow B^2$. For any fixed m, we can find an ε such that

$$P\{|X_m - \mu_m| > \varepsilon\sigma_{(n)}\} > P\{|X_m - \mu_m| > \varepsilon B\} > 0.$$

For any $n \geq m$

$$A_n(\varepsilon) = \frac{1}{\sigma^2_{(n)}} \sum_{k=1}^{n} \int_{|x-\mu_k| \geq \varepsilon\sigma_{(n)}} (x-\mu_k)^2 dF_k(x)$$

$$\geq \varepsilon^2 \sum_{k=1}^{n} P\{|X_k - \mu_k| \geq \varepsilon\sigma_{(n)}\} \geq \varepsilon^2 P\{|X_m - \mu_m| \geq \varepsilon B\} > 0$$

and hence condition (9.8.2) does not hold.

► **Example 9.8.7** *Using the central limit theorem, show that*

$$\lim_{n\to\infty} e^{-n} \sum_{k=0}^{n} \frac{n^k}{k!} = \frac{1}{2}.$$

Let $\{X_n\}$ be a sequence of Poisson variables each with parameter $\lambda = 1$. Then $\sum_{i=1}^{n} X_i$ is Poisson with parameter n. Again by CLT, $\sum_{i=1}^{n} X_i$ is asymptotically $N(n, \sqrt{n})$. Therefore,

$$e^{-n} \sum_{k=0}^{n} \frac{n^k}{k!} = P\left(\sum_{i=1}^{n} X_i - n \leq 0\right) = \frac{1}{\sqrt{2\pi}} \int_{-\infty}^{0} e^{-\frac{z^2}{2}} dz = \frac{1}{2}.$$

Exercise

1. Let X_n a sequence of independent random variables, each uniform over (0, 1). Let

$$Y_n = \frac{\min_{1\leq i\leq n} X_i}{n}.$$

Show that $Y_n \xrightarrow{L} Z$ where Z is an exponential r.v. with mean 1.

2. Let X be a $N(0, 1)$-variable. Define the sequence of r.v.'s $\{X_n\}$ such that

$$X_{2m+1} = X(m = 0, 1, \cdots), \quad X_{2m} = -X(m = 1, 2, \cdots).$$

Show that $X_n \xrightarrow{L} X$; however, X_n does not converge in probability to X.

3. Let $\{F_n\}$ be a sequence of d.f.'s such that

$$F_n(x) = \begin{cases} 0, & x < 0 \\ 1 - \frac{1}{n}, & 0 \leq x < n \\ 1, & n \leq x. \end{cases}$$

Show that $F_n \xrightarrow{w} F$ where

$$F(x) = \begin{cases} 0, & x < 0 \\ 1, & x \geq 0. \end{cases}$$

4. Let $F_n(x)$ be a sequence of negative binomial distributions with parameters (r_n, q_n) with $r_n \to \infty, q_n \to 0$ in such a way that $r_n q_n \to \lambda$. Show that $F_n(x)$ converges in distribution to Poisson distribution.

5. Show that if $X_n \xrightarrow{ms} X$ then $E(X_n) \to E(X)$.

6. Let $\{X_n\}$ be a sequence of iid variables, each uniform over $(0, \theta)$. Consider the variables $Y_n, Z_n = nY_n$ where $Y_n = \min_{i=1,\cdots,n} X_i$. Find the limiting distributions of Y_n and Z_n.

7. Let $\{X_n\}$ be a sequence of independent r.v.'s and p.m.f. of X_n is

$$P(X_n = -n-4) = \frac{1}{n+4}, \quad P(X_n = -1) = 1 - \frac{4}{n+4}, \quad P(X_n = n+4) = \frac{3}{n+4}.$$

Show that $p \lim_{n\to\infty} X_n = -1$.

8. Let $g(y)$ be a continuous function and let $\{X_n\}$ be a sequence of r.v.'s such that $X_n \xrightarrow{P} X$. Show that $g(X_n) \xrightarrow{P} g(X)$.

9. Let $\{F_n\}$ be a sequence of d.f.'s such that

$$F_n(x) = \begin{cases} 0, \ x < \theta + \frac{1}{n} \\ 1, \ x \geq \theta + \frac{1}{n}. \end{cases}$$

Show that the corresponding sequence of random variables $X_n \xrightarrow{L} X$ where X is a r.v. degenerate at $X = \theta$.

10. Show that the sequence of characteristic functions $\Psi_n(t) = \frac{(\sin nt)}{nt}$ converges to a limit function $\Psi(t)$. However, $F_n(x)$ does not converge to a limiting distribution. Explain why.

11. $\{X_n\}$ is a sequence of independent r.v.'s. Determine if they obey WLLN and/or central limit law.

   (a) $P\{X_k = 2^k\} = P\{X_k = -2^k\} = \frac{1}{2}$.

   (b) $P\{X_k = 2^k\} = P\{X_k = -2^k\} = \frac{1}{2^{2k+1}}, P\{X_k = 0\} = 1 - \frac{1}{2^{2k}}$.

   (c) $P\{X_k = 2^k\} = P\{X_k = -2^k\} = \frac{1}{2^{k+1}}, P\{X_k = 1\} = P\{X_k = -1\} = \frac{1}{2}(1 - 2^{-k})$.

   (d) $P\{X_k = 2^k\} = P\{X_k = -2^k\} = \frac{1}{2^{k+1}}, P\{X_k = 1\} = P\{X_k = -1\} = \frac{1}{2}\left(1 - \frac{1}{2^k}\right)$.

   (e) $P\left\{X_k = \frac{1}{2^k}\right\} = P\left\{X_k = -\frac{1}{2^k}\right\} = \frac{1}{2}$.

   (f) $P\{X_k = 2^{-k}\} = P\{X_k = -2^{-k}\} = \frac{1}{2^{k+1}}, P\{X_k = 1\} = P\{X_k = -1\} = \frac{1}{2}(1 - 2^{-k})$.

12. Let $\{X_k\}$ be iid with $P(X_k = 1) = P(X_k = -1) = \frac{1}{2}$. Define $Y_k = \sum_{j=1}^{k} \frac{X_j}{2^j}$. Show that $Y_k \xrightarrow{L} Z$ where $Z \sim$ uniform $(0,1)$.

13. Let $\{X_n\}$ be a sequence of independently distributed $N\left(0, \frac{\sigma}{\sqrt{n}}\right)$-variables. Show that $X_n \xrightarrow{L} X$ where X is a degenerate r.v. at $x = 0$.

14. Check if the SLLN holds for the following sequence of independent r.v.'s.

    (a) $P\{X_k = \pm 2^k\} = \frac{1}{2}$.
    (b) $P\{X_k = \pm k\} = \frac{1}{2\sqrt{k}}$, $P\{X_k = 0\} = 1 - \frac{1}{\sqrt{k}}$.
    (c) $P\{X_k = \pm 2^{k+1}\} = 2^{-(k+3)}, P(X_k = 0) = 1 - 2^{-(k+2)}$.

15. Show that a sequence $\{X_n\}$ is a.s. convergent iff it is a.s. mutually convergent i.e., iff $X_m - X_n \xrightarrow{a.s.} 0$.

16. Prove that if $\Sigma P(|X_{n+1} - X_n| \geq \varepsilon_n) < \infty$ where $\Sigma \varepsilon_n$ is a convergent sum of positive terms, then $\{X_n\}$ is a.s. convergent.

17. Let ϕ be a positive symmetric convex function. Then prove that under assumptions of theorem 9.4.5

$$\begin{aligned} P(A) &= P(|S_j| < u_j, j = 1, \cdots, n) \\ &\geq 1 - \sum_{1}^{n} \frac{E[\phi(S_j)] - E[\phi(S_{j-1})]}{\phi(u_j)} \end{aligned}$$

where we set $\phi(S_0) = 0$. In particular,

$$P(A) \geq 1 - 2^{2-r} \sum_{1}^{n} \frac{E(|X_j|^r)}{|u_j|^r}$$

for any r in [1, 2]. Here $S_n = \sum_{1}^{n} X_i$.

[*Hints:* As in the proof of theorem 9.4.5 consider a form

$$T = \Sigma(\alpha_j - \alpha_{j+1})\phi(S_j) \text{ with } \alpha_j = \phi(u_j)^{-1}$$

and use the fact that by Jensen's inequality

$$E[\phi(S_j)|X_1, \cdots, X_i] \geq \phi(S_i) \geq \phi(u_i)(j \geq i)$$

if the values of $X_1, X_2, \cdots, X_i$ imply the occurrence of B_i.]

18. Let $Y_1, Y_2, \cdots$ be iid random variables. Define $X_n = Y_n/n$. Then clearly $X_n \xrightarrow{p} 0$. Show that $X_n \xrightarrow{a.s.} 0$ iff $E|Y_1| < \infty$.

19. Let $X_1, X_2, \cdots$ be a sequence of independent r.v.'s. Then show that $X_n \xrightarrow{a.s.} 0 \iff \sum_{n=1}^{\infty} P\{|X_n| > \varepsilon\} < \infty$ for all $\varepsilon > 0$.

Chapter 10

Some Further Properties of a Multivariate Distribution

10.1 Introduction

We considered in chapter 7 general characteristics of probability distributions on R^n and two special distributions, namely, multinomial distribution and bivariate normal distribution. We consider in this chapter some further characteristics based on multivariate distributions. Section 10.2 considers expectation and covariance of random vectors, section 10.3, expectation and variance of quadratic forms in random vectors. Section 10.4 deals with rank of a multivariate distribution, section 10.5 the concept of a concentration ellipsoid.

We already considered in chapter 7 simple correlation coefficient between two variables and the regression coefficient of one variable on the other in the line of regression of the dependent variable on the independent variable. We shall consider here p variables, the regression equation of one (dependent) variable on the other (independent) variables, —the linear multiple regression equation, partial regression coefficients and multiple and partial correlation coefficients. These are considered in section 10.6 through 10.10.

Lastly in section 10.11 we consider some exact and large sample formulae for standard errors of sample moments. Section 10.12 considers large sample formulae for variance of functions of sample moments.

10.2 Expectation and Covariance of Random Vectors

We consider in this section expectation of a random vector and covariance between two random vectors.

Let $\boldsymbol{Z} = ((Z_{ij}))$ be a $m \times n$ matrix of random variables

$$Z_{ij}, \ i = 1, \cdots, m; \ j = 1, \cdots, n.$$

Definition 10.2.1 *The expectation of a random matrix is the matrix of expectation of the corresponding random variables. Thus*

$$E[Z] = \Big((E(Z_{ij}))\Big).$$

The following theorems are stated without proof. The proofs are easy.

Theorem 10.2.1 *If A, B, C are matrices of constants, of order $l \times m, n \times p$ and $l \times p$ respectively and X is a random matrix of order $m \times n$, then*

$$E[AXB + C] = AE(X)B + C.$$

Theorem 10.2.2 *If $\mathbf{X}$, $\mathbf{Y}$ are each $m \times 1$ vectors of random variables and A, B are each $l \times m$ matrices of constants, then*

$$E[A\mathbf{X} + B\mathbf{Y}] = AE(\mathbf{X}) + BE(\mathbf{Y}).$$

In particular, for scalars a, b

$$E[a\mathbf{X} + b\mathbf{Y}] = aE(\mathbf{X}) + bE(\mathbf{Y}).$$

Definition 10.2.2 *If $\mathbf{X}$, $\mathbf{Y}$ are respectively $m \times 1$ and $n \times 1$ vectors of random variables, then the matrix of covariance between $\mathbf{X}$, $\mathbf{Y}$ is*

$$C(\mathbf{X}, \mathbf{Y}) = ((\text{Cov}(X_i, Y_j))).$$

Theorem 10.2.3 $C(\mathbf{X}, \mathbf{Y}) = E(\mathbf{X} - E(\mathbf{X}))(\mathbf{Y} - E(\mathbf{Y}))'$.

Definition 10.2.3 *When $\mathbf{X}=\mathbf{Y}, C(\mathbf{X}, \mathbf{X})$ is the dispersion matrix of $\mathbf{X}$, written as $D(\mathbf{X})$,*

$$D(\mathbf{X}) = ((\text{Cov}(X_i, X_j))) = \begin{bmatrix} V(X_1) & \text{Cov}(X_1, X_2) & \cdots & \text{Cov}(X_1, X_m) \\ \cdot & \cdot & \cdots & \cdot \\ \text{Cov}(X_m, X_1) & \text{Cov}(X_m, X_2) & \cdots & V(X_m) \end{bmatrix}.$$

Clearly, $D(\mathbf{X})$ is a symmetric matrix.

Lemma 10.2.1 *If $\mathbf{a}$ is a vector of constants:*

(a) $D(\mathbf{X} - \mathbf{a}) = D(\mathbf{X})$;

(b) $E[(\mathbf{X} - \mathbf{a})(\mathbf{X} - \mathbf{a})'] = D(\mathbf{X}) + (E(\mathbf{X}) - \mathbf{a})(E(\mathbf{X}) - \mathbf{a})'$.

Theorem 10.2.4 *If $\mathbf{X}, \mathbf{Y}$ are random vectors of orders $m \times 1, n \times 1$ respectively and A, B are matrices of constants of orders $l \times m, p \times n$ respectively, then*

$$C(A\mathbf{X}, B\mathbf{Y}) = AC(\mathbf{X}, \mathbf{Y})B'.$$

Corollary 10.2.1 $D(A\mathbf{X}) = AD(\mathbf{X})A'$.

Theorem 10.2.5 *If* $\mathbf{X}$, $\mathbf{Y}$, $\mathbf{U}$, $\mathbf{V}$ *are each* $n \times 1$ *random vectors, then for all real numbers* a, b, c, d

$$C(a\mathbf{X} + b\mathbf{Y},\ c\mathbf{U} + d\mathbf{V}) = acC(\mathbf{X},\mathbf{U}) + adC(\mathbf{X},\mathbf{V}) + bcC(\mathbf{Y},\mathbf{U}) + bdC(\mathbf{Y},\ \mathbf{V}).$$

Corollary 10.2.2 $D(a\mathbf{X} + b\mathbf{Y}) = a^2D(\mathbf{X}) + b^2D(\mathbf{Y}) + 2abC(\mathbf{X},\mathbf{Y})$.

Theorem 10.2.6 *If* $\mathbf{X}$ *is a vector of random variables such that no element of* $\mathbf{X}$ *is a linear combination of the remaining elements (i.e., three exists no vector of constants* $\mathbf{a}$ *and scalar* b *such that* Prob. $(a'\mathbf{X} = b) = 1$, *then* $D(\mathbf{X})$ *is a positive definite* (p.d.) *matrix.*

Proof: For any vector of constants $\mathbf{c}$,

$$V(\mathbf{c}'\mathbf{X}) = \mathbf{c}'D(\mathbf{X})\mathbf{c} \geq 0,$$

equality holds iff $P\{\mathbf{c}'\mathbf{X} = \text{constant}\} = 1$ or $\mathbf{c} = \mathbf{0}$. The former condition, however, cannot hold.

▶ **Example 10.2.1** *Let* $\mathbf{X}$, $\mathbf{Y}$ *be* $m \times 1$ *and* $n \times 1$ *vectors of random variables and let* $D(\mathbf{X})$ *be a p.d. matrix. Then there exists a* $n \times m$ *matrix* $\boldsymbol{A}$ *for which*

$$C(\mathbf{X},\mathbf{Y} - \boldsymbol{A}\mathbf{X}) = \mathbf{0}$$

Proof:

$$\begin{aligned} C(\mathbf{X},\mathbf{Y} - \boldsymbol{A}\mathbf{X}) &= C(\mathbf{X},\mathbf{Y}) - C(\mathbf{X},\boldsymbol{A}\mathbf{X}) = C(\mathbf{X},\mathbf{Y}) - D(\mathbf{X})\boldsymbol{A}' = \mathbf{0} \\ &\Rightarrow \boldsymbol{A} = C(\mathbf{Y},\mathbf{X})(D(\mathbf{X}))^{-1}, \end{aligned}$$

since $D(\mathbf{X})$ is non-singular.

■ NOTE 10.2.1 *A typical element of* $D(\boldsymbol{X})$ *will often be denoted as* $\sigma_{ij} = \rho_{ij}\sigma_i\sigma_j$ *where* ρ_{ij} *is the correlation coefficient between* X_i *and* X_j. *In general,* $-1 \leq \rho_{ij} \leq 1$, *though in certain cases,* ρ *is restricted to smaller interval.*

For example, suppose $D(\mathbf{X})$ has the intraclass correlation structure

$$D(\boldsymbol{X}) = \sigma^2 \begin{bmatrix} 1 & \rho & \cdots & \rho \\ \rho & 1 & \cdots & \rho \\ \cdot & \cdot & \cdots & \cdot \\ \rho & \rho & \cdots & 1 \end{bmatrix} = \sigma^2 \left[(1-\rho)\boldsymbol{I} + \rho \mathbf{1}_p \mathbf{1}_p'\right]$$

where $\boldsymbol{I}$ is a $p \times p$ identity matrix and $\mathbf{1}_p = (1, \cdots, 1)'$. If $D(\mathbf{X})$ is to be a p.d. covariance matrix, one necessary (but not sufficient) condition is that $|D(\mathbf{X})| > 0$.

Now

$$|D(\mathbf{X})| = \sigma^{2p}(1-\rho)^p \left[1 + \frac{\rho p}{1-\rho}\right].$$

Hence, for $|D(\mathbf{X})| > 0$ it is necessary than $\rho > -\frac{1}{p-1}$.

A correlation matrix is a matrix of correlation coefficients, often denoted as $\boldsymbol{R}$. Thus

$$\boldsymbol{R}(\mathbf{X}) = \begin{bmatrix} 1 & \rho_{12} & \cdots & \rho_{1p} \\ \cdots & \cdots & \cdots & \cdots \\ \rho_{p1} & \rho_{p2} & \cdots & 1 \end{bmatrix}.$$

Clearly,

$$D(\mathbf{X}) = \sqrt{\boldsymbol{D}}\, \boldsymbol{R}(\mathbf{X})\sqrt{\boldsymbol{D}}$$

where $\boldsymbol{D} = \text{Diag}\,(\sigma_{11}, \cdots, \sigma_{pp})$.

10.3 Expectation and Variance of Quadratic Forms of Random Vectors

Theorem 10.3.1 *Let $\mathbf{X} = (X_1, \cdots, X_n)'$ be a random vector and $\boldsymbol{A} = ((a_{ij}))$ be a $n \times n$ symmetric matrix of constants. Let $E(\mathbf{X}) = \boldsymbol{\mu}$ and $D(\mathbf{X}) = \boldsymbol{\Sigma} = ((\sigma_{ij}))$. Then*

$$E(\mathbf{X}'\boldsymbol{A}\mathbf{X}) = tr(\boldsymbol{A\Sigma}) + \boldsymbol{\mu}'\boldsymbol{A}\boldsymbol{\mu}$$

where $tr(\boldsymbol{B})$ denotes trace of the matrix $\boldsymbol{B}$.

Proof:

$$\begin{aligned} E(\mathbf{X}'\boldsymbol{A}\mathbf{X}) &= E\left[(\mathbf{X}-\boldsymbol{\mu})'\boldsymbol{A}(\mathbf{X}-\boldsymbol{\mu}) + \boldsymbol{\mu}'\boldsymbol{A}\mathbf{X} + \mathbf{X}'\boldsymbol{A}\boldsymbol{\mu} - \boldsymbol{\mu}'\boldsymbol{A}\boldsymbol{\mu}\right] \\ &= E\left[(\mathbf{X}-\boldsymbol{\mu})'\boldsymbol{A}(\mathbf{X}-\boldsymbol{\mu})\right] + \boldsymbol{\mu}'\boldsymbol{A}\boldsymbol{\mu} \\ &= \sum_i \sum_j a_{ij}\sigma_{ij} + \boldsymbol{\mu}'\boldsymbol{A}\boldsymbol{\mu} = tr\,(\boldsymbol{A\Sigma}) + \boldsymbol{\mu}'\boldsymbol{A}\boldsymbol{\mu}. \end{aligned}$$

Corollary 10.3.1 *For any $n \times 1$ vector of constants,* b,

$$E\left[(\mathbf{X}-\mathbf{b})'\boldsymbol{A}(\mathbf{X}-\mathbf{b})\right] = tr(\boldsymbol{A\Sigma}) + (\boldsymbol{\mu}-\mathbf{b})'\boldsymbol{A}(\boldsymbol{\mu}-\mathbf{b}).$$

Corollary 10.3.2 *If $\boldsymbol{\Sigma} = \sigma^2 I_n$, then $tr(\boldsymbol{A\Sigma}) = \sigma^2 tr\boldsymbol{A}$. In this case,*

$$E(\mathbf{X}'\boldsymbol{A}\mathbf{X}) = \sigma^2 \text{ (sum of coefficients of } X_i^2) + \boldsymbol{\mu}'\boldsymbol{A}\boldsymbol{\mu}.$$

▸ **Example 10.3.1** *The r.v.'s $X_1, \cdots, X_n$ are iid with mean μ and variance σ^2. Find expected value of*

$$Q = \sum_{j=1}^{n-1}(X_j - X_{j+1})^2.$$

Here

$$D(\mathbf{X}) = \sigma^2 I_n;\ Q = \sum_{i=1}^{n} X_i^2 - X_1^2 - X_n^2 - 2\sum_{j=1}^{n-1} X_j X_{j+1};\ tr\boldsymbol{A} = 2n-2.$$

Hence,

$$E(Q) = 2\sigma^2(n-1).$$

▶ **Example 10.3.2** *The r.v.'s $X_1, \cdots, X_n$ have a common mean μ, common variance σ^2 and common covariance $\rho\sigma^2$. Find expected value of $Q = \Sigma(X_i - \bar{X})^2$.*

Here $Q = \mathbf{X}'\boldsymbol{A}\mathbf{X}$ when $\boldsymbol{A} = ((\delta_{ij} - \frac{1}{n})), \delta_{ij} = 1(0)$ if $i = (\neq)j$. Also

$$\boldsymbol{A\Sigma} = \sigma^2(1-\rho)\boldsymbol{A}.$$

$E(Q) = tr(\boldsymbol{A\Sigma})$ (the other term vanishes) $= \sigma^2(1-\rho)(n-1)$.

Theorem 10.3.2 *Let $X_1, \cdots, X_n$ be n independent r.v.'s with means $\mu_1, \cdots, \mu_n$, common variance μ_2 and common third order and fourth order central moments μ_3 and μ_4 respectively. Let $\boldsymbol{A}$ be any $n \times n$ symmetric matrix and* **a** *be the column vector consisting of the diagonal elements of $\boldsymbol{A}$. Then*

$$V(\mathbf{X}'\boldsymbol{A}\mathbf{X}) = (\mu_4 - 3\mu_2^2)\mathbf{a}'\mathbf{a} + 2\mu_2^2 tr(\boldsymbol{A}^2) + 4\mu_2\boldsymbol{\mu}'\boldsymbol{A}^2\boldsymbol{\mu} + 4\mu_3\boldsymbol{\mu}'\boldsymbol{A}\boldsymbol{\mu}$$

where $\boldsymbol{\mu} = (\mu_1, \cdots, \mu_n)'$.

Proof: Omitted.

Corollary 10.3.3 *If X_i's are normally distributed,*

$$\mu_3 = 0,\ \mu_4 = 3\mu_2^2 \quad \textit{and} \quad V(\mathbf{X}'\boldsymbol{A}\mathbf{X}) = 2\mu_2^2 tr(\boldsymbol{A}^2) + 4\mu_2\mu'\boldsymbol{A}^2\mu.$$

If further, $\boldsymbol{\mu} = \mathbf{0}$ and $\mu_2 = \sigma^2$ then

$$V(\mathbf{X}'\boldsymbol{A}\mathbf{X}) = 2\sigma^4 tr(\boldsymbol{A}^2).$$

10.4 Rank of a Distribution

Definition 10.4.1 Rank of a distribution. *Rank of a p-variate distribution is defined to be the rank of its dispersion matrix $\boldsymbol{\Sigma}$. If the rank of $\boldsymbol{\Sigma}$ is p, the distribution is said to be non-singular, otherwise it is singular.*

Theorem 10.4.1 *If rank of distribution of $\boldsymbol{\xi} = (\xi_1, \cdots, \xi_p)'$ is $r(\leq p)$, we can always find r uncorrelated r.v.'s $\zeta_1, \cdots, \zeta_r$ (say) each with zero mean and unit s.d. such that*

$$\left.\begin{array}{l} \xi_1 - m_1 = c^{11}\zeta_1 + \cdots + c^{1r}\zeta_r \\ \cdots\cdots\cdots\cdots\cdots\cdots\cdots\cdots\cdots\cdots \\ \xi_p - m_p = c^{p1}\zeta_1 + \cdots + c^{pr}\zeta_r \end{array}\right\} \tag{10.4.1}$$

hold with probability unity, where $m_i = E(\xi_i), c^{ij}$'s are constants and rank $((c^{ij}))$ is r. The column vectors of $((c^{ij}))$ are linearly independent.

Proof: Since rank $(\boldsymbol{\Sigma}) = r$, it is possible to find a non-singular matrix $\boldsymbol{C}$ such that

$$\boldsymbol{C\Sigma C'} = \text{Diag}\,(1, \cdots, 1, 0, \cdots, 0) = \Lambda \ \text{(say)}.$$

Let $\eta = \boldsymbol{C}\xi$. Then

$$V(\eta_i) = \begin{cases} 1 \text{ for } i = 1, \cdots, r \\ 0 \text{ for } i = r+1, \cdots, p. \end{cases}$$

This implies η_i is a constant for $i = r+1, \cdots, p$. Now $\xi = \boldsymbol{C}^{-1}\eta$. Thus

$$\xi_i = c^{i1}\eta_1 + c^{i2}\eta_2 + \cdots + c^{ip}\eta_p, \ \ i = 1, \cdots, p$$

with probability one. Taking expectation of both sides

$$m_i = c^{i1}E(\eta_1) + \cdots + c^{ip}E(\eta_p), \ \ i = 1, \cdots, p.$$

Therefore

$$\xi_i - m_i = c^{i1}(\eta_1 - E(\eta_1)) + \cdots + c^{ir}(\eta_r - E(\eta_r)), \ \ i = 1, \cdots, p.$$

Taking $\zeta_j = \eta_j - E(\eta_j)$ $(j = 1, \cdots, r)$ the result follows.

Definition 10.4.2 *Consider the p-dimensional space R^p and a point $\mathbf{x} = (x_1, \cdots, x_p)'$ in that space. Consider also a set of $(p-q)$ independent non-homogeneous equations*

$$a_{i0} + a_{i1}x_1 + \cdots + a_{ip}x_p = 0, \ \ i = 1, \cdots, p-q, \ \ q < p \qquad (10.4.2)$$

in $(x_1, \cdots, x_p)$. The set of points for which these $(p-q)$ equations are simultaneously satisfied is said to be a q-flat in R^p. We assume that the equations (10.4.2) are consistent.

Theorem 10.4.2 *The joint distribution of $\boldsymbol{\xi} = (\xi_1, \cdots, \xi_p)'$ would have rank r iff the point $(\xi_1, \cdots, \xi_p)'$ belongs with probability one to a r-flat in R^p, but does not belong with unit probability to a flat of lesser dimensions.*

In particular, if the distribution is non-singular, then the points $(\xi_1, \cdots, \xi_p)'$ would belong to a p flat and not less.

Proof: As rank of distribution of ξ is r we can always find r uncorrelated r.v.'s $\zeta_1, \cdots, \zeta_r$ such that (10.4.1) holds with probability unity. The column vectors of the matrix $((c^{ij}))$ in (10.4.1) are linearly independent. Let $(a_{i1}, \cdots, a_{ip})'$ $(i = 1, \cdots, p-r)$ be $(p-r)$ independent vectors each orthogonal to columns of the matrix $((c^{ij}))$. Then

$$a_{i1}(\xi_1 - m_1) + a_{i2}(\xi_2 - m_2) + \cdots + a_{ip}(\xi_p - m_p) = 0, \ \ i = 1, \cdots, p-r$$

with probability one. Therefore, with probability unity, $(\xi_1, \cdots, \xi_p)'$ satisfy $(p-r)$ linearly independent equations.

Since it is not possible to find more than $(p-r)$ linearly independent column vectors each of which is orthogonal to column vectors $(c^{1j}, c^{2j}, \cdots, c^{pj})$, $j = 1, \cdots, r$, $(\xi_1, \cdots, \xi_p)'$ can not satisfy more that $(p-r)$ linearly independent non-homogeneous equations. Hence $\boldsymbol{\xi}$ does not lie with probability one to a flat of less than r dimensions in R^p.

10.5 Concentration Ellipsoid of a Multivariate Probability Distribution

Consider first a univariate random variable X with mean μ and variance σ^2. Suppose we want to find a uniform distribution over a certain range in R^1 such that its mean is μ and variance is σ^2. Let the random variable X' be uniformly distributed over $(\mu-k\sigma, \mu+k\sigma)$ such that $E(X')=\mu, V(X')=\sigma^2$. It is easy to check that $k=\sqrt{3}$. Therefore, a uniform distribution over $(\mu-\sqrt{3}\sigma, \mu+\sqrt{3}\sigma)$ has the same mean and variance as X. The interval $(\mu-\sqrt{3}\sigma, \mu+\sqrt{3}\sigma)$ may, therefore, be taken as a geometrical representation of the concentration (opposite of dispersion) of the original distribution.

Consider now the case of a bivariate probability distribution of (X, Y) with means μ_X, μ_Y, variances σ_X^2, σ_Y^2 and covariance σ_{XY}. It is assumed that $\rho^2<1$. Here we want to find a closed curve in R^2 such that a random vector (X', Y') which is uniformly distributed over the region bounded by this curve has the same means, variances and covariance as the above bivariate distribution of X, Y. Since the problem cannot be solved in general for any closed curve, we confine ourselves to the case of ellipses in two-dimensions.

Consider the non-negative quadratic form

$$Q(x', y')=a_{11}(x'-\mu_x)^2+2a_{12}(x'-\mu_X)(y'-\mu_Y)+a_{22}(y'-\mu_Y)^2.$$

Suppose (X', Y') have a uniform distribution over the region bounded by the ellipse

$$Q(x', y')=k^2.$$

Now

$$\iint\limits_{Q(x',y')\le k^2} dx'\,dy'=\frac{\pi k^2}{\sqrt{|\boldsymbol{A}|}}$$

where $\boldsymbol{A}=\begin{bmatrix} a_{11} & a_{12} \\ a_{21} & a_{22}\end{bmatrix}$.

[In general, $\int\limits_{\boldsymbol{Z}'\boldsymbol{A}\boldsymbol{Z}\le k^2}\cdots\int \prod_{i=1}^{p} dZ_i=\dfrac{\pi^{p/2}k^p}{\sqrt{|\boldsymbol{A}|}\,\overline{|}\frac{p}{2}+1}$ where $\boldsymbol{Z}=(Z_1,\cdots,Z_p)'$ and $\boldsymbol{A}$ is a $p\times p$ matrix]

Hence, the p.d.f. of (X', Y') is

$$f(x', y')=\frac{\sqrt{|\boldsymbol{A}|}}{\pi k^2}.$$

It can be shown that

$$E(X')=\mu_X, \qquad E(Y')=\mu_Y;$$

$$V(X')=\frac{k^2}{4}\cdot\frac{a_{22}}{|\boldsymbol{A}|}=\frac{k^2}{4}a^{11}, \qquad V(Y')=\frac{k^2}{4}\cdot a^{22};$$

$$\operatorname{Cov}(X', Y')=-\frac{k^2}{4}\cdot\frac{a_{12}}{|\boldsymbol{A}|}=\frac{k^2}{4}a^{12}$$

where $((a^{ij}))=\boldsymbol{A}^{-1}$.

It can be shown that the required choices are

$$k^2 = 4, \ \ a^{11} = \sigma_X^2, \ a^{22} = \sigma_Y^2, \ a^{12} = \sigma_{XY},$$

i.e.,

$$a_{11} = \sigma^{XX} = \frac{1}{\sigma_X^2(1-\rho^2)}, \ a_{22} = \sigma^{YY} = \frac{1}{\sigma_Y^2(1-\rho^2)}, \ a_{12} = \sigma^{XY} = -\frac{\rho}{\sigma_X\sigma_Y(1-\rho^2)}$$

where $\sigma^{XX}, \sigma^{XY}, \sigma^{YY}$ are the elements of inverse of the matrix

$$\begin{bmatrix} \sigma_{XX} & \sigma_{XY} \\ \sigma_{XY} & \sigma_{YY} \end{bmatrix}.$$

It follows, therefore, a two-dimensional uniform distribution over the region enclosed by the ellipse

$$\frac{1}{1-\rho^2}\left[\frac{(x'-\mu_X)^2}{\sigma_X^2} - \frac{2\rho(x'-\mu_X)(y'-\mu_Y)}{\sigma_X\sigma_Y} + \frac{(y'-\mu_Y)^2}{\sigma_Y^2}\right] = 4 \qquad (10.5.1)$$

has the same means, same variances and same covariance as the original distribution. The ellipse (10.5.1) is called the ellipse of concentration of the distribution of (X, Y). When two bivariate distributions with the same means are such that the concentration ellipse of one lies wholly within the concentration ellipse of the other, then the former distribution may be said to have a greater concentration than the later.

Consider now the case of p variables. Let the random vector $\mathbf{X} = (X_1, \cdots, X_p)'$ have a p-variate joint distribution with mean vector $\boldsymbol{\mu} = (\mu_1, \cdots, \mu_p)'$ and dispersion matrix $\boldsymbol{\Sigma} = ((\sigma_{ij}))$, which is supposed to be positive definite. We want to find a p-dimensional ellipsoid $(\mathbf{Z}-\boldsymbol{\mu})'\boldsymbol{A}(\mathbf{Z}-\boldsymbol{\mu}) \leq k^2$ where $\boldsymbol{A} = ((a_{ij}))$ is such that a random vector $(Z_1, \cdots, Z_p)' = \boldsymbol{Z}$ which is uniformly distributed over this ellipsoid has the same mean $\boldsymbol{\mu}$ and same disperson matrix $\boldsymbol{\Sigma}$.
It can be shown that

$$E(Z_i) = \mu_i, \ V(Z_i) = \frac{k^2}{p+2} \cdot \frac{A_{ii}}{|\mathbf{A}|} \ \ (i = 1, \cdots, p)$$

$$\text{Cov}(Z_i, Z_j) = \frac{k^2}{p+2} \cdot \frac{A_{ij}}{|\mathbf{A}|} \ \ (i \neq j = 1, \cdots, p)$$

where A_{ij} is the cofactor of a_{ij} in $|\boldsymbol{A}|$. For the choice

$$k^2 = p + 2, \ \ \frac{A_{ij}}{|\mathbf{A}|} = \sigma_{ij}, \text{ i.e., } \mathbf{A} = \boldsymbol{\Sigma}^{-1}$$

the mean vector and disperson matrix of $(Z_1, \cdots, Z_p)'$ coincide with the mean vector and dispersion matrix of $(X_1, \cdots, X_p)'$. Hence, a uniform distribution over the domain enclosed by the ellipsoid

$$(\mathbf{x}-\boldsymbol{\mu})'\boldsymbol{\Sigma}^{-1}(\mathbf{x}-\boldsymbol{\mu}) = p + 2 \qquad (10.5.2)$$

has the required property of having the same means, variances and covariances as the original distribution. The ellipsoid (10.5.2) is called the ellipsoid of concentration for the given distribution. When two p-variate distributions having the same means are such that the concentration ellipsoid of one of them is enclosed wholly within that of the other then the former distribution may be considered as having more concentration that the later.

10.6 Conditional Distributions

We have already considered in details in sections 7.2 and 7.5 the notion of conditional distribution and expectation of a conditional distribution. However, for completeness, we recall these definitions here.

We are interested in a set of variables some of which are independent and the rest dependent on the set of independent variables. Our aim is to study the conditional distribution of dependent variables when the independent variables are kept fixed. This means that the independent variables are given specified values in the discrete case and are allowed to vary in small infinitesimal intervals in the continuous case.

Consider p variables $X_1, \cdots, X_p$ of which $X_{q+1}, \cdots, X_p$ are independent and $X_1, \cdots, X_q$ depend on them ($q < p$). Let $f(x_1, \cdots, x_p)$ be the joint p.d.f. of $(X_1, \cdots, X_p)$. The conditional probability distribution function

$$P\{X_1 \le x_1, \cdots, X_q \le x_q | x_{q+i} \le X_{q+i} \le x_{q+i} + \delta x_{q+i},\ i = 1, \cdots, p-q\}$$

$$= \frac{\int_{-\infty}^{x_1} \cdots \int_{-\infty}^{x_q} f(x_1, \cdots, x_p) dx_1, \cdots, dx_q (dx_{q+1} \cdots dx_p)}{\int_{-\infty}^{\infty} \cdots \int_{-\infty}^{\infty} f(x_1, \cdots, x_q x_{q+1}, \cdots, x_p) dx_1, \cdots, dx_q (dx_{q+1}, \cdots, dx_p)}$$

$$= \frac{\int_{-\infty}^{x_1} \cdots \int_{-\infty}^{x_q} f(x_1, \cdots, x_p) dx_1, \cdots, dx_q}{f_{q+1, \cdots, p}(x_{q+1}, \cdots, x_p)}$$

$$= \int_{-\infty}^{x_1} \cdots \int_{-\infty}^{x_q} f(x_1, \cdots, x_q | x_{q+1}, \cdots, x_p) dx_1, \cdots, dx_q \text{ (say)} \tag{10.6.1}$$

where $f(x_1, \cdots, x_q | x_{q+1}, \cdots, x_p)$ is the conditional density function of $X_1, \cdots, X_q$ given $X_{q+1}, \cdots, X_p$.

Consider now a function $g(x_1, \cdots, x_q)$ of $X_1, \cdots, X_q$. Then

$$E\{g(X_1, \cdots, X_q) | X_{q+1}, \cdots, X_p\}$$

$$= \frac{\int_{-\infty}^{\infty} \cdots \int_{-\infty}^{\infty} g(x_1, \cdots, x_q) f(x_1, \cdots, x_p) dx_1, \cdots, dx_q}{f_{q+1, \cdots, p}(x_{q+1}, \cdots, x_p)}$$

$$= \int_{-\infty}^{\infty} \int_{-\infty}^{\infty} g(x_1, \cdots, x_q) f(x_1, \cdots, x_q | x_{q+1}, \cdots, x_p) dx_1, \cdots, dx_q. \tag{10.6.2}$$

10.7 Multiple Regression

Suppose that X_1 is the dependent variable and $X_2, \cdots, X_p$ are all independent variables. To study the dependence of X_1 on $X_2, \cdots, X_p$ one should consider the conditional distribution of X_1 on $X_2, \cdots, X_p$. However, in case, the conditional mean exists, a simple (though often inadequate) method of investigation is to study the conditional expectation of the distribution of X_1 on $X_2, \cdots, X_p$.

The conditional expectation of X_1 given $X_2, \cdots, X_p$ is

$$E\{X_1 | x_i \leq X_i \leq x_i + dx_i;\ i = 2, \cdots, p\} = \frac{\int_{-\infty}^{\infty} x_1 f(x_1, \cdots, x_p) dx_1}{\int_{-\infty}^{\infty} f(x_1, \cdots, x_p) dx_1}$$
$$= m_1(x_2, \cdots, x_p) \text{ (say)}.$$

The locus of the point $(m_1, x_2, \cdots, x_p)$ for all possible values of $(x_2, \cdots, x_p)$ is the regression surface for X_1 on $X_2, \cdots, X_p$ and has the equation

$$x_1 = m_1(x_2, \cdots, x_p). \tag{10.7.1}$$

The equation (10.7.1) is the equation of true multiple regression of X_1 on $X_2, \cdots, X_p$. The quantity x_1 on the left-hand side of (10.7.1) is the predicted value of X_1 when $X_i = x_i$ $(i = 2, \cdots, p)$. Consider all such functions $g(X_2, \cdots, X_p)$ which may claim to predict the value of X_1 on given $X_2, \cdots, X_p$. A measure of accuracy of prediction is given by

$$E\{X_1 - g(X_2, \cdots, X_p)\}^2. \tag{10.7.2}$$

Now,

$$\begin{aligned} & E\{X_1 - g(X_2, \cdots, X_p)\}^2 \\ & = E\{X_1 - m_1(X_2, \cdots, X_p) + m_1(X_2, \cdots, X_p) - g(X_2, \cdots, X_p)\}^2 \\ & = E\{X_1 - m_1(X_2, \cdots, X_p)\}^2 + E\{m_1(X_2, \cdots, X_p) - g(X_2, \cdots, X_p)\}^2 \end{aligned}$$

the product term vanishes since

$$\begin{aligned} & E\left[\{X_1 - m_1(X_2, \cdots, X_p)\}\{m_1(X_2, \cdots, X_p) - g(X_2, \cdots, X_p)\}\right] \\ & = E_{X_2, \cdots, X_p}\left\{E_{X_1|X_2, \cdots, X_p}(X_1 - m_1)(m_1 - g)\right\} \\ & = E_{X_2, \cdots, X_p}\left\{(m_1 - g)E_{X_1|X_2, \cdots, X_p}(X_1 - m_1)\right\} \\ & = E_{X_2, \cdots, X_p}\{(m_1 - g)(m_1 - m_1)\} = 0. \end{aligned}$$

Hence,

$$E\{X_1 - g(X_2, \cdots, X_p)\}^2 \geq E\{X_1 - m_1(X_2, \cdots, X_p)\}^2 \tag{10.7.3}$$

equality holds iff the funciton g is identical with m_1. Of all the functions, therefore, that may claim to predict X_1 given $(X_2, \cdots, X_p)$ the function $m_1(X_2, \cdots, X_p)$ is the best in the sense that the expectation of sum of squares of deviations from the true value is minimum.

Instead of studying the true regression of X_1 on $X_2, \cdots, X_p$ which often would be very complicated we shall consider linear regression equation of X_1 on $X_2, \cdots, X_p$. We shall, therefore, consider a linear function

$$\alpha + \beta_2 X_2 + \cdots + \beta_p X_p$$

which gives the best prediction of X_1 in the sense that

$$\psi = E\{X_1 - \alpha - \beta_2 X_2 - \cdots - \beta_p X_p\}^2 \tag{10.7.4}$$

is minimum where the expectation is taken with respect to joint distribution of $(X_1, \cdots, X_p)$. If the true regression is linear, the minimisation of (10.7.4) gives the true regression, otherwise it gives the best fit linear regression of X_1 on $X_2, \cdots, X_p$, because minimising (10.7.4) is equivalent to minimising

$$E\{m_1(X_2, \cdots, X_p) - \alpha - \beta_2 X_2 - \cdots - \beta_p X_p\}^2. \tag{10.7.5}$$

We shall, henceforth, assume

$$E(X_1 | x_2, \cdots, x_p) = \alpha + \beta_2 x_2 + \cdots + \beta_p x_p \tag{10.7.6}$$

so that the true regression of X_1 on $X_2, \cdots, X_p$ is linear. The quantity α is the conditional mean of X_1 when $x_2 = \cdots = x_p = 0$. On the other hand β_j is the amount by which the conditional mean increases for a unit increment in the value of x_j, the other independent variables $x_2, \cdots, x_{j-1}, x_{j+1}, x_p$, being held fixed. β_j or rather β_{1j} or $\beta_{1j \cdot 23 \cdots p}$ is the *partial regression coefficient* of X_1 on X_j for fixed $X_2, \cdots, X_{j-1}, X_{j+1}, \cdots, X_p$. We shall assume throughout that $\sigma_{ij} = \text{Cov}(X_i, X_j)$ exists for all $i \neq j = 1, \cdots, p$. Also let $\mu_i = E(X_i)$, $i = 1, \cdots, p$.

10.7.1 Determination of Coefficients

We shall now determine the coefficients $\alpha, \beta_2, \cdots, \beta_p$ in (10.7.5). Assume for simplicity that the joint distribution of $(X_1, \cdots, X_p)$ is absolutely continuous. Hence,

$$\int_{-\infty}^{\infty} x_1 f(x_1 | x_2, \cdots, x_p) dx_1 = \alpha + \beta_2 x_2 + \cdots + \beta_p x_p. \tag{10.7.7}$$

Multiplying both sides of (10.7.7) by the joint probability density function of $X_2, \cdots, X_p$, say, $g(x_2, \cdots, x_p)$ and integrating over $x_2, \cdots, x_p$, we have

$$E(X_1) = \alpha + \beta_2 E(X_2) + \beta_3 E(X_3) + \cdots + \beta_p E(X_p),$$

i.e.,

$$\mu_1 = \alpha + \beta_2 \mu_2 + \beta_3 \mu_3 + \cdots + \beta_p \mu_p. \tag{10.7.8}$$

Similarly multiplying both sides of (10.7.7) by $x_j g(x_2, \cdots, x_p)$ and integrating over $x_2, \cdots, x_p$ $(j = 2, \cdots, p)$ one gets the $(p-1)$ equations

$$E(X_1X_2) = \alpha\mu_2 + \beta_2 E(X_2^2) + \beta_3 E(X_2X_3) + \cdots + \beta_p E(X_2X_p) \quad (10.7.9)$$

$$E(X_1X_3) = \alpha\mu_3 + \beta_2 E(X_2X_3) + \beta_3 E(X_3^2) + \cdots + \beta_p E(X_3X_p) \quad (10.7.10)$$

$$\cdots\cdots\cdots\cdots\cdots\cdots\cdots\cdots\cdots\cdots\cdots\cdots$$

$$E(X_1X_p) = \alpha\mu_p + \beta_2 E(X_2X_p) + \beta_3 E(X_3X_p) + \cdots + \beta_p E(X_p^2). \quad (10.7.11)$$

Substituting the value of α from (10.7.8) in (10.7.9), $\cdots$ (10.7.11), respectively, and on simplifications one gets the $(p-1)$ equations

$$\begin{bmatrix} \sigma_{22} & \sigma_{23} & \cdots & \sigma_{2p} \\ \sigma_{32} & \sigma_{33} & \cdots & \sigma_{3p} \\ \cdots & \cdots & \cdots & \cdots \\ \sigma_{p2} & \sigma_{p3} & \cdots & \sigma_{pp} \end{bmatrix} \begin{bmatrix} \beta_2 \\ \beta_3 \\ \vdots \\ \beta_p \end{bmatrix} = \begin{bmatrix} \sigma_{21} \\ \sigma_{31} \\ \vdots \\ \sigma_{p1} \end{bmatrix}. \quad (10.7.12)$$

We shall first show that the equation (10.7.12) are always solvable. For this we show that

$$\text{rank} \begin{bmatrix} \sigma_{22} & \cdots & \sigma_{2p} \\ \cdots & \cdots & \cdots \\ \sigma_{p2} & \cdots & \sigma_{pp} \end{bmatrix} = \text{rank} \begin{bmatrix} \sigma_{22} & \cdots & \sigma_{2p} & \sigma_{21} \\ \cdots & \cdots & \cdots & \vdots \\ \sigma_{p2} & \cdots & \sigma_{pp} & \sigma_{p1} \end{bmatrix}. \quad (10.7.13)$$

Consider the marginal distribution of $(X_2, \cdots, X_p)$ and suppose that its rank is $r(\leq p-1)$. Then we can write $X_2, \cdots, X_p$ as a linear function of r uncorrelated variables, say, $Y_1, \cdots, Y_r$, with probability one. Thus,

$$\begin{bmatrix} X_2 - \mu_2 \\ \vdots \\ X_p - \mu_p \end{bmatrix} = \boldsymbol{C} \begin{bmatrix} Y_1 \\ \vdots \\ Y_r \end{bmatrix} \text{ with probability one.} \quad (10.7.14)$$

Hence

$$\begin{bmatrix} \sigma_{22} & \cdots & \sigma_{2p} \\ \cdots & \cdots & \cdots \\ \sigma_{p2} & \cdots & \sigma_{pp} \end{bmatrix} = \boldsymbol{C}D(\mathbf{Y})\boldsymbol{C}'$$

where $D(\mathbf{Y})$ is the dispersion matrix of $\boldsymbol{Y} = (Y_1, \cdots, Y_r)'$. Now

$$\begin{bmatrix} \sigma_{21} \\ \vdots \\ \sigma_{p1} \end{bmatrix} = E \begin{bmatrix} X_2 - m_2 \\ \vdots \\ X_p - m_p \end{bmatrix} (X_1 - m_1) = E\{\boldsymbol{C}\mathbf{Y}(X_1 - m_1)\}$$

$$= \boldsymbol{C} \begin{bmatrix} \text{Cov}(Y_1, X_1) \\ \vdots \\ \text{Cov}(Y_r, X_1) \end{bmatrix} = \boldsymbol{C}\gamma \text{ (say).} \quad (10.7.15)$$

Ience, from (10.7.13) it is required to prove

$$\text{rank}\, \boldsymbol{CC'} = \text{rank}\, (\boldsymbol{CC'}, \boldsymbol{C\gamma}). \tag{10.7.16}$$

Iow,

$$\text{rank}\, (\boldsymbol{CC'},\ \ \boldsymbol{C\gamma}) \geq \text{rank}\, (\boldsymbol{CC'}) \tag{10.7.17}$$

ince $\boldsymbol{CC'}$ is a submatrix of $(\boldsymbol{CC'}, \boldsymbol{C\gamma})$. Again,

$$\text{rank}\, (\boldsymbol{CC'}, \boldsymbol{C\gamma}) = \text{rank}\, \{\boldsymbol{C}(\boldsymbol{C'}, \boldsymbol{\gamma})\} \leq \text{rank}\, \boldsymbol{C} = \text{rank}\, (\boldsymbol{CC'}). \tag{10.7.18}$$

Combining (10.7.17) and (10.7.18), (10.7.16) follows.

We assume that $\boldsymbol{\Sigma} = ((\sigma_{ij}; i, j = 1, \cdots, p))$ is non-singular. Hence the joint istribution of $X_1, \cdots, X_p$ is non-singular. Let ξ_1 be the predicted value of X_1 on the asis of the regression equation of X_1 on $X_2, \cdots, X_p$. Therefore, in the present case

$$E(X_1 | X_2, \cdots, X_p) = \xi_1.$$

Ience from (10.7.6)

$$\xi_1 = \alpha + \beta_2 x_2 + \cdots + \beta_p x_p. \tag{10.7.19}$$

ubtracting (10.7.8) from (10.7.19), we have

$$\xi_1 - \mu_1 = \beta_2(x_2 - \mu_2) + \cdots + \beta_p(x_p - \mu_p). \tag{10.7.20}$$

Ience from (10.7.12) we have the normal equations

$$\begin{aligned}
&\sigma_{22}\beta_2 + \sigma_{23}\beta_3 + \cdots + \sigma_{2p}\beta_p - \sigma_{21} = 0\\
&\sigma_{32}\beta_2 + \sigma_{33}\beta_3 + \cdots + \sigma_{3p}\beta_p - \sigma_{31} = 0\\
&\cdots\cdots\cdots\cdots\cdots\cdots\cdots\cdots\cdots\cdots\cdots\\
&\sigma_{p2}\beta_2 + \sigma_{p3}\beta_3 + \cdots + \sigma_{pp}\beta_p - \sigma_{p1} = 0\\
&(x_2 - \mu_2)\beta_2 + (x_3 - \mu_3)\beta_3 + \cdots + (x_p - \mu_p)\beta_p - (\xi_1 - \mu_1) = 0,
\end{aligned} \tag{10.7.21}$$

.e.,

$$\begin{bmatrix} \sigma_{22} & \sigma_{23} & \cdots & \sigma_{2p} & \sigma_{21} \\ \sigma_{32} & \sigma_{33} & \cdots & \sigma_{3p} & \sigma_{31} \\ \cdots & \cdots & \cdots & \cdots & \cdots \\ \sigma_{p2} & \sigma_{p3} & \cdots & \sigma_{pp} & \sigma_{p1} \\ x_2 - \mu_2 & x_3 - \mu_3 & \cdots & x_p - \mu_p & \xi_1 - \mu_1 \end{bmatrix} \begin{bmatrix} \beta_2 \\ \beta_3 \\ \vdots \\ \beta_p \\ -1 \end{bmatrix} = 0. \tag{10.7.22}$$

Assuming that we are not dealing with a case in which all the variables are mutually ncorrelated (when all the β_j's will be zero) we have, therefore,

$$\begin{bmatrix} \sigma_{22} & \sigma_{23} & \cdots & \sigma_{2p} & \sigma_{21} \\ \cdots & \cdots & \cdots & \cdots & \cdots \\ \sigma_{p2} & \sigma_{p3} & \cdots & \sigma_{pp} & \sigma_{p1} \\ x_2 - \mu_2 & x_3 - \mu_3 & \cdots & x_p - \mu_p & \xi_1 - \mu_1 \end{bmatrix} = \boldsymbol{0},$$

when the corresponding determinant is zero, i.e.,

$$\begin{vmatrix} \xi_1 - \mu_1 & x_2 - \mu_2 & \cdots & x_p - \mu_p \\ \sigma_{21} & \sigma_{22} & \cdots & \sigma_{2p} \\ \cdots & \cdots & \cdots & \cdots \\ \sigma_{p1} & \sigma_{p2} & \cdots & \sigma_{pp} \end{vmatrix} = 0, \tag{10.7.23}$$

i.e.,

$$(\xi_1 - \mu_1)\Sigma_{11} + (x_2 - \mu_2)\Sigma_{12} + \cdots + (x_p - \mu_p)\Sigma_{1p} = 0 \tag{10.7.24}$$

where Σ_{ij} is the cofactor of σ_{ij} in $\mathbf{\Sigma}$. The equation (10.7.24) gives

$$\xi_1 - \mu_1 = -\frac{\Sigma_{12}}{\Sigma_{11}}(x_2 - \mu_2) - \cdots - \frac{\Sigma_{1p}}{\Sigma_{11}}(x_p - \mu_p). \tag{10.7.25}$$

Comparing with the form (10.7.19) we have

$$\beta_j = \beta_{1j \cdot 23 \cdots (j-1)(j+1)} p = -\frac{\Sigma_{1j}}{\Sigma_{11}} \tag{10.7.26}$$

$\beta_j = \beta_{1j\cdot 23\cdots(j-1)(j+1)\cdots p} = \beta_{1j}$ is the partial regression coefficient of X_1 on X_j for fixed values of $X_2, \cdots, X_{j-1}, X_{j+1}, \cdots, X_p$. Also, from (10.7.19)

$$\alpha = \mu_1 - \beta_2\mu_2 - \cdots - \beta_p\mu_p. \tag{10.7.27}$$

Hence the regression equation of X_1 on $X_2, \cdots, X_p$ is given by

$$\xi_1 = \hat{X}_1 = \mu_1 - \frac{\Sigma_{12}}{\Sigma_{11}}(x_2 - \mu_2) - \frac{\Sigma_{13}}{\Sigma_{11}}(x_3 - \mu_3) - \cdots - \frac{\Sigma_{1p}}{\Sigma_{11}}(x_p - \mu_p). \tag{10.7.28}$$

Let $\boldsymbol{R}$ be the correlation matrix

$$\boldsymbol{R} = \begin{bmatrix} 1 & \rho_{11} & \rho_{12} & \cdots & \rho_{1p} \\ \rho_{21} & 1 & \rho_{23} & \cdots & \rho_{2p} \\ \cdots & \cdots & \cdots & \cdots & \cdots \\ \rho_{p1} & \rho_{p2} & \rho_{p3} & \cdots & 1 \end{bmatrix}.$$

Since

$$\begin{aligned} \sigma_{ij} &= \rho_{ij}\sigma_i\sigma_j \\ \Sigma_{11} &= \sigma_2^2 \cdots \sigma_p^2 R_{11} \\ \Sigma_{1j} &= \sigma_1\sigma_2^2 \cdots \sigma_{j-1}^2\sigma_j\sigma_{j+1}^2 \cdots \sigma_p^2 R_{1j}, \end{aligned}$$

so that

$$\beta_j = -\frac{R_{1j}}{R_{11}}\frac{\sigma_1}{\sigma_j}$$

where R_{ij} is the cofactor of ρ_{ij} in $|\boldsymbol{R}|$. Therefore, the regression equation of X_1 on $X_2, \cdots, X_p$ is

$$\xi_1 = \hat{X}_1 = \mu_1 - \frac{R_{12}}{R_{11}} \cdot \frac{\sigma_1}{\sigma_2}(x_2 - \mu_2) - \cdots - \frac{R_{1p}}{R_{11}} \cdot \frac{\sigma_1}{\sigma_p}(x_p - \mu_p). \tag{10.7.29}$$

■ NOTE 10.7.1 *In the above derivation we have assumed that the true regression of* X_1 *on* $X_2, \cdots, X_p$ *which is* $E(X_1|X_2, \cdots, X_p)$ *is linear in* $X_2, \cdots, X_p$ *(vide 10.7.6). If the true regression is not linear but we want to find a linear function of* $X_2, \cdots, X_p$ *that gives the best prediction of* X_1 *in the sense of minimising* (10.7.4), *then also we would get the regression equation* (10.7.28) *or* (10.7.29). *To find this we have to differentiate* ψ *with respect to* $\alpha, \beta_2, \cdots, \beta_p$ *and take expectation.*

We shall denote $\xi_1 = \hat{X}_1$ as $\hat{X}_{1(23\cdots p)}$, the predicted value of X_1 by the linear multiple regression equation of X_1 on $X_2, \cdots, X_p$. Therefore,

$$X_1 - \hat{X}_{1(23\cdots p)} = X_{1\cdot 23\cdots p} \tag{10.7.30}$$

is the part of X_1 that remains unpredicted by its linear regression equation on $X_2, \cdots, X_p$ and is called residual of X_1. Clearly

$$E(\hat{X}_{1(23\cdots p)}) = \mu_1. \tag{10.7.31}$$

$$\begin{aligned} V(\hat{X}_{1(23\cdots p)}) &= V(\beta_{12}X_2 + \cdots + \beta_{1p}X_p) \text{ (vide 10.7.19)} \\ &= \sum_{ij=2}^{p}\sum \beta_{1i}\beta_{1j}\sigma_{ij} = \sum_{i=2}^{p} \beta_{1i} \sum_{j=2}^{p} \beta_{1j}\sigma_{ij} \\ &= \sum_{i=2}^{p} \beta_{1i}\sigma_{1i} \text{ (from normal equations (10.7.21),} \qquad (10.7.32) \\ &= -\sum_{2}^{p} \sigma_{1i}\frac{\Sigma_{1i}}{\Sigma_{11}} = -\left[\frac{\sigma_{12}\Sigma_{12} + \cdots + \sigma_{1p}\Sigma_{1p}}{\Sigma_{11}}\right] \\ &= \frac{\sigma_{11}\Sigma_{11} - |\Sigma|}{\Sigma_{11}} = \sigma_{11} - \frac{|\Sigma|}{\Sigma_{11}}. \qquad (10.7.33) \end{aligned}$$

Again

$$E(X_{1\cdot 23\cdots p}) = E(X_1 - \hat{X}_{1(23\cdots p)}) = 0$$
$$V(X_{1\cdot 23\cdots p}) = V(X_1) + V(\hat{X}_1) - 2\,\mathrm{Cov}(X_1, \hat{X}_1).$$

Now

$$\mathrm{Cov}(X_1, \hat{X}_1) = \mathrm{Cov}(X_1, \beta_{12}X_2 + \cdots + \beta_{1p}X_p) = \sum_{i=2}^{p} \beta_{1i}\sigma_{1i} = V(\hat{X}_1).$$

Hence,

$$V(X_{1\cdot 23\cdots p}) = \sigma^2_{1\cdot 23\cdots p} \text{ (say)} = \sigma_{11} - V(\hat{X}_1) = \frac{|\Sigma|}{\Sigma_{11}} \tag{10.7.34}$$

Therefore,

$$V(X_1) = V(\hat{X}_1) + V(X_{1\cdot 23\cdots p}).$$

Hence,

$$\mathrm{Cov}(\hat{X}_1, X_{1\cdot 23\cdots p}) = 0. \tag{10.7.35}$$

Again, for $j = 2, 3, \cdots, p$

$$\begin{aligned}\text{Cov}(X_j, X_{1\cdot 23\cdots p}) &= \text{Cov}(X_j, X_1 - \alpha - \beta_{12}X_2 \cdots - \beta_{1p}X_p)\\ &= \sigma_{1j} - \beta_{12}\sigma_{2j} \cdots - \beta_{1p}\sigma_{pj}\\ &= 0 \text{ (by normal equation 10.7.21)}\end{aligned}$$

Hence, Residual $X_{1\cdot 23\cdots p}$ is uncorrelated with those variables whose subscripts occur as secondary subscripts of the residuals.

10.8 Multiple Correlation

We may be interested in knowing the accuracy of the predicting formulae $\hat{X}_{1(23\cdots p)}$. This is measured by the absolute value of the correlation coefficient between X_1 and $\hat{X}_{1(23\cdots p)}$ and this correlation coefficient is called the multiple correlation coefficient of X_1 on $(X_2, \cdots, X_p)$ and is denoted by $\rho_{1\cdot 2\cdots p}$.

Again, recalling that in the bivariate case, the absolute value of the correlation between X and Y, $|\rho_{XY}|$ equals the simple correlation coefficient between X (the independent variable) and Y_X (the predicted value of Y by its linear regression on X), we may analogously interpret that $\rho_{1\cdot 23\cdots p}$ measures the multiple correlation of X_1 on $(X_2, \cdots, X_p)$.

$$\rho_{1\cdot 23\cdots p} = \frac{\text{Cov}(X_1, \hat{X}_{1(23\cdots p)})}{\sqrt{V(X_1)V(\hat{X}_{1(23\cdots p)})}} = \frac{V(\hat{X}_1)}{\sqrt{\sigma_{11}V(\hat{X}_1)}} \tag{10.8.1}$$

$$= \sqrt{\frac{V(\hat{X}_1)}{\sigma_{11}}} = \sqrt{1 - \frac{|\boldsymbol{\Sigma}|}{\sigma_{11}\Sigma_{11}}} \tag{10.8.2}$$

$$= \sqrt{1 - \frac{|\boldsymbol{R}|}{R_{11}}}. \tag{10.8.3}$$

From (10.8.3)

$$\rho^2_{1\cdot 23\cdots p} = 1 - \frac{|\boldsymbol{R}|}{R_{11}}. \tag{10.8.4}$$

Another interpretation of $\rho_{1\cdot 23\cdots p}$ is as follows. Accuracy (rather inaccuracy) of a prediction formulae is measured by

$$E(X_1 - X_{1(23\cdots p)})^2 = V(X_{1\cdot 23\cdots p}) = \sigma^2_{1\cdot 23\cdots p}.$$

To make this measure unit-free we divide it by σ_{11}.

Again, this is not a direct measure, since higher is its value the lower is the accuracy. A unit-free measure of accuracy of the prediction formulae is, therefore,

$$1 - \frac{\sigma^2_{1\cdot 23\cdots p}}{\sigma_{11}} = \rho^2_{1\cdot 23\cdots p}. \tag{10.8.5}$$

Again

$$|\boldsymbol{R}| = R_{11} - \sum_{ij=2}^{p}\sum \rho_{1i}\rho_{1j}R_{11\cdot 1j}$$

when $R_{11\cdot 1j}$ is the cofactor of ρ_{ij} in R_{11} which is the determinant obtained by omitting the first row and first column of $|\boldsymbol{R}|$. Hence,

$$\rho^2_{1\cdot 23\cdots p} = \frac{\sum_{ij=2}^{p}\sum \rho_{1i}\rho_{1j}R_{11\cdot ij}}{R_{11}}. \tag{10.8.6}$$

It follows from (10.8.1) that $\rho_{1\cdot 23\cdots p}$ is necessarily non-negative and its value varies between 0 and 1. $\rho_{1\cdot 23\cdots p} = 0$ means $\rho_{12} = \rho_{13} = \cdots = \rho_{1p} = 0$. The correlation $\rho_{1\cdot 23\cdots p} = 1$ means $\sigma^2_{1\cdot 23\cdots p} = 0$, i.e., the residual $X_{1\cdot 23\cdots p}$ is zero with probability one, i.e., observed value of X_1 and its predicted value $\hat{X}_{1(23\cdots p)}$ coincide with probability one and X_1 is a linear function of $X_2, \cdots, X_p$.

It may be noted that the multiple correlation coefficient may be interpreted as the maximum value of simple correlation that may exist between X_1 and a linear function of $X_2, \cdots, X_p$, say,

$$X = a + b_2X_2 + \cdots + b_pX_p \tag{10.8.7}$$

or variations in values of $b_2, \cdots, b_p$ (the value of 'a' does not affect this correlation).

$$\text{Corr.}\,(X_1, X) = \frac{\sum_{j=2}^{p} b_j\sigma_{1j}}{\sqrt{\sigma_{11}\sum_{j=2}^{p}\sum_{k=2}^{p} b_jb_k\sigma_{jk}}} \tag{10.8.8}$$

It is clear that the correlation remains unchanged if b_j is replaced by cb_j (c an arbitrary constant), $j = 2, \cdots, p$. We assume that b_j's are such that they satisfy the relation

$$\sum_{j=2}^{p}\sum_{k=2}^{p} b_jb_k\sigma_{jk} = 1. \tag{10.8.9}$$

Our problem is, therefore, to maximise (10.8.8) subject to the condition (10.8.9). The maximising equations are

$$\frac{\partial}{\partial b_k}\left[\sum_{j=2}^{p} b_j\sigma_{1j} + \lambda\sum_{j=2}^{p}\sum_{k=2}^{p} b_jb_k\sigma_{jk} - 1\right] = 0, \quad k = 2, \cdots, p.$$

This gives

$$\sigma_{1k} = -\lambda\sum_{j=2}^{p} b_j\sigma_{jk}, \quad k = 2, \cdots, p.$$

Comparing with the first $(p-1)$ normal equations in (10.7.21), it, therefore, follow that b_j must be proportional to β_j, i.e., $b_j = k\beta_j (j = 2, \cdots, p)$ where k must be positiv for the correlation between X_1 and X to be a maximum. Hence,

$$\max \text{Corr}.(X_1, X) = \text{Corr}.(X_1, X_{1\cdot 23\cdots p}) = \rho_{1\cdot 23\cdots p}.$$

10.9 Partial Correlation

One may be interested in studying the degree to which two variables, say, X_1 and X_2 ar related, when the influence of the other variables, say, $X_3, \cdots, X_p$ are eliminated fron both of them. This is done by measuring the partial correlation between X_1 and X eliminating the effects of the variables $X_3, \cdots, X_p$. Thus, partial correlation between X and X_2 eliminating the effects of $X_3, \cdots, X_p$ is

$$\text{Corr}(X_{1\cdot 34\cdots p}, X_{2\cdot 34\cdots p}) = \frac{\text{Cov}(X_{1\cdot 34\cdots p}, X_{2\cdot 34\cdots p})}{\sqrt{V(X_{1\cdot 34\cdots p})V(X_{2\cdot 34\cdots p})}} \tag{10.9.1}$$

where $X_{1\cdot 34\cdots p} = X_1 - \hat{X}_{1(34\cdots p)}$, i.e., residual of X_1 after its predicted value obtained b its linear regression on $X_3, \cdots, X_p$ has been eliminated. Now

$$\begin{aligned}
&\text{Cov}(X_{1\cdot 34\cdots p}, X_{2\cdot 34\cdots p})\\
&= \text{Cov}(X_{1\cdot 34\cdots p}, X_2 - \text{ a linear function of } X_3, \cdots, X_p)\\
&= \text{Cov}(X_{1\cdot 24\cdots p}, X_2) \text{ (Since } \text{Cov}(X_{1\cdot 34\cdots p}, X_j) = 0,\ j = 3, \cdots, p)\\
&= \text{Cov}\{X_1 - \alpha - \beta_{13}X_3 - \cdots - \beta_{1p}X_p, X_2\}\\
&= \sigma_{21} - \beta_{13}\sigma_{23} - \cdots - \beta_{1p}\sigma_{2p}
\end{aligned} \tag{10.9.2}$$

where

$$\beta_{1j} = \beta_{1j\cdot 34\cdots j-1\, j+1\cdots p} = -\frac{\Sigma_{22\cdot 1j}}{\Sigma_{22\cdot 11}} \tag{10.9.3}$$

where $\Sigma_{22\cdot 1j}$ is the cofactor of σ_{1j} in Σ_{22} which is the determinant obtained by omittin the second row and second column of $\mathbf{\Sigma}$. Now consider

$$|\mathbf{\Sigma}^*| = \begin{vmatrix} \sigma_{21} & \sigma_{22} & \cdots & \sigma_{2p} \\ \sigma_{11} & \sigma_{12} & \cdots & \sigma_{1p} \\ \sigma_{31} & \sigma_{32} & \cdots & \sigma_{3p} \\ \cdots & \cdots & \cdots & \cdots \\ \sigma_{p1} & \sigma_{p2} & \cdots & \sigma_{pp} \end{vmatrix} = -|\mathbf{\Sigma}|.$$

Cofactor σ_{12} in $|\mathbf{\Sigma}|$ = cofactor of σ_{12} in $|\mathbf{\Sigma}^*|$, i.e., $\Sigma_{12} = -\Sigma_{12}^*$ or $\Sigma_{12}^* = -\Sigma_{12}$ where

$$\Sigma_{12}^* = \begin{bmatrix} \sigma_{21} & \sigma_{23} & \cdots & \sigma_{2p} \\ \sigma_{31} & \sigma_{33} & \cdots & \sigma_{3p} \\ \cdots & \cdots & \cdots & \cdots \\ \sigma_{p1} & \sigma_{p3} & \cdots & \sigma_{pp} \end{bmatrix};$$

Let us denote cofactor of σ_{ij} in Σ_{12}^* as $\Sigma_{12\cdot ij}^*$. It is seen that cofactor of σ_{1p} in $\Sigma_{22} = \Sigma_{22\cdot 1p}$ = cofactor of σ_{2p} in Σ_{12}^*, i.e., $\Sigma_{22\cdot 1p} = \Sigma_{12\cdot 2p}^*$.

Hence,

$$\Sigma_{22\cdot 11} = \Sigma_{12\cdot 21}^*; \quad \Sigma_{22\cdot 13} = \Sigma_{12\cdot 23}^*.$$

Therefore,

$$\begin{aligned}\operatorname{Cov}(X_{1\cdot 3\cdots p}, X_{2\cdot 3\cdots p}) &= \frac{\sigma_{21}\Sigma_{22\cdot 11} + \sigma_{23}\Sigma_{22\cdot 13} + \sigma_{24}\Sigma_{22\cdot 14} + \cdots + \sigma_{2p}\Sigma_{22\cdot 1p}}{\Sigma_{22\cdot 11}} \\ &= \frac{\sigma_{21}\Sigma_{12\cdot 21}^* + \sigma_{23}\Sigma_{12\cdot 23}^* + \sigma_{24}\Sigma_{21\cdot 24}^* + \cdots + \sigma_{2p}\Sigma_{12\cdot 2p}^*}{\Sigma_{22\cdot 11}} \\ &= \frac{\Sigma_{12}^*}{\Sigma_{22\cdot 11}} = -\frac{\Sigma_{12}}{\Sigma_{22\cdot 11}} = -\frac{R_{12}}{R_{22\cdot 11}}\sigma_1\sigma_2 \end{aligned} \tag{10.9.4}$$

when $R_{22\cdot 11}$ is the cofactor of ρ_{11} in the determinant obtained by deleting the second row and second column of R.

Similarly,

$$V(X_{1\cdot 34\cdots p}) = \frac{\Sigma_{22}}{\Sigma_{22\cdot 11}} = \frac{R_{22}}{R_{22\cdot 11}}\sigma_1^2 \tag{10.9.5}$$

$$V(X_{2\cdot 34\cdots p}) = \frac{\Sigma_{11}}{\Sigma_{11\cdot 22}} = \frac{R_{11}}{R_{11\cdot 22}}\sigma_2^2 \tag{10.9.6}$$

when $R_{11\cdot 22}$ is defined similarly as $R_{22\cdot 11}$.

Now $\Sigma_{11\cdot 22} = \Sigma_{22\cdot 11}$. Also $R_{11\cdot 22} = R_{22\cdot 11}$. Therefore,

$$\begin{aligned}\operatorname{Corr.}(X_{1\cdot 34\cdots p}, X_{2\cdot 34\cdots p}) &= \rho_{12\cdot 34\cdots p} \text{ (say)} \\ &= -\frac{\Sigma_{12}/\Sigma_{22\cdot 11}}{\sqrt{\dfrac{\Sigma_{11}}{\Sigma_{11\cdot 22}}\cdot\dfrac{\Sigma_{22}}{\Sigma_{22\cdot 11}}}} = -\frac{\Sigma_{12}}{\sqrt{\Sigma_{11}\Sigma_{22}}} = -\frac{R_{12}}{\sqrt{R_{11}R_{22}}}. \end{aligned} \tag{10.9.7}$$

We now enquire into the relations between partial regression coefficient $\beta_{12\cdot 34\cdots p}$ (or $\beta_{21\cdot 34\cdots p}$) and $\rho_{12\cdot 34\cdots p}$.

We have

$$\beta_{12\cdot 34\cdots p} = -\frac{R_{12}}{R_{11}}\cdot\frac{\sigma_1}{\sigma_2} = -\frac{R_{12}}{\sqrt{R_{11}R_{22}}}\cdot\frac{\sqrt{\dfrac{R_{22}}{R_{11\cdot 22}}\sigma^2}}{\sqrt{\dfrac{R_{11}}{R_{22\cdot 11}}\sigma_2^2}} = \rho_{12\cdot 34\cdots p}\frac{\sigma_{1\cdot 34\cdots p}}{\sigma_{2\cdot 34\cdots p}}. \tag{10.9.8}$$

Similarly, the regression coefficient $\beta_{21\cdot 34\cdots p}$ of X_2 on X_1, given $X_3, \ldots, X_p$ is

$$\beta_{21\cdot 34\cdots p} = \rho_{12\cdot 34\cdots p}\frac{\sigma_{2\cdot 34\cdots p}}{\sigma_{1\cdot 34\cdots p}}. \tag{10.9.9}$$

Clearly,

$$\beta_{12\cdot 34\cdots p}\beta_{21\cdot 34\cdots p} = \rho_{12\cdot 34\cdots p}^2. \tag{10.9.10}$$

10.10 A Few Further Results

(a) We have

$$
\begin{aligned}
&\operatorname{Cov}(X_{1\cdot 234\cdots p}, X_{2\cdot 34\cdots p}) \\
&= \operatorname{Cov}[\{X_1 - \mu_1 - \beta_{12\cdot 34\cdots p}(X_2 - \mu_2) - \beta_{13\cdot 24\cdots p}(X_3 - \mu_3) - \cdots \\
&= -\beta_{1p\cdot 23\cdots p-1}(X_p - \mu_p)\}, X_{2\cdot 34\cdots p}] \\
&= \operatorname{Cov}\{X_1, X_{2\cdot 34\cdots p}\} - \beta_{12\cdot 34\cdots p}\operatorname{Cov}(X_2, X_{2\cdot 34\cdots p}) \\
&= -\beta_{13\cdot 24\cdots p}\operatorname{Cov}(X_3, X_{2\cdot 34\cdots p}) - \cdots - \beta_{1p\cdot 23\cdots p-1}\operatorname{Cov}(X_p, X_{2\cdot 34\cdots p}) \\
&= \operatorname{Cov}\{X_{1\cdot 34\cdots p} + \text{ linear function of } X_3, \cdots, X_p, X_{2\cdot 34\cdots p}\} \\
&\qquad -\beta_{12\cdot 34\cdots p}\operatorname{Cov}\{X_{2\cdot 34\cdots p} + \text{linear function of } X_3, \cdots, X_p, X_{2\cdot 34\cdots p}\} \\
&= \rho_{12\cdot 34\cdots p}\sigma_{1\cdot 34\cdots p}\sigma_{2\cdot 34\cdots p} - \beta_{12\cdot 34\cdots p}\sigma^2_{2\cdot 34\cdots p} \\
&= \rho_{12\cdot 34\cdots p}\sigma_{1\cdot 34\cdots p}\sigma_{2\cdot 34\cdots p} - \rho_{12\cdot 34\cdots p}\left(\frac{\sigma_{1\cdot 34\cdots p}}{\sigma_{2\cdot 34\cdots p}}\right)\cdot\sigma^2_{2\cdot 34\cdots p} = 0.
\end{aligned}
$$

Thus $X_{1\cdot 34\cdots p}$ and $X_{2\cdot 34\cdots p}$ are uncorrelated.

(b) Consider

$$
\begin{aligned}
V(X_{1\cdot 23\cdots p}) &= \sigma^2_{1\cdot 23\cdots p} = \operatorname{Cov}(X_{1\cdot 23\cdots p}, X_{1\cdot 23\cdots p}) \\
&= \operatorname{Cov}\{(X_1 - \mu_1) + \text{ a linear function of } X_2, X_3, \cdots, p, X_{1\cdot 23\cdots p}\} \\
&= \operatorname{Cov}(X_1, X_{1\cdot 23\cdots p}) \\
&= \operatorname{Cov}\{X_{1\cdot 23\cdots p-1} + \text{ a linear function of } X_2, \cdots, X_{p-1}, X_{1\cdot 23\cdots p}\} \\
&= \operatorname{Cov}\{X_{1\cdot 23\cdots p-1}, X_{1\cdot 23\cdots p}\} \\
&= \operatorname{Cov}[X_{1\cdot 23\cdots p-1}, \{(X_1 - \mu_1) - \beta_{1p\cdot 23\cdots p-1}(X_p - \mu_p) \\
&\qquad - \text{a linear combination of } X_2, \cdots, X_{p-1}\}] \\
&= \operatorname{Cov}(X_{1\cdot 23\cdots p-1}, X_1) - \beta_{1p\cdot 23\cdots p-1}\operatorname{Cov}(X_{1\cdot 23\cdots p-1}, X_p) \\
&= \operatorname{Cov}(X_{1\cdot 23\cdots p-1}, X_{1\cdot 23\cdots p-1}) - \beta_{1p\cdot 23\cdots p-1}\operatorname{Cov}(X_{1\cdot 23\cdots p-1}, X_{p\cdot 23\cdots p-1}) \\
&= \sigma^2_{1\cdot 23\cdots p-1} - \beta_{1p\cdot 23\cdots p-1}\rho_{1p\cdot 23\cdots p-1}\sigma_{1\cdot 23\cdots p-1}\sigma_{p\cdot 23\cdots p-1} \\
&= \sigma^2_{1\cdot 23\cdots p-1} - \rho_{1p\cdot 23\cdots p-1}\frac{\sigma_{1\cdot 23\cdots p-1}}{\sigma_{p\cdot 23\cdots p-1}}\rho_{1p\cdot 23\cdots p-1}\sigma_{1\cdot 23\cdots p-1}\sigma_{p\cdot 23\cdots p-1} \\
&= \sigma^2_{1\cdot 23\cdots p-1}(1 - \rho^2_{1p\cdot 23\cdots p-1}). \qquad (10.10.1)
\end{aligned}
$$

Similarly

$$\sigma^2_{1\cdot 23\cdots p-1} = \sigma^2_{1\cdot 23\cdots p-2}(1 - \rho^2_{1p-1\cdot 23\cdots p-2}).$$

Therefore,

$$
\begin{aligned}
\sigma^2_{1\cdot 23\cdots p} &= \sigma^2_{1\cdot 23\cdots p-2}(1 - \rho^2_{1p-1\cdot 23\cdots p-2})(1 - \rho^2_{1p\cdot 23\cdot p-1}) \\
&= \cdots\cdots\cdots\cdots\cdots\cdots\cdots\cdots \\
&= \sigma^2_1(1 - \rho^2_{12})(1 - \rho^2_{13\cdot 2})(1 - \rho^2_{14\cdot 23})\cdots(1 - \rho^2_{1p\cdot 23\cdots p-1}). \qquad (10.10.2)
\end{aligned}
$$

Again

$$\rho^2_{1\cdot 23\cdots p} = 1 - \frac{\sigma^2_{1\cdot 23\cdots p}}{\sigma^2_1}$$

so that

$$\sigma^2_{1\cdot 23\cdots p} = \sigma^2_1(1 - \rho^2_{1\cdot 23\cdots p}). \tag{10.10.3}$$

Hence, from (10.10.2) and (10.10.3)

$$1 - \rho^2_{1\cdot 23\cdots p} = (1 - \rho^2_{12})(1 - \rho^2_{13\cdot 2}) \cdots (1 - \rho^2_{1p\cdot 23\cdots p-1}). \tag{10.10.4}$$

In this way such $(p-1)!$ relations may be proved.

If we introduce a $(p+1)$th variable X_{p+1}, an additional factor will appear.

$$1 - \rho^2_{1\cdot 23\cdots p+1} = (1 - \rho^2_{1\cdot 23\cdots p})(1 - \rho^2_{1p+1\cdot 23\cdots p}).$$

Since,

$$\begin{aligned} 1 - \rho^2_{1\cdot 23\cdots p+1} &\leq 1 - \rho^2_{1\cdot 23\cdots p}, \\ \rho^2_{1\cdot 23\cdots p+1} &\geq \rho^2_{1\cdot 23\cdots p}. \end{aligned} \tag{10.10.5}$$

Hence by introducing an additional predicting variable we can only improve the efficacy of the predicting formulae. If any total or partial correlation coefficient is unity, the multiple correlation coefficient will also be so.

(c) Expression for the partial correlation coefficient of a higher order in terms of those of lower order.

We have,

$$\begin{aligned}
&\text{Cov}\{X_{1\cdot 34\cdots p}, X_{2\cdot 34\cdots p}\} \\
&= \text{Cov}\{X_{1\cdot 34\cdots p}, (X_2 - \alpha' - \beta_{23\cdot 4\cdots p}X_3 - \beta_{24\cdot 3\cdots p}X_4 \cdots - \beta_{2p\cdot 34\cdots p-1}X_p)\} \\
&= \text{Cov}(X_{1\cdot 34\cdots p}, X_2) \\
&= \text{Cov}(X_{1\cdot 34\cdots p}, X_{2\cdot 34\cdots p-1} + \text{ a linear combination of } X_3, X_4, \cdots, X_{p-1}\} \\
&= \text{Cov}\{X_1 - \mu_1 - \beta_{13\cdot 4\cdots p}(X_3 - \mu_3) - \beta_{14\cdot 3\cdots p}(X_4 - \mu_4) - \cdots \\
&\qquad - \beta_{1p\cdot 34\cdots p-1}(X_p - \mu_p), X_{2\cdot 34\cdots p-1}\} \\
&= \text{Cov}\{X_1, X_{2\cdot 34\cdots p-1}\} - \beta_{1p\cdot 34\cdots p-1}\,\text{Cov}\{X_p, X_{2\cdot 34\cdots p-1}\} \\
&= \text{Cov}\{X_{1\cdot 34\cdots p-1}, X_{2\cdot 34\cdots p-1}\} - \beta_{1p\cdot 34\cdots p-1}\,\text{Cov}\{X_{p\cdot 34\cdots p-1}, X_{2\cdot 34\cdots p-1}\} \\
&= \rho_{12\cdot 34\cdots p-1}\sigma_{1\cdot 34\cdots p-1}\sigma_{2\cdot 34\cdots p-1} - \rho_{1p\cdot 34\cdots p-1}\frac{\sigma_{1\cdot 34\cdots p-1}}{\sigma_{p\cdot 34\cdots p-1}} \\
&\qquad \rho_{2p\cdot 34\cdots p-1}\sigma_{2\cdot 34\cdots p-1}\sigma_{p\cdot 34\cdots p-1} \\
&= \rho_{12\cdot 34\cdots p-1}\sigma_{1\cdot 34\cdots p-1}\sigma_{2\cdot 34\cdots p-1} - \rho_{1p\cdot 34\cdots p-1} \\
&\qquad \rho_{2p\cdot 34\cdots p-1}\sigma_{1\cdot 34\cdots p-1}\sigma_{2\cdot 34\cdots p-1}.
\end{aligned} \tag{10.10.6}$$

Now

$$\sigma^2_{1\cdot 34\cdots p} = \sigma^2_{1\cdot 34\cdots p-1}(1 - \rho^2_{1p\cdot 34\cdots p-1}) \text{ (by 10.10.1)}$$

Similarly

$$\sigma^2_{2\cdot 34\cdots p} = \sigma^2_{2\cdot 34\cdots p-1}(1-\rho^2_{2p\cdot 34\cdots p-1}).$$

Hence, from (10.10.6)

$$\rho_{12\cdot 34\cdots p}\sigma_{1\cdot 34\cdots p-1}\sigma_{2\cdot 34\cdots p-1}\sqrt{(1-\rho^2_{1p\cdot 34\cdots p-1})(1-\rho^2_{2p\cdot 34\cdots p-1})}$$
$$= (\rho_{12\cdot 34\cdots p-1} - \rho_{1p\cdot 34\cdots p-1}\rho_{2p\cdot 34\cdots p-1})\sigma_{1\cdot 34\cdots p-1}\sigma_{2\cdot 34\cdots p-1}.$$

Hence,

$$\rho_{12\cdot 34\cdots p} = \frac{\rho_{12\cdot 34\cdots p-1} - \rho_{1p\cdot 34\cdots p-1}\rho_{2p\cdot 34\cdots p-1}}{\sqrt{(1-\rho^2_{1p\cdot 34\cdots p-1})(1-\rho^2_{2p\cdot 34\cdots p-1})}}. \tag{10.10.7}$$

■ NOTE 10.10.1 *A partial correlation coefficient in which s variables have been eliminated, i.e., there are s secondary suffixes is called a partial correlation coefficient of sth order.*

(d) Partial regression coefficients in terms of the same of the lower order

$$\begin{aligned}
&\text{Cov}(X_{1\cdot 34\cdots p}, X_{2\cdot 34\cdots p}) \\
&= \beta_{12\cdot 34\cdots p}\sigma^2_{2\cdot 34\cdots p} \qquad (10.10.8) \\
&= \rho_{12\cdot 34\cdots p}\sigma_{1\cdot 34\cdots p}\sigma_{2\cdot 34\cdots p} \\
&= \frac{\rho_{12\cdot 34\cdots p-1} - \rho_{1p\cdot 34\cdots p-1}\rho_{2p\cdot 34\cdots p}}{\sqrt{(1-\rho^2_{1p\cdot 34\cdots p-1})(1-\rho^2_{2p\cdot 34\cdots p-1})}} \cdot \\
&\quad \sigma_{1\cdot 34\cdots p-1}\sigma_{2\cdot 34\cdots p-1}\sqrt{(1-\rho^2_{1p\cdot 34\cdots p-1})(1-\rho^2_{2p\cdot 34\cdots p-1})} \\
&\qquad \text{(by relation 10.10.7 and 10.10.1)} \\
&= \beta_{12\cdot 34\cdots p-1}\sigma^2_{2\cdot 34\cdots p-1} - \beta_{1p\cdot 34\cdots p-1}\sigma_{p\cdot 34\cdots p-1}\beta_{p2\cdot 34\cdots p-1}\frac{\sigma^2_{2\cdot 34\cdots p-1}}{\sigma_{p\cdot 34\cdots p-1}} \\
&= (\beta_{12\cdot 34\cdots p-1} - \beta_{1p\cdot 34\cdots p-1}\beta_{p2\cdot 34\cdots p-1})\sigma^2_{2\cdot 34\cdots p-1}. \qquad (10.10.9)
\end{aligned}$$

Now

$$\sigma^2_{2\cdot 34\cdots p} = \sigma^2_{2\cdot 34\cdots p-1}(1-\rho^2_{2p\cdot 34\cdots p-1}).$$

Hence, from (10.10.8) and (10.10.9),

$$\begin{aligned}
\beta_{12\cdot 34\cdots p} &= \frac{\beta_{12\cdot 34\cdots p-1} - \beta_{1p\cdot 34\cdots p-1}\beta_{p2\cdot 34\cdots p-1}}{1-\rho^2_{2p\cdot 34\cdots p-1}} \\
&= \frac{\beta_{12\cdot 34\cdots p-1} - \beta_{1p\cdot 34\cdots p-1}\beta_{p2\cdot 34\cdots p-1}}{1-\beta_{2p\cdot 34\cdots p-1}\beta_{p2\cdot 34\cdots p-1}} \qquad (10.10.10)
\end{aligned}$$

▶ **Example 10.10.1** *For* $p = 3$, *show that* ρ_{23} *lies between the limits*

$$\rho_{12}\rho_{13} \pm \sqrt{1-\rho^2_{12}-\rho^2_{13}+\rho^2_{12}\rho^2_{13}}.$$

Solution: By relation (10.10.7),

$$\rho_{12\cdot3} = \frac{\rho_{12} - \rho_{13}\rho_{23}}{\sqrt{(1-\rho_{12}^2)(1-\rho_{23}^2)}}. \tag{i}$$

Since

$$\rho_{12\cdot3}^2 \le 1$$

$$(\rho_{12} - \rho_{13}\rho_{23})^2 \le (1-\rho_{12}^2)(1-\rho_{23}^2)$$

$$\text{or } \rho_{12}^2 + \rho_{13}^2 + \rho_{23}^2 - 2\rho_{12}\rho_{13}\rho_{23} \le 1.$$

If we take ρ_{12}, ρ_{13} as known, this gives the required limits for ρ_{23}.

▶ **Example 10.10.2** *Suppose that $\rho_{ij} = \rho \ \forall \ i \neq j = 1, \cdots, p$. Show that for such a correlation matrix, every partial correlation of sth order (i.e., in which s variables have been eliminated) is $\frac{\rho}{1+s\rho}$ and the multiple correlation coefficient of one variate with the other $(p-1)$ variates is given by*

$$\rho^2 = 1 - (1-\rho)\frac{[1+(p-1)\rho]}{1+(p-2)\rho}.$$

Solution: (a) Using the relation (i) in example 10.10.1,

$$\rho_{12\cdot3} = \frac{\rho}{1+\rho}.$$

Again

$$\rho_{12\cdot34} = \frac{\rho_{12\cdot4} - \rho_{13\cdot4}\rho_{23\cdot4}}{\sqrt{(1-\rho_{13\cdot4}^2)(1-\rho_{23\cdot4}^2)}} = \frac{\rho}{1+2\rho}.$$

Thus the result holds for partial correlations of order 1 and 2.

Assuming that every partial correlation coefficient of order r is $\frac{\rho}{1+r\rho}$, a partial correlation coefficient of order $(r+1)$ is seen to $\rho/\{1+(r+1)\rho\}$. Hence the result follows by induction.

(b) We have,

$$1 - \rho^2 = \frac{|\boldsymbol{R}|}{R_{11}}.$$

Here

$$|\boldsymbol{R}| = \begin{vmatrix} 1 & \rho & \cdots & \rho \\ \rho & 1 & \cdots & \rho \\ \cdot & \cdot & \cdots & \cdot \\ \rho & \rho & \cdots & 1 \end{vmatrix} = [1+(p-1)\rho](1-\rho)^{p-1}.$$

Also

$$R_{11} = [1+(p-2)\rho](1-\rho)^{p-2}.$$

Hence the result.

▶ **Example 10.10.3** *Let $X_1, \cdots, X_k$ be distributed as multinomial with parameters $(n, p_1, \cdots, p_k)$. Find the regression equation of X_1 on $X_2, \cdots, X_k$.*

Since the conditional distribution of X_1 given $x_2 = X_2, \cdots, X_k = x_k$ is binomial with parameters

$$n - x_2 - \cdots - x_k = n' \text{ (say)}$$

$$\text{and } \; p_1/(1 - p_2 - \cdots - p_k) = p_1' \text{ (say) (vide 7.6.5)}$$

we have

$$E(X_1|x_2, \cdots, x_n) = n'p_1'; \quad V(X|x_2, \cdots, x_k) = n'p_1'(1 - p_1').$$

Therefore, the regression equation of X_1 (any one variable) on the remaining variables is linear. The variance of the conditional distribution is also a linear function of the independent variables.

▶ **Example 10.10.4** (***Logistic Regression***). *Assume that the response variable Y_x is of binary character (e.g., success or failure) taking values 0 and 1, so that $E(Y_x) = P(Y_x = 1) = \pi(x)$. Here x is an auxiliary variable and $\pi(x)$ is the true regression of y on x. The assumption*

$$\pi(x) = a + bx$$

can be valid only in a limited sense since we must have $0 < \pi(x) < 1 \; \forall \; x$. Instead consider the odds ratio $\frac{\pi(x)}{1-\pi(x)}$ with range $(0, \infty)$ and its logarithm with range $(-\infty, \infty)$. Thus, consider the model

$$\log \frac{\pi(x)}{1 - \pi(x)} = a + bx$$

$$\textit{or equivalently} \quad \pi(x) = \frac{e^{a+bx}}{1 + e^{a+bx}}. \tag{10.10.11}$$

The right hand side of (10.10.11) is called the logistic function and the above regression logistic regression. If for $x = x_i$ we have N_i observations on y, of which n_i observations result in successes $(i = 1, \cdots, m)$, the likelihood of the data is

$$L = \prod_{i=1}^{m} \binom{N_i}{n_i} [\pi(x_i)]^{n_i} [1 - \pi(x_i)]^{N_i - n_i}$$

$$= \left(\prod_{i=1}^{m} \binom{N_i}{n_i} \right) \frac{\exp\{a \sum n_i + b \sum n_i x_i\}}{\prod_{i=1}^{m} (1 + e^{a+bx_i})^{N_i}}.$$

The maximum likelihood estimates (see chapter 16) of a, b can be obtained by numerical iteration.

10.11 Large Sample Formulae for Standard Errors of Sample Moments

In this section we consider formulae for variances and covariances of sample moments, which are valid only in large samples. These formulae are valid for any distribution for which the relevant moments exist.

Let $X_1, \cdots, X_n$ be a random sample of size n from a distribution $f(x, \theta)$. Let

$$m'_r = \frac{1}{n}\Sigma x_i^r, \quad m_r = \frac{1}{n}\sum_{i=1}^{n}(x_i - m'_1)^r,$$

$$\mu'_r = E(X^r), \quad \mu_r = E(X - \mu'_1)^r.$$

Hence,

$$E(m'_r) = \frac{1}{n}\Sigma E(X^r) = \mu'_r. \tag{10.11.1}$$

$$\begin{aligned}
V(m'_r) &= E(m'_r - \mu'_r)^2 = E\left(\frac{1}{n}\Sigma X_i^r - \mu'_r\right)^2 \\
&= \frac{1}{n^2}E\left[(\Sigma X_i^r)^2 - 2n\mu'_r(\Sigma X_i^r) + n^2\mu_r'^2\right] \\
&= \frac{1}{n^2}E\left[(\Sigma X_i^r)^2 - n^2\mu_r'^2\right] = \frac{1}{n^2}E\left[\Sigma X_i^{2r} + \sum\sum_{i\neq j} X_i^r X_j^r - n^2\mu_r'^2\right] \\
&= \frac{1}{n^2}\left[n\mu'_{2r} + n(n-1)E(X_i^r)E(X_j^r)\right] - \mu_r'^2 \quad \text{[since } X_i\text{'s are independent]} \\
&= \frac{1}{n^2}\left[n\mu'_{2r} + n(n-1)\mu_r'^2\right] - \mu_r'^2 = \frac{1}{n}(\mu'_{2r} - \mu_r'^2).
\end{aligned} \tag{10.11.2}$$

$$\begin{aligned}
\text{Cov}(m'_r, m'_s) &= E\left[(m'_r - \mu'_r)(m'_s - \mu'_s)\right] = E\left[\left(\frac{1}{n}\sum X_i^r - \mu'_r\right)\left(\frac{1}{n}\sum X_i^s - \mu'_s\right)\right] \\
&= \frac{1}{n^2}E(\Sigma X_i^{r+s}) + \frac{1}{n^2}\left(\sum\sum_{i\neq j} X_i^r X_j^s\right) \\
&\qquad - \frac{1}{n}\mu'_r E(\Sigma X_i^s) - \frac{1}{n}\mu'_s E(\Sigma X_i^r) + E(\mu'_r\mu'_s) \\
&= \frac{1}{n^2}\cdot n\mu'_{r+s} + \frac{1}{n^2}\cdot n(n-1)\mu'_r\mu'_s - 2\mu'_r\mu'_s + \mu'_r\mu'_s \\
&= \frac{1}{n}(\mu'_{r+s} - \mu'_r\mu'_s).
\end{aligned} \tag{10.11.3}$$

$$\begin{aligned}
E(m_2) &= E\left[\frac{1}{n}\Sigma(X_i - m'_1)^2\right] = \frac{1}{n}E\left[\Sigma X_i^2 - 2m'_1\Sigma X_i + nm_1'^2\right] \\
&= \frac{1}{n}E\left[\Sigma X_i^2 - \frac{2}{n}(\Sigma X_j)(\Sigma X_i) + \frac{1}{n}(\Sigma X_i)^2\right]
\end{aligned}$$

$$= \frac{1}{n} E\left[\Sigma X_i^2 - \frac{1}{n}\Sigma X_i^2 - \frac{2}{n}\sum_{i \neq j}\sum X_i X_j\right]$$

$$= \mu_2' - \frac{1}{n}\mu_2' - n(n-1)\mu_1'^2 - \frac{n-1}{n}(\mu_2' - \mu_1'^2) = \left(1 - \frac{1}{n}\right)\mu_2. \tag{10.11.4}$$

$$V(m_2) = E(m_2^2) - \left(\frac{n-1}{n}\right)^2 \mu_2^2. \tag{10.11.5}$$

Now $$m_2^2 = \left[\frac{1}{n}\Sigma x_i^2 - \frac{1}{n^2}(\Sigma x_i)^2\right]^2 = \frac{1}{n^2}(\Sigma x_i^2)^2 - \frac{2}{n^3}(\Sigma x_i^2)(\Sigma x_i)^2 + \frac{1}{n^4}(\Sigma X_i)^4$$

$$= \frac{1}{n^2}\left(\Sigma x_i^4 + 2\sum_{i \neq j}\sum x_i^2 x_j^2\right) - \frac{2}{n^3}(\Sigma x_i^2)\left(\Sigma x_i^2 + 2\sum_{i<j}\sum x_i x_j\right)$$

$$+\frac{1}{n^4}\left(\Sigma x_i^4 + 4\sum_{i \neq j}\sum x_i^3 x_j + 6\sum_{i \neq j}\sum x_i^2 x_j^2 + 12\sum_{i \neq j \neq k}\sum\sum x_i^2 x_j x_k\right.$$

$$\left. +24 \sum_{i \neq j \neq k \neq e}\sum\sum\sum x_i x_j x_k x_e\right).$$

Hence,

$$E(m_2^2) = \frac{1}{n}\mu_4 + \frac{n-1}{n}\mu_2^2 - \frac{2}{n^2}\mu_4 - \frac{2(n-1)}{n^2}\mu_2^2 + \frac{\mu_4}{n^3} + \frac{3(n-1)}{n^3}\mu_2^2, \text{ taking } \mu_1' = 0$$

$$= \frac{(n-1)^2}{n^3}\mu_4 + \frac{(n-1)(n+1)(n-3)}{n^3}\mu_2^2. \tag{10.11.6}$$

Therefore,

$$V(m_2) = \frac{(n-1)^2}{n^3}(\mu_4 - \mu_2^2) + \frac{2(n-1)}{n^3}\mu_2^2 \text{ [from (10.11.5) and (10.11.6)].} \tag{10.11.7}$$

Hence,

$$V(m_2) = \frac{\mu_4 - \mu_2^2}{n} + 0\left(\frac{1}{n^2}\right).$$

For normal population,

$$V(m_2) \simeq \frac{3\sigma^4 - \sigma^4}{n} = \frac{2\sigma^4}{n}.$$

Defining

$$s_2^2 = \frac{n}{n-1}m_2,$$

$$V(s_2^2) = \frac{n^2}{(n-1)^2}V(m_2) = \frac{1}{n}(\mu_4 - \mu_2^2) + \frac{2}{n(n-1)}\mu_2^2 = \frac{1}{n}(\mu_4 - 3\mu_2^2) + \frac{2}{n-1}\mu_2^2$$

$$= \frac{2\sigma^4}{(n-1)} \text{ for normal population.}$$

Defining the cumulants $k_r (r = 1, 2, \cdots)$

$$k_1 = \mu_1', \quad k_2 = \mu_2, \quad k_3 = \mu_3, \quad k_4 = \mu_4 - 3\mu_2^2,$$

$$\begin{aligned} V(m_2) &= \frac{(n-1)^2}{n^3}\left(k_4 + 3\mu_2^2 - \frac{(n-3)}{(n-1)}\mu_2^2\right) \\ &= \frac{(n-1)^2}{n^3}\left(k_4 + \frac{2n}{n-1}k_2^2\right). \end{aligned} \tag{10.11.8}$$

In the same way,

$$\operatorname{Cov}(m_2, m_1') = \frac{(n-1)\mu_3}{n^2}. \tag{10.11.9}$$

The formula (10.11.1)–(10.11.9) are exact. For finding expectations, variances and covariances of higher order central moments we have to apply large sample formulae for functions of several statistics. The relevant theorem is stated below.

Theorem 10.11.1 *If $T_1, \cdots, T_k$ are k statistics calculated from a sample of size n and if each of T_i is distributed about mean ξ_i with variance of order $\frac{1}{n}$ then asymptotically,*

$$E[f(T_1, \cdots, T_k)] = f(\xi_1, \cdots, \xi_k)$$

and

$$\operatorname{Var}(f(T_1, \cdots, T_k)) = \sum_{i=1}^{k}\left(\frac{\partial f}{\partial \xi_i}\right)^2 V(T_i) + \sum\sum_{i \neq j}\left(\frac{\partial f}{\partial \xi_i}\right)\left(\frac{\partial f}{\partial \xi_j}\right)\operatorname{Cov}(T_i, T_j) \tag{10.11.10}$$

where $f(T_1, \cdots, T_k)$ is a continuous function of $T_1, \cdots, T_k$ possessing continuous partial derivatives and $\frac{\partial f}{\partial \xi_i} = \left.\frac{\partial f}{\partial T_i}\right]_{(\xi_1, \cdots, \xi_k)}$, etc.

Proof: By Taylor's theorem,

$$\begin{aligned} f(T_1, \cdots, T_k) = f(\xi_1, \cdots, \xi_k) &+ \sum_i \left[(T_i - \xi_i)\frac{\partial f}{\partial \xi_i}\right] \\ &+ \frac{1}{2!}\left[\sum_i (T_i - \xi_i)^2 \frac{\partial^2 f}{\partial \xi_i^2} + \sum\sum_{i \neq j}(T_i - \xi_i)(T_j - \xi_j)\frac{\partial^2 f}{\partial \xi_i \partial \xi_j}\right] + \cdots. \end{aligned}$$

Since $V(T_i)$ is of order $\frac{1}{n}$, we have $(T_i - \xi_i)$ of order $\frac{1}{\sqrt{n}}$. Hence neglecting terms of order $\frac{1}{n}$ and lower,

$$f(T_i, \cdots, T_k) = f(\xi_1, \cdots, \xi_k) + \sum\left[(T_i - \xi_i)\frac{\partial f}{\partial \xi}\right].$$

Therefore, to this order of approximation,

$$\begin{aligned}
E\left[f(T_1,\cdots,T_k)-f(\xi_i,\cdots,\xi_j)\right]^2 &= V(f)\\
&= E\left[\sum(T_i-\xi_i)\frac{\partial f}{\partial\xi_i}\right]^2\\
&= E\left[\sum(T_i-\xi_i)^2\left(\frac{\partial f}{\partial\xi_i}\right)^2+\sum\sum_{i\neq j}(T_i-\xi_i)(T_j-\xi_j)\frac{\partial f}{\partial\xi_i}\cdot\frac{\partial f}{\partial\xi_j}\right]\\
&= \sum\left(\frac{\partial f}{\partial\xi_i}\right)^2 V(T_i)+\sum\sum_{i\neq j}\frac{\partial f}{\partial\xi_i}\cdot\frac{\partial f}{\partial\xi_j}\,\mathrm{Cov}(T_i,\,T_j).
\end{aligned}$$

Corollary 10.11.1 *If $f_1 = f_1(T_1,\cdots,T_k)$ and $f_2 = f_2(T_1,\cdots,T_k)$ be two functions of $T_1,\cdots,T_k$, then asymptotically,*

$$f_j(T_1,\cdots,T_k)=f_j(\xi_1,\cdots,\xi_k)+\sum_{i=1}^{k}\left[(T_i-\xi_j)\frac{\partial f_j}{\partial\xi_i}\right],\quad j=1,2.$$

Hence,

$$\begin{aligned}
\mathrm{Cov}(f_1,f_2) &= E\left[\sum_{i=1}^{k}(T_i-\xi_i)\frac{\partial f_1}{\partial\xi_i}\right]\left[\sum_{i=1}^{k}(T_i-\xi_i)\frac{\partial f_2}{\partial\xi_i}\right]\\
&= \sum_{i=1}^{k}\frac{\partial f_1}{\partial\xi_i}\cdot\frac{\partial f_2}{\partial\xi_i}V(T_i)+\sum\sum_{i\neq j=1}^{k}\frac{\partial f_1}{\partial\xi_i}\cdot\frac{\partial f_2}{\partial\xi_j}\,\mathrm{Cov}(T_i,T_j). \qquad (10.11.11)
\end{aligned}$$

Corollary 10.11.2 *If T is statistic based on a sample of size n and if T is distributed about mean ξ with variance of order $\frac{1}{n}$ then asymptotically,*

$$E(f(T))=f(\xi);\;\; \mathrm{Var}(f(T))=\left(\frac{\partial f}{\partial\xi}\right)^2 V(T) \qquad (10.11.12)$$

where $f(T)$ is a continuous function of T possessing continuous partial derivatives.

□ **Remark 10.11.1** *If, further, $(T_1,\cdots,T_k)$ are assumed to have a joint k-variate normal distribution then the distribution of $f(T_1,\cdots,T_k)$ is asymptotically univariate normal. This result has been stated in lemma* 20.6.1.

We now consider large sample formulae for variances and covariances of central moments.

We have

$$\begin{aligned}
m_r &= \frac{1}{n}\sum(x_i-m_1')^r\\
&= m_r'-\binom{r}{1}m_{r-1}'m_1'+\cdots+(-1)^i\binom{r}{i}m_{r-i}'(m_1')^i\\
&\quad+\cdots+(-1)^r(m_1')^r. \qquad (10.11.13)
\end{aligned}$$

We note that if we define

$$m'_r(A) = \frac{1}{n}\sum_{i=1}^{n}(x_i - A)^r$$

then

$$m_r = m'_r(A) - \binom{r}{1} m'_{r-1}(A)m'_1(A) + \cdots + (-1)^i \binom{r}{i} m'_{r-i}(A)(m'_1(A))^i$$
$$+ \cdots + (-1)^r (m'_1(A))^r.$$

Thus the relation (10.11.13) holds even if the origin of measurement of X values is shifted. Without any loss of generality we, therefore, take $A = E(X)$ and assume that X values have been measured about the mean so that $E(X) = \mu'_1 = \mu = 0$. Under this assumption $\mu'_s = \mu_s$ $(s = 1, 2, \cdots)$. Now (10.11.13) can be written as

$$m_r = f(m'_r, m'_{r-1}, \cdots, m'_1) = f(\mu'_r, \mu'_{r-1}, \cdots, \mu'_1) + \sum_{i=1}^{r}(m'_i - \mu'_i)\frac{\partial f}{\partial \mu'_i} + \cdots$$

where $\displaystyle \frac{\partial f}{\partial \mu'_i} = \left.\frac{\partial f}{\partial m'_i}\right]_{\substack{m'_k = \mu'_k \\ k=1,2,\cdots}}$

$$= f(\mu'_r, \mu'_{r-1}, \cdots, \mu'_1) + (m'_r - \mu'_r) - (m'_1 - \mu'_1)r\mu_{r-1}$$
$$= \mu_r + (m'_r - \mu'_r) - (m'_1 - \mu'_1)r\mu_{r-1}.$$

Since m'_r is distributed about the mean μ'_r with variance $0\left(\frac{1}{n}\right)$,

$$E(m_r) = \mu_r + 0\left(\frac{1}{n}\right).$$

Again, large sample variance of m_r is

$$V(m_r) = E(m_r - \mu_r)^2 = E\left(m_r - f(\mu'_r, \mu'_{r-1}, \cdots, \mu'_1)\right)^2$$
$$= E\left[(m'_r - \mu'_r) - r(m'_1 - \mu'_1)\mu_{r-1}\right]^2$$
$$= E\left[(m'_r - \mu'_r)^2 + r^2\mu^2_{r-1}(m'_1 - \mu'_1)^2 - 2r\mu_{r-1}(m'_1 - \mu'_1)(m'_r - \mu'_r)\right]$$
$$= V(m'_r) + r^2\mu^2_{r-1}V(m'_1) - 2r\mu_{r-1}\operatorname{Cov}(m'_1, m'_r)$$
$$= \tfrac{1}{n}\left[(\mu_{2r} - \mu_r^2) + r^2\mu^2_{r-1}(\mu_2 - \mu_1^2) - 2r\mu_{r-1}(\mu_{r+1} - \mu_r\mu_1)\right]$$
$$= \tfrac{1}{n}\left[\mu_{2r} - \mu_r^2 + r^2\mu_2\mu^2_{r-1} - 2r\mu_{r-1}\mu_{r+1}\right]. \tag{10.11.14}$$

$$\operatorname{Cov}(m_r, m_s) = E\left[\{m_r - f(\mu'_r, \mu'_{r-1}, \cdots, \mu'_1)\}\{m_s - f(\mu'_s, \mu'_{s-1}, \cdots, \mu'_1)\}\right]$$
$$= E\left[\{(m'_r - \mu'_r) - r(m'_1 - \mu'_1)\mu_{r-1}\}\{(m'_s - \mu'_s) - s(m'_1 - \mu'_1)\mu_{s-1}\}\right]$$

$$= E[(m'_r - \mu'_r)(m'_s - \mu'_s) - r\mu_{r-1}(m'_1 - \mu'_1)(m'_s - \mu'_s)$$
$$-s\mu_{s-1}(m'_1 - \mu'_1)(m'_r - \mu'_r) + rs\mu_{r-1}\mu_{s-1}(m'_1 - \mu'_1)^2]$$
$$= \text{Cov}(m'_r, m'_s) - r\mu_{r-1}\,\text{Cov}(m'_s, m'_1) - s\mu'_{s-1}\,\text{Cov}(m'_r, m'_1)$$
$$+rs\mu_{r-1}\mu_{s-1}V(m'_1)$$
$$= \tfrac{1}{n}[\mu_{r+s} - \mu_r\mu_s - r\mu_{r-1}(\mu_{s+1} - \mu_s\mu_1) - s\mu_{s-1}(\mu_{r+1} - \mu_r\mu_1)$$
$$+rs\mu_{r-1}\mu_{s-1}(\mu_2 - \mu_1'^2)]$$
$$= \tfrac{1}{n}[\mu_{r+s} - \mu_r\mu_s - r\mu_{r-1}\mu_{s+1} - s\mu_{s-1}\mu_{r+1} + rs\mu_{r-1}\mu_{s-1}\mu_2]. \quad (10.11.15)$$

Again,

$$\text{Cov}(m_r, m'_s) = E\left[\{m_r - f(\mu_r, \mu_{r-1}, \cdots, \mu_1)\}\{m'_s - \mu'_s\}\right]$$
$$= E\left[\{(m'_r - \mu'_r) - r(m'_1 - \mu'_1)\mu_{r-1}\}\{m'_s - \mu'_s\}\right]$$
$$= \text{Cov}(m'_r, m'_s) - r\mu_{r-1}\,\text{Cov}(m'_1, m'_s)$$
$$= \frac{1}{n}(\mu'_{r+s} - \mu'_r\mu'_s - r\mu_{r-1}\mu'_{s+1}) \text{ (since } \mu'_1 = 0). \quad (10.11.16)$$

Note that the expressions (10.11.14)–(10.11.16) are valid only in large sample.

Standard errors of bivariate moments

In the same way it can be proved that for bivariate moments,

$$\text{Var}(m'_{r,s}) = \frac{1}{n}\left(\mu'_{2r,2s} - \mu'^{2}_{r,s}\right). \quad (10.11.17)$$

$$\text{Cov}\left(m'_{r,s}, m'_{u,v}\right) = \frac{1}{n}\left(\mu'_{r+u,s+v} - \mu'_{r,s}\mu'_{u,v}\right). \quad (10.11.18)$$

$$\text{Var}(m_{r,s}) = \frac{1}{n}(\mu_{2r,2s} - \mu^2_{r,s} + r^2\mu_{2,0}\mu^2_{r-1,s} + s^2\mu_{0,2}\mu^2_{r,s-1}$$
$$+2rs\mu_{1,1}\mu_{r-1,s}\mu_{r,s-1} - 2r\mu_{r+1,s}\mu_{r-1,s}$$
$$-2s\mu_{r,s+1}\mu_{r,s-1}). \quad (10.11.19)$$

$$\text{Cov}(m_{r,s}, m_{u,v}) = \frac{1}{n}(\mu_{r+u,s+v} - \mu_{r,s}\mu_{u,v} +$$
$$+ru\mu_{2,0}\mu_{r-1,s}\mu_{u-1,v} + sv\mu_{0,2}\mu_{r,s-1}\mu_{u,v-1} + rv\mu_{1,1}\mu_{r-1,s}\mu_{u,v-1}$$
$$+su\mu_{1,1}\mu_{r,s-1}\mu_{u-1,v} - u\mu_{r+1,s}\mu_{u-1,v}$$
$$-v\mu_{r,s+1}\mu_{u,v-1} - r\mu_{r-1,s}\mu_{u+1,v} - s\mu_{r,s-1}\mu_{u,v+1}). \quad (10.11.20)$$

► **Example 10.11.1** *Find the sampling variance of the following statistics:* (a) *Sample Standard Deviation;* (b) *Sample Coefficient of Variation;* (c) *Sample Estimate of* β_1, *viz.* $b = \frac{m_3^2}{m_2^3}$.

Solution:

(a) It is required to find $V(\sqrt{m_2})$. Here

$$\left(\frac{d\sqrt{m_2}}{dm_2}\right)_{\mu_2} = \frac{1}{2\sqrt{\mu_2}}.$$

Hence,

$$V(\sqrt{m_2}) = \left(\frac{1}{4\mu_2}\right)V(m_2) = \frac{\mu_4-\mu_2^2}{4n\mu_2} \text{ [by formula (10.11.12)]}$$

$$= \frac{\sigma^2}{2n} \text{ for a normal population.}$$

(b) The coefficient of variation (c.v.)

$$V = \frac{100\sqrt{m_2}}{m_1'},\quad E(V) \simeq \frac{100\sqrt{\mu_2}}{\mu_1'} = C \text{ (population c.v.)}$$

$$\log V = 2 + \frac{1}{2}\log m_2 - \log m_1'.$$

Taking logarithmic differentiation,

$$\frac{dV}{V} = \frac{1}{2}\cdot\frac{dm_2}{m_2} - \frac{dm_1'}{m_1'},$$

i.e.,

$$\frac{V-E(V)}{V} = \frac{1}{2}\cdot\frac{m_2-E(m_2)}{m_2} - \frac{(m_1'-E(m_1'))}{m_1'}.$$

Squaring and taking expectation, we have the approximate relation

$$\frac{\text{Var}(V)}{C^2} = \frac{1}{4}\frac{\text{Var}(m_2)}{\mu_2^2} + \frac{\text{Var}(m_1')}{\mu_1'^2} - \frac{\text{Cov}(m_2,m_1')}{\mu_2\mu_1'}$$

$$= \frac{1}{n}\left(\frac{\mu_4-\mu_2^2}{4\mu_2^2} + \frac{\mu_2}{\mu_1'^2} - \frac{\mu_3}{\mu_2\mu_1'}\right).$$

For a normal population,

$$\text{Var}(V) \simeq \frac{C^2}{n}\left(\frac{1}{2}+\frac{\mu_2}{\mu_1'^2}\right) = \frac{C^2}{2n}\left(1+\frac{2C^2}{100}\right) \simeq \frac{C^2}{2n}.$$

(c) $$\text{Var}(b_1) \simeq \left(\frac{2\mu_3}{\mu_2^3}\right)^2 V(m_3) + \left(-\frac{3\mu_3^2}{\mu_2^4}\right)V(m_2) + 2\left(-\frac{6\mu_3^3}{\mu_2^7}\right)\text{Cov}(m_3,m_2)$$

$$= \left(\frac{2\mu_3}{\mu_2^3}\right)^2\cdot\frac{\mu_6-\mu_3^2+9\mu_2^3-6\mu_2\mu_4}{n} + \left(-\frac{3\mu_3^2}{\mu_2^4}\right)^2\cdot\frac{\mu_4-\mu_2^2+4\mu_2\mu_1^2-4\mu_1\mu_3}{n}$$

$$+2\left(-\frac{6\mu_3^3}{\mu_2^7}\right)\cdot\frac{\mu_5-\mu_3\mu_2-2\mu_1\mu_4-3\mu_2\mu_3+6\mu_2\mu_1\mu_2}{n}$$

$$= 4\beta_1\left(\frac{\beta_4-\beta_1+9-6\beta_2}{n}\right) + 9\beta_1^2\left(\frac{\beta_2-1}{n}\right) - 12\beta_1\left(\frac{\beta_3-4\beta_1}{n}\right)$$

where

$$\beta_{2n+1} = \mu_3\mu_{2n+3}/\mu_2^{n+3} \quad \text{and} \quad \beta_{2n} = \mu_{2n+2}/\mu_2^{n+1}.$$

Hence,

$$\text{Var}(b_1) = \beta_1(4\beta_4 - 24\beta_2 + 36 + 9\beta_1\beta_2 - 12\beta_3 + 35\beta_1)/n.$$

For normal population

$$\text{Var}(b_1) = 6\beta_1/n.$$

▶ **Example 10.11.2** *Given a random sample $X_1, \cdots, X_n$ of size n, show that*

$$\text{(a)} \quad E\left\{\frac{1}{n}\sum_{i=1}^{n}|X_i - \mu_1'|\right\} = \delta; \quad \text{(b)} \quad E\left\{\frac{1}{n}\sum_{i=1}^{n}|X_i - \mu_1'|^2\right\} = \sigma^2.$$

when μ_1', δ, σ^2 are the population mean, mean deviation and variance respectively. Henc show that if d is the sample mean deviation,

$$\text{Var}(d) \simeq \frac{\sigma^2 - \delta^2}{n}.$$

Deduce that for a normal population,

$$\text{Var}(d) \simeq \frac{\sigma^2}{n}\left(1 - \frac{2}{n}\right).$$

Solution: Let d_α be the sample mean deviation about α,

$$d_\alpha = \frac{1}{n}\sum_{i=1}^{n}|X_i - \alpha|.$$

Then clearly

$$E(d_\alpha) = E|X - \alpha| = \delta_\alpha \text{ (say)}$$

$$E\left\{\frac{1}{n}\sum_{i=1}^{n}|X_i - \alpha|^2\right\} = E(X - \alpha)^2$$

since $|X_i - \alpha|, i = 1, \cdots, n$ are independently and identically distributed.

$$V(d_\alpha) = \frac{1}{n}V(|X - \alpha|) = \frac{1}{n}\left[E|X - \alpha|^2 - (E|X - \alpha|)^2\right]$$

$$= \frac{1}{n}\left[E|X - \mu_1' + \mu_1' - \alpha|^2 - \delta_\alpha^2\right] = \frac{1}{n}\left[\sigma^2 + (\alpha - \mu_1')^2 - \delta_\alpha^2\right].$$

When $\alpha = \mu_1'$

$$V(d_\alpha) = \frac{1}{n}[\sigma^2 - \delta^2].$$

For a normal population, mean deviation about mean,

$$\delta = \sigma\sqrt{\frac{2}{\pi}} \text{ (vide section 6.7).}$$

▶ **Example 10.11.3** *Find the standard error of sample correlation coefficient*

$$r = \frac{\sqrt{m_{11}}}{\sqrt{m_{20}m_{02}}}$$

where m_{11}, m_{20}, m_{02} are sample covariances, variances of X and Y respectively.

By theorem 10.11.1,

$$\begin{aligned} V(r) = & \left(\frac{\partial r}{\partial m_{11}}\right)_E^2 V(m_{11}) + \left(\frac{\partial r}{\partial m_{20}}\right)_E^2 V(m_{20}) + \left(\frac{\partial r}{\partial m_{02}}\right)_E^2 V(m_{02}) \\ & + 2\left(\frac{\partial r}{\partial m_{11}}\right)_E \left(\frac{\partial r}{\partial m_{20}}\right)_E \operatorname{Cov}(m_{11}, m_{20}) \\ & + 2\left(\frac{\partial r}{\partial m_{20}}\right)_E \left(\frac{\partial r}{\partial m_{02}}\right)_E \operatorname{Cov}(m_{20}, m_{02}) \\ & + 2\left(\frac{\partial r}{\partial m_{11}}\right)_E \left(\frac{\partial r}{\partial m_{02}}\right)_E \operatorname{Cov}(m_{11}, m_{02}) \end{aligned}$$

when the differentials are to be evaluated at the expected values. Now

$$V(m_{11}) \simeq \frac{\mu_{22} - \mu_{11}^2}{n}; \qquad V(m_{20}) \simeq \frac{\mu_4^2 - \mu_{20}^2}{n}.$$

$$\operatorname{Cov}(m_{11}, m_{20}) \simeq \frac{\mu_{31} - \mu_{11}\mu_{20}}{n}; \ \operatorname{Cov}(m_{11}, m_{02}) \simeq \frac{\mu_{13} - \mu_{11}\mu_{02}}{n}$$

On simplification,

$$V(r) \simeq \frac{\rho^2}{4m}\left[\frac{\mu_{40}}{\mu_{20}^2} + \frac{\mu_{04}}{\mu_{22}^2} + \frac{2\mu_{22}}{\mu_{20}\mu_{02}} + \frac{4\mu_{22}}{\mu_{11}^2} - \frac{4\mu_{31}}{\mu_{11}\mu_{20}} - \frac{4\mu_{13}}{\mu_{11}\mu_{02}}\right].$$

For a normal population, the above expression reduces to

$$V(r) \simeq \frac{(1-\rho^2)^2}{n-1}.$$

10.12 Function of Moments

In this section, we shall consider large sample formulae for expectation and variance of a function of sample raw moments. As is expected the results will be similar to those stated in theorem 10.11.1.

Let $X_1, \cdots, X_n$ be a random sample from a population with p.d.f. $f(x, \theta)$. Define

$$a_r = \frac{1}{n}\sum_{j=1}^{n} X_j^r = m'_r \ \text{ and } \ \nu_r = E(X^r) = \mu'_r.$$

I. Consider a function of moment statistic $H(a_1, \cdots, a_k)$ having the following properties:

(a) In some neighbourhood of the point $a_1 = \nu_1, \cdots, a_k = \nu_k$, the function $H(a_1, \cdots, a_k)$ is continuous and possesses continuous derivatives of first and second order with respect to $a_i (i = 1, \cdots, k)$.

(b) For all possible values of x_i's, $|H| < c \cdot n^p$ where c and p are non-negative constants. Then

$$E(H) = H_\circ + 0\left(\frac{1}{n}\right) \tag{10.12.1}$$

$$V(H) = \sum_i \sum_j H_i H_j \operatorname{Cov}(a_i, a_j) + 0\left(\frac{1}{n^{3/2}}\right) \tag{10.12.2}$$

where

$$H_\circ = H(\nu_1, \cdots, \nu_k) \text{ and } H_i = \left.\frac{\partial H}{\partial a_i}\right]_{(\nu_1, \cdots, \nu_k)}, \quad i = 1, \cdots, k.$$

Proof: (a) Let $P(S)$ denote the probability that $(X_1, \cdots, X_n) \in$ set S, where $S \in R^n$. It is known that

$$E(a_i - \nu_i)^{2r} = 0\left(\frac{1}{n^r}\right) = \frac{A}{n^r} \text{ (say)}. \tag{10.12.3}$$

Applying Theorem 3.10.1,

$$P\left\{(a_i - \nu_i)^{2r} \geq \varepsilon^{2r}\right\} < \frac{A}{n^r \varepsilon^{2r}} \tag{10.12.4}$$

for any $\varepsilon > 0$, when A is a constant independent of n and ε. (10.2.4) can be written as

$$P\{|a_i - \nu_i| \geq \varepsilon\} < \frac{A}{n^r \varepsilon^{2r}} \tag{10.12.5}$$

which is true for all $i = 1, \cdots, k$. Denote by Z be the set of points for which $|a_i - \nu_i| < \varepsilon$ $\forall\, i = 1, \cdots, k$. If Z^* is the complementary set, it is known by Boole's inequality (Theorem 1.3.3) that

$$P(Z^*) < \frac{kA}{n^r \varepsilon^{2r}} \tag{10.12.6}$$

and

$$P(Z) > 1 - \frac{kA}{n^r \varepsilon^{2r}}. \tag{10.12.7}$$

Now

$$E(H) = \int_Z H\, dP + \int_{Z^*} H\, dP.$$

By condition (b),

$$\left|\int_{Z^*} H\, dP\right| < C \cdot n^p \int_{Z^*} dP < \frac{C \cdot n^p \cdot kA}{n^r \varepsilon^{2r}} \text{ [from (10.12.6)]} \tag{10.12.8}$$

Choosing $r > p+1$, it follows, therefore,

$$E(H) = \int_Z H\,dP + 0\left(\frac{1}{n}\right). \qquad (10.12.9)$$

If ε is sufficiently small, we have, by condition (a), for any point in the set Z,

$$H(a_1, \cdots, a_k) = H(\nu_1, \cdots, \nu_k) + \sum_i (a_i - \nu_i) H_i + R \qquad (10.12.10)$$

where

$$R = \frac{1}{2} \sum_i \sum_j (a_i - \nu_i)(a_j - \nu_j) H_{ij}^{(1)},$$

$$H_{ij}^{(1)} = \frac{\partial^2 H(a_1, \cdots, a_k)}{\partial a_i \partial a_j}$$

evaluated at some intermediate point between $(a_1, \cdots, a_k)$ and $(\nu_1, \cdots, \nu_k)$. Hence

$$\int_Z H\,dP = H_\circ P(Z) + \sum_i H_i \int_Z (a_i - \nu_i)\,dP + \int_Z R\,dP. \qquad (10.12.11)$$

Now

$$H_\circ P(Z) = H_\circ(1 - P(Z^*)) = H_\circ - 0(n^{-r}) = H_\circ + 0(n^{-1}), \qquad (10.12.12)$$

since $r > p + 1 \geq 1$.

Again,

$$\int_Z (a_i - \nu_i)\,dP = E(a_i - \nu_i) - \int_{Z^*} (a_i - \nu_i)\,dP = 0 - \int_{Z^*} (a_i - \nu_i)\,dP. \qquad (10.12.13)$$

Now

$$\begin{aligned}\left|\int_{Z^*} (a_i - \nu_i)\,dP\right| &\leq \left[\int_{Z^*} (a_i - \nu_i)^2\,dP \int_{Z^*} dP\right]^{1/2} \\ &\leq \left[E(a_i - \nu_i)^2 P(Z^*)\right]^{1/2} \\ &= 0\left(n^{-\frac{1+r}{2}}\right) \text{ [because } P(Z^*) = 0\left(\tfrac{1}{n^r}\right), E(a_i - \nu_i)^2 = 0\left(\tfrac{1}{n}\right)].\end{aligned}$$

Hence,

$$\int_Z (a_i - \nu_i)\,dP = 0\left(\frac{1}{n}\right), \quad i = 1, \cdots, k.$$

Since $H_1, \cdots, H_k$ are independent of n,

$$\sum H_i \int_Z (a_i - \nu_i)\,dP = 0\left(\frac{1}{n}\right). \qquad (10.12.14)$$

Finally, by condition (a), the derivative $H_{ij}^{(1)}$'s are bounded in the neighbourhood of $(\nu_1, \cdots, \nu_k)$ and it then follows in the same way that the third term in the right-hand side of (10.12.11) is of $0\left(\frac{1}{n}\right)$. Hence, from (10.12.9) and (10.12.11), (10.12.1) follows.

(b) Consider

$$E(H-H_\circ)^2 = \int_Z (H-H_\circ)^2 dP + \int_{Z^*} (H-H_\circ)^2\, dP$$
$$= \int_Z (H-H_\circ)^2\, dP + 0(n^{-\frac{3}{2}}) \qquad (10.12.15)$$

since

$$\int_{Z^*} (H-H_\circ)^2\, dP \le C' \cdot n^{2p} \int_{Z^*} dP \le C' \cdot n^{2p} \cdot \frac{kA}{\varepsilon^{2r} n^r} \le C'' n^{2p-r}$$

and we choose $r > 2p + \frac{3}{2}$.

Again,

$$\int_Z (H-H_\circ)^2\, dP = \int_Z \left\{ \sum_i H_i(a_i - \nu_i) + R \right\}^2 dP \text{ [by expansion (10.12.10)]}$$
$$= \sum_i \sum_j H_i H_j \int_Z (a_i - \nu_i)(a_j - \nu_j)\, dP$$
$$+2\sum_i H_i \int_Z (a_i - \nu_i) R\, dP + \int_Z R^2\, dP. \qquad (10.12.16)$$

In the same way as before it can be shown that

$$\sum_i \sum_j H_i H_j \int_Z (a_i - \nu_i)(a_j - \nu_j)\, dP = \sum_i \sum_j H_i H_j \operatorname{Cov}(a_i, a_j) + 0\left(n^{-\frac{7}{4}}\right). \qquad (10.12.17)$$

$$\left[\because \int_Z (a_i - \nu_i)(a_j - \nu_j)\, dP = \operatorname{Cov}(a_i, a_j) - \int_{Z^*} (a_i - \nu_j)(a_j - \nu_j)\, dP \right.$$

$$\left| \int_{Z^*} (a_i - \nu_i)(a_j - \nu_j)\, dP \right| \le \left[\int_{Z^*} \{(a_i - \nu_i)(a_j - \nu_j)\}^2\, dP \int_{Z^*} dP \right]^{1/2}$$
$$\le \left[\left\{ \int_{Z^*} (a_i - \nu_i)^4\, dP \right\} \left\{ \int_{Z^*} (a_j - \nu_j)^4\, dP \right\} \right]^{1/4} \left[\int_{Z^*} dP \right]^{1/2}$$
$$\le \left[E(a_i - \nu_i)^4 E(a_j - \nu_j)^4 \right]^{1/4} 0\left(n^{-\frac{r}{2}}\right)$$
$$\left. \le 0\left(\frac{1}{n}\right) 0\left(n^{-\frac{3}{4}}\right) \text{ as } r > 2p + \tfrac{3}{2} \ge \tfrac{3}{2} \right].$$

Again,

$$\sum H_i \int_Z (a_i - \nu_i) R\, dP = 0\left(n^{-\frac{3}{2}}\right) \qquad (10.12.18)$$

[Consider one term in the left-hand side of (10.12.18) :

$$\left|H_1 \int_Z (a_1 - \nu_1)^3 H_{11}^{(1)}\, dP\right| < E(|a_1 - \nu_1|^3)$$

$$\leq \left\{E|a_1 - \nu_1|^4\right\}^{3/4} \quad \text{[because } \alpha_r^{1/r} \leq \alpha_{r+1}^{1/(r+1)}, \text{ where } \alpha_r = E|X - \nu_1|^r \text{ (vide theorem 3.4.3)]}$$

$$= 0\left(n^{-\frac{3}{2}}\right).]$$

Similarly, it can be shown that

$$\int_Z R^2\, dP = 0(n^{-2}) \tag{10.12.19}$$

Again,

$$V(H) = E(H - E(H))^2 = E(H - H_\circ)^2 + [H_\circ - E(H)]^2 + 0\left(\frac{1}{n^2}\right)^2. \tag{10.12.20}$$

Combining (10.12.15)–(10.12.20), (10.12.2) follows.

□ **Remark 10.12.1** *One important multivariate distribution is the multivariate normal distribution. This has been considered in details in chapter* 13.

Exercise

1. (a) If $\boldsymbol{X}, \boldsymbol{Y}$ are $m \times 1$ and $n \times 1$ random vectors and **a, b** are $m \times 1, n \times 1$ vector of constants, prove that

$$C[\boldsymbol{X} - \mathbf{a}, \boldsymbol{Y} - \mathbf{b}] = C[\boldsymbol{X}, \boldsymbol{Y}].$$

   (b) Prove that

$$C[\boldsymbol{X}, \boldsymbol{Y}] = E[\boldsymbol{X}\boldsymbol{Y}'] - (E(\boldsymbol{X}))(E(\boldsymbol{Y}))'.$$

2. Let $\boldsymbol{X} = (X_1, \cdots, X_n)'$ be a vector of random variables and let

$$Y_1 = X_1, Y_i = X_i - X_{i-1}, \quad i = 2, \cdots, n.$$

   If the Y_i's are mutually independent random variables each with unit variance, find $D(\boldsymbol{X})$.

3. If $X_1, \cdots, X_n$ are random variables each with variance σ^2, and $X_{i+1} = \rho X_i + a$ $(i = 1, \cdots, n)$ where a and ρ are constants, find $D(\boldsymbol{X})$. [$\text{Cov}(X_i, X_j) = \sigma^2 \rho^{|i-j|}$]

4. If $\boldsymbol{A}$ is a symmetric matrix and $\boldsymbol{X}$ is a random vector, prove that

$$E(\boldsymbol{X}'\boldsymbol{A}\boldsymbol{X}) = tr[\boldsymbol{A}E(\boldsymbol{X}\boldsymbol{X}')].$$

5. If $X_1, \cdots, X_n$ are mutually independent r.v.'s with common mean μ and variance $\sigma_1^2, \cdots, \sigma_n^2$ respectively, prove that

$$E\left[\frac{1}{n(n-1)}\sum_{i=1}^{n}(X_i - \bar{X})^2\right] = V(\bar{X}).$$

6. The r.v.'s $X_1, \cdots, X_n$ have a common mean μ, a common variance σ^2 and a common correlation coefficient ρ. Find $V(\bar{X})$. Show that

$$E\left\{\sum_{i=1}^{n}(X_i - \bar{X})^2\right\} = (n-1)(1-\rho)\sigma^2.$$

7. Let $X_1, \cdots, X_n$ be i.i.d. $N(\mu, \sigma^2)$ variables. Define

$$S^2 = \sum_{i=1}^{n}(X_i - \bar{X})^2/(n-1), \quad Q = \sum_{i=1}^{n-1}(X_{i+1} - X_i)^2/2(n-1).$$

(a) Prove that $V(S^2) = 2\sigma^4/(n-1)$.
(b) Show that $E(Q) = \sigma^2$.

8. Show that if in a p-variate distribution all the correlations $\rho_{ij} (i \neq j)$ are equal to ρ, then

$$\rho \geq -\frac{1}{p-1}.$$

9. Show that

$$1 - \rho_{1p\cdot 23\cdots p-1}^2 = \frac{1-\rho_{1\cdot 23\cdots p}^2}{1-\rho_{1\cdot 23\cdots (p-1)}^2}.$$

10. Show that if $\rho_{1\cdot 23\cdots p}$ is zero, partial correlation coefficients of all orders vanish. Show also that $\rho_{1\cdot 23\cdots p}^2 \leq 1$.

11. Let X_1, X_2, X_3 be three variates each with mean zero and variance unity. Find the correlation coefficient between $X_1 + X_3$ and $X_2 + X_3$ in terms of ρ_{12}, ρ_{13} and ρ_{23} and show that it is equal to

(a) $\dfrac{\rho_{12}+1}{2}$ if $\rho_{13} = \rho_{23} = 0$; (b) $\dfrac{\rho_{12}+3}{4}$ if $\rho_{13} = \rho_{23} = 1$.

12. Prove that the necessary and sufficient conditions for the coincidence of three regression planes for a bivariate distribution is

$$\rho_{12}^2 + \rho_{13}^2 + \rho_{23}^2 - 2\rho_{12}\rho_{13}\rho_{23} = 1.$$

13. Show that $\rho_{1\cdot 23} \geq \rho_{12}$.

14. Show that

$$1 - \rho^2_{1\cdot 23\cdots p} = \left(1 - \rho^2_{12}\right)\left(1 - \rho^2_{13\cdot 2}\right)\cdots\left(1 - \rho^2_{1p\cdot 23\cdots p-1}\right).$$

Hence show that if $\rho_{1\cdot 23\cdots p}$ is zero, X_1 is uncorrelated with each of the other variables.

15. If the random vector $\boldsymbol{X} = (X_1, \cdots, X_4)'$ have covariance matrix

$$\sum = \begin{bmatrix} \sigma^2 & \sigma_{12} & \sigma_{13} & \sigma_{14} \\ \sigma_{12} & \sigma^2 & \sigma_{14} & \sigma_{13} \\ \sigma_{13} & \sigma_{14} & \sigma^2 & \sigma_{12} \\ \sigma_{14} & \sigma_{13} & \sigma_{12} & \sigma^2 \end{bmatrix}$$

show that the four multiple correlation coefficients between one variable and the other three are equal.

16. Suppose $\boldsymbol{X}_{(p\times 1)}$ follows Multinomial $(n; \pi_1, \cdots, \pi_p)$ distribution. Show that the distribution of $\boldsymbol{X}$ is singular. Also prove that the regression of X_1 on $(X_2, \cdots, X_q)$ $(q < p)$ is linear and conditional variance of X_1 given $(X_2, \cdots, X_q)$ depends on the conditioning variables.

Show that

$$\rho_{12.34\cdots k} = -\left[\pi_1\pi_2/(1 - \pi_1 - \pi_3 - \cdots - \pi_k)(1 - \pi_2 - \pi_3 - \cdots - \pi_k)\right]^{1/2}$$

for $k = 3, 4, \cdots, p$.

17. Show that in large samples from a symmetric distribution

(a) $V(m'_{2r}) = V(m_{2r})$;

(b) $\text{Cov}(m'_1, m_r) = 0, \quad r = 2, 4, \cdots$;

(c) $\text{Cov}(m_{2r}, m_{2s}) = \text{Cov}(m'_{2r}, m'_{2s})$;

(d) $\text{Cov}(m_r, m_{r+1}) = 0$.

[*Hints:* Use the fact that for a symmetric distribution all central moments of odd order vanish.]

18. Prove the following results:

(a) $V(m_3) = \frac{1}{n}[\mu_6 - \mu_3^2 - 6\mu_4\mu_2 + 9\mu_2^3]$;

(b) $V(m_4) = \frac{1}{n}[\mu_8 - \mu_4^2 - 8\mu_5\mu_3 + 16\mu_2\mu_3^2]$;

(c) $V(k_4) = V(m_4) - 12\mu_2\,\text{Cov}(m_4, m_2) + 36\mu_2^2 V(m_2) = \frac{1}{n}[\mu_8 - 12\mu_6\mu_2 - 8\mu_5\mu_3 - \mu_4^2 + 48\mu_4\mu_2^2 + 64\mu_3^2\mu_2 - 36\mu_2^4]$ when k_4 is the sample cumulant of order 4;

(d) $V(b_2) = \frac{1}{n}[\beta_6 - 6\beta_4\beta_2 + 4\beta_2^3 - \beta_2^2 + 16\beta_2\beta_1 - 8\beta_3 + 16\beta_1]$ when $\beta_{2n+1} = \mu_3\mu_{2n+3}/\mu_2^{n+3}$, $\beta_{2n} = \mu_{2n+2}/\mu_2^{n+1}$;

(e) $m'_r \xrightarrow{a.s.} \mu'_r$; $m_r \xrightarrow{a.s.} \mu_r$.

19. Let $b_k = \frac{1}{n}\sum_{i=1}^{n}(X_i - \mu)^k$. Show that

(a) $E(b_k) = \mu_k$;

(b) For $\mu_{2k} < \infty$, $\text{Var}(b_k) = \frac{\mu_{2k} - \mu_k^2}{n}$;

(c) $b_k \xrightarrow{a.s.} \mu_k$.

20. Show that in R^k, the random vector $\boldsymbol{X}_n = (X_{n1}, \cdots, X_{nk})'$ converges in distribution to the random vector $\boldsymbol{X}$ iff each linear combination of the components of $\boldsymbol{X}_n$ converges in distribution to the same linear combination of the components of $\boldsymbol{X}$. **(Cramer and Wald, 1936)**

21. Let $\boldsymbol{X}_1, \boldsymbol{X}_2, \cdots$ and $\boldsymbol{X}$ be random k-vectors defined on a probability space and let g be a vector-valued Borel function defined on R^k. Suppose that g is continuous with P_X-probability unity. Then, show that

(a) $\boldsymbol{X}_n \xrightarrow{a.s.} \boldsymbol{X}$ implies $g(\boldsymbol{X}_n) \xrightarrow{a.s.} g(\boldsymbol{X})$;

(b) $\boldsymbol{X}_n \xrightarrow{P} \boldsymbol{X}$ implies $g(\boldsymbol{X}_n) \xrightarrow{P} g(\boldsymbol{X})$;

(c) $\boldsymbol{X}_n \xrightarrow{L} \boldsymbol{X}$ implies $g(\boldsymbol{X}_n) \xrightarrow{L} g(\boldsymbol{X})$.

Hence, prove that

(a) if $X_n \xrightarrow{L} N(0,1)$, then $X_n^2 \xrightarrow{L} \chi^2_{(1)}$;

(b) if $(X_n, Y_n) \xrightarrow{L} N(0, I)$, then $X_n/Y_n \xrightarrow{L}$ Cauchy.

22. *Continuation.* Suppose that the k-vector $\boldsymbol{X}_n$ converges to the k-vector $\boldsymbol{X}$ almost surely or in probability or in distribution. Let $\boldsymbol{A}_{m\times k}$ and $\boldsymbol{B}_{k\times k}$ be matrices. Then, show that $\boldsymbol{A}\boldsymbol{X}_n \to \boldsymbol{A}\boldsymbol{X}$ and $\boldsymbol{X}'_n\boldsymbol{B}\boldsymbol{X}_n \to \boldsymbol{X}'\boldsymbol{B}\boldsymbol{X}$ in the given mode of convergence.

23. Let $\mu'_{2k} < \infty$. Show that the random vector $\sqrt{n}(m'_1 - \mu'_1, \cdots, m'_k - \mu'_k)'$ converges in distribution to k-variate normal distribution with mean vector $(0, \cdots, 0)'$ and covariance matrix $((\sigma_{ij}))_{k\times k}$ where $\sigma_{ij} = \mu'_{i+j} - \mu'_i\mu'_j$.

24. (***Slutsky's Theorem***). Show that if $X_n, Y_n, \cdots, Z_n$ are random variables converging in probability to $a, b, \cdots, c$ respectively, then any rational function $g(X_n, Y_n, \cdots, Z_n)$ converges in probability to the constant $g(a, b, \cdots, c)$, provided the later is finite. Also, prove that for any $k > 0$, $g^k(X_n, Y_n, \cdots, Z_n) \xrightarrow{P} g^k(a, b, \cdots, c)$.

Chapter 11

A Few Basic Distributions

1.1 Introduction

ı this chapter we shall consider three basic distributions,—χ^2, t and F distributions. ection 11.2 considers χ^2 distribution, section 11.3 student's t and section 11.4, nedecor's F. Both the central and non-central distributions have been considered. istributions of different statistics useful for some common tests of significance have een considered in section 11.5. Sections 11.6, 11.7, 11.8 consider distributions of order atistics, sample cumulative distribution functions and sample quantiles respectively.

1.2 The Chi-Square Distribution

et $X_1, \cdots, X_n$ be n independent r.v.'s distributed as $N(\mu_1, \sigma_1), \cdots, N(\mu_n, \sigma_n)$ respec-vely. Then the distribution of the r.v.

$$\chi^2 = \sum_{i=1}^{n} \left(\frac{X_i - \mu_i}{\sigma_i}\right)^2$$

called a χ^2 distribution with n d.f. We shall derive the distribution of χ^2.

he joint distribution of $X_1, \cdots, X_n$ is

$$f(\mathbf{x}|\boldsymbol{\mu}, \boldsymbol{\sigma})\, dx = \frac{1}{(2\pi)^{n/2} \prod_{i=1}^{n} \sigma_i} \exp\left[-\frac{1}{2}\sum_{i=1}^{n}\left(\frac{x_i - \mu_i}{\sigma_i}\right)^2\right] \prod_{i=1}^{n} dx_i$$

here $\mathbf{x} = (x_1, \cdots, x_n)$, $\mu = (\mu_1, \cdots, \mu_n)$, $\boldsymbol{\sigma} = (\sigma_1, \cdots, \sigma_n)$.

utting $(x_i - \mu_i)/\sigma_i = z_i$,

$$f(\mathbf{z})\, dz = \frac{1}{(2\pi)^{n/2}} \exp\left[-\frac{1}{2}\sum_{i=1}^{n} z_i^2\right] \prod_{i=1}^{n} dz_i.$$

We are interested in the probability that

$$\chi^2 = \sum_{i=1}^{n} Z_i^2$$

will lie in an infinitesimal interval

$$\left(\chi^2 - \frac{1}{2}d\chi^2,\ \chi^2 + \frac{1}{2}d\chi^2\right).$$

Consider the poler transformation

$$(z_1, \cdots, z_n) \to (\chi, \theta_1, \cdots, \theta_{n-1})$$

such that

$$\begin{aligned}
z_1 &= \chi \cos\theta_1 & 0 < \chi < \infty \\
z_2 &= \chi \sin\theta_1 \cos\theta_2 & 0 < \theta_i < \pi,\ i = 1, \cdots, n-2 \\
&\vdots & 0 < \theta_{n-1} < 2\pi \\
z_{n-1} &= \chi \sin\theta_1 \sin\theta_2 \cdots \sin\theta_{n-2} \cos\theta_{n-1} \\
z_n &= \chi \sin\theta_1 \sin\theta_2 \cdots \sin\theta_{n-2} \sin\theta_{n-1}.
\end{aligned}$$

The Jacobian of the transformation is

$$|J| = \chi^{n-1} \sin^{n-2}\theta_1 \sin^{n-3}\theta_2 \cdots \sin\theta_{n-2}.$$

The p.d.f. of the joint distribution of $\chi, \theta_1, \theta_2, \cdots, \theta_{n-1}$ is therefore,

$$\frac{1}{(2\pi)^{n/2}} \exp\left[-\frac{1}{2}\chi^2\right] \chi^{n-1} \sin^{n-2}\theta_1 \sin^{n-3}\theta_2 \cdots \sin\theta_{n-2} d\chi\, d\theta_1 \cdots\ d\theta_{n-2}\, d\theta_{n-1}.$$

Hence the probability that χ will lie within an infinitesimal interval of length $d\chi$ is

$$\frac{1}{(2\pi)^{n/2}} e^{-\chi^2/2} \chi^{n-1}\, d\chi \int_0^{\pi} \sin^{n-2}\theta_1\, d\theta_1 \int_0^{\pi} \sin^{n-3}\theta_2\, d\theta_2$$

$$\times \int_0^{\pi} \sin^{n-4}\theta_3\, d\theta_3 \cdots \int_0^{\pi} \sin\theta_{n-2}\, d\theta_{n-2} \int_0^{2\pi} d\theta_{n-1}.$$

Now

$$\int_0^{\pi} \sin^p\theta\, d\theta = 2\int_0^{\pi/2} \sin^p\theta d\theta = \frac{\Gamma\left(\frac{p+1}{2}\right)\Gamma\left(\frac{1}{2}\right)}{\Gamma\left(\frac{p+2}{2}\right)}.$$

[Recall that

$$B(m, n) = 2\int_0^{\pi/2} \sin^{2m-1}\theta \cos^{2n-1}\theta\, d\theta$$

vide (6.5.2)].

ence,

$$P\left\{\chi - \frac{1}{2}d\chi \le \chi \le \chi + \frac{1}{2}d\chi\right\} = \frac{1}{(2\pi)^{n/2}}e^{-\frac{1}{2}\chi^2}\chi^{n-1}\,d\chi\frac{\left(\sqrt{\frac{1}{2}}\right)^{n-2}\cdot 2\pi}{\sqrt{\frac{n}{2}}}$$

$$= \frac{1}{2^{\frac{n}{2}-1}\sqrt{\frac{n}{2}}}e^{-\frac{1}{2}\chi^2}\chi^{n-1}d\chi \quad (11.2.1)$$

$$= \frac{1}{2^{\frac{n}{2}}\sqrt{\frac{n}{2}}}e^{-\frac{1}{2}\chi^2}(\chi^2)^{\frac{n}{2}-1}d\chi^2. \quad (11.2.2)$$

11.2.1) gives the p.d.f. of χ and (11.2.2) the p.d.f. of χ^2. It follows from (6.4.12) that $^2/2$ is a Gamma variable with parameter $p = n/2$. The properties of a χ^2 distribution, herefore, follow from those of a Gamma distribution. Thus, for χ^2 distribution

$$\mu_1' = n, \qquad \mu_2' = n(n+2),\ \mu_r' = 2^r\frac{\sqrt{\frac{n}{2}+r}}{\sqrt{\frac{n}{2}}},\ \mu_2 = 2n,\ \mu_3 = 8n,\ \mu_4 = 12n(n+4)$$

$$\sqrt{\beta_1} = \frac{2\sqrt{2}}{\sqrt{n}},\ \beta_2 = 3 + \frac{12}{n}.$$

The m.g.f. is $M(t) = (1-2t)^{-n/2}, t < \frac{1}{2}$.

As $n \to \infty$, the standardised χ^2-variable, $(\chi^2 - n)/\sqrt{2n}$ converges in law to a standard $N(0,1)$ variable (example 9.2.9). We shall often write $\chi^2_{(n)}$ to denote a χ^2 random variable with n.d.f.

The $\chi^2_{(n)}$ distribution is tabulated for values of $n = 1, 2, \cdots, 30$ (table A.5 in the Appendix). For $n > 30$, it is possible to use normal approximation.

It follows that (Fisher, 1922) if X follows $\chi^2_{(n)}$, then

$$\lim_{n\to\infty} P\left\{\sqrt{2X} \le z + \sqrt{2n-1}\right\} = \Phi(z)$$

when $\Phi(z)$ is the area to the left of the ordinate at z of a standard normal variate.

It follows from the additivity property of a Gamma variable (theorem 6.4.1) that the χ^2-distribution has the following reproductive property. However, a different proof is given below.

Theorem 11.2.1 (***Additivity property of Central*** χ^2). *If χ_1^2, χ_2^2 are two independent χ^2 variables with n_1, n_2 d.f. respectively, then $\chi_1^2 + \chi_2^2$ follows a χ^2 with (n_1+n_2) d.f.*

Proof: The joint distribution of χ_1, χ_2 (where χ_i is the positive square root of $\chi_i^2, i = 1, 2$) is

$$f(\chi_1)f(\chi_2)d\chi_1 d\chi_2 = \frac{1}{2^{\frac{n_1+n_2-4}{2}}\sqrt{\frac{n_1}{2}}\sqrt{\frac{n_2}{2}}}e^{-\frac{1}{2}(\chi_1^2+\chi_2^2)}\chi_1^{n_1-1}\chi_2^{n_2-1}d\chi_1 d\chi_2.$$

Put

$$\chi_1 = \chi \cos\theta, \quad \chi_2 = \chi \sin\theta, \quad 0 < \theta < \frac{\pi}{2}.$$

Thus

$$\chi^2 = \chi_1^2 + \chi_2^2, \quad |J| = \chi.$$

The joint distribution of χ^2 and θ is

$$\begin{aligned} g(\chi^2, \theta) d\chi^2\, d\theta &= \frac{1}{2^{(n_1+n_2-4)/2}\,\Gamma\frac{n_1}{2}\,\Gamma\frac{n_2}{2}} \exp[-\chi^2/2]\chi^{n_1+n_2-1}(\cos\theta)^{n_1-1}(\sin\theta)^{n_2-1} d\chi\, d\theta \\ &= \left[\frac{1}{2^{(n_1+n_2)/2}\,\Gamma\frac{n_1+n_2}{2}} e^{-\chi^2/2}(\chi^2)^{(n_1+n_2)/2-1} d\chi^2\right] \\ &\quad \left[\frac{2}{B\left(\frac{n_1}{2}, \frac{n_2}{2}\right)}(\cos\theta)^{n_1-1}(\sin\theta)^{n_2-1}\, d\theta\right]. \end{aligned} \quad (11.2.3)$$

The marginal distribution of χ^2 is obtained by integrating (11.2.3) with respect t θ over $(0, \frac{\pi}{2})$. It follows by formulae (6.5.2) that the integral over the second term i (11.2.3) has value 1. Hence the variable $\chi^2 = \chi_1^2 + \chi_2^2$ follows a χ^2 distribution wit $(n_1 + n_2)$ d.f.

Corollary 11.2.1 *The marginal distribution of θ is*

$$\frac{2}{B\left(\frac{n_1}{2}, \frac{n_2}{2}\right)} \cos^{n_1-1}\theta \sin^{n_2-1}\theta\, d\theta.$$

Define

$$\frac{\chi_1^2/n_1}{\chi_2^2/n_2} = F, \text{ i.e., } F = \frac{n_2 \cot^2\theta}{n_1}.$$

Thus

$$\sin^2\theta = \left(1 + \frac{n_1 F}{n_2}\right)^{-1}, \quad \cos^2\theta = \frac{n_1 F/n_2}{1 + n_1 F/n_2}, \quad dF = \frac{n_2}{n_1} \operatorname{cosec}^2\theta\, d\theta.$$

The distribution of F is, therefore,

$$\begin{aligned} f(F) dF &= \frac{2}{B\left(\frac{n_1}{2}, \frac{n_2}{2}\right)} \left\{\frac{n_1 F/n_2}{1 + \frac{n_1 F}{n_2}}\right\}^{\frac{n_1-1}{2}} \times \frac{1}{\left(\frac{n_1 F}{n_2} + 1\right)^{\frac{n_2-1}{2}}} \times \frac{dF}{\frac{n_2}{n_1}\left(1 + \frac{n_1 F}{n_2}\right)} \\ &= \frac{1}{B\left(\frac{n_1}{2}, \frac{n_2}{2}\right)} \frac{\left(\frac{n_1 F}{n_2}\right)^{\frac{n_1}{2}-1} d\left(\frac{n_1}{n_2}F\right)}{\left(1 + \frac{n_1 F}{n_2}\right)^{\frac{n_1+n_2}{2}}}, \quad 0 \le F < \infty. \end{aligned} \quad (11.2.4)$$

so that $n_1 F/n_2$ follows a Beta-distribution of the second type [vide (6.5.7)] wit parameters $\left(\frac{n_1}{2}, \frac{n_2}{2}\right)$. The distribution (11.2.4) is called Snedecor's F-distribution wit (n_1, n_2) d.f. and would be studied in details in section 11.4.

Corollary 11.2.2 *If $\chi_1^2, \cdots, \chi_r^2$ are independent r.v.'s such that χ_j^2 follows chi-square with n_j d.f. then the sum $\chi^2 = \sum_{j=1}^{r} \chi_j^2$ follows χ^2 with $n = \sum_{j=1}^{r} n_j$ d.f.*

We consider another property of a χ^2-distribution. Suppose $X_i (i = 1, \cdots, n)$ are independent standard normal variables. We are interested in the conditional distribution of $\sum_{i=1}^{n} X_i^2$ when the r.v.'s satisfy $m(< n)$ independent linear homogeneous constraints given by

$$\left.\begin{array}{ll} a_{11}x_1 + a_{12}x_2 + \cdots + a_{1n}x_n & = 0 \\ a_{21}x_1 + a_{22}x_2 + \cdots + a_{2n}x_n & = 0 \\ \cdots\cdots\cdots\cdots\cdots\cdots\cdots\cdots & \\ a_{m1}x_1 + a_{m2}x_2 + \cdots + a_{mn}x_n & = 0 \end{array}\right\}. \tag{11.2.5}$$

Theorem 11.2.2 *The conditional distribution of $\chi^2 = \sum_{i=1}^{n} X_i^2$ under the condition (11.2.5) is a χ^2-distribution with $(n - m)$ d.f.*

Proof: The conditions (11.2.5) may be expressed as

$$\begin{array}{ll} l_{11}x_1 + l_{12}x_2 + \cdots + l_{1n}x_n & = 0 \\ \cdots\cdots\cdots\cdots\cdots\cdots\cdots\cdots & \\ l_{m1}x_1 + l_{m2}x_2 + \cdots + l_{mn}x_n & = 0 \end{array} \tag{11.2.6}$$

where $\mathbf{l}_i = (l_{i1}, l_{i2}, \cdots, l_{in})'(i = 1, \cdots, m)$ are m mutually orthogonal and unitary vectors. Thus

$$\boldsymbol{L}_1\mathbf{x} = 0 \tag{11.2.7}$$

where

$$\boldsymbol{L}_1 = \begin{bmatrix} \boldsymbol{l}_1' \\ \vdots \\ \boldsymbol{l}_m' \end{bmatrix},$$

$\mathbf{x} = (x_1, \cdots, x_n)'; \boldsymbol{L}_1\boldsymbol{L}_1' = \boldsymbol{I}_m, \boldsymbol{I}_p$ denoting a unit matrix of order $p \times p$.

Let us make the transformation $(x_1, \cdots, x_n) \to (y_1, \cdots, y_n)$ such that

$$\begin{array}{ll} y_1 & = l_{11}x_1 + \cdots + l_{1n}x_n \\ \cdots\cdots & \cdots\cdots\cdots\cdots\cdots\cdots \\ y_m & = l_{m1}x_1 + \cdots + l_{mn}x_n \\ y_{m+1} & = l_{m+11}x_1 + \cdots + l_{m+1n}x_n \\ \cdots\cdots & \cdots\cdots\cdots\cdots\cdots\cdots \\ y_n & = l_{n1}x_1 + \cdots + l_{nn}x_n \end{array} \tag{11.2.8}$$

where the vectors

$$\mathbf{l}_{m+1} = (l_{m+11}, \cdots, l_{m+1n})', \cdots, \mathbf{l}_n = (l_{n1}, \cdots, l_{nn})'$$

are each unitary, mutually orthogonal and are also orthogonal to $\mathbf{l}'_1, \cdots, \mathbf{l}'_m$. Thus

$$\mathbf{y} = L\mathbf{x} \tag{11.2.9}$$

where

$$L = \begin{bmatrix} L_1 \\ L_2 \end{bmatrix}, \quad L_2 = \begin{bmatrix} \mathbf{l}'_{m+1} \\ \vdots \\ \mathbf{l}'_n \end{bmatrix}, \quad L_2 L'_2 = I_{(n-m)}, \quad L_1 L'_2 = \mathbf{0}, \quad \mathbf{y} = (y_1, \cdots, y_n)'.$$

Clearly,

$$L'L = [L'_1 L'_2] \begin{bmatrix} L_1 \\ L_2 \end{bmatrix} = \begin{bmatrix} I_m & 0 \\ 0 & I_{(n-m)} \end{bmatrix} = I_n.$$

Under the transformation (11.2.7), $Y_1, \cdots, Y_n$ are independent standard normal variates. Also

$$\sum_{i=1}^{n} y_i^2 = \mathbf{y}'\mathbf{y} = \mathbf{x}'L'L\mathbf{x} = \sum_{i=1}^{n} x_i^2.$$

The problem, therefore, reduces to that of finding the conditional distribution of $\sum_{i=1}^{n} Y_i^2$ where $Y_i (i = 1, \cdots, n)$ satisfy the conditions

$$Y_1 = 0, \quad \cdots, \quad Y_m = 0 \tag{11.2.10}$$

with probability one, by virtue of (11.2.6) and (11.2.8). This problem is, therefore same as finding the unconditional distribution of $\sum_{i=m+1}^{n} Y_i^2$ where Y_i's are independent standard normal variables. Thus

$$\chi^2 = \sum_{i=m+1}^{n} Y_i^2 = \sum_{i=1}^{n} X_i^2$$

subject to the constraint (11.2.6), follows $\chi^2_{(n-m)}$.

► **Example 11.2.1** *Show that if n is even,*

$$P = \frac{1}{2^{n/2}\Gamma(n/2)} \int_x^\infty \exp(-\chi^2/2)\chi^{n-1} d\chi$$

$$= \exp(-\chi^2/2) \left[1 + \frac{\chi^2}{2} + \cdots + \frac{\chi^{n-1}}{2.4.6\cdots(n-2)}\right].$$

Hence the value of P for given χ^2 can be derived from tables of Poisson exponential limit.

Solution: See theorem 6.4.2. Put $\chi^2/2 = u$.

The above distributions of χ^2 [as given in (11.2.2)] is called the central χ^2-distribution. We now consider non-central χ^2-distributions.

Definition 11.2.1 *Let $X_1, \cdots, X_n$ be independent normal r.v.'s with $E(X_i) = \mu_i$ and $V(X_i) = \sigma^2$ $(i = 1, \cdots, n)$. Let $Y = \sum_{i=1}^{n} X_i^2/\sigma^2$. Then the r.v. Y is said to follow a non-central chi-squared distribution with non-centrality parameter (n.c.p.) $\lambda = \sum_{i=1}^{n} \mu_i^2/\sigma^2$ and n.d.f. We will write $Y \sim \chi^2(n, \lambda)$.*

We shall find the distribution of $\chi^2(1, \lambda)$.

Theorem 11.2.3 *Let $X \sim N(\mu, \sigma^2)$. Then the p.d.f. of $Y = X^2/\sigma^2$ is*

$$f(y) = e^{-\lambda/2} \sum_{j=0}^{\infty} \left(\frac{\lambda}{2}\right)^j \cdot \frac{1}{j!} \cdot \frac{1}{2^{\frac{2j+1}{2}} \Gamma\left(\frac{2j+1}{2}\right)} e^{-y/2} y^{j-\frac{1}{2}}, \quad 0 \le y < \infty \tag{11.2.11}$$

where $\lambda = \mu^2/\sigma^2$. (11.2.11) is the p.d.f. of a $\chi^2(1, \lambda)$ variable.

Proof: The distribution of X is

$$g(x)\, dx = \frac{1}{\sigma\sqrt{2\pi}} \exp\left[-\frac{1}{2\sigma^2}(x - \mu)^2\right], \quad -\infty < x < \infty.$$

Let us put $z = x^2$. Corresponding to values of z in each interval on the axis $\mathbb{R}^1[0, \infty)$, x will lie in two intervals on the real line $\mathbb{R}^1(-\infty, \infty)$ and hence the chance that z will lie in an infinitesimal interval $\left(z - \frac{1}{2}\, dz,\ z + \frac{1}{2}\, dz\right)$ is equal to the sum of the chances that x will lie in $\left(-x - \frac{1}{2}\, dx,\ -x + \frac{1}{2}\, dx\right)$ and in $\left(x - \frac{1}{2}\, dx,\ x + \frac{1}{2}\, dx\right)$.
Therefore, probability differential of Z is

$$\begin{aligned}
f(z)\, dz &= \frac{1}{\sigma\sqrt{2\pi}} \left[e^{-\frac{1}{2\sigma^2}(\sqrt{z}-\mu)^2} + e^{-\frac{1}{2\sigma^2}(-\sqrt{z}-\mu)^2}\right] \frac{dz}{2\sqrt{z}} \\
&= \frac{1}{\sigma\sqrt{2\pi}} e^{-\frac{1}{2\sigma^2}(z+\mu^2)} \frac{dz}{2\sqrt{z}} \left[e^{\mu\sqrt{z}/\sigma^2} + e^{-\mu\sqrt{z}/\sigma^2}\right] \\
&= \frac{1}{\sigma\sqrt{2\pi}} e^{-\frac{1}{2\sigma^2}(z+\mu^2)} \cosh\left(\frac{\mu\sqrt{z}}{\sigma^2}\right) z^{-1/2}\, dz \quad \left[\because \cosh x = \frac{e^x + e^{-x}}{2}\right] \\
&= \frac{1}{\sqrt{2\pi}} e^{-\frac{1}{2\sigma^2}(z+\mu^2)} \left(\frac{z}{\sigma^2}\right)^{-1/2} d\left(\frac{z}{\sigma^2}\right) \sum_{j=0}^{\infty} \left(\frac{\mu}{\sigma}\sqrt{\frac{z}{\sigma^2}}\right)^{2j} \cdot \frac{1}{2j!}.
\end{aligned} \tag{11.2.12}$$

Let us put $\mu^2/\sigma^2 = \lambda, z/\sigma^2 = y$. Then (11.2.12) gives the probability differential of Y as

$$f(y)\, dy = \frac{1}{\sqrt{2\pi}} e^{-(\lambda+y)/2} y^{-\frac{1}{2}} \sum_{j=0}^{\infty} \frac{(\lambda y)^j}{2j!}\, dy, \quad y > 0. \tag{11.2.13}$$

Now

$$\Gamma(2j+1) = \frac{\Gamma\left(\frac{2j+1}{2}\right)}{\sqrt{\pi}} 2^{2j}\,\Gamma(j+1) \quad \text{[by the relation (6.4.6)].}$$

Hence (11.2.13) is

$$e^{-(\lambda+y)/2} y^{-\frac{1}{2}} \sum_{j=0}^{\infty} \frac{(\lambda y)^j}{\Gamma\left(\frac{2j+1}{2}\right) 2^{2j+\frac{1}{2}}\,\Gamma(j+1)}\, dy$$

$$= e^{-\lambda/2} \sum_{j=0}^{\infty} \left(\frac{\lambda}{2}\right)^j \cdot \frac{1}{j!} \cdot \frac{e^{-y/2}}{2^{(2j+1)/2}\,\Gamma\left(\frac{2j+1}{2}\right)} \cdot y^{(2j-1)/2}\, dy, \quad y > 0 \tag{11.2.14}$$

$$= e^{-\lambda/2} \sum_{j=0}^{\infty} \left(\frac{\lambda}{2}\right)^j \cdot \frac{1}{j!} \phi(y|2j+1, 0)\, dy \tag{11.2.15}$$

where $\phi(y|p, 0)$ is the p.d.f. of a central χ^2 with p.d.f.

The p.d.f. of a $\chi^2(n, \lambda)$ variable can be shown to be (for a proof see theorem 11.2.15)

$$f_n(y) = \frac{1}{\sqrt{\pi} 2^{n/2}} \exp\left\{-\frac{1}{2}(\lambda + y)\right\} y^{(n-2)/2} \sum_{j=0}^{\infty} \frac{(\lambda y)^j}{2j!} \frac{\Gamma\left(j+\frac{1}{2}\right)}{\Gamma\left(j+\frac{1}{2n}\right)}, \quad y > 0 \tag{11.2.16}$$

$$= e^{-\lambda/2} \sum_{j=0}^{\infty} \left(\frac{\lambda}{2}\right)^j \frac{1}{j!} \frac{e^{-y/2}}{2^{(2j+n)/2}\,\Gamma\left(\frac{2j+n}{2}\right)} y^{(2j+n-2)/2}, \quad y > 0 \tag{11.2.17}$$

$$= e^{-\lambda/2} \sum_{j=0}^{\infty} \left(\frac{\lambda}{2}\right)^j \cdot \frac{1}{j!} \phi(y|2j+n, 0). \tag{11.2.18}$$

For $\lambda = 0$, (11.2.18) reduces to the central $\chi^2_{(n)}$ p.d.f. given in (11.2.2).
The m.g.f. of $\chi^2(n, \lambda)$ is

$$M(t) = E\left(e^{t\Sigma_{i=1}^n X_i^2/\sigma^2}\right) = \prod_{i=1}^{n} E\left(e^{tX_i^2/\sigma^2}\right)$$

where $X_1, \cdots, X_n$ are independent and X_i is $N(\mu_i, \sigma^2)$.
Now

$$E\left(e^{tX_i^2/\sigma^2}\right) = \frac{1}{\sigma\sqrt{2\pi}} \int_{-\infty}^{\infty} e^{tx_i^2/\sigma^2 - (x_i-\mu_i)^2/2\sigma^2}\, dx_i, \quad t < \frac{1}{2}$$

$$= \frac{1}{\sqrt{1-2t}} \exp\left\{\frac{t\mu_i^2}{\sigma^2(1-2t)}\right\}, \quad t < \frac{1}{2} \tag{11.2.19}$$

Hence,

$$M(t) = (1-2t)^{-n/2} \exp\left\{\frac{t}{1-2t}\lambda\right\}, \quad t < \frac{1}{2}. \tag{11.2.20}$$

The mean and variance of $\chi^2(n,\lambda)$ can be easily calculated.

$$E(Y) = \frac{1}{\sigma^2}E\left(\sum_{i=1}^{n} X_i^2\right) = \frac{n\sigma^2 + \sum_1^n \mu_i^2}{\sigma^2} = n + \lambda. \tag{11.2.21}$$

$$V(Y) = \frac{1}{\sigma^4}V\left(\sum_{1}^{n} X_i^2\right) = \frac{1}{\sigma^4}\left\{\sum_{i=1}^{n} E(X_i^4) - \sum_{i=1}^{n}\left(E(X_i^2)\right)^2\right\}$$

$$= \frac{1}{\sigma^4}\left\{\sum_{i=1}^{n}\left(3\sigma^4 + 6\sigma^2\mu_i^2 + \mu_i^4\right) - \sum_{i=1}^{n}\left(\sigma^2 + \mu_i^2\right)^2\right\} = 2(n + 2\lambda). \tag{11.2.22}$$

It readily follows from (11.2.20) that if $Y_1, \cdots, Y_k$ are independent and $Y_i \sim \chi^2(n_i, \lambda_i)$ then $\Sigma_i Y_i$ follows $\chi^2(\Sigma n_i, \Sigma\lambda_i)$. A direct proof is given in theorem 11.2.6.

Theorem 11.2.4 *If we have two independent variables X, Y where $X \sim \chi^2_{(r)}$ and $Y \sim \chi^2(1,\lambda)$ then $Z = X + Y$ follows a $\chi^2(r+1,\lambda)$.*

Proof: The joint distribution of X and Y is

$$\frac{1}{2^{\frac{r}{2}}\Gamma\frac{r}{2}} e^{-x/2} x^{\frac{r}{2}-1}\, dx \sum_{j=0}^{\infty}\left(\frac{\lambda}{2}\right)^j \cdot \frac{1}{j!} \cdot \frac{e^{-\frac{1}{2}(y+\lambda)} y^{(2j+1)/2-1}}{2^{(2j+1)/2}\Gamma\frac{2j+1}{2}}\, dy. \tag{11.2.23}$$

Let us make the polar transformation

$$X = Z\cos^2\theta, \quad Y = Z\sin^2\theta; \quad |J| = Z.$$

The joint distribution of Z and θ is

$$g(Z,\theta) = \frac{e^{-\lambda/2}}{2^{\frac{r}{2}}\Gamma\frac{r}{2}} e^{-z/2}\left(z\cos^2\theta\right)^{r/2-1} \sum_{j=0}^{\infty}\left(\frac{\lambda}{2}\right)^j \cdot \frac{1}{j!}$$

$$\times \frac{1}{2^{(2j+1)/2}\Gamma\frac{2j+1}{2}} (z\sin^2\theta)^{(2j-1)/2} z\, dz\, d\theta$$

$$= \left[e^{-\lambda/2}\sum_{j=0}^{\infty}\left(\frac{\lambda}{2}\right)^j \cdot \frac{1}{j!} \cdot \frac{1}{2^{(2j+r+1)/2}\Gamma\frac{2j+r+1}{2}} e^{-z/2} z^{(2j+r-1)/2}\right]$$

$$\left[\frac{1}{B\left(\frac{r}{2}, \frac{2j+1}{2}\right)}\left(\cos^2\theta\right)^{r/2-1}\left(\sin^2\theta\right)^{(2j-1)/2}\right] \tag{11.2.24}$$

which shows that Z and θ are independently distributed. Z follows $\chi^2(r+1,\lambda)$.

Theorem 11.2.5 *If $X_1, \cdots, X_n$ are independent r.v.'s with $X_i \sim N(\mu_i, \sigma^2)$ then $Y = \sum_{i=1}^{n} X_i^2/\sigma^2$ follows $\chi^2(n,\lambda)$ where n.c.p. $\lambda = \sum_{i=1}^{p}\mu_i^2/\sigma^2$ and the distribution of Y is given in (11.2.18).*

Proof: The joint probability distribution of $X_1, \cdots, X_n$ is

$$f(\mathbf{x})\,d\mathbf{x} = \frac{1}{(\sqrt{2\pi}\sigma)^n} \exp\left[-\frac{1}{2\sigma^2}\sum_{i=1}^{n}(x_i - \mu_i)^2\right] \prod dx_i. \tag{11.2.25}$$

Consider the orthogonal transformation $(x_1, \cdots, x_n) \to (y_1, \cdots, y_n)$.

$$y_1 = \frac{(\mu_1 x_1 + \cdots + \mu_n x_n)}{\sqrt{\sum_{i=1}^{n} \mu_i^2}} \tag{11.2.26}$$

$$y_j = (l_{j1}x_1 + \cdots + l_{jn}x_n), \; j = 2, \cdots, n$$

where the vectors $\mathbf{l}_j = (l_{j1}, \cdots, l_{jn})'(j = 2, \cdots, n)$ are each unitary, mutually orthogonal and orthogonal to $l_1 = (\mu_1/\Sigma\mu_i^2, \cdots, \mu_n/\Sigma\mu_i^2)'$. Then $\Sigma x_i^2 = \Sigma y_i^2$ and $|J| = 1$.

The joint distribution of $Y_1, \cdots, Y_n$ is

$$f(\mathbf{y})\,d\mathbf{y} = \frac{1}{(\sqrt{2\pi}\sigma)^n} \exp\left[-\frac{1}{2\sigma^2}\Sigma\left(y_i^2 - 2y_1\sqrt{\Sigma\mu_i^2} + \mu_i^2\right)\right]\prod dy_i$$

$$= \frac{1}{\left(\sqrt{2\pi}\sigma\right)^n} \exp\left[-\frac{1}{2\sigma^2}\left(y_1 - \sqrt{\Sigma\mu_i^2}\right)^2 - \frac{1}{2\sigma^2}\Sigma_2^n\, y_i^2\right]\prod dy_i \tag{11.2.27}$$

which shows that $Y_1^2/\sigma^2 \sim \chi^2(1, \lambda), \;\; \lambda = \Sigma\mu_i^2/\sigma^2$ and $\Sigma_2^n Y_i^2/\sigma^2 \sim \chi^2_{(n-1)}$.

Hence, by theorem 11.2.4, $\Sigma_1^n Y_i^2/\sigma^2 = \Sigma_1^n X_i^2/\sigma^2$ follows $\chi^2(n, \lambda)$.

Corollary 11.2.3 *If $X_1, \cdots, X_n$ are independent and $X_i \sim N(\mu_i, \sigma_i^2)$ then $\Sigma_1^n X_i^2/\sigma_i^2$ follows $\chi^2(n, \lambda)$ where $\lambda = \Sigma_1^n \mu_i^2/\sigma_i^2$.*

Proof: Consider orthogonal transformation

$$(x_1/\sigma_1, \cdots, x_n/\sigma_n) \to (y_1, \cdots, y_n), \;\; y_i = \sum_1^n \frac{\mu_i x_i}{\sigma_i} \Bigg/ \sqrt{\sum_1^n \frac{\mu_i^2}{\sigma_i^2}}, y_2, \cdots, y_n$$

are suitably chosen. Then the joint p.d. of $(Y_1, \cdots, Y_n)$ is obtained from $(X_1, \cdots, X_n)$ as

$$\frac{1}{\left(\sqrt{2\pi}\sigma\right)^n} \exp\left[-\frac{1}{2}\left\{\left(y_1 - \sqrt{\sum\frac{\mu_i^2}{\sigma_i^2}}\right)^2 + \sum_2^n y_i^2\right\}\right]\prod dy_i$$

which shows that

$$\sum Y_i^2 = \sum \frac{X_i^2}{\sigma_i^2} \sim \chi^2(n, \lambda).$$

Theorem 11.2.6 (*Additivity property of non-central* χ^2). *If Y_1, Y_2 are independent and $Y_i \sim \chi^2(r_i, \lambda_i)$ then $Y_1 + Y_2 \sim \chi^2(r_1 + r_2, \lambda_1 + \lambda_2)$.*

Proof: The joint probability distribution of Y_1, Y_2 is

$$f(y_1, y_2)\, dy_1\, dy_2 = e^{-(\lambda_1+\lambda_2)/2} \left[\sum_{j=0}^{\infty} \left(\frac{\lambda_1}{2}\right)^j \cdot \frac{1}{j!} \cdot \frac{1}{2^{(2j+r_1)/2}\Gamma\frac{2j+r_1}{2}} e^{-y_1/2} (y_1)^{(2j+r)/2-1}\right]$$
$$\left[\sum_{k=0}^{\infty} \left(\frac{\lambda_2}{2}\right)^k \cdot \frac{1}{k!} \cdot \frac{1}{2^{(2k+r_2)/2}\Gamma\frac{2k+r_2}{2}} e^{-y_2/2} (y_2)^{(2k+r_2)/2-1}\right] dy_1\, dy_2. \tag{11.2.28}$$

Since this is a product of two series of positive terms we can write it as a series of double summation.

$$e^{-(\lambda_1+\lambda_2)/2} \sum_{j=0}^{\infty}\sum_{k=0}^{\infty} \left(\frac{\lambda_1}{2}\right)^j \left(\frac{\lambda_2}{2}\right)^k \cdot \frac{1}{j!k!}$$
$$\phi(y_1|2j + r_1, 0)\phi(y_2|2k + r_2, 0)\, dy_1\, dy_2.$$

Hence, by additivity property of central χ^2, the probability distribution of Y is

$$\begin{aligned} f(y)\, dy &= e^{-(\lambda_1+\lambda_2)/2} \sum_{j=0}^{\infty}\sum_{k=0}^{\infty} \left(\frac{\lambda_1}{2}\right)^j \left(\frac{\lambda_2}{2}\right)^k \frac{1}{j!k!} \phi\left(y|2(j+k) + (r_1 + r_2), 0\right) dy \\ &= e^{-(\lambda_1+\lambda_2)/2} \sum_{t=0}^{\infty} \phi\left(y|2t + r_1 + r_2, 0\right) \frac{1}{t!}\, dy \sum\sum_{j,k:j+k=t} (j+k)! \left(\frac{\lambda_1}{2}\right)^j \left(\frac{\lambda_2}{2}\right)^k \frac{1}{j!k!} \\ &= e^{-(\lambda_1+\lambda_2)/2} \sum_{t=0}^{\infty} \phi(y|2t + r_1 + r_2, 0) \left(\frac{\lambda_1}{2} + \frac{\lambda_2}{2}\right)^t \cdot \frac{1}{t!}\, dy \end{aligned} \tag{11.2.29}$$

which is $\chi^2(r_1 + r_2, \lambda_1 + \lambda_2)$ distribution.

Theorem 11.2.7 *If $Q_i \cap \chi^2_{(r_i)}, i = 1, 2$ for $r_1 > r_2$ and if $Q = Q_1 - Q_2$ is statistically independent of Q_2 then $Q \cap \chi^2_{(r_1-r_2)}$.*

Proof: The m.g.f. of Q_1 is

$$E\left(t^{tQ_1}\right) = (1 - 2t)^{-r_1/2} = E\left[e^{t(Q+Q_2)}\right] = E\left[e^{tQ}\right](1 - 2t)^{-r_2/2}.$$

Hence,

$$E\left[e^{tQ}\right] = (1 - 2t)^{-(r_1-r_2)/2}.$$

We state below without proof Cochran's theorem on partitioning of χ^2.

Theorem 11.2.8 *If $X_i (i = 1, \cdots, n)$ are independently distributed $N(0,1)$ variable and if*

$$\sum_{i=1}^{n} X_i^2 = Q_1(\mathbf{X}) + \cdots + Q_k(\mathbf{X})$$

where $Q_j(\mathbf{X}) = \mathbf{X}'\boldsymbol{Q}_j\mathbf{X}$ is a positive definite or positive semidefinite quadratic form in $\mathbf{X} = (\mathbf{X}_1, \cdots, \mathbf{X}_n)'$ and if rank $(\boldsymbol{Q}_j) = r_j (j = 1, \cdots, k)$ then a necessary and sufficient condition that Q_j's will be independently distributed as central χ^2 with r_j d.f. is that

$$n = \sum_{j=1}^{k} r_j.$$

11.3 Student's *t*-distribution

We now consider the distribution of Student's t statistic which arises in sampling from a normal population.

Definition 11.3.1 *Let $X \sim N(0,1)$ and $Y^2 \sim \chi^2_{(n)}$ and let X and Y^2 be independent. Then the statistic*

$$T = \frac{X}{\sqrt{Y^2/n}} \tag{11.3.1}$$

is said to have a Student's t-distribution with n.d.f., $T \sim t_n$.

We now derive the probability distribution of T.
The joint p.d.f. of X and $Y = +\sqrt{Y^2}$ is

$$f(X,Y) = \frac{1}{\sqrt{2\pi}2^{(n-2)/2}\Gamma\frac{n}{2}} \exp\left[-\frac{1}{2}(x^2+y^2)\right] y^{n-1}, \quad -\infty < x < \infty, \; 0 < y < \infty.$$

Let us make the polar transformation

$$(x,y) \to (r,\theta), \; x = r\cos\theta, \; y = r\sin\theta, \; 0 < r < \infty, \; 0 < \theta < \pi.$$

Then $x^2 + y^2 = r^2, J = r$. Hence the p.d.f. of r and θ is

$$g_1(r,\theta) = \frac{1}{2^{(n-1)/2}\Gamma\frac{n}{2}\sqrt{\pi}} \exp\left[-\frac{1}{2}r^2\right] r^n \sin^{n-1}\theta.$$

The marginal p.d.f. of θ is, therefore,

$$g(\theta) = C \cdot \sin^{n-1}\theta, \; 0 < \theta < \pi \tag{11.3.2}$$

where C is a suitable constant and is

$$\frac{1}{B\left(\frac{n}{2}, \frac{1}{2}\right)}$$

[by (6.5.2)]. Now

$$t = \frac{x}{\sqrt{\frac{y}{n}}} = \sqrt{n}\cot\theta.$$

Hence

$$\theta = \cot^{-1}(t/\sqrt{n}), \quad \sin^{n-1}\theta = \left(1 + \frac{t^2}{n}\right)^{-(n-1)/2}.$$

Therefore from (11.3.2), the p.d.f. of T is obtained as

$$f_n(t) = \frac{1}{B\left(\frac{1}{2}, \frac{n}{2}\right)} \cdot \frac{1}{\sqrt{n}\left(1 + \frac{t^2}{n}\right)^{(n+1)/2}}, \quad -\infty < t < \infty. \tag{11.3.3}$$

Note that T^2/n is a beta variable of the second type with parameters $\left(\frac{1}{2}, \frac{n}{2}\right)$.

The p.d.f. of T is symmetric about $t = 0$. Also $f_n(t) \to 0$ as $t \to \pm\infty$. For large n the t-distribution is close to the normal distribution. For small n, however, there is considerable difference between the two distributions.

We shall denote $t_{n,\alpha}$ to denote the upper $\alpha\%$ point of a t-distribution, i.e.,

$$P\{T_n > t_{n,\alpha}\} = \alpha.$$

Since the distribution is symmetric in t, therefore,

$$P\{|T_n| > t_{n,\alpha/2}\} = \alpha.$$

Also, obviously,

$$t_{n,1-\alpha} = -t_{n,\alpha}.$$

It follows that for small n,

$$P\{|T| \geq t_0\} > P\{|\tau| > t_0\}$$

where $\tau \sim N(0,1)$. This means that for small n, t-distribution contains more probability mass at the tails than a standard normal distribution.

The table A.6 in the appendix gives values of $t_{n,\alpha}$ for different values of n and α.

It is seen that for $n = 1, T$ is a Cauchy r.v. We shall, therefore, assume that $n > 1$.

It is easy to prove that for this distribution $E(T^r)$ exists for $r < n$ and $E(T^r) = 0$ for $r(< n)$ odd and

$$E(T^r) = \frac{\Gamma\left(\frac{r+1}{2}\right)\Gamma\left(\frac{n-r}{2}\right)}{\Gamma\left(\frac{1}{2}\right)\Gamma\left(\frac{n}{2}\right)} n^{r/2} \text{ for } r(< n) \text{ even.}$$

Hence

$$E(T) = 0 \text{ and } E(T^2) = V(T) = n/(n-2), \quad n > 2.$$

Definition 11.3.2 *If $X \cap N(\mu, \sigma^2)$ and $Y^2/\sigma^2 \cap \chi^2_{(n)}$ and X and Y are independent, then*

$$T = \frac{X}{\sqrt{Y^2/n}}$$

is said to follow a non-central t-distribution with d.f. n and non-centrality parameter $\lambda = \mu/\sigma$. The p.d.f. of a non-central t-distribution is given by

$$f_n(t, \lambda) = \frac{n^{\frac{n}{2}} e^{-\lambda^2/2}}{\sqrt{\pi}\,\Gamma(n/2)\,(n+t^2)^{(n+1)/2}} \sum_{j=0}^{\infty} \Gamma\left(\frac{n+j+1}{2}\right)\left(\frac{\lambda^j}{j!}\right)\left(\frac{2t^2}{n+t^2}\right)^{j+2}. \quad (11.3.4)$$

For $\lambda = 0$, (11.3.4) reduces to the p.d.f. of a central t with n.d.f.

Non-central t-distribution is not symmetric about $t = 0$. If $\lambda > 0$, the mode is shifted towards the right hand side while if $\lambda < 0$, the mode is shifted towards the left.

11.4 Snedecor's F-distribution

Definition 11.4.1 *If χ_1^2 and χ_2^2 are independent (central) χ^2 r.v.'s with m, n d.f. respectively, then the r.v. $F = (\chi_1^2/m)/(\chi_2^2/n)$ is said to have Snedecor's F-distribution with (m, n) d.f. $F \sim F(m, n)$.*

We now derive the distribution of $F(m, n)$.

Theorem 11.4.1 *The probability distribution of F is*

$$g(F)dF = \frac{1}{B\left(\frac{m}{2}, \frac{n}{2}\right)} \frac{\left(\frac{m}{n}F\right)^{m/2-1}\left(\frac{m}{n}dF\right)}{\left(1+\frac{m}{n}F\right)^{(m+n)/2}}, \quad F > 0.$$

So that $\frac{m}{n}F$ follows a Beta-distribution of the second type with parameters $\left(\frac{m}{2}, \frac{n}{2}\right)$.

Consider the marginal p.d.f. of θ as obtained in (11.2.3)

$$f(\theta)\, d\theta = \frac{1}{B\left(\frac{m}{2}, \frac{n}{2}\right)} \cos\theta^{m-1} \sin^{n-1}\theta\, d\theta, \;\; 0 < \theta < \pi.$$

We have

$$F = \frac{n}{m}\cot^2\theta.$$

'herefore,

$$f(F)dF = \frac{1}{B\left(\frac{m}{2},\frac{n}{2}\right)}\left\{\frac{mF/n}{1+\frac{mF}{n}}\right\}^{(m-1)/2}\cdot\frac{\left(\frac{m}{n}\right)}{\left(\frac{mF}{n}+1\right)^{(n-1)/2}}\frac{dF}{\left(1+\frac{mF}{n}\right)}$$

$$= \frac{1}{B\left(\frac{m}{2},\frac{n}{2}\right)}\frac{\left(\frac{m}{n}F\right)^{\frac{m}{2}-1}d\left(\frac{m}{n}F\right)}{\left(1+\frac{m}{n}F\right)^{\frac{m+n}{2}}},\quad F>0. \tag{11.4.1}$$

] **Remark 11.4.1** *If* $m = 1$, *then* $F = t_n^2$ *so that* $F(1,n)$ *and* t_n^2 *have the same istribution.*

] **Remark 11.4.2** *If* $F \sim F(m,n)$ *then* $\frac{1}{F} \sim F(n,m)$.

] **Remark 11.4.3** *If* X *is a Cauchy r.v. with parameter* $\theta = 0$, *so that its p.d.f. is*

$$f(x) = \frac{1}{\pi(1+x^2)},$$

hen X^2 *is a* $F(1,1)$ *variable.*

] **Remark 11.4.4** *Let* $F_{m,n,\alpha}$ *denote the upper* $\alpha\%$ *point of* $F_{m,n}$ *distribution,* $P\{F_{(m,n)} > F_{m,n;\alpha}\} = \alpha$. *It follows from remark* 11.4.2 *that*

$$F_{m,n;\alpha} = \frac{1}{F_{n,m;(1-\alpha)}}.$$

The table A.7 *in the appendix gives values of* $F_{m,n;\alpha}$ *for selected values of* m, n *and* α.

Theorem 11.4.2 *Let* $X \sim F(m,n)$. *Then for any positive integer* $k < \frac{n}{2}$

$$E(X^k) = \left(\frac{n}{m}\right)^k \cdot \frac{\Gamma\left(\frac{m}{2}+k\right)\Gamma\left(\frac{n}{2}-k\right)}{\Gamma\left(\frac{m}{2}\right)\Gamma\left(\frac{n}{2}\right)}. \tag{11.4.2}$$

Proof: We have

$$\int_0^\infty F^{\frac{m}{2}+k-1}\left(1+\frac{m}{n}F\right)^{-(m+n)/2} dF$$

$$= \left(\frac{n}{m}\right)^{k+(m/2)}\int_0^1 y^{k+(m/2)-1}(1-y)^{(n/2)-k-1}\,dy \tag{11.4.3}$$

vhere $y = \frac{m}{n}F/\left\{1+\frac{m}{n}F\right\}$. The integral in (11.4.3) converges for $\frac{n}{2} - k > 0$.

In particular,

$$E(X) = \frac{n}{m} \cdot \frac{\frac{m}{2}}{\frac{n}{2}-1} = \frac{n}{n-2}, \; n > 2 \tag{11.4.4}$$

$$E(X^2) = \left(\frac{n}{m}\right)^2 \cdot \frac{m(m+2)}{(n-2)(n-4)}, \; n > 4. \tag{11.4.5}$$

Hence,

$$V(X) = \frac{2n^2(m+n-2)}{m(n-2)^2(n-4)}, \; n > 4. \tag{11.4.6}$$

□ **Remark 11.4.5** *If $X \sim F(m,n)$ then*

$$Y = \frac{1}{1+\frac{m}{n}X}$$

is a beta-variable of the first kind with parameters

$$a = \frac{m+n}{2}+1, \; b = \frac{m}{2}.$$

Fisher's Z-Distribution

The statistic $Z = \frac{1}{2}\log_e F$ is known as Fisher's Z-statistic. Its distribution is obtained from (11.4.1) as

$$dP = \frac{2m^{m/2}n^{n/2}e^{mz}\,dz}{B\left(\frac{m}{2},\frac{n}{2}\right)\left(me^{2z}+n\right)^{(m+n)/2}}, \; -\infty < z < \infty.$$

The distribution (11.4.1) is a central F-distribution. We now consider non-central F-distribution.

Definition 11.4.2 *If χ_1^2 follows a $\chi^2(m,\lambda)$ and χ_2^2 a central $\chi^2_{(n)}$ then the statistic $F = \frac{\chi_1^2/m}{\chi_2^2/n}$ follows a non-central F-distribution with (m,n) d.f. and non-centrality parameter (ncp)λ. $F \sim F(m,n;\lambda)$.*

The joint probability distribution of χ_1^2 and χ_2^2 is

$$f(\chi_1^2,\chi_2^2)d\chi_1^2 d\chi_2^2 = \frac{1}{2^{\frac{n}{2}}\Gamma\frac{n}{2}} e^{-\frac{1}{2}\chi_2^2}\left(\chi_2^2\right)^{\frac{n}{2}-1} d\chi_2^2 e^{-\lambda/2}$$

$$\sum_{j=0}^{\infty}\left(\frac{\lambda}{2}\right)^j \frac{1}{j!} \cdot \frac{1}{2^{(2j+m)/2}\Gamma\frac{2j+m}{2}} e^{-\frac{1}{2}\chi_1^2}\left(\chi_1^2\right)^{(2j+m)/2-1} d\chi_1^2.$$

Let

$$\chi_1^2/\chi_2^2 = F'; \; |J| = \frac{\partial\left(\chi_1^2,\chi_2^2\right)}{\partial\left(F',\chi_2^2\right)} = \chi_2^2.$$

The joint probability distribution of F' and χ_2^2 is

$$g\left(F',\chi_2^2\right)dF'd\chi_2^2 = e^{-\lambda/2}\sum_{j=0}^{\infty}\left(\frac{\lambda}{2}\right)^j\cdot\frac{1}{j!}\cdot\frac{1}{2^{n/2}\Gamma\frac{n}{2}}e^{-\frac{1}{2}\chi_2^2(F'+1)}$$

$$\left(\chi_2^2\right)^{\frac{n}{2}-1}\frac{1}{2^{(2j+m)/2}\Gamma\frac{2j+m}{2}}\left(\chi_2^2F'\right)^{(2j+m)/2-1}\chi_2^2d\chi_2^2dF'$$

$$= e^{-\lambda/2}\sum_{j=0}^{\infty}\left(\frac{\lambda}{2}\right)^j\cdot\frac{1}{j!}\cdot\frac{1}{2^{\frac{n}{2}}\Gamma\frac{n}{2}}e^{-\frac{1}{2}\chi_2^2(F'+1)}\left(\chi_2^2\right)^{\frac{2j+m+n}{2}-1}$$

$$\frac{(1+F')^{(2j+m+n)/2-1}}{2^{(2j+m)/2}\Gamma\frac{2j+m}{2}}\frac{(F')^{(2j+m)/2-1}}{(1+F')^{(2j+m+n)/2}}d\left\{(1+F')\chi_2^2\right\}dF$$

$$= e^{-\lambda/2}\sum_{j=0}^{\infty}\left(\frac{\lambda}{2}\right)^j\frac{1}{j!}\cdot\frac{1}{B\left(\frac{2j+m}{2},\frac{n}{2}\right)}\frac{(F')^{(2j+m)/2-1}}{(1+F')^{(2j+m+n)/2}}dF'$$

$$\left[\frac{1}{2^{(2j+m+n)/2}\Gamma\frac{2j+m+n}{2}}\right.$$

$$e^{-\chi_2^2(F'+1)/2}\left\{\chi_2^2(1+F')\right\}^{(2j+m+n)/2-1}$$

$$\left.d\left\{(1+F')\chi_2^2\right\}\right]. \tag{11.4.7}$$

The distribution of F' is obtained by integrating (11.4.7) over χ_2^2. The integral of the quantity inside the curly bracket over the region $0 \le \chi_2^2 < \infty$ is unity. Hence the probability element of $F = \frac{n}{m}F'$ is

$$f(F)\,dF = e^{-\lambda/2}\sum_{j=0}^{\infty}\left(\frac{\lambda}{2}\right)^j\cdot\frac{1}{j!}\cdot\frac{1}{B\left(\frac{2j+m}{2},\frac{n}{2}\right)}$$

$$\frac{\left(\frac{m}{n}F\right)^{(2j+m)/2-1}}{\left(1+\frac{m}{n}F\right)^{(2j+m+n)/2}}d\left(\frac{m}{n}F\right),\quad F>0. \tag{11.4.8}$$

If $\lambda = 0$, (11.4.8) reduces to central $F(m,n)$-p.d.f. given in (11.4.1).
It follows that

$$E(F) = \frac{n(m+\lambda)}{m(n-2)},\quad n>2 \tag{11.4.9}$$

$$V(F) = \frac{2n^2}{m^2(n-4)(n-2)^2}\left[(m+\lambda)^2+(n-2)(m+2\lambda)\right],\quad n>4. \tag{11.4.10}$$

Theorem 11.4.3 *If χ_1^2 is a $\chi^2(m,\lambda_1)$ and χ_2^2 is a independent $\chi^2(n,\lambda_2)$ then the distribution of*

$$(\chi_1^2/m)/(\chi_2^2/n)$$

is a non-central $F(m,n;\lambda_1+\lambda_2)$.

Proof: The joint distribution of χ_1^2, χ_2^2 is

$$\begin{aligned} f(\chi_1^2,\chi_2^2)d\chi_1^2 d\chi_1^2 &= \left[e^{-\lambda_1/2}\sum_{j=0}^{\infty}\left(\frac{\lambda_1}{2}\right)^j \cdot \frac{1}{j!}\phi\left(\chi_1^2|2j+m,0\right)d\chi_1^2\right] \\ &\quad \left[e^{-\lambda_2/2}\sum_{k=0}^{\infty}\left(\frac{\lambda_2}{2}\right)^k \cdot \frac{1}{k!}\phi\left(\chi_2^2|2k+n,0\right)d\chi_2^2\right] \\ &= e^{-(\lambda_1+\lambda_2)/2}\sum_{j=0}^{\infty}\sum_{k=0}^{\infty}\left(\frac{\lambda_1}{2}\right)^j\left(\frac{\lambda_2}{2}\right)^k\frac{1}{j!k!} \\ &\quad \phi\left(\chi_1^2|2j+m,0\right)\phi\left(\chi_2^2|2k+n,0\right)d\chi_1^2 d\chi_2^2. \end{aligned} \tag{11.4.11}$$

The probability element of $F' = \chi_1^2/\chi_2^2$ is obtained by integrating (11.4.11) over the region

$$F' - \frac{1}{2}dF' \le F' \le F' + \frac{1}{2}dF'.$$

This gives

$$\begin{aligned} g(F')dF' = e^{-(\lambda_1+\lambda_2)/2}\sum_{j=0}^{\infty}\sum_{k=0}^{\infty}\left(\frac{\lambda_1}{2}\right)^j\left(\frac{\lambda_2}{2}\right)^k\frac{1}{j!k!} \\ \frac{1}{B\left(\frac{2j+m}{2},\frac{2k+n}{2}\right)}\frac{(F')^{(2j+m)/2-1}}{(1+F')^{(2j+2k+m+n)/2}}dF' \end{aligned} \tag{11.4.12}$$

by theorem 11.4.1.

The expression (11.4.12) reduces to

$$e^{-(\lambda_1+\lambda_2)/2}\sum_{t=0}^{\infty}\frac{(F')^{(2t+m)/2-1}}{B\left(\frac{2t+m}{2},\frac{n}{2}\right)(1+F')^{(2t+m+n)/2}}\left(\frac{\lambda_1+\lambda_2}{2}\right)^t\cdot\frac{1}{t!}.$$

11.5 Tests of Significance

We now consider the distributions of several statistic which are used for testing hypotheses regarding normal populations. The theory of testing hypotheses will be considered in details in chapters 17 and 18. The statistic used for testing would often involve distributions of sample mean $\bar{x}$ and sample variance

$$s^2 = \frac{1}{(n-1)}\sum_{i=1}^{n}(x_i - \bar{x})^2$$

of a sample from a normal population. These distributions will be derived in chapter 12. For the time being we shall assume that these distributions are known and the statistic used for testing these hypotheses (which are also supported by common sense) are also known. We shall in this section consider the distributions of these statistics under the null hypothesis by using the three distributions, χ^2, t and F.

11.5.1 Single Sample Problems

(**a.1**) Let $(X_1, \cdots, X_n)$ be a random sample from a $N(\mu, \sigma^2)$ population. We want to test the null hypothesis $H_\circ(\mu = \mu_0)$.

(i) Suppose σ^2 is known. The test statistic is $\tau = \sqrt{n}(\bar{x} - \mu_0)/\sigma$ where

$$\bar{x} = \frac{1}{n}\sum_{i=1}^{n} x_i.$$

Under $H_\circ, \tau$ follows $N(0,1)$ distribution.

(ii) σ^2 is unknown. An unbiased estimate of σ^2 is

$$s^2 = \sum_{i=1}^{n}(x_i - \bar{x})^2/(n-1).$$

Here the test statistic is

$$t = \sqrt{n}(\bar{x} - \mu_0)/s = \left[\sqrt{n}(\bar{x} - \mu_0)/\sigma\right]/\sqrt{s^2/\sigma^2}.$$

Since $\sqrt{n}(\bar{x} - \mu_0)/\sigma$ follows a $N(0,1)$ distribution and

$$\sum_{i=1}^{n}(x_i - \bar{x})^2/\sigma^2$$

follows a $\chi^2_{(n-1)}$ distribution under $H_\circ$ and since $\bar{x}$ and s^2 are independently distributed (vide section 12.2) it follows by theorem 11.3.1 that t follows a student's t-distribution with $(n-1)$ d.f. under $H_\circ$.

(**a.2**) Suppose we want to test the null hypothesis $H_\circ(\sigma^2 = \sigma_0^2)$ against

$$H_1(\sigma^2 < \sigma_0^2) \quad \text{or} \quad H_2(\sigma^2 > \sigma_0^2) \quad \text{or} \quad H(\sigma^2 \neq \sigma_0^2).$$

The test statistic is

$$\chi^2 = \frac{(n-1)s^2}{\sigma_0^2}.$$

Under $H_\circ, \chi^2$ follows $\chi^2_{(n-1)}$ (section 12.2).

11.5.2 Two Sample Problems

(b.1) Testing the equality of two population means. Let

$$x_{1i}(i = 1, \cdots, n_1), \ \ x_{2j}(j = 1, \cdots, n_2)$$

be two independent samples drawn from $N(\mu_1, \sigma^2), N(\mu_2, \sigma^2)$ populations respectively where σ^2 is unknown. Let $\bar{x}_1, \bar{x}_2$ be the sample means and

$$s_1^2 = \sum_{i=1}^{n_1}(x_{1i} - \bar{x}_1)^2/(n_1 - 1), \ \ s_2^2 = \sum_{j=1}^{n_2}(x_{2j} - \bar{x}_2)^2/(n_2 - 1)$$

the sample variances.
Suppose we want to test $H_0(\mu_1 = \mu_2)$.
The test statistic is

$$t = \frac{(\bar{x}_1 - \bar{x}_2) - (\mu_1 - \mu_2)}{\sqrt{s^2\left(\dfrac{1}{n_1} + \dfrac{1}{n_2}\right)}} \tag{11.5.1}$$

where

$$s^2 = \frac{(n_1 - 1)s_1^2 + (n_2 - 1)s_2^2}{\nu} = \frac{\Sigma_i(x_{1i} - \bar{x}_1)^2 + \Sigma_j(x_{2j} - \bar{x}_2)^2}{\nu}, \tag{11.5.2}$$

and $\nu = n_1 + n_2 - 2$.

The statistic s^2 is the pooled estimate of σ^2. The quantity $(n_1 - 1)s_1^2/\sigma^2$ and $(n_2 - 1)s_2^2/\sigma^2$ are independent $\chi^2_{(n_1-1)}, \chi^2_{(n_2-1)}$ variates respectively (section 12.2). Therefore, by the additivity property of χ^2 variates, $(n_1 - 1)s_1^2/\sigma^2 + (n_2 - 1)s_2^2/\sigma^2$ is $\chi^2_{(\nu)}$ and hence $E(s^2) = \sigma^2$. Now

$$\frac{t^2}{\nu} = \frac{\left[(\bar{x}_1 - \bar{\mu}_1) - (\bar{x}_2 - \mu_2)\right]^2/\sigma^2\left(\dfrac{1}{n_1} + \dfrac{1}{n_2}\right)}{\left[(n_1 - 1)s_1^2 + (n_2 - 1)s_2^2\right]/\sigma^2} \tag{11.5.3}$$

The sample mean $\bar{x}_i$ follows

$$N(\mu_i, \sigma^2/n_i), \ \ (i = 1, 2).$$

Since the samples are independent,

$$(\bar{x}_1 - \bar{x}_2) \cap N\left(\mu_1 - \mu_2, \sigma^2\left(\frac{1}{n_1} + \frac{1}{n_2}\right)\right).$$

Hence

$$\{(\bar{x}_1 - \bar{x}_2) - (\mu_1 - \mu_2)\}/\sigma\sqrt{\frac{1}{n_1} + \frac{1}{n_2}} \cap N(0, 1).$$

Therefore, the numerator of (11.5.3) follows $\chi^2_{(1)}$. Again since in sampling from normal populations, sample means and sample variances are independently distributed, the enominator of (11.5.3) is distributed independently of the numerator and is a $\chi^2_{(1)}$. herefore, $t^2/(n_1+n_2-2)$ is a beta variate of the second type with parameters $\left(\frac{1}{2}, \frac{n_1+n_2-2}{2}\right)$. Therefore, the statistic t follows student's t distribution with (n_1+n_2-2) .f.

One frequently samples from a bivariate normal population $N(\mu_1, \mu_2, \sigma_1, \sigma_2, \rho)$. It required to test $H_o(\mu_1 = \mu_2)$. Let (X_i, Y_i) $(i = 1, \cdots, n)$ be a sample from this opulation.

Then $D_i = (X_i - Y_i)$ follows univariate $N(\mu_1 - \mu_2, \sigma^2)$ where $\sigma^2 = \sigma_1^2 + \sigma_2^2 - 2\rho\sigma_1\sigma_2$. he test statistic

$$t = \frac{\sqrt{n}\bar{D}}{s_D}$$

here $s_D^2 = \sum_{i=1}^{n}(D_i - \bar{D})^2/(n-1)$ follows Student's t with $(n-1)$ d.f. under H_o. This called 'paired t'.

b.2) Testing the equality of two population variances

Let $x_{1i}(i = 1, \cdots, n_1), x_{2j}(j = 1, \cdots, n_2)$ be two independent samples from $N(\mu_1, \sigma_1^2), N(\mu_2, \sigma^2)$ populations respectively, when μ_i's are unknown. We want to test he hypothesis $\mu_0(\sigma_1^2 = \sigma_2^2)$. The test statistic is $F = \frac{s_1^2}{s_2^2}$. Consider

$$F' = \frac{\dfrac{(n_1-1)s_1^2}{\sigma_1^2}\Big/(n_1-1)}{\dfrac{(n_2-1)s_2^2}{\sigma_2^2}\Big/(n_2-1)} = F\left(\frac{\sigma_2^2}{\sigma_1^2}\right).$$

Under null hypothesis $\sigma_1^2 = \sigma_2^2$ and $F' = F$. If H_o is true, numerator of F' follows $\chi^2_{(n_1-1)}$ and denominator is an independent $\chi^2_{(n_2-1)}$. Hence F follows a $F_{(n_1-1),(n_2-1)}$.

Suppose again we want to test $H'_o(\sigma_1^2 = \theta\sigma_2^2), \theta \neq 1$. Here the test statistic is $F' = F/\theta$. Under H'_o, F' follows Snedector's $F_{(n_1-1),(n_2-1)}$. Its probability distribution is

$$\frac{1}{B\left(\frac{n_1-1}{2}, \frac{n_2-1}{2}\right)} \frac{\left(\frac{n_1-1}{n_2-1}\cdot\frac{F}{\theta}\right)^{\frac{n_1-1}{2}-1}}{\left(1+\frac{n_1-1}{n_2-1}\cdot\frac{F}{\theta}\right)^{\frac{n_1+n_2-2}{2}}} d\left(\frac{n_1-1}{n_2-1}\cdot\frac{F}{\theta}\right). \tag{11.5.4}$$

1.5.3 K-Sample Problems

uppose we have k independent samples $x_{ij}(j = 1, \cdots, n_i)$ of size n_i, drawn from k normal opulations with mean $\mu_i(i = 1, \cdots, k)$. We are interested in testing the following null ypotheses

(i) $H_{01}(\mu_1 = \mu_2 = \cdots = \mu_k|$ values of $\sigma_1, \sigma_2, \cdots \sigma_k$, are known)

(ii) $H_{02}(\mu_1 = \mu_2 = \cdots = \mu_k|\sigma_1 = \cdots = \sigma_k = \sigma_0$ but the value is unknown).

(i) Since $\bar{x}_i \cap N(\mu_i, \sigma_i), n_i(\bar{x}_i - \bar{\mu}_i)^2/\sigma_i^2 \cap \chi^2_{(1)}$ and hence

$$\sum_1^k n_i(\bar{x}_i - \mu_i)^2/\sigma_i^2 \cap \chi^2_{(k)}.$$

If H_{01} is true, the maximum likelihood estimate (m.l.e.) of the common mean valu μ is

$$\hat{\mu} = \left(\sum_{i=1}^k n_i\bar{x}_i/\sigma_i^2\right) \Big/ \left(\sum_{i=1}^k n_i/\sigma_i^2\right).$$

(The m.l.e. has been considered in chapter 16.) We shall show that under H_{o1},

$$\sum_{i=1}^k \frac{n_i(\bar{x}_i - \hat{\mu})^2}{\sigma_i^2}$$

follows $\chi^2_{(k-1)}$.

Proof: The joint probability distribution of $\bar{x}_1, \cdots, \bar{x}_k$ under $H_{o1}(\mu_1 = \cdots = \mu_k = \mu$ is

$$\frac{\sqrt{\prod_{i=1}^k n_i}}{(2\pi)^{k/2}\prod_{i=1}^k \sigma_i} \exp\left[-\frac{1}{2}\sum_{i=1}^k \frac{(\sqrt{n_i}\bar{x}_i - \sqrt{n_i}\mu)^2}{\sigma_i^2}\right] \prod_{i=1}^k d\bar{x}_i$$

Let us put $y_i = \sqrt{n_i}\bar{x}_i/\sigma_i$. Then

$$f(y_1, \cdots, y_n)\Pi\, dy_i = C\exp\left[-\frac{1}{2}\sum_{i=1}^k\left(y_i - \frac{\sqrt{n_i}}{\sigma_i}\mu\right)^2\right]\prod_{i=1}^k dy_i.$$

where $C = (2\pi)^{-k/2}$. Let us make the orthogonal transformation $(y_1, \cdots, y_n) -$ $(z_1, \cdots, z_n)$ where

$$z_1 = \left(\sum_{i=1}^k \sqrt{n_i}y_i/\sigma_i\right)\Big/\sqrt{\sum_1^k \frac{n_i}{\sigma_i^2}}$$

and $z_2, \cdots, z_n$ are chosen suitably. Therefore,

$$\begin{aligned} f(z_1, \cdots, z_n)\Pi\, dz_i &= C\exp\left[-\frac{1}{2}\left\{\sum_1^k z_i^2 - 2\mu z_1\sqrt{\sum\frac{n_i}{\sigma_i^2}} + \mu^2\sum_1^k\frac{n_i}{\sigma_i^2}\right\}\right]\prod_1^k dz_i \\ &= C\exp\left[-\frac{1}{2}\left\{\left(z_1 - \mu\sqrt{\sum\frac{n_i}{\sigma_i^2}}\right)^2 + \sum_2^k z_i^2\right\}\right]\prod dz_i \end{aligned}$$

which shows that $Z_2, \cdots, Z_k$ are each independent $N(0,1)$ variables. Hence

$$\sum_2^k Z_i^2 \cap \chi^2_{(k-1)}.$$

ow

$$\sum_{2}^{k} z_i^2 = \sum_{1}^{k} z_i^2 - z_1^2 = \sum_{1}^{k} y_i^2 - z_1^2 = \sum_{1}^{k} \frac{n_i \bar{x}_i^2}{\sigma_1^2} - \frac{\left(\sum_{1}^{k} \frac{n_i \bar{x}_i}{\sigma_i}\right)^2}{\sum \frac{n_i}{\sigma_i^2}}$$

$$= \sum_{1}^{k} \frac{n_i \bar{x}_i^2}{\sigma_i^2} - \left(\hat{\mu}\sqrt{\sum \frac{n_i}{\sigma_i^2}}\right)^2 = \sum_{1}^{k} \frac{n_i(\bar{x}_i - \hat{\mu})^2}{\sigma_i^2}.$$

Again

$$z_1 = \sum \frac{n_i \bar{x}_i}{\sigma_i^2} \Big/ \sqrt{\sum \frac{n_i}{\sigma_i^2}} \cap N\left(\mu\sqrt{\sum \frac{n_i}{\sigma_i^2}}, 1\right).$$

Hence

$$\hat{\mu} \cap N\left(\mu, \frac{1}{\sum \frac{n_i}{\sigma_i^2}}\right).$$

If the null hypothesis H_{01} does not hold then

$$\sum_{1}^{k} \frac{n_i(\bar{x}_i - \hat{\mu})^2}{\sigma_i^2} \cap \chi^2\left(k-1, \lambda = \sum_{1}^{k} \frac{n_i(\mu_i - \bar{\mu})^2}{\sigma_i^2}\right)$$

where

$$\bar{\mu} = \left(\sum_{1}^{k} n_i \mu_i / \sigma_i^2\right) \Big/ \left(\sum_{1}^{k} n_i / \sigma_i^2\right).$$

(ii) Under the null hypothesis H_{02} the m.l.e. of μ is

$$\widetilde{\mu} = \left(\sum_{i=1}^{k} n_i \bar{x}_i / \sigma^2\right) \Big/ \left(\sum_{1}^{k} n_i / \sigma^2\right) = \bar{\bar{x}}.$$

In this case $\sum_1^k n_i(\bar{x}_i - \widetilde{\mu})^2/\sigma^2$ follows $\chi^2_{(k-1)}$.

Again $(n_i - 1)s_i^2/\sigma^2 \cap \chi^2_{(n_i-1)}$ where $s_i^2 = \sum (x_{ij} - \bar{x}_i)^2/(n_i - 1)$.

Hence, $\sum_{1}^{k} (n_i - 1)s_i^2/\sigma^2 \cap \chi^2_{n-k}$ where $n = \sum_{i=1}^{k} n_i$. The test statistic is

$$F = \frac{\sum_{i=1}^{k} n_i(\bar{x}_i - \tilde{\mu})^2/(k-1)}{\sum_{i=1}^{k} (n_i - 1)s_i^2/(n-k)} \tag{11.5.5}$$

which follows $F_{((k-1),(n-k))}$ under $H_\circ$.

We give below an analytic proof of this result.

In a single sample the joint probability distribution of $\bar{x}_i$ and s_i is (vide section 12.2

$$f(\bar{x}_i, s_i)d\bar{x}_i ds_i = \text{Const.}\exp\left[-\frac{1}{2\sigma^2}\left\{n_i(\bar{x}_i-\mu_i)^2+(n_i-1)s_i^2\right\}\right]s_i^{n_i-2}d\bar{x}_i ds_i.$$

Since the samples are drawn independently, the joint probability distribution of $(\bar{x}_1, \cdots, \bar{x}_k)$, $(s_1, \cdots, s_k)$ is

$$\prod_{i=1}^{k} f(\bar{x}_i, s_i)d\bar{x}_i ds_i = \text{Const.}\exp\left[-\frac{1}{2\sigma^2}\sum_{i=1}^{k}\left\{n_i(\bar{x}_i-\mu_i)^2+(n_i-1)s_i^2\right\}\right]$$

$$\prod_{i=1}^{k} s_i^{n_i-2}\prod_{i=1}^{k} d\bar{x}_i ds_i$$

Let $y_i = \sqrt{n_i}\bar{x}_i$. We make the orthogonal transformation $(y_1, \cdots, y_k) \to (z_1, \cdots, z_k)$ where

$$\begin{aligned}
z_1 &= \frac{\sum\sqrt{n_i}y_i}{\sum\sqrt{n_i}}\\
z_2 &= \frac{\sqrt{n_1 n_2}y_1 - n_1 y_2}{\sqrt{n_1(n_1+n_2)}}\\
&\cdots\cdots\cdots\cdots\cdots\cdots\\
z_k &= \frac{(\sqrt{n_1}y_1+\cdots+\sqrt{n_{k-1}}y_{k-1})\sqrt{n_k}y_k - (n_1+\cdots+n_{k-1})y_k}{\sqrt{(n_1+\cdots+n_{k-1})(n_1+\cdots+n_k)}}.
\end{aligned}$$

The joint probability distribution of $(y_1, \cdots, y_k)$ and $(s_1, \cdots, s_k)$ is, therefore,

$$\text{Const.}\exp\left[-\frac{1}{2\sigma^2}\left\{(z_1-\tilde{\mu}\sqrt{\Sigma n_i})^2+\sum_{2}^{k} z_i^2+\sum(n_i-1)s_i^2\right\}\right]\prod_{i=1}^{k} s_i^{n_i-2}\prod_{i=1}^{k} dz_i ds_i.$$

We introduce now two sets of poder transformations;

$$\begin{aligned}
z_2 &= r\cos\theta_1\\
z_3 &= r\sin\theta_1\cos\theta_2\\
z_{k-1} &= r\sin\theta_1\sin\theta_2\cdots\sin\theta_{k-3}\cos\theta_{k-2}\\
z_k &= r\sin\theta_1\sin\theta_2\cdots\sin\theta_{k-3}\sin\theta_{k-2}\\
|J| &= r^{k-2}\sin^{k-3}\theta_1\sin^{k-4}\theta_2\cdots\sin\theta_{k-3}\\
\text{and } \sqrt{n_i-1}s_1 &= u\cos\phi_1\\
\sqrt{n_2-1}s_2 &= u\sin\phi_1\cos\phi_2\\
&\cdots\cdots\cdots\cdots\cdots\\
\sqrt{n_{k-1}-1}s_{k-1} &= u\sin\phi_1\sin\phi_2\cdots\sin\phi_{k-2}\cos\phi_{k-1}\\
\sqrt{n_k-1}s_k &= u\sin\phi_1\sin\phi_2\cdots\sin\phi_{k-2}\sin\phi_{k-1}\\
|J| &= u^{k-1}\sin^{k-2}\phi_1\sin^{k-3}\phi_2\cdots\sin\phi_{k-2}.
\end{aligned}$$

The transformed probability element is, therefore,

$$\text{Const.}\exp\left[-\frac{1}{2\sigma^2}\left\{\left(z_1-\widetilde{\mu}\sqrt{n_i}\right)^2-(r^2+u^2)\right\}\right]\prod_{i=1}^{k}(u\sin\phi_1\cdots\sin\phi_{i-1}\cos\phi_i)^{n_i-2}$$

$$\times r^{k-2}\sin^{k-3}\theta_1\cdots\sin\theta_{k-3}u^{k-1}\sin^{k-2}\phi_1\cdots\sin\phi_{k-2}\,dr\,d\theta_1$$

$$\cdots d\theta_{k-2}\,du\,d\phi_1\cdots d\phi_{k-1}.$$

Integrating over z_1 and the anguler variables $\theta_1,\cdots,\theta_{k-2}$, $\phi_1,\cdots,\phi_{k-1}$ we get the probability distribution of r and u as

$$g(r,u)\,dr\,du=\text{Const.}\exp\left[-\frac{1}{2\sigma^2}(r^2+u^2)\right]r^{k-2}u^{n-k-1}\,dr\,du.$$

Let us put $F=(r^2/k)/\left(u^2/(n-k)\right)$ so that

$$r=\sqrt{F}.u.\sqrt{\frac{k-1}{n-k}}.$$

Hence

$$g_1(F,u)dF\,du=\text{Const.}\exp\left[-\frac{u^2}{2\sigma^2}\left(1+F.\frac{k-1}{n-k}\right)\right]$$

$$F^{(k-3)/2}u^{n-2}dF\,du.$$

Integrating over u throughout the range $(0,\infty)$ the marginal distribution of F is found as

$$g(F)dF=\text{Const.}F^{(k-3)/2}dF\Big/\left(1+\frac{k-1}{n-k}F\right)^{\frac{n-1}{2}}$$

$$=\frac{1}{B\left(\frac{k-1}{2},\frac{n-k}{2}\right)}\cdot\frac{\left(\frac{k-1}{n-k}F\right)^{(k-3)/2}d\left(\frac{k-1}{n-k}F\right)}{\left(1+\frac{k-1}{n-k}F\right)^{(n-1)/2}}$$

which is a central $F_{(k-1),(n-k)}$ distribution. Now

$$u^2=\sum(n_i-1)s_i^2,\quad r^2=\sum n_i\left(\bar{x}_i-\bar{\bar{x}}\right)^2.$$

Hence the proof of the result (11.5.5).

The total sum of squares can be partitioned in the analysis of variance table as follows:

Sources of variation	$d.f.$	Sum of squares (SS)	Mean Squares $(= SS/(d.f.))$	F
Between samples	$k-1$	$\sum_{i=1}^{k} n_i(\bar{x}_i - \overline{\overline{x}})^2$	V_B	V_B/V_W
Within samples	$n-k$	$\sum_i \sum_j (x_{ij} - \bar{x}_i)^2$	V_W	
Total $S.S.$	$n-1$	$\sum_i \sum_j (x_{ij} - \overline{\overline{x}})^2$		

Clearly V_B is the numerator of (11.5.5) while V_W is its denominator. It is easy to check that under this normal model

$$E(V_B) = \sigma^2 + \frac{\sum n_i(\mu_i - \bar{\mu})^2}{k-1} \quad \text{when } \bar{\mu} = \sum n_i\mu_i / \sum n_i;$$
$$E(V_W) = \sigma^2.$$

If the null hypothesis is true, the value of V_B/W_W is expected to be near to unity. Hence a high value of F should lead to the rejection of the hypothesis that μ_i's are all equal.

If the null hypothesis is not true V_B follows a $\chi^2(k-1, \lambda)$ where

$$\lambda = \sum n_i(\mu_i - \bar{\mu})^2/(k-1)$$

and hence the statistic F will follow a non-central $F(k-1, n-k, \lambda)$.

11.6 Order Statistics

Let $X_1, \cdots, X_n$ denote a random sample of size n from a c.d.f. $F_X(x)$. Let X's be rearranged in order of magnitude from the least to the greatest denoted by $X_{(1)}, X_{(2)}, \cdots, X_{(n)}$. The statistics $(X_{(1)}, \cdots, X_{(n)})$ is called the order statistics corresponding to the sample $(X_1, \cdots, X_n)$, $X_{(r)}$ [the rth smallest observation] being called the rth order statistic.

The order statistics are not independent even if the original sample observations are so, because, if $X_{(j)} \geq y$, then $X_{(j+1)} \geq y$. For a continuous distribution, one can always take $X_{(1)} < X_{(2)} < \cdots < X_{(n)}$. We have already obtained the marginal distribution of $X_{(1)} = \min_i X_i$ and $X_{(n)} = \max_i X_i$ in section 8.6. We will now find the marginal sampling distribution of an arbitrary order statistic.

11.6.1 Exact sampling distribution of order statistics

Let $\boldsymbol{X} = (X_1, \cdots, X_n)$ be a random sample from a continuous p.d.f. $f_X(x)$ and $(X_{(1)}, \cdots, X_{(n)})$ be the order statistic corresponding to it. Since given the ordered sample $(X_{(1)}, \cdots, X_{(n)})$, there are $n!$ samples $\boldsymbol{X}$ that give rise to this ordered sample, and since the observations $X_1, \cdots, X_n$ are independent, the joint p.d.f. of $X_{(1)}, \cdots, X_{(n)}$ is

$$g_n(x_{(1)}, \cdots, x_{(n)}) = n! \prod_{i=1}^{n} f(x_{(i)}) \text{ for } x_{(1)} < \cdots < x_{(n)}. \tag{11.6.1}$$

The joint p.d.f. of any $k(< n)$ order statistics, say, $X_{(r_1)} < X_{(r_2)} < \cdots < X_{(r_k)}$ is obtained from (11.6.1) by integrating out the remaining $(n-k)$ order statistics. This is given by

$$\begin{aligned} g_k(x_{(r_1)}, \cdots, x_{(r_k)}) = & \frac{n!}{(r_1-1)!(r_2-r_1-1)!(r_3-r_2-1)!\cdots(n-r_k)!} \Big[F(x_{(r_1)})\Big]^{r_1-1} \\ & \Big[F(x_{(r_2)}) - F(x_{(r_1)})\Big]^{r_2-r_1-1} \Big[F(x_{(r_3)}) - F(x_{(r_2)})\Big]^{r_3-r_2-1} \cdots \\ & \cdots \Big[1 - F(x_{(r_k)})\Big]^{n-r_k} f(x_{(r_1)}) \cdots f(x_{(r_k)}) \\ & \text{for } x_{(r_1)} < \cdots < x_{(r_k)} \end{aligned} \tag{11.6.2}$$

$$\begin{aligned} = & \frac{n!}{\prod_{i=1}^{k+1}(r_i - r_{i-1} - 1)!} \prod_{i=1}^{k+1} \Big[F(x_{(r_i)}) - F(x_{(r_{i-1})})\Big]^{r_i - r_{i-1}-1} \prod_{i=1}^{k} f(x_{(r_i)}) \\ & \text{for } x_{(r_1)} < \cdots < x_{(r_k)} \end{aligned} \tag{11.6.3}$$

when $r_0 = 0, r_{k+1} = n+1$ and $x_{(r_0)} = -\infty, x_{(r_{k+1})} = \infty$.

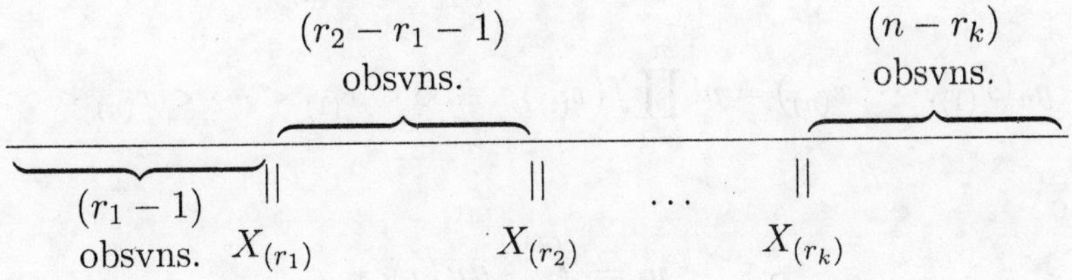

Fig. 11.1

In particular, the joint p.d.f. of two order statistic $X_{(r)}$ and $X_{(s)}$ is given by

$$\begin{aligned} g_2(x_{(r)}, x_{(s)}) = & \frac{n!}{(r-1)!(s-r-1)!(n-s)!} \Big[F(x_{(r)})\Big]^{r-1} \\ & \Big[F(x_{(s)}) - F(x_{(r)})\Big]^{s-r-1} \Big[1 - F(x_{(s)})\Big]^{n-s} f(x_{(r)}) f(x_{(s)}) \\ & \text{for } x_{(r)} < x_{(s)} \end{aligned} \tag{11.6.4}$$

The marginal p.d.f. of the rth order statistic $X_{(r)}$ is given by

$$\begin{aligned} g(x_{(r)}) &= \frac{n!}{(r-1)!(n-r)!}\left[F(x_{(r)})\right]^{r-1}\left[1-F(x_{(r)})\right]^{n-r} f(x_{(r)}) \\ &= \binom{n}{r} r \left[F(x_{(r)})\right]^{r-1}\left[1-F(x_{(r)})\right]^{n-r} f(x_{(r)}) \end{aligned} \tag{11.6.5}$$

The mean of the rth order statistic $X_{(r)}$ is

$$E(X_{(r)}) = \frac{n!}{(r-1)!(n-r)!}\int_{-\infty}^{\infty} y\,[F(y)]^{r-1}\,[1-F(y)]^{n-r}\, f(y)\,dy. \tag{11.6.6}$$

The variance of $X_{(r)}$ is $E(X_{(r)}^2) - E(X_{(r)})^2$.

The product moments of the order statistics are obtained from (11.6.3) as follows:

$$E\left(X_{(r_1)}^{\alpha_1}\cdots X_{(r_k)}^{\alpha_k}\right) = \frac{n!}{\prod_{i=1}^{k+1}(r_i - r_{i-1} - 1)!}\iiint_{-\infty<y_1<\cdots<y_k<\infty} \prod_{i=1}^{k+1}[F(y_i) - F(y_{i-1})]^{r_i - r_{i-1} - 1} \prod_{j=1}^{k}\left\{y_j^{\alpha_j} f(y_j)\,dy_j\right\}. \tag{11.6.7}$$

We now consider an important property of the order statistic. This property is very useful in the theory of non-parametric inference considered in chapter 22.

Theorem 11.6.1 *The distribution of area under the density function between any two ordered observations is independent of the form of the density function.*

Proof: From (11.6.1), the joint p.d.f. of $X_{(1)}, \cdots, X_{(n)}$ is

$$g_n(x_{(1)}, \cdots, x_{(n)}) = n!\prod_{i=1}^{n} f(x_{(i)}),\quad -\infty < x_{(1)} < \cdots < x_{(n)} < \infty.$$

Let us write

$$u_i = \int_{-\infty}^{x(i)} dF(x). \tag{11.6.8}$$

Then the joint probability distribution of u_i's $(i = 1, \cdots, n)$ is given by

$$g(u_1, u_2, \cdots, u_n)du_1\cdots du_n = n!\,du_1\cdots\, du_n,\quad 0 \le u_1 < u_2 < \cdots < u_n \le 1. \tag{11.6.9}$$

The density function of $(u_1, u_2, \cdots, u_n)$ enables us to find the distribution of areas under $f(x)$ between any two pairs of ordered sample observations. For example, suppose we want to find the distribution of the area, say v, of $f(x)$ between $x_{(1)}$ and $x_{(n)}$.

$$v = F(x_{(n)}) - F(x_{(1)}) = u_n - u_1;\quad u_n = u_1 + v$$

For fixed $v, 0 < u_1 < 1 - v$ and $v < u_n < 1$. The joint density of (u_1, u_n) is obtained rom (11.6.9) by integrating over $u_2, \cdots, u_{n-1}$.

$$h(u_1, u_n) = \int_{u_1}^{u_n} \int_{u_1}^{u_{n-1}} \cdots \int_{u_1}^{u_3} n! \prod_{i=2}^{n-1} du_i = n! \int_{u_1}^{u_n} \cdots \int_{u_1}^{u_4} (u_3 - u_1) \prod_{i=3}^{n-1} du_i$$

$$= n! \int_{u_1}^{u_n} \cdots \int_{u_1}^{u_5} \frac{(u_4 - u_1)^2}{2} \prod_{i=4}^{n-1} du_i = n! \frac{(u_n - u_1)^{n-2}}{(n-2)!}, \quad 0 \leq u_1 \leq u_n \leq 1. \tag{11.6.10}$$

The joint distribution of u_1 and $v (= u_n - u_1)$ is, therefore, given by

$$K(u_1, v)\, du_1 dv = n(n-1) v^{n-2}\, du_1 dv. \tag{11.6.11}$$

Hence the marginal distribution of v is

$$\begin{aligned} dF(v) &= dv \int_0^{1-v} n(n-1) v^{n-2}\, du_1 \\ &= n(n-1) v^{n-2} (1 - v), \quad 0 < v < 1 \end{aligned} \tag{11.6.12}$$

which is a Beta distribution with parameter $(n - 1, 2)$.

Consider the more general case. First of all we find the joint distribution of u_r and $u_s (s > r)$. This is

$$dF = du_r\, du_s \int_R n!\, du_1 \cdots du_{r-1}\, du_{r+1} \cdots du_{s-1}\, du_{s+1} \cdots du_n \tag{11.6.13}$$

where the range of integration R is:

$$\begin{aligned} &0 \leq u_1 \leq u_2 \leq \cdots \leq u_{r-1} \\ &\leq u_r - \frac{1}{2} du_r \leq u_r \leq u_r + \frac{1}{2} du_r \leq u_{r+1} \leq \cdots \leq u_{s-1} \\ &\leq u_s - \frac{1}{2} du_s \leq u_s \leq u_s + \frac{1}{2} du_s \leq u_{s+1} \leq \cdots \leq u_n \leq 1. \end{aligned}$$

For finding the value of (11.6.13) proceed as follows. Consider first

$$\begin{aligned} &\int_0^{u_r} \cdots \int_{u_{r-3}}^{u_r} \int_{u_{r-2}}^{u_r} du_1 \cdots du_{r-1}; \\ &= \int_0^{u_r} \cdots \int_{u_{r-3}}^{u_r} du_1 \cdots du_{r-2} (u_r - u_{r-2}) \text{ [integrating over } u_{r-1}] \\ &= \int_0^{u_r} \cdots \int_{u_{r-4}}^{u_r} \frac{(u_r - u_{r-3})^2}{2!} du_1 \cdots du_{r-3} \text{ [integrating over } u_{r-2}]. \end{aligned}$$

Proceeding in this way, one ultimately gets

$$\int_0^{u_r} \frac{(u_r - u_1)^{r-2}}{(r-2)!}\, du_1 = \frac{u_r^{r-1}}{(r-1)!}. \quad (11.6.1$$

Similarly for the second set consider integration over $u_{r+1}, \cdots, u_{s-1}$ and for this se the value of the integral is $\frac{(u_s-u_r)^{s-r-1}}{(s-r-1)!}$.

For the third set, the integration is over $u_{s+1}, \cdots, u_n$ and the integral becom $\frac{(1-u_s)^{n-s}}{(n-s)!}$.

Combining the joint distribution of u_r and u_s is

$$n! \frac{u_r^{r-1}}{(r-1)!} \frac{(u_s - u_r)^{s-r-1}}{(s-r-1)!} \frac{(1-u_s)^{n-s}}{(n-s)!}\, du_r\, du_s. \quad (11.6.1$$

The same distribution can also be obtained alternatively as follows.

Probability that $(r-1)$ values $u_1, \cdots, u_{r-1}$ are less than $u_r - \frac{1}{2}du_r$ one value betwee $\left(u_r - \frac{1}{2}du_r, u_r + \frac{1}{2}du_r\right)$, $(s-r-1)$ observations between $u_r + \frac{1}{2}du_r$, $u_s - \frac{1}{2}du_s$, or value between $u_s - \frac{1}{2}du_s$, $u_s + \frac{1}{2}du_s$ and $(n-s)$ values between $u_s + \frac{1}{2}du_s$ and 1 is give by (11.6.15). Let $w = u_s - u_r$, $s > r$.

From (11.6.15), the distribution of w is given by

$$dF(w) = \frac{n!}{(s-r-1)!(n-s+r)!} w^{s-r-1}(1-w)^{n-s+r}\, dw \quad (11.6.1$$

which is a Beta distribution with parameters $(s-r, n-s+r+1)$. The p.d.f. (11.6.1 is independent of $f(x)$.

Again, from (11.6.9),

$$\begin{aligned} E(u_i) &= \int_0^1 \int_0^{u_n} \cdots \int_0^{u_2} n! u_i\, du_1 \cdots du_n \\ &= \int_0^1 \frac{n!}{(i-1)!(n-i)!} u_i^i (1-u_i)^{n-i}\, du_i \qquad \text{(using the technique by which (11.6.15) was obtained)} \\ &= \frac{n!}{(i-1)!(n-i)!} B(i+1, n-i+1) = \frac{i}{n+1}. \end{aligned} \quad (11.6.1$$

Therefore, $E(u_i) - E(u_{i-1})$ = expected area under $f(x)$ between any two successive ordered observations is $i/(n+1) - (i-1)/(n+1) = 1/(n+1)$.

Thus, on the average n ordered observations divide the area under $f(x)$ into $(n+1$ equal parts of area $1/(n+1)$ each. This expected value of the area is independent o $f(x)$.

11.6.2 Distribution of functions of order statistics

We shall now consider distribution of certain functions of the order statistic.

(a) Sample median, sample range and sample mid range.

The sample median is defined to be middle order statistic if n is odd and the average of the two middle order statistics if n is even.

Thus, if $n = 2k + 1$, where k is some positive integer, then $X_{(k+1)}$ is the sample median. Its distribution can be found from (11.6.5). If $n = 2k$, the sample median is $\left[X_{(k)} + X_{(k+1)}\right]/2$, the distribution of which can be obtained by a transformation starting with the joint p.d.f. of $X_{(k)}$ and $X_{(k+1)}$, as given in (11.6.4).

For finding the p.d.f. of R consider the joint p.d.f. of $X_{(1)}, X_{(n)}$.

$$g_2(x_{(1)}, x_{(n)}) = n(n-1)\left[F(x_{(n)}) - F(x_{(1)})\right]^{n-2} f(x_{(1)})f(x_{(n)}), \quad x_{(1)} < x_{(n)}. \tag{11.6.18}$$

Let us transform $\left(X_{(1)}, X_{(n)}\right) \to (Y, R)$ when $X_{(1)} = Y, X_{(n)} = Y + R \cdot |J| = 1$.

The p.d.f. of R can be obtained from the joint p.d.f. of Y and R by integrating out Y. This is given by

$$g_R(r) = n(n-1)\int_{-\infty}^{\infty} [F(y+r) - F(y)]^{n-2} f(y+r)f(y)\, dy. \tag{11.6.19}$$

The distribution function of R is

$$G_R(r) = n\int_{-\infty}^{\infty} [F(y+r) - F(y)]^{n-1} f(y)\, dy. \tag{11.6.20}$$

Consider now the case when the p.d.f. of X is defined over the finite range (a, b). Since $Y = X_{(n)} - R$ cannot exceed the value $b - r$ for the value of R fixed at r, the p.d.f. of R is given from (11.6.19) as

$$g_R(r) = n(n-1)\int_{a}^{b-r} [F(y+r) - F(y)]^{n-2} f(y+r)f(y)\, dy.$$

Other statistics of interest are the mth extremes, that is the pair $X_{(m)}, X_{(n-m+1)}$ which are respectively the mth value from the bottom and the mth value from the top in the set of observations, absolute extreme = max $\left[X_{(n)}, -X_{(1)}\right]$ and extremal quotient $Q = X_{(n)}/(-X_{(1)})$. If the original distribution is symmetrical, the distributions of mth extremes will also be symmetrical. Each of the statistics $X_{(1)}$ and $X_{(n)}$ are called extreme value statistic.

Order statistics are very useful in problems of censored data. A typical situation arises in connection with life-testing experiments, in which a fixed number n of items are put on test and the experiment is terminated as soon as a prescribed number r have failed. The observed lifetimes are thus $X_{(1)}^{(n)} \leq \cdots \leq X_{(r)}^{(n)}$, whereas the lifetimes $X_{(r+1)}^{(n)} \leq \cdots \leq X_{(n)}^{(n)}$

remain unobserved. For further details on properties and applications of order statistics in life testing and reliability problems, the reader may refer to Mann, Schafer and Singpurwalla (1974), Gupta and Panchapakesan (1974), among others.

As an illustration of the use of order statistics, we present results from the theory of outliers suggested by Neyman and Scott (1971). Intuitively, an outlier is an observation that is sufficiently far away from the remaining observations so as to invoke suspicion that this particular observation obeyed other probability distribution than that obeyed by the remaining observations.

Definition 11.6.1 *A sample $x_1, \cdots, x_n$ contains an (r, n) right outlier, if*

$$x_{(n)} - x_{(n-1)} > r\left(x_{(n-1)} - x_{(1)}\right), \quad n \geq 3, r > 0. \tag{11.6.21}$$

Theorem 11.6.2 *A random sample $X_1, \cdots, X_n$ from the distribution with cdf F and density f will contain an (r, n) outlier with probability*

$$\pi(r, n; F) = n(n-1)\int_{-\infty}^{\infty}\int_{x}^{\infty}\left[F\left(\frac{y+rx}{r+1}\right) - F(x)\right]^{n-2} f(y)f(x)dy\,dx. \tag{11.6.22}$$

Proof: Suppose $x_{(1)} = x$ and $x_{(n)} = y(> x)$. The sample will contain an (r, n) outlier, if (11.6.21) holds, i.e., if

$$x < X_{(n-1)} < \frac{y+rx}{r+1}.$$

This event occurs if $(n-2)$ elements $X_{(2)}, \cdots, X_{(n-1)}$ of the sample lie between x and $\frac{y+rx}{r+1}$, which occurs with probability $\left[F\left(\frac{y+rx}{r+1}\right) - F(x)\right]^{n-2}$. Multiplying this by the joint density of $(X_{(1)}, X_{(n)})$ and integrating, we obtain (11.6.22).

► **Example 11.6.1** *A random sample of size n is taken from a population with p.d.f. $f(x) = e^{-x}, x \geq 0$. Find the p.d.f. of range R. Also show that $E(R) = 1+\frac{1}{2}+\frac{1}{3}+\cdots+\frac{1}{n-1}$.*

Solution: Here $F(y+r) - F(y) = e^{-y}(1-e^{-r})$.
From (11.6.19)

$$g_R(r) = (n-1)e^{-r}(1-e^{-r})^{n-2}$$

$$E(R) = (n-1)\int_0^{\infty} re^{-r}(1-e^{-r})^{n-2}\,dr.$$

Putting $1 - e^{-r} = z$ and using $\log(1-z) = z + \frac{z^2}{2} + \frac{z^3}{3} + \cdots$,

$$E(R) = (n-1)\int_0^1\left(z + \frac{z^2}{2} + \frac{z^3}{3} + \cdots\right)z^{n-2}\,dz \quad \text{(the series is uniformly convergent in (0, 1))}$$

$$= (n-1)\left[\frac{1}{n} + \frac{1}{2}\cdot\frac{1}{n+1} + \frac{1}{3}\cdot\frac{1}{n+2} + \cdots\right]$$

$$= \left(1 - \frac{1}{n}\right) + \left(\frac{1}{2} - \frac{1}{n+1}\right) + \left(\frac{1}{3} - \frac{1}{n+2}\right) + \cdots = 1 + \frac{1}{2} + \frac{1}{3} + \cdots + \frac{1}{n-1}.$$

Example 11.6.2 *A finite population consists of N distinct values* $x_1 < x_2 < \cdots < x_N$. *sample of size n is drawn from this population by simple random sampling without placement. Find the p.m.f. of the sample rth order statistic* $X_{(r)}$.

Possible values of $X_{(r)}$ are $x_r, x_{r+1}, \cdots, x_{N-n+r}$ since there are $(r-1)$ values in the mple which are less than $X_{(r)}$ and $(n-r)$ values in the sample which are greater than $_{(r)}$. Thus $X_{(r)}$ takes the value $x_m (m = r, r+1, \cdots, N-n+r)$ if there are exactly $(r-$ values in the sample which are less than x_m and $(n-r)$ values in the sample which e greater than x_m and this probability is

$$P\{X_{(r)} = x_m\} = \frac{\binom{m-1}{r-1}\binom{N-m}{n-r}}{\binom{N}{n}}, \quad m = r, r+1, \cdots, \quad N-n+r.$$

milarly,

$$P\{X_{(r)} = x_m, X_{(s)} = x_t, r < s, m < t\} = \frac{\binom{m-1}{r-1}\binom{t-m-1}{s-r-1}\binom{N-t}{n-s}}{\binom{N}{n}},$$
$$m = r, r+1, \cdots, t-s+r; \quad t = s, s+1, \cdots, N-n+s.$$

Example 11.6.3 *Consider a finite population of N units with the variable y taking ılue* Y_i *on unit* $i(=1, \cdots, N)$. *Let* $Y_{(1)} < Y_{(2)} < \cdots < Y_{(N)}$ *be the ordered values for the pulation.*

The (t/N)th quantile is $Y_{(t)}$ where t is an integer in $[1, N]$.

A simple random sample of size n is drawn from the population and let $_{1)} < y_{(2)} \cdots < y_{(n)}$ be the ordered values in the sample. Now

$$P\left(y_{(k)} \le Y_{(t)} \le y_{(r)}\right) = P\left(y_{(k)} \le Y_{(t)}\right) - P\left(y_{(r)} \le Y_{(t-1)}\right).$$

gain,

$$P\left(y_{(k)} \le Y_{(t)}\right) = P\left(\bigcup_{i=k}^{n} A_i\right),$$

here A_i is the event "exactly i sample elements have values of y with $y \le Y_{(t)}$". Thus

$$P\left(y_{(k)} \le Y_{(t)}\right) = \sum_{i=\max(k,t+n-N)}^{\min(t,n)} \binom{t}{i}\binom{N-t}{n-i} \Big/ \binom{N}{n}$$

nd $$P\left(y_{(k)} \le Y_{(t)} \le y_{(r)}\right) = \left\{\sum_{i=k}^{n}\binom{t}{i}\binom{N-t}{n-i} - \sum_{i=r}^{n}\binom{t-1}{i}\binom{N-t+1}{n-i}\right\} \Big/ \binom{N}{n}$$
$$= \left\{\sum_{i=k}^{r-1}\binom{t}{i}\binom{N-t}{n-i} + \binom{t-1}{r-1}\binom{N-t}{n-r}\right\} \Big/ \binom{N}{n}.$$

(Sedransk and Meyer, 1978)

11.6.3 Asymptotic distribution of order statistics

Let

$$X_{(1)}^{(n)}, X_{(2)}^{(n)}, \cdots, X_{(n)}^{(n)}$$

be the order statistic for a sample of size n. For any given p, $0 < p < 1$, let $p^{(n)}$ be su that $np^{(n)}$ is an integer and $p^{(n)}$ is approximately equal to p. We shall find the asympto distribution of $X_{(np^{(n)})}^{(n)}$ as n tends to ∞. This is given in the following theorem.

Theorem 11.6.3 *Let $X_1, X_2, \cdots$ be i.i.d. r.v.'s with common p.d.f. $f(\cdot)$ and cumulat d.f. $F(\cdot)$. Assume that $F(x)$ is strictly monotone for $0 < F(x) < 1$. For a giv $p(0 < p < 1)$, let ξ_p be the pth quantile i.e., $F(\xi_p) = p$. Let $p^{(n)}$ be such that $np^{(n)}$ an integer and $p^{(n)}$ is close to p. Assume that $n|p^{(n)} - p|$ is bounded. Then $X_{(np^{(n)})}^{(n)}$ asymptotically normally distributed with mean ξ_p and variance $p(1-p)/n[f(\xi_p)]^2$.*

We shall give here the proof of this theorem for the particular case $p = \frac{1}{2}$. T theorem with a slight modification is restated in Theorem 11.8.3 and a proof has be given.

Theorem 11.6.4 *For large n, the sample median $\tilde{x}$ is approximately normally d tributed with mean μ_e, the population median and variance $[4nf(\mu_e)^2]^{-1}$.*

Proof: Let $f(\mu_e) = y_0$ and $\displaystyle\int_{-\infty}^{x} f(t)\,dx = F(x)$.

Expanding $F(\tilde{x})$ around the point $F(\mu_e)$,

$$\begin{aligned} F(\tilde{x}) &= F(\mu_e) + (\tilde{x} - \mu_e)F'(\mu_e) + \frac{(\tilde{x} - \mu_e)^2}{2!}F''(\mu_e) + \cdots \\ &= \frac{1}{2} + (\tilde{x} - \mu_e)y_0 + \frac{(\tilde{x} - \mu_e)^2}{2!}f'(\mu_e) + \cdots \\ \therefore\quad f(\tilde{x}) &= y_0 + (\tilde{x} - \mu_e)f'(\mu_e) + \cdots. \end{aligned}$$

For a symmetrical distribution $f'(\mu_e)$ will be zero and for a moderately asymmetri distribution $f'(\mu_e)$ will be small. Now the p.d.f. of sample median $\tilde{x}$ is

$$g(\tilde{x}) = \frac{(2m+1)!}{m!m!}[F(\tilde{x})]^m[1 - F(\tilde{x})]^m f(\tilde{x})$$

when $n = 2m + 1$. For large m, using Stirling's approximation for factorials,

$$\begin{aligned} g(\tilde{x}) \simeq{} & \frac{\sqrt{2\pi}e^{-(2m+1)}(2m+1)^{(2m+3/2)}}{\left(\sqrt{2\pi}e^{-m}m^{m+\frac{1}{2}}\right)^2} \\ & \times \left[\frac{1}{2} + (\tilde{x} - \mu_e)y_0 + \cdots\right]^m \left[\frac{1}{2} - (\tilde{x} - \mu_e)y_0 + \cdots\right]^m y_0 \end{aligned}$$

$$= \frac{e^{-1}}{\sqrt{2\pi}} \frac{2^{2m+3/2}\left(m+\frac{1}{2}\right)^{2m+3/2}}{m^{2m+1}}$$

$$\times \left[\frac{1}{2} + (\tilde{x} - \mu_e)y_0 + \cdots\right]^m \left[\frac{1}{2} - (\tilde{x} - \mu_e)y_0 + \cdots\right]^m y_0$$

$$\simeq \frac{e^{-1}}{\sqrt{2\pi}} \sqrt{m} \cdot 2^{2m+3/2}(e) \left(\frac{1}{4}\right)^m \left[1 - 4m(\tilde{x} - \mu_e)^2 y_0^2 - \cdots\right] y_0$$

$$\left(\text{since } \lim_{m\to\infty} \left(1 + \tfrac{1}{2m}\right)^{2m+3/2} = e\right)$$

$$= \frac{1}{\sqrt{2m}} \left(\frac{1}{2\sqrt{2m}y_0}\right)^{-1} e^{-4m(\tilde{x}-\mu_e)^2 y_0^2}$$

$$= \frac{1}{\sqrt{2\pi}} \cdot \frac{1}{\sigma} e^{-(\tilde{x}-\mu_e)^2/2\sigma^2} \quad \text{when } \sigma^2 = \frac{1}{8my_0^2} \sim \frac{1}{4ny_0^2}.$$

Hence the proof.

Further properties of order statistics have been considered in chapter 22.

1.7 Sample Cumulative Distribution Function

et $X_1, \cdots, X_n$ be a random sample from the distribution $F(\cdot)$. The sample cumulative istribution function $F_n(x)$ is defined as

$$F_n(x) = \frac{1}{n} \text{ (no. of observations } X_j \leq x). \tag{11.7.1}$$

or fixed x, $F_n(x)$ is a statistic since it depends on the sample.

heorem 11.7.1 *Let $X_1, \cdots, X_n$ be a random sample from $F(\cdot)$. Then the p.m.f. of $F_n(x)$ is*

$$P\left[F_n(x) = \frac{k}{n}\right] = \binom{n}{k} (F(x))^k (1 - F(x))^{n-k}$$

.e., $F_n(x)$ follows a Binomial distribution with parameters $(n, F(x))$.

Proof: Let $Z_i = 1(0)$ if $X_i \leq x$ (otherwise). Here $P(Z_i = 1) = F(x)$. Also $Z_1, \cdots, Z_n$ re independent r.v.'s. Therefore, $\sum_{i=1}^{n} Z_i$ has a Binomial distribution $B(n, F(x))$. Again

$$P\left(\sum_{i=1}^{n} Z_i = k\right) = P\left(F_n(x) = \frac{k}{n}\right).$$

Ience the proof.

Theorem 11.7.2 (*Glivanko-Cantelli theorem*). *Let $F_n(x)$ be as defined above. The*

$$\sup_{-\infty<x<\infty} |F_n(x) - F(x)| \to 0$$

with probability 1, as $n \to \infty$. Thus, $F_n(x)$ converges to $F(x)$ for every x. Moreover, t convergence is uniform.

For a proof, the reader is referred to Fisz (1963, p. 391).

11.8 The Sample Quantiles

Corresponding to a sample $\{X_1, \cdots, X_n\}$ of observations on F, the sample pth quanti is defined as the pth quantile of the sample distribution function F_n, that is, $F_n^{-1}(p$ We denote it as Z_{pn} or simply Z_p. Clearly $Z_p \simeq X^{(n)}_{(np^{(n)})}$.
Actually,

$$\begin{aligned} Z_{pn} &= X^{(n)}_{(np)} \quad \text{if } np \text{ is an integer} \\ &= X^{(n)}_{([np]+1)} \quad \text{if } np \text{ is not an integer.} \end{aligned}$$

We state below the following properties without proof.

Theorem 11.8.1 *Let $0 < p < 1$. If ξ_p is the unique solution x of $F(x-) \leq p \leq F(x$ then $Z_{pn} \xrightarrow{a.s.} \xi_p$.*

Theorem 11.8.2 *Let $0 < p < 1$. Suppose that ξ_p is the unique solution x of $F(x-)$ $p \leq F(x)$. Then for every $\varepsilon > 0$,*

$$\begin{aligned} P\left(|Z_{pn} - \xi_p| > \varepsilon\right) &\leq 2e^{-2n\delta_\varepsilon^2} \quad \textit{for all } n, \\ \textit{where} \quad \delta_\varepsilon &= \min\left\{F(\xi_p + \varepsilon) - p, p - F(\xi_p - \varepsilon)\right\}. \end{aligned}$$

Corollary 11.8.1 *Under the assumptions of the above theorem, for every $\varepsilon > 0$,*

$$P\left(\sup_{m \geq n} |Z_{pm} - \xi_p| > \varepsilon\right) \leq \frac{2}{1-\rho_\varepsilon}\rho_\varepsilon^n \quad \textit{for all } n,$$

where $\rho_\varepsilon = \exp(-2\delta_\varepsilon^2)$ and δ_ε is as defined above.

We shall now consider the asymptotic distribution of Z_p.

Theorem 11.8.3 *If in some neighbourhood of the population quantile ξ_p, $f(x)$ continuous and has a continuous derivative and if $f(\xi_p) > 0$, the distribution of samp quantile Z_p is asymptotically*

$$N\left(\xi_p, \frac{p(1-p)}{nf^2(\xi_p)}\right).$$

Proof: Let $r = [np]+1$ or np according as np is not an integer or so. Then $Z_p = X_{(r)}^{(n)}$. The p.d.f. of Z_p is

$$f_{rn}(x) = \frac{n!}{(r-1)!(n-r)!}[F(x)]^{r-1}[1-F(x)]^{n-r}f(x).$$

Let

$$Y_{rn} = \sqrt{\frac{n}{pq}}f(\xi_p)(Z_p - \xi_p)$$

where $q = 1 - p$.
The p.d.f. of Y_{rn} is

$$\begin{aligned}
g_{rn}(y) &= \sqrt{\frac{pq}{n}}\cdot\frac{1}{f(\xi_p)}\cdot\frac{n!}{(r-1)!(n-r)!}\left[F\left\{\xi_p + \sqrt{\frac{pq}{n}}\cdot\frac{y}{f(\xi_p)}\right\}\right]^{r-1} \\
&\quad \left[1 - F\left\{\xi_p + \sqrt{\frac{pq}{n}}\cdot\frac{y}{f(\xi_p)}\right\}\right]^{n-r} f\left(\xi_p + \sqrt{\frac{pq}{n}}\cdot\frac{y}{f(\xi_p)}\right) \\
&= \left\{\sqrt{\frac{pq}{n}}\cdot\frac{n!}{(r-1)!(n-r+1)!}\,p^{r-1}q^{n-r+1} \times \frac{n-r+1}{q}\right\} \\
&\quad \left\{\frac{f\left(\xi_p + \sqrt{\frac{pq}{n}}\cdot\frac{y}{f(\xi_p)}\right)}{f(\xi_p)}\right\}\left\{\left(\frac{F(x)}{p}\right)^{r-1}\left(\frac{1-F(x)}{q}\right)^{n-r}\right\} \qquad (11.8.1) \\
&= \{A_{n1}\}\{A_{n2}\}\{A_{n3}\} \quad \text{(say)},
\end{aligned}$$

where A_{n1} is the quantity in the first curly bracket of (11.8.1) and similarly for A_{n2}, A_{n3}. From Stirling's approximation for factorials ($n! \simeq \sqrt{2\pi n}\,e^{-n}n^n$) (vide (6.4.9))

$$\begin{aligned}
\frac{n!}{(r-1)!(n-r+1)!} &\simeq \frac{n^n\sqrt{2\pi n}}{\sqrt{2\pi(r-1)}\sqrt{2\pi(n-r+1)}(r-1)^{(r-1)}(n-r+1)^{n-r+1}} \\
&\simeq \frac{1}{\sqrt{2\pi pqn}}\cdot\frac{1}{p^{r-1}q^{n-r+1}}
\end{aligned}$$

using $r - 1 \simeq np$. Using this result,

$$A_{n1} \simeq \frac{1}{\sqrt{2\pi}}\cdot\frac{n-r+1}{n}\cdot\frac{1}{q}.$$

Hence,

$$\lim_{n\to\infty} A_{n1} = \frac{1}{\sqrt{2\pi}}.$$

By the assumption that $f(x)$ is continuous at $x = \xi_p$,

$$\lim_{n\to\infty} A_{n2} = 1.$$

Again, expanding the function $F(x)$ in the neighbourhood of the point $x = \xi_p$,

$$\begin{aligned} F(x) &= F(\xi_p) + (x - \xi_p) f(\xi_p) + o(x - \xi_p) \\ &= p + \sqrt{\frac{pq}{n}} \cdot y + o\left[\sqrt{\frac{pq}{n}} \cdot \frac{y}{f(\xi_p)}\right] \\ &= p + y\sqrt{\frac{pq}{n}}\,[1 + Q(n, y)], \end{aligned}$$

where the function $Q(n, y) \to 0$ as $n \to \infty$.

Let

$$\begin{aligned} a &= \sqrt{\frac{q}{p}}\,[1 + Q(n, y)] \\ b &= \sqrt{\frac{p}{q}}\,[1 + Q(n, y)]. \end{aligned}$$

Then

$$A_{n3} = \left(1 + \frac{ay}{\sqrt{n}}\right)^{r-1} \left(1 - \frac{by}{\sqrt{n}}\right)^{n-r}.$$

Hence,

$$\log A_{n3} = (r - 1) \log \left(1 + \frac{ay}{\sqrt{n}}\right) + (n - r) \log \left(1 - \frac{by}{\sqrt{n}}\right).$$

Since $a/\sqrt{n}$, $b/\sqrt{n}$ tend to zero as $n \to \infty$, expanding log-functions and simplifying

$$\begin{aligned} \log A_{n3} &= \frac{(r-1)a - (n-r)b}{\sqrt{n}} y - \frac{(r-1)a^2 + (n-r)b^2}{n} \cdot \frac{y^2}{2} + n\,o\left(\frac{1}{n}\right) \\ &= B_{n1} - B_{n2} + n\,o\left(\frac{1}{n}\right) \quad \text{(say)}, \end{aligned} \tag{11.8.2}$$

where the first term in (11.8.2) is B_{n1}. Substituting for a and b

$$B_{n1} = \left[\left(\frac{r}{n} - p\right)\sqrt{n} - \frac{q}{\sqrt{n}}\right] \frac{1 + Q(n, y)}{\sqrt{pq}} \quad \to 0 \text{ as } \quad n \to \infty.$$

Similarly,

$$\lim_{n \to \infty} B_{n2} = \lim_{n \to \infty} \left[\frac{rq}{np} + \left(1 - \frac{r}{n}\right)\frac{p}{q} - \frac{q^2}{n}\right] [1 + Q(n, y)]^n = 1.$$

Also $n\,o\left(\frac{1}{n}\right) \to 0$ as $n \to \infty$. Thus, $\lim\limits_{n \to \infty} A_{n3} = e^{-y^2/2}$.

Combining,

$$\lim_{n \to \infty} g_{rn}(y) = \frac{1}{\sqrt{2\pi}} e^{-y^2/2}.$$

Thus limiting distribution of Y_{rn} is $N(0, 1)$ and hence the limiting distribution of Z is as stated.

The following is the multivariate generalisation of theorem 11.8.3.

Theorem 11.8.4 *Let $0 < p_1 < \cdots < p_k < 1$. Suppose that F has a density f in neighbourhoods of $\xi_{p1}, \cdots, \xi_{p_k}$ and that f is positive and continuous at $\xi_{p1}, \cdots, \xi_{p_k}$. Let $Z_{p1}, \cdots, Z_{p_k}$ be the corresponding sample quantiles. Then $(Z_{p1}, \cdots, Z_{p_k})'$ is asymptotically normal with mean vector $(\xi_{p1}, \cdots, \xi_{p_k})$ and covariance matrix $((\frac{\sigma_{ij}}{n}))$ when*

$$\sigma_{ij} = \frac{p_i(1-p_j)}{f(\xi_{pi})f(\xi_{pj})} \quad \text{for } i \le j$$

and $\sigma_{ij} = \sigma_{ji}$ for $i > j$.

(Multivariate normal distribution has been considered in Chapter 13).

Estimation of semi-interquartile range

As noted in section 3.8, semi-interquartile range

$$Q = \frac{1}{2}\left(\xi_{3/4} - \xi_{1/4}\right)$$

is a measure of dispersion of the distribution f. A natural estimator of Q is

$$\hat{Q} = \frac{1}{2}\left(\hat{\xi}_{3/4} - \hat{\xi}_{1/4}\right).$$

By Theorem 11.8.4, it follows that $\hat{Q}$ is asymptotically normal

$$N\left(Q, \frac{1}{64n}\left(\frac{3}{f^2(\xi_{3/4})} - \frac{2}{f(\xi_{1/4})f(\xi_{3/4})} + \frac{3}{f^2(\xi_{1/4})}\right)\right).$$

For sampling from a normal $N(\mu, \sigma^2)$ population, $\hat{Q}$ is asymptotically normal

$$N\left(0.6745\sigma, \frac{(0.7867)^2\sigma^2}{n}\right).$$

Exercise

1. Obtain the distributions of a) $u = \frac{1}{X^2}$, b) $v = \frac{X^2}{1+X^2}$.

2. If $X_1, \cdots, X_4$ are i.i.d. $N(0,1)$ variables obtain the distribution of

   (a) $u = \frac{X_1^2+X_2^2}{X_1^2+X_2^2+X_3^2}$;

   (b) $v = \frac{X_1+X_2+X_3}{\left[\frac{1}{2}(X_1-X_3)^2+\frac{1}{6}(X_1-2X_2+X_3)^2+X_4^2\right]^{1/2}}$.

3. Let X_1, X_2 be a random sample from exponential distribution $f(x) = e^{-x}, 0 < x < \infty$. Show that

   (a) $Z = X_1/X_2$ has an F distribution;

   (b) $(X_1 + X_2)$ and X_1/X_2 are independently distributed.

4. (a) Show that for $\nu > 2$, the variance of t distribution with ν d.f. is $\nu/(\nu-2)$.
   (b) Show that for t distribution with $\nu(>4)$ d.f.,
   (i) $\mu_4 = \frac{3\nu^2}{(\nu-2)(\nu-4)}$; (ii) $\beta_2 \left(= \frac{\mu_4}{\mu_2^2}\right) = 3 + \frac{6}{\nu-4}$.
   [*Hints:* Put $1 + \frac{t^2}{\nu} = \frac{1}{u}$.]

5. Let X_i $(i = 1, \cdots, n)$ be i.i.d. $N(0, \sigma^2)$ variables. Let $Y_i = \sum_{j=1}^{n} C_{ij} X_j$, where the vectors $\mathbf{C}_i = (C_{i1}, \cdots, C_{in})'$ $(i = 1, \cdots, p)$ are mutually orthogonal. Show that

$$\left(\sum_1^n X_i^2 - \sum_1^p Y_i^2\right) \Big/ \sigma^2 \cap \chi^2_{(n-p)}.$$

6. Show that the mode of the F distribution with $n_1(\geq 2), n_2$ d.f. is given by $F = \frac{n_2(n_1-2)}{n_1(n_2+2)}$ and is always less than unity. Show also that the two points of inflexion which exist for $n_1 \geq 4$ are equidistant from the mode.

7. Let the variables $X_i (i = 1, 2, 3)$ be i.i.d. $N(0, 1)$ variables. Obtain from suitable statistical tables, the probabilities,
   (a) $P\{3X_1 + 7X_2 + 5X_3 \geq 6\}$; (b) $P\{X_1^2 + 2X_2^2 + X_3^2 \geq 7\}$; (c) $P\{9X_1^2 - 4X_2^2 - X_3^2 \leq 0\}$.

8. If S_1 and S_2 are the standard deviations of independent random samples of sizes $n_1 = 61$ and $n_2 = 31$ from normal populations with $\sigma_1^2 = 12, \sigma_2^2 = 18$, find $P(S_1^2/S_2^2 > 1.16)$.

9. The claim that the variance of a normal population is $\sigma^2 = 4$ is to be rejected if the variance of a random sample of size 9 exceeds 7.7535. What is the probability that this claim will be rejected even though $\sigma^2 = 4$?

10. Let Y_1, Y_2 be independent χ^2 variables, each with n d.f. Show that $\sqrt{n}(Y_1 - Y_2)/2\sqrt{Y_1 Y_2}$ has the t distribution with n d.f. and is independent of $Y_1 + Y_2$.

11. Using the Stirling's formula (6.4.9) show that when $\nu \to \infty$, the t distribution approaches the standard normal distribution.

12. Show that for $\nu_2 > 2$, the mean of the F_{ν_1,ν_2} distribution is $\frac{\nu_2}{\nu_2-2}$, making use of the definition of $F = (\chi_1^2/\nu_1)/(\chi_2^2/\nu_2)$ and the fact that $E\left(\frac{1}{\chi_2^2}\right) = \frac{1}{\nu_2-2}$.

13. (a) Integrating by parts, show that if t is a positive integer

$$I = \frac{1}{t!}\int_\lambda^\infty e^{-x} x^t \, dx = \frac{1}{t!}\lambda^t e^{-\lambda} + \int_\lambda^\infty \frac{1}{(t-1)!} e^{-x} x^{t-1} \, dx$$

   and by continuing the process, show that $I = e^{-\lambda}\left(1 + \lambda + \frac{\lambda^2}{2!} + \cdots + \frac{\lambda^t}{t!}\right)$.

Hence, show that the probabilities, that in random sampling the value of χ^2 with ν (an even number) d.f., will exceed χ_0^2 is obtained from this expression by putting $\lambda = \frac{1}{2}\chi_0^2$ and $t = (\nu - 2)/2$.
Hence or otherwise show that the values of $P(\chi^2 \geq \chi_0^2)$ for a given value of χ_0^2 can be obtained from the tables of Poisson distribution.

(b) For χ^2 distribution of $2(\nu + 1)$ d.f., prove that

$$P(\chi^2 \geq \chi_0^2) = e^{-\lambda} \sum_{1}^{\nu} \frac{\lambda^r}{r!} = \frac{1}{\nu!} \int_{\lambda}^{\infty} e^{-y} y^{\nu} \, dy$$

where $\lambda = \chi_0^2/2$. Explain the uses of this result.

[*Hints:* vide Theorem 6.4.2.]

14. Let $F \cap F_{m,n}$. The r.v. $Z = \frac{1}{2} \log_e F$ is known as Fisher's Z-statistic. Find the p.d.f. of Z.

15. Show that for large n, $\sqrt{2\chi^2}$ is normally distributed about mean $\sqrt{2n-1}$ with variance unity.

[*Hints:* $\sqrt{2\chi^2} \leq \sqrt{2n-1} + z \;\Leftrightarrow\; \chi^2 \leq n - \frac{1}{2} + z\sqrt{2n-1} + \frac{z^2}{2}$;

As $n \to \infty$, the last expression tends to $n + z\sqrt{2n}$. Therefore,

$$P\left[\sqrt{2\chi^2} - \sqrt{2n-1} \leq z\right] \sim P\left[\frac{\chi^2 - n}{\sqrt{2n}} \leq z\right] = \Phi(z),$$

since by the CLT, $(\chi^2 - n)/\sqrt{2n} \sim N(0, 1)$ as $n \to \infty$.]

16. Let $X_1, \cdots, X_n$ be a random sample from $N(\mu, \sigma^2)$. Define

$$\bar{X}_k = \sum_{1}^{k} X_i/k, \quad \bar{X}_{n-k} = \sum_{k+1}^{n} X_i/(n-k), \bar{X} = \sum_{1}^{n} X_i/n,$$

$$S_k^2 = \sum_{1}^{k} (X_i - \bar{X}_k)^2/(k-1), \quad S_{n-k}^2 = \sum_{k+1}^{n} (X_i - \bar{X}_{n-k})^2/(n-k-1),$$

$$S^2 = \sum_{1}^{n} (X_i - \bar{X})^2/(n-1).$$

Find the distribution of

(a) $\left[(k-1)S_k^2 + (n-k-1)S_{n-k}^2\right]/\sigma^2$,

(b) $(\bar{X}_k + \bar{X}_{n-k})/2$,

(c) S_k^2/S_{n-k}^2. **[Mood, Graybill and Boes, 1974]**

17. If $X_i (i = 1, \cdots, n)$ are a random sample from $N(\mu, \sigma^2)$, show that

$$\sqrt{\frac{n}{n-1}}(X_1 - \bar{X}) \Big/ \sqrt{\{(n-1)S^2 - \frac{n}{n-1}(X_1 - \bar{X})^2\}/(n-2)},$$

follows t distribution with $(n-2)$ d.f.

18. Let a sample of size n_1 from $N(\mu_1, \sigma_1^2)$ have sample variance S_1^2 and anothe independent sample of size n_2 from $N(\mu_2, \sigma_2^2)$ have sample mean $\bar{X}_2$ and varianc S_2^2. Find the joint density of $T_1 = \frac{\sqrt{n_2}(\bar{X} - \mu_2)}{S_2}$ and $T_2 = \frac{S_1^2}{S_2^2}$.

19. Let $F \sim F_{m,n}$. Show that $Y = \left(1 + \frac{m}{n}F\right)^{-1} \sim B\left(\frac{n}{2}, \frac{m}{2}\right)$.

Show also that for $x > 0$, $P\{F \leq x\} = 1 - P\left\{Y \leq \left(1 + \frac{m}{n}x\right)^{-1}\right\}$.

20. The variates $X_1, \cdots, X_n$ are independently distributed uniformly over $(0, 1)$ a $dF = dx$. Show that $-2 \log_e \left(\prod_1^n X_i\right)$ is distributed $\chi^2_{(2n)}$.

21. The r.v.'s $X_1, \cdots, X_n$ are mutually independent, $X_i \cap N(\mu_i, \sigma_i^2)$. Let $\bar{X}$ = $\left(\Sigma_i X_i/\sigma_i^2\right)\left(\Sigma_i 1/\sigma_i^2\right)^{-1}$. Show that $\Sigma_i (X_i - \bar{X})^2/\sigma_i^2 \cap \chi^2_{(n-1)}$.

22. Show that the $F_{(4,4)}$ distribution is given by

$$g_F(f) = \begin{bmatrix} 6f(1+f)^{-4}, & f > 0 \\ 0, & \text{otherwise} \end{bmatrix}$$

and use this density to find the probability that for independent random sample of size 5 from normal populations with the same variance, S_1^2/S_2^2 will take on value less than $\frac{1}{2}$ or greater than 2.

23. If $X_i (i = 1, \cdots, n)$ are a random sample from $N(0, 1)$, show that

$$Y_1 = X_1 \Big/ \sqrt{\sum_2^n X_i^2},\ Y_2 = X_2 \Big/ \sqrt{\sum_3^n X_i^2}, \cdots, Y_{n-1} = X_{n-1}\Big/\sqrt{X_n^2},$$

$$Y_n = \sum_i^n X_i^2$$

are independently distributed and determine their distributions.

24. Verify that if F has F_{ν_1, ν_2} distribution and $\nu_2 \to \infty$, the distribution of $Y = \nu_1 F$ approaches the $\chi^2_{(\nu_1)}$.

25. Assuming Stirling's approximating formula $n! \sim e^{-n} n^n \sqrt{2\pi n}$ prove that for n sufficiently large and $h (> 0)$ fixed, $\overline{|n+h} \sim n^h \overline{|(n)}$.

Hence, show that if X has a Beta distribution $f_X(x) = \frac{1}{B(p,q)}x^{p-1}(1-x)^{q-1}$, $(0 \leq X \leq 1)$, then for large p and relatively small q, a first approximation gives $-(2p)\log_e X \cap \chi^2_{(2q)}$.

Also prove that as an improved approximation $-(2p+q-1)\log_e X \cap \chi^2_{(2p)}$.

[*Hints:* Put $x = e^{-y}$. Then the distribution of Y is

$$\frac{1}{B(p,q)}e^{-py}y^{q-1}\left[1 - \frac{y}{2} + \frac{y^2}{6} - \cdots\right]^{q-1} dy \quad (0 \leq y < \infty)$$

$$\sim \frac{p^q}{\lceil(q)}e^{-py}y^{q-1}\,dy;$$

The improved estimation is obtained by considering the p.d.f. of Y as $\propto e^{-py}y^{q-1}e^{-(q-1)y/2}$.]

26. If X, Y are independent random variables such that X is $N(0,1)$ and Y is $\chi^2_{(n)}$, prove that $T = \frac{X\sqrt{n}}{\sqrt{Y}}$ has Student's $t_{(n)}$ distribution. Assuming that for sufficiently large n,

$$\frac{\lceil\{(n+1)/2\}}{\sqrt{(n/2)}\lceil(n/2)} \sim \left(1 - \frac{1}{4n}\right)$$

derive Bartlett's approximation that $\left(n - \frac{1}{2}\right)\log_e\left[1 + \frac{T^2}{n}\right]$ is distributed as $\chi^2_{(1)}$.

[*Hints:*

$$f_T(t)dt = \frac{dt}{\sqrt{n}B(n/2,1/2)\left(1+\frac{t^2}{n}\right)^{(n+1)/2}}, \quad -\infty < t < \infty.$$

Hence p.d.f. of $W = 1 + \frac{T^2}{n}$ is

$$\frac{1}{B(n/2,1/2)}w^{(n-2)/2}(1-w)^{-1/2}dw, \quad 0 \leq w \leq 1$$

and the distribution of $Y = n\log W$ is proportional to

$$e^{-y/2}y^{-1/2}\left[1 - \frac{y}{2n} + 0(n^{-2})\right] dy \simeq e^{-(n+1)y/2n}y^{-1/2}.]$$

27. Let $X \cap N(0,\sigma_1^2), Y \cap N(0,\sigma_2^2)$. Find the probability distribution of $U = X/Y$. Hence, prove that if $(X_1,\cdots,X_n)$ and $(Y_1,\cdots,Y_n)$ are independent samples from $N(0,\sigma_1^2), N(0,\sigma_2^2)$ respectively, then the statistics

$$V = \sum_1^n X_i \Big/ \sum_1^n Y_i \text{ and } W = \frac{1}{n}\sum_1^n (X_i/Y_i)$$

have identical distributions.

[*Hints:* The joint p.d.f. of U, Y is

$$\frac{1}{2\pi\sigma_1\sigma_2}\exp\left\{-\frac{y^2}{2}\left(\frac{u^2}{\sigma_1^2} + \frac{1}{\sigma_2^2}\right)\right\}|y|\,dy\,du, \quad -\infty < u, y < \infty.$$

Integration over y gives p.d.f. of U as

$$f(u)\,du = \frac{\mu}{\pi} \cdot \frac{du}{u^2+\mu^2}, \quad \mu = \sigma_1/\sigma_2, (-\infty < u < \sigma). \qquad \text{(i}$$

Since $\Sigma_1^n X_i, \Sigma_1^n Y_i$ are independent $N(0, n\sigma_1^2), N(0, n\sigma_2^2)$ respectively, the p.d.f. of V is same a that of U.

Let $r_i = x_i/y_i$. From (i), the jt. p.d.f. of $R_1, \cdots, R_n$ is

$$\left(\frac{\mu}{\pi}\right)^n \prod_{i=1}^{n} \frac{dr_i}{(\mu^2+r_i^2)} \quad (-\infty < r_i < \infty). \qquad \text{(ii}$$

Let $Z_i = R_i (i = 1, \cdots, n-1), Z_n = \sum_1^n R_i$. Integrating out successively $Z_{n-1}, Z_{n-2}, \cdots, Z_1$ b the method of partial fractions the p.d.f. of Z_n is obtained from (ii) as

$$\frac{\mu}{\pi} \frac{n\,dz_n}{(n\mu)^2+z_n^2}, \quad (-\alpha < z_n < \infty,].$$

28. Let $(X_1, \cdots, X_m)$ be a random sample from $N(\mu_1, \sigma_1^2)$ and let $(Y_1, \cdots, Y_n)$ be random sample from $N(\mu_2, \sigma_2^2), m < n$. Scheffé's test for the difference $\mu_1 - \mu_2$ use a statistic

$$T = \sqrt{m(m-1)}(\bar{X} - \bar{Y}) / \left[\sum_{j=1}^{m} (U_j - \bar{U})^2\right]^{1/2}$$

when

$$\bar{x} = \sum_1^m x_i/m, \quad \bar{y} = \sum_1^n y_i/n, u_j = x_j - \sqrt{\frac{m}{n}} y_j \quad (j = 1, \cdots, m),$$

$$\bar{u} = \sum_1^m u_j/m.$$

Obtain the distribution of T assuming $\mu_1 = \mu_2$.

[*Hints:* Show that $E\left\{\frac{1}{m}\sum_{j=1}^{m}(U_j - \bar{U})^2\right\} = \sigma_1^2/m + \sigma_2^2/n, T \cap t_{(m-1)}$.]

29. Using (11.6.4), show that the joint p.d.f. of sample range $R = X_{(n)} - X_{(1)}$ an sample mid-range $T = \left[X_{(1)} + X_{(n)}\right]/2$ is given by

$$g_{R,T}(r,t) = n(n-1)\left[F\left(t+\frac{r}{2}\right) - F\left(t-\frac{r}{2}\right)\right]^{n-2} f\left(t-\frac{r}{2}\right) f\left(t+\frac{r}{2}\right), r > 0.$$

30. A sample of size n is drawn from the uniform density $f(x) = \frac{1}{\theta}, 0 \leq n \leq \theta$ Show that the p.d.f. of range R is $g_R(r) = (n-1)\theta^{-n} r^{n-2}(\theta - r)$ and $E(R) =$ $(n-1)\theta/(n+1)$.

31. A sample of size n is drawn from the uniform density $f(x) = 1, 0 \le n \le 1$. Show that the probability that the range will exceed 0.5 is $1 - \frac{n+1}{2^n}$. Find also the least value of n for which the Prob. {range will exceed 0.90} > 0.95.

32. Show that the distribution of the range in samples of n from the distribution

$$dF = 2(1-x)\,dx, \quad 0 \le x \le 1$$

is $$g_R(r) = \frac{n(n-1)r^{n-2}}{2}\left[\frac{2r^{n+1}}{n^2-1} + \frac{(2-r)^{n+1}}{n+1} - \frac{r^2(1-r)^{n-1}}{n-1}\right].$$

33. Let $X_1, \cdots, X_n$ be a random sample from a uniform distribution $(\mu - \sqrt{3}\sigma, \mu + \sqrt{3}\sigma)$. Show that the p.d.f. of the range R is

$$f_R(r) = \frac{n(n-1)}{(2\sqrt{3}\sigma)^n} r^{n-2}(2\sqrt{3}\sigma - r), \quad 0 < r < 2\sqrt{3}\sigma$$

and $$E(R) = \frac{2\sqrt{3}\sigma(n-1)}{n+1}.$$

34. Find $\pi(r, n, F_\theta)$ (defined in (11.6.22)) when F_θ is the uniform distribution on $(0, \theta]$. Find $\pi(r, n, \mathcal{F})$ for the family $\mathcal{F}$ of all uniform distributions on $[0, \theta], 0 < \theta < \infty$.

35. Show that in random samples of size $(2m+1)$ from an exponential distribution with p.d.f., $f(x;\theta) = \frac{1}{\theta}e^{-x/\theta}, \quad 0 < x < \infty, \ \theta > 0$ the sampling distribution of the median is

$$g(\tilde{x}) = \begin{cases} \frac{(2m+1)!}{m!m!\theta} e^{-\tilde{x}(m+1)/\theta}\left[1 - e^{-\tilde{x}/\theta}\right]^m & \text{for } \tilde{x} > 0. \\ 0 \text{ elsewhere.} \end{cases}$$

36. Let y_1, y_2 $(y_1 < y_2)$ be the order statistics of a random sample of size two from $N(\mu, \sigma^2)$ where σ^2 is unknown. Show that $P(y_1 < \mu < y_2) = \frac{1}{2}$. Compute the expected value of the random length $y_1 - y_2$.

37. A random sample of size n is drawn from the exponential population, $dF = \frac{1}{\theta}e^{-x/\theta}$, $0 \le x \le \infty$. Show that the rth raw moment μ_r' of $X_{(m)}$ is $\frac{(-1)^r\theta^r}{B(m,n-m+1)} \frac{d^r}{dn^r} B(m, n-m+1)$.

Hence that

$$\mu_1' = \frac{1}{\theta}\left(\frac{1}{n} + \frac{1}{n-1} + \cdots + \frac{1}{n-m+1}\right)$$

$$\mu_2^1 = \frac{1}{\theta^2}\left[\frac{1}{n^2} + \frac{1}{(n-1)^2} + \cdots + \frac{1}{(n-m+1)^2}\right].$$

[*Hints:*

$$\mu_r' = \frac{1}{B(m, n-m+1)} \int_0^1 x^r F^{m-1}(1-F)^{n-m+1}\,dF$$

where $F = 1 - e^{-x/\theta}$. Differentiate under the integral sign with respect to n. Use

$$\frac{d}{d\alpha}\log\lceil(\alpha) = \lim_{n\to\infty}\left(\log n - \sum_{j=0}^{n-1}\frac{1}{\alpha+j}\right)$$

$$\left.\frac{d^2}{d\alpha^2}\log\lceil(\alpha) = \sum_{j=0}^{\infty}\frac{1}{(\alpha+j)^2}.\right]$$

[Kendall, 1954]

38. Let $X_1, \cdots, X_n$ be a random sample of size n drawn from the rectangular population $R(0, a)$. Find the joint distribution of $U = \frac{1}{2}(X_{(n)} - X_{(1)})$, $V = \frac{1}{2}(X_{(n)} + X_{(1)})$. Hence, derive the marginal distribution of V. Prove that V is an unbiased estimate of the population mean and that its efficiency, as compared to the sample mean $\bar{x}$ of the n random observations is $\frac{(n+1)(n+2)}{6n} > 1$ for $n \geq 3$.

   Also show that the standardised random variable $W = \frac{V-E(V)}{\sqrt{\text{Var}(V)}}$ has, as $n \to \infty$, the limiting Laplace distribution $\frac{1}{\sqrt{2}}e^{-\sqrt{2}|w|}dw(0 < w < \infty)$, whereas the standardised sample mean $\bar{x}$ has a limiting unit normal distribution. In view of this limiting distributional behaviour, comment on the use of v as an estimate of the mean of the rectangular population. **[Rahman, 1967]**

39. It is assumed that the life-times of electric bulbs follow an exponential distribution with an unknown mean λ^{-1}. To estimate λ^{-1} a sample of n bulbs is taken and the life-times of the first $r(< n)$ bulbs failed, say, $X_{(1)} < X_{(2)} < \cdots < X_{(r)}$ are observed.

   (a) Prove that $X_{(i)} - X_{(i-1)}$, $i = 1, 2, \cdots, r(X_{(0)} = 0)$ are independently distributed.

   (b) Also show that

   $$(X_{(1)} + \cdots + X_{(r)})/r + (n-r)X_{(r)}/r$$

   has the minimum variance among all the linear unbiased estimators of λ^{-1} based on $X_{(1)}, \cdots, X_{(r)}$ (i.e., estimators which are linear functions of $X_{(1)}, \cdots, X_{(r)}$).

40. Let $X_{(1)}, \cdots, X_{(n)}$ be an ordered sample from

$$f(x) = \frac{(k+1)x^k}{a^{(k+1)}}, \quad 0 \leq x \leq a$$

   where a, k are positive integers. Find the joint distribution of $X_{(i)} (i = 1, \cdots, n)$. Hence, prove that for $i > j$ and r, s positive integers

$$E\left(X_{(i)}^2 X_{(j)}^s\right) = \frac{a^{r+s}n(k+1)}{\{n(k+1)+s+r\}} \cdot \frac{\lceil(n)\lceil\left(j+\frac{s}{k+1}\right)\lceil\left(i+\frac{s+r}{k+1}\right)}{\lceil(j)\lceil\left(i+\frac{s}{k+1}\right)\lceil\left(n+\frac{s+r}{k+1}\right)},$$

$$\text{and hence, } \operatorname{Cov}\left(X_{(j)}, X_{(n)}\right) = \frac{n(k+1)a^2}{\{1+n(k+1)\}\{2+n(k+1)\}} \times \frac{\Gamma(n)\,\Gamma[j+\{1/(k+1)\}]}{\Gamma(j)\,\Gamma[n+\{1/(k+1)\}]}.$$

41. Find the condition for the statistic $Y = \sum_{i=1}^{n} C_i X_{(i)}$, where C_i's are constants, to be unbiased for the population mean. Hence, assuming k to be known, show that $Y^* = \frac{n(k+1)+1}{n(k+2)} X_{(n)}$ is the best linear unbiased estimate of the population mean and that its efficiency as compared to the sample mean $\bar{x}$ is $\frac{n(k+1)+2}{k+3} > 1$ for $n \geq 2$. Also, prove that the standardised random variable $U = \frac{Y* - E(Y*)}{\sqrt{V(Y^*)}}$ has, as $n \to \infty$, the limiting distribution $e^{u-1}\, du \ (-\infty < u \leq 1)$.

42. Let $Y_{(1)} < Y_{(2)} < \cdots < Y_{(n)}$ be the order statistic in a random sample of size n drawn from a distribution with p.d.f.

$$f_\theta(x) = \frac{1}{\theta}(0), \quad 0 < x < \theta \text{ (otherwise)}.$$

(a) Show that the correlation coefficient between $Y_{(1)}$ and $Y_{(n)}$ tends to zero as $n \to \infty$.

(b) Show that $Y_{(n)}$ and $\frac{Y_{(1)}}{Y_{(n)}}$ are independently distributed.

(c) Show that for any $k = 1, 2, \cdots$

$$E\left(\frac{Y_{(1)}}{Y_{(n)}}\right)^k = \frac{E(Y_{(1)}^k)}{E(Y_{(n)}^k)}.$$

43. Let $X_1, \cdots, X_n$ be i.i.d. with d.f.

$$F(y) = \begin{cases} y^\theta, \ 0 < y < 1, \ \theta > 0 \\ 0, \ \text{otherwise} \end{cases}$$

Show that $X_{(i)}/X_{(n)}, i = 1, 2, \cdots, n-1$, and $X_{(n)}$ are independent.

44. Let $X_1, \cdots, X_n$ be a random sample from $f(x) = 1, 0 < x < 1$.

Show that $Y_1 = X_{(1)}/X_{(2)}, Y_2 = X_{(2)}/X_{(3)}, \cdots, Y_{n-1} = X_{(n-1)}/X_{(n)}$ and $Y_n = X_{(n)}$ are independent. Find the p.d.f. of $Y_1, \cdots, Y_n$.

45. Let $Y_1 < \cdots < Y_n$ be the order statistic of a sample from the strictly increasing continuous distribution function $F(x)$. Show that

(a) $P[F(Y_n) - F(Y_1) > \beta] = 1 - n\beta^{n-1} + (n-1)\beta^n$;

(b) $P[Y_r < Z_p] = I_p(r, n-r+1)$

where $I_p(a, b)$ is the incomplete beta function and $F(Z_p) = p$.

46. From a population of N cards marked $1, 2, \cdots, N$, a simple random sample of size n is drawn without replacement.

Show that the p.m.f. of the largest observation $X_{(n)}$ is

$$P\left[X_{(n)} = r\right] = \binom{r-1}{n-1} \Big/ \binom{N}{n}, \quad r = n, n+1, \cdots, N.$$

Show also that

$$E(X_{(n)}) = \frac{n(N+1)}{n+1}$$

$$V(X_{(n)}) = \frac{n(N+1)(N-n)}{(n+1)^2(n+2)}.$$

Show that if $n = 2m+1$, the sample median $X_{(m+1)}$ has the p.m.f.

$$P(X_{(m+1)} = r) = \binom{r-1}{m}\binom{N-r}{m} \Big/ \binom{N}{n}, \quad r = m+1, \cdots, N-m$$

and $E(X_{(m+1)}) = \dfrac{N+1}{2}$, $V(X_{(m+1)}) = \dfrac{(N+1)(N-n)}{4(n+2)}$.

47. (a) Show that in odd samples of size n from a rectangular population of unit range, the sample variance of the distribution of median is given exactly by $\frac{1}{4(n+2)}$.

(b) Derive an expression for the standard error of the sample median in samples from a population with density $\theta e^{-\theta x}, x > 0$.

48. Show that the distribution of range in samples of size three from a normal population with unit variance is given by

$$g_R(r) = \frac{6}{\pi\sqrt{2}} e^{-r^2/4} \int_0^{r/\sqrt{6}} e^{-z^2/2}\, dz.$$

[*Hints:* From the result of problem (30)

$$g_R(r) = \frac{6}{(\sqrt{2\pi})^3} \int_{-\infty}^{\infty} \left[\int_{(t-\frac{r}{2})}^{(t+\frac{r}{2})} e^{-z^2/2}\, dz\right] \exp\left[-\frac{1}{2}\left\{\left(t-\frac{r}{2}\right)^2 + \left(t+\frac{r}{2}\right)^2\right\}\right] dt$$

where $t = [x_{(1)} + x_{(3)}]/2$.]

49. Show that for the rectangular distribution

$$f(x) = \frac{1}{\theta}, \quad \alpha - \frac{\theta}{2} \le x \le \alpha + \frac{\theta}{2}; \quad E(X_{(r)}) = \frac{\theta r}{n+1}\left(\alpha - \frac{\theta}{2}\right).$$

50. Show that the distribution of the largest observation $x_{(n)}$ of the sample drawn from the population

$$dF = \frac{2x}{2\theta+1}\,dx,\ \ 0 \le x \le \theta+1 \ \text{ is } \ \frac{2nX_{(n)}}{(2\theta+1)^n}(x_{(n)}^2 - \theta^2)^{n-1}\,dx_{(n)}.$$

51. Let $X_{(1)}, \cdots, X_{(n)}$ be the order statistic corresponding to the random sample $(X_1, \cdots, X_n)$ drawn from the distribution $F_X(x)$. Show that

$$E(X_{(p+1)} - X_{(p)}) = D_{n.p} = \frac{n}{p}\int_{-\infty}^{\infty} F(x)^{n-p}(1-F(x))^p\,dx.$$

Show also that

$$\binom{n}{p} D_{n-1,p-1} - \frac{n-p+1}{p} D_{n,p-1} = D_{n,p}.$$

[Kendall, 1954]

52. (a) Let (X_1, X_2, X_3) be a random sample from $f(x) = \theta e^{-x\theta}$, $x \ge 0$, $\theta > 0$. Show that $Y_1 = X_{(3)} - X_{(2)}$ and $Y_2 = X_{(2)}$ are independent.

(b) Let $(X_1, \cdots, X_n)$ be a random sample from $f(x) = \theta e^{-x\theta}$, $x \ge 0, \theta > 0$,

(i) Show that $X_{(r)}$ and $X_{(s)} - X_{(r)}$ are independent for any $s > r$.

(ii) Find the p.d.f. of $X_{(r+1)} - X_{(r)}$.

(iii) Let $Z_1 = nX_{(1)}, Z_2 = (n-1)(X_{(2)} - X_{(1)}), Z_3 = (n-2)\ (X_{(3)} - X_{(2)})$, $\cdots$, $Z_n = (X_{(n)} - X_{(n-1)})$. Show that $(Z_1, \cdots, Z_n)$ and $(X_1, \cdots, X_n)$ are identically distributed. **[Rohatgi, 1976]**

53. Show that for the uniform distribution over $(0, 1)$, the correlation coefficient between the two order statistic $X_{(r)}, X_{(s)}, r < s$ is $\sqrt{\frac{r(n-s+1)}{s(n-r+1)}}$.

54. If $X_{(1)} < \cdots < X_{(n)}$ are the ordered values of a random sample of size n from a population having a continuous distribution function $F(x)$ and

$$F_n(x) = \begin{cases} 0 & \text{if } x < x_{(1)} \\ \dfrac{r}{n} & \text{if } X_{(r)} \le x < X_{(r+1)} \qquad (r = 1, 2, \cdots, n-1). \end{cases}$$

Find the distribution of $F_n(x)$ and prove that for every $\varepsilon > 0$,

$$P\left[|F_n(x) - F(x)| < \varepsilon\right] \to 1 \ \text{ as } \ n \to \infty.$$

Chapter 12

Sampling Distributions arising out of Univariate and Bivariate Normal Distributions

12.1 Introduction

We consider in this chapter sampling distributions of some statistics arising from univariate and bivariate normal distributions. Section 12.2 considers distributions of sample mean and sample variance in sampling from a univariate normal distribution. Sampling distributions of sample means, sample variance, sample covariance, sample correlation and sample regression coefficient in sampling from a bivariate normal population are considered in section 12.3.

12.2 Sampling Distributions arising from Univariate Normal Distributions

Let $X_1, \cdots, X_n$ be a random sample from a $N(\mu, \sigma^2)$ population. We are interested in the sampling distribution of the sample mean,

$$\bar{X} = \frac{1}{n}\sum_{i=1}^{n} X_i$$

and sample variance

$$S^2 = \frac{1}{n-1}\sum_{i=1}^{n}(X_i - \bar{X})^2.$$

It will be seen in chapter 15 that $\bar{X}$ and S^2 are unbiased estimators of μ and σ^2 respectively, when both of these parameters are unknown.

The joint p.d.f. of $X_1, \cdots, X_n$ is

$$f(x_1, \cdots, x_n) = \frac{1}{(\sigma\sqrt{2\pi})^n} \exp\left[-\frac{1}{2\sigma^2}\sum_{i=1}^{n}(x_i-\mu)^2\right], \quad -\infty < x_i < \infty. \tag{12.2.1}$$

Let us now make the one-to-one transformation $(x_1, \cdots, x_n) \to (y_1, \cdots, y_n)$ such that

$$\begin{aligned} y_1 &= \frac{x_1-\mu}{\sqrt{n}} + \cdots + \frac{x_n-\mu}{\sqrt{n}} \\ y_2 &= l_{21}x_1 + \cdots + l_{2n}x_n \\ &\cdots\cdots\cdots\cdots \\ &\cdots\cdots\cdots\cdots \\ y_n &= l_{n1}x_1 + \cdots + l_{nn}x_n \end{aligned}$$

where the vectors $\boldsymbol{l}_i' = (l_{i1}, \cdots, l_{in})$ are unitary, mutually orthogonal and each orthogonal to the vector

$$\left(\frac{1}{\sqrt{n}}, \cdots, \frac{1}{\sqrt{n}}\right)'.$$

Hence,

$$\sum_{i=1}^{n} y_i^2 = \sum_{i=1}^{n}(x_i-\mu)^2.$$

Also $|J| = 1$. Hence the joint p.d.f. of $y_1, \cdots, y_n$ is

$$g(y_1, \cdots, y_n) = \frac{1}{(\sigma\sqrt{2\pi})^n} \exp\left[-\frac{1}{2\sigma^2}\sum_{i=1}^{n} y_i^2\right], \quad -\infty < y_i < \infty, \quad i = 1, \cdots, n. \tag{12.2.2}$$

Hence, $Y_1, \cdots, Y_n$ are each independent $N(0, \sigma^2)$ variables. Now

$$Y_1 = \sqrt{n}(\bar{X} - \mu) \cap N(0, \sigma^2).$$

Therefore, the sample mean $\bar{X}$ is normally distributed about mean μ and variance

$$\sigma^2/n(-\infty < \bar{X} < \infty).$$

Again,

$$\sum_{i=2}^{n} y_i^2 = \sum_{1}^{n} y_i^2 - y_1^2 = \sum_{1}^{n}(x_i-\mu)^2 - n(\bar{x}-\mu)^2 = \sum_{i=1}^{n}(x_i-\bar{x})^2 = (n-1)s^2.$$

But

$$\sum_{i=2}^{n} Y_i^2/\sigma^2$$

follows a χ^2 distribution with $(n-1)$ d.f. Hence $(n-1)S^2/\sigma^2$ follows a central $\chi^2_{(n-1)}$. The p.d.f. of S^2 is

$$\frac{[(n-1)/\sigma^2]^{(n-1)/2}}{2^{(n-1)/2}\overline{\left|\left(\frac{n-1}{2}\right)\right.}} \exp\left[-\frac{(n-1)S^2}{2\sigma^2}\right] (S^2)^{(n-3)/2}, \; 0 < S^2 < \infty. \tag{12.2.3}$$

The p.d.f. of the sample standard deviation S is

$$\frac{[(n-1)/\sigma^2]^{(n-1)/2}}{2^{(n-3)/2}\overline{\left|\left(\frac{n-1}{2}\right)\right.}} \exp\left[-\frac{(n-1)S^2}{2\sigma^2}\right] S^{n-2}, \; 0 < S < \infty. \tag{12.2.4}$$

Since Y_i's are independent variables, it follows that $\bar{X}$ and S^2 are independently distributed random variables.

It has already been proved (section 11.3) that the statistic

$$T = \frac{\sqrt{n}(\bar{X}-\mu)}{S}$$

follows Students' $t_{(n-1)}$ distribution.

12.3 Sampling Distributions of some Statistics arising out of Bivariate Normal Population

We shall consider here sampling distribution of sample means, sample variances and sample correlation coefficient obtained from a sample from a bivariate normal population. For this we first consider the following lemma.

12.3.1 A Lemma

Lemma 12.3.1 *The volume of a n-dimensional hypersphere intersected by a k-dimensional flat is*

$$J = \int \prod_{i=1}^{n} dx_i$$

$$s^2 - \frac{1}{2}\,ds^2 \le s^2 = \sum_1^n x_i^2 \le s^2 + \frac{1}{2}\,ds^2 \tag{12.3.1}$$

$$w_j - \frac{1}{2}\,dw_j \le w_j = \sum_{i=1}^{n} l_{ji}x_i \le w_j + \frac{1}{2}\,dw_j, \;\; j = 1, \cdots, k < n.$$

This is given by (12.3.9).

Proof: First make a linear transformation

$$\boldsymbol{x} = \begin{bmatrix} x_1 \\ \vdots \\ x_n \end{bmatrix} \to \boldsymbol{W} = \begin{bmatrix} \boldsymbol{W}_{1(k\times 1)} \\ \boldsymbol{W}_2 \end{bmatrix} = \begin{bmatrix} \boldsymbol{L}_{1(k\times n)} \\ \boldsymbol{L}_{2(\overline{n-k}\times n)} \end{bmatrix} \boldsymbol{x} \tag{12.3.2}$$

where $\boldsymbol{L}_1 = (l_{ji})_{\substack{j=1,\cdots,k \\ i=1,\cdots,n}}$ is completely known. Choose $\boldsymbol{L}_2$ such that

$$\boldsymbol{L}_2\boldsymbol{L}_2' = I_{(n-k)}, \quad \boldsymbol{L}_1\boldsymbol{L}_2' = \boldsymbol{0}.$$

Jacobian of transformation is

$$|\boldsymbol{L}^{-1}| = \frac{1}{\sqrt{|\boldsymbol{A}|}}$$

where $\boldsymbol{A} = \boldsymbol{L}_1\boldsymbol{L}_1'$.

Now

$$s^2 = x'x = W_1'(\boldsymbol{L}_1L_1')^{-1}\boldsymbol{W}_1 + \boldsymbol{W}_2'\boldsymbol{W}_2 = \sum\sum_{ij=1}^{k} a^{ij}w_iw_j + \sum_{t=k+1}^{n} w_t^2 \text{ where } ((a^{ij})) = \boldsymbol{A}^{-1} \tag{12.3.3}$$

By this transformation the integral (12.3.1) becomes

$$J = \int \frac{1}{\sqrt{|\boldsymbol{L}_1\boldsymbol{L}_1'|}} \prod_{j=1}^{n} dw_j$$

$$s^2 - \sum\sum_{ij=1}^{k} a^{ij}w_iw_j - \frac{1}{2}ds^2 \le \sum_{t=k+1}^{n} w_t^2 \le s^2 - \sum\sum_{ij=1}^{k} a^{ij}w_iw_j + \frac{1}{2}ds^2$$

$$w_j - \frac{1}{2}dw_j \le w_j \le w_j + \frac{1}{2}dw_j, \quad j = 1, \cdots, k \tag{12.3.4}$$

$$= \frac{1}{\sqrt{|\boldsymbol{L}_1\boldsymbol{L}_1'|}} \prod_{j=1}^{k} dw_j \int \prod_{t=k+1}^{n} dw_t$$

$$s^2 - \sum\sum_{ij=1}^{k} a^{ij}w_iw_j - \frac{1}{2}ds^2 \le \sum_{k+1}^{n} w_t^2 \le s^2 - \sum\sum_{ij=1}^{k} a^{ij}w_iw_j + \frac{1}{2}ds^2$$

$$= \frac{1}{\sqrt{|\boldsymbol{L}_1\boldsymbol{L}_1'|}} \prod_{j=1}^{k} dw_j \times \text{contents of the spherical shell of dimension } (n-k) \text{ and of radius}$$

$$T = \left\{ s^2 - \sum\sum_{ij=1}^{k} a^{ij}w_iw_j \right\}^{1/2} \text{ and thickness } ds^2. \tag{12.3.5}$$

Now the volume of a p-dimensional sphere of radius R is

$$\int_{\Sigma_{i=1}^{p} x_i^2 \le R^2} \prod_{i=1}^{p} dx_i = \frac{\pi^{p/2}}{\Gamma(\frac{p}{2}+1)}(R^2)^{p/2}. \tag{12.3.6}$$

ılso the contents of the spherical shell of this sphere of thickness dR^2 is

$$\int\limits_{R^2-\frac{1}{2}dR^2\le\Sigma_{i=1}^p x_i^2\le R^2+\frac{1}{2}dR^2} \prod dx_i = \frac{\pi^{p/2}}{\Gamma(\frac{p}{2})}(R^2)^{\frac{p}{2}-1}\,dR^2. \tag{12.3.7}$$

Hence (12.3.5) is

$$\frac{1}{\sqrt{|\boldsymbol{L}_1\boldsymbol{L}_1'|}}(T^2)^{(n-k-2)/2}\frac{\pi^{(n-k)/2}}{\Gamma\left(\frac{n-k}{2}\right)}\,ds^2\prod_{j=1}^{k}dw_j. \tag{12.3.8}$$

Now

$$T^2 = s^2-\sum\sum_{ij=1}^{k}|A_{ij}|w_iw_j/|\boldsymbol{A}| \text{ where } |A_{ij}| \text{ is the cofactor of } a_{ij} \text{ in } |\boldsymbol{A}|$$

$$= \frac{1}{|\boldsymbol{A}|}\left\{S^2|\boldsymbol{A}|-\sum\sum_{ij=1}^{k}|A_{ij}|w_iw_j\right\} = \frac{1}{|\boldsymbol{A}|}\begin{vmatrix} s^2 & w_1 & w_2 & \cdots & w_k \\ w_1 & a_{11} & a_{12} & \cdots & a_{1k} \\ \vdots & & & & \\ w_k & a_{k1} & a_{k2} & \cdots & a_{kk}\end{vmatrix}$$

$$= \frac{1}{|\boldsymbol{L}_1\boldsymbol{L}_1'|}\begin{vmatrix} S^2 & \boldsymbol{W}_1' \\ \boldsymbol{W}_1 & \boldsymbol{L}_1\boldsymbol{L}_1'\end{vmatrix}.$$

Hence (12.3.8) is

$$J = \frac{1}{|\boldsymbol{L}_1\boldsymbol{L}_1'|^{(n-k-1)/2}}\begin{vmatrix} S^2 & \boldsymbol{W}_1' \\ \boldsymbol{W}_1 & \boldsymbol{L}_1\boldsymbol{L}_1'\end{vmatrix}^{(n-k-2)/2}\frac{\pi^{(n-k)/2}}{\Gamma\left(\frac{n-k}{2}\right)}\,ds^2\prod_{j=1}^{k}dw_j. \tag{12.3.9}$$

12.3.2 The distribution of sample means, sample variances and sample covariances of a sample from a bivariate normal population

Let (x_i, y_i) $(i = 1, \cdots, n)$ be a random sample from a bivariate normal population

$$N_2(\mu_1, \mu_2, \sigma_1, \sigma_2, \rho).$$

The joint-distribution of sample observations is

$$\prod_{i=1}^{n} f(x_i, y_i)\,dx_i\,dy_i = \frac{1}{\left(2\pi\sigma_1\sigma_2\sqrt{1-\rho^2}\right)^n}\exp\left[-\frac{1}{2(1-\rho^2)}\times\sum_{i=1}^{n}\left\{\frac{(x_i-\mu_1)^2}{\sigma_1^2}\right.\right.$$

$$\left.\left.-2\rho\frac{(x_i-\mu_1)(y_i-\mu_2)}{\sigma_1\sigma_2}+\frac{(y_i-\mu_2)^2}{\sigma_2^2}\right\}\right]\prod_{i=1}^{n}dx_i\,dy_i. \tag{12.3.10}$$

Let

$$\bar{x} = \sum_1^n x_i/n, \qquad \bar{y} = \sum_1^n y_i/n, \qquad S_{11} = \sum_1^n (x_i - \bar{x})^2,$$
$$S_{22} = \sum_1^n (y_i - \bar{y})^2, \; S_{12} = \sum_1^n (x_i - \bar{x})(y_i - \bar{y}). \tag{12.3.11}$$

We introduce the orthogonal transformations

$$\boldsymbol{u} = \boldsymbol{L}\boldsymbol{x}, \;\; \boldsymbol{v} = \boldsymbol{L}\boldsymbol{y}, \tag{12.3.12}$$

where

$$\boldsymbol{x} = (x_1, \cdots, x_n)', \; \boldsymbol{y} = (y_1, \cdots, y_n)', \; \boldsymbol{u} = (u_1, \cdots, u_n)', \; \boldsymbol{v} = (v_1, \cdots, v_n)',$$
$$u_1 = \sqrt{n}\bar{x}, \qquad v_1 = \sqrt{n}\bar{y}, \qquad u_2, \cdots, u_n, \qquad v_2, \cdots, v_n$$

being chosen suitably so that $\boldsymbol{L}$ is an orthogonal matrix

$$\boldsymbol{L}\boldsymbol{L}' = I.$$

Now

$$\sum (x_i - \mu_1)^2/\sigma_1^2 = \left\{S_{11} + n(\bar{x} - \mu_1)^2\right\}/\sigma_1^2 = \frac{S_{11} + (u_1 - \sqrt{n}\mu_1)^2}{\sigma_1^2}$$
$$= \frac{\sum_2^n u_i^2 + (u_1 - \sqrt{n}\mu_1)^2}{\sigma_1^2}, \tag{12.3.13}$$

since

$$S_{11} = \sum_1^n x_i^2 - n\bar{x}^2 = \sum_1^n u_i^2 - u_1^2 = \sum_2^n u_i^2.$$

Similarly,

$$\sum_1^n (y_i - \mu_2)^2/\sigma_2^2 = \frac{\sum_2^n v_i^2 + (v_1 - \sqrt{n}\mu_2)^2}{\sigma_2^2}. \tag{12.3.14}$$

Also

$$\frac{\sum (x_i - \bar{x})(y_i - \bar{y})}{\sigma_1\sigma_2} = \frac{S_{12} + n(\bar{x} - \mu_1)(\bar{y} - \mu_2)}{\sigma_1\sigma_2}$$
$$= \frac{\sum_2^n u_i v_i + (u_1 - \sqrt{n}\mu_1)(v_1 - \sqrt{n}\mu_2)}{\sigma_1\sigma_2} \tag{12.3.15}$$

since

$$S_{12} = \sum x_i y_i - n\bar{x}\bar{y} = \sum_1^n u_i v_i - u_1 v_1 = \sum_2^n u_i v_i.$$

Hence, from (12.3.10) the joint distribution of $(u_i, v_i)(i = 1, \cdots, n)$ is

$$\prod_{i=1}^{n} f(u_i, v_i)\, du_i\, dv_i = \left[\frac{1}{2\pi\sigma_1\sigma_2\sqrt{1-\rho^2}} \exp\left\{-\frac{1}{2(1-\rho^2)} \times \left(\frac{(u_1-\sqrt{n}\mu_1)^2}{\sigma_1^2} - 2\rho\frac{(u_1-\sqrt{n}\mu_1)(u_2-\sqrt{n}\mu_2)}{\sigma_1\sigma_2} + \frac{(v_1-\sqrt{n}\mu_2)^2}{\sigma_2^2}\right)\right\} du_1\, dv_1\right]\left[\frac{1}{(2\pi\sigma_1\sigma_2\sqrt{1-\rho^2})^{n-1}} \times \exp\left\{-\frac{1}{2(1-\rho^2)}\left(\sum_2^n \frac{u_i^2}{\sigma_1^2} - 2\rho\sum_2^n \frac{u_i v_i}{\sigma_1\sigma_2} + \sum_2^n \frac{v_i^2}{\sigma_2^2}\right)\right\} \prod_{i=2}^{n} du_i\, dv_i\right] \tag{12.3.16}$$

(12.3.16) shows that $(\bar{X}, \bar{Y})$ (where $\bar{x} = u_1/\sqrt{n}, \bar{y} = v_1/\sqrt{n}$) follows a bivariate normal distribution

$$N_2(\mu_1, \mu_2, \sigma_1/\sqrt{n}, \sigma_2/\sqrt{n}, \rho)$$

and is distributed independently of $S_{ij}(i, j = 1, 2)$ (note that $S_{11} = \sum_2^n u_i^2$, etc.). The distribution of $(\bar{X}, \bar{Y})$ is obtained from the expression in the first [ ] in the right-hand side of (12.3.16).

To find the distribution of S_{11}, S_{12}, S_{22} consider the joint distribution of $(u_i, v_i)(i = 2, \cdots, n)$ given by the expression in the second [ ] in the right-hand side (12.3.16). This is to be integrated over the region

$$R_1 : \begin{cases} S_{11} - \frac{1}{2}\, dS_{11} \le \sum_2^n u_i^2 \le S_{11} + \frac{1}{2}\, dS_{11} \\ S_{22} - \frac{1}{2}\, dS_{22} \le \sum_2^n u_i v_i \le S_{22} + \frac{1}{2}\, dS_{22} \\ S_{12} - \frac{1}{2}\, dS_{12} \le \sum_2^n v_i^2 \le S_{12} + \frac{1}{2}\, dS_{12}. \end{cases} \tag{12.3.17}$$

In the domain R_1, each of the quantities S_{11}, S_{12}, S_{22} varies within a infinitesimal interval so that the integrand can be taken to be a constant over this range and can be brought outside the integral sign. Hence

$$dF(S_{11}, S_{12}, S_{22}) = \frac{1}{\left[2\pi\sigma_1\sigma_2\sqrt{1-\rho^2}\right]^{n-1}} \exp\left[-\frac{1}{2(1-\rho^2)}\left\{\frac{S_{11}}{\sigma_1^2} - 2\rho\frac{S_{12}}{\sigma_1\sigma_2} + \frac{S_{22}}{\sigma_2^2}\right\}\right] \int_{R_1} \prod_2^n du_i\, dv_i. \tag{12.3.18}$$

Now

$$\int_{R_1} \prod_2^n du_i\, dv_i = \left[\int\limits_{S_{22}-\frac{1}{2}\, dS_{22} \le \Sigma_2^n v_i^2 \le S_{22}+\frac{1}{2}\, dS_{22}} dv_2 \cdots dv_n \right] \int\limits_{\substack{S_{11}-\frac{1}{2}\, dS_{11} \le \Sigma_2^n u_i^2 \le S_{11}+\frac{1}{2}\, dS_{11} \\ S_{12}-\frac{1}{2}\, dS_{12} \le \Sigma_2^n u_i v_i \le S_{12}+\frac{1}{2}\, dS_{12}}} du_2 \cdots du_n. \tag{12.3.19}$$

In the second integral in the right-hand side of (12.3.19), v_i's are considered as constants and hence

$$\sum_2^n u_i v_i$$

is a linear function of u_i's. Applying the result (12.3.9) of lemma 12.3.1, (12.3.19) is

$$\left[\int\limits_{S_{22}-\frac{1}{2}\, S_{22} \le \Sigma_2^n v_i^2 \le S_{22}+\frac{1}{2}\, dS_{22}} \prod_{i=2}^n dv_2 \right] \cdot \frac{\pi^{(n-2)/2}}{\Gamma\left(\frac{n-2}{2}\right) S_{22}^{(n-3)/2}} \begin{vmatrix} S_{11} & S_{12} \\ S_{21} & S_{22} \end{vmatrix}^{(n-4)/2} dS_{11}\, dS_{22} \tag{12.3.20}$$

Again, applying the result (12.3.7), the integral in [ ] in (12.3.20) is

$$\frac{\pi^{(n-1)/2}}{\Gamma\left(\frac{n-1}{2}\right)} S_{22}^{(n-3)/2}\, dS_{22}. \tag{12.3.21}$$

Therefore, combining (12.3.19), (12.3.20) and (12.3.21), the joint distribution of S_{11}, S_{12}, S_{22} is

$$dF(S_{11}, S_{12}, S_{22}) = \frac{\pi^{(2n-3)/2}}{(2\pi\sigma_1\sigma_2\sqrt{1-\rho^2})^{n-1}\Gamma\left(\frac{n-1}{2}\right)\Gamma\left(\frac{n-2}{2}\right)} \exp\left[-\frac{1}{2(1-\rho^2)}\left\{\frac{S_{11}}{\sigma_1^2} - 2\rho\frac{S_{12}}{\sigma_1\sigma_2} + \frac{S_{22}}{\sigma_2^2}\right\}\right] (S_{11}S_{22} - S_{12}^2)^{(n-4)/2}\, dS_{11}\, dS_{12}\, dS_{22}. \tag{12.3.22}$$

Transforming

$$S_{11} = (n-1)s_1^2, \quad S_{12} = (n-1)r s_1 s_2, \quad S_{22} = (n-1)s_2^2,$$

$$J\left(\frac{S_{11}, S_{12}, S_{22}}{s_1, s_2, r}\right) = 4(n-1)^3 s_1^2 s_2^2$$

the joint distribution of s_1, s_2, r is obtained from (12.3.22) as

$$dF(s_1, s_2, r) = \frac{\pi^{n-\frac{3}{2}} 4(n-1)^{(n-1)}}{(2\pi\sigma_1\sigma_2\sqrt{1-\rho^2})^{n-1}\Gamma\left(\frac{n-1}{2}\right)\Gamma\left(\frac{n-2}{2}\right)} \exp\left[-\frac{n-1}{2(1-\rho^2)}\left\{\frac{s_1^2}{\sigma_1^2} - 2\rho r\frac{s_1 s_2}{\sigma_1\sigma_2} + \frac{s_2^2}{\sigma_2^2}\right\}\right] s_1^{n-2} s_2^{n-2}(1-r^2)^{(n-4)/2}\, ds_1\, ds_2\, dr. \tag{12.3.23}$$

Now using the duplicating formula (6.4.6) for a gamma function

$$\frac{1}{\Gamma\left(\frac{n-1}{2}\right)\Gamma\left(\frac{n-2}{2}\right)} = \frac{2^{n-3}}{\Gamma(n-2)\sqrt{\pi}}. \tag{12.3.24}$$

Hence (12.3.23) can be written as

$$dF(s_1, s_2, r) = \frac{(n-1)^{(n-1)}}{(\sigma_1\sigma_2\sqrt{1-\rho^2})^{n-1}\pi\Gamma(n-2)} \exp\left[-\frac{n-1}{2(1-\rho^2)}\left\{\frac{s_1^2}{\sigma_1^2} - 2\rho r\frac{s_1 s_2}{\sigma_1\sigma_2} + \frac{s_2^2}{\sigma_2^2}\right\}\right] s_1^{(n-2)} s_2^{(n-2)} (1-r^2)^{(n-4)/2}\, ds_1\, ds_2\, dr. \tag{12.3.25}$$

12.3.3 Distribution of sample correlation coefficient, r

CASE (a) : Suppose $\rho = 0$. In this case, the joint distribution of s_1, s_2, r is obtained from (12.3.25) as

$$dF(s_1, s_2, r) = \text{Const.}\exp\left[-\frac{n-1}{2}\left\{\frac{s_1^2}{\sigma_1^2} + \frac{s_2^2}{\sigma_2^2}\right\}\right] s_1^{n-2} s_2^{n-2} (1-r^2)^{(n-4)/2}\, ds_1 ds_2 dr.$$

Integrating over s_1 and s_2, the marginal distribution of r is obtained as

$$dF(r) = \text{Const.}(1-r^2)^{(n-4)/2}\, dr = \frac{1}{B\left(\frac{n-2}{2}, \frac{1}{2}\right)}(1-r^2)^{(n-4)/2}\, dr, \quad -1 \le r \le 1 \tag{12.3.26}$$

which shows that r^2 is $B\left(\frac{n-2}{2}, \frac{1}{2}\right)$ variable.

It may be noted that if we put

$$t = r\sqrt{(n-2)}/\sqrt{(1-r^2)}$$

then the distribution of t is obtained from (12.3.26) as

$$f(t|\rho = 0) = \frac{1}{B\left(\frac{n-2}{2}, \frac{1}{2}\right)\sqrt{(n-2)}} \cdot \frac{dt}{\left(1 + \frac{t^2}{n-2}\right)^{(n-1)/2}}, \quad -\infty < t < \infty \tag{12.3.27}$$

which shows that t follows a Student's t distribution with $(n-2)$ d.f. For testing a null hypothesis $H_o(\rho = 0)$ one may use the statistic t.

CASE (b) : Suppose $\rho \neq 0$. Here, put

$$\xi = s_1 s_2/\sigma_1\sigma_2, \quad e^z = (s_1/\sigma_1)/(s_2/\sigma_2)$$

so that

$$z = \log_e\left(\frac{s_1\sigma_2}{s_2\sigma_1}\right), \quad \xi e^z = s_1^2/\sigma_1^2, \quad \xi e^{-z} = s_2^2/\sigma_2^2, \quad -\infty \le z \le \infty \text{ and } 0 \le \xi \le \infty.$$

Also

$$\frac{1}{|J|}=\left|\frac{\partial(\xi,z,r)}{\partial(s_1,s_2,r)}\right|=\frac{2}{\sigma_1\sigma_2}.$$

Hence exponent in (12.3.23) is

$$-\frac{n-1}{2(1-\rho^2)}\xi(e^z+e^{-z}-2\rho r)=-\frac{n-1}{1-\rho^2}\xi(\cosh z-\rho r).$$

Hence, from (12.3.23),

$$dF(\xi,z,r)=\frac{(n-1)^{(n-1)}}{2\pi\overline{|(n-2)}\left\{\sigma_1\sigma_2\sqrt{1-\rho^2}\right\}^{n-1}}$$

$$\exp\left[-\frac{n-1}{1-\rho^2}\xi(\cosh z-\rho r)\right]\xi^{n-2}(1-r^2)^{(n-4)/2}\,d\xi\,dz\,dr.$$

$$\therefore\ dF(z,r)=\frac{(n-1)^{(n-1)}}{2\pi\overline{|(n-2)}\,(1-\rho^2)^{(n-1)/2}}(1-r^2)^{(n-4)/2}\,dz\,dr$$

$$\int_0^\infty\exp\left[-\frac{n-1}{1-\rho^2}(\cosh z-\rho r)\right]\xi^{n-2}\,d\xi$$

$$=\frac{\overline{|(n-1)}}{2\pi\overline{|(n-2)}}\cdot\frac{(1-\rho^2)^{(n-1)/2}(1-r^2)^{(n-4)/2}}{(\cosh z-\rho r)^{n-1}}\,dr\,dz. \tag{12.3.28}$$

Therefore, marginal distribution of r is obtained by integrating (12.3.28) over $-\infty\le z<\infty$. Now, it is known,

$$\int_0^\infty\frac{dz}{\cosh z+\cos\theta}=\frac{\theta}{\sin\theta}. \tag{12.3.29}$$

Differentiating both sides of (12.3.29) with respect to $\cos\theta$, $(n-2)$ times,

$$\frac{d^{n-2}}{d(\cos\theta)^{n-2}}\int_0^\infty\frac{dz}{\cosh z+\cos\theta}=\frac{d^{n-2}}{d(\cos\theta)^{n-2}}\left(\frac{\theta}{\sin\theta}\right) \tag{12.3.30}$$

$$\text{or,}\ \overline{|(n-1)}\int_0^\infty\frac{dz}{(\cosh z+\cos\theta)^{n-1}}=\frac{d^{n-2}}{d(-\cos\theta)^{n-2}}\left(\frac{\theta}{\sin\theta}\right).$$

Substituting $\cos\theta=-\rho r$,

$$\int_{-\infty}^\infty\frac{dz}{(\cosh z-\rho r)^{n-1}}=\frac{2}{\overline{|(n-1)}}\frac{d^{n-2}}{d(\rho r)^{(n-2)}}\left(\frac{\theta}{\sin\theta}\right)$$

$$=\frac{2}{\overline{|(n-1)}}\frac{d^{n-2}}{d(\rho r)^{(n-2)}}\left(\frac{\cos^{-1}(-\rho r)}{\sqrt{1-r^2\rho^2}}\right). \tag{12.3.31}$$

Hence from (12.3.28)

$$dF(r)=\frac{(1-\rho^2)^{(n-1)/2}}{\pi\overline{|(n-2)}}(1-r^2)^{(n-4)/2}\frac{d^{n-2}}{d(\rho r)^{n-2}}\left\{\frac{\cos^{-1}(-\rho r)}{\sqrt{1-r^2\rho^2}}\right\}dr. \tag{12.3.32}$$

12.3.4 Distribution of sample regression coefficient in sampling from a bivariate normal population

We consider in this subsection the distribution of sample regression coefficient.

The sample regression coefficient is defined as

$$b = \frac{S_{12}}{S_{11}} = r\frac{s_2}{s_1}.$$

Its population analogue is $\beta = \rho\frac{\sigma_2}{\sigma_1}$. Define

$$\begin{aligned} A &= \frac{S_{11}}{2\sigma_1^2(1-\rho^2)}, \quad & B &= \frac{S_{22}}{2\sigma_2^2(1-\rho^2)}, \quad & C &= \frac{S_{12}}{2\sigma_1\sigma_2(1-\rho^2)}, \\ 0 &\le A \le \infty, & 0 &\le B \le \infty, & -\infty &\le C \le \infty. \end{aligned} \tag{12.3.33}$$

The joint distribution of A, B, C is obtained from $dF(S_{11}, S_{12}, S_{22})$ in (12.3.22) as

$$dF(A,B,C) = K\exp[-(A-2\rho C+B)](AB-C^2)^{(n-4)/2}\,dA\,dB\,dC. \tag{12.3.34}$$

We now use the transformation

$$u = A, \quad v = C/A, \quad w = B - C^2/A. \tag{12.3.35}$$

Hence $C = uv, \quad B = w + uv^2$

$$\frac{1}{|J|} = J\left(\frac{u,v,w}{A,B,C}\right) = \frac{1}{A}.$$

The joint distribution of (u,v,w) is, therefore,

$$\begin{aligned} dF(u,v,w) &= \text{Const.}\exp\left[-\left\{u(1-\rho^2) + w + u(v-\rho)^2\right\}\right](uw)^{(n-4)/2}u\,du\,dv\,dw \\ &= \left[\text{Const.}e^{-u\{(1-\rho^2)+(v-\rho)^2\}}u^{\frac{n-2}{2}}\,du\,dv\right]\left[e^{-w}w^{(n-4)/2}\,dw\right] \\ &= dF(u,v)\,dF(w). \end{aligned} \tag{12.3.36}$$

This shows that w is independently distributed of u and v, but u, v are not mutually independent.

The marginal distribution of v is obtained by integrating $dF(u,v)$ over u

$$dF(v) = \frac{K\,dv}{\{(1-\rho^2)+(v-\rho)^2\}^{n/2}}, \quad -\infty \le v \le \infty \tag{12.3.37}$$

K, a constant. Letting

$$v - \rho = \sqrt{1-\rho^2}\tan\theta, \quad dv = \sqrt{1-\rho^2}\sec^2\theta\,d\theta.$$

From (12.3.37),

$$I = K \int_{-\pi/2}^{\pi/2} (1-\rho^2)^{-(n-1)/2} \cos^{n-2}\theta d\theta$$

which gives

$$K = \frac{(1-\rho^2)^{(n-1)/2}\Gamma\left(\frac{n}{2}\right)}{\Gamma\left(\frac{\pi}{2}\right)\Gamma\left(\frac{n-1}{2}\right)}. \tag{12.3.38}$$

Again,

$$v = \frac{C}{A} = \frac{\sigma_1}{\sigma_2} \cdot b.$$

Therefore, from (12.3.37) and (12.3.38)

$$dF(b) = \frac{\Gamma\left(\frac{n}{2}\right)\sigma_2^{n-1}(1-\rho^2)^{(n-1)/2}}{\sqrt{\pi}\Gamma\left(\frac{n-1}{2}\right)\sigma_1^{n-1}} \cdot \frac{db}{\left\{\frac{\sigma_2^2}{\sigma_1^2}(1-\rho^2) + \left(b - \rho\frac{\sigma_2}{\sigma_1}\right)^2\right\}^{n/2}}. \tag{12.3.39}$$

The distribution depends on three parameters σ_1, σ_2, ρ. This is a Pearsonian type II distribution, symmetrical about $\beta = \rho\frac{\sigma_2}{\sigma_1}$ and tends to normality fairly rapidly. We now consider the distribution of the statitistic

$$t = \frac{(b-\beta)\sqrt{S_{11}}}{\sqrt{\frac{S_{22} - bS_{12}}{n-2}}} \tag{12.3.40}$$

which is used for testing the null hypothesis $H_\circ(\beta = \beta_0)$.

Consider the distribution of A, B, C given in (12.3.34). Use the transformation $(A, B, C) \to (A, D, C)$ where

$$D = \frac{C - \rho A}{\sqrt{AB - C^2}}; \text{ hence } B = \frac{(C-\rho A)^2}{AD^2} + \frac{C^2}{A}, \tag{12.3.41}$$

$$dB = -\frac{2(C-\rho A)^2}{AD^3} dD, \quad (AB - C^2)^{(n-4)/2} = \frac{(C-\rho A)^{n-4}}{D^{n-4}},$$

$$-(A - 2\rho C + B) = -A\left(1 + \frac{\rho^2}{D^2}\right) + \left(2\rho C - \frac{C^2}{A}\right)\left(1 + \frac{1}{D^2}\right).$$

Therefore,

$$dF(A, D, C) = \text{Const.} e^{-A\left(1+\frac{\rho^2}{D^2}\right)} (AD^{n-1})^{-1} e^{\frac{D^2+1}{D^2}\left(2\rho C - \frac{C^2}{A}\right)} (C - \rho A)^{n-2}\, dA\, dD\, dC. \tag{12.3.42}$$

Integrating over the range of C one gets

$$dF(A,D) = e^{-A\left(1+\frac{\rho^2}{D^2}\right)}(AD^{n-1})^{-1}\,dAdD\int_{-\infty}^{\infty} e^{-\frac{Q}{A}(C^2-2\rho CA)}(C-\rho A)^{(n-2)}\,dC \tag{12.3.43}$$

where $Q = (D^2+1)/D^2$. The integral in (12.3.43) is found to be Const. $e^{A\rho^2 Q}$ by substituting $C - \rho A = Z$

$$\begin{aligned} dF(A,D) &= \text{Const.}e^{-A\left(1+\frac{\rho^2}{D^2}\right)+A\rho^2 Q}\cdot(AD^{n-1})^{-1}\left(\frac{D^2 A}{D^2+1}\right)^{(n-1)/2} dAdD \\ &= \text{Const.}e^{-A(1-\rho^2)}A^{(n-3)/2}\frac{dD\,dA}{(1+D^2)^{(n-1)/2}}. \end{aligned} \tag{12.3.44}$$

Hence, A, D are independently distributed

$$dF(D) = \text{Const.}\frac{dD}{(1+D^2)^{(n-1)/2}}.$$

Now

$$t = D\sqrt{n-2}.$$

Hence,

$$dF(t) = \frac{1}{\sqrt{n-2}B\left(\frac{n-2}{2},\frac{1}{2}\right)}\frac{dt}{\left(1+\frac{t^2}{n-2}\right)^{(n-1)/2}},\quad -\infty < t < \infty \tag{12.3.45}$$

which is a student's t with $(n-2)$ d.f.

12.3.5 Alternative derivation

In deriving the distribution (12.3.45) of the statistic t it was assumed that the joint distribution of x and y is bivariate normal. We shall now show that the distribution (12.3.45) holds good even when (x, y) does not follow bivariate normal. We assume only that the conditional distribution of y given x is normal with mean given as a linear function of x and variance σ^2 (a constant).

We assume that $y_1, \cdots, y_n$ is a random sample from the conditional distribution of y given x which is

$$f(y|x)\,dy = \frac{1}{\sigma\sqrt{2\pi}}\exp\left[-\frac{1}{2\sigma^2}(y-\alpha-\beta x)^2\right]dy. \tag{12.3.46}$$

Hence the joint distribution of y_i given $x_i, i = 1, \cdots, n$ is

$$\begin{aligned} &f(y_1,\cdots,y_n|x_1,\cdots,x_n)\Pi\,dy \\ &= \frac{1}{(\sigma\sqrt{2\pi})^n}\exp\left[-\frac{1}{2\sigma^2}\sum_{i=1}^{n}\{y_i-\alpha'-\beta(x_i-\bar{x})\}^2\right]\prod_{i=1}^{n} dy_i \end{aligned} \tag{12.3.47}$$

where $\alpha' = \alpha + \beta\bar{x}$.

Consider orthogonal transformation $(y_1, \cdots, y_n) \to (u_1, \cdots, u_n)$ when

$$u_1 = \frac{\Sigma y_i}{\sqrt{n}} = \sqrt{n}\bar{y},\ u_2 = \frac{(x_1-\bar{x})y_1 + \cdots + (x_n - \bar{x})y_n}{\sqrt{\Sigma(x_i-\bar{x})^2}}$$

$$= \frac{S_{12}}{\sqrt{S_{11}}} = b\sqrt{S_{11}}, \tag{12.3.48}$$

and $u_3, \cdots, u_n$ are chosen suitably. Hence,

$$\sum_1^n y_i^2 = \sum_1^n u_i^2,$$

$$\sum_1^n \{y_i - \alpha' - \beta(x_i - \bar{x})\}^2$$

$$= \sum_1^n y_i^2 + n\alpha'^2 + \beta^2 \sum (x_i - \bar{x})^2 - 2\alpha' \sum_i^n y_i - 2\beta \sum_1^n y_i(x_i - \bar{x})$$

$$= \sum_1^n u_i^2 + n\alpha'^2 + \beta^2 \sum (x_i-\bar{x})^2 - 2\alpha'\sqrt{n}u_1 - 2\beta u_2 \sqrt{\Sigma(x_i-\bar{x})^2}$$

$$= \sum_3^n u_i^2 + (u_1 - \sqrt{n}\alpha')^2 + \left(u_2 - \beta\sqrt{\Sigma(x_i-\bar{x})^2}\right)^2.$$

Therefore,

$$dF(u_1, \cdots, u_n | x_1, \cdots, x_n)$$

$$= \frac{1}{(\sigma\sqrt{2\pi})^n} \exp\left[-\frac{1}{2\sigma^2}\left\{(u_1 - \sqrt{n}\alpha')^2 + (u_2 - \beta\sqrt{S_{11}})^2 + \sum_3^n u_i^2\right\}\right] \prod du_i. \tag{12.3.49}$$

This show that

$$(u_1 - \sqrt{n}\alpha')/\sigma,\ \left(u_2 - \beta\sqrt{S_{11}}\right)/\sigma,\ u_3/\sigma, \cdots, u_n/\sigma$$

are independently normally distributed. Hence

$$\sum_3^n u_i^2/\sigma^2 \cap \chi^2_{(n-2)}.$$

Hence

$$\frac{u_2 - \beta\sqrt{S_{11}}}{\sigma} \Bigg/ \sqrt{\frac{\sum_3^n u_i^2}{\sigma^2(n-2)}} \cap t_{(n-2)}. \tag{12.3.50}$$

Now

$$\sum_3^n u_i^2 = \sum_1^n (y_i - \bar{y})^2 - u_2^2 = S_{22} - bS_{12}.$$

Therefore,

$$\frac{(b-\beta)\sqrt{S_{11}}}{\sqrt{\frac{S_{22}-bS_{12}}{n-2}}} \cap t_{(n-2)}. \tag{12.3.51}$$

Again

$$t' = \frac{u_1 - \sqrt{n}\alpha'}{\sigma} \Bigg/ \sqrt{\frac{\sum_3^n u_i^2}{\sigma^2(n-2)}} = \frac{\sqrt{n}(\bar{y}-\alpha')}{\sqrt{\frac{S_{22}-bS_{12}}{n-2}}} \cap t_{(n-2)}. \tag{12.3.52}$$

The statistic t' is used for testing $H_\circ(\alpha' = \alpha'_0)$.

Note that if (x, y) follows a bivariate normal distrtibution then the conditional distribution of y given x follows the distribution (12.3.46). However, the latter distribution may follow otherwise also. Hence the assumption of the distribution (12.3.46) is more general than the assumption that (x, y) follows bivariate normal.

► **Example 12.3.1** *If $(X_1, \cdots, X_n)$ is a random sample from $N(\mu, \sigma^2)$ population, find the distribution of $Z = \frac{X_1 - \bar{X}}{S}$, where $\bar{X}$ and S are sample mean and sample s.d., respectively.*

$$dF(x_1, \cdots, x_n) = \frac{1}{(\sigma\sqrt{2\pi})^n} \exp\left[-\frac{1}{2\sigma^2}\sum_{i=1}^n (x_i-\mu)^2\right] \prod_{i=1}^n dx_i.$$

Consider Helmert's transformation $(x_1, \cdots, x_n) \to (y_1, \cdots, y_n)$.

$$\begin{aligned} y_1 &= \frac{\Sigma x_i}{\sqrt{n}} = \sqrt{n}\bar{x}, \\ y_2 &= \frac{(n-1)x_1 - x_2 - \cdots - x_n}{\sqrt{n(n-1)}}, \\ y_3 &= \frac{(n-2)x_2 - x_3 - \cdots - x_n}{\sqrt{(n-1)(n-2)}}, \\ &\cdots\cdots\cdots\cdots\cdots\cdots, \\ y_n &= \frac{x_{n-1} - x_n}{\sqrt{1.2}}. \end{aligned}$$

This is an orthogonal transformation and hence $|J| = 1, \Sigma x_i^2 = \Sigma y_i^2$.

$$dF(y_1, \cdots, y_n) = \frac{1}{(\sigma\sqrt{2\pi})^n} \exp\left[-\frac{1}{2\sigma^2}\left\{(y_1 - \sqrt{n}\mu)^2 + \sum_2^n y_i^2\right\}\right] \prod dy_i. \tag{i}$$

Thus Y_1 is a $N(\sqrt{n}\mu, \sigma)$ variable and is distributed independently of

$$\chi^2 = \sum_2^n y_i^2/\sigma^2$$

which is a $\chi^2_{(n-1)}$ variable. Let us now introduce polar transformation,

$$\begin{aligned}
y_2 &= r\cos\theta_1, \\
y_3 &= r\sin\theta_1\cos\theta_2, \\
&\cdots\cdots\cdots\cdots\cdots\cdots\cdots\cdots, \\
y_{n-1} &= r\sin\theta_1\sin\theta_2\cdots\sin\theta_{n-3}\cos\theta_{n-2}, \\
y_n &= r\sin\theta_1\sin\theta_2\cdots\sin\theta_{n-3}\sin\theta_{n-2}, \\
|J| &= r^{n-2}\sin^{n-3}\theta_1\sin^{n-4}\theta_2\cdots\sin\theta_{n-3};\ r^2 = \sum_2^n y_i^2.
\end{aligned}$$

Hence,

$$dF(r, \theta_1, \cdots, \theta_{n-2}) = \text{Const.}e^{-r^2/2\sigma^2} r^{n-2}\sin^{n-3}\theta_1\cdots\sin\theta_{n-3}\,dr\,d\theta_1\cdots d\theta_{n-2}$$

Hence,

$$dF(\theta_1) = \text{Const.}\sin^{n-3}\theta_1 d\theta_1 = \frac{1}{B\left(\frac{n-2}{2}, \frac{1}{2}\right)}\sin^{n-3}\theta_1 d\theta_1. \qquad \text{(ii)}$$

Now

$$r^2 = (n-1)s^2; \quad \frac{y_2}{\sqrt{(n-1)s^2}} = \cos\theta_1; \quad \frac{y_2\sqrt{(n-1)}}{\sqrt{n}} = x_1 - \bar{x};$$

$$\frac{y_2}{\sqrt{(n-1)s^2}} = \frac{x_1-\bar{x}}{s}\,\frac{\sqrt{n}}{n-1} = z\cdot\frac{\sqrt{n}}{n-1}; \quad \sin\theta_1 = \left\{1 - \frac{nz^2}{(n-1)^2}\right\}^{1/2}.$$

From (ii)

$$dF(z_1) = \frac{\sqrt{n}}{(n-1)B\left(\dfrac{n-2}{2}, \dfrac{1}{2}\right)}\left\{1 - \frac{nz^2}{(n-1)^2}\right\}^{(n-4)/2} dz, \quad -\infty \leq z \leq \infty.$$

► **Example 12.3.2** *Let $(x_i, y_i)(i = 1, \cdots, n)$ be a random sample from a bivariate normal population $N_2(\mu_1, \mu_2, \sigma_1, \sigma_2, \rho)$. Find the distribution of s_1^2/s_2^2 in the non-null case, i.e. when $\rho \neq 0$.*

We note that if $\rho = 0$, distributions of $(n-1)s_1^2/\sigma_1^2$ and $(n-1)s_2^2/\sigma_2^2$ are $\chi^2_{(n-1)}$ so that

$$s_1^2/s_2^2 \cap \frac{\sigma_1^2}{\sigma_2^2}F_{(n-1,n-1)}.$$

Consider the distribution of s_1, s_2, r given in (12.3.25)

$$dF(s_1, s_2, r) = \text{Const.} \exp\left[-\frac{(n-1)}{2(1-\rho^2)}\left\{\frac{s_1^2}{\sigma_1^2} + \frac{s_2^2}{\sigma_1^2}\right\}\right] s_1^{n-2} s_2^{n-2}$$

$$\exp\left[\frac{(n-1)}{(1-\rho^2)} r\rho \frac{s_1 s_2}{\sigma_1 \sigma_2}\right] (1-r^2)^{(n-4)/2}\, ds_1\, ds_2\, dr \qquad (12.3.53)$$

$$= \text{Const.} \exp\left[-\frac{n-1}{2(1-\rho^2)}\left\{\frac{s_1^2}{\sigma_1^2} + \frac{s_2^2}{\sigma_2^2}\right\}\right] s_1^{n-2} s_2^{n-2}\, ds_1\, ds_2$$

$$\left[\sum_{i=0}^{\infty} \frac{1}{i!}\left\{\frac{n-1}{(1-\rho^2)} r\rho \frac{s_1 s_2}{\sigma_1 \sigma_2}\right\}^i (1-r^2)^{(n-4)/2}\, dr\right]. \qquad (12.3.54)$$

To find the joint distribution of s_1, s_2 we are to integrate (12.3.54) with respect to r over $(-1, 1)$. Since the above expression is a series of positive terms, term by term integration is possible. Now

$$\int_{-1}^{1} r^i (1-r^2)^{(n-4)/2}\, dr = 0 \text{ if } i \text{ is odd.} \qquad (12.3.55)$$

If i is even, say, $i = 2j$, the integral (12.3.55) is

$$2\int_0^1 r^{2j}(1-r^2)^{(n-4)/2}\, dr = \int_0^1 (r^2)^{j-\frac{1}{2}}(1-r^2)^{(n-4)/2} d(r^2)$$

$$= B\left(\frac{2j+1}{2}, \frac{n-2}{2}\right). \qquad (12.3.56)$$

Hence, the joint distribution of s_1^2, s_2^2 is

$$dF(s_1^2, s_2^2) = \text{Const.} \exp\left[-\frac{n-1}{2(1-\rho^2)}\left\{\frac{s_1^2}{\sigma_1^2} + \frac{s_2^2}{\sigma_2^2}\right\}\right] s_1^{n-2} s_2^{n-2}\, ds_1\, ds_2$$

$$\left[\sum_{j=0}^{\infty} \frac{1}{(2j)!}\left\{\frac{(n-1)\rho}{(1-\rho^2)} \frac{s_1 s_2}{\sigma_1 \sigma_2}\right\}^{2j} B\left(\frac{2j+1}{2}, \frac{n-2}{2}\right)\right]$$

$$= \text{Const.} \sum_{j=0}^{\infty} A_j \exp\left[-\frac{n-1}{2(1-\rho^2)}\left\{\frac{s_1^2}{\sigma_1^2} + \frac{s_2^2}{\sigma_2^2}\right\}\right]$$

$$(s_1^2)^{(n-3)/2+j}(s_2^2)^{(n-3)/2+j}\, ds_1^2\, ds_2^2 \qquad (12.3.57)$$

where

$$A_j = \frac{\overline{\left|\left(\frac{2j+1}{2}\right)\right.}}{\overline{|(2j+1)}\,\overline{\left|\left(\frac{n+2j-1}{2}\right)\right.}}.$$

We change from (s_1^2, s_2^2) to (F, s_2^2) where $F = s_1^2/s_2^2$. $|J| = s_2^2$. Hence

$$dF(F, s_2^2) = \text{Const.} \sum_{j=0}^{\infty} A_j \exp\left[-\frac{(n-1)s_2^2}{2(1-\rho^2)}\left\{\frac{F}{\sigma_1^2} + \frac{1}{\sigma_2^2}\right\}\right]$$
$$(s_2^2)^{n-2+2j} F^{(n-3)/2+j}\, ds_2^2\, dF.$$

Therefore,

$$dF(F) = \text{Const.} \sum_{j=0}^{\infty} A_j F^{(n-3)/2+j}\, dF$$
$$\int_0^{\infty} \exp\left[-\frac{(n-1)s_2^2}{2(1-\rho^2)}\left\{\frac{F\sigma_2^2 + \sigma_1^2}{\sigma_1^2\sigma_2^2}\right\}(s_2^2)^{n-2+2j}\, ds_2^2\right]. \qquad (12.3.58)$$

The integral is

$$\left\{\frac{2(1-\rho^2)\sigma_1^2\sigma_2^2}{(n-1)(F\sigma_2^2+\sigma_1^2)}\right\}^{n+2j-1} \Gamma(n+2j-1).$$

Hence, (12.3.58) is

$$\text{Const.} \sum_{j=0}^{\infty} A_j \Gamma(n+2j-1) \cdot \frac{F^{(n-3)/2+j}\, dF}{(F\sigma_2^2+\sigma_1^2)^{(n+2j-1)}} \cdot \left\{\frac{2(1-\rho^2)\sigma_1^2\sigma_2^2}{n-1}\right\}^{n+2j-1}. \qquad (12.3.59)$$

Using the relation

$$\frac{\Gamma(2j+1)}{\Gamma\left(\frac{2j+1}{2}\right)} = \frac{2^{2j}\Gamma(j+1)}{\sqrt{\pi}}$$

(12.3.59) simplifies to

$$dF(F) = \text{Const.}\sqrt{\pi} \sum_{j=0}^{\infty} \frac{\Gamma(n+2j-1)}{\Gamma\left(\frac{n+2j-1}{2}\right) 2^{2j}\Gamma(j+1)} \frac{F^{(n-3)/2+j}\, dF}{(F\sigma_2^2+\sigma_1^2)^{n+2j-1}} \times$$
$$\left\{\frac{2(1-\rho^2)\sigma_1^2\sigma_2^2}{n-1}\right\}^{n+2j-1}, \qquad (12.3.60)$$

The distribution (12.3.60) depends on the nuisance parameter ρ. Hence for testing $H_\circ(\sigma_1^2 = \sigma_2^2)$, we cannot use the statistic $F = s_1^2/s_2^2$ as its distribution (12.3.60) depends on the unknown parameter ρ.

For testing $H_\circ(\sigma_1^2 = \sigma_2^2)$ on the basis of a sample drawn from a $N_2(\mu_1, \mu_2, \sigma_1^2, \sigma_2^2, \rho$ population we consider instead modified variables

$$U = X + Y, \; V = X - Y.$$

Then

$$V(U) = \sigma_1^2 + \sigma_2^2 + 2\rho\sigma_1\sigma_2,$$

$$V(V) = \sigma_1^2 + \sigma_2^2 - 2\rho_1\sigma_2,$$

$$\text{Cov}(U, V) = \sigma_1^2 - \sigma_2^2.$$

If $\sigma_1^2 = \sigma_2^2$, $\text{Cov}(U, V) = 0$ and hence $\rho_{U,V} = 0$. Therefore, $H_\circ\,(\sigma_1^2 = \sigma_2^2)$ is equivalent to $H_\circ'\,(\rho_{U,V} = 0)$. Under null hypothesis $H_\circ'$, $r_{uv}\sqrt{n-2}/\sqrt{1-r_{uv}^2}$ follows Student's $t_{(n-2)}$ [vide (12.3.27)]. This is known as Pitman's test.

▶ **Example 12.3.3** *Let $(x_i, y_i)(i = 1, \cdots, n)$ be a sample from a bivariate normal $N_2(\mu_1, \mu_2, \sigma_1, \sigma_2, \rho)$ population. Find the distribution of $F = (s_1/\sigma_1)/(s_2/\sigma_2)$ and hence show that under $H_\circ(\sigma_1 = \sigma_2)$ the distribution of $F' = s_1/s_2$ is independent of the common value of σ_1 and σ_2 but involves the population correlation coefficient ρ.*

The joint distribution of s_1, s_2 and r is given by (12.3.25)

$$dF(s_1, s_2, r) = \text{Const.}\exp\left[-\frac{n-1}{2(1-\rho^2)}\left\{\frac{s_1^2}{\sigma_1^2} + \frac{s_2^2}{\sigma_2^2} - 2r\rho\frac{s_1 s_2}{\sigma_1\sigma_2}\right\}\right]$$

$$s_1^{(n-2)} s_2^{(n-2)} (1-r^2)^{(n-4)/2}\, ds_1\, ds_2\, dr. \tag{12.3.61}$$

Make transformations $(s_1, s_2, r) \to (F, s_2, r)$, $|J| = s_2\sigma_1/\sigma_2$.

$$dF(F, s_2^2, r) = \text{Const.}\exp\left[-\frac{(n-1)s_2^2}{2(1-\rho^2)\sigma_2^2}\left\{F^2 - 2r\rho F + 1\right\}\right]$$

$$(s_2^2)^{(n-2)} F^{n-2} (1-r^2)^{(n-4)/2} F\, ds_2^2\, dr. \tag{12.3.62}$$

Integrating out s_2^2 over $(0, \infty)$ we get the joint distribution of F and r as,

$$dF(F, r) = \text{Const.}(1-r^2)^{(n-4)/2} F^{n-2}\, dF\, dr$$

$$\int_0^\infty \exp\left[-\frac{(n-1)s_2^2}{2(1-\rho^2)\sigma_2^2}\left\{F^2 - 2r\rho F + 1\right\}\right] (s_2^2)^{(n-2)} d(s_2^2) \tag{12.3.63}$$

$$= \frac{\text{Const.}(1-r^2)^{(n-4)/2} F^{n-2}\, dF\, dr}{(F^2 - 2r\rho F + 1)^{(n-1)}}. \tag{12.3.64}$$

Now, use the transformation,

$$r = u(\lambda + \mu) - (1-u)(\lambda - \mu) \text{ where } \lambda = 1 + F^2,\ \ \mu = 2\rho F,$$
$$d = u(\lambda + \mu) + (1-u)(\lambda - \mu). \tag{12.3.65}$$

Hence,

$$1 - r^2 = \frac{4u(1-u)(\lambda^2 - \mu^2)}{d^2}, \tag{12.3.66}$$

$$1 - 2\rho F r + F^2 = \lambda - \mu r = (\lambda^2 - \mu^2)/d$$

and

$$\frac{dr}{du} = \frac{2(\lambda^2 - \mu^2)}{d^2}.$$

Therefore,

$$dF(u, F) = \frac{\text{Const.} F^{(n-2)} u^{(n-4)/2} (1-u)^{(n-4)/2}}{(\lambda^2 - \mu^2)^{n/2}} d\, du\, dF.$$

Note that $r = -1 \Rightarrow u = 0, r = 1, \Rightarrow u = 1$. Hence,

$$dF(F) = \frac{\text{Const.} F^{(n-2)}}{(\lambda^2 - \mu^2)^{n/2}} dF \left[\int_0^1 u^{(n-4)/2} (1-u)^{(n-4)/2} \{u(\lambda + \mu) + (\lambda - \mu)(1-u)\}\, du \right]$$

$$= \frac{\text{Const.} F^{(n-2)}}{(\lambda^2 - \mu^2)^{n/2}} dF$$

$$\left[(\lambda + \mu) \int_0^1 u^{(n-2)/2} (1-u)^{(n-4)/2}\, du + (\lambda - \mu) \int_0^1 u^{(n-4)/2} (1-u)^{(n-2)/2}\, du \right]$$

$$= \frac{\text{Const.} F^{(n-2)}}{(\lambda^2 - \mu^2)^{n/2}} dF \left[(\lambda + \mu) B\left(\frac{n}{2}, \frac{n-2}{2}\right) + (\lambda - \mu) B\left(\frac{n}{2}, \frac{n-2}{2}\right) \right]$$

$$= \frac{\text{Const.} F^{(n-2)}}{(\lambda^2 - \mu^2)^{n/2}} \lambda B\left(\frac{n}{2}, \frac{n-2}{2}\right) dF$$

$$= \text{Const.} \frac{\lambda F^{(n-2)}\, dF}{(\lambda^2 - \mu^2)^{n/2}}.$$

But

$$\lambda^2 - \mu^2 = (1 + F^2)^2 \left[1 - \frac{4\rho^2 F^2}{(1 + F^2)^2} \right].$$

Hence,

$$dF(F) = \text{Const.} \frac{F^{n-2}}{(1 + F^2)^{n-1}} \left[1 - \frac{4\rho F^2}{(1 + F^2)^2} \right]^{-\frac{n}{2}}. \tag{12.3.67}$$

Under $H_\circ(\sigma_1 = \sigma_2)$, F becomes F' and as such the distribution of F' is obtained by putting $F = F'$ in (12.3.67). The distribution (12.3.67) does not involve the common value of the population s.d. but involves ρ.

▶ **Example 12.3.4** *(X, Y) follows a bivariate normal distribution $N_2(\mu_1, \mu_2, \sigma_1, \sigma_2, \rho)$. Find the distribution of $T = aX + bY$, where a, b are given constants.*

$$f(x,y)\,dx\,dy = \frac{1}{2\pi\sigma_2\sigma_2\sqrt{1-\rho^2}}\exp\left[-\frac{1}{2(1-\rho^2)}\left\{\frac{(x-\mu_1)^2}{\sigma_1^2}\right.\right.$$

$$\left.\left.-2\rho\frac{(x-\mu_1)(y-\mu_2)}{\sigma_1\sigma_2}+\frac{(y-\mu_2)^2}{\sigma_2^2}\right\}\right]dx\,dy. \qquad \text{(i)}$$

Put

$$x' = x-\mu_1,\ \ y' = y-\mu_2,\ \ T' = ax'+by',\ \ R = cx'+dy',$$

where c, d satisfy the equations

$$eT'^2 + R^2 = \frac{x'^2}{\sigma_1^2} - 2\rho\frac{x'y'}{\sigma_1\sigma_2} + \frac{y'^2}{\sigma_2^2}$$

$$\text{or, } e(a^2x'^2 + b^2y'^2 + 2abx'y') + (c^2x'^2 + d^2y'^2 + 2c\,dx'y')$$

$$= \frac{x'^2}{\sigma_1^2} - 2\rho\frac{x'y'}{\sigma_1\sigma_2} + \frac{y'^2}{\sigma_2^2}.$$

Equating coefficient of $x'^2, y'^2, x'y'$ we get

$$\frac{1}{\sigma_1^2} - ea^2 = c^2 \quad \cdots\text{(ii)} \qquad \therefore -\frac{\rho}{\sigma_1\sigma_2} - eab = cd \quad \cdots\text{(iii)} \qquad \therefore \frac{1}{\sigma_2^2} - eb^2 = d^2.\cdots\text{(iv)}$$

From (ii) and (iv)

$$c^2d^2 = \left(\frac{1}{\sigma_1^2} - ea^2\right)\left(\frac{1}{\sigma_2^2} - eb^2\right) = \frac{1}{\sigma_1^2\sigma_2^2} + e^2a^2b^2 - \frac{1}{\sigma_1^2}eb^2 - \frac{1}{\sigma_2^2}ea^2 \qquad \text{(v)}$$

and from (iii)

$$c^2d^2 = \frac{\rho^2}{\sigma_1^2\sigma_2^2} + e^2a^2b^2 + \frac{2e\rho ab}{\sigma_1\sigma_2}. \qquad \text{(vi)}$$

Equating (v) and (vi), one gets

$$e = \frac{1-\rho^2}{a^2\sigma_1^2 + b^2\sigma_2^2 + 2\rho\sigma_1\sigma_2 ab}. \qquad \text{(vii)}$$

Again,

$$J\left(\frac{x', y'}{T', R}\right) = \frac{1}{ad-bc}.$$

Also

$$(ad-bc)^2 = \frac{a^2\sigma_1^2 + b^2\sigma_2^2 + 2\rho\sigma_1\sigma_2}{\sigma_1^2\sigma_2^2}.$$

Hence, from (i)

$$dF(T,R) = \frac{1}{2\pi\sigma_1\sigma_2\sqrt{1-\rho^2}}$$

$$\exp\left[-\frac{1}{2(1-\rho^2)}(eT'^2+R^2)\right] \times \frac{\sigma_1\sigma_2 dT'\, dR}{\sqrt{a^2\sigma_1^2+b^2\sigma_2^2+2\rho\sigma_1\sigma_2 ab}}$$

$$= \frac{1}{\sqrt{2\pi}\sqrt{a^2\sigma_1^2+b^2\sigma_2^2+2\rho\sigma_1\sigma_2 ab}}$$

$$\exp\left[-\frac{eT'^2}{2(1-\rho^2)}\right] dT' \times \frac{1}{\sqrt{2\pi}} \exp\left[-\frac{R^2}{2(1-\rho^2)}\right] d\left(\frac{R}{\sqrt{1-\rho^2}}\right).$$

Integrating over R

$$dF(T') = \frac{1}{\sqrt{2\pi(a^2\sigma_1^2+b^2\sigma_2^2+2ab\rho\sigma_1\sigma_2)}}$$

$$\exp\left[-\frac{(ax'+by')^2}{2(a^2\sigma_1^2+b^2\sigma_2^2+2ab\rho\sigma_1\sigma_2)}\right] dT.$$

Now

$$T' = T - (a\mu_1 + b\mu_2).$$

Hence,

$$T \cap N\left(a\mu_1 + b\mu_2, a^2\sigma_1^2+b^2\sigma_2^2+2ab\rho\sigma_1\sigma_2\right).$$

[**Note :** The problem can easily be worked out by calculating the moment generating function of T.]

Exercise

1. Let $X_1, \cdots, X_n$ be a random sample from $N(\mu, \sigma^2)$.

   (a) Find the conditional density of X_1 given $X_2 = x_2, \cdots, X_n = x_n$ and then set $X_1 = n\bar{X} - X_2 \cdots - X_n$, and use the transformation technique to find the conditional density of $\bar{X}$ given $X_2 = x_2 = \cdots = X_n = x_n$;

   (b) Find the joint density of $\bar{X}, X_2, \cdots, X_n$ by multiplying the conditional density of $\bar{X}$ obtained in part (a) by the joint density of $X_2, X_3, \cdots, X_n$ and show that

$$g(x_2, \cdots, x_n|\bar{x}) = \sqrt{n}\left(\frac{1}{\sigma\sqrt{2\pi}}\right)^{n-1} e^{-(n-1)s^2/2\sigma^2},$$

$$-\infty < x_i < \infty, \;\; i = 2, 3, \cdots, n;$$

(c) Show that the conditional moment-generating function of $(n-1)S^2/\sigma^2$ given $\bar{X} = \bar{x}$ is

$$E\left[\exp\left(\frac{(n-1)S^2}{\sigma^2}\right)\Big|\bar{x}\right] = (1-2t)^{(n-1)/2} \text{ for } t < \tfrac{1}{2}.$$

Since this result is free of $\bar{x}$, it follows that $\bar{X}$ and S^2 are independent; it also shows that $(n-1)S^2/\sigma^2$ has a $\chi^2_{(n-1)}$ distribution. **[Freund, 1992]**

2. Show that if s^2 is an estimate of variance σ^2 of normal population,

$$E(s) = \sigma\frac{\Gamma\left(\frac{n}{2}\right)}{\Gamma\left(\frac{n-1}{2}\right)}\sqrt{\frac{2}{n}} = \sigma + 0\left(\frac{1}{n}\right)$$

$$V(s) = \left[\frac{n-1}{n} - \frac{\left(\Gamma\left(\frac{n}{2}\right)\right)^2}{\left(\Gamma\left(\frac{n-1}{2}\right)\right)^2}\cdot\frac{2}{n}\right]\sigma^2 = \frac{\sigma^2}{n} + 0\left(\frac{1}{n^2}\right).$$

3. Let s_i^2 be the sample variance obtained from a sample of size n_i from $N(\mu_i, \sigma_i^2)$ population $(i = 1, 2)$. Show that the distribution of $F = s_1^2\sigma_2^2/s_2^2\sigma_1^2$ depends only on n_1, n_2. Show also that the mean of this distribution does not involve n_1. Show that when $n_1 = 2$,

$$\int_{F_0}^{\infty} f_F(f)\,df = p \text{ gives } F_0 = \frac{n_2}{2}\left(p^{-2/n_2} - 1\right)$$

where $f_F(f)$ is the p.d.f. of F distribution.

4. A random sample of n observations $(x_1, y_1), \cdots, (x_n, y_n)$ is drawn from a $N_2(\mu_1, \mu_2, \sigma_1, \sigma_2, \rho)$ population. Prove that

$$Q^2 = \frac{1}{1-\rho^2}\sum_{i=1}^{n}\left[\frac{(x_i-\mu_1)^2}{\sigma_1^2} - 2\rho\frac{(x_i-\mu_1)(y_i-\mu_2)}{\sigma_1\sigma_2} + \frac{(y_i-\mu_2)^2}{\sigma_2^2}\right].$$

is distributed as $\chi^2_{(2n)}$.

[*Hints:* Transform: $u_i = (x_i-\mu_1)/\sigma_1, v_i = (y_i-\mu_2)/\sigma_2\ (i = 1, \cdots, n)$; set $w_i = (v_i - \rho u_i)/\sqrt{1-\rho^2}$. Hence $Q^2 = \sum_{i=1}^{n}(u_i + w_i^2) \cap \chi^2_{(2n)}$.]

5. Let $(X, Y) \cap N_2(0, 0, \sigma_1, \sigma_2, \rho)$ distribution. Show that

$$U = \frac{X}{\sigma_1} + \frac{Y}{\sigma_2};\quad V = \frac{X}{\sigma_1} - \frac{Y}{\sigma_2}$$

are independent normal variates with variance $2(1+\rho)$ and $2(1-p)$ respectively. If r is the correlation coefficient between U and V in a random samples of size n from this N_2 population, show that r^2 follows $B_1\left(\frac{1}{2}, \frac{n}{2} - 1\right)$.

6. If $(X_1, \cdots, X_n)$ is a random sample from $N(\mu, \sigma^2)$ population and

$$Z = (x_1 - \bar{x}) \Big/ \sqrt{\sum_1^n (x_i - \bar{x})^2},$$

show that

$$t = \frac{z\sqrt{n-2}}{\sqrt{n-1-z^2}}$$

has a Student's $t_{(n-2)}$ distribution.

7. Let $(X_i, Y_i)(i = 1, \cdots, n)$ be a random sample form $N_2(\mu_1, \mu_2, \sigma^2, \sigma^2, \rho)$. Let $U_i = X_i + Y_i$, $V_i = X_i - Y_i$ $(i = 1, \cdots, n)$, $\bar{U} = \Sigma U_i/n$, $\bar{V} = \Sigma V_i/n$, $s_u^2 = \Sigma(U_i - \bar{U})^2/(n-1)$, $s_v^2 = \Sigma\,(V_i - \bar{V})^2/(n-1)$. Show that $F = (1-\rho)\ s_u^2/(1+\rho)\ s_v^2 \cap F_{(n-1, n-1)}$.

8. Show that in sampling from $N(\mu, \sigma^2)$ population, the p.d.f. of s^2 is maximum when $s^2 = \sigma^2$ as $n \to \infty$ but the p.d.f. of s is not maximum when $s = \sigma$ as $n \to \infty$.

9. Let r be the sample correlation coefficient based on n independent observations from a bivariate normal population with correlation coefficient ρ. Show that

$$\sqrt{n}(r - \rho) \xrightarrow{L} N\left(0, \frac{(1-\rho^2)^2}{n}\right).$$

Chapter 13

Multivariate Normal Distribution and Related Distributions

13.1 Introduction

We consider in this chapter multivariate normal distribution and its properties. Section 13.3 considers some results on quadratic forms of random variables which have joint multinormal distribution. Multivariate Central Limit Theorem is considered in section 13.4. Multivariate t distribution, Spherical and Elliptic distributions and multivariate cumulants are considered in sections 13.6, 13.8 and 13.9 respectively. The Dirichlet distribution, though not related to normal distribution is considered due to its usefulness, in section 13.7.

13.2 Multivariate Normal Distribution

Let $\boldsymbol{X}' = (X_1, \cdots, X_p)$ be a random vector of p random variables. Let $\boldsymbol{\mu} = (\mu_1, \cdots, \mu_p)'$ be a vector of real constants and $\boldsymbol{A}$ be a $p \times p$ real, symmetric and positive definite (p.d.) matrix.

Definition 13.2.1 *The r.v.'s $X_1, \cdots, X_p$ have a joint multivariate normal distribution if their joint probability density can be written in the form*

$$f_{\boldsymbol{X}'}(x_1, \cdots, x_p) = C. \quad \exp\left[-\frac{1}{2}(\boldsymbol{x}-\boldsymbol{\mu})'\boldsymbol{A}(\boldsymbol{x}-\boldsymbol{\mu})\right] \qquad (13.2.1)$$

where $\boldsymbol{x} = (x_1, \cdots, x_p)'$ and C is a suitable constant. In other words, $\boldsymbol{X}$ follows a joint p-variate normal distribution, if its density can be written in the form

$$f_{\boldsymbol{X}'}(x_1, \cdots, x_p) = C. \quad \exp\left[-\frac{1}{2}(\textit{positive definite quadratic form in } \boldsymbol{x})\right]. \qquad (13.2.2)$$

It is obvious that C must be a function of $\boldsymbol{\mu}$ and $\boldsymbol{A}$. In order to determine the constant C we consider the moment generating function (m.g.f.) of $\boldsymbol{X}$ about $(0, \cdots, 0)$. The m.g.f. is

$$M_{\boldsymbol{X}}(t) = E(e^{\boldsymbol{t}'\boldsymbol{x}})$$

where $\boldsymbol{t}' = (t_1, \cdots, t_p)$, a set of arbitrary real numbers.

Theorem 13.2.1 *The m.g.f. of the distribution in* (13.2.1) *is given by*

$$M(t_1, \cdots, t_p) = \exp\left\{\boldsymbol{t}'\boldsymbol{\mu} + \frac{1}{2}\boldsymbol{t}'\boldsymbol{A}^{-1}\boldsymbol{t}\right\}. \tag{13.2.3}$$

Proof:

$$M(\boldsymbol{t}) = C\int_{-\infty}^{\infty} \cdots \int_{-\infty}^{\infty} \exp\left[-\frac{1}{2}(\boldsymbol{x}-\boldsymbol{\mu})'\boldsymbol{A}(\boldsymbol{x}-\boldsymbol{\mu}) + \boldsymbol{t}'\boldsymbol{x}\right] d\boldsymbol{x}$$

Let $x_i - \mu_i = y_i, i = 1, \cdots, p$. Hence,

$$M(\boldsymbol{t}) = Ce^{\boldsymbol{t}'\boldsymbol{\mu}}\int_{-\infty}^{\infty} \cdots \int_{-\infty}^{\infty} \exp\left[-\frac{1}{2}\boldsymbol{y}'\boldsymbol{A}\boldsymbol{y} + \boldsymbol{t}'\boldsymbol{y}\right] d\boldsymbol{y}. \tag{13.2.4}$$

Since $\boldsymbol{A}$ is p.d., we can write $\boldsymbol{A} = \boldsymbol{H}'\boldsymbol{H}$ where $\boldsymbol{H}$ is non-singular.
Transform $\boldsymbol{z} = \boldsymbol{H}\boldsymbol{y}; |J| = 1/|\boldsymbol{H}| = (\sqrt{|\boldsymbol{A}|})^{-1}$. Hence (13.2.4) reduces to

$$M(\boldsymbol{t}) = \frac{C}{|\boldsymbol{H}|}e^{\boldsymbol{t}'\boldsymbol{\mu}}\int_{-\infty}^{\infty} \cdots \int_{-\infty}^{\infty} \exp\left[-\frac{1}{2}\boldsymbol{z}'\boldsymbol{z} + \boldsymbol{t}'\boldsymbol{H}^{-1}\boldsymbol{z}\right] d\boldsymbol{z} \tag{13.2.5}$$

$$= \frac{C}{|\boldsymbol{H}|}e^{\boldsymbol{t}'\boldsymbol{\mu}}\int_{-\infty}^{\infty} \cdots \int_{-\infty}^{\infty} \exp\left[-\frac{1}{2}\sum_{i=1}^{p}(z_i^2 + 2b_iz_i)\right] \Pi\, dz_i \tag{13.2.6}$$

where $\boldsymbol{b}' = t'\boldsymbol{H}^{-1}$.

$$= \frac{C}{|\boldsymbol{H}|}\exp\left(\boldsymbol{t}'\boldsymbol{\mu} + \frac{1}{2}\boldsymbol{b}'\boldsymbol{b}\right)\int_{-\infty}^{\infty} \cdots \int_{-\infty}^{\infty} \exp\left[-\frac{1}{2}(z_i + b_i)^2\right] \Pi\, dz_i$$

$$= \frac{C}{|\boldsymbol{H}|}(2\pi)^{p/2}\exp\left\{\boldsymbol{t}'\boldsymbol{\mu} + \frac{1}{2}\boldsymbol{b}'\boldsymbol{b}\right\}.$$

Now $\boldsymbol{b}'\boldsymbol{b} = \boldsymbol{t}'(\boldsymbol{H}'\boldsymbol{H})^{-1}\boldsymbol{t} = \boldsymbol{t}'\boldsymbol{A}^{-1}\boldsymbol{t}$. Hence,

$$M(\boldsymbol{t}) = \frac{C}{\sqrt{|\boldsymbol{A}|}}(2\pi)^{p/2}\exp\left\{\boldsymbol{t}'\boldsymbol{\mu} + \frac{1}{2}\boldsymbol{t}'\boldsymbol{A}^{-1}\boldsymbol{t}\right\}. \tag{13.2.7}$$

Since $M(\boldsymbol{0}) = 1$,

$$C = \sqrt{|\boldsymbol{A}|}(2\pi)^{-p/2}. \tag{13.2.8}$$

Hence the proof.

The joint density function (13.2.1) is, therefore,

$$f_{\boldsymbol{X}}(\boldsymbol{x}') = \frac{\sqrt{|\boldsymbol{A}|}}{(2\pi)^{p/2}} \exp\left\{-\frac{1}{2}(\boldsymbol{x}-\boldsymbol{\mu})'\boldsymbol{A}(\boldsymbol{x}-\boldsymbol{\mu})\right\}. \tag{13.2.9}$$

We now find the values of $\boldsymbol{\mu}$ and $\boldsymbol{A}$ in terms of moments of the distribution. From (13.2.3).

$$\begin{aligned} E(\boldsymbol{X}) &= \boldsymbol{\mu}, \\ D(\boldsymbol{X}) &= \boldsymbol{\Sigma}\,(\text{say}) = ((\sigma_{ij})) = \boldsymbol{A}^{-1}. \end{aligned} \tag{13.2.10}$$

Hence,

$$f_{\boldsymbol{X}}(\boldsymbol{x}') = \frac{1}{\sqrt{|\boldsymbol{\Sigma}|}(2\pi)^{p/2}} \exp\left\{-\frac{1}{2}(\boldsymbol{x}-\boldsymbol{\mu})'\boldsymbol{\Sigma}^{-1}(\boldsymbol{x}-\boldsymbol{\mu})\right\}. \tag{13.2.11}$$

We shall denote in this case $\boldsymbol{X} \cap N_p(\boldsymbol{\mu}, \boldsymbol{\Sigma})$.

■ NOTE 13.2.1 *The m.g.f.* $M(\boldsymbol{t})$ *in* (13.2.3) *shows, that all cumulants and cross-cumulants of order higher than two of the distribution are zero.*

■ NOTE 13.2.2 *The density function* (13.2.1) *is constant whenever the quadratic form in the exponent is so, so that it is constant on the ellipsoid*

$$\left\{\boldsymbol{x} : (\boldsymbol{x}-\boldsymbol{\mu})'\boldsymbol{\Sigma}^{-1}(\boldsymbol{x}-\boldsymbol{\mu}) = k\right\} \tag{13.2.12}$$

in R^p, *for every* $k > 0$. *This ellipsoid has centre* $\boldsymbol{\mu}$ *and* $\boldsymbol{\Sigma}$ *determines its shape and orientation.*

■ NOTE 13.2.3 *If* $\boldsymbol{A}$ *is only positive semi-definite (i.e.,* $|\boldsymbol{A}| = 0$*), the joint distribution of* $(X_1, \cdots, X_p)$ *is called a singular multinormal distribution.*

It follows from (13.2.3) that $M(0, \cdots, t_i, 0, \cdots, 0) = \exp\left(t_i\mu_i + \frac{t_i^2}{2}\sigma_{ii}\right)$ is the m.g.f. of $X_i (i = 1, \cdots, n)$. Thus each $X_i \cap N(\mu_i, \sigma_{ii}), i = 1, \cdots, p$.

For $i \neq j$, the joint m.g.f. of X_i, X_j is

$$M(0, \cdots, 0, t_i, 0, \cdots, 0, t_j, 0, \cdots, 0) = \exp\left\{t_i\mu_i + t_j\mu_j + \frac{1}{2}(\sigma_{ii}t_i^2 + 2\sigma_{ij}t_it_j + t_j^2\sigma_{jj})\right\}.$$

This is the m.g.f. of a bivariate normal distribution with means μ_i, μ_j, variance σ_{ii}, σ_{jj} and covariance $\sigma_{ij} = \rho_{ij}\sigma_i\sigma_j$ where ρ_{ij} is the correlation coefficient between X_i and $X_j (i \neq j)$.

It follows similarly that the marginal distribution of any subset $(X_1, \cdots, X_r)$ (say) of $\boldsymbol{X}(r < p)$ is a r-variate normal with mean vector $(\mu_1, \cdots, \mu_r)'$ and dispersion matrix $((\sigma_{ij}; i, j = 1, \cdots, r))$. (vide corollary 13.2.2).

If $\boldsymbol{\Sigma} = \text{Diag}(\sigma_{ii}, i = 1, \cdots, p)$ then $X_1, \cdots, X_p$ are all mutually independent, $X_i \cap N(\mu_i, \sigma_{ii})$.

We now consider several properties of a N_p-distribution.

Theorem 13.2.2 *If $\boldsymbol{X} \cap N_p(\boldsymbol{\mu}, \boldsymbol{\Sigma})$ and $\boldsymbol{C}$ is a $m \times p$ matrix with rank m, then*

$$\boldsymbol{CX} \cap N_m(\boldsymbol{C\mu}, \boldsymbol{C\Sigma C'}). \tag{13.2.13}$$

Proof: Let $\boldsymbol{Y} = \boldsymbol{CX}$. Then for every real $\boldsymbol{t}$

$$\begin{aligned} E\{\exp(\boldsymbol{t'Y})\} &= E\{\exp(\boldsymbol{t'CX})\} = E\{\exp(\boldsymbol{S'X})\} \text{ where } \boldsymbol{S'} = \boldsymbol{t'C} \\ &= \exp\left(\boldsymbol{S'\mu} + \frac{1}{2}\boldsymbol{S'\Sigma S}\right) = \exp\left[\boldsymbol{t'C\mu} + \frac{1}{2}\boldsymbol{t'}(\boldsymbol{C\Sigma C'})\boldsymbol{t}\right]. \end{aligned} \tag{13.2.14}$$

Since $\boldsymbol{C\Sigma C'}$ is positive definite, (13.2.14) is the m.g.f. of a $N_m(\boldsymbol{C\mu}, \boldsymbol{C\Sigma C'})$ distribution. The result follows from the uniqueness of m.g.f..

Corollary 13.2.1 *Any linear function of $X_1, \cdots, X_p$ is a univariate normal variable. In particular, every marginal distribution is a univariate normal distribution.*

Corollary 13.2.2 *The marginal distribution of any subset of the elements of $\boldsymbol{X}$ is also multivariate normal.*

Proof: Suppose $\boldsymbol{Y} = (X_1, \cdots, X_m)$ $(m < p)$. Put $\boldsymbol{C} = (I_m, \boldsymbol{O})$.

Corollary 13.2.3 *Suppose $\boldsymbol{X} \cap N_p(\mu 1_p, \boldsymbol{\Sigma})$ where $\boldsymbol{\Sigma} = \sigma^2 I_p$. Consider $\boldsymbol{Y} = \boldsymbol{LX}$ where $\boldsymbol{L}$ is an orthogonal matrix. Then $\boldsymbol{Y} \cap N_p(\mu \boldsymbol{L}, \boldsymbol{\Sigma})$.*

The corollary states that mutually independent normal variables with the same variance remains mutually independent normal variables with the same variance under orthogonal transformation. The above invariance property is a characterisation of a normal distribution.

► **Example 13.2.1** *Let $X_1, \cdots, X_n$ be mutually independent $N(\mu, \sigma^2)$ variables. Prove that $\bar{X}$ is statistically independent of*

$$Q = \sum_{i=1}^{n}(X_i - \bar{X})^2/\sigma^2 \quad \text{and} \quad Q \cap \chi^2_{(n-1)}.$$

Proof: Let $Y_i = (X_i - \mu)/\sigma$. Then Y_i follows $N(0,1)$ and $\boldsymbol{Y}, N_p(0, I)$. Consider an orthogonal transformation (Helmert's transformation)

$$\begin{aligned} Z_1 &= \frac{Y_1 + \cdots + Y_n}{\sqrt{n}}, \\ Z_2 &= \frac{Y_1 - Y_2}{\sqrt{2.1}}, \\ Z_3 &= \frac{Y_1 + Y_2 - 2Y_3}{\sqrt{3.2}}, \\ &\cdots\cdots\cdots\cdots\cdots\cdots, \\ Z_n &= \frac{Y_1 + \cdots + Y_{n-1} - (n-1)Y_n}{\sqrt{n(n-1)}}. \end{aligned}$$

By corollary (13.2.3), $Z_1, \cdots, Z_n$ are independent normal variables with the same ariance unity. Also $E(Z_i) = 0, i = 1, \cdots, n$.

Jow

$$\sum_{i=1}^{n} Z_i^2 = \sum_{i=1}^{n} Y_i^2 = n\bar{Y}^2 + \sum_{i=1}^{n} (Y_i - \bar{Y})^2 \tag{13.2.15}$$

Hence, $Q = \Sigma_{i=1}^n (Y_i - \bar{Y})^2 = \Sigma_2^n Z_i^2$.

ince Z_1 is independent of $Z_2, \cdots, Z_n$, it is independent of $\sum_2^n Z_i^2 = Q$.

Also $Z_i^2 \cap \chi^2_{(i)}$ so that $Q \cap \chi^2_{(n-1)}$.

NOTE 13.2.4 *We have noted that the marginal distribution of each component of* $\boldsymbol{X}$ *s univariate normal. The converse is not true in general. (Obviously, the converse is lso true if the components of* $\boldsymbol{X}$ *are all independent and normal.)*

For example, suppose the r.v.'s X_1, X_2 have joint distribution function

$$F(X_1, X_2) = \Phi(x_1)\Phi(x_2)[1 + \alpha(1 - \Phi(x_1))(1 - \Phi(x_2)]$$

where $|\alpha| \leq |, \Phi(x)$ is the distribution function of the standard normal distribution. It an be shown that the marginal distributions of X_1 and X_2 are standard normal (see lso exercise 20).

Theorem 13.2.3 *The r.v.* $\boldsymbol{X} = (X_1, \cdots, X_p)'$ *has a p-variate normal distribution ff every linear function of* $\boldsymbol{X}$, $\boldsymbol{l}'\boldsymbol{X} = l_1X_1 + \cdots + l_pX_p$ *follows a univariate normal listribution.*

Proof: Suppose $\boldsymbol{l}'\boldsymbol{X}$ has a univariate normal distribution for any $\boldsymbol{l}$. Then, the m.g.f. of $\boldsymbol{l}'\boldsymbol{X}$ is

$$\Phi(t) = \exp\left\{bt + \frac{1}{2}t^2\sigma^2\right\}$$

where $b = E(\boldsymbol{l}'\boldsymbol{X}) = \boldsymbol{l}'\boldsymbol{\mu}$ (say), $\sigma^2 = V(\boldsymbol{l}'\boldsymbol{X})$ and t is any real number. Thus

$$\Phi(t) = \exp\left\{\boldsymbol{l}'\boldsymbol{\mu}t + \frac{1}{2}t^2\boldsymbol{l}'\boldsymbol{\Sigma}\boldsymbol{l}\right\}.$$

Let $t = 1$. Then

$$\Phi(1) = \exp\left\{\boldsymbol{l}'\boldsymbol{\mu} + \frac{1}{2}\boldsymbol{l}'\boldsymbol{\Sigma}\boldsymbol{l}\right\} = M(l_1, \cdots, l_p)$$

which is the m.g.f. of a $N_p(\boldsymbol{\mu}, \boldsymbol{\Sigma})$ distribution. The converse follows from corollary (13.2.1).

■ NOTE 13.2.5 *Note that for the results of theorem (13.2.3) it is necessary that ever linear combination* $\boldsymbol{l}'\boldsymbol{X}$ *is univariate normal and not just* $X_1, \cdots, X_n$. *The relevance c note (13.2.4) will now be clear.*

Theorem (13.2.3) also states that the multivariate normal distribution is the onl multivariate distribution which has the property that every linear combination $\boldsymbol{l}'\boldsymbol{X}$ is univariate normal.

Corollary 13.2.4 *If* $\boldsymbol{X} = (X_1, X_2)'$ *and the marginal distributions of* X_1 *and* X_2 *ar both univariate normal, then* $\boldsymbol{X}$ *has a bivariate normal distribution iff* $l_1X_1 + l_2X_2$ *univariate normal for all real* l_1 *and* l_2.

Theorem 13.2.4 *Let* $\boldsymbol{X} \cap N_p(\boldsymbol{\mu}, \boldsymbol{\Sigma})$ *and suppose that* $\boldsymbol{X} = \begin{bmatrix} \boldsymbol{X}^{(1)} \\ \boldsymbol{X}^{(2)} \end{bmatrix}$ *where* $\boldsymbol{X}^{(1)}$ *has r elements,* $m < p$. *Then* $\boldsymbol{X}^{(1)}, \boldsymbol{X}^{(2)}$ *are statistically independent iff* $C\left[\boldsymbol{X}^{(1)}, \boldsymbol{X}^{(2)}\right] = \mathbf{0}$.

Proof: If $\boldsymbol{X}^{(1)}, \boldsymbol{X}^{(2)}$ are independent, then all pairs of elements $X_i^{(1)}, X_j^{(2)}$ are sc Hence,

$$\operatorname{Cov}\left(X_i^{(1)}, X_j^{(2)}\right) = 0 \quad \text{and} \quad C\left(\boldsymbol{X}^{(1)}, \boldsymbol{X}^{(2)}\right) = \mathbf{0}.$$

Conversely, if $C\left(\boldsymbol{X}^{(1)}, \boldsymbol{X}^{(2)}\right) = \boldsymbol{\Sigma}_{12} = \mathbf{0}$, then

$$\boldsymbol{\Sigma} = \begin{bmatrix} \boldsymbol{\Sigma}_{11} & \boldsymbol{\Sigma}_{12} \\ \boldsymbol{\Sigma}_{21} & \boldsymbol{\Sigma}_{22} \end{bmatrix} = \begin{bmatrix} \boldsymbol{\Sigma}_{11} & \mathbf{0} \\ \mathbf{0} & \boldsymbol{\Sigma}_{22} \end{bmatrix}.$$

Hence,

$$|\boldsymbol{\Sigma}| = |\boldsymbol{\Sigma}_{11}| \cdot |\boldsymbol{\Sigma}_{22}|; \quad \boldsymbol{\Sigma}^{-1} = \begin{bmatrix} \boldsymbol{\Sigma}_{11}^{-1} & \mathbf{0} \\ \mathbf{0} & \boldsymbol{\Sigma}_{22}^{-1} \end{bmatrix}.$$

Therefore,

$$f_{\boldsymbol{X}'}(\boldsymbol{x}') = \left\{\frac{1}{(2\pi)^{m/2}\sqrt{\boldsymbol{\Sigma}_{11}}} \exp\left[-\frac{1}{2}\left(\boldsymbol{x}^{(1)} - \boldsymbol{\mu}^{(1)}\right)' \boldsymbol{\Sigma}_{11}^{-1}\left(\boldsymbol{x}^{(1)} - \boldsymbol{\mu}^{(1)}\right)\right]\right\}$$

$$\left\{\frac{1}{(2\pi)^{(p-m)/2}\sqrt{|\boldsymbol{\Sigma}_{22}|}} \exp\left[-\frac{1}{2}\left(\boldsymbol{x}^{(2)} - \boldsymbol{\mu}^{(2)}\right)' \boldsymbol{\Sigma}_{22}^{-1}\left(\boldsymbol{x}^{(2)} - \boldsymbol{\mu}^{(2)}\right)\right]\right\}$$

so that $\boldsymbol{X}^{(1)}, \boldsymbol{X}^{(2)}$ are independent.

Corollary 13.2.5 *If* $\boldsymbol{X}' = \left(\boldsymbol{X}^{(1)'}, \boldsymbol{X}^{(2)'}, \cdots, \boldsymbol{X}^{(r)'}\right)$ *and* $C\left(\boldsymbol{X}^{(i)}, \boldsymbol{X}^{(j)}\right) = \mathbf{0}$ *for eac* $i \neq j$ $(= 1, \cdots, r)$ *and if* $\boldsymbol{X} \cap N_p(\boldsymbol{\mu}, \boldsymbol{\Sigma})$ *then* $\boldsymbol{X}^{(i)}$*'s are mutually independent and nc just pair-wise independent.*

Corollary 13.2.6 *If $X \cap N_p(\mu, \sigma^2 I)$, then $a'X$, $b'X$ are independent iff $a'b = 0$.*

Proof: The proof follows from theorem (13.2.4), observing that

$$\text{Cov}(a'X, b'Y) = \sigma^2 a'b = 0.$$

Theorem 13.2.5 *Suppose $X \cap N_p(\mu, \sigma^2 I_p)$ and let $U = AX, V = BX$. Let A_1 represent the linearly independent row vectors of A and let $U_1 = A_1 X$. Suppose $C(U, V) = 0$. Then*

(a) U_1 is independent of $V'V$.

(b) $U'U$ is independent of $V'V$.

Proof: $C(U, V) = AD(X)B' = \sigma^2 AB' = 0$. Hence, every row of A is orthogonal to every row of B. Let B_1 represent the linearly independent rows of B. Let $C = \begin{bmatrix} A_1 \\ B_1 \end{bmatrix}$.

The rows of C are linearly independent. Hence, CX has a multivariate normal distribution.

Let $V_1 = B_1 X$. $C(U_1, V_1) = \sigma^2 A_1 B_1' = 0$.

Hence U_1, V_1 are independent (theorem (13.2.4)).

Without any loss of generality assume $B = \begin{bmatrix} B_1 \\ B_2 \end{bmatrix}$.

Since rows of B_2 are linearly dependent on those of B_1, let $B_2 = HB_1$.

Hence,

$$V = BX = \begin{bmatrix} I \\ H \end{bmatrix} B_1 X = MV_1 \quad \text{(say);}$$

$$V'V = V_1' M' M V_1.$$

Since U_1 is independent of V_1, it is independent of $V'V$. Similarly U can be expressed as $U = LU_1$ and since V_1 is independent of U_1, $V'V$ is independent of $U'U$.

▶ **Example 13.2.2** *Suppose X_i's $(i = 1, \cdots, n)$ are i.i.d. $N(0, \sigma^2)$. Show that $\bar{X}$ and $Q = \Sigma_{i=1}^n (X_i - \bar{X})^2$ are statistically independent.*

$X \cap N_n(0, \sigma^2 I)$.

Let $\bar{X} = U, V = X - \bar{X} 1_n$.

Now, $\text{Cov}(\bar{X}, V_i) = 0, \ i = 1, \cdots, n$.

Hence, $\text{Cov}(\bar{X}, V) = 0$.

Therefore, $\bar{X}$ and $V'V$ are statistically independent.

Theorem 13.2.6 *Let $\boldsymbol{X} \cap N_p(\boldsymbol{\mu}, \sigma^2 I_n)$ and $\boldsymbol{P}$ be a non-negative symmetric matrix of rank r. Then $Q = (\boldsymbol{X} - \boldsymbol{\mu})'\boldsymbol{P}(\boldsymbol{X} - \boldsymbol{\mu})/\sigma^2$ is distributed as $\chi^2_{(r)}$ iff $\boldsymbol{P}$ is a idempotent matrix.*

Proof: Suppose $\boldsymbol{P}$ is idempotent. Since $\boldsymbol{P}^2 = \boldsymbol{P}, \boldsymbol{P}$ has r eigenvalues equal to unity and the rest $(n-r)$ eigenvalues equal to zero. Hence, there exists an orthogonal matrix $\boldsymbol{T}$ such that

$$\boldsymbol{T}'\boldsymbol{P}\boldsymbol{T} = \text{Diag}(\lambda_1, \cdots, \lambda_n) = \wedge \text{ (say)},$$

where λ_i's are eigenvalues of $\boldsymbol{P}$.
Let $\boldsymbol{Y} = \boldsymbol{T}'(\boldsymbol{X} - \boldsymbol{\mu})$.
Then $\boldsymbol{Y} \cap N_n(\boldsymbol{0}, \sigma^2 I_n)$.
Therefore, $Q = \boldsymbol{Y}'\wedge\boldsymbol{Y}/\sigma^2 \cap \chi^2_{(r)}$.
Conversely, suppose $Q \cap \chi^2_{(r)}$. Then

$$E(e^{tQ}) = (1-2t)^{-r/2}. \tag{13.2.16}$$

Since, $\boldsymbol{P}$ is symmetric there exists a orthogonal matrix $\boldsymbol{S}$ such that $\boldsymbol{S}'\boldsymbol{P}\boldsymbol{S} = \wedge$.
Let $\boldsymbol{X} - \boldsymbol{\mu} = \boldsymbol{S}\boldsymbol{Y}$.
Therefore, $Q = \boldsymbol{\Sigma}_1^n \lambda_i Y_i^2/\sigma^2$.
Now

$$E[\exp(tQ)] = E\left[\exp\left(t\boldsymbol{\Sigma}_1^n \lambda_i Y_i^2/\sigma^2\right)\right].$$

But $\boldsymbol{Y} = \boldsymbol{S}'(\boldsymbol{X} - \boldsymbol{\mu}) \cap N_n(\boldsymbol{0}, \sigma^2 I_n)$ so that Y_i's are independent $N(0, \sigma^2)$. Hence $E(\exp(t\lambda_i Y_i^2/\sigma^2)) = (1 - 2\lambda_i t)^{-1/2}$. Therefore, from (13.2.16)

$$(1-2t)^{-r/2} = \prod_{i=1}^{n} (1 - 2t\lambda_i)^{-1/2}$$

identically in t for sufficiently small $|t|$. By the uniqueness property of polynomial roots we must have r λ_i's equal to unity and the rest λ_i's equal to zero. Hence, $\boldsymbol{P}^2 = \boldsymbol{P}$ and rank $(\boldsymbol{P}) = r$.

► **Example 13.2.3** *Let $\boldsymbol{X} \cap N_n(\boldsymbol{0}, I_n)$, $\boldsymbol{X}'\boldsymbol{X} = \boldsymbol{X}'\boldsymbol{A}\boldsymbol{X} + \boldsymbol{X}'\boldsymbol{B}\boldsymbol{X}$ and $\boldsymbol{X}'\boldsymbol{A}\boldsymbol{X} \cap \chi^2_{(r)}$; then*

$$\boldsymbol{X}'\boldsymbol{B}\boldsymbol{X} \cap \chi^2_{(n-r)}.$$

$\boldsymbol{A}$ is symmetric idempotent of rank r. Also $\boldsymbol{B} = I_n - \boldsymbol{A}$ is symmetric and $\boldsymbol{B}^2 = \boldsymbol{B}$ so that $\boldsymbol{B}$ is idempotent. Hence, $\boldsymbol{X}'\boldsymbol{B}\boldsymbol{X} \cap \chi^2_{(p)}$. Now

$$p = \text{rank } \boldsymbol{B} = \text{trace } \boldsymbol{B} = n - \text{trace } \boldsymbol{A} = n - \text{rank } \boldsymbol{A} = n - r.$$

[If $\boldsymbol{P}$ is a symmetric idempotent matrix (i.e., a projection matrix), trace $\boldsymbol{P}$ = rank $\boldsymbol{P}$.]

Example 13.2.4 *If $\boldsymbol{X} \cap N_n(\boldsymbol{0}, I_n)$ prove that*

$$\sum_{i=1}^{n}(X_i - \bar{X})^2 \cap \chi^2_{(n-1)}.$$

We have

$$\sum_{i=1}^{n}(X_i - \bar{X})^2 = \boldsymbol{X}'\boldsymbol{A}\boldsymbol{X}$$

where

$$\boldsymbol{A} = \left(\delta_{ij} - \frac{1}{n}\right), \quad \delta_{ij} = 1(0)$$

if $i = (\neq)j$. Hence, $\boldsymbol{A}$ is idempotent. Also rank $(\boldsymbol{A})$ = trace $\boldsymbol{A} = n - 1$.

Theorem 13.2.7 *If the $p \times 1$ random vectors $\boldsymbol{X}$ and $\boldsymbol{Y}$ are independent and $\boldsymbol{X} + \boldsymbol{Y}$ has p-variate normal distribution, then both $\boldsymbol{X}, \boldsymbol{Y}$ are normal.*

Proof: Omitted.

We have considered two characterising properties of normal distributions in corollary (13.2.3) and theorem (13.2.3). Other characterisations of univariate and multivariate normal distributions are due to Lukacs (1956), Laha (1957), Rao (1969, 1972, 1973), Kingman and Graybill (1970), Patil and Boswell (1970), Anderson (1971).

We now consider the conditional joint distribution of $X_1, \cdots, X_r$ given $X_{r+1}, \cdots, X_p$. We partition the matrix $\boldsymbol{A}(= \boldsymbol{\Sigma}^{-1})$ at the rth row and column as

$$\boldsymbol{A} = \begin{bmatrix} \boldsymbol{A}_{11} & \boldsymbol{A}_{12} \\ \boldsymbol{A}_{21} & \boldsymbol{A}_{22} \end{bmatrix}.$$

(Note that $\boldsymbol{A}_{21} = \boldsymbol{A}'_{12}$). Consider the $p \times p$ matrix

$$\boldsymbol{C} = \begin{bmatrix} I_r & \boldsymbol{0} \\ -\boldsymbol{A}_{21}\boldsymbol{A}_{11}^{-1} & I_{(p-r)} \end{bmatrix};$$

$$\boldsymbol{C}\boldsymbol{A}\boldsymbol{C}' = \begin{bmatrix} \boldsymbol{A}_{11} & \boldsymbol{0} \\ \boldsymbol{0} & \boldsymbol{A}_{22} - \boldsymbol{A}_{21}\boldsymbol{A}_{11}^{-1}\boldsymbol{A}_{12} \end{bmatrix}. \tag{13.2.17}$$

Transform $(\boldsymbol{x} - \boldsymbol{\mu})' = \boldsymbol{y}'\boldsymbol{C}$.

Now

$$(\boldsymbol{x} - \boldsymbol{\mu})'\boldsymbol{A}(\boldsymbol{x} - \boldsymbol{\mu}) = \boldsymbol{y}'\boldsymbol{C}\boldsymbol{A}\boldsymbol{C}'\boldsymbol{y} = \boldsymbol{y}'_{(1)}\boldsymbol{A}_{11}\boldsymbol{y}_{(1)} + \boldsymbol{y}'_{(2)}\boldsymbol{D}\boldsymbol{y}_{(2)}$$

where

$$\boldsymbol{y}'_{(1)} = (y_1, \cdots, y_r), \ \boldsymbol{y}'_{(2)} = (y_{r+1}, \cdots, y_p) \text{ and } \boldsymbol{D} = \boldsymbol{A}_{22} - \boldsymbol{A}_{21}\boldsymbol{A}_{11}^{-1}\boldsymbol{A}_{12}.$$

The joint density function of $\boldsymbol{Y} = (Y_1, \cdots, Y_p)'$ [where $(\boldsymbol{X} - \boldsymbol{\mu})' = \boldsymbol{Y}'\boldsymbol{C}$] is, therefore,

$$\boldsymbol{f}_{\boldsymbol{Y}'}\left(\boldsymbol{y}'_{(1)}, \boldsymbol{y}'_{(2)}\right) = \left\{\frac{\sqrt{|\boldsymbol{A}_{11}|}}{(2\pi)^{r/2}} \exp\left(-\frac{1}{2}\boldsymbol{y}'_{(1)}\boldsymbol{A}_{11}\boldsymbol{y}_{(1)}\right)\right\}$$

$$\left\{\frac{\sqrt{|\boldsymbol{D}|}}{(2\pi)^{(p-r)/2}} \exp\left(-\frac{1}{2}\boldsymbol{y}'_{(2)}\boldsymbol{D}\boldsymbol{y}_{(2)}\right)\right\} \qquad (13.2.18)$$

since

$$|\boldsymbol{A}| = |\boldsymbol{A}_{11}| \cdot |\boldsymbol{D}| \text{ (from (13.2.17), since } |\boldsymbol{C}| = 1).$$

It follows from (13.2.18) that the random vectors $\boldsymbol{Y}'_{(1)} = (Y_1, \cdots, Y_r)$ and $\boldsymbol{Y}'_{(2)} = (Y_{r+1}, \cdots, Y_p)$ are independent of each other and each has a joint multinormal distribution. Again

$$\boldsymbol{Y}'_{(2)} = \boldsymbol{X}'_{(2)} - \boldsymbol{\mu}'_2; \ \ \boldsymbol{Y}'_{(1)} = \left(\boldsymbol{X}_{(1)} - \boldsymbol{\mu}_{(1)}\right)' + \left(\boldsymbol{X}_{(2)} - \boldsymbol{\mu}_{(2)}\right)' \boldsymbol{A}_{21}\boldsymbol{A}_{11}^{-1}.$$

Therefore,

(a) $(X_{r+1}, \cdots, X_p)'$ have a joint multinormal distribution with mean vector $(\mu_{r+1}, \cdots$ and variance-covariance matrix

$$\left(\boldsymbol{A}_{22} - \boldsymbol{A}_{21}\boldsymbol{A}_{11}^{-1}\boldsymbol{A}_{12}\right)^{-1}.$$

(b) The conditional joint distribution of $\boldsymbol{X}_{(1)} = (X_1, \cdots, X_r)'$ given $(X_{r+1}, \cdots, X_p)$ is multinormal with expected values

$$\boldsymbol{\mu}_{(1)} - \boldsymbol{A}_{11}^{-1}\boldsymbol{A}_{12}(\boldsymbol{X}_{(2)} - \boldsymbol{\mu}_{(2)}) \qquad (13.2.19)$$

and variance-covariance matrix $\boldsymbol{A}_{11}^{-1}$. Formulae (13.2.19) shows that the regression of each of $X_1, \cdots, X_r$ on the set $\boldsymbol{X}'_{(2)} = (X_{r+1}, \cdots, X_p)$ is linear and the conditional distribution are homoscedastic (since $\boldsymbol{A}_{11}^{-1}$ does not depend on $\boldsymbol{X}_{(2)}$).

Special Cases:

r=1: Consider the conditional distribution of X_1 given $X_2, \cdots, X_p$

$$E(X_1|X_2, \cdots, X_p) = \mu_1 - \sum_{j=2}^{p}(x_j - \mu_j)\frac{\sigma^{1j}}{\sigma^{11}}; \ \ V(X_1|X_2, \cdots, X_p) = \frac{1}{\sigma^{11}}.$$

This means that the true regression equation of X_1 on $X_2, \cdots, X_p$ is linear.
The quantity

$$1 - \frac{V(X_1|X_2, \cdots, X_p)}{V(X_1)} = \rho^2_{1.2\cdots p}$$

where $\rho_{1.2\cdots p}$ is the multiple correlation coefficient of X_1 on $X_2, \cdots, X_p$.

Hence,

$$\rho^2_{1\cdot 2\cdots p} = 1 - \frac{1}{\sigma^{11}\sigma_{11}}.$$

$p = 2$: Here

$$E(X_1|X_3,\cdots,X_p) = \mu_1 - \frac{\sigma^{22}}{H}\sum_{j=3}^{p}(X_j-\mu_j)\sigma^{j1} + \frac{\sigma^{21}}{H}\sum_{j=3}^{p}(X_j-\mu_j)\sigma^{j2},$$

$$E(X_2|X_3,\cdots,X_p) = \mu_2 - \frac{\sigma^{11}}{H}\sum_{j=3}^{p}(X_j-\mu_j)\sigma^{j2} + \frac{\sigma^{12}}{H}\sum_{j=3}^{p}(X_j-\mu_j)\sigma^{j1}$$

where $H = \sigma^{11}\sigma^{22} - (\sigma^{12})^2$.

Also,

$$D\left(\begin{pmatrix} X_1 \\ X_2 \end{pmatrix}\middle| X_3,\cdots,X_p\right) = \begin{bmatrix} \sigma^{11} & \sigma^{12} \\ \sigma^{21} & \sigma^{22} \end{bmatrix}^{-1}$$

$$\text{Corr}(X_1, X_2|X_3,\cdots,X_p) = -\frac{\sigma^{12}}{\sqrt{\sigma^{11}\sigma^{22}}}.$$

The correlation $(X_1, X_2|X_3,\cdots,X_p)$ is called the partial correlation between X_1, X_2 eliminating the effects of $X_3,\cdots,X_p$ and is denoted as $\rho_{12\cdot 34\cdots p}$. For a multivariate normal population this is independent of values of $X_3,\cdots,X_p$.

It has been shown that if $X_1,\cdots,X_p$ have a joint p-variate multinormal distribution, then

$$P_r\left[\bigcap_{j=1}^{p}(|X_j-\mu_j|\le C_j)\right] \ge \prod_{j=1}^{p} P_r\left[|X_j-\mu_j|\le C_j\right].$$

The standardised multivariate normal distribution has mean vector $\mathbf{0}$ and dispersion matrix $\boldsymbol{R}$ where $\boldsymbol{R}(=((\rho_{ij})))$ is the correlation matrix of $\boldsymbol{X}$. Hence the density function of a standardised multinormal vector $\boldsymbol{X}$ is

$$f_{\boldsymbol{X}}(\boldsymbol{x}) = \frac{1}{(2\pi)^{p/2}\sqrt{|\boldsymbol{R}|}}\exp\left[-\frac{1}{2}\boldsymbol{x}'\boldsymbol{R}^{-1}\boldsymbol{x}\right].$$

Generalising the univariate probability integral $\Phi(z)$, we define,

$$\Phi_p(z_1,\cdots,z_p;\boldsymbol{R}) = P_r\left[\bigcap_{j=1}^{p}(X_j\le z_j)\right]$$

$$= \frac{1}{(2\pi)^{p/2}\sqrt{|\boldsymbol{R}|}}\int_{-\infty}^{z_1}\cdots\int_{-\infty}^{z_p}\exp\left[-\frac{1}{2}\boldsymbol{x}'\boldsymbol{R}^{-1}\boldsymbol{x}\right]d\boldsymbol{x}.$$

13.3 Some Results on Quadratic Forms

We consider in this section some results on quadratic forms in multinormal variables.

Theorem 13.3.1 *If* $\boldsymbol{X}$ *is* $N_p(\boldsymbol{\mu}, \boldsymbol{\Sigma})$ *and* $\boldsymbol{\Sigma}$ *is non-singular, then*

(a) $(\boldsymbol{X}-\boldsymbol{\mu})'\boldsymbol{\Sigma}^{-1}(\boldsymbol{X}-\boldsymbol{\mu}) \cap \chi^2_{(p)}$;

(b) $\boldsymbol{X}'\boldsymbol{\Sigma}^{-1}\boldsymbol{X} \cap \chi^2(p, \lambda)$ where $\lambda = \boldsymbol{\mu}'\boldsymbol{\Sigma}^{-1}\boldsymbol{\mu}$.

Proof: (a) Let $\boldsymbol{\Sigma} = \boldsymbol{C}\boldsymbol{C}'$ where $\boldsymbol{C}$ is non-singular. Put $\boldsymbol{U} = \boldsymbol{C}^{-1}(\boldsymbol{X}-\boldsymbol{\mu})$ so that $\boldsymbol{U}$ is $N_p(\boldsymbol{0}, I_p)$. Also

$$(\boldsymbol{X}-\boldsymbol{\mu})'\boldsymbol{\Sigma}^{-1}(\boldsymbol{X}-\boldsymbol{\mu}) = \boldsymbol{U}'\boldsymbol{U}$$

which is a sum of squares of p-independent $N(0,1)$ variables and hence is $\chi^2_{(p)}$.

(b) Let $\boldsymbol{V} = \boldsymbol{C}^{-1}\boldsymbol{X}$. Hence

$$\boldsymbol{V} \cap N_p(\boldsymbol{C}^{-1}\boldsymbol{\mu}, I_p) \text{ and } \boldsymbol{X}'\boldsymbol{\Sigma}^{-1}\boldsymbol{X} = \boldsymbol{V}'\boldsymbol{V}$$

which follows $\chi^2(p, \lambda)$ where $\lambda = (\boldsymbol{C}^{-1}\boldsymbol{\mu})'(\boldsymbol{C}^{-1}\boldsymbol{\mu}) = \boldsymbol{\mu}'\boldsymbol{\Sigma}^{-1}\boldsymbol{\mu}$.

■ NOTE 13.3.1 [1] *The theorem is important in testing the hypothesis about the mea[n] vector of a normal population when* $\boldsymbol{\Sigma}$ *is known.*

Suppose that $\boldsymbol{X}_1, \cdots, \boldsymbol{X}_n$ is a random sample from $N_p(\boldsymbol{\mu}, \boldsymbol{\Sigma})$. Then $\bar{\boldsymbol{X}} = (\bar{X}_1, \cdots, \bar{X}$ where

$$\bar{X}_j = \frac{1}{n}\sum_{i=1}^{n} X_{ji} (j = 1, \cdots, p)$$

follows $N_p(\boldsymbol{\mu}, n^{-1}\boldsymbol{\Sigma})$ by results in section 14.2.

Suppose we want to test $H_o(\boldsymbol{\mu} = \boldsymbol{\mu}_0)$ against $H(\boldsymbol{\mu} \neq \boldsymbol{\mu}_0)$. If the null hypothesis true, the statistic

$$W = n(\bar{\boldsymbol{X}} - \boldsymbol{\mu}^0)'\boldsymbol{\Sigma}^{-1}(\bar{\boldsymbol{X}} - \boldsymbol{\mu}^0) \cap \chi^2_{(p)}.$$

A critical region for H_o against H would be to reject H_o if

$$W > \chi^2_{p,(1-\alpha)}$$

where $\chi^2_{p,(1-\alpha)}$ is the upper $100(1-\alpha)\%$ point of $\chi^2_{(p)}$.

If the null hypothesis is not true, $(\bar{\boldsymbol{X}} - \boldsymbol{\mu}^0) \cap N_p(\boldsymbol{\mu} - \boldsymbol{\mu}^0, n^{-1}\boldsymbol{\Sigma})$. Hence, by (th[e] second part of) theorem 13.3.1, W follows $\chi^2(p, \lambda)$ where $\lambda = n(\boldsymbol{\mu} - \boldsymbol{\mu}^0)'\boldsymbol{\Sigma}^{-1}(\boldsymbol{\mu} - \boldsymbol{\mu}^0$
The power of the test is

$$\gamma(\lambda) = P\left\{\chi^2(p, \lambda) > \chi^2_{p,1-\alpha}\right\}$$

which can be found from the table of $\chi^2(p, \lambda)$ distribution function.

[1] Should be read along with chapter 17.

Theorem 13.3.2 *If $\boldsymbol{X}$ follows $N_p(\boldsymbol{\mu}, I_p)$ and $\boldsymbol{B}$ is a $p\times p$ symmetric matrix then $\boldsymbol{X}'\boldsymbol{B}\boldsymbol{X}$ follows non-central χ^2 iff $\boldsymbol{B}$ is idempotent. In this case, d.f. of χ^2 is r = Rank $(\boldsymbol{B})$ and the non-centrality parameter $\lambda = \boldsymbol{\mu}'\boldsymbol{B}\boldsymbol{\mu}$.*

Proof: Omitted.

Corollary 13.3.1 *If $\boldsymbol{X} \cap N_p(\boldsymbol{\mu}, \boldsymbol{\Sigma})$ where $\boldsymbol{\Sigma}$ is non-singular and $\boldsymbol{X}, \boldsymbol{\mu}, \boldsymbol{\Sigma}$ are partitioned as*

$$\boldsymbol{X} = \begin{bmatrix} \boldsymbol{X}_1 \\ \boldsymbol{X}_2 \end{bmatrix}, \quad \boldsymbol{\mu} = \begin{bmatrix} \boldsymbol{\mu}_1 \\ \boldsymbol{\mu}_2 \end{bmatrix}, \quad \boldsymbol{\Sigma} = \begin{bmatrix} \boldsymbol{\Sigma}_{11} & \boldsymbol{\Sigma}_{12} \\ \boldsymbol{\Sigma}_{21} & \boldsymbol{\Sigma}_{22} \end{bmatrix}$$

where $\boldsymbol{X}_1$ and $\boldsymbol{\mu}_1$ are $r \times 1$ and $\boldsymbol{\Sigma}_{11}$ is $r \times r$, then

$$Q \equiv (\boldsymbol{X} - \boldsymbol{\mu})'\boldsymbol{\Sigma}^{-1}(\boldsymbol{X} - \boldsymbol{\mu}) - (\boldsymbol{X}_1 - \boldsymbol{\mu}_1)'\boldsymbol{\Sigma}^{-1}(\boldsymbol{X}_1 - \boldsymbol{\mu}_1)$$

follows $\chi^2_{(p-r)}$.

Proof: Let $\boldsymbol{\Sigma} = \boldsymbol{C}\boldsymbol{C}'$ where $\boldsymbol{C}$ is non-singular. Partition $\boldsymbol{C}$

$$\boldsymbol{C} = \begin{bmatrix} C_1 \\ C_2 \end{bmatrix}$$

where $\boldsymbol{C}_1$ is $(r \times p)$ so that $\boldsymbol{\Sigma}_{11} = \boldsymbol{C}_1\boldsymbol{C}_1'$. Let $\boldsymbol{U} = \boldsymbol{C}^{-1}(\boldsymbol{X} - \boldsymbol{\mu})$.
Then

$$\boldsymbol{U} \cap N_p(\boldsymbol{0}, I_p) \text{ and } \boldsymbol{X}_1 - \boldsymbol{\mu}_1 = \boldsymbol{C}_1\boldsymbol{U}.$$

Then

$$Q = \boldsymbol{U}'\boldsymbol{U} - \boldsymbol{U}'\boldsymbol{C}_1'(\boldsymbol{C}_1\boldsymbol{C}_1')^{-1}\boldsymbol{C}_1\boldsymbol{U} = \boldsymbol{U}'[I_p - \boldsymbol{C}_1'(\boldsymbol{C}_1\boldsymbol{C}_1')^{-1}\boldsymbol{C}_1]\boldsymbol{U} = \boldsymbol{U}'(I_p - \boldsymbol{P})\boldsymbol{U}$$

where $\boldsymbol{P} = \boldsymbol{C}_1'(\boldsymbol{C}_1\boldsymbol{C}_1')^{-1}\boldsymbol{C}_1$. Now $\boldsymbol{P}$ is symmetric and idempotent.

Hence $\boldsymbol{B} = (I_p - \boldsymbol{P})$ is also so. Therefore, the result follows by theorem 13.3.2.

The theorem 13.3.2 is generalised to theorems 13.3.3 and 13.3.4 below.

Theorem 13.3.3 *Suppose $\boldsymbol{X} \cap N_p(\boldsymbol{0}, \boldsymbol{\Sigma})$. A necessary and sufficient condition for $\boldsymbol{X}'\boldsymbol{B}\boldsymbol{X}$ to have a χ^2-distribution is $\boldsymbol{\Sigma}\boldsymbol{B}\boldsymbol{\Sigma}\boldsymbol{B}\boldsymbol{\Sigma} = \boldsymbol{\Sigma}\boldsymbol{B}\boldsymbol{\Sigma}$, in which case the d.f. of χ^2 is r = rank $(\boldsymbol{B}\boldsymbol{\Sigma})$.*

Theorem 13.3.4 *If $\boldsymbol{X}$ is $N_p(\mu, \Sigma)$, where $\boldsymbol{\Sigma}$ is non-singular and $\boldsymbol{B}$ is a $p\times p$ symmetric matrix, then $\boldsymbol{X}'\boldsymbol{B}\boldsymbol{X}$ is $\chi^2(r, \lambda)$ where r = rank $(\boldsymbol{B})$, $\lambda = \boldsymbol{\mu}'\boldsymbol{\Sigma}\boldsymbol{\mu}$, iff $\boldsymbol{B}\boldsymbol{\Sigma}$ is idempotent.*

Proof: Let $\boldsymbol{C}$ be a non-singular $p \times p$ matrix such that $\boldsymbol{C}\boldsymbol{\Sigma}\boldsymbol{C}' = I_p$. Put $\boldsymbol{Y} = \boldsymbol{C}\boldsymbol{X}$. Then $\boldsymbol{X}'\boldsymbol{B}\boldsymbol{X} = \boldsymbol{Y}'\boldsymbol{C}^{-1'}\boldsymbol{B}\boldsymbol{C}^{-1}\boldsymbol{Y}$, where $\boldsymbol{Y} \cap N_p(\boldsymbol{C}\boldsymbol{\mu}, I_p)$. It follows from theorem 13.3.2 that $\boldsymbol{X}'\boldsymbol{B}\boldsymbol{X}$ is non-central χ^2 iff $\boldsymbol{C}^{-1'}\boldsymbol{B}\boldsymbol{C}^{-1}$ is idempotent. It is, therefore, sufficient to show that this is so iff $\boldsymbol{B\Sigma}$ is idempotent: If $\boldsymbol{B\Sigma}$ is idempotent, $\boldsymbol{B} = \boldsymbol{B\Sigma B} = \boldsymbol{BC}^{-1}\boldsymbol{C}^{-1'}\boldsymbol{B}$; hence,

$$\boldsymbol{C}^{-1'}\boldsymbol{BC}^{-1} = (\boldsymbol{C}^{-1'}\boldsymbol{BC}^{-1})(\boldsymbol{C}^{-1'}\boldsymbol{BC}^{-1})$$

so that $\boldsymbol{C}^{-1'}\boldsymbol{BC}^{-1}$ is idempotent. If $\boldsymbol{C}^{-1'}\boldsymbol{BC}^{-1}$ is idempotent then

$$\boldsymbol{C}^{-1'}\boldsymbol{BC}^{-1} = \boldsymbol{C}^{-1'}\boldsymbol{BC}^{-1}\boldsymbol{C}^{-1'}\boldsymbol{BC}^{-1} = \boldsymbol{C}^{-1'}\boldsymbol{B\Sigma BC}^{-1}$$

so that $\boldsymbol{B} = \boldsymbol{B\Sigma B}$ and hence $\boldsymbol{B\Sigma}$ is idempotent.

13.4 Multivariate Central Limit Theorem

We now consider multivariate central limit theorem due to Cramer (1946) and Anderson (1974).

Theorem 13.4.1 *Let $\boldsymbol{X}_1, \boldsymbol{X}_2, \cdots$ be a sequence of independently and identically distributed m-dimensional random vectors with mean $\boldsymbol{\mu}$ and covariance matrix $\boldsymbol{\Sigma}$. Let*

$$\bar{\boldsymbol{X}}_n = \frac{1}{n}\sum_{i=1}^{n} \boldsymbol{X}_i, \quad n \geq 1.$$

Then, as $n \to \infty$, the asymptotic distribution of

$$\sqrt{n}(\bar{\boldsymbol{X}}_n - \boldsymbol{\mu}) = n^{-\frac{1}{2}}\sum_{i=1}^{n}(\boldsymbol{X}_i - \boldsymbol{\mu})$$

is $N(\boldsymbol{0}, \boldsymbol{\Sigma})$.

Proof: Let

$$\boldsymbol{Y}_n = n^{-\frac{1}{2}}\sum_{i=1}^{n}(\boldsymbol{X}_i - \boldsymbol{\mu}).$$

By the continuity theorem of the characteristic function (theorem 9.2.4) it suffices to show that $\phi_n(\boldsymbol{t})$, the characteristic function of $\boldsymbol{Y}_n$ converges to

$$\exp\left(-\frac{1}{2}\boldsymbol{t}'\boldsymbol{\Sigma t}\right),$$

the characteristic function of $N_m(\boldsymbol{0}, \boldsymbol{\Sigma})$ distribution. Now the characteristic function of $\boldsymbol{t}'\boldsymbol{Y}_n$, when $\boldsymbol{t} \in R^m$, is

$$f_n(\alpha, \boldsymbol{t}) = E\left[\exp(i\alpha \boldsymbol{t}'\boldsymbol{Y}_n)\right]$$

considered as a function of $\alpha \in R^1$. Again,

$$\boldsymbol{t}'\boldsymbol{Y}_n = n^{-\frac{1}{2}}\sum_{j=1}^{n}(\boldsymbol{t}'\boldsymbol{X}_j - \boldsymbol{t}'\boldsymbol{\mu}).$$

Now $t'X_j - t'\mu$ is random variable in R^1. It, therefore, follows by the univariate entral limit theorem (theorem 9.8.1) that as $n \to \infty$, the asymptotic distribution of $'Y_n$ is $N(0, t'\Sigma t)$ and hence, as $n \to \infty$

$$f_n(\alpha, t) \to \exp\left(-\frac{1}{2}\alpha^2 t'\Sigma t\right)$$

or all t and α. Putting $\alpha = 1$,

$$\phi_n(t) = f_n(1, t) \to \exp\left(-\frac{1}{2}t'\Sigma t\right),$$

s $n \to \infty$. Hence the proof.

When the random vectors X_i are independently but not identically distributed a ultivariate extention of Theorem 9.8.3 is as follows.

heorem 13.4.2 *Let $\{X_i\}$ be independent random vectors with means $\{\mu_i\}$ and ispersion matrices $\{\Sigma_i\}$ and distribution function $\{F_i\}$. Suppose that*

$$\frac{\Sigma_1 + \cdots + \Sigma_n}{n} \to \Sigma, \quad \text{as } n \to \infty$$

nd that

$$\frac{1}{n}\sum_{i=1}^{n} \int_{\|x_i - \mu_i\| \geq \varepsilon\sqrt{n}} \|x_i - \mu_i\|^2 \cdot dF_i(x_i) \to 0, \quad \text{as } n \to \infty, \quad \text{for each } \varepsilon > 0.$$

Then $\frac{1}{n}\sum_{i=1}^{n} X_i$ *is asymptotically normal* $N\left(\frac{1}{n}\sum_{i=1}^{n}\mu_i, \frac{1}{n}\Sigma\right)$.

Here $\|z\| = \sqrt{\sum_{i=1}^{p} z_i^2}$ *where* $z = (z_1, \cdots, z_p)'$.

The theorem 13.4.3 below follows as an application of the central limit theorem 13.4.1. Consider the following notation. If T is a $p \times q$ matrix

$$T = (t_1, t_2, \cdots, t_q)$$

here t_i is $p \times 1 (i = 1, \cdots, q)$, then

$$\text{vec}\,(T) = \begin{bmatrix} t_1 \\ t_2 \\ \vdots \\ t_q \end{bmatrix}$$

hich is $pq \times 1$.

In the following theorem when we talk about the asymptotic normality of a random atrix T we mean the asymptotic normality of vec (T).

Theorem 13.4.3 *Let $\boldsymbol{X}_1, \boldsymbol{X}_2, \cdots$ be a sequence of independently and identically dis tributed $m \times 1$ random vectors with finite fourth moments and mean $\boldsymbol{\mu}$ and covarianc matrix $\boldsymbol{\Sigma}$ and let*

$$\boldsymbol{A}(n) = \sum_{i=1}^{n} (\boldsymbol{X}_i - \bar{\boldsymbol{X}}_n)(\boldsymbol{X}_i - \bar{\boldsymbol{X}}_n)'$$

where

$$\bar{\boldsymbol{X}}_n = \frac{1}{n}\sum_{i=1}^{n} \boldsymbol{X}_i.$$

Then the asymptotic distribution of

$$\boldsymbol{T}(n) = n^{-\frac{1}{2}}(\boldsymbol{A}(n) - n\boldsymbol{\Sigma})$$

is normal with mean zero and covariance matrix

$$\boldsymbol{V} = \text{Cov}\left[\text{vec}\left((\boldsymbol{X}_1 - \boldsymbol{\mu})(\boldsymbol{X}_1 - \boldsymbol{\mu})'\right)\right].$$

Corollary 13.4.1 *Let $\boldsymbol{S} = (n-1)^{-1}\boldsymbol{A}(n)$. Under the conditions of theorem 13.4.3, th asymptotic distribution of $\boldsymbol{U}(n) = \sqrt{n}(\boldsymbol{S}(n) - \boldsymbol{\Sigma})$ is normal with mean $\boldsymbol{0}$ and covarianc matrix $\boldsymbol{V}$.*

It may be noted that this asymptotic normal distribution is singular, because $\boldsymbol{V}$ i singular. This is due to the fact that $\boldsymbol{V}$ is $m^2 \times m^2$ covariance matrix in the asymptoti distribution of vec $(\boldsymbol{T}(n))$ or vec $(\boldsymbol{U}(n))$ and, because $\boldsymbol{T}(n)$ and $\boldsymbol{U}(n)$ are symmetri these vectors have repeated elements.

□ **Remark 13.4.1** *Given an underlying distribution for the $\boldsymbol{X}_i$, it is rather tedious t find the elements of the asymptotic covariance matrix, since this involves finding all th fourth order mixed moments of the distribution. However, the calculations are straigh forward if the sampling is from a $N_m(\boldsymbol{\mu}, \boldsymbol{\Sigma})$ distribution. In this case, the elements the asymptotic covariance matrix are given by* $\text{Cov}(u_{ij}(n), u_{ke}(n)) = \sigma_{ik}\sigma_{je} + \sigma_{ie}\sigma_{jk}$.

We note below an asymptotic property of multivariate normal distribution.

Theorem 13.4.4 *Let $\boldsymbol{U}(n) = (U_1(n), \cdots, U_m(n))'$ be a random vector (depending c sample size n) and $\boldsymbol{b} = (b_1, \cdots, b_m)'$ be a fixed vector. Assume $\sqrt{n}(\boldsymbol{U}(n) - \boldsymbol{b})$ asymptotically normal $N_m(\boldsymbol{0}, \boldsymbol{T})$.*

Let $\boldsymbol{V} = f(\boldsymbol{U})$ be a function of $\boldsymbol{U}$ with first and second order derivatives in th neighbourhood of $\boldsymbol{U} = \boldsymbol{b}$. Let $\partial f(\boldsymbol{U})/\partial U_i]_{\boldsymbol{U}=\boldsymbol{b}}$ be the ith component of $\boldsymbol{\Phi}_{\boldsymbol{b}}$. Then th limiting distribution of

$$\sqrt{n}\,[f(\boldsymbol{U}) - f(\boldsymbol{b})] \text{ is } N\left(0, \boldsymbol{\Phi}_{\boldsymbol{b}}'\boldsymbol{T}\boldsymbol{\Phi}_{\boldsymbol{b}}\right).$$

The theorem is proved in Cramer (1946, p.366)

13.5 Matrix Normal Distribution

Let $\boldsymbol{X}(n \times p)$ be a matrix whose n rows $\boldsymbol{x}_1', \cdots, \boldsymbol{x}_n'$ are independently distributed as $N_p(\boldsymbol{\mu}, \Sigma)$. Then $\boldsymbol{X}$ has the *matrix normal distribution* and represents a normal sample from $N_p(\boldsymbol{\mu}, \Sigma)$. The p.d.f. of $\boldsymbol{X}$ is

$$f(\boldsymbol{X}) = |2\pi\boldsymbol{\Sigma}|^{-n/2} \exp\left\{-\frac{1}{2}\sum_{r=1}^{n}(\boldsymbol{x}_r - \boldsymbol{\mu})'\boldsymbol{\Sigma}^{-1}(\boldsymbol{x}_r - \boldsymbol{\mu})\right\}$$

$$= |2\pi\boldsymbol{\Sigma}|^{-n/2} \exp\left\{-\frac{1}{2} tr\left[\boldsymbol{\Sigma}^{-1}(\boldsymbol{X} - \mathbf{1}\boldsymbol{\mu}')'(\boldsymbol{X} - \mathbf{1}\boldsymbol{\mu}')\right]\right\}.$$

where $\mathbf{1}$ is a $n \times 1$ vectors of 1's.

13.6 The Multivariate t-Distribution

Let $\boldsymbol{X}$ be a $m \times 1$ random vector; then $\boldsymbol{X}$ has a t-distribution with d degrees of freedom, mean vector $\boldsymbol{\mu}$ and precision matrix $\boldsymbol{T}$ if the density of $\boldsymbol{X}$ is

$$f(\boldsymbol{x}|d, \boldsymbol{\mu}, \boldsymbol{T}) \propto \left[1 + \frac{(\boldsymbol{x} - \boldsymbol{\mu})'\boldsymbol{T}(\boldsymbol{x} - \boldsymbol{\mu})}{d}\right]^{-(d+m)/2}, \quad \boldsymbol{x} \in R^m$$

where $d > 0$ and $\boldsymbol{T}$ is $m \times m$ symmetric positive definite matrix.

It can be shown that when $m = 1$, X has a univariate t-distribution as defined earlier and that $\sqrt{T}(x - \mu)$ has Student's t-distribution with d-degree of freedom. An interesting property of the multivariate t-distribution is that if $m \geq 2$, the components of $\boldsymbol{X}$ cannot be independent but they can be uncorrelated since the dispersion matrix of $\boldsymbol{X}$ is

$$D(\boldsymbol{X}) = \frac{d}{d-2}\boldsymbol{T}^{-1}, \quad d > 2.$$

The mean vector of $\boldsymbol{X}$ is $E(\boldsymbol{X}) = \boldsymbol{\mu}$ and the marginal distribution of $\boldsymbol{X}^{(1)}$ is also a t with d degrees of freedom, location parameter $\boldsymbol{\mu}^{(1)}$ and precision matrix $\boldsymbol{T}_{11} - \boldsymbol{T}_{12}\boldsymbol{T}_{22}^{-1}\boldsymbol{T}_{21}$, where

$$\boldsymbol{T} = \begin{pmatrix} \boldsymbol{T}_{11} & \boldsymbol{T}_{12} \\ \boldsymbol{T}_{21} & \boldsymbol{T}_{22} \end{pmatrix}$$

and $\boldsymbol{T}_{11}$ is of order m_1. Here $\boldsymbol{X}^{(1)}$ consists of the first m_1 components of $\boldsymbol{X}$ and $\boldsymbol{\mu}$ is partitioned as

$$\boldsymbol{\mu} = \begin{pmatrix} \boldsymbol{\mu}^{(1)} \\ \boldsymbol{\mu}^{(2)} \end{pmatrix}$$

where $\boldsymbol{\mu}^{(1)}$ is $m_1 \times 1$.

The conditional distribution of $\boldsymbol{X}^{(1)}$ given $\boldsymbol{X}^{(2)}$ is also a t where the conditiona mean vector of $\boldsymbol{X}^{(1)}$ is

$$E\left[\boldsymbol{X}^{(1)}|\boldsymbol{X}^{(2)} = \boldsymbol{x}_2\right] = \boldsymbol{\mu}^{(1)} - \boldsymbol{T}_{11}^{-1}\boldsymbol{T}_{12}\left[\boldsymbol{x}_2 - \boldsymbol{\mu}^{(2)}\right],$$

and the conditional precision matrix is

$$\frac{(d+m_2)\boldsymbol{T}_{11}}{d + [\boldsymbol{x}_2 - \boldsymbol{\mu}^{(2)}]'\left[\boldsymbol{T}_{22} - \boldsymbol{T}_{21}\boldsymbol{T}_{11}^{-1}\boldsymbol{T}_{12}\right][\boldsymbol{x}_2 - \boldsymbol{\mu}^{(2)}]}.$$

Unlike the normal distribution, the precision matrix of the conditional distributio of $\boldsymbol{X}^{(1)}$ given $\boldsymbol{X}^{(2)}$ depends on the value of the conditioning variable.

13.7 The Dirichlet Distribution

This distribution is not related to multivariate normal distribution. However, th distribution is noted due to its usefulness.

A multivariate generalisation of the univariate beta distribution is the Dirichle distribution.

The random vector $\boldsymbol{X} = (X_1, \cdots, X_p)'$ follows a p-variate Dirichlet distributio $D(\gamma_1, \cdots, \gamma_p; \gamma_{p+1})$ (with parameters $\gamma_1, \cdots, \gamma_{p+1}$) if its p.d.f. is given by

$$f_{\boldsymbol{X}'}(\boldsymbol{x}) = \frac{\Gamma(\gamma_1 + \cdots + \gamma_{p+1})}{\Gamma(\gamma_1)\cdots\Gamma(\gamma_{p+1})} x_1^{\gamma_1 - 1}\cdots x_p^{\gamma_p - 1}(1 - x_1 - \cdots - x_p)^{\gamma_{p+1}-1}$$

whenever,

$$x_k \geq 0 \quad (k = 1, \cdots, p), \ \sum_{k=1}^{p} x_k \leq 1,$$

$$\gamma_k > 0, \quad k = 1, \cdots, p+1.$$

The moments about the origin are given by

$$E(X_1^{r_1}\cdots X_p^{r_p}) = \frac{\Gamma(\gamma_1 + r_1)\cdots\Gamma(\gamma_p + r_p)\,\Gamma(\gamma_1 + \cdots + \gamma_{p+1})}{\Gamma(\gamma_1)\cdots\Gamma(\gamma_p)\,\Gamma(\gamma_1 + \cdots + \gamma_{p+1} + r_1 + \cdots + r_p)}.$$

Thus

$$E(X_k) = \frac{\gamma_k}{\gamma_1 + \cdots + \gamma_{p+1}}, \qquad k = 1, \cdots, p$$

$$V(X_k) = \frac{\gamma_k(\gamma_1 + \cdots + \gamma_{p+1} - \gamma_k)}{(\gamma_1 + \cdots + \gamma_{p+1})^2(\gamma_1 + \cdots + \gamma_{p+1} + 1)}, \qquad k = 1, \cdots, p$$

$$\text{Cov}(X_j, X_k) = -\frac{\gamma_j\gamma_k}{(\gamma_1 + \cdots + \gamma_{p+1})^2(\gamma_1 + \cdots + \gamma_{p+1} + 1)}, \qquad (j \neq k = 1, \cdots, p).$$

Again if $\boldsymbol{X} = (X_1, \cdots, X_p)' \cap D(\gamma_1, \cdots, \gamma_p; \gamma_{p+1})$ then

$$\boldsymbol{X}^{(1)} = (X_1, \cdots, X_q)'(q < p) \cap D(\gamma_1, \cdots, \gamma_q; \gamma_{q+1} + \cdots + \gamma_{p+1}).$$

It follows that if the r.v. Y_k has a univariate gamma density $f_G(\gamma_k) = e^{-y} y^{\gamma_k - 1} / \overline{|}(\gamma_k)$ $(y > 0, \gamma_k > 0)$ $(k = 1, \cdots, p+1)$ and Y_k's are independent then the variable $\boldsymbol{X} = (X_1, \cdots, X_p)'$ where

$$X_k = \frac{Y_k}{Y_1 + \cdots + Y_{p+1}}, \quad k = 1, \cdots, p$$

follows $D(\gamma_1, \cdots, \gamma_p; \gamma_{p+1})$.

[Another distribution, multivariate Poisson distribution has been considered in exercise 24 of this chapter].

13.8 Spherical and Elliptic Distributions

Definition 13.8.1 *A $p \times 1$ random vector $\boldsymbol{X}$ is said to have a spherical distribution if $\boldsymbol{X}$ and $\boldsymbol{HX}$ have the same distribution for all $p \times p$ orthogonal matrices $\boldsymbol{H}$.*

If $\boldsymbol{X}$ has a spherical distribution with a density function then the density function depends on $\boldsymbol{x}$ only through the value of $\boldsymbol{x}'\boldsymbol{x}$. Some examples are:

(a) the p-variate normal distribution $N_p(\mathbf{0}, \sigma^2 I_p)$,

(b) the ε-contaminated normal distribution with density function

$$(1-\varepsilon) \cdot \frac{1}{(2\pi)^{p/2}} \exp\left(-\frac{1}{2}\boldsymbol{x}'\boldsymbol{x}\right) + \varepsilon \cdot \frac{1}{(2\pi\sigma^2)^{p/2}} \exp\left(-\frac{1}{2}\boldsymbol{x}'\boldsymbol{x}/\sigma^2\right) \qquad (13.8.1)$$

where $0 \le \varepsilon < 1$.

The class of some spherical distributions may be generated as follows. Let $X_1, \cdots, X_p, Z(> 0)$ be random variables such that for given Z, $X_1, \cdots, X_p$ have independent $N(0, Z)$ distribution. If Z has a distribution function G, then the joint density of $X_1, \cdots, X_p$ is

$$f(x_1, \cdots, x_p) = \int_0^\infty \frac{1}{\left(\sqrt{2\pi z}\right)^p} \exp\left(-\frac{1}{2z}\sum_{i=1}^p x_i^2\right) dG(z). \qquad (13.8.2)$$

Clearly, the densities given in (13.8.2) are spherical. The class of such spherical distributions formed by varying G is called the class of compound normal distributions. If, in particular, Z takes the values 1 and σ^2 with probabilities $(1-\varepsilon)$ and ε respectively, $\boldsymbol{X}$ has the ε-contaminated normal distribution given in (13.8.1) above. Also if $\frac{n}{Z}$ has $\chi^2_{(n)}$, then $\boldsymbol{X}$ has the p-variate t-distribution given in section 13.6.

Definition 13.8.2 *The $p \times 1$ random vector $\boldsymbol{X}$ is said to have an elliptical distribution with parameters $\boldsymbol{\mu}$ and $\boldsymbol{V} = ((v_{ij}))$ if its density function is of the form*

$$C_p \cdot |\boldsymbol{V}|^{-\frac{1}{2}} h\left((\boldsymbol{x}-\boldsymbol{\mu})'\boldsymbol{V}^{-1}(\boldsymbol{x}-\boldsymbol{\mu})\right) \tag{13.8.3}$$

for some function $h(\cdot)$, when $\boldsymbol{V}$ is positive definite and C_p is a normalising constant. The class of elliptic distribution will be written as $\varepsilon_p(\boldsymbol{\mu}, \boldsymbol{V})$ and a member of this class as $E_p(\boldsymbol{\mu}, \boldsymbol{V})$. Clearly, if $\boldsymbol{X}$ is $N_p(\boldsymbol{\mu}, \boldsymbol{V})$ then $\boldsymbol{X}$ is $E_p(\boldsymbol{\mu}, \boldsymbol{V})$. If $\boldsymbol{X}$ is $E_p(\boldsymbol{0}, I_p)$, then obviously $\boldsymbol{X}$ has a spherical distribution. Also if $\boldsymbol{Y}$ has a p-variate spherical distribution with a density function and $\boldsymbol{X} = \boldsymbol{C}\boldsymbol{Y} + \boldsymbol{\mu}$, where $\boldsymbol{C}$ is a non-singular $p \times p$ matrix, then $\boldsymbol{X}$ is $E_p(\boldsymbol{\mu}, \boldsymbol{V})$, where $\boldsymbol{V} \equiv \boldsymbol{C}\boldsymbol{C}'$.

If $\boldsymbol{X}$ is $E_p(\boldsymbol{\mu}, \boldsymbol{V})$, then the characteristic function of $\boldsymbol{X}$ is

$$\phi(\boldsymbol{t}) = E(e^{i\boldsymbol{t}'\boldsymbol{X}}) = e^{i\boldsymbol{t}'\boldsymbol{\mu}}\psi(\boldsymbol{t}'\boldsymbol{V}\boldsymbol{t}) \tag{13.8.4}$$

for some function ψ.

Again,

$$E(\boldsymbol{X}) = \boldsymbol{\mu}, \; D(\boldsymbol{X}) = \alpha\boldsymbol{V} \text{ where } \alpha = -2\psi'(0) \tag{13.8.5}$$

provided the respective moments exist. It follows, therefore, that all distributions in the class $\varepsilon_p(\boldsymbol{\mu}, \boldsymbol{V})$ have the same mean $\boldsymbol{\mu}$ and same correlation matrix $((\rho_{ij}))$ where

$$\rho_{ij} = \frac{\alpha v_{ij}}{\sqrt{(\alpha v_{ii})(\alpha v_{ij})}} = \frac{v_{ij}}{\sqrt{v_{ii}v_{ij}}}.$$

where $\boldsymbol{V} = ((v_{ij}))$. It follows from (13.8.4), that all marginal distributions of an elliptical variate are elliptical and all marginal density functions of dimension $r < p$ have the same functional form. For example, if

$$\boldsymbol{X} = \begin{bmatrix} \boldsymbol{X}_1 \\ \boldsymbol{X}_2 \end{bmatrix}, \; \boldsymbol{\mu} = \begin{bmatrix} \boldsymbol{\mu}_1 \\ \boldsymbol{\mu}_2 \end{bmatrix}, \; \boldsymbol{V} = \begin{bmatrix} \boldsymbol{V}_{11} & \boldsymbol{V}_{12} \\ \boldsymbol{V}_{21} & \boldsymbol{V}_{22} \end{bmatrix}$$

where $\boldsymbol{X}_1$ and $\boldsymbol{\mu}_1$ are $r \times 1$ and $\boldsymbol{V}_{11}$ is $r \times r$, then the characteristic function of $\boldsymbol{X}_1$ obtained from (13.8.4) by putting $\boldsymbol{t} = (\boldsymbol{t}_1', \boldsymbol{0}')'$, where $\boldsymbol{t}_1$ is $r \times 1$, is

$$\exp(i\boldsymbol{t}_1'\boldsymbol{\mu}_1)\psi(\boldsymbol{t}_1'\boldsymbol{V}_{11}\boldsymbol{t}_1)$$

which is the characteristic function of a random vector with an $E_r(\boldsymbol{\mu}_1, \boldsymbol{V}_{11})$ distribution. It follows that if any marginal distribution is normal, then $\boldsymbol{X}$ is normal, for the characteristic function of $\boldsymbol{X}$ has the same functional form as the characteristic function of the marginal distribution, viz., normal.

It has been shown (theorem 13.2.4), that if $\boldsymbol{X}$ is $N_p(\boldsymbol{\mu}, \boldsymbol{\Sigma})$ and $\boldsymbol{\Sigma}$ is diagonal, then $\boldsymbol{X}_1, \cdots, \boldsymbol{X}_p$ are all independent. The following theorem characterises the p-variate normal distribution within the class $\varepsilon_p(\boldsymbol{\mu}, \boldsymbol{V})$ through independence of variables and diagonal structure of $\boldsymbol{V}$.

Theorem 13.8.1 *Let $\boldsymbol{X}$ be $E_p(\boldsymbol{\mu}, \boldsymbol{V})$, when $\boldsymbol{V}$ is diagonal. Then if $X_1, \cdots, X_p$ are independent, $\boldsymbol{X}$ is $N_p(\boldsymbol{\mu}, \boldsymbol{V})$.*

We now consider some properties of conditional distribution of $\boldsymbol{X}$. We have seen that if $\boldsymbol{X}$ is $N_p(\boldsymbol{\mu}, \boldsymbol{\Sigma})$, then the conditional expectation of a subvector of $\boldsymbol{X}$ given the remaining components is linear in the fixed variables, and the conditional covariance matrix does not depend on the fixed variables. The first of these properties hold for Elliptic distributions also.

Theorem 13.8.2 *If $\boldsymbol{X}$ is $E_p(\boldsymbol{\mu}, \boldsymbol{V})$ and if $\boldsymbol{X}, \boldsymbol{\mu}, \boldsymbol{V}$ are partitioned as*

$$\boldsymbol{X} = \begin{bmatrix} \boldsymbol{X}_1 \\ \boldsymbol{X}_2 \end{bmatrix}, \quad \boldsymbol{\mu} = \begin{bmatrix} \boldsymbol{\mu}_1 \\ \boldsymbol{\mu}_2 \end{bmatrix}, \quad \boldsymbol{V} = \begin{bmatrix} \boldsymbol{V}_{11} & \boldsymbol{V}_{12} \\ \boldsymbol{V}_{21} & \boldsymbol{V}_{22} \end{bmatrix}$$

where $\boldsymbol{X}_1, \boldsymbol{\mu}_1$ are $r \times 1$ and $\boldsymbol{V}_{11}$ is $r \times r$, then, provided they exist,

$$E(\boldsymbol{X}_1|\boldsymbol{X}_2) = \boldsymbol{\mu} + \boldsymbol{V}_{12}\boldsymbol{V}_{22}^{-1}(\boldsymbol{X}_1 - \boldsymbol{\mu}_1)$$

$$\mathrm{Cov}(\boldsymbol{X}_1|\boldsymbol{X}_2) = g(\boldsymbol{X}_2)\left(\boldsymbol{V}_{11} - \boldsymbol{V}_{12}\boldsymbol{V}_{22}^{-1}\boldsymbol{V}_{22}\right)$$

for some function g. Moreover, the conditional distribution of $\boldsymbol{X}_1$ given $\boldsymbol{X}_2$ is k-variate normal. It can be shown that if the conditional covariance matrix of $\boldsymbol{X}_1$ given $\boldsymbol{X}_2$ does not depend on $\boldsymbol{X}_2$, then $\boldsymbol{X}$ must be normal. This property, therefore, characterises the multivariate normal distribution in the class of elliptical distributions.

An interesting property of a spherically distributed vector $\boldsymbol{X}$ is that a transformation to polar co-ordinates yields angles and a radius which are all independently distributed, with the angles having the same distributions for all $\boldsymbol{X}$.

Theorem 13.8.3 *If $\boldsymbol{X}$ is $E_p(0, I_p)$, with density function $C_p h(\boldsymbol{x}\boldsymbol{x}')$, and*

$$\begin{aligned}
X_1 &= r \sin\theta_1 \sin\theta_2 \cdots \sin\theta_{p-2} \sin\theta_{p-1}, \\
X_2 &= r \sin\theta_1 \sin\theta_2 \cdots \sin\theta_{p-2} \cos\theta_{p-1}, \\
X_3 &= r \sin\theta_1 \sin\theta_2 \cdots \cos\theta_{p-2}, \\
&\cdots\cdots\cdots\cdots\cdots\cdots\cdots\cdots\cdots \\
X_{p-1} &= r \sin\theta_1 \cos\theta_2, \\
X_p &= r \cos\theta_1
\end{aligned}$$

$(r > 0, 0 \le \theta_i \le \pi, i = 1, \cdots, p-2, 0 \le \theta_{p-1} \le 2\pi)$ then $r, \theta_1, \cdots, \theta_{p-1}$ are all independent, the distributions of $\theta_1, \cdots, \theta_{p-1}$ are the same for all $\boldsymbol{X}$, with θ_k having p.d.f. proportional to $\sin^{p-1-k}\theta_k$ and $r^2 = \boldsymbol{X}'\boldsymbol{X}$ having a distribution.

$$f_{r^2}(y) = \frac{C_p \pi^{p/2}}{\Gamma\left(\frac{p}{2}\right)} y^{p/2-1} h(y) \quad (y > 0)$$

where C_p is a constant of proportionality.

Theorem 13.8.4 *Let* $\boldsymbol{X}$ *have a p-variate spherical distribution with* $P(\boldsymbol{X} = \boldsymbol{0}) = 0$.

(a) If $W = \dfrac{\boldsymbol{\alpha}'\boldsymbol{X}}{\|\boldsymbol{X}\|}$ where $\boldsymbol{\alpha} \in R^p$, $\boldsymbol{\alpha}'\boldsymbol{\alpha} = 1$, then

$$Y = \frac{\sqrt{(p-1)W}}{\sqrt{(1-W^2)}}$$

has the t_{p-1} distribution.

(b) If $\boldsymbol{B}$ is a $p \times p$ symmetric idempotent matrix of rank k then

$$Z = \frac{\boldsymbol{X}'\boldsymbol{B}\boldsymbol{X}}{\|\boldsymbol{X}\|^2}$$

has the beta distribution with parameters $\frac{k}{2}, \frac{(p-k)}{2}$.

An interesting example of this theorem where normality assumption can be droppe is as follows. It is known that if $(X_1, \cdots, X_n)$ is a random sample from a $N(\mu, \sigma^2$ population, then under the null hypothesis $H_\circ(\mu = 0)$

$$t = \frac{\sqrt{n}\bar{X}}{S} \text{ where } S^2 = \frac{1}{n-1}\sum_{i=1}^{n}(X_i - \bar{X})^2$$

follows Student's $t_{(n-1)}$. For this result to hold it is enough, therefore, to assume th vector

$$\boldsymbol{X} = (X_1, \cdots, X_n)'$$

has a spherical distribution with $P(\boldsymbol{X} = \boldsymbol{0}) = 0$ (since here $\alpha = \left(\frac{1}{\sqrt{n}}, \cdots, \frac{1}{\sqrt{n}}\right)' \in R^n$).

The spherical and elliptic distributions are of considerable interest in the studie of robust procedures where multivariate normality assumption is replaced by th assumption of such distributions.

13.9 Multivariate Cumulants

Let $\boldsymbol{X}$ be a p-variate distribution with characteristic function $\phi(t)$ and suppose fo simplicity that the moments exist. The characteristic function of X_j is $\phi_j(t_j) = \phi(t$ where

$$\boldsymbol{t} = (0, \cdots, 0, t_j, \theta, \cdots, 0)',$$

and the cumulants of X_j are the coefficient $k_r^{(j)}$ in the

$$\log \phi_j(t_j) = \sum_{r=0}^{\infty} k_r^{(j)} \frac{(it_j)^r}{r!}.$$

The first four cumulants of X_j in terms of moments are:

$$\begin{aligned}
k_1^{(j)} &= \mu_1^{(j)}, \\
k_2^{(j)} &= \mu_2^{(j)} - \left(\mu_1^{(j)}\right)^2 = \mathrm{Var}(X_j), \\
k_3^{(j)} &= \mu_3^{(j)} - 3\mu_1^{(j)}\mu_2^{(j)} + 2\left(\mu_1^{(j)}\right)^3, \\
k_4^{(j)} &= \mu_4^{(j)} - 4\mu_1^{(j)}\mu_3^{(j)} - 3\left(\mu_2^{(j)}\right)^2 + 12\mu_2^{(j)}\left(\mu_1^{(j)}\right)^2 - 6\left(\mu_1^{(j)}\right)^4.
\end{aligned}$$

The skewness $\gamma_1^{(j)}$ and curtosis $\gamma_2^{(j)}$ of distribution of X_j are

$$\gamma_1^{(j)} = \frac{k_3^{(j)}}{\left(k_2^{(j)}\right)^{3/2}}, \quad \gamma_2^{(j)} = \frac{k_4^{(j)}}{\left(k_2^{(j)}\right)^2}.$$

If X_j has a normal distribution, all cumulants $k_r^{(j)}$ of order $r > 2$ are all zero.

The cumulants of a joint distribution are defined in a similar way. For example, denoting the joint characteristic function of X_j and X_k as $\phi_{j,k}(t_j, t_k)$, the cumulants of their joint distribution are the coefficients $k_{r_1,r_2}^{j,k}$ in

$$\log \phi_{j,k}(t_j, t_k) = \sum\sum_{r_1,r_2=0}^{\infty} k_{r_1,r_2}^{j,k} \frac{(it_j)^{r_1}(it_k)^{r_2}}{r_1!r_2!}$$

[Clearly, $k_{1,1}^{j,k} = \mathrm{Cov}(X_j, X_k)$]. The cumulants of the joint distribution of $X_1, \cdots, X_p$ are the coefficients $k_{r_1,r_2,\cdots,r_p}^{1,2,\cdots,p}$ in

$$\log \phi(\boldsymbol{t}) = \sum \cdots \sum_{r_1,\cdots,r_p=0}^{\infty} k_{r_1,r_2,\cdots,r_p}^{1,2,\cdots,p} \frac{(it_i)^{r_1}(it_2)^{r_2}\cdots(it_p)^{r_p}}{r_1!r_2!\cdots r_p!}.$$

If $\boldsymbol{X}$ is normal, all cumulants for which $\Sigma r_i > 2$ are zero.

If $\boldsymbol{X}$ has p-variate elliptical distribution $E_m(\boldsymbol{\mu}, \boldsymbol{V})$ with characteristic function

$$\phi(\boldsymbol{t}) = e^{i\boldsymbol{\mu}'\boldsymbol{t}}\psi(\boldsymbol{t}'\boldsymbol{V}\boldsymbol{t}),$$

covariance matrix

$$\boldsymbol{\Sigma} = -2\psi'(0)\boldsymbol{V} = ((\sigma_{ij}))$$

and finite fourth moments, then it follows that

(1) the marginal distributions of $X_j (j = 1, \cdots, p)$ all have zero skewness and the same kurtosis

$$\gamma_2^{(j)} = \frac{k_r^{(j)}}{\left(k_2^{(j)}\right)^2} = \frac{3\left[\psi''(0) - \psi'(0)\right]^2}{\psi'(0)^2} = 3k \text{ (say) } (j = 1, \cdots, p).$$

(2) all fourth-order cumulants are determined by this kurtosis parameter k as

$$k_{1111}^{ijkl} = k\left(\sigma_{ij}\sigma_{kl} + \sigma_{ik}\sigma_{jl} + \sigma_{il}\sigma_{jk}\right).$$

□ **Remark 13.9.1** *We have noted that the asymptotic distribution of a sample covariance matrix is normal with a covariance matrix depending on the fourth-order moments of the underlying distribution. The result (2) above shows that if we are sampling from an elliptical distribution, these moments have simple forms.*

Exercise

1. Suppose $\boldsymbol{X} = (X_1, X_2, X_3)' \cap N_3(\boldsymbol{\mu}, \boldsymbol{\Sigma})$ where

$$\boldsymbol{\mu} = \begin{bmatrix} 2 \\ 1 \\ 2 \end{bmatrix}, \quad \boldsymbol{\Sigma} = \begin{bmatrix} 2 & 1 & 1 \\ 1 & 3 & 0 \\ 1 & 0 & 1 \end{bmatrix}.$$

Find the joint distribution of $Y_1 = X_1 + X_2 + X_3$ and $Y_2 = X_1 - X_2$.

[*Ans.* N_2 with mean $\begin{bmatrix} 5 \\ 1 \end{bmatrix}$ and dispersion matrix $\begin{bmatrix} 10 & 0 \\ 0 & 3 \end{bmatrix}$].

2. Let $X_1, \cdots, X_n$ be independent $N(0,1)$ variables. Find the joint moment generating function of $\bar{X}, X_1 - \bar{X}, X_2 - \bar{X}, \cdots, X_n - \bar{X}$ and hence deduce that $\bar{X}$ and

$$\sum_{i=1}^{n}(X_i - \bar{X})^2$$

are statistically independent. **[Hogg and Craig, 1970]**

3. Suppose $\boldsymbol{X} \cap N_n(\boldsymbol{\mu}, \boldsymbol{\Sigma})$ and consider the partition

$$\boldsymbol{X} = \begin{bmatrix} \boldsymbol{X}_1 \\ \boldsymbol{X}_2 \end{bmatrix}, \quad \boldsymbol{\mu} = \begin{bmatrix} \boldsymbol{\mu}_1 \\ \boldsymbol{\mu}_2 \end{bmatrix}, \quad \boldsymbol{\Sigma} = \begin{bmatrix} \boldsymbol{\Sigma}_{11} & \boldsymbol{\Sigma}_{12} \\ \boldsymbol{\Sigma}_{21} & \boldsymbol{\Sigma}_{22} \end{bmatrix}$$

where $\boldsymbol{X}_1$ and $\boldsymbol{\mu}_1$ are $p \times 1$ variables, $\boldsymbol{\Sigma}_{11}$ is a $p \times p$ matrix ($p < n$). If $\boldsymbol{Y} = \boldsymbol{X}_1 - \boldsymbol{\Sigma}_{12}\boldsymbol{\Sigma}_{22}^{-1}\boldsymbol{X}_2$, prove that $C(\boldsymbol{Y}, \boldsymbol{X}_2) = \boldsymbol{0}$ and deduce that $\boldsymbol{Y}, \boldsymbol{X}_2$ are independent. Hence, prove that the conditional distribution of $\boldsymbol{X}_1$ given $\boldsymbol{X}_2 = \boldsymbol{x}_2$ is

$$N_p\left(\boldsymbol{\mu}_1 + \boldsymbol{\Sigma}_{12}\boldsymbol{\Sigma}_{22}^{-1}(\boldsymbol{x}_2 - \boldsymbol{\mu}_2), \boldsymbol{\Sigma}_{11} - \boldsymbol{\Sigma}_{12}\boldsymbol{\Sigma}_{22}^{-1}\boldsymbol{\Sigma}_{21}\right).$$

4. Let $\boldsymbol{X} \cap N_n(\boldsymbol{\mu}, \boldsymbol{\Sigma})$ where $\sigma_{ii} = \sigma^2 (\forall\, i)$ and $\sigma_{ij} = (1-\rho)\sigma^2 (\forall\, i \neq j)$. Prove that $\bar{X}$ is statistically independent of

$$\sum_{i=1}^{n}(X_i - \bar{X})^2.$$

5. If $\boldsymbol{X}$ and $\boldsymbol{Y}$ are independent p-variate multinormal random vectors, prove that $a\boldsymbol{X} + b\boldsymbol{Y}$ is also multivariate normal where a, b are given constants.

6. Let $\boldsymbol{X} \cap N_n(\boldsymbol{0}, \boldsymbol{\Sigma})$ where $\boldsymbol{\Sigma} = \sigma^2[(1-\rho)I_n + \rho\boldsymbol{1}\boldsymbol{1}'], (0 < \rho < 1)$. Using Helmert's transformation, prove that $\bar{X}$ is independent of

$$Q = \sum_{i=1}^{n}(X_i - \bar{X})^2/\sigma^2(1-\rho)$$

and that $Q \cap \chi^2_{(n-1)}$.

7. Let $\boldsymbol{Y} \cap N_n(0, \boldsymbol{I}_n)$ and let $\boldsymbol{X} = \boldsymbol{A}\boldsymbol{Y}, \boldsymbol{U} = \boldsymbol{B}\boldsymbol{Y}, \boldsymbol{V} = \boldsymbol{C}Y$, where $\boldsymbol{A}, \boldsymbol{B}$ and $\boldsymbol{C}$ are all $r \times n$ matrices of rank $r(< n)$. If $C(\boldsymbol{X}, \boldsymbol{U}) = \boldsymbol{0}$ and $C(\boldsymbol{X}, \boldsymbol{V}) = \boldsymbol{0}$, prove that $\boldsymbol{X}$ is independent of $\boldsymbol{U} + \boldsymbol{V}$.

8. Let $\boldsymbol{X}_1, \boldsymbol{X}_2$ be two random p-vectors and let $\boldsymbol{X} = \begin{bmatrix} \boldsymbol{X}_1 \\ \boldsymbol{X}_2 \end{bmatrix}$. If $\boldsymbol{X}$ has a multivariate normal distribution with mean vector $\boldsymbol{0}$ and dispersion matrix

$$\begin{bmatrix} \boldsymbol{A} & \boldsymbol{B} \\ \boldsymbol{B}' & \boldsymbol{C} \end{bmatrix},$$

when $\boldsymbol{A}, \boldsymbol{B}, \boldsymbol{C}$ are all $p \times p$ matrices, show that a necessary and sufficient condition for $\boldsymbol{X}_1 + \boldsymbol{X}_2$ to be independent of $\boldsymbol{X}_1 - \boldsymbol{X}_2$ is $\boldsymbol{A} = \boldsymbol{C}, \boldsymbol{B} = \boldsymbol{B}'$.

9. Let $\boldsymbol{y}_\alpha \cap N_p(a_\alpha\boldsymbol{\mu}, \boldsymbol{\Sigma}), \alpha = 1, \cdots, n$ be independent. Show that $Q = \sum_{\alpha=1}^{n} \boldsymbol{y}_\alpha \boldsymbol{y}'_\alpha$ and $\boldsymbol{z}\boldsymbol{z}'$ are independently distributed where

$$\boldsymbol{z} = \sum_{\alpha=1}^{n} \frac{a_\alpha \boldsymbol{y}_\alpha}{\sqrt{\sum_{\alpha=1}^{n} a_\alpha^2}}.$$

10. If X is $N_p(\boldsymbol{\mu}, \boldsymbol{\Sigma})$ prove that
    (a) $E(\boldsymbol{X}'\boldsymbol{A}\boldsymbol{X}) = tr(\boldsymbol{A}\boldsymbol{\Sigma}) + \boldsymbol{\mu}'\boldsymbol{A}\boldsymbol{\mu}$;
    (b) $V(\boldsymbol{X}'\boldsymbol{A}\boldsymbol{X}) = 2[tr(\boldsymbol{A}\boldsymbol{\Sigma}\boldsymbol{A}\boldsymbol{\Sigma}) + 2\boldsymbol{\mu}'\boldsymbol{A}\boldsymbol{\Sigma}\boldsymbol{A}\boldsymbol{\mu}]$.

11. Suppose that $\boldsymbol{X} \cap N_3(0, \boldsymbol{\Sigma})$ where

$$\boldsymbol{\Sigma} = \begin{bmatrix} 1 & \sigma_{12} & \sigma_{13} \\ \sigma_{21} & 1 & \sigma_{23} \\ \sigma_{31} & \sigma_{32} & 1 \end{bmatrix}.$$

Show that

$$P(X_1 > 0, X_2 > 0, X_3 > 0) = \frac{1}{8} + \frac{1}{4\pi}\left(\sin^{-1}\sigma_{12} + \sin^{-1}\sigma_{13} + \sin^{-1}\sigma_{23}\right).$$

12. Suppose that the vector $\begin{bmatrix} Y \\ \boldsymbol{X} \end{bmatrix}$ when Y is $1 \times 1, \boldsymbol{X}$ is $(p-1) \times 1$ has mean vector $\mu \mathbf{1}$, and covariance matrix
$$\begin{bmatrix} \sigma_{11} & \boldsymbol{\sigma}'_{12} \\ \boldsymbol{\sigma}_{12} & \boldsymbol{\Sigma}_{22} \end{bmatrix}$$
where $\sigma_{11} = V(Y), \boldsymbol{\Sigma}_{22} = \text{Cov}(\boldsymbol{X})$. Find the coefficient vector $\boldsymbol{\alpha}$ of a linear function $\boldsymbol{\alpha}'\boldsymbol{X}$ which minimises $V(Y - \boldsymbol{\alpha}'\boldsymbol{X})$ subject to the condition $E(\boldsymbol{\alpha}'\boldsymbol{X}) = E(Y)$.

13. Let $\boldsymbol{X} = (X_1, \cdots, X_p)'$ be distributed as $N_p(\boldsymbol{\mu}, \boldsymbol{\Sigma})$. Show that if $\boldsymbol{\mu} = \mathbf{1}_p$,
$$\boldsymbol{\Sigma} = \begin{bmatrix} 1 & 1 & 1 & \cdots & 1 & 1 \\ 1 & 2 & 2 & \cdots & 2 & 2 \\ 1 & 2 & 3 & \cdots & 3 & 3 \\ \cdots & \cdots & \cdots & \cdots & \cdots & \cdots \\ \cdots & \cdots & \cdots & \cdots & \cdots & \cdots \\ 1 & 2 & 3 & \cdots & p-1 & p-1 \\ 1 & 2 & 3 & \cdots & p-1 & p \end{bmatrix}$$
the random variable
$$R = (X_2 - X_1)^2 + \cdots + (X_p - X_{p-1})^2$$
has a chi-square distribution.

14. Show that if $\boldsymbol{X}_n \xrightarrow{L} N(\boldsymbol{\mu}, \boldsymbol{\Sigma})$ and if $\boldsymbol{C}$ be a $m \times k$ matrix then $\boldsymbol{C}\boldsymbol{X}_n \xrightarrow{L} N(\boldsymbol{C}\boldsymbol{\mu}, \boldsymbol{C}\boldsymbol{\Sigma}\boldsymbol{C}')$.

15. Derive the characteristic function of the multivariate normal distribution with mean vector $\boldsymbol{\mu} = \mathbf{0}$ and dispersion matrix $\boldsymbol{\Sigma} = ((\sigma_{ij}))$. Show that
$$E(X_i X_j X_k X_l) = \sigma_{ij}\sigma_{kl} + \sigma_{il}\sigma_{jk} + \sigma_{ik}\sigma_{jl}.$$

16. Let $\boldsymbol{A}$ and $\boldsymbol{B}$ be two real symmetric matrices of order $n \times n$ and let $X_1, \cdots, X_n$ be normal variates with zero mean and unit variances. Show that it is a necessary and sufficient condition for $\boldsymbol{X}'\boldsymbol{A}\boldsymbol{X}$ and $\boldsymbol{X}'\boldsymbol{B}\boldsymbol{X}$ to be distributed independently as chi-squared that $\boldsymbol{AB} = 0$ and $\boldsymbol{A}^2 = \boldsymbol{A}, \boldsymbol{B}^2 = \boldsymbol{B}$.

17. Show that if rows of $\boldsymbol{X}$ are $i.i.d$ $N_p(\boldsymbol{\mu}, \boldsymbol{\Sigma})$ and if $\boldsymbol{C}_1, \cdots, \boldsymbol{C}_k$ are symmetric matrices, then $\boldsymbol{X}'\boldsymbol{C}_1\boldsymbol{X}, \cdots, \boldsymbol{X}'\boldsymbol{C}_k\boldsymbol{X}$ are jointly independent if $\boldsymbol{C}_r\boldsymbol{C}_s = \mathbf{0}$ for all $r \neq s$.

18. Let $\boldsymbol{X}'\boldsymbol{A}\boldsymbol{X}, \boldsymbol{X}'\boldsymbol{B}\boldsymbol{X}$ be two quadratic forms in independent normal variates $X_1, \cdots, X_n$ each of unit variance. Show that their joint moment generating function is given by $|I - \theta_1\boldsymbol{A} - \theta_2\boldsymbol{B}|^{-\frac{1}{2}}$ where $\theta_1 = 2t_1, \theta_2 = 2t_2$. Hence derive a theorem (due to Cochran) that it is necessary and sufficient for the two forms to be independent is that
$$|I - \theta_1\boldsymbol{A} - \theta_2\boldsymbol{B}| = |I - \theta_1\boldsymbol{A}||I - \theta_2\boldsymbol{B}|$$
identically. **[Kendall, 1954].**

19. Suppose that X is $N_3(\mathbf{0}, \mathbf{\Sigma})$ when

$$\mathbf{\Sigma} = \begin{bmatrix} 1 & \rho & 0 \\ \rho & 1 & \rho \\ 0 & \rho & 1 \end{bmatrix}.$$

Is there a value of ρ for which $X_1 + X_2 + X_3$ and $X_1 - X_2 - X_3$ are independent?

20. Show that for the following bivariate distributions, the marginal distributions of X_1 and X_2 are normal.

(a) $f(x_1, x_2) = \frac{1}{2\pi} \exp\left[-\frac{1}{2}(x_1^2 + x_2^2)\right] \left\{1 + x_1 x_2 \exp\left[-\frac{1}{2}(x_1^2 + x_2^2)\right]\right\}, -\infty < x_1, x_2 < \infty.$

(b) $f(x_1, x_2) = \frac{1}{2\pi} \exp\left[-\frac{1}{2}(x_1^2 + x_2^2)\right] \left(1 - \frac{x_1 x_2}{(1+x_1^2)(1+x_2^2)}\right).$

21. Suppose two variables X, Y follow a bivariate normal distribution $N(\boldsymbol{\mu}, \mathbf{\Sigma})$ where

$$\boldsymbol{\mu} = \begin{bmatrix} 6 \\ 2 \end{bmatrix}, \quad \mathbf{\Sigma} = \begin{bmatrix} 9 & 3 \\ 3 & 4 \end{bmatrix}.$$

Find (a) $E(Y|X), E(X|Y)$, (b) $V(Y|X), V(X|Y)$, (c) $P\{X \le 5\}$, (d) $P\{2 \le X \le 5, 1 \le Y \le 3\}$.

22. Let $\boldsymbol{X} \cap N(\boldsymbol{\mu}, \mathbf{\Sigma}), |\mathbf{\Sigma}| > 0$. It is found that first q variables form a group, so $\boldsymbol{X}$ is partitioned as

$$\boldsymbol{X} = \begin{bmatrix} \boldsymbol{X}^{(1)} \\ \boldsymbol{X}^{(2)} \end{bmatrix}, \quad \boldsymbol{X}^{(1)} : q \times 1.$$

Find the distribution of $\boldsymbol{AX}^{(1)} + \boldsymbol{BX}^{(2)}$, where $\boldsymbol{A} : q \times q$ and $\boldsymbol{B} : q \times (p-q), 2q < p$.

23. Suppose $\boldsymbol{X} = (X_1, X_2, X_3)'$ follow $N_3(\boldsymbol{\mu}, \mathbf{\Sigma})$ where

$$\boldsymbol{\mu} = \begin{bmatrix} 3 \\ 10 \\ 8 \end{bmatrix}, \quad \mathbf{\Sigma} = \begin{bmatrix} 1 & 3 & 1 \\ 3 & 16 & 2 \\ 1 & 2 & 4 \end{bmatrix}.$$

Find the distribution of (a) X_1, for given (X_1, X_3), (b) (X_1, X_2) for given X_3 and (c) X_3, for given X_1 and X_2.

24. *A multivariate Poisson distribution.* Let $u_0, u_1, \cdots, u_p$ be independent Poisson variables with parameters $\lambda_0, \lambda_1 - \lambda_0, \cdots, \lambda_p - \lambda_0$, respectively. Write down the joint

distribution of $x_i = u_i + u_0 (i = 1, \cdots, p)$ and show that the marginal distributio of $x_1, \cdots, x_p$ are all Poisson. For $p = 2$, show that the p.m.f. is given by

$$f(x, y) = \exp(-\lambda_1 - \lambda_2 + \lambda_0) \frac{a^x b^y}{x!y!} \sum_{r=0}^{s} \frac{x^{(r)} y^{(r)} \lambda_0^r}{a^r b^r r!}$$

where $s = \min(x, y)$, $a = \lambda_1 - \lambda_0$, $b = \lambda_2 - \lambda_0$, $\lambda_1 > \lambda_0 > 0$, $\lambda_2 > \lambda_0 > 0$, and $x^{(r)} = x(x-1)\cdots(x-r+1)$. Furthermore,

$$E(y|x) = b + \left(\frac{\lambda_0}{\lambda_1}\right) x, \quad \text{Var}(y|x) = b + \left\{\frac{a\lambda_0}{\lambda_1^2}\right\} x$$

[Mardia, et al., 197

25. *Fre'chet* (1951) *inequality.* Let X, Y be random variables with c.d.f. $H(x, y)$ an marginal c.d.f's $F(x)$ and $G(y)$. Show that

$$\max(F + G - 1, 0) \leq H \leq \min(F, G).$$

26. Let $\boldsymbol{X}_j = (X_{j1}, \cdots, X_{jnj})' \cap N_{n_j}(\mu \mathbf{1}_{n_j}, \tau^2 I_{n_j} + \sigma^2 \mathbf{1}_{n_j} \mathbf{1}'_{n_j})$ where I_p is a identit matrix of order p and $\mathbf{1}_q = (1, \cdots, 1)'$. Suppose we have m such random vector $\boldsymbol{X}_j (j = 1, \cdots, m)$. Let

$$\bar{X}_j = \sum_{i=1}^{n_j} X_{ji}/n_j, \quad \bar{X}_{00} = \sum_{j=1}^{m} n_j \bar{X}_j / n_T$$

where

$$n_T = \sum_{j=1}^{m} n_j, \quad MSB = \sum_{j=1}^{m} n_j (\bar{X}_j - \bar{X}_{00})^2/(m-1),$$

$$MSW = \sum_{j=1}^{m} \sum_{i=1}^{n_j} (X_{ji} - \bar{X}_j)^2/(n_T - m).$$

Show that

$$E(MSW) = \tau^2, \quad V(MSW) = 2\tau^4/(n_T - m)$$

$$E(MSB) = \tau^2 + \left(n_T - \left(\sum_{j=1}^{m} n_j^2\right) \Big/ n_T\right) \frac{\sigma^2}{(m-1)}$$

$$V(MSB) = \frac{2}{(m-1)^2} \left[\sum_{j=1}^{m} \sigma_j^4 + \left(\sum_{j=1}^{m} n_j \sigma_j^2\right)^2 \Big/ n_T^2 - 2 \sum_{j=1}^{m} n_j \sigma_j^4 / n_T\right]$$

where $\sigma_j^2 = \tau^2 + \sigma^2 n_j (i \leq j \leq m)$.

[*Hints:* $\sum_{i=1}^{n_j}(X_{ji}-\bar{X}_j)^2 \cap \chi^2_{(n_j-1)} \cdot \tau^2$, since $(X_{ji}-\bar{X}_j) \cap N(0,\tau^2)$. Again $\sqrt{n}(\bar{X}_j-\mu)$ are mutually independent $N(0,\sigma_j^2)$ variables.

Let $Z_j = \sqrt{n}(\bar{X}_j-\mu)/\sigma_j$, $\boldsymbol{Z} = (Z_1,\cdots,Z_m)'$, $\boldsymbol{A} = \boldsymbol{D} - \boldsymbol{uu'}$ where $\boldsymbol{D} = \text{Diag}(\sigma_1^2,\cdots,\sigma_m^2)$, $\boldsymbol{u}' = \frac{1}{\sqrt{nT}}(\sqrt{n}_1\sigma_1,\cdots,\sqrt{n_m}\sigma_m)$. Then $MSB = \boldsymbol{Z'AZ}/(m-1)$.

Now $E(\boldsymbol{Z'AZ}) = tr(\boldsymbol{A})$. Also $V(\boldsymbol{Z'AZ}) = 2tr(\boldsymbol{A}^2)$.] **[Ghosh and Meeden, 1986]**

27. Consider the model $Y_i = \theta + \varepsilon_i (i = 1,\cdots,N)$ where $\theta, \varepsilon_1,\cdots,\varepsilon_N$ are independent r.v.'s with $\theta \cap N(\mu,\sigma^2)$ and ε_i's are i.i.d. $N(0,\tau^2)$. Show that the joint probability distribution of $\boldsymbol{Y} = (Y_1,\cdots,Y_N)$ given μ,σ^2,τ^2 is $N_N(\mu\mathbf{1}_N, \tau^2 I_N + \sigma^2\mathbf{1}_N\mathbf{1}_N')$. Show also that the joint conditional distribution of $Y_{n+1},\cdots,Y_N$ given that $Y_i = y_i (i = 1,\cdots,n)$ is $(N-n)$ variate-normal with mean vector $\frac{M\mu+n\bar{y}}{M+n}\mathbf{1}_{N-n}$ and dispersion matrix $\tau^2[I_{N-n} + \mathbf{1}_{N-n}\mathbf{1}_{N-n}'/(M+n)]$ where $M = \tau^2/\sigma^2$ and $\bar{y} = \frac{1}{n}\Sigma_{i=1}^n y_i$. Hence, show that

$$E\left(\sum_{i=1}^N Y_i | Y_i = y_i,\ i = 1,\cdots,n\right) = n\bar{y} + (N-n)\{B\mu + (1-B)\bar{y}\}$$

where $B = \frac{M}{M+n}$.

28. Let X have an m-variate elliptical t-distribution with n d.f. and the parameters $\boldsymbol{\mu}, \boldsymbol{V}$, i.e.; $\boldsymbol{X}$ has density function

$$\frac{\Gamma\left(\frac{n+m}{2}\right)}{\Gamma\left(\frac{n}{2}\right)(n\pi)^{m/2}} |\boldsymbol{V}|^{-\frac{1}{2}} \cdot \frac{1}{\left[1 + \frac{1}{n}(\boldsymbol{x}-\boldsymbol{\mu})'\boldsymbol{V}^{-1}(\boldsymbol{x}-\boldsymbol{\mu})\right]^{(n+m)/2}}.$$

(a) Show that $E(\boldsymbol{X}) = \boldsymbol{\mu}$, $\text{Cov}(\boldsymbol{X}) = \frac{n}{n-2}\boldsymbol{V} (n > 2)$.

(b) If $\boldsymbol{X} = (\boldsymbol{X}_1', \boldsymbol{X}_2')'$, when $\boldsymbol{X}_1$ is $k \times 1$, show that the marginal distribution of $\boldsymbol{X}_1$ is k-variate elliptical t.

(c) Show that for this distribution, the Kurtosis parameter k defined in Section 13.9 is $k = 2/(n-4)$. **[Muirhead, 1982]**

29. If $\boldsymbol{X}$ is $N_m(\boldsymbol{\mu}, \boldsymbol{\Sigma})$, where $\boldsymbol{\Sigma}$ is p.d., $\boldsymbol{A}$ is an $m \times m$ symmetric matrix and $\boldsymbol{B}$ is an $r \times m$ matrix, prove that $X'\boldsymbol{AX}$ and $\boldsymbol{BX}$ are independent iff $\boldsymbol{B\Sigma A} = 0$.

Chapter 14

Distributions arising out of Multivariate Normal Distributions

14.1 Introduction

We consider in this chapter sampling distributions of statistic obtained from samples from p-variate normal distributions.

In section 14.2 we consider sampling distribution of sample means and joint distribution of sample variances and covariances (so called Wishart distribution). Sections 14.3, 14.4 deal, respectively, with distribution of partial regression coefficients and multiple correlation coefficient. Section 14.5 considers distribution of intraclass correlation coefficient, section 14.6 that of the generalised variance. Hotelling's generalized T^2-statistic has been considered in details in chapter 18.

14.2 Distribution of Sample Mean, Sample Variances and Covariances

Suppose a random sample of size n is drawn from a p-variate normal distribution,

$$X = (X_1, X_2, \cdots, X_p)' \cap N_p(\boldsymbol{\mu}, \boldsymbol{\Sigma})$$

where $\boldsymbol{\mu} = (\mu_1, \cdots, \mu_p)'$, $\boldsymbol{\Sigma} = ((\sigma_{ij}))$. Let the matrix of observations be

$$\boldsymbol{X} = \begin{bmatrix} x_{11} & x_{21} & \cdots & x_{p1} \\ x_{12} & x_{22} & \cdots & x_{p2} \\ & & & \\ x_{1n} & x_{2n} & \cdots & x_{pn} \end{bmatrix} = (\boldsymbol{X}_1 \ \boldsymbol{X}_2 \ \cdots \ \boldsymbol{X}_p) \text{ (say)} \tag{14.2.1}$$

where $\boldsymbol{X}_i = (x_{i1} \ x_{i2} \ \cdots \ x_{in})'$ and $x_{i\lambda}$ denote the λth observation on variable X_i. Let

$$\bar{x}_i = \sum_{\lambda=1}^{n} x_{i\lambda}/n,$$

sample mean of X_i,

$$s_{ij} = \frac{1}{(n-1)} S_{ij} = \frac{1}{n-1} \sum_{\lambda=1}^{n} (x_{i\lambda} - \bar{x}_i)(x_{j\lambda} - \bar{x}_j),$$

the sample covariance between X_i and X_j. We are required to find the joint distribution of $\bar{x}_i$'s and s_{ij}'s.

The joint distribution of n sample observation $x_{i\lambda}$ $(i = 1, \cdots, p; \lambda = 1, \cdots, n)$ is

$$dF(x_{i\lambda}, i = 1, \cdots, p; \lambda = 1, \cdots, n)$$
$$= \frac{1}{(2\pi)^{np/2} |\boldsymbol{\Sigma}|^{n/2}} \exp \left[-\frac{1}{2} \sum_{\lambda=1}^{n} \sum_{ij=1}^{p} \sum \sigma^{ij} (x_{i\lambda} - \mu_i)(x_{j\lambda} - \mu_j) \right] \prod_{\lambda=1}^{n} \prod_{i=1}^{p} dx_{i\lambda} \tag{14.2.2}$$

where $((\sigma^{ij})) = \boldsymbol{\Sigma}^{-1}$. We make an orthogonal transformation from $\boldsymbol{X}$ to $\boldsymbol{U}$ such that

$$\boldsymbol{U} = \begin{bmatrix} u_{11} & u_{21} & \cdots & u_{p1} \\ u_{12} & u_{22} & \cdots & u_{p2} \\ \cdots & \cdots & \cdots & \cdots \\ u_{1n} & u_{2n} & \cdots & u_{pn} \end{bmatrix} = \boldsymbol{LX} \tag{14.2.3}$$

when $u_{i1} = \sqrt{n}\bar{x}_i$ $(i = 1, \cdots, p)$. The jacobian of transformation is unity.

Hence,

$$\sum_{\lambda=1}^{n} x_{i\lambda}^2 = \sum_{\lambda=1}^{n} u_{i\lambda}^2, \sum_{\lambda=1}^{n} x_{i\lambda} x_{j\lambda} = \sum_{\lambda=1}^{n} u_{i\lambda} u_{j\lambda}, \quad i \neq j,$$

$$\sum_{\lambda=1}^{n} (x_{i\lambda} - \mu_i)^2 = \sum_{\lambda=2}^{n} u_{i\lambda}^2 + (u_{i1} - \sqrt{n}\mu_i)^2,$$

$$\sum_{\lambda=1}^{n} (x_{i\lambda} - \mu_i)(x_{j\lambda} - \mu_j) = \sum_{\lambda=2}^{n} u_{i\lambda} u_{j\lambda} + (u_{i1} - \sqrt{n}\mu_i)(u_{ji} - \sqrt{n}\mu_j).$$

Hence, the joint distribution of $u_{i\lambda}$'s is

$$dF(u_{i\lambda}, \lambda = 1, \cdots, n; i = 1, \cdots, p)$$
$$= \frac{1}{(2\pi)^{np/2} |\boldsymbol{\Sigma}|^{n/2}} \exp \left[-\frac{1}{2} \sum_i \sum_j \sigma^{ij} \left\{ (u_{i1} - \sqrt{n}\mu_i)(u_{j1} - \sqrt{n}\mu_j) + \sum_{\lambda=2}^{n} u_{i\lambda} u_{j\lambda} \right\} \right]$$
$$\prod_{i=1}^{p} du_{i1} \prod_{\lambda=2}^{n} du_{i\lambda}$$
$$= \left\{ \frac{n^{p/2}}{(2\pi)^{p/2} |\boldsymbol{\Sigma}|^{1/2}} \exp \left[-\frac{n}{2} \sum_i \sum_j \sigma^{ij} (\bar{x}_i - \mu_i)(\bar{x}_j - \mu_j) \right] \prod_{i=1}^{p} d\bar{x}_i \right\}$$
$$\left\{ \frac{1}{(2\pi)^{(n-1)p/2} |\boldsymbol{\Sigma}|^{(n-1)/2}} \exp \left[-\frac{1}{2} \sum_{\lambda=2}^{n} \sum_{ij=1}^{p} \sum \sigma^{ij} u_{i\lambda} u_{j\lambda} \prod_{i=1}^{p} \prod_{\lambda=2}^{n} du_{i\lambda} \right] \right\} \tag{14.2.4}$$

(14.2.4) shows that the joint distribution of sample means $\bar{x}_i$'s is independent of the joint distribution of $u_{i\lambda}$'s $(i = 1, \cdots, p; \lambda = 2, \cdots, n)$ and that the mean vector $\bar{\boldsymbol{x}} = (\bar{x}_1, \cdots, \bar{x}_p)'$ is distributed as $N_p(\mu, \boldsymbol{\Sigma}/n)$.

The joint distribution of $u_{i\lambda}$'s $(i = 1, \cdots, p; \lambda = 2, \cdots, n)$ is given by the term in the last { }ed portion of (14.2.4). Now,

$$\begin{aligned}\sum_{\lambda=2}^{n} u_{i\lambda}u_{j\lambda} &= \sum_{\lambda=1}^{n} u_{i\lambda}u_{j\lambda} - u_{i1}u_{j1} = \sum_{\lambda=1}^{n} x_{i\lambda}x_{j\lambda} - n\bar{x}_i\bar{x}_j \\ &= \sum_{\lambda=1}^{n}(x_{i\lambda} - \bar{x}_i)(x_{j\lambda} - \bar{x}_j) = S_{ij}.\end{aligned}$$

Hence,

$$\begin{aligned}&dF(u_{i\lambda}, i = 1, \cdots, p; \lambda = 2, \cdots, n) \\ &= \frac{1}{(2\pi)^{(n-1)p/2}|\boldsymbol{\Sigma}|^{(n-1)/2}} \exp\left[-\frac{1}{2}\sum_{ij=1}^{p}\sum \sigma^{ij}S_{ij}\right] \prod_{i=1}^{p}\prod_{\lambda=2}^{n} du_{i\lambda}.\end{aligned} \tag{14.2.5}$$

In order to obtain the joint distribution of S_{ij}'s we are to integrate (14.2.5) over the region

$$R : S_{ij} - \frac{1}{2}dS_{ij} \le S_{ij} = \sum_{\lambda=2}^{n} u_{i\lambda}u_{j\lambda} \le S_{ij} + \frac{1}{2}dS_{ij}, \;\; i, j = 1, \cdots, p. \tag{14.2.6}$$

In the domain, R, the quantity S_{ij} varies within an infinitesimal range,

$$\left(S_{ij} - \frac{1}{2}dS_{ij}, \;\; S_{ij} + \frac{1}{2}dS_{ij}\right);$$

as such the whole of the non-differentiable part of (14.2.5) can be regarded as a constant in the domain. Thus the joint distribution of S_{ij}'s is

$$\begin{aligned}&dF(S_{ij}, i, j = 1, \cdots, p) \\ &= \frac{1}{(2\pi)^{(n-1)p/2}|\boldsymbol{\Sigma}|^{(n-1)/2}} \exp\left[-\frac{1}{2}\sum_{i}\sum_{j}\sigma^{ij}S_{ij}\right] \int_R \prod_{i=1}^{p}\prod_{\lambda=2}^{n} du_{i\lambda}.\end{aligned} \tag{14.2.7}$$

Let

$$V = \int_R \prod_{1}^{p}\prod_{2}^{n} du_{i\lambda} = \int_{R_p} \prod_{i=1}^{p-1}\prod_{\lambda=2}^{n} du_{i\lambda} \int_Q du_{p2} \cdots du_{pn} \tag{14.2.8}$$

$$\begin{aligned}\text{where } Q : \Big\{ &S_{pp} - \frac{1}{2}dS_{pp} \le \sum_{\lambda=2}^{n} u_{p\lambda}^2 \le S_{pp} + \frac{1}{2}dS_{pp} \\ &S_{pj} - \frac{1}{2}dS_{p_j} \le \sum_{\lambda=2}^{n} u_{p\lambda}u_{j\lambda} \le S_{p_j} + \frac{1}{2}dS_{p_j}, j = p-1, p-2, \cdots, 1\Big\}\end{aligned}$$

and

$$R_p : \ S_{ij} - \frac{1}{2}\, dS_{ij} \leq S_{ij} \leq S_{ij} + \frac{1}{2}\, dS_{ij}, \ \ i, j = 1, \cdots, p-1$$

Now we recall the result of Lemma 12.3.1 for evaluating (14.2.8)

$$\int \prod_{i=1}^{n} dx_i$$

$$S^2 - \frac{1}{2}\, dS^2 \ \leq \sum_{1}^{n} x_i^2 \leq S^2 + \frac{1}{2}\, dS^2$$

$$w_j - \frac{1}{2}\, dw_j \ \leq w_j = \sum_{i=1}^{n} l_{ji} x_i \leq w_j + \frac{1}{2}\, dw_j, j = 1, \cdots, k, \ (k < n)$$

$$= \frac{\pi^{(n-k)/2}}{\left|\left(\frac{n-k}{2}\right)\right.} \begin{vmatrix} S^2 & \boldsymbol{W}_1' \\ \boldsymbol{W}_1 & \boldsymbol{L}_1 \boldsymbol{L}_1' \end{vmatrix}^{\frac{n-k-2}{2}} |\boldsymbol{L}_1 \boldsymbol{L}_1'|^{-\frac{(n-k-1)}{2}}\, dS^2\, dw_1 \cdots dw_k \quad (14.2.9)$$

and

$$\boldsymbol{W}_1' = (w_1, w_2, \cdots, w_k)$$

$$\boldsymbol{L}_1 = (l_{ji}, \ \ j = 1, \cdots, k; \ i = 1, \cdots, n).$$

Here,

$$\boldsymbol{L}_1 = \begin{bmatrix} u_{p-12} & u_{p-13} & \cdots & u_{p-1n} \\ u_{p-22} & u_{p-23} & \cdots & u_{p-2n} \\ & & & \vdots \\ u_{12} & u_{13} & \cdots & u_{1n} \end{bmatrix}$$

$$\boldsymbol{L}_1 \boldsymbol{L}_1' = \begin{bmatrix} S_{p-1p-1} & S_{p-1p-2} & \cdots & S_{p-11} \\ S_{p-2p-1} & S_{p-2p-2} & \cdots & S_{p-21} \\ \cdot & \cdot & \cdots & \cdot \\ S_{1p-1} & S_{1p-2} & \cdots & S_{11} \end{bmatrix}$$

$$\boldsymbol{W}_1' = [S_{pp-1} \ \ S_{pp-2} \ \ \cdots \ \ S_{p_1}]. \quad (14.2.10)$$

Therefore, from (14.2.8),

$$V = \left[\int_{R_p} \prod_{i=1}^{p-1} \prod_{\lambda=2}^{n} du_{i\lambda} \right] \frac{\pi^{(n-p)/2}}{\left|\left(\frac{n-p}{2}\right)\right.} \begin{vmatrix} S_{pp} & S_{pp-1} & \cdots & S_{p1} \\ \cdot & \cdot & \cdots & \cdot \\ S_{1p} & S_{1p-1} & \cdots & S_{11} \end{vmatrix}^{(n-p-2)/2}$$

$$\begin{vmatrix} S_{p-1p-1} & \cdots & S_{p-11} \\ \cdot & \cdot & \vdots \\ S_{1p-1} & \cdots & S_{11} \end{vmatrix}^{-(n-p-1)/2} dS_{1p} \cdots dS_{pp}. \quad (14.2.11)$$

The integral in the [ ] in (14.2.11) can be evaluated similarly by repeated application of the results of the lemma in (14.2.9). Thus

$$V = \frac{\pi^{\frac{1}{2}\{(n-p)+(n-p+1)+\cdots+(n-1)\}}}{\prod_{i=1}^{p}\Gamma\left(\frac{n-i}{2}\right)}|\boldsymbol{S}|^{(n-p-2)/2}\prod\prod_{i<j}^{p} dS_{ij}. \tag{14.2.12}$$

Therefore, from (14.2.7),

$$\begin{aligned} dF(S_{ij}; i,j=1,\cdots,p) &= \frac{\exp\left(-\frac{1}{2}\Sigma_i\Sigma_j\sigma^{ij}S_{ij}\right)}{(2\pi)^{(n-1)p/2}|\boldsymbol{\Sigma}|^{(n-1)/2}}V \\ &= \frac{\exp\left(-\frac{1}{2}\sum\sum_{ij=1}^{p}\sigma^{ij}S_{ij}\right)|\boldsymbol{S}|^{(n-p-2)/2}}{2^{(n-1)p/2}\pi^{p(p-1)/4}|\boldsymbol{\Sigma}|^{(n-1)/2}\prod_{i=1}^{p}\Gamma\left(\frac{n-i}{2}\right)}\prod\prod_{i<j=1}^{p} dS_{ij} \end{aligned} \tag{14.2.13}$$

where $|\boldsymbol{S}| = |S_{ij}; i,j=1,\cdots,p|$. Writing

$$s_{ij} = S_{ij}/(n-1), \quad \prod_{i<j=1}^{p} dS_{ij} = (n-1)^{p(p+1)/2}\prod\prod_{i<j=1}^{p} ds_{ij}, \; |\boldsymbol{S}| = (n-1)^p|\boldsymbol{s}|$$

where $\boldsymbol{s} = ((s_{ij}, i,j=1,\cdots,p))$.

Again,

$$tr\left(\boldsymbol{\Sigma}^{-1}\boldsymbol{s}\right) = \sum_i\sum_j \sigma^{ij}s_{ij}.$$

Hence the joint distribution of s_{ij}'s is

$$\begin{aligned} dF(s_{ij}, i<j=1,\cdots,p) &= \frac{\left(\frac{n-1}{2}\right)^{(n-1)p/2}}{\pi^{p(p-1)/4}|\boldsymbol{\Sigma}|^{(n-1)/2}\prod_{i=1}^{p}\Gamma\left(\frac{n-i}{2}\right)} e^{-\frac{n-1}{2}tr\left(\boldsymbol{\Sigma}^{-1}\boldsymbol{s}\right)} \\ &\quad |\boldsymbol{s}|^{(n-p-2)/2}\prod_{i<j=1}^{p} dF(s_{ij}). \end{aligned} \tag{14.2.14}$$

If we define $s_{ij} = S_{ij}/n$, from (14.2.13),

$$\begin{aligned} dF(s_{ij}, i<j=1,\cdots,p) &= \frac{\left(\frac{n}{2}\right)^{np/2}}{\pi^{p(p-1)/4}|\boldsymbol{\Sigma}|^{(n-1)/2}\prod_{i=1}^{p}\Gamma\left(\frac{n-i}{2}\right)} e^{-\frac{n}{2}tr\left(\boldsymbol{\Sigma}^{-1}\boldsymbol{s}\right)} \\ &\quad |\boldsymbol{s}|^{(n-p-2)/2}\prod_{i<j} dF(s_{ij}). \end{aligned} \tag{14.2.15}$$

This ((14.2.14) or (14.2.15)) is known as Wishart distribution with $(n-1)$ d.f. and denoted as

$$W(p, n-1, \boldsymbol{\Sigma}, \boldsymbol{s}).$$

We can write the expression (14.2.13) as Wishart distribution

$$W(p, n-1, \boldsymbol{\Sigma}, \boldsymbol{S}).$$

In the univariate case, $p = 1, \boldsymbol{\Sigma} = \sigma^2, S^2 = \sum_{i=1}^{n}(x_i - \bar{x})^2$ and the p.d.f. of S^2 is, from (14.2.13),

$$\frac{e^{-S^2/2\sigma^2}}{2^{(n-1)/2}\sigma^{(n-1)}\left\lceil\left(\frac{n-1}{2}\right)\right.} S^{(n-3)/2}\, dS^2.$$

Putting $v = S^2/\sigma^2$, we obtain the density function of v as

$$\frac{1}{2^{(n-1)/2}\left\lceil\frac{n-1}{2}\right.} e^{-v/2} v^{(n-1)/2-1}\, dv$$

which is $\chi^2_{(n-1)}$. Hence $W(1, n-1, \sigma^2)/\sigma^2$ follows a $\chi^2_{(n-1)}$ distribution.

14.2.1 Characteristic Function of Wishart Distribution

Let us introduce a real $p \times p$ symmetric matrix $\boldsymbol{\theta} = ((\theta_{ij}))$ corresponding to $((S_{ij}))$. The characteristic function of the distribution is

$$\Phi(\boldsymbol{\theta}) = E\left(\exp\left\{i\sum_{ij=1}^{p}\sum \theta_{ij} S_{ij}\right\}\right) \tag{14.2.16}$$

$$= E\left(\exp\left\{i\sum_i\sum_j \theta_{ij}\sum_{\lambda=2}^{n} u_{i\lambda}u_{j\lambda}\right\}\right)$$

$$= \prod_{\lambda=2}^{n}\left[E\left(\exp\left\{i\sum_i\sum_j \theta_{ij}u_{i\lambda}u_{j\lambda}\right\}\right)\right]$$

$$= \left[E\left(\exp\left\{i\sum_{ij=1}^{p}\sum \theta_{ij}u_{i\lambda}u_{j\lambda}\right\}\right)\right]^{n-1} = [\Phi(\lambda)]^{n-1} \text{ (say)} \tag{14.2.17}$$

since the observations are independent and $(u_{1\lambda}, \cdots, u_{p\lambda})'(\lambda = 2, \cdots, n)$ are each independent multivariate normal. Now

$$\Phi(\lambda) = \int_{-\infty}^{\infty}\cdots\int_{-\infty}^{\infty} \exp\left\{i\sum_{ij=1}^{p}\sum \theta_{ij}u_{i\lambda}u_{j\lambda}\right\} \cdot \frac{1}{(2\pi)^{p/2}|\boldsymbol{\Sigma}|^{1/2}}$$

$$\exp\left\{-\frac{1}{2}\sum_{ij=1}^{p}\sum \sigma^{ij}u_{i\lambda}u_{j\lambda}\right\}\prod_{i=1}^{p} du_{i\lambda} \text{ [from (14.2.4)]}$$

$$= \frac{1}{(2\pi)^{p/2}|\Sigma|^{1/2}} \int_{-\infty}^{\infty} \cdots \int_{-\infty}^{\infty} \exp\left[-\frac{1}{2}\sum\sum_{ij=1}^{p}(\sigma^{ij} - 2i\theta_{ij})u_{i\lambda}u_{j\lambda}\right] \prod_{i=1}^{p} du_{i\lambda}$$

$$= \frac{1}{(2\pi)^{p/2}|\Sigma|^{1/2}} \int_{-\infty}^{\infty} \cdots \int_{-\infty}^{\infty} \exp\left[-\frac{1}{2}\boldsymbol{u}'(\boldsymbol{\Sigma}^{-1} - 2i\boldsymbol{\theta})\boldsymbol{u}\right] \prod_{i=1}^{p} du_{i\lambda} \qquad (14.2.18)$$

where $\boldsymbol{u} = (u_{1\lambda}, \cdots, u_{p\lambda})'$. Now $\boldsymbol{\Sigma}$ is a positive definite matrix and $\boldsymbol{\theta}$ is a real symmetric matrix. Hence, there exists a non-singular matrix $\boldsymbol{A}$ such that $\boldsymbol{A}\boldsymbol{\Sigma}^{-1}\boldsymbol{A}' = I$ and $\boldsymbol{A}\boldsymbol{\theta}\boldsymbol{A}' = \boldsymbol{D}$ where $\boldsymbol{D}$ is the real diagonal matrix $\boldsymbol{D} = \operatorname{diag}(d_1, \cdots, d_p)$ and d_i's are the roots of the equation $|\boldsymbol{\theta} - d\boldsymbol{\Sigma}^{-1}| = 0$.

Consider the transformation, $\boldsymbol{u} = \boldsymbol{A}\boldsymbol{v}$; $|J| = |\boldsymbol{A}| = \sqrt{|\boldsymbol{\Sigma}|}$,

$$\boldsymbol{u}'(\boldsymbol{\Sigma}^{-1} - 2i\boldsymbol{\theta})\boldsymbol{u} = \boldsymbol{v}'\boldsymbol{A}'(\boldsymbol{\Sigma}^{-1} - 2i\boldsymbol{\theta})\boldsymbol{A}\,\boldsymbol{v} = \boldsymbol{v}'I\boldsymbol{v} - 2i\boldsymbol{v}'\boldsymbol{D}\boldsymbol{v}$$

$$= \sum_{j=1}^{p} v_j^2(1 - 2id_j).$$

Therefore, (14.2.18)

$$= \frac{1}{(2\pi)^{p/2}} \prod_{j=1}^{p} \int_{-\infty}^{\infty} e^{-\frac{1}{2}v_j^2(1-2id_j)}\, dv_j$$

$$= \frac{1}{(2\pi)^{p/2}} \prod_{j=1}^{p} \frac{(2\pi)^{p/2}}{(1 - 2id_j)^{1/2}} = \frac{1}{\prod_{j=1}^{p}(1 - 2id_j)^{1/2}}. \qquad (14.2.19)$$

Now,

$$\prod_{j=1}^{p}(1 - 2id_j) = |I - 2i\boldsymbol{D}| = |\boldsymbol{A}\boldsymbol{\Sigma}^{-1}\boldsymbol{A}' - 2i\boldsymbol{A}\boldsymbol{\theta}\boldsymbol{A}'|$$

$$= |\boldsymbol{A}||\boldsymbol{\Sigma}^{-1} - 2i\boldsymbol{\theta}||\boldsymbol{A}'| = |\boldsymbol{A}|^2|\boldsymbol{\Sigma}^{-1} - 2i\boldsymbol{\theta}|.$$

Therefore, (14.2.19)

$$= \frac{1}{\sqrt{|\boldsymbol{\Sigma}|}\sqrt{|\boldsymbol{\Sigma}^{-1} - 2\boldsymbol{\theta}i|}}.$$

Therefore, the characteristic function of the distribution of S_{ij}'s is obtained from (14.2.17) as

$$\Phi(\boldsymbol{\theta}) = \frac{|\boldsymbol{\Sigma}^{-1}|^{(n-1)/2}}{|\boldsymbol{\Sigma}^{-1} - 2i\boldsymbol{\theta}|^{(n-1)/2}}. \qquad (14.2.20)$$

14.2.2 Reproductive Property of Wishart Distribution

Let $\xi^{(1)}, \cdots, \xi^{(k)}$ be independently distributed as $W(p, n_1 - 1, \boldsymbol{\Sigma}, \boldsymbol{S}_1) \cdots, W(p, n_k - 1$ $\boldsymbol{\Sigma}, \boldsymbol{S}_k)$ respectively. Then

$$\xi = \sum_{j=1}^{k} \xi^{(j)}$$

is distributed as $W(p, n - k, \boldsymbol{\Sigma}, \boldsymbol{S})$ where

$$n = \sum_{1}^{k} n_j, \quad \boldsymbol{S} = \sum_{j=1}^{k} \boldsymbol{S}_j.$$

Proof: The characteristic function of $\xi^{(j)}$ is

$$\frac{|\boldsymbol{\Sigma}^{-1}|^{(n_j-1)/2}}{|\boldsymbol{\Sigma}^{-1} - 2i\boldsymbol{\theta}|^{(n_j-1)/2}} = \Phi_j \text{ (say)}.$$

By virtue of the independence of $\xi^{(j)}$'s, the characteristic function of ξ is

$$\prod_{j=1}^{k} \Phi_j = \frac{|\boldsymbol{\Sigma}^{-1}|^{(n-k)/2}}{|\boldsymbol{\Sigma}^{-1} - 2i\boldsymbol{\theta}|^{(n-k)/2}}$$

which is the characteristic function of $W(p, n - k, \boldsymbol{\Sigma}, \boldsymbol{S})$. Hence the theorem.

In order that the result holds, the $\xi^{(j)}$'s must be of the same dimension so that the sum ξ is defined and each variable must have the same parameter $\boldsymbol{\Sigma}$.

14.2.3 Some Other Properties of Wishart Distribution

(a) If $\boldsymbol{A}$ is $W(p, n, \boldsymbol{\Sigma})$ and $\boldsymbol{M}$ is $k \times p$ of rank k, then $\boldsymbol{MAM'}$ is $W(k, n, \boldsymbol{M\Sigma M'})$.

Proof: Let $\boldsymbol{A} = \sum_{\alpha=1}^{n} \boldsymbol{Z}_\alpha \boldsymbol{Z}'_\alpha$ when $\boldsymbol{Z}_\alpha$ is distributed as independent $N_p(\boldsymbol{0}, \boldsymbol{\Sigma})$.

Then $\boldsymbol{Y}_\alpha = \boldsymbol{MZ}_\alpha$ is distributed as $N_k(\boldsymbol{0}, \boldsymbol{M\Sigma M'})$.

Therefore,

$$\sum_{\alpha=1}^{n} \boldsymbol{Y}_\alpha \boldsymbol{Y}'_\alpha = \sum_{\alpha=1}^{n} \boldsymbol{MZ}_\alpha \boldsymbol{Z}'_\alpha \boldsymbol{M'} = \boldsymbol{M} \sum_{\alpha=1}^{n} \boldsymbol{Z}_\alpha \boldsymbol{Z}'_\alpha \boldsymbol{M'}$$

$$= \boldsymbol{MAM'}$$

is distributed $W(k, n, \boldsymbol{M\Sigma M'})$.

(b) Consider a transformation $\boldsymbol{A} = \boldsymbol{CBC'}$ where $\boldsymbol{C}$ is a non-singular $p \times p$ matrix.

If $\boldsymbol{A}$ is distributed as Wishart $W(p, n, \boldsymbol{\Sigma})$ then $\boldsymbol{B}$ is distributed as $W(p, n, \boldsymbol{\Phi})$ where

$$\boldsymbol{\Phi} = \boldsymbol{C}^{-1} \boldsymbol{\Sigma} \boldsymbol{C}'^{-1}.$$

Proof: Let

$$A = \sum_{\alpha=1}^{n} Z_\alpha Z'_\alpha$$

when Z_α is distributed as independent $N_p(\mathbf{0}, \Sigma)$.

Then $Y_\alpha = C^{-1} Z_\alpha$ is distributed as

$$N_p(\mathbf{0}, (C^{-1})\Sigma(C^{-1})') = N_p(\mathbf{0}, \Phi).$$

Again

$$B = C^{-1} A (C')^{-1} = C^{-1} \sum_{\alpha=1}^{n} Z_\alpha Z'_\alpha (C')^{-1} = \sum_{\alpha=1}^{n} (C^{-1} Z_\alpha)(C^{-1} Z_\alpha)' = \sum_{\alpha=1}^{n} Y_\alpha Y'_\alpha.$$

Therefore,

$$B \cap W(p, n, \Phi).$$

(c) Let $A \cap W(p, n, \Sigma)$. Let also A, Σ be partitioned as

$$A = \begin{bmatrix} A_{11} & A_{12} \\ A_{21} & A_{22} \end{bmatrix}, \quad \Sigma = \begin{bmatrix} \Sigma_{11} & \Sigma_{12} \\ \Sigma_{21} & \Sigma_{22} \end{bmatrix}$$

where A_{11}, Σ_{11} are each $q \times q$.
Then

$$A_{11} \cap W(q, n, \Sigma_{11}).$$

Proof: A is distributed as

$$\sum_{\alpha=1}^{n} Z_\alpha Z'_\alpha$$

where Z_α is independent $N_p(\mathbf{0}, \Sigma)$. Partition Z_α as

$$Z_\alpha = \begin{bmatrix} Z_\alpha^{(1)} \\ Z_\alpha^{(2)} \end{bmatrix}$$

when $Z_\alpha^{(1)}$ is $q \times 1$. Then the $Z_\alpha^{(1)}$'s are independent $N_q(\mathbf{0}, \Sigma_{11})$.
Hence A_{11} is distributed as

$$\sum_{\alpha=1}^{n} Z_\alpha^{(1)} Z_\alpha^{(1)\prime},$$

which follows $W(p, n, \Sigma_{11})$.

(d) Let A and Σ be partitioned into $(p_1, p_2, \cdots, p_q)$ rows and $(p_1, p_2, \cdots, p_q)$ columns $(p_1 + \cdots + p_q = p)$.

$$A = \begin{bmatrix} A_{11} & \cdots & A_{1q} \\ \cdot & \cdots & \\ A_{q1} & \cdots & A_{qq} \end{bmatrix}, \quad \sum = \begin{bmatrix} \Sigma_{11} & \cdots & \Sigma_{1q} \\ \cdot & \cdots & \\ \Sigma_{q1} & \cdots & \Sigma_{qq} \end{bmatrix}.$$

A_{ij} is of order $p_i \times p_j (i, j = 1, \cdots, q)$.

If $\boldsymbol{\Sigma}_{ij} = \mathbf{0}$ for $i \neq j$ and if $\boldsymbol{A} \cap W(p, n, \boldsymbol{\Sigma})$ then $A_{11}, \cdots, A_{qq}$ are independent distributed and $A_{jj} \cap W(p_j, n, \boldsymbol{\Sigma}_{jj})$.

Proof: $\boldsymbol{A}$ is distributed as

$$\sum_{\alpha=1}^{n} \boldsymbol{Z}_\alpha \boldsymbol{Z}'_\alpha$$

where $\boldsymbol{Z}_\alpha$'s are independently distributed as $N_p(\mathbf{0}, \boldsymbol{\Sigma})$. Let $\boldsymbol{Z}_\alpha$ be partitioned as

$$\boldsymbol{Z}_\alpha = \begin{bmatrix} \boldsymbol{Z}_\alpha^{(1)} \\ \vdots \\ \boldsymbol{Z}_\alpha^{(q)} \end{bmatrix}$$

when $\boldsymbol{Z}_\alpha^{(j)}$ is of order $(p_j \times 1)$. Since $\Sigma_{ij} = \mathbf{0}$, $\boldsymbol{Z}_\alpha^{(i)}$ is independent of $\boldsymbol{Z}_\alpha^{(j)}$. Also

$$\boldsymbol{A}_{ii} = \sum_{\alpha=1}^{n} \boldsymbol{Z}_\alpha^{(i)} \boldsymbol{Z}_\alpha^{(i)\prime}$$

is independent of

$$\boldsymbol{A}_{jj} = \sum_{\alpha=1}^{n} \boldsymbol{Z}_\alpha^{(j)\prime} \boldsymbol{Z}_\alpha^{(j)\prime}.$$

The result now follows from the previous result.

(e) If $\boldsymbol{A}$ is $W(p, n, \boldsymbol{\Sigma})$ and $\boldsymbol{Y}$ is any $p \times 1$ random vector which is independent of $\boldsymbol{A}$ wit $P(\boldsymbol{Y} = \mathbf{0}) = 0$, then $\boldsymbol{Y}'\boldsymbol{A}\boldsymbol{Y}/\boldsymbol{Y}'\boldsymbol{\Sigma}\boldsymbol{Y}$ is $\chi^2_{(n)}$ and is independent of $\boldsymbol{Y}$.

Proof: In the result (a), put $\boldsymbol{M} = \boldsymbol{Y}'(1 \times p)$. Then conditional on $\boldsymbol{Y}, \boldsymbol{Y}'\boldsymbol{A}\boldsymbol{Y}$ follow

$$W(1, n, \boldsymbol{Y}'\boldsymbol{\Sigma}\boldsymbol{Y}).$$

Therefore,

$$\boldsymbol{Y}'\boldsymbol{A}\boldsymbol{Y}/\boldsymbol{Y}'\boldsymbol{\Sigma}\boldsymbol{Y}$$

is $\chi^2_{(n)}$. Since this distribution does not depend on $\boldsymbol{Y}$, it is also the condition distribution of $\boldsymbol{Y}'\boldsymbol{A}\boldsymbol{Y}/\boldsymbol{Y}'\boldsymbol{\Sigma}\boldsymbol{Y}$.

Corollary 14.2.1 *If $\bar{\boldsymbol{x}}$ and $\mathbf{s} = ((s_{ij}))$ where $s_{ij} = \frac{1}{(n-1)}\Sigma(x_{i\lambda} - \bar{x}_i)(x_{j\lambda} - \bar{x}_j)$, are th mean and covariance matrix formed from a sample of size n from $N_p(\boldsymbol{\mu}, \boldsymbol{\Sigma})$ distributior then $(n-1)\bar{\boldsymbol{X}}'\mathbf{s}\bar{\boldsymbol{X}}/\bar{\boldsymbol{X}}'\boldsymbol{\Sigma}\bar{\boldsymbol{X}}$ follows $\chi^2_{(n-1)}$ and is independent of $\bar{\boldsymbol{X}}$.*

(We are using the symbol $\bar{\boldsymbol{X}}$ to denote the random variable corresponding to $\bar{\boldsymbol{x}}$ whil we shall often use the same symbol s_{ij} to denote a random variable as well its realise value).

Proof: It has been found that $\bar{\boldsymbol{X}}$ and $\boldsymbol{S}$ are independent. Also $\boldsymbol{S}$ is

$$W(p, n-1, \boldsymbol{\Sigma}).$$

Hence the result follows from (e).

Another related statistic, Wilk's statistic is discussed in Section 18.11.

Example 14.2.1 *Find the distribution of sample correlation coefficients r_{ij}'s $(i < j = \cdots, p)$ when $\rho_{ij} = 0 (i \neq j = 1, \cdots, p)$ in sampling from a p-variate normal distribution.*

Let us draw a sample of size $n, (x_{1\lambda}, x_{2\lambda}, \cdots, x_{p\lambda})$ $(\lambda = 1, \cdots, n)$ from a p-variate normal distribution $N_p(\boldsymbol{\mu}, \boldsymbol{\Sigma})$ when $\boldsymbol{\Sigma} = \text{Diag}\,(\sigma_{ii}, i = 1, \cdots, p)$. It has been found that the joint distribution of S_{ij}'s

$$\left(S_{ij} = \sum_{\lambda=1}^{n} (x_{i\lambda} - \bar{x}_i)(x_{j\lambda} - \bar{x}_j), \right.$$
$$\left. \bar{x}_i = \frac{1}{n} \sum_{\lambda=1}^{n} x_{i\lambda}, i, j = 1, \cdots, p \right)$$

is a Wishart distribution, $W(p, n-1, \boldsymbol{\Sigma})$,

$$dF(S_{ij}, i < j = 1, \cdots, p) = \frac{|S|^{(n-p-2)/2} \exp\left(-\frac{1}{2} \sum_i \sum_j \sigma^{ij} S_{ij} \right)}{2^{(n-1)p/2} \pi^{p(p-1)/4} |\boldsymbol{\Sigma}|^{(n-1)/2} \prod_{i=1}^{p} \Gamma\left(\frac{n-i}{2}\right)} \prod_{i<j=1}^{p}\prod dS_{ij}. \quad (14.2.21)$$

Here $|\boldsymbol{\Sigma}| = \prod_{i=1}^{p} \sigma_{ii}$. Transform

$$S_{ij} = \sqrt{S_{ii} S_{jj}}\, r_{ij}, i < j \quad (14.2.22)$$
$$S_{ii} = S_{ii}. \quad (14.2.23)$$

Jacobian of transformation is

$$|J| = \prod_{i=1}^{p} S_{ii}^{(p-1)/2}.$$

Hence, from (14.2.21), the joint density of S_{ii}'s and r_{ij}'s is

$$\frac{\left|\sqrt{S_{ii} S_{jj}}\, r_{ij}\right|^{(n-p-2)/2}}{2^{(n-1)p/2} \pi^{p(p-1)/4} \left(\prod_{i=1}^{p} \sigma_{ii} \right)^{(n-1)/2} \prod_{i=1}^{p} \Gamma\left(\frac{n-i}{2}\right)} \exp\left(-\frac{1}{2} \sum_{i=1}^{p} \frac{S_{ii}}{\sigma_{ii}} \right) \prod_{i=1}^{p} S_{ii}^{(p-1)/2}$$

$$= \frac{|r_{ij}|^{(n-p-2)/2}}{\pi^{p(p-1)/4} \prod_{i=1}^{p} \Gamma\left(\frac{n-i}{2}\right)} \prod_{i=1}^{p} \left\{ \frac{S_{ii}^{(n-1)/2-1} \exp\left(-\frac{1}{2} \frac{S_{ii}}{\sigma_{ii}} \right)}{2^{(n-1)/2} \sigma_{ii}^{(n-1)/2}} \right\} \quad (14.2.24)$$

since

$$\left|\sqrt{S_{ii} S_{jj}}\; r_{ij}\right| = \left(\prod_{i=1}^{p} S_{ii} \right) |r_{ij}|.$$

Writing $S_{ii}/\sigma_{ii} = u_i$ and integrating (14.2.24) over u_i, the ith integral in the produc term is

$$\int_0^\infty \frac{S_{ii}^{(n-1)/2-1} \exp\left(-\frac{1}{2}\frac{S_{ii}}{\sigma_{ii}}\right)}{2^{(n-1)/2}\sigma_{ii}^{(n-1)/2}} dS_{ii} = \int_0^\infty u_i^{(n-1)/2-1} e^{-u_i}\, du_i = \left\lceil\left(\frac{n-1}{2}\right)\right. .$$

Hence, from (14.2.24), the distribution of r_{ij}'s is

$$dF(r_{ij}, i < j = 1, \cdots, p) = \frac{\left\{\left\lceil\left(\frac{n-1}{2}\right)\right.\right\}^p |r_{ij}|^{(n-p-2)/2}}{\pi^{p(p-1)/4} \prod_{i=1}^{p} \left\lceil\left(\frac{n-i}{2}\right)\right.} \prod_{i<j=1}^{p}\prod dr_{ij}. \tag{14.2.25}$$

14.3 Distribution of Sample Partial Regression Coefficients

Consider a set of $(p+1)$ variables $Y, X_1, \cdots, X_p$ and let n observations be taken a random on each of these $(p+1)$ variables. The matrix of observations is

$$\begin{bmatrix} y_1 & x_{11} & x_{21} & \cdots & x_{p1} \\ \cdots & \cdots & \cdots & \cdots & \cdots \\ \cdots & \cdots & \cdots & \cdots & \cdots \\ y_n & x_{1n} & x_{2n} & \cdots & x_{pn} \end{bmatrix}, \tag{14.3.1}$$

y_λ being the λth observation on Y and $x_{i\lambda}$ the λth observation on X_i ($i = 1, \cdots, p; \lambda = 1, \cdots, n$). The equation of regression of Y on $X_1, \cdots, X_p$ is

$$\hat{Y} = E(Y|x_1, \cdots, x_p). \tag{14.3.2}$$

Suppose that the regression equation is linear in $x_1, \cdots, x_p$ so that the expected valu of Y corresponding to the observations $(x_{1\lambda}, \cdots, x_{p\lambda})$ is

$$\begin{aligned} \hat{Y}_\lambda &= \alpha' + \beta_1 x_{1\lambda} + \beta_2 x_{2\lambda} + \cdots + \beta_p x_{p\lambda} \\ &= \alpha + \beta_1(x_{1\lambda} - \bar{x}_1) + \beta_2(x_{2\lambda} - \bar{x}_2) + \cdots + \beta_p(x_{p\lambda} - \bar{x}_p) \end{aligned} \tag{14.3.3}$$

where $\alpha, \beta_1, \cdots, \beta_p$ are unknown constants. The quantities β_j's are called the partia regression coefficients of Y on X_j eliminating the effects of $X_1, \cdots, X_{j-1}, X_{j+1}, \cdots, X_p$.

We assume that the conditional distribution of Y given $X_1, \cdots, X_p$ is normal with mean given in (14.3.3) and a constant (independent of x) variance σ^2. Hence the joint conditional distribution of $Y_1, \cdots, Y_n$ given $x_{i\lambda}$'s ($\lambda = 1, \cdots, n; i = 1, \cdots, p$) is

$$\begin{aligned} dF(y_1, \cdots, y_n) = \frac{1}{(\sqrt{2\pi}\sigma)^n} \exp\Bigg[-\frac{1}{2\sigma^2}\sum_\lambda \{y_\lambda - \alpha - \beta_1(x_{1\lambda} - \bar{x}_1) - \cdots \\ - \beta_p(x_{p\lambda} - \bar{x}_p)\}^2 \Bigg] \prod_{\lambda=1}^{n} dy_\lambda. \end{aligned} \tag{14.3.4}$$

Note that the above assumption is less restrictive than the assumption $(Y, X_1, \cdots, X_p)'$ follow a $(p+1)$-variate normal distribution; because if $(Y, X_1, \cdots, X_p)$ follow a N_{p+1} then (14.3.4) holds. However, (14.3.4) may hold even if $(Y, X_1, \cdots, X_p)$ does not follow a N_{p+1}. Define

$$\bar{y} = \frac{1}{n}\sum y_\lambda, \quad S_{ij} = \sum_\lambda (x_{i\lambda} - \bar{x}_i)(x_{j\lambda} - \bar{x}_j), \quad i, j = 1, \cdots, p$$

$$\begin{aligned} S_{iy} &= \sum_{\lambda=1}^{n} (x_{i\lambda} - \bar{x}_i)(y_\lambda - \bar{y}) \qquad (14.3.5) \\ z_\lambda &= y_\lambda - \alpha - \beta_1(x_{1\lambda} - \bar{x}_1) - \cdots - \beta_p(x_{p\lambda} - \bar{x}_p) \\ S_{iz} &= \sum_{\lambda=1}^{n} (z_\lambda - \bar{z})(x_{i\lambda} - \bar{x}_i) = \sum_{\lambda=1}^{n} (x_{i\lambda} - \bar{x}_i) z_\lambda \\ &= S_{yi} - \beta_1 S_{1i} - \beta_2 S_{2i} - \cdots - \beta_p S_{pi}. \end{aligned}$$

The sample regression coefficients $(b_1, \cdots, b_p)' = \boldsymbol{b}$ is the solution of the normal equations

$$\begin{bmatrix} S_{11} & \cdots & S_{1p} \\ & & \vdots \\ S_{p1} & \cdots & S_{pp} \end{bmatrix} \begin{bmatrix} b_1 \\ \vdots \\ b_p \end{bmatrix} = \begin{bmatrix} S_{1y} \\ \vdots \\ S_{py} \end{bmatrix}$$

$$\text{or,} \quad \boldsymbol{Sb} = ((S_{iy})) \qquad (14.3.6)$$

$$\text{or,} \quad \boldsymbol{b} = (\boldsymbol{S})^{-1}((S_{iy})) = \boldsymbol{C}((S_{iy})) \text{ (say)}$$

where $\boldsymbol{C} = ((c_{ij}, \ i, j = 1, \cdots, p)) = (\boldsymbol{S})^{-1}$.

The joint distribution of $\boldsymbol{z} = (z_1, \cdots, z_n)'$ given $x_{i\lambda}$'s, is, therefore,

$$dF(\boldsymbol{z}') = \frac{1}{(\sqrt{2\pi}\sigma)^n} \exp\left[-\frac{1}{2\sigma^2}\sum_\lambda z_\lambda^2\right] \prod_{i=1}^{n} dz_i. \qquad (14.3.7)$$

Consider the transformation $(z_1, \cdots, z_n) \to (u_1, \cdots, u_n)$ such that

$$u_1 = \frac{z_1 + \cdots + z_n}{\sqrt{n}} = \sqrt{n}\bar{z} \qquad (14.3.8)$$

$$\begin{bmatrix} u_2 \\ \vdots \\ u_{p+1} \end{bmatrix} = \begin{bmatrix} C_{11} & \cdots & C_{1p} \\ & & \\ C_{p1} & \cdots & C_{pp} \end{bmatrix} \begin{bmatrix} S_{1z} \\ \vdots \\ S_{pz} \end{bmatrix} \qquad (14.3.9)$$

and $u_{p+2}, \cdots, u_n$ are chosen in such a way that they are orthogonal among themselves and also to $u_1, \cdots, u_{p+1}$. Note that the whole transformation $(z_1, \cdots, z_n) \to (u_1, \cdots, u_n)$ is not generally orthogonal as $u_2, \cdots, u_{p+1}$ are not generally orthogonal among themselves. Now,

$$((S_{iz})) = \begin{bmatrix} S_{1z} \\ \vdots \\ S_{pz} \end{bmatrix} = \begin{bmatrix} x_{11} - \bar{x}_1 & \cdots & x_{1n} - \bar{x}_1 \\ \cdots & & \\ x_{p1} - \bar{x}_p & \cdots & x_{pn} - \bar{x}_p \end{bmatrix} \begin{bmatrix} z_1 \\ \vdots \\ z_n \end{bmatrix} = \boldsymbol{X}^0\boldsymbol{z} \text{ (say)}. \qquad (14.3.10)$$

Hence, from (14.3.9)

$$(u_2, \cdots, u_{p+1})' = C((S_{iz})) = CX^0 z. \tag{14.3.11}$$

We can, therefore, write the transformation as

$$\boldsymbol{u} = \begin{bmatrix} u_1 \\ u_2 \\ \vdots \\ u_{p+1} \\ u_{p+2} \\ \vdots \\ u_n \end{bmatrix} = \begin{bmatrix} \boldsymbol{u}_1 \\ \boldsymbol{u}_2 \\ \boldsymbol{u}_3 \end{bmatrix} = \begin{bmatrix} \boldsymbol{L}_1 \\ \boldsymbol{L}_2 \\ \boldsymbol{L}_3 \end{bmatrix} \boldsymbol{z} = \boldsymbol{Lz} \tag{14.3.12}$$

where

$$\boldsymbol{L}_1 = \left(\frac{1}{\sqrt{n}}, \cdots, \frac{1}{\sqrt{n}}\right)$$
$$\boldsymbol{L}_2 = CX^0$$

and $\boldsymbol{L}_3$ is such that

$$\boldsymbol{L}_3\boldsymbol{L}_3' = I, \;\; \boldsymbol{L}_3\boldsymbol{L}_1' = \boldsymbol{0}, \;\; \boldsymbol{L}_3\boldsymbol{L}_2' = \boldsymbol{0}. \tag{14.3.13}$$

Clearly,

$$\boldsymbol{L}_1\boldsymbol{L}_1' = 1,$$

$$\boldsymbol{L}_2\boldsymbol{L}_1' = \boldsymbol{C}\begin{bmatrix} x_{11} - \bar{x}_1 & \cdots & x_{1n} - \bar{x}_1 \\ \\ x_{p1} - \bar{x}_p & & x_{pn} - \bar{x}_p \end{bmatrix}\begin{bmatrix} \frac{1}{\sqrt{n}} \\ \vdots \\ \frac{1}{\sqrt{n}} \end{bmatrix}$$

$$= \boldsymbol{C}\begin{bmatrix} \frac{1}{\sqrt{n}}\ \Sigma_\lambda\ (x_{1\lambda} - \bar{x}_1) \\ \vdots \\ \frac{1}{\sqrt{n}}\ \Sigma_\lambda\ (x_{p\lambda} - \bar{x}_p) \end{bmatrix} = \boldsymbol{C0} = \boldsymbol{0}.$$

Now

$$\boldsymbol{L}_2\boldsymbol{L}_2' = \boldsymbol{C}\boldsymbol{X}^0\boldsymbol{X}^{0'}\boldsymbol{C} = \boldsymbol{CSC}' = \boldsymbol{C}.$$
$$\therefore\ (\boldsymbol{L}_2\boldsymbol{L}_2')^{-1} = \boldsymbol{S}.$$

Again, $\boldsymbol{U} = \boldsymbol{LZ}$ i.e., $\boldsymbol{Z} = \boldsymbol{L}^{-1}\boldsymbol{U}$. Hence,

$$\sum_1^n Z_\lambda^2 = \boldsymbol{Z}'\boldsymbol{Z} = \boldsymbol{U}'(\boldsymbol{L}^{-1})'\boldsymbol{L}^{-1}\boldsymbol{U} = \boldsymbol{U}'(\boldsymbol{LL}')^{-1}\boldsymbol{U}.$$

again,

$$\boldsymbol{LL'} = \begin{bmatrix} \boldsymbol{L_1L_1'} & 0 & 0 \\ 0 & \boldsymbol{L_2L_2'} & 0 \\ 0 & 0 & \boldsymbol{L_3L_3'} \end{bmatrix} = \begin{bmatrix} 1 & 0 & 0 \\ 0 & \boldsymbol{S}^{-1} & 0 \\ 0 & 0 & I \end{bmatrix}.$$

$$\therefore\ (\boldsymbol{LL'})^{-1} = \begin{bmatrix} 1 & 0 & 0 \\ 0 & \boldsymbol{S} & 0 \\ 0 & 0 & I \end{bmatrix}.$$

$$\therefore\ \sum_\lambda Z_\lambda^2 = (U_1'\ U_2'\ U_3') \begin{bmatrix} 1 & 0 & 0 \\ 0 & \boldsymbol{S} & 0 \\ 0 & 0 & I \end{bmatrix} \begin{bmatrix} U_1 \\ U_2 \\ U_3 \end{bmatrix}$$

$$= \boldsymbol{U_1'U_1 + U_2'SU_2 + U_3'U_3}$$

$$= U_1^2 + \sum_{ij=2}^{p+1}\sum S_{ij}U_iU_j + \sum_{p+2}^{n} U_i^2$$

where suffixes of S_{ij} run from $2, \cdots, p+1$; it is to be understood that new $S_{ij} = S_{i-1j-1}$ of (14.3.6)).

In the transformation $\boldsymbol{U = LZ}$, the Jacobian of transformation is $|\boldsymbol{L}^{-1}|$. We have

$$|\boldsymbol{LL'}| = |\boldsymbol{S}|^{-1}. \quad \therefore\ |\boldsymbol{S}| = |\boldsymbol{LL'}|^{-1} = |\boldsymbol{L}^{-1}|^2. \quad \therefore\ |\boldsymbol{L}^{-1}| = \sqrt{|\boldsymbol{S}|}.$$

The joint distribution of $u_1, \cdots, u_n$ given $x_1, \cdots, x_n$ is, therefore, by (14.3.7),

$$\frac{\sqrt{|\boldsymbol{S}|}}{(\sigma\sqrt{2n})^n} \exp\left[-\frac{1}{2\sigma^2}\left\{u_1^2 + \sum_{i,j=2}^{p+1}\sum S_{ij}u_iu_j + \sum_{i=p+2}^{n} u_i^2\right\}\right] \prod_{i=1}^{n} du_i. \tag{14.3.14}$$

Thus we see that $u_1, (u_2, \cdots, u_{p+1}), (u_{p+2}, \cdots, u_n)$ are independently distributed. This shows that distribution of u_1 given $x_1, \cdots, x_n$ is $\frac{1}{\sigma\sqrt{2\pi}}e^{-u_1^2/2\sigma^2}\,du_1$. Again

$$u_1 = \frac{1}{\sqrt{n}}\sum_{\lambda=1}^{n}\{y_\lambda - \alpha - \beta_1(x_{1\lambda} - \bar{x}_1) - \cdots - \beta_p(x_{p\lambda} - \bar{x}_p)\} = \sqrt{n}(\bar{y} - \alpha).$$

Also $du_1 = \sqrt{n}d\bar{y}$.

Hence, distribution of $\bar{Y}$ for given $x_1, \cdots, x_n$ is

$$\frac{\sqrt{n}}{\sigma\sqrt{2\pi}}e^{-\frac{n}{2\sigma^2}(\bar{y}-\alpha)^2}\,d\bar{y} \tag{14.3.15}$$

i.e., conditional distribution of $\bar{Y}$ for given x's is $N\left(\alpha, \frac{\sigma^2}{n}\right)$, where $\alpha' = \alpha - \beta_1\bar{x}_1 - \cdots - \beta_p\bar{x}_p$. Again from (14.3.14) we see that the u_i is $N(0, \sigma^2)$, $i = p+2, \cdots, n$ and are independent of $u_1, \cdots, u_{p+1}$. Hence

$$\chi^2 = \sum_{i=p+2}^{n} u_i^2/\sigma^2$$

follows a Central χ^2 with $(n-p-1)$ d.f.

Also, from (14.3.14), the joint distribution of $u_2, \cdots, u_{p+1}$ is

$$\frac{\sqrt{|S|}}{(\sigma\sqrt{2\pi})^p} \exp\left(-\frac{1}{2\sigma^2}\sum_{ij=2}^{p+1}\sum S_{ij}u_iu_j\right) du_2 \cdots du_{p+1} \qquad (14.3.16)$$

so that $(u_2, \cdots, u_{p+1})'$ follows a p-variate normal distribution with mean vector zero an dispersion matrix $\sigma^2((S_{ij}))^{-1} = ((C_{ij}\sigma^2))$.
Now,

$$\sum_{ij=2}^{p+1}\sum S_{ij}u_iu_j = \boldsymbol{u}_2'(S_{ij})\boldsymbol{u}_2.$$

Again,

$$\begin{aligned}\boldsymbol{U}_2 &= \boldsymbol{L}_2\boldsymbol{Z} = \boldsymbol{C}\boldsymbol{X}^0(\boldsymbol{Y} - \alpha\boldsymbol{1}_n - \boldsymbol{X}^{0'}\boldsymbol{\beta}) \\ \text{or, } \boldsymbol{U}_2 &= \boldsymbol{C}\boldsymbol{X}^0\boldsymbol{Y} - \alpha\boldsymbol{C}\boldsymbol{X}^0\boldsymbol{1}_n - \boldsymbol{C}\boldsymbol{S}\boldsymbol{\beta} \\ &= ((S_{ij}))^{-1}((S_{iy})) - \alpha((C_{ij}))\boldsymbol{0} - ((C_{ij}))((S_{ij}))\boldsymbol{\beta} \\ &= ((S_{ij}))^{-1}((S_{iy})) - ((S_{ij}))^{-1}((S_{ij}))\boldsymbol{\beta} = \boldsymbol{b} - \boldsymbol{\beta}.\end{aligned}$$

$$\therefore \begin{bmatrix} u_2 \\ \vdots \\ u_{p+1}\end{bmatrix} = \begin{bmatrix} b_1 - \beta_1 \\ \vdots \\ b_p - \beta_p\end{bmatrix}.$$

Hence,

$$\begin{aligned}\boldsymbol{u}_2'(S_{ij})\boldsymbol{u}_2 &= (b_1 - \beta_1 \cdots b_p - \beta_p)\begin{bmatrix} S_{11} & \cdots & S_{1p} \\ \vdots & & \\ S_{p1} & \cdots & S_{pp}\end{bmatrix}\begin{bmatrix} b_1 - \beta_1 \\ \vdots \\ b_p - \beta_p\end{bmatrix} \\ &= \sum_i\sum_j S_{ij}(b_i - \beta_i)(b_j - \beta_j).\end{aligned}$$

Hence, the joint distribution of $b_1, \cdots, b_p$ given x's is

$$\frac{\left|\frac{S_{ij}}{\sigma^2}\right|^{1/2}}{(2\pi)^{p/2}} \exp\left(-\frac{1}{2}\sum_{i,j=1}^{p}\sum \frac{S_{ij}}{\sigma^2}(b_i - \beta_i)(b_j - \beta_j)\right) db_1 \cdots db_p \qquad (14.3.17)$$

which is a p-variate normal distribution with mean vector $\boldsymbol{\beta}$ and dispersion matrix

$$\left(\left(\frac{S_{ij}}{\sigma^2}\right)\right)^{-1} = \sigma^2((C_{ij})).$$

Again

$$\begin{aligned}\sum_{\lambda=1}^{n} z_\lambda^2 &= \boldsymbol{u}'(\boldsymbol{L}\boldsymbol{L}')^{-1}\boldsymbol{u} = u_1^2 + \sum_{ij=1}^{p}\sum S_{ij}u_{i+1}u_{j+1} + \sum_{p+2}^{n} u_i^2 \\ &= n(\bar{y} - \alpha)^2 + \sum_{ij=1}^{p}\sum S_{ij}(b_i - \beta_i)(b_j - \beta_j) + \sum_{p+2}^{n} u_i^2.\end{aligned} \qquad (14.3.18)$$

We shall now show that $u_i, i = p+2, \cdots, n$ do not involve $\alpha, \beta_1, \cdots, \beta_p$.

$$U_3 = L_3 Z = L_3 Y - L_3 \alpha \mathbf{1} - L_3 X^{0'} \beta.$$

Now $L_1 = \frac{1}{\sqrt{n}} \mathbf{1}'$. As $L_3 L_1' = \mathbf{0}$, $\sqrt{n} L_3 L_1' = \mathbf{0}$ or $L_3(\sqrt{n} L_1)' = \mathbf{0}$ or $L_3 \mathbf{1} = \mathbf{0}$.
Again $L_3 L_2' = \mathbf{0}$ or $L_3 X^{0'} C = \mathbf{0}$; $\therefore$ $L_3 X^{0'} = \mathbf{0}$, since $((C_{ij})) \neq \mathbf{0}$ $\therefore$ $L_3 X^{0'} \beta = \mathbf{0}$.
Hence,

$$U_3 = L_3 Y$$

so that $u_{p+2}, \cdots, u_n$ do not involve $\alpha, \beta_1, \cdots, \beta_p$. Hence (14.3.18) is an identity in y's and u's for any value of α and β. So putting $\alpha = 0, \beta_i = 0, i = 1, \cdots, p$ in (14.3.18) we get

$$\sum_{\lambda=1}^{n} y_\lambda^2 = u_1^2 + \sum\sum_{ij=1}^{p} S_{ij} b_i b_j + \sum_{i=p+2}^{n} u_i^2 = n\bar{y}^2 + \sum\sum_{ij=1}^{p} S_{ij} b_i b_j + \sum_{i=p+2}^{n} u_i^2.$$

$$\therefore \sum_{i=p+2}^{n} u_i^2 = \sum_{\lambda=1}^{n} y_\lambda^2 - n\bar{y}^2 - \sum\sum_{ij=1}^{p} S_{ij} b_i b_j = \sum_{\lambda=1}^{n} (y_\lambda - \bar{y})^2 - \sum\sum_{ij=1}^{p} S_{ij} b_i b_j.$$

$$= TSS - SS_{\text{regression}} = \text{Residual } SS = S_e^2. \qquad (14.3.19)$$

Hence

$$\sum_{i=p+2}^{n} u_i^2 / \sigma^2 = \chi^2$$

with $(n-p-1)$ d.f. or $\frac{S_e^2}{\sigma^2}$ follows a central $\chi^2_{(n-p-1)}$. Further, $S_e^2/(n-p-1) = s_e^2$ gives an unbiased estimate of σ^2.

To test $H_\circ(\alpha = \alpha_0)$: under $H_\circ$, $\sqrt{n}(\bar{y} - \alpha_0)/\sigma$ is $N(0,1)$ and S_e^2/σ^2 is central $\chi^2_{(n-p-1)}$. Hence under $H_\circ$,

$$t = \{\sqrt{n}(\bar{y} - \alpha_0)/\sigma\} / \sqrt{\chi^2/(n-p-1)} = \frac{\sqrt{n}(\bar{y} - \alpha_0)}{\sqrt{S_e^2/(n-p-1)}} = \frac{\sqrt{n}(\bar{y} - \alpha_0)}{s_e}$$

which follows a central t-distribution with $(n-p-1)$ d.f. This distribution of t is independent of that of X's so that whatever be the distribution of $X_1, \cdots, X_p$ the above test is valid.

To test $H_\circ(\beta_i = \beta_i^0)$; we have,

$$f(\boldsymbol{b}) d\boldsymbol{b} = \frac{\left|\frac{S_{ij}}{\sigma^2}\right|^{1/2}}{(2\pi)^{p/2}} \exp\left[-\frac{1}{2\sigma^2} \sum\sum_{ij=1}^{p} S_{ij}(b_i - \beta_i)(b_j - \beta_j)\right] \prod_{i=1}^{p} db_i.$$

This is a multivariate normal distribution. Now if we integrate over $b_1, \cdots, b_{i-1}$, $b_{i+1}, \cdots, b_p$, we get the marginal distribution of b_i as $N(\beta_i, \sigma^2 S^{ii})$. Hence under $H_\circ$, b_i is

$N(\beta_i^0, \sigma^2 S^{ii})$ and since $\sum\limits_{i=p+2}^{n} u_i^2/\sigma^2$ is always a central $\chi^2_{(n-p-1)}$, under $H_\circ$, we have

$$t = \{(b_i - \beta_i^0)/\sigma\sqrt{S^{ii}}\} \Big/ \sqrt{\sum_{i=p+2}^{n} u_i^2/\sigma^2(n-p-1)} = \frac{b_i - \beta_i^0}{s_e\sqrt{S^{ii}}} \text{ follows } t_{(n-p-1)}.$$

To test regarding a linear function of β's, say

$$H_\circ : \sum_{i=1}^{p} l_i\beta_i = \gamma_0 :$$

$\Sigma l_i b_i$ being a linear function of p-variables $b_1, \cdots, b_p$ whose joint distribution is p-variate normal, distribution of $\Sigma l_i b_i$ is univariate normal with mean $\Sigma l_i\beta_i$ and variance

$$\sigma^2 \sum_i \sum_j l_i l_j S^{ij}.$$

Hence as a test criterion we use

$$t = \frac{\Sigma l_i b_i - \gamma_0}{s_e\sqrt{\sum\limits_i \sum\limits_j l_i l_j S^{ij}}} = \frac{\Sigma l_i(b_i - \beta_i)}{s_e\sqrt{\sum\limits_i \sum\limits_j l_i l_j S^{ij}}}$$

which follows a student's $t_{(n-p-1)}$.

To test $H_\circ : l_0\alpha + \Sigma l_i\beta_i = \delta_0$. Here $l_0\bar{y} + \Sigma l_i b_i$ is

$$N\left(l_0\alpha + \Sigma l_i\beta_i,\ \sigma^2\left(\frac{l_0^2}{n} + \sum_{ij=1}^{p}\sum l_i l_{\bar{y}} S^{ij}\right)\right);$$

hence, under $H_\circ$,

$$\frac{l_0\bar{y} + \sum l_i b_i - \delta_0}{\sigma\sqrt{\dfrac{l_0^2}{n} + \sum\limits_{ij=1}^{p}\sum l_i l_j S^{ij}}} \sim N(0,1).$$

Therefore, to test the hypothesis we use the test statistic

$$t = \frac{l_0\bar{y} + \sum l_i b_i - \delta_0}{s_e\sqrt{\dfrac{l_0^2}{n} + \sum\limits_{ij=1}^{p}\sum l_i l_j S^{ij}}}$$

with d.f. $(n-p-1)$.

To test $H_\circ : \beta_1 = \cdots = \beta_p = 0$ i.e., $H_\circ : \rho_{y.12\cdots p} = 0$.

Now

$$dF(\boldsymbol{b}) = \frac{\left|\frac{S_{ij}}{\sigma^2}\right|^{1/2}}{(2\pi)^{p/2}} \exp\left\{-\frac{1}{2\sigma^2}\sum_{ij=1}^{p}\sum(b_i-\beta_i)(b_j-\beta_j)S_{ij}\right\}\prod_{i=1}^{p} db_i.$$

$$\therefore \quad \frac{1}{\sigma^2}\sum_{ij=1}^{p}\sum(b_i-\beta_i)(b_j-\beta_j)S_{ij} \sim \text{Central } \chi^2_{(p)}.$$

Hence under $H_\circ$,

$$\frac{1}{\sigma^2}\sum_{ij=1}^{p}\sum b_i b_j S_{ij} \text{ is } \chi^2_{(p)}. \quad \therefore \quad F = \frac{\sum_{ij=1}^{p}\sum b_i b_j S_{ij}/p}{S_e^2/(n-p-1)}$$

follows central $F_{p,n-p-1}$.

Again, square of the sample multiple correlation coefficient,

$$R^2_{y.12\cdots p} = \frac{S.S. \text{ due to regression}}{\text{Total } S.S.} = \frac{\sum_i\sum_j b_i b_j S_{ij}}{S_{yy}}.$$

$$\therefore \quad \sum_i\sum_j b_i b_j S_{ij} = R^2_{y.12\cdots p}S_{yy}.$$

$$\therefore \quad F = \frac{R^2_{y.12\cdots p}/p}{(1-R^2_{y.12\cdots p})/(n-p-1)} \text{ with } (p, n-p-1) \text{ d.f.}$$

To test $H_\circ : \beta_i = 0$ i.e., $H_\circ : \rho_{yi.12\cdots(i-1)(i+1)\cdots p} = 0$.

The test statistic is

$$t = \frac{b_i}{s_e\sqrt{S^{ii}}} = \frac{r_{yi.12\cdots(i-1)(i+1)\cdots p}}{\sqrt{1-r^2_{yi.12\cdots\overline{i-1}\,\overline{i+1}\cdots p}}}\sqrt{n-p-1}$$

which follows a central $t_{(n-p-1)}$.

To test $H_\circ(\hat{Y}_\xi = \eta)$ i.e. the expected value of Y corresponding to a given set of values of $\boldsymbol{\xi} = (\xi_1, \cdots, \xi_p)$ of $\boldsymbol{X} = (\boldsymbol{X}_1, \cdots, \boldsymbol{X}_p)$ is η [vide (14.3.3)].

Consider the statistic $\tilde{Y}_\xi = \bar{y} + b_1(\xi_1 - \bar{x}_1) + \cdots + b_p(\xi_p - \bar{x}_p)$. Under $H_\circ$, $E(\tilde{Y}_\xi) = \eta$

$$\begin{aligned} V(\tilde{Y}_\xi) &= V(\bar{y}) + V\left\{\sum_{i=1}^{p} b_i(\xi_i - \bar{x}_i)\right\} \\ &= \frac{\sigma^2}{n} + \sigma^2\sum_{ij=1}^{p}\sum\{(\xi_i - \bar{x}_i)(\xi_j - \bar{x}_j)\}S^{ij} \\ &= \sigma^2\left[\frac{1}{n} + \sum_{ij=1}^{p}\sum(\xi_i - \bar{x}_i)(\xi_j - \bar{x}_j)S^{ij}\right]. \end{aligned}$$

Under $H_\circ$,

$$\frac{\tilde{Y}_\xi - \eta}{\sigma\sqrt{\frac{1}{n} + \sum_i \sum_j (\xi_i - \bar{x}_i)(\xi_j - \bar{x}_j)S^{ij}}} \sim N(0,1)$$

and we use it as our test criterion.

If however σ^2 is unknown we replace it by its unbiased estimate s_e^2 and the statistic

$$t = \frac{\tilde{Y}_\xi - \eta}{s_e\sqrt{\frac{1}{n} + \sum_i \sum_j (\xi_i - \bar{x}_i)(\xi_j - \bar{x}_j)S^{ij}}}$$

follows a Student's t-distribution with $(n-p-1)$ d.f.

In the above case the distribution of the statistic t is a conditional one for fixed $X_1, \cdots, X_p$. However, since the distribution of t is independent of x's the statistic t follows student's t, whatever the distribution of $X_1, \cdots, X_n$. However when $X_1 = \bar{x}_1, \cdots, X_p = \bar{x}_p$ the estimate $\hat{Y}$ has the smallest variance; because in that case the estimate of Y is $\bar{y}$ and its variance is $\frac{\sigma^2}{n}$. In all cases

$$V(\tilde{Y}_x) = \frac{\sigma^2}{n} + \sigma^2 \sum \sum_{ij=1}^{p} (x_i - \bar{x}_i)(x_j - \bar{x}_j)S^{ij}$$

is a p.d. quadratic form.

Now we consider the test statistic under alternative hypotheses.

(a) for testing $H_\circ : \alpha = 0$, $t = \frac{\sqrt{n}\bar{y}}{s_e}$. It follows non-central t distribution under alternative hypothesis with non-centrality parameter $\lambda = \frac{\sqrt{n}\alpha}{\sigma}$ (for given $x_1, \cdots, x_p$). Since λ does not involve x's in any way, the unconditional distribution of t (for given x's) remains non-central t under alternative hypothesis whatever the distribution of x's.

(b) for testing $H_\circ : \beta_i = 0$ use $t = \frac{b_i}{s_e\sqrt{S^{ii}}}$. Under alternative hypothesis it follows non-central t with non-centrality parameter $\lambda = \beta_i/\sigma\sqrt{S^{ii}}$. However, λ involves x's (in S^{ii}). Therefore, the unconditional distribution of t will not remain a non-central t under alternative hypothesis.

(c) for testing $H_\circ : \beta_1 = \cdots = \beta_p = 0$ or $H_\circ : \rho_{y.12\cdots p} = 0$ use,

$$F = \frac{\sum_i \sum_j b_i b_j S_{ij}/p}{s_e^2};$$

under alternative hypothesis, denominator remains central χ^2, but the numerator becomes non-central χ^2 with non-centrality parameter

$$\lambda = \frac{1}{\sigma^2} \sum_i \sum_j \beta_i \beta_j S_{ij}.$$

Hence under alternative hypothesis, $\frac{R^2}{1-R^2} \cdot \frac{n-p-1}{p}$ is a non-central F with non-centrality parameter λ. This is a conditional distribution. The unconditional distribution will evidently be different from this non-central F, since λ involves x's.

Example 14.3.1 *If $\boldsymbol{X} = (X_1, \cdots, X_p)'$ follows $N_p(\boldsymbol{m}, \boldsymbol{\Sigma})$ then show that distribution of $\Sigma_i \, \Sigma_j \, \sigma^{ij} \, X_i \, X_j$ is a non-central $\chi^2_{(p)}$, if $\boldsymbol{m} \neq \boldsymbol{0}$.*

The distribution of $\boldsymbol{X}$ is

$$\frac{1}{(2\pi)^{p/2}\sqrt{1\Sigma 1}} \exp\left[-\frac{1}{2}\sum_i\sum_j \sigma^{ij}(x_i - m_i)(x_j - m_j)\right] \prod dx_i. \tag{i}$$

Since $\boldsymbol{\Sigma} = ((\sigma_{ij}))$ is p.d. matrix there exists $\boldsymbol{L}$ such that $\boldsymbol{L}'\boldsymbol{\Sigma}\boldsymbol{L} = \boldsymbol{I}_{p\times p}$.
Let us consider the transformation $\boldsymbol{X} = \boldsymbol{L}^{-1}\boldsymbol{Y}$; let $\boldsymbol{m} = \boldsymbol{L}^{-1}\boldsymbol{\xi}$.
Hence exponent in (i) is

$$\begin{aligned}
-\frac{1}{2}(\boldsymbol{X} - \boldsymbol{m})'\boldsymbol{\Sigma}^{-1}(\boldsymbol{X} - \boldsymbol{m}) &= -\frac{1}{2}(\boldsymbol{Y} - \boldsymbol{\xi})'(\boldsymbol{L}^{-1})'\boldsymbol{\Sigma}^{-1}\boldsymbol{L}^{-1}(\boldsymbol{Y} - \boldsymbol{\xi}) \\
&= -\frac{1}{2}(\boldsymbol{Y} - \boldsymbol{\xi})'(\boldsymbol{L}'\boldsymbol{\Sigma}\boldsymbol{L})^{-1}(\boldsymbol{Y} - \boldsymbol{\xi}) \\
&= -\frac{1}{2}(\boldsymbol{Y} - \boldsymbol{\xi})'(\boldsymbol{Y} - \boldsymbol{\xi}) \\
&= -\frac{1}{2}\Sigma(y_i - \xi_i)^2.
\end{aligned}$$

Also $d\boldsymbol{x} = |\boldsymbol{L}^{-1}|\, d\boldsymbol{y}$. Hence joint distribution of $\boldsymbol{Y}$ is

$$\begin{aligned}
dF(\boldsymbol{y}') &= \frac{1}{(2\pi)^{p/2}\,|\boldsymbol{\Sigma}|^{1/2}} e^{-\frac{1}{2}\Sigma(y_i - \xi_i)^2} \frac{1}{|\boldsymbol{L}|}\, d\boldsymbol{y} \\
&= \frac{1}{(2\pi)^{p/2}} e^{-\frac{1}{2}\Sigma(y_i - \xi_i)^2}\, d\boldsymbol{y} \quad [\because \ \boldsymbol{L}\boldsymbol{\Sigma}\boldsymbol{L}' = I].
\end{aligned}$$

Hence Y_i's $(i = 1, \cdots, p)$ are independent $N(\xi_i, 1)$. So $\sum_1^p Y_i^2$ is distributed as non-central $\chi^2_{(p)}$ with the non-centrality parameter (n.c.p.)

$$\lambda = \sum_1^p \xi_i^2.$$

Now

$$\begin{aligned}
\sum_1^p Y_i^2 &= \boldsymbol{Y}'\boldsymbol{Y} = \boldsymbol{X}'\boldsymbol{\Sigma}^{-1}\boldsymbol{X} = \sum_i\sum_j \sigma^{ij} X_i X_j \\
\text{and } \lambda &= \boldsymbol{\xi}'\boldsymbol{\xi} = \boldsymbol{m}'\boldsymbol{\Sigma}^{-1}\boldsymbol{m} = \sum_i\sum_j \sigma^{ij} m_i m_j.
\end{aligned}$$

Hence

$$\sum_i \sum_j \sigma^{ij} X_i X_j$$

follows non-central $\chi^2_{(p)}$ with n.c.p.

$$\sum_i \sum_j \sigma^{ij} m_i m_j.$$

14.4 Distribution of Sample Multiple Correlation Coefficient

We consider here the distribution of sample multiple correlation coefficient.

Case 1. Population multiple correlation coefficient $\rho_{y.12\cdots p} = 0$.

Testing $\beta_1 = \beta_2 = \cdots = \beta_p = 0 \equiv H_\circ : \rho^2_{y.12\cdots p} = 0$.

Under $H_\circ$, $\frac{1}{\sigma^2} \sum_i \sum_j b_i b_j S_{ij}$ is a central $\chi^2_{(p)}$ and S_e^2/σ^2 is always a central $\chi^2_{(n-p-1}$ and these two $\chi^2 s$ are independent. Hence under $H_\circ$,

$$F = \frac{\sum_i \sum_j S_{ij} b_i b_j / p}{S_e^2/(n-p-1)} = \sum_i \sum_j \frac{S_{ij} b_i b_j}{S_e^2} \cdot \frac{n-p-1}{p}$$

is a central $F_{(p,n-p-1)}$.

Now distribution of a central F is

$$dF = \frac{1}{B\left(\frac{p}{2}, \frac{n-p-1}{2}\right)} \frac{\left(\frac{p}{n-p-1} F\right)^{\frac{p}{2}-1}}{\left(1 + \frac{p}{n-p-1} F\right)^{\frac{n-1}{2}}} d\left(\frac{p}{n-p-1} F\right). \tag{14.4.1}$$

Again,

$$F = \sum_i \sum_j \frac{S_{ij} b_i b_j}{S_e^2} \cdot \frac{n-p-1}{p} = \frac{R^2 S_{yy}}{(1-R^2) S_{yy}} \frac{n-p-1}{p}.$$

$$\frac{dF}{dR} = \frac{n-p-1}{p} \cdot \frac{1}{(1-R^2)^2}.$$

Hence,

$$dF(R^2 | \rho_{y.12\cdots p} = 0)$$

$$= \frac{1}{B\left(\frac{p}{2}, \frac{n-p-1}{2}\right)} \frac{\left(\frac{R^2}{1-R^2}\right)^{\frac{p}{2}-1}}{\left(1 + \frac{R^2}{1-R^2}\right)^{\frac{n-1}{2}}} \cdot \frac{p}{n-p-1} \cdot \frac{n-p-1}{p} \cdot (1-R^2)^{-2} dR^2$$

$$= \frac{1}{B\left(\frac{p}{2}, \frac{n-p-1}{2}\right)} (R^2)^{\frac{p}{2}-1} (1-R^2)^{\frac{n-p-1}{2}-1} dR^2. \tag{14.4.2}$$

This is the conditional distribution of R^2 when $X_1, \cdots, X_p$ are fully fixed. Since it does not involve x's it is also the unconditional distribution of R^2.

Case 2. $\rho_{y.12\cdots p} \neq 0$.

The joint distribution of $\bar{y}, b_1, \cdots, b_p, S_e^2$ is given by

$$\frac{\sqrt{n}}{\sigma\sqrt{2\pi}} e^{-\frac{n}{2\sigma^2}(\bar{y}-\alpha)^2} d\bar{y} \times \frac{\sqrt{|S_{ij}|}}{(\sigma^2\pi)^{p/2}} \exp\left(-\frac{1}{2\sigma^2}\sum_i\sum_j S^{ij}(b_i-\beta_i)(b_j-\beta_j)\right)\prod_{i=1}^{p} db_i$$

$$\times \frac{1}{2^{\frac{n-p-1}{2}}\left\lceil\frac{n-p-1}{2}\right.} e^{-\frac{1}{2}\frac{S_e^2}{\sigma^2}} \left(\frac{S_e^2}{\sigma^2}\right)^{\frac{n-p-3}{2}} d\left(\frac{S_e^2}{\sigma^2}\right).$$

When $H_o(\rho_{y.12\cdots p} = 0)$ is not tenable, $\chi_1^2 = \frac{\Sigma\Sigma S_{ij} b_i b_j}{\sigma^2}$ is a non-central (n.c.) $\chi^2_{(p)}$ and non-centrality parameter

$$\theta = \sum_i\sum_j S_{ij}\beta_i\beta_j/\sigma^2,$$

$\frac{S_e^2}{\sigma^2}$ is always a central $\chi^2_{(n-p-1)} = \chi_2^2$ (say). χ^2's are independent of each other. Hence,

$$dF\left(\chi_1^2, \chi_2^2\right) = e^{-\frac{\theta}{2}} \sum_{i=0}^{\alpha}\left(\frac{\theta}{2}\right)^i \frac{1}{i!}\frac{1}{2^{\frac{2i+p}{2}}\left\lceil\frac{2i+p}{2}\right.} e^{-\frac{1}{2}\chi_1^2}\left(\chi_1^2\right)^{\frac{2i+p-2}{2}}$$

$$d\chi_1^2 \cdot \frac{1}{2^{\frac{n-p-1}{2}}\left\lceil\frac{n-p-1}{2}\right.} e^{-\frac{1}{2}\chi_2^2}\left(\chi_2^2\right)^{\frac{n-p-1}{2}-1} d\chi_2^2$$

$$= e^{-\frac{\theta}{2}} \sum_{i=0}^{\infty}\frac{1}{i!}\left(\frac{\theta}{2}\right)^i \frac{1}{2^{(2i+n-1)/2}\left\lceil\frac{p}{2}+i\right.\left\lceil\frac{n-p-1}{2}\right.} e^{-\frac{1}{2}\left(\chi_1^2+\chi_2^2\right)}$$

$$\left(\chi_1^2\right)^{\frac{p}{2}+i-1}\left(\chi_2^2\right)^{\frac{n-p-1}{2}-1} d\chi_1^2\, d\chi_2^2. \qquad (14.4.3)$$

Now by definition,

$$R^2 = \frac{SS \text{ due to regression}}{TSS} = \frac{\sum_i\sum_j S_{ij}b_ib_j}{\Sigma(y_i-\bar{y})^2}$$

$$= \frac{\sum_i\sum_j S_{ij}b_ib_j}{S_e^2 + \sum_i\sum_j S_{ij}b_ib_j} = \frac{\chi_1^2}{\chi_1^2+\chi_2^2} \qquad (14.4.4)$$

and

$$\rho^2 = \frac{SS \text{ due to regression in the population}}{TSS \text{ in the population}}$$

$$= \frac{\Sigma\Sigma S_{ij}\beta_i\beta_j/(n-1)}{\sigma^2 + \frac{\Sigma\Sigma S_{ij}\beta_i\beta_j}{n-1}} \quad \text{(note that } x\text{'s are fixed)}$$

$$= \frac{\theta/(n-1)}{1+\theta/(n-1)}. \quad \therefore \quad \theta = \frac{\rho^2(n-1)}{1-\rho^2}; \quad \rho = 0 \Rightarrow \theta = 0.$$

To find the distribution of R^2 we are to integrate (14.4.3). Since the series is uniformly convergent one can perform integration term by term. The ith term is

$$e^{-\theta/2}\left(\frac{\theta}{2}\right)^i \frac{1}{i!}\frac{1}{2^{i+\frac{n-1}{2}}\,\overline{\left|\frac{p}{2}+i\right.}\,\overline{\left|\frac{n-p-1}{2}\right.}}$$

$$\cdot e^{-\frac{1}{2}(\chi_1^2+\chi_2^2)}(\chi_1^2)^{\frac{p}{2}+i-1}(\chi_2^2)^{\frac{n-p-3}{2}}\,d\chi_1^2\,d\chi_2^2. \qquad (14.4.5)$$

Now

$$R^2 = \frac{\chi_1^2}{\chi_1^2+\chi_2^2}. \quad \therefore \quad \chi_1^2(1-R^2) = R^2\chi_2^2.$$

$$J\left(\frac{\chi_1^2, \chi_2^2}{R^2, \chi_2^2}\right) = \begin{vmatrix} \chi_2^2/(1-R^2)^2 & 0 \\ 0 & 1 \end{vmatrix} = \frac{\chi_2^2}{(1-R^2)^2}.$$

Hence, after transformation (14.4.5) becomes

$$\frac{e^{-\theta/2}}{2^{i+\frac{n-1}{2}}\,\overline{\left|\frac{p}{2}+i\right.}\,\overline{\left|\frac{n-p-1}{2}\right.}\,i!}\left(\frac{\theta}{2}\right)^i e^{-\frac{1}{2}\left(\frac{R^2}{1-R^2}+1\right)\chi_2^2}.$$

$$\left(\frac{R^2}{1-R^2}\chi_2^2\right)^{\frac{p}{2}+i-1}\left(\chi_2^2\right)^{\frac{n-p-3}{2}}\frac{\chi_2^2}{(1-R^2)^2}\,d\chi_2^2\,dR^2$$

$$= \frac{e^{-\theta/2}}{2^{i+\frac{n-1}{2}}\,\overline{\left|\frac{p}{2}+i\right.}\,\overline{\left|\frac{n-p-1}{2}\right.}}\left(\frac{\theta}{2}\right)^i \frac{1}{i!}e^{-\frac{\chi_2^2}{2(1-R^2)}}\left[\frac{1}{1-R^2}\right]^{\frac{p}{2}+i+1}$$

$$(R^2)^{\frac{p}{2}+i-1}(\chi_2^2)^{\frac{n-3}{2}+i}\,dR^2\,d\chi_2^2.$$

Integrating over χ_2^2 we get the ith term in the distribution of R^2 as

$$\frac{e^{-\theta/2}(R^2)^{\frac{p}{2}+i-1}}{2^{\frac{n-1}{2}+i}\,\overline{\left|\frac{p}{2}+i\right.}\,\overline{\left|\frac{n-p-1}{2}\right.}}\left(\frac{\theta}{2}\right)^i\frac{1}{i!}\left(\frac{1}{1-R^2}\right)^{\frac{p}{2}+i-1}dR^2$$

$$\int_0^\infty e^{-\frac{\chi_2^2}{2(1-R^2)}}(\chi_2^2)^{\frac{n-1}{2}+i-1}\,d\chi_2^2$$

$$= \frac{e^{-\theta/2}}{2^{\frac{n-1}{2}+i}\Gamma\left(\frac{p}{2}+i\right)\Gamma\left(\frac{n-p-1}{2}\right)}\left(\frac{\theta}{2}\right)^i \frac{1}{i!}(R^2)^{\frac{p}{2}+i-1}$$

$$\times \frac{1}{(1-R^2)^{\frac{p}{2}+i+1}}\, dR^2 \times \frac{\Gamma\left(\frac{n-1}{2}+i\right)}{\left[\frac{1}{2(1-R^2)}\right]^{\frac{n-1}{2}+i}}$$

$$= \frac{e^{-\theta/2}\Gamma\left(\frac{n-1}{2}+i\right)}{\Gamma\left(\frac{p}{2}+i\right)\Gamma\left(\frac{n-p-1}{2}\right)}\left(\frac{\theta}{2}\right)^i \frac{1}{i!}(R^2)^{\frac{p}{2}+i-1}(1-R^2)^{\frac{n-p-3}{2}}\, dR^2$$

$$= \frac{e^{-\theta/2}}{B\left(\frac{p}{2}+i, \frac{n-p-1}{2}\right)}\left(\frac{\theta}{2}\right)^i \frac{1}{i!}(R^2)^{\frac{p}{2}+i-1}(1-R^2)^{\frac{n-p-3}{2}}\, dR^2.$$

Hence from (14.4.3)

$$dF(R^2/\theta) = \sum_{i=0}^{\infty} \frac{e^{-\theta/2}}{B\left(\frac{p}{2}+i, \frac{n-p-1}{2}\right)}\left(\frac{\theta}{2}\right)^i \frac{1}{i!}(R^2)^{\frac{p}{2}+i-1}(1-R^2)^{\frac{n-p-1}{2}-1}\, dR^2 \qquad (14.4.6)$$

hen

$$\theta = \frac{1}{\sigma^2}\sum_i \sum_j S_{ij}\beta_i\beta_j.$$

This is the conditional distribution when x's are kept fixed.

Suppose $(X_1, \cdots, X_p)'$ follow $N_p(\boldsymbol{m}, \Sigma)$. In order to get unconditional distribution we re to multiply $dF(R^2/\theta)$ by the joint distribution of X's and integrate the product over X's. But the conditional distribution of R^2 depends on X's only through θ. So we can nultiply $dF(R^2/\theta)$ by the distribution of θ and integrate over θ to get the unconditional istribution of R^2. Let

$$\boldsymbol{X} = \begin{bmatrix} x_{11} & x_{21} & \cdots & x_{p1} \\ \cdot & \cdot & \cdots & \cdot \\ \cdot & \cdot & \cdots & \cdot \\ \cdot & \cdot & \cdots & \cdot \\ x_{1n} & x_{2n} & \cdots & x_{pn} \end{bmatrix}.$$

We apply an orthogonal transformation transforming

$$\begin{bmatrix} x_{i1} \\ x_{i2} \\ \vdots \\ x_{in} \end{bmatrix} \text{ to } \begin{bmatrix} u_{i1} \\ u_{i2} \\ \vdots \\ u_{in} \end{bmatrix}$$

vhen $u_{i1} = \sqrt{n}\bar{x}_i, u_{i\lambda}$'s, $\lambda = 2, \cdots, n$ so chosen so as to make the whole set of ransformations orthogonal $(i = 1, \cdots, p)$.

Therefore,

$$\sum_{\lambda=1}^{n}(x_{i\lambda}-m_i)(x_{j\lambda}-m_j) = \sum x_{i\lambda}x_{j\lambda} - m_i\sum x_{j\lambda} - m_j\sum x_{i\lambda} + nm_im_j$$

$$= \sum u_{i\lambda}u_{j\lambda} - \sqrt{n}m_i u_{j1} - \sqrt{n}m_j u_{i1} + nm_i m_j$$

$$= \sum_{\lambda=2}^{n} u_{i\lambda}u_{j\lambda} + (u_{i1} - \sqrt{n}m_i)(u_{j1} - \sqrt{n}m_j)$$

$$= \sum_{\lambda=2}^{n} u_{i\lambda}u_{j\lambda} + n(\bar{x}_i - m_1)(\bar{x}_j - mj).$$

$$\therefore \sum_{\lambda=2}^{n} u_{i\lambda}u_{j\lambda} = \sum_{\lambda=1}^{n}(x_{i\lambda} - m_i)(x_{j\lambda} - m_j) - n(\bar{x}_i - m_i)(\bar{x}_j - m_j)$$

$$= \sum_{\lambda=1}^{n}(x_{i\lambda} - \bar{x}_i)(x_{j\lambda} - \bar{x}_j) = S_{ij}.$$

$$\therefore \theta = \frac{1}{\sigma^2}\sum_i\sum_j S_{ij}\beta_i\beta_j = \frac{1}{\sigma^2}\sum_{\lambda=2}^{n}\sum_{ij=1}^{p} u_{i\lambda}u_{j\lambda}\beta_i\beta_j$$

$$= \frac{1}{\sigma^2}\sum_{\lambda=2}^{n}\left\{\sum_{i=1}^{p} u_{i\lambda}\beta_i\right\}^2. \qquad (14.4.7)$$

Now each $u_{i\lambda}(i = 1, \cdots, p;\ \lambda = 2, \cdots, n)$ is univariate normal [since $u_{i\lambda}$ is a linea function of $x_{i\lambda}$].

$$\therefore \sum_{i=1}^{p} u_{i\lambda}\beta_i, \ \ \lambda = 2, \cdots, n$$

is univariate normal with $E(\Sigma u_{i\lambda}\beta_i) = 0$ and

$$V\left(\sum_i u_{i\lambda}\beta_i\right) = \sum_i\sum_j \sigma_{ij}\beta_i\beta_j. \quad \therefore \quad \frac{\sum_{\lambda=2}^{n}\left(\sum_{i=1}^{p} u_{i\lambda}\beta_i\right)^2}{\sum_i\sum_j \sigma_{ij}\beta_i\beta_j} = \text{Central } \chi^2_{(n-1)}. \qquad (14.4.8)$$

But l.h.s. of (14.4.8) is

$$\frac{\theta\sigma^2}{\rho^2\sigma_{yy}} = \frac{\theta}{\rho^2\sigma_{yy}\sigma^{yy}} \quad \text{[Compare with (10.7.34)].}$$

Again,

$$\frac{1}{\sigma_{yy}\sigma^{yy}} = (1 - \rho^2) \quad \text{[Compare with (10.8.5)].}$$

$\therefore\ \frac{\theta(1-\rho^2)}{\rho^2}$ is a central $\chi^2_{(n-1)}$ having distribution

$$\frac{1}{2^{\frac{n-1}{2}}\overline{\left|\frac{n-1}{2}\right.}} e^{-\theta(1-\rho^2)/2\rho^2}\left\{\frac{\theta(1-\rho^2)}{\rho^2}\right\}^{\frac{n-1}{2}-1} d\left\{\frac{\theta(1-\rho^2)}{\rho^2}\right\}. \qquad (14.4.9)$$

Iultiplying $dF(R^2/\theta)$ of (14.4.6) by (14.4.9) we get

$$dF(R^2,\theta) = \sum_{i=0}^{\infty} e^{-\theta/2} \frac{1}{B\left(\frac{p}{2}+i, \frac{n-p-1}{2}\right)} \left(\frac{\theta}{2}\right)^i \frac{1}{i!} (R^2)^{\frac{p}{2}+i-1} (1-R^2)^{\frac{n-p-1}{2}-1}$$
$$\times \frac{1}{2^{\frac{n-1}{2}} \overline{\left| \frac{n-1}{2} \right.}} e^{-\frac{\theta(1-\rho^2)}{2\rho^2}} \left\{\frac{\theta(1-\rho^2)}{\rho^2}\right\}^{\frac{n-1}{2}-1} d\left\{\frac{\theta(1-\rho^2)}{\rho^2}\right\}.$$

We have to integrate it over the range of θ viz. 0 to ∞.

Due to uniform convergence of the series term by term integration is permissible. 'he ith term is

$$\frac{1}{2^{\frac{n-1}{2}} \overline{\left| \frac{n-1}{2} \right.}} \frac{1}{B\left(\frac{p}{2}+i, \frac{n-p-1}{2}\right)} \frac{1}{i!} (R^2)^{\frac{p}{2}+i-1} (1-R^2)^{\frac{n-p-1}{2}-1} \, dR^2 \times I \tag{14.4.10}$$

'here

$$I = \int_0^\infty e^{-\theta/2} \left(\frac{\theta}{2}\right)^i e^{-\frac{\theta(1-\rho^2)}{2\rho^2}} \left\{\frac{\theta(1-\rho^2)}{\rho^2}\right\}^{\frac{n-1}{2}-1} d\left\{\frac{\theta(1-\rho^2)}{\rho^2}\right\}. \tag{14.4.11}$$

Iow

$$I = \left(\frac{1}{2}\right)^i (1-\rho^2)^{\frac{n-1}{2}} \left(\frac{1}{\rho}\right)^{\frac{n-1}{2}} \frac{\overline{\left| \frac{n-1}{2}+i \right.}}{\left(\frac{1}{2}\rho^2\right)^{\frac{n-1}{2}+i}}$$
$$= 2^{-i} (1-\rho^2)^{\frac{n-1}{2}} (\rho^2)^{-\frac{n-1}{2}} 2^{\frac{n-1}{2}+i} (\rho^2)^{\frac{n-1}{2}+i} \overline{\left| \frac{n-1}{2}+i \right.}$$
$$= 2^{\frac{n-1}{2}} (1-\rho^2)^{\frac{n-1}{2}} (\rho^2)^i \overline{\left| \left(\frac{n-1}{2}+i\right) \right.}. \tag{14.4.12}$$

Ience we have

$$dF(R^2) = \sum_{i=0}^{\infty} \frac{1}{2^{\frac{n-1}{2}} \overline{\left| \frac{n-1}{2} \right.}} \frac{1}{i!} \frac{1}{B\left(\frac{p}{2}+i, \frac{n-p-1}{2}\right)} (R^2)^{\frac{p}{2}+i-1}$$
$$(1-R^2)^{\frac{n-p-1}{2}-1} 2^{\frac{n-1}{2}} (\rho^2)^i (1-\rho^2)^{\frac{n-1}{2}} \overline{\left| \frac{n-1}{2}+i \right.} \, dR^2$$
$$= \frac{1}{\overline{\left| \frac{n-1}{2} \right.} \overline{\left| \frac{n-p-1}{2} \right.}} (1-\rho^2)^{\frac{n-1}{2}} (1-R^2)^{\frac{n-p-3}{2}} (R^2)^{\frac{p}{2}-1}$$
$$\sum_{i=0}^{\infty} \frac{\overline{\left| \frac{n-1}{2}+i \right.} \overline{\left| \frac{n-1}{2}+i \right.}}{\overline{\left| \frac{p}{2}+i \right.}} \frac{(R^2)^i}{i!} (\rho^2)^i \, dR^2$$
$$= \frac{1}{\overline{\left| \frac{n-1}{2} \right.} \overline{\left| \frac{n-p-1}{2} \right.}} (1-\rho^2)^{\frac{n-1}{2}} (1-R^2)^{\frac{n-p-3}{2}} (R^2)^{\frac{p}{2}-1} \, dR^2$$
$$\left[\sum_{i=0}^{\infty} \frac{\left\{\left(\frac{n-1}{2}+i-1\right) \cdots \left(\frac{n-1}{2}+1\right) \left(\frac{n-1}{2}\right) \left(\overline{\left| \frac{n-1}{2} \right.}\right)\right\}^2}{\left\{\left(\frac{p}{2}+i-1\right) \cdots \left(\frac{p}{2}+1\right) \frac{p}{2} \overline{\left| \frac{p}{2} \right.}\right\}} \frac{(R^2\rho^2)^i}{i!} \right]$$

whence the quantity within [ ] is a confluent hypergeometric series 2^{F_1}

$$\left[2^{F_1}(a,c,b,x) = 1 + \frac{ac}{b}x + \frac{a(a+1)c(c+1)}{b(b+1)}\frac{x^2}{2!} + \cdots \right.$$
$$\left. + \frac{a(a+1)\cdots(a+i-1)c(c+1)\cdots(c+i-1)}{b(b+1)\cdots(b+i-1)}\frac{x^i}{i!} + \cdots \right].$$

14.5 Sampling Distribution of Intraclass Correlation Coefficient

Suppose we have a sample from a $N_p(m\mathbf{1}, \mathbf{\Sigma})$ population where $\mathbf{\Sigma} = ((\sigma_{ij}))$ is such th

$$\rho_{ij} = \rho\sigma^2 \quad \forall\, i \neq j = 1, \cdots, p.$$

Then

$$|\Sigma| = \sigma^{2p} \begin{vmatrix} 1 & \rho & \cdots & \rho \\ \cdot & \cdot & \cdots & \cdot \\ \rho & \rho & \cdots & 1 \end{vmatrix} = \sigma^{2p} \begin{vmatrix} 1+(p-1)\rho & \rho & \cdots & \rho \\ 1+(p-1)\rho & 1 & \cdots & \rho \\ \cdots & \cdot & \cdots & \cdot \\ 1+(p-1)\rho & \rho & \cdots & 1 \end{vmatrix}$$

$$= \sigma^{2p}\{1+(p-1)\rho\} \begin{vmatrix} 1 & \rho & \cdots & \rho \\ \rho & 1 & \cdots & \rho \\ \cdot & \cdot & \cdots & \cdot \\ \rho & \rho & \cdots & 1 \end{vmatrix} = \sigma^{2p}\{1+(p-1)\rho\}(1-\rho)^{p-1}.$$

The cofactor for each diagonal element is

$$\sigma^{2(p-1)}\{1+(p-2)\rho\}(1-\rho)^{p-2}.$$

The cofactor for each non-diagonal element can be put in the form

$$-\sigma^{2(p-1)} \begin{vmatrix} \rho & \rho & \cdots & \rho \\ \rho & 1 & \cdots & \rho \\ \rho & \rho & \cdots & 1 \end{vmatrix},$$

either directly or after interchanging rows and columns and hence, is

$$\left[-\sigma^{2(p-1)}(1-\rho)^{p-2}\rho\right].$$

Therefore,

$$\sigma^{ii} = \frac{1+(p-2)\rho}{D}, \quad \sigma^{ij} = -\frac{\rho}{D}$$

where $D = \sigma^2(1+\overline{p-1}\rho)(1-\rho)$.

The density function of $(X_1, \cdots, X_p)$ is

$$f(x_1, \cdots, x_p) = \frac{1}{(2\pi)^{p/2}\sqrt{|\Sigma|}} \exp\left[-\frac{1}{2}\sum_i\sum_j \sigma^{ij}(x_i - m)(x_j - m)\right]. \qquad (14.5.1)$$

Now,

$$\begin{aligned}
&\sum_i\sum_j \sigma^{ij}(x_i - m)(x_j - m) \\
&= \frac{1+(p-2)\rho}{D}\sum_i (x_i - m)^2 - \frac{\rho \sum\sum_{i\neq j}(x_i - m)(x_j - m)}{D} \\
&= \frac{1}{\sigma^2(1-\rho)}\sum_{i=1}^{p}(x_i - m)^2 - \frac{p^2\rho}{\sigma^2(1-\rho)\{1+(p-1)\rho\}}(\bar{x} - m)^2.
\end{aligned}$$

Hence the p.d.f. (14.5.1) reduces to

$$\frac{1}{(2\pi)^{p/2}\left[\sigma^{2p}\{1+(p-1)\rho\}(1-\rho)^{p-1}\right]^{1/2}}$$
$$\exp\left[-\frac{1}{2\sigma^2(1-\rho)}\left\{\sum_{i=1}^{p}(x_i - m)^2 - \frac{p^2\rho}{1+(p-1)\rho}(\bar{x} - m)^2\right\}\right]. \qquad (14.5.2)$$

Let us now consider two linear functions of x's as $L_1 = \Sigma l_i x_i, L_2 = \Sigma l_i' x_i$. We shall prove that if L_1, L_2 are orthogonal linear functions and if at least one of them is orthogonal to $\bar{x}$, then L_1, L_2 are independently normally distributed. As $(X_1, \cdots, X_p)'$ follow N_p, the joint distribution of L_1, L_2 is bivariate normal. To show that they are independently distributed is to show that $\text{Cov}(L_1, L_2) = 0$. Now

$$\begin{aligned}
E\sum_i\sum_j l_i l_j'(X_i - m)(X_j - m) &= \rho\sigma^2 \sum\sum_{i\neq j} l_i l_j' + \sigma^2 \sum_i l_i l_i' \\
&= \sigma^2 \sum_i l_i l_i' - \rho\sigma^2 \sum_i l_i l_i' = 0 \;\; (\text{since } \Sigma_i l_i l_i' = 0).
\end{aligned}$$

We now consider orthogonal transformation $(x_1, \cdots, x_p) \to (u_1, \cdots, u_p)$ where $u_1 = \sqrt{p}\bar{x}$, $u_2, \cdots, u_p$ are suitably chosen. Then by the above result $u_2, \cdots, u_p$ are independently distributed. Also

$$\begin{aligned}
E(u_1) &= \sqrt{p}m, \\
E(u_i) &= 0, \;\; i \geq 2; \\
\sum_i (x_i - m)^2 &= (u_1 - \sqrt{p_1}m)^2 + \sum_2^p u_i^2 \\
p^2(\bar{x} - m)^2 &= p(u_1 - \sqrt{p}m)^2.
\end{aligned}$$

The exponential part of the density function in (14.5.2) simplifies to

$$\exp\left[-\frac{1}{2\sigma^2}\left\{\frac{(u_1 - m\sqrt{p})^2}{1+(p-1)\rho} + \frac{1}{1-\rho}\sum_2^p u_i^2\right\}\right].$$

Consider now n independent samples from this population

$$(x_{11}, \cdots, x_{p1})$$
$$\vdots$$
$$(x_{1n}, \cdots, x_{pn}),$$

each sample being considered as a family. Compute

$$\bar{x} = \bar{y} = \frac{(p-1)\sum_i\sum_j x_{ij}}{np(p-1)},$$

$$s_x^2 = s_y^2 = \frac{(p-1)\sum_i\sum_j (x_{ij} - \bar{x})^2}{np(p-1)}.$$

Now

$$\sum_{j=1}^n \sum_{i \neq i'=1}^p \sum (x_{ij} - \bar{x})(x_{i'j} - \bar{x}) = \sum_{j=1}^n \sum_{i=1}^p (x_{ij} - \bar{x}) \left[\sum_k (x_{kj} - \bar{x}) - (x_{ij} - \bar{x})\right]$$

$$= \sum_{j=1}^n \left[\sum_i (x_{ij} - \bar{x})\right]^2 - \sum_{j=1}^n \sum_{i=1}^p (x_{ij} - \bar{x})^2$$

$$= p^2 \sum_j (\bar{x}_j - \bar{x})^2 - \sum_{j=1}^n \sum_{i=1}^p (x_{ij} - \bar{x})^2.$$

The intraclass correlation coefficient is defined as

$$r_{\text{intra}} = r = \frac{p^2 \sum_{j=1}^n (\bar{x}_j - \bar{x})^2 - \sum_{i=1}^p \sum_{j=1}^n (x_{ij} - \bar{x})^2}{(p-1)\sum_i \sum_j (x_{ij} - \bar{x})^2}.$$

Defining

$$s_b^2 = p\sum_{j=1}^n (\bar{x}_j - \bar{x})^2/(n-1),$$

$$s_w^2 = \sum_i \sum_j (x_j - \bar{x}_i)^2/n(p-1),$$

$$r = \frac{p(n-1)s_b^2 - \{(n-1)s_b^2 + n(p-1)s_w^2\}}{(p-1)\{(n-1)s_b^2 + n(p-1)s_w^2\}}$$

$$= \frac{(n-1)s_b^2 - ns_w^2}{(n-1)s_b^2 + n(p-1)s_w^2} = \frac{(n-1)F' - n}{(n-1)F' + n(p-1)}$$

where

$$F' = \frac{n}{n-1} \cdot \frac{1+(p-1)r}{1-r} = \frac{s_b^2}{s_w^2}.$$

We now make orthogonal transformation

$$(x_{1\lambda}, x_{2\lambda}, \cdots, x_{p\lambda}) \to (u_{1\lambda}, u_{2\lambda}, \cdots, u_{p\lambda}) \quad (\lambda = 1, \cdots, n)$$

where $u_{1\lambda} = \sqrt{p}\bar{x}_\lambda$; $E(U_{1\lambda}) = \sqrt{p}m$.
Also $U_{1\lambda}, U_{2\lambda}, \cdots, U_{p\lambda}$ are each independently normally distributed. Again,

$$V(\sqrt{p}\bar{X}_\lambda) = p\left[\frac{\sigma^2}{p} + \frac{p-1}{p}\rho\sigma^2\right] = \sigma^2[1+(p-1)\rho].$$

Again $U_{11}, U_{12}, \cdots, U_{1n}$ are each independently normally distributed.
Therefore,

$$\sum_{\lambda=1}^{n} \frac{(U_{1\lambda} - \bar{U}_1)^2}{\sigma^2(1+\overline{p-1}\rho)} \cap \chi^2_{(n-1)}$$

where

$$\bar{U}_1 = \frac{1}{n}\sum_{\lambda=1}^{n} U_{1\lambda}.$$

Also

$$\sum_{i=2}^{p}\sum_{\lambda=1}^{n} U_{i\lambda}^2/\sigma^2(1-\rho)$$

follows independent $\chi^2_{n(p-1)}$ d.f.

$$\left[E\sum_{i=2}^{p} U_{i\lambda}^2 = E\left\{\sum_{i=1}^{p} X_{i\lambda}^2 - p\bar{X}_\lambda^2\right\} = \sum_{i=1}^{p}\left\{V(X_{i\lambda}) + m^2\right\} - p\left\{V(\bar{X}_\lambda) + m^2\right\}\right.$$
$$= p\sigma^2 - \sigma^2\{1+(p-1)\rho\} = \sigma^2[p-1-(p-1)\rho]$$
$$\left. = \sigma^2(p-1)(1-\rho);\ \text{hence } E(U_{i\lambda}^2) = \sigma^2(1-\rho)\right].$$

Again

$$\sum_{\lambda=1}^{n}(u_{1\lambda} - \bar{u}_1)^2 = p\sum_{\lambda=1}^{n}(\bar{x}_i - \bar{x})^2 = s_b^2(n-1).$$

Also

$$\sum_{\lambda=1}^{n}\sum_{i=2}^{p}\frac{u_{i\lambda}^2}{\sigma^2(1-\rho)} = \frac{\sum_{\lambda=1}^{n}\sum_{i=1}^{p}(x_{i\lambda} - \bar{x}_i)^2}{\sigma^2(1-\rho)} = \frac{n(p-1)s_w^2}{\sigma^2(1-\rho)}.$$

Thus,

$$\frac{s_b^2}{\sigma^2(1+\overline{p-1}\rho)} \div \frac{s_w^2}{\sigma^2(1-\rho)} = F'\frac{1-\rho}{1+\overline{p-1}\rho}$$

follows central $F_{(n-1),n(p-1)}$. The distribution of F' is

$$dF(F') = \frac{\Gamma\left(\frac{np-1}{2}\right)}{\Gamma\left(\frac{n-1}{2}\right)\Gamma\left(\frac{n(p-1)}{2}\right)} \frac{\left\{\dfrac{(n-1)(1-\rho)}{n(p-1)(1+\overline{p-1}\rho)}\right\}^{(n-1)/2} F'^{(n-3)/2}\, dF'}{\left\{1+\dfrac{(n-1)F'(1-\rho)}{n(p-1)\left\{1+(p-1)\rho\right\}}\right\}^{(n-p-1)/2}}.$$

Replacing F' by $\frac{n\{1+(p-1)r\}}{(n-1)(1-r)}$ one gets the distribution of intraclass correlation coefficient r as

$$dF(r) = \frac{\Gamma\left(\frac{np-1}{2}\right)}{\Gamma\left(\frac{n-1}{2}\right)\Gamma\left(\frac{n(p-1)}{2}\right)}$$

$$\frac{\left\{\frac{1-\rho}{1+(p-1)\rho}\cdot\frac{1}{p-1}\right\}^{(n-1)/2}\left\{\frac{1+r(p-1)}{1-r}\right\}^{(n-3)/2} \frac{p}{(1-r)^2}\,dr}{\left\{1+\frac{(n-1)}{n(p-1)}\cdot\frac{1-\rho}{1+(p-1)\rho}\cdot\frac{n}{n-1}\cdot\frac{1+\overline{p-1}r}{1-r}\right\}^{(np-1)/2}}.$$

When $\rho = 0$,

$$dF(r) = \frac{\Gamma\left(\frac{np-1}{2}\right)}{\Gamma\left(\frac{n-1}{2}\right)\Gamma\left(\frac{n(p-1)}{2}\right)} \frac{(p-1)^{n(p-1)/2}}{p^{(np-3)/2}}\{1+(p-1)r\}^{(n-3)/2}(1-r)^{n(p-1)/2-1}\,dr.$$

Putting $1+\overline{p-1}r = Z$, distribution of Z is

$$\beta_1\left(\frac{n-1}{2}, \frac{n(p-1)}{2}\right).$$

14.6 Sampling Distribution of the Generalised Variance

Suppose we have a sample of size n on each of p variables $X_1, \cdots, X_p$, $x_{i\lambda}$ being the λth observation on the variable X_i, $i = 1, \cdots, p$. Define

$$S_{ij} = \sum_{\lambda=1}^{n}(x_{i\lambda}-\bar{x}_i)(x_{j\lambda}-\bar{x}_j), \quad \bar{x}_i = \frac{1}{n}\sum_{\lambda=1}^{n} x_{i\lambda}, \quad \boldsymbol{S} = ((S_{ij}))_{p\times p}.$$

The sample generalised variance is defined as

$$|\boldsymbol{s}| = \left|\frac{1}{n-1}\boldsymbol{S}\right|.$$

Assume that the conditional distribution of X_1 given $X_2, \cdots, X_p$ is $N(\alpha+\beta_2 x_2+\cdots+\beta_p x_p; 1)$.

First let us find the sampling distribution of

$$\frac{|S_{ij}|}{|S_{ij\cdot 1}|} = \frac{1}{S^{11}} \tag{14.6.1}$$

where

$$|S_{ij\cdot 1}| = \begin{vmatrix} S_{22} & \cdots & S_{2p} \\ & & \vdots \\ S_{p2} & \cdots & S_{pp} \end{vmatrix}. \tag{14.6.2}$$

The vector of sample regression coefficients $\boldsymbol{b} = (b_2, \cdots, b_p)'$ is defined as

$$\begin{bmatrix} S_{22} & S_{23} & \cdots & S_{2p} \\ & & & \vdots \\ S_{p2} & S_{p3} & \cdots & S_{pp} \end{bmatrix} \begin{bmatrix} b_2 \\ \vdots \\ b_p \end{bmatrix} = \begin{bmatrix} S_{12} \\ \vdots \\ S_{1p} \end{bmatrix}, \text{ i.e., } ((\boldsymbol{S}_{ij\cdot 1}))\boldsymbol{b} = ((S_{1j})) \text{ (say).} \tag{14.6.3}$$

Consider

$$\begin{aligned} &\sum_{\lambda=1}^{n} \{x_{1\lambda} - \bar{x}_1 - b_2(x_{2\lambda} - \bar{x}_2) - \cdots - b_p(x_{p\lambda} - \bar{x}_p)\}^2 \\ &= S_{11} - b_2 S_{12} - \cdots - b_p S_{1p} \\ &= S_{11} - ((S_{ij}))'((S_{ij\cdot 1}))^{-1}((S_{1j})) \\ &= S_{11} - \sum_{ij=2}^{p}\sum S_{1i}S_{1j}S^{ij\cdot 1}. \end{aligned} \tag{14.6.4}$$

Again

$$S^{ij\cdot 1} = \frac{T_{ij\cdot 1}}{|S_{ij\cdot 1}|},$$

where $T_{ij\cdot 1}$ is the cofactor of $S_{ij\cdot 1}$ in $((S_{ij\cdot 1}))$.
Therefore,

$$\begin{aligned} (14.6.4) &= S_{11} - \frac{1}{|S_{ij\cdot 1}|}\sum_{ij=2}^{p}\sum S_{1i}S_{1j}T_{ij\cdot 1} \\ &= \frac{1}{|S_{ij\cdot 1}|}\left\{ S_{11}|S_{ij\cdot 1}| - \sum_{ij=2}^{p}\sum S_{1i}S_{1j}T_{ij\cdot 1} \right\} = \frac{|S_{ij}|}{|S_{ij\cdot 1}|} = \frac{1}{S^{11}}. \end{aligned} \tag{14.6.5}$$

Assume also that X_1 (measured about its mean) is $N(0,1)$.

The joint distribution of $X_{1\lambda} - \bar{X}_1$, (where $\bar{X}_1 = \frac{1}{n}\sum_{\lambda=1}^{n} X_{1\lambda}$), $\lambda = 1, \cdots, n$ is

$$f(x_{1\lambda}, \lambda = 1, \cdots, n) = \frac{1}{\left(\sqrt{2\pi}\right)^n} \exp\left[-\frac{1}{2}\sum_{\lambda=1}^{n}(x_{1\lambda} - \bar{x}_1)^2\right]. \tag{14.6.6}$$

Now

$$\boldsymbol{b} = ((S_{ij\cdot 1}))^{-1}((S_{1j})) = ((S_{ij\cdot 1}))^{-1} \begin{bmatrix} x_{21} - \bar{x}_2 & \cdots & x_{2n} - \bar{x}_2 \\ \vdots & & \\ x_{p1} - \bar{x}_p & \cdots & x_{pn} - \bar{x}_p \end{bmatrix} \begin{bmatrix} x_{11} - \bar{x}_1 \\ \vdots \\ x_{1n} - \bar{x}_1 \end{bmatrix}$$

$$= (S_{ij\cdot 1})^{-1} \begin{bmatrix} \boldsymbol{x}'_2 \\ \vdots \\ \boldsymbol{x}'_p \end{bmatrix} \boldsymbol{x}_1 = \boldsymbol{L}_1 \boldsymbol{x}_1 \text{ (say)}.$$

$$\boldsymbol{L}_1 \boldsymbol{L}'_1 = ((S_{ij\cdot 1}))^{-1}.$$

Transform

$$(x_{11} - \bar{x}_1, \cdots, x_{1n} - \bar{x}_1)' \to (b_2, \cdots, b_p, u_1, \cdots, u_{n-p+1})'$$

$$= \begin{bmatrix} \boldsymbol{b} \\ \boldsymbol{u} \end{bmatrix} = \begin{bmatrix} \boldsymbol{L}_1 \\ \boldsymbol{L}_2 \end{bmatrix} \boldsymbol{x}_1 = \boldsymbol{L}\boldsymbol{x}_1$$

where $\boldsymbol{L}_2$ is such that $\boldsymbol{L}_2 \boldsymbol{L}'_1 = \boldsymbol{0}, \boldsymbol{L}_2 \boldsymbol{L}'_2 = I$.
Then

$$S_{11} = \boldsymbol{X}'_1 \boldsymbol{X}_1 = (\boldsymbol{b}'\boldsymbol{u}')(\boldsymbol{L}\boldsymbol{L}')^{-1} \begin{bmatrix} \boldsymbol{b} \\ \boldsymbol{u} \end{bmatrix}$$

$$= \boldsymbol{b}'((S_{ij\cdot 1}))\boldsymbol{b} + \boldsymbol{u}'\boldsymbol{u}.$$

Jacobian of transformation is $|J| = |\boldsymbol{L}^{-1}| = \sqrt{|S_{ij.1}|}$. From (14.6.6), the joint distribution of $\boldsymbol{b}$ and $\boldsymbol{u}$ is

$$\left\{ \frac{\sqrt{|S_{ij.1}|}}{\left(\sqrt{2\pi}\right)^{p-1}} \exp\left[-\frac{1}{2}\boldsymbol{b}'((S_{ij\cdot 1}))\boldsymbol{b}\right] \prod_{i=2}^{p} db_i \right\}$$
$$\left\{ \frac{1}{(2\pi)^{(n-p+1)/2}} \exp\left[-\frac{1}{2} \sum_{i=1}^{n-p+1} u_i^2\right] \prod_{i=1}^{n-p+1} du_i \right\}.$$

Hence, $U_1, \cdots, U_{n-p+1}$ follow independent $N(0,1)$ distribution and hence $\sum_{1}^{n-p+1} U_i^2$ follows $\chi^2_{(n-p+1)}$. Now,

$$\sum_{1}^{n-p+1} U_i^2 = \boldsymbol{X}'_1 \boldsymbol{X}_1 - \boldsymbol{b}'((S_{ij\cdot 1}))\boldsymbol{b}$$

$$= S_{11} - \boldsymbol{b}'((S_{1j})) = \frac{1}{S^{11}} \text{ [from (14.6.5)]}.$$

Hence $\frac{1}{S^{11}}$ follows a central $\chi^2_{(n-p+1)}$ distribution.

14.6.1 Suppose we have a sample of size n from $N(0, \Sigma)$ population. We shall show that the sample generalised variance has a sampling distribution which is the product of independent central χ^2-distributions.

Let $\boldsymbol{X} = (X_1, \cdots, X_p)'$ follow $N_p(0, \Sigma)$ distribution. A sample of size n is drawn from this distribution, $x_{i\lambda}$ being the λth observation on $X_i (\lambda = 1, \cdots, n; i = 1, \cdots, p)$. Assume without any loss of generalites that the observations are measured against their means so that $\sum_{\lambda} x_{i\lambda} = 0$. Let $X = ((x_{i\lambda}))_{n \times p}$

$$S_{ij}(x) = \sum_{\lambda=1}^{n} x_{i\lambda} x_{j\lambda}, \quad s_{ij}(x) = S_{ij}(x)/n.$$

From $\boldsymbol{x}_\lambda = (x_{1\lambda}, \cdots, x_{p\lambda})'$ we pass on to $\boldsymbol{y}_\lambda = (y_{1\lambda}, \cdots, y_{p\lambda})'$ such that

$$\begin{bmatrix} x_{1\lambda} \\ \vdots \\ x_{p\lambda} \end{bmatrix} = \boldsymbol{B} \begin{bmatrix} y_{1\lambda} \\ \vdots \\ y_{p\lambda} \end{bmatrix} \quad \lambda = 1, \cdots, n$$

where $\boldsymbol{B}$ is such that $\boldsymbol{B}'\boldsymbol{\Sigma}^{-1}\boldsymbol{B} = I$. Then

$$\boldsymbol{x}_\lambda' \boldsymbol{\Sigma}^{-1} \boldsymbol{x}_\lambda = \boldsymbol{y}_\lambda' \boldsymbol{B}' \boldsymbol{\Sigma}^{-1} \boldsymbol{B} \boldsymbol{y}_\lambda = \boldsymbol{y}_\lambda' \boldsymbol{y}_\lambda = \sum_{i=1}^{p} y_{i\lambda}^2.$$

Henceforth, we shall write $S_{ij}(y) = S_{ij}$. Now

$$|S_{ij}| = S_{11} \frac{\begin{vmatrix} S_{11} & S_{12} \\ S_{21} & S_{22} \end{vmatrix}}{S_{11}} \cdot \frac{\begin{vmatrix} S_{11} & S_{12} & S_{13} \\ S_{21} & S_{22} & S_{23} \\ S_{31} & S_{32} & S_{33} \end{vmatrix}}{\begin{vmatrix} S_{11} & S_{12} \\ S_{21} & S_{22} \end{vmatrix}} \cdots \frac{\begin{vmatrix} S_{11} & \cdots & S_{1p} \\ \cdot & \cdots & \cdot \\ S_{p1} & \cdots & S_{pp} \end{vmatrix}}{\begin{vmatrix} S_{11} & \cdots & S_{1p-1} \\ \cdot & \cdots & \cdot \\ S_{p-11} & \cdots & S_{p-1\,p-1} \end{vmatrix}}. \tag{14.6.7}$$

Consider the sample regression equation of Y_p on $Y_1, \cdots, Y_{p-1}$ viz. $\tilde{y}_p = b_1 y_1 + \cdots + b_{p-1} y_{p-1}$ where $\tilde{y}_p$ is the predicted value of Y_p for given values of $y_1, \cdots, y_{p-1}$.

The sample partial regression coefficients $b_1, \cdots, b_{p-1}$ are given by

$$\begin{bmatrix} b_1 \\ \vdots \\ b_{p-1} \end{bmatrix} = \begin{bmatrix} S_{11} & \cdots & S_{1p-1} \\ \\ S_{p-11} & \cdots & S_{p-1\,p-1} \end{bmatrix}^{-1} \begin{bmatrix} S_{1p} \\ \vdots \\ S_{p-1\,p} \end{bmatrix} = (S_{ij \cdot p})^{-1} \begin{bmatrix} \boldsymbol{Y}'_{(1)} \\ \vdots \\ \boldsymbol{Y}'_{(p-1)} \end{bmatrix} \boldsymbol{Y}_{(p)} = \boldsymbol{L}_1 \boldsymbol{Y}_{(p)}, \tag{14.6.8}$$

where $\boldsymbol{Y}'_{(i)} = (Y_{i1}, \cdots, Y_{in})$ and $(S_{ij\cdot p})$ is the $(p-1) \times (p-1)$ matrix with elements S_{ij} in which variable Y_p has been excluded.

Hence $\boldsymbol{L}_1 \boldsymbol{L}_1' = (S_{ij \cdot p})^{-1}$.

Now the joint distribution of $(Y_{p1}, \cdots, Y_{pn})'$ is

$$\frac{1}{(2\pi)^{n/2}} \exp\left[-\frac{1}{2}\boldsymbol{y}'_{(p)}\boldsymbol{y}_{(p)}\right] \prod_{\lambda=1}^{n} dy_{p\lambda}. \tag{14.6.9}$$

We apply the transformation

$$\begin{bmatrix} b_1 \\ \vdots \\ b_{p-1} \\ u_1 \\ \vdots \\ u_{n-p+1} \end{bmatrix} = \begin{bmatrix} \boldsymbol{b} \\ \boldsymbol{u} \end{bmatrix} = \begin{bmatrix} \boldsymbol{L_1} \\ \boldsymbol{L_2} \end{bmatrix} \boldsymbol{Y}_{(p)} = \boldsymbol{LY}_{(p)}$$

where $\boldsymbol{L}_2$ is such that $\boldsymbol{L_2L_1'} = \boldsymbol{0}, \boldsymbol{L_2L_2'} = I$. Hence

$$\boldsymbol{y}'_{(p)}\boldsymbol{y}_{(p)} = [\boldsymbol{b}'\boldsymbol{u}'](\boldsymbol{LL}')^{-1}\begin{bmatrix} \boldsymbol{b} \\ \boldsymbol{u} \end{bmatrix} = \boldsymbol{b}'((S_{ij\cdot p}))\boldsymbol{b} + \boldsymbol{u}'\boldsymbol{u}.$$

$|J| = \sqrt{|S_{ij\cdot p}|}$.

The joint distribution of $\boldsymbol{b}$ and $\boldsymbol{u}$ is

$$\frac{\sqrt{|S_{ij\cdot p}|}}{(2\pi)^{(p-1)/2}} \exp\left[-\frac{1}{2}\boldsymbol{b}'((S_{ij\cdot p}))\boldsymbol{b}\right] \prod_{1}^{p-1} db_i \times \frac{1}{(2\pi)^{(n-p+1)/2}}$$

$$\exp\left[-\frac{1}{2}\sum_{i=1}^{n-p+1} u_i^2\right] \prod_{i=1}^{n-p-1} du_i. \tag{14.6.10}$$

This shows that $\sum_{i=1}^{n-p+1} u_i^2$ follows $\chi^2_{(n-p+1)}$. Again

$$\begin{aligned} \sum_{i=1}^{n-p+1} u_i^2 &= \boldsymbol{y}'_{(p)}\boldsymbol{y}_{(p)} - \boldsymbol{b}'((S_{ij.p}))\boldsymbol{b} \\ &= S_{pp} - (S_{1p} \cdots S_{p-1\,p})((S_{ij\cdot p}))^{-1}\begin{bmatrix} S_{1p} \\ \vdots \\ S_{p-1\,p} \end{bmatrix} \\ &= \frac{|S_{ij}|}{|S_{ij\cdot p}|} \quad \text{[by (14.6.5)]}. \end{aligned} \tag{14.6.11}$$

Hence

$$|S_{ij}|/|S_{ij\cdot p}| \cap \chi^2_{(n-p+1)}.$$

Note that this χ^2 is related to Y_p only and its values are independent of values of $Y_1, \cdots, Y_{p-1}$.

Similarly considering the sample regression equation of Y_{p-1} on $Y_1, \cdots, Y_{p-2}$ we see that

$$\frac{|S_{ij\cdot p}|}{|S_{ij\cdot p(p-1)}|} \cap \chi^2_{(n-p+2)},$$

and this is related only to Y_{p-1} and is independent of the remaining variables.

Proceeding in this way we see ultimately that S_{11} is distributed as $\chi^2_{(n)}$ and is related only to Y_1. Hence from (14.6.7)

$$|S_{ij}| \cap \chi^2_{(n)}\chi^2_{(n-1)} \cdots \chi^2_{(n-p+1)}.$$

Returning to the original variables $X_1, \cdots, X_p$, note that

$$|S_{ij}(x)| = |\boldsymbol{X}'\boldsymbol{X}| = |\boldsymbol{B}|^2|\boldsymbol{Y}'\boldsymbol{Y}| = |\boldsymbol{\Sigma}||S_{ij}(y)|;$$
$$|s_{ij}(x)| = n^{-p}|S_{ij}(x)| = n^{-p}|\boldsymbol{\Sigma}||S_{ij}(y)|.$$

Hence,

$$|s_{ij}(x)| \cap n^{-p}|\boldsymbol{\Sigma}|\chi^2_{(n)}\chi^2_{(n-1)} \cdots \chi^2_{(n-p+1)}$$

where all the χ^2's are independent.

▶ **Example 14.6.1** *A sample of size n is drawn from a $N_p(\boldsymbol{m}, \boldsymbol{\Sigma})$ population. Show that the distribution of sample generalised variance is the product of several independently distributed χ^2 variables multiplied by a constant.*

Let

$$S_{ij}(x) = \sum_{\lambda=1}^{n}(x_{i\lambda} - \bar{x}_i)(x_{j\lambda} - \bar{x}_j)$$
$$s_{ij}(x) = S_{ij}(x)/(n-1).$$

We transform from $\boldsymbol{x}_\lambda$ to $\boldsymbol{y}_\lambda$ such that

$$\begin{bmatrix} x_{1\lambda} \\ \vdots \\ x_{p\lambda} \end{bmatrix} = \boldsymbol{B}\begin{bmatrix} y_{1\lambda} \\ \vdots \\ y_{p\lambda} \end{bmatrix}, \quad \lambda = 1, \cdots, n$$
$$\boldsymbol{B}'\boldsymbol{\Sigma}^{-1}\boldsymbol{B} = I.$$

Then $\boldsymbol{y}_\lambda$ is distributed as $N_p(\boldsymbol{\mu}, I)$ when $\boldsymbol{\mu} = B^{-1}\boldsymbol{m}$. Let

$$S_{ij}(y) = \sum_{\lambda=1}^{n}(y_{i\lambda} - \bar{y}_i)(y_{j\lambda} - \bar{y}_j) = S_{ij}.$$

Now,

$$|S_{ij}| = S_{11}\frac{\begin{vmatrix} S_{11} & S_{12} \\ S_{21} & S_{22} \end{vmatrix}}{S_{11}} \cdot \frac{\begin{vmatrix} S_{11} & S_{12} & S_{13} \\ S_{21} & S_{22} & S_{23} \\ S_{31} & S_{32} & S_{33} \end{vmatrix}}{\begin{vmatrix} S_{11} & S_{12} \\ S_{21} & S_{22} \end{vmatrix}} \cdots \times \frac{|S_{ij}|}{|S_{ij\cdot p}|}.$$

Consider the sample regression equation of y_p on $y_1, \cdots, y_{p-1}$. This is given by

$$\tilde{y}_p = b_1(y_1 - \bar{y}_1) + \cdots + b_{p-1}(y_{p-1} - \bar{y}_{p-1})$$

where $\bar{y}_i = \frac{1}{n}\Sigma_\lambda y_{i\lambda}$ and $\tilde{y}_p$ is the predicted value of y_p on $y_1, \cdots, y_{p-1}$. The coefficient $b_1, \cdots, b_{p-1}$ are given by

$$\boldsymbol{b} = \begin{bmatrix} b_1 \\ \vdots \\ b_{p-1} \end{bmatrix} = \begin{bmatrix} S_{11} & \cdots & S_{1p-1} \\ \cdot & \cdots & \cdot \\ S_{p-11} & \cdots & S_{p-1\,p-1} \end{bmatrix}^{-1} \begin{bmatrix} S_{1p} \\ \vdots \\ S_{p-1\,p} \end{bmatrix} = \boldsymbol{L}_2 \boldsymbol{y}_{(p)} \text{ (say)}$$

where $\boldsymbol{y}_{(p)} = (y_{p1}, y_{p2}, \cdots, y_{pn})'$ so that $\boldsymbol{L}_2\boldsymbol{L}_2' = (S_{ij\cdot p})^{-1}$.

The joint distribution of $y_{p\lambda}$'s $(\lambda = 1, \cdots, n)$ is

$$\frac{1}{\left(\sqrt{2\pi}\right)^n} \exp\left[-\frac{1}{2}\sum_{\lambda=1}^{n}(y_{p\lambda} - \mu_p)^2\right] \prod_{\lambda=1}^{n} dy_{p\lambda}.$$

Apply transformation

$$\boldsymbol{u} = \begin{bmatrix} u_1 \\ \cdots\cdots \\ b_1 \\ \vdots \\ b_{p-1} \\ \cdots\cdots \\ u_1 \\ \vdots \\ u_{n-p} \end{bmatrix} = \begin{bmatrix} u_1 \\ \boldsymbol{b} \\ \boldsymbol{u}_3 \end{bmatrix} = \begin{bmatrix} \boldsymbol{L}_1 \\ \boldsymbol{L}_2 \\ \boldsymbol{L}_3 \end{bmatrix} \boldsymbol{y}_{(p)} = \boldsymbol{L}\boldsymbol{y}_{(p)}$$

where $\boldsymbol{L}_1 = \left(\frac{1}{\sqrt{n}}, \cdots, \frac{1}{\sqrt{n}}\right)$, $\boldsymbol{L}_2$ defined as above and $\boldsymbol{L}_3$ is such that

$$\boldsymbol{L}_3\boldsymbol{L}_1' = \boldsymbol{0}, \quad \boldsymbol{L}_3\boldsymbol{L}_2' = \boldsymbol{0}, \quad \boldsymbol{L}_3\boldsymbol{L}_3' = I.$$

Then

$$\begin{aligned}
\sum_{\lambda=1}^{n}(y_{p\lambda} - \mu_p)^2 &= \left(\boldsymbol{y}_{(p)} - \mu_p\boldsymbol{1}\right)'\left(\boldsymbol{y}_{(p)} - \mu_p\boldsymbol{1}\right) \\
&= (\boldsymbol{u} - \mu_p\boldsymbol{L}\boldsymbol{1})'\left(\boldsymbol{L}\boldsymbol{L}'\right)^{-1}(\boldsymbol{u} - \mu_p\boldsymbol{L}\boldsymbol{1}) \\
&= \left(u_1 - \sqrt{n}\mu_p\right)^2 + \boldsymbol{b}'\left(S_{ij\cdot p}\right)\boldsymbol{b} + \boldsymbol{u}_3'\boldsymbol{u}_3 \\
|J| = |\boldsymbol{L}|^{-1} &= \sqrt{|S_{ij\cdot p}|}.
\end{aligned}$$ (i

Hence

$$\boldsymbol{u}_3'\boldsymbol{u}_3 = \sum_{1}^{n-p} u_i^2 \cap \chi^2_{(n-p)}.$$

Again, putting $\mu_p = 0$ in indentity relation (i) we see that

$$\sum_1^{n-p} u_i^2 = \sum_\lambda (y_{p\lambda} - \bar{y}_p)^2 - \boldsymbol{b}'(S_{ij\cdot p})\boldsymbol{b} = S_{pp} - \boldsymbol{b}'(S_{ij\cdot p})\boldsymbol{b} = |S_{ij}|/|S_{ij\cdot p}|.$$

Hence $|S_{ij}|/|S_{ij\cdot p}| \cap \chi^2_{(n-p)}$. Also this χ^2 is independent of the conditionality variables $Y_1, \cdots, Y_{p-1}$.

Similarly, considering the sample regression of Y_{p-1} on $Y_1, \cdots, Y_{p-2}$ we see that $|S_{ij\cdot p}|/\ |S_{ij} \cdot \overline{p-1},\ p|$ is distributed as an independent $\chi^2_{(n-p-1)}$.

Proceeding in this way we see that $S_{11} \cap \chi^2_{(n-1)}$. Also all these chi-squares are independent. Again $((S_{ij})) = (\boldsymbol{Y} - \bar{\boldsymbol{Y}})'(\boldsymbol{Y} - \bar{\boldsymbol{Y}})$ when

$$\bar{\boldsymbol{Y}} = \begin{pmatrix} \bar{y}_1 & \cdots & \bar{y}_p \\ \cdot & \cdots & \cdot \\ \bar{y}_1 & \cdots & \bar{y}_p \end{pmatrix}_{(n\times p)}.$$

$$\begin{aligned} |S_{ij}(x)| &= |(\boldsymbol{X} - \bar{\boldsymbol{X}})'(\boldsymbol{X} - \bar{\boldsymbol{X}})| = |\boldsymbol{B}((S_{ij}))\boldsymbol{B}'| \\ &= |\boldsymbol{B}|^2|S_{ij}| = |\boldsymbol{\Sigma}|\chi^2_{(n-1)}\chi^2_{(n-2)}\cdots\chi^2_{(n-p)}. \end{aligned} \qquad \text{(ii)}$$

Hence, from (ii), the generalised variance

$$|s_{ij}(x)| = (n-1)^{-p}|S_{ij}(x)| \cap (n-1)^{-p}|\boldsymbol{\Sigma}| \prod_{i=1}^{p} \chi^2_{(n-i)}.$$

▶ **Example 14.6.2** *Find the moments of the distribution of the generalised variance.*

It is known that

$$\begin{aligned} E\left(\chi^2_{(r)}\right)^h &= \frac{2^h}{\left\lceil\left(\frac{r}{2}\right)\right.} \int_0^\infty e^{-\chi^2/2} \left(\frac{\chi^2}{2}\right)^{(r+2h-1)/2} d(\chi^2/2) \\ &= \frac{2^h \left\lceil\left(\frac{r}{2}+h\right)\right.}{\left\lceil\left(\frac{r}{2}\right)\right.}. \end{aligned}$$

Hence the hth order moment of the generalised variance is

$$\begin{aligned} E\left(|s_{ij}|^h\right) &= (n-1)^{-hp}|\boldsymbol{\Sigma}|^h E\left\{\prod_{i=1}^{p} \chi^2_{n-i}\right\} \\ &= (n-1)^{-hp}|\boldsymbol{\Sigma}|^h \prod_{i=1}^{p} \frac{2^h \left\lceil\left(\frac{n-i}{2}+h\right)\right.}{\left\lceil\left(\frac{(n-i)}{2}\right)\right.} \\ &= \left\{\frac{2}{(n-1)^p}\right\}^h |\boldsymbol{\Sigma}|^h \prod_{i=1}^{p} \frac{\left\lceil\left(\frac{n-i}{2}+h\right)\right.}{\left\lceil\left(\frac{n-i}{2}\right)\right.}. \end{aligned}$$

► **Example 14.6.3** *Show that in sampling from $N_p(\mathbf{0},\mathbf{\Sigma})$, $|s_{ij}|$ is distributed asymptotically normally.*

We have

$$E\left(\chi^2_{(n-i)}\right) = n-i;\ \ V\left(\chi^2_{(n-i)}\right) = 2(n-i).$$

Hence

$$\frac{\chi^2_{(n-i)}-(n-i)}{\sqrt{2(n-i)}} \sim N(0,1)$$

asymptotically

$$\text{i.e., } u_i^{(n)} = \frac{\chi^2_{(n-i)}}{n-i} \to N\left(1,\frac{2}{n}\right) \ \text{ as } n\to\infty.$$

Now

$$|s_{ij}| = n^{-p}|\mathbf{\Sigma}|\prod_{i=0}^{p-1}\chi^2_{(n-i)}.$$

Hence

$$\frac{|s_{ij}|}{|\mathbf{\Sigma}|} = \prod_{i=0}^{p-1}\frac{\chi^2_{(n-i)}}{n} = u_0^{(n)}u_1^{(n)}\cdots u_{p-1}^{(n)} \to N\left(1,\frac{2p}{n}\right)$$

$$\text{i.e., } \left(\frac{|s_{ij}|}{|\mathbf{\Sigma}|}-1\right) \to N\left(0,\frac{2p}{n}\right)$$

$$\text{or, } \sqrt{n}\left\{\frac{|s_{ij}|}{|\mathbf{\Sigma}|}-1\right\} \to N(0,2p).$$

► **Example 14.6.4** *The joint p.d.f. of $2n$ random variables $X_1,\cdots,X_n,Y_1,\cdots,Y_n$ is given by*

$$\text{Const.}\exp\left[-\frac{1}{2}\sum\sum_{ij=1}^{n}a_{ij}\left\{(x_i-y_i)(x_j-y_j)+y_iy_j\right\}\right]$$

where $\boldsymbol{A}=((a_{ij}))$ is a positive definite matrix.

Find the distribution of

$$\sqrt{\sum_i\sum_j a_{ij}x_ix_j}.$$

We have

$$dF(x_i,y_i;i=1,\cdots,n) = \text{Const.}\exp\left[-\frac{1}{2}\sum_i\sum_j a_{ij}y_iy_j\right]$$

$$\exp\left[-\frac{1}{2}\sum_i\sum_j a_{ij}(x_i-y_i)(x_j-y_j)\right]\Pi\, dx_i\, dy_i. \qquad \text{(i}$$

If we keep $y_1, \cdots, y_n$ fixed and integrate over x's we get the joint distribution of $y_1, \cdots, y_n$.

$$dF(y_1, \cdots, y_n) = \text{Const.} \exp\left[-\frac{1}{2}\sum_i\sum_j a_{ij}y_iy_j\right] \Pi\, dy_i$$

$$\int_{-\infty}^{\infty} \cdots \int_{-\infty}^{\infty} \exp\left[-\frac{1}{2}\sum\sum_{ij=1}^{n} a_{ij}(x_i - y_i)(x_j - y_j)\right] \Pi\, dx_i. \quad \text{(ii)}$$

The expression under the integral sign represents the p.d.f. of a multivariate normal distribution with mean vector $\boldsymbol{y} = (y_1, \cdots, y_n)'$ and dispersion matrix $\boldsymbol{A}^{-1} = ((a^{ij}))$ where $A = ((a_{ij}))$. Hence the integral is $\frac{(2\pi)^{n/2}}{\sqrt{|\boldsymbol{A}|}}$.

Therefore,

$$\begin{aligned} dF(y_1, \cdots, y_n) &= \text{Const.} \exp\left[-\frac{1}{2}\sum\sum_{ij=1}^{n} a_{ij}y_iy_j\right] \prod_{i=1}^{n} dy_i \\ &= \text{Const.} \exp\left[-\frac{1}{2}\boldsymbol{y}'\boldsymbol{A}\boldsymbol{y}\right] \prod_{i=1}^{n} dy_i. \end{aligned} \quad \text{(iii)}$$

Let us make transformation $\boldsymbol{y} \to \boldsymbol{z} = (z_1, \cdots, z_n)$ such that $\boldsymbol{y} = \boldsymbol{C}\boldsymbol{z}$ when $\boldsymbol{C}'\boldsymbol{A}\boldsymbol{C} = I$.

Therefore,

$$\boldsymbol{y}'\boldsymbol{A}\boldsymbol{y} = \boldsymbol{z}'\boldsymbol{z} = \sum_{i=1}^{n} z_i^2; \quad |J| = |\boldsymbol{C}| = \frac{1}{\sqrt{|\boldsymbol{A}|}}.$$

Hence,

$$dF(z_1, \cdots, z_n) = \text{Const.} e^{-\sum_1^n z_i^2/2} \Pi\, dz_i \quad \text{(iv)}$$

which shows that

$$\sum_1^n z_i^2 \cap \chi^2_{(n)}.$$

Therefore

$$\sum_i\sum_j a_{ij}y_iy_j = \chi_1^2 \cap \chi^2_{(n)}.$$

Therefore, the conditional distribution of $(x_1, \cdots, x_n)$ given $(y_1, \cdots, y_n)$ is

$$\begin{aligned} dF(x_1, \cdots, x_n | y_1, \cdots, y_n) &= \text{Const.} \exp\left[-\frac{1}{2}\left\{\sum_i\sum_j a_{ij}(x_i - y_i)(x_j - y_j)\right\}\right] \prod_{i=1}^{n} dx_i \\ &= \text{Const.} \exp\left[-\frac{1}{2}(\boldsymbol{x} - \boldsymbol{y})'\boldsymbol{A}(\boldsymbol{x} - \boldsymbol{y})\right] \prod_{i=1}^{n} dx_i. \end{aligned} \quad \text{(v)}$$

Consider the transformation

$$\boldsymbol{x} = \boldsymbol{C}\boldsymbol{w}, \quad \boldsymbol{w} = (w_1, \cdots, w_n)', \quad \boldsymbol{y} = \boldsymbol{C}\boldsymbol{z}, \tag{v}$$

so that

$$\begin{aligned}(\boldsymbol{x}-\boldsymbol{y})'\boldsymbol{A}(\boldsymbol{x}-\boldsymbol{y}) &= (\boldsymbol{w}-\boldsymbol{z})'\boldsymbol{C}'\boldsymbol{A}\boldsymbol{C}(\boldsymbol{w}-\boldsymbol{z})\\ &= (\boldsymbol{w}-\boldsymbol{z})'(\boldsymbol{w}-\boldsymbol{z}); \quad \text{also } x'\boldsymbol{A}x = \sum_{i=1}^{n} w_i^2.\end{aligned}$$

The transformed density is

$$dF(w_1, \cdots, w_n | z_1, \cdots, z_n) = \text{Const.}\exp\left[-\frac{1}{2}\sum_{i=1}^{n}(w_i - z_i)^2\right]\prod_{i=1}^{n} dz_i$$

which shows that $\sum_{i=1}^{n} w_i^2$ is conditionally distributed as non-central χ^2 with n d.f. an n.c.p.

$$\sum_{i=1}^{n} z_i^2 = \sum_i \sum_j a_{ij} y_i y_j = \lambda.$$

Therefore, distribution of

$$\sum\sum_{ij=1}^{n} a_{ij} x_i x_j = \chi_2^2$$

given $(y_1, \cdots, y_n)$ is a $\chi^2(n, \lambda)$.

$$dF\left(\chi_2^2 | y_1, \cdots, y_n\right) = e^{-\lambda/2}\sum_{i=0}^{\infty}\left(\frac{\lambda}{2}\right)^i \frac{1}{i!}\frac{e^{-\chi_2^2/2}\left(\chi_2^2\right)^{\frac{n}{2}+i-1}}{2^{\frac{n}{2}+i}\Gamma\left(\frac{n}{2}+i\right)} d\chi_2^2. \tag{vi}$$

The distribution of χ_2^2 depends on $y_1, \cdots, y_n$ only through λ. Hence (vii) is th conditional distribution of χ_2^2 given λ. Again $\lambda = \chi_1^2 \cap \chi_{(n)}^2$. Therefore,

$$\begin{aligned}dF\left(\chi_2^2, \lambda\right) &= dF\left(\chi_2^2 | \lambda\right) dF(\lambda)\\ &= \frac{1}{2^{\frac{n}{2}}\Gamma\left(\frac{n}{2}\right)}\sum_{i=0}^{\infty}\frac{e^{-\chi_2^2/2}\left(\chi_2^2\right)^{\frac{n}{2}+i-1}}{i!\,2^{n/2+2i}\Gamma\left(\frac{n}{2}+i\right)} e^{-\lambda}\lambda^{\frac{n}{2}+i-1} d\chi_2^2\, d\lambda.\end{aligned} \tag{vii}$$

The unconditional distribution of χ_2^2 is obtained by integrating (viii) with respect to over $(0, \lambda)$. As the series is uniformly convergent, term by term integration is permissibl Now

$$\int_0^{\infty} e^{-\lambda}\lambda^{\frac{n}{2}+i-1}\, d\lambda = \Gamma\left(\frac{n}{2}+i\right).$$

lence,

$$dF\left(\chi_2^2\right) = \frac{1}{2^{\frac{n}{2}}\Gamma\left(\frac{n}{2}\right)} e^{-\chi_2^2/2}\left(\chi_2^2\right)^{\frac{n}{2}-1}\sum_{i=0}^{\infty}\frac{\left(\chi_2^2\right)^i\Gamma\left(\frac{n}{2}+i\right)d\chi_2^2}{i!\,2^{\frac{n}{2}+2i}\Gamma\left(\frac{n}{2}+i\right)}$$

$$= \frac{1}{2^n\Gamma\left(\frac{n}{2}\right)} e^{-\chi_2^2/2}\left(\chi_2^2\right)^{\frac{n}{2}-1}\sum_{i=0}^{\infty}\frac{\left(\chi_2^2/4\right)^i}{i!}\,d\chi_2^2$$

$$= \frac{1}{2^n\Gamma\left(\frac{n}{2}\right)} e^{-\chi_2^2/4}\left(\chi_2^2\right)^{\frac{n}{2}-1} d\chi_2^2.$$

lence the distribution of $\chi_2 = \sqrt{\sum_i\sum_j a_{ij}x_ix_j}$ is

$$dF\left(\chi_2\right) = \frac{1}{2^{n-1}\Gamma\left(\frac{n}{2}\right)} e^{-\chi_2^2/4}\left(\chi_2^2\right)^{\frac{n-1}{2}} d\chi.$$

Exercise

1. Suppose that $\boldsymbol{A}$ is $W(m, n, \boldsymbol{\Sigma}), n > m - 1$ where

$$\boldsymbol{\Sigma} = \sigma^2\begin{bmatrix} 1 & \rho & \cdots & \rho \\ \rho & 1 & \cdots & \rho \\ \vdots \\ \rho & \rho & \cdots & 1\end{bmatrix}.$$

Show that

$$\boldsymbol{\Sigma}^{-1} = \frac{1}{\sigma^2(1-\rho)}I_m - \frac{\rho}{\sigma^2(1-\rho)[1+(m-1)\rho]}\mathbf{1}\mathbf{1}'$$

and

$$|\boldsymbol{\Sigma}| = (\sigma^2)^m(1-\rho)^{m-1}[1+(m-1)\rho].$$

Show that the maximum likelihood estimate of σ^2 and ρ are

$$\widehat{\sigma^2} = \frac{tr(\boldsymbol{A})}{mn} \text{ and } \hat{\rho} = \frac{\mathbf{1}'\boldsymbol{A}\mathbf{1} - tr\boldsymbol{A}}{(m-1)tr\boldsymbol{A}} = \frac{2\sum\sum_{i<j=1}^{m} a_{ij}}{(m-1)tr\boldsymbol{A}}.$$

[*Hints:* Try this problem after reading Section 16.2.]

2. Show that in Wishart distribution (14.2.14),

$$E(s_{ij}) = \sigma_{ij}$$

$$E(s_{ij}s_{ke}) = \sigma_{ij}\sigma_{ke} + \frac{1}{n-1}\{\sigma_{ik}\sigma_{je} + \sigma_{ie}\sigma_{jk}\};$$

hence, show that

$$\text{Cov}(s_{ij}, s_{ke}) = \frac{1}{n-1}\{\sigma_{ik}\sigma_{je} + \sigma_{ie}\sigma_{jk}\}.$$

3. Let $\boldsymbol{A} \cap W(p, n, \boldsymbol{\Sigma})$. Let also $\boldsymbol{A}, \boldsymbol{\Sigma}$ be partitioned as

$$\boldsymbol{A} = \begin{bmatrix} \boldsymbol{A}_{11} & \boldsymbol{A}_{12} \\ \boldsymbol{A}_{21} & \boldsymbol{A}_{22} \end{bmatrix}, \quad \boldsymbol{\Sigma} = \begin{bmatrix} \boldsymbol{\Sigma}_{11} & \boldsymbol{\Sigma}_{12} \\ \boldsymbol{\Sigma}_{21} & \boldsymbol{\Sigma}_{22} \end{bmatrix}$$

where $\boldsymbol{A}_{11}, \boldsymbol{\Sigma}_{11}$ are each $q \times q$. Suppose $\boldsymbol{\Sigma}_{12} = \boldsymbol{0}$. Show that $\boldsymbol{A}_{11}$ and $\boldsymbol{A}_{22}$ ar[e] independent and there distributions are $W(q, n, \boldsymbol{\Sigma}_{11}), W(p-q, n, \boldsymbol{\Sigma}_{22})$ respectivel[y]

4. The matrix $\boldsymbol{S}$ of corrected sum of squares and products of p variables based o[n] a sample of $\nu + 1$ observations from a p-dimensional normal distribution wit[h] dispersion matrix $\boldsymbol{\Sigma}$ follows the Wishart distribution $W(p, \nu + 1, \boldsymbol{\Sigma})$.

   Work out $E(|\boldsymbol{S}|^t)$ and reduce it to its simplest form when $p = 2$ and $\boldsymbol{\Sigma} = I$.

5. The dispersion matrix of a bivariate normal distribution has elements $\sigma_{11} = 16.90$ $\sigma_{12} = 22.01$ and $\sigma_{22} = 51.72$. A sample of $n = 50$ pairs of observations yielded th[e] mean values $\bar{x}_1 = 2.54$ and $\bar{x}_2 = -1.72$ in the sample. Are these data consisten[t] with the hypothesis that in the population two mean values are zeroes?

6. (a) Write down in the simplest form the joint probability density function of [p] random variables $X_1, \cdots, X_p$ which follow a multivariate normal distributio[n] in which all the mean values are equal to μ, all the variances are equal to σ[²] and all the covariances are equal $\rho\sigma^2$.

   (b) Show that $\rho \geq -\frac{1}{p-1}$.

   [*Hints:* Consider the variance of $X_1 + \cdots + X_p$.]

   (c) If $p \geq 5$ and Y_1, Y_2, Y_3, Y_4 are four normalised mutually orthogonal linea[r] compounds of $X_1, X_2, \cdots, X_p$ each orthogonal to $X_1 + X_2 + \cdots + X_p$, wor[k] out the probability distribution of

$$F = \frac{Y_1^2 + Y_2^2}{Y_3^2 + Y_4^2}.$$

7. Let $\boldsymbol{X} = (X_1, \cdots, X_p)' \cap N_p(\boldsymbol{\mu}, \boldsymbol{\Sigma})$.

   (a) Find the distribution of the $(p+1)$-dimensional random vector $\Big(X_1, \cdots, X_p$ $\sum_{j=1}^{p} l_j X_j\Big)'$ where l_j are given constants.

   (b) Find the conditional joint distribution of $X_1, \cdots, X_p$ given $l_1X_1 + \cdots + l_pX_p =$ c, c being a constant.

(c) Show that the covariance between X_i, X_j conditional upon $l_1X_1+\cdots+l_pX_p = c$, is

$$\sigma_{ij} - \frac{\left(\sum_{k=1}^{p} l_k\sigma_{ik}\right)\left(\sum_{k=1}^{p} l_k\sigma_{jk}\right)}{\sum_{m=1}^{p}\sum_{k=1}^{p} l_k l_m \sigma_{km}}$$

for $i, j = 1, \cdots, p$.

8. If a symmetric matrix $\boldsymbol{S} = ((S_{ij})),\ \ i, j = 1, \cdots, p$ has a Wishart distribution with m degrees of freedom based on a matrix $\boldsymbol{\Sigma}$, find the distribution of the matrix $((S_{ij})),\ i, j = 1, \cdots, r < p$ with its parameters.

9. Let $\boldsymbol{X}, \boldsymbol{Y}, \boldsymbol{Z}$ be each a $p \times 1$ random vector; let $(\boldsymbol{X}', \boldsymbol{Y}', \boldsymbol{Z}')'$ have a $3p$-dimensional random distribution with mean vector $(\boldsymbol{\lambda}', \boldsymbol{\mu}', \boldsymbol{\nu}')$ and non-singular dispersion matrix $\boldsymbol{\Sigma}$ where $E(\boldsymbol{X}) = \boldsymbol{\lambda}$, $E(\boldsymbol{Y}) = \boldsymbol{\mu}$, $E(\boldsymbol{Z}) = \boldsymbol{\nu}$. A random sample of size n is drawn from this population. Derive a suitable test for the hypothesis

$$H_1 : 3\lambda = \boldsymbol{\mu} + 2\boldsymbol{\nu}$$

against

$$H_2 : 3\lambda \neq \boldsymbol{\mu} + 2\boldsymbol{\nu}.$$

10. Show that if $\boldsymbol{M} \sim \boldsymbol{W}(p, m, \boldsymbol{\Sigma})$ then (i) diagonal submatrices of $\boldsymbol{M}$ themselves have a Wishart distribution (ii) $\boldsymbol{\Sigma}^{-1/2}\boldsymbol{M}\boldsymbol{\Sigma}^{-1/2} \sim \boldsymbol{W}(p, m, \boldsymbol{I})$.

11. Let $\boldsymbol{X}(n \times p) = (\boldsymbol{x}_1', \cdots, \boldsymbol{x}_n')'$ where $\boldsymbol{x}_i'(p \times 1)(i = 1, \cdots, n)$ is an independent sample from $N_p(\mu, \boldsymbol{\Sigma})$. Let $\boldsymbol{C}(n \times n)$ be a symmetric matrix. Show that

   (a) $\boldsymbol{X}'\boldsymbol{C}\boldsymbol{X}$ has the same distribution as a weighted sum of independent Wishart $\boldsymbol{W}(p, 1, \boldsymbol{\Sigma})$ matrices, where the weights are eigenvalues of $\boldsymbol{C}$;

   (b) $\boldsymbol{X}'\boldsymbol{C}\boldsymbol{X}$ has a Wishart distribution iff $\boldsymbol{C}$ is idempotent, in which case $\boldsymbol{X}'\boldsymbol{C}\boldsymbol{X} \sim \boldsymbol{W}(p, r, \boldsymbol{\Sigma})$ where $r = tr\ \boldsymbol{C} = \text{rank}\,\boldsymbol{C}$.

12. Let $\boldsymbol{M} \sim \boldsymbol{W}(p, m, \boldsymbol{\Sigma})$, where

$$\boldsymbol{M} = \begin{bmatrix} \boldsymbol{M}_{11} & \boldsymbol{M}_{12} \\ \boldsymbol{M}_{21} & \boldsymbol{M}_{22} \end{bmatrix}$$

and $\boldsymbol{M}_{11}$ is a $a \times a$ matrix $(a < m)$. Show that

   (a) $\boldsymbol{M}_{22.1} = \boldsymbol{M}_{22} - \boldsymbol{M}_{21}\boldsymbol{M}_{11}^{-1}\boldsymbol{M}_{12}$ has the $\boldsymbol{W}(p, m-a, \boldsymbol{\Sigma}_{22.1})$ distribution and is independent of $(\boldsymbol{M}_{11}, \boldsymbol{M}_{12})$, and

   (b) if $\boldsymbol{\Sigma}_{12} = \boldsymbol{0}$, then $\boldsymbol{M}_{22} - \boldsymbol{M}_{22.1} = \boldsymbol{M}_{21}\boldsymbol{M}_{11}^{-1}\boldsymbol{M}_{12}$ has the $\boldsymbol{W}(p, a, \boldsymbol{\Sigma}_{22})$ distribution and $\boldsymbol{M}_{21}\boldsymbol{M}_{11}^{-1}\boldsymbol{M}_{12}$, $\boldsymbol{M}_{11}$, and $\boldsymbol{M}_{22.1}$ are jointly independent. (Here, $\boldsymbol{\Sigma}_{11}$ is defined similarly as $\boldsymbol{M}_{11}$).

13. Let $M \sim W(p, m, \Sigma)$, $m > p$. Show that

    (a) the ratio $a'\Sigma^{-1}a/a'M^{-1}a$ has the $\chi^2_{(m-p+1)}$ distribution for any fixed p-vector a, and, in particular, $\sigma^{ii}/m^{ii} \sim \chi^2_{(m-p+1)}$ for $i = 1, \cdots, p$;

    (b) m^{ii} is independent of all the elements of M except m_{ii}.

14. Let $M \sim W(p, m, \Sigma)$, $m \geq p$. Show that $|M|$ is $|\Sigma|$ times p-independent chi-square random variables with degrees of freedom $m, m-1, \cdots, m-p+1$. Hence, note that if $\Sigma > 0$, then $M > 0$ with probability one.

Chapter 15

Theory of Point Estimation

15.1 Introduction

We consider in this chapter the basic concepts of the theory of point estimation. Let $X_1, \cdots, X_n$ be n random variables having a joint p.d.f. which belongs to a family $\{f_\theta(x_1, \cdots, x_n), \theta \in \Theta\}$, where Θ is the parameter space. Our purpose is to estimate the value of θ or any function $\psi(\theta)$ of θ on the basis of observations $\boldsymbol{x} = (x_1, \cdots, x_n)'$ of the random vector $\boldsymbol{X} = (X_1, \cdots, X_n)'$. In doing so, the form of the joint distribution $f_\theta(\boldsymbol{x})$ is, of course, assumed known. In this chapter we develop properties of a good estimator.

In sections 15.2, 15.3 we introduce the concepts of unbiasedness, consistency and mean square consistency. Section 15.4 considers the concepts of sufficiency and minimal sufficiency. The concept of completeness of a family of distributions is considered in section 15.5. Section 15.6 considers uniformly minimum variance unbiased estimation. This section also considers consistent asymptotically normal estimates. Different lower bounds for variance of estimators are considered in section 15.7. Lastly we consider criteria for comparison among several competing statistics for estimating a parameter vector $\boldsymbol{\theta}$ in the multivariate set up. The last two sections consider robust estimators and estimators based on resampling methods respectively.

Definition 15.1.1 *Any function of the random variables $X_1, \cdots, X_n$ that are being observed, say, $t_n(X_1, \cdots, X_n)$ is called a statistic. Clearly, a statistic is a random variable. If it is used to estimate a unknown parameter θ of the distribution, it is called an estimator. A particular value of the estimator, say, $t_n(x_1, \cdots, x_n)$ is called an estimate of θ.*

We shall, however, often use the terms *estimator* and *estimate* somewhat loosely, their actual implications being clear from the context.

15.2 Unbiasedness

Definition 15.2.1 ***Unbiasedness.*** *Let X be a r.v. with p.d.f. $f_\theta(x)$, $-\infty < x < \infty$, $\theta \in \Theta$.*

A statistic $t(X)$ is said to be unbiased for θ if

$$E_\theta\left[t(X)\right] = \int_{-\infty}^{\infty} t(x) f_\theta(x)\, dx = \theta \;\; \forall\, \theta \in \Theta. \tag{15.2.1}$$

Any function $\psi(\theta)$ for which a function $t(X)$ exists satisfying $E_\theta\left[t(X)\right] = \psi(\theta) \;\; \forall\, \theta$ is called an estimable function.

Definition 15.2.2 ***Bias of an estimator.*** *Bias of an estimator $t_n(\mathbf{X})$ when $\mathbf{X} = (X_1, \cdots, X_n)'$ in estimating θ is*

$$b_n(t_n, \theta) = E_\theta\left[t_n(\mathbf{X}) - \theta\right]. \tag{15.2.2}$$

■ NOTE 15.2.1 (15.2.1) *requires that t be integrable (summable) for every $\theta \in \Theta$ i.e., $E_\theta(t)$ exists. The definition can be easily extended to the case where θ is a vector of parameters and t is a random vector.*

Definition 15.2.3 *An estimator $t_n(\mathbf{X})$ is asymptotically unbiased estimator of θ if*

$$\lim_{n \to \infty} b_n(t_n, \theta) = 0 \;\; \forall\, \theta \in \Theta. \tag{15.2.3}$$

Definition 15.2.4 *The quantity $E_\theta\left(t(X) - \theta\right)^2$ is called the mean square error (m.s.e.) of $t(X)$ about θ.*

$$\begin{aligned} MSE(t(X)) &= E_\theta\left[t(X) - \theta\right]^2 \\ &= E_\theta\left[t(X) - E_\theta(t(X))\right]^2 + \left[E_\theta(t(X)) - \theta\right]^2 \\ &= V_\theta(t(X)) + \left[\text{Bias}\ (t(X), \theta)\right]^2 \end{aligned} \tag{15.2.4}$$

where $V_\theta(t(X))$ is the variance of t. If t is unbiased for θ, $MSE(t)$ reduces to $V(t)$.

In the terminology of decision theory, if a squared error loss function is assumed so that the loss in estimating θ by $t(X)$ is $\left(t(X) - \theta\right)^2$, the risk function which is defined as the expectation of the loss function is the variance of $t(X)$ if t is unbiased for θ.

▶ **Example 15.2.1** *Let X be a binomial r.v. with parameters n and θ. Consider the estimator*

$$t_n(X) = \frac{X + \frac{\sqrt{n}}{2}}{n + \sqrt{n}}.$$

Since

$$E(X) = n\theta, \;\; E_\theta\left[t_n(X)\right] = \frac{n\theta + \frac{\sqrt{n}}{2}}{n + \sqrt{n}} \neq \theta.$$

Hence t_n is a biased estimator.

However, bias of $t_n(X)$, $b(t_n(X), \theta) = E\left[t_n(X) - \theta\right] = \frac{\frac{1}{2} - \theta}{\sqrt{n} + 1} \to 0$ as $n \to \infty$. Thus, $t(X)$ is asymptotically unbiased.

▶ **Example 15.2.2** *An unbiased estimator may sometimes be absurd. Let X be a r.v. having a Poisson p.m.f., $P_\lambda(X = x) = e^{-\lambda}\frac{\lambda^x}{x!}, x = 0, 1, 2, \cdots$. Suppose we want to estimate $\psi(\lambda) = e^{-3\lambda}$.*

Consider the statistic $t(X) = (-2)^X$.

$$E_\lambda\left[t(X)\right] = e^{-\lambda}\sum_{x=0}^{\infty}(-2)^x \cdot \frac{\lambda^x}{x!} = e^{-3\lambda}. \tag{15.2.5}$$

However, the statistic $t(X)$ is absurd as an estimator, because $\psi(\lambda)$ is always positive, while $t(X)$ is sometimes negative.

▶ **Example 15.2.3** *Unbiased estimates may not always exist. Consider a Bernoulli r.v. with parameter p. Let $\psi(p) = p^2$. For an unbiased estimator $t(X)$ we must have*

$$E_p(t) = pt(1) + (1-p)t(0) = p^2 \ \forall\, p \in [0, 1] \tag{15.2.6}$$
$$i.e.\ p^2 - p\{t(1) - t(0)\} - t(0) = 0 \ \forall\, p \in [0, 1].$$

This is not possible, because if a convergent power series $\sum_0^\infty C_n z^n$ vanishes for all values of z lying in a closed interval, the coefficients $C_0, C_1, \cdots,$ must be identically zero (vide lemma 15.7.1). Hence no unbiased estimator of p^2 exists (see also exercise 4 in section 15.3).

▶ **Example 15.2.4** $X_1, \cdots, X_n$ *is a random sample from the population*

$$f(x) = \begin{cases} e^{-(x-\theta)} & \text{for } x > \theta \\ 0 & \text{otherwise.} \end{cases}$$

Note that

$$E(X) = \int_\theta^\infty xe^{-(x-\theta)} = 1 + \theta,$$

so that $E(\bar{X}) = E(X) = 1 + \theta$. Thus $\bar{X}$ is a biased estimator of θ.
However, $E(\bar{X} - 1) = \theta$.

▶ **Example 15.2.5** *Let $(X_1, \cdots, X_n)$ be a random sample from uniform distribution*

$$f_\theta(x) = \begin{cases} \frac{1}{\theta}, \ 0 < x < \theta \\ 0, \ \text{otherwise.} \end{cases}$$

Consider $X_{(n)} = \max_i X_i$. The p.d.f. of $X_{(n)}$ is

$$g_\theta(x_{(n)}) = \begin{cases} \frac{n}{\theta}\left[\int_0^{x(n)} \frac{dx}{\theta}\right]^{n-1} = \frac{n}{\theta^n}x_{(n)}^{n-1}, \ 0 < x_{(n)} < \theta \\ 0 \qquad\qquad \text{otherwise.} \end{cases}$$

$E(X_{(n)}) = \frac{n}{n+1}\theta$. Thus $X_{(n)}$ is not unbiased for θ, though it is asymptotically so. However, $\frac{n+1}{n}X_{(n)}$ is unbiased for θ. Check that $2\bar{X}$ is also unbiased for θ.

► **Example** 15.2.6 *Consider the geometric distribution* $P(N = x) = q^{x-1}p$, $x = 1, 2, \cdots$ $q = 1 - p$, $0 < p < 1$. *Suppose we wish to find functions of N alone,* $\psi(N)$ *say, which are unbiased for* p.

$$E_p[\psi(N)] = p \ \forall\, p \text{ i.e., } \sum_{x=1}^{\infty} \psi(x)pq^{x-1} = p$$

$$\text{i.e., } \sum_{x=1}^{\infty} \psi(x)q^{x-1} = 1 \ \forall\, p.$$

Now it is known that the two power series $\sum_0^\infty a_j z^j$ and $\sum_0^\infty b_j z^j$ have the same value for every z in some interval iff $a_j = b_j \ \forall\, j = 0, 1, 2, \cdots$. Hence we must have $\psi(1) = 1$. Therefore, the estimator is

$$\psi(N) = \begin{cases} 1 \text{ if } N = 1 \\ 0 \text{ if } N > 1. \end{cases}$$

The function $\psi(N)$ is absurd as an estimator of p.
Hence for geometric distribution no sensible estimator of p based on N alone exists

15.2.1 Correction for bias

In case $E(t)$ is a complicated function of θ, say, $E(t) = \theta + b(\theta)$, Quenouille (1956) proposed a method of correcting for bias in a general class of situations. Suppose t_n is a biased but consistent (see definition 15.3.1) estimator of θ and

$$E(t_n - \theta) = \frac{a_1}{n} + \frac{a_2}{n^2} + \frac{a_3}{n^3} + \cdots. \tag{15.2.7}$$

Consider a new estimator

$$t'_n = nt_n - (n-1)\bar{t}_{n-1} \tag{15.2.8}$$

where

$$\bar{t}_{n-1} = \frac{1}{n}\sum_{i=1}^{n} t_{n-1}(i),$$

the statistic $t_{n-1}(i)$ is based on all the observations except x_i.

$$E(t'_n) = \theta - \frac{a_2}{n^2} - \frac{a_2 + a_3}{n^3} - \cdots.$$

Thus t'_n is also consistent but biased to order n^{-2}.
Similarly,

$$t''_n = \left\{n^2 t'_n - (n-1)^2 t'_{n-1}\right\} / \left\{n^2 - (n-1)^2\right\}$$

will be consistent but biased only to order n^{-3}.
This is known as method of *jack-knifing*.

▶ **Example 15.2.7** *For Binomial (n, p) population, an intuitive estimator of p^2 is $t_n = (x/n)^2$ where x is the number of successes. Now, t_{n-1} takes value $(x/(n-1))^2$ if an observation with failure is omitted and $\left(\frac{x-1}{n-1}\right)^2$ if an observation with success is omitted. Hence,*

$$\bar{t}_{n-1} = \frac{1}{n}\left[(n-x)\left(\frac{x}{n-1}\right)^2 + x\left(\frac{x-1}{n-1}\right)^2\right] = \frac{x^2(n-2)+x}{n(n-1)^2}.$$

Hence

$$t'_n = nt_n - (n-1)\bar{t}_{n-1} = \frac{x(x-1)}{n(n-1)}.$$

It can be verified that t'_n is exactly unbiased for p^2.

If $t(X)$ is unbiased for $\theta, \psi(t)$ is not, in general, unbiased for $\psi(\theta)$ unless $\psi(\theta)$ is a linear function. This is one of the difficulties associated with the concept of unbiasedness. Another difficulty is that unbiased estimators are not necessarily unique (example 15.2.5). Again, if t_1, t_2 are two unbiased estimators of θ, any convex combination $\lambda t_1 + (1-\lambda)t_2, 0 < \lambda < 1$ is also unbiased for θ. Hence, we have, here, an infinite number of unbiased estimators for θ. One, therefore, has to find some criteria for differentiating among these competing estimators. We shall consider this point in details in section 15.6.

15.3 Consistency

It is intuitively clear that for t_n to be a good estimator the difference $|t_n - \theta|$ should be as small as possible. However, t_n is a random variable and has its own sampling distribution whose range may be infinitely large. We will be, therefore, satisfied if the sampling distribution of t_n becomes more and more concentrated around θ as the sample size n increases. This means that for each fixed $\theta \in \Theta$, the probability

$$P_\theta\left[|t_n - \theta| \le \varepsilon\right] \tag{15.3.1}$$

for any given $\varepsilon(> 0)$ should be an increasing function of n. This idea leads to the concept of consistency as a criterion of a good estimator. Hence we have.

Definition 15.3.1 *A statistic t or rather the sequence $\{t_n\}$ is said to be consistent for θ if t_n converges in probability to $\theta (t_n \xrightarrow{P} \theta)$ as $n \to \infty$ for each fixed $\theta \in \Theta$. Thus t is said to be consistent if for every fixed $\theta \in \Theta$ and every pair of positive quantities ε and η, however, small, it is possible to find an n_0, depending on ε and η, such that*

$$P_\theta\left[|t_n - \theta| < \varepsilon\right] > 1 - \eta \tag{15.3.2}$$

whenever $n \ge n_0(\varepsilon, \eta)$.

If such statistics are used the accuracy of the estimate increases with the increase in the value of n. It is to be noted that consistency is a large sample property as it is concerned with the behaviour of an estimator as the sample size becomes infinitely large. Moreover, one should speak of consistency of a sequence of estimators rather than one point estimator.

Further, if there exists a consistent estimate t_n we may construct a number of other consistent estimates. For example, $\frac{a+n}{b+n}t_n, t_n + \frac{a}{n}$ where a, b are finite constants are also consistent.

▶ **Example 15.3.1** *Consider a random sample of size n from a Bernoulli population with parameter p. It is known that by Bernoulli's* WLLN *the sample mean*

$$\bar{X}_n \left(= \frac{1}{n} \sum_{i=1}^{n} X_i \right) \xrightarrow{P} p$$

for every fixed $p \in [0, 1]$. *Hence* $\bar{X}$ *is consistent for p.*

▶ **Example 15.3.2** *Let* $\{X_n\}$ *be a sequence of iid random variables having a finite moment of rth order (r, a positive integer). Then by Khintchen's weak law of large numbers (Section 9.6)*

$$\frac{1}{n}\sum_{i=1}^{n} X_i^k \xrightarrow{P} E(X^k) \text{ for } 1 \le k \le r.$$

Hence the kth sample moment is a consistent estimator of the population moment.

▶ **Example 15.3.3** *Let* $\{X_n\}$ *be a random sample from a distribution with finite mean and variance,* μ *and* σ^2 *respectively. It is known by Chebychev's* WLLN *(Section 9.6)* $X_n \xrightarrow{P} \mu$. *Hence the sample mean* $\bar{X}$ *is a consistent estimator of* μ.

By Khintchen's WLLN, $\bar{X}_n \xrightarrow{P} \mu$ even if the variance does not exist. Hence $\bar{X}$ is a consistent estimator of μ when the mean of the distribution exists.

▶ **Example 15.3.4** *Let* $\{X_n\}$ *be a sequence of iid* $N(\mu, \sigma^2)$ *r.v.'s. If*

$$s_n^2 = \frac{1}{n-1}\sum_{i=1}^{n} (X_i - \bar{X}_n)^2$$

then it is known that

$$(n-1)s_n^2/\sigma^2 \sim \chi^2_{n-1}.$$

Hence

$$E\left(s_n^2/\sigma^2\right) = 1, \quad V\left(s_n^2/\sigma^2\right) = 2/(n-1).$$

It follows by Chebychev's inequality,

$$P\left\{|s_n^2 - \sigma^2| > \varepsilon\right\} \le \frac{V(s_n^2)}{\varepsilon^2} = \frac{2\sigma^4}{(n-1)\varepsilon^2} \to 0 \text{ as } n \to \infty.$$

Thus $s^2 \xrightarrow{P} \sigma^2$. Hence s^2 is consistent for σ^2.

▶ **Example 15.3.5** *For the population in example* 15.2.4 *show that the statistic* $X_{(1)} = \min_i X_i$ *is a consistent estimator of* θ.

The p.d.f. of $X_{(1)}$ is

$$g(x_{(1)}) = n \cdot e^{-(x_{(1)}-\theta)} \left[\int_{x_{(1)}}^{\infty} e^{-(x-\theta)}\, dx \right]^{n-1} = n\, e^{-n(x_{(1)}-\theta)}.$$

for $x_{(1)} > \theta$ and $g(x_{(1)}) = 0$ otherwise. Hence, $E(X_{(1)}) = \theta + \frac{1}{n}$ so that $X_{(1)}$ is asymptotically unbiased for θ.

$$\begin{aligned} P\left[|X_{(1)} - \theta| < \varepsilon\right] &= P\left[\theta < X_{(1)} < \theta + \varepsilon\right] \\ &= \int_{\theta}^{\theta+\varepsilon} n\, e^{-n(x_{(1)}-\theta)}\, dx_{(1)} \\ &= 1 - e^{-n\varepsilon} \to 1 \text{ as } n \to \infty. \end{aligned}$$

Hence $X_{(1)}$ is consistent for θ.

▶ **Example 15.3.6** *Let* $X_1, \cdots, X_n$ *be a random sample from the uniform distribution*

$$f(x, \theta) = \begin{cases} 1/\theta, & 0 \le x \le \theta \\ 0\ , & \textit{elsewhere.} \end{cases}$$

Examine the consistences of the estimators $T_1 = \max_i X_i$, $T_2 = \min X_i + \max X_i$, $T_3 = (n+1) \min X_i$, $T_4 = 2\bar{X}$ *for estimating* θ.

We have

$$\begin{aligned} P(T_1 \le t) &= P(X_1 \le t, \cdots, X_n \le t) \\ &= \left(\frac{t}{\theta}\right)^n. \end{aligned}$$

Therefore, for $0 < \varepsilon < \theta$,

$$\begin{aligned} P\left[|T_1 - \theta| \le \varepsilon\right] &= P\left[T_1 \ge \theta - \varepsilon\right] \\ &= 1 - \left(\frac{\theta - \varepsilon}{\theta}\right)^n \to 1 \text{ as } n \to \infty. \end{aligned}$$

Here T_1 is consistent for θ.

Again,

$$P\{\min_i X_i > \varepsilon\} = \prod_{i=1}^{n} P(X_i > \varepsilon) = \left(1 - \frac{\varepsilon}{\theta}\right)^n \to 0.$$

Thus $\min_i X_i \xrightarrow{P} 0$.

Consequently,

$$T_2 = T_1 + \min_i X_i \xrightarrow{P} \theta.$$

Thus T_2 is consistent for θ.
Again,

$$\begin{aligned} P\{|T_3 - \theta| \le \varepsilon\} &= P\{|(n+1)\min X_i - \theta| \le \varepsilon\} \\ &= P\left\{\frac{\theta-\varepsilon}{n+1} \le \min X_i \le \frac{\theta+\varepsilon}{n+1}\right\} \\ &= F\left(\frac{\theta+\varepsilon}{n+1}\right) - F\left(\frac{\theta-\varepsilon}{n+1}\right) \qquad \text{(i)} \end{aligned}$$

where $F(\cdot)$ is the cdf of $\min\limits_i X_i$ and is given in section 8.6. Thus (i)

$$\begin{aligned} &= 1 - \left(1 - \frac{\theta+\varepsilon}{\theta(n+1)}\right)^n - \left(1 - \left(1 - \frac{\theta-\varepsilon}{\theta(n+1)}\right)^n\right) \\ &= \left(1 - \frac{\theta-\varepsilon}{\theta(n+1)}\right)^n - \left(1 - \frac{\theta+\varepsilon}{\theta(n+1)}\right)^n \\ &\to e^{-(\theta-\varepsilon)/\theta} - e^{-(\theta+\varepsilon)/\theta} < 1. \end{aligned}$$

Thus $\lim\limits_{n\to\infty} P\{|T_3 - \theta| > \varepsilon\} > 0$, which shows that T_3 is not a consistent estimator of θ.
Again, by the law of large numbers,

$$\bar{X} \xrightarrow{P} E(X_1) = \frac{\theta}{2}$$

so that

$$T_4 = 2\bar{X} \xrightarrow{P} \theta.$$

Therefore, T_4 is consistent for θ.

Theorem 15.3.1 *Let $\{t_n\}$ be a sequence of estimates such that for every $\theta \in \Theta$, the expectation and variance of t_n exist and $E(t_n) \to \theta$ and $V(t_n) \to 0$ as $n \to \infty$. Then t_n is consistent for θ.*

Proof: We have, by Chebychev's inequality

$$\begin{aligned} P\{|t_n - \theta| > \varepsilon\} &< \frac{E(t_n - \theta)^2}{\varepsilon^2} \\ &= \frac{V(t_n) + (E(t_n) - \theta)^2}{\varepsilon^2} \to 0 \text{ as } n \to \infty. \end{aligned}$$

Example 15.3.7 *Let $X_1, \cdots, X_n$ be a random sample from a continuous distribution ith p.d.f. f_0. Let μ_e be the median of X and let $1/f(\mu_e)$ be finite. If $\tilde{X}$ be the sample edian, then*

$$E(\tilde{X}_n) = \mu_e + 0\left(\frac{1}{n}\right)$$

$$V(\tilde{X}_n) = \frac{1}{4n[f(\mu_e)]^2} + 0\left(\frac{1}{n}\right)$$

vide Theorem 11.6.3). Hence, $E(\tilde{X}_n) \to \mu_e$ and $V(\tilde{X}_n) \to 0$ for every $\theta \in \Theta$. Thus $\tilde{X}_n$ s consistent for μ_e.

Example 15.3.8 *Let $\{X_n\}$ be a sequence of iid random variables for which moment f order $2r$ about $0, \mu'_{2r}$ exists.*

Let m'_r, m_r, μ_r denote, respectively, the sample raw and central moment and opulation central moment. Then

$$E(m'_r) = \mu'_r, \quad V(m'_r) = \frac{\mu'_{2r} - \mu'^2_r}{n};$$

ence m'_r is consistent for μ'_r.

Since μ'_{2r} exists so does μ_{2r}. Again,

$$E(m_r) = \mu_r + 0\left(\frac{1}{n}\right)$$

$$V(m_r) = \frac{1}{n}\left\{\mu_{2r} - \mu_r^2 - 2r\mu_{r-1}\mu_{r+1} + r^2\mu^2_{r-1}\mu_2\right\} + 0\left(\frac{1}{n^2}\right).$$

ence, $E(m_r) \to \mu_r$, $V(m_r) \to 0$ for every $\theta \in \Theta$ and m_r is consistent for μ_r.

The following theorem gives an invariance property of a consistent estimator.

heorem 15.3.2 *Let t be a consistent estimator for θ and let $\psi(\theta)$ be a continuous nction of θ. Then $\psi(t)$ is a consistent estimator of $\psi(\theta)$.*

Proof: Since t is consistent for θ, given $\varepsilon(> 0)$ and $\eta(> 0)$, however small, we can nd, for each $\theta \in \Theta$, an $n_0(\varepsilon, \eta)$ such that

$$P\{|t_n - \theta| \leq \varepsilon\} > 1 - \eta$$

henever $n \geq n_0$.

Again, since $\psi(\theta)$ is a continuous function, for each $\theta \in \Theta$, given $\varepsilon_1 > 0$, we can noose $\varepsilon > 0$ such that

$$|t_n - \theta| \leq \varepsilon \Rightarrow |\psi(t_n) - \psi(\theta)| \leq \varepsilon_1.$$

ence for each $\theta \in \Theta$

$$P\left[|\psi(t_n) - \psi(\theta)| < \varepsilon_1\right] \geq P\left[|t_n - \theta| \leq \varepsilon\right] > 1 - \eta$$

henever $n \geq n_0(\varepsilon, \eta)$. Hence $\psi(t_n)$ converges stochastically to $\psi(\theta)$ for each $\theta \in \Theta$.

15.3.1 Generalisation to multiparametric case

Let the probability density function of the r.v. X be $f_X(\theta_1, \cdots, \theta_k)$ involving parameters $\theta_1, \cdots, \theta_k$. Consider k functionally unrelated statistic $t_{1n}, t_{2n}, \cdots, t_{kn}$. Le $\boldsymbol{\theta} = (\theta_1, \cdots, \theta_k)'$, $t_n = (t_{1n}, \cdots, t_{kn})'$ and $\theta \in \Theta$ where $\theta \in R^k$.

The sequence $\{\boldsymbol{t}_n\}$ will be consistent for $\boldsymbol{\theta}$ if $\boldsymbol{t}_n$ converges in probability to $\boldsymbol{\theta}$ as $n \to \infty$ for each fixed $\boldsymbol{\theta} \in \Theta$, i.e., for every fixed $\boldsymbol{\theta}$ and every pair of positive quantities ε and η however, small, it is possible to find as $n_0(\varepsilon, \eta)$ such that

$$P\left[|\boldsymbol{t}_n - \boldsymbol{\theta}| > \boldsymbol{\varepsilon}\right] < \eta \tag{15.3.1}$$

whenever $n \geq n_0(\varepsilon, \eta)$, where $\boldsymbol{\varepsilon} = (\varepsilon, \cdots, \varepsilon)'$.

Theorem 15.3.3 *A sufficient condition for $\mathbf{t}_n$ to be consistent for $\boldsymbol{\theta}$ is that each t_i should be consistent for $\theta_i (i = 1, \cdots, k)$.*

Proof: If t_i is consistent for θ_i then

$$P\left\{|t_{in} - \theta_i| > \varepsilon\right\} < \eta/k, \quad \text{for } n > n_{0i}(\varepsilon, \eta),$$

$i = 1, \cdots, k$. Let $n_0 = \max\limits_{i=1,\cdots,k} n_{0i}$.

Therefore, $P\left\{|t_{in} - \theta_i| > \varepsilon\right\} < \eta/k$ for $n > n_0(\varepsilon, \eta)$ holds for all $i = 1, \cdots, k$.

Let S_i denote the event $\{|t_{in} - \theta_i| > \varepsilon\}$, the complimentary event being $S_i^* = \{|t_{in} - \theta_i| \leq \varepsilon\}$. Now, by Boole's inequality,

$$P\left(\bigcup_{i=1}^{k} S_i\right) \leq \sum_{i=1}^{k} P(S_i).$$

Hence,

$$P\left(\bigcap_{i=1}^{k} S_i^*\right) = 1 - P\left(\bigcup_{i=1}^{k} S_i\right) \geq 1 - \sum_{i=1}^{k} P(S_i) \geq 1 - \eta$$

whenever $n \geq n_0(\varepsilon, \eta)$. Hence $\boldsymbol{t}_n$ is consistent for $\boldsymbol{\theta}$.

■ NOTE 15.3.1 *We have seen that t is consistent for θ, if $t_n \xrightarrow{P} \theta$. Similarly, one ca define strong consistency of a sequence of estimators if t_n converges almost surely to θ i.e., $t_n \xrightarrow{a.s.} \theta$. One can, also, define consistency in the rth mean if $t_n \xrightarrow{r} \theta$. In this treatise whenever we say than t_n is consistent we shall mean weak consistency i.e., $t_n \xrightarrow{P} \theta$.*

Definition 15.3.2 *A statistic $t_n(X_1, \cdots, X_n)$ is consistent in mean square (or consistent in quadratic mean) if*

$$E_\theta\left(t_n - \theta\right)^2 \to 0 \ \text{ as } n \to \infty \ \forall\, \theta \in \Theta.$$

Theorem 15.3.4 *A statistic $t_n(X_1, \cdots, X_n)$ is consistent in mean square iff* (i) *$V_\theta(t_n) \to 0$ as $n \to \infty$ and* (ii) *t_n is asymptotically unbiased.*

Proof: $E_\theta\left(t_n - \theta\right)^2 = V_\theta(t_n) + \left[b_n(t_n, \theta)\right]^2 \to 0$ iff both (i) and (ii) holds.

Theorem 15.3.5 *If $E_\theta(t_n - \theta)^2 \to 0\ \forall\, \theta$ then $t_n \xrightarrow{P} \theta\ \forall\, \theta$.*

Proof: By Chebychev's inequality, for any $\varepsilon > 0$

$$P\left[|t_n - \theta| \le \varepsilon\right] \ge 1 - \frac{E_\theta (t_n - \theta)^2}{\varepsilon^2} \to 1 \text{ as } n \to \infty \text{ (by the condition of the theorem).}$$

We now consider the following theorem, given without proof, on consistency and unbiasedness of sample mean and median.

Theorem 15.3.6 (***Dudewicz*, 1976, *p*.181-84**). *Let $X_1, \cdots, X_n$ be independent r.v.'s, each with d.f. $F(x|\theta), \theta \in \Theta$. Then, supposing, $E_\theta(X_1)$ exists,*

(i) $\bar{X}$ is unbiased and consistent for $E_\theta(X_1)$;

(ii) Sample median $\tilde{X}$ is a consistent (not generally unbiased) estimator of population median μ_e.

(iii) $\tilde{X}$ is an unbiased estimator of μ_e if X_i has a p.d.f. that is symmetric about μ_e.

▶ **Example 15.3.9** *If $X_1, \cdots, X_n$ are iid $N(\mu, \sigma^2)$, both $\bar{X}$ and $\tilde{X}$ are unbiased and consistent for μ.*

▶ **Example 15.3.10** *Let $X_1, \cdots, X_n$ be a random sample from the distribution $F(x)$. The sample cumulative distribution function $F_n(x)$ has a binomial distribution with parameters $(n, F(x))$ (vide section 11.7). Here $F_n(x)$ is a unbiased estimator of $F(x)$ with variance*

$$\text{Var}[F_n(x)] = \frac{F(x)[1 - F(x)]}{n} \to 0$$

as $n \to \infty$. Thus $F_n(x)$ is consistent in mean square and hence weakly consistent for $F(x)$. Furthermore, in a direct application of the Strong Law of Large Numbers.

$$F_n(x) \xrightarrow{a.s.} F(x).$$

It follows from the Lindeberg-Levy CLT, that for each fixed x, $-\infty < x < \infty$, $F_n(x)$ is asymptotically normal $\left(F(x), \frac{F(x)(1-F(x))}{n}\right)$.

Exercise

1. Let $X_1, \cdots, X_n$ be iid Bernoulli (p) r.v.'s and let

$$S_n = \sum_{i=1}^{n} X_i.$$

   (a) Show that $S_n(S_n - 1)/n(n-1)$ is unbiased for p^2 and $S_n(S_n-1)(S_n-2)/n(n-1)(n-2)$ is unbiased for p^3. Examine the consistency of these estimators.

(b) Show that $S_n(n-S_n)/\{n(n-1)\}$ is a unbiased estimator of pq (when $q = 1-p$) and has a smaller variance than the estimator $\frac{S_n}{n}\left(1-\frac{S_n}{n}\right)$.

(c) Find an unbiased estimator of pq^2. Show that $\frac{S_n}{n}\left(1-\frac{S_n}{n}\right)^2$ is not unbiased for pq^2 and its bias tends to zero as $n \to \infty$.

2. Let $Z_1, \cdots, Z_n$ be iid r.v.'s. Let c be a fixed quantity $(-\infty < c < \infty)$. Define

$$X_i = \begin{cases} 1, \text{ if } Z_i \geq c \\ 0, \text{ otherwise.} \end{cases}$$

Let $P = \text{Prob}\ (Z_i \geq c)$, $Q = 1 - P$, and $S_n = \sum_{i=1}^{n} X_i$.

Show that the statistic $t_n = \frac{S_n}{n}\left(1-\frac{S_n}{n}\right)$ is not unbiased for PQ, though it is consistent for the same.

[*Hints.* Show that $E\left\{\frac{n}{n-1}t_n\right\} = PQ$. Also show that S_n/n is consistent in mean square to P. Then apply theorem 15.3.5.]

3. Let $(X_1, \cdots, X_n)$ be iid Poisson (λ) variables. Show that $t_n = \bar{X}_n^2 - \bar{X}_n$ is a biased estimator of λ^2. Find its bias $b_n(\lambda)$ and hence find a unbiased estimator of λ^2.

4. Show that for a Poisson (λ) distribution, there is no exactly unbiased estimator of $1/\lambda$.

5. Let X be a truncated Poisson r.v. with

$$P(X = r) = \frac{e^{-\lambda}\lambda^r}{r!(1-e^{-\lambda})}, r = 1, 2, \cdots$$

λ being an unknown parameter. A random sample of size n contains n_r observations having the value r. Show that

$$\hat{\lambda} = \sum_{r=2}^{\infty} r n_r / n$$

is an unbiased estimate of λ.

6. Let X be a random variable having a hypergeometric distribution

$$P_N(X = K) = \frac{\binom{M}{K}\binom{N-M}{n-K}}{\binom{N}{n}}, \qquad K = 0, 1, \cdots, n;\ \ N = M, M+1, \cdots.$$

Show that there does not exist a unbiased estimator of N.

7. Let $X_1, \cdots, X_n$ be iid r.v.'s with $E(X_i) = \theta$ and $E|X_i|^2 < \infty$. Show that $t(X_1, \cdots, X_n) = 2\sum_{i=1}^{n} iX_i/n(n+1)$ is a consistent estimator of μ.

8. Show that if t is a unbiased estimator of θ, then t^2 is a biased estimator of θ^2 but if t is a consistent estimator of θ, then t^2 is also a consistent estimator of θ^2.

9. If $S^2 = \frac{1}{(n-1)}\sum_{i=1}^{n}(X_i - \bar{X})^2$ is the variance of a random sample from an infinite population with finite variance σ^2 show that $E(S^2) = \sigma^2$. Also show that S is not unbiased for σ.

10. Let $X_1, \cdots, X_n$ be independent r.v.'s each with p.d.f. $f(x) = \lambda e^{-\lambda x}, 0 \le x < \infty$, $\lambda > 0$. Show that $1/\bar{X}$ is a biased estimator of λ. Compute its bias. Based on this find a unbiased estimator of λ.

11. Let X have p.d.f. $f(x;\theta) = \frac{2(x-\theta)}{(1-\theta)^2}, \theta < x < 1, 0 < \theta < 1$. Show that $E(X - \theta) = \frac{2}{3}(1-\theta)$. Hence find a unbiased estimator of θ based on a sample of size one. If $(X_1, \cdots, X_n)$ be a random sample of size n from the above population, find a function of $\bar{X}_n$ which is unbiased for θ and find its bias.

12. Let $(X_1, \cdots, X_n)$ be a random sample from a gamma (α, β) population

$$f(x;\alpha,\beta) = \begin{cases} \dfrac{1}{\beta^\alpha \overline{|\alpha}}\, e^{-x/\beta} x^{\alpha-1}, & 0 \le x < \infty, \beta > 0, \alpha > 1 \\ 0 & \text{otherwise} \end{cases}$$

where value of α is known. Show that the statistic $t_n = \bar{X}_n/\alpha$ is unbiased and mean-square consistent for β.

Show also that the statistic $\sum_{i=1}^{n} X_i^2/n\alpha(\alpha+1)$ is unbiased and consistent for β^2.

13. Let $(X_1, \cdots, X_n)$ be a random sample from $N(0, \sigma^2)$. Show that some multiple of $\sum_{1}^{n}|X_i|$ is consistent for σ^2.

14. Let $(X_1, \cdots, X_n)$ be a random sample from a rectangular distribution $f(x;\theta) = \frac{1}{\theta}$, $0 < x < \theta$, $\theta > 0$. Show that the geometric mean

$$t(\mathbf{X}) = \left(\prod_{i=1}^{n} X_i\right)^{1/n}$$

is a consistent estimator of θe^{-1}.

15. Prove that the sample mean from a Cauchy population

$$f(x;\theta) = \frac{1}{\pi\{1 + (x-\theta)^2\}}$$

is not a consistent estimator of θ. However, sample median is a consistent estimator of θ.

16. Let $X_1, \cdots, X_n$ be a random sample from Laplace p.d.f.

$$f(x;\theta) = \frac{1}{2}e^{-|x-\theta|}, \quad -\infty < x < \infty, \quad -\infty < \theta < \infty.$$

Let $X_{(1)} \leq \cdots \leq X_{(n)}$ be the order statistic. Show that $t_n = \frac{X_{(1)}+X_{(n)}}{2}$ is unbiased for θ.

17. Let $X_1, \cdots, X_n$ be iid r.v.'s $N(\mu, 1)$. Find a unbiased estimator of μ^2 that is a function of $\bar{X}$.

[*Hints:* Consider $\bar{X}_n^2 - \frac{1}{n}$.]

18. Let $(X_1, \cdots, X_n)$ be a random sample from a $N(\mu, \mu)$ population for some $\mu > 0$ Find a consistent estimator of μ^2.

[*Hints:* Since $E(\bar{X}) = \mu, E(s^2) = \mu$, consider $t_n = \bar{X}_n s_n^2$.]

19. A sample of n independent observations $x_1, \cdots, x_n$ is taken from a normal population (μ, σ^2). Show that

$$t = \frac{1}{2(n-1)} \sum_{i=1}^{n} (x_{i+1} - x_i)^2$$

is an unbiased estimator of σ^2 and the efficiency of t with respect to the usual estimate

$$s^2 = \frac{1}{(n-1)} \sum_{i=1}^{n} (x_i - \bar{x})^2, \quad \bar{x} = \frac{1}{n} \sum_{i=1}^{n} x_i, \quad \text{is } \frac{2(n-1)}{3n-4}.$$

[*Hints:* Efficiency of an estimator t_1 with respect to another estimator t_2 is $V(t_2)/V(t_1)$.]

20. Show that in sampling from a $N(\theta, \sigma)$ population, the estimators

$$t_1 = \sqrt{\frac{1}{2}\sum_{i=1}^{n} X_i^2}\left(\frac{\Gamma\left(\frac{n}{2}\right)}{\Gamma\left(\frac{n+1}{2}\right)}\right),$$

$$t_2 = \sqrt{\frac{1}{2}\sum_{i=1}^{n} (X_i - \bar{X})^2}\left(\frac{\Gamma\left(\frac{n-1}{2}\right)}{\Gamma\left(\frac{n}{2}\right)}\right)$$

are both unbiased for σ.

21. Let $(X_1, \cdots, X_n)$ be a random sample from $N(\mu, \sigma^2)$ population, $(\sigma^2 < \infty)$. Show that the sample variance s^2 is unbiased for σ^2. Show that, in general, the kth sample moment m_k is unbiased for the kth population moment μ_k, when it exists. Show that s is not a unbiased estimator of σ and bias of s is

$$b(\sigma, s) = \sigma\left[\sqrt{\frac{2}{n-1}}\,\Gamma\left(\frac{n}{2}\right)\left\{\Gamma\left(\frac{n-1}{2}\right)\right\}^{-1} - 1\right].$$

22. Let S and T be independent statistics with a joint cdf $F(t;\sigma,\tau)\,G(s;\sigma)$ depending on the parameters $(\sigma,\tau)\in\Theta_1\times\Theta_2$. Let $b(\sigma,T)$ be an unbiased estimator of $\{h(\sigma,\tau),\tau\in\Theta_2\}$ for each $\sigma\in\Theta_1$ and let $a(S,\tau)$ be an unbiased estimator of $\{b(\sigma,\tau),\sigma\in\Theta_1\}$ for each $\tau\in\Theta_2$. Then show that $a(S,T)$ is an unbiased estimator of $\{h(\sigma,\tau),(\sigma,\tau)\in\Theta_1\times\Theta_2\}$.

.5.4 Sufficiency, Minimal Sufficiency

'he concept of sufficiency is of paramount importance in mathematical and applied tatistics. After the data have been collected the statistician is confronted with the ask of making some statistically valid conclusion about the parameter of interest. It vould, indeed, be immensely helpful if some summarisation of raw data is possible vhich preserves all the available information about the population. Fisher (1921, 1925) ıtroduced the concept of sufficiency as the one containing all the relevant information. Jeyman (1935) gave a method of finding sufficient statistic. (The term 'information' ere means the relevant knowledge about the parameter. The technical definition of nformation' is given in definition 15.6.6.)

)efinition 15.4.1 *Let $\boldsymbol{X}=(X_1,\cdots,X_n)$ be a sample from a family of distribution unctions $\mathcal{F}=\{F_\theta,\theta\in\Theta\}$. A statistic $T=T(\boldsymbol{X})$ is said to be sufficient for the family or simply for θ, iff the conditional distribution of $\boldsymbol{X}$ given $T=t$ is independent of θ provided that the conditional distribution is defined almost everywhere for each θ).*

The statistic T and the parameter θ may be vectors in which case $\boldsymbol{T}=(T_1,\cdots,T_k)'$ s said to be jointly sufficient for $\boldsymbol{\theta}=(\theta_1,\cdots,\theta_k)'$.

NOTE 15.4.1 *The definition is, however, not quite satisfactory, since the conditional 'istribution may not be defined always. A more rigorous definition is beyond the scope f the present treatise. However, if the r.v.'s are all discrete or all continuous, this 'efinition is quite sufficient.*

Example 15.4.1 *Let $X_i (i=1,\cdots,n)$ be independent r.v.'s each having the p.d.f.* $P(X_i=1)=p$, $P(X_i=0)=1-p$, $0<p<1$, $i=1,\cdots,n$.

Here

$$T(\boldsymbol{X})=\sum_{i=1}^{n}X_i$$

s sufficient for p, because the conditional probability

$$P(X_1=x_1,\cdots,X_n=x_n|T=t)=\frac{P\{(X_1=x_1,\cdots,X_n=x_n)\cap(T=t)\}}{P(T=t)}$$

$$=\begin{cases}\frac{p^{\Sigma_1^n x_i}(1-p)^{n-\Sigma_1^n x_i}}{\binom{n}{t}p^t(1-p)^{n-t}}=\frac{1}{\binom{n}{t}}, & \text{if } \Sigma_{i=1}^n x_i=t\\ 0, & \text{otherwise.}\end{cases}$$

ı any case the conditional probability is independent of p.

► **Example 15.4.2** *Let $X_i (i = 1, \cdots, n)$ be independent r.v.'s each having p.d.f.,*

$$f(x; \lambda) = e^{-\lambda}\lambda^x/x!, \quad \lambda > 0, \quad x_i = 0, 1, 2, \cdots; \quad i = 1, \cdots, n.$$

Then $T(\boldsymbol{X}) = \sum_{i=1}^{n} X_i$ is sufficient for λ, because the p.d.f. of T is

$$P(T = t) = e^{-n\lambda}\frac{(n\lambda)^t}{t!}, \quad t = 0, 1, 2, \cdots$$

and hence

$$P\{X_1 = x_1, \cdots, X_n = x_n | T = t\} = \frac{P\{(X_1 = x_1, \cdots, X_n = x_n) \cap (T = t)\}}{P(T = t)}$$

$$= \begin{cases} \dfrac{e^{-\lambda n}\lambda^{\Sigma_1^n x_i}/\Pi_1^n x_i!}{e^{-\lambda n}(n\lambda)^t/t!} = \dfrac{t!}{n^t \Pi_{i=1}^n x_i!}, & \text{if } \Sigma_1^n x_i = t \\ 0, & \text{otherwise.} \end{cases}$$

In any case the conditional distribution is independent of λ.

► **Example 15.4.3** *Let $X_i (i = 1, \cdots, n)$ be independent r.v.'s each having the discre rectangular distribution $P(X_i = k) = \frac{1}{N}$, $k = 1, 2, \cdots, N$, $N \in Z$, the set of positi integers. Let $X_{(1)} < X_{(2)} < \cdots < X_{(n)}$ be the order statistic. Then $T = X_{(n)}$ is sufficie for the parameter N.*

The p.d.f. of T is

$$P[T(\boldsymbol{X}) = t] = P\left[X_{(n)} \le t\right] - P\left[X_{(n)} < t\right], \qquad t = 1, \cdots, n$$

$$= \left(\frac{t}{N}\right)^n - \left(\frac{t-1}{N}\right)^n.$$

Hence,

$$P[X_1 = x_1, \cdots, X_n = x_n | T(\boldsymbol{x}) = t]$$

$$= \frac{P\left[(X_1 = x_{(1)}, \cdots, X_{(n)} = x_n) \cap X_{(n)} = t\right]}{P(X_{(n)} = t)}$$

$$= \frac{1/N^n}{\left(\frac{t}{N}\right)^n - \left(\frac{t-1}{N}\right)^n} = \frac{1}{t^n - (t-1)^n},$$

which is independent of N.

An alternative definition of sufficiency which follows readily from definition 15.4. may be stated as follows.

Definition 15.4.2 *Let $X_i (i = 1, \cdots, n)$ be a random sample from a population havi p.d.f. $f(x; \theta)$, $\theta \in \Theta$. The statistic $T_1(\boldsymbol{X})$ is called a sufficient statistic for θ iff f any other statistic $T_2(\boldsymbol{X}), \cdots, T_n(\boldsymbol{X})$ the conditional p.d.f. $\chi(t_2, \cdots, t_n | t_1)$ of $T_2, \cdots, T$ given $T_1 = t_1$ is independent of θ.*

Proof: Let us transfrom from $(X_1, \cdots, X_n)$ to $(T_1, \cdots, T_n)$. Suppose the transformation is one-to-one and let $X_i = W_i(\boldsymbol{T})$, $i = 1, \cdots, n$ where $\boldsymbol{T} = (T_1, \cdots, T_n)$.

If T_1 is sufficient for θ, by definition 15.4.1,

$$f(x_1, \cdots, x_n; \theta) = g(t_1, \theta) h(x_1, \cdots, x_n) \tag{15.4.1}$$

where $h(x_1, \cdots, x_n)$ is independent of θ.

The joint p.d.f. of $\boldsymbol{T}$ is, therefore,

$$\begin{aligned} \phi(t_1, \cdots, t_n; \theta) &= |J| f\left[w_1(\boldsymbol{t}), \cdots, w_n(\boldsymbol{t}); \theta\right] \\ &= |J| g(t_1, \theta) h\left[w_1(\boldsymbol{t}), \cdots, w_n(\boldsymbol{t})\right] \end{aligned} \tag{15.4.2}$$

where $|J|$ is the jacobian of transformation, $\boldsymbol{t} = (t_1, \cdots, t_n)'$ when the r.v.'s are of the continuous type and is

$$\phi(t_1, \cdots, t_n; \theta) = g(t_1, \theta) h\left[w_1(\boldsymbol{t}), \cdots, w_n(\boldsymbol{t})\right] \tag{15.4.3}$$

for random variables of the discrete type. If the transformation is m to one ($m \geq 2$), the right hand member of (15.4.2) or (15.4.3) is the sum of m terms of this form. In any case, the marginal distribution of T_1 is

$$\begin{aligned} &\int_{-\infty}^{\infty} \cdots \int_{-\infty}^{\infty} \phi(t_1, \cdots, t_n; \theta)\, dt_2 \cdots dt_n \\ &= g(t_1, \theta) \int_{-\infty}^{\infty} \cdots \int_{-\infty}^{\infty} h\left[w_1(\boldsymbol{t}), \cdots, w_n(\boldsymbol{t})\right] |J| \prod_2^n dt_i \\ &= g_1(t_1, \theta) \quad \text{(say)} \end{aligned}$$

in the continuous case and

$$g(t_1, \theta) \sum_{t_2} \cdots \sum_{t_n} h\left[w_1(\boldsymbol{t}), \cdots, w_n(\boldsymbol{t})\right] = g_1(t_1, \theta)$$

in the discrete case. In any case, the conditional p.d.f. of $T_2, \cdots, T_n$ given $T_1 = t_1$, $k(t_2, \cdots, t_n | t_1)$ is independent of θ, provided the marginal p.d.f. $g_1(t_1, \theta) > 0$. Hence the proof.

In particular, if for the statistic T_1 and any other statistic T_2 the condition

$$\phi(t_1, t_2; \theta) = \phi_1(t_1; \theta) \phi_2(t_1, t_2) \tag{15.4.4}$$

where $\phi(t_1, t_2; \theta)$ is the joint p.d.f. of T_1 and T_2 is satisfied whatever the other statistic T_2, then T_1 is a sufficient statistic. Here ϕ_1 contains t_1 and θ but not t_2; ϕ_2 may contain t_1 and t_2 but not θ.

If (15.4.4) is satisfied, marginal distribution of T_1 is

$$\begin{aligned} \int_{\text{over } t_2} \phi(t_1, t_2)\, dt_2 &= \phi_1(t_1; \theta) \int_{\text{over } t_2} \phi_2(t_1, t_2)\, dt_2 \\ &= \phi_1(t_1; \theta) \phi_3(t_1) = \psi(t_1; \theta) \quad \text{(say)}. \end{aligned}$$

Hence the conditional p.d.f. of T_2 given $T_1 = t_1$ is

$$\chi(t_2|t_1) = \frac{\phi(t_1, t_2; \theta)}{\phi(t_1; \theta)} = \frac{\phi_2(t_1, t_2)}{\phi_3(t_1)},$$

which is independent of θ. Hence, T_1 is sufficient for θ.

The concept of sufficiency means in a nutshell that the statistic should summarise a the information about the population parameter. No other statistic should throw an further light on the population than the one provided by the sufficient statistic. (Th phrase 'information' has been used here in a non-technical sense. A technical definitio of 'information' has been given in definition 15.6.6 and section 16.2).

■ NOTE 15.4.2 *If T is sufficient for θ we need only concentrate on T for getting all th information that the sample contains about θ. In practice there will be several sufficie statistic for a family of distributions and the question arises as to which one should used in a given situation. The matter will be investigated subsequently.*

The procedure of guessing a statistic T and then checking to see whether T sufficient by definition 15.4.1 or the procedure of combining a large number of statistic to test whether T is sufficient by definition 15.4.2 is tedius. The following theorem give necessary and sufficient conditions for a statistic T to be sufficient.

Fisher-Neyman Factorisation Theorem

Theorem 15.4.1 *Let $X_1, \cdots, X_n$ be all discrete or continuous. The statistic T will sufficient for θ iff the joint p.m.f. or p.d.f. of $\mathbf{X} = (X_1, \cdots, X_n)'$ is expressible in th form.*

$$f_{\mathbf{X}}(x_1, \cdots, x_n; \theta) = g_T(t; \theta)h(x_1, \cdots, x_n) \tag{15.4.5}$$

where the first factor in the right hand side depends on θ and on $x_1, \cdots, x_n$ only throug t (which is a particular value of T), while the second factor is independent of θ (θ an T both may be vectors).

Proof: (a) **Discrete case:**

(i) **Necessity:** Let T be sufficient for θ. Then by definition, $P\{\mathbf{X} = \boldsymbol{x}|\ T = t\}$ independent of θ. We may write

$$\begin{aligned} P_\theta\{\mathbf{X} = \boldsymbol{x}\} &= P_\theta\{\mathbf{X} = \boldsymbol{x}, T(\mathbf{X}) = t\} \\ &= P_\theta\{T(\mathbf{X}) = t\}P\{\mathbf{X} = \boldsymbol{x}|T(\mathbf{X}) = t\} \end{aligned}$$

provided the conditional probability $P\{\mathbf{X} = \boldsymbol{x}|T(\mathbf{X}) = t\}$ is well defined. For valu of $\mathbf{X}$ for which $P_\theta\{\mathbf{X} = \boldsymbol{x}\} > 0$ for some θ, define $h(\boldsymbol{x}) = P\{\mathbf{X} = \boldsymbol{x}|T(\mathbf{X}) = t\}$ ar $g_T(t; \theta) = P_\theta\{T(\mathbf{X}) = t\}$ when (15.4.5) follows.

For values of $\boldsymbol{X}$ for which $P_\theta\{\boldsymbol{X} = \boldsymbol{x}\} = 0$ for all θ, define $h(\boldsymbol{x}) = 0$. Hence (15.4.5) holds.

(ii) **Sufficiency:** Suppose that (15.4.5) holds. Then for fixed t,

$$\begin{aligned} P_\theta(T(\boldsymbol{X}) = t) &= \sum_{\{\boldsymbol{x}:T(\boldsymbol{x})=t\}} P_\theta(\boldsymbol{X} = \boldsymbol{x}) = \sum_{\{\boldsymbol{x}:T(\boldsymbol{x})=t\}} g_T(T(\boldsymbol{x});\theta)h(\boldsymbol{x}) \\ &= g_T(t;\theta) \sum_{\{\boldsymbol{x}:T(\boldsymbol{x})=t\}} h(\boldsymbol{x}). \end{aligned} \tag{15.4.6}$$

Suppose that $P_\theta(T = t) > 0$ for some θ. Then

$$P_\theta\{\boldsymbol{X} = \boldsymbol{x}|T = t\} = \frac{P_\theta\{(\boldsymbol{X} = \boldsymbol{x}) \cap (T = t)\}}{P_\theta(T = t)} = \begin{cases} 0, & \text{if } T(\boldsymbol{X}) \neq t \\ \frac{P_\theta(\boldsymbol{X}=\boldsymbol{x})}{P_\theta(T=t)}, & \text{if } T(\boldsymbol{X}) = t. \end{cases}$$

Thus if $T(\boldsymbol{X}) = t$,

$$\frac{P_\theta(\boldsymbol{X} = \boldsymbol{x})}{P_\theta(T = t)} = \frac{h(\boldsymbol{x})}{\sum_{\{x:T(\boldsymbol{x})=t\}} h(\boldsymbol{x})} \quad \text{by (15.4.6).}$$

In any case the conditional probability is independent of θ. Hence, the proof.

(b) **Continuous case:** We shall assume that the statistic T is continuous with continuous partial derivatives with respect to $X_1, \cdots, X_n$ except for a set of values of A of $\boldsymbol{X}$ such that $P_\theta(A) = 0$ for all values of $\theta \in \Theta$.

(i) **Necessity:** Let the statistic T_1 be sufficient for θ. Consider any $(n-1)$ additional statistic $T_2, \cdots, T_n$ which are algebraically unrelated to T_1. Then the conditional distribution of $(T_2, \cdots, T_n)$ given T_1 must be independent of θ. That is, the joint p.d.f. of $\boldsymbol{T} = (T_1, \cdots, T_n)'$ is

$$\phi_{\boldsymbol{T}}(t_1, \cdots, t_n; \theta) = \psi_{T_1}(t_1;\theta)\chi(t_2 \cdots t_n|t_1) \tag{15.4.7}$$

where the first factor on the r.h.s is the marginal p.d.f. of T_1 and the second factor giving the conditional p.d.f. of $T_2, \cdots, T_n$ given $T_1 = t_1$ is independent of θ.

We now derive from ϕ the joint p.d.f. of $X_1, \cdots, X_n$. Assume that $t_1, t_2, \cdots, t_n$ admit of continuous partial derivatives w.r.t. $x_1, \cdots, x_n$. Then the joint p.d.f. of $\boldsymbol{X}$ is

$$f_{\boldsymbol{X}}(\boldsymbol{x};\theta) = \phi\left(t_1(\boldsymbol{x}), t_2(\boldsymbol{x}), \cdots, t_n(\boldsymbol{x});\theta\right) \times \left|J\left(\frac{t_1, \cdots, t_n}{x_1, \cdots, x_n}\right)\right| \tag{15.4.8}$$

assuming that the transformation $(x_1, \cdots, x_n) \to (t_1, \cdots, t_n)$ is one-to-one. Because of (15.4.7) this takes the form

$$\psi(t_1,\theta)\chi\left[t_2(\boldsymbol{x}), \cdots, t_n(\boldsymbol{x})|t_1(\boldsymbol{x})\right] \times |J| = g(t_1,\theta)h(x_1, \cdots, x_n) \quad \text{(say)}$$

where $g(t_1,\theta) = \psi(t_1,\theta)$ and $f(\boldsymbol{x})$ is independent of θ. Hence the form (15.4.5).

If the transformation $\boldsymbol{t} \to \boldsymbol{x}$ is m-to-one ($m \geq 2$), then also $f_{\boldsymbol{X}}(\boldsymbol{x};\theta)$ will be of the same form, because $f_{\boldsymbol{X}}(\boldsymbol{x};\theta)$ will be sum of m terms each of the form (15.4.8) corresponding to the set of values $(t_1, \cdots, t_n)$ leading to the same set of values of $(x_1, \cdots, x_n)$. Hence the necessity of the factorisation form (15.4.5).

(ii) **Sufficiency :** Suppose the factorisation form (15.4.5) holds. Then the joint p.d.f. of $(T_1, \cdots, T_n)$, where $(T_2, \cdots, T_n)$ is any set of statistic unrelated to T_1, is

$$f(x_1(\boldsymbol{t}), x_2(\boldsymbol{t}), \cdots, x_n(\boldsymbol{t}); \theta) \times \left| J\left(\frac{x_1, \cdots, x_n}{t_1, \cdots, t_n}\right)\right| = g(t_1, \theta) h\Big[x_1(\boldsymbol{t}), \cdots, x_n(\boldsymbol{t})\Big] \times |J|$$

where $\boldsymbol{t} = (t_1, \cdots, t_n)'$, if the transformation $(x_1, \cdots, x_n) \to (t_1, \cdots, t_n)$ is one-to-one and is the sum of m such expressions with the same value of $g(t_1, \theta)$ in case the transformation is m-to-one ($m \geq 2$). In either case, the joint p.d.f. of T is of the form

$$\phi_{\boldsymbol{T}}(\boldsymbol{t}) = g_1(t_1, \theta) C(t_1, \cdots, t_n). \tag{15.4.9}$$

The marginal p.d.f. of T_1 is obtained by integrating (15.4.9) over $t_2, \cdots, t_n$ and is of the form

$$\begin{aligned} & g_1(t_1, \theta) \int_{-\infty}^{\infty} \cdots \int_{-\infty}^{\infty} C(t_1, \cdots, t_n) \prod_{i=2}^{n} d\,t_i \\ &= g_1(t_1, \theta) d(t_1) = \psi(t_1, \theta) \ \text{(say)}. \end{aligned}$$

Hence the conditional distribution of $T_2, \cdots, T_n$ given $T_1 = t_1$ is

$$\frac{c(t_1, \cdots, t_n)}{d(t_1)} = \chi(t_2, \cdots, t_n | t_1),$$

which is independent of θ. Since this is true for any set of statistics which are functionally unrelated to T_1, T_1 is sufficient for θ.

■ NOTE 15.4.3 *The factorisation theorem holds if θ is a vector of parameters and T is a random vector and we say that T is jointly sufficiently of $\boldsymbol{\theta}$. Even if θ is a scalar, T may be vector* (*exercise* 6 *of this section*). *If $\boldsymbol{\theta}$ and $\boldsymbol{T}$ are vectors of the same dimension and if $\boldsymbol{T}$ is sufficient for $\boldsymbol{\theta}$, it does not follow that the jth component of $\boldsymbol{T}$ is sufficient for the jth component of $\boldsymbol{\theta}$.* (*Example* 15.4.7, *Case* (c)).

The following theorem holds readily from *F.N.* theorem.

Theorem 15.4.2 *If the statistic T is sufficient for θ, any monotonic function $\xi(T)$ will also be sufficient for θ.*

Proof: Let $Y = \xi(T)$. Since $\xi(T)$ is monotonic, $T = \xi^{-1}(Y)$ exists. By the condition of sufficiency, from (15.4.5)

$$\begin{aligned} f_{\boldsymbol{X}}(x_1, \cdots, x_n; \theta) &= g(t; \theta) h(x_1, \cdots, x_n) \\ &= g(\xi^{-1}(y), \theta) h(x_1, \cdots, x_n) \\ &= \psi(y, \theta) h(x_1, \cdots, x_n). \end{aligned}$$

Hence Y is sufficient for θ.

We now use F.N. theorem to detect sufficient statistics in some examples.

▶ **Example 15.4.4** *Let $X_i(i = 1, \cdots, n)$ be a random sample from a continuous uniform distribution $R(0, \theta)$. The p.d.f. of $\boldsymbol{X}$ is*

$$f(\boldsymbol{x};\theta) = \begin{cases} \dfrac{1}{\theta^n}, & \text{if } x_i \in [0,\theta], i = 1, \cdots, n; \theta \in R^1 \\ 0, & \text{otherwise.} \end{cases}$$

Let $X_{(1)} = \min\limits_i X_i, X_{(n)} = \max\limits_i X_i$. Define a function $d(a, b)$ such that

$$d(a,b) = \begin{cases} 1, & \text{if } a \le b \\ 0, & \text{otherwise.} \end{cases} \tag{i}$$

Hence

$$f(\boldsymbol{x};\theta) = \left\{\frac{d(x_{(n)},\theta)}{\theta^n}\right\} d(0, x_{(1)}).$$

Therefore, by F. N. criterion, $X_{(n)}$ is sufficient for θ.

Again, using the definition 15.4.1 directly,

$$f_\theta(\boldsymbol{x}|X_{(n)} = x_{(n)}) = \theta^{-n} / \frac{n}{\theta}\left[\int_0^{x_{(n)}} \frac{dt}{\theta}\right]^{n-1} = \frac{1}{nx_{(n)}^{n-1}}$$

is independent of θ. Hence $X_{(n)}$ is sufficient for θ.

The result can easily be generalised where $f(x;\theta) = Q(\theta)H(x), 0 < x < \theta$ and $\int_0^\theta H(x)\,dx = [Q(\theta)]^{-1}$. In this case

$$\begin{aligned} f(\boldsymbol{x},\theta) &= [Q(\theta)]^n \prod_{i=1}^n H(x_i) d(x_{(n)},\theta) d(0, x_{(1)}) \\ &= \left\{[Q(\theta)]^n\, d(x_{(n)},\theta)\right\}\left\{\prod_{i=1}^n H(x_i) d(0, x_{(1)})\right\}; \end{aligned}$$

hence $X_{(n)}$ is sufficient for θ.

▶ **Example 15.4.5** *Let $X_i, i = 1, \cdots, n$ be a random sample from a rectangular population with p.d.f.*

$$f_X(x;\alpha,\beta) = \begin{cases} \frac{1}{\beta-\alpha}, & \beta \le X \le \alpha \\ 0, & \textit{otherwise.} \end{cases}$$

The parameter space, $\Theta = \{-\infty < \alpha < \beta < \infty\}$. Let $X_{(1)} < X_{(2)} < \cdots < X_{(n)}$ be the order statistic. The joint p.d.f.

$$f_{\boldsymbol{X}}(\boldsymbol{x};\alpha,\beta) = \begin{cases} \frac{1}{(\beta-\alpha)^n}, & \alpha \le x_i \le \beta, \ i = 1,\cdots,n \\ 0, & \text{otherwise} \end{cases}$$

$$= \frac{1}{(\beta-\alpha)^n} d(\alpha, x_{(1)}) d(x_{(n)}, \beta)$$

where $d(a,b)$ is as defined at (i) of example 15.4.4.

Hence writing $T_1 = X_{(1)}, T_2 = X_{(n)}$;

$$g(t_1, t_2; \alpha, \beta) = d(\alpha, x_{(1)}) d(x_{(n)}, \beta)/(\beta-\alpha)^n, h(x_1, \cdots, x_n) = 1,$$

$(X_{(1)}, X_{(n)})$ is jointly sufficient for (α, β) by F.N. theorem.

▶ **Example 15.4.6** *Let $X_i (i = 1, \cdots, n)$ be a random sample from the distribution $f(x;\theta) = e^{-(x-\theta)}, x > \theta, -\infty < \theta < \infty$.*

The p.d.f. of $X_{(1)}$ is

$$g(x_{(1)};\theta) = \int_{x_{(1)}}^{\infty}\int_{x_{(2)}}^{\infty}\cdots\int_{x_{(n-1)}}^{\infty} n! \exp\left[-(x_{(1)}-\theta)-(x_{(2)}-\theta)-\cdots -(x_{(n)}-\theta)\right] dx_{(n)}\, dx_{(n-1)} \cdots dx_{(2)}$$

$$= n\, e^{-n(x_{(1)}-\theta)}, \theta < x_{(1)} < \infty.$$

The joint p.d.f. of $\boldsymbol{X}$ is

$$f_{\boldsymbol{X}}(\boldsymbol{x};\theta) = \exp\left[-\sum_{i=1}^{n}(x_i - \theta)\right]$$

$$= \left\{n \exp\left[-n(x_{(1)}-\theta)\right]\right\}\left\{\frac{1}{n}\exp\left[-(x_{(2)}+\cdots+x_{(n)}-(n-1)x_{(1)})\right]\right\}$$

so that $X_{(1)}$ is sufficient for θ.

▶ **Example 15.4.7** *Let $X_i (i = 1, \cdots, n)$ be a random sample from a $N(\mu, \sigma^2)$ population. We consider the sufficient statistic for μ or σ^2 or both according as the other parameter is known or unknown.*

Case (a). μ unknown, σ^2 known.

$$f_{\boldsymbol{X}}(\boldsymbol{x};\mu,\sigma^2) = \frac{1}{\left(\sigma\sqrt{2\pi}\right)^n}\exp\left[-\frac{1}{2\sigma^2}\sum_{i=1}^{n}(x_i-\mu)^2\right] \quad \text{(i)}$$

$$= \frac{1}{(\sigma\sqrt{2\pi})^n}\cdot\exp\left[-\frac{n(\bar{x}-\mu)^2}{2\sigma^2} - \sum_{i=1}^{n}(x_i-\bar{x})^2\Big/2\sigma^2\right], \quad \bar{x} = \frac{1}{n}\sum_{i=1}^{n} x_i \quad \text{(ii)}$$

$$= g_{\bar{X}}(\bar{x},\mu) h(x_1,\cdots,x_n)$$

when

$$g_{\bar{X}}(\bar{x}, \mu) = \frac{\sqrt{n}}{\sigma\sqrt{2\pi}} \exp\left[-\frac{n(\bar{x}-\mu)^2}{2\sigma^2}\right],$$

$$h(x_1, \cdots, x_n) = K \cdot \exp\left[-\frac{1}{2\sigma^2} \cdot \sum_{i=1}^{n} (x_1 - \bar{x})^2\right],$$

K, a suitable constant. Hence $\bar{X}$ is sufficient for μ when σ is known.

Case (b). μ known, σ^2 unknown. The joint p.d.f. (i) can be written as

$$f_{\mathbf{X}}(\boldsymbol{x}; \mu, \sigma^2) = g(T, \sigma^2)h(x_1, \cdots, x_n)$$

where

$$T = \sum_{i=1}^{n} (X_i - \mu)^2, \;\; g(t; \sigma^2) = \left(\sigma\sqrt{2\pi}\right)^{-n} \exp(-t/2\sigma^2), \;\; h(x_1, \cdots, x_n) = 1.$$

Therefore, T and

$$\frac{T}{n} = \frac{1}{n}\sum_{i=1}^{n} (X_i - \mu)^2$$

is sufficient for σ^2. In fact T/σ^2 is a $\chi^2_{(n)}$ variable.

Case (c). Both μ, σ^2 unknown. Writing

$$S^2 = \sum_{i=1}^{n} (x_i - \bar{x})^2/(n-1),$$

the joint p.d.f. in (ii) can be written as

$$f(x; \mu, \sigma^2) = g(\bar{x}, s^2; \mu, \sigma^2)h(x_1, \cdots, x_n)$$

where $g(\bar{x}, s^2; \mu, \sigma^2)$ is given by the right hand side of (ii) and $h(\boldsymbol{x}) = 1$. Hence $(\bar{X}, S^2)$ are jointly sufficient for (μ, σ^2) where

$$S^2 = \sum_{i=1}^{n} (X_i - \bar{X})^2/(n-1).$$

An equivalent sufficient statistic is $\left(\sum_{i=1}^{n} X_i, \sum_{i=1}^{n} X_i^2\right)$. Note that $\bar{X}$ is not sufficient for μ if σ^2 is unknown and S^2 is not sufficient for σ^2 if μ is unknown.

Families of Distributions Admitting Sufficient Statistic

Koopman (1956) investigated the classes of distributions admitting sufficient statistic. Before considering his theorem we consider a few definitions.

Definition 15.4.3 *Let $\mathcal{F} = \{f_\theta; \theta \in \Theta\}$ be a family of p.d.f.'s (or p.m.f.'s). Assume that the set $S = \{x; f_\theta(x) > 0\}$ is independent of θ. The family of distributions $\mathcal{F}$ is a one-parameter exponential family if there exists real-valued functions $A(\theta)$ and $B(\theta)$ defined on Θ and Borel measurable functions $v(x)$ and $C(x)$ defined on R^1 such that*

$$f_\theta(x) = \exp\{A(\theta)v(x) + B(\theta) + C(x)\}. \tag{15.4.10}$$

where $\frac{dA}{d\theta} \neq 0$ for all $\theta \in \Theta$ and $\{v(x), 1\}$ are linearly independent over support S.

▶ **Example 15.4.8** *Let $X \cap N_p(\mu, \Sigma)$. Show that the sample mean vector $\bar{X}$ and the sample covariance matrix $\boldsymbol{s} = ((s_{ij}))$ where $s_{ij} = \frac{1}{n}\sum\limits_{r=1}^{n}(X_{ri} - \bar{X}_i)(X_{rj} - \bar{X}_j)$, X_{ri} being the rth observation of variable X_i, are jointly sufficient for (μ, Σ).*

From (18.6.10)

$$\begin{aligned}\log f(X; \mu, \Sigma) = & -\frac{np}{2}\log(2\pi) - \frac{n}{2}\log|\Sigma| - \frac{1}{2}\text{tr}\,\Sigma^{-1}\boldsymbol{s} \\ & -\frac{n}{2}\text{tr}\,\Sigma^{-1}(\bar{X} - \mu)(\bar{X} - \mu)'\end{aligned}$$

which shows that $(\bar{X}, \boldsymbol{s})$ is jointly sufficient for (μ, Σ).

▶ **Example 15.4.9** *Let $X \cap N(\mu, \sigma^2)$.*

Case (a). σ^2 known, μ unknown, $\mu \in R_1$.

$$f_\mu(x) = \frac{1}{\sigma\sqrt{2\pi}}\exp\left\{\frac{\mu x}{\sigma^2} - \frac{\mu^2}{2\sigma^2} - \frac{x^2}{2\sigma^2}\right\}.$$

Here,

$$\begin{aligned} A(\mu) &= \mu/\sigma^2, \;\; v(x) = x, \;\; B(\mu) = -\mu^2/2\sigma^2, \\ C(x) &= -\frac{x^2}{2\sigma^2} - \ln(\sqrt{2\pi}\sigma). \end{aligned}$$

Case (b). μ known, σ^2 unknown, $\sigma^2 > 0$.

$$f_{\sigma^2}(x) = \exp\left\{-\log(\sigma\sqrt{2\pi}) - \frac{(x-\mu)^2}{2\sigma^2}\right\}.$$

Here,

$$\begin{aligned} A(\sigma^2) &= -\frac{1}{2\sigma^2}, \;\; v(x) = (x-\mu)^2, \;\; B(\sigma^2) = -\log(\sigma\sqrt{2\pi}), \\ C(x) &= 0. \end{aligned}$$

▶ **Example 15.4.10** *Let* $X \cap$ *Poisson* $(\lambda), \lambda > 0$.

$$P(X = x) = \exp\{-\lambda + x \log \lambda - \log x!\}$$

is an exponential family. Here $A(\lambda) = \log \lambda,\ v(x) = x,\ B(\lambda) = -\lambda,\ C(x) = -\log x!$.

It can be similarly shown that Binomial, Gamma (with only one parameter unknown) belong to one parameter exponential family.

Definition 15.4.4 *Let* $\mathcal{F} = \{f_{\boldsymbol{\theta}}; \boldsymbol{\theta} \in \Theta\}$ *be a family of p.d.f.'s (or p.m.f.'s). Assume that the set* $S = \{x; f_{\boldsymbol{\theta}}(x) > 0\}$ *is independent of* $\boldsymbol{\theta}$. *The family* $\mathcal{F}$ *is a k-parameter exponential family if there exist real-valued functions* $A_i (i = 1, \cdots, k)$, B *defined on* Θ *and Borel-measurable functions* $v_i(x)(i = 1, \cdots, k), C(x)$ *defined on* R^1 *such that*

$$f_{\boldsymbol{\theta}}(x) = \exp\left\{\sum_{i=1}^{k} A_i(\boldsymbol{\theta}) v_i(x) + B(\boldsymbol{\theta}) + C(x)\right\} \tag{15.4.11}$$

where the Jacobian $\left|\dfrac{\partial(A_1, \cdots, A_k)}{\partial(\theta_1, \cdots, \theta_k)}\right| \neq 0$ and $\{v_1, \cdots, v_k\}$ are linearly independent over S.

▶ **Example 15.4.11** *Let* $X \cap N(\mu, \sigma^2)$, *both* μ, σ^2 *are unknown.* $\Theta = \{(\mu, \sigma^2), \mu \in R_1, \sigma^2 > 0\}$.

$$f_{\mu,\sigma^2}(x) = \exp\left[-\frac{x^2}{2\sigma^2} + \frac{\mu x}{\sigma^2} - \frac{1}{2}\left(\frac{\mu^2}{\sigma^2} + \log(2\pi\sigma^2)\right)\right] \tag{15.4.12}$$

is a 2-parameter exponential family with

$$\begin{aligned} A_1(\mu, \sigma^2) &= -\frac{1}{2\sigma^2}, v_1(x) = x^2 \\ A_2(\mu, \sigma^2) &= \frac{\mu}{\sigma^2}, v_2(x) = x \\ B(\mu, \sigma^2) &= -\frac{1}{2}\left(\frac{\mu^2}{\sigma^2} + \log(2\pi\sigma^2)\right) \\ C(x) &= 0. \end{aligned}$$

It can be shown that the Gamma distributions, Beta distributions when both the parameters of these distributions are unknown are 2-parameter exponential families. Uniform distribution over (α, β) is not an exponential family.

Definition 15.4.5 *If* $A(\theta)$ *is a continuous function of* θ *and* $C(x), v'(x)(\not\equiv 0)$ *are continuous functions of* x *then the p.d.f.* (15.4.10) *is said to belong to the Koopman-Pitman (KP) class of p.d.f.'s and if the set* $\{x : f_\theta(x) > 0\}$ *is independent of* θ *we say that we have a regular case.*

Similar definitions extend to the multiparameter case in definition 15.4.4.

The p.d.f.'s in examples 15.4.8, 15.4.9 *are regular cases of* KP *class.*

Theorem 15.4.3 *Let $X_i, i = 1, \cdots, n$ be a random sample from the distribution having a p.d.f. of the form* (15.4.10) *which is a regular case of the* KP *class. Then $\sum_{i=1}^{n} v(X_i)$ is a sufficient statistic for θ.*

Proof: Transform from

$$(X_1, \cdots, X_n) \to (Y_1, \cdots, Y_n)$$

where

$$Y_1 = \sum_{i=1}^{n} v(X_i), Y_j = X_j, j = 2, \cdots, n.$$

The p.d.f. of Y_1 is

$$g_\theta(y_1) = \int_{-\infty}^{\infty} \cdots \int_{-\infty}^{\infty} e^{A(\theta)y_1 + nB(\theta) + \Sigma_{i=1}^{n} C(w_i(y))} \left| J\left(\frac{x_1, \cdots, x_n}{y_1, \cdots, y_n}\right) \right| \prod_{i=2}^{n} dy_i$$

where $x_i = w_i(\boldsymbol{y}), i = 1, \cdots, n$, $\boldsymbol{y} = (y_1, \cdots, y_n)'$, assuming without loss of generality that the transformation is $1:1$.

$$= R(y_1)\exp[A(\theta)y_1 + nB(\theta)], \theta \in \Theta$$

when $R(y_1)$ is independent of θ and the set $\{y_1 : g_\theta(y_1) > 0\}$ is independent of θ. Hence,

$$f_\theta(\boldsymbol{x}) = g_\theta(y_1)\left\{e^{\Sigma c(x_i)}/R(y_1)\right\}$$

when by F.N. theorem, Y_1 is sufficient for θ.

► **Example 15.4.12** *For normal distribution with σ^2 known, $\sum_{i=1}^{n} X_i = \bar{X}$ is sufficient for μ; when μ is known, $\sum_{i=1}^{n}(X_i - \mu)^2$ is sufficient for σ^2.*

We now consider Koopman's theorem giving the form of the distribution admitting sufficient statistic.

Theorem 15.4.4 *Let $X_1, \cdots, X_n$ be independent r.v.'s with the comman p.d.f. $f_\theta(x)$ such that*

$$\prod_{i=1}^{n} f_\theta(x_i) = g_\theta(t)h(x, \cdots, x_n)$$

and f, g are differentiable with respect to θ and x_i's and the range of x_i's do not depend on θ. Then

$$\prod_{i=1}^{n} f_\theta(x_i) = h(x_1, \cdots, x_n)\exp\left[A(\theta)K(t) + B(\theta)\right]. \tag{15.4.13}$$

Proof:

$$\prod_{i=1}^{n} f_\theta(x_i) = g(t;\theta)h(x_1,\cdots,x_n).$$

Therefore,

$$\sum_{i=1}^{n} \frac{\partial \log f_\theta(x_i)}{\partial \theta} = \frac{\partial \log g(t;\theta)}{\partial \theta} = K(t;\theta). \tag{15.4.14}$$

Since (15.4.14) is true for all θ, we choose an arbitrary θ_0 to obtain

$$r = \sum_{i=1}^{n} v(x_i) = K(t;\theta_0) = K(t)$$

when

$$v(x_i) = \left.\frac{\partial \log f_\theta(x_i)}{\partial \theta}\right]_{\theta=\theta_0}.$$

Assuming that $K(t)$ is a monotonic function of t so that $K^{-1}(t)$ exists we obtain $= K^{-1}(r)$ and define

$$q(r,\theta) = K(K^{-1}(r),\theta).$$

Differentiating (15.4.14) w.r.t. x_i,

$$\frac{\partial^2}{\partial x_i \partial \theta} \log f_\theta(x_i) = \frac{\partial q(r;\theta)}{\partial r} \cdot \frac{\partial r}{\partial x_i}.$$

Therefore,

$$\left[\frac{\partial^2}{\partial x_i \partial \theta} \log f_\theta(x_i)\right] \Big/ \frac{\partial r}{\partial x_i} = \frac{\partial q(r;\theta)}{\partial r}. \tag{15.4.15}$$

Now on the left hand side of (15.4.15), the numerator is a function of x_i and θ while the denominator is a function of x_i only. The right hand side is a function of θ and r only and r is a symmetric function of $x_1,\cdots,x_n$. Thus (15.4.14) must be a function of θ alone, say

$$\frac{\partial q(r,\theta)}{\partial r} = c_1(\theta)$$

and so

$$q(r,\theta) = rc_1(\theta) + c_2(\theta)$$

$$\text{or,}\quad K(t,\theta) = K(t,\theta_0)c_1(\theta) + c_2(\theta) = K(t)c_1(\theta) + c_2(\theta)$$

$$\text{or,}\quad \frac{\partial \log g(t;\theta)}{\partial \theta} = K(t)c_1(\theta) + c_2(\theta).$$

Therefore,

$$g(t;\theta) = \exp\left[A(\theta)K(t) + B(\theta)\right]$$

and so (15.4.13) follows.

More insight into the implication of sufficiency can be obtained by considering the left hand side of (15.4.15) as follows.

$$\frac{\partial^2}{\partial x_i \partial \theta} \log f_\theta(x_i) = c_1(\theta)\frac{\partial r}{\partial x_i} = c_1(\theta)\frac{\partial v(x_i)}{\partial x_i}$$

and so

$$\frac{\partial \log f_\theta(x_i)}{\partial \theta} = c_1(\theta)v(x_i) + c_2(\theta).$$

Therefore,

$$\log f_\theta(x_i) = A(\theta)v(x_i) + B(\theta) + w(x_i)$$

and so

$$\begin{aligned} f_\theta(x_i) &= h(x_i)\exp\left[A(\theta)v(x_i) + B(\theta)\right] \\ &= \exp\left[A(\theta)v(x_i) + B(\theta) + C(x_i)\right]. \end{aligned} \tag{15.4.16}$$

Hence

$$f(x_1, \cdots, x_n; \theta) = \exp\left[A(\theta)\sum_{i=1}^{n} v(x_i) + B_1(\theta) + \sum_{i=1}^{n} C(x_i)\right]$$

when $B_1(\theta) = nB(\theta)$. Note that $\sum_{i=1}^{n} v(x_i)$ is a sufficient statistic. (15.4.16) is precisely the form of the exponential distributions discussed in definition 15.4.3.

The result can be generalised to the case where $\boldsymbol{\theta}$ is a vector, $\boldsymbol{\theta} = (\theta_1, \cdots, \theta_K)'$. The family of distributions that admits K jointly sufficient statistic for $\boldsymbol{\theta}$ will have the following Koopman-Pitman form:

$$f_{\boldsymbol{\theta}}(x) = \exp\left[\sum_{j=1}^{k} A_j(\boldsymbol{\theta})v_j(x) + B(\boldsymbol{\theta}) + C(x)\right]. \tag{15.4.17}$$

We have already given examples of regular cases of KP family of distributions and identified the relevant sufficient statistics.

15.4.1 Minimal Sufficient Statistic

We now consider the concept of minimal sufficient statistic. It is usual to call the hypersurface of observed points $(x_1, \cdots, x_n)$ where a sufficient statistic t has a specified value as a contour of t [Fig. 15.1]. The principle of sufficiency states that it is enough

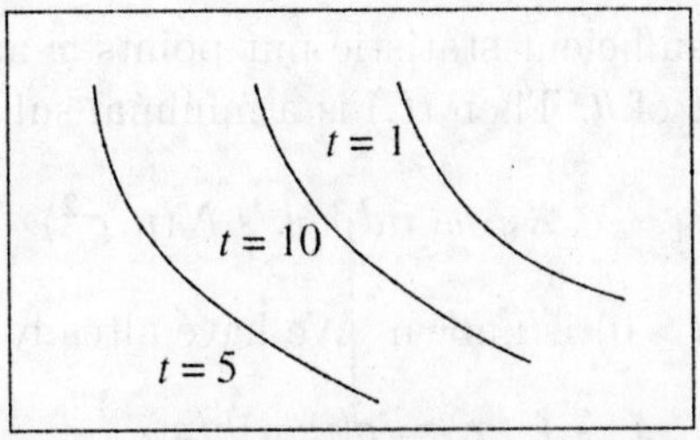

Fig. 15.1 Contours of a Sufficient Satistic

to know in which contour of t our observed point lies to make inferance about θ. Further specification of the co-ordinates of the observed point $(x_1, \cdots, x_n)$ throws no light on the value of θ. Consequently any two observed points which lie on the same contour of t give the same information about θ.

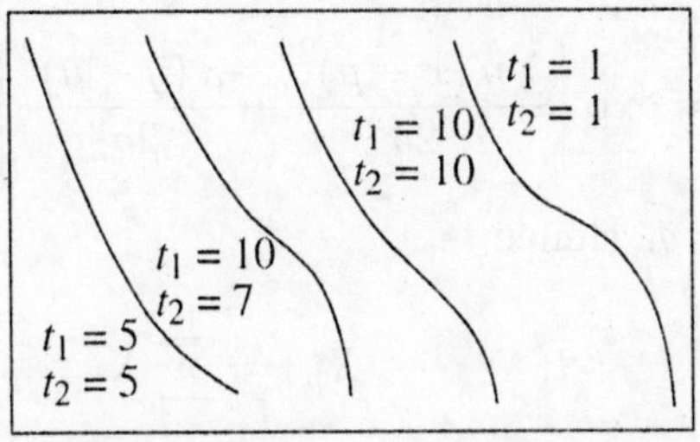

Fig. 15.2 Contours of two Sufficient Satistic (t_1, t_2)

Suppose there are two sufficient statistic t_1, t_2 with their centours shown in Fig. 15.2. Obviously we should prefer t_1 to t_2 because it unites more points and is till sufficient. Generalising the concept, among a number of sufficient statistic we should choose one, t_0 (say), whose power of summarising data is maximum i.e., whose every contour $\{t_0 = k\}$ covers the contours $\{t = k_i\}, i = 1, 2, \cdots$ of every other sufficient statistic t for each k. Such a statistic is called a minimal sufficient statistic.

Definition 15.4.6 *A statistic $T(X_1, \cdots, X_n)$ is a minimal sufficient statistic if T is a sufficient statistic and is a function of every other sufficient statistic, i.e.,*

$$T(X_1, \cdots, X_n) = \phi_t(t(X_1, \cdots, X_n)) \tag{15.4.18}$$

for every sufficient statistic t.

A minimal sufficient statistic, however, may not always exist. For finding a minimal sufficient statistic consider

Lemma 15.4.1 *For two points $\boldsymbol{x} = (x_1, \cdots, x_n), \boldsymbol{y} = (y_1, \cdots, y_n)$ the ratio $f_\theta(\boldsymbol{x})/f_\theta(\boldsymbol{y})$ is independent of θ if $t(\boldsymbol{x}) = t(\boldsymbol{y})$.*

Proof: By Neyman's factorisation theorem

$$\frac{f_\theta(\boldsymbol{x})}{f_\theta(\boldsymbol{y})} = \frac{g(t(\boldsymbol{x}); \theta)h(\boldsymbol{x})}{g(t(\boldsymbol{y}); \theta)h(\boldsymbol{y})} = \frac{h(\boldsymbol{x})}{h(\boldsymbol{y})}$$

which is independent of θ.

For finding a minimal sufficient statistic put points $\boldsymbol{x}$ and $\boldsymbol{y}$ on the same contour $f_\theta(\boldsymbol{x})/f_\theta(\boldsymbol{y})$ is independent of θ. Then $t(.)$ is a minimal sufficient statistic.

▶ **Example 15.4.13** *Let $X_1, \cdots, X_n$ be iid r.v.'s $N(\mu, \sigma^2)$.*

Case (a). Suppose $\sigma^2 (> 0)$ is known. We have already seen (example 15.4.7) that

$$f_\mu(\boldsymbol{x}) = g(\bar{x}; \mu) h(\boldsymbol{x})$$

where

$$g(\bar{x}; \mu) = \frac{\sqrt{n}}{\sigma\sqrt{2\pi}} \exp\left[-\frac{n(\bar{x}-\mu)^2}{2\sigma^2}\right]$$

so that $\bar{X}$ is a sufficient statistic. Now

$$\frac{f_\mu(\boldsymbol{x})}{f_\mu(\boldsymbol{y})} = \exp\left[-\frac{n(\bar{x}-\mu)^2}{2\sigma^2} + \frac{n(\bar{y}-\mu)^2}{2\sigma^2}\right] \cdot \frac{h(\boldsymbol{x})}{h(\boldsymbol{y})}$$

is independent of μ iff $\bar{x} = \bar{y}$. Hence

$$t(X_1, \cdots, X_n) = \sum_{i=1}^{n} X_i$$

is minimal sufficient for μ.

Case (b). Suppose both μ, σ^2 are unknown. We have already seen (example 15.4.7 that

$$f_{\mu,\sigma^2}(\boldsymbol{x}) = g(\bar{x}, s^2; \mu, \sigma^2) h(\boldsymbol{x})$$

where

$$g(\bar{x}, s^2; \mu, \sigma^2) = \frac{1}{\left(\sqrt{2\pi}\sigma\right)^n} \exp\left[-\frac{n(\bar{x}-\mu)^2}{2\sigma^2} - \frac{ns^2}{2\sigma^2}\right],$$

$$s^2 = \sum_{i=1}^{n} (x_i - \bar{x})^2 / (n-1)$$

so that

$$\left(\sum_{i=1}^{n} X_i, \sum_{i=1}^{n} X_i^2\right)$$

is jointly sufficient for (μ, σ^2). Now $f_{\mu,\sigma^2}(\boldsymbol{x})/f_{\mu,\sigma^2}(\boldsymbol{y})$ is independent of (μ, σ^2) iff

$$\left\{\sum_{i=1}^{n} x_i = \sum_{i=1}^{n} y_i, \sum_{i=1}^{n} x_i^2 = \sum_{i=1}^{n} y_i^2\right\}.$$

Hence the statistic $t(X_1, \cdots, X_n) = \left(\sum_{i=1}^{n} X_i, \sum_{i=1}^{n} X_i^2\right)$ is jointly minimal sufficient fo (μ, σ^2).

▶ **Example 15.4.14** (*Dudewicz and Mishra, 1988*). *Let X be a single observation from the probability function*

$$f_\theta(x) = \begin{cases} \theta^2 & \text{if } x = -1, 3 \\ \frac{1}{2} - \frac{\theta^2}{2} & \text{if } x = 0 \\ -\frac{\theta^2}{2} + \frac{\theta}{2} & \text{if } x = 2, 4 \\ \frac{1}{2} - \frac{\theta^2}{2} - \theta & \text{if } x = 1 \end{cases}$$

when θ is an unknown number between 0 and $\left(\sqrt{2}-1\right)$. Find a minimal sufficient statistic for θ.

X itself is a sufficient statistic for θ. To find a minimal sufficient statistic we put on the same contour a set of points such that, for every pair of points (x, y) in the set, $f_\theta(x)/f_\theta(y)$ is independent of θ. We, therefore, partition the sample space into the sets $\{-1, 3\}$, $\{0\}$, $\{2, 4\}$, $\{1\}$ and a minimal sufficient statistic is

$$t(X) = \begin{cases} K_1 & \text{if } X = -1 \text{ or } 3 \\ K_2 & \text{if } X = 0 \\ K_3 & \text{if } X = 2 \text{ or } 4 \\ K_4 & \text{if } X = 1 \end{cases}$$

where K_1, K_2, K_3, K_4 are any set of distinct constants. The p.m.f. of $t(x)$ is

$$p_{t(X)}(K) = P\left[t(X) = K\right] = \begin{cases} 2\theta^2 & \text{if } K = K_1 \\ \frac{1}{2} - \frac{\theta^2}{2} & \text{if } K = K_2 \\ \theta - \theta^2 & \text{if } K = K_3 \\ \frac{1}{2} - \frac{\theta^2}{2} - \theta & \text{if } K = K_4. \end{cases}$$

▶ **Example 15.4.15** *Consider a sample of size 2 from Cauchy p.d.f.*

$$f_\theta(x) = \frac{1}{\pi\{1 + (x-\theta)^2\}}, \quad -\infty < x < \infty.$$

Show that no non-trivial sufficient statistic exists.

Suppose there exists a non-trivial sufficient statistic t. Then for some two points

$$\boldsymbol{x} = (x_1, x_2), \quad \boldsymbol{y} = (y_1, y_2)$$

we must have $t(\boldsymbol{x}) = t(\boldsymbol{y})$. By lemma 15.4.1, we must have $f_\theta(\boldsymbol{x})/f_\theta(\boldsymbol{y})$ independent of θ. Now

$$\frac{f_\theta(\boldsymbol{x})}{f_\theta(\boldsymbol{y})} = \frac{\dfrac{1}{\{1+(x_1-\theta)^2\}\{1+(x_2-\theta)^2\}}}{\dfrac{1}{\{1+(y_1-\theta)^2\}\{1+(y_2-\theta)^2\}}} = P(x_1, x_2, y_1, y_2)$$

or, $\{1+(x_1-\theta)^2\}\{1+(x_2-\theta)^2\}P(x_1, x_2, y_1, y_2) = \{1+(y_1-\theta)^2\}\{1+(y_2-\theta)^2\}$ (i) should hold for all θ. (i) reduces to

$$a\theta^4 + b\theta^3 + c\theta^2 + d\theta + e = 0 \ \forall \, \theta. \tag{ii}$$

The right hand side of (ii) is a polynomial of degree 4 and its value cannot be zero for all θ. Hence there cannot exist any non-trivial sufficient statistic.

However, it can be shown that the order statistic (defined below) is minimal sufficient.

15.4.2 Sufficiency of General Order Statistic

Let $X_1, \cdots, X_n$ be n independent observations from $f_\theta(x)$ and $X_{(1)} \leq X_{(2)} \leq \cdots \leq X_{(n)}$ be the observations arranged in the increasing order of magnitude. The statistic $t(X_1, \cdots, X_n) = (X_{(1)}, \cdots, X_{(n)})$ is called the general order statistic (vide Section 11.6).

Lemma 15.4.2 *If $g(x_1, \cdots, x_n)$ is a symmetric function it can be written as a function of the order statistic.*

$$\begin{aligned} g(x_1, \cdots, x_n) &= h\left[t(x_1, \cdots, x_n)\right] \\ &= h\left[x_{(1)}, \cdots, x_{(n)}\right] \end{aligned}$$

and conversely, if $g(x_1, \cdots, x_n)$ can be written as a function of the order statistic, then it is a symmetric function.

Proof: Since $g(x_1, \cdots, x_n)$ has the same value for all the $n!$ points $(x_{i_1}, \cdots, x_{i_n})$, it must be a function of the order statistic.

Conversely, if $g(x_1, \cdots, x_n) = h(x_{(1)}, \cdots, x_{(n)})$, then since $g(x_1, \cdots, x_n)$ have the same value for all the $n!$ points $(x_{i_1}, \cdots, x_{i_n})$, it must be a symmetric function.

Theorem 15.4.5 *If $X_1, \cdots, X_n$ are iid continuous r.v.'s then the order statistic*

$$t(X_1, \cdots, X_n) = (X_{(1)}, \cdots, X_{(n)})$$

is a sufficient statistic.

Proof: For a proof see Theorem 22.2.1.

Exercise

1. Show that if t_1 is sufficient for θ and t_2 is any other estimator (not a function of t_1), the joint p.d.f. of t_1, t_2 can be written as $\phi(t_1, t_2; \theta) = \phi_1(t_1; \theta)\phi_2(t_1, t_2)$.

   Explain the importance of this result.

   [*Hints:* $\ln f(x_1, \cdots, x_n; \theta) = g(t_1, \theta)h(x_1, \cdots, x_n)$ transform from $(x_1, \cdots, x_n) \rightarrow (t_1, t_2, x_3, \cdots, x_n)$.]

2. Show that if t is sufficient for θ, $\frac{\partial \log f(x_1, \cdots, x_n; \theta)}{\partial \theta}$ is a function of t and θ only.

3. Suppose (X_1, X_2) is a random sample from a Bernoulli population $p^x(1-p)^x$, $x = 0, 1$, $0 < p < 1$. Show that (i) $X_1 - X_2$ is not sufficient for p; (ii) X_2 is not sufficient for p.

4. Prove the result (ii) in exercise 3 in general. If $(X_1, \cdots, X_n)$ is a random sample from a population in the family $\mathcal{F} = \{F_\theta; \theta \in \Theta\}$, then any statistic U in which an element X_i is missing cannot be a sufficient statistic.

5. Show that in sampling from a rectangular population $f(x;\theta) = \frac{1}{\theta}, 0 \le x \le \theta$, the sample total $\sum_{i=1}^{n} X_i$ is not sufficient for p. Find an unbiased estimate of θ^α (α known) based on the sufficient statistic.

6. Let $X_1, \cdots, X_n$ be a random sample from a p.d.f.

$$f(x;\theta) = \begin{cases} \frac{1}{\theta}, \; x \in \left[-\frac{\theta}{2}, \frac{\theta}{2}\right], \theta > 0 \\ 0, \text{ otherwise.} \end{cases}$$

Show that $(X_{(1)}, X_{(n)})$ is sufficient for θ when $X_{(1)} = \min X_i$, $X_{(n)} = \max X_i$.

7. Let X be a random variable with p.d.f.

$$f(x;\theta) = \tfrac{1}{2}, \quad \theta \le x \le \theta + 2$$

where θ is any real number. Prove that there does not exist any sufficient statistic for θ.

8. Let X_1 and X_2 be independent discrete random variables on $\{1, 2, \cdots, N\}$ where N is unknown. Obtain the conditional distribution of (X_1, X_2) given $T = \max(X_1, X_2)$. Hence show that T is sufficient for N but $X_1 + X_2$ is not.

9. Let $X_i (= 1, \cdots, n)$ be a random sample from a distribution

$$f(x; N_1, N_2) = \frac{1}{N_2 - N_1}, \quad n = N_1 + 1, N_1 + 2, \cdots, N_2$$

and $N_1, N_2 (N_1 < N_2)$ are integers. Find a sufficient statistic for (i) N_1 when N_2 is known (ii) N_2 when N_1 is known (iii) (N_1, N_2).

10. Prove that in a random sample of size n from a geometric distribution pq^{x-1}, $q = 1 - p$, $0 < p < 1$, $x = 1, 2, \cdots$ the sample total $\sum_{i=1}^{n} X_i$ is sufficient for p.

11. Prove that $U = \sum_{i=1}^{n} X_i$ is a sufficient statistic for the parameter θ in a random sample of size n from an exponential population $f_X(x;\theta) = \theta e^{-\theta x}, x > 0$.

12. There are n white balls and $(N-n)$ black balls in an urn where N is a known quantity, but n is unknown. Show that the total number of white balls obtained in a simple random sample without replacement of r balls from the population $(r < n, r < N-n)$ is a sufficient statistic from n.

13. Show that for the double-exponential distribution $f(x;\theta) = \frac{1}{2}\exp(-|x-\theta|)$, $-\infty < \theta < \infty$, there does not exist any single statistic which is sufficient for θ. The whole set of observation $(X_1, \cdots, X_n)$ or the order statistic $(X_{(1)}, \cdots, X_{(n)})$ is, of course, sufficient for θ.

14. Let $X_1, \cdots, X_n$ be independent r.v.'s with common p.d.f. $N(0, \sigma^2)$. Prove that $\sum_1^n X_i^2$ is a sufficient statistic for σ^2.

15. Show that if μ is known, the estimator

$$S = \sqrt{\frac{1}{n}\sum_{i=1}^{n}(X_i-\mu)^2}$$

is sufficient for σ in samples of size n from a $N(\mu, \sigma)$ population.

16. Let X_1, X_2 be independent r.v.'s with X_i following Poisson (m_i), $m_i > 0, i = 1, 2$. Prove that (X_1, X_2) is a sufficient statistic for (m_1, m_2).

17. Let X_1, X_2 be independent r.v.'s with X_i following a Bernoulli distribution $b(n, p_i)$, $0 < p_i < 1$, $i = 1, 2$. Prove that (X_1, X_2) is a sufficient statistic for (p_1, p_2).

18. Let $X_i (i = 1, \cdots, n)$ be a random sample from the log-normal distribution

$$f(x, \mu, \sigma) = \frac{1}{x\sigma\sqrt{2\pi}}\exp\left\{-\frac{1}{2\sigma^2}(\log x - \mu)^2\right\}, \quad x > 0.$$

Find a sufficient statistic for (i) μ when σ is known (ii) σ when μ is known (iii) (μ, σ).

19. Let $X_i (i = 1, \cdots, n)$ be a random sample from a distribution with p.d.f.

$$f(x, \theta_1, \theta_2) = \begin{cases} \frac{1}{\theta_2}\exp\left[-\frac{x-\theta_1}{\theta_2}\right], & \theta_1 < x < \infty, \quad 0 < \theta_i < \infty, \quad i = 1, 2 \\ 0, & \text{otherwise.} \end{cases}$$

Show that $\left(X_{(1)}, \sum_{i=1}^{n} X_i\right)$ are jointly sufficient for (θ_1, θ_2).

20. Let $X_1, \cdots, X_n$ be a random sample from a gamma distribution

$$f(x, \theta, p) = \frac{\theta^p}{\lceil p} e^{-\theta x} x^{p-1}, \;\; 0 < x < \infty, \;\; \theta > 0, \;\; p > 0.$$

Show that (i) $\left(\sum_i X_i, \prod_i X_i\right)$ are jointly sufficient for (θ, p) (ii) if θ is known, $\prod_i X_i$ is sufficient for p (iii) if p is known $\sum_i X_i$ is sufficient for θ.

21. For each of the following cases find sufficient statistics

(a) $f(x; \theta) = Q(\theta) e^{-x/\theta}, x = \theta, \theta + 1, \theta + 2, \cdots, \theta > 0.$

(b) $f(x; \theta, p) = qp^{x-\theta}, q = 1 - p, 0 < p < 1, x = \theta, \theta + 1, \cdots.$

when (a) p is known, θ unknown (b) θ is known, p unknown (c) p, θ both unknown.

22. Let W be a n-dimensional Euclidian space and let F be all the class of Borel subsets of W. μ is a σ-finite measure on F, w is a subset of r-dimensional Euclidian space and $f(x; \theta), \theta \in w$, is a generalised p.d.f. with respect to μ such that

$$f(\boldsymbol{x}; \boldsymbol{\theta}) = K(\boldsymbol{\theta}) h(\boldsymbol{x}) \exp \sum_{i=1}^{r} \theta_i t_i(\boldsymbol{x}), \;\; \boldsymbol{x} \in W$$

where $h, t_1, \cdots, t_r$ are real-valued Borel measurable functions. Prove that $\boldsymbol{T} = (T_1, \cdots, T_r)$ is a sufficient statistic for $\boldsymbol{\theta} = (\theta_1, \cdots, \theta_r)$.

23. Let $X_1, \cdots, X_n$ be independent random variables with common p.d.f.

$$f(x; \boldsymbol{\theta}) = \exp\left[-x^{2n} + \theta_1 x + \theta_2 x^2 + \cdots + \theta_n x^n\right]$$

where $\boldsymbol{\theta} = (\theta_1, \cdots, \theta_n)$ is a point in the n-dimensional Euclidian space. Prove that $(X_{(1)}, \cdots, X_{(n)})$ is a sufficient statistic for $\boldsymbol{\theta}$ where $X_{(i)}$ is the ith order statistic.

24. Let $\boldsymbol{X}_1, \cdots, \boldsymbol{X}_n$ be a random sample from the multivariate normal distribution

$$f(\boldsymbol{X}; \boldsymbol{\mu}, \wedge) = \frac{\sqrt{|\wedge|}}{(2\pi)^{k/2}} \exp\left[-\frac{1}{2}(\boldsymbol{x} - \boldsymbol{\mu})' \wedge (\boldsymbol{x} - \boldsymbol{\mu})\right]$$

where $\boldsymbol{x} = (x_1, \cdots, x_k)'$, $\boldsymbol{\mu} = (\mu_1, \cdots, \mu_k)'$, $-\infty < \mu_i < \infty$, $i = 1, \cdots, k$, $\wedge = ((\lambda_{ij}))$, $i, j = 1, \cdots, k$, any positive definite matrix.

Find a sufficient statistic for $(\boldsymbol{\mu}, \wedge)$.

25. Consider the above problem when $\wedge = \text{Diag}(\lambda_{ii}, i = 1, \cdots, k)$.

26. Let $X_i (i = 1, \cdots, n)$ be a random sample from a Beta distribution

$$f(x; \alpha, \beta) = \frac{1}{B(\alpha, \beta)} x^{\alpha-1}(1-x)^{\beta-1}, \;\; 0 < x < 1, \;\; \alpha > 0, \;\; \beta > 0.$$

Find sufficient statistic for (i) α if β is known (ii) β if α is known (iii) (α, β) if both are unknown.

27. Let $X_1, \cdots X_n$ be r.v.'s taking the values 0 or 1 such that

$$P_{A,N}(X_1 = 1) = A/N, \;\; P_{A,N}\{X_{k+1} = 1 | X_1 = x_1, \cdots, X_k = x_k\} = \frac{A - \Sigma_1^k x_i}{N-k}$$

where $k = 1, \cdots, n-1, \sum_{i=1}^{k} A_i \leq A, \sum_{i=1}^{n} (1 - x_i) \leq N - A, \, n = 1, \cdots, N.$

Let $\Omega = \{(A, N) = \cdots A = 0, 1, \cdots, N; N = n, n+1, \cdots, \}$.

Prove that $T = \sum_{i=1}^{n} X_i$ is sufficient for Ω.

28. Let $\{X_i\}, \, i = 1, \cdots, n$ be a random sample from the distribution

$$f(x; a, b) = \begin{cases} \frac{1}{2b}, & a - b < x < a + b \\ 0, & \text{elsewhere,} \end{cases}$$

where $-\infty < a < \infty, -\infty < b < \infty$. Show that the statistic $T = (X_{(1)}, X_{(n)})$, where $X_{(1)} = \min X_i, X_{(n)} = \max X_i$, is jointly sufficient for (a, b).

29. Let X be a r.v. with p.d.f. of a regular case of the Koopman – Pitman class (15.4.10). Show that $E(v(x)) = -\frac{B'(\theta)}{A'(\theta)}$ by differentiating both sides of the equality

$$\int_{-\infty}^{\infty} e^{A(\theta)v(x) + B(\theta) + C(x)} \, dx = 1$$

with respect to θ. Find $V(v(x))$ by a second differentiation.

30. Let (X, Y) be a bivariate random vector with joint distribution given as follows. $P[X = 1] = p$; $P[X = 0] = q = 1 - p$; $P[Y = 1 | X = 1] = p$, $P[Y = 0 | X = 1] = q$; $P[Y = 1 | X = 0] = P[Y = 0 | X = 0] = \frac{1}{2}$. $p \in [0, 1]$. Find a minimal sufficient statistic for p.

[*Hints:* Clearly (X, Y) is a sufficient statistic for p. To find a minimal sufficient statistic note that the joint distribution of (X, Y) is

$$P_{X,Y}(x, y) = \begin{cases} q/2 \text{ if } (x, y) = (0, 0) \text{ or } (0, 1) \\ pq \;\; \text{if } (x, y) = (1, 0) \\ p^2 \;\; \text{if } (x, y) = (1, 1). \end{cases}$$

Proceed as in example 15.4.13.]

31. Let X be a single observation from the distribution

$$f_\theta(x) = \begin{cases} \frac{\theta}{4} & \text{if } x = -2 \\ \frac{\theta}{3} & \text{if } x = 1 \\ \frac{1-2\theta}{3} & \text{if } x = 0, 3, 12 \\ \theta^2 + \frac{5\theta}{12} & \text{if } x = 36 \\ \theta - \theta^2 & \text{if } x = 60 \end{cases}$$

where $0 < \theta < \frac{1}{2}$. Find a minimal sufficient statistic for θ.

32. Let $X_1, \cdots, X_n$ be i.i.d. $N(\theta, \theta^2)$ variables. Find a minimal sufficient statistic for θ.

33. Let $X_1, \cdots, X_n$ be a random sample of size n from the discrete uniform distribution $f_\theta(x) = \frac{1}{\theta}$, $0 < x < \theta$ ($= 0$ otherwise).

Show that $X_{(n)} = \max_i X_i$ is minimal sufficient for θ.

34. Let $X_1, \cdots, X_n$ be a random sample from the p.d.f. $f_\theta(x) = \frac{\theta}{(1+x)^{1+\theta}}$, $0 < x < \infty$, $\theta > 0$. Show that $\prod_{i=1}^{n}(1 + X_i)$ is minimal sufficient statistic for θ.

35. Let $X_1, \cdots, X_n$ be a random sample of size n from a lognormal distribution with p.d.f. $f_{\mu,\sigma^2}(x) = \exp[-(\ln(x) - \mu)^2 / 2\sigma^2]/(x\sigma\sqrt{2\pi}), 0 < x < \infty$.

Find the joint sufficient statistic for (μ, σ^2).

36. Let $X_1, \cdots, X_n$ be a random sample from the "displaced Laplace" distribution $f_\theta(x) = \frac{1}{2} e^{-|x-\theta|}$, $-\infty < x < \infty$, $-\infty < \theta < \infty$. Show that the order statistic $(X_{(1)}, \cdots, X_{(n)})$ when $X_{(1)} \leq \cdots \leq X_{(n)}$ is minimal sufficient statistic.

37. Let $X_1, \cdots, X_n$ be a random sample of size n from the "displaced exponential" p.d.f. $f_{\alpha,\beta}(x) = e^{-(x-\alpha)/\beta}$, $\alpha < x < \infty$. Let $Y_1 \leq Y_2 \cdots \leq Y_n$ be the order statistic.

Show that (Y_1, T) when $T = \sum_{i=1}^{n} (X_i - Y_1)$ is jointly minimal sufficient statistic for (α, β).

38. Let $X_1, \cdots, X_n$ be a random sample of size n from $N(0, \sigma^2)$ distribution. If $n = 1$, show that X_1^2 is minimal sufficient for σ^2. If $n > 1$, find a sufficient statistic for σ^2.

39. (a) Let $X_1,\cdots,X_n$ be independence r.v.'s each having Bernoulli (p) distribution $0 \leq p \leq 1$. Show that $\sum_{i=1}^{n} X_i$ is a minimal sufficient statistic.

(b) Show that for the three-parameter distribution

$$dF(x) = \frac{1}{\Gamma(p)}\left(\frac{x-\alpha}{\sigma}\right)^{p-1} \exp\left[-\left(\frac{x-\alpha}{\sigma}\right)\right]\frac{dx}{\sigma}, \quad p, \sigma > 0, \alpha < x < \infty$$

there are sufficient statistic for p and σ individually when the other two parameters are known. Hence, show that there are sufficient statistic for p and σ jointly if α is known. Show also that if σ is known and $p = 1$, there is a sufficient statistic for α.

40. Let X be a Poisson variable with unknown parameter $\lambda \geq 0$. Show that no non-trivial sufficient statistic exists.

[*Hints:* $f_\lambda(x)/f_\lambda(y) = (y!\lambda^{x-y})/x!$ which is independent of λ iff $x = y$. Hence X is a minimal sufficient statistic for λ. X is also a trivial sufficient statistic. Since a minimal sufficient statistic is a function of every other sufficient statistic, no non-trivial sufficient statistic exists.]

41. Prove that if we have n independent observations from the Cauchy distribution no non-trivial sufficient statistic exists.

42. Suppose we have a random sample

$$X = (X_1, \cdots, X_n;\ Y_1, \cdots, Y_m)$$

when $X_i \sim N(\mu, \sigma_1^2)$ $(i = 1, \cdots, n)$, $Y_j \sim N(\mu, \sigma_2^2)$ $(j = 1, \cdots, m)$, $-\infty < \mu < \infty$, $\sigma_k^2 > 0$ $(k = 1, 2)$. Show that

$$T = \left(\sum_{i=1}^{n} X_i,\ \sum_{i=1}^{n} X_i^2,\ \sum_{j=1}^{m} Y_j,\ \sum_{j=1}^{m} Y_j^2\right)$$

is a sufficient statistic for $(\mu, \sigma_1^2, \sigma_2^2)$ but not a complete sufficient statistic.

43. Show that the distribution

$$dF(x) = \frac{\theta_1\, dx}{\pi\left\{\theta_1^2 + (x-\theta_2)^2\right\}}, \quad -\infty < x < \infty$$

does not possess a single sufficient statistic for either parameter if the other is known or a pair of jointly sufficient statistic if both are unknown.

15.5 Completeness

Definition 15.5.1 *A family $\mathcal{F} = \{f_\theta; \theta \in \Theta\}$ of probability distributions is complete if*

$$\int_{-\infty}^{\infty} \Psi(x) f_\theta(x)\, dx = 0 \ \forall\, \theta \in \Theta \tag{15.5.1}$$

implies

$$\Psi(x) = 0 \ \ a.e.\ \mathcal{F}$$

i.e. $\Psi(x) = 0$ *everywhere except for the null sets* N *for which* $f_\theta(x) = 0 \ \forall\, \theta$ *whenever* $x \in N$. *This is also stated as the distribution of the random variable* X *is complete or* X *is complete for* θ.

Definition 15.5.2 *A statistic* T *is complete if the p.d.f. of* T *is complete over the space* $(\mathcal{T}, \beta)$ *where* $\mathcal{T}$ *is the space of values of* T *and* β *is the* σ*-field generated by the Borel sets of* $\mathcal{T}$. *This is also said as* T *is complete for* $\mathcal{F}$ *where* $f_\theta(x)$ *is the p.d.f. of* X *from which p.d.f. of* T *has been obtained.*

Theorem 15.5.1 *If* T *is complete for* $\mathcal{F} = \{f_\theta(x), \theta \in \Theta\}$ *then any statistic depending on* X *only through* T *is complete.*

Proof: Let $T^* = g(T)$. Suppose there is a function h satisfying

$$E_\theta[h(T^*)] = 0 \ \forall\, \theta \in \Theta. \tag{15.5.2}$$

Then

$$E_\theta[h\{g(T)\}] = 0 \ \forall\, \theta \in \Theta. \tag{15.5.3}$$

By completeness property of T, (15.5.3) implies $h\{g(T)\} = 0$ *a.e.* i.e., $h(T^*) = 0$ *a.e.*$\mathcal{F}$. This completes the proof.

Before proceeding to give a few illustrative examples we recall two results which will be very useful.

Lemma 15.5.1 *If the infinite series* $\sum_{j=0}^{\infty} a_j z^j$ *converges to zero for all* z *in a given interval then* $a_0 = a_1 = a_2 = \cdots = 0$.

Definition 15.5.3 ***The Laplace transformation.*** *If the integral* $\int_0^\infty e^{-st} f(s)\, ds$ *converges for some values of* t, *then the integral*

$$\Phi(t) = \int_0^\infty e^{-st} f(s)\, ds \tag{15.5.4}$$

is called the Laplace unilateral transform of f. *The integral*

$$\Phi_1(t) = \int_{-\infty}^{\infty} e^{-st} f(s)\, ds \tag{15.5.5}$$

when it converges for some value of t *is called the bilateral transform of* f. *Since*

$$\int_{-\infty}^{\infty} e^{-st} f(s)\, ds = \int_0^\infty e^{-st} f(s)\, ds + \int_0^\infty e^{st} f(-s)\, ds$$

a bilateral transform can be studied by studying two unilateral transforms in one of which the variable t *has been replaced by* $-t$.

The Laplace transform Φ of a continuous function is unique. This is called the unicity property of Laplace transform. For further details see Wedder (1946).

Definition 15.5.4 *The integral*

$$\Psi(t) = \int_0^{\infty} s^{t-1} f(s)\, dt$$

when it converges for some value of t is called the Mellin transformation of $f(s)$.

▶ **Example 15.5.1** *Let $P = \{p_\theta = b(n;\theta), 0 \leq \theta \leq 1\}$ be the family of Binomial distributions corresponding to n Bernoulli trials with constant probability of success θ, $\theta \in [0,1]$.*

$$p_\theta(X = x) = \binom{n}{x} \theta^x (1-\theta)^{n-x}, \quad x = 0, 1, \cdots, n.$$

The family P is complete. We have to show that

$$E_\theta\left(\Psi(x)\right) = \sum_{x=0}^{n} \binom{n}{x} \theta^x (1-\theta)^{n-x} \Psi(x) = 0 \;\; \forall\, \theta \tag{15.5.6}$$

implies $\Psi(x) = 0$ for $x = 0, 1, \cdots, n$. Now (15.5.6) implies

$$\sum_{x=0}^{n} a(x) \Phi^x = 0 \tag{15.5.7}$$

where $a(x) = \binom{n}{x}\Psi(x), \Phi = \frac{\theta}{1-\theta}$. The left hand side of (15.5.7) is a polynomial in Φ of degree n. Hence by lemma 15.5.1 it vanishes iff $a(x) = 0$ i.e., $\Psi(x) = 0$ for $x = 0, 1, \cdots, n$.

We have seen before that if $X_i (i = 1, \cdots, n)$ is a random sample from a Bernoulli distribution with parameter θ then the total number of successes,

$$X = \sum_{i=1}^{n} X_i$$

is sufficient for θ. Hence X is a complete sufficient statistic for θ.

▶ **Example 15.5.2** *Let $P = \{p_\theta(x) = e^{-\theta}\theta^x/x!, x = 0, 1, 2, \cdots; \theta > 0\}$ be a family of Poisson distributions. The family P is complete.*

We have to show that

$$E_\theta[\psi(x)] = e^{-\theta} \sum_{x=0}^{\infty} \psi(x) \frac{\theta^x}{x!} = 0 \text{ for } \theta > 0$$

implies that $\psi(x) = 0$ for $x = 0, 1, 2, \cdots$.

Since $\psi(x)/x!$ is the coefficient of θ^x in the power series expansion of zero it must be zero for $x = 0, 1, 2, \cdots$, i.e., $\psi(x) = 0$ a.e. P. Hence X is a complete statistic.

We have seen that if $X_i, i = 1, \cdots, n$ be a random sample from a Poisson (θ) distribution then

$$T = \sum_{i=1}^{n} X_i$$

is sufficient for θ. Again, T follows Poisson $(n\theta)$. Thus, T is complete sufficient for θ.

▶ **Example 15.5.3** *Let*

$$f = \{f_\theta(x) = \left(\sqrt{2\pi}\right)^{-1} \exp\left(-\frac{(x-\theta)^2}{2}\right), \quad -\infty < x < \infty, -\infty < \theta < \infty\}.$$

Then f is complete. We have to prove that

$$E_\theta(\psi(x)) = \frac{1}{\sqrt{2\pi}} \int_{-\infty}^{\infty} \psi(x) \exp\left(-\frac{(x-\theta)^2}{2}\right) dx = 0 \ \forall\, \theta \tag{15.5.8}$$

mplies $\psi(x) = 0$ a.e. f.
15.5.8) can be written as

$$\int_{-\infty}^{\infty} \psi(x) e^{-x^2/2} e^{\theta x}\, dx = 0 \ \forall\, \theta. \tag{15.5.9}$$

The right hand side of (15.5.9) is the bilateral transform of the function $\psi(x)e^{-x^2/2}$. Hence by the unicity property of Laplace transform

$$\psi(x)e^{-x^2/2} = 0$$

.e., $\psi(x) = 0$ a.e. Lebesgue measure. Hence f is a complete family.

Example 15.5.4 *The family of chi-squared distributions*

$$f(x;p) = \begin{cases} \dfrac{1}{2^p \lceil p} e^{-x/2} x^{p-1}, & x > 0, p \text{ a positive integer } \geq 1 \\ 0, & \text{otherwise} \end{cases}$$

s complete, because

$$E_p(\psi(x)) = \frac{1}{2^p \lceil p} \int_0^{\infty} e^{-x/2} \psi(x) x^{p-1}\, dx = 0$$

mplies $\psi(x)x^{p-1} = 0$ for every x by the unicity property of the Laplace transformation. Hence $\psi(x) = 0$ a.e. Lebesgue and so X is complete.

▶ **Example 15.5.5** *Let X be a r.v. with Cauchy distribution*

$$f_\theta(x) = \frac{K}{\pi\left\{1 + \left(\frac{x}{\theta}\right)^2\right\}}, \quad |x| < A, \ -\infty < \theta < \infty$$

where K is a suitable constant. X is not complete. Let $\psi(x)$ be an odd function of x [i.e., $\psi(-x) = -\psi(x)$].

Suppose that

$$E_\theta[\psi(X)] = \frac{K}{\pi} \int_{-\infty}^{\infty} \frac{\psi(x)\, dx}{1 + (x/\theta)^2} = 0 \ \forall\, \theta.$$

This does not imply that $\psi(x) = 0$ a.e. Labesgue.

► **Example 15.5.6** *Let X be a r.v. with a uniform distribution over $[\theta, \theta+1]$, where is any real number. The family of probability distributions is not complete.*

Consider a function $\psi(x)$ such that

$$\psi(x) = \begin{cases} 1 \text{ for } N < x \leq N + \frac{1}{2} \\ -1 \text{ for } N + \frac{1}{2} < x \leq N + 1 \end{cases}$$

for every integral N. Then if $\theta < N \leq \theta + \frac{1}{2}$,

$$E_\theta[\psi(X)] = \int_\theta^{\theta+1} \psi(x)\, dx = \int_\theta^N \psi(x)\, dx + \int_N^{N+\frac{1}{2}} \psi(x)\, dx + \int_{N+\frac{1}{2}}^{\theta+1} \psi(x)\, dx$$
$$= -(N-\theta) + \frac{1}{2} - \left(\theta + 1 - N - \frac{1}{2}\right) = 0.$$

It can be shown similarly that $E_\theta[\psi(X)] = 0$ even if $\theta + \frac{1}{2} < N \leq \theta + 1$. Hence $\psi(x$ is a non-zero function with $E(\psi(X)) = 0$.

► **Example 15.5.7** *Let $X_i (i = 1, \cdots, n)$ be a random sample from the population wit p.m.f $P(X_i = K) = N^{-1}, K = 1, \cdots, N, N$ being any positive integer. Then $T = \max_i X$ is a complete statistic.*

The p.m.f. of T is

$$P(T = t) = \left(\frac{t}{N}\right)^n - \left(\frac{t-1}{N}\right)^n.$$

We have to show that

$$E_N(\psi(T)) = \frac{1}{N^n} \sum_{t=1}^{N} \psi(t)\,[t^n - (t-1)^n] = 0, N = 1, 2, \cdots \quad (15.5.1($$

implies that $\psi(t) = 0$ for $t = 1, 2, \cdots, N$.

For $N = 1$, (15.5.10) $\Rightarrow \psi(t) = 0$.

For $N = 2$, (15.5.10) $\Rightarrow \psi(1) = 0, \psi(2) = 0$

and so on. Hence, the family of distributions of T is complete. T is a complete sufficie statistic.

► **Example 15.5.8** *Let $X_i (i = 1, \cdots, n)$ be a random sample from the continuo uniform distribution*

$$f_\theta(x) = \begin{cases} \frac{1}{\theta},\ 0 < x < \theta,\ 0 < \theta < \infty \\ 0,\ \textit{elsewhere}. \end{cases}$$

The family of distributions induced by $T = \max_i X_i$ is complete. The p.d.f. of T is

$$g_\theta(t) = \begin{cases} nt^{n-1}/\theta^n, & 0 < t < \theta \\ 0, & \text{otherwise.} \end{cases}$$

For a measurable function $\psi(T)$ of T,

$$E_\theta[\psi(T)] = 0 \ \forall\, \theta \ \Rightarrow \ \int_0^\theta \psi(t)t^{n-1}\,dt = 0 \ \forall\, \theta.$$

Differentiating both sides with respect to θ,

$$\psi(\theta)\theta^{n-1} = 0 \ \forall\, \theta \ \text{ or } \ \psi(\theta) = 0 \ \forall\, \theta. \tag{15.5.11}$$

Since $0 < \theta < \infty$, (15.5.11) implies $\psi(t) = 0 \ \forall\, t \in (0, \theta)$ and $\forall\, \theta \in (0, \infty)$. Hence $\psi(t) = 0$ a.e. $\{g_\theta(t), 0 < \theta < 1\}$. Hence T is complete. T is a complete sufficient statistic.

▶ **Example 15.5.9** *Consider the normal distribution $N(\mu, \theta)$ when μ is known but the variance θ is unknown, $0 < \theta < \infty$. We have seen that*

$$T = \sum_{i=1}^{n} (X_i - \mu)^2$$

is sufficient for θ. The p.d.f. of T is

$$g_\theta(t) = \frac{1}{\theta 2^{n/2}\Gamma\frac{n}{2}} e^{-t/2\theta}\left(\frac{t}{2\theta}\right)^{\frac{n}{2}-1}, \quad 0 < t < \infty$$

$$E_\theta(\psi(T)) = 0 \ \forall\, \theta \ \Rightarrow \ \int_0^\infty \psi(t)e^{-t/2\theta}t^{\frac{n}{2}-1}\,dt = 0 \ \forall\, \theta. \tag{15.5.12}$$

The left hand side of equation (15.5.12) is the unilateral Laplace transform of $\psi(t)t^{(n-2)/2}$. Hence by the unicity property of this transformation

$$\psi(t)t^{(n-2)/2} = 0 \ \forall\, t \in (0, \infty)$$
$$\text{i.e., } \psi(t) = \text{ a.e. } \{g_\theta(t), \theta > 0\}.$$

Hence T is a complete sufficient statistic.

The following theorem stated without proof gives a large class of probability distributions for which a complete sufficient statistic exists.

Theorem 15.5.2 *Let $\{f_{\boldsymbol\theta} : \theta \in \Theta\}$ be a k-parameter exponential family*

$$f_{\boldsymbol\theta}(x_1, \cdots, x_n) = \exp\left\{\sum_{j=1}^{k} A_j(\boldsymbol\theta)v_j(x) + B(\boldsymbol\theta) + C(x)\right\} \tag{15.5.13}$$

where $\boldsymbol{x} = (x_1, \cdots, x_n)$, $\boldsymbol{\theta} = (\theta_1, \cdots, \theta_k) \in \Theta$, *an interval in* $R^k, v_1, \cdots, v_k, C$ *are defin on* $R^n, k \leq n$. *Let* $\mathbf{A} = (A_1, \cdots, A_k)$ *and suppose that range of* A *contains an open s in* R^k. *Then*

$$(v_1(\boldsymbol{x}), \cdots, v_k(\boldsymbol{x}))$$

is a complete sufficient statistic.

If one has completeness for a class of probability distributions, one can, in some case infer completeness for a larger class. The following theorem states this property.

Theorem 15.5.3 *Let* $\mathcal{F} = \{F_\theta, \theta \in w\}$ *be a complete family of distributions and* $w \in \Omega$. *Then* $\mathcal{F}_1 = \{F_\theta, \theta \in \Omega\}$ *is also complete provided none of the added probabili measures assign positive probability to sets having zero probability for each measure of*

Definition 15.5.5 *The family* $\mathcal{F} = \{F_\theta, \theta \in \Theta\}$ *is boundedly complete if*

$$\int \psi(x) dF_\theta(x) = 0 \ \forall \ \theta \in \Theta$$

and $\psi(x)$ *bounded and real implies* $\psi(x) = 0$ *a.e.* $\mathcal{F}$.

Theorem 15.5.4 *If a family* $\mathcal{F}$ *of probability measures is complete, then it is bounded complete.*

Proof: Follows easily.

A family of probability measures can be boundedly complete without being complet

► **Example 15.5.10** *Consider the family of p.d.f.'s* $f = \{f_\theta(x) = (1-\theta)^2\theta^x,\ x = 1, \cdots; = \theta$ *for* $x = -1; = 0$ *otherwise;* $0 < \theta < 1\}$. *Then*

$$E_\theta(\psi(X)) = 0 \ \forall \ \theta \in \Theta \tag{15.5.1}$$

implies

$$\theta\psi(-1) + \sum_{x=0}^{\infty} \psi(x)(1-\theta)^2\theta^x = 0.$$

$$\therefore \sum_{x=0}^{\infty} \psi(x)\theta^x = -\psi(-1)\theta(1-\theta)^{-2} = -\psi(-1)\sum_{x=0}^{\infty} x\theta^x$$

and hence

$$\sum_{x=0}^{\infty} [\psi(x) + \psi(-1)x]\theta^x = 0 \tag{15.5.1}$$

(15.5.15) implies that $\psi(x) + \psi(-1)x = 0$ for $x = -1, 0, 1, \cdots$. Hence, either $\psi(-1)$ not zero and $\psi(x)$ is unbounded or $\psi(x) = 0$ for $x = -1, 0, 1, \cdots$.

Hence if $\psi(x)$ is bounded and satisfies (15.5.14) then $\psi(x) = 0$ a.e. f and thus f boundedly complete. Alternatively, if $\psi(-1) \neq 0$ and $\psi(x) = -x\psi(-1)$ then (15.5.14) satisfied but $\psi(x) \neq 0$ a.e. f and so f is not complete.

heorem 15.5.5 (**Lehmann (1959), p.132-3**). *Let $\boldsymbol{X} = (X_1, \cdots, X_n)'$ be a random ector in the exponential class i.e., with density*

$$f_{\boldsymbol{\theta}}(x_1, \cdots, x_n) = \exp\left[\sum_{j=1}^{k} A_j(\boldsymbol{\theta})T_j(\boldsymbol{x}) + B(\boldsymbol{\theta}) + C(\boldsymbol{x})\right] \tag{15.5.16}$$

here $\boldsymbol{x} = (x_1, \cdots, x_n)'$, for all $\boldsymbol{\theta} = (\theta_1, \cdots, \theta_k)' \in \Theta \subseteq R_k$. Then $\boldsymbol{T} = (T_1(\boldsymbol{x}), \cdots, T_k(\boldsymbol{X}))'$ a minimal sufficient statistic for $\boldsymbol{\theta} = (\theta_1, \cdots, \theta_k)'$. $\boldsymbol{T}$ is also complete as long as Θ ontains all θ in some k-dimensional rectangle.

Example 15.5.11 *We have already seen a few cases of exponential family of distribu-ons in examples* 15.5.8–15.5.11. *Thus, if $X_1, \cdots, X_n$ be independent r.v.'s each $N(\mu, \sigma^2)$ ith μ, σ^2 unknown, then $T(X_1, \cdots, X_n) = (\Sigma X_i, \Sigma X_i^2)$ is a complete minimal sufficient tatistic for (μ, σ^2). If $X_1, \cdots, X_n$ are independent r.v.'s each with the same distribution hich is either Bernoulli, Poisson or Gamma with unknown parameters the respective omplete sufficient statistic are as follows.*

Bernoulli: ΣX_i for p

Poisson: ΣX_i for λ

Gamma: $\left[\frac{1}{\beta^{\alpha+1}\overline{|}(\alpha+1)} x^{\alpha} e^{-x/\beta},\ 0 \le x < \infty\right]$; $(\Sigma X_i, \Sigma \ln(X_i))$ for (α, β).

heorem 15.5.6 *Let $X_1, \cdots, X_n$ be independent r.v.'s each with distribution function , where F is unknown and F belongs to the class of all continuous d.f.'s. Then the rder statistic $(X_{(1)}, \cdots, X_{(n)})$ is a complete sufficient statistic for F.*

5.5.1 Ancillary Statistic

efinition 15.5.6 *A statistic $u(X_1, \cdots, X_n)$ is called an ancillary statistic of first order f $E_\theta[u(X_1, \cdots, X_n)]$ is a constant independent of θ. The statistic $u(X_1, \cdots, X_n)$ is called n ancillary statistic for θ if the distribution function of $u(X_1, \cdots, X_n)$ does not depend n θ.*

Thus, an ancillary statistic does not contain any information about θ.

heorem 15.5.7 (**Basu, 1955**). *Let $t(X_1, \cdots, X_n)$ be a complete sufficient statistic nd $u(X_1, \cdots, X_n)$ be an ancillary statistic. Then t and u are independently distributed.*

Example 15.5.12 *Let X_1, X_2 be independent $N(\mu, \sigma^2)$ r.v.'s with μ unknown, σ^2 nown. Let $t = X_1 + X_2$, $u = X_1 - X_2$. Then $u(X_1, X_2) \sim N(0, 2\sigma^2)$ and its distribution is ndependent of μ. Hence u is an ancillary statistic for μ. Since t is a complete sufficient tatistic distributions of t and u are independent.*

► **Example 15.5.13** *Let $X_1, \cdots, X_n$ be iid r.v.'s each distributed uniformly on $[0, \theta]$. L Y_1 \le Y_2 \le \cdots \le Y_n$ denote the corresponding order statistic. Here Y_n and Y_1/Y_n a independently distributed. Y_n is complete sufficient. It is, therefore, enough to show th distribution of Y_1/Y_n is independent of θ.*

$$\begin{aligned} F_{Y_1/Y_n}(z) &= P(Y_1 \le zY_n) \\ &= \int_0^\theta P(Y_1 \le zy) f_{Y_n}(y)\, dy \\ &= \int_0^\theta \left[1 - \left(1 - \frac{yz}{\theta}\right)^n\right] \frac{ny^{n-1}}{\theta^n}\, dy \\ &= 1 - n\int_0^1 (1 - zx)^n x^{n-1}\, dx \ \ (\textit{putting } x = y/\theta). \end{aligned}$$

which is independent of θ.

Exercise

1. Let X be a r.v. with exponential density $f_\theta(x) = \theta e^{-\theta x}, x > 0, \theta > 0$. Show tha distribution of X is complete.

2. Let $\{X_i\}$ be a random sample from $f_\theta(x) = e^{-(x-\theta)}$, $\theta < x < \infty, -\infty < \theta <$ ∝ Show that $X_{(1)} = \min\limits_i X_i$ is complete sufficient for θ.

3. Let X be a random sample from $N(0, \sigma^2)$. Show that X^2 is complete sufficient fc σ^2. Show also that X is sufficient but not complete.

4. Let X have, for every $\theta \in \Theta$, a distribution which is symmetrical about 0. Sho that if $\psi(x)$ is an odd function of x such that $E_\theta[\psi(x)]$ exists for every θ, then th family of distributions is not complete.

   Hence, show that the family $\{f_\theta(x) = \left(\theta\sqrt{2\pi}\right)^{-1} \exp(-x^2/2\theta^2), -\infty < x <$ ∝ $0 < \theta < \infty\}$ is not complete.

5. (a) Show that the family of hypergeometric distributions $H(N, n, M)$,

   $$f_M(x) = \binom{M}{x}\binom{N-M}{n-x}, \quad M = 0(1)N$$

   where N, n are fixed positive integers are complete.

   (b) Suppose now n and M are fixed and N belongs to the set of positive intege greater than $\max(M, n)$. Is X complete?

6. Show that the family of rectangular distributions $f = \{f_\theta(x) = 1, \theta < x < \theta + 1; = 0$ otherwise; $-\infty < \theta < \infty\}$ is not (even) boundedly complete.

   [*Hints:* Consider $\psi(x)$ as a periodic function with period 1 and such that $\int_0^1 \psi(t)\, dt = 0$. Then $E_\theta[\psi(X)] = 0 \Rightarrow \int_0^1 \psi(t)\, dt = 0$; however, $\psi(x) \neq 0$ a.e. f.]

7. Let $X_i (i = 1, \cdots, n)$ be a random sample from $N(\mu, \sigma^2)$ where both (μ, σ^2) are unknown, $\Theta = \{(\mu, \sigma^2) : -\infty < \mu < \infty, \sigma^2 > 0\}$. Show that $\boldsymbol{T} = (T_1, T_2)$ where $T_1 = \sqrt{n}\bar{X}$, $T_2 = \sum_{i=1}^{n} (X_i - \bar{X})^2$ is jointly sufficient and complete for (μ, σ^2).

8. Let X be a r.v. with p.d.f. $f_N(x) = (N+1)x^N, 0 \leq x \leq 1$, N belonging to the set of non-negative integers. Show that $T(X) = X$ is complete.

9. (a) Let X be a sample from $f(x;\theta) = \frac{1}{\theta}, 0 < x < \theta$, $0 < \theta < \infty$. Show that distribution of X is complete.

   (b) Suppose X follows a continuous uniform distribution

$$f(x;\theta) = \begin{cases} \frac{1}{2}\theta, & -\theta < x < \theta \\ 0, & \text{otherwise} \end{cases}$$

   $-\infty < \theta < \infty$. Show that the family of distributions is not complete.

   (c) Let the p.d.f. of X be $f(x;\theta) = Q(\theta)M(x)$, $0 < x < \infty, 0 < \theta < \infty, M(x) (> 0)$ being a continuous function of x. Show that the family of p.d.f.'s of $X_{(n)} (= \max_i X_i)$ is complete.

10. Let $\{X_i\}$ be independent random variables with common p.d.f.

$$f_{\alpha,\beta}(x) = \begin{cases} \frac{1}{\beta-\alpha}, & \alpha \leq x \leq \beta \\ 0, & \text{otherwise} \end{cases}$$

   where α, β are real numbers such that $\alpha < \beta$. Consider the statistic $T = X_{(n)} - X_{(1)}$ where $X_{(n)} = \max_i X_i, X_{(1)} = \min_i X_i$. Is T complete or boundedly complete?

11. Is the family of distribution of order statistic in sampling from a Poisson distribution complete?

12. Let X_1, X_2 be independent r.v.'s where X_i is binomial $b(n_i, p_i) (i = 1, 2)$ and $(p_1, p_2) \in \omega$ the interior of the unit square. Prove that $T = (X_1, X_2)$ is complete.

13. Let $\{X_{ij}, j = 1, \cdots, n_i\}$ be a random sample from

$$N(\mu_i, \sigma_i^2)(i = 1, 2), \quad \Omega = \{(\mu_1, \mu_2, \sigma_1^2, \sigma_2^2); \ -\infty < \mu_i < \infty, \sigma_i^2 > 0, i = 1, 2\}.$$

   Prove that

$$T = \left(\sum_{j=1}^{n_1} X_{1j}, \ \sum_{j=1}^{n_2} X_{2j}, \ \sum_{j=1}^{n_1} X_{1j}^2, \ \sum_{j=1}^{n_2} X_{2j}^2 \right)$$

   is complete.

14. Let $\{X_i\}$ be independent random samples from $N(\mu, \sigma^2)$. Construct complet sufficient statistic for the parametric upper half plane $\{\mu > 0, \sigma^2 > 0\}$.

[*Hints:* Use theorem 15.5.2.]

15. Let $\{X_i\}$ be a sample from $N(\alpha\sigma, \sigma^2)$ where α is a known real number. Show tha the statistic

$$T = \left(\sum_1^n X_i, \ \sum_1^n X_i^2 \right)$$

is sufficient for σ, but the family of distributions of T is not complete.

16. Let $\{X_i\}$ be a random sample from the uniform distribution

$$f_\theta(x) = \begin{cases} 1, \ \theta - \frac{1}{2} < x < \theta + \frac{1}{2} \\ 0, \text{ otherwise,} \end{cases}$$

$-\infty < \theta < \infty$. Show that the statistic $T = (\min X_i, \max X_i)$ is sufficient for θ, bu not complete.

17. Let X_1, X_2 be a random sample of size 2 from the exponential p.d.f. $f_\theta(x) =$ $e^{-x/\theta}$, $0 < x < \infty$. Show that $t_1 = X_1 + X_2$, $t_2 = \frac{X_1}{X_1+X_2}$ are independent r.v.'s

[*Hints:* Use Basu's theorem 15.5.7.]

Generalise this result to prove the following. If $X_1, \cdots, X_n$ is a random sample c size n from the above p.d.f., show that $t_1 = S_n, t_2 = X_i/S_n$, where $S_n = \sum_{i=1}^n X_i$ ar (for any fixed $i = 1, \cdots, n$) independent r.v.'s.

18. Let $\boldsymbol{N} = (N_1, \cdots, N_k)$ have a distribution with parameters $\theta_1, \cdots, \theta_k$ such tha $\theta_i > 0 \ \forall \ i$ and $\sum_{i=1}^k \theta_i = 1$ with p.m.f.

$$P\{N_1 = n_1, \cdots, N_k = n_k\} = \begin{cases} \dfrac{n!}{\prod_{i=1}^k n_i!}, \text{ if } n_i \geq 0 \ \forall \ i \text{ and } \Sigma n_i = n \\ 0, \quad \text{otherwise} \end{cases}$$

where n is a given fixed integer. Show that $\boldsymbol{N}$ is a complete sufficient statistic fc $\boldsymbol{\theta} = (\theta_1, \cdots, \theta_k)$.

15.6 Uniformly Minimum Variance Unbiased Estimation Consistent Asymptotically Normal Estimators

We have seen in Section 15.2 that for the same parameter there may exist an infinit number of unbiased estimators. We need, therefore, some criteria for choosing a optimum one among these estimators. Clearly the one whose scatter around θ is smaller would be preferable. A measure of dispersion of an unbiased estimator t is given by its variance $V_\theta(t) = E_\theta(t-\theta)^2$. We examine in this section the concepts based on the minimisation of variance of an unbiased estimator.

We consider in sub-section 15.6.1 the concept of uniformly minimum variance unbiased (UMVU) estimation and different theorems for obtaining UMVU-estimators and investigate their properties.

In sub-section 15.6.2 we consider modal-unbiased and median-unbiased estimation. The subsection 15.6.3 deals with consistent asymptotically normal (CAN) estimators.

15.6.1 Uniformly Minimum Variance Unbiased (UMVU) Estimation, Locally Best Unbiased Estimation

Let X be a r.v. with density function $f_\theta(x), \theta \in \Theta$. Let u_θ be the class of unbiased estimates t of θ such that $E_\theta t^2 < \infty$.

Definition 15.6.1 *An estimator $t_0 \in u_\theta$ is said to be locally best unbiased estimator of θ at θ_0 if*

$$E_{\theta_0}\left(t_0(X)-\theta_0\right)^2 \le E_{\theta_0}\left(t(X)-\theta_0\right)^2 < \infty \tag{15.6.1}$$

for all $t \in u_\theta$.

Definition 15.6.2 *An estimator $t^* \in u_\theta$ is said to be uniformly minimum variance unbiased estimator (UMVUE) of θ if*

$$E_\theta\left(t^*(X)-\theta\right)^2 \le E_\theta\left(t(X)-\theta\right)^2$$

for all $\theta \in \Theta$ and every $t \in u_\theta$.

Clearly, if a UMVUE exists, it is the most desirable one if we restrict ourselves to the class of unbiased estimators. Sometimes, the precision of an estimator is measured in terms of its mean square error (mse) $E_\theta\left(t(X)-\theta\right)^2$. An estimator t_1 is said to be at least as good as another estimator t_2 in the mse-sense if

$$E_\theta\left(t_1-\theta\right)^2 \le E_\theta\left(t_2-\theta\right)^2 \;\; \forall\, \theta \in \Theta. \tag{15.6.2}$$

A uniformly minimum mse-estimator does not exist unless we restrict ourselves to certain class of estimators.

▶ **Example 15.6.1** *Let $X_1, \cdots, X_n$ be a random sample from a Rectangular population $(0,\theta)$:*

$$f(x,\theta) = \frac{1}{\theta}(0), \quad 0 \le x \le \theta \text{ (otherwise).}$$

Consider the five estimators for estimating $\theta : T_1 = X_{(n)}$,
$T_2 = \frac{n+1}{n}X_{(n)}$, $T_3 = X_{(1)} + X_{(n)}$, $T_4 = (n+1)X_{(1)}$,
$T_5 = 2\bar{X}$ *when* $X_{(1)} = \min\limits_i X_i$, $X_{(n)} = \max\limits_i X_i$.

As shown in example 15.3.6
$$E(T_1) = \frac{n}{n+1} \cdot \theta.$$
Also,
$$E(T_1^2) = \frac{n}{n+2}\theta^2.$$
Thus,
$$V(T_1) = \frac{n}{(n+1)^2(n+2)}\theta^2$$
$$\text{MSE } (T_1) = V(T_1) + [\text{Bias } (T_1, \theta)]^2$$
$$= \frac{2\theta^2}{(n+1)(n+2)}.$$

T_2 is unbiased for θ.
$$\text{MSE } (T_2) = V(T_2) = \frac{\theta^2}{n(n+2)}.$$
Considering T_3, note that
$$E(X_{(1)}) = \frac{\theta}{n+1}.$$
Hence T_3 is unbiased for θ.
$$V(T_3) = V(X_{(1)}) + V(X_{(n)}) + 2\,\text{Cov}(X_{(1)}, X_{(n)}).$$
The joint density of $X_{(1)}$ and $X_{(n)}$ is given from (11.6.4) as
$$f_{X_{(1)},X_{(n)}}(u,t) = \frac{n(n-1)}{\theta^n}(t-u)^{n-2}, \quad 0 \le u < t \le \theta.$$
Therefore,
$$\text{Cov}(X_{(1)}, X_{(n)}) = \frac{n(n-1)}{\theta^2}\int_0^\theta\int_0^t tu(t-u)^{n-2}du\,dt - \frac{n\theta^2}{(n+1)^2}$$
$$= \frac{\theta^2}{n+2} - \frac{n\theta^2}{(n+1)^2} = \frac{\theta^2}{(n+1)^2(n+2)}.$$
Therefore,
$$V(T_3) = \frac{2\theta^2}{(n+1)(n+2)}$$
T_4 is unbiased.
$$V(T_4) = (n+1)^2 V(X_{(1)}) = \frac{n}{n+2}\theta^2.$$
T_5 is unbiased.
$$V(T_5) = V\left(\frac{2\bar{X}}{n}\right) = \frac{4}{n^2}\cdot nV(X_i) = \frac{\theta^2}{3n}. \quad \left(\because V(X_i) = \frac{\theta^2}{12}\right)$$

It follows from above that for $n > 1$ the best of the five suggested estimators is T_2. To determine the best estimator in the class $T = \alpha T_1$, where α is a suitably chosen constant, we consider

$$E(T) = \alpha E(T_1) = \frac{\alpha n}{n+1}\theta.$$

Therefore,

$$B(T,\theta) = E(T) - \theta = \frac{\alpha n - (n+1)}{n+1}\theta$$

$$V(T) = \alpha^2 V(T_1) = \frac{\alpha^2 n}{(n+1)^2(n+2)}\theta^2.$$

$$\begin{aligned}\text{MSE }(T) &= V(T) + [\text{Bias }(T,\theta)]^2 \\ &= \left[\frac{\alpha^2 n + (n+2)(\alpha^2 n^2 - 2\alpha n(n+1) + (n+1)^2}{(n+1)^2(n+2)}\right]\theta^2\end{aligned}$$

Differentiating w.r.t. α we find the value of α that minimises MSE (T) as

$$\alpha = \frac{n+2}{n+1}.$$

For this value of α.

$$\text{MSE }(T) = \frac{\theta^2}{(n+1)^2} < \frac{\theta^2}{n(n+1)} = \text{MSE }(T_2).$$

Thus, $\frac{n+2}{n+1}T_1$ is the best estimator in the class αT_1 for $n > 1$. The same conclusion also holds for $n = 1$ when MSE $\left(\frac{3}{2}X_1\right) = \frac{\theta^2}{4}$.

Definition 15.6.3 *Let $X_1, \cdots, X_n$ be n independent random variables with mean μ and variance $\sigma_k^2 (k = 1, \cdots, n)$. Then $\sum_{i=1}^{n} C_i X_i$, when $C_1, \cdots, C_n$ are constants with $\sum_{i=1}^{n} C_i = 1$, is a linear unbiased estimator of μ. An estimator in this class with minimum variance in this class is called a best linear unbiased estimater (BLUE) of μ.*

We now state and prove the Rao-Blackwell theorem which gives a technique for obtaining from a given unbiased estimator $\delta(X)$, a new unbiased estimator $t(\delta(X))$ of θ such that $t(\delta)$ has uniformly smaller variance then δ.

Theorem 15.6.1 ***Rao-Blackwell theorem* (*Rao*, 1945; *Blackwell*, 1947).** *Let $\mathcal{F} = \{F_\theta, \theta \in \Theta\}$ be a family of probability distributions and let δ be any statistic in U where U is the non-empty class of all unbiased estimators of θ with $E_\theta(\delta^2) < \infty$. Let t be a sufficient statistic for $\mathcal{F}$. Then the conditional expectation*

$$E_\theta[\delta(X)|t(X)] = h(t) \tag{15.6.3}$$

is unbiased for θ. Furthermore,

$$E_\theta[h(t)-\theta]^2 \le E_\theta[\delta(X)-\theta]^2 \;\forall\, \theta \in \Theta \tag{15.6.4}$$

equality holds iff

$$\delta(X) = h[t(X)]$$

a.e. $\mathcal{F}$ *i.e.,* $P_\theta[\delta = E(\delta|t)] = 1 \;\forall\, \Theta$ *except for null sets.*

Proof: Since t is a sufficient statistic, the conditional probability distribution of $\delta(X)$ given t is independent of θ and hence $h(t)$ is not a function of θ. Therefore, $h(t)$ is a statistic defined over the space of t,

$$E_\theta[h(t)] = E_\theta[E_\theta\{\delta(X)|t\}] = E_\theta(\delta(X)) = \theta.$$

It is, therefore, necessary and sufficient to show

$$E_\theta[h(t)^2] \le E_\theta[\delta(X)^2] \;\forall\, \theta \in \Theta.$$

But

$$E_\theta[\delta(X)^2] = E_\theta\left[E\{\delta(X)^2|t\}\right].$$

Hence it is sufficient to show that

$$[E\{\delta(X)|t\}]^2 \le E\left\{\delta(X)^2|t\right\}.$$

By the Cauchy-Schwartz inequality

$$E\left\{\delta(X)^2|t\right\}E\{1|t\} \ge [E\{\delta(X)|t\}]^2$$

and hence (15.6.4) follows. The equality in (15.6.4) holds iff

$$\begin{aligned}
& E_\theta\left[E^2\{\delta(X)|t\}\right] = E_\theta[\delta(X)^2] \\
\text{i.e.,}\quad & E_\theta[E\{\delta(X)^2|t\}] - E_\theta[E^2\{\delta(X)|t\}] = 0 \\
\text{i.e.,}\quad & E_\theta[\mathrm{Var}(\delta(X)|t)] = 0 \\
& \text{i.e., iff } \mathrm{Var}(\delta(X)|t) = 0 \\
& \text{i.e., iff } \delta(X) = E(\delta(X)|t) = h(t) \text{ a.e. } \mathcal{F}.
\end{aligned}$$

Theorem 15.6.2 *If t is a complete sufficient statistic for θ then there exists at most one estimator $h(t)$ such that $E\{h(t)\} = \theta$.*

Proof: If possible let there be two unbiased estimates $h_1(t), h_2(t)$. Then

$$E_\theta\{h_1(t) - h_2(t)\} = 0 \;\forall\, \theta \in \Theta.$$

Since t is a complete sufficient statistic $h_1(t) - h_2(t) = 0$ a.e. (θ).

Theorem 15.6.3 (***Lehmann and Scheffe,* 1950**). *If t is complete sufficient statistic for θ and if there exists an unbiased estimator g of θ, then there exists a unique* UMVUE *of θ and it is given by $E\{g|t\}$.*

Proof: By theorem 15.6.1, $E\{g|t\}$ is unbiased for θ with $V\{E(g|t)\} \leq V\{t\} \quad \forall\, \theta$. Also $E\{g|t\} = h(t)$ is a function of t. By theorem 15.6.1, $h(t)$ is the unique unbiased estimator of θ. Hence $h(t)$ is the unique UMVUE of θ.

Corollary 15.6.1 *If t is a complete sufficient statistic then any function $h(t)$ with finite second moment is a* UMVUE *of its expectation.*

Corollary 15.6.2 *If t is a complete sufficient statistic, then there exists a one-to-one correspondence between the elements of the class of estimable perametric functions and those of class of functions of sufficient statistic. This property is sometimes referred to as saying that the class of estimates based on sufficient statistic is complete.*

Theorem 15.6.4 *Let U be the non-empty class of all unbiased estimator t of θ with $E_\theta(t^2) < \infty \quad \forall\, \theta \in \Theta$. Let U_0 be the corresponding class of all unbiased estimator ν of 0 i.e., $E_\theta(\nu) = 0$, $E_\theta(\nu^2) < \infty \quad \forall\, \theta \in \Theta$. Then an estimator $t^* \in U$ is* UMVUE *of θ iff*

$$E_\theta(t^*\nu) = 0 \;\forall\, \theta \in \Theta \text{ and } \forall\, \nu \in U_0. \tag{15.6.5}$$

Proof: Since both $E_\theta(t^2)$ and $E_\theta(\nu^2)$ exist, $E_\theta(t^*\nu)$ exists.

Necessity : Suppose that $t^* \in U$ is UMVUE but $E_{\theta_0}(t^*\nu_0) \neq 0$ for some $\theta_0 \in \Theta$ and for some $\nu_0 \in U_0$. Now $t^* + C\nu_0 \in U$ for all real constants C.

If $E_{\theta_0}(\nu_0^2) = 0$ then, since $\nu_0 \in U_0, P(\nu_0 = 0) = 1$. In this case $E_{\theta_0}(t^*\nu_0) \neq 0$ cannot hold. We must have, therefore, $E_{\theta_0}(\nu_0^2) > 0$. Choose

$$C_0 = -\frac{E_{\theta_0}(t^*\nu_0)}{E_{\theta_0}(\nu_0^2)}.$$

Therefore,

$$E_{\theta_0}(t^* + C_0\nu_0)^2 = E_{\theta_0}(t^{*2}) - \frac{\{E_{\theta_0}(t^*\nu_0)\}^2}{E_{\theta_0}(\nu_0^2)} < E_{\theta_0}(t^{*2}). \tag{15.6.6}$$

Since both t^* and $t^* \pm C_0\nu_0 \in U$,

$$V_{\theta_0}(t^* + C_0\nu_0) < V_{\theta_0}(t^*)$$

which is a contradiction. Hence we must have (15.6.5).

Sufficiency : Suppose (15.6.5) holds for some $t^* \in U$, $\forall\, \nu \in U_0$ and $\forall\, \theta \in \Theta$. Let $t \in U$. Then $t^* - t \in U_0$ and hence

$$E_\theta\{t^*(t^* - t)\} = 0 \;\forall\, \theta; \text{ i.e., } E_\theta(t^{*2}) = E_\theta(t^*t). \tag{15.6.7}$$

Now, by, Cauchy-Schwartz inequality

$$E_\theta(t^*t) \leq \sqrt{E_\theta(t^{*2})E_\theta(t^2)}. \tag{15.6.8}$$

Hence from (15.6.7) and (15.6.8)

$$E_\theta(t^{*2}) \leq E_\theta(t^2) \quad \text{or} \quad V_\theta(t^*) \leq V_\theta(t) \; \forall\, \theta \in \Theta. \tag{15.6.9}$$

Since t is arbitrary, t^* is UMVUE of θ.

Theorem 15.6.5 *Let U be the non-empty class of all unbiased estimators t of θ. Then if there exists a* UMVUE *of θ, it must be unique.*

Proof: If possible, let there be two UMVUE's of θ, t_1, t_2 with common variance V. Consider a new estimator

$$t = \frac{t_1 + t_2}{2}.$$

Evidently $t \in U$.

$$V(t) = \frac{1}{4}[V(t_1) + V(t_2) + 2\,\mathrm{Cov}(t_1, t_2)] \leq V \tag{15.6.10}$$

since $\mathrm{Cov}(t_1, t_2) \leq \sqrt{V(t_1)V(t_2)} = V$ by Cauchy-Schwartz inequality. Hence $V(t) = V$ and this holds when

$$t_1 - \theta = k(\theta)(t_2 - \theta) \;\text{ a.e. } (\theta). \tag{15.6.11}$$

Multiplying both sides by $(t_1 - \theta)$ and taking expectation,

$$V(t_1) = k(\theta)\,\mathrm{Cov}(t_1, t_2)$$
$$\text{i.e.}\quad k(\theta) = 1.$$

Hence $t_1 = t_2$ a.e. (θ).

Theorem 15.6.6 *Suppose T^* is the most efficient estimate of θ and T is another estimate whose efficiency is e. The correlation coefficient between T^* and T is $\sqrt{e}$.*

Proof: Consider a new estimator $T' = KT + (1 - K)T^*$.

$$\begin{aligned} V(T') &= K^2 V(T) + (1-K)^2 V(T^*) + 2K(1-K)\rho_{T,T^*}\sqrt{V(T)V(T^*)} \\ &= K^2 \frac{V(T^*)}{e} + (1-K)^2 V(T^*) + 2K(1-K)\rho_{T,T^*}\frac{V(T^*)}{\sqrt{e}} \\ &= V(T^*)\left[K^2\left(\frac{1}{e} + 1 - \frac{2\rho}{\sqrt{e}}\right) + 2K\left(\frac{\rho}{\sqrt{e}} - 1\right) + 1\right] \geq 0. \end{aligned}$$

We choose K sufficiently small so that the sign of the addition of the first two terms in the curly bracket is determined by that of the second. If $\frac{\rho}{\sqrt{e}} - 1$ is a positive quantity we choose K as a small negative quantity and vice versa so that the sign of the second term is negative.

By such a choice of $K, V(T^*) \geq V(T')$ which is contrary to our assumption.

Hence $\rho/\sqrt{e} - 1$ must be zero.

■ NOTE 15.6.1 *We know that for normal $N(\mu, \sigma)$ population both sample mean $\bar{X}$ and sample median $\tilde{X}$ is unbiased for μ with variance $\frac{\sigma^2}{n}$ and $\frac{\sigma^2\pi}{2n}$ respectively. Also $\bar{X}$ is* MVU-*estimator.*

Hence correlation coefficient between $\bar{X}$ and $\tilde{X}$ is $\sqrt{\frac{2}{\pi}} = \sqrt{0.637}$.

Theorem 15.6.7 *Consider two estimators T_1 and T_2 with efficiency e_1, e_2 respectively. Then*

$$\sqrt{e_1e_2} - \sqrt{(1-e_1)(1-e_2)} \le \rho(T_1, T_2) \le \sqrt{e_1e_2} + \sqrt{(1-e_1)(1-e_2)}.$$

Proof: Consider a new estimator $T = (1-K)T_1 + KT_2$.

$$V(T) = (1-K)^2V(T_1) + K^2V(T_2) + 2K(1-K)\rho_{T_1,T_2}\sqrt{V(T_1)V(T_2)}.$$

Let V_0 be the minimum variance. Then

$$V(T) = (1-K)^2\frac{V_0}{e_1} + k^2\frac{V_0}{e_2} - 2K(1-K)\rho\frac{V_0}{\sqrt{e_1e_2}} \ge 0$$

$$\Rightarrow \quad K^2\left(\frac{1}{e_1} + \frac{1}{e_2} - \frac{2\rho}{\sqrt{e_1e_2}}\right) - 2K\left(\frac{1}{e_1} - \frac{\rho}{\sqrt{e_1e_2}}\right) + \left(\frac{1}{e_1} - 1\right) \ge 0$$

$$\Rightarrow \quad \left(\frac{1}{e_1} - \frac{\rho}{\sqrt{e_1e_2}}\right)^2 \ge 4\left(\frac{1}{e_1} + \frac{1}{e_2} - \frac{2\rho}{\sqrt{e_1e_2}}\right)\left(\frac{1}{e_1} - 1\right)$$

$$\Rightarrow \quad \left(\frac{\rho}{\sqrt{e_1e_2}} - 1\right)^2 \ge \left(\frac{1}{e_1} - 1\right)\left(\frac{1}{e_2} - 1\right)$$

from which the desired result follows.

Corollary 15.6.3 *When $e_1 = e_2 = 1, \rho_{T_1,T_2} = 1$ providing the uniqueness of the most efficient unbiased estimator.*

If $e_1 = e$, $e_2 = 1$, $\rho = \sqrt{e}$.

Corollary 15.6.4 *If $e_1 = e_2 = e$ then we have*

$$2e - 1 \le \rho \le 1.$$

▶ **Example 15.6.2** *Let $X_1, \cdots, X_n$ be a random sample from a uniform distribution*

$$P(X_i = k) = \begin{cases} \frac{1}{N}, & k = 1, 2, \cdots, N \\ 0, & otherwise. \end{cases}$$

Find a UMVUE *of N.*

It has been proved before that $t(\boldsymbol{X}) = \max_{i=1,\cdots,n} X_i$ is a complete sufficient statistic for this distribution.
Consider $\delta(X) = 2X - 1$ when X is any random observation from the population.

$$E(\delta(X)) = N. \quad \left[\because \; E(X) = \frac{N+1}{2}\right]$$

Now probability distribution of $t(\boldsymbol{X})$ is

$$\begin{aligned} P_N[t(\boldsymbol{X}) = y] &= P_N[t(\boldsymbol{X}) \leq y] - P_N[t(\boldsymbol{X}) \leq y-1] \\ &= \frac{y^n - (y-1)^n}{N^n}. \end{aligned}$$

complete for N. (vide example 15.5.7).
Now the conditional probability

$$\begin{aligned} &P_N[X_1 = x_1, \cdots, X_n = x_n | t(\boldsymbol{X}) = y] \\ &= \frac{P_N[X_1 = x_1, \cdots, X_n = x_n, t(\boldsymbol{X}) = y]}{P_N[t(\boldsymbol{X}) = y]} \\ &= \begin{cases} \frac{1/N^n}{[y^n-(y-1)^n]/N^n} = \frac{1}{y^n-(y-1)^n}, & \text{if } t(\boldsymbol{x}) = y \\ 0, & \text{otherwise.} \end{cases} \end{aligned}$$

Conditional probability distribution of any random observation X_1 given $t(X) = y$

$$\begin{aligned} P[X_1 = x | t(\boldsymbol{X}) = y] &= \frac{P(X_1 = x_1, t(\boldsymbol{X}) = y)}{P(t(\boldsymbol{X}) = y)} \\ &= \frac{y^{n-1} - (y-1)^{n-1}}{y^n - (y-1)^n} \quad \text{for } x_1 = 1, 2, \cdots, y-1 \\ &= \frac{y^{n-1}}{y^n - (y-1)^n} \quad \text{if } x_1 = y. \end{aligned}$$

$$\begin{aligned} E(\delta(X)|t(X) = y) &= \frac{y^{n-1} - (y-1)^{n-1}}{y^n - (y-1)^n} \sum_{1}^{y-1} (2x - 1) + \frac{y^{n-1}(2y-1)}{y^n - (y-1)^n} \\ &= \frac{y^{n+1} - (y-1)^{n+1}}{y^n - (y-1)^n}. \end{aligned}$$

which is UMVUE of N.

► **Example 15.6.3** *Let $X_1, \cdots, X_n$ be iid Poisson(λ) variables. Find a* UMVUE *of μ.*

Since $T = \sum_{i=1}^{n} X_i$ is a complete sufficient statistic for λ, and since

$$\bar{X} = \frac{1}{n} \sum_{i=1}^{n} X_i,$$

a function of T, is unbiased for λ, $\bar{X}$ is the unique UMVUE for λ.

Alternatively, suppose that we do not know that $\bar{X}$ is unbiased for λ and we only know that

$$T = \sum_{i=1}^{n} X_i$$

is complete sufficient for λ. We consider a function $t(X_1, \cdots, X_n) = X_1$ (say) which is known to be unbiased for λ. Then $E(X_1|T)$ is the unique UMVUE of λ.

Now,

$$\begin{aligned} & P(X_1 = i|T = t), \ 0 \le i \le t \\ &= \frac{P(X_1 = i, T = t)}{P(T = t)} = \frac{P(X_1 = i)P(X_2 + \cdots + X_n = t - i)}{P(T = t)} \\ &= \left[\left\{e^{-\lambda}\frac{\lambda^i}{i!}\right\}\left\{e^{-(n-1)\lambda}\frac{((n-1)\lambda)^{t-i}}{(t-i)!}\right\}\right]\left[e^{-n\lambda}\frac{(n\lambda)^t}{t!}\right]^{-1} \\ &= \frac{t!}{i!(t-i)!}\left(\frac{n-1}{n}\right)^{t-i}\left(\frac{1}{n}\right)^i \end{aligned} \qquad \text{(a)}$$

which is a Binomial distribution $B\left(t, \frac{1}{n}\right)$. Hence

$$E\left(X_1|T = t\right) = \frac{t}{n}.$$

Therefore, $\frac{T}{n} = \bar{X}$ is UMVUE of λ.

▶ **Example 15.6.4** *$X_1, \cdots, X_n$ are iid Poisson (λ). Find a* UMVUE *of $g(\lambda) = e^{-\lambda}(1+\lambda)$.*

We note that $P(X = 0) + P(X = 1) = e^{-\lambda} + \lambda e^{-\lambda} = g(\lambda)$. If we can find some function of $T = \sum_{i=1}^{n} X_i$ which is unbiased for $g(\lambda)$, then that function will be UMVUE of $g(\lambda)$.

Consider a function $\psi(X)$ such that

$$\psi(X_i) = \begin{cases} 1, \text{ if } X_i = 0 \text{ or } 1 \\ 0, \text{ otherwise.} \end{cases}$$

Then

$$E\left(\frac{1}{n}\sum_{i=1}^{n}\psi(X_i)\right) = g(\lambda).$$

However,

$$h'(X_1, \cdots, X_n) = \frac{1}{n}\sum_{i=1}^{n}\psi(X_i)$$

is not a function of complete sufficient statistic $T = \sum_{i=1}^{n} X_i$. Therefore, we have to find $E(h'|T)$.

However, we take a simpler function $h(X_1, \cdots, X_n) = \psi(X_1)$ which is unbiased for $g(\lambda)$ and find $E[\psi(X_1)|T = t]$.

Now,

$$P(\psi(X_1) = 0|T = 0) = P(X_1 > 1|T = 0) = 0$$
$$P(\psi(X_1) = 1|T = 0) = P(X_1 = 0|T = 0) = 1.$$

Hence,

$$E(\psi(X_1)|T = 0) = 1. \quad \text{(i)}$$

Similarly, it can be shown that

$$E(\psi(X_1)|T = 1) = 1. \quad \text{(ii)}$$

when $T = t > 1$,

$$\begin{aligned} P(\psi(X_1) = 1|T = t) &= P(X_1 = 0 \text{ or } 1|T = t) \\ &= P(X_1 = 0|T = t) + P(X_1 = 1|T = t) \\ &= \left(1 - \frac{1}{n}\right)^t + t\left(\frac{n-1}{n}\right)^{t-1}\left(\frac{1}{n}\right) \quad \text{[using the result (a) of previous example]} \\ &= \frac{(n-1)^{t-1}(n-1+t)}{n^t} = \phi(t) \text{ (say)}. \end{aligned}$$

Hence

$$E(\psi(X_1)|T = t) = \phi(t). \quad \text{(iii)}$$

Combining (i), (ii) and (iii), the UMVUE of $g(\lambda)$ is,

$$h(t) = E[\psi(X_1, \cdots, X_n)|T = t] = \begin{cases} 1 & \text{if } t = 0 \text{ or } 1 \\ \phi(t) & \text{if } t > 1. \end{cases}$$

► **Example 15.6.5** *Let $X_1, \cdots, X_n$ be iid r.v.'s each having exponential density $f_\mu(x) = \mu e^{-\mu x}$. Show that the MVU-estimate of the cumulative distribution function $F_X(x)$ is*

$$1 - \{1 - \min(x, T)/T\}^n$$

when $T = \sum_{i=1}^{n} X_i$.

Consider the statistic U which is an unbiased estimator of $F_X(x)$:

$$U = \begin{cases} 1 \text{ if } X_1 \le x \\ 0 \text{ otherwise.} \end{cases}$$

Now for this density T is a complete sufficient statistic. By Rao-Blackwellisation technique $V = E(U|T)$ is an UMVUE. To compute this we have to find the conditional distribution of X_1 given $T = t$. This is

$$P(X_1 = x|T = t) = \frac{f_{X_1}(x;\mu) f_{T-X_1}(t-x;\mu)}{f_{X_1,\cdots,X_n}(t;\mu)}, \qquad 0 \le x \le t$$

It follows that

$$\begin{aligned} E(U|T) &= 1 \times P(X_1 \le x|T) + 0 \times P(X_1 > x|T) \\ &= \int_0^x n\left(1-\frac{u}{t}\right)^{n-1} du. \end{aligned}$$

$$V = E(U|T) = P(U = 1|T) = \begin{cases} 1, & \text{if } x \ge T \\ 1-(1-x/T)^n, & \text{if } x < T. \end{cases}$$

15.6.2 Modal Unbiased and Median Unbiased Estimators

Definition 15.6.4 ***Modal-unbiased estimate.*** *Let $X_i (i = 1,\cdots,n)$ be iid r.v.'s with common p.d.f. $f(x;\theta)$ and let $t(X_1,\cdots,X_n)$ be a statistic such that the mode of the density function of t is θ. Then $t(X_1,\cdots,X_n)$ is said to be a modal-unbiased estimate of θ.*

Definition 15.6.5 ***Median-unbiased estimate.*** *Let $X_i (i = 1,\cdots,n)$ be iid r.v.'s with common p.d.f. $f(x;\theta)$ and let $t(X_1,\cdots,X_n)$ be a statistic such that the median of the density function of t is θ. Then $t(X_1,\cdots,X_n)$ is called a median-unbiased estimate of θ.*

▶ **Example 15.6.6** *Let $\{X_i, i = 1,\cdots,n\}$ be independent random samples from a continuous uniform distribution*

$$f(x;\theta) = \begin{cases} 1/\theta, & 0 < x \le \theta \\ 0, & \textit{otherwise.} \end{cases}$$

Consider $Y = \max_i X_i$. The density function of Y is

$$\phi(y;\theta) = \frac{ny^{n-1}}{\theta^n}, \quad 0 < y \le \theta.$$

The mode of this distribution is θ. Hence, y is a modal unbiased estimate of θ. A modal-unbiased estimate may not be unique.

▶ **Example 15.6.7** *Let $X_1,\cdots,X_n$ be independent random variables each having exponential density*

$$f(x;\theta) = \frac{1}{\theta} e^{-x/\theta}, \quad x > 0, \theta > 0.$$

For this distribution, the statistic

$$Y_1 = X_{(n)}/\log n \text{ where } X_{(n)} = \max_i X_i$$

$$Y_2 = S_n/(n-1) \text{ where } S = \sum_{i=1}^{n} X_i$$

are both modal-unbiased for θ.

The p.d.f. of $X_{(n)}$ is

$$g(x_{(n)}, \theta) = \frac{n}{\theta}(1 - e^{-x_{(n)}/\theta})^{n-1} e^{-x_{(n)}/\theta}, \quad x_{(n)} > 0, \theta > 0.$$

Hence the p.d.f. of Y_1 is

$$g(y_1, \theta) = \frac{n \log n}{\theta}\left[1 - (n^{-1/\theta})^{y_1}\right]^{n-1}\left(n^{-1/\theta}\right)^{y_1}. \tag{i}$$

We have to show that (i) is maximum when $y_1 = \theta$. For this we note that

$$g(y_1, \theta) = \frac{n \log n}{\theta}(1 - m^{y_1})^{n-1} m^{y_1} \tag{ii}$$

where $m = n^{-1/\theta}$. Differentiating (ii) with respect to y_1 and equating it to zero gives

$$m^{y_1} = \frac{1}{n} = m^{\theta}.$$

Hence $y_1 = \theta$. It can be easily verified that second derivative of (ii) at $y_1 = \theta$ is negative and hence Y_1 is modal unbiased estimate of θ.

The p.d.f. of S_n is

$$h(s_n, \theta) = \frac{s_n^{n-1}}{\overline{|(n)}\, \theta^n} e^{-s_n/\theta}, \quad s_n > 0, \theta > 0$$

and hence the p.d.f. of $Y_2 = S_n/(n-1)$ is

$$h(y_2; \theta) = \frac{(n-1)^n}{\overline{|(n)}\, \theta^n} e^{-(n-1)y_2/\theta} y_2^{n-1}, \quad y_2 > 0, \theta > 0.$$

It can be shown that

$$\frac{dh(y_2, \theta)}{dy_2} = 0$$

gives $y_2 = \theta$ as the solution. Again the second derivative of $h(y_2, \theta)$ at $y_2 = \theta$ is negative. Hence Y_2 is a modal-unbiased estimate of θ.

Thus a modal-unbiased estimate is not always unique.

15.6.3 Consistent Asymptotically Normal (CAN) Estimators

We have already discussed unbiasedness and consistency properties of an estimator and considered how to choose an estimator (UMVU-estimator) among a class of unbiased estimators. We consider further about how to choose among a class of rival estimators.

An estimator t_1 may be said to be more concentrated around θ than another estimator t_2 if

$$P[|t_1 - \theta| < C] \geq P[|t_2 - \theta| < C]$$

holds for all $C(> 0)$ with strict inequality holding for at least one C. Clearly t_1 will be preferable to t_2 as an estimator of θ.

In general a "uniformly closer to θ" estimator t in a class of estimators will not exist. However, if t_1, t_2 are normally distributed around θ, a simple result exists.

Theorem 15.6.8 *If t_1, t_2 are respectively $N(\theta, \sigma_1^2)$, $N(\theta, \sigma_2^2)$-variables than t_1 is more concentrated about θ if $\sigma_1^2 \leq \sigma_2^2$.*

Proof:

$$[P|t_i - \theta| \leq C] = P\left[-\frac{C}{\sigma_i} < \frac{t_i - \theta}{\sigma_i} < \frac{C}{\sigma_i}\right] = 2\Phi\left(\frac{C}{\sigma_i}\right) - 1$$

when $\Phi(x)$ is the area under $N(0, 1)$ density to the left of x. Hence

$$P[|t_1 - \theta| \leq C] \geq P[|t_2 - \theta| \leq C]$$

$$\iff \frac{C}{\sigma_1} \geq \frac{C}{\sigma_2} \iff \sigma_1 \leq \sigma_2.$$

Thus, if the competing estimators t_1, t_2 both have normal distribution one with the lower variance will be preferable. We have seen that under general regularity conditions many statistic $t_n(X_1, \cdots, X_n)$ tend to be distributed normally around mean θ as $n \to \infty$ (see the C. L. Theorems). Hence among such statistics one should use one with the smallest variance. We formalise these concepts.

Definition 15.6.6 *If an estimator $t_n(X_1, \cdots, X_n)$ is such that $\sqrt{n}(t_n - \theta)$ is asymptotically normal $N(0, J^2(\theta))$ for some function $J^2(\theta)$, then it is called a consistent asymptotically normal* (CAN) *estimator.*

The quantity $I(\theta) = \frac{1}{J^2(\theta)}$ is called Fisher's information about θ contained in the estimator.

▶ **Example 15.6.8** *Let $X_1, \cdots, X_n$ be a random sample from a distribution with mean μ, variance σ^2 and Kurtosis $\beta_2(= \mu_4/\sigma^4) > 1$.*

Consider $\hat{\mu} = \bar{X}_n$, $\hat{\sigma}^2 = S_n^2 = \frac{1}{n-1}\sum_{i=1}^{n}(X_i - \bar{X}_n)^2$.

Now by weak law of large numbers,

$$\hat{\mu} \xrightarrow{P} \mu, \quad \hat{\sigma}^2 \xrightarrow{P} \sigma^2.$$

Also by CLT

$$\sqrt{n}(\hat{\mu}-\mu) \xrightarrow{L} N(0,\sigma^2), \ \sqrt{n}(\hat{\sigma}^2-\sigma^2) \xrightarrow{L} N\left(0,(\beta_2-1)\sigma^4\right).$$

Thus $\hat{\mu}$ is a CAN estimator with $J^2=\sigma^2$ and $\hat{\sigma}^2$ is a CAN estimator with $J^2=(\beta_2-1)\sigma^4$ Fisher's information about μ in $\hat{\mu}$ is $1/\sigma^2$ and that about σ^2 in $\hat{\sigma}^2$ is $1/(\beta_2-1)\sigma^4$.

► **Example 15.6.9** *Let $X_1,\cdots,X_n$ be iid Bernoulli (p)-variables. Consider $\hat{p}=\bar{X}_n$ Now by* WLLN

$$\sqrt{n}(\bar{X}_n-p) \xrightarrow{L} N(0,pq).$$

Hence $\hat{p}$ is a CAN estimator with Fisher's information $1/pq$.

Definition 15.6.7 *If t_1,t_2 are two* CAN *estimators for θ with Fisher's informatio $1/J_1^2(\theta)$, $1/J_2^2(\theta)$ then t_1 is better than t_2 in the sense of being asymptotically mor concentrated about θ if $J_1^2(\theta)\le J_2^2(\theta) \ \ \forall\ \theta\in\Theta$ (with strictly inequality for at leas one θ).*

Definition 15.6.8 *The asymptotic relative efficiency of a* CAN *estimator t_2 with $J_2^2(\theta$ with respect to a* CAN *estimator t_1 with $J_1^2(\theta)$ is*

$$e_\theta(t_1,t_2)=\frac{J_1^2(\theta)}{J_2^2(\theta)}.$$

► **Example 15.6.10** *Let $X_1,\cdots,X_n$ be iid $N(\mu,\sigma^2)$. Then both $\bar{X}_n$ and sample media $\bar{X}_{me}$ are* CAN *with $J_1^2(\mu)=\sigma^2$ and $J_2^2(\mu)=\frac{\pi\sigma^2}{2}$. Hence $e_\mu(t_1,t_2)=\frac{2}{\pi}=0.64<1$.*

$\bar{X}_n$ can (in large sample) estimate μ as efficiently with $0.64n$ observation as $\bar{X}_{me}$ wit n observations.

Estimation of population mean by sample median $\bar{X}_{me}$ in preference to sample mea $\bar{X}$ is, however, sometimes done due to the reasons that some X_i's are often affected b observational errors or errors of measurement and $\bar{X}_n$ is more affected than $\bar{X}_{me}$.

► **Example 15.6.11** *Theorem* 11.6.3 *shows that if $X_1,\cdots,X_n$ are iid with d.f. $F(x)$ an p.d.f. $f(x)$ and if $f(x)$ is continuous and positive at $x=\xi_p$ the sample percentile Z_p i a* CAN *estimator of population percentile ξ_p with $J^2(\xi_p)=\frac{pq}{f^2(\xi_p)}$.*

Thus, sample median is CAN estimator for population median μ_e with informatio $I(\mu_e)=4f^2(\mu_e)$.

Definition 15.6.9 *An estimator $t_n(X_1,\cdots,X_n)$ of θ is a best asymptotically norm (*BAN*) estimator of θ if t_n is asymptotically normal, is consistent and most efficien for θ.*

Theorem 15.6.9 *If two most efficient estimators t_n and t_n' tend in the limit to b distributed in the bivariate normal form, then $\rho(t_n,t_n')=1$.*

Proof: The limiting joint distribution of t_n, t'_n is

$$dF = \frac{1}{2\pi v_n\sqrt{1-\rho^2}} \exp\left[-\frac{1}{2v_n(1-\rho^2)}\left\{(t_n-\theta)^2 - 2\rho(t_n-\theta)(t'_n-\theta) + (t'_n-\theta)^2\right\}\right] dt_n\, dt'_n$$

when Var $(t_n) = \text{Var}(t'_n) = v_n$.

Consider new estimators

$$u_n = \frac{1}{2}(t_n + t'_n), \quad u'_n = \frac{1}{2}(t_n - t'_n).$$

Then the joint distribution of u_n and u'_n is

$$dF = \text{Const.} \exp\left[-\frac{1}{2v_n(1-\rho^2)}\left\{2(1-\rho)(u_n-\theta)^2 + 2(1+\rho)u_n'^2\right\}\right] du_n\, du'_n.$$

Hence u_n, u'_n are independent with Var $(u_n) = \left(\frac{1+\rho}{2}\right) v_n$.

Since t_n, t'_n are most efficient we must have

$$\text{Var}(u_n) \geq v_n \quad \text{i.e.,} \quad \rho \geq 1.$$

Hence $\rho = 1$. Thus in the limit all most efficient estimators tend to be equivalent.

Exercise

1. Let $X_1, \cdots, X_n$ be independent $N(\mu, \sigma^2)$, variable, $-\infty < x < \infty$, $-\infty < \mu < \infty$, $\sigma > 0$.

   Find UMVUE of μ. Find UMVUE of the 60th percentile.

2. Let t_1, t_2 be two unbiased estimators of θ with variances σ_1^2, σ_2^2 and correlation ρ. Find the best unbiased linear combination of t_1, t_2 along with its variance.

3. (a) Let $\{X_i\}, i = 1, \cdots, n$ be a random sample from the Bernoulli distribution $p^x(1-p)^{1-x}$, $x = 0, 1$. Find the MVUE of (i) $7p$ (ii) $5p - 1$. [use the fact that $\bar{X}$ is complete and sufficient for p].

   (b) Find the MVUE of λ^3 where λ is a Poisson parameter.

4. Let X be a random variable having rectangular distribution $P(X = k) = \frac{1}{N}, k = 1, \cdots, N$; $\Theta = \{N | N \geq M; M > 1\}$. (i) Characterise all unbiased estimates of N (ii) Find a UMVUE of N.

5. Let $X_1, \cdots, X_n$ be independent samples from the power series distribution

$$P(X = x) = C(x)\theta^x / f(\theta), \;\; x = 0, 1, 2, \cdots; \theta > 0, C(x) > 0,$$

$$C(0) = 1, f(\theta) = \sum_0^{\infty} C(x)\theta^x.$$

Let

$$a(t, n) = \sum \prod_{i=1}^{n} C(x_i)$$

where Σ is over all non-negative integral values of $x_1, \cdots, x_n$ subject to th condition $x_1 + \cdots + x_n = t$.

Prove that for any positive integer m,

$$U_m(t) = \begin{cases} 0, & \text{for } t > m \\ \dfrac{a(t-m, n)}{a(t, n)}, & \text{for } t \geq m \end{cases}$$

where $t = \sum_{i=1}^{n} X_i$, is UMVUE of θ^m. **[Roy and Mitra, 195**

6. Show that if t_1 is an efficient estimator and t_2 another consistent estimator of then covariance of t_1 and $(t_2 - t_1)$ is asymptotically zero. Hence, show that if th estimators are jointly normally distributed we may regard the variance of t_2 – as composed of two independent parts, one being the variation of $(t_1 - \theta)$ and th other a component due to inefficiency of estimation. **[Kendall and Stuart, 196**

7. Let $X_1, \cdots, X_n$ be independent random samples from $N(\mu, \sigma)$ with both μ, σ unknown. find the minimum-variance-unbiased estimate of the proportion p,

$$p = \frac{1}{\sqrt{2\pi}\sigma} \int_c^{\infty} e^{-(x-\mu)^2/2\sigma^2}\, dx.$$

[Kolmogorov, 195

8. Let X be a r.v. having a Normal $N(\mu, \sigma)$ p.d.f. Show that X is a modal-unbiase estimate of μ.

9. Let $X_i (i = 1, \cdots, n)$ be independent r.v.'s with common p.d.f.

$$f(x, \theta) = \frac{1}{\theta} e^{-x/\theta}, \;\; x > 0, \theta > 0.$$

(a) Show that $Y_1 = S_n/n$ is a mean-unbiased estimate of θ and $Y_2 = S_n/(n -$ is a modal-unbiased estimate of θ where $S_n = \sum_{i=1}^{n} X_i$.

(b) Considering the loss function

$$L(\hat{\theta}, \theta) = \left(\frac{1}{\hat{\theta}} - \frac{1}{\theta}\right)^2$$

where $\hat{\theta}$ is an estimate of θ, show that the risk (expected loss) of Y_2 is uniformly smaller than that of Y_1.

(c) For the loss function

$$L(\hat{\theta}, \theta) = \frac{\hat{\theta}}{\theta} + \frac{\theta}{\hat{\theta}} - 2$$

show that both Y_1 and Y_2 have the same expected loss, $1/(n-1)$.

10. Let $X_i (i = 1, \cdots, n)$ be a random sample from $N(\mu, \sigma^2)$.
Find the UMVUE of (i) $3\mu + 5\sigma^2$ (ii) $\mu^2 - 5\sigma^2$.
[use the fact that $(\bar{X}, S^2)$ are jointly sufficient and complete.]

11. Consider the Weibull distribution

$$f(x; c, \alpha) = \alpha c^{\alpha} x^{\alpha-1} e^{-(cx)^{\alpha}}, \quad x > 0, \ c > 0, \ \alpha \geq 2 \text{ but known.}$$

Show that

$$Y_1 = \frac{1}{\sqrt{\left(1 - \frac{1}{\alpha}\right) X}}$$

is a mean-unbiased estimate of c, while

$$Y_2 = \left(\frac{\alpha+1}{\alpha}\right)^{1/\alpha} \cdot \frac{1}{X}$$

is a modal-unbiased of c. For the loss function $L(\hat{c}, c) = \left(\frac{1}{\hat{c}} - \frac{1}{c}\right)^2$, where $\hat{c}$ is an estimate of c, show that Y_2 has a uniformly smaller risk than Y_1.

12. Let X be a r.v. with p.d.f. $f(x; \theta)$ and let $Y = g(X)$. Let $\phi(y)$ be the p.d.f. of Y such that

(a) $\frac{d\phi(y)}{dy} = 0$ at $y = \theta$;

(b) $\frac{d^2\phi(y)}{dy^2} < 0$ at $y = \theta$;

(c) g is a one-to-one transform from x to y and from y to x.
Show that the solution of the following differential equations are modal unbiased estimates of θ;

(d) $f(x; \theta)\frac{d^2x}{dy^2} + \frac{df(x;\theta)}{dx}\left(\frac{dx}{dy}\right)^2 = 0$ at $y = \theta$;

(e) $\frac{d^2f(x;\theta)}{dx^2}\left(\frac{dx}{dy}\right)^3 + 3\left(\frac{df(x;\theta)}{dx}\right)\left(\frac{dx}{dy}\right)\left(\frac{d^2x}{dy^2}\right) + f(x, \theta)\frac{d^3x}{dy^3} < 0.$

[*Hints:* The left hand side of relations (d) and (e) are $\frac{d\phi(y)}{\alpha y}$ and $\frac{d^2\phi(y)}{dy^2}$ respectively.]

13. Let $X_i (i = 1, \cdots, n = 2m + 1)$ be iid r.v.'s with negative exponential density

$$f(x, \theta) = \frac{1}{\theta} e^{-x/\theta}, \quad x > 0, \theta > 0.$$

Let $X_{(j)}$ be the jth order statistic, $X_{(1)} < X_{(2)} < \cdots < X_{(n)}$. Show that

$$Z_1 = \frac{2m+1}{\log_e 2} X_{(1)}$$

is median-unbiased estimate of θ with

$$E(Z_1 - \theta)^2 = \theta^2 \frac{4m^2(2 - \log_e 2) + 8m + 1 + (m \log_e 2 - 1)^2}{m^2 \log_e^2 2}.$$

Hence show that

$$E_\theta (Y - \theta)^2 < E_\theta (Z_1 - \theta)^2 \text{ for all } 0 < \theta < \infty$$

and all $m \geq 2$ when Y is the modal-unbiased estimator $\sum_{i=1}^{n} X_i / 2m$.

Show also that the estimator $Z_{m+1} = X_{(m+1)} / \log_e 2$ is also median-unbiased.

[*Hints:* $X_{(1)}$ has a negative-exponential distribution with p.d.f.

$$f_{X_{(1)}}(x) = \begin{cases} \frac{2m+1}{\theta} \exp\left[-\frac{2m+1}{\theta} x\right], & 0 \leq x \leq \infty \\ 0, & \text{otherwise.]} \end{cases}$$

14. Let X be a r.v. with truncated Poisson p.m.f.

$$P(X = k) = \frac{e^{-\lambda} \lambda^k}{k!(1 - e^{-\lambda})}, \quad k = 1, 2, \cdots; \lambda > 0.$$

Prove that the only unbiased estimate of $(1 - e^{-\lambda})$ is 0 for X an odd integer and 2 for X an even integer.

15. Let $X_1, X_2, \cdots, X_n$ be iid r.v.s with common p.d.f. $f(x - \theta)$, $-\infty < x < \infty$, $-\infty < \theta < \infty$. State and prove the conditions that

$$t(x_1, \cdots, x_n) = \frac{\int \theta \prod_{i=1}^{n} f(x_i - \theta)\, dx_i}{\int \prod_{i=1}^{n} f(x_i - \theta) dx_i}$$

has a minimum variance-unbiased estimate among the class of all estimators $\psi(x_1, \cdots, x_n)$ with the translation property i.e., estimators satisfying

$$\psi(x_1 + a, \cdots, x_n + a) = \psi(x_1, \cdots, x_n) + a$$

where a is any real number.

16. Let $X_1, \cdots, X_n$ be iid $N(\mu, \sigma^2)$ variables, both μ and σ^2 being unknown. Find UMVUE's of μ and σ^2. $[\bar{x}, s^2 = \Sigma(x_i - \bar{x})^2/(n-1)]$

17. Let $X_1, \cdots, X_n$ be iid Bernoulli (p) variables. Show that there does not exist any UMVUE of $1/p$.

[*Hints:* Show that no function of $\sum_{i=1}^{n} X_i$ (complete sufficient statistic for p) is unbiased for $1/p$.]

18. If t is a most efficient estimator and t' a less efficient estimator with efficiency e and if the correlation coefficient of t and t' is ρ, show by considering the estimator

$$t'' = (1 + e - 2\rho\sqrt{e})^{-1}[(1 - \rho\sqrt{e})t + (e - \rho\sqrt{e})t']$$

that (i) $\rho = \sqrt{e}$ (ii) Cov $(t, t' - t) = 0$.

19. (a) If T_1 and T_2 be the MVUE's of $g_1(\theta)$ and $g_2(\theta)$ respectively, then show that $a_1T_1 + a_2T_2$ is the MVUE of $a_1g_1(\theta) + a_2g_2(\theta)$. Show also that T_1T_2 is the MVUE of $E(T_1T_2)$.

(b) If T be the MVUE of $g(\theta)$, then show that $\rho_k(T) = a_0 + a_1T + a_2T^2 + \cdots + a_kT^k \quad (k \neq 0)$ is the MVUE of $E[\rho_k(T)]$.

20. Obtain two unbiased estimators of the parameter p of the Gamma distribution

$$f_p(x) = \frac{1}{\overline{|p}} e^{-x} x^{p-1}, \quad 0 \leq x < \infty.$$

Hence obtain infinitely many unbiased estimators of p and state which is the most preferable one.

[*Hints:* Both sample mean and variance are unbiased for p.]

21. (x_1, x_2) is a random sample of size 2 from the distribution

$$f_\theta(x) = \frac{1}{\theta} e^{-x\theta}, \quad x \geq 0, \theta > 0.$$

Show that $t_1 = x_1 + x_2$ is sufficient and $t_2 = x_2$ is unbiased for θ. Compute $E(t_2|t_1)$ and verify Rao-Blackwell theorem.

22. Let $X_1 = 2, X_2 = 1, X_3 = 2, X_4 = 1, X_5 = 1$ be independent observations on the number of deaths in car accidents with at least one death. Assume that the observations are from a Poisson distribution with parameter λ *truncated* at zero. Find UMVUE of λ.

23. Let $\{f_\theta(x), \theta \in R_1\}$ be a family of p.d.f.'s and let $g(\theta)$ be an estimable parametric function. Show that the UMVUE of $g(\theta)$ based on a sample $x_1, \cdots, x_n$ is a symmetric function of $x_1, \cdots, x_n$.

24. Obtain the UMVUE of μ^2 in $N(\mu, 1), -\infty < \mu < \infty$ based on a sample of size n.

25. $X_1, \cdots, X_n$ are iid Poisson with parameter λ. Find the UMVUE of $P(X = 0) = e^{-\lambda}$.

15.7 Different Lower Bounds for Variance of Estimators

15.7.1 Rao-Cramer Lower Bound

In Section 15.6 we were concerned with minimisation of variance of unbiased estimators. We considered the class of CAN estimators and were also concerned with the minimisation of variance of estimators in this class. The question arises if there exists a lower bound to the mean square error (variance) of estimators of a parameter θ or a function $\psi(\theta)$ for a family of distributions $F_\theta, \theta \in \Theta$. This problem was first investigated in the celebrated Rao-Cramer [RC] (Rao, 1945; Cramer, 1946) inequality (the inequality was first given in M. Fréchet, 1937).

Theorem 15.7.1 *Let $X_1, \cdots, X_n$ be r.v.'s having the joint p.d.f. $f_\theta(x_1, \cdots, x_n), \theta \in \Theta$. If the statistic $t(X_1, \cdots, X_n)$ is such that $E_\theta(t(X_1, \cdots, X_n)) = \psi(\theta) \quad \forall\ \theta$, and if the following regularity conditions hold, then*

$$E_\theta(t(\boldsymbol{X}) - \psi(\theta))^2 \geq \frac{[\psi'(\theta)]^2}{E_\theta\left(\frac{\partial \log_e f_\theta(\boldsymbol{X})}{\partial \theta}\right)^2} \tag{15.7.1}$$

where $\psi'(\theta) = \frac{d\psi(\theta)}{d\theta}$.

The regularity conditions (of Wolfowitz (1947)) are:

(i) θ lies in a non-degenerate open interval Θ in the real line; Θ may be infinite;
(ii) $\partial f_\theta(\boldsymbol{x})/\partial\theta$ exists $\forall\ \theta \in \Theta$;
(iii) $\int f_\theta(\boldsymbol{x})\, d\boldsymbol{x}$ can be differentiated with respect to θ under the integral sign;
(iv) $\int t(\boldsymbol{x}) f_\theta(\boldsymbol{x})\, d\boldsymbol{x}$ can be differentiated under the integral sign;
(v) $E_\theta\left[\frac{\partial \log f_\theta(\boldsymbol{x})}{\partial\theta}\right]^2$ exists and is positive $\forall\ \theta \in \Theta$.

Proof: Consider two random variables $U = t - E(t)$, $V = \frac{\partial \log f_\theta(\boldsymbol{x})}{\partial\theta} - E\left(\frac{\partial \log f_\theta(\boldsymbol{x})}{\partial\theta}\right)$. By Cauchy-Schwartz inequality,

$$\begin{aligned}&\left[E_\theta\left\{(t - E(t))\frac{\partial \log f_\theta(\boldsymbol{x})}{\partial\theta}\right\}\right]^2\\ &\quad \leq E_\theta(t - E(t))^2 E_\theta\left\{\frac{\partial \log f_\theta(\boldsymbol{x})}{\partial\theta} - E_\theta\left(\frac{\partial \log f_\theta(\boldsymbol{x})}{\partial\theta}\right)\right\}^2.\end{aligned} \tag{15.7.2}$$

Now

$$\begin{aligned}E\left(\frac{\partial \log f_\theta(\boldsymbol{x})}{\partial\theta}\right) &= \int_w \frac{\partial \log f_\theta(\boldsymbol{x})}{\partial\theta} f_\theta(\boldsymbol{x})\, dx\\ &= \int_w \frac{1}{f}\frac{\partial f}{\partial\theta} f(\boldsymbol{x})\, d\boldsymbol{x} = \frac{\partial}{\partial\theta}\int_w f(\boldsymbol{x})\, d\boldsymbol{x} = \frac{\partial}{\partial\theta}(1) = 0\end{aligned} \tag{15.7.3}$$

where w is the sample space of $\boldsymbol{x}$.

Again,

$$E\left(t\frac{\partial \log f_\theta(\boldsymbol{x})}{\partial \theta}\right) = E\left(t\frac{f'}{f}\right)\left(\text{ where } f' = \frac{\partial f}{\partial \theta}\right) = \int_w tf'\,d\boldsymbol{x}.$$

Now

$$\int_w tf(\boldsymbol{x})\,d\boldsymbol{x} = \psi(\theta);\ \text{hence}$$

$$\int_w tf'(\boldsymbol{x})\,dx = \psi'(\theta). \tag{15.7.4}$$

Therefore, from (15.7.2), (15.7.3), (15.7.4), the inequality (15.7.1) follows.

Corollary 15.7.1 *If $E(t) = \psi(\theta) = \theta + b(\theta)$ so that t is an estimator of θ with bias $b(\theta)$, a function of θ, then*

$$E(t-\psi(\theta))^2 = E(t-\theta)^2 - (b(\theta))^2$$

$$\psi'(\theta) = 1 + b'(\theta), \quad \textit{where } b'(\theta) = \frac{db(\theta)}{d\theta}$$

and the inequality (15.7.1) *reduces to*

$$E(t-\theta)^2 \geq b^2(\theta) + \frac{(1+b'(\theta))^2}{E_\theta\left(\frac{\partial \log_e f_\theta(x)}{\partial \theta}\right)^2}. \tag{15.7.5}$$

The inequality in (15.7.5) gives the lower bound to the mean square error (mse) (around θ) of any biased estimator t of θ, whose bias is $b(\theta)$, a function of θ. For the class of estimators whose bias is a fixed function $b(\theta)$ of θ, (15.7.5) gives a lower bound to the mse of t.

Corollary 15.7.2 *If t is unbiased for θ,* (15.7.1) *reduces to*

$$V_\theta(t) \geq \frac{1}{E_\theta\left(\frac{\partial \log_e f_\theta(\boldsymbol{x})}{\partial \theta}\right)^2} \tag{15.7.6}$$

Corollary 15.7.3 *Suppose that $\int f_\theta(\boldsymbol{x})\,d\boldsymbol{x}$ is differentiable with respect to θ, under the sign of integration, not only once, but twice. Now*

$$\frac{\partial \log_e f_\theta(\boldsymbol{x})}{\partial \theta} = \frac{1}{f}\frac{\partial f}{\partial \theta};$$

$$\textit{hence } \frac{\partial^2 \log f_\theta(\boldsymbol{x})}{\partial \theta^2} = \frac{1}{f}\frac{\partial^2 f}{\partial \theta^2} - \frac{1}{f^2}\left(\frac{\partial f}{\partial \theta}\right)^2;$$

$$\textit{hence } E_\theta\left(\frac{\partial^2 \log f_\theta(\boldsymbol{x})}{\partial \theta^2}\right) = -E_\theta\left[\frac{1}{f}\frac{\partial f}{\partial \theta}\right]^2 = -E_\theta\left[\frac{\partial \log f}{\partial \theta}\right]^2$$

$$\textit{since } E_\theta\left(\frac{1}{f}\frac{\partial^2 f}{\partial \theta^2}\right) = \int_w \frac{\partial^2 f}{\partial \theta^2}\,d\boldsymbol{x} = \frac{\partial^2}{\partial \theta^2}\int_w f\,d\boldsymbol{x} = 0.$$

In this case, the inequality (15.7.1) reduces to

$$E_\theta(t-\psi(\theta))^2 \geq -\frac{(\psi'(\theta))^2}{E_\theta\left(\frac{\partial^2 \log f_\theta(\boldsymbol{x})}{\partial\theta^2}\right)}. \tag{15.7.7}$$

Corollary 15.7.4 *In case $X_1, \cdots, X_n$ are iid r.v.'s with common p.d.f. $f_\theta(x)$, then $f_\theta(x_1, \cdots, x_n) = \prod_{i=1}^{n} f_\theta(x_i)$ and*

$$\begin{aligned}
E_\theta\left(\frac{\partial \log f_\theta(\boldsymbol{x})}{\partial\theta}\right)^2 &= E_\theta\left(\sum_{i=1}^{n}\frac{\partial \log f_\theta(x_i)}{\partial\theta}\right)^2 \\
&= \sum_{i=1}^{n} E_\theta\left(\frac{\partial \log f_\theta(x_i)}{\partial\theta}\right)^2 \quad (\textit{since } X_i, X_j, i \neq j, \textit{ are independent}) \\
&= nE_\theta\left(\frac{\partial \log f_\theta(x)}{\partial\theta}\right)^2. \qquad (15.7.8)
\end{aligned}$$

Similarly it can be shown that

$$E_\theta\left(\frac{\partial^2 \log f_\theta(\boldsymbol{x})}{\partial\theta^2}\right) = nE_\theta\left(\frac{\partial^2 \log f_\theta(x)}{\partial\theta^2}\right). \tag{15.7.9}$$

The inequalities (15.7.1), (15.7.5), (15.7.7) can be modified accordingly.

■ NOTE 15.7.1 *The quantity $I_\theta = E_\theta\left(\frac{\partial \log_e f_\theta(X)}{\partial\theta}\right)^2$ is called Fisher's information about θ contained in the observation X.*

■ NOTE 15.7.2 *The* RC*-inequality remains valid even when the r.v.'s $X_1, \cdots, X_n$ are all discrete. The underlying assumptions and proofs are similar, only the multiple integrals are to be replaced by the appropriate multiple sums.*

■ NOTE 15.7.3 *The cases in which regularity conditions (i)–(v) are satisfied may be called regular estimation cases. An unbiased estimator t of $\psi(\theta)$ for which the* RC*-lower bound is attained is called minimum variance bound unbiased estimator (*MVBUE*) or best regular unbiased estimator (*BRUE*) or the most efficient unbiased estimator.*

Obviously, even in a regular estimation case a MVB estimator may not exist. Therefore, the variance of a UMVU estimator, whose variance is the least attainable variance in the class of unbiased estimators, may exceed the RC-bound even if we restrict ourselves to the regular cases.

Again, in a non-regular case, the variance of a MVU-estimator may be lower than the one given by RC-bound (which, however, does not exist in this case) [vide example 15.7.5].

□ **Remark 15.7.1** *The regularity conditions are not generally satisfied if the set of all possible values of X depends on the values of unknown parameter θ. Generally it is easy to check the regularlity conditions* (i), (ii) *and* (v). *To check conditions* (iii) *and* (iv) *one can use the following result of Cramer.*

Let us assume that (i), (ii) and (v) hold. Suppose that for each $\theta_0 \in \Theta$, there are numbers a and b and integrable functions G and H such that $\theta_0 \in (a, b) \subset \Theta$ and such that

$$\left|\frac{\partial f_\theta(\boldsymbol{x})}{\partial \theta}\right| < G(\boldsymbol{x}), \theta \in (a, b);$$

$$\left|t(\boldsymbol{x})\frac{\partial f_\theta(\boldsymbol{x})}{\partial \theta}\right| < H(\boldsymbol{x}), \theta \in (a, b).$$

Then (iii) and (iv) are satisfied provided

$$\int_w G(\boldsymbol{x})\, dx < \infty, \quad \int_w H(\boldsymbol{x})\, dx < \infty.$$

[For situation where regularity conditions hold, even when the range of X depends on θ, see exercise 13 at the end of this chapter].

▶ **Example 15.7.1** *Let X be normally distributed with mean θ and known variance σ^2, $\theta \in \Theta = (-\infty, \infty)$.*

$$f_\theta(x) = \frac{1}{\sqrt{2\pi}} e^{-(x-\theta)^2/2\sigma^2}$$

$$\log f_\theta(x) = \text{Const.} - \frac{(x-\theta)^2}{2\sigma^2}$$

$$\frac{\partial \log f_\theta(x)}{\partial \theta} = \frac{(x-\theta)}{\sigma^2}; \quad E\left(\frac{\partial \log f_\theta(x)}{\partial \theta}\right) = \frac{1}{\sigma^2}.$$

Hence minimum variance bound is $\frac{\sigma^2}{n}$.

Consider $t = \bar{X}$; the condition (iv) can be shown to hold. It is known that $\bar{X}$ is unbiased for θ with $V(\bar{X}) = \sigma^2/n$. Hence $\bar{X}$ is MVB-estimator of θ.

▶ **Example 15.7.2** *Let $X_1, \cdots, X_n$ be iid $N(\mu, \sigma^2)$ with μ known. We want to estimate $\sigma^2, \Theta = (0, \infty)$. Here the regularity conditions hold.* (a) *Estimation of $\sigma^2 = \theta$ (say).*

$$f_\theta(x) = \frac{1}{\sqrt{2\pi\theta}} \exp[-(x-\mu)^2/2\theta]$$

$$\log f_\theta(x) = \text{Const.} - \frac{\log \theta}{2} - \frac{(x-\mu)^2}{2\theta}$$

$$\frac{\partial \log f_\theta(x)}{\partial \theta} = \frac{(x-\mu)^2}{2\theta^2} - \frac{1}{2\theta}$$

$$\frac{\partial^2 \log f_\theta(x)}{\partial \theta^2} = -\frac{(x-\mu)^2}{\theta^3} + \frac{1}{2\theta^2}.$$

Hence

$$E_\theta\left[\frac{\partial^2 \log f_\theta(x)}{\partial\theta^2}\right] = -\frac{E_\theta(X-\mu)^2}{\theta^3} + \frac{1}{2\theta^2} = -\frac{1}{2\theta^2}.$$

Therefore CR-lower bound to variance of an unbiased estimator of θ is $\frac{2\theta^2}{n}$.
Consider the statistic

$$S^2 = \sum_{i=1}^{n}(X_i-\mu)^2/n.$$

We have,

$$E_\theta(S^2) = \theta, \;\; V_\theta(S^2) = \frac{2\theta^2}{n}.$$

Thus S^2 is an MVBU estimator of $\theta = \sigma^2$.

(b) Estimation of $\sigma = \sqrt{\theta}$ when $\theta = \sigma^2$. Let $\psi(\theta) = \sqrt{\theta}$. $\psi'(\theta) = \frac{1}{2\sqrt{\theta}}$. Hence CR-lower bound to variance of an unbiased estimator of σ is $\frac{\theta}{2n}$.

Consider the statistic $S = \sqrt{\sum_{i=1}^{n}(X_i-\mu)^2/n} = \frac{\chi_{(n)}\sqrt{\theta}}{\sqrt{n}}$.

Now it is known that for a $\chi^2_{(n)}$ distribution

$$E(\chi_{(n)}) = \sqrt{2}\,\Gamma\!\left(\frac{n+1}{2}\right)\Big/\Gamma\!\left(\frac{n}{2}\right).$$

Hence

$$E_\theta(S) = \sqrt{2}\cdot\frac{\Gamma\left(\frac{n+1}{2}\right)}{\Gamma\left(\frac{n}{2}\right)}\cdot\sqrt{\frac{\theta}{n}}.$$

Hence an unbiased estimator of $\sqrt{\theta} = \sigma$ is

$$S' = \sqrt{\frac{n}{2}}\,\frac{\Gamma\left(\frac{n}{2}\right)}{\Gamma\left(\frac{n+1}{2}\right)}\cdot S.$$

$$\begin{aligned}
\text{Var}(S') &= \text{Var}\left(\sqrt{\frac{\theta}{2}}\cdot\frac{\Gamma\left(\frac{n}{2}\right)}{\Gamma\left(\frac{n+1}{2}\right)}\cdot\chi_{(n)}\right)\\
&= \frac{\theta}{2}\left\{\frac{\Gamma\left(\frac{n}{2}\right)}{\Gamma\left(\frac{n+1}{2}\right)}\right\}^2\left[n - 2\left\{\frac{\Gamma\left(\frac{n+1}{2}\right)}{\Gamma\left(\frac{n}{2}\right)}\right\}^2\right]\\
&= \theta\left[\frac{n}{2}\left\{\frac{\Gamma\left(\frac{n}{2}\right)}{\Gamma\left(\frac{n+1}{2}\right)}\right\}^2 - 1\right].
\end{aligned}$$

Therefore, S' is not a MVB-estimator.

Efficiency of $S' = \frac{\theta/2n}{\text{Var}(S')}$.

Efficiency increases steadily with increasing n. For $n \geq 10$, efficiency is almost equal to one.

Alternative estimators of σ :

(i) $S_0' = \sqrt{\frac{n-1}{2}} \cdot \frac{\Gamma\left(\frac{n-1}{2}\right)}{\Gamma\left(\frac{n}{2}\right)} \cdot S_0$, where $S_0^2 = \frac{n}{n-1} S^2$, is unbiased for σ.

$$\text{Var}(S_0') = \theta \left[\frac{n-1}{2} \left\{ \frac{\Gamma\left(\frac{n-1}{2}\right)}{\Gamma\left(\frac{n}{2}\right)} \right\}^2 - 1 \right].$$

Hence efficiency of $S_0' = \frac{\theta/2n}{\text{Var}(S_0')}$, which is always lower than that of S'.

(ii) Consider $S_1 = \frac{1}{n} \sum_i |X_i - \mu|$, the sample mean deviation about μ. We have, for a $N(\mu, \sigma^2)$, variable,

$$E(|X - \mu|) = \sigma \sqrt{\frac{2}{\pi}}.$$

Hence an unbiased estimator of $\sigma = \sqrt{\theta}$ is

$$S_1' = \sqrt{\frac{\pi}{2}} \cdot S_1.$$

$$\begin{aligned} \text{Var}(S_1') &= \frac{\pi}{2} \cdot \frac{1}{n^2} \sum_i \text{Var}(|X_i - \mu|) \\ &= \frac{\pi}{2} \cdot \frac{1}{n^2} \cdot n \left(\theta - \frac{2}{\pi}\theta \right) \quad \left[\because \ \text{Var}\,|X_i - \mu| = E|X_i - \mu|^2 - [E|X_i - \mu|]^2 \right] \\ &= \frac{\theta}{2n}(\pi - 2). \end{aligned}$$

Hence S_1' is also not an MVB estimator.

Efficiency of $S_1' = \frac{\theta/2n}{\text{Var}(S_1')} = \frac{1}{\pi - 2} = 0.8760$, a constant for all n.

Example 15.7.3 *Let X have p.m.f. $f_\theta(x) = \binom{n}{x} \theta^x (1-\theta)^{n-x}$, where n is a known positive integer, $x = 0, 1, 2, \cdots, n$ and $\Theta = (0, 1)$. Regularity conditions* (i), (ii) *and* (v) *can easily be checked to be satisfied. Consider condition* (iv). *Since*

$$E(t(x)) = \sum_{x=0}^{n} t(x) f_\theta(x) = \sum_{x=0}^{n} t(x) \binom{n}{x} \theta^x (1-\theta)^{n-x}$$

is a polynomial in θ, it can be differentiated.

Similarly (iii) can be seen to hold.

$E_\theta\left[\frac{\partial \log f_\theta(X)}{\partial\theta}\right]^2 = \frac{n}{\theta(1-\theta)}$. Hence RC-lower bound of variance to unbiased estimator of θ is $\frac{\theta(1-\theta)}{n}$.

Since $p = \frac{X}{n}$ is unbiased for θ and $V_\theta(p) = \frac{\theta(1-\theta)}{n}, p$ is an MVB estimator of θ.

► **Example 15.7.4** *X has p.m.f.* $f_\theta(x) = e^{-\theta}\frac{\theta^x}{x!}$, $x = 0, 1, 2, \cdots, \theta > 0$; $p_\theta(\boldsymbol{x}) = e^{-n\theta}\frac{\theta^{\Sigma x_i}}{\Pi X_i!}$.

It can be easily checked that it is a regular estimation case.

$$\frac{\partial \log p_\theta(\boldsymbol{x})}{\partial\theta} = \frac{n}{\theta}(\bar{x} - \theta), \text{ when } \bar{x} = \Sigma x_i/n.$$

$$E\left(\frac{\partial \log p_\theta(\boldsymbol{x})}{\partial\theta}\right)^2 = \frac{n}{\theta}.$$

Hence RC-bound to variance of unbiased estimator of θ is θ/n. Since $\bar{X}$ is unbiased for θ with $V_\theta(\bar{X})$ is $\theta/n, \bar{X}$ is MVBUE of θ.

Consider another unbiased estimator

$$t_1 = \frac{1}{2(n-1)}\sum_{i=1}^{n-1} X_i + \frac{1}{2}X_n;\ V(t_1) = \frac{n\theta}{4(n-1)} > \frac{\theta}{n}.$$

► **Example 15.7.5** *Consider Poisson distribution as in example* 15.7.4 *above. Suppos we want to estimate* $g(\theta) = e^{-\theta}(1+\theta)$. *Consider the statistic* $t = \sum_{i=1}^{n}\psi(X_i)$ *where*

$$\psi(X_i) = \begin{cases} 1 \text{ if } X_i = 0 \text{ or } 1 \\ 0 \text{ otherwise.} \end{cases}$$

$$E(\psi(X_i)) = g(\theta). \text{ Hence } E(t) = g(\theta)$$

$$V(t) = \frac{1}{n}V(\psi(X_1)) = e^{-\theta}(1+\theta)[1 - e^{-\theta}(1+\theta)]/n.$$

RC-*lower bound is* $(g'(\theta))^2/nI_\theta = \frac{\theta^3 e^{-2\theta}}{n} > V(t)$.

► **Example 15.7.6** *X has p.d.f.* $f_\theta(x) = \frac{1}{\theta}(0)$ *if* $0 \le x \le \theta$ *(otherwise),* $\theta \in (0, \infty)$. *Her* $\partial f_\theta(x)/\partial\theta$ *exists if* $\theta \ne x$ *but does not exist if* $\theta = x$. *Hence condition* (ii) *is not satisfied*

Consider the unbiased estimator $t' = \frac{n+1}{n}X_{(n)}$ *where* $X_{(n)} = \max_i X_i$. $V(t') = \frac{\theta^2}{n(n+2)}$

Also $t'' = 2\bar{X}$ *is unbiased for* θ *with* $V(t'') = \frac{\theta^2}{3n}$.

RC-lower bound of variance of unbiased estimator of t is $V(t) = \theta^2/n$. Clearly, both $V(t'), V(t'')$ are $< \frac{\theta^2}{n}$.

However, the inequality (15.7.1) does not hold in this non-regular case.

(See example (15.7.14), where a lower bound to variance has been obtained)

Conditions for equality in RC inequality

If the equality in RC-inequality (15.7.11) holds then necessarily $\psi'(\theta) \neq 0$.
Equality in RC-inequality is satisfied by an unbiased estimator t of $\psi(\theta)$ iff

$$\frac{\partial \log f_\theta(\boldsymbol{x})}{\partial \theta} \propto (t - \psi(\theta)) \ \ a.e.$$

i.e., if

$$\frac{\partial \log f_\theta(\boldsymbol{x})}{\partial \theta} = c_n(\theta)(t - \psi(\theta)) \ \ a.e. \tag{15.7.10}$$

for some constant $c_n(\theta)$ i.e., $\partial \log f_\theta(\boldsymbol{x})/\partial\theta$ is linearly related to $t - \psi(\theta)$. (15.7.10) implies

$$\log f_\theta(\boldsymbol{x}) = \int c_n(\theta)(t - \psi(\theta))\, d\theta + k_n(\boldsymbol{x})$$

$$\text{or,}\ \ f_\theta(\boldsymbol{x}) = \exp\left[\int c_n(\theta)(t - \psi(\theta))\, d\theta\right] h_n(x) \tag{15.7.11}$$

$$= \exp\left[t(\boldsymbol{x})K(\theta) - Q(\theta) + \lambda_n(\boldsymbol{x})\right] \tag{15.7.12}$$

where $k_n(\boldsymbol{x})$ and hence $h_n(\boldsymbol{x})$, $\lambda_n(\boldsymbol{x})$ are independent of θ but may depend on $\boldsymbol{x}$ and

$$K(\theta) = \int c_n(\theta)\, d\theta, Q(\theta) = \int \psi(\theta) c_n(\theta)\, d\theta.$$

(15.7.11) shows that $t(\boldsymbol{x})$ is a sufficient statistic for $\psi(\theta)$. Hence a MVBU estimator must be a sufficient statistic.

Evaluation of $c_n(\theta)$: Multiplying both sides of (15.7.10) by $(t - \psi(\theta))$ and taking expectation

$$E\left\{(t(\boldsymbol{x}) - \psi(\theta))\frac{\partial \log f_\theta(\boldsymbol{x})}{\partial \theta}\right\} = c_n(\theta)E(t - \psi(\theta))^2.$$

Now

$$E\left\{(t(\boldsymbol{x}) - \psi(\theta))\frac{\partial \log f_\theta(\boldsymbol{x})}{\partial \theta}\right\} = \int (t - \psi(\theta))\frac{1}{f}\frac{\partial f}{\partial \theta} f\, d\boldsymbol{x}$$

$$= \frac{\partial}{\partial \theta}\int (t - \psi(\theta))\, f\, d\boldsymbol{x} = \psi'(\theta).$$

Therefore,

$$c_n(\theta) = \frac{\psi'(\theta)}{\underset{\theta}{\mathrm{Var}}(t(\boldsymbol{x}))}. \tag{15.7.13}$$

In particular, if $\psi(\theta) = \theta$, $c(\theta) = 1/\underset{\theta}{\mathrm{Var}}(t(x))$.

We have already seen that the BRUE t of $\psi(\theta)$ is sufficient and satisfies (15.7.10). Let us examine if the converse is also true.

If t is sufficient then the conditional distribution given t is independent of θ. Using (15.7.10)

$$E_\theta\left\{\frac{\partial \log f_\theta(\boldsymbol{x})}{\partial\theta}\right\}^2 = c_n(\theta)^2 \operatorname*{Var}_\theta(t(x)); \qquad (15.7.14)$$

using (15.7.13)

$$\operatorname*{Var}_\theta(t(x)) = \frac{(\psi'(\theta))^2}{E_\theta\left\{\frac{\partial \log f_\theta(x)}{\partial\theta}\right\}^2}$$

which show that t is the BRUE. Hence

Theorem 15.7.2 *A necessary and sufficient condition for an estimator t to be* BRUE *is that it is sufficient and satisfies* (15.7.10).

Corollary 15.7.5 *If $\{f_\theta, \theta \in \Theta\}$ is a one-parameter exponential family (given in* (15.4.10)), *and $Q(\theta)$ has a continuous non-vanishing derivative in Θ then $t(\boldsymbol{X})$ is* BRUE *of its expectation.*

► **Example 15.7.7** (i) *Considering example* 15.7.1, $\frac{\partial \log f(x)}{\partial\theta} = \frac{n(\bar{x}-\theta)}{\sigma^2}$ *which is of the form* (15.7.10).

Hence $\bar{X}$ is a sufficient statistic and BRUE.

(ii) In example 15.7.2 for estimation of $\theta = \sigma^2$,

$$\frac{\partial \log f(\boldsymbol{x})}{\partial\theta} = \frac{n}{2\theta^2}\left[\sum_{i=1}^{n}(x_i-\mu)^2/n - \theta\right].$$

Hence

$$s^2 = \frac{1}{n}\sum_{i=1}^{n}(X_i-\mu)^2$$

is a sufficient statistic and BRUE for θ.

If, however, one wants to estimate $\sigma = \sqrt{\theta} = \psi(\theta)$, then there exists no t for which representation in the form (15.7.10) is feasible. Hence no MVBUE of σ exists.

(iii) For example 15.7.4, $\frac{\partial \log f(\boldsymbol{x})}{\partial\theta} = \frac{n}{\theta}(\bar{x}-\theta)$.

Hence $\bar{X}$ is sufficient and BRUE.

(iv) For example 15.7.5, investigating into the condition (15.7.10) for equality in RC-inequality, we note

$$\frac{\partial \log f_\theta(\boldsymbol{x})}{\partial\theta} = \frac{n}{\theta}(\bar{x}-\theta) = c_n(\theta)(t-g(\theta))$$

implies that

$$t = \frac{n}{\theta c_n(\theta)}\bar{x} + \left(e^{-\theta}(1+\theta) - \frac{n}{c_n(\theta)}\right)$$

There is no choice of $c_n(\theta)$ for which t is independent of θ. Hence in this case there is no MVBU of $g(\theta)$.

▸ **Example 15.7.8** *Show that in samples from*

$$dF = \frac{1}{\Gamma(p)\,\theta^p} x^{p-1} e^{-x/\theta}\, dx, \quad p > 0, \quad 0 \le x < \infty,$$

the MVB *estimator of θ for fixed p is $\bar{X}/p$, with variance θ^2/np, while if $\theta = 1$, that of $\frac{\partial}{\partial p}\log\Gamma(p)$ is $\frac{1}{n}\sum_{i=1}^{n}\log X_i$ with variance $\left\{\frac{\partial^2 \log\Gamma(p)}{n\partial p^2}\right\}$.*

Here,

$$\frac{\partial}{\partial\theta}\log f_{\theta,p}(\boldsymbol{x}) = c_n(\theta)(t-\theta)$$

where $c_n(\theta) = \frac{np}{\theta^2}, t = \frac{\bar{x}}{p}$;

$$\operatorname*{Var}_{\theta}(t) = \frac{\psi'(\theta)}{c_n(\theta)} = \frac{\theta^2}{np} \quad [\because \ \psi(\theta) = \theta].$$

If

$$\theta = 1, \quad \frac{\partial \log f(\boldsymbol{x})}{\partial p} = g_n(p)(t' - \phi(\theta))$$

when

$$t' = \frac{1}{n}\sum_{i=1}^{n}\log x_i, \quad \phi(p) = \frac{\partial \log\Gamma(p)}{\partial p}, \quad g_n(p) = n$$

$$\operatorname{Var}(t') = \frac{\phi'(p)}{g_n(p)} = \frac{\partial^2 \log\Gamma(p)}{n\partial p^2}.$$

Example 15.7.9 *Let $X_1, \cdots, X_n$ be iid r.v.'s, each distributed as a Cauchy variable*

$$f_\theta(x) = \frac{1}{\pi[1+(x-\theta)^2]}, \quad -\infty < \theta < \infty.$$

Here

$$\frac{\partial \log f_\theta(x)}{\partial\theta} = \frac{2(x-\theta)}{1+(x-\theta)^2};$$

hence

$$\frac{\partial \log f_\theta(x_1, \cdots, x_n)}{\partial\theta} = 2\sum_i \frac{(x_i-\theta)}{1+(x_i-\theta)^2}.$$

It is found that there exists no statistic $t(x_1, \cdots, x_n)$ for which it is possible to write this derivative in the form (15.7.10). Hence there does not exist any unbiased estimator of θ which attains RC lower bound.

Rao-Cramer Inequality in Multiparameter Case

Lemma 15.7.1 (*Lehmann and Hodges*). *Let* $\boldsymbol{A} = (A_1, \cdots, A_m)'$ *and* $\boldsymbol{B} = (B_1, \cdots, B_n)'$ *be two sets of r.v.'s each being linearly independent. Let the cavariance matrix of* $\boldsymbol{A}$ *be* $\boldsymbol{C} = ((c_{ij}))$ *and of* $\boldsymbol{B}$ *be* $\boldsymbol{D} = ((d_{ij}))$ *and that between* $\boldsymbol{A}$ *and* $\boldsymbol{B}$ *be* $\boldsymbol{H} = ((h_{ij}))$*. Then* $\boldsymbol{D} - \boldsymbol{H}'\boldsymbol{C}^{-1}\boldsymbol{H}$ *is p.s.d.*

Proof: We have to show that for any vector $\boldsymbol{L} = (l_1, \cdots, l_n)'$

$$\boldsymbol{L}'(\boldsymbol{D} - \boldsymbol{H}'\boldsymbol{C}^{-1}\boldsymbol{H})\boldsymbol{L} \geq 0.$$

Assume without loss of generality $E(\boldsymbol{A}) = \boldsymbol{O}, E(\boldsymbol{B}) = \boldsymbol{O}$.
Now by Cauchy Schwartz inequality,

$$\left[\text{Cov}\left(\sum_{i=1}^{m} u_i A_i, \sum_{j=1}^{n} l_j B_j\right)\right]^2 \leq \text{Var}\left(\sum_{i=1}^{m} u_i A_i\right) \text{Var}\left(\sum_{j=1}^{n} l_j B_j\right)$$

where u_i's are elements of some vector $\boldsymbol{U}$. That is

$$(\boldsymbol{U}'\boldsymbol{H}\boldsymbol{L})^2 \leq (\boldsymbol{U}'\boldsymbol{C}\boldsymbol{U})(\boldsymbol{L}'\boldsymbol{D}\boldsymbol{L}).$$

If we set $\boldsymbol{U} = \boldsymbol{C}^{-1}\boldsymbol{H}\boldsymbol{L}$

$$\begin{aligned}(\boldsymbol{L}'\boldsymbol{H}'\boldsymbol{C}^{-1}\boldsymbol{H}\boldsymbol{L})^2 &\leq (\boldsymbol{L}'\boldsymbol{H}'\boldsymbol{C}^{-1}\boldsymbol{C}\boldsymbol{C}^{-1}\boldsymbol{H}\boldsymbol{L})(\boldsymbol{L}'\boldsymbol{D}\boldsymbol{L}) \\ &= (\boldsymbol{L}'\boldsymbol{H}'\boldsymbol{C}^{-1}\boldsymbol{H}\boldsymbol{L})(\boldsymbol{L}'\boldsymbol{D}\boldsymbol{L}).\end{aligned}$$

Since the A's are linearly independent, the matrix $\boldsymbol{C}$ is positive definite and hence $\boldsymbol{L}'\boldsymbol{H}'\boldsymbol{C}^{-1}\boldsymbol{H}\boldsymbol{L}$ is also a positive definite quadratic form. Hence

$$\boldsymbol{L}'\boldsymbol{H}'\boldsymbol{C}^{-1}\boldsymbol{H}\boldsymbol{L} \leq \boldsymbol{L}'\boldsymbol{D}\boldsymbol{L}.$$

Since $\boldsymbol{D}$ is positive semidefinite, it follows that for all $\boldsymbol{L}$

$$\boldsymbol{L}'(\boldsymbol{D} - \boldsymbol{H}'\boldsymbol{C}^{-1}\boldsymbol{H})\boldsymbol{L} \geq 0.$$

► **Example 15.7.10** *Suppose* $n = 1$; *here* D *and* $\boldsymbol{H}'\boldsymbol{C}\boldsymbol{H}$ *are real numbers. Hence*

$$D = \sigma_B^2 \geq \boldsymbol{H}'\boldsymbol{C}^{-1}\boldsymbol{H}.$$

Theorem 15.7.3 *Let* $A_1, \cdots, A_m$ *be* m *linearly independent r.v.'s, each with mean zero and covariance matrix* $\boldsymbol{C} = ((c_{ij}))$*. If* B *is some r.v. such that* $E(A_1B) = 1$ *and* $E(A_kB) = 0$, $k = 2, 3, \cdots, m$, *then*

$$\sigma_B^2 \geq \frac{\textit{Cofactor of } c_{11}}{|\boldsymbol{C}|}.$$

Proof: Here $\boldsymbol{H} = (1, 0, \cdots, 0)'$, $\boldsymbol{H}'\boldsymbol{C}^{-1}\boldsymbol{H} = (1, 0, \cdots, 0)\boldsymbol{C}^{-1}(1, 0, \cdots, 0)'$ is the leading element of $\boldsymbol{C}^{-1}$ and hence the result follows from the lemma.

▸ **Example 15.7.11** *Let $X_1, \cdots, X_n$ be n iid r.v.'s each with common p.d.f. $f_{\theta_1,\cdots,\theta_k}(x)$. Find a lower bound to the variance of an unbiased estimator of θ_1. Assume that the regularity conditions of Wolfwitz are satisfied.*

We have

$$\int \prod_{j=1}^{n} f_{\theta_1,\cdots,\theta_k}(x_j)\, dx_j = 1.$$

Differentiating both sides with respect to θ_i gives, since differentiation under the integration sign is permissible,

$$\int \left\{ \sum_{j=1}^{n} \frac{\partial f_{\boldsymbol{\theta}}(x_j)}{\partial \theta_i} \prod_{l(\neq j)=1}^{n} f_{\boldsymbol{\theta}}(x_l) \right\} dx_1 \cdots dx_n = 0$$

$$\text{i.e.,} \quad \int \left\{ \sum_{j=1}^{n} \frac{1}{f_{\boldsymbol{\theta}}(x_j)} \frac{\partial f_{\boldsymbol{\theta}}(x_j)}{\partial \theta_i} \right\} \prod_{r=1}^{n} f_{\boldsymbol{\theta}}(x_r)\, dx_r = 0$$

$$\text{i.e.,} \quad \int \left\{ \sum_{j=1}^{n} \frac{\partial \log f_{\boldsymbol{\theta}}(x_j)}{\partial \theta_i} \right\} \prod_{r=1}^{n} f_{\boldsymbol{\theta}}(x_r)\, dx_r = 0. \tag{15.7.15}$$

If

$$A_i = \sum_{j=1}^{n} \frac{\partial \log f_{\boldsymbol{\theta}}(x_j)}{\partial \theta_i}$$

(15.7.15) means

$$E(A_i) = 0, \;\; i = 1, \cdots, k. \tag{15.7.16}$$

Let $t(X_1, \cdots, X_n)$ be an unbiased estimator of θ_1. Then

$$\int t \prod_{j=1}^{n} f_{\boldsymbol{\theta}}(x_j)\, dx_j = \theta_1. \tag{15.7.17}$$

Differentiating both sides with respect to θ_i gives

$$\int t \sum_{j=1}^{n} \frac{\partial \log f_{\boldsymbol{\theta}}(x_j)}{\partial \theta_i} \prod_{r=1}^{n} f_{\boldsymbol{\theta}}(x_r)\, dx_r = \begin{cases} 1 \text{ if } i = 1 \\ 0 \text{ if } i = 2, \cdots, k. \end{cases}$$

$$\text{or,} \quad E(A_1 t) = 1, \;\; E(A_i t) = 0, i = 2, \cdots, k.$$

We assume

$$C_{ie} = E(A_i A_e) = n\alpha_{ie} = nE\left(\frac{\partial \log f_{\boldsymbol{\theta}}(X)}{\partial \theta_i} \frac{\partial \log f_{\boldsymbol{\theta}}(X)}{\partial \theta_e} \right).$$

Hence by theorem 15.7.3,

$$\sigma_t^2(\theta_1, \cdots, \theta_k) \geq \frac{\text{cofactor of } \alpha_{11}}{n|\alpha_{ie}|}$$

where $|\alpha_{ie}| = |\boldsymbol{\alpha}|, \boldsymbol{\alpha} = ((\alpha_{ij}))_{k\times k}$.

▶ **Example 15.7.12** *Let $X_1, \cdots, X_n$ be independent r.v.'s with p.d.f.*

$$f_{\theta_1,\theta_2}(X) = \frac{1}{\sqrt{2\pi\theta_2}} e^{-(x-\theta_1)^2/2\theta_2}, \quad -\infty < x < \infty, -\infty < \theta_1 < \infty, \theta_2 > 0.$$

Let us find a lower bound to the variance of unbiased estimator t of θ_1. It is found that

$$E\left(\frac{\partial^2 \log f_{\theta_1,\theta_2}(X)}{\partial^2\theta_1}\right) = -\frac{1}{\theta_2}, \quad E\left(\frac{\partial^2 \log f_{\theta_1,\theta_2}(X)}{\partial\theta_1\partial\theta_2}\right) = 0,$$

$$E\left(\frac{\partial^2 \log f_{\theta_1,\theta_2}(X)}{\partial\theta_2^2}\right) = -\frac{1}{2\theta_2^2}.$$

Then

$$\sigma_t^2(\theta_1, \theta_2) \geq \frac{\left|-\frac{1}{2\theta_2^2}\right|}{n\begin{vmatrix} -\frac{1}{\theta_2} & 0 \\ 0 & -\frac{1}{2\theta_2^2} \end{vmatrix}} = \frac{\theta_2}{n}.$$

15.7.2 Bhattacharyya Bounds

Bhattacharyya (1946) generalised Rao-Crammer inequality.

Theorem 15.7.4 *Let $X_1, \cdots, X_n$ be r.v.'s with p.d.f. $f_\theta(x_1, \cdots, x_n)$, $\theta \in \Theta$. We want to find a lower bound to the variance of an unbiased estimator $t(X_1, \cdots, X_n)$ of θ. Assume that the following regularity conditions (similar to those in* RC *inequality) hold.*

(a) Θ is a non-degenerate open interval in the real line;

(b) For almost all $x_1, \cdots, x_n$ and all $\theta \in \Theta$

$$\partial^i f_\theta(x)/\partial\theta^i, \quad i = 1, \cdots, k$$

exist, the exceptional set if any have probability measure zero.

(c) $\frac{\partial^i}{\partial\theta^i} \int_w f_\theta(x)\, dx = \int_w \frac{\partial^i}{\partial\theta^i} f_\theta(x)\, dx, \quad i = 1, \cdots, k;$

(d) $\frac{\partial^i}{\partial\theta^i} \int_w t f_\theta(x)\, dx = \int_w t \frac{\partial^i}{\partial\theta^i} f_\theta(x)\, dx, \quad i = 1, \cdots, k;$

(e) If $c_{ij} = \mathrm{Cov}_\theta(A_i, A_j)(i, j = 1, \cdots, k)$ where

$$A_i = \frac{1}{f_\theta(x)} \frac{\partial^i}{\partial\theta^i} f_\theta(x),$$

then the determinant $|c_{ij}| > 0 \ \forall\ \theta$.

Now

$$\int_w f_\theta(x)\, dx = 1.$$

Differentiating both sides with respect to θ, i times (since differentiation under integration sign is permissible), we get

$$\int \frac{1}{f_\theta(\boldsymbol{x})}\left\{\frac{\partial^i}{\partial\theta^i}(f_\theta(\boldsymbol{x}))\right\} f_\theta(\boldsymbol{x})\,d\boldsymbol{x} = 0 \tag{15.7.18}$$

i.e., $E(A_i) = 0$.

Again,

$$\int t(x_1, \cdots, x_n) f_\theta(\boldsymbol{x})\,d\boldsymbol{x} = \theta. \tag{15.7.19}$$

Differentiating both sides with respect to θ, k times,

$$E(A_1 t) = 1, \;\; E(A_i t) = 0, \;\; i = 2, \cdots, k.$$

Then

$$\text{Var}(t) \geq L_k = \frac{\boldsymbol{C}_{11}}{|\boldsymbol{C}|} \tag{15.7.20}$$

where $\boldsymbol{C}_{11}$ is the cofactor of c_{11} in $|\boldsymbol{C}| = |\boldsymbol{C}_k|$.

Proof: Since the dispersion matrix of $(t, A_1, \cdots, A_k)$ is positive semi-definite,

$$\begin{vmatrix} V(t) & E(A_1 t) & \cdots & E_1(A_k t) \\ E(A_1 t) & V(A_1) & \cdots & E(A_k A_1) \\ & & & \\ E(A_k t) & E(A_k A_1) & \cdots & V(A_k) \end{vmatrix} \geq 0$$

i.e.,

$$\begin{vmatrix} V(t) & 1 & 0 & \cdots & 0 \\ 1 & c_{11} & c_{12} & \cdots & c_{1k} \\ 0 & c_{21} & c_{22} & \cdots & c_{2k} \\ \cdots & \cdots & \cdots & \cdots & \cdots \\ 0 & c_{k1} & c_{k2} & \cdots & c_{kk} \end{vmatrix} = V(t)|\boldsymbol{C}_k| - \text{ cofactor of } c_{11} \text{ in } |\boldsymbol{C}| \geq 0.$$

Hence the result (15.7.20).

▶ **Example 15.7.13** *Consider that X has p.d.f. $f_\theta(x) = \theta(1-\theta)^x, x = 0, 1, 2, \cdots$.*

$$\begin{aligned} \frac{\partial f_\theta(x)}{\partial\theta} &= \frac{1}{\theta} f(x) - x f(x-1) \\ \frac{\partial^2 f_\theta(x)}{\partial\theta^2} &= \frac{2x}{\theta} f(x-1) + x(x-1) f(x-2) \\ \frac{\partial^3 f_\theta(x)}{\partial\theta^3} &= \frac{3x(x-1)}{6} f(x-2) - x(x-1)(x-2) f(x-3). \end{aligned}$$

Here

$$A_1 = \frac{1}{f(x)}\frac{\partial f}{\partial \theta} = -\frac{1}{1-\theta}\left(x - \frac{1-\theta}{\theta}\right)$$

$$A_2 = \frac{1}{f(x)}\frac{\partial^2 f}{\partial \theta^2} = \frac{x^2}{(1-\theta)^2} - \frac{[1+(1-\theta)]x}{\theta(1-\theta)^2}$$

$$E(X) = \frac{\theta'}{\theta}, E(X^2) = \frac{\theta' + \theta'^2}{\theta^2}$$

$$E(X^3) = \frac{\theta' + 4\theta'^2 + \theta'^3}{\theta^3}$$

$$E(X^4) = \frac{\theta' + 11\theta'^2 + 11\theta'^3 + \theta'^4}{\theta^4}$$

$$V(A_1) = \frac{1}{\theta^2\theta'}, \text{ Cov}(A_1, A_2) = -\frac{2}{\theta^3\theta'}$$

$$V(A_2) = \frac{4(1+\theta')}{\theta^4\theta'^2} \quad \text{where } \theta' = 1-\theta.$$

Hence RC-lower bound $L_1 = \frac{1}{V(A_1)} = \theta^2(1-\theta)$.

Again,

$$|C_2| = \begin{vmatrix} \frac{1}{\theta^2\theta'} & -\frac{2}{\theta^3\theta'} \\ -\frac{2}{\theta^3\theta'} & \frac{4(1+\theta')}{\theta^4\theta'^2} \end{vmatrix} = \frac{4}{\theta^6\theta'^3}.$$

Bhattacharyya lower bound

$$L_2 = \frac{[1+\theta']\theta^6\theta'^3}{\theta^4\theta'^2} = \theta^2(1-\theta)[1+(1-\theta)] = \theta^2(1-\theta)(2-\theta).$$

We have already seen that for f_θ, the only unbiased estimater of θ is

$$u(X) = \begin{cases} 1 \text{ if } X = 0 \\ 0 \text{ otherwise} \end{cases}$$

with variance $V(u(X)) = \theta(1-\theta), u(X)$ is, therefore, the MVU-estimator of θ.

Note that $L_1 < L_2 < \theta(1-\theta)$.

Bhattacharyya inequality provides us with a sequence of increasing lower bound tending to the limit $\theta(1-\theta)$.

15.7.3 Chapman-Robbins-Kiefer Lower Bound

This inequality does not require the regularity conditions of RC-inequality.

Theorem 15.7.5 *Consider n r.v.'s $X_1, \cdots, X_n$ having a joint p.d.f(or p.m.f.) $f_\theta(\boldsymbol{x})$, $\theta \in \Theta$ when $\Theta \in R^1$. Assume that if $\theta_1 \neq \theta \in \Theta$, then $f_\theta \neq f_{\theta_1}$ in the sense that*

$$\mu[\boldsymbol{x} : f_\theta(\boldsymbol{x}) \neq f_{\theta_1}(\boldsymbol{x})] > 0$$

where μ is a Labesgue measure. Assume also that there exists a $\theta_1 \in \Theta$ such that $\theta_1 \neq \theta$ and

$$W_\theta = \{\boldsymbol{x} : f_\theta(\boldsymbol{x}) > 0\} \supset W_{\theta_1} = \{\boldsymbol{x} : f_{\theta_1}(\boldsymbol{x}) > 0\}.$$

Let t be an unbiased estimator of $\psi(\theta)$ with $E_\theta(t^2) < \infty \ \forall \, \theta \in \Theta$. Then

$$\operatorname*{Var}_\theta(t) \geq \sup_{\{\theta_1 | \theta_1 \neq \theta; W_{\theta_1} \subset W_\theta\}} \frac{[\psi(\theta_1) - \psi(\theta)]^2}{E_\theta\{(f_{\theta_1} - f_\theta)/f_\theta\}^2}. \tag{15.7.21}$$

Proof: Define

$$g_\theta(\boldsymbol{x}) = \begin{cases} \frac{f_{\theta_1}(\boldsymbol{x}) - f_\theta(\boldsymbol{x})}{f_\theta(\boldsymbol{x})} & \text{if } f_\theta(\boldsymbol{x}) > 0 \\ 0 & \text{otherwise.} \end{cases}$$

Therefore,

$$E_\theta[g_\theta(\boldsymbol{X})] = \int_{W_\theta} \frac{f_{\theta_1}(\boldsymbol{x}) - f_\theta(\boldsymbol{x})}{f_\theta(\boldsymbol{x})} f_\theta(\boldsymbol{x}) \, d\boldsymbol{x} = 0.$$

Also

$$E_\theta[t(\boldsymbol{X}) g_\theta(\boldsymbol{X})] = \int_{W_\theta} t(\boldsymbol{x}) \frac{f_{\theta_1}(\boldsymbol{x}) - f_\theta(\boldsymbol{x})}{f_\theta(\boldsymbol{x})} f_\theta(\boldsymbol{x}) \, d\boldsymbol{x} = \psi(\theta_1) - \psi(\theta).$$

Using the Cauchy-Schwartz inequality,

$$\operatorname*{Cov}_\theta^2 \{t(\boldsymbol{X}), g(\boldsymbol{X})\} \leq \operatorname*{Var}_\theta(t(\boldsymbol{X})) \operatorname*{Var}_\theta\left(\frac{f_{\theta_1}(\boldsymbol{X})}{f_\theta(\boldsymbol{X})} - 1\right).$$

Thus,

$$\operatorname*{Var}_\theta(t(\boldsymbol{X})) \geq \frac{[\psi(\theta_1) - \psi(\theta)]^2}{E_\theta\left[\frac{f_{\theta_1}(\boldsymbol{X})}{f_\theta(\boldsymbol{X})} - 1\right]^2}.$$

Hence the result.

In the discrete case, the integrals have to be replaced by sums only.

□ **Remark 15.7.2** *Sometimes (15.7.21) is expressible in an alternative form. Let θ and $\theta + h$ $(h \neq 0)$ be any two distinct values in Θ such that $W_{\theta+h} \subset W_\theta$ and suppose $\psi(\theta) = \theta$. Then*

$$\operatorname*{Var}_\theta(t) \geq \sup_h \frac{h^2}{E_\theta\left\{\left(\frac{f_{\theta+h}(\boldsymbol{X})}{f_\theta(\boldsymbol{X})} - 1\right)^2\right\}} \tag{15.7.22}$$

where the supremum is taken over all $h(\neq 0)$ such that $W_{\theta+h} \subset W_\theta$.

□ **Remark 15.7.3** *Inequality* (15.7.21) *applies even if the parameter space* Θ *is discrete, where* RC *inequality does not hold.*

▸ **Example 15.7.14** *Let* X *be a r.v. with p.d.f.*

$$f_\theta(x) = \begin{cases} \frac{1}{\theta}, & 0 \le x \le \theta \\ 0, & otherwise. \end{cases}$$

We have seen in example that RC-regularity conditions do not hold in this case. Let $\psi(\theta)=\theta$. If $\theta_1 < \theta, W_{\theta_1} \subset W_\theta$.

$$E_\theta \left[\frac{f_{\theta_1}(X)}{f_\theta(X)}\right]^2 = \int_0^{\theta_1} \frac{\theta}{\theta_1^2}\, dx \ [\text{since } W_{\theta_1} \subset W_\theta] = \frac{\theta}{\theta_1}.$$

Hence

$$\operatorname*{Var}_\theta(t) \ge \sup_{\{\theta_1 : \theta_1 < \theta\}} \frac{(\theta_1 - \theta)^2}{\left(\frac{\theta}{\theta_1} - 1\right)} = \sup_{\{\theta_1 : \theta_1 < \theta\}} \theta_1(\theta - \theta_1) = \frac{\theta^2}{4},$$

for any unbiased estimator $t(X)$ of θ.

We have already seen that X is a complete sufficient statistic and $2X$ is unbiased for θ so that $t'(X) = 2X$ is UMVUE.

$$\operatorname*{Var}_\theta(2X) = \frac{\theta^2}{3} > \frac{\theta^2}{4}.$$

The CRK inequality is not achieved by any unbiased estimator of θ.

▸ **Example 15.7.15** X *has a discrete rectangular p.m.f.*

$$P(X = k) = \begin{cases} \frac{1}{N}, & k = 1, 2, \cdots, N \\ 0, & elsewhere. \end{cases}$$

Let $\theta = \{N : N \ge M, M(> 1) \text{ given}\}$.
Take $\psi(N) = N$. Note that for $N \ne N_1 \in \Theta \subset R^1$,

$$W_N = \{1, \cdots, N\} \supset W_{N_1} = \{1, \cdots, N_1\} \text{ if } n_1 < N.$$

Also $P_N(x) \ne P_{N_1}(x)$ for $N \ne N_1$. Thus, CRK-conditions hold (although RC-conditions do not). Hence for any unbiased estimator $t(X)$,

$$\operatorname*{Var}_N(t) \ge \sup_{\{N_1 : N_1 < N\}} \frac{(N - N_1)^2}{\operatorname*{Var}_N(P_{N_1}/P_N)}.$$

Now

$$\frac{P_{N_1}(x)}{P_N(x)} = \begin{cases} \frac{N}{N_1}, & \text{for } x = 1, 2, \cdots, N_1(< N) \\ 0, & \text{otherwise.} \end{cases}$$

Hence

$$E_N\left(\frac{P_{N_1}(x)}{P_N(x)}\right)^2 = \frac{N}{N_1}; \ \underset{N}{\operatorname{Var}}(P_{N_1}(x)/P_N(x)) = \frac{N}{N_1} - 1, \ N > N_1.$$

Therefore,

$$\underset{N}{\operatorname{Var}}(t) \geq \sup_{N_1<N} \{N_1(N-N_1)\}.$$

Now, for integral y,

$$\phi_N(y) = \frac{y(N-y)}{(y-1)(N-y+1)} > 1 \text{ iff } y < \frac{N+1}{2}.$$

Therefore, $\phi_N(y)$ increases so long as $y < \frac{N+1}{2}$ and decreases if $y > \frac{N+1}{2}$. Therefore, $\phi_N(y)$ attains its maximum at $y = \left[\frac{N+1}{2}\right]$. Hence,

$$\underset{N}{\operatorname{Var}}(t) \geq \left[\frac{N+1}{2}\right]\left\{N - \left[\frac{N+1}{2}\right]\right\} = V_N \text{ (say)}$$

where $[x]$ is the largest integer contained in x.

As noted before X is a complete sufficient statistic and $(2X-1)$ is an unbiased and hence UMVUE of N.

$$V_N(2X-1) = \frac{N^2-1}{3} > V_N \text{ for } N > 2.$$

15.7.4 Efficiency of Regular Estimators

We now introduce the concept of efficiency.

Definition 15.7.1 *Let t_1, t_2 be two unbiased estimators of $\psi(\theta)$, with $E(t_i^2) < \infty \ \forall \ \theta \in \Theta$, $i = 1, 2$. Then efficiency of t_1 with respect to t_2 is*

$$e_\theta(t_1|t_2) = \frac{V_\theta(t_1)}{V_\theta(t_2)};$$

t_1 is more efficient than t_2 if

$$e_\theta(t_1|t_2) < 1.$$

In the case of regular estimators it is usual to calculate the efficiency of an estimator t with respect to the BRUE or MVBUE when it exists.

Definition 15.7.2 *In a regular estimation case, the efficiency of an unbiased regular estimator $t(X_1, \cdots, X_n) = t_n(\mathbf{X})$ of θ is*

$$e_\theta[t_n(\mathbf{X})] = \frac{1/E_\theta\left(\frac{\partial \log f_\theta(\mathbf{X})}{\partial\theta}\right)^2}{V_\theta[t_n(\mathbf{X})]} = \frac{1/nI_\theta(X)}{V_\theta[t_n(\mathbf{X})]}.$$

Clearly $0 \leq e_\theta[t_n(\boldsymbol{X})] \leq 1$. If $e_\theta(t_n(\boldsymbol{X})) = 1, t_n(\boldsymbol{X})$ is the most efficient estimator and BRUE of θ. Clearly $t_n(\boldsymbol{X})$ then attains the RC lower bound.

We have already examined in sub-section 15.7.1 the situations when an estimator becomes most efficient estimator or BRUE for θ.

Definition 15.7.3 *In any regular estimation case, the asymptotic efficiency of an unbiased regular estimator* $t(X_1, \cdots, X_n) = t_n(\boldsymbol{X})$ *(say) is*

$$\lim_{n\to\infty} e_\theta(t_n(X)).$$

■ NOTE 15.7.4 *The definitions* 15.7.2 *and* 15.7.3 *relate to regular estimators only. The concepts similar to that in definition* 15.7.1 *and* 15.7.3, *namely, that of 'asymptotic relative efficiency' was developed in case of* CAN *estimators.*

▶ **Example 15.7.16** *In case of* $N(\mu, \sigma)$ *population sample mean* $\bar{X}_n$ *is* BRUE *for* μ *with variance* σ^2/n. *Also sample median* $\tilde{X}_n$ *is unbiased with variance* $\pi\sigma^2/2n$. *Hence efficiency* $e_\mu(\tilde{X}_n) = \frac{2}{\pi}$. *This is also its asymptotic efficiency.*

15.8 Criteria for Comparison among Several Competing Statistics for Estimating a Parameter θ in the Multivariate set up

Let $\boldsymbol{t} = (t_1, \cdots, t_p)'$ be a statistic for estimating $\boldsymbol{\theta} = (\theta_1, \cdots, \theta_p)'$.

Definition 15.8.1 ***Multivariate Efficiency.*** *If two unbiased estimators* $\boldsymbol{t}_1, \boldsymbol{t}_2$ *have variance-covariance matrices* $\boldsymbol{\delta}_1, \boldsymbol{\delta}_2$ *respectively, then* $\boldsymbol{t}_1$ *is more efficient than* $\boldsymbol{t}_2 (\boldsymbol{t}_1 \succ \boldsymbol{t}_2)$ *if* $\boldsymbol{\delta}_2 - \boldsymbol{\delta}_1$ *is a p.s.d. matrix.*

Definition 15.8.2 ***Multivariate MSE.*** *For any statistic* $\boldsymbol{t}$,

$$\begin{aligned} MSE(\boldsymbol{t}) &= E(\boldsymbol{t}-\boldsymbol{\theta})'(\boldsymbol{t}-\boldsymbol{\theta}) = E(\boldsymbol{t}-E(\boldsymbol{t})+E(\boldsymbol{t})-\boldsymbol{\theta})'(\boldsymbol{t}-E(\boldsymbol{t})+E(\boldsymbol{t})-\boldsymbol{\theta}) \\ &= E(\boldsymbol{t}-E(\boldsymbol{t}))'(\boldsymbol{t}-E(\boldsymbol{t})) + (E(\boldsymbol{t})-\boldsymbol{\theta})'(E(\boldsymbol{t})-\boldsymbol{\theta}) \\ &= tr[V(\boldsymbol{t})] + [\text{bias }(\boldsymbol{t})]'[\text{bias }(\boldsymbol{t})] \\ &= \sum_{i=1}^{p} V(t_i) + \sum_{i=1}^{p} (\text{bias }(t_i))^2 \end{aligned} \tag{15.8.1}$$

where $V(\cdot)$ *denotes dispersion matrix of* $(\cdot)$.

In statistical decision theory $E(\boldsymbol{t}-\boldsymbol{\theta})'(\boldsymbol{t}-\boldsymbol{\theta}) = \sum_{i=1}^{k} E(t_i - \theta_i)^2$ is called the expected squared loss function or *risk function* w.r.t. squared loss function.

If we define Eucledian length of a vector $\boldsymbol{a}$ by $\|\boldsymbol{a}\| = (\boldsymbol{a}'\boldsymbol{a})^{1/2}$, then MSE $(\boldsymbol{t}) = E\|\boldsymbol{t}-\boldsymbol{\theta}\|^2$, the expected squared Eucledian distance between $\boldsymbol{t}$ and $\boldsymbol{\theta}$. Thus an estimator with lower MSE will be close to true parameter.

Definition 15.8.3 ***Weighted MSE.*** *The weighted MSE of* $\boldsymbol{t}$ *with the p.s.d. matrix* $\boldsymbol{W}$ *is defined as*

$$W\,\text{MSE}(\boldsymbol{t}) = E(\boldsymbol{t}-\boldsymbol{\theta})'\boldsymbol{W}(\boldsymbol{t}-\boldsymbol{\theta}).$$

Definition 15.8.4 ***Matrix MSE.*** *This is defined as*

$$Mtx\,\text{MSE}(\boldsymbol{t}) = E(\boldsymbol{t}-\boldsymbol{\theta})(\boldsymbol{t}-\boldsymbol{\theta})'.$$

This can also be written as

$$\begin{aligned} Mtx\,\text{MSE}(\boldsymbol{t}) &= E(\boldsymbol{t}-E(\boldsymbol{t})+E(\boldsymbol{t})-\boldsymbol{\theta})(\boldsymbol{t}-E(\boldsymbol{t})+E(\boldsymbol{t})-\boldsymbol{\theta})' \\ &= E(\boldsymbol{t}-E(\boldsymbol{t}))(\boldsymbol{t}-E(\boldsymbol{t}))' + (E(\boldsymbol{t})-\boldsymbol{\theta})(E(\boldsymbol{t})-\boldsymbol{\theta})' \\ &= V(\boldsymbol{t}) + [\text{Bias }(\boldsymbol{t})][\text{Bias }(\boldsymbol{t})]'. \end{aligned}$$

Note that

$$tr\{Mtx\,\text{MSE}(\boldsymbol{t})\} = \text{MSE}(\boldsymbol{t}) = E(\boldsymbol{t}-\boldsymbol{\theta})'(\boldsymbol{t}-\boldsymbol{\theta}).$$

Multivariate MSE-criterion

Consider two competing estimators $\boldsymbol{t}_1$ and $\boldsymbol{t}_2$ of θ. If the difference $\Delta = Mtx\,\text{MSE}(\boldsymbol{t}_1) - Mtx\,\text{MSE}(\boldsymbol{t}_2)$ is p.d., then $\boldsymbol{t}_2$ is preferred to $\boldsymbol{t}_1 (\boldsymbol{t}_2 \succ \boldsymbol{t}_1)$. To check up positive-definiteness, it is easier in practice to check whether the scalar $\boldsymbol{\eta}'\boldsymbol{\Delta\eta} > 0 \;\forall\; \boldsymbol{\eta} \neq \boldsymbol{0}$ where $\boldsymbol{\eta}$ is any non-zero column vector. Also, if $\boldsymbol{\eta}'\boldsymbol{\Delta\eta} > 0 \;\forall\; \boldsymbol{\eta} \neq \boldsymbol{0}$, then $\text{MSE}(\boldsymbol{\eta}'\boldsymbol{t}_1) > \text{MSE}(\boldsymbol{\eta}'\boldsymbol{t}_2) \;\forall\; \boldsymbol{\eta} \neq \boldsymbol{0}$ and vice-versa.

Hence, checking up if $\boldsymbol{\eta}'\boldsymbol{\Delta\eta} > 0$ may be taken as equivalent to checking if $\boldsymbol{t}_2 \succ \boldsymbol{t}_1$.

A measure of efficiency of $\boldsymbol{t}_2$ wrt $\boldsymbol{t}_1$ may be $\frac{|D(\boldsymbol{t}_1)|}{|D(\boldsymbol{t}_2)|}$, where $D(\cdot)$ denotes the covariance matrix of $(\cdot)$.

To compare the performance of $\boldsymbol{t}_1$ and $\boldsymbol{t}_2$ one may use the spectral norm of the difference between $D(\boldsymbol{t}_1)$ and $D(\boldsymbol{t}_2)$.

$$d = \|D(\boldsymbol{t}_1) - D(\boldsymbol{t}_2)\|_2^2$$

where $\|\boldsymbol{A}\|_2$ denotes the spectral norm of $\boldsymbol{A}$ (see e.g., Horn and Johnson, 1985), i.e.,

$$\|\boldsymbol{A}\|_2 = \sup_{\substack{\boldsymbol{X}\in R^p \\ \|\boldsymbol{X}\|=1}} \|\boldsymbol{AX}\|.$$

For any covariance matrix $\boldsymbol{A}$,

$$\begin{aligned} \|\boldsymbol{A}\|_2^2 = \sup_{\substack{\boldsymbol{X}\in R^p \\ \|\boldsymbol{X}\|=1}} \|\boldsymbol{AX}\|^2 &= \sup_{\substack{\boldsymbol{X}\in R^p \\ \|\boldsymbol{X}\|=1}} \|\boldsymbol{XA'AX}\| \\ &= \lambda_1(\boldsymbol{A'A}) = (\lambda_1(\boldsymbol{A}))^2 \end{aligned}$$

where $\lambda(\boldsymbol{A})$ is the largest eigenvalue of $\boldsymbol{A}$. The measure d is thus related to the largest eigenvalue of the difference between the two covariance matrices. •

15.9 Robust Estimators

In this section we shall consider two methods of obtaining estimators whose properties would relatively remain unaffected by deviaions from the assumed models. Such estimators are called robust.

As an example, consider the problem of estimating the mean μ of a distribution. We know that the sample mean $\bar{X}$ is an unbiased estimator of μ. It is also consistent and efficient. These properties are, however, valid under the specific assumption about the underlying distribution. If the actual distribution differs in some respect from the assumed one, $\bar{X}$ may not have these desirable properties.

A type of deviation from the model is called *contamination*. Such deviations may be caused by the occurrence of outliers or gross errors, i.e., observations that follow a distribution different from the one assumed in the model, the difference consisting in the mass of outlier distribution being concentrated around much larger (or smaller) values than those typically encountered in the model.

Suppose that the postulated model is $f(x, \theta)$ while the actual sample is taken from the distribution

$$\varphi(x, \theta) = (1 - \varepsilon)f(x, \theta) + \varepsilon g(x, \theta) \tag{15.9.1}$$

where most of the mass of $g(x, \theta)$ is far away from the range of typical values of $f(x, \theta)$.

It follows that the sample mean $\bar{X}$ is very sensitive to deviation from a model of the type described in (15.9.1). Various suggestions for its remedy led to the theory of robust estimation. Here we mention two of these procedures, L-estimators and M-estimators.

15.9.1 *L*-estimators

L-estimators are constructed primarily for estimation of location parameters.

Let $X_1, \cdots, X_n$ be a random sample from the distribution $f(x, \theta)$ and let

$$Y_{1,n} \le Y_{2,n} \le \cdots \le Y_{n,n} \tag{15.9.2}$$

denote the order statistics for the sample. A L-statistic is a linear combination of the order statistic,

$$T = \sum_{k=1}^{n} \alpha_{k,n} Y_{k,n} \tag{15.9.3}$$

where $\alpha_{k,n}$ $(k = 1, \cdots, n)$ is an array of coefficients. The class of L-estimators contains many well-known estimators.

The choice $\alpha_{k,n} = 1/n \; \forall \; k = 1, \cdots, n$, gives $T = \bar{X}$. The choice $\alpha_{k,n} = 0 \; \forall \; k \ge 2$, $\alpha_{1,n} = 1$ or $\alpha_{n,n} = 1$, $\alpha_{k,n} = 0 \; \forall \; k < n$, gives two extreme order statistics, $Y_{1,n} = \min(X_1, \cdots, X_n)$ and $Y_{n,n} = \max(X_1, \cdots, X_n)$ respectively. In a similar way one can obtain sample quantities. The choice $\alpha_{[3n/4]+1,n} = 1$, $\alpha_{[n/4]+1,n} = -1$, $\alpha_{k,n} = 0$ for the remaining k gives the interquartile range.

The most important L-estimators are the trimmed and windsorized means, defined as follows.

Definition 15.9.1 *Let $0 < \alpha < \frac{1}{2}$. Then the α-trimmed mean is*

$$U_{n,\alpha} = \frac{1}{n - 2[n\alpha]} \sum_{k=[n\alpha]+1}^{n-[n\alpha]} Y_{k,n}. \tag{15.9.4}$$

Definition 15.9.2 *Let $0 < \alpha < \frac{1}{2}$. The α-windsorized mean is*

$$V_{n,\alpha} = \frac{1}{n} \left\{ [n\alpha] Y_{[n\alpha]+1,n} + \sum_{k=[n\alpha]+1}^{n-[n\alpha]} Y_{k,n} + [n\alpha] Y_{n-[n\alpha],n} \right\} \tag{15.9.5}$$

Thus, α-trimming consists of rejecting from the sample the fraction α of the largest and the fraction α of the smallest observations and taking the average of the remaining observations.

The α-windsorizing consists of replacing each observation in the lowest α-fraction by the sample quantile of order α and each observation in the upper α-fraction by the sample quantile of order $(1-\alpha)$ and keeping the other observations unchanged.

The windsorized mean is then calculated as the mean of the windsorized sample.

Clearly, the purpose of trimming (or windsorizing) is to eliminate (or reduce) the effect of outliers in the sample.

The main area of research in the theory of L-estimators is to define the notion of optimality and then to determine the optimum value of α at which the mean value should be trimmed or windsorized and establishing their asymptotic properties.

15.9.2 *M*-estimators

Let $g(x, u)$ be a function of two arguments. Given a sample $x_1, \cdots, x_n$ from the distribution $f(x, \theta)$, one takes an estimate of θ the solution of the equation

$$\sum_{k=1}^{n} g(x_k, u) = 0. \tag{15.9.6}$$

Such estimators are often obtained by solving an approximate minimization problem,

$$\sum_{k=1}^{n} \rho(x_k, u) = \min! \tag{15.9.7}$$

Here $g(x, \theta) = \partial\rho(x, \theta)/\partial\theta$. The function $\rho(x, u)$ can be thought of as some distance function. The sum of distances from all the sample points is $\sum_{k=1}^{n} \rho(x_k, u)$. As an estimate of θ we choose a point u^* that minimizes the sum $\sum_{k=1}^{n} \rho(x_k, u)$. In this sense u^* is the point closest to the sample, with closeness being expressed by the function ρ.

Note that the choice $\rho(x, \theta) = -\log f(x, \theta)$ gives the ordinary maximum likelihood estimate.

If we take appropriate functions ρ and g, we can also obtain different variants of least square estimators. Similarly, trimmed or windsorized means can be obtained by appropriate choice of the functions.

If we are interested in estimating the location parameter, we minimise

$$\sum_{k=1}^{n} \rho(x_k - u)$$

or obtain solution of the equation

$$\sum_{k=1}^{n} g(x_k - u) = 0. \tag{15.9.8}$$

The equation (15.9.8) can be written as

$$\sum_{k=1}^{n} \omega_k(x_k - u) = 0$$

where

$$\omega_k = \frac{g(x_k - u)}{x_k - u}.$$

This gives a formal representation of u as a weighted sum

$$u = \frac{\sum_{1}^{n} \omega_k x_k}{\sum_{1}^{n} \omega_k}$$

with weights depending on the sample.

The main direction of research of M-estimators is to study its asymptotic properties under some general assumptions on functions g or ρ and distribution of X.

15.10 Estimators Based on Resampling Methods

We shall now mention briefly some recent methods for improving estimators. These methods require extensive computations and are applicable now with the availability of fast computers.

The first method, called *jack-knifing* is applied to reduce the bias of the estimator. Let $X_1, \cdots, X_n$ be a random sample from a distribution with cdf F. The function F is unknown and the objective is to estimate some parameter θ which depends on $F, \theta = g(F)$. Some examples are the mean $\int x\, dF(x)$, the variance $\int [x - E(X)]^2 dF(x)$

tc. Generally, g is a function defined on a class of cdf's (eg. for all cdf's where mean xist, etc.) and values of g are real numbers.

As estimators we consider the sample counterparts of the corresponding parameters. Given the sample $x_1, \cdots, x_n$ we consider the discrete probability distribution which gives nass $1/n$ to each of the points $x_1, \cdots, x_n$. Let F_n be the cdf of such a distribution. We now take $g(F_n)$ as an estimate of θ. When the points $X_1, \cdots, X_n$ are taken as random variables, $g(F_n)$ becomes random leading to an estimate $\hat{\theta}$, with value $g(F_n)$ for the sample $x_1, \cdots, x_n$.

Example 15.10.1 *If θ is the mean, then $g(F_n) = \int x\, dF_n(x) = \frac{1}{n}\sum_{i=1}^{n} x_i$ is the sample mean $\bar{x}$. For estimation of the variance, the estimate is $g(F_n) = \int (x-\bar{x})^2 dF_n(x) = \frac{1}{n}\sum_{i=1}^{n}(x_i-\bar{x})^2 = s^2$.*

In general, the estimator $\hat{\theta}$ is biased, with bias

$$E_F(\hat{\theta}) - \theta = F_F[g(F_n)] - g(F) \tag{15.10.1}$$

We now remove the sample point x_i from the sample and let $F_{n-1}^{(i)}$ be the sample cdf based on the remaining units in the sample. We now calculate the value of an estimator at $F_{n-1}^{(i)}$, obtaining

$$\hat{\theta}^{(i)} = g\left(F_{n-1}^{(i)}\right), \quad i = 1, \cdots, n. \tag{15.10.2}$$

Let

$$\bar{\hat{\theta}} = \frac{1}{n}\sum_{i=1}^{n} \hat{\theta}^{(i)} \tag{15.10.3}$$

Consider the expression

$$\hat{B} = (n-1)(\bar{\hat{\theta}} - \hat{\theta}) \tag{15.10.4}$$

It is seen that under assumptions (15.10.6) below, $\hat{B}$ is a good estimate of the bias of $\hat{\theta}$, so that we may improve $\hat{\theta}$ by considering the jack-knife estimator of θ, defined as

$$\begin{aligned}\hat{\theta}_J &= \hat{\theta} - (n-1)(\bar{\hat{\theta}} - \hat{\theta}) \\ &= n\hat{\theta} - (n-1)\bar{\hat{\theta}}\end{aligned} \tag{15.10.5}$$

Assume that for each sample size m, the expectation of the statistic $g(F_m) = \hat{\theta}(X_1, \cdots, X_m)$ can be expressed as

$$E[g(F_m)] = \theta + \frac{a_1}{m} + \frac{a_2}{m^2} + \cdots \tag{15.10.6}$$

where $a_1, a_2, \cdots$ are constants depending on F only (not on m). Thus, the bias of $\hat{\theta}$ is of the order of $1/n$, where n is the sample size,

Now,

$$E(\bar{\hat{\theta}}) = \frac{1}{n}\sum_{i=1}^{n} E(\hat{\theta}^{(i)}) = \frac{1}{n}\sum_{i=1}^{n} E\left\{g\left(F_{n-1}^{(i)}\right)\right\}$$
$$= E\left[g(F_{n-1})\right] \tag{15.10.7}$$

by assumption (15.10.6). Therefore, by (15.10.5),

$$E(\hat{\theta}_J) = nE(\hat{\theta}) - (n-1)E(\bar{\hat{\theta}})$$
$$= n\left[\theta + \frac{a_1}{n} + \frac{a_2}{n^2} + \cdots\right] - (n-1)\left[\theta + \frac{a_1}{n-1} + \frac{a_2}{(n-1)^2} + \cdots\right]$$
$$= \theta - \frac{a_2}{n(n-1)} - \frac{(2n-1)a_3}{n^2(n-1)^2} - \cdots \tag{15.10.8}$$

The bias of $\hat{\theta}_J$ is, therefore, of order $1/n^2$.

In most cases, the bias of $\hat{\theta}$ is unknown, $a_1, a_2, \cdots$, in (15.10.6) being hard to find out. The jack-knife estimation has reduced bias.

► **Example 15.10.2** *Suppose we want to estimate θ in the uniform distribution on $[0, \theta]$. The functional g is of the form*

$$g(F) = \inf\{t : F(t) = 1\}.$$

Hence $g(F) = \theta$. If $X_1, \cdots, X_n$ is a sample from F, and F_n is the empirical cdf $g(F_n) = \max(X_1, \cdots, X_n) = X_{n,n}$. Therefore, $E_F[g(F_n)] = E_F(X_{n,n}) = \frac{n}{n+1}\theta$. Hence, expansion (15.10.6) takes the form

$$E_F[g(F_m)] = \frac{m}{m+1}\theta = \frac{\theta}{1+\frac{1}{m}}$$
$$= \theta - \frac{\theta}{m} + \frac{\theta}{m^2} - \frac{\theta}{m^3} + \cdots$$

so that $a_i = (-1)^i\theta$.

The estimator $\hat{\theta}^{(i)}$ based on the sample with X_i removed, is equal to either $X_{n,n}$ or $X_{n-1,n}$, the latter being the case only if the largest element of the sample is removed. Hence,

$$\bar{\hat{\theta}} = \frac{n-1}{n}X_{n,n} + \frac{1}{n}X_{n-1,n}.$$

The jack-knife estimate is, therefore,

$$\hat{\theta}_j = nX_{n,n} - (n-1)\bar{\hat{\theta}}$$
$$= nX_{n,n} - (n-1)\left\{\frac{n-1}{n}X_{n,n} + \frac{1}{n}X_{n-1,n}\right\}$$
$$= \frac{2n-1}{n}X_{n,n} - \frac{n-1}{n}X_{n-1,n}.$$

Since

$$E(X_{n,n}) = \frac{n}{n+1}\theta, \quad E(X_{n-1,n}) = \frac{n-1}{n+1}\theta,$$

$$E(\hat{\theta}_J) = \frac{n^2}{n(n+1)}\theta = \theta - \frac{\theta}{n(n+1)}.$$

Thus, the jack-knife estimator is biased; but the bias is much smaller than the bias of the estimator $X_{n,n}$.

Another example has been given in subsection 15.2.1.

Efron (1982) introduced the *bootstrap* sample. If $x_1, \cdots, x_n$ is the original sample, we can consider the bootstrap random variable, say $X^{(B)}$, defined as

$$P\left\{X^{(B)} = x_i\right\} = \frac{1}{n}, \quad i = 1, \cdots, n.$$

If $X_1^{(B)}, \cdots, X_m^{(B)}$ form a bootstrap sample, this means $X_1^{(B)}, \cdots, X_m^{(B)}$ are independently and identically distributed with the same distribution (15.10.9). This means that we are sampling the elements of the original sample with replacement.

The bootstrap samples are most often used to provide information about the variance of the estimators.

Exercise

1. Let $X_1, \cdots, X_n$ be independent r.v.'s with common p.d.f. $P(X_i = 1) = p$, $P(X_i = 0) = 1 - p$, $0 \le p \le 1$, $i = 1, \cdots, n$. Let $t(X_1, \cdots, X_n)$ be a unbiased estimator of p. Find a lower-bound to the variance of t and check the regularity conditions.

2. $X_1, \cdots, X_n$ are n iid r.v.'s each having an exponential density $f_\theta(x) = \frac{1}{\theta}\exp[-x/\theta]$. Find the total scale $S_\theta = \partial \log f_\theta(x_1, \cdots, x_n)/\partial\theta$ and the information contained in the sample about θ, $E(-\partial S_\theta/\partial\theta)$. *Ans:* information $= \frac{n}{\Theta^2}$]

3. Let $X_1, \cdots, X_n$ be a random sample of size n from a p.d.f. $f_\theta(x) = \theta/(1+x)^{(1+\theta)}$, if $x > 0 (= 0$, otherwise), for some $\theta > 0$. Find RC-lower bound to variance of unbiased estimator of (i) $1/\theta$ (ii) e^θ (iii) θ^2.

4. X is a random variable assuming values $0, 1, 2, 3$ with probabilities $\theta, \theta^2, \theta^3$, $1 - \theta - \theta^2 - \theta^3$ respectively. Find the RC-lower bound to the variance of an unbiased estimator of θ.

5. Let $X_1, \cdots, X_n$ be iid Bernoulli (p-) variables. Show that $t_n = S_n(n - S_n)(n - 1 - S_n)$, where $S_n = \Sigma_{i=1}^n X_i$, is an unbiased estimator of pq^2 and find RC-lower bound to the variance of this estimator.

6. Let $X_1, \cdots, X_n$ be iid r.v.'s from a distribution $f_\theta(x) = (1-\theta) + \frac{\theta}{2\sqrt{x}}$, $0 < x < 1$ $0 < \theta < 1$. Find the lower bound for $E_\theta(\bar{X}_n - \theta)^2$. [Note that $\bar{X}_n$ is a biase estimator of θ].

7. Let $X_1, \cdots, X_n$ be iid with common p.d.f. $f_\theta(x) = \exp[-(x-\theta) - \exp(x-\theta)]$ Find RC lower bound for the variance of an unbiased estimator of (i) θ, (ii) θ^2, (iii $\theta^2 + \theta$, (iv) e^θ.

8. Show that if $X_1, \cdots, X_n$ are iid with p.d.f. $f_\theta(x) = \exp\{A(\theta)B(x) + C(\theta) + D(x)\}$ then $T = \Sigma_{i=1}^n B(X_i)/n$ attains the Rao-Cramer lower bound for an unbiase estimator of $E(T)$.

   [*Hints:* Show that since $T(x_1, \cdots, x_n)$ is a linear function of $\partial \log f_\theta(x_1, \cdots, x_n)/\partial\theta$, equality hold in RC-inequality.]

9. In a random sample of size n from the Poisson population with parameter λ, th frequency of the value zero is r. Set up a consistent estimate of λ based on alone. Assuming n to be large, find the variance and hence, the efficiency of you estimate.

10. Let $X_1, X_2, \cdots, X_n$ be iid r.v.'s with common p.d.f. $f_\theta(x) = e^{-(x-\theta)}(0)$, $x \geq \theta >$ (otherwise). Find the Chapman-Robbins-Kiefer lower bound of variance of a unbiased estimator of θ.

11. Let $X_1, \cdots, X_n$ be n independent random variables such for any i,

$$P(X_i = r) = p(1-p)^r, \quad r = 0, 1, \cdots; 0 < p < 1.$$

   Find a sufficient statistic for p and hence obtain an unbiased estimator of $1/p$. Als obtain the variance of the estimator and compare it with the Rao-Cramer lowe bound for the unbiased estimator of $1/p$.

12. Let $X_1, \cdots, X_n$ be iid r.v.'s with p.d.f. $f_{\theta_1,\theta_2}(x) = \frac{1}{\sqrt{2\pi\theta_2}} \exp\left[-(x-\theta_1)^2/2\theta_2\right]$ $-\infty < x < \infty$; $-\infty < \theta_1 < \infty$; $0 < \theta_2 < \infty$.

   Let $t(X_1, \cdots, X_n)$ be a unbiased estimator of θ_2. Find the Rao-Cramer lower boun to $V(t)$.

13. (a) Show that Rao-Cramer inequality holds good for a distribution $f_\theta(x)$ whos range (a, b) depends on θ, provided that $E\left(\frac{\partial \log p_\theta(\boldsymbol{x})}{\partial\theta}\right) = 0$, where $p_\theta(\boldsymbol{x}) = f_\theta(x_1, \cdots, x_n)$ holds. Show that this is so iff $f_\theta(a) = f_\theta(b) = 0$.

   If in addition $\left.\frac{\partial f_\theta(x)}{\partial\theta}\right]_{x=a} = \left.\frac{\partial f_\theta(x)}{\partial\theta}\right]_{x=b} = 0$, then $E\left[\left(\frac{\partial \log p_\theta(\boldsymbol{x})}{\partial\theta}\right)^2\right] = -E\left(\frac{\partial^2 \log p_\theta(\boldsymbol{x})}{\partial\theta^2}\right)$ and the MVB for $\psi(\theta)$ can be written in the form

$$\operatorname*{Var}_\theta(t) \geq -\frac{\{\psi'(\theta)\}^2}{E_\theta\left[\frac{\partial^2 \log p_\theta}{\partial\theta^2}\right]}.$$

(b) Apply this result to show that the MVB holds for the estimation of θ in

$$dF(x) = \frac{1}{\Gamma(p)}(x-\theta)^{p-1}\exp\{-(x-\theta)\}\,dx, \quad 0 \le x < \infty;\ p > 2$$

and is equal to $(p-2)/n$, but is not attainable since there is no single-sufficient statistic for θ. **[Kendall and Stuart, 1961]**

14. Let $X_1, \cdots, X_n$ be iid r.v.'s with common mean μ and variance σ^2. Find the efficiency of the unbiased estimate $2\Sigma_{i=1}^{n} iX_i/n(n+1)$ relative to $\bar{X}_n$.

15. Based on a random sample of size n from the distribution with p.d.f.

$$f(x,\alpha) = \frac{1}{\Gamma(\alpha)} e^{-x} x^{\alpha-1}, x \ge 0, \alpha > 0.$$

Obtain an estimator of $\frac{\partial \log \Gamma(\alpha)}{\partial x}$ which attain the RC-lowerbound.

16. Let $X_1, \cdots, X_n$ be a random sample from a Gamma distribution

$$f_{\alpha \cdot p}(x) = \begin{cases} \frac{1}{p^{\alpha+1}\Gamma(\alpha+1)} x^{\alpha} e^{-x/p}, & 0 \le x < \infty \\ 0, & \text{otherwise.} \end{cases}$$

Let t_{1n} be an unbiased estimator α where p is known and t_{2n} be an unbiased estimator of p where α is known. Find the Rao-Cramer lower bound for the variance of these estimators.

17. Let X be a r.v. with continuous p.d.f. $f_\theta(x)$ such that the function is specified when θ is specified. Let again θ be a r.v. with distribution $\xi(\theta)$ defined over a non-degenerate range Θ.

Let $t = t(x)$ be an estimate of θ functionally independent of θ. Let $\psi(\theta) = E(t|\theta)$. Assume

(a) $\frac{\partial f_\theta(x)}{\partial \theta}$ exists for all $\theta \in \Theta$;

(b) $\left|\frac{\partial f_\theta(x)}{\partial \theta}\right| < G_1(x)$;

(c) G_1 and tG_1 are integrable over the entire real line R^1, the range of x, which is independent of θ.

Prove that $\text{Var}(t-\theta)E\left[E\left\{\left(\frac{\partial \log f_\theta(x)}{\partial \theta}\right)^2 \Big| \theta\right\}\right] \ge E^2\left(\frac{\partial \psi(\theta)}{\partial \theta}\right)$. **[Gart, 1959]**

18. Find the CRK lower bound for the variance of an unbiased estimate of θ, based on a sample of size n from the distribution

(a) $f_\theta(x) = \begin{cases} e^{-(x-\theta)}, & \theta < x < \infty; \\ 0, & \text{otherwise.} \end{cases}$

(b) $N(\mu, 1)$.

19. Compute CRK lower bound for the variance of an estimator of θ based on X wit $P_\theta[X=0]=1-\theta$, $P_\theta[X=1]=\theta, 0<\theta<1$.

20. If $f(x,\theta)=\dfrac{\theta^p}{\Gamma(p)}e^{-\theta x}x^{p-1}$, where p is known and θ is unknown show that ther exists an unbiased estimate of $1/\theta$ based on a sample of size one, whose varianc is equal to the CR-lower bound.

Chapter 16

Different Methods of Estimation

16.1 Introduction

n this chapter we consider different methods of parametric estimation. The method f maximum likelihood is considered in section 16.2, the method of moments in section 6.3, the method of minimum chi-square in section 16.4 and the method of least squares n section 16.5.

16.2 The Method of Maximum Likelihood

he method of maximum likelihood is one of the oldest and most important methods of stimation in estimation theory. It was first used by Gauss in developing the theory of east squares and subsequently re-introduced by Fisher in (1912).

Let $X_1, \cdots, X_n$ be the observed values of a set of random variables $X_1, \cdots, X_n$ having joint p.d.f. $f_\theta(x_1, \cdots, x_n)$, $\theta \in \Theta$. The function $f_\theta(x_1, \cdots, x_n)$ may be regarded as a unction of θ for a given set of values $(x_1, \cdots, x_n)$. When regarded as a function of , the expression $f_\theta(x_1, \cdots, x_n)$ is referred to as the likelihood function (L.F.) of θ, $(\theta|x_1, \cdots, x_n)$ and expresses the probability that the value of the random variable θ is for given values of observations $x_1, \cdots, x_n$. The maximum likelihood estimate (m.l.e.) f θ is that value of θ, within the admissible range of values of θ, which makes the kelihood function a maximum, i.e., the m.l.e. of θ is the number $\hat{\theta}$, if it exists, such hat $L(\hat{\theta}|x_1, \cdots, x_n) > L(\theta'|x_1, \cdots, x_n)$ whatever be θ', any other value in Θ.

Ordinarily the parameter θ may be regarded as continuous and in this case the etermination of m.l.e. becomes simple. Assuming that (a) the likelihood is a positive ifferentiable function of θ and (b) the maximum of the likelihood does not occur on he boundary of the interval in R^1 of all admissible values of θ, which may be infinite in oth the directions, the stationary values of the likelihood function within the interval re given by the roots of the equation

$$\frac{\partial L(\theta|x_1, \cdots, x_n)}{\partial \theta} = 0.$$

A sufficient condition that any of these values, say, $\hat{\theta}$ be a real maximum is

$$\left.\frac{\partial^2 L(\theta|x_1,\cdots,x_n)}{\partial\theta^2}\right]_{\theta=\hat{\theta}} < 0.$$

Since $\log L$ attains its maximum value for the same value of θ as L it is usual t maximise $\log L$ in lieu of L. Therefore, we shall seek solution of

$$\frac{\partial \log L(\theta|x_1,\cdots,x_n)}{\partial\theta} = 0 \tag{16.2.1}$$

subject to the condition

$$\frac{\partial^2 \log L(\theta|x_1,\cdots,x_n)}{\partial\theta^2} < 0. \tag{16.2.2}$$

If there are more than one solution satisfying (16.2.1) and (16.2.2) the maximum c these solutions is to be taken.

We shall ignore any solution which is independent of the observations, i.e. an constant solution. The equation (16.2.1) is generally referred to as Likelihood equatio (L.E.). If the observations are iid,

$$f_\theta(x_1,\cdots,x_n) = \prod_{i=1}^{n} f_\theta(x_i)$$

where $f_\theta(x)$ is the common p.d.f. and here

$$\log L(\theta) = \sum_{i=1}^{n} \log f_\theta(x).$$

The method holds even if all the variables $X_1,\cdots,X_n$ are discrete and in this cas the density function is to be replaced by p.m.f.

When more than one parameter are involved, the m.l.e.'s are defined in an analogeou manner. Thus if a sample has the p.d.f. $f_{\boldsymbol{\theta}}(x_1,\cdots,x_n)$ where $\boldsymbol{\theta} = (\theta_1,\cdots,\theta_k) \in \Theta \subset R^k$ this can be regarded as the likelihood of $\boldsymbol{\theta}$ when sample observations $x_1,\cdots,x_n$ are giver The m.l.e.'s are the numbers $\hat{\theta}_1,\cdots,\hat{\theta}_k$, if such a set exists, which maximises f as function of $\boldsymbol{\theta}$. If the likelihood function does not have a maxima on the boundary of se Θ, the maximum of the L.F. is obtained by the solution of

$$\frac{\partial \log L}{\partial\theta_i} = 0,\quad i = 1,\cdots,k$$

subject to the condition that the matrix

$$\left.\left(\left(\frac{\partial^2 \log L}{\partial\theta_i\partial\theta_j}\right)\right)_{i,j=1,\cdots,r}\right]_{\boldsymbol{\theta}=\hat{\boldsymbol{\theta}}}$$

is negative-definite.

☐ **Remark 16.2.1** *If assumptions* (a) *and* (b) *do not hold, the m.l.e. cannot be obtained by solving the likelihood equation* (16.2.1).

▶ **Example 16.2.1** *$X_1, \cdots, X_n$ are iid Bernoulli (p), $p \in [0,1]$. Find m.l.e. of p.*

Likelihood

$$L(p|x_1, \cdots, x_n) = p^{\sum_{i=1}^{n} x_i} (1-p)^{n - \sum_{i=1}^{n} x_i}$$

when $x_i = 1(0)$ if a success (failure) occurs.

$$\log L = \sum_{i=1}^{n} x_i \log p + \left(n - \sum_{i=1}^{n} x_i\right) \log(1-p)$$

$$\frac{d \log L}{dp} = 0 \text{ gives } \hat{p} = \frac{1}{n} \sum_{i=1}^{n} x_i = \bar{x}.$$

Example 16.2.2 *$X_1, \cdots, X_n$ are iid Poisson (λ). Find m.l.e. of λ.*

$$L = e^{-n\lambda} \frac{\lambda^{\sum_{i=1}^{n} x_i}}{\prod_{i=1}^{n} x_i!}; \quad \log L = -n\lambda + \sum_{i=1}^{n} x_i \log \lambda + \text{Const.}$$

$$\frac{\partial \log L}{\partial \lambda} = 0 \text{ gives } \hat{\lambda} = \frac{1}{n} \sum_{i=1}^{n} x_i = \bar{x}.$$

Example 16.2.3 *Let $X_1, \cdots, X_n$ be a random sample from a Binomial (m, p) population with m known, $0 \le p \le 1$. Find m.l.e. of p. Here the likelihood is*

$$L(p|m, x_i, i = 1, \cdots, n) = \prod_{i=1}^{n} \binom{m}{x_i} p^{x_i} (1-p)^{m - x_i}, \qquad x_i = 0, 1, 2, \cdots, m.$$

Assume $\sum_{i=1}^{n} x_i > 0$

$$\log L = \sum_{i=1}^{n} x_i \log p + \left(nm - \sum_{i=1}^{n} x_i\right) \log(1-p) + \text{Const.}$$

$$\frac{d \log L}{dp} = 0 \text{ gives } \frac{\sum_{i=1}^{n} x_i}{p} = \frac{mn - \sum_{i=1}^{n} x_i}{1-p} = 0 \text{ i.e., } \quad \hat{p} = \frac{\sum_{i=1}^{n} x_i}{mn} = \frac{\bar{x}}{m}.$$

To show that $\hat{p}$ actually provides a supremum of the likelihood consider

$$\frac{d^2 \log L}{dp^2} = -\frac{\sum_{i=1}^{n} x_i}{p^2} - \frac{mn - \sum_{i=1}^{n} x_i}{(1-p)^2}$$
$$< 0, \text{ if } 0 < \sum_{i=1}^{n} x_i \le nm.$$

Hence $\hat{p} = \bar{x}/m$ is m.l.e. of p if

$$\sum_{i=1}^{n} x_i > 0.$$

If

$$\sum_{i=1}^{n} x_i = 0, \quad L(p) = (1-p)^{mn}$$

which becomes maximum if $p = 0$.

Thus, whatever be the value of

$$\sum_{i=1}^{n} x_i, \quad \hat{p} = \bar{x}/m \text{ is m.l.e. of } p.$$

By theorem 16.2.3, m.l.e. of pq is $\frac{\bar{x}}{n}\left(1 - \frac{\bar{x}}{n}\right)$.

► **Example 16.2.4** *Let X_1, X_2 be a random sample from a population*

$$f_\theta(x) = \frac{2}{\theta^2}(\theta - x), \quad 0 < x < \theta.$$
$$L = \frac{4}{\theta^4}(\theta - x_1)(\theta - x_2)$$
$$\frac{d \log L}{d\theta} = -\frac{4}{\theta} + \frac{1}{\theta - x_1} + \frac{1}{\theta - x_2} = 0$$
$$\Rightarrow \quad \hat{\theta} = \frac{3(x_1 + x_2) + \sqrt{9(x_1 - x_2)^2 + 4x_1x_2}}{4}.$$

16.2.1 Properties of Maximum Likelihood Solution

Theorem 16.2.1 *If a sufficient statistic t for θ exists, the m.l.e. will be a function of the sufficient statistic.*

Proof: Since t is sufficient, by factorisation theorem

$$f_\theta(\boldsymbol{x}) = \phi_1(t, \theta)\phi_2(\boldsymbol{x})$$

hence

$$\log L(\theta| = \boldsymbol{x}) = \log \phi_1(t,\theta) + \log \phi_2(\boldsymbol{x})$$

$$\left.\frac{d\log L(\theta)}{d\theta}\right]_{\hat\theta} = 0 \text{ gives } \left.\frac{d\log \phi_1(t,\theta)}{d\theta}\right]_{\hat\theta} = 0. \tag{16.2.3}$$

Since $\phi_1(t,\theta)$ involves $x_1, \cdots, x_n$ and equation (16.2.3) is satisfied only for $\theta = \hat\theta$, solution of

$$\frac{d\log \phi_1(t,\theta)}{d\theta} = 0$$

will also involve t. Hence $\hat\theta$ is an explict function of t.

□ **Remark 16.2.2** *Theorem* 16.2.1 *does not state that a m.l.e. will itself be a sufficient statistic although this will be the case in general.*

Theorem 16.2.2 *Suppose that the regularity conditions of Rao-Cramer inequality are satisfied and θ belongs to an open interval Θ in the real line. If a* MVB *estimator t exists (which attains the RC lower bound), the likelihood equation has a unique solution $\hat\theta = t$ that maximises the likelihood.*

Proof: If t be a MVB estimator,

$$\frac{\partial \log f_\theta(x_1, \cdots, x_n)}{\partial\theta} = c(\theta)(t-\theta) \text{ with probability one.} \tag{16.2.4}$$

Also

$$\left.\frac{\partial \log f_\theta(x_1, \cdots, x_n)}{\partial\theta}\right]_{\hat\theta} = 0 \tag{16.2.5}$$

if it is to satisfy the m.l. equation.

From (16.2.4) and (16.2.5) it follows that the likelihood equation has the unique solution $\hat\theta = t$ since $c(\theta)$ being a function of θ cannot be zero.

Again, from (16.2.4)

$$\frac{\partial^2 \log f_\theta}{\partial\theta^2} = c'(\theta)(t-\theta) - c(\theta)$$

$$\left.\frac{\partial^2 \log f_\theta(x)}{\partial\theta^2}\right]_{\hat\theta} = -c(\hat\theta) = -E_\theta\left[\frac{\partial \log f_\theta(x)}{\partial\theta}\right]^2_{\hat\theta} < 0. \text{ [by (15.11.3)]}$$

Hence the solution $\hat\theta = t$ determines the maximum of the likelihood.

□ **Remark 16.2.3** *In theorem* 16.2.2 *we assumed the differentiability of $c(\theta)$ and the existence of second order partial derivative $\partial^2 \log f_\theta(x)/\partial\theta^2$. If the conditions of theorem* 16.2.2 *are satisfied, the most efficient estimator is necessarily m.l.e. However, the converse is not necessarily true, i.e., every m.l.e. is not necessarily a* MVB *estimator.*

For example, in sampling from $N(\mu, \sigma^2)$ population,

$$\hat{\sigma}^2 = \frac{1}{n}\sum_{i=1}^{n}(X_i - \bar{X})^2$$

is the MLE of σ^2; however $\hat{\sigma}^2$ is not a MVBE, because $V(\hat{\sigma}^2) = 2(n-1)\sigma^4/n^2$ which doe not attain the CR-bound $2\sigma^4/n$.

Theorem 16.2.3 (***Invariance Property***). *If $\hat{\theta}$ is the m.l.e. of θ, then $u(\hat{\theta})$ is the m.l.e of $u(\theta)$ provided $u(\theta)$ is some single-valued function of θ.*

Proof: Since $u(\theta)$ is a single valued function of θ, we may write the inverse functio as $\theta = v(u)$.

Now

$$\partial \log f_\theta(x_1, \cdots, x_n)/\partial[u(\theta)] = \frac{\partial \log f_\theta(x_1, \cdots, x_n)}{\partial\theta} \cdot \frac{\partial\theta}{\partial[u(\theta)]} = 0$$

at the m.l.e. $\theta = \hat{\theta}$, since $\partial[u(\theta)]/\partial\theta \neq 0$.

Hence

$$\frac{\partial \log f_\theta(x_1, \cdots, x_n)}{\partial u} = 0$$

at the point $\hat{\theta} = v(\hat{u})$, where $\hat{u}$ is that value of u for which the likelihood function i maximum. Hence $\hat{u} = u(\hat{\theta})$ is the m.l.e. of $u(\theta)$.

■ NOTE 16.2.1 *If $u(\theta)$ is a many-to-one function then also the theorem* 16.2.3 *hold (the proof is omitted).*

Bias of the m.l.e. For a finite population m.l.e. is a biased one. However, sligh modifications may do away with the bias.

Theorem 16.2.4 *The maximum likelihood estimate is unique.*

Proof: If possible, let there be two estimates θ_1^*, θ_2^*. Hence

$$\partial \log f/\partial\theta]_{\theta_1^*} = 0, \quad \partial \log f/\partial\theta]_{\theta_2^*} = 0.$$

Then, by Roll's theorem $\partial^2 \log f/\partial\theta^2 = 0$ at some point in (θ_1^*, θ_2^*) which is quit impossible.

Large Sample Properties

The maximum likelihood solution attempts to find the mode of the likelihood $L(\theta|x$ $\cdots, x_n)$ over variations in values of θ in Θ. In estimation theory, the mode is usuall inferior to either mean or median and so it is often found that the m.l.e. has poor sma sample properties and is often replaced by methods such as minimax method. For large sample, however, the mode tends to mean and median, if they exist and the m.l.e. has many optimum large sample properties. In fact, the resulting estimate is a consistent, asymptotically normal and asymptotically efficient estimate of θ for large sample sizes.

The following are the asymptotically optimum properties of a m.l.e.

Under certain regularity conditions it can be shown that

(i) (Kulldorff, 1957) the likelihood equation has a solution which converges in probability to the true value θ_0;

(ii) a consistent solution of the likelihood equation corresponds to the maximum of the likelihood.

(iii) (Kulldorff, 1957) a consistent solution of the likelihood equation is asymptotically normally distributed about the true value θ and is asymptotically most efficient.

(iv) the likelihood equation has one and only one consistent solution (Huzurbazar, 1948).

(v) the solution of the likelihood equation which corresponds to the absolute maximum of the likelihood is necessarily consistent (Wald, 1949).

All these results are true with probability tending to unity as the sample size tends to infinity.

If, however, the sample values are not independent or the number of parameters is not fixed, regardless of the sample size, then the above classical asymptotic results may not hold (see, for instance, Cox and Hinkley (1974), Chapter 9).

Theorem 16.2.5 *Under certain regularity conditions (as stated below), a consistent solution of the likelihood equation always exists. The conditions are*

(i) *For almost every* x, $\dfrac{\partial \log f_\theta(x)}{\partial \theta}, \dfrac{\partial^2 \log f_\theta(x)}{\partial \theta^2}$ *exist* $\forall\ \theta$ *belonging to an interval* Θ_0 *containing the true value* θ_0 *and are bounded by integrable functions.*

(ii) $\displaystyle\int_{-\infty}^{\infty} \frac{\partial \log f_\theta(x)}{\partial \theta} f_\theta(x)\, dx = 0 \quad \forall\, \theta \in \Theta.$

[This is different from the assumption that the differentiation under the integration sign is permissible].

(iii) $\displaystyle\int_{-\infty}^{\infty} \left.\frac{\partial^2 \log f_\theta(x)}{\partial \theta^2}\right]_{\theta'} f_\theta(x)\, dx < -k^2$ *for all* (θ, θ') *belonging to* Θ_0 *and is independent of* (θ, θ').

Proof: We have

$$\frac{1}{n}\frac{\partial \log L}{\partial \theta} = \frac{1}{n}\frac{\partial \log L}{\partial \theta}\bigg]_{\theta_0} + \frac{1}{n}(\theta - \theta_0)\frac{\partial^2 \log L}{\partial \theta^2}\bigg]_{\theta'} \quad \text{where } \theta' = \theta_0 + \lambda(\theta - \theta_0), 0 < \lambda < 1.$$

Now

$$\frac{1}{n}\frac{\partial \log L}{\partial \theta}\bigg]_{\theta_0} = \frac{1}{n}\sum_{i=1}^{n}\frac{\partial \log f(x_i)}{\partial \theta}\bigg]_{\theta_0} = B_0 \text{ (say)} \xrightarrow{P} 0 \tag{16.2.6}$$

by Khintchen's WLLN and by assumption (ii).

Similarly

$$\frac{1}{n}\frac{\partial^2 \log L}{\partial \theta^2}\bigg]_{\theta'} = \frac{1}{n}\sum_{i=1}^{n}\frac{\partial^2 \log f(x_i)}{\partial \theta}\bigg]_{\theta'} = B_1 \text{ (say)}$$

$$\xrightarrow{P} \int_{-\infty}^{\infty}\frac{\partial^2 \log f(x_i)}{\partial \theta^2}\bigg]_{\theta'} f_\theta(x)\,dx = -k_1^2 \text{ (say)}. \tag{16.2.7}$$

Hence, from (16.2.6) and (16.2.7),

$$P\left\{|B_0| > \delta^2\right\} < \frac{\varepsilon}{2} \text{ for } n > n_0(\delta) \text{ (say)},$$

$$P\left\{B_1 > -k_1^2 + \lambda^2\right\} < \frac{\varepsilon}{2} \text{ for } n > n_1(\lambda) \text{ (say)} \tag{16.2.8}$$

for δ, λ sufficiently small.

Let $n' = \max(n_0, n_1)$. Hence for $n > n'$ both the inequalities in (16.2.8) hold.

Hence for $n > n'$ $P\left\{|B_0| < \delta^2, B_1 < -k_1^2\right\} \geq 1 - \varepsilon$ by Boole's inequality.

Now

$$\frac{1}{n}\frac{\partial \log L}{\partial \theta} = B_0 + (\theta - \theta_0)B_1 = B_0 + B_1\delta \text{ (say)}, \tag{16.2.9}$$

where $\delta = \theta - \theta_0$, a sufficiently small quantity.

Again for sufficiently large n, $|B_0| < \delta^2$ and $B_1 < -k^2$ with probability tending t unity. The sign of the right hand side of (16.2.9) can be made to depend on those c B_1, δ.

For δ positive it is negative (since B_1 is negative) and δ negative it is positive.

Hence we can find a solution $\theta^* \in [\theta_0 - \delta, \theta_0 + \delta]$ of $\frac{\partial \log L}{\partial \theta} = 0$ which tends to the tru value θ_0 (the true value) as $n \to \infty$.

Theorem 16.2.6 *A consistent solution of the likelihood equation is asymptoticall normal and is asymptotically efficient.*

Proof: In addition to the assumptions made in the previous theorem we make th following assumptions.

(iv) $\dfrac{\partial^2 \log f_\theta(x)}{\partial \theta^2}$ is continuous in θ.

(v) $\int f_\theta(x)\,dx$ is differentiable with respect to θ under the integral sign, not only once but twice.

Let θ^* be a consistent solution of the L.E. Then by (16.2.9)

$$B_0 + (\theta^* - \theta_0)B_1 = 0. \tag{16.2.10}$$

Now

$$B_1 = \frac{1}{n}\sum \frac{\partial^2 \log f}{\partial \theta^2}\bigg]_{\theta'} = \left\{\frac{1}{n}\sum \frac{\partial^2 \log f}{\partial \theta^2}\bigg]_{\theta'} - \frac{1}{n}\sum \frac{\partial^2 \log f}{\partial \theta^2}\bigg]_{\theta_0}\right\} + \frac{1}{n}\sum \frac{\partial^2 \log f}{\partial \theta^2}\bigg]_{\theta_0}. \tag{16.2.11}$$

For n sufficiently large we can take $|\theta' - \theta_0| < \eta$ where η is a pre-assigned small quantity and therefore by continuity assumption (iv)

$$\left|\left\{\frac{1}{n}\sum \frac{\partial^2 \log f}{\partial \theta^2}\bigg]_{\theta'} - \frac{1}{n}\sum \frac{\partial^2 \log f}{\partial \theta^2}\bigg]_{\theta_0}\right\}\right| < \eta'. \tag{16.2.12}$$

Therefore, for sufficiently large n,

$$B_1 \to \frac{1}{n}\sum \frac{\partial^2 \log f}{\partial \theta^2}\bigg]_{\theta_0}. \tag{16.2.13}$$

Therefore if $\frac{\partial \log L}{\partial \theta}\Big]_{\theta^*} = 0$, then for sufficiently large n, it follows from (16.2.10) and (16.2.13) that

$$\theta^* - \theta_0 = \frac{-\frac{1}{n}\sum_1^n \frac{\partial \log f(x_i)}{\partial \theta}\Big]_{\theta_0}}{B_1}. \tag{16.2.14}$$

We now define,

$$E\left(\frac{\partial^2 \log f}{\partial \theta^2}\right)_{\theta_0} = \int_{-\infty}^{\infty}\left[\frac{1}{f}\frac{\partial^2 f}{\partial \theta^2} - \left(\frac{1}{f}\frac{\partial f}{\partial \theta}\right)^2\right]_{\theta_0} f_\theta(x)\,dx = -E\left(\frac{\partial \log f}{\partial \theta}\right)^2\bigg]_{\theta_0} = -k_0^2 \text{ (say)} \tag{16.2.15}$$

(16.2.14) can be written as

$$\sqrt{n}k_0(\theta^* - \theta_0) = \frac{\frac{\sqrt{n}}{k_0}\frac{1}{n}\sum \frac{\partial \log f_\theta(x_i)}{\partial \theta}\Big]_{\theta_0}}{-B_1/k_0^2}. \tag{16.2.16}$$

Clearly $B_1 \xrightarrow{P} -k_0^2$ by Khintchen's WLLN. Therefore, denominator of (16.2.16) converges in probability to 1. Again by Lindeberg and Levy CLT.

$$\frac{1}{n}\sum_{i=1}^{n} \frac{\partial \log f_\theta(x_i)}{\partial \theta}\bigg]_{\theta_0}.$$

is asymptotically normal with mean 0 and s.d. $k_0/\sqrt{n}$ and consequently

$$\frac{1}{k_0\sqrt{n}}\sum \frac{\partial \log f_\theta(x_i)}{\partial\theta}\Bigg]_{\theta_0}$$

is asymptotically $N(0,1)$. Hence by *Slutsky's theorem* (Cramer, 1946), $\sqrt{n}k_0(\theta^*-\theta_0) \sim N(0,1)$ i.e.,

$$\theta^* \sim N\left(\theta_0, \frac{1}{nk_0^2}\right).$$

We have seen in section 16.11 that the MVB estimator in the class of all regular estimators has the variance

$$\frac{1}{nE\left(\frac{\partial \log f}{\partial\theta}\right)^2\Big]_{\theta_0}}$$

and an estimator which attains this bound asymptotically is asymptotically most efficient. A consistent solution θ^* of the likelihood equation is therefore, asymptotically most efficient.

■ NOTE 16.2.2 *Sometimes the conditions of theorem* 16.2.6 *are not satisfied, till the m.l.e. becomes consistent and asymptotically normal.*

▶ **Example 16.2.5** (***Kulldorff*** **1957**). $\{X_i, i = 1, \cdots, n\}$ *is a random sample from* $N(0,\sigma^2)$. *Here m.l.e.* $\hat\sigma^2 = \frac{1}{n}\Sigma X_i^2$, $E(X^2)=\sigma^2, V(X^2) = 2\sigma^4$; *also* $E_{\sigma^2}\left\{\frac{\partial \log f_{\sigma^2}(X)}{\partial\sigma^2}\right\}^2 = \frac{1}{2\sigma^4}$.

By Central Limit Theorem,

$$\sqrt{n}(\hat\sigma^2-\sigma^2) \xrightarrow{L} N(0, 2\sigma^4).$$

Now

$$\frac{\partial^3 \log f_{\sigma^2}}{\partial^3(\sigma^2)} = -\frac{1}{(\sigma^2)^3} + \frac{3x^2}{(\sigma^2)^4} \to \infty \text{ as } \sigma^2 \to 0.$$

and is not bounded in $0 < \sigma^2 < \infty$. Hence the condition (iv) is not satisfied.

Kulldorff (1957) proved theorems 16.2.5 and 16.2.6 without using condition (iv).

Theorem 16.2.7 *A consistent solution of the likelihood equation is unique.*

Proof: If possible, suppose likelihood equation has more than one consistent solution and let θ_1^*, θ_2^* be two such solutions. It follows then from Roll's theorem that the equation $\frac{d^2L}{d\theta^2} = 0$ has at least one solution, say θ_3^* lying between θ_1^* and θ_2^*. Since both θ_1^*, θ_2^*

converge in probability to $\theta_0, \theta_3^* \xrightarrow{P} \theta_0$, i.e., θ_3^* is a consistent estimator. Hence by corollary 16.2.1, we have

$$P\left\{\left.\frac{d^2L}{d\theta^2}\right]_{\theta_3^*} < 0\right\} \to 1 \text{ as } n \to \infty.$$

This is a contradiction with the finding $\left.\frac{d^2L}{d\theta^2}\right]_{\theta_3^*} = 0$.

Hence the L.E. has a unique consistent solution.

Theorem 16.2.8 *Under regularity conditions of theorem* 16.2.6 *a consistent solution of the likelihood equation corresponds to the maximum of the likelihood.*

Let θ^* be any consistent solution of the L.E. By the Mean Value Theorem

$$\left.\frac{1}{n}\frac{\partial^2 L}{\partial\theta^2}\right]_{\theta^*} - \left.\frac{1}{n}\frac{\partial^2 L}{\partial\theta^2}\right]_{\theta_0} = \left.\frac{1}{n}\frac{\partial^3 L}{\partial\theta^3}\right]_{\theta'} (\theta^* - \theta_0) \tag{16.2.17}$$

where θ' lies between θ^* and θ_0.

Now

$$\left.\frac{1}{n}\frac{\partial^2 L}{\partial\theta^2}\right]_{\theta_0} = \frac{1}{n}\sum_{i=1}^{n}\left.\frac{\partial^2 \log f_\theta(x_i)}{\partial\theta^2}\right]_{\theta_0} \xrightarrow{P} E\left(\frac{\partial^2 \log f_\theta(X)}{\partial\theta^2}\right)_{\theta_0} = -I_{\theta_0}(\boldsymbol{X})$$

by Khintchen's WLLN.

Again

$$\left|\frac{1}{n}\frac{\partial^3 L}{\partial\theta^3}\right| \leq \frac{1}{n}\sum_{i=1}^{n}\left|\frac{\partial^3}{\partial\theta^3}\log f_\theta(x_i)\right| < \frac{1}{n}\sum_{i=1}^{n} M(x_i)$$

where it is assumed $\left|\frac{\partial^3 \log f_\theta(x_i)}{\partial\theta^3}\right| < M(x_i)$.

By Khintchen's WLLN,

$$\frac{1}{n}\sum_{i=1}^{n} M(x_i) \xrightarrow{P} E(M(X)).$$

Hence if B is some number $> E(M(X))$,

$$P\left\{\frac{1}{n}\sum_{i=1}^{n} M(x_i) < B\right\} \to 1 \text{ as } n \to \infty$$

and consequently

$$P\left\{\frac{1}{n}\frac{\partial^3 L}{\partial\theta^3} < B\right\} \to 1 \text{ as } n \to \infty, \text{ whatever } \theta.$$

Hence from (16.2.17),

$$\frac{1}{n}\frac{\partial^2 L}{\partial \theta^2}\bigg]_{\theta^*} \xrightarrow{P} -I_{\theta_0} \text{ as } n \to \infty.$$

We have, therefore, for any arbitrarily small positive number ε,

$$P\left\{\frac{1}{n}\frac{\partial^2 L}{\partial \theta^2}\bigg]_{\theta^*} < -I_{\theta_0} + \varepsilon\right\} \to 1 \text{ as } n \to \infty. \tag{16.2.18}$$

Since $-I_{\theta_0}$ is a negative quantity ($I(\theta)$ being positive $\forall\,\theta$), (16.2.18) means

$$P\left\{\frac{1}{n}\frac{\partial^2 L}{\partial \theta^2}\bigg]_{\theta^*} < 0\right\} \to 1 \text{ as } n \to \infty. \tag{16.2.19}$$

Therefore, the probability that L is maximum at θ^*, approaches unity as $n \to \infty$.

Corollary 16.2.1 *If θ^* is any consistent solution (not necessarily the solution of the* L.E.*), then*

$$P\left\{\frac{1}{n}\frac{\partial^2 L}{\partial \theta^2}\bigg]_{\theta^*} < 0\right\} \to 1 \textit{ as } n \to \infty.$$

► **Example 16.2.6** *Consider* $f_\theta(x) = 1 + \frac{2x-1}{\sigma}\cos\theta + \frac{3x^2-1}{\sigma}\sin\theta \quad (0 \le x \le 1;\ 0 < \theta \le 100\pi)$.

The distribution satisfies all the regularity conditions for the existence of a consistent solution.

Since the L.F. is continuous in θ and is periodic with period 2π, the L.E. has at least one solution between $2\pi r$ and $2\pi(r+1), r = 0, 1, \cdots, 49$. Hence there are at least 50 solutions of the L.E.

Of these solutions only one solution is consistent, namely, the one which tends to the true value θ_0 (of θ) as $n \to \infty$. Others are inconsistent, because if $\theta^* \xrightarrow{P} \theta_0$, then $\theta^* + 2\pi r \xrightarrow{P} \theta_0 + 2\pi r$. Therefore, $\theta^* + 2\pi r$ is inconsistent if $r \neq 0$.

Obviously the likelihood attains maximum at some of these inconsistent solutions.

For the class of distributions for which there exists a unique solution of the likelihood, the solution must be consistent because a consistent solution of the likelihood always exists, provided some regularity conditions are satisfied (Theorem 16.2.7). Hence for such distributions the likelihood equation has a unique maxima (hence also an absolute maxima).

16.2.2 MLE of One-Parameter Exponential Family

Consider the one-parameter exponential family

$$\varepsilon_\theta = \left\{ f_\theta(x) = \exp\{C(\theta) + Q(\theta)t(x) + h(x)\}, \theta \in \Theta \subset R^1 \right\}. \tag{16.2.20}$$

Assume:

(i) the first two derivatives of $C(\theta)$ and $Q(\theta)$ exist and are continuous;

(ii) $I_\theta(X) = E_\Theta \left[\frac{\partial \log L}{\partial \theta}\right]^2$ exists and is positive.

Here

$$\frac{\partial \log L}{\partial \theta} = 0 \Rightarrow C'(\theta) + Q'(\theta)\sum_{i=1}^{n} t(x) = 0 \Rightarrow \frac{1}{n}\sum_{i=1}^{n} t(x) = -\frac{C'(\theta)}{Q'(\theta)}. \tag{16.2.21}$$

The following results hold under assumptions (i) and (ii) above.

Theorem 16.2.9 *There exists a solution of the* L.F. *iff* $t(X)$ *and* $-\frac{C'(\theta)}{Q'(\theta)}$ *have the same range of values.*

If part. Suppose $t(X) \in (\alpha, \beta)$. Then

$$\frac{1}{n}\sum_{i=1}^{n} t(x_i) \in (\alpha, \beta).$$

For any value of t, say, t_0 we can always find a value of θ, say, θ_0 for which $t_0(x) = -\frac{C'(\theta_0)}{Q'(\theta_0)}$. Hence θ_0 is the m.l.e. Here, obviously

$$-\frac{C'(\theta)}{Q'(\theta)} \in (\alpha, \beta).$$

Only if part. Suppose there exists a solution of the L.E. Suppose $t(x) \in (\alpha, \beta)$. Hence for $\theta = \hat{\theta}$,

$$\frac{\partial \log f}{\partial \theta} = C'(\theta) + Q'(\theta)t(X) = 0. \tag{16.2.22}$$

Taking expectation of both sides

$$E_\theta[t(X)] = -\frac{C'(\theta)}{Q'(Q)} \tag{16.2.23}$$

where $C'(\theta), Q'(\theta)$ are functions of the solutions θ of the L.E.
From (16.2.22)

$$t(X) = -\frac{C'(\hat{\theta})}{Q'(\hat{Q})} \tag{16.2.24}$$

where $\hat{\theta}$ is the solution of L.E.

Since $t(X) \in (\alpha, \beta)$, it is clear from (16.2.23) and (16.2.24) that $-C'(\theta)/Q'(\theta) \in (\alpha, \beta)$.

Theorem 16.2.10 *If $f \in \varepsilon_\theta$, any solution of the* L.F., *if it exists, gives the unique m.l.e.*

Proof: (a) Any solution of the L.E. provides a maximum of the L.F.
Any solution $\hat{\theta}$ of L.E. satisfies (16.2.22) and hence $t(X)$ satisfies (16.2.24).
It is enough to show that $\left.\frac{\partial^2 \log L}{\partial \theta^2}\right]_{\hat{\theta}} < 0$. Now

$$\left.\frac{\partial^2 \log L}{\partial \theta^2}\right]_{\theta=\hat{\theta}} = C''(\hat{\theta}) + Q''(\hat{\theta})t(\boldsymbol{X}) = C''(\hat{\theta}) - Q''(\hat{\theta})\frac{C'(\hat{\theta})}{Q'(\hat{\theta})}. \tag{16.2.25}$$

Now

$$\begin{aligned} -I_\theta(\boldsymbol{X}) = E_\theta\left(\frac{\partial^2 \log L}{\partial \theta^2}\right) &= E_\theta[C''(\theta) + Q''(\theta)t(\boldsymbol{X})] \\ &= C''(\theta) - Q''(\theta)\frac{C'(\theta)}{Q'(\theta)}. \end{aligned} \tag{16.2.26}$$

Hence from (16.2.25)

$$\left.\frac{\partial^2 \log L}{\partial \theta^2}\right]_{\theta=\hat{\theta}} = \left.-I_\theta(\boldsymbol{X})\right]_{\theta=\hat{\theta}} < 0,$$

since $I_\theta(\boldsymbol{X})$, the information contained in $\boldsymbol{X}$ about θ is > 0.

(b) Solution of the L.F., if it exists, is unique.

If possible, let there be two solutions $\hat{\theta}_1, \hat{\theta}_2$. Then by (a), both $\hat{\theta}_1, \hat{\theta}_2$ provide local maxima of the L.F. Hence there exists another solution of the L.E., $\hat{\theta}_3 \in (\hat{\theta}_1, \hat{\theta}_2)$ at which the L.F. is maximum. This is impossible by (a). Hence the solution if it exists is unique.

(a) and (b) together proves the theorem.

□ **Remark 16.2.4** *When $f \in \varepsilon_\theta, t(x) = -\frac{C'(\hat{\theta})}{Q'(\hat{\theta})}$. Hence there exists a $1 : 1$ correspondence between the m.l.e. and the sufficient statistic. Again $t(X)$ is a complete sufficient statistic. Hence* MVUE *of θ can be obtained by correcting $\hat{\theta}$ for bias.*

□ **Remark 16.2.5** *When $f_\theta \in \varepsilon_\theta$,*

$$E_\theta\left(\frac{\partial^2 \log L}{\partial \theta^2}\right) = \left.\frac{\partial^2 \log L}{\partial \theta^2}\right]_{\hat{\theta}=\theta}.$$

As was seen in theorem 16.2.6, the large-sample variance of m.l.e. is given by $-1/E\left(\frac{\partial^2 \log L}{\partial \theta^2}\right)$. When $f_\theta \in \varepsilon_\theta$,

$$V(\hat{\theta}) = -1/\left.\frac{\partial^2 \log L}{\partial \theta^2}\right]_{\hat{\theta}=\theta}.$$

16.2.3 BAN-property of the MLE when θ is Vector-Valued

Let $X_1, \cdots, X_n$ be iid random variables with common p.d.f. $f_\theta(x), \theta = (\theta_1, \cdots, \theta_q) \subset \Theta \in R^q$, when Θ is an open subset of R^q. We assume that

(i) all the regularity conditions to ensure the consistency of m.l.e. hold;

(ii) $\dfrac{\partial \log f}{\partial \theta_i}, \dfrac{\partial^2 \log f}{\partial \theta_i \partial \theta_j}$ exist and are continuous for almost all x and $\forall\, \theta$;

(iii) $J_{ij}(\theta) = E_\theta \left[\dfrac{\partial \log f_\theta(X)}{\partial \theta_i} \cdot \dfrac{\partial \log f_\theta(X)}{\partial \theta_j} \right] = -E_\theta \left[\dfrac{\partial^2 \log f_\theta(X)}{\partial \theta_i \partial \theta_j} \right]$

and $E_\theta \left[\dfrac{\partial \log f_\theta(X)}{\partial \theta_i} \right] = 0 \ \forall\, \theta$;

(iv) $\boldsymbol{J}(\theta) = ((J_{ij}(\theta))) = n\boldsymbol{j}(\theta)$ exists and is p.d. $\forall\, \theta$.

Theorem 16.2.11 *Under the regularity conditions (i)—(iv) above the estimator $\hat{\theta}$ of θ is* BAN *estimator whenever it is obtained through* L.E.

Proof: Let the true value of θ be $\theta_0 = (\theta_{10}, \cdots, \theta_{q0})$.
Then by Taylor expansion

$$\left.\frac{\partial \log L}{\partial \theta_i}\right]_\theta = \left.\frac{\partial \log L}{\partial \theta_i}\right]_{\theta_0} + \sum_{j=1}^{q} (\theta_j - \theta_{j0}) \left.\frac{\partial^2 \log L}{\partial \theta_i \partial \theta_j}\right]_{\theta_1}$$

where $\theta_1 = \theta_0 + \lambda(\theta - \theta_0), \lambda < 1$.

Let $\hat{\theta} = (\hat{\theta}_1, \cdots, \hat{\theta}_q)$ be the m.l.e. of θ. Then

$$\left.\frac{\partial \log L}{\partial \theta_i}\right]_{\theta=\hat{\theta}} = \left.\frac{\partial \log L}{\partial \theta_i}\right]_{\theta_0} + \sum_{j=1}^{q} (\hat{\theta}_j - \theta_{j0}) \left.\frac{\partial^2 \log L}{\partial \theta_i \partial \theta_j}\right]_{\theta_1} = 0 \tag{16.2.27}$$

$$\Rightarrow \sum_{j=1}^{q} (\hat{\theta}_j - \theta_{j0}) \left.\frac{\partial^2 \log L}{\partial \theta_i \partial \theta_j}\right]_{\theta_1} = - \left.\frac{\partial \log L}{\partial \theta_i}\right]_{\boldsymbol{\theta}_0}, i = 1, \cdots, q. \tag{16.2.28}$$

Writing

$$\boldsymbol{A}_n(\theta) = \left(\left(\frac{1}{n} \frac{\partial^2 \log L}{\partial \theta_i \partial \theta_j} \right) \right)$$

$$\boldsymbol{b}_n(\theta) = \frac{1}{\sqrt{n}} \left(\frac{\partial \log L}{\partial \theta_1}, \cdots, \frac{\partial \log L}{\partial \theta_q} \right)'$$

(16.2.28) means

$$\sqrt{n}\boldsymbol{A}_n(\theta_1)(\hat{\boldsymbol{\theta}} - \boldsymbol{\theta}_0) = -\boldsymbol{b}_n(\theta_0). \tag{16.2.29}$$

Now

$$\left.\frac{1}{n}\frac{\partial^2 \log L}{\partial\theta_i \partial\theta_j}\right]_{\theta_0} = \left.\frac{1}{n}\sum_{k=1}^{n}\frac{\partial^2 \log f_\theta(x_k)}{\partial\theta_i\partial\theta_j}\right]_{\theta_0} \xrightarrow{P} E\left(\frac{\partial^2 \log f_\theta(X)}{\partial\theta_i\partial\theta_j}\right)_{\theta_0} = -j_{ij}(\theta_0). \quad (16.2.30)$$

Since $\hat{\theta}$ is the m.l.e. of $\theta, \hat{\theta} \xrightarrow{P} \theta_0$.
Again θ_1 lies on the line segment joining θ_0 and $\hat{\theta}$ implies

$$\theta_1 \xrightarrow{P} \theta_0. \quad (16.2.31)$$

Hence from (16.2.30),

$$\left.\frac{1}{n}\frac{\partial^2 \log L}{\partial\theta_i\partial\theta_j}\right]_{\theta_1} \xrightarrow{P} -j_{ij}(\theta_0) \text{ i.e., } \boldsymbol{A}_n(\theta_1) \xrightarrow{P} -\boldsymbol{j}(\theta_0). \quad (16.2.32)$$

Again,

$$\boldsymbol{b}_n(\theta_0) = \sqrt{n}\begin{bmatrix} \frac{1}{n}\sum \frac{\partial \log f_\theta(x_i)}{\partial\theta_1} \\ \vdots \\ \frac{1}{n}\sum \frac{\partial \log f_\theta(x_i)}{\partial\theta_q}\end{bmatrix}_{\theta_0} = \frac{\sqrt{n}}{n}\Sigma \boldsymbol{y}_i \text{ (say)}$$

where

$$\boldsymbol{y}_i = \left(\frac{\partial \log f_\theta(x_i)}{\partial\theta_1}, \cdots, \frac{\partial \log f_\theta(x_i)}{\partial\theta_q}\right)'_{\theta_0}.$$

$\boldsymbol{y}_i$'s are iid with $E_\theta(\boldsymbol{y}_i), = \boldsymbol{0}$ and dispersion matrix of $\boldsymbol{y}_i$ is

$$D_{\theta_0}(\boldsymbol{y}_i) = \left(\left(E_\theta\left[\frac{\partial \log f_\theta(x_i)}{\partial\theta_j}\frac{\partial \log f_\theta(x_i)}{\partial\theta_k}\right]\right)\right) = \boldsymbol{j}(\theta_0).$$

Hence by the multivariate central limit theorem,

$$\sqrt{n}\bar{\boldsymbol{y}} \to \boldsymbol{Z} \sim N_q(0, \boldsymbol{j}(\theta_0))$$
$$\Rightarrow \boldsymbol{b}_n(\theta_0) \xrightarrow{L} N_q(0, \boldsymbol{j}(\theta_0)).$$

Since $\boldsymbol{j}(\theta_0)$ is assumed to be p.d. for large n, $\boldsymbol{A}_n(\theta_1)$ can be taken as non-singular. From (16.2.29), $\sqrt{n}(\hat{\theta} - \theta_0) = -\boldsymbol{A}_n^{-1}(\theta_1)\boldsymbol{b}_n(\theta_0)$. Also $-\boldsymbol{A}_n^{-1}(\theta_1) \xrightarrow{P} \boldsymbol{j}^{-1}(\theta_0)$ (by 16.2.32). Hence $-\boldsymbol{A}_n^{-1}(\theta_1)\boldsymbol{b}_n(\theta_0) \xrightarrow{L} N_q\left(0, \boldsymbol{j}^{-1}(\theta_0)\right)$. Hence the asymptotic variance of $\hat{\theta}$ is $\boldsymbol{j}^{-1}(\theta_0)/n = \boldsymbol{J}^{-1}(\theta_0)$. Hence $\hat{\theta}$ is BAN-estimator.

Corollary 16.2.2 *If we are to estimate* $\boldsymbol{g}(\theta) = (g_1(\theta), \cdots, g_r(\theta))'$, *let*

$$\wedge_{r\times q} = \left(\left(\frac{\partial g_i(\theta)}{\partial\theta_j}\right)\right)_{\substack{i=1,\cdots,r \\ j=1,\cdots,q}} \quad \textit{exist.}$$

Since $\sqrt{n}(\hat{\theta}-\theta) \xrightarrow{L} N_q\left(0, j^{-1}(\theta)\right)$

$$\sqrt{n}\begin{bmatrix} g_1(\hat{\theta})-g_1(\theta) \\ \vdots \\ g_r(\hat{\theta})-g_r(\theta) \end{bmatrix} \xrightarrow{L} N_r\left(0, \wedge j^{-1}(\theta)\wedge'\right).$$

The asymptotic variance of $g(\hat{\theta})$ is

$$\frac{\wedge j(\theta)\wedge'}{n}.$$

$g(\hat{\theta})$ is also BAN.

16.2.4 Maximum Likelihood Estimation in case of Multiparameter Exponential Family

Consider the multiparameter exponential family of p.d.f.'s $\varepsilon_\theta = \{f_\theta, \theta = (\theta_1, \cdots, \theta_q) \in \Theta \subset R^q\}$ when

$$f_\theta = \exp\left\{C(\theta) + \sum_{j=1}^{q} Q_j(\theta)t_j(x) + h(x)\right\}.$$

We assume that (i) the first two derivatives of $Q_j(\theta), j = 1, \cdots, k$ and $C(\theta)$ exist and are continuous (ii) the matrix $\boldsymbol{J}(\theta) = ((J_{ij}(\theta)))$ where

$$J_{ij}(\theta) = -E_\theta\left[\frac{\partial^2 \log L}{\partial\theta_i\partial\theta_j}\right]$$

exists and is p.d.

Here the L.E. is given by

$$\frac{\partial C(\theta)}{\partial\theta_j} + \sum_{j=1}^{q} \frac{\partial Q_j(\theta)}{\partial\theta_j} t_j(x) = 0, \quad j = 1, \cdots, q. \tag{16.2.33}$$

Theorem 16.2.12 *If $f \in \varepsilon_\theta$, any solution of the* L.E. *provides a maximum of the likelihood.*

Proof: Let $\hat{\theta} = (\hat{\theta}_1, \cdots, \hat{\theta}_q)$ be a solution of the L.E.

Then (16.2.33) holds for $\theta = \hat{\theta}$.

It is enough to show that $\left(\left(\frac{\partial^2 \log L}{\partial\theta_j\partial\theta_s}\right)\right)_{\hat{\theta}}$ is negative definite.

$$\left.\frac{\partial^2 \log L}{\partial\theta_j\partial\theta_s}\right]_{\hat{\theta}} = \left.\frac{\partial^2 C(\theta)}{\partial\theta_j\partial\theta_s}\right]_{\hat{\theta}} + \sum_{j=1}^{k} \left.\frac{\partial^2 Q_j(\theta)}{\partial\theta_j\partial\theta_s}\right]_{\hat{\theta}} t_j(x). \tag{16.2.34}$$

Again, taking expectation over (16.2.33) for m.l.e. of θ,

$$\frac{\partial C(\theta)}{\partial \theta_j} + \sum_{j=1}^{k} \frac{\partial \theta_j(\theta)}{\partial \theta_j} E_\theta[t_j(X)] = 0. \tag{16.2.35}$$

Now

$$-J_{js}(\theta) = E_\theta\left(\frac{\partial^2 \log L}{\partial \theta_j \partial \theta_s}\right) = \frac{\partial^2 C(\theta)}{\partial \theta_j \partial \theta_s} + \sum_{j=1}^{k} \frac{\partial^2 Q_j(\theta)}{\partial \theta_j \partial \theta_s} E_\theta(t_j(X)). \tag{16.2.36}$$

By eliminating $E_\theta[t_j(X)]$ from (16.2.36) using (16.2.35) and replacing θ by $\hat{\theta}$ one get value of the right hand side of (16.2.36) as is obtained from (16.2.34) by eliminating $t_j(x$ using (16.2.33). Hence

$$-J_{js}(\hat{\theta}) = \left.\frac{\partial^2 \log L}{\partial \theta_j \partial \theta_s}\right]_{\theta=\hat{\theta}}.$$

Since $\boldsymbol{J}$ is p.d., the result follows.

16.2.5 Successive approximation to maximum likelihood estimators

Sometimes the likelihood equation is very complicated and its explicit algebraic solutio is not available. It is to be solved numerically in that case by the method of iteration using as a starting value the observed value of some consistent (but inefficient) estimato which can be easily computed. In large samples, such an estimator will tend to be fairl close to the m.l.e. and higher is its efficiency, the greater is the closeness.

Let t' be an inefficient estimator with efficiency e and t a most efficient estimator Let $\delta = t' - t$. Then

$$\begin{aligned} V_\theta(\delta) &= V_\theta(t') + V_\theta(t) - 2\,\text{Cov}_\theta(t', t) \\ &= \frac{1-e}{e} V_\theta(t) \quad (\text{since } \rho(t, t') = \sqrt{e}). \end{aligned}$$

If t' is nearly most efficient, $1 - e$ will be small, and $V(\delta)$ will also be small. Henc δ is expected to be small.

If t is the m.l.e. of θ, $\left.\frac{\partial \log L}{\partial \theta}\right]_{\theta=t} = 0$.

Now

$$\left.\frac{\partial \log L}{\partial \theta}\right]_{\theta=t'} = \left.\frac{\partial \log L}{\partial \theta}\right]_{\theta=t} + (t'-t)\left.\frac{\partial^2 \log L}{\partial \theta^2}\right]_{\theta=t} + \begin{array}{l}\text{terms of higher order in } (t'-t) \\ \text{[which are negligible} \\ \text{since } \delta \text{ is small]}\end{array}$$

$$= (t'-t)\left.\frac{\partial^2 \log L}{\partial \theta^2}\right]_{\theta=t}. \tag{16.2.37}$$

For large n,

$$V_\theta(t) \simeq -\frac{1}{\left.\frac{\partial^2 \log L}{\partial\theta^2}\right]_{\theta=t}}$$

hence,

$$\left.\frac{\partial \log L}{\partial\theta}\right]_{\theta=t'} = \frac{t-t'}{V_\theta(t)};$$

therefore, we take

$$t'' = t' + V_\theta(t)\left[\frac{\partial \log L}{\partial\theta}\right]_{\theta=t'} \tag{16.2.38}$$

as an efficient estimator of t to a better order of approximation. In (16.2.37), if necessary, $V_\theta(t)$ is to be estimated by replacing θ by t'.

The process may be repeated and further approximations are:

$$t''' = t'' + V_\theta(t)\left.\frac{\partial \log L}{\partial\theta}\right]_{\theta=t''},$$

$$t^{(iv)} = t''' + V_\theta(t)\left.\frac{\partial \log L}{\partial\theta}\right]_{\theta=t'''}.$$

▶ **Example 16.2.7** *Suppose we want to find the m.l.e of θ in the Cauchy population*

$$f_\theta(x) = \frac{1}{\pi[1+(x-\theta)^2]}, \quad -\infty < x < \infty.$$

The sample mean is not consistent for θ but the sample median t' is and its large sample-variance is $\pi^2/4n$. The likelihood equation is

$$\frac{\partial \log L}{\partial\theta} = -2\sum_i \frac{(x_i-\theta)}{1+(x_i-\theta)^2} = 0.$$

This equation is difficult to solve. Now the large-sample variance of the m.l.e. $\hat\theta$ is

$$-\left[nE\left(\frac{\partial^2 \log f}{\partial\theta^2}\right)\right]^{-1}.$$

Here

$$\int_{-\infty}^{\infty}\frac{\partial^2 \log f}{\partial\theta^2} f\,dx = \frac{1}{\pi}\int_{-\infty}^{\infty}\frac{2(x-\theta)^2-2}{\{1+(x-\theta)^2\}^3}\,dx = \frac{4}{\pi}\int_0^{\infty}\frac{x^2-1}{(1+x^2)^3}\,dx = -\frac{1}{2}$$

so that

$$V_\theta(\hat\theta) = \frac{2}{n}.$$

Consider therefore,

$$t'' = t' + V_\theta(\hat\theta)\left(\frac{\partial \log L}{\partial\theta}\right)_{\theta=t'} = t' - \frac{4}{n}\sum_{i=1}^{n}\frac{(x_i-t')}{1+(x_i-t')^2}$$

which is an improved estimator over t'.

▶ **Example** 16.2.8 (***Kendall and Stuart* II 1961**). *Consider a multinomial distribution with probabilities* $p_1 = (2+\theta)/4$, $p_2 = p_3 = (1-\theta)/4$, $p_4 = \theta/4$ *(where the parameter* $\theta \in [0,1]$*) and the respective frequencies* n_1, n_2, n_3, n_4 $(\Sigma n_i = n)$.

The likelihood

$$L(\theta|n_1, \cdots, n_4) \propto (2+\theta)^{n_1}(1-\theta)^{n_2+n_3}\theta^{n_4}$$

$$\frac{\partial \log L}{\partial \theta} = 0 \Rightarrow n\theta^2 + \{2(n_2+n_3) + n_4 - n_1\}\theta - 2n_4 = 0. \tag{i}$$

Its solution gives $\hat{\theta}$.

$$\operatorname{Var}\hat{\theta} \sim -\frac{1}{E\left(\frac{\partial^2 \log L}{\partial \theta^2}\right)} = \frac{2\theta(1-\theta)(2+\theta)}{n(1+2\theta)}. \tag{ii}$$

Suppose we wish to find $\hat{\theta}$ iteratively. A simple inefficient estimator to start with is

$$t = \frac{n_1 + n_4 - (n_2+n_3)}{n}.$$

t is consistent and $V(t) = \frac{(1-\theta^2)}{n}$. Efficiency of t is

$$\frac{2\theta(1-\theta)(2+\theta)}{(1-\theta^2)(1+2\theta)} = \frac{2\theta(2+\theta)}{(1+\theta)(1+2\theta)}.$$

Suppose $n_1 = 1997$, $n_2 = 906$, $n_3 = 904$, $n_4 = 32$ (Fisher 1959, Ch. 9). Then $t = 0.0570$.

$$\left(\frac{\partial \log L}{\partial \theta}\right)_{\theta=t} = -387.1713,$$

$$\left.\operatorname{Var}\hat{\theta}\right]_{\theta=t} = 0.0000517 \text{ [from (ii)]}.$$

First iteration gives, from (16.2.38),

$$\hat{\theta}_1 = 0.0370.$$

For second iteration,

$$\left(\frac{\partial \log L}{\partial \theta}\right)_{\theta=\hat{\theta}_1} = -34.315,$$

$$\left.\operatorname{Var}\hat{\theta}\right]_{\theta=\hat{\theta}_1} = 0.0000352 \text{ and hence } \hat{\theta}_2 = 0.0358.$$

The value of m.l.e., $\hat{\theta}$, obtained by solving (i), for the above data is 0.0357.

■ NOTE 16.2.3 *Other common iterations for finding m.l.e. by successively numerical approximations are:*

$$\text{(a) } t'' = t' - \left[\left(\frac{\partial^2}{\partial\theta^2}\log L\right)^{-1}\right]_{\theta=t'}\left(\frac{\partial}{\partial\theta}\log L\right)_{\theta=t'};$$

$$\text{(b) } t'' = t - \frac{k}{n}\left(\frac{\partial}{\partial\theta}\log L\right)_{\theta=t'} \quad \text{for some suitable } k.$$

Kale (1961) investigated the convergence of several iterative processes.

▶ **Example 16.2.9** *Let $X_1, \cdots, X_n$ be iid $N(\mu, \sigma^2)$. Find m.l.e. of μ and σ^2 when both are unknown.*

$$L(\mu, \sigma^2|x_1, \cdots, x_n) = (2\pi\sigma^2)^{-n/2}\exp\left[-\sum_{i=1}^{n}\frac{(x_i-\mu)^2}{2\sigma^2}\right]$$

$$\log L = -\sum_{i=1}^{n}\frac{(x_i-\mu)^2}{2\sigma^2} - \frac{n}{2}\log\sigma^2 - \frac{n}{2}\log 2\pi$$

$$\frac{\partial \log L}{\partial\mu} = \frac{1}{\sigma^2}\sum_{i=1}^{n}(x_i-\mu) = 0$$

$$\frac{\partial \log L}{\partial\sigma^2} = \sum_{i=1}^{n}\frac{(x_i-\mu)^2}{2\sigma^4} - \frac{n}{2\sigma^2} = 0.$$

Solving these equations, we obtain maximum-likelihood estimates

$$\mu^* = \frac{1}{n}\sum_{i=1}^{n}X_i = \bar{X}, \tag{i}$$

$$\sigma^{*2} = \frac{1}{n}\sum_{i=1}^{n}(X_i-\bar{X})^2 = s'^2. \tag{ii}$$

These estimates are also obtained by the method of least squares (vide section 16.5). This is a particular example of the fact that the m.l. estimate in the case of normal variables is the same as the least square estimate. Note that while $\bar{X}$ is unbiased for μ, the estimate σ^{*2} is biased, a common feature of m.l. estimates.

■ NOTE 16.2.4 *If σ^2 is know, m.l.e. of μ is $\bar{X}$. If μ is known m.l.e. of σ^2 is $\frac{1}{n}\Sigma(X_i - \mu)^2 = s_0^2$ (say).*

■ NOTE 16.2.5 $\frac{\partial^2 \log L}{\partial\mu^2} = -\frac{n}{\sigma^2}$, $\frac{\partial^2 \log L}{\partial(\sigma^2)^2} = -\frac{n}{2\sigma^4}$, $\frac{\partial^2 \log L}{\partial\mu\partial\sigma^2} = -\frac{n}{\sigma^4}(\bar{x}-\mu)$. *Hence the inverse of the dispersion matrix $\boldsymbol{V}$ in large sample is*

$$\boldsymbol{V}^{-1} = \begin{bmatrix} n/\sigma^2 & 0 \\ 0 & n/2\sigma^4 \end{bmatrix}.$$

■ NOTE 16.2.6 *If we known μ, the m.l.e. of σ^2 is $\hat{\sigma}'^2 = \frac{1}{n}\Sigma_{i=1}^{n}(x_i-\mu)^2$. Suppose we estimate μ by $\hat{\mu} = \bar{x}$ and then want to maximise the likelihood for this given value of μ viz., $\bar{x}$. The solution will of course not be $s'^2 = \frac{1}{n}\Sigma_{i=1}^{n}(x_i-\bar{x})^2$, because this is the problem of maximisation of the likelihood over a restricted parameter space $\{(\bar{x},\sigma^2) : \sigma^2 > 0\}$.*

Now the likelihood function of μ, σ^2 is

$$L(\mu,\sigma^2|x_1,\cdots,x_n) = \left[\frac{\sqrt{n}}{\sigma\sqrt{2\pi}}\exp -\left\{\frac{n}{2\sigma^2}(\bar{x}-\mu)^2\right\}\right]$$
$$\left[\frac{n^{\frac{n-1}{2}}}{\left|\left(\frac{n-1}{2}\right)2^{\frac{n-3}{2}}\sigma\right.}\left(\frac{s'}{\sigma}\right)^{n-2}\exp\left(-\frac{ns'^2}{2\sigma^2}\right)\right]$$

If we maximise the likelihood in this form for simultaneous variations in μ and σ we obtain solutions (i) and (ii). If on the other hand $\bar{x}$ has a fixed value the distribution of s'^2 becomes χ^2 with $(n-1)$ d.f. The likelihood of σ^2 is then

$$L(\sigma^2|x_1,\cdots,x_n;\bar{x}) \propto \frac{s'^{n-2}}{\sigma^{n-1}}\exp\left(-\frac{ns'^2}{2\sigma^2}\right)$$

and is maximised for $\hat{\sigma}^2 = s^2 = \frac{ns'^2}{n-1} = s^2 = \frac{1}{n-1}\Sigma_{i=1}^{n}(x_i-\bar{x})^2$.

The difference between two solutions $(\bar{x}, s'^2)$ and $(\bar{x}, s^2)$ is very subtle. Both lead to the maximisation of the likelihood, —the first set for simultaneous variations of μ and σ^2 and the second over the step-wise maximisation, —initially over variations in μ and then over variations in σ^2 for fixed value of $\bar{x}$. However, in large sample, the difference between two solutions is unimportant.

▶ **Example 16.2.10** *Observations n in number are made on a random variable with p.d.f. $f_\theta(x) = \theta\, e^{-\theta x}$, $\theta > 0$. The exact value of each observation less than A is recorded. Suppose these values are $x_1,\cdots,x_m$. The remaining $(n-m)$ observations exceed A. Find m.l.e. of θ.*

The probability that an observation exceeds A

$$P_A = \int_A^\infty \theta e^{-\theta x}\, dn = e^{-\theta A}.$$

Likelihood of θ is

$$L(\theta) = \left[\prod_{j=1}^{m}\theta e^{-\theta x_j}\right](e^{-\theta A})^{n-m},$$

$$\frac{\partial \log L}{\partial\theta} = \sum_{j=1}^{m}\frac{\partial}{\partial\theta}\log(\theta e^{-\theta x_j}) + \frac{\partial}{\partial\theta}[-\theta A(n-m)] = \frac{m}{\theta} - \sum_{j=1}^{m}x_j - A(n-m).$$

The m.l.e. is

$$\hat{\theta} = \frac{m}{\sum_{j=1}^{m} x_j + A(n-m)}.$$

▶ **Example 16.2.11** *$X = (X_1, \cdots, X_p)'$ follows p-variate normal with mean vector $\mu = \mu(1, \cdots, 1)'$ and dispersion matrix $\Sigma = ((\sigma_{ij}))$.*

Find m.l.e. of μ when σ_{ij}'s are known.

The likelihood of μ given a single observation $(x_1, \cdots, x_p)$ is

$$L(\mu|\sigma_{ij}, x_k, i, j = 1, \cdots, p; k = 1, \cdots, p) = \frac{1}{(2\pi)^{p/2}\sqrt{|\Sigma|}} \exp\left[-\frac{1}{2}\sum_i\sum_j (x_i - \mu)\sigma^{ij}(x_j - \mu)\right],$$

where $((\sigma^{ij})) = \Sigma^{-1}$.

$$\log L = \text{Const.} - \frac{1}{2}\sum_i\sum_j (x_i - \mu)\sigma^{ij}(x_j - \mu),$$

$$\frac{\partial \log L}{\partial \mu} = \frac{1}{2}\sum_i\sum_j \sigma^{ij}[(x_i - \mu) + (x_j - \mu)].$$

Now

$$\sum_i\sum_j \sigma^{ij}(x_i - \mu) = \sum_i (x_i - \mu)\sum_j \sigma^{ij} = \sum_i k_i(x_i - \mu), \text{ where } \sum_j \sigma^{ij} = k_i.$$

Hence

$$\frac{\partial \log L}{\partial \mu} = \sum_{i=1}^{p} k_i(x_i - \mu) = 0; \quad \hat{\mu} = \sum_{i=1}^{p} k_i x_i \Big/ \sum_{i=1}^{p} k_i.$$

In particular if $X_1, \cdots, X_p$ are p independent normally distributed variables such that $X_i \sim N(\mu, \sigma_i^2)$,

$$\Sigma^{-1} = \text{Diag}\left(\frac{1}{\sigma_1^2}, \cdots, \frac{1}{\sigma_p^2}\right), \quad k_i = \frac{1}{\sigma_i^2}, \quad \hat{\mu} = \left(\sum_{i=1}^{p} x_i/\sigma_i^2\right) \Big/ \left(\sum_{i=1}^{p} 1/\sigma_i^2\right).$$

▶ **Example 16.2.12** *Let $X_1, \cdots, X_n$ be iid r.v.'s having Laplace density*

$$f_\theta(x) = \tfrac{1}{2}\exp[-|x - \theta|].$$

$$\text{Likelihood} = \text{Const.} \exp\left[-\sum_{i=1}^{n} |x_i - \theta|\right].$$

Maximisation of likelihood is equivalent to minimisation of $\sum_{i=1}^{n}|x_i-\theta|$. Now $\sum_{i=1}^{n}|x_i-\theta|$ is the mean deviation about θ which is least if θ is the sample median. Hence m.l.e. of θ is sample median. In large sample, $V(\hat{\theta}) = \frac{1}{n}$.

The regularity conditions of theorem 16.2.6 do not hold here so that the result on the $V(\hat{\theta})$ cannot be applied. However, note that

$$\frac{\partial \log f_\theta(x)}{\partial \theta} = \begin{cases} 1 \text{ if } x > \theta \\ -1 \text{ if } x < \theta \end{cases}$$

(does not exist for $x = \theta$).
Therefore, taking

$$\begin{aligned} E\left(\frac{\partial \log f_\theta(x)}{\partial \theta}\right)^2 &= \lim_{C\to\infty}\left[\int_{-\infty}^{C} + \int_{C}^{\infty}\right]\left(\frac{\partial \log f_\theta(x)}{\partial \theta}\right)^2 f_\theta(x)\,dx \\ &= \lim_{C\to\infty}\left[\int_{-\infty}^{C} + \int_{C}^{\infty}\right] 1.f_\theta(x)\,dx = 1 \end{aligned}$$

we have

$$E\left(\frac{\partial \log L}{\partial \theta}\right)^2 = nE\left(\frac{\partial \log f}{\partial \theta}\right)^2 = n.$$

Hence, the MVB for an unbiased estimator of θ is $\frac{1}{n}$. Therefore, $\hat{\theta}$ is asymptotically most efficient.

► **Example 16.2.13** *Let $X_1, \cdots, X_n$ be iid r.v.'s each having exponential density $f_\theta(x) =$* $\exp\,[-(x-\theta)]$, $x \geq \theta$

$$f_\theta(x_1, \cdots, x_n) = \exp\left[-\sum_{i=1}^{n}(x_i - \theta)\right].$$

Here $\frac{d \log L}{d\theta} = 0$ does not lead to any meaningful equation. The method of differentiation does fail. We have to maximise the likelihood by other means.

To maximise the likelihood $f_\theta(x)$ is to minimise $(x-\theta)$. Since x is $\geq \theta$, $(x-\theta)$ will be minimum if θ is the minimum of the sample observations. Therefore, m.l.e. of θ is $x_{(1)}$ where $x_{(1)} = \min\limits_i x_i$.
Alternative Method :

$$f_\theta(x) = e^{-(x-\theta)} I_{[\theta,\infty]}(x)$$

where I is an indicator function such that $I_{[\theta,\infty)}(x) = 1(0)$ if $x \in [\theta, \infty)$ (otherwise).
Hence

$$L(\theta|x_1, \cdots, x_n) = e^{-(\Sigma x_i - n\theta)} I_{[\theta,\infty)}(x_{(1)}) \cdots I_{[\theta,\infty)}(x_{(n)}).$$

$L(\theta)$ is not differentiable with respect to θ. To find the m.l.e. of θ, we note that

$$L(\theta) = e^{-\Sigma x_i + n\theta} I_{[\theta,\infty)}(x_{(1)}).$$

Thus $L(\theta)$ is the product of an increasing function of θ and an indicator function which is 1 for $\theta \leq x_{(1)}$ (0 otherwise). Hence $L(\theta)$ is maximised at $\hat{\theta} = x_{(1)}$.

▶ **Example 16.2.14** *$X_1, \cdots, X_n$ are iid each having p.d.f. (extreme-value distribution)*

$$f_{\alpha,\beta}(x) = \alpha \exp\left[-\alpha(x-\beta) - \exp\{-\alpha(x-\beta)\}\right], -\infty < x < \infty.$$

Here

$$\log L = n\log\alpha - \alpha\sum_{i=1}^{n}(x_i - \beta) - \sum_{i=1}^{n}\exp\{-\alpha(x_i-\beta)\}$$

$$\frac{\partial \log L}{\partial \alpha} = \frac{n}{\alpha} - \sum_{i=1}^{n}(x_i-\beta) + \sum_{i=1}^{n}[(x_i-\beta)\exp\{-\alpha(x_i-\beta)\}]$$

$$\frac{\partial \log L}{\partial \beta} = n\alpha - \alpha\sum_{i=1}^{n}\exp\{-\alpha(x_i-\beta)\}.$$

Setting these two equations to zero, m.l.e.'s of α, β are given by

$$(\hat{\alpha})^{-1} = \bar{x} - \frac{\sum_{i=1}^{n} x_i e^{-\hat{\alpha}x_i}}{\sum_{i=1}^{n} e^{-\hat{\alpha}x_i}}; \quad e^{-\hat{\alpha}\hat{\beta}} = \frac{1}{n}\sum_{i=1}^{n} e^{-\hat{\alpha}x_i}.$$

▶ **Example 16.2.15** *Sample of size n is drawn from a rectangular population*

$$f_{\alpha,\beta}(x) = \frac{1}{\beta-\alpha}, \quad \alpha < x < \beta.$$

Likelihood $= L(\alpha,\beta|x_1,\cdots,x_n) = \frac{1}{(\beta-\alpha)^n}$.

$$\frac{\partial \log L}{\partial \alpha} = 0 \quad \Rightarrow \quad \frac{n}{\beta-\alpha} = 0$$

$$\frac{\partial \log L}{\partial \beta} = 0 \quad \Rightarrow \quad \frac{n}{\beta-\alpha} = 0.$$

Hence at least one of α or β must be infinite. Actually for these values of α or β, likelihood function is a minimum. Again to be an estimate the solution of the l.h. equations must depend on the sample values. The l.h. equations, therefore, do not have a solution here, which maximise the likelihood.

We shall therefore, maximise the likelihood in a different way. It is clear that the likelihood $[f_{\alpha,\beta}(x)]$ can be made as large as we please when $\beta-\alpha$ can be made as small as possible.

Let $x_{(1)} = \min_i x_i$, $x_{(n)} = \max_i x_i$. Clearly α can not be made larger than $x_{(1)}$ and β cannot be made smaller than $x_{(n)}$. Hence the smallest value for $\beta-\alpha$ is $x_{(n)} - x_{(1)}$ i.e., the m.l.e.'s are $\hat{\alpha} = x_{(1)}, \hat{\beta} = x_{(n)}$.

► **Example 16.2.16** *In a truncated binomial distribution $B_t(m, \pi|1)$ where the observation $x = 0$ is excluded, p.m.f of X is*

$$f_\pi(x) = \frac{\binom{m}{x}\pi^x(1-\pi)^{m-x}}{1-(1-\pi)^m}, \quad x = 1, 2, \cdots, m.$$

A random sample of size n is drawn from the above population. Find m.l.e of π when $m=2$.

Here likelihood

$$L(\pi|\boldsymbol{x}) = \frac{\text{Const.}\pi^{\Sigma x_i}(1-\pi)^{2n-\Sigma x_i}}{(2\pi-\pi^2)^n}$$

$$\frac{d\log L}{d\pi} = \frac{n\bar{x}}{\pi} + \frac{n(\bar{x}-2)}{1-\pi} - \frac{n}{\pi} + \frac{n}{2-\pi} = 0$$

$$\Rightarrow \quad \hat{\pi} = 2\left(1-\frac{1}{\bar{x}}\right).$$

► **Example 16.2.17** *X has a p.d.f.*

$$f_{\alpha,\beta,p}(x) = \frac{1}{\beta\overline{|(p)}}\left(\frac{x-\alpha}{\beta}\right)^{p-1} \exp\left[-\left(\frac{x-\alpha}{\beta}\right)\right], \quad \alpha \le x < \infty, \ \alpha > 0, \ p > 2.$$

Here

$$\log L = -np\log\beta - n\log\beta - n\log\overline{|(p)} + (p-1)\sum_{i=1}^{n}\log(x_i-\alpha) - \sum_{i=1}^{n}\left(\frac{x_i-\alpha}{\beta}\right)$$

$$\frac{\partial \log L}{\partial \alpha} = -(p-1)\sum\frac{1}{(x_i-\alpha)} + \frac{n}{\beta} = 0, \tag{i}$$

$$\frac{\partial \log L}{\partial \beta} = -\frac{np}{\beta} + \frac{1}{\beta^2}\sum_{i=1}^{n}(x_i-\alpha) = 0, \tag{ii}$$

$$\frac{\partial \log L}{\partial p} = -n\log\beta - n\frac{d}{dp}\log\overline{|(p)} + \sum_{i=1}^{n}\log(x_i-\alpha) = 0. \tag{iii}$$

Solving (i), (ii), (iii) one gets m.l.e.'s.

The inverse of the dispersion matrix V is

$$V^{-1} = \begin{bmatrix} \frac{1}{\beta^2(p-2)} & \frac{1}{\beta^2} & \frac{1}{\beta(p-1)} \\ \frac{1}{\beta^2} & \frac{p}{\beta^2} & \frac{1}{\beta} \\ \frac{1}{\beta(p-1)} & \frac{1}{\beta} & \frac{d^2}{dp^2}\log\Gamma(p) \end{bmatrix}.$$

This gives

$$V(\hat{\alpha}) = \frac{n^2}{|V^{-1}|\beta^2}\left\{p\frac{d^2}{dp^2}\log\Gamma(p) - 1\right\},$$

$$V(\hat{\beta}) = \frac{n^2}{|V^{-1}|\beta^2}\left\{\frac{1}{p-2}\frac{d^2}{dp^2}\log\Gamma(p) - \frac{1}{(p-1)^2}\right\},$$

$$V(\hat{p}) = \frac{2n^2}{|V^{-1}|(p-2)\beta^4},$$

where $|V^{-1}| = \frac{n^3}{(p-2)\beta^4}\left\{2\frac{d^2\log\Gamma(p)}{dp^2} - \frac{2}{p-1} + \frac{1}{(p-1)^2}\right\}$.

For large p, using Stirling's approximation for $\Gamma(p+1)$, $V(\hat{p}) \simeq \frac{6}{n}\{(p-1)^3 + \frac{1}{5}(p-1)\}$.

▶ **Example 16.2.18** *Let $X_1, \cdots, X_n$ be a random sample from a rectangular distribution*

$$f_\theta(x) = \begin{cases} \frac{1}{2\theta}, & -\theta \le x \le \theta, \quad \theta > 0 \\ 0, & \textit{otherwise.} \end{cases}$$

Here

$$L(\theta|x_1, \cdots, x_n) = \begin{cases} \left(\frac{1}{2\theta}\right)^n, & \text{if } -\theta \le x_i \le \theta, \quad i = 1, \cdots, n \\ 0, & \text{otherwise.} \end{cases}$$

Thus $L(\theta)$ will be maximum for the smallest possible θ that keeps $L(\theta) > 0$. But $x_i \le |\theta|$ iff $|x_i| < \theta$. Hence, $\theta \ge (|x_1|, |x_2|, \cdots, |x_n|)$. Thus m.l.e. of θ is $\max_i(|x_i|)$.

■ NOTE 16.2.7 *If X is a random sample from a rectangular distribution $f_\theta(x) = \frac{1}{2\theta}$, $a - \theta \le x \le a + \theta$, $\theta > 0$, 'a' known, similar analysis shows that m.l.e. of θ is* $\max_i(|x_i - a|)$.

▶ **Example 16.2.19** *Suppose $X_1, \cdots, X_n$ is a random sample from a rectangular distribution $f_\theta(x) = \frac{1}{2\theta}, a - \theta \le x \le a + \theta, \theta > 0$, a unknown but θ known; $a \in (-\infty, \infty)$. The p.d.f is zero otherwise.*

$$L(a|x_1, \cdots, x_n) = \begin{cases} \frac{1}{(2\theta)^n}, & \textit{if } \boldsymbol{x} \in S(a, \theta) \\ 0, & \textit{otherwise.} \end{cases}$$

where $S(a, \theta) = \{\boldsymbol{x} : a - \theta \le x_i \le a + \theta, i = 1, \cdots, n\}$.

Therefore, any a satisfying $x \in S(a,\theta)$ makes the likelihood maximum. Now

$$\begin{aligned} & a-\theta \le x_i \le a+\theta \\ \text{iff} \quad & x_i-\theta \le a \le x_i+\theta. \end{aligned}$$

Hence any a satisfying $\max_i x_i - \theta \le \min_i x_i + \theta$ is a m.l.e of 'a'. Note that m.l.e. is not unique in this case.

► **Example 16.2.20** *Let $X_1, \cdots, X_n$ be a random sample from the rectangular population*

$$f_{(a,\theta)}(x) = \begin{cases} \frac{1}{2\theta}, \ a-\theta \le x \le a+\theta, \ \ a, \theta \text{ both} \\ \qquad \text{unknown, } a \in (-\infty, \infty), \theta > 0 \\ 0, \ \text{otherwise.} \end{cases}$$

Here

$$L(a,\theta|x_1,\cdots,x_n) = \begin{cases} \frac{1}{(2\theta)^n}, & \text{if } a-\theta \le x_i \le a+\theta, \ \ i=1,\cdots,n \\ 0, & \text{otherwise.} \end{cases}$$

for any fixed (a,θ). Here $L(a,\theta)$ is maximised by the smallest θ for which, for all $i = 1, \cdots, n$,

$$a-\theta \le x_i \le a+\theta. \tag{i}$$

From (i) $\theta \ge a - \min_i x_i$ and $\theta \ge \max_i x_i - a$.

$$\therefore \quad \theta \ge \max\left[-\min_i(x_i-a), \max_i(x_i-a)\right].$$

Therefore,

$$\theta = \begin{cases} -\min_i(x_i-a), \text{ if } \max_i(x_i-a) \le -\min_i(x_i-a) \\ \max_i(x_i-a), \ \text{ if } -\min_i(x_i-a) \le \max_i(x_i-a) \end{cases}$$

i.e.,

$$\theta = \begin{cases} -\min_i x_i + a, \text{ if } a \ge \left(\max_i x_i + \min_i x_i\right)/2 \\ \max_i x_i - a, \ \text{ if } a \le \left(\max_i x_i + \min_i x_i\right)/2. \end{cases}$$

In any case,

θ is minimised for

$$a = \frac{\max_i x_i + \min_i x_i}{2} = \hat{a} \text{ (say).}$$

For this value of a,

$$\theta = \left(\max_i x_i - \min_i x_i\right)/2 = \hat{\theta} \text{ (say).}$$

$(\hat{a}, \hat{\theta})$ is m.l.e. of (a,θ).

▶ **Example 16.2.21** *Let $X_1, \cdots, X_n$ be independent r.v's each having exponential distribution*

$$f_{(\theta_1,\theta_2)}(a) = \begin{cases} \frac{1}{\theta_1} e^{-(x-\theta_2)/\theta_1}, & \theta_2 \le x < \infty \\ 0, & x < \theta_2 \end{cases}$$

where $-\infty < \theta_2 < \infty$, $\theta_1 > 0$, both θ_1, θ_2 being unknown. Find m.l.e. of (θ_1, θ_2), say $(\hat{\theta}_1, \hat{\theta}_2)$. Find also $E(\hat{\theta}_1)$, $E(\hat{\theta}_2)$.

$$\log L(\theta_1, \theta_2 | x_1, \cdots, x_n) = -n \log \theta_1 - \frac{1}{\theta_1} \sum_{i=1}^{n} (x_i - \theta_2)$$

$$\frac{\partial \log L}{\partial \theta_1} = 0 \quad \Rightarrow \quad -\frac{n}{\theta_1} + \frac{n(\bar{x} - \theta_2)}{\theta_1^2} = 0, \tag{i}$$

$$\frac{\partial \log L}{\partial \theta_2} = 0 \quad \Rightarrow \quad \frac{n}{\theta_1} = 0. \tag{ii}$$

The second equation gives $\theta_1 = \infty$ which is clearly inadmissible and in this case the first equation becomes absurd. The method of differentiation to maximise the likelihood, therefore, fails. We have to maximise the likelihood by other means. Note that

$$L(\theta_1, \theta_2) = \begin{cases} \frac{1}{\theta_1^n} e^{-\sum\limits_{i=1}^{n} (x_i - \theta_2)/\theta_1}, & \text{if } \min\limits_i x_i \ge \theta_2 \\ 0, & \text{otherwise.} \end{cases}$$

Now, for any fixed θ_1, $L(\theta_1, \theta_2)$ is maximised if θ_2 is chosen as large as possible. Since $\theta_2 \le x_i \ \forall\, i$,the largest possible value of θ_2 is $\min\limits_i x_i = x_{(1)}$ (say). Hence $\hat{\theta}_2 = x_{(1)}$.

Equation (i) now gives $\hat{\theta}_1 = \bar{x} - x_{(1)}$.
Again,

$$\frac{\partial^2}{\partial \theta_1^2} \log L(\theta_1, \hat{\theta}_2) = \frac{n}{\theta_1^2} - \frac{2n(\bar{x} - \hat{\theta}_2)}{\theta_1^3} < 0$$

$$\text{iff} \quad \theta_1 < 2(\bar{x} - x_{(1)}) = 2\hat{\theta}_1.$$

Hence $\hat{\theta}_1$ maximises the likelihood $L(\theta_1, \hat{\theta}_2)$ and is the m.l.e. of θ_1. Note that the likelihood $L(\theta_1, \theta_2)$ has no terminal maximum near the boundary values of θ_1, since $L(\theta_1, \theta_2) \to 0$ as $\theta_1 \to 0$ or ∞.
It follows that $E(X_1) = E(\bar{X}) = \theta_1 + \theta_2$.
Again the distribution function of $X_{(1)} = \min\limits_i X_i$ is

$$F_{X_{(1)}}(x) = P[X_{(1)} \le x] = 1 - (1 - P(X_1 \le x))^n$$
$$= 1 - \exp\left[-n\left(\frac{x - \theta_2}{\theta_1}\right)\right].$$

Therefore, p.d.f. of $X_{(1)}$ is

$$f_{X_{(1)}}(x) = \begin{cases} \frac{n}{\theta_1} \exp\left[-n\left(\frac{x-\theta_2}{\theta_1}\right)\right], & x \ge \theta_2 \\ 0, & \text{otherwise.} \end{cases}$$

Hence

$$E(X_{(1)}) = \int_{\theta_2}^{\infty} x f(x)\, dx = \theta_2 + \frac{\theta_1}{n}.$$

Thus,

$$E(\hat{\theta}_1) = \theta_1 \left(1 - \frac{1}{n}\right), \quad E(\hat{\theta}_2) = \theta_2 + \frac{\theta_1}{n}.$$

► **Example** 16.2.22 (*Kendall and Stuart,* 1961). *Let X_i be a random sample from the p.d.f. $f_\theta(x) = \frac{1}{\theta}$, $k\theta \le x \le (k+1)\theta$, $k(>0)$ known, θ unknown.*

Here $(x_{(1)}, x_{(n)})$ where $x_{(1)} = \min x_i, x_{(n)} = \max x_i$ are jointly sufficient for θ. However, m.l.e. of θ is $\hat{\theta} = x_{(n)}/(k+1)$. Note that $\hat{\theta}$ is a function of $x_{(n)}$ alone.

$$V(\hat{\theta}) = \frac{V}{(k+1)^2} \quad \text{(say)}.$$

Consider another estimator $\hat{\hat{\theta}} = x_{(1)}/k$. $V(\hat{\hat{\theta}}) = \frac{V}{k^2}$. $x_{(1)}$ and $x_{(n)}$ are asymptotically independently distributed.

Consider the estimator $\tilde{\theta} = a\hat{\theta} + (1-a)\hat{\hat{\theta}}$ when a is a suitable constant in $[0, 1]$.

$$V(\tilde{\theta}) = V\left[\frac{a^2}{(k+1)^2} + \frac{(1-a)^2}{k^2}\right] \quad \text{(in large sample)},$$

which is minimised for $a = \frac{(k+1)^2}{k^2+(k+1)^2}$, when $V(\tilde{\theta})$ is

$$V_0(\tilde{\theta}) = \frac{V}{k^2 + (k+1)^2}.$$

Hence efficiency of $\hat{\theta}$ with respect to $\tilde{\theta}$ is

$$\frac{V_0(\tilde{\theta})}{V(\hat{\theta})} = \frac{(k+1)^2}{k^2 + (k+1)^2} < 1.$$

For large k, $V(\hat{\theta}) \simeq 2V(\tilde{\theta})$.

► **Example** 16.2.23 *Find the most general form of the distribution differentiable in θ for which the sample mean is the maximum likelihood estimator.*

Here, a solution of

$$\frac{\partial \log L}{\partial \theta} = \sum_{i=1}^{n} \frac{1}{f(x_i)} \frac{\partial f(x_i)}{\partial \theta} = 0 \text{ is } \theta = \bar{x}$$

so that

$$\sum_{i=1}^{n} (x_i - \theta) = 0.$$

This is true for all x and θ and hence

$$\frac{1}{f}\frac{\partial f}{\partial \theta} = (x-\theta)k(\theta) \tag{i}$$

when $k(\theta)$ is independent of x but may depend upon θ. Suppose $k(\theta) = \frac{\partial^2 \psi}{\partial \theta^2}$. Therefore, integrating (i) with respect to θ,

$$\log f = \int (x-\theta)\frac{\partial^2 \psi}{\partial \theta^2}d\theta = (x-\theta)\frac{\partial \psi}{\partial \theta} + \psi(\theta) + \xi(x)$$

when $\xi(x)$ is an arbitrary function of x. Hence

$$f = k\exp\left[(x-\theta)\frac{\partial \psi}{\partial \theta} + \psi(\theta) + \xi(x)\right]. \tag{ii}$$

In particular, if,

$$\psi(\theta) = \frac{1}{2}\theta^2, \ \ \xi(x) = -\frac{1}{2}x^2,$$

the expression (ii) becomes the p.d.f. of a normal distribution.

▶ **Example 16.2.24** *For the bivariate normal distribution*

$$f(x,y) = \frac{1}{2\pi\sigma_1\sigma_2\sqrt{1-\rho^2}}$$
$$\exp\left[-\frac{1}{2(1-\rho^2)}\left\{\left(\frac{x-\mu_1}{\sigma_1}\right)^2 - 2\rho\left(\frac{x-\mu_1}{\sigma_1}\right)\left(\frac{x-\mu_2}{\sigma_2}\right) + \left(\frac{y-\mu_2}{\sigma_2}\right)^2\right\}\right],$$
$$-\infty \le x, \ \ y \le \infty, \ \ -\infty < \mu_1, \ \ \mu_2 < \infty, \ \ \sigma_1, \ \ \sigma_2 > 0, \ \ |\rho| < 1$$

likelihood equations are

$$\frac{\partial \log L}{\partial \mu_1} = \frac{n}{\sigma_1(1-\rho^2)}\left\{\frac{(\bar{x}-\mu_1)}{\sigma_1} - \rho\frac{(\bar{y}-\mu_2)}{\sigma_2}\right\} = 0. \tag{i}$$

$$\frac{\partial \log L}{\partial \mu_2} = \frac{n}{\sigma_2(1-\rho^2)}\left\{\frac{(\bar{y}-\mu_2)}{\sigma_2} - \rho\frac{(\bar{x}-\mu_1)}{\sigma_1}\right\} = 0. \tag{ii}$$

$$\frac{\partial \log L}{\partial \sigma_1^2} = -\frac{1}{2\sigma_1^2(1-\rho^2)}\left\{n(1-\sigma^2) - \frac{\Sigma(x_i-\mu_1)^2}{\sigma_1^2} + \rho\frac{\Sigma(x_i-\mu_1)(y_i-\mu_2)}{\sigma_1\sigma_2}\right\} = 0. \tag{iii}$$

$$\frac{\partial \log L}{\partial \sigma_2^2} = -\frac{1}{2\sigma_2^2(1-\rho^2)}\left\{n(1-\rho^2) - \frac{\Sigma(y_i-\mu_2)^2}{\sigma_2^2} + \rho\frac{\Sigma(x_i-\mu_1)(y_i-\mu_2)}{\sigma_1\sigma_2}\right\} = 0. \tag{iv}$$

$$\frac{\partial \log L}{\partial \rho} = \frac{1}{1-\rho^2}\left\{n\rho - \frac{1}{1-\rho^2}\left[\rho\left(\frac{\Sigma(x_i-\mu_1)^2}{\sigma_1^2} + \frac{\Sigma(y_i-\mu_2)^2}{\sigma_2^2}\right)\right.\right.$$
$$\left.\left. -(1+\rho^2)\frac{\Sigma(x_i-\mu_1)(y_i-\mu_2)}{\sigma_1\sigma_2}\right]\right\} = 0. \tag{v}$$

Case 1. Suppose $\mu_1, \mu_2, \sigma_1, \sigma_2$ are known, ρ can be estimated by solving (v) [vide exercise 25]

Case 2. Suppose μ_1, μ_2 are known and σ_1^2, σ_2^2, ρ are to be estimated. We have to solve (iii)—(v). The equations imply,

$$n(1-\rho^2) = \frac{\Sigma(x_i-\mu_1)^2}{\sigma_1^2} - \rho\frac{\Sigma(x_i-\mu_1)(y_i-\mu_2)}{\sigma_1\sigma_2}, \tag{vi}$$

$$n(1-\rho^2) = \frac{\Sigma(y_i-\mu_2)^2}{\sigma_2^2} - \rho\frac{\Sigma(x_i-\mu_1)(y_i-\mu_2)}{\sigma_1\sigma_2}, \tag{vii}$$

$$n(1-\rho^2) = \frac{\Sigma(x_i-\mu_1)^2}{\sigma_1^2} + \frac{\Sigma(y_i-\mu_2)^2}{\sigma_2^2} - \frac{(1+\rho^2)\Sigma(x_i-\mu_1)(y_i-\mu_2)}{\rho\sigma_1\sigma_2}. \tag{viii}$$

(vi) + (vii) − (viii) gives

$$n(1-\rho^2) = \frac{1-\rho^2}{\rho}\frac{\Sigma(x_i-\mu_1)(y_i-\mu_2)}{\sigma_1\sigma_2} \quad \text{or,} \quad \rho = \frac{\Sigma(x_i-\mu_1)(y_i-\mu_2)}{n\sigma_1\sigma_2}.$$

Substituting in (vi) and (vii) gives

$$\hat{\sigma}_1^2 = \frac{1}{n}\Sigma(x_i-\mu_1)^2, \quad \hat{\sigma}_2^2 = \frac{1}{n}\Sigma(y_i-\mu_2)^2 \quad . \tag{ix}$$

when $\hat{\rho} = \Sigma(x_i-\mu_1)(y_i-\mu_2)/n\hat{\sigma}_1\hat{\sigma}_2$.

Case 3. Suppose all the five parameters are unknown.
Equations (i) and (ii) reduce to

$$\frac{\bar{x}-\mu_1}{\sigma_1} = \rho\frac{\hat{y}-\mu_2}{\sigma_2} \tag{x}$$

$$\frac{\bar{y}-\mu_2}{\sigma_2} = \rho\frac{\hat{x}-\mu_1}{\sigma_1} \tag{xi}$$

(x) and (xi) hold iff

$$\hat{x}-\mu_1 = \bar{y}-\mu_2 = 0. \tag{xii}$$

These conditions coupled with the solutions (ix), which are the solutions of (iii)—(v) give the m.l.e.'s as

$$\begin{aligned} &\hat{\mu}_1 = \bar{x}, \quad \hat{\mu}_2 = \bar{y}, \quad \hat{\sigma}_1^2 = \frac{1}{n}\Sigma(x_i-\bar{x})^2, \\ &\hat{\sigma}_2^2 = \frac{1}{n}\Sigma(y_i-\bar{y})^2, \quad \rho = \frac{\Sigma(x_i-\bar{x})(y_i-\bar{y})}{n\hat{\sigma}_1\hat{\sigma}_2}. \end{aligned} \tag{xiii}$$

We now consider the large sample dispersion matrices of the m.l.e.'s

Case (i) When estimating ρ alone, $\hat{\rho}$ is not sufficient.
Its large sample variance is $V(\hat{\rho}) = \frac{(1-\rho^2)^2}{n(1+\rho^2)}$ [vide exercise 25].

Case (ii) μ_1, μ_2 are known. Here $\boldsymbol{t} = (\hat{\sigma}_1^2, \hat{\sigma}_2^2, \hat{\rho})$ given in (ix) are jointly sufficient. The inverse of the large-sample dispersion matrix of $\boldsymbol{t}$, $\boldsymbol{V}_3^{-1}$ (say) is

$$\boldsymbol{V}_3^{-1} = -\left(\frac{\partial^2 \log L}{\partial\theta_r \partial\theta_s}\right)_{\hat{\boldsymbol{\theta}}=\boldsymbol{\theta}}$$

$$= \frac{n}{(1-\rho^2)}\begin{pmatrix} \frac{2-\rho^2}{4\sigma_1^4} & -\frac{\rho^2}{4\sigma_1^4\sigma_2^2} & -\frac{\rho}{2\sigma_1^2} \\ -\frac{\rho^2}{4\sigma_1^2\sigma_2^2} & \frac{2-\rho^2}{4\sigma_2^4} & -\frac{\rho}{2\sigma_2^2} \\ -\frac{\rho}{2\sigma_1^2} & -\frac{\rho}{2\sigma_2^2} & \frac{1+\rho^2}{1-\rho^2} \end{pmatrix}. \tag{xiv}$$

Here

$$\boldsymbol{V}_3 = \frac{1}{n}\begin{pmatrix} 2\sigma_1^4 & 2\rho^2\sigma_1^2\sigma_2^2 & \rho(1-\rho^2)\sigma_1^2 \\ 2\rho^2\sigma_1^2\sigma_2^2 & 2\sigma_2^4 & \rho(1-\rho^2)\sigma_2^2 \\ \rho(1-\rho^2)\sigma_1^2 & \rho(1-\rho^2)\sigma_2^2 & (1-\rho^2)^2 \end{pmatrix}. \tag{xv}$$

Case (iii) All the parameters are unknown. Here $\boldsymbol{t}_1 = (\hat{\mu}_1, \hat{\mu}_2, \hat{\sigma}_1^2, \hat{\sigma}_2^2, \hat{\rho})$ is jointly sufficient.

$$\frac{\partial^2 \log L}{\partial \mu_i^2} = -\frac{n}{\sigma_i^2(1-\rho^2)}, \quad i = 1, 2$$

$$\frac{\partial^2 \log L}{\partial \mu_1 \partial \mu_2} = \frac{n\rho}{\sigma_1\sigma_2(1-\rho^2)},$$

$$\frac{\partial^2 \log L}{\partial \mu_i \partial \sigma_j^2} = 0 = \frac{\partial^2 \log L}{\partial \mu_i \partial \rho}, \quad i, j = 1, 2$$

at $\hat{\boldsymbol{\theta}} = \boldsymbol{\theta}$. Writing $\boldsymbol{V}_2^{-1}$ as the inverse of the dispersion matrix of $\hat{\mu}_1, \hat{\mu}_2$ and $\boldsymbol{V}_5^{-1}$ as the same for all the five parameters

$$\boldsymbol{V}_2^{-1} = \frac{n}{(1-\rho^2)}\begin{pmatrix} \frac{1}{\sigma_1^2} & -\frac{\rho}{\sigma_1\sigma_2} \\ -\frac{\rho}{\sigma_1\sigma_2} & \frac{1}{\sigma_2^2} \end{pmatrix}$$

$$\boldsymbol{V}_5^{-1} = \begin{bmatrix} \boldsymbol{V}_2^{-1} & \mathbf{0} \\ \mathbf{0} & \boldsymbol{V}_3^{-1} \end{bmatrix}.$$

Hence $\boldsymbol{V}_5 = \begin{bmatrix} \boldsymbol{V}_2 & \mathbf{0} \\ \mathbf{0} & \boldsymbol{V}_3 \end{bmatrix}$ where $\boldsymbol{V}_2 = \frac{1}{n}\begin{bmatrix} \sigma_1^2 & \rho\sigma_1\sigma_2 \\ \rho\sigma_1\sigma_2 & \sigma_2^2 \end{bmatrix}$ and $\boldsymbol{V}_3$ is given in (xv).

▶ **Example 16.2.25** *Let $X \sim$ Bernoulli (p), $p \in [0, 1/2]$. Here likelihood $L(p|x) = p^x(1-p)^{1-x}, x = 0, 1$. $dL(p)/dp = 0$ gives $\hat{p} = x$, but since x is either 0 or 1 and since $p \in [0, 1/2]$, the differentiation of likelihood gives an inadmissible solution.*

Now

$$L(p) = \begin{cases} p & \text{if } x = 1 \\ 1 - p & \text{if } x = 0. \end{cases}$$

Clearly $L(p)$ is maximised for $p = \frac{1}{2}$ at $x = 1$ and for $p = 0$ at $x = 0$. Hence, $\hat{p} = \frac{1}{2}$ for $x = 1$ and $\hat{p} = 0$ for $x = 0$. Thus, the m.l.e. of p is

$$\hat{p}(X) = \frac{X}{2}.$$

$E(\hat{p}(X)) = \frac{p}{2}$ so that $\hat{p}$ is a biased estimator. Also mean square error of $\hat{p}$ is

$$E_p(\hat{p} - p)^2 = \frac{p}{4}.$$

► **Example 16.2.26** *Consider*

$$f(x;\theta) = \begin{cases} \frac{k}{\theta^k} x^{k-1}, & \textit{if } 0 \le x \le \theta \\ 0, & \textit{if } x > \theta \end{cases} \qquad k > 0,\ 0 \le \theta \le 1.$$

Show that the m.l.e. is consistent.

Here range of X depends on θ. Hence the regularity conditions of theorem 16.2.6 are not satisfied. $X_{(1)} = \min\limits_i X_i$ is the m.l.e. of θ. $X_{(1)}$ is also a sufficient statistic for θ. The p.d.f. of $X_{(1)}$ is

$$f(x_{(1)};\theta) = \frac{nk}{\theta^{kn}} x_{(1)}^{kn-1},$$

$$E(X_{(1)}) = \frac{nk}{nk+1}\theta,$$

$$V\left(\frac{nk+1}{nk} X_{(1)}\right) = \frac{\theta^2}{(nk+2)nk} \to 0 \text{ as } n \to \infty.$$

Hence

$$\frac{nk+1}{nk} X_{(1)}, X_{(1)}$$

are consistent estimates of θ.

► **Example 16.2.27** *Find the m.l.e. of the parameters of a multivariate normal distribution.*

Let $x_{i\lambda}$, $i = 1, \cdots, p$; $\lambda = 1, \cdots, n$ be a random sample of size n from the p-parameter normal density

$$f = \frac{1}{(2\pi)^{p/2}\sqrt{|\Sigma|}} \exp\left[-\frac{1}{2}\sum_1^p \sum_1^p \sigma^{ij}(x_i - m_i)(x_j - m_j)\right]$$

where $\Sigma = ((\sigma_{ij}))$, $\Sigma^{-1} = ((\sigma^{ij})) = A = ((a_{ij}))$.

The likelihood function is

$$L(m_1, \cdots, m_p, \sigma_{11}, \cdots, \sigma_{pp} | x_{ij}, i = 1, \cdots, p; j = 1, \cdots, n)$$
$$= \frac{|A|^{n/2}}{(2\pi)^{np/2}} \exp\left[-\frac{1}{2}\sum_{\lambda=1}^{n}\sum_{i,j=1}^{p} a_{ij}(x_{i\lambda} - m_i)(x_{j\lambda} - m_j)\right]$$

$$\log L = \frac{n}{2}\log|A| - \frac{1}{2}\sum_{\lambda=1}^{n}\sum_{i,j=1}^{p} a_{ij}(x_{i\lambda} - m_i)(x_{j\lambda} - m_j) + \text{Const.}$$

$$\frac{\partial \log L}{\partial m_i} = \frac{1}{2}\sum_{\lambda=1}^{n}\sum_{j=1}^{p} a_{ij}(x_{j\lambda} - m_j) = 0,$$

$$\text{or,} \quad \sum_{j=1}^{p} a_{ij}(\bar{x}_j - m_j) = 0 \quad \left(\text{where } \bar{x}_j = \frac{1}{n}\sum_{\lambda} xj\lambda\right), i = 1, \ldots, p$$

$$\text{or,} \quad A(\bar{\boldsymbol{X}} - \boldsymbol{m}) = \boldsymbol{0} \tag{i}$$

where $\bar{\boldsymbol{x}} = (\bar{x}_1, \cdots, \bar{x}_p)'$, $\boldsymbol{m} = (m_1, \cdots, m_4)$. Since A is non-singular, $\bar{\boldsymbol{X}} - \boldsymbol{m} = \boldsymbol{0}$ i.e., m.l.e. of $\boldsymbol{m}$ is $\bar{\boldsymbol{x}}$.

Assume for the time-being that $a_{ij} \neq a_{ji}$.

$$\frac{\partial \log L}{\partial a_{ij}} = \frac{n}{2|A|}A_{ij} - \frac{1}{2}\sum_{\lambda=1}^{n}(x_{i\lambda} - \hat{m}_i)(x_{j\lambda} - \bar{m}_j) = 0 \tag{ii}$$

$$\text{i.e.,} \quad \frac{n}{2}a^{ij} - \frac{1}{2}\sum_{\lambda=1}^{n}(x_{i\lambda} - \bar{x}_i)(x_{j\lambda} - \bar{x}_j) = 0.$$

Therefore,

$$\hat{\sigma}_{ij} = \frac{1}{n}\sum_{\lambda=1}^{n}(x_{i\lambda} - \bar{x}_i)(x_{j\lambda} - \bar{x}_j).$$

16.3 The Method of Moments

Estimation by the method of moments was introduced by Karl Pearson (1900).

Definition 16.3.1 *Let $X_1, \cdots, X_n$ be iid with common p.d.f. $f_\theta(x), \theta \subset \Theta \subset R^1$. Suppose the moments $\mu_1', \cdots, \mu_r'$ exist when $\mu_r' = E(X^r)$ and we are interested in estimating the parameters*

$$\theta_i = h_i(\mu_1', \mu_2', \cdots, \mu_r'), \quad i = 1, \cdots, r \tag{16.3.1}$$

where h is a known numerical function. The method of moments estimator (MME) of θ_i is

$$t_i(X_1, \cdots, X_n) = h_i(m_1', m_2', \cdots, m_r') \tag{16.3.2}$$

where
$$m'_r = \frac{1}{n}\sum_{k=1}^{n} x_k^r.$$

In particular the MME of μ'_i is $m'_i (i = 1, \cdots, r)$.

■ NOTE 16.3.1 *By WLLN, $m'_j \xrightarrow{P} \mu'_j$. Thus for estimating the population moments, the* MME*'s are consistent and unbiased. Again the estimates are in this case asymptotically normally distributed.*

If there is a one-to-one correspondence between the parameters $(\theta_1, \cdots, \theta_r)$ and the moments $(\mu'_1, \cdots, \mu'_r)$ and the functions $h_i(\mu'_1, \cdots, \mu'_r)$ are continuous in $\mu'_1, \cdots, \mu'_r$, then the estimates $t_i (i = 1, \cdots, r)$ given in (16.3.2) are consistent estimates of θ's.

Generally the MM-estimators will be less efficient than m.l.e.'s. The two methods will give identical estimates only if the parent population is of a particular form. In this case m.l.e.'s must be obtained as linear functions of sample moments. Hence we must have

$$\frac{\partial \log L}{\partial \theta_j} = C_0 + C_1 m'_1 + C_2 m'_2 + C_3 m'_3 + \cdots$$
$$\text{or,}\quad f_{\boldsymbol{\theta}}(x) = \exp[b_0 + b_1 x + b_2 x^2 + \cdots] \tag{16.3.3}$$

where b's are functions of $\boldsymbol{\theta} = (\theta_1, \theta_2, \cdots)$.

► **Example 16.3.1** $f_\theta(x) = \frac{1}{\theta} e^{-x/\theta}, 0 < x < \infty$. *The p.d.f. is of the form* (16.3.3). *Hence the* MME*'s and m.l.e.'s must be identical. The m.l.e. of θ is $\bar{x}$. Again $\mu'_1 = \theta$ so that* MME *is $\hat{\theta} = \bar{x}$.*

► **Example 16.3.2** $f_\theta(x) = \frac{1}{\sigma\sqrt{\pi}} \exp\left[-(x-\mu)^2/2\sigma^2\right],\ -\infty < x < \infty$. *The p.d.f. is of the form* (16.3.3). *Here also both* MME *and m.l.e. coincide. For normal distribution $E(X) = \mu$; $\mu_2 = E(X-\mu)^2 = \sigma^2$. Hence equating the first two sample central moments with the population central moments,*

$$\hat{\mu} = \bar{x}, \quad \hat{\mu}_2 = \frac{1}{n}\sum_{i=1}^{n}(x_i - \bar{x})^2.$$

The method of moments estimation sometimes yield estimates easily in comparison with maximum likelihood estimates.

For example, consider the example 16.2.17. The m.l.e.'s were obtained there very tediously.

Consider now the method of moments estimators. By equating sample moments to the appropriate moments, we find

$$\mu'_1 = \alpha + \beta p,$$
$$\mu_2 = \beta^2 p,$$
$$\mu_3 = 2\beta^3 p,$$

so that sample coefficient of skewness.

$$b_1 = \frac{m_3^2}{m_2^3} = \frac{4}{\hat{\hat{p}}} \tag{i}$$

whatever $\alpha, \beta, \hat{\hat{p}}$ being the MME of p.

Now

$$\text{Var}(b_1) = \frac{\beta_1}{n}\{4\beta_4 - 24\beta_2 + 36 + 9\beta_1\beta_2 - 12\beta_3 + 35\beta_1\}$$

For the present distribution

$$\text{Var}(b_1) = \frac{\beta_1^2}{n}\frac{6(p+1)(p+5)}{p}.$$

Hence from (i),

$$\text{Var}(\hat{\hat{p}}) \simeq \left[\frac{\partial}{\partial b_1}\left(\frac{4}{b_1}\right)\right]^2_{b_1=\beta} V(b) \quad \text{(by formula 7.3.19)}$$

$$\simeq \frac{p^4}{16}\,\text{Var}(b) = \frac{6p(p+1)(p+5)}{n}.$$

The efficiency of $\hat{\hat{p}}$ with respect to the m.l.e. $\hat{p}$ for large p is

$$\frac{\text{Var}(\hat{p})}{\text{Var}(\hat{\hat{p}})} = \frac{\left\{(p-1)^3 + \frac{1}{5}(p-1)\right\}}{p(p+1)(p+5)} < 1.$$

16.4 Method of Minimum Chi-Square (MCS)

The method is only appropriate to data classified into a finite number of groups. Suppose we have a sample of size n from a multinomial population of k mutually exclusive and exhaustive groups with probabilities $p_1, \cdots, p_k$, the corresponding observed frequencies being $f_1, \cdots, f_k (\Sigma p_k = 1, \Sigma f_k = n)$. Suppose p_i's are functions of l parameters $\boldsymbol{\theta} = (\theta_1, \cdots, \theta_l)$ so that $p_i = p_i(\boldsymbol{\theta})$, $i = 1, \cdots, k$. The expected frequency in the ith class is $np_i(\boldsymbol{\theta}) = m_i(\boldsymbol{\theta})$ (say).

A measure of discrepancy between the observed frequencies and the corresponding set of expected frequencies is given by the Pearsonain Chi-Square, defined as

$$\chi^2(\boldsymbol{\theta}) = \sum_i [f_i - m_i(\boldsymbol{\theta})]^2 / m_i(\theta) = \sum_i f_i^2 / m_i(\theta) - n. \tag{16.4.1}$$

Clearly $\chi^2(\boldsymbol{\theta})$ is a function of $\boldsymbol{\theta}$. The minimum χ^2-estimate (MCS) of $\boldsymbol{\theta}$ is that value of $\boldsymbol{\theta}$ which minimises $\chi^2(\boldsymbol{\theta})$. The estimates are obtained by solving the simultaneous equations,

$$\frac{\partial \chi^2}{\partial \theta_j} = 0, \quad j = 1, \cdots, l \tag{16.4.2}$$

i.e.,
$$2\sum_i \frac{(f_i - m_i(\boldsymbol{\theta}))}{m_i(\boldsymbol{\theta})} \frac{\partial m_i(\boldsymbol{\theta})}{\partial \theta_j} - \sum_i \left(\frac{f_i - m_i(\boldsymbol{\theta})}{m_i(\boldsymbol{\theta})}\right)^2 \frac{\partial m_i(\boldsymbol{\theta})}{\partial \theta_j} = 0$$

or,
$$-\sum_{i=1}^{k} \frac{f_i^2}{m_i^2} \frac{\partial m_i}{\partial \theta_j} = 0. \tag{16.4.3}$$

Consider now the likelihood

$$L(p_1, \cdots, p_k | f_1, \cdots, f_k) = \frac{n!}{\prod_i f_i!} \prod_i p_i^{f_i}$$

$$\log L = \text{Const.} + \sum_i f_i \log p_i.$$

The log-likelihood equations are

$$\frac{\partial \log L}{\partial \theta_j} = \sum_i \frac{f_i}{p_i} \frac{\partial p_i}{\partial \theta_j} = 0. \tag{16.4.4}$$

Now we can write (16.4.3) as

$$\sum_i \frac{f_i^2 - m_i^2}{m_i^2} \frac{\partial mi}{\partial \theta_j} = 0 \quad \left(\text{since } \sum_i \frac{\partial m_i}{\partial \theta_j} = 0\right)$$

or,
$$\sum_i \frac{(f_i - m_i)(f_i + m_i)}{m_i^2} \frac{\partial m_i}{\partial \theta_j} = 0. \tag{16.4.5}$$

In large sample $f_i + m_i \simeq 2m_i$ and (16.4.5) reduces to

$$2\sum_i \frac{f_i - m_i}{m_i} \frac{\partial m_i}{\partial \theta_j} = 0 \text{ or } \sum_i \frac{f_i}{m_i} \frac{\partial m_i}{\partial \theta_j} = 0 \text{ or, } \sum_i \frac{f_i}{p_i} \frac{\partial p_i}{\partial \theta_j} = 0.$$

Hence, for large n, the minimum χ^2 equations and the likelihood equations are identical, providing identical estimates.

Hence the MCS-estimates share all the asymptotic properties of m.l.-estimates and are, therefore, BAN estimates. Rao (1957) investigated the asymptotic properties of MCS-estimates when there is only a single parameter θ.

Theorem 16.4.1 *The minimum χ^2-estimate is consistent.*

Developing

$$\sum_i \frac{f_i^2}{m_i^2} \frac{dm_i}{d\theta}$$

around the true value θ_0

$$n\sum_i \frac{f_i^2}{m_i^2}\frac{dp_i}{d\theta} = n\sum_i \frac{f_i^2}{m_i^2}\frac{dp_i}{d\theta}\Bigg]_{\theta_0} + n(\theta-\theta_0)\sum_i A_i(\theta)]_{\theta'} \quad \text{(say)}, \tag{16.4.6}$$

where

$$A_i(\theta) = \frac{f_i^2}{m_i^2}\frac{\partial^2 p_i}{\partial\theta^2} - 2\frac{f_i^2 n}{m_i^3}\left(\frac{dp_i}{d\theta}\right)^2, \quad \theta' = \theta_0 + \lambda(\theta-\theta_0), \quad 0<\lambda<1.$$

Since

$$\frac{f_i}{n} \xrightarrow{P} p_i, \qquad \sum_{i=1}^{k}\frac{f_i^2}{m_i^2}\frac{dp_i}{d\theta}\Bigg]_{\theta_0} \xrightarrow{P} \sum_{i=1}^{k}\frac{dp_i}{d\theta}\Bigg]_{\theta_0} = 0.$$

If $|\theta-\theta_0| < \delta$, then with proper assumption of continuity, $\Sigma_i A_i]_{\theta'}$ can be made as close as possible to $\Sigma_i A_i\theta]_{\theta_0}$.

Now

$$\sum_i A_i(\theta)\Bigg]_{\theta_0} \xrightarrow{P} -2\sum_i \frac{1}{p_i}\left(\frac{dp_i}{d\theta}\right)^2\Bigg]_{\theta_0}. \tag{16.4.7}$$

Hence

$$n\sum_i \frac{f_i^2}{m_i^2}\frac{dp_i}{d\theta} \xrightarrow{P} -2n(\theta-\theta_0)\sum_i \frac{1}{p_i}\left(\frac{dp_i}{d\theta}\right)^2\Bigg]_{\theta_0}, \tag{16.4.8}$$

so that the left-hand side of (16.4.6)converges asymptotically to a value which is negative if $\theta > \theta_0$ and positive if $\theta < \theta_0$. Hence if $\theta \in [\theta_0-\delta, \theta_0+\delta]$ there is at least one value of θ for which $n\Sigma_i \frac{f_i^2}{m_i}\frac{dp_i}{d\theta} = 0$ with probability $> (1-\eta)$. Hence with probability tending to unity, the minimum χ^2-equation has a solution which converges to θ_0.

Theorem 16.4.2 *The minimum χ^2-estimate is unique.*

Proof: If possible suppose there are two solutions $\hat\theta_1, \hat\theta_2$.

Therefore,

$$\sum_i \frac{f_i^2}{m_i^2}\frac{dp_i}{d\theta}\Bigg]_{\hat\theta_1} = 0, \quad \sum_i \frac{f_i^2}{m_i^2}\frac{dp_i}{d\theta}\Bigg]_{\hat\theta_2} = 0.$$

Hence by Roll's theorem

$$\frac{d}{d\theta}\left\{\sum_i \frac{f_i^2}{m_i^2}\frac{dp_i}{d\theta}\right\}\Bigg]_{\hat\theta'} = \sum_i A_i(\theta)\Bigg]_{\hat\theta'} = 0$$

where $A_i(\theta)$ is given in (16.4.6) and $\hat\theta_1 < \hat\theta' < \hat\theta_2$.

But as shown in (16.4.7) $\Sigma_i A(\hat\theta') \xrightarrow{P}$ a negative quantity. Thus, we come to a contradiction. Hence, $\hat\theta_1$, $\hat\theta_2$ cannot be different.

Theorem 16.4.3 *A solution of the χ^2-square equation* (16.4.3) *corresponds to the minimum of the χ^2-square.*

$$\frac{d^2\chi^2}{d\theta} = -\sum_{i=1}^{k} \frac{f_i^2}{m_i^2}\frac{dm_i}{d\theta},$$

$$\frac{d^2\chi^2}{d\theta^2} = -n\sum_{i=1}^{k} A_i(\theta). \tag{16.4.9}$$

Now as shown in (16.4.7) the right hand side of (16.4.9)

$$\xrightarrow{P} 2n\sum_i \frac{1}{p_i}\left(\frac{dp_i}{d\theta}\right)^2.$$

Hence $\left.\frac{d^2\chi^2}{d\theta^2}\right]_{\hat{\theta}} \xrightarrow{P}$ a positive quantity.

Since minimum χ^2-estimate is asymptotically equivalent to m.l.e., the minimum χ^2 estimate is asymptotically normal and efficient.

16.4.1 Modified MCS method

In the minimum χ^2-method some difficulty arises due to $p_i(\theta)$ occurring in the denominator. In the modified MCS-method f_i is taken in the denominator. Here one minimises

$$\chi'^2 = \sum_i \frac{(f_i - np_i)^2}{f_i} = \sum_i \frac{n^2p_i^2}{f_i} - n. \tag{16.4.10}$$

Now the likelihood

$$L = \frac{n!}{\prod_i f_i!}\prod_i p_i^{f_i} = \frac{n!}{\prod_i f_i!}\prod_i\left(\frac{np_i}{f_i}\right)^{f_i}\prod_i\left(\frac{f_i}{n}\right)^{f_i} \tag{16.4.11}$$

$$\log L = \sum_i f_i \log\left(\frac{np_i}{f_i}\right) + k$$

where k is a constant. For large sample, we may write

$$np_i = f_i + a_i\sqrt{n} \tag{16.4.12}$$

where a_i's are small compared to f_i and $\sum_i a_i = 0$. Hence

$$\log L = k + \sum_i f_i \log\left(1 + \frac{a_i\sqrt{n}}{f_i}\right)$$

$$= k + \sum_i a_i\sqrt{n} - \sum_i \frac{a_i^2 n}{2f_i} + \sum_i \frac{a_i^3 n^{3/2}}{3f_i^2} - \cdots$$

$$= k - \frac{1}{2}\sum_i \frac{(np_i - f_i)^2}{f_i} + O\left(n^{-1/2}\right). \tag{16.4.13}$$

Therefore, if we neglect terms of order $O\left(n^{-1/2}\right)$, the maximisation of $\log L$ becomes equivalent to minimisation of χ'^2. Thus, for large n the estimate obtained by minimisation of χ'^2 method would be approximately equivalent to the m.l.e.

However, difficulty arises in defining χ'^2 if there are empty cells. If $f_i = 0, p_i^{f_i} = 1$ so that contribution of this cell to $\log L$ is zero. However, writing np_i in the form (16.4.12) would mean a contribution of

$$\sum_i a_i\sqrt{n} = n\sum_i p_i = A, \text{ (say)},$$

(the summation being over all the empty cells), from these cells to $\log L$. Hence we should have

$$\log L = k - \frac{1}{2}\left\{\sum_i \frac{(np_i - f_i)^2}{f_i} + 2A\right\} + O\left(n^{-1/2}\right) \tag{16.4.14}$$

the summation being over all occupied cells.

Hence χ'^2 would be given by

$$\chi'^2 = \sum_i \frac{(np_i - f_i)^2}{f_i} + 2A$$

the summating being taken over all occupied cells.

The modified minimum chi-square estimators have been shown to be a Best Asymptotically Normal (BAN) estimators by Neyman (1949).

The M.L.-method, the MCS method and the modified MCS method all have the some asymptotic properties. From the point of view of computational convenience, no clear cut choice among the three procedures is feasible in general. Sometimes the ML-procedure, sometimes the MCS-procedure or modified MCS-procedure is advantageous over others. The MCS-procedures suffer from the further draw-back that the data must be grouped in several disjoint classes. If the variable is continuous, this seems to be an undesirable restriction on this estimation procedure. Again, determination of p_i's in terms of θ's sometimes may pose problems.

► **Example 16.4.1** (***Kendall and Stuart,* 1961**). *Consider Poisson distribution* $P_\theta(x=j) = e^{-\theta}\frac{\theta^j}{j!}, j = 0, 1, 2, \cdots$. *Here* $E(f_j) = m_j = np_j$ *where* $p_j = P_\theta(X = j)$

$$\frac{dp_j}{d\theta} = -e^{-\theta}\frac{\theta^j}{j!} + e^{-\theta}\frac{\theta^{j-1}}{(j-1)!} = e^{-\theta}\frac{\theta}{j!}\left[\frac{j}{\theta} - 1\right] = p_j\left[\frac{j}{\theta} - 1\right].$$

The minimising equation (16.4.3) is, therefore, dropping the factor $\frac{1}{n}$,

$$\sum_j \frac{f_j^2}{p_j}\left(1-\frac{j}{\theta}\right)=0. \tag{16.4.15}$$

This equation can be solved by using an iteration method of solution similar to that used for m.l.e. in section 16.2.5.

Expanding (16.4.15) as a Taylor series in θ about $\bar{x}$ (the m.l.e. of θ), which is taken as the initial trial value,

$$\sum_j \frac{f_j^2}{p_j}\left(1-\frac{j}{\theta}\right)=\sum_j \frac{f_j^2}{\pi_j}\left(1-\frac{j}{\bar{x}}\right)+(\theta-\bar{x})\sum_j\left\{\frac{j}{\bar{x}^2}+\left(1-\frac{j}{\bar{x}}\right)^2\right\}\frac{f_j^2}{\pi_j}, \tag{16.4.16}$$

where $\pi_j = e^{-\bar{x}}\bar{x}^j/j!$.

Equating (16.5.16) to zero, by (16.5.15), we have

$$\theta-\bar{x}=\bar{x}\left[\frac{\sum_j \frac{f_j^2}{\pi_j}(j-\bar{x})}{\sum_j \frac{f_j^2}{\pi_j}\{j+(j-\bar{x})^2\}}\right] \tag{16.4.17}$$

from which $\hat{\theta}^{(1)}$, the first estimate of θ can be found. The successive estimates can be found by the process of iteration.

In the above example $\bar{x}$ is not the MCS-estimate. It is known that for Poisson distribution $\bar{x}$ is a sufficient statistic and is a m.l.e. The example shows that MCS-method does not necessarily yield a sufficient statistic, even if it exists.

Similarly, one can also find the modified MCS estimator which is a root of the equation

$$\sum_j \frac{p_j^2}{f_j}\left(1-\frac{j}{\theta}\right)=0. \tag{16.4.18}$$

The equation (16.4.18) can also be solved iteratively as above, starting from an initial solution $\bar{x}$.

16.5 Method of Least Squares

Suppose we have n observations $(x_1, y_1), \cdots, (x_n, y_n)$ on two variables X and Y of which Y may be considered as a dependent variable whose values are determined by those of X. Suppose we want to fit a curve.

$$y=\phi(x; b_0, b_1, \cdots) \tag{16.5.1}$$

to the (x, y) values giving y as a function of x, where $b_0, b_1, \cdots$ are constants to be suitably determined. A measure of the goodness of fit of the curve ϕ to y-values is given by

$$SSE = \sum_i \{y_i - \phi(x_i; b_0, b_1, \cdots)\}^2, \tag{16.5.2}$$

$y_i - \phi(x_i) = e_i$ is the amount of error committed in estimating y_i by $\phi(x_i)$. Clearly lower is the value of the error sum of squares (SSE) better is the fit of the curve ϕ to the y-values. The method of least square estimates of constants $b_0, b_1, \cdots$ are those values of $b_0, b_1, \cdots$, (say) for which the SSE is minimum.

Suppose now $\phi(x, b_0, \cdots)$ is the expression for conditional expectation of Y given x. Then the equation (16.5.1) is called the regression equation of Y on x. Here we determine the best fit regression equation of Y on x by the method of least squares.

If we assume that the errors e_i are independently normally distributed with $(0, \sigma^2)$ then the likelihood function

$$L(b_0, b_1, \cdots, |y_i, x_i, \sigma^2) = \text{Const.} \exp\left[-\frac{1}{2\sigma^2}\sum_{i=1}^{n}\{y_i - \phi(x_i, b_0, b_1, \cdots)\}^2\right]. \tag{16.5.3}$$

Here maximising L amounts to minimising

$$\sum_{i=1}^{n}\{y_i - \phi(x_i; b_0, b_1, \cdots)\}^2.$$

Hence, in this case m.l.e.'s and l.s.e.'s of $b_0, b_1, \cdots$, are equivalent.

In case e_i's are independent $N(0, \sigma_i^2)$ variables, maximising L is equivalent to minimising

$$\sum_{i=1}^{n}\{y_i - \phi(x_i; b_0, b_1, \cdots)\}^2/\sigma_i^2 \tag{16.5.4}$$

which is the sum of squares of residuals each weighted by the inverse of its variance. This is called the weighted least square method.

▶ **Example 16.5.1** *Suppose $E(Y|x) = \phi(x) = b_0 + b_1 x + b_2 x^2 + \cdots + b_p x^p, n > p+1$ the pth degree polynomial regression equation of Y on X. Here we are to minimise*

$$S = \sum_{i=1}^{n}(y_i - b_0 - b_1 x_i - b_2 x_i^2 - \cdots - b_p x_i^p)^2$$

with respect to $b_0, b_1, \cdots, b_p$. The minimising equations are

$$\frac{\partial S}{\partial b_j} = 2\sum_{i=1}^{n} x_i^j (y_i - b_0 - b_1 x_i - \cdots - b_p x_i^p)^2 = 0$$

$$i.e., \quad \sum_{i=1}^{n} x_i^j y_i = b_0 \sum_{i=1}^{n} x_i^j + b_1 \sum_{i=1}^{n} x_i^{j+1} + \cdots + b_p \sum_{i=1}^{n} x_i^{j+p}, \quad j = 0, 1, \cdots, p$$

from which $b_0, b_1, \cdots, b_p$ can be solved.

► **Example 16.5.2** *Consider $(p+1)$ variables $Y, X_1, \cdots, X_p$ with observations $(y_i, x_{1i}$ $\cdots, x_{pi})$ $i = 1, \cdots, n$. Suppose $\phi(x_1, \cdots, x_n) = E(Y|x_1, \cdots, x_n) = b_0 + b_1x_1 + b_2x_2 + \cdots$ b_px_p, the multiple regression equation of Y on $x_1, \cdots, x_p$. The least square estimates o $b_0, b_1, \cdots, b_p$ are obtained by minimising*

$$S = \sum_{i=1}^{n}(y_i - b_0 - b_1x_{1i} - b_2x_{2i} - \cdots - b_px_{p_i})^2.$$

The normal equations are

$$\frac{\partial S}{\partial b_0} = 0 \quad \Rightarrow \quad \sum_{i=1}^{n} e_i = 0 \qquad (16.5.5.o)$$

$$\cdots \quad \cdots \quad \cdots \quad \cdots$$

$$\frac{\partial S}{\partial b_j} = 0 \quad \Rightarrow \quad \sum_{i=1}^{n} e_ix_{ji} = 0 \quad (j = 1, \cdots, p) \qquad (16.5.5.j)$$

(16.5.5.o) *to* (16.5.5.j) *give respectively,*

$$\sum_{i=1}^{n} y_i = nb_0 + b_1\sum_{i=1}^{n} x_{1i} + \cdots + b_p\sum_{i=1}^{n} x_{pi}$$

$$\cdots \quad \cdots \quad \cdots \quad \cdots$$

$$\sum_{i=1}^{n} y_ix_{ji} = b_0\sum_{i} x_{ji} + b_1\sum_{1} x_{1i}x_{ji} + \cdots + b_p\sum_{i} x_{ji}x_{pi}, \quad (j = 1, \cdots, p)$$

from which $b_0, b_1, \cdots, b_p$ can be solve out.

The method of least squares can be used even when we do not know the probabilit density function of the distribution being sampled and is particularly convenient whe the sample values are independent. In many experimental situations it is realisticall supposed that the observations are composed from a known number of fixed effect together with an experimental error. Often these fixed effects can be represented a a linear combination of unknown parameters. Also the errors are often independentl distributed with constant variances. For this case, the least-squares estimators are als the linear unbiased estimators with minimum variance. When it is further known th the random variables are normally distributed, then the least-square estimators of fixe effects coincide with the maximum likelihood estimators. (For further discussion on lea squares theory the reader may refer to Chapter 21).

Exercise

1. Let $X_i (i = 1, \cdots, n)$ be a random sample of size n from the geometric distributic $f_p(x) = p(1-p)^{x-1}, x = 1, 2, \cdots; 0 < p < 1$. Find the m.l.e. of p.

   Suppose a randomly selected sample of size 5 gives the following data: $X_1 =$ $X_2 = 7$, $X_3 = 6$, $X_4 = 5$, $X_3 = 9$ (i.e., in the first sample, the first trial resul in failure and the second trial in success, etc.) Based on these data compute th m.l.e. of $P(X \geq 4)$.

2. Samples are drawn at random from a Bernoulli (p)-population [$P(X = 1) = p, P(X = 0) = 1 - p, 0 \le p \le 1$] until the sum of the sampled observations is n. Suppose that m samples are drawn to obtain the sample total n. Show that the p.m.f. of m is $\binom{m-1}{n-1} p^n(1-p)^{m-n}$, $m = n, n+1, \cdots$, and that the m.l.e. of p is n/m.

3. From a box containing N identical balls marked $1, \cdots, N, M$ balls are drawn one after another without replacement. Let $X_1, \cdots, X_M$ denote the numbers marked on the sampled balls. Find the m.l.e. and UMVUE of N. Compare their m.s.e.'s.

4. Let $X_1, \cdots, X_m$ be independent random variables from Binomial $B(n_1, p), \cdots,$ $B(n_m, p)$ distributions respectively. Find the m.l.e. of p.

5. For a random sample of size n corresponding to the pmf

$$f(x;\theta) = \begin{cases} \theta^x(1-\theta)^{1-x}, & x = 0, 1; 0 < \theta < 1 \\ 0 \quad , & \text{otherwise} \end{cases}$$

find an unbiased estimator of the variance of the maximum likelihood estimator of θ. Justify why it is not possible to find an unbiased estimator of $\frac{1}{\theta}$.

6. A random sample of size n is drawn from a Poisson (λ)-population. Find an m.l.e. of $e^{-\lambda}$. Find an unbiased estimator of $e^{-\lambda}$.

7. A random sample of size n is drawn from a multinomial population with parameters $(p_1, \cdots, p_k), \Sigma p_i = 1$. Show that the m.l.e.'s of $p_1, \cdots, p_k$ are $\bar{x}_1, \cdots, \bar{x}_k$ where $\bar{x}_i = \frac{1}{n}\Sigma_{j=1}^n x_{ij}$ and that the variance covariance matrix of the estimators is $\left(\left(\frac{1}{n}(\delta_{ij}p_i - p_ip_j)\right)\right)$ where $\delta_{ij} = 1(0)$ if $i = j$ (otherwise).

8. A random sample of size n is drawn from the following population:

$$f_{\theta,c}(x) = \begin{cases} \theta, & \text{if } x = 1 \\ (1-\theta)/(c-1), & \text{if } x = 2, 3, \cdots, c \\ 0, & \text{otherwise} \end{cases}$$

where $\theta(\varepsilon[0,1])$ and c $(= 2, 3, \cdots)$ are unknown.

Find m.l.e.'s of θ and c.

[*Hints:* $L(\theta, c|x_1, \cdots, x_n) = \frac{\theta^s(1-\theta)^{n-s}}{(c-1)^{n-s}}$ if $x_i \le c$

$(i = 1, \cdots, n)$ where s is the number of x_i's that equal one.

Case 1. $s < n$: here $(c-1)^{n-s}$ is minimised for $c = \max_i x_i = x_{(n)}$ (say); hence, $\hat{c} = x_{(n)}$. Again $\theta^s(1-\theta)^{n-s}$ is maximised for $\hat{\theta} = s/n$.

Case 2. $s = n$: here $L(\theta, c) = \theta^n$, a constant $\forall$ c. Since $\theta \in [0, 1]$, $\hat{\theta} = 1$. Again any admissible value of c maximises the likelihood. The m.l.e. of c is not unique].

9. A random sample of size n is drawn from a population with p.d.f. $f_\theta(x) = Kx^\theta(1-x)$, $0 \le x \le 1$ where K is a constant. Show that the m.l.e. of θ is $\hat{\theta} = \frac{-(3G+2)+\sqrt{G^2+4}}{2G}$ where $G = \frac{1}{n}\ln\left(\prod_{i=1}^{n} x_i\right)$.

[*Hints:* $f_\theta(x) = (\theta+1)(\theta+2)x^\theta(1-x)$, when $\theta > -1$].

10. In a sample survey for estimating the size of fish-population in a lake, a random sample of c fishes are captured and then released in the water after tagging. After sufficient time have elapsed for their thorough mixing with the untagged fishes, n fishes are recaptured and the number of tagged fishes x is counted. Find the m.l.e. of N, the total number of fishes in the lake, assuming no change in the size of the fish-population during the experiment.

[*Hints:* $L(N|x) = \binom{c}{x}\binom{N-c}{n-x}/\binom{N}{n}$; $L(N)/L(N-1) < 1$ iff $(N-n)(N-c) < N(N-c-n+x)$ i.e. iff $N > nc/x$. Hence $\hat{N} = nc/x$]

11. Find the m.l.e. of θ based on a sample of size n from the rectangular population over $[\theta, 2\theta]$. If $x_{(1)} = \min x_i, x_{(n)} = \max x_i$, show that $t = [2x_{(n)} + x_{(1)}]/5$ is more efficient than the m.l.e.

12. Let $X_1, \cdots, X_n$ be a sample from a Rectangular population over $\left[\theta - \frac{1}{2}, \theta + \frac{1}{2}\right]$. Show that every statistic $t(X_1, \cdots, X_n)$ satisfying $\max x_i - \frac{1}{2} \le t(X_1, \cdots, X_n) \le \min x_i + \frac{1}{2}$ is an m.l.e. of θ.

13. $X_1, \cdots, X_n$ are independent random variables with the following distributions. Find m.l.e.'s of the parameters as indicated. Are the m.l.e.'s MVU?

(a) $P(X_i = 0) = 1 - p, P(X_i = 1) = p$ where $p \in \left[\frac{1}{4}, \frac{3}{4}\right]$.

(b) $N(\theta, \theta^2), \theta \in (0, \infty)$.

(c) $f_\theta(x) = \frac{\theta}{1-\theta}x^{(2\theta-1)/(1-\theta)}, 0 < x \le 1, \frac{1}{2} < \theta < 1$.

(d) $f_\theta(x) = \frac{x}{\theta^2}\exp\left\{-\frac{x^2}{\theta^2}\right\}$, $x > 0, \theta > 0$ (Reyleigh distribution).

(e) $f_{\alpha,\beta}(x) = \frac{\alpha\beta^\alpha}{x^{\alpha+1}}$, $\alpha > 0, x \ge \beta > 0$ (Pareto distribution).

(f) $f_{c,a,\mu}(x) = \frac{c}{a}\left(\frac{x-\mu}{a}\right)^{c-1}\exp\left[-\left(\frac{x-\mu}{a}\right)^2\right], x > \mu$ (Weibull distribution).

(g) $f_{\theta,\mu}(x) = \theta^{-1}e^{-\frac{(x-\mu)}{\theta}}\exp\left[-e^{-\frac{(x-\mu)}{\theta}}\right], \theta > 0$ (Extreme Value distribution).

(h) $f_{\mu,\sigma}(x) = \frac{e^{-(x-\mu)/\sigma}}{\sigma\left[1+e^{-\frac{(x-\mu)}{\sigma}}\right]^2}, -\infty < x < \infty, \sigma > 0$ (Logistic distribution).

(i) $f_\theta(x) = \theta(1-x)^{\theta-1}, 0 \le x \le 1, \theta > 1$.

(j) $f_\theta(x) = (\theta\ \theta_1)x^{\theta_1-1}\exp\left[-\theta x^{\theta_1}\right], x > 0, \theta_1(> 0)$ known.

(k) $f_{b,c}(x) = \frac{\left(\frac{x}{b}\right)^{c-1} e^{-\frac{x}{b}}}{b[(c-1)!]}$, $0 \le x \le \infty, b > 0, c$ (integer) > 0. [Erlang distribution].

(l) A Cauchy distribution.

(m) $f_\theta(x) = \frac{\theta^{m+1} x^m e^{-\theta x}}{\Gamma(m+1)}$ for fixed (known) $m; \theta, x > 0$.

14. Show that the m.l.e. of the parameter α and β of the distribution

$$f_{\alpha,\beta}(x) = \frac{1}{\Gamma(\beta)} \left(\frac{\beta}{\alpha}\right)^{\beta} \exp\left[-\frac{\beta x}{\alpha}\right] x^{\beta-1}, \quad 0 < x < \infty$$

are

$$\bar{x}, 1 \Big/ \sum_{i=1}^{n} \left(2 \log \bar{x} - \frac{1}{n} \sum_{i=1}^{n} \log x_i\right).$$

Find also the variance and covariance of the estimators in large samples. Assume that for large β, $\frac{d}{d\beta} \log \Gamma(\beta) = \log \beta - \frac{1}{2\beta}$.

15. Let $X_1, \cdots, X_n$ be iid gamma variables $f_{\alpha,\beta}(x) = \frac{1}{\Gamma(\alpha)\, \beta^\alpha} e^{-x/\beta} x^{\alpha-1}$, $x > 0$, $\alpha > 0$, $\beta > 0$.

Find the m.l.e. of (α, β).

[*Hints:*

$$\ln L(\alpha, \beta) = -n \ln(\Gamma(\alpha)) - n\alpha \ln(\beta) - \sum_{i=1}^{n} x_i/\beta + (\alpha - 1) \sum \ln(x_i)$$

$$\frac{\partial \ln L(\alpha, \beta)}{\partial \alpha} = 0 \quad \Rightarrow \quad \frac{n\Gamma(\alpha)'}{\Gamma(\alpha)} + n \ln(\beta) - \sum \ln(x_i) = 0 \qquad \text{(i)}$$

$$\frac{\partial \ln L(\alpha, \beta)}{\partial \beta} = 0 \quad \Rightarrow \quad n\beta\alpha - \Sigma x_i = 0. \qquad \text{(ii)}$$

(ii) $\Rightarrow \hat{\beta} = \bar{x}/\hat{\alpha}$. (i) implies

$$\frac{n\Gamma(\hat{\alpha})'}{\Gamma(\hat{\alpha})} + n \ln(\bar{x}/\hat{\alpha}) - \sum \ln(x_i) = 0 \qquad \text{(iii)}$$

$\left[\text{where } \Gamma(m)' = \int_0^\infty x^{m-1} \ln(x) e^{-x}\, dx\right]$.

Simplifying (iii),

$$\frac{\Gamma(\hat{\alpha})'}{\Gamma(\hat{\alpha})} - \ln(\hat{\alpha}) - \ln(R) = 0 \qquad \text{(iv)}$$

where

$$R = \left[\left(\prod_{i=1}^{n} x_i\right)^{1/n}\right] \Big/ \bar{x}.$$

The solution $\hat{\alpha}$ of equation (iv) can be obtained numerically.

Wilks, Gnanadesikan and Huyett (1962) computed the relationship between $\hat{\alpha}$ and $1/(1 - R)$.]

16. Let $X_1, \cdots, X_n$ be iid as an exponential distribution $f_\theta(x) = \theta e^{-\theta x}, 0 < x < \infty$, $\theta > 0$. Find the m.l.e. of θ and show that it is consistent and asymptotically normal.

17. Let $X_i (i = 1, \cdots, n)$ be a random sample from the truncated Laplace distribution

$$f_\theta(x) = \frac{1}{2(1-e^{-\theta})} e^{-|x|}, \quad |x| < \theta, \theta > 0.$$

Find an m.l.e. of θ. Is the estimator unbiased? Consistent?.

18. A random sample of size n is drawn from the lognormal distribution

$$f_{\mu,\sigma}(x) = (\sigma\sqrt{2\pi})^{-1} x^{-1} \exp\left[-\frac{1}{2\sigma^2}(\log x - \mu)^2\right], \quad x > 0.$$

Find the m.l.e. of μ, σ^2, $E(X)$ and $V(X)$.

$$\left[\hat{\mu} = \bar{Y}, \hat{\sigma}^2 = \frac{1}{n}\sum_{i=1}^{n}(Y_i - \bar{Y})^2, \hat{E}(X) = \exp\left[\bar{Y} + \sum_{i=1}^{n}(Y_i - \bar{Y})^2/(2n)\right]\right.$$

$$\left.\hat{V}(X) = \left[\exp\{2\bar{Y} + \Sigma(Y_i - \bar{Y})^2/n\}\right]\left[\exp\left\{\Sigma(Y_i - \bar{Y})^2/n\right\} - 1\right] \text{ where } Y_i = \ln X_i\right]$$

19. $\{X_i, i = 1, \cdots, n\}$ is a random sample from the Beta distribution,

$$f_{\alpha,\beta}(x) = \frac{1}{B(\alpha,\beta)} x^{\alpha-1}(1-x)^{\beta-1}, \quad 0 < x < 1, \beta (> 0)$$

known, but $\alpha (> 0)$ unknown.

Find the m.l.e. of α if $\beta = 1, 2$.

20. Let $\{X_i, i = 1, \cdots, n\}$ be a random sample of size n from a population with p.d.f. $f_\theta(x) = x\, e^{-x/\theta}, \theta > 0, x > 0$. Find the m.l.e. of θ and calculate its variance. Show that $\sum_{i=1}^{n} x_i^2/3n$ is an unbiased estimate of $V(X)$ and $\bar{x}/2$ is a biased estimate.

21. (a) Let $\{X_i, i = 1, \cdots, n\}$ be a random sample from $N(\mu, \sigma^2)$ population and independently $\{Y_i, i = 1, \cdots, m\}$ be a random sample from $N(\mu, \gamma\sigma^2)$ population $(\gamma > 0)$. Find the m.l.e. of γ when μ, σ^2 are (i) known (ii) all unknown.

(b) Independent samples of sizes n_1, n_2 are taken from two normal population with equal means μ and variance $\lambda\sigma^2$ and σ^2 respectively. Find the maximum likelihood estimate of μ and show that its large sample variance is

$$\frac{\sigma^2}{\left(\frac{n_1}{\lambda} + n_2\right)}.$$

Hence show that the unbiased estimator

$$t = \frac{n_1\bar{x}_1 + n_2\bar{x}_2}{n_1 + n_2}$$

has efficiency $\lambda(n_1+n_2)^2/(n_1\lambda+n_2)(n_1+n_2\lambda)$ which attains the value 1 iff $\lambda = 1$.

22. Let $\{X_i, i = 1, \cdots, m\}$ be a random sample from $N(\mu, \sigma_1^2)$ distribution and independently $\{Y_i, i = 1, \cdots, n\}$ be a random sample from $N(\mu, \sigma_2^2)$ distribution. Find m.l.e.'s of $\mu, \sigma_1^2, \sigma_2^2$. Find variance of these estimators.

23. Let $f_i(x)$ be a p.d.f. with mean μ_i, variance σ_i^2 $(i = 1, 2)$. Consider a random sample of size 2 from the mixture distribution $f_\theta(x) = \theta f_1(x) + (1-\theta) f_2(x)$. If $\mu_i, \sigma_i^2 (i = 1, 2)$ are all known, find m.l.e. of θ.

24. Let (X, Y) follow bivariate normal $N_2(\mu_1, \mu_2, \sigma_1^2, \sigma_2^2, \rho)$.

    Suppose m observations are available on the pair (X_i, Y_i), $i = 1, \cdots, m$ while the other $(n - m)$ observations are available only on X. Find the m.l.e. of $\mu_1, \mu_2, \sigma_1, \sigma_2, \rho$.

    [*Hints:* Joint distribution $f_{X,Y}(x, y|\mu_1, \mu_2, \sigma_1^2, \sigma_2^2, \rho) = f_X(x|\mu_1, \sigma_1^2) f_{Y|X}(y|\eta_x, \sigma_2^2(1-\rho^2))$ (i) where $\eta_x = \mu_2 + \rho\frac{\sigma_2}{\sigma_1}(x - \mu_1)$ and $f(\bullet|a, b^2)$ denotes a normal density with mean a, variance b^2.

    For estimating μ_1, σ_1^2 we can use all the n available X_i values while for the other parameters we have also to use the incomplete data on Y_i. The joint distribution depends on five parameters $\mu_1, \sigma_1^2, \mu_2' = \mu_2 - \rho\frac{\sigma_2}{\sigma_1}$, $\rho' = \rho\frac{\sigma_2}{\sigma_1}$, $\sigma_2'^2 = \sigma_2^2(1-\rho^2)$.

    We first maximise the likelihood (i) with respect to μ_1, σ_1^2 and then with respect to $\mu_2', \rho', \sigma_2'^2$.]

25. Let $\{(X_i, Y_i), i = 1, \cdots, n\}$ be a sample of size n from a standard bivariate normal population. Show that the m.l.e. of ρ is a root of the equation

$$n\rho^3 - \rho^2\Sigma xy + \rho\left(\Sigma x^2 + \Sigma y^2 - n\right) - \Sigma xy = 0.$$

    By calculating $E\left(\frac{\partial^2 \log L}{\partial \rho^2}\right)$ show that asymptotically, $V(\hat{\rho}) = \frac{(1-\rho^2)^2}{n(1+\rho^2)}$.

26. Suppose $\boldsymbol{X}_1, \cdots, \boldsymbol{X}_n$ are each iid N_p $(\boldsymbol{\mu}, \boldsymbol{\Sigma})$. Show that

$$\bar{\boldsymbol{X}} = \frac{1}{n}\sum_{i=1}^{n} \boldsymbol{X}_i \text{ and } \boldsymbol{S} = \frac{1}{(n-1)}\sum_{i=1}^{n}(\boldsymbol{X}_i - \bar{\boldsymbol{X}})(\boldsymbol{X}_i - \bar{\boldsymbol{X}})'$$

    are unbiased for $\boldsymbol{\mu}$ and $\boldsymbol{\Sigma}$ respectively. Prove also that $\left(\bar{\boldsymbol{X}}, \frac{(n-1)}{n}\boldsymbol{S}\right)$ are joint m.l.e.'s of $\boldsymbol{\mu}$ and $\boldsymbol{\Sigma}$ respectively. Find the m.l.e. of $\boldsymbol{\sigma}$ when it is known that $\boldsymbol{\mu} = \boldsymbol{\mu}_0$ (given).

27. Show that the most general form of a distribution for which the sample geometric mean $G = \left\{ \prod_{i=1}^{n} X_i \right\}^{1/n}$ is the m.l.e. of θ has the p.d.f. (p.m.f.)

$$f_\theta(x) = \left(\tfrac{x}{\theta}\right)^{\theta A'(\theta)} \exp\left[A(\theta) + B(x)\right].$$

28. Show that the most general form of a distribution for which the sample harmonic mean

$$H = n \Big/ \sum_{i=1}^{n} \frac{1}{X_i}$$

is the m.l.e. of θ has the p.d.f. (p.m.f.) $f_\theta(x) = \exp\left[\frac{1}{x}\{\theta A'(\theta) - A(\theta)\} - A'(\theta) + B\right.$

29. Let $Y_1 < \cdots < Y_n$ be the order statistic of a random sample of size n from the population with density

$$\sigma^{-1} \exp\left(-\frac{x-\mu}{\sigma}\right),$$

where μ and σ are both unknown. Further, let $L(\mu, \sigma)$ be the likelihood function. Writing $nS = \sum_{i=1}^{n} (Y_i - Y_1)$, show that for all (μ, σ),

$$L(Y_1, S) \geq L(\mu, \sigma)$$

and comment. Also show that, using the first principle, Y_1 is a consistent estimator of μ.

30. Suppose the r.v. X has a Binomial (n, p) distribution where n is known, $0 \leq p \leq 1$. Show that the MCS-estimator of p is identical with the m.l.e. for any n. Also show that if X is not 0 or n, the modified MCS estimator is also identical with the m.l.e.

31. A random sample of size n is drawn from a Poisson (λ) population and the frequency n_0 for $X = 0$ and n_1 for $X \geq 1$ are recorded separately. Obtain the m.l.e. $\hat{\lambda}$ of λ and its large sample variance. Show that the efficiency of $\hat{\lambda}$ with respect to the usual m.l.e. $\hat{\hat{\lambda}}$ when the observations are fully recorded is $\lambda/(e^\lambda - 1)$. Similarly, if n_0, n_1, n_2 denote respectively the frequencies for $X = 0, X = 1, X \geq 2$ respectively, show that the m.l.e. of λ is

$$\tilde{\lambda} = \frac{\lambda[1-(1+\lambda)e^{-\lambda}]}{ne^{-\lambda}(1-e^{-\lambda})(1-\lambda)^2}.$$

Hence show that for sufficiently small λ,

$$V(\tilde{\lambda}) \simeq \tfrac{\lambda}{2} \, V(\hat{\lambda}).$$

32. Let X be a random variable denoting the time for the failure of an item (say, an electric bulb). If time is measured in discrete periods, assume that the p.m.f. of X is

$$P_\theta[X = k] = \theta^{k-1}(1-\theta), \quad k = 1, 2, \cdots, 0 < \theta < 1.$$

Suppose that we only record the time of failure if $k \leq r$, otherwise only note that the item has lived at least r time periods. Thus, we observe only the random variables $Y_1, \cdots, Y_n$ where $Y_i = k$ if the ith item fails at time $k (k = 1, \cdots, r)$ and $Y_i = r + 1$ if the item survives at least $r + 1$ time periods. The r.v.'s $Y_1, \cdots, Y_n$ are, therefore, iid with p.m.f.

$$f_\theta(k) = P_\theta(X = k), k = 1, \cdots, r,$$
$$f_\theta(r+1) = 1 - P_\theta(X \leq r) = \theta^r.$$

If M is the number of elements with $Y_i = r + 1$, show that the m.l.e. of θ based on $(Y_1, \cdots, Y_n)$ is

$$\hat{\theta} = \frac{\sum_{i=1}^{n} Y_i - n}{\sum_{i=1}^{n} Y_i - M}.$$

33. In an animal breeding experiment, four distinct kinds of progeny were observed with frequencies n_1, n_2, n_3, n_4 $(\Sigma n_i = N)$. The corresponding expected frequencies are $\frac{N}{4}(2+p)$, $\frac{N}{4}(1-p)$, $\frac{N}{4}(1-p)$, $\frac{N}{4}p$ where p is an unknown parameter. Find $\hat{p}$, the m.l.e. of p and show that its large-sample variance is $\frac{2p(1-p)(2+p)}{N(1+2p)}$.

34. A random sample of size n drawn from a population with three kinds of elements in Hardy-Weinberg proportion $\theta^2, 2\theta(1-\theta), (1-\theta)^2$ gives the frequencies as n_1, n_2, n_3 respectively. Find the m.l.e. of θ. $[\hat{\theta} = \frac{2n_1+n_2}{2n}]$

35. Observed cell-frequencies and the corresponding probabilities are given below for a coupling intercross, where θ is the recombination fraction of two linked genes.

Class	Frequency	Probability
1	125	$\frac{1}{4}(3 - 2\theta + \theta^2)$
2	18	$\frac{1}{4}(2\theta - \theta^2)$
3	20	$\frac{1}{4}(2\theta - \theta^2)$
4	34	$\frac{1}{4}(1 - 2\theta + \theta^2)$.

Find the m.l.e. of θ.

36. Describe the method of scoring for simultaneous estimation of k parameters by the method of maximum likelihood.

Let X have a multinomial distribution with probabilities

$$\left\{\frac{1}{4}(2+\theta^2),\ \frac{1}{4}(1-\theta^2),\ \frac{1}{4}(1-\theta^2),\ \frac{1}{4}\theta^2\right\}.$$

Obtain the first iterate in the method of scoring with an initial estimate $\hat{\theta}_0$ for θ.

37. Let $X_1, \cdots, X_n$ be a sample from $N(\mu, \sigma^2)$. Find the MME's of μ and σ^2.

38. Let $X_1, \cdots, X_n$ be a sample from Gamma population

$$\frac{1}{\left|(\alpha)\,\beta^\alpha\right.} e^{-x/\beta} x^{\alpha-1}, \quad 0 < x < \infty, \quad \alpha > 0, \quad \beta > 0.$$

Find MME's of (α, β).

39. Let $X_1, \cdots, X_n$ be a random sample from the lognormal p.d.f.

$$f_{(\mu,\sigma)}(x) = (\sigma\sqrt{2\pi})^{-1} x^{-1} \exp\left\{-\frac{1}{2\sigma^2}(\log x - \mu)^2\right\}, \quad x > 0.$$

Show that the MME's are

$$\hat{\mu} = \frac{1}{2}\log\left[\frac{x^4}{s^2+\bar{x}^2}\right], \quad \hat{\sigma}^2 = \log\left(\frac{\bar{x}^2+s^2}{\bar{x}^2}\right)$$

where s^2 is the sample variance.

[*Hints:* $E(X) = e^{\mu+\frac{\sigma^2}{2}}$, $E(X^2) = \left(e^{\mu+\sigma^2/2}\right)^2 e^{\sigma^2}$. Equating $E(X)$ with $\bar{X}$ and $E(X^2)$ with $\frac{1}{n}\Sigma X_i^2$,

$$\hat{\sigma}^2 = \ln\left(\frac{\Sigma X_i^2}{n\bar{X}^2}\right), \quad \hat{\mu} = \ln\left(\sqrt{n}\bar{X}^2 / \sqrt{\Sigma X_i^2}\right).$$

Alternatively, since $\ln(X_i)$ is $N(\mu, \sigma^2)$, based on the random sample $\ln(X_1), \ln(X_2), \cdots, \ln(X_n)$, the MME's are

$$\hat{\hat{\mu}} = \frac{1}{n}\sum \ln(X_i), \quad \hat{\hat{\sigma}}^2 = \frac{1}{n}\Sigma\left(\ln(X_i) - \hat{\hat{\mu}}\right)^2.]$$

40. Let $X_1, \cdots, X_m$ be a random sample from Binomial $B(n, p)$ population. Show that the method of moment estimators for p and n are given by

$$\hat{p} = \frac{\bar{x}}{t_2(x_1, \cdots, x_n)}, \quad \hat{n} = t_2(x_1, \cdots, x_n) = \frac{\bar{x}^2}{\bar{x} - \frac{1}{N}\Sigma(x_i - \bar{x})^2}$$

Show that $\bar{x} \xrightarrow{P} np$, but $t_2(X_1, \cdots, X_n) \not\xrightarrow{P} n$.

41. Let $X_1, \cdots, X_n$ be a random sample from Rectangular (a, b)-population. By considering the first two moments, show that MME's of a and b, respectively, are

$$\hat{a} = \bar{x} - \sqrt{3\frac{(n-1)}{n}s^2}; \quad \hat{b} = \bar{x} + \sqrt{\frac{3(n-1)}{n}s^2}$$

where $s^2 = \frac{1}{(n-1)}\sum_{i=1}^{n}(x_i - \bar{x})^2$.

42. Find MME for the Laplace distribution

$$f_\theta(x) = \frac{1}{2\theta} e^{-|x|/\theta}, \quad -\infty < x < \infty, \theta > 0.$$

[*Hints:* Here $E(X) = 0$. Hence $E(X) = \hat{X}$ does not yield an estimate of θ.

Consider, therefore, $E(X^2) = 2\theta^2$. The equation $E(X^2) = \frac{1}{n}\Sigma X_i^2$ yields

$$\hat{\theta} = \sqrt{\frac{1}{2n}\sum_1^n X_i^2}.$$

Note that m.l.e. of θ is $\frac{1}{n}\sum |X_i|$.]

43. Let $X_1, \cdots, X_n$ be independent r.v.'s each with p.d.f. $f_\theta(x) = \frac{1}{2\theta}, -\theta < x < \theta$, $\theta > 0$. Find MME of θ and show that it is a biased estimator.

44. Let $X_1, \cdots, X_n$ be a random sample of size n from the p.d.f. $f_\theta(x) = \frac{\theta}{x^2}, \theta \le x < \infty$, $\theta > 0$. Find MME of θ.

45. Let $X_1, \cdots, X_n$ be a random sample of size n from the p.d.f.

$$f_{(\alpha,\beta)}(x) = \frac{1}{\Gamma(\beta)}\left(\frac{\beta}{\alpha}\right)^\beta e^{-(\beta/\alpha)x} x^{\beta-1}, \quad 0 < x < \alpha.$$

Find the MME's of α and β.

46. Let $X_1, \cdots, X_n$ be a random sample from Beta population

$$\frac{1}{B(\alpha,\beta)} x^{\alpha-1}(1-x)^{\beta-1}, \quad 0 < x < 1, \quad \alpha > 0, \quad \beta > 0.$$

Find the MME's of (α, β).

47. Consider the following lines model

$$\boldsymbol{Y} = \boldsymbol{X\beta} + \boldsymbol{e}$$

when

$$\boldsymbol{Y} = (\boldsymbol{Y}'_1, \cdots, \boldsymbol{Y}'_q)'$$
$$\boldsymbol{X} = [\boldsymbol{X}'_1, \cdots, \boldsymbol{X}'_1]',$$
$$\boldsymbol{e} = (\boldsymbol{e}'_1, \cdots, \boldsymbol{e}'_q),$$
$$\boldsymbol{\beta} = [\beta_1, \cdots, \beta_p]'.$$

$\boldsymbol{Y}_i$ is $n_i \times 1$, $\boldsymbol{X}_i$ is $n_i \times p$, β is a $p \times 1$ vector of unknown parameters, $\boldsymbol{e}_i$ is $n_i \times$ 1
$n = \sum_{i=1}^{q} n_i$, $\boldsymbol{X}$ is a matrix of known elements, and $\boldsymbol{e}$ is a random vector having n-variate normal distribution with mean vector $E(\boldsymbol{e}) = \boldsymbol{0}$ and dispersion matrix

$$\sum = \begin{bmatrix} \sigma_1^2 \boldsymbol{I} n_1 & \boldsymbol{0} & \cdots & \boldsymbol{0} \\ \boldsymbol{0} & \sigma_2^2 \boldsymbol{I} n_2 & \cdots & \boldsymbol{0} \\ \boldsymbol{0} & \boldsymbol{0} & \cdots & \sigma_q^2 \boldsymbol{I} n_q \end{bmatrix},$$

$\boldsymbol{I}_k$ being a unit matrix of order k. Assume $n_i \geq (p+2)\ \forall\, i = 1, \cdots, q$.

Show that the m.l. estimator of $\boldsymbol{\beta}$ and σ_i^2 are

$$\hat{\boldsymbol{\beta}} = \left[\sum_{i=1}^{q} \hat{\sigma}_i^{-2} \boldsymbol{X}_i' \boldsymbol{X}_i\right]^{-1} \left[\sum_{i=1}^{q} \hat{\sigma}_i^{-2} \boldsymbol{X}_i' \boldsymbol{Y}_i\right],$$

$$\hat{\sigma}_i^2 = \frac{1}{n_i}(\boldsymbol{Y}_i - \boldsymbol{X}_i \hat{\boldsymbol{\beta}})'(\boldsymbol{Y}_i - \boldsymbol{X}_i \hat{\boldsymbol{\beta}}) . \ (i = 1, \cdots, q).$$

Chapter 17

Tests of Hypotheses I

17.1 Introduction

In this chapter we consider tests of statistical hypotheses. Let $X_1, \cdots, X_n$ be a random sample from a population distribution F_θ, $\theta \in \Theta$, where the functional form of F_θ is known, except perhaps, for the parameter θ. Thus, for example, we may have a random sample from $N(\mu, 1)$ population where the value of μ is unknown. One may be interested in examining the validity of the assertion that the value of μ lies in a certain known range, say, (μ_1, μ_2) on the basis of the sample drawn from the population. A problem of this type is usually referred to as a problem of testing of hypotheses. In this chapter we will develop the fundamental concepts of Neyman-Pearson theory of testing of hypotheses. Sections 17.2, 17.3 give respectively some basic concepts and optimality criteria. Section 17.4 considers testing a simple hypothesis against a simple alternative. Section 17.5 deals with the relation between the most powerful test and sufficient statistics for the same problem. Section 17.6 considers the randomised tests. Testing of simple and composite null hypotheses against composite alternative hypotheses have been considered in Sections 17.7, 17.8. Section 17.9 narrates some properties of class of UMP tests.

17.2 Some Basic Notions

Definition 17.2.1 *A parametric hypothesis is a statement about the unknown parameter θ. It is usually referred to as the null hypothesis $H_o : \theta \in \Theta_0$ where $\Theta_0 \subset \Theta$. The statement $H_1 : \theta \in \Theta_1$ is usually referred to as the alternative hypothesis. Our task is to test H_o against H_1. Clearly θ may be a vector of parameters $\boldsymbol{\theta} = (\theta_1, \cdots, \theta_q)'$ where $\boldsymbol{\theta} \in \boldsymbol{\Theta}$.*

Definition 17.2.2 *If Θ_0 contains only one point, say, θ_0, the hypothesis H_o is said to be a simple one. Otherwise, i.e., if Θ_0 contains more than one point, the hypothesis H_o is said to be a composite one. Similar definition holds also for alternative hypotheses.*

Clearly, under a simple hypothesis the p.d.f. of a random variable X is completely specified.

▶ **Example 17.2.1** *Suppose $X \sim N(\mu, \sigma^2)$ where σ^2 is known. The hypothesis $H_o(\mu = \mu_0)$ is a simple hypothesis. The hypotheses $H(\mu \leq \mu_0)$, $H(\mu > \mu_0)$ are composit hypotheses.*

If σ^2 is also unknown, $H_o(\mu = \mu_0)$ is a composite hypothesis, because, here $\Theta = \{(\mu, \sigma^2), -\infty < \mu < \infty, \sigma^2 > 0\}$ and $\Theta_0 = \{\mu = \mu_0, \sigma^2 > 0\}$ contains infinitely many points. Again, $H_o(\mu = \mu_0, \sigma^2 = \sigma_0^2)$ is a simple hypothesis.

Often we are interested in testing a simple hypothesis $H_o(\theta = \theta_0)$ against alternative composite hypothesis $H_1(\theta \neq \theta_0)$ called a both-sided alternative or one-sided composite alternatives $H_1(\theta < \theta_0)$, $H_2(\theta > \theta_0)$.

The problem of testing a hypothesis H_o against an alternative H_1 may be described as follows. Given the sample observations, $\boldsymbol{x} = (x_1, \cdots, x_n)'$, we make a decision which will lead to either acceptance or rejection of H_o. The sample space $\mathcal{X}_n$ of the random vector $\boldsymbol{X} = (X_1, \cdots, X_n)'$ is divided into two disjoint subsets ω and $\omega^c = \mathcal{X}_n - \omega$ such that H_o is rejected if $\boldsymbol{x} \in \omega$ and is accepted if $\boldsymbol{x} \in \omega^c$. The region ω is called the critical region and ω^c the region of acceptance. Such a test is called a non-randomised test of H_o against H_1.

Let $\delta(\boldsymbol{x})$ denote a function denoting the probability of rejecting the null hypothesis H_o when $\boldsymbol{x}$ is the sample observation. Clearly, for a non-randomised test

$$\delta(\boldsymbol{x}) = \begin{cases} 1 \text{ if } \boldsymbol{x} \in \omega \\ 0 \text{ if } \boldsymbol{x} \in \omega^c. \end{cases} \tag{17.2.1}$$

Definition 17.2.3 *Every Borel-measurable mapping ϕ of $R^n \rightarrow [0, 1]$ is called a test function.*

Clearly, $\delta(\boldsymbol{x})$ is a test function. Some other examples of a test function are $\phi(\boldsymbol{x}) = 0 \ \forall \ \boldsymbol{x} \in R^n$; $\phi(\boldsymbol{x}) = 1 \ \forall \ \boldsymbol{x} \in R^n$; $\phi(\boldsymbol{x}) = \alpha \ \forall \ \boldsymbol{x} \in R^n$ where $0 < \alpha < 1$ Clearly, an indicator function $I_A(\boldsymbol{x})$ of some Borel-measurable subset A of R^n is a test function. In (17.2.1), $\delta(\boldsymbol{x})$ is the indicator function of the set ω. (An indicator function has been defined in definition 2.2.3).

Definition 17.2.4 *A non-randomised test for testing H_o against H_1 is a test-function $\delta(\boldsymbol{x})$ defined for all $\boldsymbol{x} \in \mathcal{X}_n$ such that $\delta(\boldsymbol{x})$ is as in (17.2.1). Clearly, the critical region ω depends on the test function δ. If δ changes, $\omega(\delta)$ will be different. In a non-randomised test if $\boldsymbol{x}$ is observed we either accept H_o or reject it with probability unity.*

Definition 17.2.5 *A randomised test for testing H_o against H_1 is a test function $\delta(\boldsymbol{x})$ defined for all $\boldsymbol{x} \in \mathcal{X}_n$ such that $0 \leq \delta(\boldsymbol{x}) \leq 1 \ \forall \ \boldsymbol{x}$ (but not like (17.2.1)). If we observe $\boldsymbol{x}$ we make a Bernoulli experiment with probability of success $\delta(\boldsymbol{x})$. If a success occurs we reject H_o; otherwise we accept it.*

Randomised tests will be needed in general only if X is a discrete r.v. For the time being, we shall consider non-randomised tests only. In the problem of testing of hypotheses, clearly the true value of θ remains unknown. We are only aiming at testing whether the observation $\boldsymbol{x}$ supports our assertion $\theta \in \Theta_0$ i.e., $\boldsymbol{x}$ is a random sample from the p.d.f. $f_\theta(\boldsymbol{x}), \theta \in \Theta_0$. Hence, the acceptance (rejection) of $H_\circ$ on the basis of $\boldsymbol{x}$ does not necessarily imply that $H_\circ$ is true (false). Therefore, we may reject (by mistake) $H_\circ$ when it is, in fact, true; or we may accept (by mistake) $H_\circ$ when it is false. In both the cases we commit some errors. The first type of error (error in rejecting $H_\circ$ when it is true) is called the '*error of type* I'. The second type of error (error in accepting $H_\circ$ when it is false) is called the '*error of type* II'.

For simplicity, we consider that the null hypothesis is a simple one, $H_\circ(\theta = \theta_0)$ so that $\Theta_0 = \{\theta_0\}$.
In this case, Prob. {type I error}:

$$\begin{aligned} P_{\theta_0}(\omega) &= P\{\boldsymbol{x} \in \omega | H_\circ\} = \int_\omega f_{\theta_0}(\boldsymbol{x}) d\boldsymbol{x} \\ &= \int \delta(\boldsymbol{x}) f_{\theta_0}(\boldsymbol{x}) d\boldsymbol{x} = E_{\theta_0}[\delta(\boldsymbol{x})]. \end{aligned} \tag{17.2.2}$$

Prob. {type II error}:

$$P\{\boldsymbol{x} \in \omega^c | H_1\} = P_\theta(\omega^c) = \int_{\omega^c} f_\theta(\boldsymbol{x}) d\boldsymbol{x}, \ \ \theta \in \Theta_1. \tag{17.2.3}$$

The probability of rejecting a true $H_\circ$ (type I error) depends on the test function δ and the value θ_0 and is called the *level of significance* of test or the *size* of the critical region ω.

The probability of rejecting a null hypothesis $H_\circ$ when it is false i.e., when $\theta \in \bar{\Theta}_0 = \Theta - \{\theta_0\}$, is

$$\begin{aligned} P_\theta(x \in \omega) &= 1 - P_\theta(x \in \omega^c), \\ &= 1 - P_\theta(\omega^c), \ \ \theta \in \bar{\Theta}_0 \\ &= \gamma_\theta(\omega), \ \ \theta \in \bar{\Theta}_0 \end{aligned}$$

$\gamma_\theta(\omega)$ is called the *power of the test* ω. Clearly, $\gamma_\theta(\omega)$ depends on the test function δ and the value of the parameter $\theta \in \bar{\Theta}_0$. Note that $\gamma_\theta(\omega) = E_\theta[\delta(\boldsymbol{x})]$ where $\theta \in \bar{\Theta}_0$.
For $\theta \in \Theta_1 (\subseteq \bar{\Theta}_0)$,

$$\gamma_\theta(\omega) = 1 - P_\theta(\omega^c) = 1 - \text{Probability of error of type II.} \tag{17.2.4}$$

Clearly, the probabilities of type II error, $P_\theta(\omega^c)$ depends on the test function δ and the value of the parameter $\theta \in \Theta_1$. The quantity $P_\theta(\omega^c) = \beta_\theta(\omega)$ is also called the value of the *Operating Characteristic* (OC) *function* of the test ω at the point θ.

The above definition of level of significance or size of a test does not apply when $H_\circ$ is a composite hypothesis, since $P_\theta(\omega)$ depends on $\theta \in \Theta_0$. In this case, the size of the critical region or level of signifiance of the test or propability of type I error is defined as

$$\max_{\theta \in \Theta_0} P_\theta(\omega). \tag{17.2.5}$$

Ideally, one would like to find a critical region for which the probability of both types of errors are zero, or are as small as possible. However, with a fixed number of random variables $X_1, \cdots, X_n$, i.e., for a non-sequential test it is not possible to minimise both these probabilities at the same time. A test that minimises the probability of type error, in fact, maximises the probability of error of the other type. The procedure used in practice is to limit the probability of type I error to some pre-assigned small value α (usually $0 \cdot 05$ or $0 \cdot 01$) and to minimise the probability of type II error (β_θ) subject to this constraint.

Note that if δ_1, δ_2 are two tests of the same size α and λ is a real number, $0 < \lambda < 1$ then $\lambda\delta_1 + (1-\lambda)\delta_2$ is also a test function of size α and it follows that the class of all test functions Δ_α is convex.

► **Example 17.2.2** *Let $X_1, \cdots, X_n$ be a random sample from $N(\mu, 1)$ population with μ unknown, $\mu \in \Theta = R^1$. We want to test $H_o(\mu = \mu_0)$ when $\mu_0 = 0.8$ against the alternative hypothesis $H_1(\mu \neq \mu_0)$. Here $\Theta_0 = \{\mu_0\}, \Theta_1 = \{\mu : \mu \in R^1 - \{\mu_0\}\}$. Thus H_o is a simple hypothesis and H_1 is a composite hypothesis. We shall confine to non randomised tests. We, therefore, have to find ω so that by looking at the observation x we can decide whether $\boldsymbol{x} \in \omega$ (i.e., H_o is to be rejected) or $\boldsymbol{x} \subset \omega^c$ (i.e., H_o is to be accepted).*

It is reasonable to use a sufficient statistic, preferably, a minimal sufficient statistic for μ. It is known that $\sum_{i=1}^{n} X_i$ is a minimal sufficient statistic for μ (example 15.5.12). Again, $\bar{X}$ is a UMVUE for μ. Hence, it is reasonable to base our decision on comparing the value of $\bar{X}$ with θ_0.

Suppose we confine to tests with levels of signifiance $\alpha = 0 \cdot 05$.

Let us consider two possible tests.

(a) H_o is rejected if $|\bar{X} - \mu_0| > a$ and is accepted otherwise. Hence, test function $\delta_1(\boldsymbol{x})$ is:

$$\delta_1(\boldsymbol{x}) = \begin{cases} 1 \text{ if } |\bar{x} - \mu_0| > a \\ 0 \text{ otherwise.} \end{cases}$$

(b) H_o is rejected if $\bar{X} < \mu_0 - b$. Hence, test function $\delta_2(\boldsymbol{x})$ is:

$$\delta_2(\boldsymbol{x}) = \begin{cases} 1 \text{ if } \bar{x} < \mu_0 - b \\ 0 \text{ otherwise.} \end{cases}$$

In both the cases a, b are to be set such that size of the critical region is α.

(i) **For test** (a): the level of significance is

$$\begin{aligned} P_{\mu_0}(\omega) &= P_{\mu_0}[|\bar{X} - \mu_0| > a] = 1 - P_{\mu_0}[|\bar{X} - \mu_0| \leq a] \\ &= 1 - \int_{\mu_0 - a}^{\mu_0 + a} \sqrt{\frac{n}{2\pi}} \exp\left[-\frac{n}{2}(\bar{x} - \mu_0)^2\right] d\bar{x} \\ &= 1 - \int_{-a\sqrt{n}}^{a\sqrt{n}} \frac{1}{\sqrt{2\pi}} e^{-\tau^2/2} d\tau \quad [\text{writing } \sqrt{n}(\bar{x} - \mu_0) = \tau] \\ &= 2[1 - \Phi(a\sqrt{n})] \qquad \text{(i)} \end{aligned}$$

Equating (i) with $0 \cdot 05$,

$$a = 1.96/\sqrt{n}.$$

Consider now the power of the test.

$$\begin{aligned} \gamma_1(\mu) &= P_\mu[\boldsymbol{x} \in \omega], \quad \mu \neq \mu_0 \\ &= P_\mu[|\bar{X} - \mu_0| > a], \quad \mu \neq \mu_0 \\ &= 1 - P_\mu[|\bar{X} - \mu_0| \leq a] \\ &= 1 - \int_{\mu_0 - a}^{\mu_0 + a} \sqrt{\frac{n}{2\pi}} \exp\left[-\frac{n}{2}(\bar{x} - \mu)^2\right] d\bar{x} \\ &= 1 - [\Phi((a + \mu_0 - \mu)\sqrt{n}) - \Phi((-a + \mu_0 - \mu)\sqrt{n})] \qquad \text{(ii)} \end{aligned}$$

where $a = 1.96/\sqrt{n}$.

We note that $\gamma_1(\mu)$ is a function of $\mu \in \Theta_1$. If $\mu = \mu_0$, $\gamma_1(\mu_0) = 0 \cdot 05 (= \alpha)$.

Table 17.1 *Showing the power function of the tests for* $n = 100$.

μ	$\gamma_1(\mu)$	$\gamma_2(\mu)$
.30	.9988	.9996
.35	.9945	.9979
.40	.9793	.9909
.45	.9382	.9686
.50	.8508	.9131
.55	.7995	.8051
.60	.5160	.6406
.65	.3231	.4443
.70	.1700	.2611
.75	.0790	.1271
.80	.0500	.0500
.85	.0790	.0162
.90	.1700	.0041
1.0	.3231	.0008
1.5	.5160	0

The power-curve $\gamma_1(\mu)$ is symmetric around μ_0, since $\gamma_1(\mu_0+\lambda)=\gamma_1(\mu_0-\lambda)$ for all λ. Again, for $\mu_2>\mu_1(>\mu_0)$, $\gamma_1(\mu_2)>\gamma_1(\mu_1)$. Again, $\lim\limits_{\mu\to\pm\infty}\gamma_1(\mu)=1$. The power function of the test $\gamma_1(\mu)$ has been shown in fig. 17.1 for $n=100$.

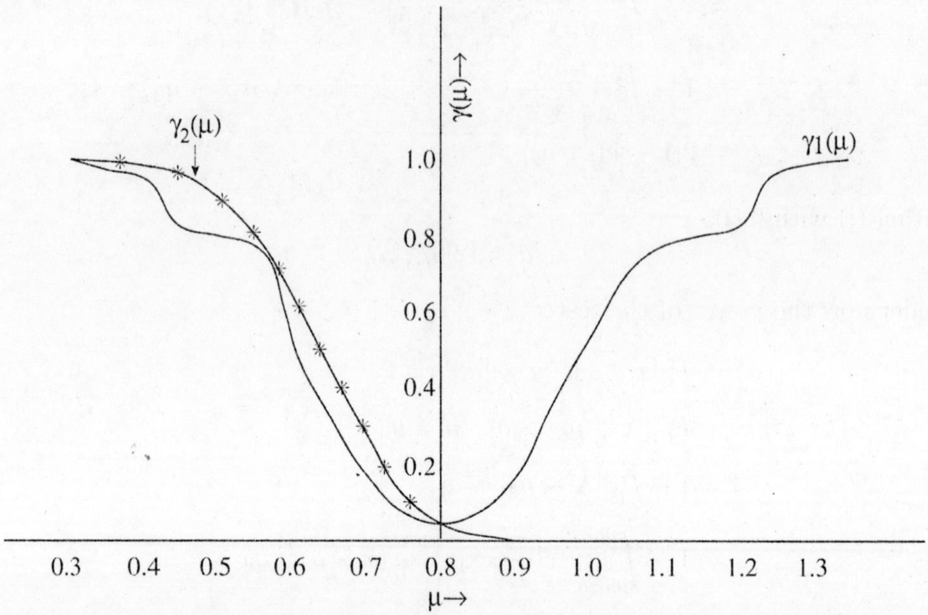

Fig. 17.1 Showing the power curves for two tests

(ii) **For test** (b): The size of the test is

$$P_{\mu_0}[\bar{X}<\mu_0-b]=\int_{-\infty}^{\mu_0-b}\sqrt{\frac{n}{2\pi}}\exp\left[-n(\bar{x}-\mu_0)^2/2\right]d\bar{x}$$
$$=\Phi(-b\sqrt{n})=0.05.$$

This gives $b=1.64/\sqrt{n}$. The power of the test,

$$\gamma_\mu=P_\mu[\bar{X}<\mu_0-b],\quad \mu\neq\mu_0$$
$$=\int_{-\infty}^{\mu_0-b}\sqrt{\frac{n}{2\pi}}\exp\left[-n(\bar{x}-\mu)^2\right]d\bar{x}$$
$$=\Phi(\sqrt{n}(\mu_0-b-\mu))=\Phi(\sqrt{n}(\mu_0-\mu)-1.64)=\gamma_2(\mu)\ \text{(say)}.$$

It is clear that if $\mu_2<\mu_1<\mu_0$, then $\gamma_2(\mu_1)<\gamma_2(\mu_2)$; also $\gamma_2(\mu_0)=0\cdot05$; $\lim\limits_{\mu\to-\infty}\gamma_2(\mu)=1$. The power function $\gamma_2(\mu)$ has been plotted in fig. 17.1 for $n=100$.

From the figures it is seen that both the tests have the same level. For $\mu<\mu_0$, $\gamma_2(\mu)>\gamma_1(\mu)\ \forall\ \mu$. Thus, for testing $H_o(\mu=\mu_0)$ against $H(\mu<\mu_0)$, we should use the test (b) (and not test (a)) as this test is more powerful than the test (a) for all $\mu<\mu_0$ i.e., test (b) has more power of rejecting H_o when it is false than the other test whatever be the true value of the parameter $\mu < \mu_0$. If, however, the alternative hypothesis is $H_1(\mu \neq \mu_0)$, then we may use test (a) as it gives high probability of rejecting $H_\circ$ when it is false and some other value $\mu(\neq \mu_0)$ is true value of the parameter.

Note that in this case the continuity of the r.v. X (i.e., the absolute continuity of the distribution function of X) allows us to achieve any size α.

17.3 Same Optimality Criteria

17.3.1 A Few Definitions

We consider a few definitions concerning optimality of tests in different situations.

Definition 17.3.1 *The critical region ω_0 is said to be most powerful* (MP) *for testing $H_\circ(\theta = \theta_0)$ against $H(\theta = \theta_1)$ among all size-α tests, if*

$$P_{\theta_0}(\omega_0) = \alpha, \tag{17.3.1}$$

$$P_{\theta_1}(\omega_0) \geq P_{\theta_1}(\omega) \tag{17.3.2}$$

for all ω satisfying (17.3.1).

Definition 17.3.2 *The critical region ω_0 is said to be uniformly most powerful* (UMP) *for testing $H_\circ : \theta = \theta_0$ against $H : \theta \in \Theta_1$ if*

$$P_{\theta_0}(\omega_0) = \alpha \tag{17.3.3}$$

$$P_{\theta}(\omega_0) \geq P_{\theta}(\omega) \ \forall\, \theta \in \Theta_1 \tag{17.3.4}$$

whatever be the other region ω satisfying (17.3.3).

Definition 17.3.3 *A critical region ω_0 is said to be unbiased for testing $H_\circ : \theta = \theta_0$ against $H : \theta \in \Theta_1$ if*

$$P_{\theta}(\omega_0) \geq P_{\theta_0}(\omega_0) \ \forall\, \theta \in \Theta_1. \tag{17.3.5}$$

The unbiasedness is a very desirable property of a test. If a test ω is biased, then, for some $\theta = \theta_1 \in \Theta_1$,

$$P_{\theta_1}(\omega) < P_{\theta_0}(\omega)$$
$$\text{i.e., } 1 - P_{\theta_1}(\omega) > 1 - P_{\theta_0}(\omega).$$

Therefore, for such a region ω, probability of accepting $H_\circ$ when it is false ($\theta = \theta_1$) is greater than probability of accepting $H_\circ$ when it is true. This is a very undesirable feature and so we should try to avoid biased tests.

It will be seen that for many alternatives, specially both-sided alternatives no UMP test generally exists. In such situations one may consider a UMP test among the class of unbiased tests. Such a test is called a uniformly most powerful unbiased (UMPU) test.

Definition 17.3.4 *A critical region ω_0 is said to be a* UMPU *critical region for testing $H_o(\theta=\theta_0)$ against $H(\theta \in \Theta_1)$ if*

$$P_{\theta_0}(\omega_0) = \alpha, \tag{17.3.6}$$

$$P_\theta(\omega_0) \geq \alpha \ \forall\, \theta \in \Theta_1, \tag{17.3.7}$$

$$P_\theta(\omega_0) \geq P_\theta(\omega) \ \forall\, \theta \in \Theta_1 \tag{17.3.8}$$

whatever be the other critical region ω, satisfying (17.3.6) *and* (17.3.7).

When θ is a single parameter and Θ is a real interval, we may replace the conditions (17.3.7) and (17.3.8) by alternative conditions expressed in terms of derivatives. Assume that for every measurable set ω in the sample space $\mathcal{X}_n$, the function $P_\theta(\omega)$ is twice differentiable with respect to θ, the second derivative being continuous in the neighbourhood of θ_0.

Definition 17.3.5 *A critical region ω_0 is said to be a type A critical region for testing H_o $(\theta = \theta_0)$ against $H(\theta \in \Theta_1)$ if*

$$P_{\theta_0}(\omega_0) = \alpha, \tag{17.3.9}$$

$$\left.\frac{dP_\theta(\omega_0)}{d\theta}\right]_{\theta_0} = 0, \tag{17.3.10}$$

$$\left.\frac{d^2P_\theta(\omega_0)}{d\theta^2}\right]_{\theta_0} \geq \left.\frac{d^2P_\theta(\omega)}{d\theta^2}\right]_{\theta_0} \tag{17.3.11}$$

whatever be the other critical region ω satisfying the conditions (17.3.9) and (17.3.10).

To study the implications of conditions (17.3.10) and (17.3.11) expand $P_\theta(\omega)$ around $\theta = \theta_0$.

$$P_\theta(\omega) = P_{\theta_0}(\omega) + (\theta - \theta_0)\left.\frac{dP_\theta(\omega)}{d\theta}\right]_{\theta_0} + \frac{(\theta-\theta_0)^2}{2}\left.\frac{d^2P_\theta(\omega)}{d\theta^2}\right]_{\theta_0} + \varepsilon$$

$$= \alpha + \frac{(\theta-\theta_0)^2}{2}\left.\frac{d^2P_\theta(\omega)}{d\theta^2}\right]_{\theta_0} + \varepsilon \tag{17.3.12}$$

using conditions (17.3.9) and (17.3.10), when $\varepsilon \to 0$ as $\theta \to \theta_0$. If $|\theta - \theta_0|$ is small, (17.3.11), therefore, implies that

$$P_\theta(\omega_0) \geq P_\theta(\omega) (\geq \alpha).$$

Thus, a type A critical region ω_0 is a MP critical region for testing $H_o(\theta = \theta_0)$ against alternative $H(\theta = \theta_1)$ provided θ_1 is close to θ_0. Thus, (in case Θ is an interval), a type A critical region may be called as a locally most powerful critical region of size α.

It should, however, be noted that one is generally interested in alternatives θ_1 which differ markedly from θ_0 and not in those which are close to θ_0. Thus, a type A critical region which ensures locally most powerfulness is mostly of theoretical interest.

The local optimality of a type A region occurs from the condition (17.3.11). Replacing this condition by the one which ensures maximum power for all $\theta \neq \theta_0$ we get a type A_1 critical region.

Definition 17.3.6 *A critical region ω_0 is said to be a type A_1 critical region of size α for testing $H_\circ(\theta = \theta_0)$ against $H(\theta \neq \theta_0)$ if*

$$P_{\theta_0}(\omega_0) = \alpha, \tag{17.3.13}$$

$$\left.\frac{dP_\theta(\omega_0)}{d\theta}\right]_{\theta_0} = 0, \tag{17.3.14}$$

$$P_\theta(\omega_0) \geq P_\theta(\omega), \theta \neq \theta_0 \tag{17.3.15}$$

whatever be the other critical region ω satisfying (17.3.13) *and* (17.3.14).

In many cases a type A_1 region will also be a UMPU region. However, this is not always so, because the conditions (17.3.7) and (17.3.14) are not identical. In many cases a UMPU region may be such that the power function is not differentiable at $\theta = \theta_0$.

In the next section we consider derivation of a MP critical region for testing a simple hypothesis against a simple alternative.

17.4 Testing a Simple $H_\circ$ against a Simple Alternative

We consider the case when Θ contains only two points θ_0 and θ_1 and where the null hypothesis is $H_\circ(\theta = \theta_0)$ and the alternative hypothesis is $H(\theta = \theta_1)$. Our problem is to find a critical region $\omega(X_1, \cdots, X_n)$ based on the sample $(X_1, \cdots, X_n)$ such that its power is maximum among all the critical regions of a pre-assigned size α for testing $H_\circ$ against H_1. Such a critical region is the best critical region of size α. This is given by the following lemma due to Neyman and Pearson (1933).

Theorem 17.4.1 ***Neyman-Pearson Lemma.*** *Let $(X_1, \cdots, X_n)'$ be a random sample from a distribution with p.d.f. $f(x;\theta)$. Suppose the parameter space $\Theta = \{\theta_0, \theta_1\}$. Let*

$$L(\theta) = L(\boldsymbol{x}|\theta) = \prod_{i=1}^{n} f(x_i;\theta)$$

be the likelihood function of the sample $\boldsymbol{x} = (x_1, \cdots, x_n)'$ when the value of the parameter is θ. Then any critical region ω of the form

$$\omega = \{(x_1, \cdots, x_n)' : L(\theta_1)/L(\theta_0) \geq k\} \tag{17.4.1}$$

for some $0 \leq k \leq \infty$ has the property that

$$\int_\omega L(\boldsymbol{x}|\theta_1)d\boldsymbol{x} \geq \int_{\omega'} L(\boldsymbol{x}|\theta_1)d\boldsymbol{x} \tag{17.4.2}$$

where $dx = \prod_{i=1}^{n} dx_i$ and ω' is any critical region such that

$$\int_{\omega'} L(\boldsymbol{x}|\theta_0)dx \le \int_{\omega} L(\boldsymbol{x}|\theta_0)dx. \tag{17.4.3}$$

If $k = \infty$, the critical region

$$\omega = \{\boldsymbol{x} : L(\theta_0) = 0\} \tag{17.4.4}$$

is most powerful among all critical regions of size zero.

Proof: We have

$$\omega = (\omega \cap \omega') \cup (\omega \cap \bar{\omega}')$$
$$\text{and } \omega' = (\omega' \cap \omega) \cup (\omega' \cap \bar{\omega})$$

where the set $\bar{A}$ denotes $(\mathcal{X} - A)$, $\mathcal{X}$ denoting the sample space of $(X_1, \cdots, X_n)'$.

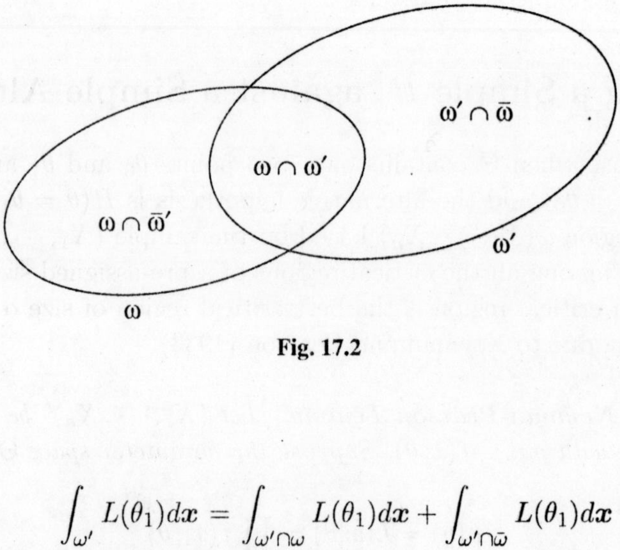

Fig. 17.2

Hence

$$\int_{\omega'} L(\theta_1)dx = \int_{\omega'\cap\omega} L(\theta_1)dx + \int_{\omega'\cap\bar{\omega}} L(\theta_1)dx \tag{17.4.5}$$

$$\text{and } \int_{\omega} L(\theta_1)dx = \int_{\omega\cap\omega'} L(\theta_1)dx + \int_{\omega\cap\bar{\omega}'} L(\theta_1)dx. \tag{17.4.6}$$

Subtracting (17.4.6) from (17.4.5),

$$\int_{\omega'} L(\theta_1)dx - \int_{\omega} L(\theta_1)dx = \int_{\omega'\cap\bar{\omega}} L(\theta_1)dx - \int_{\omega\cap\bar{\omega}'} L(\theta_1)dx. \tag{17.4.7}$$

Note that for $\boldsymbol{x} \in w' \cap \bar{\omega} \subseteq \omega'$, $L(\theta_1) \le kL(\theta_0)$ and for $\boldsymbol{x} \in \omega \cap \bar{\omega}' \subseteq \omega$, $L(\theta_1) \ge kL(\theta_0)$.

Hence, from (17.4.7),

$$\begin{aligned}
P_{\theta_1}(\omega') - P_{\theta_1}(\omega) &\le k\int_{\omega'\cap\bar{\omega}} L(\theta_0)\prod_{i=1}^{n} dx_i - k\int_{\omega\cap\bar{\omega}'} L(\theta_0)\prod_{i=1}^{n} dx_i \\
&= k\left[\int_{\omega'} L(\theta_0)\prod_{i=1}^{n} dx_i - \int_{\omega} L(\theta_0)\prod_{i=1}^{n} dx_i\right] \\
&\le 0 \text{ (by (17.4.3)).}
\end{aligned}$$

Consider now $k = \infty$, and tests of size zero. The critical region ω has now size zero. If critical region ω' has size zero, then $\omega' \subseteq \omega$. Hence

$$\int_{\omega'} L(\theta_1)\prod_{i=1}^{n} dx_i \le \int_{\omega} L(\theta_1)\prod_{i=1}^{n} dx_i$$

$$\text{i.e., } P_{\theta_1}(\omega') \le P_{\theta_1}(\omega).$$

■ NOTE 17.4.1 *The theorem identifies the most powerful test in a class of non-randomised tests only. The theorem does not refer to the class of randomised tests.*

The test function corresponding to (17.4.1) is

$$\delta(\boldsymbol{x}) = \begin{cases} 1 \text{ if } L(\boldsymbol{x}|\theta_1) \ge kL(\boldsymbol{x}|\theta_0) \\ 0 \text{ otherwise.} \end{cases}$$

■ NOTE 17.4.2 *Any point $\boldsymbol{x}$ for which $L(\theta_1|\boldsymbol{x}) \neq 0$ but $L(\theta_0|\boldsymbol{x}) = 0$ yields a ratio $L(\theta_1|\boldsymbol{x})/L(\theta_0|\boldsymbol{x})$ which exceeds every finite k and hence is included in every critical region. Set of such points has zero probability under $H_\circ$ and is the best critical region* (BCR) *of size zero. The* BCR *of size $\alpha(\neq 0)$ for $H_\circ$ against H_1 includes sets of all such points with $L(\theta_0|\boldsymbol{x}) = 0$, as the omission of such points may decrease the power of the test (since $L(\theta_1|\boldsymbol{x}) \neq 0$ for such points) (vide example* 17.4.8*).*

■ NOTE 17.4.3 *The Neyman-Pearson theorem* 17.4.1 *does not require the sample values to be independently and identically distributed. The distributions specified under $H_\circ$ and H_1 need not even belong to the same family (example* 17.4.7*). What is needed is that distributions must be completely specified under $H_\circ$ and H_1. Again θ may be a vector of parameters $\boldsymbol{\theta} = (\theta_1, \cdots, \theta_k)'$.*

■ NOTE 17.4.4 *It can be seen that the test given in* (17.4.1) *or* (17.4.4) *is unique (except on a null set), i.e., if δ is a* MP *test of size α of $H_\circ$ against H_1, it must be of the form* (17.4.1) *or* (17.4.4)*, except perhaps for a set A with $P_{\theta_0}(A) = P_{\theta_1}(A) = 0$. [For a proof see necessity part of theorem* 17.8.1]

■ NOTE 17.4.5 *It can be shown that for the family of Neyman-Pearson tests, the larger the value of α, the smaller the value of β = Probability (type II error).*

Theorem 17.4.2 *If ω_0 be an* MP *test of size α for testing $H_\circ : \theta = \theta_0$ against $H_1 : \theta = \theta_1$, then it is necessarily unbiased.*

Proof: If ω_0 be an MP region of size α, then

$$\int_{\omega_0} L(\boldsymbol{x}|\theta_0)d\boldsymbol{x} = \int_{\{\boldsymbol{x}:L(\boldsymbol{x}|\theta_1)\geq kL(\boldsymbol{x}|\theta_0)\}} L(\boldsymbol{x}|\theta_0)d\boldsymbol{x} = \alpha$$

$$\text{and} \int_{\omega_0} L(\boldsymbol{x}|\theta_1)d\boldsymbol{x} = \int_{\{\boldsymbol{x}:L(\boldsymbol{x}|\theta_1)\geq kL(\boldsymbol{x}|\theta_0)\}} L(\boldsymbol{x}|\theta_1)d\boldsymbol{x}$$

$$> k\int_{\{\boldsymbol{x}:L(\boldsymbol{x}|\theta_1)\geq kL(\boldsymbol{x}|\theta_0)\}} L(\boldsymbol{x}|\theta_0)d\boldsymbol{x} = k\alpha. \tag{17.4.8}$$

Also

$$1 - \int_{\omega_0} L(\boldsymbol{x}|\theta_1)d\boldsymbol{x} = \int_{\{\boldsymbol{x}:L(\boldsymbol{x}|\theta_1)<kL(\boldsymbol{x}|\theta_0)\}} L(\boldsymbol{x}|\theta_1)d\boldsymbol{x}$$

$$< k\int_{\{\boldsymbol{x}:L(\boldsymbol{x}|\theta_1)<kL(\boldsymbol{x}|\theta_0)\}} L(\boldsymbol{x}|\theta_0)d\boldsymbol{x} = k(1-\alpha). \tag{17.4.9}$$

If $k \geq 1$, then (17.4.8) gives

$$\int_{\omega_0} L(\boldsymbol{x}|\theta_1)d\boldsymbol{x} > \alpha.$$

If $k < 1$, then (17.4.9) gives

$$1 - \int_{\omega_0} L(\boldsymbol{x}|\theta_1)d\boldsymbol{x} < 1 - \alpha$$

which also implies

$$\int_{\omega_0} L(\boldsymbol{x}|\theta_1)d\boldsymbol{x} > \alpha.$$

Hence the critical region ω_0 is unbiased.

Corollary 17.4.1 *If ω_0 is a* UMP *region of size α for testing $H_\circ : \theta = \theta_0$ against $H : \theta \in \Theta_1$ then also the above proof holds (replacing θ_1 by $\theta \in \Theta_1$) and a* UMP *test for $H_\circ : \theta = \theta_0$ against $H : \theta \in \Theta_1$ is necessarily unbiased.*

► **Example 17.4.1** *Let $(X_1, \cdots, X_n)'$ be a random sample from the exponential distribution $f(x;\theta) = \theta e^{-\theta x}, 0 < x < \infty, \theta > 0$. Consider $H_\circ(\theta = \theta_0)$, $H_1(\theta = \theta_1)$ where $\theta_1 > \theta_0$. Here*

$$L(\boldsymbol{x}|\theta) = \theta^n \exp\left[-\theta\sum_1^n x_i\right].$$

By NP-theorem, the most powerful critical region of size α consists of those sample points $\boldsymbol{x}$ for which

$$L(\boldsymbol{x}|\theta_1)/L(\boldsymbol{x}|\theta_0) \geq k \text{ i.e., } \frac{\theta_1^n \exp\left[-\theta_1\Sigma x_i\right]}{\theta_0^n \exp\left[-\theta_0\Sigma x_i\right]} \geq k$$

$$\text{or, } \exp\left[(\theta_0 - \theta_1)\Sigma x_i\right] \geq k\left(\frac{\theta_0}{\theta_1}\right)^n.$$

Since $\theta_1 > \theta_0$, $\exp[(\theta_0 - \theta_1)\Sigma x_i]$ is a decreasing function of Σx_i and hence it is greater than $k(\theta_0/\theta_1)^n$ when $\sum_1^n x_i$ is less than some constant. A BCR is, therefore, of the form

$$\omega = \left\{ \boldsymbol{x} : \sum_{i=1}^{n} x_i < c \right\}. \tag{i}$$

The size of ω is

$$P\left\{ \sum_1^n X_i < c | \theta = \theta_0 \right\}.$$

Note that $Y = \sum_1^n X_i$ has a gamma distribution with parameters (n, θ). The p.d.f. of Y is

$$g(y; \theta) = \frac{\theta(\theta y)^{n-1}}{(n-1)!} \exp(-\theta y), \quad 0 \le y < \infty.$$

Hence size of ω is

$$\int_0^c g(y; \theta_0) dy = \int_0^{\theta_0 c} \left\{ z^{n-1} \exp(-z)/(n-1)! \right\} dz \text{ where } z = \theta_1 y.$$

Equating this to a known value α, value of c can be found out.

The power of the test is

$$\begin{aligned} P\left\{ \sum_1^n X_i < c | \theta = \theta_1 \right\} &= \int_0^c g(y; \theta_1) dy \\ &= \int_0^{\theta_1 c} \left\{ z^{n-1} \exp(-z)/(n-1)! \right\} dZ. \end{aligned}$$

We note that so long $\theta_1 > \theta_0$, the test ω remains of the form (i). The value of the constant c depends only on θ_0 and α. Hence the test (i) is uniformly most powerful critical region for testing $H_\circ(\theta = \theta_0)$ against the composite hypothesis $H_1(\theta > \theta_0)$.

Example 17.4.2 *$(X_1, \cdots, X_n)'$ is a random sample from the Poisson distribution with parameter θ. Consider $H_\circ(\theta = \theta_0)$, $H_1(\theta = \theta_1)$, $\theta_1 < \theta_0$.*

$$\frac{L(\boldsymbol{x}|\theta_1)}{L(\boldsymbol{x}|\theta_0)} = \frac{\prod_1^n \theta_1^{x_i} \exp(-n\theta_1)}{\prod_1^n \theta_0^{x_i} \exp(-n\theta_0)} = (\theta_1/\theta_0)^{\Sigma x_i} \exp[n(\theta_0 - \theta_1)].$$

Since $\theta_1 < \theta_0$, the ratio is a decreasing function of Σx_i. Hence the most powerful test for $H_\circ$ against H_1 is

$$L(\theta_1)/L(\theta_0) > k \Rightarrow \sum_1^n x_i < c. \tag{i}$$

The size of the test is

$$P_{\theta_0}\left[\sum_1^n X_i < c\right] = 1 - \sum_{y=c+1}^{\infty} (n\theta_0)^y \exp(-n\theta_0)/y! \qquad \text{(ii)}$$

assuming c to be an integer. The power of the test is $P_{\theta_1}\left[\sum_1^n X_i < c\right]$.

Since the distribution of the test statistic $\sum_1^n X_i$ is discrete, it is not always possible to find a value of c for which a pre-assigned size α is attained. To attain this, some randomisation at the boundary values of c may be required.

Suppose $n = 10$, $\theta_0 = 0.6$, $\alpha = 0 \cdot 05$. From tables of Poisson distribution

$$P\left[\sum_1^{10} X_i < 2|\theta = 0.6\right] = 0 \cdot 017$$

$$P\left[\sum_1^{10} X_i < 3|\theta = 0.6\right] = 0 \cdot 062.$$

Hence, there is no discrete value of c for which the size $0 \cdot 05$ is exactly attained. We choose a quantity λ such that

$$P\left[\sum_1^{10} X_i < 2|\theta = 0 \cdot 6\right] + \lambda P\left[\sum_1^{10} X_i = 2|\theta = 0 \cdot 6\right] = 0 \cdot 05$$

or, $\quad 0 \cdot 017 + 0 \cdot 033\lambda = 0 \cdot 05$

or, $\quad \lambda = 0 \cdot 733.$

Consider now the following randomisation test. If $\sum_1^{10} X_i < 2$, reject H_o. If $\sum_1^{10} X_i > 2$ accept H_o. If $\sum_1^{10} X_i = 2$ reject H_o with probability $\lambda = 0 \cdot 733$ (and accept H_o with probability $1 - \lambda = 0 \cdot 267$).

The critical function corresponding to this critical region is

$$\delta(x) = \begin{cases} 1 & \text{if } \sum_1^{10} X_i < 2 \\ 0 & \text{if } \sum_1^{10} X_i > 2 \\ 0 \cdot 733 & \text{if } \sum_1^{10} X_i = 2. \end{cases}$$

The power of the test when $\theta_1 = 0 \cdot 5$ is

$$P\left[\sum_1^{10} X_i < 2|\theta_1 = 0 \cdot 5\right] + 0 \cdot 733P\left[\sum_1^{10} X_i = 2|\theta_1 = 0 \cdot 5\right] = 0 \cdot 102305.$$

Note that the test statistic does not depend on θ_1 so long as $\theta_1 < \theta_0$. Hence the critical region (i) is UMP for H_o against the composite hypothesis $H_1(\theta < \theta_0)$.

▶ **Example 17.4.3** *Let $(X_1, \cdots, X_n)'$ be a random sample from $N(\mu, 1)$ distribution. Consider $H_o(\mu = \mu_0)$, $H_1(\mu = \mu_1)$, $\mu_1 \neq \mu_0$.*

$$\begin{aligned} L(\mu) &= (2\pi)^{-n/2} \exp\left[-\frac{1}{2}\sum_{i=1}^{n}(x_i - \mu)^2\right] \\ &= (2\pi)^{-n/2} \exp\left[-\frac{n}{2}\{s^2 + (\bar{x} - \mu)^2\}\right] \end{aligned}$$

where $s^2 = \sum_{i=1}^{n}(x_i - \bar{x})^2/n$. *For the* BCR,

$$\begin{aligned} \frac{L(\mu_1)}{L(\mu_0)} &= \exp\left[\frac{n}{2}\{(\bar{x} - \mu_0)^2 - (\bar{x} - \mu_1)^2\}\right] \\ &= \exp\left[\frac{n}{2}\{(\mu_1 - \mu_0)2\bar{x} + (\mu_0^2 - \mu_1^2)\}\right] \geq k_\alpha \end{aligned}$$

$$\text{or,} \quad (\mu_1 - \mu_0)\bar{x} \geq \frac{1}{2}(\mu_1^2 - \mu_0^2) + \frac{1}{n}\log k_\alpha.$$

Hence, if $\mu_1 > \mu_0$, the BCR is

$$\bar{x} \geq \frac{1}{2}(\mu_0 + \mu_1) + \log k_\alpha/\{n(\mu_1 - \mu_0)\}. \tag{i}$$

If $\mu_1 < \mu_0$, the BCR is

$$\bar{x} \leq \frac{1}{2}(\mu_0 + \mu_1) + \log k_\alpha/\{n(\mu_1 - \mu_0)\}. \tag{ii}$$

It is seen, therefore, that given μ_0, μ_1 and α, the BCR is determined by the value of $\bar{x}$ alone. This is what is expected since $\bar{x}$ is the MVB estimate and sufficient statistic for μ.

To obtain a most powerful test of size α for testing against $\mu_1 > \mu_0$, we, therefore, determine from (i) a quantity $\bar{x}_\alpha$ such that

$$\int_{\bar{x}_\alpha}^{\infty} \sqrt{\frac{n}{2\pi}} \exp\left[-\frac{n}{2}(\bar{x} - \mu_0)^2\right] d\bar{x} = \alpha.$$

This gives,

$$\bar{x}_\alpha = \mu_0 + \tau_\alpha/\sqrt{n}$$

where $\Phi(-\tau_\alpha) = \alpha$, Φ being the cumulative probability integral of a standard normal variate. The BCR is

$$\{\boldsymbol{x} : \bar{x} \geq \mu_0 + \tau_\alpha/\sqrt{n}\}.$$

Since the test does not depend on the actual value of μ_1 so long as $\mu_1 > \mu_0$, the test is UMP for one-sided alternatives $H(\mu = \mu_1 > \mu_0)$.
The power of the test at $\mu = \mu_1$ is

$$\int_{\bar{x}_\alpha}^{\infty} \frac{\sqrt{n}}{\sqrt{2\pi}} \exp\left[-\frac{n}{2}(\bar{x}-\mu_1)^2\right] d\bar{x} = \Phi(\sqrt{n}(\mu_1-\mu_0)-\tau_\alpha).$$

The power is a monotonic function of both n and $(\mu_1 - \mu_0)$.
Similar results follow for testing $H_\circ(\mu = \mu_0)$ against $H(\mu_1 < \mu_0)$.

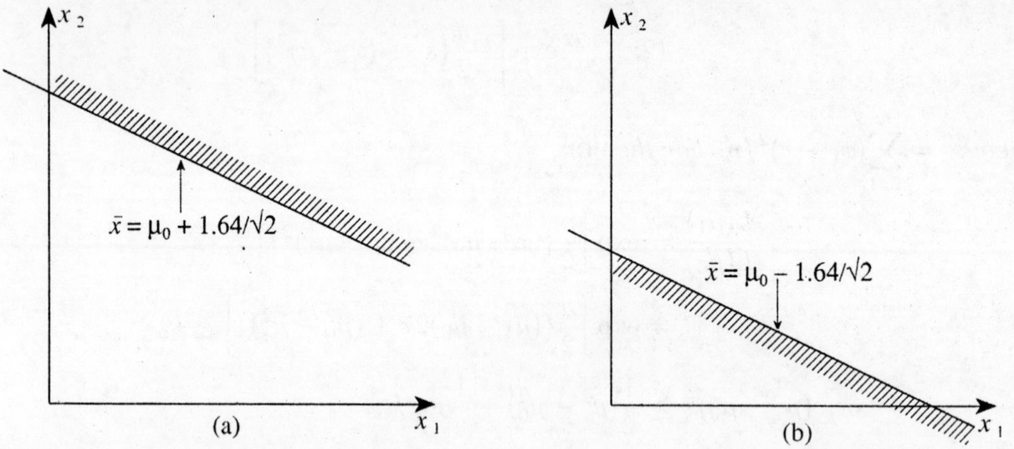

Fig. 17.3 The shaded portion in (a) [(b)] shows that critical region of size 0.05 for testing $H_\circ$ ($\mu = \mu_0$) against $H(\mu > \mu_0)$ [$H(\mu < \mu_0)$] for a normal population with $\sigma^2 = 1$

► **Example 17.4.4** *Let* $(X_1, \cdots, X_n)'$ *be a random sample from the* $N(\mu, \sigma^2)$ *distribution. Test* $H_\circ(\mu = \mu_0, \sigma = \sigma_0)$ *against* $H_1(\mu = \mu_1, \sigma = \sigma_1)$.

$$\frac{L(\mu_1, \sigma_1)}{L(\mu_0, \sigma_0)} = \left(\frac{\sigma_0}{\sigma_1}\right)^n e^A$$

where

$$A = \frac{1}{2}\left\{\frac{\Sigma(x_i-\mu_0)^2}{\sigma_0^2} - \frac{\Sigma(x_i-\mu_1)^2}{\sigma_1^2}\right\}.$$

The BCR depends on A. The BCR is $A > k'$ or ignoring the constants,

$$(1/\sigma_0^2 - 1/\sigma_1^2)\sum x_i^2 + 2\left(\mu_1/\sigma_1^2 - \mu_0/\sigma_0^2\right)\sum x_i > k. \tag{i}$$

It is clear that there will not generally be a common BCR for different members of H_1 even if we limit ourselves by $\sigma_1 < \sigma_0$ and $\mu_1 < \mu_0$ or similar restrictions.

Case 1. Suppose $\sigma_0 = \sigma_1$. In this case, (i) reduces to $(\mu_1 - \mu_0)\bar{x} > c$; if $\mu_1 > \mu_0$, this implies $\bar{x} > t$. The size of the test is

$$P_{\mu_0,\sigma_0}[\bar{X} > t] = P_{\mu_0,\sigma_0}\left[\frac{\sqrt{n}(\bar{X}-\mu_0)}{\sigma_0} > \frac{\sqrt{n}(t-\mu_0)}{\sigma_0}|H_\circ\right] = \alpha.$$

Hence the critical region is

$$\{\boldsymbol{x} : \bar{x} > \mu_0 + \tau_\alpha\sigma_0/\sqrt{n}\} \qquad \text{(ii)}$$

where τ_α is as defined in example 17.4.3. The power of the test is

$$P_{\mu_1,\sigma_0}\left[\bar{X} > \mu_0 + \tau_\alpha\frac{\sigma_0}{\sqrt{n}}\right] = P_{\mu_1,\sigma_0}\left[\frac{\sqrt{n}(\bar{X}-\mu_1)}{\sigma_0} > \frac{\sqrt{n}(\mu_0-\mu_1)}{\sigma_0} + \tau_\alpha\right].$$

When H_1 is true, $\sqrt{n}(\bar{X}-\mu_1)/\sigma_0$ is a standard normal variate, when the power can be calculated.

Since the critical region (ii) does not depend on μ_1, the proposed test is UMP for $H_\circ$ against $H_1 : (\mu > \mu_0)$. It can be shown that the power of the above test increases to 1 as $n \to \infty$.

The case $\mu_1 < \mu_0$ can be similarly worked out.

Case 2. Suppose $\mu_0 = \mu_1$. Here the condition of equation (i) depends on

$$\left(\frac{1}{\sigma_0^2} - \frac{1}{\sigma_1^2}\right)\sum_1^n (x_i - \mu_0)^2.$$

If $\sigma_0 < \sigma_1$, the BCR is

$$\left\{\boldsymbol{x} : \sum_1^n (x_i - \mu_0)^2 > k'\right\}.$$

The size of the test is

$$P_{\mu_0,\sigma_0}\left\{\sum_1^n (X_i - \mu_0)^2 > k'\right\} = \alpha.$$

Therefore, $k' = \sigma_0^2\chi^2_{\alpha,n}$ where $\chi^2_{\alpha,n}$ is the $100(1-\alpha)\%$ point of the $\chi^2_{(n)}$ distribution, since, under $H_\circ$,

$$\sum_1^n (X_i - \mu_0)^2/\sigma_0^2 \sim \chi^2_{(n)}.$$

The power of the test is

$$P_{\mu_0,\sigma_1}\left[\sum_1^n (X_i - \mu_0)^2 > \sigma_0^2\chi^2_{n,\alpha}\right] = P_{\mu_0,\sigma_1}\left[\sum_1^n (X_i - \mu_0)^2/\sigma_1^2 > (\sigma_0/\sigma_1)^2\chi^2_{n,\alpha}\right].$$

Under H_1, $\sum_1^n (X_i - \mu_0)^2/\sigma_1^2$ has $\chi^2_{(n)}$ distribution when the power can be calculated.

In fact this test is UMP for $H_\circ$ against $H_1(\sigma > \sigma_0)$. For the case $\sigma < \sigma_0$ the UMP critical region is

$$\omega_0 = \left\{ \boldsymbol{x} : \sum_i (x_i - \mu_0)^2/\sigma_0^2 < \chi^2_{n,1-\alpha} \right\}.$$

► **Example 17.4.5** *Let $(X_1, \cdots, X_n)'$ be a random sample from a $N(\mu, \sigma^2)$ distribution with μ known. Find a UMP critical region of size α for testing $H_\circ(\sigma^2 = \sigma_0^2)$ against $H(\sigma^2 > \sigma_0^2)$.*

This case reduces to the case (2) of example 17.4.4 (vide Fig 17.4).

For the above examples, the BCR lies in the tail of the distribution of the statistic. The following example shows that this is not the case always.

► **Example 17.4.6** *X is a single distribution from Cauchy population*

$$dF(x) = \frac{dx}{\pi\{1 + (x - \theta)^2\}}, \quad -\infty \leq x \leq \infty.$$

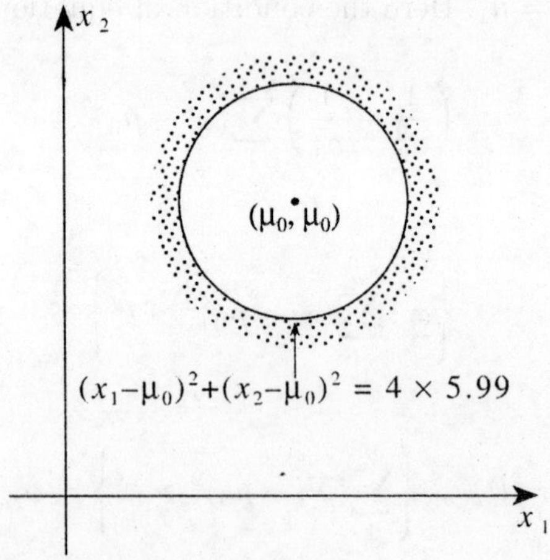

Fig. 17.4 Shows UMP critical region of size 0.05 for testing $H_\circ$ ($\sigma^2 = 4$) against $H(\sigma^2 > 4)$ for $n = 2$.

To test $H_\circ(\theta = 0)$ against $H_1(\theta = 1)$. The BCR is given by

$$\frac{L(\theta_1)}{L(\theta_0)} = \frac{1 + x^2}{1 + (x - 1)^2} \geq k_\alpha \tag{i}$$

i.e.,

$$x^2(k_\alpha - 1) + 2x + (k_\alpha - 2) \geq 0. \tag{ii}$$

The form of the BCR depends on the value of α. If, for example, $k_\alpha = 1$, (ii) reduces to $x \geq \frac{1}{2}$. If $k_\alpha = 0.5$, (ii) reduces to

$$x^2 - 4x + 3 \leq 0 \text{ or, } (x-2)^2 \leq 1$$

which holds if $1 \leq x \leq 3$.

▶ **Example 17.4.7** *Suppose X (a sample of size one) has $N(0,1)$ density $f(x)$ under H_o. The alternative hypothesis H_1 has a double exponential density,*

$$g(x) = \frac{1}{2}\exp(-|x|), \quad -\infty < x < \infty.$$

Note that the density function is completely specified under both H_o and H_1. Using NP theorem the most powerful critical region is

$$\lambda(x) = \frac{g(x)}{f(x)} \geq k$$

$$\text{i.e., } \lambda(x) = \sqrt{\frac{\pi}{2}} \cdot \exp\left\{-|x| + \frac{x^2}{2}\right\} \geq k.$$

Now,

$$\frac{x^2}{2} - |x| = \frac{1}{2}\left[(|x|-1)^2 - 1\right].$$

Hence, $\lambda(x)$ is an increasing function of $|(|x|-1)|$. Therefore,

$$\lambda(x) \geq k \;\Rightarrow\; ||x|-1| \geq k'.$$

The BCR is of the form

$$w = \{x : |x| \leq k_1 \text{ or, } |x| \geq k_2\}.$$

The size of the test is

$$P_f(\omega) = \int_{|x|\leq k_1} (1/\sqrt{2\pi})e^{-x^2/2}\,dx + \int_{|x|\geq k_2} (1/\sqrt{2\pi})e^{-x^2/2}\,dx$$
$$= 2[1 + \Phi(k_1) - \Phi(k_2)].$$

The power of the test is

$$P_g(\omega) = \int_{|x|\leq k_1} \frac{1}{2}e^{-|x|}\,dx + \int_{|x|\geq k_2} \frac{1}{2}e^{-|x|}\,dx = e^{-k_2} + 1 - e^{-k_1}.$$

▶ **Example 17.4.8** *Let $X_1, \cdots, X_n$ be a random sample from the uniform distribution $f(x;\theta) = \frac{1}{\theta}$. Consider $H_o(\theta = \theta_0)$ against $H_1(\theta = \theta_1 > \theta_0)$. Here*

$$L(\theta) = \begin{cases} \frac{1}{\theta^n}, & \text{if } \max x_i < \theta \\ 0, & \text{otherwise.} \end{cases}$$

Therefore,

$$\frac{L(\theta_1)}{L(\theta_0)} = \begin{cases} \infty, & \text{if } \theta_0 < \max x_i < \theta \\ \left(\frac{\theta_0}{\theta_1}\right)^n, & \text{if } 0 < \max x_i < \theta_0. \end{cases}$$

Hence for any $k < \left(\frac{\theta_0}{\theta_1}\right)^n$, every sample point $\boldsymbol{x}$ satisfies

$$\frac{f(\boldsymbol{x};\theta_1)}{f(\boldsymbol{x};\theta_0)} > k.$$

Any point $\boldsymbol{x}$ for which $\theta_0 < \max x_i < \theta_1$ has zero probability under $H_\circ$ and is the BCR of size zero and hence is included in any BCR of size α for $H_\circ$ against H_1 (vide note 17.4.2).

Consider now any critical region of size α. Since for this distribution $\max\limits_i X_i$ is sufficient for θ, we should use this statistic. Now

$$\begin{aligned} P\left[0 < \max_i X_i \le c|H_\circ\right] &= P[0 < X_i \le c \ \forall\, i = 1, \cdots, n|H_\circ] \\ &= \{P[0 < X_i \le c|H_\circ]\}^n = (c/\theta_0)^n. \end{aligned} \qquad \text{(i)}$$

If (i) is equal to α,

$$c = \theta_0 \alpha^{1/n}.$$

Hence, a BCR of size α is

$$\omega = \left\{\boldsymbol{x} : \boldsymbol{x} \in \left(0 < \max_i x_i \le \theta_0\alpha^{1/n}\right) \cup \left(\theta_0 < \max_i x_i < \theta_1\right)\right\}.$$

The power of the test is

$$\begin{aligned} &P_{\theta_1}\left\{0 < \max_i X_i \le \theta_0\alpha^{1/n}\right\} + P_{\theta_1}\left\{\theta_0 < \max_i X_i < \theta_1\right\} \\ &= \left(\frac{\theta_0\alpha^{1/n}}{\theta_1}\right)^n + 1 - \left(\frac{\theta_0}{\theta_1}\right)^n = 1 - \left(\frac{\theta_0}{\theta_1}\right)^n (1-\alpha). \end{aligned}$$

► **Example 17.4.9** *Let $(X_1, \cdots, X_n)'$ be a random sample from the exponential distribution*

$$f(x;\theta) = e^{-(x-\theta)}, \ \ \theta \le x \le \infty.$$

For this distribution $X_{(1)} = \min\limits_i X_i$ is sufficient for θ.

Suppose $H_\circ : \theta = \theta_0$ and $H_1 : \theta = \theta_1 > \theta_0$.

Now,

$$\frac{L(\theta_1|\boldsymbol{x})}{L(\theta_0|\boldsymbol{x})} = \begin{cases} 0, & \text{if } \theta_0 \le x_{(1)} < \theta_1 \\ e^{n(\theta_1-\theta_0)} (> 1), & \text{if } x_{(1)} \ge \theta_1. \end{cases}$$

For any $k \leq e^{n(\theta_1-\theta_0)}$ any sample observation $(x_1, \cdots, x_n)'$ with $x_{(1)} \geq \theta_1$ satisfies $L(\theta_1)/L(\theta_0) \geq k$ and hence is the BCR of the same size.

We note that the choice of $k(\leq e^{n(\theta_1-\theta_0)})$ does not determine the size of the critical regions in this case. For this we have to utilise the distribution of the sufficient statistic $X_{(1)}$.

$$\begin{aligned} P_{\theta_0}\left[X_{(1)} - \theta_1 \geq c\right] &= P_{\theta_0}\left[X_i - \theta_1 \geq c \ \forall\, i\right] \\ &= \left[\int_{c+\theta_1}^{\infty} e^{-(x-\theta_0)}\, dx\right]^n \\ &= e^{-n(c+\theta_1-\theta_0)} = \alpha \end{aligned}$$

implies $c = -\frac{1}{n}\log\alpha + \theta_0 - \theta_1$.

Hence the BCR of size α is

$$\left\{\boldsymbol{x} : x_{(1)} \geq \theta_0 - \frac{1}{n}\log\alpha\right\}.$$

Consider now the alternative hypothesis $H_2 : \theta \neq \theta_0$. Here

$$\frac{L(\theta_1|\boldsymbol{x})}{L(\theta_0|\boldsymbol{x})} = \begin{cases} 0, & \text{if } \theta_0 \leq x_{(1)} < \theta_1 \quad \text{(i)} \\ e^{n(\theta_1-\theta_0)}(<1), & \text{if } x_{(1)} \geq \theta_0 > \theta_1 \quad \text{(ii)} \\ e^{n(\theta_1-\theta_0)}(>1), & \text{if } x_{(1)} \geq \theta_1 > \theta_0 \quad \text{(iii)} \\ \infty, & \text{if } \theta_1 \leq x_{(1)} < \theta_0. \quad \text{(iv)} \end{cases}$$

Obviously, (i) and (iii) relate to the case $\theta_0 < \theta_1$ and (ii) and (iv) to the case $\theta_1 < \theta_0$. For the case $\theta_1 < \theta_0$, any $\boldsymbol{x}$ with $\theta_1 \leq x_{(1)} < \theta_0$ is a BCR of size 0. For the case $\theta_1 > \theta_0$, the critical region of size α is $\boldsymbol{x}$ with $\boldsymbol{x}_{(1)} \geq \theta_0$. Hence the BCR is given by

$$\left\{\boldsymbol{x} : \boldsymbol{x} \in \{(x_{(1)} - \theta_0) < 0\} \cup \{(x_{(1)} - \theta_0) > c\}\right\}.$$

The event $(x_{(1)} - \theta_0) < 0$ has probability zero under H_o. For the second set c is so determined that the probability of the event $X_{(1)} - \theta_0 > c$ is α under H_o.

17.5 Most Powerful Tests and Sufficient Statistic

If both H_o and H_1 specify the distributions from the same family $\{F_\theta, \theta \in \Theta\}$ and if a sufficient statistic t for θ exists, it follows from the factorisation theorem of the sufficient statistic that

$$\frac{L(\theta_1|\boldsymbol{x})}{L(\theta_0|\boldsymbol{x})} = \frac{g(t;\theta_1)}{g(t;\theta_0)}.$$

Hence, it follows from (17.4.1) that the BCR is a function of the sufficient statistic However, the BCR will not be of the form $t \geq c_\alpha$ or $t \leq c_\alpha$ always. Example 17.4.6 is an example. The BCR will be of this form if $g(t;\theta_1)/g(t;\theta_0)$ is a monotonic function of t for $\theta_1 \neq \theta_0$.

If the family $\{f_\theta, \theta \in \Theta\}$ admits a sufficient statistic, one can restrict attention to tests based on sufficient statistic.

If δ is a test function and T is a sufficient statistic, then $\psi(X) = E\{\delta(\boldsymbol{X})|T\} \in [0,1]$ and itself is a test function.

Clearly,

$$E_\theta[\psi(\boldsymbol{X})] = E_\theta[\delta(\boldsymbol{X})]$$

and hence $\psi(\boldsymbol{X})$ and $\delta(\boldsymbol{X})$ have the same power function (and size).

17.6 Most Powerful Randomized Tests

The following version of NP theorem appliable to randomized tests also gives a most powerful test for $H_\circ(\theta = \theta_0)$ against $H_1(\theta = \theta_1)$.

Theorem 17.6.1 *Consider $H_\circ(\theta = \theta_0)$ and $H_1(\theta = \theta_1)$. Any test of the form*

$$\delta(\boldsymbol{x}) = \begin{cases} 1, \text{ if } L(\theta_1|\boldsymbol{x}) > kL(\theta_0|\boldsymbol{x}) \\ \lambda, \text{ if } L(\theta_1|\boldsymbol{x}) = kL(\theta_0|\boldsymbol{x}) \\ 0, \text{ if } L(\theta_1|\boldsymbol{x}) < kL(\theta_0|\boldsymbol{x}) \end{cases} \tag{17.6.1}$$

for some $k(>0)$ and $0 \leq \lambda \leq 1$ is MP *of its size for testing $H_\circ$ against H_1.*

If $k = \infty$, the test

$$\delta(\boldsymbol{x}) = \begin{cases} 1, \text{ if } L(\theta_0|\boldsymbol{x}) = 0 \\ 0, \text{ if } L(\theta_0|\boldsymbol{x}) > 0 \end{cases} \tag{17.6.2}$$

is MP of size 0 for testing $H_\circ$ against H_1.

The quantity λ is so determined that the size of the test is a pre-assigned quantity α.

Proof: Let $\phi(\boldsymbol{x})$ be any test function satisfying $E_{\theta_0}\{\phi(\boldsymbol{x})\} \leq E_{\theta_0}\{\delta(\boldsymbol{x})\}$. The probability of rejecting $H_\circ$ is, therefore,

$$\int \cdots \int \phi(\boldsymbol{x})L(\theta|\boldsymbol{x})d\boldsymbol{x} = P(\theta) \tag{17.6.3}$$

where $P(\theta)$ is the power function of ϕ. Consider then, for any positive k, the difference

$$P(\theta_1) - kP(\theta_0) = \int \cdots \int \phi(\boldsymbol{x})[L(\theta_1|\boldsymbol{x}) - kL(\theta_0|\boldsymbol{x})]d\boldsymbol{x}. \tag{17.6.4}$$

The integral will be maximised if for each $\boldsymbol{x}$ such that

$$L(\theta_1|\boldsymbol{x}) > kL(\theta_0|\boldsymbol{x})$$

we assign $\phi(\boldsymbol{x})$ the value 1 and for each $\boldsymbol{x}$ such that

$$L(\theta_1|\boldsymbol{x}) < kL(\theta_0|\boldsymbol{x})$$

we assign $\phi(\boldsymbol{x})$ the value 0. Samples for which

$$L(\theta_1|\boldsymbol{x}) = kL(\theta_0|\boldsymbol{x})$$

contribute nothing to the value of the integral in (17.6.4) and the value of ϕ is immaterial. Suppose it is λ. Denote the critical function with such values as $\delta(\boldsymbol{x})$ and let its power function be $\gamma(\theta)$. We have, therefore, by construction,

$$\begin{aligned}\gamma(\theta_1) - k\gamma(\theta_0) &\geq P(\theta_1) - kP(\theta_0)\\ \text{or,}\quad \gamma(\theta_1) - P(\theta_1) &\geq k[\gamma(\theta_0) - P(\theta_0)].\end{aligned}$$

Hence

$$\gamma(\theta_0) \geq P(\theta_0) \quad \Rightarrow \quad \gamma(\theta_1) > P(\theta_1).$$

Therefore, $\delta(\boldsymbol{x})$ is the most powerful test of its size, since

$$P(\theta_0) = E_{\theta_0}(\phi(\boldsymbol{x})) \leq \gamma(\theta_0) = E_{\theta_0}(\delta(\boldsymbol{x})).$$

If $k = \infty$, any test of size zero must vanish on the set $\{\boldsymbol{x} : L(\theta_0|\boldsymbol{x}) > 0\}$, as is evident from (17.6.1).

$$\gamma(\theta_1) - P(\theta_1) = \int_{\{\boldsymbol{x}:L(\theta_0|\boldsymbol{x})=0\}} (1 - \phi(\boldsymbol{x}))L(\theta_1|\boldsymbol{x})d\boldsymbol{x} \geq 0. \tag{17.6.5}$$

Hence the test given in (17.6.2) is the most powerful test of size 0.

To prove the last part we need to restrict ourselves to the case when $0 < \alpha \leq 1$, since the MP test with $\alpha = 0$ has been considered in the second part. The size of the test $\delta(\boldsymbol{x})$ is

$$\begin{aligned}E_{\theta_0}\{\delta(\boldsymbol{x})\} &= P_{\theta_0}\{L(\theta_1|\boldsymbol{x}) > kL(\Theta_0|\boldsymbol{x})\} + \lambda P_{\theta_0}\{L(\theta_1|\boldsymbol{x}) = kL(\theta_0|\boldsymbol{x})\}\\ &= 1 - P_{\theta_0}\{L(\theta_1|\boldsymbol{x}) \leq kL(\theta_0|\boldsymbol{x})\} + \lambda P_{\theta_0}\{L(\theta_1|\boldsymbol{x}) = kL(\theta_0|\boldsymbol{x})\}.\end{aligned} \tag{17.6.6}$$

Since $P_{\theta_0}\{L(\theta_0|\boldsymbol{x}) = 0\} = 0$, we can rewrite (17.6.6) as

$$\begin{aligned}E_{\theta_0}\{\delta(\boldsymbol{x})\} &= 1 - P_{\theta_0}\left\{\frac{L(\theta_1|\boldsymbol{x})}{L(\theta_0|\boldsymbol{x})} \leq k\right\} + \lambda P_{\theta_0}\left\{\frac{L(\theta_1|\boldsymbol{x})}{L(\theta_0|\boldsymbol{x})} = k\right\}\\ &= \alpha \text{ (by requirement).}\end{aligned} \tag{17.6.7}$$

The equation (17.6.7) has to be solved for λ. (17.6.7) gives

$$P_{\theta_0}\left\{\frac{L(\theta_1|\boldsymbol{x})}{L(\theta_0|\boldsymbol{x})} \le k\right\} - \lambda P_{\theta_0}\left\{\frac{L(\theta_1|\boldsymbol{x})}{L(\theta_0|\boldsymbol{x})} = k\right\} = 1-\alpha. \tag{17.6.8}$$

Now

$$P_{\theta}\left\{\frac{L(\theta_1|\boldsymbol{x})}{L(\theta_0|\boldsymbol{x})} \le k\right\}$$

is a distribution function so that it is a non-decreasing, a right continuous function of k. The quantity

$$P_{\theta_0}\left\{\frac{L(\theta_1|\boldsymbol{x})}{L(\theta_0|\boldsymbol{x})} = k\right\}$$

is saltus at k. If there exists a non-negative quantity k_0 such that

$$P_{\theta_0}\left\{\frac{L(\theta_1|\boldsymbol{x})}{L(\theta_0|\boldsymbol{x})} \le k_0\right\} = 1-\alpha$$

then take $\lambda = 0$; otherwise, there exists a quantity k_0 such that

$$P_{\theta_0}\left\{\frac{L(\theta_1|\boldsymbol{x})}{L(\theta_0|\boldsymbol{x})} < k_0\right\} < 1-\alpha < P_{\theta_0}\left\{\frac{L(\boldsymbol{x}|\theta_1)}{L(\boldsymbol{x}|\theta_0)} \le k_0\right\}.$$

In this case, take

$$\lambda = \frac{P_{\theta_0}\left\{\frac{L(\theta_1|\boldsymbol{x})}{L(\theta_0|\boldsymbol{x})} \le k_0\right\} - (1-\alpha)}{P_{\theta_0}\left\{\frac{L(\theta_1|\boldsymbol{x})}{L(\theta_0|\boldsymbol{x})} = k_0\right\}}. \tag{17.6.9}$$

With this choice of λ, the size α is exactly attained.

It should be noted that if $X_1, \cdots, X_n$ are continuous random variables, the MP randomised test for testing $H_\circ$ against H_1 will generally be identical with the MP non-randomised test. This is because $L(\theta_1|\boldsymbol{x})/L(\theta_0|\boldsymbol{x})$ will also be a continuous random variable and, therefore, $P[L(\theta_1|\boldsymbol{x}) = kL(\theta_0|\boldsymbol{x})]$ will be zero. Even when $X_1, \cdots, X_n$ are discrete, a MP critical region of size α (if it exists) and a MP randomised test of size α will be identical. However, it will often turn out that no MP critical region of exact size α exists. In such cases we should take instead a slightly lower level for which a MP critical region exists and make use of the randomisation at the boundary. The MP randomised test leads to an increase in power, while still attaining the desired level α.

Example 17.4.2 gives an example of a UMP randomised critical region of size α for testing $H_\circ(\theta = \theta_0)$ against $H(\theta < \theta_0)$ for a Poisson population.

17.7 Testing a Simple Hypothesis (or a One-sided Composite Hypothesis) against a One-side Alternative

Consider the case when $H_\circ$ is simple but H_1 is composite. Suppose, for instance, $H_\circ$ is $\theta = \theta_0$, H_1 is $\theta \ne \theta_0$. For each of the alternative hypotheses $H_{1t} : \theta = \theta_t$ (the null hypothesis being $H_\circ$ in each case) we find the BCR ω_t for a fixed t. If ω_t does not depend on θ_t, this would be the UMP critical region for H_1. However, a UMP critical region does not generally exist unless we restrict the class of alternatives. In examples 17.4.2–17.4.5 we obtained UMP critical regions for the alternatives of the form $H(\theta < \theta_0)$ and $H(\theta > \theta_0)$. We introduce here the concept of monotone likelihood ratio MLR for a distribution. It will be seen that the distribution having MLR provide UMP tests for $H_\circ(\theta = \theta_0)$ against $H(\theta > \theta_0)$ [or $H(\theta < \theta_0)$].

Definition 17.7.1 *Let $\{f_\theta, \theta \in \Theta\}$ be a family of p.d.f.'s (or p.m.f.'s), $\Theta \subseteq R^1$ and let $L(\theta|\boldsymbol{x})$ denote the likelihood for $\boldsymbol{x}$, a random sample of size n from the distribution $f_\theta, \theta \in \Theta$. The family $\{f_\theta\}$ has a monotone likelihood ratio* (MLR) *in the statistic $t(\boldsymbol{x})$ if for $\theta_1 < \theta_2$, whenever $f_{\theta_1}, f_{\theta_2}$ are distinct, the ratio $L(\theta_2|\boldsymbol{x})/L(\theta_1|\boldsymbol{x})$ is a monotonic (either non-decreasing or non-increasing) function of $t(\boldsymbol{x})$ for the set of values of $\boldsymbol{x}$ for which at least one of $L(\theta_1|\boldsymbol{x})$ and $L(\theta_2|\boldsymbol{x}) > 0$.*

▶ **Example 17.7.1** *The one-parameter exponential family*

$$L(\theta|\boldsymbol{x}) = \exp\left[Q(\theta)t(\boldsymbol{x}) + S(\boldsymbol{x}) + D(\theta)\right]$$

where $Q(\theta)$ is non-decreasing, has an MLR *in $t(\boldsymbol{x})$.*

Proof: For $\theta_2 > \theta_1$, $Q(\theta_2) > Q(\theta_1)$ and hence

$$\frac{L(\theta_2|\boldsymbol{x})}{L(\theta_1|\boldsymbol{x})} = \exp\{t(\boldsymbol{x})(Q(\theta_2) - Q(\theta_1)) + (D(\theta_2) - D(\theta_1))\}$$

is non-decreasing in $t(\boldsymbol{x})$.

The one-parameter exponential family includes normal, binomial, poisson, gamma (one parameters fixed), beta (one parameter fixed) distributions.

Example 17.7.2 *Let $X_1, \cdots, X_n$ be a random sample from Rectangular $(0, \theta)$ distribution, $\theta > 0$. The likelihood of $\boldsymbol{x}$ is*

$$L(\theta|\boldsymbol{x}) = \begin{cases} \frac{1}{\theta^n}, & \text{if } 0 \leq x_{(n)} \leq \theta \\ 0, & \text{otherwise.} \end{cases}$$

where $x_{(n)} = \max_i x_i$.

Let $\theta_2 > \theta_1$.

$$\frac{L(\theta_2|\boldsymbol{x})}{L(\theta_1|\boldsymbol{x})} = \frac{(1/\theta_2^n)I_{[\max x_i \leq \theta_2]}}{(1/\theta_1^n)I_{[\max x_i \leq \theta_1]}} = t(\boldsymbol{x})\left(\frac{\theta_1}{\theta_2}\right)^n.$$

Now,

$$Q(\boldsymbol{x}) = \frac{I_{[x_{(n)} \leq \theta_2]}}{I_{[x_{(n)} \leq \theta_1]}} = 1 \text{ if } x_{(n)} \leq \theta_1$$

$$= \infty \text{ if } x_{(n)} \in [\theta_1, \theta_2].$$

Define $Q(\boldsymbol{x}) = \infty$ if $x_{(n)} > \theta_2$. Thus $L(\theta_2|\boldsymbol{x})/L(\theta_1|\boldsymbol{x})$ is a non-decreasing function of $t(\boldsymbol{x}) = x_{(n)}$ and the family of rectangular densities on $[0, \theta]$ has a MLR in $x_{(n)}$.

Theorem 17.7.1 *Suppose that X is a r.v. with distribution function $F(x|\theta), \theta \in \Theta$. Assume that $F(x|\theta)$ has a monotone likelihood ratio in $t(\boldsymbol{x})$. Consider $H_\circ(\theta = \theta_0)$ and $H_1(\theta > \theta_0)$. A* UMP *non-randomised test of size α for testing $H_\circ$ against H_1 is*

$$\delta(\boldsymbol{x}) = \begin{cases} 1, & \text{if } t(\boldsymbol{x}) \geq k \\ 0, & \text{otherwise} \end{cases} \tag{17.7.1}$$

where k is such that

$$E_{\theta_0}[\delta(\boldsymbol{x})] = \alpha. \tag{17.7.2}$$

Proof: Consider the problem of testing $H_\circ : \theta = \theta_0$ against $H_1' : \theta = \theta_1$ for $\theta_1 > \theta_0$. Both $H_\circ$ and H_1' are simple. By NP-theorem a MP test of size α for testing $H_\circ$ against H_1' is

$$\delta(\boldsymbol{x}) = \begin{cases} 1 \text{ if } L(\theta_1|\boldsymbol{x})/L(\theta_0|\boldsymbol{x}) \geq k \\ 0, \text{ otherwise.} \end{cases} \tag{17.7.3}$$

Since $F(x|\theta)$ has a MLR in $t(\boldsymbol{x})$, (17.7.3) can be written as

$$\delta(\boldsymbol{x}) = \begin{cases} 1, & \text{if } t(\boldsymbol{x}) \geq c \\ 0, & \text{otherwise} \end{cases} \tag{17.7.4}$$

where c is such that the condition (17.7.2) is satisfied. It is seen that the test (17.7.4) is the same regardless of the choice of θ_1 so long as $\theta_1 > \theta_0$. Hence a test satisfying (17.7.4) and (17.7.2) is UMP for $H_\circ$ against $H_1(\theta > \theta_0)$.

Corollary 17.7.1 *If H_1 is replaced by $H_1^* : \theta_1 < \theta_0$ then the* UMP *test satisfies (17.7.2) and*

$$\delta(\boldsymbol{x}) = \begin{cases} 1, & \text{if } t(\boldsymbol{x}) \leq c \\ 0, & \text{otherwise.} \end{cases}$$

Suppose now we want to test one-sided null hypothesis $H_\circ(\theta \leq \theta_0)$ against the one sided alternative $H_1(\theta > \theta_0)$ [or its dual $H_\circ' : \theta \geq \theta_0$ against $H_1'(\theta < \theta_0)$]. In general, it is not possible to find UMP test for this problem. The MP test of $H_\circ(\theta \leq \theta_0)$ against the alternative $H(\theta = \theta_1(> \theta_0))$ would depend on θ_1 and hence cannot be UMP. However, if the distribution $F(x|\theta)$ admits of a MLR, then there exists a UMP test for $H_\circ(\theta \leq \theta_0)$ against $H_1(\theta > \theta_0)$ (or its dual).

Theorem 17.7.2 *If the likelihood $L(\theta|\boldsymbol{x})$ is such that whenever $\theta_1 < \theta_2$, $L(\theta_2|\boldsymbol{x})/L(\theta_1|\boldsymbol{x})$ is monotone in some function $t(\boldsymbol{x})$ of the observations $\boldsymbol{x}$, then there exists a* UMP *test of $H_\circ(\theta \leq \theta_0)$ against $H(\theta > \theta_0)$ based on the statistic $t(\boldsymbol{x})$.*

Proof: Let us denote a critical function $\delta_0(X)$ such that

$$\delta_0(\boldsymbol{x}) = \begin{cases} 1, \text{ if } L(\theta_2|\boldsymbol{x})/L(\theta_1|\boldsymbol{x}) > k \\ 0, \text{ if } L(\theta_2|\boldsymbol{x})/L(\theta_1|\boldsymbol{x}) < k \\ \pi, \text{ if } L(\theta_2|\boldsymbol{x})/L(\theta_1|\boldsymbol{x}) = k \end{cases} \tag{17.7.5}$$

i.e.,

$$\delta_0(\boldsymbol{x}) = \begin{cases} 1, \text{ if } t(\boldsymbol{x}) > t_0 \\ 0, \text{ if } t(\boldsymbol{x}) < t_0 \\ \pi, \text{ if } t(\boldsymbol{x}) = t_0. \end{cases}$$

Then

$$\begin{aligned} E\{\delta_0(X)|\theta_2\} = \alpha(\theta_2) \text{ (say)} &= \int_{t(\boldsymbol{x})>t_0(\boldsymbol{x})} L(\theta_2|\boldsymbol{x})d\boldsymbol{x} + \pi \int_{t(\boldsymbol{x})=t_0(\boldsymbol{x})} L(\theta_2|\boldsymbol{x})d\boldsymbol{x} \\ &= \int_{L(\theta_2|\boldsymbol{x})>kL(\theta_1|\boldsymbol{x})} L(\theta_2|\boldsymbol{x})d\boldsymbol{x} \\ &\quad + \pi \int_{L(\theta_2|\boldsymbol{x})=kL(\theta_1|\boldsymbol{x})} L(\theta_2|\boldsymbol{x})d\boldsymbol{x} \\ &> k \int_{L(\theta_2|\boldsymbol{x})>kL(\theta_1|\boldsymbol{x})} L(\theta_1|\boldsymbol{x})d\boldsymbol{x} \\ &\quad + k\pi \int_{L(\theta_2|\boldsymbol{x})=kL(\theta_1|\boldsymbol{x})} L(\theta_1|\boldsymbol{x})d\boldsymbol{x} \\ &= kE\{\delta_0(X)|\theta_1\} = k\alpha(\theta_1). \end{aligned} \tag{i}$$

Similarly, it can be shown that

$$1 - \alpha(\theta_2) \le k(1 - \alpha(\theta_1)). \tag{ii}$$

Hence,

$$\frac{1-\alpha(\theta_2)}{\alpha(\theta_2)} \le \frac{k(1-\alpha(\theta_1))}{k\alpha(\theta_1)} \text{ or, } \alpha(\theta_1) < \alpha(\theta_2) \text{ for } \theta_1 < \theta_2.$$

Thus $E(\delta_0|\theta)$ is a monotonically increasing function of θ. Hence if t_0 and π are so chosen that $E(\delta_0|\theta_0) = \alpha$, then $E(\delta_0|\theta) < \alpha$ for $\theta < \theta_0$ and δ_0 is of size α for testing $H_o(\theta \le \theta_0)$.

Again according to NP lemma among all tests for which $E(\delta|\theta_0) \le \alpha$, the critical function δ_0' which maximises $E(\delta_0'|\theta')$ for any $\theta' > \theta_0$ is given by

$$\delta_0' = \begin{cases} 1, \text{ if } L(\boldsymbol{x}|\theta')/L(\boldsymbol{x}|\theta_0) > k \text{ i.e., if } t(\boldsymbol{x}) > t_0 \\ 0, \text{ if } L(\boldsymbol{x}|\theta')/L(\boldsymbol{x}|\theta_0) < k \text{ i.e., if } t(\boldsymbol{x}) < t_0 \\ \pi, \text{ if } L(\boldsymbol{x}|\theta')/L(\boldsymbol{x}|\theta_0) = k \text{ i.e., if } t(\boldsymbol{x}) = t_0. \end{cases}$$

$L(\boldsymbol{x}|\theta')/L(\boldsymbol{x}|\theta_0)$ being a monotonically increasing function of $t(\boldsymbol{x})$ for $\theta' > \theta_0$ where π, t_0 are uniqualy determined from the condition $E(\delta_0'|\theta_0) = \alpha$. It follows from our choice of δ_0 that δ_0 and δ_0' are the same.

Thus taking any δ of size α i.e., $E(\delta|\theta) \le \alpha$ for $\theta \le \theta_0$ so that $E(\delta|\theta_0) \le \alpha$, we have from NP-lemma and the remarks above that $E(\delta_0|\theta) \ge E(\delta|\theta)$ if $\theta > \theta_0$. Therefore, δ_0 is UMP size α-test for testing $H_o(\theta \le \theta_0)$ against $H(\theta > \theta_0)$.

Theorem 17.7.3 *If the density function is such that it has monotone likelihood ratio, the power function of the test is a monotonically increasing function of* θ.

Proof: By assumption, $L(\boldsymbol{x}|\theta)/L(\boldsymbol{x}|\theta')$ is a monotonically increasing function of a statistic $t(\boldsymbol{x})$ for $\theta > \theta'$. Define a critical function

$$\delta_0(\boldsymbol{x}) = \begin{cases} 1, \text{ if } L(\boldsymbol{x}|\theta)/L(\boldsymbol{x}|\theta') > k \text{ i.e., if } t(\boldsymbol{x}) > t_0 \\ 0, \text{ if } L(\boldsymbol{x}|\theta)/L(\boldsymbol{x}|\theta') < k \text{ i.e., if } t(\boldsymbol{x}) < t_0 \\ \pi, \text{ if } L(\boldsymbol{x}|\theta)/L(\boldsymbol{x}|\theta') = k \text{ i.e., if } t(\boldsymbol{x}) = t_0. \end{cases}$$

Then,

$$E_\theta(\delta_0(\boldsymbol{x})) = \int_{t(\boldsymbol{x})>t_0} L(\boldsymbol{x}|\theta)d\boldsymbol{x} + \pi\int_{t(\boldsymbol{x})=t_0} L(\boldsymbol{x}|\theta)d\boldsymbol{x}$$

$$> k\left[\int_{t(\boldsymbol{x})>t_0} L(\boldsymbol{x}|\theta')d\boldsymbol{x} + \pi\int_{t(\boldsymbol{x})=t_0} L(\boldsymbol{x}|\theta')d\boldsymbol{x}\right] = kE_{\theta'}(\delta_0(\boldsymbol{x}))$$

i.e., $\alpha(\theta) > k\alpha(\theta')$; $(\theta > \theta')$.

Similarly, it can be shown that

$$1 - \alpha(\theta) \leq k[1 - \alpha(\theta')].$$

Hence

$$\frac{1-\alpha(\theta)}{\alpha(\theta)} \leq \frac{1-\alpha(\theta')}{\alpha(\theta')} \quad \text{or,} \quad \alpha(\theta) \geq \alpha(\theta').$$

Thus, $E(\delta_0|\theta)$ is a monotonically increasing function of θ. Hence to test $H_o(\theta \leq \theta_0)$ against $H(\theta > \theta_0)$ if we choose $\theta_2 > \theta_1 > \theta_0$, we have $E(\delta_0|\theta_2) > E(\delta_0|\theta_1)$ showing that $E(\delta_0|\theta)$ is a monotonically increasing function of θ for $\theta > \theta_0$. Hence the proof.

If the density function of a r.v. X admits of a sufficient statistic T, the density function of X is given as

$$f_\theta(x) = C(\theta)e^{\phi(\theta)T(x)}$$

where ϕ is strictly monotonic. Therefore, for $\theta' > \theta$

$$\frac{f_{\theta'}(x)}{f_\theta(x)} = [C(\theta')/C(\theta)]\exp[\phi(\theta') - \phi(\theta)]T(x).$$

Hence $f_{\theta'}(x)/f_\theta(x)$ is a monotonically increasing function of $T(x)$ for $\theta' > \theta$. Thus $f_\theta(x)$ has monotone likelihood ratio and as such there exists a UMP test for $H_o(\theta \leq \theta_0)$ against $H(\theta > \theta_0)$.

If a sufficient statistic t for θ exists,

$$\frac{L(\boldsymbol{x}|\theta_2)}{L(\boldsymbol{x}|\theta_1)} = \frac{h(t|\theta_2)g(x_1, \cdots, x_n)}{h(t|\theta_1)g(x_1, \cdots, x_n)} = \frac{h(t|\theta_2)}{h(t|\theta_1)}.$$

Theorem 17.7.2 may be applied if $h(t|\theta_2)/h(t|\theta_1)$ is monotonic in t.

▶ **Example 17.7.3** *Let $(X_1, \cdots, X_n)'$ be a random sample from Poisson distribution with parameter λ. Here*

$$\frac{L(\boldsymbol{x}|\lambda_2)}{L(\boldsymbol{x}|\lambda_1)} = \left(\frac{\lambda_2}{\lambda_1}\right)^{\Sigma x_i} \exp[n(\lambda_1 - \lambda_2)].$$

For every $\lambda_1 < \lambda_2$ this is an increasing function of $\sum_1^n x_i$. Hence there is a uniformly most powerful test of $H_\circ : \lambda \leq \lambda_0$ against $H_1 : \lambda > \lambda_0$ based on the statistic $\sum_1^n x_i$.

▶ **Example 17.7.4** *Let $X_1, \cdots, X_n$ be a random sample from an exponential distribution $f(x;\theta) = \frac{1}{\theta}\, e^{-x/\theta}$, $0 < x < \infty$, $\theta > 0$. Consider $H_\circ : \theta \leq \theta_0$ against $H : \theta > \theta_0$.*

First consider $H'_\circ : \theta = \theta_0$ against $H' : \theta = \theta_1 (> \theta_0)$. By NP lemma, a MP test is

$$\delta(\boldsymbol{x}) = \begin{cases} 1, \text{ if } \sum_1^n x_i \geq k(\theta_0) \\ 0, \text{ otherwise,} \end{cases} \tag{i}$$

where k is such that $E_{\theta_0}[\delta(\boldsymbol{x})] = \alpha$. Since $\sum_1^n X_i$ follows a Gamma distribution with parameter (θ_0, n) under $H'_\circ$,

$$\frac{1}{\theta_0^n \overline{|n}} \int_{k'(\theta_0)}^{\infty} e^{-x/\theta_0} x^{n-1}\, dx = \alpha.$$

It is found that k' does not depend on θ. Hence the test is UMP for $H'_\circ$ against H.

Consider now $H^*_\circ : \theta = \theta^* (\leq \theta_0)$ against H, for which we can find an UMP. Let us find $k'(\theta^*)$ such that

$$\sup_{\theta^* \subset (0,\theta_0)} \frac{1}{\theta_0^n \overline{|n}} \int_{k'(\theta^*)}^{\infty} e^{-x/\theta_0} x^{n-1}\, dx = \alpha.$$

In this case all the tests $H^*_\circ(\theta^*)$ against H have sizes $\leq \alpha$. It can be checked that the supremum is attained at the boundary point i.e., when $\theta = \theta_0$. Now each of the tests $H^*_\circ(\theta^*)$ against H is of the same form (i). Hence the test of $H'_\circ(\theta = \theta_0)$ against H may be taken as the UMP test of size α for the composite $H_\circ : \theta \leq \theta_0$ against $H : \theta > \theta_0$.

Note that the exponential distribution possesses MLR and hence the result follows by theorem 17.7.2.

▶ **Example 17.7.5** *In example 17.4.4 we obtained an UMP test of size α for $H'_\circ : \mu = \mu_0$ versus $H : \mu > \mu_0$ where $(X_1, \cdots, X_n)'$ is a random sample from $N(\mu, \sigma^2)$ distribution. If $H_\circ$ is composite, $H_\circ : \mu \leq \mu_0$ then for any $\mu^* \subset (-\infty, \mu_0]$ we can find a UMP (size α) test for $H^*_\circ(\mu = \mu^*)$ against H. Determine k' such that*

$$\sup_{\mu^* \subset (-\infty, \mu_0].} P_{\mu^*}\left(\bar{X} > k'(\mu^*)\right) = \alpha.$$

It can be seen that the supremum is attained at $\mu = \mu_0$. Hence the UMP test for $H'_\circ : \mu = \mu_0$ against H can be taken as the UMP (size α) test for $H_\circ : \mu \leq \mu_0$ against $H : \mu > \mu_0$.

17.8 Testing a Hypothesis for a Single Parameter for Two-sided Alternatives

We have already considered tests for hypotheses $H_\circ(\theta \leq \theta_0)$ against $H(\theta > \theta_0)$. We now consider testing $\theta_1 \leq \theta \leq \theta_2$ against the two sided alternatives $H(\theta < \theta_1 \text{ or } \theta > \theta_2)$. Generally, a UMP test does not exist for each cases.

In example 17.4.3 we noted that no UMP critical region existed for the both-sided alternative $H(\theta \neq \theta_0)$.

We prove below that for a simple $H_\circ(\theta = \theta_0)$ no UMP test exists in general for two-sided alternative $H(\theta \neq \theta_0)$ under regularity conditions which include the condition that $\partial L/\partial\theta$ is continuous in θ.

Expanding the likelihood $L(\theta_1|x)$ around θ_0

$$L(\theta_1|x) = L(\theta_0|x) + (\theta_1 - \theta_0)\frac{\partial L(\theta|x)}{\partial\theta}\bigg]_{\theta^*} \tag{17.8.1}$$

where $\theta^* \in (\theta_0, \theta_1)$. For the BCR we must have

$$\frac{L(\theta_1|x)}{L(\theta_0|x)} = 1 + \frac{(\theta_1 - \theta_0)L'(\theta^*|x)}{L(\theta_0|x)} \geq K_\alpha(\theta_1). \tag{17.8.2}$$

Clearly,

$$K_\alpha(\theta_0) = 1 \tag{17.8.3}$$

identically. Expanding $K_\alpha(\theta_1)$ in a Taylor series about θ_0,

$$K_\alpha(\theta_1) = 1 + (\theta_1 - \theta_0)K'_\alpha(\theta^{**}) \tag{17.8.4}$$

where θ^{**} lies in the interval (θ_0, θ_1). (17.8.2), (17.8.3), (17.8.4) give, for all $\boldsymbol{x}$ in the BCR,

$$(\theta_1 - \theta_0)\left\{\frac{L'(\theta^*|\boldsymbol{x})}{L(\theta_0|\boldsymbol{x})} - K'_\alpha(\theta^{**})\right\} \geq 0. \tag{17.8.5}$$

Consider now points $\tilde{x}$ on the boundary of BCR. For such points

$$\frac{L(\theta_1|\tilde{\boldsymbol{x}})}{L(\theta_0|\tilde{\boldsymbol{x}})} = K_\alpha(\theta_1)$$

so that

$$K'_\alpha(\theta^{**}) = \frac{L'(\theta^{**}|\tilde{\boldsymbol{x}})}{L(\theta_0|\tilde{\boldsymbol{x}})}.$$

Hence, from (17.8.5), we have, for points $\boldsymbol{x}$ in the BCR

$$(\theta_1 - \theta_0)\left\{\frac{L'(\theta^*|\boldsymbol{x})}{L(\theta_0|\boldsymbol{x})} - \frac{L'(\theta^{**}|\tilde{\boldsymbol{x}})}{L(\theta_0|\tilde{\boldsymbol{x}})}\right\} \geq 0 \tag{17.8.6}$$

(17.8.6) holds for all $\boldsymbol{x}$, $\tilde{\boldsymbol{x}}$ in the BCR and all $\theta_1 (\neq \theta_0)$.

Since $(\theta_1 - \theta_0)$ changes sign (17.8.6) must, therefore, be zero.

Again, for points $\boldsymbol{x}$ outside the BCR we must have

$$(\theta_1 - \theta_0)\left\{\frac{L'(\theta^*|\boldsymbol{x})}{L(\theta_0|\boldsymbol{x})} - \frac{L(\theta^{**}|\tilde{\boldsymbol{x}})}{L(\theta_0|\tilde{\boldsymbol{x}})}\right\} \leq 0. \tag{17.8.7}$$

The similar argument shows that

$$\frac{L'(\theta^*|\boldsymbol{x})}{L(\theta_0|\boldsymbol{x})} - \frac{L'(\theta^{**}|\tilde{\boldsymbol{x}})}{L(\theta_0|\tilde{\boldsymbol{x}})} = 0 \tag{17.8.8}$$

for points $\boldsymbol{x}$, inside or outside the BCR, when θ^*, θ^{**} are points in (θ_0, θ_1). By virtue of the continuity of L' in θ, therefore,

$$\frac{L'(\theta_0|\boldsymbol{x})}{L(\theta_0|\boldsymbol{x})} = \left.\frac{\partial \log L(\theta_0|\boldsymbol{x})}{\partial \theta}\right]_{\theta=\theta_0} = \text{constant} \tag{17.8.9}$$

(17.8.9) is the necessary condition for the existence of a both-sided BCR. It cannot be satisfied for the distributions with

$$E\left(\frac{\partial \log L}{\partial \theta}\right) = 0$$

(e.g., for distributions with range independent of θ) unless the constant in (17.8.9) is zero (because $E\left(\frac{\partial \log L}{\partial \theta}\right) = 0$ with the condition (17.8.9) implies that $\partial \log L/\partial\theta]_{\theta=\theta_0} = 0$ identically).

Example (17.4.9) gives an instance where a two-sided BCR exists. For this distribution $\partial \log L/\partial \theta = n$ so that (17.8.9) is satisfied.

There are, however, distributions for which UMPU tests exist. For studying this problem we require an extension of Neyman-Pearson theorem.

Theorem 17.8.1 ***Extended Neyman-Pearson Lemma.*** *Let $\xi_0, \xi_1, \cdots, \xi_m$ be non-negative integrable functions of $x_1, \cdots, x_n$ and ω be regions of R^n (Borel sets) such that*

$$\int_\omega \xi_i \, d\boldsymbol{x} = \alpha_i \quad (i = 1, \cdots, m). \tag{17.8.10}$$

Let ω_0 be one of these regions satisfying

$$\omega_0 = \left\{\boldsymbol{x} : \xi_0 > \sum_{i=1}^m k_i \, \xi_i\right\} \tag{17.8.11}$$

where the constants k_i's are so chosen as to satisfy (17.8.10). *Then*

$$\int_{\omega_0} \xi_0\, dx \geq \int_{\omega} \xi_0\, dx. \tag{17.8.12}$$

Conversely, if any region ω_0' satisfies (17.8.10) *and* (17.8.11), *then it is identical with ω_0 except perhaps for a set of points with Labesgue measure* 0.

Proof: **Sufficiency:**

$$\begin{aligned}\int_{\omega_0} \xi_0 dx - \int_{\omega} \xi_0 dx &= \int_{\omega_0 \cap \omega} \xi_0\, dx + \int_{\omega_0 \cap \omega^c} \xi_0\, dx - \int_{\omega \cap \omega_0} \xi_0\, dx - \int_{\omega \cap \omega_0^c} \xi_0\, dx \\ &= \int_{\omega_0 \cap \omega^c} \xi_0\, dx - \int_{\omega \cap \omega_0^c} \xi_0\, dx.\end{aligned} \tag{17.8.13}$$

Now, $\omega_0 \cap \omega^c \subseteq \omega_0$. Hence, by (17.8.11)

$$\int_{\omega_0 \cap \omega^c} \xi_0\, dx \geq \sum_{i=1}^{m} k_i \int_{\omega_0 \cap \omega^c} \xi_i\, dx. \tag{17.8.14}$$

Again, $\omega \cap \omega_0^c \not\subset \omega_0$. Hence

$$\int_{\omega \cap \omega_0^c} \xi_0\, dx \leq \sum_{i=1}^{m} k_i \int_{\omega \cap \omega_0^c} \xi_i\, dx. \tag{17.8.15}$$

Using (17.8.14) and (17.8.15), we have, from (17.8.13),

$$\begin{aligned}\int_{\omega_0} \xi_0\, dx - \int_{\omega} \xi_0\, dx &\geq \sum_{i=1}^{m} k_i \left[\int_{\omega_0 \cap \omega^c} \xi_i\, dx - \int_{\omega \cap \omega_0^c} \xi_i\, dx\right] \\ &= \sum_{i=1}^{m} k_i \left[\int_{\omega_0} \xi_i\, dx - \int_{\omega} \xi_i\, dx\right] = 0 \text{ by (17.8.10).}\end{aligned}$$

Necessity: Suppose ω_0' be one of the regions ω, satisfying (17.8.10), such that

$$\int_{\omega_0'} \xi_0 dx \geq \int_{\omega} \xi_0 dx$$

for all ω. Now

$$\int_{\omega_0'} \xi_0\, dx = \int_{\omega_0} \xi_0\, dx. \tag{17.8.16}$$

We shall show that $\omega_0' \cap \omega_0^c$ and $\omega_0 \cap \omega_0'^c$ have Lebesgue measure 0. If not, then

$$\int_{\omega_0 \cap \omega_0'^c} \xi_0\, dx \geq \sum_{i} k_i \int_{\omega_0 \cap \omega_0'^c} \xi_i\, dx \tag{17.8.17}$$

while

$$\int_{\omega_0^c \cap \omega_0'} \xi_0 \, d\boldsymbol{x} \le \sum_i k_i \int_{\omega_0^c \cap \omega_0'} \xi_i \, d\boldsymbol{x}.$$

Now

$$\int_{\omega_0 \cap \omega_0'^c} \xi_i \, d\boldsymbol{x} = \int_{\omega_0^c \cap \omega_0'} \xi_i \, d\boldsymbol{x} \;\; \forall \, i,$$

because ω_0' satisfies (17.8.10). Hence (17.8.17) implies

$$\int_{\omega_0 \cap \omega_0'^c} \xi_0 \, d\boldsymbol{x} > \int_{\omega_0^c \cap \omega_0'} \xi_0 \, d\boldsymbol{x}$$

which implies

$$\int_{\omega_0} \xi_0 d\boldsymbol{x} > \int_{\omega_0'} \xi_0 d\boldsymbol{x},$$

a contraction. Hence the proof.

■ NOTE 17.8.1 *Theorem 17.8.1 gives* MP *critical region for* $H_\circ(\theta = \theta_0)$ *against* $H(\theta = \theta_1)$ *for* $m = 1$, $\xi_0 = L(\boldsymbol{x}|\theta_1)$, $\xi_1 = L(\boldsymbol{x}|\theta_0)$. *Clearly, the theorem gives non-randomised tests only.*

■ NOTE 17.8.2 *Suppose we want to test* $H_\circ(\theta = \theta_1$ *or* $\theta_2)$ *against alternative* $H(\theta = \theta^*)$. *To find an optimal test for this problem put* $m = 2$, $\xi_0 = L(\boldsymbol{x}|\theta^*)$, $\xi_1 = L(\boldsymbol{x}|\theta_1)$, $\xi_2 = L(\boldsymbol{x}|\theta_2)$. *The most powerful critical region among all critical regions for which probability of type I error at* θ_i *is* $\alpha_i (i = 1, 2)$ *is given by the set* ω_0 *given by*

$$\omega_0 : \{\boldsymbol{x} : L(\theta^*|\boldsymbol{x}) \ge k_1 L(\theta_1|\boldsymbol{x}) + k_2 L(\theta_2|\boldsymbol{x})\}$$

where k_1, k_2 are determined from the conditions

$$P_{\theta_i}(\omega_0) = \alpha_i, \;\; i = 1, 2.$$

The corresponding test function is

$$\delta(\boldsymbol{x}) = \begin{cases} 1, \text{ if } L(\theta^*|\boldsymbol{x}) \ge k_1 L(\theta_1|\boldsymbol{x}) + k_2 L(\theta_2|\boldsymbol{x}) \\ 0, \text{ otherwise.} \end{cases}$$

If the critical region ω_0 is independent of θ^* for all θ^* lying in a certain region $\Theta^* \subset \Theta$, then the test is UMP for all $\theta \in \Theta^*$.

■ NOTE 17.8.3 *Suppose we want to test* $H_\circ(\theta = \theta_1, \cdots, \theta_m)$ *against* $H(\theta = \theta^*)$. *Here put* $\xi_i = L(\boldsymbol{x}|\theta_i), i = 1, \cdots, m$ *and* $\xi_0 = L(\boldsymbol{x}|\theta^*)$. *The* MP *critical region is*

$$\omega_0 : \left\{ \boldsymbol{x} : L(\boldsymbol{x}|\theta^*) \ge \sum_{i=1}^{m} k_i L(\boldsymbol{x}|\theta_i) \right\} \tag{17.8.18}$$

where the constants k_i*'s are to be suitably chosen so that*

$$P_{\theta_i}(\omega_0) = \alpha_i, \;\; i = 1, \cdots, m. \tag{17.8.19}$$

$$\begin{aligned}\int_{\omega_0} L(\boldsymbol{x}|\theta^*)d\boldsymbol{x} &= 1 - \int_{\omega_0^c} L(\boldsymbol{x}|\theta^*)d\boldsymbol{x} \\ &\geq 1 - \sum_{i=1}^{m} k_i \int_{\omega_0^c} L(\boldsymbol{x}|\theta_i)d\boldsymbol{x} \\ &= 1 - \sum_{i=1}^{m} k_i \left\{1 - \int_{\omega_0} L(\boldsymbol{x}|\theta_i)d\boldsymbol{x}\right\}. \end{aligned} \tag{17.8.20}$$

Also it follows from (17.8.18) that

$$\int_{\omega_0} L(\boldsymbol{x}|\theta^*)d\boldsymbol{x} \geq \sum_{i=1}^{m} k_i \int_{\omega_0} L(\boldsymbol{x}|\theta_i)d\boldsymbol{x}. \tag{17.8.21}$$

Therefore, if

$$\begin{aligned}\int_{\omega_0} L(\boldsymbol{x}|\theta_i)d\boldsymbol{x} &= \text{constant } \forall\, i \\ &= \int_{\omega_0} L(\boldsymbol{x}|\theta_1)d\boldsymbol{x} \text{ (say)},\end{aligned}$$

then, it follows from (17.8.21)

$$\int_{\omega_0} L(\boldsymbol{x}|\theta^*)d\boldsymbol{x} \geq \left(\sum_{i=1}^{m} k_i\right) \int_{\omega_0} L(\boldsymbol{x}|\theta_1)d\boldsymbol{x}. \tag{17.8.22}$$

Also, it follows from (17.8.20) that

$$\int_{\omega_0} L(\boldsymbol{x}|\theta^*)d\boldsymbol{x} \geq 1 - \left(\sum_{i=1}^{m} k_i\right)\left[1 - \int_{\omega_0} L(\boldsymbol{x}|\theta_1)d\boldsymbol{x}\right]. \tag{17.8.23}$$

Therefore, whether

$$\sum_{i=1}^{m} k_i \geq 1 \quad \text{or,} \quad \sum_{i=1}^{m} k_i < 1,$$

(17.8.22) and (17.8.23) imply

$$\int_{\omega_0} L(\boldsymbol{x}|\theta^*)d\boldsymbol{x} \geq \int_{\omega_0} L(\boldsymbol{x}|\theta_1)d\boldsymbol{x}. \tag{17.8.24}$$

Thus, if the probability of type I error at $\theta = \theta_i$ is kept constant $(i = 1, \cdots, m)$, the critical region ω_0 given by (17.8.18) is unbiased for testing $H_\circ(\theta = \theta_i, i = 1, \cdots, m)$against $H(\theta = \theta^*)$.

■ NOTE 17.8.4 *Clearly observations in note 17.4.3 apply also to the derivation of the* MP *critical region* (17.8.18).

■ NOTE 17.8.5 *The test function corresponding to* (17.8.18) *is a non-randomised one. This would be sufficient if* $(X_1, \cdots, X_n)'$ *have a joint continuous distribution function. If there are some discrete variables, it may not be possible to attain the size condition* (17.8.19) *exactly always. In such situations one would have to randomise at the boundary points. This can be done following theorem* 17.6.1 *and example* 17.4.2.

When no UMP critical region of size α for testing $H_\circ : \theta = \theta_0$ against $H : \theta \neq \theta_0$ exists, we should look for UMPU critical regions of the same size. However, this will not be an easy task. We first consider construction of type A and type A_1 regions.

In constructing a type A region we make the following assumptions. For every ω,

$$\frac{\partial^i P_\theta(\omega)}{\partial \theta^i} \quad (i = 1, 2)$$

exists (the second derivative being continuous) in the neighbourhood of $\theta = \theta_0$ and differentiation under integration sign is permissible so that

$$\frac{\partial^i}{\partial \theta^i} \int_\omega L(\boldsymbol{x}|\theta) d\boldsymbol{x} = \int_\omega \frac{\partial^i}{\partial \theta^i} L(\boldsymbol{x}|\theta) d\boldsymbol{x} \;\; (i = 1, 2). \tag{17.8.25}$$

Let us write

$$\left.\frac{\partial}{\partial \theta} L(\boldsymbol{x}|\theta)\right]_{\theta=\theta_0} = L'(\boldsymbol{x}|\theta_0), \quad \left.\frac{\partial^2}{\partial \theta^2} L(\boldsymbol{x}|\theta)\right]_{\theta=\theta_0} = L''(\boldsymbol{x}|\theta_0).$$

Theorem 17.8.2 *A type A critical region* ω_0 *for testing* $H_\circ : \theta = \theta_0$ *against* $H : \theta \neq \theta_0$ *is given by*

$$\omega_0 = \{\boldsymbol{x} : L''(\boldsymbol{x}|\theta_0) \geq k_1 L(\boldsymbol{x}|\theta_0) + k_2 L'(\boldsymbol{x}|\theta_0)\} \tag{17.8.26}$$

when k_1, k_2 *are constants such that*

$$\int_{\omega_0} L(\boldsymbol{x}|\theta) d\boldsymbol{x} = \alpha, \tag{17.8.27}$$

$$\int_{\omega_0} L'(\boldsymbol{x}|\theta_0) d\boldsymbol{x} = 0. \tag{17.8.28}$$

Proof: In Theorem 17.8.1, put $m = 2$, $\xi_0 = L''(\boldsymbol{x}|\theta)$, $\xi_1 = L(\boldsymbol{x}|\theta_0)$, $\xi_2 = L'(\boldsymbol{x}|\theta_0)$, $\alpha_1 = \alpha$, $\alpha_2 = 0$. By virtue of the assumption (17.8.25), (17.8.28) is the condition (17.3.10); (17.8.12) implies

$$\left.\frac{d^2 P_\theta(\omega_0)}{d\theta^2}\right]_{\theta_0} \geq \left.\frac{d^2 P_\theta(\omega)}{d\theta^2}\right]_{\theta_0}$$

which is the condition (17.3.11); (17.8.27) gives the desired size. Hence the proof.

Corollary 17.8.1 *Let us denote*

$$\frac{\partial}{\partial\theta}\log L(\boldsymbol{x}|\theta)\bigg]_{\theta=\theta_0} = \phi,$$

$$\frac{\partial^2}{\partial\theta^2}\log L(\boldsymbol{x}|\theta)\bigg]_{\theta=\theta_0} = \phi'. \tag{17.8.29}$$

Hence,

$$L'(\boldsymbol{x}|\theta_0) = \phi L(\boldsymbol{x}|\theta_0),$$

$$L''(\boldsymbol{x}|\theta_0) = (\phi' + \phi^2)L(\boldsymbol{x}|\theta_0). \tag{17.8.30}$$

Consider the special case where

$$\phi' = a + b\phi \tag{17.8.31}$$

for some a, b. The type A region may then be defined as

$$\omega_0 = \left\{\boldsymbol{x} : a + b + \phi^2 > k_1 + k_2\phi\right\}$$

$$\text{or, } \omega_0 = \left\{\boldsymbol{x} : a' + b'\phi + \phi^2 > 0\right\}.$$

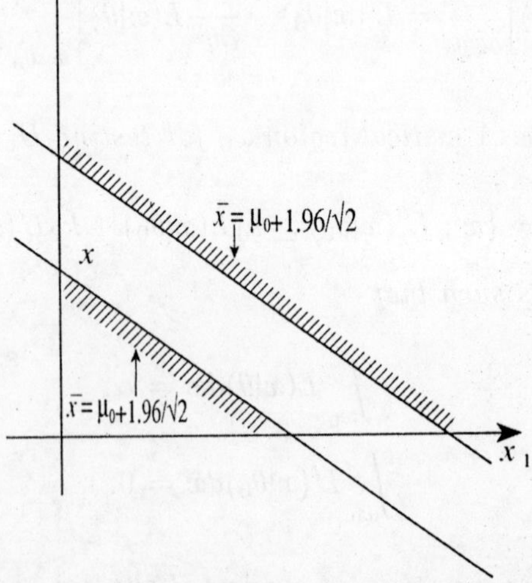

Fig. 17.5 The shaded portion shows UMPU critical region of size 0.05 for testing H_o ($\mu = \mu_0$) against H ($\mu \neq \mu_0$) for a normal population of size 2 with $\sigma^2 = 1$

Now, the second derivative of the function $\zeta(\phi) = a' + b'\phi + \phi^2 > 0$.

Hence the function has only one minimum and no maximum. Writing $\zeta(\phi) = (\phi - c_1)$ $(\phi - c_2)$, the type A region is given by

$$\omega_0 = \{\boldsymbol{x} : \phi < c_1 \cup \phi > c_2\} \tag{17.8.32}$$

where c_1, c_2 are constants such that the conditions (17.8.27) and

$$\int_{\omega_0} \phi L(\boldsymbol{x}|\theta_0)d\boldsymbol{x} = 0 \tag{17.8.33}$$

are satisfied.

► **Example 17.8.1** *Obtain a type A critical region of size α for testing $H_\circ : \theta = \theta_0$ against $H : \theta \neq \theta_0$ for a sample from $N(\theta, \sigma^2)$ population with a known value of σ^2.*

Here $\phi = \dfrac{n(\bar{x}-\theta_0)}{\sigma^2}$ where $\bar{x} = \dfrac{1}{n}\sum_{i=1}^{n} x_i$. Also $\phi' = -n/\sigma^2 = a + b\phi$ where $a = -n/\sigma^2$, $b = 0$. Thus, type A critical region is

$$\omega_0 : \{\boldsymbol{x} : \bar{x} < d_1 \cup \bar{x} > d_2\}$$

where d_1, d_2 are such that (17.8.27), (17.8.28) are satisfied. Let $g(\bar{x};\theta_0)$ denote the marginal p.d.f. of $\bar{x}$. Then these conditions are

$$\int_{-\infty}^{d_1} g(\bar{x};\theta_0)d\bar{x} + \int_{d_2}^{\infty} g(\bar{x};\theta_0)d\bar{x} = \alpha. \tag{i}$$

$$\int_{-\infty}^{d_1} \phi g(\bar{x};\theta_0)d\bar{x} + \int_{d_2}^{\infty} \phi g(\bar{x};\theta_0)d\bar{x} = 0. \tag{ii}$$

(i) implies

$$\frac{1}{\sqrt{2\pi}} \int_{\sqrt{n}(d_1-\theta_0)/\sigma}^{\sqrt{n}(d_2-\theta_0)/\sigma} e^{-y^2/2}dy = 1 - \alpha. \tag{iii}$$

(ii) implies

$$\int_{\sqrt{n}(d_1-\theta_0)/\sigma}^{\sqrt{n}(d_2-\theta_0)/\sigma} ye^{-y^2/2}dy = 0$$

$$\Rightarrow \sqrt{n}(d_1-\theta_0)/\sigma = -\sqrt{n}(d_2-\theta_0)/\sigma. \tag{iv}$$

Hence,

(iii) implies

$$\frac{1}{\sqrt{2\pi}} \int_{-\sqrt{n}(d_2-\theta_0)/\sigma}^{\sqrt{n}(d_2-\theta_0)/\sigma} e^{-y^2/2}dy = 1 - \alpha$$

$$\Rightarrow \sqrt{n}(d_2-\theta_0)/\sigma = \tau_{\alpha/2} \;\Rightarrow\; d_2 = \theta_0 + \tau_{\alpha/2}\frac{\sigma}{\sqrt{n}}.$$

Hence type A critical region is

$$\omega_0 : \left\{\boldsymbol{x} : \sqrt{n}|\bar{x}-\theta_0|/\sigma > \tau_{\alpha/2}\right\}.$$

► **Example 17.8.2** *Obtain a type A critical region of size α for testing $H_\circ : \theta = \theta_0$ against $H : \theta \neq \theta_0$ for a $N(\mu, \theta)$ population where the value of μ is known.*

Here

$$\phi = -\frac{n}{2\theta_0} + \frac{\sum_i (x_i - \mu)^2}{2\theta_0^2},$$

$$\phi' = a + b\phi$$

where $a = -n/2\theta_0^2$, $-2/\theta_0$.

Type A region is

$$\begin{aligned}\omega_0 &= \{\boldsymbol{x} : \phi < c_1 \cup \phi > c_2\} \\ &= \left\{\boldsymbol{x} : \sum_i (x_i - \mu)^2/\theta_0 < d_1 \cup \sum_i (x_i - \mu)^2/\theta_0 > d_2\right\}\end{aligned}$$

where d_1, d_2 satisfy (17.8.27) and (17.8.33) respectively. Since $\sum_i (x_i - \mu)^2/\theta_0$ follow $\chi^2_{(n)}$ under $H_\circ$, these two conditions are equivalent to

$$\frac{1}{2^{n/2}\overline{|\frac{n}{2}}} \int_{d_1}^{d_2} e^{-y/2} y^{\frac{n-2}{2}} dy = 1 - \alpha \qquad \text{(i}$$

and

$$\frac{1}{2^{n/2}\overline{|n/2}} \int_{d_1}^{d_2} (y - n) e^{-y/2} y^{(n-2)/2} dy = 0. \qquad \text{(ii}$$

The second condition reduces to

$$\frac{1}{2^{n/2}\overline{|n/2}} e^{-d_1/2} d_1^{n/2} = \frac{1}{2^{n/2}\overline{|n/2}} e^{-d_2/2} d_2^{n/2}. \qquad \text{(iii}$$

Hence the critical values $d_1 = \chi_1^2$, $d_2 = \chi_2^2$ are such that

$$e^{-\chi_1^2/2} (\chi_1^2)^{n/2} = e^{-\chi_2^2/2} (\chi_2^2)^{n/2} \qquad \text{(iv}$$

and

$$P\left\{\chi^2_{(n)} < \chi_1^2\right\} + P\left\{\chi^2_{(n)} > \chi_2^2\right\} = \alpha. \qquad \text{(v}$$

The values χ_1^2, χ_2^2 are to be solved from equations (iv) and (v) for which method c successive approximation may be used.

We now consider construction of type A_1 regions. For this we assume that $P_\theta(\omega)$ i differentiable with respect to θ for every critical region ω in the neighbourhood of θ_0 an that

$$\frac{\partial}{\partial\theta} \int_\omega L(\boldsymbol{x}|\theta) d\boldsymbol{x} = \int_\omega L'(\boldsymbol{x}|\theta) d\boldsymbol{x}. \qquad (17.8.34)$$

Theorem 17.8.3 *A type A_1 critical region of size α for testing $H_o(\theta = \theta_0)$ against $H(\theta \neq \theta_0)$ is given by*

$$\omega_0 = \{\boldsymbol{x} : L(\boldsymbol{x}|\theta) > k_1 L(\boldsymbol{x}|\theta_0) + k_2 L'(\boldsymbol{x}|\theta_0)\} \tag{17.8.35}$$

where k_1, k_2 are constants such that

$$\int_{\omega_0} L(\boldsymbol{x}|\theta_0) d\boldsymbol{x} = \alpha, \tag{17.8.36}$$

$$\int_{\omega_0} L'(\boldsymbol{x}|\theta_0) d\boldsymbol{x} = 0. \tag{17.8.37}$$

Proof: In theorem 17.8.1, put $m = 2$, $\xi_0 = L(\boldsymbol{x}|\theta)$, $\xi_1 = L(\boldsymbol{x}|\theta_0)$, $\xi_2 = L'(\boldsymbol{x}|\theta_0)$, $\alpha_1 = \alpha$, $\alpha_2 = 0$. By virtue of our assumptions (17.8.35) the result follows.

▶ **Example 17.8.3** *Consider $H_o(\theta = \theta_0)$ versus $H_1(\theta \neq \theta_0)$ for a normal population with known σ^2. Here*

$$L(\boldsymbol{x}|\theta) > k_1 L(\boldsymbol{x}|\theta_0) + k_2 L'(\boldsymbol{x}|\theta_0)$$

$$\Rightarrow \exp\left[-\sum (x_i - \theta)^2/2\sigma^2\right] > k_1 \exp\left[-\sum (x_i - \theta_0)^2/2\sigma^2\right] + \frac{k_2}{\sigma^2}\sum_i (x_i - \theta_0)$$

$$\exp\left[-\sum_i (x_i - \theta_0)^2/2\sigma^2\right]$$

$$\Rightarrow e^{c\bar{x}} > a + b\bar{x},$$

where $c = n(\theta - \theta_0)/\sigma^2$ and a, b are suitable constants such that (17.8.36) *and* (17.8.37) *hold. Let*

$$\phi(\bar{x}) = e^{c\bar{x}} - a - b\bar{x};$$

$$\phi'(\bar{x}) = ce^{c\bar{x}} - b;$$

$$\phi''(\bar{x}) = c^2 e^{c\bar{x}} > 0 \ \forall \ \bar{x}, \text{ since } c \neq 0.$$

Hence the function $\phi(\bar{x})$ has only one minimum and no maximum. Therefore,

$$e^{c\bar{x}} > a + b\bar{x} \quad \Leftrightarrow \quad \bar{x} < d_1 \text{ or } \bar{x} > d_2$$

where $\phi(\bar{x}) = (\bar{x} - d_1)(\bar{x} - d_2)$. Therefore,

$$\omega_0 : \{\boldsymbol{x} : \bar{x} < d_1 \cup \bar{x} > d_2\}$$

when

$$\int_{d_1}^{d_2} g(\bar{x}; \theta_0) d\bar{x} = 1 - \alpha \tag{i}$$

and

$$\int_{\omega_0} L'(\boldsymbol{x}|\theta_0) d\boldsymbol{x} = 0 \text{ or, } \int_{d_1}^{d_2} (\bar{x} - \theta_0) g(\bar{x}; \theta_0) d\bar{x} = 0, \tag{ii}$$

$g(\bar{x}; \theta_0)$ being the marginal p.d.f. of $\bar{x}$ when $\theta = \theta_0$.

Comparing with example 17.8.1, it is seen that ω_0 satisfies the same conditions as the type A region of size α and as such for this problem both type A region and type A_1 region coincide.

For this case the UMPU region of size α is also the same region.

► **Example 17.8.4** *The following example shows relations between* MP *test for* $H_o(\theta_1 = \theta_{10},\ \theta_2 = \theta_{20})$ *and* UMP *tests for* $H_{0i}(\theta_i = \theta_{i0})$ $(i = 1, 2)$ *against the relevant hypotheses.*

X_1, X_2 are independently distributed as $N(\theta_1, 1), N(\theta_2, 1)$ respectively. Consider the simple hypotheses $H_o(\theta_1 = 0, \theta_2 = 0)$ against $H_1(\theta_1 = \theta_1', \theta_2 = \theta_2')$. By the NP theorem 17.4.1, the MP test of H_o against H_1 is based on the statistic $\theta_1' X_1 + \theta_2' X_2$. In particular, if $\theta_1' = 1, \theta_2' = -1$, the MP test is based on large values of $x_1 - x_2$. Since the distribution of $(X_1 - X_2)$ is $N(\theta_1 - \theta_2, 2)$, it can be shown that 97.5% critical region of H_o is

$$\left\{x : x_1 - x_2 > (1.29)\sqrt{2}\right\} \tag{i}$$

and the power of this test when $\theta_1 = 1,\ \theta_2 = -1$ is

$$P\left\{X_1 - X_2 > 1.29\sqrt{2}|\theta_1 = 1, \theta_2 = -1\right\} \simeq 0.55.$$

Consider now the distributions and the respective hypotheses separately. X_1 is distributed as $N(\theta_1, 1)$. The UMP test of $H_o(\theta_1 = 0)$ against $H_{11}(\theta_1 > 0)$ of size 0.05 has critical region $\{x_1 : x_1 > 1.64\}$. The power of this test when $\theta_1 = 1$ is $P\{X_1 > 1.64|\theta_1 = 1\} = 0.261$.

X_2 is distributed independently of X_1. The UMP test of size 0.05 for $H_{02}(\theta_2 = 0)$ against $H_{21}(\theta_2 < 0)$ is $\{x_2 : x_2 < -1.64\}$. The power of this test when $\theta_2 = -1$ is again 0.261.

The probability of rejecting at least one of $H_{01}(\theta_1 = 0)$ and $H_{02}(\theta_2 = 0)$ when both are true is $1 - (1 - 0.05)^2 = 0.0975$.

The probability of rejecting at least one of $H_{01}(\theta_1 = 0)$ and $H_{02}(\theta_2 = 0)$ when $\theta_1 = 1$, $\theta_2 = -1$ is $1 - (1 - 0.261)^2 = 0.454$ (the MP test (i) has power approximately equal to 0.55).

Consider now UMPU critical region of size α for testing $H_{01}(\theta_1 = 0)$ against the two-sided alternative H_{12} $(\theta_1 \neq 0)$, based on the distribution $N(\theta_1, 1)$ of X_1. Its power against some particular value θ_1 is $\gamma_1(\theta_1) \geq \alpha$. Similarly, if X_2 is distributed independently of X_1 as $N(\theta_2, 1)$, the power of any UMPU test for H_{02} $(\theta_2 = 0)$ against H_{22} $(\theta_2 \neq 0)$ is $\gamma_2(\theta_2) \geq \alpha$. The overall probability of rejecting at least one of $\theta_1 = 0$ and $\theta_2 = 0$ when the same pair $(\theta_1,\ \theta_2)$ is true is $1 - (1 - \gamma_1\ (\theta_1))\ (1 - \gamma_2(\theta_2)) \geq 1 - (1 - \alpha)^2 =$ prob. of rejecting at least one of the hypotheses, $\theta_1 = 0$ and $\theta_2 = 0$ when $(\theta_1 = 0, \theta_2 = 0)$. Hence the combined test is also unbiased. However, by constructing separate tests at levels α_1, α_2 such that $1 - (1 - \alpha_1)(1 - \alpha_2) = 1 - (1 - \alpha)^2$ we have another unbiased test at the same level, but with a different power function.

For example, 5% tests for $H_{0i}(\theta_i = 0)$ against $H_{i2}(\theta_i \neq 0)$ is $|X_i| \geq 1.96 (i = 1, 2)$. The probability $P[|X_1 \geq 1.96$ or $|X_2| \geq 0.196|\theta_1 = 1, \theta_2 = -1] \simeq 0.31(\geq 1-0.9025 = 0.0975)$.

Suppose, now, whatever x_1 we accept $\theta_1 = 0$, i.e., $\alpha_1 = 0$. In this case, the power against any $\theta_1 \neq 0$ is also zero.

For testing $H_{02}(\theta_2 = 0)$ against $H_{22}(\theta_2 \neq 0)$ we find α_2 such that $1-(1-\alpha_2) = 0.0975$. This gives $\alpha_2 = 0.0975$. In this case $P_r[|X_2| \geq d|\theta_2 = -1] \simeq 0.26$ where d satisfies $P_r[|X_2| \geq d|\theta_2 = 0] = 0.0975$. This power of the critical region $\{|X_2| \geq d$, whatever $X_1|\theta_1 = 1, \theta_2 = -1] \simeq 0.26$.

Thus, if X_1, X_2 are independently distributed and are both members of one-parameter exponential families, then there does not exist a UMPU test of $H_\circ(\theta_1 = \theta_1', \theta_2 = \theta_2')$ against $H(\theta_1 \neq \theta_1', \theta_2 \neq \theta_2')$. There exist a infinite number of combinations of critical regions for testing for individual parameters. One can make a selection by picking out a symmetrical acceptance region.

17.9 Properties of Class of UMP Tests

Definition 17.9.1 *For testing $H_\circ(\theta \leq \theta_0)$ against $H(\theta > \theta_0)$ if there are two tests $\delta_1(X)$ and $\delta_2(X)$ then $\delta_1(X)$ is said to be uniformly better than $\delta_2(X)$ if $E(\delta_1(X)|\theta) \geq E(\delta_2(X)|\theta)$ for all $\theta \geq \theta_0$ and $E(\delta_1(X)|\theta) \leq E(\delta_2(X)|\theta)$ for all $\theta \leq \theta_0$) with strict inequality holding for at least one θ.*

Definition 17.9.2 *A test $\delta(X)$ is said to be admissible in a class Δ, if there exists no other test which is uniformly better than $\delta(X)$.*

Definition 17.9.3 *A class of tests is said to be complete if given any test outside this class there exists at least one test inside this class which is uniformly better than this.*

Definition 17.9.4 *A class of tests is said to be essentially complete if given any test outside this class there exists at least one test inside this class which is uniformly better than or equivalent to it.*

Definition 17.9.5 *A class of tests is said to be minimal complete if it is complete and contains no subclass of it which is complete.*

Definition 17.9.6 *A class of tests is said to be essentially minimal complete if it is essentially complete and contains no subclass of it which is essentially complete.*

Theorem 17.9.1 *If the class of admissible tests is complete it must be minimal complete.*

Theorem 17.9.2 *A minimal complete class of tests consists of only admissible tests.*

The proof of theorems 1 and 2 are easy and are omitted.

In section 17.7 we have seen that a UMP test exists for $H_\circ(\theta \leq \theta_0)$ against $H(\theta > \theta_0)$ if the p.d.f. admits of a monotone likelihood ratio. We now consider some properties of the class of UMP tests for such hypotheses.

Theorem 17.9.3 *Suppose that the joint density function $L(\theta|\boldsymbol{x})$ has* ML*-ratio and let the null hypothesis be $H_\circ(\theta \leq \theta_0)$ against $H(\theta > \theta_0)$. Then the class of* UMP *tests corresponding to different levels of significance α is essentially complete.*

Proof: Let $\delta(X)$ be a critical function corresponding to any size α-test outisde the class of UMP tests. We have to show that there must exist at least one test having critical function $\delta(X)$ within the class of UMP tests which is uniformly better than or equivalent to it where $E(\delta_0(X)|\theta_0) = E(\delta(X)|\theta_0) = \alpha(\theta_0)$.

If it is not so, we must have $E(\delta|\theta) < E(\delta_0|\theta)$ for at least one $\theta < \theta_0$ and/or $E(\delta|\theta) > E(\delta_0|\theta)$ for at least one $\theta > \theta_0$. Suppose that for $\theta'(< \theta_0)$, $E(\delta|\theta) = \alpha' < E(\delta_0|\theta') = \alpha_0'$.

Consider the MP test of $H_\circ'(\theta = \theta')$ against $H_1'(\theta = \theta_0)$ of size α'. Since δ does not belong to the class of UMP tests of size α', for testing $H_\circ'$ against H_1' there must exist at least one δ_0' of size α' which is more powerful than δ i.e.,

$$E(\delta_0'|\theta_0) > E(\delta|\theta_0) = \alpha(\theta_0).$$

Since for testing a simple hypothesis against a simple alternative power increases with size, we must have

$$E(\delta_0'|\theta_0) < E(\delta_0|\theta_0) = \alpha(\theta_0).$$

Hence a contradiction. Therefore, the class of UMP tests for distributions having ML ratio is complete.

■ NOTE 17.9.1 *If the set $\{x : L(\theta|\boldsymbol{x}) > 0\}$ is independent of θ, then the above class is essentially minimal complete.*

In order to show that the class of UMP tests is essentially minimal complete we have to show that it contains no subclass which is also complete. If possible, let there be two tests $\delta_1(X)$ and $\delta_2(X)$ within the above-mentioned class of UMP tests such that one, say, $\delta_1(X)$, is uniformly better than the other $\delta_2(X)$. Then suppose that

$$E(\delta_1|\theta_1) = \alpha_1(\theta_1) < E(\delta_2|\theta_1) = \alpha_2(\theta_1) \text{ for } \theta_1 \leq \theta_0 \tag{17.9.1}$$

$$E(\delta_1|\theta_2) = \gamma_1(\theta_2) > E(\delta_2|\theta_2) = \gamma_2(\theta_2) \text{ for } \theta_2 > \theta_0. \tag{17.9.2}$$

Consider the test of $H_\circ'(\theta = \theta_1)$ against $H_1'(\theta = \theta_2)$. We have $\alpha_1(\theta_1) < \alpha_2(\theta_1)$. Because of the monotonicity of the LR, δ_1, δ_2 may be considered as MP tests of $H_\circ'$ against H_1'. Also the density function is non-vanishing for any θ. Since for testing a simple hypothesis against a simple alternative, power increases with size, we must have $\gamma_1(\theta_2) < \gamma_2(\theta_1)$ which contradicts (17.9.2). Similarly, it can be seen that if (17.9.2) holds (17.9.1) is violated. Hence the result follows by theorem 17.9.3.

Theorem 17.9.4 *Every* UMPU *test which is unique is admissible.*

Let $\delta_0(X)$ be a unique UMPU test and let $\delta(X)$ be any other test for testing $H_\circ(\theta \le \theta_0)$ against $H(\theta > \theta_0)$. If possible let $\delta(X)$ be uniformly better than $\delta_0(X)$. Let sizes corresponding to $\delta(X)$ and $\delta_0(X)$ be $\alpha(\theta)$ and $\alpha_0(\theta)$ and powers be $\gamma(\theta)$ and $\gamma_0(\theta)$ respectively. Then we shall have

$$\alpha(\theta) \le \alpha_0(\theta), \text{ for } \theta \le \theta_0$$
$$\gamma(\theta) \ge \gamma_0(\theta), \text{ for } \theta \ge \theta_0$$

with strict inequality holding for some θ. Since $\delta_0(X)$ is unbiased.

$$\gamma_0(\theta) \ge \alpha_0(\theta_0) \text{ for } \theta > \theta_0.$$

Hence

$$\gamma(\theta) \ge \gamma_0(\theta) \ge \alpha_0(\theta_0) \ge \alpha(\theta_0) \text{ for } \theta > \theta_0$$

so that $\delta(X)$ is also unbiased and it is then UMP having greater power than $\delta_0(X)$. This is not possible, neither $\delta(X)$ can be equivalent to $\delta_0(X)$ since $\delta_0(X)$ is unique UMPU-test. Therefore, $\delta_0(X)$ is admissible.

Exercise

1. Let $X_1, \cdots, X_m$ be a random sample from Bernoulli distribution $f(x;\theta) = \theta^x(1-\theta)^{1-x}$, $x = 0, 1$. Consider $H_\circ(\theta = \theta_0)$, $H_1(\theta = \theta_1)$ where $\theta_1 < \theta_0$. Find the MP test. Find UMP test for $H_\circ$ against $H(\theta < \theta_0)$. [$\Sigma x_i < c$]

2. In each of the following cases find the MP test of its size. Assume that a random sample of size n has been drawn in each case.

   (a) $f(x;\theta) = \theta \cdot x^{-2}, 0 < \theta \le x < \infty; H_\circ(\theta = \theta_0), H_1(\theta = \theta_1), \theta_1 > \theta_0;$

   (b) $f(x;\theta) = \theta(1-\theta)^{x-1}, x = 1, 2, \cdots; 0 < \theta < 1; H_\circ(\theta = \theta_0), H_1(\theta = \theta_1); \theta_1 > \theta_0;$

   (c) $f(x;\theta) = 2\theta x + 2(1-\theta)(1-x), 0 < x < 1; \theta \in [0,1]. H_\circ(\theta = \theta_0), H(\theta = \theta_1).$

3. Let X be a sample of size one from the Cauchy distribution $f(x;\theta) = \frac{1}{\pi\{1+(x-\theta)^2\}}$, $-\infty < x < \infty$.

   Show that for testing $H_\circ(\theta = -1)$ against $H(\theta = 0)$, $L(x|0)/L(x|-1) > 2$ implies $0 \le x \le 2$.

4. Show that for a single value from Cauchy population $f(x|\theta_2)/f(x|\theta_1)$ is not monotonic in x.

5. For a sample of size one test the null hypothesis that X follows $N(0,1)$ against the alternative hypothesis that X has a Cauchy distribution $f(x;1) = \frac{1}{\pi(1+x^2)}$, $-\infty < x < \infty$.

   Show that the power of the test is $1 - \frac{2}{\pi}\tan^{-1}(\tau_{\alpha/2})$.

6. Let $(X_1, \cdots, X_{10})'$ be a random sample from a Bernoulli (p) distribution, $0 \leq p \leq 1$. Suggest from preliminary considerations a test for the null hypothesis $H_\circ \left(p = \frac{1}{2}\right)$ against the alternative $H\left(p \neq \frac{1}{2}\right)$ when the size of the test is $0 \cdot 05$.

7. (a) Let $X_i (i = 1, \cdots, n)$ be a random sample from a distribution with density

$$f(x; a, b) = \begin{cases} \frac{1}{a} \exp\left[-\frac{x-b}{a}\right] & \text{if } x \geq b \\ 0, & \text{otherwise.} \end{cases}$$

Determine the UMP test for testing $H_\circ : b = b_0$ versus $H_1 : b \neq b_0$ when $a (> 0)$ is assumed known.

(b) Let $X_1, \cdots, X_n$ be jointly distributed as a n-variate normal with

$$E(X_1) = n\theta, \quad \theta > 0,$$
$$E(X_j) = 0, \quad j > 1$$

and the dispersion matrix

$$\boldsymbol{V} = \begin{bmatrix} n - 1 + \theta^2 & -1 & 0 & \cdots & -1 \\ -1 & 1 & 0 & \cdots & 0 \\ -1 & 0 & 1 & \cdots & 0 \\ \cdots & \cdots & \cdots & \cdots & \cdots \\ \cdots & \cdots & \cdots & \cdots & \cdots \\ -1 & 0 & 0 & \cdots & 1 \end{bmatrix}.$$

Determine a sufficient statistic for θ. Discuss the existence of a UMP test for $H_\circ : \theta = \theta_0 (> 0)$ versus $H_1 : \theta = \theta_1 (> 0)$.

8. Let $X_1, \cdots, X_m$ be a random sample from the log-normal distribution

$$f(x; \theta) = \frac{1}{\sqrt{2\pi}\theta x} \exp\left[-\frac{1}{2}\left(\frac{\log x}{\theta}\right)^2\right], \quad 0 < x < \infty, \theta > 0.$$

Find the UMP test for $H_\circ : \theta = \theta_0$ against $H : \theta > \theta_0$. $[\Sigma (\log x_i)^2 > k]$

9. Let $X_1, \cdots, X_n$ be a random sample from the gamma distribution $f(x; p, \lambda) = \frac{1}{\lambda^p \Gamma p}$ $e^{-x/\lambda}\, x^{p-1}, x > 0$ with known parameter p and unknown parameter λ. Find the UMP test of $H_\circ (\lambda = \lambda_0)$ against $H(\lambda > \lambda_0)$. Show that the power curve is a non-decreasing function of λ. $[\Sigma x_i < k]$

10. Find a UMP test of size α for $H_\circ : \theta \leq \theta_0$ against $H_1 : \theta > \theta_0$ based on a sample of size n for the following families of p.d.f.'s. Find the expression for power function in each case.

(a) $X \sim N(\theta, 1), -\infty < x < \infty, -\infty < \theta < \infty;$

(b) $f_\theta(x) = \theta x^{-(\theta+1)}, 1 \leq x < \infty, \theta > 0;$

(c) $f_\theta(x) = \frac{1}{\theta}e^{-x/\theta}$, $x > 0$, $\theta > 0$;

(d) $f_\theta(x) = \frac{1}{\theta}x^{(1-\theta)/\theta}$, $0 < x < 1$;

(e) $f_\theta(x) = \frac{1}{\theta}x^{\theta-1}$, $0 < x < 1$, $\theta > 0$.

11. Let X be a single observation from a probability function $p(x)$ which is positive on the non-negative integers and zero elsewhere. For testing

$$H_\circ : p(x) = \frac{e^{-1}}{x!}$$

against

$$H_1 : p(x) = \frac{1}{2^{x+1}}.$$

Consider the following test:

$$w : \left\{x : 2^x/x! \leq \frac{e}{2}\right\}.$$

(a) Show, using the first principle, that the test is unbiased.

(b) Show that the test is most powerful.

12. Let $X_1, \cdots, X_n$ be a random sample from $N(\mu, \sigma^2)$ distribution. Show that for testing $H_\circ(\mu = \mu_0, \sigma = \sigma_0)$ against $H(\mu = \mu_1, \sigma = \sigma_1 > \sigma_0)$, the BCR consists of sufficiently large values of

$$\sum_1^n \left[x_i - \left(\frac{\mu_0\sigma_1^2 - \mu_1\sigma_0^2}{\sigma_1^2 - \sigma_0^2}\right)\right]^2.$$

13. Let $X_1, \cdots, X_n$ be independent random variables where X_i follows $N(\mu, \sigma_i^2)$ distribution with known $\sigma_i^2 (i = 1, \cdots, n)$. Derive UMP test of size α for $H_\circ(\mu = \mu_0)$ against $H(\mu > \mu_0)$. Is there any UMP test of size α for $H_\circ$ against two sided alternative $H'(\mu \neq \mu_0)$?

14. Let X have a hypergeometric distribution

$$P\{X = x|N, M\} = \frac{\binom{M}{x}\binom{N-M}{n-x}}{\binom{N}{n}}, \quad x = 0, 1, \cdots, M,$$

where N is known and M is unknown. Show that the distribution has a MLR in x. Hence, show that a UMP test of $H_\circ(M \leq M_0)$ against $H(M > M_0)$ is given by

$$\delta(x) = \begin{cases} 1, \text{ if } x > k \\ \lambda, \text{ if } x = k \\ 0, \text{ if } x < k \end{cases}$$

where k, λ are determined from the condition $E_{M_0}(\delta(x)) = \alpha$.

15. Show that there does not exist any UMP test of size α for testing $H_o(\sigma^2 = \sigma_0^2)$ versus $H(\sigma^2 \neq \sigma_0^2)$, based on a random sample of size n from a $N(\mu, \sigma^2)$ population with μ known. Also show how the power function of the MP test for $H_o : \sigma^2 = \sigma_0^2$ versus $H_1 : \sigma^2 = \sigma_1^2 (> \sigma_0^2)$ changes as a function of $c = \sigma_1^2/\sigma_0^2$.

16. Using the idea of MLR obtain a UMP (level α) test for $H_o : \sigma^2 = \sigma_0^2$ versus $H_1 : \sigma^2 > \sigma_0^2$ on the basis of a sample drawn from a $N(\mu, \sigma^2)$ population.

    Find an expression for the OC curve and show that the test is unbiased.

17. Let $X_1, \cdots, X_n$ be a random sample from uniform distribution

$$f(x;\theta) = \begin{cases} 1, \text{ if } \theta \leq x \leq \theta + 1 \\ 0, \text{ otherwise.} \end{cases}$$

    Obtain UMP tests for $H_o(\theta = \theta_0)$ against one-sided alternatives $H'(\theta > \theta_0)$ and $H''(\theta < \theta_0)$. Show that there does not exist any UMP test for H_o against two-sided alternative $H(\theta \neq \theta_0)$.

18. Let $X_1, \cdots, X_n$ be a sample of size n from the p.m.f.

$$P_M(x) = \frac{1}{M} \quad \text{for } x = 1, 2, \cdots, M, \quad M = 1, 2, \cdots.$$

    Show that the test

$$\delta(\boldsymbol{x}) = \begin{cases} 1, \text{ if } \max\limits_{1 \leq i \leq n} x_i > M_0 \\ \alpha, \text{ if } \max\limits_{1 \leq i \leq n} x_i \leq M_0 \end{cases}$$

    is UMP of size α for testing $H_o(M \leq M_0)$ against $H(M > M_0)$.

19. Let $X_1, \cdots, X_n$ be a random sample from the continuous rectangular distribution $R(0, \theta)$, $\theta > 0$. Show that the test

$$\delta(\boldsymbol{x}) = \begin{cases} 1, \text{ if } \max\limits_{1 \leq i \leq n} x_i > \theta_0 \\ \alpha, \text{ if } \max\limits_{1 \leq i \leq n} x_i \leq \theta_0 \end{cases}$$

    is UMP of size α for testing $H_o(\theta \leq \theta_0)$ against $H(\theta > \theta_0)$.

20. Let $Y_1 < Y_n$ be the extreme statistics in a random sample of size n from a rectangular distribution over $(\theta, \theta + 1)$. To test $H_o : \theta = 0$ against $\theta > 0$, the following test was used:

    Reject H_o iff $Y_1 > 1$ or $Y_n > c$, where c is a constant.

    (a) Determine c so that the test with have size α.

    (b) Find the power function of the test and show that the test is unbiased.

21. Let X be a random sample of size one. Consider test of the null hypothesis $H_o : X$ is a sample from a uniform $U(0,1)$ distribution against the alternative hypothesis $H : X$ is a sample from a beta $B(2,1)$ distribution

$$f(x) = 2x, \;\; 0 \leq x \leq 1.$$

Find a MP test of size $0 \cdot 05$.

22. Let $X_1, \cdots, X_n$ be a random sample from the $B(\alpha, \beta)$ distribution

$$f(x|\alpha,\beta) = \frac{1}{B(\alpha+1,\beta+1)} x^{\alpha}(1-x)^{\beta}, \;\; 0 \leq x \leq 1.$$

Consider testing of $H_o : \alpha = \alpha_0$ versus $H : \alpha > \alpha_0$ at level $0 \cdot 05$.

Show that the ratio $L(\boldsymbol{x}|\alpha_2)/L(\boldsymbol{x}|\alpha_1)$ increases with $\prod_{i=1}^{n} x_i$, when $\alpha_2 - \alpha_1 (> 0)$ is kept fixed. Hence, show that a UMP test has critical function $\delta(\boldsymbol{x})$ such that

$$\delta(\boldsymbol{x}) = \begin{cases} 1, \text{ if } \prod_{i=1}^{n} x_i \geq k \\ 0, \text{otherwise} \end{cases}$$

and $E_{\alpha_0}[\delta(\boldsymbol{x})] = 0 \cdot 05$.

23. Let $X_1, \cdots, X_n$ be a random sample from a uniform $U(0, \theta)$ population. (a) Show that for testing $H_o(\theta = \theta_0)$ against $H(\theta > \theta_0)$ any critical region ω for which $P_{\theta_0}(\omega) = \alpha$ and which is such that $\boldsymbol{x} \in \omega$ if $x_{(n)} > \theta_0$ is UMP at level α. (b) Show that for testing $H_o(\theta = \theta_0)$ against $H'(\theta \neq \theta_0)$, a unique UMP test exists and is given by

$$\omega : \left\{ \boldsymbol{x} | x_{(n)} > \theta_0 \cup x_{(n)} < \theta_0 \sqrt[n]{\alpha} \right\}$$

when $x_{(n)} = \max_i x_i$.

24. (a) Suppose $X_1, \cdots, X_n$ are independent r.v.'s each having uniform $U(0, \theta)$ distribution, $\theta > 0$. Show that a MP test for $H_o : \theta = 1$ against $H_1 : \theta = 2$ of size $\cdot 05$ is

$$\delta(\boldsymbol{x}) = \begin{cases} 0 \cdot 05, \text{ if } x_{(1)} \geq 0 \;\text{ and }\; x_{(n)} \leq 1 \\ 1, \quad \text{otherwise} \end{cases}$$

where $x_{(n)}, x_{(1)}$ denote, respectively, the maximum and the minimum value of the observations.

Show also that the power function of the test is

$$\gamma(\theta) = \begin{cases} 0 \cdot 05, & \text{if } 0 < \theta \leq 1 \\ 1 - \frac{0 \cdot 95}{\theta^n}, & \text{for } \theta > 1. \end{cases}$$

For testing the same hypothesis, consider a test based on the sufficient statistic $X_{(n)} = Y$. Show that the p.d.f. of Y is

$$g(y;\theta) = ny^{n-1}/\theta^n, \;\; 0 \le y \le \theta.$$

Hence, show that the MP test of size $0 \cdot 05$ of $H_\circ$ against H_1 is

$$\delta_1(y) = \begin{cases} 1, \text{ if } y > c \\ 0, \text{ otherwise,} \end{cases}$$

where $c = \sqrt[n]{0 \cdot 95}$. The power function of this test is

$$\gamma_1(\theta) = \begin{cases} 0, & \text{if } \theta \le (0 \cdot 95)^{1/n} \\ 1 - \frac{0\cdot 95}{\theta^n}, & \text{if } \theta > (0 \cdot 95)^{1/n}. \end{cases}$$

Observe that the former test is the MP among tests based on X, while the later test is the MP among tests based on Y. Observe also that in the sense of Neyman-Pearson theorem both the tests are equivalent since they have the same level $(0\cdot 05)$ and the same power at $\theta = 2$.

25. Let $X_1, \cdots, X_{30}$ be a random sample from a Poisson (λ) distribution. It is desired to test $H_\circ\{\lambda \le 0 \cdot 2 \text{ or } \lambda \ge 0 \cdot 8\}$ versus $H_1\{\lambda \in (0 \cdot 2, 0 \cdot 8)\}$. A suggested test is

$$\delta(\boldsymbol{x}) = \begin{cases} 1, \text{ if } c_1 < \Sigma_{i=1}^{30} x_i < c_2 \\ \lambda_1, \text{ if } \Sigma x_i = c_1 \\ \lambda_2, \text{ if } \Sigma x_i = c_2 \\ 0, \text{ otherwise.} \end{cases}$$

Using the Normal approximation to Poisson, find c_1, c_2 such that

$$E_{\lambda=0.2}\{\delta(\boldsymbol{X})\} = E_{\lambda=0.8}\{\delta(\boldsymbol{X})\} = 0.05.$$

Use the exact Poisson distribution to find an $\alpha = .05$ test and $c_1, c_2, \gamma_1, \gamma_2$ such that

$$E_{\lambda=0.2}\{\delta(\boldsymbol{X})\} = \alpha = E_{\lambda=0.8}\{\delta(\boldsymbol{X})\}.$$

26. Let $(X_1, \cdots, X_n)'$ be a random sample from a $N(\mu, \sigma^2)$ population where σ^2 is known. Show that the UMP test of size α for $H_\circ(\mu = \mu_0)$ against $H(\mu > \mu_0)$ is actually uniformly least powerful for testing $H_\circ$ against $H'(\mu < \mu_0)$ (and vice-versa).

27. Find a type A_1 critical region of size α for testing $H_\circ : \sigma^2 = \sigma_0^2$ against $H(\sigma^2 \ne \sigma_0^2)$ for a $N(\mu, \sigma^2)$ population with μ known. Show that the critical region coincides with the type A region of the same size.

Chapter 18

Testing of Hypothesis II

18.1 Introduction

We consider in this chapter testing of composite hypotheses by the criteria of similar region-most powerful tests, and likelihood ratio statistic. Invariance tests and some properties of consistent tests are also considered. The concepts of similar regions is considered in section 18.2. Section 18.3 considers construction of most powerful similar regions. The likelihood ratio test criterion and its asymptotic properties are considered in sections 18.4 and 18.5. Section 18.6 considers Hotelling's generalised "student" test. The hypothesis of equality of mean vectors and dispersion matrices of K multinormal populations are considered respectively in sections 18.7 and 18.8. The next section considers the hypothesis of equality of multinormal populations. The hypothesis of independence of sets of variables in a multinormal population is considered in section 18.10. Wilk's criteria are considered in section 18.11. Principle of invariance is dealt with in Section 18.12. The last section considers monotonicity of power functions.

18.2 Composite Hypotheses

As stated in Section 17.2 that a null hupothesis $H_\circ(\theta \in \Theta_0)$ is said to be composite if Θ_0 contains more than one point θ. Thus, under a composite hypothesis the joint distribution of $(X_1, \cdots, X_n)'$ is not completely specified. For testing such a hypothesis against an alternative hypothesis $H(\theta \in \Theta_1)$ at level α (which is the probability of rejecting $H_\circ$ when it is true) we should consider critical regions ω such that

$$P_\theta(\omega) \leq \alpha \text{ for all } \theta \in \Theta_0. \tag{18.2.1}$$

This means that the test function $\delta(\boldsymbol{x})$ should be such that

$$E_\theta[\delta(\boldsymbol{x})] \leq \alpha \ \forall\, \theta \in \Theta_0. \tag{18.2.2}$$

In particular one may consider critical regions ω such that

$$P_\theta(\omega) = \alpha \ \forall\, \theta \in \Theta_0.$$

Definition 18.2.1 *A test $\delta(\boldsymbol{x})$ is said to be α-similar on a subset $\Theta^* \subset \Theta$ if*

$$E_\theta[\delta(\boldsymbol{x})] = \gamma_\theta(\delta) = \alpha \ \forall\, \theta \in \Theta^*.$$

A test is said to be similar on a subset Θ^* if it is α-similar on Θ^* for some $\alpha, 0 \le \alpha \le 1$.

Clearly, if $\delta(\boldsymbol{x}) = \alpha$ for all $\boldsymbol{x}$ with probability one for all $\theta \in \Theta^*$, $\delta(\boldsymbol{x})$ is α-similar on Θ^*. Thus, there exists at least one similar test on every Θ^*.

Definition 18.2.2 *A critical region ω is said to α-similar on Θ^* if the corresponding test $\delta(\boldsymbol{x})$ is α-similar on Θ^*. A critical region is said to be similar to the sample space $\mathcal{X}$ [because $P_\theta(\mathcal{X}) = 1$ (a constant) $\forall\, \theta \in \Theta^*$] with respect to $\theta \in \Theta^*$ or more briefly, a similar region if it is α-similar for at least one α.*

In the case of testing a composits hypothesis $H_\circ(\theta \in \Theta_0)$ at level α, one has to, therefore, find all α-similar critical regions and choose from them one α-similar region which is best from the consideration of power of the test.

For finding similar regions one may use the sufficient statistic for θ. Let T be a sufficient statistic for $\theta \in \Theta_0$ and let $\delta_\omega(\boldsymbol{x})$ be a non-randomised test function corresponding to the critical region ω such that

$$\delta_\omega(\boldsymbol{x}) = \begin{cases} 1, & \text{if } \boldsymbol{x} \in \omega \\ 0, & \text{otherwise.} \end{cases}$$

Theorem 18.2.1 *If ω be such that for a given value t of the sufficient statistic T for $\theta \in \Theta_0$, the conditional expectation*

$$E_\theta(\delta_\omega(\boldsymbol{x})|T = t) = \alpha \ \forall\, \theta \in \Theta_0 \tag{18.2.3}$$

then $\delta_\omega(\boldsymbol{x})$ is a size-α similar test and ω is a size α similar critical region.

Proof:

$$\begin{aligned} E_\theta(\delta_\omega(\boldsymbol{x})) &= E_{T,\theta}[E_\theta(\delta_\omega(\boldsymbol{x})|T = t)] \\ &= E_\theta(\alpha) = \alpha \ \forall\, \theta \in \Theta_0. \end{aligned}$$

Therefore,

$$P_\theta(\omega) = \alpha \ \forall\, \theta \in \Theta_0.$$

We have shown that (18.2.3) is a sufficient condition for ω to be similar. This does not, however, guarantee that there are not other similar critical regions not based on the sufficient statistic. We shall show that if the sufficient statistic T is boundedly complete for all $\theta \in \Theta_0$ then (18.2.3) is also a necessary condition for all size-α similar regions for $H_\circ(\theta \in \Theta_0)$. The concept of bounded completeness has been explained in section 15.7.

Definition 18.2.3 *A (non-randomised) test with critical region ω is said to of Neyman structure with respect to a sufficient statistic T for $\theta \in \Theta_0$ if $E(\delta_\omega|T)$ is a constant $\forall\, \theta \in \Theta_0$. In particular, the test $\delta_\omega(\boldsymbol{x})$ in (18.2.3) has Neyman structure.*

Theorem 18.2.2 *If the family of distributions of sufficient statistic T for $\theta \in \Theta_0$, $P^T = \{p_\theta(T), \theta \in \Theta_0\}$ is boundedly complete, then every (non-randomised) size-α similar test δ will have a Neyman-structure with respect to the sufficient statistic T.*

Proof: Since δ is a size-α similar test

$$E_\theta[\delta_\omega(\boldsymbol{x})] = P_\theta(\omega) = \alpha \;\forall\, \theta \in \Theta_0$$

$$\text{i.e., } E_\theta[\delta_\omega(\boldsymbol{x}) - \alpha] = 0 \;\forall\, \theta \in \Theta_0$$

$$\text{i.e., } E_{\theta,T}[E_\theta\{(\delta_\omega(\boldsymbol{x}) - \alpha)|T = t\}] = 0 \;\forall\, \theta \in \Theta$$

$$\text{i.e., } E_{\theta,T}[\psi_\theta(t)] = 0 \;\forall\, \theta \in \Theta_0.$$

Now since $\delta_\omega(\boldsymbol{x}) - \alpha \in [-\alpha, 1-\alpha]$, its conditional expected value $\psi_\theta(t)$ is also a bounded function of t. By the property of bounded completeness of P^T,

$$\psi_\theta(t) = 0 \text{ a.e. } \forall\, \theta \in \Theta_0$$

$$\text{i.e., } E_\theta\{\delta_\omega(\boldsymbol{x})|T = t\} = \alpha \text{ a.e. } \forall\, \theta \in \Theta.$$

Hence the test $\delta_\omega(\boldsymbol{x})$ has Neyman structure with respect to T.

18.3 Construction of Most Powerful Similar Regions

Suppose the parameters vector $\boldsymbol{\theta} = (\theta_1, \cdots, \theta_m)'$ and $H_\circ$ specifies the values of first r parameters, $H_\circ(\theta_1 = \theta_{10}, \theta_2 = \theta_{20}, \cdots, \theta_r = \theta_{r0})$ while the remaining $(m-r)$ parameters $\theta_{r+1}, \theta_{r+2}, \cdots, \theta_m$ remain unspecified. Such a hypothesis is said to be a hypothesis with $(m-r)$ d.f.

Let us denote by $\boldsymbol{\theta}_s$ the set of parameters $(\theta_1, \cdots, \theta_r)'$ whose values are specified by $H_\circ$ and $\boldsymbol{\theta}_t = (\theta_{r+1}, \cdots, \theta_m)'$ the set of parameters with unspecified values. The parameters $\theta_{r+1}, \cdots, \theta_m$ are also called nuisance parameters. Thus, the null hypothesis is $H_\circ(\boldsymbol{\theta}_s = \boldsymbol{\theta}_s^0)$.

Suppose the statistic T is a boundedly complete sufficient statistic for $\boldsymbol{\theta}_t$ when $H_\circ$ holds. The results in theorems 18.2.1 and 18.2.2 then imply every size-α similar region satisfies (18.2.3). Hence every similar region of size-α is composed of parts of the contours of the sample space defined by $T = t$, for varying t, the (conditional) probability content of each such part being α under $H_\circ$.

Consider, therefore,the contour for which value of T is a constant, say, t. Since T is sufficient for $\boldsymbol{\theta}_t$ under $H_\circ$, the conditional distribution of $(X_1, \cdots, X_n)'$ given T will be independent of $\boldsymbol{\theta}_t$ and will depend on $\boldsymbol{\theta}_s$ alone. Thus, on this contour the composite $H_\circ$

with $\boldsymbol{\theta}_t$ unspecified is reduced to a simple $H_\circ$ with $\boldsymbol{\theta}_s$ specified. If T is also sufficient for $\boldsymbol{\theta}_t$ when H_1 holds, the composite H_1 is also reduced to a simple H_1 on the contour $T = t$. (It follows that in this case, the power as well as the size of any critical region on a contour with t constant, will be independent of θ). If, however, T is not sufficient for $\boldsymbol{\theta}_t$ when H_1 holds, then we consider H_1 as a class of simple alternatives to the simple alternative $H_\circ$ (since T is sufficient for $\boldsymbol{\theta}_t$ under $H_\circ$).

Therefore, by keeping t constant, we reduce the problem of testing a composite hypothesis $H_\circ(\boldsymbol{\theta}_s = \boldsymbol{\theta}_s^0, \boldsymbol{\theta}_t)$ to that of testing a simple hypothesis $H_\circ(\boldsymbol{\theta}_s = \boldsymbol{\theta}_s^0)$ against a simple alternative (class of simple alternatives constituting H_1). The methods of last chapter, based on the Neyman-Pearson lemma may now be used to seek a BCR (or common BCR) for $H_\circ$ against H_1. If there is a BCR for each fixed value t of T for, $H_\circ$ (now a simple hypothesis) against a H_1 (also a simple hypothesis) then by combining these regions for varying t, we get a MP test for $H_\circ$ against H_1. If this test remains most powerful against a class of alternative hypotheses constituting the composite alternative hypothesis, then it is a UMP similar test for the original problem.

► **Example 18.3.1** *Let X_1, X_2 be independently distributed $N(\theta_1, 1), N(\theta_2, 1)$ respectively. It is required to test $H_\circ(\theta_1 = \theta_2)$ against $H(\theta_1 > \theta_2)$. Note that both $H_\circ, H$ are composite hypotheses.*

Let us reparametrize as $\phi_1 = \theta_1 - \theta_2$, $\phi_2 = \theta_1 + \theta_2$. Then the hypotheses are $H_\circ(\phi_1 = 0)$, $H(\phi_1 > 0)$ with ϕ_2 as a nuisance parameter. The likelihood is

$$\begin{aligned} f(x_1, x_2; \phi_1, \phi_2) &= \frac{1}{2\pi} \exp\left[-\frac{1}{2}\left\{[x_1 - (\phi_1 + \phi_2)/2]^2 + [x_2 - (\phi_2 - \phi_1)/2]^2\right\}\right] \\ &= \frac{1}{2\pi} \exp\left[-\frac{1}{2}\left\{x_1^2 + x_2^2 - (x_1 + x_2)\phi_2 - (x_1 - x_2)\phi_1 + (\phi_2^2 + \phi_1^2)/2\right\}\right]. \end{aligned}$$

For known values of ϕ_1, the statistic $T = X_1 + X_2$ is sufficient for ϕ_2.

Hence the conditional distribution of (X_1, X_2) given T is independent of ϕ_2 under $H_\circ$. Therefore, on the contour $T = x_1 + x_2$, the problem becomes one of testing a simple $H_\circ(\phi_1 = 0)$ against a simple alternative $H(\phi_1 > 0)$. Instead of restricting to the conditional distribution of (X_1, X_2) given T one can restrict to the conditional distribution of $(X_1 - X_2)$ given T. It follows that a critical region with size 5% is $\{(x_1, x_2) : x_1 - x_2 \geq 1.64\sqrt{2}\}$.

■ NOTE 18.3.1 *Suppose the null hypothesis is $H_\circ(\theta_1 = \cdots = \theta_r, \theta_{r+1}, \cdots, \theta_m)$, where the values of the parameters whose equality is to be tested is not specified as in the case of the hypothesis $H'_\circ(\theta_1 = \theta_{10}, \cdots, \theta_r = \theta_{r\circ}, \theta_{r+1}, \cdots, \theta_m)$. By suitable reparametrization $H_\circ$ can often be written in the form $H'_\circ$. For example, for $H_\circ(\theta_1 = \theta_2 = \theta_3)$, the reparametrisation*

$$\phi_1 = (\theta_1 - \theta_2), \;\; \phi_2 = (\theta_1 + \theta_2 - 2\theta_3), \;\; \phi_3 = (\theta_1 + \theta_2 + \theta_3)$$

implies $H_\circ$ is same as $H'_\circ(\phi_1 = 0, \phi_2 = 0, \phi_3)$ where ϕ_3 becomes a nuisance parameter.

▸ **Example 18.3.2** *Test $H_o(\mu = \mu_0)$ against $H_1(\mu = \mu_1)$ for a normal population $N(\mu, \sigma)$.*

The hypothesis H_o has one d.f., the parameter σ^2 being unspecified. The joint p.d.f. of $(X_1, \cdots, X_n)'$ under H_o, is

$$f(\boldsymbol{x}|\mu_0, \sigma^2) = \frac{1}{(\sigma\sqrt{2\pi})^n} \exp\left[-\sum_i (x_i - \mu_0)^2/2\sigma^2\right].$$

This shows that the statistic

$$T = \sum_i (X_i - \mu_0)^2$$

is sufficient of σ^2 under H_o, but not otherwise. Also the statistic T is complete. All similar regions for H_o therefore, consist of fractions α of contours with fixed values t of T.

Since the conditional distribution of $(X_1, \cdots, X_n)'$ given $T = t$ is independent of σ, the problem becomes that of testing a simple hypothesis $H_o'(\mu = \mu_0, \sigma = \sigma_0)$ against the composite hypothesis $H_1(\mu = \mu_1, \sigma)$ on a contour with $T = t$, where σ_0 is arbitrary. Consider, in particular, testing

$$H_o'(\mu = \mu_0, \sigma = \sigma_0) \text{ against } H_1'(\mu = \mu_1, \sigma = \sigma_1)$$

where H_1' is a member of the class $H_1(\mu = \mu_1, \sigma)$.

The MP critical region for H_o' against H_1' is on the contour $T = t$

$$\omega_0 = \{\boldsymbol{x} : f(\boldsymbol{x}|\mu_1, \sigma_1) > k(t) f(\boldsymbol{x}|\mu_0, \sigma_0)\} \tag{i}$$

where $k(t)$ is such that the conditional probability that $\boldsymbol{x}$ lies in ω_0 given that $T = t$ is α.

Taking logarithm of both sides, the inequality $f(\boldsymbol{x}|\mu_1, \sigma_1) > k(t) f(\boldsymbol{x}|\mu_0, \sigma_0)$ reduces to

$$(\mu_1 - \mu_0)(\bar{x} - \mu_0) > k_1(t) \tag{ii}$$

when k_1 is a constant containing no function of x except t. Thus, the BCR consists of large values of $\bar{x}$ if $\mu_1 - \mu_0 > 0$ and of small values of $\bar{x}$ if $\mu_1 - \mu_0 < 0$.

Case 1. $\mu_1 > \mu_0$.

Here condition (ii) implies

$$(\bar{x} - \mu_0) > k_2(t) \text{ or, } \sqrt{n}(\bar{x} - \mu_0)/\sqrt{t} > k_3(t).$$

Hence,

$$\omega_0 = \{\boldsymbol{x} : \sqrt{n}(\bar{x} - \mu_0)/\sqrt{t} > k_3(t)\} \tag{iii}$$

when $k_3(t)$ is so determined that $P_{\theta_0}\{\omega_0|T = t\} = \alpha$.

Now $\sqrt{n}(\bar{X}-\mu_0)/\sqrt{T}$ and T are independently distributed so that the conditiona distribution of $\sqrt{n}(\bar{X}-\mu_0)/\sqrt{T}$ given $T=t$ is the same as the marginal distribution c $\sqrt{n}(\bar{X}-\mu_0)/\sqrt{T}$. As such $k_3(t)$ is independent of t.
Now

$$\frac{\sqrt{n}(\bar{X}-\mu_0)}{\sqrt{T}} = \frac{\sqrt{n}(\bar{X}-\mu_0)}{\{n(\bar{X}-\mu_0)^2+\sum_i (X_i-\bar{X})^2\}^{1/2}} = \frac{t}{\sqrt{t^2+n-1}}$$

when

$$t=\sqrt{n}(\bar{X}-\mu_0)/s, \quad s=\frac{1}{n-1}\sum_{i=1}^{n}(x_i-\bar{x})^2$$

is a student's t statistic with $(n-1)$ d.f. Therefore, (iii) implies

$$\omega_0=\{\boldsymbol{x}: t>k_4\}$$

when k_4 is such that $P_{\theta_0}(\omega_0)=\alpha$. Hence k_4 is the upper α-point of the t distributio with $n-1$ d.f. i.e., $k_4=t_{n-1,\alpha}$. Thus,

$$\omega_0: \{\boldsymbol{x}: \sqrt{n}(\bar{x}-\mu_0)/s > t_{\alpha,n-1}\}.$$

Since this is independent of σ^2 and σ_1^2, the critical region ω_0 is MP similar region o size α for testing $H_\circ$ against H_1. Again, since ω_0' is the same for all $\mu_1>\mu_0$, it is UMI similar region of size α for testing the composite null hypothesis $H_\circ: (\mu=\mu_0)$ agains the composite alternative $H_1: (\mu=\mu_1)$ so long as $\mu_1>\mu_0$.

Since this is independent of σ_0^2 and σ_1^2, the critical region ω_0 is MP similar regior of size α for testing $H_\circ$ against H_1, when $\mu_1>\mu_0$. Again since ω_0 is the same for al $\mu_1>\mu_0$, it is UMP for testing $H_\circ(\mu=\mu_0)$ against the one sided composite alternativ $H(\mu_1>\mu_0)$.

Case 2. $\mu_1<\mu_0$.

It can be similarly shown that the UMP size-α similar region for testing $H_\circ(\mu=\mu_0$ against $H(\mu_1<\mu_0)$ is

$$\omega_0: \{\boldsymbol{x}: \sqrt{n}(\bar{x}-\mu_0)/s < -t_{\alpha,n-1}\}.$$

► **Example 18.3.3** *Test the null hypothesis* $H_\circ$ $(\sigma^2=\sigma_0^2)$ *against the alternativ hypothesis* $H_1(\sigma^2=\sigma_1^2)$ *for a normal* $N(\mu,\sigma^2)$ *population.*

The hypothesis $H_\circ$ has one d.f., the parameter μ being unspecified. The join distribution of $(X_1,\cdots,X_n)'$ under $H_\circ$ is

$$f(\boldsymbol{x}|\mu,\sigma_0)=\frac{1}{(\sigma_0\sqrt{2\pi})^n}\exp\left[-\sum_i (x_i-\mu)^2/2\sigma_0^2\right].$$

The statistic

$$\bar{X} = \sum_{i=1}^{n} X_i/n$$

is a complete sufficient statistic for μ under $H_\circ$. All similar regions for $H_\circ$, therefore, consist of fractions α of contours with fixed values $\bar{x}$ of $\bar{X}$.

Since the conditional distribution of $(X_1, \cdots, X_n)'$ given $\bar{X} = \bar{x}$ is independent of μ, the problem becomes that of testing a simple hypothesis $H'_\circ(\mu = \mu_0, \sigma = \sigma_0)$ against the composite hypothesis $H_1(\mu, \sigma = \sigma_1)$ on a contour with $\bar{X} = \bar{x}$, where μ_0 is arbitrary. Consider, in particular, testing

$$H'_\circ(\mu = \mu_0, \sigma = \sigma_0) \text{ against } H'_1(\mu = \mu_1, \sigma = \sigma_1)$$

when H'_1 is a member of the class $H_1(\mu = \mu_1, \sigma)$.

The MP critical region of size α for testing $H'_\circ$ against H'_1 on the contour $\bar{X} = \bar{x}$ is

$$\omega_0 = \{\boldsymbol{x} : f(\boldsymbol{x}|\mu_1, \sigma_1) > k(\bar{x}) f(\boldsymbol{x}|\mu_0, \sigma_0)\} \tag{i}$$

where $k(\bar{x})$ is such that the conditional probability that $\boldsymbol{x}$ lies in ω_0 given that $\bar{X} = \bar{x}$ is α. Taking logarithm of both sides, the inequality, $f(\boldsymbol{x}|\mu_1, \sigma_1) > k(\bar{x}) f(\boldsymbol{x}|\mu_0, \sigma_0)$ reduces to

$$(\sigma_1^2 - \sigma_0^2) \sum_i (x_i - \bar{x})^2 > k_1(\bar{x}) \tag{ii}$$

where k_1 is a constant containing no function of x except $\bar{x}$. Thus, the BCR consists of large values of

$$\sum_i (x_i - \bar{x})^2 \quad \text{if } \sigma_1^2 > \sigma_0^2$$

and of small values of

$$\sum_i (x_i - \bar{x})^2 \quad \text{if } \sigma_1^2 < \sigma_0^2.$$

Case 1. $\sigma_1^2 > \sigma_2^2$.

Here condition (ii) implies

$$\sum_i (x_i - \bar{x})^2 > k_2(\bar{x}),$$

$$\text{or, } \sum_i (x_i - \bar{x})^2/\sigma_0^2 > k_3(\bar{x}).$$

Hence,

$$\omega_0 = \left\{\boldsymbol{x} : \sum_i (x_i - \bar{x})^2/\sigma_0^2 > k_3(\bar{x})\right\} \tag{iii}$$

when $k_3(\bar{x})$ is so determined that $P_{\theta_0}\{\omega_0|\bar{X} = \bar{x}\} = \alpha$.

Now

$$\sum_i (X_i - \bar{X})^2/\sigma_0^2 \quad \text{and} \quad \bar{X}$$

are independently distributed so that the conditional distribution of

$$\sum_i (X_i - \bar{X})^2/\sigma_0^2 \quad \text{given } \bar{X} = \bar{x}$$

is the same as the marginal distribution of

$$\sum_i (X_i - \bar{X})^2/\sigma_0^2.$$

As such $k_3(\bar{x})$ is independent of $\bar{x}$.

Now

$$\sum_i (X_i - \bar{X})^2/\sigma_0^2$$

is distributed as χ^2 with $(n-1)$ d.f. under H_o. Therefore, (iii) implies

$$\omega_0 = \left\{ \boldsymbol{x} : \sum_i (x_i - \bar{x})^2/\sigma_0^2 > \chi^2_{(n-1),\alpha} \right\} \quad \text{(iv)}$$

when $\chi^2_{(n-1),\alpha}$ is the upper $100\alpha\%$ point of a $\chi^2_{(n-1)}$ distribution. Since ω_0 is independent of μ_0 and μ_1 it is the MP similar region of size α for testing $H_o(\sigma = \sigma_0)$ against $H_1(\sigma = \sigma_1)$ where $\sigma_1^2 > \sigma^2$. Again, since ω_0 is the same for all $\sigma_1^2 > \sigma_0^2$, it is UMP similar region of size α for testing H_o against the one-sided alternative $H(\sigma_1^2 > \sigma_0^2)$.

Case 2. $\sigma_1^2 < \sigma_0^2$.

It can be similarly shown that the UMP similar region of size α for testing H_o against the one sided alternative $H(\sigma_1^2 < \sigma_0^2)$ is

$$\omega_0 : \left\{ \boldsymbol{x} : \sum_i (x_i - \bar{x})^2/\sigma_0^2 < \chi^2_{(n-1),1-\alpha} \right\}$$

when $\chi^2_{(n-1),(1-\alpha)}$ is the lower α-point of the χ^2 distribution with $(n-1)$ d.f.

The topic of two-sided tests in the presence of nuisence parameters requires a more advanced approach. (A full treatment may be found in Lehmann (1959)).

18.4 Likelihood Ratio Tests

In the general case, we may be concerned with several parameters, some of which may be involved in the composite null and alternative hypotheses, the remaining being nuisance parameters. Some general technique is needed for testing such hypotheses. Neyman and Pearson (1928) suggested a simple method of test construction, called the *likelihood ratio tests* which is a natural extension of the test defined by (17.3.2) to a composite hypothesis. Unfortunately, it does not appear to be a best test in any general case.

We recall that the joint probability density function of the fixed sample values, regarded as a function of the parameters is called the likelihood. One school of thought maintains that any inference from a sample is to be determined solely from the likelihood function. (We have already used the likelihood function in maximum likelihood estimation.) It has been suggested that the relative support provided by the sample for two different sets of values of the parameters should be provided by the ratio of the corresponding likelihood functions. In hypothesis testing we want to know whether there is sufficient support for some set of values of parameters. One way of examining this is to compare the likelihood function at a set of parameter values with the largest value achievable by varying the parameters (subject to constraints, if any). One can then divide the parametric space into two classes, acceptable class and non-acceptable class of values of parameters according as the ratio is high or low.

Let $\boldsymbol{X} = (X_1, \cdots, X_n)'$ have a joint distribution $L(x_1, \cdots, x_n|\theta)$ depending on an unknown parameter θ. The likelihood function of θ for fixed values $(x_1, \cdots, x_n)'$ is

$$L(\theta|x_1, \cdots, x_n).$$

Suppose we want to test the simple null hypothesis $H_\circ(\theta = \theta_0)$ against the composite alternative $H(\theta \neq \theta_0)$. A measure of how far the value θ_0 of θ is supported by the observation $\boldsymbol{x} = (x_1, \cdots, x_n)'$ is given by the likelihood ratio (LR),

$$\lambda(\boldsymbol{x}) = \frac{L(\theta_0|\boldsymbol{x})}{\sup_{\theta\in\Theta} L(\theta|\boldsymbol{x})}. \tag{18.4.1}$$

A small value of λ will lead to the rejection of $H_\circ$. The size-α critical region is given

$$\{\boldsymbol{x} : \lambda(\boldsymbol{x}) \leq \lambda_0\} \tag{18.4.2}$$

where

$$P\{\lambda(\boldsymbol{x}) \leq \lambda_0|\theta_0\} = \alpha. \tag{18.4.3}$$

Note that we have here tested a simple $H_\circ$ against a composite alternative and Θ_0 together with Θ_1 form the parameter space Θ. Testing of a composite $H_\circ$ using LR test has been considered subsequently.

▶ **Example 18.4.1** *The r.v. $X_i (i = 1, \cdots, n)$ is distributed independently as $N(\theta_i, \sigma^2)$ where $\sigma^2 (> 0)$ is known, $-\infty < \theta_i < \infty$. It is required to test $H_\circ$ $(\theta_1 = \cdots = \theta_n = 0)$ against H $(\theta_i \neq 0, i = 1, \cdots, n)$.*

The likelihood is

$$L(\theta_1, \cdots, \theta_n|\boldsymbol{x}; \sigma^2) = \frac{1}{(\sqrt{2\pi}\sigma)^n} \exp\left[-\frac{1}{2}\sum_{i=1}^{n}(x_i - \theta_i)^2/\sigma^2\right].$$

Under null hypothesis, $\boldsymbol{\theta}_0 = (\theta_1 = 0, \cdots, \theta_n = 0)$

$$L(\boldsymbol{\theta}_0|\boldsymbol{x}; \sigma^2) = \frac{1}{(\sqrt{2\pi}\sigma)^2} \exp\left[-\frac{1}{2}\sum_{i=1}^{n} x_i^2/\sigma^2\right].$$

Since there is no restriction on θ_i $(i = 1, \cdots, n)$, the likelihood is maximised when $\theta_i = x_i$ $(i = 1, \cdots, n)$ and hence

$$\sup_{\theta \in \Theta} L(\theta|\boldsymbol{x}) = \frac{1}{(\sqrt{2\pi}\sigma)^n}.$$

Therefore

$$\lambda(\boldsymbol{x}) = \exp\left[-\frac{1}{2}\sum_{i=1}^{n} x_i^2/\sigma^2\right]. \quad \text{(i)}$$

We find c such that

$$P\{\lambda(\boldsymbol{x}) \le c|\theta_1 = \cdots = \theta_n = 0] = \alpha.$$

Since the ratio (i) is a decreasing function of $\sum_1^n x_i^2$, we may equivalently find c^* such that

$$P\left\{\sum_{i=1}^{n} X_i^2 > c^*|\theta_1 = \cdots = \theta_n = 0\right] = \alpha.$$

Under $H_\circ$,

$$\sum_{i=1}^{n} X_i^2/\sigma^2$$

has a $\chi^2_{(n)}$ distribution. Hence $c^* = \sigma^2\chi^2_{n,\alpha}$.

► **Example 18.4.2** *Let $X_1, \cdots, X_n$ be a random sample from a normal distribution $N(\mu, \sigma^2)$ where μ is unknown, but σ is known. Consider the simple $H_\circ(\mu = \mu_0)$.*

Case 1. Alternative hypothesis, $H(\mu \ne \mu_0)$. Here $\Theta = \{\mu : -\infty < \mu < \infty\}$.

$$L(\mu|\boldsymbol{x}) = k\exp\left[-\sum_1^n (x_i - \mu)^2/2\sigma^2\right]$$

where $k = 1/(\sqrt{2\pi}\sigma)^n$. The m.l.e. of μ, when $\mu \in \Theta$ is

$$\bar{x} = \frac{1}{n}\sum_{i=1}^{n} x_i$$

and hence

$$\sup_{\mu \in \Theta} L(\mu|\boldsymbol{x}) = k\exp\left[-\sum_1^n (x_i - \bar{x})^2/2\sigma^2\right].$$

Therefore,

$$\lambda(\boldsymbol{x}) = \frac{L(\mu_0|\boldsymbol{x})}{\sup\limits_{\mu \in \Theta} L(\mu|\boldsymbol{x})} = \exp\left[-\frac{1}{2\sigma^2}\left\{\sum_{i=1}^{n}(x_i - \mu_0)^2 - \sum_{i=1}^{n}(x_i - \bar{x})^2\right\}\right]$$

$$= \exp\left[-\frac{n}{2\sigma^2}(\bar{x} - \mu_0)^2\right].$$

Now,

$$\lambda(\boldsymbol{x}) \le \lambda_0 \Leftrightarrow -\frac{n}{2\sigma^2}(\bar{x}-\mu_0)^2 < \lambda_1$$
$$\Leftrightarrow |\tau(\bar{x})| > \lambda_2$$

where $\tau(\bar{x}) = \frac{\sqrt{n}(\bar{x}-\mu_0)}{\sigma}$.
Hence, the LR test has critical region

$$\omega = \{\boldsymbol{x} : |\tau(\bar{x})| > \lambda_2\}$$

where λ_2 is such that $P_{\mu_0}(\omega) = \alpha$. Now, under $H_\circ$, $\tau(\bar{x}) \sim N(0,1)$. Here $\lambda_2 = \tau_{\alpha/2}$ where $\tau_{\alpha/2}$ is the upper $\alpha/2$-point of the standard normal distribution.

Case 2. Alternative is $H : \mu > \mu_0$. Here $\Theta = \{\mu : \mu_0 \le \mu < \infty\}$. The m.l.e. of μ for $\mu \in \Theta$ is

$$\hat{\mu} = \begin{cases} \mu_0 \text{ if } \bar{x} < \mu_0 \\ \bar{x} \text{ if } \bar{x} \ge \mu_0. \end{cases}$$

Therefore,

$$\sup_{\mu\in\Theta} L(\mu|\boldsymbol{x}) = \begin{cases} K\exp\left[-\sum_i (x_i-\mu_0)^2/2\sigma^2\right], \text{ if } \bar{x} < \mu_0 \\ K\exp\left[-\sum_i (x_i-\bar{x})^2/2\sigma^2\right], \text{ if } \bar{x} \ge \mu_0; \end{cases}$$

and hence

$$\lambda(\boldsymbol{x}) = \begin{cases} 1, & \text{if } \bar{x} < \mu_0 \\ \exp\left[-\frac{n}{2\sigma^2}(\bar{x}-\mu_0)^2\right], & \text{if } \bar{x} \ge \mu_0. \end{cases}$$

If $\lambda(\boldsymbol{x}) = 1$ we must accept $H_\circ$. Therefore, the region $\{\boldsymbol{x} : \bar{x} < \mu_0\}$ must belong to the acceptance region. If $\bar{x} \ge \mu_0$, the critical region is $\omega_1 = \{\boldsymbol{x} : \lambda(\boldsymbol{x}) < \lambda_0\}$. Now,

$$\lambda(\boldsymbol{x}) < \lambda_0 \Leftrightarrow \frac{\sqrt{n}(\bar{x}-\mu_0)}{\sigma} > \lambda_1$$

so that

$$\omega_1 = \left\{\boldsymbol{x} : \frac{\sqrt{n}(\bar{x}-\mu_0)}{\sigma} > \lambda_1\right\}$$

where λ_1 is such that

$$P_{\mu_0}\left\{\sqrt{n}(\bar{x}-\mu_0)/\sigma > \lambda_1\right\} = \alpha.$$

Since $\sqrt{n}(\bar{X}-\mu_0)/\sigma$ has a $N(0,1)$ distribution under μ_0, $\lambda_1 = \tau_\alpha$. Thus,

$$\omega_1 = \left\{\boldsymbol{x} : \frac{\sqrt{n}(\bar{x}-\mu_0)}{\sigma} > \tau_\alpha\right\}.$$

Case 3. Alternative is $H : \mu < \mu_0$. Here $\Theta = \{\mu : -\infty < \mu \le \mu_0\}$. The m.l.e. of μ for $\mu \in \Theta$ is

$$\hat{\mu} = \begin{cases} \mu_0 \text{ if } \bar{x} > \mu_0 \\ \bar{x} \text{ if } \bar{x} \le \mu_0. \end{cases}$$

Proceeding in the same way as in case 2, the LR test has the critical region

$$\omega_2 = \left\{ \boldsymbol{x} : \frac{\sqrt{n}(\bar{x} - \mu_0)}{\sigma} < \tau_\alpha \right\}.$$

It is seen that ω is also the type A UMPU critical region and ω_1, ω_2 are also the UMP critical regions for the relevant tests.

If the null hypothesis is composite, we consider the maximum value that the likelihood can attain when the null hypothesis is true. This is reasonable, because if the ratio of the maximum to that of the overall maximum is not acceptable, then any smaller value of the likelihood should also be unacceptable. When $H_\circ$ is composite hypothesis suppose $H_\circ : \theta \in \Theta_0 (\subset \Theta)$. Here, the test is based on the likelihood ratio

$$\lambda(\boldsymbol{x}) = \frac{\sup\limits_{\theta \in \Theta_0} L(\theta|\boldsymbol{x})}{\sup\limits_{\theta \in \Theta} L(\theta|\boldsymbol{x})}. \tag{18.4.4}$$

The critical region of the test is

$$\omega = \{\boldsymbol{x} : \lambda(\boldsymbol{x}) < \lambda_0\}$$

where λ_0 is such that $P_\theta(\omega) \le \alpha$ for $\theta \in \Theta_0$.

► **Example 18.4.3** *Suppose $X_1, \cdots, X_n$ is a random sample from $N(\mu, \sigma^2)$ distribution where both μ, σ^2 are unknown. Consider the problem of testing the null hypothesis $H_\circ(\mu = \mu_0)$.*

Case 1. Alternative is $H(\mu \ne \mu_0)$. Here

$$\Theta = \left\{(\mu, \sigma^2) | -\infty < \mu < \infty, 0 < \sigma^2 < \infty\right\}$$
$$\Theta_0 = \left\{(\mu, \sigma^2) | \mu = \mu_0, 0 < \sigma^2 < \infty\right\}.$$

when $(\mu, \sigma^2) \in \Theta$.

$$L(\boldsymbol{\theta} \in \Theta|\boldsymbol{x}) = (2\pi\sigma^2)^{-n/2} \exp\left[-\sum_{i=1}^{n}(x_i - \mu)^2/2\sigma^2\right].$$

$$\log L(\boldsymbol{\theta}|\boldsymbol{x}) = -n \log\sqrt{2\pi} - \frac{n}{2}\log\sigma^2 - \frac{1}{2\sigma^2}\sum_{i=1}^{n}(x_i - \mu)^2$$

$$\frac{\partial \log L}{\partial \mu} = \frac{\sum\limits_{i=1}^{n}(x_i - \mu)}{\sigma^2}, \quad \frac{\partial \log L}{\partial(\sigma^2)} = -\frac{n}{2\sigma^2} + \frac{\sum\limits_{i=1}^{n}(x_i - \mu)^2}{2\sigma^4}.$$

'he likelihood is maximum at

$$\hat{\mu} = \bar{x}, \quad \hat{\sigma}^2 = \sum_{i=1}^{n} (x_i - \bar{x})^2/n \text{ where } \bar{x} = \frac{1}{n}\sum_{i=1}^{n} x_i.$$

lence

$$\sup_{\boldsymbol{\theta}\in\Theta} L(\mu, \sigma^2|\boldsymbol{x}) = L(\hat{\mu}, \hat{\sigma}^2|\boldsymbol{x}) = \frac{1}{(2\pi)^{n/2}} \left(\frac{n}{\sum_{i=1}^{n}(x_i - \bar{x})^2} \right)^{n/2} \exp\left(-\frac{n}{2}\right).$$

Vhen $(\mu, \sigma^2) \in \Theta_0$,

$$L(\boldsymbol{\theta} \in \Theta_0|\boldsymbol{x}) = (2\pi\sigma^2)^{-(n/2)} \exp\left[-\sum_{i=1}^{n}(x_i - \mu_0)^2/2\sigma^2\right].$$

lere μ_0 is fixed and the maximum now occurs at

$$\hat{\hat{\sigma}}^2 = \sum_{i=1}^{n}(x_i - \mu_0)^2/n.$$

Therefore,

$$\sup_{\boldsymbol{\theta}\in\Theta_0} L(\mu, \sigma^2|\boldsymbol{x}) = L(\mu_0, \hat{\hat{\sigma}}^2|\boldsymbol{x}) = \frac{1}{(2\pi)^{n/2}} \left(\frac{n}{\sum_{i=1}^{n}(x_i - \mu_0)^2} \right)^{n/2} \exp(-\frac{n}{2}).$$

The likelihood ratio is

$$\lambda(\boldsymbol{x}) = \sup_{\boldsymbol{\theta}\in\Theta_0} L / \sup_{\boldsymbol{\theta}\in\Theta} L = \left\{ \frac{\sum_{i=1}^{n}(x_i - \bar{x})^2}{\sum_{i=1}^{n}(x_i - \mu_0)^2} \right\}^{n/2}.$$

Hence, the test is: reject $H_o(\mu = \mu_0)$, when $\lambda(\boldsymbol{x}) \le c$.

Now,

$$\sum_{i=1}^{n}(x_i - \mu_0)^2 = (n-1)s^2 + n(\bar{x} - \mu_0)^2$$

where

$$s^2 = \sum_{i=1}^{n}(x_i - \bar{x})^2/(n-1).$$

$$\therefore \quad \lambda^{2/n} = \frac{1}{1+\frac{n(\bar{x}-\mu_0)^2}{(n-1)s^2}} = \frac{1}{1+t^2/(n-1)}$$

where $t = \sqrt{n}(\bar{x}-\mu_0)/s$. Thus

$$\lambda \leq \lambda_1 \iff |t| \geq t_0.$$

Under $H_\circ$, $\frac{\sqrt{n}(\bar{X}-\mu_0)}{S}$ has student's t distribution with $(n-1)$ d.f. Hence, $t_0 = t_{(n-1),\alpha/2}$ where $t_{(n-1),\alpha/2}$ is the upper 100 $(\alpha/2)$ per cent point of Student's t distribution with $(n-1)$ d.f. Therefore, the critical region is

$$\omega = \left\{ \boldsymbol{x} : \frac{\sqrt{n}(\bar{x}-\mu_0)}{s} > t_{(n-1),\alpha/2} \right\}.$$

Case 2. Alternative is $H : \mu > \mu_0$. Here

$$\Theta = \left\{(\mu,\sigma^2)|\mu_0 \leq \mu < \infty, 0 < \sigma^2 < \infty\right\}$$

$$\Theta_0 = \left\{(\mu,\sigma^2)|\mu = \mu_0, 0 < \sigma^2 < \infty\right\}.$$

For $(\mu,\sigma^2) \in \Theta$, the m.l.e.'s of μ, σ^2 are respectively,

$$\hat{\mu} = \begin{cases} \bar{x}, & \text{if } \bar{x} \geq \mu_0 \\ \mu_0, & \text{if } \bar{x} < \mu_0 \end{cases}$$

$$\hat{\sigma}^2 = \begin{cases} (n-1)s^2/n, & \text{if } \bar{x} \geq \mu_0 \\ s_0^2, & \text{if } \bar{x} < \mu_0 \end{cases}$$

where

$$s_0^2 = \sum_{i=1}^{n} (x_i - \mu_0)^2/n.$$

Thus,

$$\sup_{\boldsymbol{\theta}\in\Theta} L(\boldsymbol{\theta}|\boldsymbol{x}) = \begin{cases} \dfrac{1}{[2\pi(n-1)s^2/n]^{n/2}} \exp(-\frac{n}{2}), & \text{if } \bar{x} \geq \mu_0 \\ \dfrac{1}{[2\pi s_0^2]^{n/2}} \exp(-\frac{n}{2}), & \text{if } \bar{x} < \mu_0. \end{cases}$$

For $(\mu,\sigma^2) \in \Theta_0$, the m.l.e. of σ^2 is $\hat{\hat{\sigma}}^2 = s_0^2$. Hence

$$\sup_{\boldsymbol{\theta}\in\Theta_0} L(\boldsymbol{\theta}| = \boldsymbol{x}) = \frac{1}{(2\pi s_0^2)^{n/2}} \exp(-\frac{n}{2}).$$

The test criterion is, therefore,

$$\lambda(\boldsymbol{x}) = \begin{cases} [(n-1)s^2/ns_0^2]^{n/2}, & \text{if } \bar{x} \geq \mu_0 \\ 1, & \text{if } \bar{x} < \mu_0. \end{cases}$$

The observations $\boldsymbol{x}$ for which $\bar{x} < \mu_0$ must be included in the acceptance region. For observations with $\bar{x} \geq \mu_0$, the critical region is

$$\omega_1 = \{\boldsymbol{x} : \lambda(\boldsymbol{x}) < \lambda_0\}.$$

Now,

$$\begin{aligned}\lambda(\boldsymbol{x}) < \lambda_0 \Leftrightarrow & \; ns_0^2/s^2 > \lambda_1 \\ & \text{or, } n(\bar{x} - \mu_0)^2/s^2 > \lambda_2 \\ & \text{or, } \sqrt{n}(\bar{x} - \mu_0)/s > \lambda_3.\end{aligned}$$

Since, $\sqrt{n}(\bar{X} - \mu_0)/S$ is distributed as a Student's t with $(n-1)$ d.f. under $H_o, \lambda_3 = t_{(n-1),\alpha}$. Thus,

$$\omega_1 = \{\boldsymbol{x} : \sqrt{n}(\bar{x} - \mu_0)/s > t_{(n-1),\alpha}\}.$$

Case 3. Alternative is $H : \mu < \mu_0$. Here,

$$\begin{aligned}\Theta &= \left\{(\mu, \sigma^2) | -\infty < \mu \leq \mu_0, 0 < \sigma^2 < \infty\right\} \\ \Theta_0 &= \left\{(\mu, \sigma^2) | \mu = \mu_0, 0 < \sigma^2 < \infty\right\}.\end{aligned}$$

For $(\mu, \sigma^2) \in \Theta$, the m.l.e.'s of μ and σ^2 are

$$\hat{\mu} = \begin{cases} \bar{x}, \text{ if } \bar{x} \leq \mu_0 \\ \mu_0, \text{ if } \bar{x} > \mu_0 \end{cases}$$

$$\hat{\sigma}^2 = \begin{cases} (n-1)s^2/n, \text{ if } \bar{x} \leq \mu_0 \\ s_0^2, \quad \text{ if } \bar{x} > \mu_0. \end{cases}$$

For $(\mu, \sigma^2) \in \Theta_0$, the m.l.e. of σ^2 is s_0^2. Similar exercise gives the critical region for LR test as

$$\omega_2 = \{\boldsymbol{x} : \sqrt{n}(\bar{x} - \mu_0)/s < -t_{(n-1),\alpha}\}.$$

It is seen that ω_1, ω_2 are also the similar size-α UMP critical regions for the relevant hypotheses. The critical region ω_0 can be shown to be the similar size-α UMPU critical region.

▶ **Example 18.4.4** *Let $X_1, \cdots, X_n$ be independent r.v.'s each with exponential p.d.f.*

$$f(x|\theta, \lambda) = \frac{1}{\lambda} e^{-(x-\theta)/\lambda}, x > \theta$$

where λ is known. Consider $H_o(\theta = \theta_0)$ against $H_1(\theta > \theta_0)$. Find the LR test.

Here

$$\Theta = \{\theta; \theta_0 < \theta < \infty\}$$
$$\Theta_0 = \{\theta = \theta_0\}.$$

The p.d.f. of $(X_1, \cdots, X_n)'$ is

$$f(\boldsymbol{x}|\theta) = \begin{cases} \frac{1}{\lambda^n} e^{-\sum\limits_{i=1}^{n} (x_i - \theta)/\lambda}, & \text{if } x_{(1)} > \theta \\ 0, & \text{otherwise} \end{cases}$$

where $x_{(1)} = \min x_i$. Under $H_\circ$,

$$\max_{\theta \in \Theta_0} L(\theta|\boldsymbol{x}) = \lambda^{-n} \exp\left[-\sum_{i=1}^{n}(x_i - \theta_0)/\lambda\right], \quad x_{(1)} \geq \theta_0.$$

under Θ

$$\max_{\theta \in \Theta} L(\theta|\boldsymbol{x}) = \lambda^{-n} \exp\left[-\sum_{i=1}^{n}(x_i - \hat{\theta})/\lambda\right], \quad x_{(1)} \geq \hat{\theta}$$

when $\hat{\theta} = x_{(1)}$. Hence

$$\lambda(\boldsymbol{x}) = \frac{\exp\left[-\sum\limits_{i=1}^{n}(x_i - \theta_0)/\lambda\right] \cdot I_{[\theta_0,\infty)}(x_{(1)})}{\exp\left[-\sum\limits_{i=1}^{n}(x_i - \hat{\theta})/\lambda\right] \cdot I_{[\hat{\theta},\infty)}(x_{(1)})}$$

when $I_{[a,b)}(z) = 1(0)$ if $z \in [a, b)$ (otherwise). Now $I_{[\hat{\theta},\infty)}(x_{(1)}) = 1$. Hence

$$\lambda(\boldsymbol{x}) = \begin{cases} \exp[n(\theta_0 - \hat{\theta})/\lambda], & \text{if } x_{(1)} \geq \theta_0 \\ 0, & \text{otherwise.} \end{cases}$$

Therefore,

(a) whenever $x_{(1)} \geq \theta_0$ $\delta(\boldsymbol{x}) = \begin{cases} 1, & \text{if } \exp[n(\theta_0 - \hat{\theta})/\lambda] < C \\ 0, & \text{otherwise.} \end{cases}$;

(b) whenever $x_{(1)} < \theta_0$, $\delta(\boldsymbol{x}) = 1$ always.

Now,

$$\exp[n(\theta_0 - \hat{\theta})/\lambda] < C \Rightarrow \hat{\theta} > \theta_0 - \frac{\lambda}{n}\ln(C) = k \text{ (say)}.$$

To determine the value of k for a given size α of the test, we have

$$\begin{aligned}
\alpha &= P[\hat{\theta} > \max(k, \theta_0)|\theta_0] + P[\hat{\theta} < \theta_0|\theta_0] \\
&= P[X_{(1)} > \max(k, \theta_0)|\theta_0] + P[X_{(1)} < \theta_0|\theta_0] \\
&= \{P[X_1 > \max(k, \theta_0)|\theta_0]\}^n + \{1 - P[X_{(1)} \geq \theta_0|\theta_0]\} \\
&= \left\{\int_{\max(k,\theta_0)} \frac{1}{\lambda} e^{-(x-\theta_0)/\lambda} dx\right\}^n + 1 - \left\{\int_{\theta_0}^{\infty} \frac{1}{\lambda} e^{-(x-\theta_0)/\lambda}\right\}^n \\
&= \left(e^{\min\left(0, \frac{\theta_0 - k}{\lambda}\right)}\right)^n
\end{aligned}$$

$$\text{or,} \quad k = \theta_0 - \frac{\lambda}{n} \ln(\alpha).$$

Hence we reject $H_\circ$ if $X_{(1)} > \theta_0 - \frac{\lambda}{n} \ln(\alpha)$.

▶ **Example 18.4.5** *Likelihood ratio test for comparing the means of k normal populations with common variance.*

Let $X_{ij} (j = 1, \cdots, n_i)$ be k independent random samples of size n_i from Normal populations $N(\mu_i, \sigma^2)$ $(i = 1, \cdots, k)$, where σ^2 is a known constant. Our aim is to test the null $H_\circ(\mu_1 = \cdots = \mu_k)$ against alternative $H(\mu_i$'s are not all equal). Here

$$\Theta = \left\{(\mu_1, \cdots, \mu_k, \sigma^2)| -\infty < \mu_i < \infty, \sigma^2 > 0\right\}$$

$$\Theta_0 = \left\{(\mu_1, \cdots, \mu_k, \sigma^2)|\mu_i = \mu \,\forall\, i; \sigma^2 > 0\right\}.$$

The likelihood function is

$$\begin{aligned}
L(\boldsymbol{\theta}|\boldsymbol{x}) &= \prod_{i=1}^{k}\prod_{j=1}^{n_i}\left\{\frac{1}{\sqrt{2\pi}\sigma} \exp\left\{-(x_{ij} - \mu_i)^2/2\sigma^2\right\}\right\} \\
&= \frac{1}{(2\pi\sigma^2)^{n/2}} \exp\left[-\sum_i\sum_j (x_{ij} - \mu_i)^2/2\sigma^2\right]
\end{aligned}$$

where $n = \sum_i n_i$. For $\boldsymbol{\theta} \in \Theta$, the m.l.e.'s of the parameters are

$$\begin{aligned}
\hat{\mu}_i &= \bar{x}_i \quad (i = 1, \cdots, k) \\
\hat{\sigma}^2 &= \sum_i\sum_j (x_{ij} - \bar{x}_i)^2/n = SSW/n \text{ (say)}.
\end{aligned}$$

For $\boldsymbol{\theta} \in \Theta_0$, the m.l.e.'s are

$$\begin{aligned}
\hat{\mu} &= \sum_{i=1}^{k} n_i\bar{x}_i/n = \bar{x}, \\
\hat{\sigma}^2 &= \sum_i\sum_j (x_{ij} - \bar{x})^2/n = TSS/n \text{ (say)}.
\end{aligned}$$

Now

$$TSS = \sum_i \sum_j (x_{ij} - \bar{x})^2 = \sum_i n_i(\bar{x}_i - \bar{x})^2 + \sum_i \sum_j (x_{ij} - \bar{x}_i)^2$$
$$= SSB + SSW \text{ (say)}.$$

Hence

$$\sup_{\theta \in \Theta} L(\boldsymbol{\theta}|\boldsymbol{x}) = \left(\frac{n}{2\pi SSW}\right)^{n/2} \exp(-n/2)$$
$$\sup_{\theta \in \Theta_0} L(\boldsymbol{\theta}|\boldsymbol{x}) = \left(\frac{n}{2\pi TSS}\right)^{n/2} \exp(-n/2).$$

The LR test for H_o against H_1 is, therefore,

$$\lambda = \lambda(\boldsymbol{x}) = \left(\frac{SSW}{TSS}\right)^{n/2}.$$

Now

$$\lambda < \lambda_0 \Leftrightarrow \frac{TSS}{SSW} > k_1 \Rightarrow \frac{SSB}{SSW} > k_2$$
$$\Rightarrow \frac{SSB/(k-1)}{SSW/(n-k)} > k_3.$$

Again, $SSB/(k-1)$ which is the sum of squares between samples divided by the d.f. $(k-1)$ is the mean square between samples, MSB. Similarly sum of squares within samples (SSW) divided by $(n-k)$ gives the mean square within samples, MSW. Under H_o, the statistic MSB/MSW follows $F_{(k-1),(n-k)}$ distribution. Hence the critical region is

$$w = \{\boldsymbol{x} : MSB/MSW > F_{(k-1),(n-k);\alpha}\}.$$

This test based on partitioning the total sum of squares, TSS into two parts, SSB and SSW is called a *Analysis of Variance* test.

18.5 Asymptotic Properties of Likelihood Ratio Tests

In large samples the likelihood ratio tests have optimal asymptotic properties under the same conditions for which asymptotically normally distributed maximum likelihood estimates exist. Asymptotic distributions of the likelihood ratio tests of composite hypotheses in large samples when the null hypothesis is true were first discussed by Wilks (1938) and later considered in more details by Wald (1941a,b).

Before considering the asymptotic properties of LR test we consider a definition.

Definition 18.5.1 ***Consistent test.*** *Consider $H_\circ(\theta \in \Theta_0)$ against $H_1(\theta \in \Theta_1), \Theta = \Theta_0 \cup \Theta_1$. A sequence of tests $\{\omega_{\alpha,n}\}$ (based on a sample of size n) is said to be a consistent test of size-α for testing $H_\circ$ against H_1 if*

$$\lim_{n\to\infty} P(\omega_{\alpha,n}|\theta) = 1 \ \ \textit{for } \theta \in \Theta - \Theta_0.$$

The definition was introduced by Wald and Wolfwitz (1940). This is a natural requirement as one would expect that the power should increase with the increasing sample size.

18.5.1 Asymptotic Properties of a LR Test for a Simple Hypothesis

We first consider the asymptotic properties of LR tests for simple hypothesis $H_\circ(\theta = \theta_0)$ where the parameter θ is one-dimentional.

Asymptotic distribution of likelihood ratio in large samples

Theorem 18.5.1 *Let $(X_1, \cdots, X_n)'$ be a random sample from a distribution $F(x;\theta)$. Consider the simple null hypothesis $H_\circ(\theta = \theta_0)$, the parameter space Θ being an interval on the real line containing some (open) interval Θ_H which contains θ_0. Assume that for a sample of size n, there exists a maximum likelihood estimator $\hat{\theta}_n$ for θ_0, such that the sequence $\{\hat{\theta}_n\}$ converges in probability to θ_0. The corresponding sequence of LR tests is, therefore, given by*

$$\lambda_n = \frac{\prod_{i=1}^{n} L(x_i;\theta_0)}{\prod_{i=1}^{n} L(x_i;\hat{\theta}_n)}, \quad n = 1, 2, \cdots. \tag{18.5.1}$$

It then follows that under certain regularity conditions

$$-2\log_e \lambda_n, \quad n = 1, 2, \cdots,$$

converges in distribution to the Chi-square distribution $\chi^2_{(1)}$ when $H_\circ$ is true.

Proof: We shall assume that $F(x;\theta)$ is regular with respect to its first and second order derivatives for all θ belonging to Θ_H, i.e., $\forall\, \theta \in \Theta_H$

$$\frac{\partial}{\partial\theta}\int_{-\infty}^{\infty} dF(x;\theta) = \int_{-\infty}^{\infty}\left[\frac{\partial}{\partial\theta}\log dF(x;\theta)\right] dF(x;\theta) = 0 \tag{18.5.2}$$

$$\frac{\partial^2}{\partial\theta^2}\int_{-\infty}^{\infty} dF(x;\theta) = \int_{-\infty}^{\infty}\left[\frac{\partial^2}{\partial\theta^2}\log dF(x;\theta)\right] dF(x;\theta) + \int_{-\infty}^{\infty}\left[\frac{\partial}{\partial\theta}\log dF(x;\theta)\right]^2 dF(x;\theta) = 0. \tag{18.5.3}$$

[These conditions are same as the conditions (i), (ii) of theorem 16.2.5]

Under these conditions

$$A(\theta_0, \theta) = \int_{-\infty}^{\infty} \log dF(x;\theta) dF(x;\theta_0) \quad (18.5.$$

can be differentiated twice under the integral sign with respect to θ, giving

$$\begin{aligned} A'(\theta_0, \theta_0) &= 0 \\ A''(\theta_0, \theta_0) &= -B^2(\theta_0, \theta_0) \end{aligned} \quad (18.5.$$

where

$$B^2(\theta, \theta_1) = \int_{-\infty}^{\infty} \left[\frac{\partial \log dF(x,\theta)}{\partial \theta}\right]_{\theta=\theta_1}^2 dF(x,\theta).$$

Therefore, $A(\theta_0, \theta)$ has a relative maximum

$$A(\theta_0, \theta_0) \text{ at } \theta = \theta_0. \quad (18.5.$$

Now, in a sample $(x_1, \cdots, x_n)$ from the distribution $F(x, \theta_0)$, the quantity

$$\frac{1}{n} \sum_{i=1}^{n} \log dF(x_i; \hat{\theta}_n) \quad (18.5.$$

is a consistent estimator of $A(\theta_0, \theta_0)$. It follows by the Khintchen's law of large numbe that

$$\frac{1}{n} \sum_{i=1}^{n} \log dF(x_i; \theta_0)$$

converges in probability to $A(\theta_0, \theta_0)$. Again, if $\hat{\theta}_n \xrightarrow{P} \theta_0$, then

$$\frac{1}{n} \sum_{i=1}^{n} \log dF(x_i, \hat{\theta}_n)$$

converges in probability to $A(\theta_0, \theta_0)$.

Assume now that $\hat{\theta}_n$ is asymptotically normally distributed in accordance wi theorem 16.2.6. Since $\hat{\theta}$ converges almost certainty to θ_0 as $n \to \infty$, it follows th for an arbitrary $\varepsilon > 0$, there exists a $n(\varepsilon)$ such that for all $n > n(\varepsilon)$, the followi equality holds with probability exceeding $1 - \varepsilon$.

$$\sum_{i=1}^{n} \log dF(x_i, \theta_0) = \sum_{i=1}^{n} \log dF(x_i; \hat{\theta}_n) + \frac{1}{2} \sum_{i=1}^{n} \left[\frac{\partial^2}{\partial \theta^2} \log dF(x_i; \theta)\right]_{\theta=\theta^*} (\theta_0 - \hat{\theta}_n)^2 \quad (18.5.$$

when θ^* is a r.v. such that $|\theta_0 - \theta^*| < |\theta_0 - \hat{\theta}|$. Now, (18.5.8) can be written as

$$-2 \sum_{i=1}^{n} \log \left(\frac{\partial F(x_i; \theta_0)}{\partial F(x_i; \hat{\theta})}\right) = -\frac{1}{n} \sum_{i=1}^{n} \left[\frac{\partial^2}{\partial \theta^2} \log dF(x_i; \theta)\right]_{\theta=\theta^*} [\sqrt{n}(\theta_0 - \hat{\theta})]^2. \quad (18.5.$$

The l.h.s of (18.5.9) is $-2\log_e \lambda_n$. For $n > n(\varepsilon)$, the equality of two sides in (18.5.9) holds with probability exceeding $(1-\lambda)$. Now

$$-\frac{1}{n}\sum_{i=1}^{n}\frac{\partial^2}{\partial\theta^2}\log dF(x_i;\theta)\Bigg]_{\theta=\theta^*}$$

converges in probability to $B^2(\theta_0,\theta)$. Furthermore, $\sqrt{n}B(\theta_0,\theta_0)(\theta_0-\hat{\theta})$ converges in distribution to $N(0,1)$. Therefore, the $-2\log_e \lambda_n$ converges in distribution to the $\chi^2_{(1)}$. That is,

$$\lim_{n\to\infty} P(-2\log_e \lambda_n < \chi^2|\theta_0) = \frac{1}{\sqrt{2\pi}}\int_0^{\chi^2} u^{-1/2}e^{-u/2}\,du. \tag{18.5.10}$$

Consistency of a Likelihood Ratio Test

Consider as before the null hypothesis $H_\circ(\theta=\theta_0)$. Suppose ω_α is the region

$$\omega_\alpha\{\boldsymbol{x} : -2\log\lambda(\boldsymbol{x}) > \chi^2_{(1),\alpha}\}$$

when $\chi^2_{(1),\alpha}$ is the number for which $P\left\{\chi^2_{(1)} > \chi^2_{(1),\alpha}\right\} = \alpha$.

If $F(x;\theta)$ is regular with respect to its first and second order derivatives in Θ_0, then it is known from (18.5.10) that

$$\lim_{n\to\infty} P\{\omega_\alpha|\theta_0\} = \alpha.$$

Let $\theta_1 \neq \theta_0$ be any point in Θ_0 and consider

$$\lim_{n\to\infty} P\{\omega_\alpha|\theta_1\}. \tag{18.5.11}$$

Writing

$$\lambda(\theta) = \prod_{i=1}^{n}\frac{dF(x_i;\theta)}{dF(x_i;\hat{\theta})}, \tag{18.5.12}$$

we have

$$\lambda(\theta_0) = \lambda(\theta_1)\cdot\nu \tag{18.5.13}$$

where

$$\nu = \prod_{i=1}^{n}\frac{dF(x_i;\theta_0)}{dF(x_i;\theta_1)}. \tag{18.5.14}$$

Writing

$$-2\log\lambda(\theta_1) = u_n, \quad -2\log\nu = nv_n \tag{18.5.15}$$

the region ω_α is the set in χ_n for which

$$u_n + nv_n > \chi^2_{(1),\alpha}.$$

If the true value of θ is θ_1, then it follows by theorem 18.5.1 that u_n is a sequence of r.v.'s converging in distribution to $\chi^2_{(1)}$ distribution. From (18.5.14), we can write

$$v_n = 2\left[\frac{1}{n}\sum_{i=1}^{n}\log dF(x_i;\theta_1) - \frac{1}{n}\sum_{i=1}^{n}\log dF(x_i;\theta_0)\right]. \qquad (18.5.16)$$

It follows by Khintchen's weak law of large numbers that if θ_1 is the true value of θ,

$$\frac{1}{n}\sum_{i=1}^{n}\log dF(x_i;\theta_1) \quad \text{and} \quad \frac{1}{n}\sum_{i=1}^{n}\log dF(x_i;\theta_0)$$

converge in probability to $A(\theta_1, \theta_1)$ and $A(\theta_1, \theta_0)$ respectively. However, it follows from (18.5.6) that $A(\theta_1, \theta_1) > A(\theta_1, \theta_0)$. Therefore, the sequence of r.v.'s v_n, $\{n = 1, 2, \cdots, \}$ converges in probability to the positive number $v_0 = 2[A(\theta_1, \theta_1) - A(\theta_1, \theta_0)]$. Since when $\theta = \theta_1$, u_n converges in distribution to $\chi^2_{(1)}$, it follows that

$$\lim_{n\to\infty} P(u_n + nv_n > \chi^2_{(1),\alpha}) = 1$$

$$\text{i.e.,} \lim_{n\to\infty} P(\omega_\alpha|\theta = \theta_1) = 1$$

that is, ω_α is a consistent test for $H_\circ(\theta = \theta_0)$.

Therefore, for θ in Θ_0, the sequence of power functions $\{P(\omega_\alpha|\theta)\}, n = 1, 2, \cdots$ converges to the function

$$\gamma(\theta) = \begin{cases} 1, \text{ if } \theta = \theta_0 \\ 0, \text{ if } \theta \neq \theta_0 \end{cases}$$

which means that for sufficiently large n, ω_α is a consistent test for $H_\circ(\theta = \theta_0)$. Hence we have following theorem.

Theorem 18.5.2 *If $F(x;\theta)$ is regular with respect to its first and second order derivatives with respect to θ, then the likelihood ratio test defined by* (18.4.2) *is a consistent test for $H_\circ(\theta = \theta_0)$.*

We now consider extention to simple hypothesis with a r-dimensional parameter vector.

Assume that the r.v. X has a c.d.f. $F(x|\boldsymbol{\theta})$ where $\boldsymbol{\theta} = (\theta_1, \cdots, \theta_r)'$ is a r-dimensional vector of parameters, $\boldsymbol{\theta} \in \boldsymbol{\Theta}_r$, the parameter space. The hypothesis to be tested is the simple hypothesis $H_\circ(\boldsymbol{\theta} = \boldsymbol{\theta}_0)$ against $H(\boldsymbol{\theta} \neq \boldsymbol{\theta}_0)$. The asymptotic results in theorems 18.5.1, 18.5.2 readily extend to the case of r-dimensional vector, $\boldsymbol{\theta}$ which we state below without proof.

Theorem 18.5.3 *If $\boldsymbol{\theta}$ is a r-dimensional vector and if $F(x, \boldsymbol{\theta})$ is regular with respect to all of its first and second order θ-derivatives in a open interval $\boldsymbol{\Theta}_{rH}$ containing $\boldsymbol{\theta}$, then*

$A(\theta_0, \theta)$ as defined in (18.5.4) [where θ is replaced by $\boldsymbol{\theta}$] has a maximum of $A(\boldsymbol{\theta}_0, \boldsymbol{\theta}_0)$ at $\boldsymbol{\theta} = \boldsymbol{\theta}_0$. Again, under H_o,

$$\lim_{n\to\infty} P(-2\log\lambda < \chi^2) = \frac{1}{2^{\frac{r}{2}}\Gamma\frac{r}{2}}\int_0^{\chi^2} e^{-\frac{1}{2}u}u^{\frac{r}{2}-1}\,du$$

i.e., $-2\log\lambda$ converges in law to the $\chi^2_{(r)}$ distribution when H_o is true. Also the LR *test for H_o is a consistent test.*

18.5.2 Asymptotic Properties of a LR Test for a Composite Null Hypothesis

Consider now the composite null hypothesis $H_o(\theta_1 = \theta_{10}, \cdots, \theta_r = \theta_{ro}, \theta_{r+1}, \cdots, \theta_m)'$ of dimension $(m-r)$, when $\boldsymbol{\theta} = (\theta_1, \cdots, \theta_m)' \in \Theta_m$ the parameter space. The likelihood ratio test for such a hypothesis has been defined in the usual way in (18.4.4). In large sample, the only part of Θ_m of interest is that lying within some open r-dimensional interval Θ_{mH} containing the true parameter points $(\theta_{10}, \cdots, \theta_{ro}, \theta_{r+1}, \cdots, \theta_m)$. The following asymptotic result is due to Wilks (1938).

Theorem 18.5.4 *Suppose $(x_1, \cdots, x_n)'$ is a sample from the c.d.f. $F(x|\boldsymbol{\theta})$ where $\boldsymbol{\theta}$ is m-dimentional and $F(x|\boldsymbol{\theta})$ is regular with respect to all of its first and second derivatives for $\theta \in \Theta_{mH}$. Then, if H_o is true*

$$\lim_{n\to\infty} P(-2\log_e\lambda < \chi^2) = \frac{1}{2^{\nu/2}\Gamma\frac{\nu}{2}}\int_0^{\chi^2} e^{-\frac{u}{2}}u^{\frac{\nu}{2}-1}\,du$$

where $\nu = (m-r)$, i.e., $-2\log_e\lambda$ converges in law to the $\chi^2_{(m-r)}$ distribution when H_o is true. Also the LR *test is consistent i.e.,*

$$\lim_{n\to\infty} P(-2\log_e\lambda < \chi^2_{(m-r),r}|\theta \in \Theta_m - \Theta_{mH}) = 1.$$

18.6 Hotelling's Generalised "Students" Test

Suppose a sample O_n is drawn from a multivariate normal population with distribution

$$dF(x_1, \cdots, x_n) = \frac{\sqrt{|\boldsymbol{A}|}}{(2\pi)^{p/2}}\exp\left[-\frac{1}{2}\sum_i^p\sum_{j=1} A_{ij}(x_i-\mu_i)(x_j-\mu_j)\right]\prod_1^p dx_i \qquad (18.6.1)$$

when $\boldsymbol{A} = \boldsymbol{\Sigma}^{-1}$ [inverse of the dispersion matrix of $(X_1, \cdots, X_p)'$] is a positive definite matrix. It is desired to test the hypothesis $H_o(\mu_i = \mu_i^0,\ i = 1, \cdots, p)$. The parameter space is

$$\Theta = \begin{cases} A_{ij}\text{'s such that } \boldsymbol{A} \text{ is positive definite} \\ -\infty < \mu_i < \infty,\quad i = 1, \cdots, p \end{cases}$$

$$\Theta_0 = \Big\{ \text{The subspace of } \Theta \text{ for which } \mu_i = \mu_i^0, i = 1, \cdots, p.$$

This hypothesis is the p-variate analogue of the hypothesis that a normal population has a specified mean $\mu = \mu_0$, whatever be its variance σ^2.

Let us draw a sample of size n from the distribution (18.6.1), $X_{i\alpha}$ being the αth observation of the variable X_i. The matrix of observations is

$$\boldsymbol{X} = \begin{bmatrix} x_{11} & \cdots & x_{p1} \\ \vdots & & \\ x_{1n} & \cdots & x_{pn} \end{bmatrix}.$$

The likelihood ratio statistic is

$$\lambda(\boldsymbol{X}) = \frac{\max\limits_{\theta\in\Theta_0} L(\boldsymbol{X})}{\max\limits_{\theta\in\Theta} L(\boldsymbol{X})}. \tag{18.6.2}$$

Now,

$$L(\boldsymbol{X}) = \frac{|\boldsymbol{A}|^{n/2}}{(2\pi)^{pn/2}} \exp\left[-\frac{1}{2}\sum_{\alpha=1}^{n}\sum_{i}^{p}\sum_{j=1}^{p} A_{ij}(x_{i\alpha}-\mu_i)(x_{j\alpha}-\mu_j)\right]\prod_{i=1}^{p}\prod_{\alpha=1}^{n} dx_{i\alpha} \tag{18.6.3}$$

$$\log L = -\frac{np}{2}\log(2\pi) + \frac{n}{2}\log|\boldsymbol{A}| - \frac{1}{2}\sum_{\alpha=1}^{n}(\boldsymbol{x}_\alpha - \boldsymbol{\mu})'\boldsymbol{A}(\boldsymbol{x}_\alpha - \boldsymbol{\mu}) \tag{18.6.4}$$

where

$$\boldsymbol{x}_\alpha = (x_{1\alpha}, x_{2\alpha}, \cdots, x_{p\alpha})', \quad \boldsymbol{\mu} = (\mu_1, \cdots, \mu_p)'. \tag{18.6.5}$$

Let

$$\bar{\boldsymbol{x}} = (\bar{x}_1, \cdots, \bar{x}_p)'; \text{ where } \bar{x}_i = \frac{1}{n}\sum_{\alpha=1}^{n} x_{i\alpha}.$$

$$\boldsymbol{S} = \sum_{\alpha=1}^{n}(\boldsymbol{x}_\alpha - \bar{\boldsymbol{x}})(\boldsymbol{x}_\alpha - \bar{\boldsymbol{x}})' = \left(\sum_{\alpha=1}^{n}(x_{i\alpha}-\bar{x}_i)(x_{j\alpha}-\bar{x}_j)\right)_{p\times p}.$$

Lemma 18.6.1 *For any vector* $\boldsymbol{b} = (b_1, \cdots, b_p)'$,

$$\sum_{\alpha=1}^{n}(\boldsymbol{x}_\alpha - \boldsymbol{b})(\boldsymbol{x}_\alpha - \boldsymbol{b})' = \sum_{\alpha=1}^{n}(\boldsymbol{x}_\alpha - \bar{\boldsymbol{x}})(\boldsymbol{x}_\alpha - \bar{\boldsymbol{x}})' + n(\bar{\boldsymbol{x}} - \boldsymbol{b})(\bar{\boldsymbol{x}} - \boldsymbol{b})'. \tag{18.6.6}$$

Proof:

$$\begin{aligned}\sum_{\alpha=1}^{n}(\boldsymbol{x}_\alpha - \boldsymbol{b})(\boldsymbol{x}_\alpha - \boldsymbol{b})' &= \sum_{\alpha=1}^{n}[(\boldsymbol{x}_\alpha - \bar{\boldsymbol{x}}) + (\bar{\boldsymbol{x}} - \boldsymbol{b})][(\boldsymbol{x}_\alpha - \bar{\boldsymbol{x}}) + (\bar{\boldsymbol{x}} - \boldsymbol{b})]' \\ &= \sum_{\alpha=1}^{n}(\boldsymbol{x}_\alpha - \bar{\boldsymbol{x}})(\boldsymbol{x}_\alpha - \bar{\boldsymbol{x}})' + \sum_{\alpha=1}^{n}(\boldsymbol{x}_\alpha - \bar{\boldsymbol{x}})(\bar{\boldsymbol{x}} - \boldsymbol{b})' \\ &\quad + \sum_{\alpha=1}^{n}(\bar{\boldsymbol{x}} - \boldsymbol{b})(\boldsymbol{x}_\alpha - \bar{\boldsymbol{x}})' + n(\bar{\boldsymbol{x}} - \boldsymbol{b})(\bar{\boldsymbol{x}} - \boldsymbol{b})'. \end{aligned} \tag{18.6.7}$$

The second and the third term in (18.6.7) vanish because $\sum_{\alpha}(\boldsymbol{x}_\alpha - \bar{\boldsymbol{x}}) = \mathbf{0}$. Hence, ne lemma.

etting $\boldsymbol{b} = \boldsymbol{\mu}$,

$$\sum_{\alpha=1}^{n}(\boldsymbol{x}_\alpha - \boldsymbol{\mu})(\boldsymbol{x}_\alpha - \boldsymbol{\mu})' = \boldsymbol{S} + n(\bar{\boldsymbol{x}} - \boldsymbol{\mu})(\bar{\boldsymbol{x}} - \boldsymbol{\mu})'. \tag{18.6.8}$$

gain,

$$\begin{aligned}\sum_{\alpha=1}^{n}(\boldsymbol{x}_\alpha - \boldsymbol{\mu})'\boldsymbol{A}(\boldsymbol{x}_\alpha - \boldsymbol{\mu}) &= tr\sum_{\alpha=1}^{n}(\boldsymbol{x}_\alpha - \boldsymbol{\mu})(\boldsymbol{x}_\alpha - \boldsymbol{\mu})'\boldsymbol{A} \\ &= tr[\boldsymbol{SA} + n(\bar{\boldsymbol{x}} - \boldsymbol{\mu})(\bar{\boldsymbol{x}} - \boldsymbol{\mu})'\boldsymbol{A}] \text{ [by (18.6.8)]} \\ &= tr\boldsymbol{AS} + ntr\boldsymbol{A}(\bar{\boldsymbol{x}} - \boldsymbol{\mu})(\bar{\boldsymbol{x}} - \boldsymbol{\mu})'. \end{aligned} \tag{18.6.9}$$

ence,

$$\log L = -\frac{np}{2}\log(2\pi) + \frac{n}{2}\log|\boldsymbol{A}| - \frac{1}{2}tr\boldsymbol{AS} - \frac{n}{2}tr\boldsymbol{A}(\bar{\boldsymbol{x}} - \boldsymbol{\mu})(\bar{\boldsymbol{x}} - \boldsymbol{\mu})'. \tag{18.6.10}$$

Now $tr\boldsymbol{A}(\bar{\boldsymbol{x}} - \boldsymbol{\mu})(\bar{\boldsymbol{x}} - \boldsymbol{\mu})' = (\bar{\boldsymbol{x}} - \boldsymbol{\mu})'\boldsymbol{A}(\bar{\boldsymbol{x}} - \boldsymbol{\mu})$ and this is non-negative, since $\boldsymbol{A}$ is ositive semi-define. This is zero, iff $\bar{\boldsymbol{x}} = \boldsymbol{\mu}$.

We now consider the following lemma.

emma 18.6.2 *Let*

$$f(C) = \frac{n}{2}\log|C| - \frac{1}{2}\sum_{i}\sum_{j=1}^{p} C_{ij}d_{ij} \tag{18.6.11}$$

here $\boldsymbol{C} = ((C_{ij}))$ is positive semi-definite and where $\boldsymbol{D} = ((d_{ij}))$ is positive definite. hen the maximum of $f(\boldsymbol{C})$ is attained at $\boldsymbol{C} = n\boldsymbol{D}^{-1}$ and the maximum value is

$$f(n\boldsymbol{D}^{-1}) = \frac{np}{2}\log n - \frac{n}{2}\log|\boldsymbol{D}| - \frac{np}{2}.$$

Proof: Omitted [For details, see Anderson, (1974) p. 46–47].

Applying this lemma to (18.6.10) we see that the contribution due to the second and hird term is maximised if

$$\boldsymbol{A} = \boldsymbol{A}^* = \left(\frac{\boldsymbol{S}}{n}\right)^{-1} \tag{18.6.12}$$

here we assume that $\boldsymbol{S}$ is non-singular. It can be shown that the probability of drawing sample of size $n(> p)$ such that $\boldsymbol{S}$ is singular is zero. Thus, $\boldsymbol{S}^{-1}$ exists and $\boldsymbol{A}^*$ is positive efinite. Hence, when $\boldsymbol{\theta} \in \Theta$, the maximum likelihood estimates are

$$\boldsymbol{\mu}^* = \bar{\boldsymbol{x}} \quad \text{and} \quad \boldsymbol{A}^* = n\boldsymbol{S}^{-1}.$$

Therefore, from (18.6.4),

$$\max_{\theta\in\Theta} L(\boldsymbol{x}) = \frac{e^{-np/2}}{(2\pi)^{np/2}\left|\frac{\boldsymbol{S}}{n}\right|^{n/2}}. \tag{18.6.13}$$

When $\theta \in \Theta_0$, $\boldsymbol{\mu} = \boldsymbol{\mu}_0$.
In this case, the m.l.e. of $\boldsymbol{A}$ is $\boldsymbol{A}^{**} = n\boldsymbol{S}_0^{-1}$ where

$$\boldsymbol{S}_0 = \sum_{\alpha}(\boldsymbol{x}_\alpha - \boldsymbol{\mu}_0)(\boldsymbol{x}_\alpha - \boldsymbol{\mu}_0)'$$

and hence

$$\max_{\theta\in\Theta_0} L(\boldsymbol{X}) = \frac{e^{-np/2}}{(2\pi)^{np/2}\left|\frac{\boldsymbol{S}_0}{n}\right|^{n/2}}. \tag{18.6.14}$$

Therefore,

$$\lambda = \frac{|\boldsymbol{S}|^{n/2}}{|\boldsymbol{S}_0|^{n/2}} = \frac{|\boldsymbol{S}|^{n/2}}{|\boldsymbol{S} + n(\bar{\boldsymbol{x}} - \boldsymbol{\mu}_0)(\bar{\boldsymbol{x}} - \boldsymbol{\mu}_0)'|^{n/2}}. \tag{18.6.15}$$

Now, for $|\boldsymbol{B}| \neq 0$

$$\begin{aligned}\begin{vmatrix}\boldsymbol{B} & \boldsymbol{C}\\ \boldsymbol{D} & \boldsymbol{E}\end{vmatrix} &= \begin{vmatrix}\boldsymbol{B} & \boldsymbol{C}\\ \boldsymbol{D} & \boldsymbol{E}\end{vmatrix}\begin{vmatrix}\boldsymbol{I} & -\boldsymbol{B}^{-1}\boldsymbol{C}\\ \boldsymbol{O} & \boldsymbol{I}\end{vmatrix}\\ &= \begin{vmatrix}\boldsymbol{B} & \boldsymbol{O}\\ \boldsymbol{D} & \boldsymbol{E} - \boldsymbol{D}\boldsymbol{B}^{-1}\boldsymbol{C}\end{vmatrix} = |\boldsymbol{B}||\boldsymbol{E} - \boldsymbol{D}\boldsymbol{B}^{-1}\boldsymbol{C}|.\end{aligned} \tag{18.6.16}$$

Use of this result shows that

$$\begin{aligned}\lambda^{2/n} &= \frac{|\boldsymbol{S}|}{|\boldsymbol{S} + [\sqrt{n}(\bar{\boldsymbol{x}} - \boldsymbol{\mu}_0)][\sqrt{n}(\bar{\boldsymbol{x}} - \boldsymbol{\mu}_0)]'|} = \frac{|\boldsymbol{S}|}{\begin{vmatrix}1 & \sqrt{n}(\bar{\boldsymbol{x}} - \boldsymbol{\mu}_0)'\\ -\sqrt{n}(\bar{\boldsymbol{x}} - \boldsymbol{\mu}_0) & \boldsymbol{S}\end{vmatrix}}\\ &= \frac{1}{1 + n(\bar{\boldsymbol{x}} - \boldsymbol{\mu}_0)'\boldsymbol{S}^{-1}(\bar{\boldsymbol{x}} - \boldsymbol{\mu}_0)} = \frac{1}{1 + \frac{T^2}{n-1}}\end{aligned} \tag{18.6.17}$$

where

$$T^2 = n(n-1)(\bar{\boldsymbol{x}} - \boldsymbol{\mu}_0)'\boldsymbol{S}^{-1}(\bar{\boldsymbol{x}} - \boldsymbol{\mu}_0) \tag{18.6.18}$$

$$= n(\bar{\boldsymbol{x}} - \boldsymbol{\mu}_0)'\boldsymbol{s}^{-1}(\bar{\boldsymbol{x}} - \boldsymbol{\mu}_0) \tag{18.6.19}$$

where $\boldsymbol{s} = \boldsymbol{S}/(n-1)$ is the sample variance-covariance matrix. The statistic (18.6.19) is known as Hotelling's T^2 statistic with $(n-1)$ d.f. In the univariate case $(p = 1)$ (18.6.19) reduces to

$$t^2 = n(\bar{x} - \mu_0)^2/s^2$$

which is the square of a student's t variate. The distribution of the T^2 statistic for $\boldsymbol{\mu} = \boldsymbol{\mu}_0$ (known), was derived by Hotelling (1931). The distribution for arbitrary $\boldsymbol{\mu}$ was given by Bose and Roy (1938) and by Hsu (1938). A compact proof for the general case was given by Bowker (1960).

Theorem 18.6.1 *Let*

$$F = \left(\frac{T^2}{n-1}\right)\left(\frac{n-p}{p}\right)$$

where T^2 *is as defined in* (18.6.19). *Then* $F \cap F(p, n-p, \phi)$ *when*

$$\phi = n(\boldsymbol{\mu} - \boldsymbol{\mu}_0)'\boldsymbol{A}(\boldsymbol{\mu} - \boldsymbol{\mu}_0).$$

In general, let

$$T^2 = \boldsymbol{Y}'\boldsymbol{s}^{-1}\boldsymbol{Y}$$

where $\boldsymbol{Y}$ is distributed as $N_p(\boldsymbol{\nu}, \boldsymbol{\Sigma})$ and

$$n\boldsymbol{s} = \sum_{\alpha=1}^{n} \boldsymbol{Z}_\alpha \boldsymbol{Z}'_\alpha$$

where $\boldsymbol{Z}_\alpha$'s are independent $N_p(\boldsymbol{0}, \boldsymbol{\Sigma})$ $\alpha = 1, \cdots, n$. Also let $\boldsymbol{Z}_\alpha$ be distributed independently of $\boldsymbol{Y}$. Then

$$\frac{T^2}{n} \cdot \frac{n-p+1}{p} \cap F(p, n-p+1, \phi)$$

where $\phi = \boldsymbol{\nu}'\boldsymbol{\Sigma}^{-1}\boldsymbol{\nu}$. In particular, if $\boldsymbol{\nu} = \boldsymbol{0}$, then the above statistic follows central $F_{(p,n-p+1)}$.

Proof: Let $\boldsymbol{D}$ be a non-singular $p \times p$ matrix such that $\boldsymbol{D\Sigma D}' = I_p$. Define

$$\boldsymbol{Y}^* = \boldsymbol{DY}, \;\; \boldsymbol{\nu}^* = \boldsymbol{D\nu}, \;\; s^* = \boldsymbol{DsD}', \;\; \boldsymbol{Z}^*_\alpha = \boldsymbol{DZ}_\alpha.$$

Hence,

$$\boldsymbol{Y}^* \cap N_p(\boldsymbol{\nu}^*, I_p),$$
$$\boldsymbol{Z}^*_\alpha \cap N_p(\boldsymbol{0}, I_p).$$

Also

$$ns^* = \sum_{\alpha=1}^{n} \boldsymbol{Z}^*_\alpha \boldsymbol{Z}^{*'}_\alpha.$$

$$\boldsymbol{Y}^{*'}\boldsymbol{s}^{*^{-1}}\boldsymbol{Y}^* = \boldsymbol{Y}'\boldsymbol{D}'(\boldsymbol{DsD}')^{-1}\boldsymbol{DY} = \boldsymbol{Y}'\boldsymbol{s}^{-1}\boldsymbol{Y} = T^2.$$

Consider orthogonal transformation $\boldsymbol{Y}^* \to \boldsymbol{U} = (U_1, \cdots, U_p)'$,

$$\boldsymbol{U} = \boldsymbol{QY}^*$$

when the first row of $\boldsymbol{Q}$ is

$$q_{1j} = \frac{Y^*_j}{\sqrt{\boldsymbol{Y}^{*'}\boldsymbol{Y}^*}}, \;\; \boldsymbol{Y}^* = (Y^*_1, \cdots, Y^*_p)'.$$

Let

$$B = nQs^*Q'.$$

Hence,

$$U_1 = \sum_{j=1}^{p} q_{1j}Y_j^* = \sqrt{Y^{*'}Y^*}$$

$$U_i = \sum_{j=1}^{p} q_{ij}Y_j^* = \sum_{j=1}^{p} q_{ij}(q_{1j}\sqrt{Y^{*'}Y^*})$$

$$= \sqrt{Y^{*'}Y^*}\sum_{j=1}^{p} q_{ij}q_{1j} = 0, i = 2, \cdots, p.$$

Now,

$$T^2 = Y^{*'}s^{*^{-1}}Y^* = (Q^{-1}U)'s^{*^{-1}}(Q^{-1}U)$$
$$= nU'B^{-1}U = nb^{11}U_1^2.$$

Hence,

$$\frac{T^2}{n} = (Y^{*'}Y^*)/\left(\frac{1}{b^{11}}\right)$$

where $B^{-1} = ((b^{ij}))$. Now by the result in Section 14.6, $\frac{1}{b^{11}}$ follows a central χ^2 distribution, $\chi^2_{(n-p+1)} = \chi_1^2$ (say). Since this distribution does not depend on Q, it is also the unconditional distribution of $\frac{1}{b^{11}}$. Again, $Y^{*'}Y^*$ follows a non-central $\chi^2(p, \phi) = \chi_2^2$ (say) where $\phi = \nu^{*'}\nu^* = \nu'\Sigma^{-1}\nu$. Also, χ_1^2, χ_2^2 are independently distributed. Therefore,

$$\frac{T^2}{n}\cdot\frac{n-p+1}{p} = \frac{\chi_2^2/p}{\chi_1^2/(n-p+1)}$$

follows a non-central $\chi^2(p, n-p+1, \phi)$.

To test $H_\circ(\mu = \mu_0)$ against $H(\mu \neq \mu_0)$ one, therefore, uses the statistic (from (18.6.17)),

$$\lambda^{2/n} \leq \lambda_0$$

$$\Rightarrow \quad T^2 \geq \frac{(n-1)\cdot p}{n-p} F_{p,n-1;1-\alpha}.$$

■ NOTE 18.6.1 *A statistic relatied to T^2 is Mahalanobis' D^2-statistic,*

$$D^2 = (\bar{X}_1 - \bar{X}_2)'s^{-1}(\bar{X}_1 - \bar{X}_2)$$

and the corresponding population measure,

$$\Delta^2 = (\mu_1 - \mu_2)'\Sigma^{-1}(\mu_1 - \mu_2)$$

is called Mahalanobis' generalised distance.

18.6.1 Invariance Property of T^2-test

Suppose originally $\boldsymbol{X}$ was distributed as $N_p(\boldsymbol{\mu}, \boldsymbol{\Sigma})$ and we wanted to test $H_\circ(\boldsymbol{\mu} = \boldsymbol{\mu}^0)$. Now, if the vector of random variables is transformed as $\boldsymbol{Y} = \boldsymbol{CX}$ (where $|\boldsymbol{C}| \neq 0$) then $\boldsymbol{Y} \cap N_p(\boldsymbol{C\mu}, \boldsymbol{C\Sigma C'})$. The transformed hypothesis is $H_\circ(\boldsymbol{\mu}^* = \boldsymbol{\mu}^{0*})$ where $\boldsymbol{\mu}^* = \boldsymbol{C\mu}$ and $\mu^{0*} = \boldsymbol{C\mu}^0$. Now T^2 calculated from the transformed observations $\boldsymbol{Cx}_\alpha = \boldsymbol{y}_\alpha$ is

$$T^2(y) = n(n-1)(\bar{\boldsymbol{x}} - \boldsymbol{\mu}_0)'\boldsymbol{C}'\boldsymbol{S}(\boldsymbol{y})^{-1}\boldsymbol{C}(\bar{\boldsymbol{x}} - \boldsymbol{\mu}_0).$$

Again,

$$\boldsymbol{S}(\boldsymbol{y}) = \sum_{\alpha=1}^{n} \boldsymbol{C}(\boldsymbol{x}_\alpha - \bar{\boldsymbol{x}})(\boldsymbol{x}_\alpha - \bar{\boldsymbol{x}})'\boldsymbol{C}' = \boldsymbol{CS}(x)\boldsymbol{C}'.$$

Hence,

$$T^2(y) = n(n-1)(\bar{\boldsymbol{x}} - \boldsymbol{\mu}_0)'\boldsymbol{C}'(\boldsymbol{C}')^{-1}\boldsymbol{S}(x)^{-1}\boldsymbol{C}^{-1}\boldsymbol{C}(\bar{\boldsymbol{x}} - \boldsymbol{\mu}_0) = T^2(x).$$

Hence $T^2(y)$ calculated on the basis of the transformed observations $\boldsymbol{y}_\alpha$ is the same as $T^2(x)$ calculated on the basis of original observation $\boldsymbol{x}_\alpha$.

Optimum properties of a T^2-test

The T^2-test for $H_\circ : \boldsymbol{\mu} = \boldsymbol{\mu}_0$ has a number of optimum properties:

1. The test is unbiased.

2. The test is similar.

3. The test is admissible i.e., there exists no other test which is uniformly more powerful.

4. The power function of the test is a monotonically increasing function of $n(\boldsymbol{\mu} - \boldsymbol{\mu}^0)'\boldsymbol{\Sigma}^{-1}(\boldsymbol{\mu} - \boldsymbol{\mu}^0)$.

5. Of all the tests based on $\boldsymbol{x}_\alpha = (x_{1\alpha}, \cdots, x_{p\alpha})'$ from $N_p(\boldsymbol{\mu}, \boldsymbol{\Sigma})$ for the hypothesis $H_\circ(\boldsymbol{\mu} = \boldsymbol{\mu}^0)$ which are invariant with respect to the transformation $\boldsymbol{x}_\alpha^* = \boldsymbol{Cx}_\alpha(|\boldsymbol{C}| \neq 0)$, the T^2-test is a UMP test.

6. Of all the tests based on $\boldsymbol{x}_\alpha = (x_{1\alpha}, \cdots, x_{p\alpha})'$ from $N_p(\boldsymbol{\mu}, \boldsymbol{\Sigma})$ for the hypothesis $H_\circ(\boldsymbol{\mu} = \boldsymbol{\mu}^0)$ which have power depending on $n(\boldsymbol{\mu} - \boldsymbol{\mu}^0)'\boldsymbol{\Sigma}^{-1}(\boldsymbol{\mu} - \boldsymbol{\mu}^0)$, the T^2-test is a UMP test.

For proof of the results the reader may refer to Anderson (1974).

18.6.2 A Confidence Region for the Mean Vector

If $\boldsymbol{\mu}$ is the mean of $N_p(\boldsymbol{\mu}, \boldsymbol{\Sigma})$ then it is known that

$$P\left[n(\bar{\boldsymbol{x}} - \boldsymbol{\mu})'\boldsymbol{s}^{-1}(\bar{\boldsymbol{x}} - \boldsymbol{\mu}) \leq T_0^2(\alpha)\right] = 1 - \alpha. \tag{18.6.20}$$

Thus, if for a particular sample (18.6.20) is computed then the region

$$n(\bar{\boldsymbol{x}} - \boldsymbol{\mu})'\boldsymbol{s}^{-1}(\bar{\boldsymbol{x}} - \boldsymbol{\mu}) \leq T_0^2(\alpha) \tag{18.6.21}$$

covers the true value of $\boldsymbol{\mu}$ with probability $(1 - \alpha)$. The region (18.6.21) is the interior (including the boundary) of an ellipsoid in the p-dimensional space of $\boldsymbol{\mu}$ with centre at $\bar{\boldsymbol{x}}$ and with size and shape determined by $\boldsymbol{s}^{-1}$ and α.

18.6.3 A Two-sample Problem

Suppose $\boldsymbol{x}_1^{(i)}, \cdots, \boldsymbol{x}_{n_i}^{(i)}$ be a sample of size n_i from a $N_p(\boldsymbol{\mu}^{(i)}, \boldsymbol{\Sigma})$ population $(i = 1, 2)$. We wish to test the hypothesis $H_\circ(\boldsymbol{\mu}^{(1)} = \boldsymbol{\mu}^{(2)})$. Let, $\bar{\boldsymbol{x}}^{(i)} = (\bar{x}_1^{(i)}, \cdots, \bar{x}_p^{(i)})$ when

$$\bar{x}_j^{(i)} = \sum_{\alpha=1}^{n_i} x_{j\alpha}^{(i)}/n_i.$$

Now

$$\bar{\boldsymbol{x}}^{(i)} \cap N_p(\boldsymbol{\mu}^{(i)}, \ \boldsymbol{\Sigma}/n_i).$$

Therefore,

$$\sqrt{n_1 n_2/(n_1 + n_2)}(\bar{\boldsymbol{x}}^{(1)} - \bar{\boldsymbol{x}}^{(2)}) \cap N_p(\boldsymbol{0}, \boldsymbol{\Sigma})$$

under $H_\circ$. Again, if we define

$$\boldsymbol{s} = \frac{1}{n_1 + n_2 - 2}\left\{\sum_{\alpha=1}^{n_1}(\boldsymbol{x}_\alpha^{(1)} - \bar{\boldsymbol{x}}^{(1)})(\boldsymbol{x}_\alpha^{(1)} - \bar{x}^{(1)})' + \sum_{\alpha=1}^{n_2}(\boldsymbol{x}_\alpha^{(2)} - \bar{x}^{(2)})(\boldsymbol{x}_\alpha^{(2)} - \bar{x}^{(2)})'\right\}$$

them $(n_1 + n_2 - 2)\boldsymbol{s}$ is distributed as

$$\sum_{\alpha=1}^{n_1+n_2-2} \boldsymbol{Z}_\alpha \boldsymbol{Z}_\alpha'$$

where $\boldsymbol{Z}_\alpha$ is distributed as $N_p(\boldsymbol{0}, \boldsymbol{\Sigma})$. Therefore,

$$T^2 = \frac{n_1 n_2}{n_1 + n_2}(\bar{\boldsymbol{x}}^{(1)} - \bar{\boldsymbol{x}}^{(2)})'\boldsymbol{s}^{-1}(\bar{\boldsymbol{x}}^{(1)} - \bar{\boldsymbol{x}}^{(2)})$$

is distributed as T^2 with $(n_1 + n_2 - 2)$ d.f. The critical region is

$$T^2 \geq \frac{(n_1 + n_2 - 2)p}{n_1 + n_2 - p - 1} F_{p, n_1+n_2-p-1; \alpha}$$

with significance level α.

18.6.4 A k-sample Problem

Let $\boldsymbol{X}_\alpha^{(i)} (\alpha = 1, \cdots, n_i, i = 1, \cdots, k)$ be samples from $N_p(\boldsymbol{\mu}^{(i)}, \boldsymbol{\Sigma})$ respectively. We want to test the hypothesis

$$H_\circ : \sum_{i=1}^{k} l_i \boldsymbol{\mu}^{(i)} = \boldsymbol{\mu}$$

where $l_i (i = 1, \cdots, k)$ are given scalars and $\boldsymbol{\mu}$ is known.
Under $H_\circ$,

$$\sum_{i=1}^{k} l_i \bar{\boldsymbol{x}}^{(i)} \cap N_p \left(\boldsymbol{\mu}, \sum_{i=1}^{k} \frac{l_i^2}{n_i} \boldsymbol{\Sigma} \right)$$

where $\bar{\boldsymbol{x}}^{(i)}$ is the sample mean vector of the ith sample. Therefore,

$$c \left(\sum_{i=1}^{k} l_i \bar{\boldsymbol{x}}_i^{(i)} - \boldsymbol{\mu} \right)' \boldsymbol{\Sigma}^{-1} \left(\sum_{i=1}^{k} l_i \bar{\boldsymbol{x}}^{(i)} - \boldsymbol{\mu} \right) \cap \chi_p^2 \text{ where } c = \left(\sum_{i=1}^{k} \frac{l_i^2}{n_i} \right)^{-1}.$$

Again

$$\sum_{\alpha=1}^{n_i} (\boldsymbol{X}_\alpha^{(i)} - \bar{\boldsymbol{X}}^{(i)})(\boldsymbol{X}^{(i)} - \bar{X}^{(i)})' \cap W(p, n_i - 1, \boldsymbol{\Sigma}).$$

Therefore,

$$\sum_{i=1}^{k} \sum_{\alpha=1}^{n_i} (\boldsymbol{X}_\alpha^{(i)} - \bar{\boldsymbol{X}}^{(i)})(\boldsymbol{X}_\alpha^{(i)} - \bar{X}^{(i)})' \cap W \left(p, \sum_{i=1}^{k} n_i - k, \boldsymbol{\Sigma} \right).$$

Defining

$$\boldsymbol{s} = \frac{1}{n-k} \sum_{i=1}^{k} \sum_{\alpha=1}^{n_i} (\boldsymbol{X}_\alpha^{(i)} - \bar{\boldsymbol{X}}^{(i)})(\boldsymbol{X}_\alpha^{(i)} - \bar{\boldsymbol{X}}^{(i)})', \quad n = \sum_{i=1}^{k} n_i$$

therefore,

$$c \left(\sum_{i=1}^{k} l_i \bar{\boldsymbol{x}}_i^{(i)} - \boldsymbol{\mu} \right)' \boldsymbol{s}^{-1} \left(\sum_{i=1}^{k} l_i \bar{\boldsymbol{x}}_i^{(i)} - \boldsymbol{\mu} \right) = T^2$$

follows a T^2-distribution with $(n-k)$ d.f.

18.6.5 Testing all the Elements of μ are Equal

Let $\boldsymbol{X}_\alpha (\alpha = 1, \cdots, n)$ be a sample of size n from $N_p(\boldsymbol{\mu}, \boldsymbol{\Sigma})$. We want to test the hypothesis

$$H_\circ (\mu_1 = \cdots = \mu_p).$$

Let $\boldsymbol{C}$ be any $(p-1) \times p$ matrix of rank $(p-1)$ such that $\boldsymbol{C1} = \boldsymbol{O}$. Let $\boldsymbol{Y}_\alpha = \boldsymbol{C X}_\alpha$. Then

$$\boldsymbol{Y}_\alpha \cap N_{p-1}(\boldsymbol{C\mu}, \boldsymbol{C \Sigma C}').$$

Hence

$$\bar{\boldsymbol{Y}} \cap N_{p-1}(\boldsymbol{C\mu}, \boldsymbol{C\Sigma C}'/n).$$

Again

$$\sum_{\alpha=1}^{n}(\boldsymbol{Y}_\alpha - \bar{\boldsymbol{Y}})(\boldsymbol{Y}_\alpha - \bar{\boldsymbol{Y}})' = \boldsymbol{C}\sum_{\alpha=1}^{n}(\boldsymbol{X}_\alpha - \bar{\boldsymbol{X}})(\boldsymbol{X}_\alpha - \bar{\boldsymbol{X}})'\boldsymbol{C}' \cap W(\boldsymbol{C\Sigma C}', p-1, n-1).$$

Also, under $H_\circ, \boldsymbol{C\mu} = \boldsymbol{0}$.
Therefore, under $H_\circ$

$$n\bar{\boldsymbol{Y}}'\boldsymbol{s}^{-1}\bar{\boldsymbol{Y}} = T^2,$$

where

$$\boldsymbol{s} = \frac{1}{n-1}\sum_{\alpha=1}^{n}(\boldsymbol{Y}_\alpha - \bar{\boldsymbol{Y}})(\boldsymbol{Y}_\alpha - \bar{\boldsymbol{Y}})',$$

and T^2 is a $(p-1)$-dimensional T^2-distribution with $(n-1)$ d.f.

18.6.6 The Multivariate Behrens-Fisher Problem

Let $\boldsymbol{X}_\alpha^{(i)} (\alpha = 1, \cdots, n_i; i = 1, 2)$ be samples from $N_p(\boldsymbol{\mu}^{(i)}, \boldsymbol{\Sigma}_i)$ $(i = 1, 2)$. We want to test $H_\circ(\boldsymbol{\mu}^{(1)} = \boldsymbol{\mu}^{(2)})$.
Now,

$$\bar{\boldsymbol{X}}^{(i)} \cap N_p(\boldsymbol{\mu}^{(i)},\ \boldsymbol{\Sigma}_i/n_i).$$

Thus,

$$\bar{\boldsymbol{X}}^{(1)} - \bar{\boldsymbol{X}}^{(2)} \cap N_p(\boldsymbol{\mu}^{(1)} - \boldsymbol{\mu}^{(2)},\ \boldsymbol{\Sigma}_1/n_1 + \boldsymbol{\Sigma}_2/n_2).$$

However,

$$\sum_{i=1}^{2}\sum_{\alpha=1}^{n_i}(\boldsymbol{X}_\alpha^{(i)} - \bar{\boldsymbol{X}}^{(i)})(\boldsymbol{X}_\alpha^{(i)} - \bar{\boldsymbol{X}}^{(i)})'$$

does not have a Wishart distribution with covariance matrix given by a multiple of $\boldsymbol{\Sigma}_1/n_1 + \boldsymbol{\Sigma}_2/n_2$. Hence, the technique in subsection 18.6.3 cannot be applied here.
If $n_1 = n_2$, the T^2-test can be used obviously. Consider,

$$\boldsymbol{Y}_\alpha = \boldsymbol{X}_\alpha^{(1)} - \boldsymbol{X}_\alpha^{(2)}.$$

Under $H_\circ$, $\bar{\boldsymbol{Y}} \cap N_p(\boldsymbol{0}(\boldsymbol{\Sigma}_1 + \boldsymbol{\Sigma}_2)/n)$.
Also,

$$(n-1)\boldsymbol{s} = \sum_{\alpha=1}^{n}(\boldsymbol{Y}_\alpha - \bar{\boldsymbol{Y}})(\boldsymbol{Y}_\alpha - \bar{\boldsymbol{Y}})' \cap W(p, n-1, \boldsymbol{\Sigma}_1 + \boldsymbol{\Sigma}_2).$$

Therefore,

$$n\bar{\boldsymbol{Y}}\boldsymbol{s}^{-1}\bar{\boldsymbol{Y}}'$$

follows T^2 with $(n-1)$ d.f. under $H_\circ$.

Assume now $n_1 < n_2$. Define

$$\boldsymbol{Y}_\alpha = \boldsymbol{X}_\alpha^{(1)} - \sqrt{\frac{n_1}{n_2}}\boldsymbol{X}_\alpha^{(2)} + \frac{1}{\sqrt{n_1 n_2}}\sum_{\beta=1}^{n_1}\boldsymbol{X}_\beta^{(2)} - \frac{1}{n_2}\sum_{\gamma=1}^{n_2}\boldsymbol{X}_\gamma^{(2)}, \quad \alpha = 1, \cdots, n_1$$

$$E(\boldsymbol{Y}_\alpha) = \boldsymbol{\mu}^{(1)} - \boldsymbol{\mu}^{(2)}.$$

Also

$$E(\boldsymbol{Y}_\alpha - E(\boldsymbol{Y}_\alpha))(\boldsymbol{Y}_\beta - E(\boldsymbol{Y}_\beta))'$$

$$= E\left[(\boldsymbol{X}_\alpha^{(1)} - \boldsymbol{\mu}^{(1)}) - \sqrt{\frac{n_1}{n_2}}(\boldsymbol{X}_\alpha^{(2)} - \boldsymbol{\mu}^{(2)}) + \frac{1}{\sqrt{n_1 n_2}}\sum_{\gamma=1}^{n_1}(\boldsymbol{X}_\gamma^{(2)} - \boldsymbol{\mu}^{(2)}) - \frac{1}{n_2}\sum_{\gamma=1}^{n_2}(\boldsymbol{X}_\gamma^{(2)} - \boldsymbol{\mu}^{(2)})\right]$$

$$\left[(\boldsymbol{X}_\beta^{(1)} - \boldsymbol{\mu}^{(1)}) - \sqrt{\frac{n_1}{n_2}}(\boldsymbol{X}_\beta^{(2)} - \boldsymbol{\mu}^{(2)}) + \frac{1}{\sqrt{n_1 n_2}}\sum_{\gamma=1}^{n_1}(\boldsymbol{X}_\gamma^{(2)} - \boldsymbol{\mu}^{(2)}) - \frac{1}{n_2}\sum_{\gamma=1}^{n_2}(\boldsymbol{X}_\gamma^{(2)} - \boldsymbol{\mu}^{(2)})\right]'$$

$$= \delta_{\alpha\beta}\boldsymbol{\Sigma}_1 + \frac{n_1}{n_2}\delta_{\alpha\beta}\boldsymbol{\Sigma}_2 + \boldsymbol{\Sigma}_2\left(-\frac{2}{n_2} + \frac{2}{n_2}\sqrt{\frac{n_1}{n_2}} + \frac{n_1}{n_1 n_2} - 2\frac{n_1}{\sqrt{n_1 n_2}\, n_2} + \frac{n_2}{n_2^2}\right)$$

$$= \delta_{\alpha\beta}(\boldsymbol{\Sigma}_1/n_1 + \boldsymbol{\Sigma}_2/n_2)n_1$$

where $\delta_{\alpha\beta} = 1(0)$ if $\alpha = (\neq)\beta$. Therefore, $\bar{\boldsymbol{Y}}$'s is normal with mean $\boldsymbol{\mu}^{(1)} - \boldsymbol{\mu}^{(2)}$ and covariance matrix $(\boldsymbol{\Sigma}_1/n_1 + \boldsymbol{\Sigma}_2/n_2)$. Hence, under $H_\circ$,

$$T^2 = n_1 \bar{\boldsymbol{Y}}' \boldsymbol{s}^{-1} \bar{\boldsymbol{Y}}$$

where

$$(n_1 - 1)\boldsymbol{s} = \sum_{\alpha=1}^{n_1}(\boldsymbol{Y}_\alpha - \bar{\boldsymbol{Y}})(\boldsymbol{Y}_\alpha - \bar{\boldsymbol{Y}})'$$

follows T^2 with $(n_1 - 1)$ d.f.

Scheffé (1943) suggested this procedure in the univariate case and showed that this technique gives the shortest confidence intervals obtained by using the t-distribution.

The above technique can be generalised to a k-sample problem as follows. Let

$$\boldsymbol{X}_\alpha^{(i)}, \quad \alpha = 1, \cdots, n_i;\ i = 1, \cdots, k$$

be k samples drawn from $N(\boldsymbol{\mu}^{(i)}, \boldsymbol{\Sigma})$ respectively. Consider testing

$$H_\circ : \sum_{i=1}^{k} l_i \boldsymbol{\mu}^{(i)} = \boldsymbol{\mu}$$

where $l_1, \cdots, l_k$ are given scalars and $\boldsymbol{\mu}$ a known vector. Suppose n_i's are unequal and n_1 is the smallest among them. Define

$$\boldsymbol{Y}_\alpha = l_1 \boldsymbol{X}_\alpha^{(1)} + \sum_{i=2}^{k} l_i \sqrt{\frac{n_1}{n_i}}\left(\boldsymbol{X}_\alpha^{(i)} - \frac{1}{n_1}\sum_{\beta=1}^{n_1}\boldsymbol{X}_\beta^{(i)} + \frac{1}{\sqrt{n_1 n_i}}\sum_{\gamma=1}^{n_i}\boldsymbol{X}_\gamma^{(i)}\right).$$

Then,

$$E(\boldsymbol{Y}_\alpha) = \sum_{i=1}^{k} l_i \boldsymbol{\mu}^{(i)},$$

$$D(\boldsymbol{Y}_\alpha) = \sum_{i=1}^{k} \frac{l_i^2 n_1}{n_i} \boldsymbol{\Sigma}_i$$

and $\boldsymbol{Y}_\alpha$'s $(\alpha = 1, \cdots, n_1)$ are independently distributed. Hence

$$\bar{\boldsymbol{Y}} \sim N_p \left(\sum_{i=1}^{k} l_i \boldsymbol{\mu}^{(i)}, \sum_{i=1}^{k} \frac{l_i^2}{n_i} \boldsymbol{\Sigma}_i \right).$$

Hence

$$n_1 \bar{\boldsymbol{Y}}' \boldsymbol{s}^{-1} \bar{\boldsymbol{Y}} \cap T^2(n_1 - 1)$$

under H_o,where

$$(n_1 - 1)\boldsymbol{S} = \sum_{\alpha=1}^{n_1} (\bar{\boldsymbol{Y}}_\alpha - \bar{\boldsymbol{Y}})(\bar{\boldsymbol{Y}}_\alpha - \bar{\boldsymbol{Y}})'.$$

18.6.7 Testing that Two Sub-vectors have Equal Means

Let $\boldsymbol{X} = (\boldsymbol{X}'_{(1)} \; \boldsymbol{X}'_{(2)})' \cap N_{2p}(\boldsymbol{\mu}, \boldsymbol{\Sigma})$ where

$$\boldsymbol{\mu} = \begin{bmatrix} \boldsymbol{\mu}_1 \\ \boldsymbol{\mu}_2 \end{bmatrix}, \quad \boldsymbol{\Sigma} = \begin{bmatrix} \boldsymbol{\Sigma}_{11} & \boldsymbol{\Sigma}_{12} \\ \boldsymbol{\Sigma}_{21} & \boldsymbol{\Sigma}_{22} \end{bmatrix}$$

Assume $\boldsymbol{X}_1, \boldsymbol{\mu}_1$ are of order $p \times 1$ and $\boldsymbol{\Sigma}_{11}$ of order $p \times p$. We want to test $H_o : \boldsymbol{\mu}_1 = \boldsymbol{\mu}_2$. Then $\boldsymbol{X}_{(1)} - \boldsymbol{X}_{(2)}$ follows N_p with mean $\boldsymbol{\mu}^{(1)} - \boldsymbol{\mu}^{(2)}$ and dispersion matrix

$$\boldsymbol{\Sigma}_{11} - \boldsymbol{\Sigma}_{12} - \boldsymbol{\Sigma}_{21} + \boldsymbol{\Sigma}_{22}.$$

Hence, under H_o,

$$n(\bar{\boldsymbol{X}}_{(1)} - \bar{\boldsymbol{X}}_{(2)})' \boldsymbol{s}^{-1} (\bar{\boldsymbol{X}}_{(1)} - \boldsymbol{X}_{(2)})$$

follows T^2 with $(n-1)$ d.f., when

$$(n-1)\boldsymbol{s} = \sum_{\alpha=1}^{n} (\boldsymbol{X}_{(1)\alpha} - \boldsymbol{X}_{(2)\alpha} - \bar{\boldsymbol{X}}_{(1)} + \bar{\boldsymbol{X}}_{(2)})(\boldsymbol{X}_{(1)\alpha} - \boldsymbol{X}_{(2)\alpha} - \bar{\boldsymbol{X}}_{(1)} + \bar{\boldsymbol{X}}_{(2)})'.$$

18.7 Hypothesis of Equality of Mean Vectors of K Multinormal Populations

Suppose 0_{n_t} $(x_{i\lambda}^{(t)}, i = 1, \cdots, p; \lambda = 1, \cdots, n_t; t = 1, \cdots, K)$ are K sets of samples from K multi-(p-) variate populations,

$$\boldsymbol{X}^{(t)} = ((x_{i\lambda}^{(t)}))_{p\times n_t}$$

being the group of n_t samples drawn from the tth population $N_p(\boldsymbol{\mu}^{(t)}, \boldsymbol{\Sigma})$, where $\boldsymbol{\mu}^{(t)} = (\mu_1^{(t)}, \cdots, \mu_p^{(t)})'$ $t = 1, \cdots, K$, all the populations having the same unknown dispersion matrix $\boldsymbol{\Sigma}$. We are interested in testing the null hypothesis $H_{\circ 1}(\boldsymbol{\mu}^{(1)} = \cdots = \boldsymbol{\mu}^{(K)})$. The parametric space Θ and its subspace under $H_{\circ 1}$ are, therefore,

$$\Theta = \begin{cases} \boldsymbol{\Sigma} = ((\sigma_{ij})) \text{ such that } \boldsymbol{\Sigma} \text{ is positive definite} \\ -\infty < \mu_i^{(t)} < \infty, i = 1, \cdots, p; t = 1, \cdots, K \end{cases}$$

$$\Theta_{01} = \begin{cases} \text{Subspace of } \Theta \text{ for which } \boldsymbol{\mu}_i^{(1)} = \cdots = \boldsymbol{\mu}_i^{(K)} = \boldsymbol{\mu}_i, \\ i = 1, \cdots, p; -\infty < \mu_i < \infty. \end{cases}$$

Let

$$S_{ij}^{(t)} = \sum_{\lambda=1}^{n_t}(x_{i\lambda}^{(t)} - \bar{x}_i^{(t)})(x_{j\lambda}^{(t)} - \bar{x}_j^{(t)}), \quad \bar{S}_{ij} = \sum_{t=1}^{K} S_{ij}^{(t)}$$

$$\bar{\boldsymbol{x}}^{(t)} = (\bar{x}_1^{(t)}, \cdots, \bar{x}_p^{(t)})'$$

$$S_{ij} = \sum_{t=1}^{K}\sum_{\lambda=1}^{n_t}(x_{i\lambda}^{(t)} - \bar{x}_i)(x_{j\lambda}^{(t)} - \bar{x}_j).$$

where

$$\bar{x}_i^{(t)} = \frac{1}{n_t}\sum_{\lambda} x_{i\lambda}, \quad \bar{x}_i = \frac{\Sigma n_t \bar{x}_i^{(t)}}{n}, \quad n = \sum_t n_t.$$

The likelihood function for all the samples 0_{n_t} is

$$f(\boldsymbol{x}^{(t)}, t = 1, \cdots, K) = \frac{1}{(2\pi)^{np/2}|\boldsymbol{\Sigma}|^{n/2}} \exp\left[-\frac{1}{2}\sum_{t=1}^{K}\sum_{i}^{p}\sum_{j=1}\sum_{\lambda=1}^{n_t}(x_{i\lambda}^{(t)} - \mu_i^{(t)})\sigma^{ij}(x_{j\lambda}^{(t)} - \mu_j^{(t)})\right]\prod_{i,\lambda,t} dx_{i\lambda}^{(t)}.$$

The exponent is

$$-\frac{1}{2}\sum_{i}^{p}\sum_{j=1}\sigma^{ij}\sum_{t=1}^{K} c_{ij}^{(t)},$$

where

$$c_{ij}^{(t)} = \sum_{\lambda=1}^{n_t} (x_{i\lambda}^{(t)} - \mu_i^{(t)})(x_{j\lambda}^{(t)} - \mu_j^{(t)}).$$

The m.l.e. of the parameters under Θ are

$$\hat{\mu}_i^{(t)} = \bar{x}_i^{(t)}, \quad ((\hat{\sigma}_{ij})) = \left(\left(\frac{\bar{S}_{ij}}{n}\right)\right)$$

when the likelihood is

$$\max_{\theta \in \Theta} L(\theta | \boldsymbol{x}^{(t)}, t = 1, \cdots, K) = \frac{e^{-np/2}}{(2\pi)^{np/2} \left|\frac{\bar{S}_{ij}}{n}\right|^{n/2}}.$$

Similarly, the m.l.e. of parameters under Θ_{01} are

$$\hat{\mu}_i = \bar{x}_i, \quad ((\hat{\hat{\sigma}}_{ij})) = \left(\left(\frac{S_{ij}}{n}\right)\right)$$

and hence,

$$\max_{\theta \in \Theta_{01}} L(\theta | \boldsymbol{x}^{(t)}, t = 1, \cdots, K) = \frac{e^{-np/2}}{(2\pi)^{np/2} \left|\frac{S_{ij}}{n}\right|^{n/2}}.$$

The LR test statistic is, therefore,

$$\lambda_1 = \frac{\left|\frac{\bar{S}_{ij}}{n}\right|^{n/2}}{\left|\frac{S_{ij}}{n}\right|^{n/2}} = \frac{|\bar{S}_{ij}|^{n/2}}{|S_{ij}|^{n/2}}$$

when

$$\lambda_1^{2/n} = \frac{|\bar{S}_{ij}|}{|S_{ij}|} = Z \text{ (say)}.$$

The critical region is $\{\boldsymbol{X} : Z \le z_0\}$ where z_0 is such that the size of the test is α.

18.8 Hypothesis of Equality of Dispersion Matrices of K Multinormal Populations

Consider the K sets of samples from K p-variate populations,

$$\mathbf{X}^{(t)} = ((x_{i\lambda}^{(t)}))_{p \times n_t}, \quad t = 1, \cdots, K$$

being the sampled observations from the tth population $N_p(\boldsymbol{\mu}^{(t)}, \boldsymbol{\Sigma}^{(t)})$. We are interested in testing of null hypothesis $H_{o2}(\boldsymbol{\Sigma}^{(1)} = \cdots = \boldsymbol{\Sigma}^{(K)})$ whatever $\boldsymbol{\mu}^{(t)}$ $(t = 1, \cdots, K)$.

The space Θ is the parameter space in which $\mathbf{\Sigma}^{(t)}$ is positive definite and $\boldsymbol{\mu}^{(t)}$ any ector. The space Θ_{02} is the subspace of Θ for which $\mathbf{\Sigma}^{(1)} = \cdots = \mathbf{\Sigma}^{(K)}$. Under Θ the .l.e. of $\boldsymbol{\mu}^{(t)}$ and $\mathbf{\Sigma}^{(t)}$ are

$$\hat{\boldsymbol{\mu}}^{(t)} = \bar{\boldsymbol{x}}^{(t)}, \quad \hat{\sum}^{(t)} = \frac{1}{n_t}\boldsymbol{S}^{(t)} \quad \text{where } \boldsymbol{S}^{(t)} = ((S_{ij}^{(t)})).$$

nder Θ_{02} the m.l.e. of $\boldsymbol{\mu}^{(t)}$ remains also $\bar{\boldsymbol{x}}^{(t)}$. Under H_{o2} the likelihood function is

$$L(\boldsymbol{X}) = \frac{1}{(2\pi)^{pn/2}|\mathbf{\Sigma}|^{n/2}} \exp\left\{-\frac{1}{2}\sum_t\sum_\lambda \left(\boldsymbol{x}_\lambda^{(t)} - \bar{\boldsymbol{x}}^{(t)}\right)' \mathbf{\Sigma}^{-1}\left(\boldsymbol{x}_\lambda^{(t)} - \bar{\boldsymbol{x}}^{(t)}\right)\right\}$$

here

$$\boldsymbol{x}_\lambda^{(t)} = \left(x_{1\lambda}^{(t)}, \cdots, x_{p\lambda}^{(t)}\right)', \quad \bar{\boldsymbol{x}}^{(t)} = \left(\bar{x}_1^{(t)}, \cdots, \bar{x}_p^{(t)}\right)'.$$

y lemma 18.6.2 the m.l.e. of $\mathbf{\Sigma}$ is

$$\hat{\sum}_0 = \frac{\bar{\boldsymbol{S}}}{n} \quad \text{where } \bar{\boldsymbol{S}} = ((\bar{S}_{ij})),$$

$_{ij}$ being defined in section 18.7 and the maximum of the likelihood function is

$$\frac{1}{(2\pi)^{np/2}\left|\hat{\sum}_0\right|^{n/2}} e^{-np/2}.$$

he likelihood ratio statistic is, therefore,

$$\lambda_2 = \frac{\prod_{t=1}^{K}\left|\hat{\sum}^{(t)}\right|^{n_t/2}}{\left|\hat{\sum}_0\right|^{n/2}}$$

$$= \frac{\prod_{t=1}^{K}|\boldsymbol{S}^{(t)}|^{n_t/2}}{|\bar{\boldsymbol{S}}|^{n/2}} \cdot \frac{n^{np/2}}{\prod_{t=1}^{K} n_t^{pn_t/2}}.$$

he critical region is $\{\boldsymbol{X} : \lambda \leq \lambda_0(\alpha)\}$ where $\lambda_0(\alpha)$ is such that the size of the test is α.

8.9 Hypothesis that K Multinormal Populations are Identical

n section 18.7 we considered the null hypothesis

$$H_{o1} : \boldsymbol{\mu}^{(1)} = \cdots = \boldsymbol{\mu}^{(K)} \quad \text{given that } \mathbf{\Sigma}^{(1)} = \cdots = \mathbf{\Sigma}^{(K)}.$$

The null hypothesis considered in section 18.8 was

$$H_{o2}: \boldsymbol{\Sigma}^{(1)} = \cdots = \boldsymbol{\Sigma}^{(K)} \quad \text{whatever} \quad \boldsymbol{\mu}^{(t)} \quad (t = 1, \cdots, K).$$

Now we consider the null hypothesis

$$H_o: H_{o12}: \boldsymbol{\mu}^{(1)} = \cdots = \boldsymbol{\mu}^{(K)}, \quad \boldsymbol{\Sigma}^{(1)} = \cdots = \boldsymbol{\Sigma}^{(K)} \tag{18.9.1}$$

which is a combination of H_{o1} and H_{o2}. As in section 18.8 let $\boldsymbol{x}_\lambda^{(t)}$ be a sample from $N(\boldsymbol{\mu}^{(t)}, \boldsymbol{\Sigma}^{(t)})$ $(t = 1, \cdots, K)$. Then Θ is the unrestricted parameter space of $\{\boldsymbol{\mu}^{(t)}, \boldsymbol{\Sigma}^{(t)}\}$ $(t = 1, \cdots, K)$, where $\boldsymbol{\Sigma}^{(t)}$ is positive definite and Θ_{012} is the subspace of Θ restricted by (18.9.1).

The hypothesis H_{o2} of section 18.8 is that $\boldsymbol{\theta} \in \boldsymbol{\Theta}_{02}$ given that $\boldsymbol{\theta} \in \boldsymbol{\Theta}$. The hypothesis H_{o1} of section 8.7 is that $\boldsymbol{\theta} \in \boldsymbol{\Theta}_{01} \cap \boldsymbol{\Theta}_{02}$ given that $\boldsymbol{\theta} \in \boldsymbol{\Theta}_{02}$. The hypothesis H_{o12} is that $\theta \in \boldsymbol{\Theta}_{01} \cap \boldsymbol{\Theta}_{02}$ given then $\boldsymbol{\theta} \in \boldsymbol{\Theta}$.

Lemma 18.9.1 *Let $L(\boldsymbol{y}|\boldsymbol{\theta})$ be the likelihood of $\boldsymbol{\theta}$ based on observation $\boldsymbol{y}$, $\boldsymbol{\theta} \in \Omega$. Let H_a be the hypothesis $\boldsymbol{\theta} \in \Omega_a \subset \Omega$. Let H_b be the hypothesis $\boldsymbol{\theta} \in \Omega_b \cap \Omega_a$ given that $\boldsymbol{\theta} \in \Omega_a$. Let H_{ab} be the hypothesis $\boldsymbol{\theta} \in \Omega_a \cap \Omega_b$ given that $\boldsymbol{\theta} \in \Omega$. If $\lambda_a, \lambda_b, \lambda_{ab}$, the likelihood criteria for testing H_a, H_b, H_{ab} respectively are uniquely determined, then*

$$\lambda_{ab} = \lambda_a \lambda_b. \tag{18.9.2}$$

Proof: We have

$$\lambda_a = \frac{\sup\limits_{\boldsymbol{\theta} \in \Omega_a} L(\boldsymbol{y}|\boldsymbol{\theta})}{\sup\limits_{\boldsymbol{\theta} \in \Omega} L(\boldsymbol{y}|\boldsymbol{\theta})}$$

$$\lambda_b = \frac{\sup\limits_{\boldsymbol{\theta} \in \Omega_b \cap \Omega_a} L(\boldsymbol{y}|\boldsymbol{\theta})}{\sup\limits_{\theta \in \Omega_a} L(\boldsymbol{y}|\boldsymbol{\theta})}$$

$$\lambda_{ab} = \frac{\sup\limits_{\boldsymbol{\theta} \in \Omega_b \cap \Omega_a} L(\boldsymbol{y}|\boldsymbol{\theta})}{\sup\limits_{\boldsymbol{\theta} \in \Omega} L(\boldsymbol{y}|\boldsymbol{\theta})}.$$

Hence, equality (18.9.2) follows obviously.

For H_{o12} the likelihood criterion is therefore

$$\lambda = \lambda_1 \lambda_2 = \left(\prod_{t=1}^{K} \frac{|\boldsymbol{S}^{(t)}|^{n_t/2}}{n_t^{pn_t/2}} \right) \frac{n^{pn/2}}{|\boldsymbol{S}|^{n/2}} \tag{18.9.3}$$

where $\boldsymbol{S} = ((S_{ij})), S_{ij}$ being defined in section 18.7.

Now,

$$\begin{aligned} \boldsymbol{S} &= \sum_{t=1}^{K}\sum_{\lambda=1}^{n_t}\left(\boldsymbol{x}_\lambda^{(t)} - \bar{\boldsymbol{x}}\right)\left(\boldsymbol{x}_\lambda^{(t)} - \bar{\boldsymbol{x}}\right)' \\ &= \bar{\boldsymbol{S}} + \sum_{t=1}^{K} n_t\left(\bar{\boldsymbol{x}}^{(t)} - \bar{\boldsymbol{x}}\right)\left(\bar{\boldsymbol{x}}^{(t)}. - \bar{\boldsymbol{x}}\right)'. \end{aligned}$$

The critical region is defined by

$$\{\boldsymbol{X} : \lambda \leq \lambda(\alpha)\}$$

where $\lambda(\alpha)$ is such that the size of the critical region is α.

18.10 The Hypothesis of Independence of Sets of Variables in a Multivariate Normal Population

Let $\boldsymbol{X} = (X_1, \cdots, X_p)'$ follow a p-variate normal population $N(\boldsymbol{\mu}, \boldsymbol{\Sigma})$. Suppose the variates are grouped into r groups as follows:

$$G_1 : (X_1, \cdots, X_{p_1}), \quad G_2 : (X_{p_1+1}, \cdots, X_{p_1+p_2}), \cdots,$$
$$G_r : (X_{p_1+\cdots+p_{r-1}+1}, \cdots, X_p), \quad p = p_1 + p_2 + \cdots + p_r.$$

We want to test the hypothesis that the groups of variates are mutually independent i.e., $\sigma_{ij} = 0$ for all i, j not belonging to the same group. Let $\boldsymbol{\Sigma}^0$ be the structure of $\boldsymbol{\Sigma}$ when H_o holds. Therefore, here

Θ : space of all σ_{ij}'s for which $\boldsymbol{\Sigma} = ((\sigma_{ij}))$ is positive definite

Θ_0 : subspace of Θ for which $\boldsymbol{\Sigma} = \boldsymbol{\Sigma}^0$.

A sample $\boldsymbol{x}_\lambda = (x_{1\lambda}, \cdots, x_{p\lambda})'$ $(\lambda = 1, \cdots, n)$ is drawn from the population.

Define

$$S_{ij} = \sum_{\lambda}(x_{i\lambda} - \bar{x}_i)(x_{j\lambda} - \bar{x}_j), \quad i, j = 1, \cdots, p.$$

The maximum likelihood function for variations of the parameters over Θ is

$$\frac{e^{-np/2}}{(2\pi)^{np/2}\left|\dfrac{S_{ij}}{n}\right|^{n/2}}; \tag{18.10.1}$$

The maximum likelihood function for variations of the parameters over Θ_0 is

$$\frac{e^{-np/2}}{(2\pi)^{np/2}\left|\dfrac{S^*_{ij}}{n}\right|^{n/2}} \tag{18.10.2}$$

when $S_{ij}^* = S_{ij}(0)$ if i, j belong to the same group of variates (otherwise). Therefor $|S_{ij}^*|$ is the product of mutually exclusive principal minors

$$\prod_{u=1}^{r} |S_{i(u)j(u)}|,$$

($|S_{i(u)j(u)}|$ being the determinant of all S_{ij}'s associated with the uth group). Similarly

$$|\mathbf{\Sigma}^0| = \prod_{u=1}^{r} |\sigma_{i(u)j(u)}|.$$

The likelihood ratio for testing $H_o(\mathbf{\Sigma} = \mathbf{\Sigma}^0)$ is

$$\lambda = \frac{|S_{ij}|^{n/2}}{|S_{ij}^*|^{n/2}}. \tag{18.10.3}$$

The statistic $\lambda^{2/n} = w$ (say), may be used for testing H_o. It can be shown that

$$E(W^h) = \prod_{t=1}^{r} \prod_{i=1}^{p_t} \frac{\Gamma\left(\frac{n-i}{2}\right)}{\Gamma\left(\frac{n-i}{2} + h\right)} \left[\prod_{i=1}^{p} \frac{\Gamma\left(\frac{n-i}{2} + h\right)}{\Gamma\left(\frac{n-i}{2}\right)}\right]. \tag{18.10.4}$$

As a special case, suppose we wish to test the hypothesis that X_1 is independent of $X_2, X_3, \cdots, X_p$. In this case, $p_1 = 1, p_2 = p - 1$. The w-criterion is

$$w = \frac{|r_{ij}|}{R_{11}} = 1 - R^2$$

where R_{11} is the minor of r_{11} in $|r_{ij}|$ and R is the sample multiple correlation coefficient of X_1 on $X_2, X_3, \cdots, X_p$. The hth moment of w for this case is, from (18.10.4),

$$\frac{\Gamma\left(\frac{n-1}{2}\right)\Gamma\left(\frac{n-p}{2} + h\right)}{\Gamma\left(\frac{n-1}{2} + h\right)\Gamma\left(\frac{n-p}{2}\right)}.$$

The probability element of w would be

$$\frac{\Gamma\left(\frac{n-1}{2}\right)}{\Gamma\left(\frac{n-p}{2}\right)\Gamma\left(\frac{p-1}{2}\right)} w^{\frac{n-p}{2}-1}(1-w)^{\frac{p-1}{2}-1}\, dw.$$

Letting $w = 1 - R^2$, we easily find the distribution law of R^2 to be

$$\frac{1}{B\left(\frac{n-p}{2}, \frac{p-1}{2}\right)} (R^2)^{\frac{p-1}{2}-1}(1-R^2)^{\frac{n-p}{2}-1} d(R^2)$$

when the hypothesis of independence of X_1 and $X_2, \cdots, X_p$ is true which is equivalent to the hypothesis that the population multiple correlation coefficient $\rho_{1.2\cdots p}$ is zero.

18.11 Wilk's Criterion

In testing of different hypotheses for a normal population model $N_p(\mu, \boldsymbol{\Sigma})$ the likelihood ratio test statistic involves

$$W = \frac{|\hat{\boldsymbol{\Sigma}}_{H_\circ \cup H}|}{|\hat{\boldsymbol{\Sigma}}_{H_\circ}|}. \tag{18.11.1}$$

Since a sample variance-covariance matrix is an unbiased estimate of $\boldsymbol{\Sigma}$, W would be a ratio of the determinants of two sample variance-covariance matrices and more specifically would be of the form

$$W = \frac{|\boldsymbol{B}|}{|\boldsymbol{B}+\boldsymbol{C}|} \tag{18.11.2}$$

where $\boldsymbol{B}, \boldsymbol{C}$ are two symmetric $p \times p$ matrices, $\boldsymbol{B} = ((b_{ij}))$

$$C = ((c_{ij})), \ \ \boldsymbol{B} \sim W(p, k_1, \boldsymbol{\Sigma}), \ \ \boldsymbol{C} \sim W(p, k_2, \boldsymbol{\Sigma}),$$

$\boldsymbol{B} \overset{ind}{\sim} \boldsymbol{C}$, $k_1 \geq p$. The statistic (18.11.2) is known as Wilk's statistic with (k_1, k_2) d.f. We shall find the distribution of W. Note that we can express

$$\left.\begin{aligned} \boldsymbol{B} &= \sum_{\lambda=1}^{k_1} \boldsymbol{Z}_\lambda \boldsymbol{Z}'_\lambda \\ \boldsymbol{C} &= \sum_{\lambda=k_1+1}^{k_1+k_2} \boldsymbol{Z}_\lambda \boldsymbol{Z}'_\lambda \end{aligned}\right\} \tag{18.11.3}$$

where $\boldsymbol{Z}_\lambda = (Z_{1\lambda}, \cdots, Z_{p\lambda})'$, $\lambda = 1, \cdots, k_1 + k_2$ are independent random variables each distributed as $N_p(\boldsymbol{0}, \boldsymbol{\Sigma})$.

The conditional distribution of $Z_{p\lambda}$ given $(Z_{1\lambda}, \cdots, Z_{p-1\lambda})$ is univariate normal with mean

$$\beta_1 z_{1\lambda} + \cdots + \beta_{p-1} z_{p-1\lambda} \text{ (say)}$$

and variance $\frac{1}{\sigma^{pp}}$ where $\boldsymbol{\Sigma}^{-1} = ((\sigma^{ij}))$.

The least square estimate of $\beta_1, \cdots, \beta_{p-1}$ from the observations $\boldsymbol{Z}_\lambda$, $\lambda = 1, \cdots, k-1$ is obtained by minimising

$$\sum_{\lambda=1}^{k_1} (z_{p\lambda} - \beta_1 z_{1\lambda} - \cdots - \beta_{p-1} z_{p-1\lambda})^2 \tag{18.11.4}$$

and the minimum value of sum of squares is

$$\sum_{\lambda=1}^{k_1} (z_{p\lambda} - b_1 z_{1\lambda} - \cdots - b_{p-1} z_{p-1\lambda})^2 = \frac{|\boldsymbol{B}|_p}{|\boldsymbol{B}|_{p-1}}$$

where $|\boldsymbol{H}|_i$ denotes the determinant formed by the first i rows and first i columns of $\boldsymbol{H}$.

This ratio is distributed as χ_1^2/σ^{pp} where χ_1^2 is a central $\chi^2_{(k_1-p+1)}$ (vide section 14.6).

Similarly, if we consider the least square estimate of $\beta_1, \cdots, \beta_{p-1}$ on the basis o observations $(z_{1\lambda}, \cdots, z_{p-1\lambda})$, $\lambda = 1, \cdots, k_1 + k_2$, then denoting the estimated value of β as b_j' $(j = 1, \cdots, p-1)$,

$$\sum_{\lambda=1}^{k_1+k_2} \left(z_{p\lambda} - b_1' z_{1\lambda} - \cdots - b_{p-1}' z_{p-1\lambda}\right)^2 = \frac{|\boldsymbol{B}+\boldsymbol{C}|_p}{|\boldsymbol{B}+\boldsymbol{C}|_{p-1}}$$

is distributed as $(\chi_1^2 + \chi_2^2)/\sigma^{pp}$ where χ_2^2 is a central chi-square with k_2 d.f. and i independent of χ_1^2.

The ratio

$$\frac{|\boldsymbol{B}|_p}{|\boldsymbol{B}|_{p-1}} \Big/ \frac{|\boldsymbol{B}+\boldsymbol{C}|_p}{|\boldsymbol{B}+\boldsymbol{C}|_{p-1}}$$

is of the form $\chi_1^2/(\chi_1^2 + \chi_2^2)$ and hence is distributed as a beta $B\left(\frac{k_1-p+1}{2}, \frac{k_2}{2}\right)$ variable.

Now,

$$W = \left[\frac{|\boldsymbol{B}|_p}{|\boldsymbol{B}|_{p-1}} \div \frac{|\boldsymbol{B}+\boldsymbol{C}|_p}{|\boldsymbol{B}+\boldsymbol{C}|_{p-1}}\right]\left[\frac{|\boldsymbol{B}|_{p-1}}{|\boldsymbol{B}|_{p-2}} \div \frac{|\boldsymbol{B}+\boldsymbol{C}|_{p-1}}{|\boldsymbol{B}+\boldsymbol{C}|_{p-2}}\right] \cdots [|\boldsymbol{B}|_1 \div |\boldsymbol{B}+\boldsymbol{C}|_1]$$
$$= \omega_p \times \omega_{p-1} \times \cdots \times \omega_1 \text{ (say)}$$

where ω_i is a distributed as $B\left(\frac{k_1-i+1}{2}, \frac{k_2}{2}\right)$.

Now ω_p is distributed independently of $Z_{i\lambda}$ $(i = 1, \cdots, p-1)$ and hence is independen of ω_i $(i = 1, \cdots, p-1)$. Similarly, ω_{p-1} is distributed independently of $Z_{i\lambda}$ $(i = 1, \cdots, p-2)$, and hence is independent of ω_i $(i = 1, \cdots, p-2)$. The statistic W i the product of p independent beta random variables ω_i $(i = 1, \cdots, p)$.

It can be shown that

(1) if $k_2 = 1$, then

$$\frac{(1-W)/p}{W/(k_1-p+1)} \cap F_{(p, k_1-p+1)}.$$

(2) if $k_2 = 2$, then

$$\frac{(1-\sqrt{W})/p}{\sqrt{W}/(k_1-p+1)} \cap F_{2p, 2k_1-2p+2}.$$

(3) if $p = 1$, then W is a Beta variable and

$$\frac{(1-W)/k_1}{W/k_2} \cap F_{k_1, k_2}$$

For testing a hypothesis $H_\circ\left(\boldsymbol{\mu}^{(1)} = \cdots = \boldsymbol{\mu}^{(g)}\right)$ in sampling from g multivaria normal populations $N_p(\boldsymbol{\mu}^{(i)}, \boldsymbol{\Sigma})$, where $\boldsymbol{x}_\lambda^{(i)} = \left(x_{1\lambda}^{(i)}, \cdots, x_{p\lambda}^{(i)}\right)'$ $(\lambda = 1, \cdots, n_i)$, i = $1, \cdots, g$, the null hypothesis is rejected if the ratio of the generalized variances

$$\wedge^* = \frac{|\boldsymbol{W}|}{|\boldsymbol{B}+\boldsymbol{W}|} = \frac{\left|\sum_{i=1}^{g}\sum_{\lambda=1}^{n_i}\left(\boldsymbol{x}_\lambda^{(i)} - \bar{\boldsymbol{x}}^{(i)}\right)\left(\boldsymbol{x}_\lambda^{(i)} - \bar{\boldsymbol{x}}^{(i)}\right)'\right|}{\left|\sum_{i=1}^{g}\sum_{\lambda=1}^{n_i}\left(\boldsymbol{x}_\lambda^{(i)} - \bar{\boldsymbol{x}}\right)\left(\boldsymbol{x}_\lambda^{(i)} - \bar{\boldsymbol{x}}\right)'\right|}$$

is too small. Here

$$B = \sum_{i=1}^{g} n_i \left(\bar{x}^{(i)} - \bar{x}\right)\left(\bar{x}^{(i)} - \bar{x}\right)'$$

and the notations have obvious interpretations.

The quantity $\wedge^*$ was proposed originally by Wilks (1932). The exact distribution of $\wedge^*$ can be derived for the special cases listed in table 18.1. For other cases and large sample sizes, a modification of $\wedge^*$ due to Bartlett (1938) can be used to test H_o.

Table 18.1. *Distribution of Wilk's* $\wedge^* = |W|/|B+W|$

No. of variables	No. of groups	Sampling distribution for multivariate normal data
$p=1$	$g \geq 2$	$\left(\frac{\sum n_i - g}{g-1}\right)\left(\frac{1-\wedge^*}{\wedge^*}\right) \sim F_{g-1, \Sigma n_i - g}$
$p=2$	$g \geq 2$	$\left(\frac{\sum n_i - g - 1}{g-1}\right)\left(\frac{1-\sqrt{\wedge^*}}{\sqrt{\wedge^*}}\right) \sim F_{2(g-1), 2(\Sigma n_i - g - 1)}$
$p \geq 1$	$g=2$	$\left(\frac{\sum n_i - p - 1}{p}\right)\left(\frac{1-\wedge^*}{\wedge^*}\right) \sim F_{p, \Sigma n_i - p - 1}$
$p \geq 1$	$g=3$	$\left(\frac{\sum n_i - p - 2}{p}\right)\left(\frac{1-\sqrt{\wedge^*}}{\sqrt{\wedge^*}}\right) \sim F_{2p, 2(\Sigma n_i - p - 2)}$

Bartlett has shown that if H_o is true and $\Sigma n_i = n$ is large,

$$-\left(n - 1 - \frac{(p+g)}{2}\right)\ln \wedge^* = -\left(n - 1 - \frac{(p+g)}{2}\right)\ln\left(\frac{|W|}{|B+W|}\right) \qquad (18.11.5)$$

has approximately a chi-square distribution with $p(g-1)$ d.f. Consequently, for Σn_i large, we reject H_o at significance level α if

$$-\left(n - 1 - \frac{p+g}{2}\right)\ln\left(\frac{|W|}{|B+W|}\right) > \chi^2_{p(g-1);\alpha}.$$

18.12 Principle of Invariance and Invariant Tests

Many statistical problems show some kinds of symmetry. For such problems one should consider statistical procedures which are also symmetric in nature. For instance, suppose $X_1, \cdots, X_n$ are independent r.v.'s having p.d.f.'s $f_{\theta_1}(x_1), \cdots, f_{\theta_n}(x_n)$, respectively. For testing $H_o(\theta_1 = \cdots = \theta_n)$ against the alternative hypothesis that θ's are not all equal one should consider a test statistic which is symmetric in $x_1, \cdots, x_n$, as otherwise acceptance or rejection of the hypothesis would depend on the system of numbering (which is often arbitrary) of these variables.

Let X be a r.v. distributed according to a p.d.f. f_θ, $\theta \in \Theta$ and let g be a $1:1$ transformation of the sample space χ onto itself. Let gX be the transformed r.v. which takes the value gx when $X = x$. Suppose that the distribution of gX is $f_{\theta'}$, where θ' also lies in Θ. The element θ' of Θ associated with θ in this manner will be denoted as $\bar{g}\theta$ Thus, for any measurable subset A of χ

$$P_\theta[X : gX \in A] = P_{\bar{g}\theta}[X \in A] \quad (18.12.1)$$

$$\text{i.e., } P_\theta[g^{-1}A] = P_{\bar{g}\theta}(A). \quad (18.12.2)$$

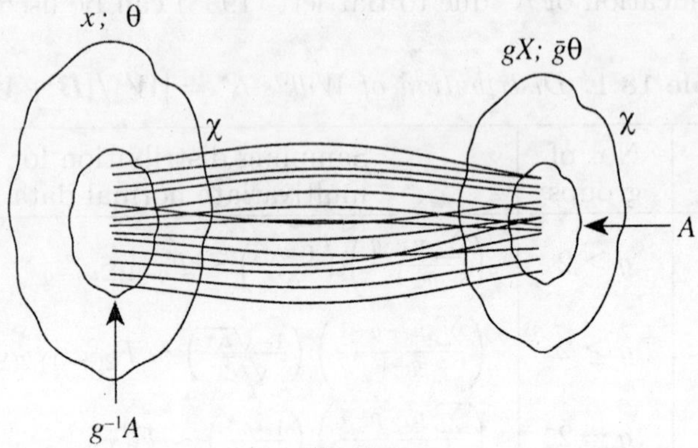

Fig. 18.1

(Fig. 18.1). The parameter set Θ remains invariant under g (or is preserved by g) if $\bar{g}\theta \in \Theta$ for all θ in Θ and if, further, for any θ' in Θ there exists a $\theta \in \Theta$ such that $\bar{g}\theta = \theta'$ These two conditions may be expressed by stating

$$\bar{g}\Theta = \Theta. \quad (18.12.3)$$

The transformation $\bar{g}$ of Θ onto itself defined in this way is $1 : 1$ provided the distributions f_θ corresponding to different values of θ are distinct.

Lemma 18.12.1 *Let g, g' be two transformations preserving Θ. Then the transformations $g'g$ and g^{-1} defined by*

$$(g'g)x = g'(gx) \quad and \quad g(g^{-1}x) = x \ \forall\, x \in \mathcal{X}$$

also preserve Θ and satisfy

$$\overline{g'g} = \bar{g}' \cdot \bar{g} \quad and \quad \overline{(g^{-1})} = (\bar{g})^{-1}. \quad (18.12.4)$$

Proof: If the distribution of X is f_θ, that of gX is $f_{\bar{g}\theta}$, and that of $(g'g)X = g'(gX)$ is $f_{\bar{g}'\cdot\bar{g}\theta}$. This establishes the first equation of (18.12.4). The proof of the second equation follows analogously.

We shall say that the problem of testing $H_o : \theta \in \Theta_0$ against $H_1 : \theta \in \Theta_1$ (where $\Theta_0 \cup \Theta_1 = \Theta$) remains invariant under a transformation g if $\bar{g}$ preserves both Θ_0 and Θ_1 so that the equation

$$\bar{g}\Theta_0 = \Theta_0 \quad (18.12.5)$$

satisfied in addition to (18.12.3). Let $\mathcal{G}$ be a class of transformations satisfying (18.12.3) d (18.12.5) and let G be the smallest class of transformations containing $\mathcal{G}$ and such at $g, g' \in G$ implies $gg' \in G$ and $g^{-1} \in G$. Then G is a group of transformations all of iich preserve both Θ and Θ_0 by lemma 18.12.1. Any class $\mathcal{G}$ of transformations leaving e problem invariant can therefore, be extended to a group G.

It follows from lemma 18.12.1 that the class of induced transformations $\bar{g}$ form a oup $\bar{G}$. The two equations (18.12.4) state that $\bar{G}$ is a homomorphism of G. When e transformation $g \in G$ preserve the sample space $\mathcal{X}$ and the induced transformation $\in \bar{G}$ preserve both Θ and Θ_0, it is usual to restrict attention to tests δ which are also mmetric i.e., which satisfy

$$\delta(gx) = \delta(x), \ \forall\, x \in \mathcal{X} \text{ and } g \in G. \tag{18.12.6}$$

A test δ satisfying (18.12.6) is said to be invariant under G.

If a problem is invariant under a group of transformations, the principle of invariance stricts attention to invariant tests. We first characterise the totality of all invariant sts.

vo points $x_1, x_2 \in \mathcal{X}$ are considered equivalent under G,

$$x_1 \sim x_2(\text{mod}G),$$

here exists a transformation $g \in G$ for which $x_2 = gx_1$. A point x traces out an orbit all transformations g of G are applied to it; that is, the orbit containing x consists of e totality of points gx with $g \in G$. From the definition of invariance it follows that a action is invariant iff it takes a constant value on each orbit.

A function T is said to be maximal invariant if it is invariant and if

$$T(x_1) = T(x_2) \text{ implies } x_2 = gx_1 \text{ for some } g \in G.$$

ieorem 18.12.1 *Let $T(x)$ be a maximal invariant statistic with respect to G. Then a cessary and sufficient condition for a statistic ϕ to be invariant is that it depends on only through $T(x)$, i.e., there exists a measurable function h for which $\delta(x) = h[T(x)]$ all x.*

Proof: If $\phi(x) = h[T(x)]$ for all x, then $\phi(g(x)) = h[T(gx)] = h[T(x)] = \phi(x)$ so that s invariant. On the other hand, if ϕ is invariant, and if $T(x_1) = T(x_2)$ then $x_2 = gx_1$ some g and therefore $\phi(x_2) = \phi(x_1)$.

Example 18.12.1 *Let $x = (x_1, \cdots, x_n)'$ and let G be the set of $n!$ permutations of the ordinates of x. Consider the set of order statistic $T(x) = (x_{(1)} \leq x_{(2)} \leq \cdots \leq x_{(n)})$. permutation of values $x_1, \cdots, x_n$ in x does not change set of values of the co-ordinates d therefore, preserve the order statistic and hence, the values $x_{(i)} i = 1, \cdots, n$. On the er hand if for two points x, x', $T(x) = T(x')$ then $x_{(i)} = x'_{(i)}$ $\forall\, i$ and hence obviously can be obtained from x through a permutation of co-ordinates. Therefore, the order tistic is maximal invariant under G.*

▶ **Example 18.12.2** *Let $x = (x_1, \cdots, x_n)$ and let G be the group of transformations*

$$gx = (x_1 + c, x_2 + c, \cdots, x_n + c), \quad c \neq 0, \quad -\infty < c < \infty.$$

Then the set of differences $y = (x_1 - x_n, x_2 - x_n, \cdots, x_{n-1} - x_n)$ is invariant under G

Consider another point $x' = (x'_1, \cdots, x'_n)$

$$gx' = (x'_1 + c, x'_2 + c, \cdots, x'_n + c).$$

$$y' = (x'_1 - x'_n, \cdots, x'_{n-1} - x'_n).$$

Suppose that

$$y = y' \text{ i.e., } x_i - x_n = x'_i - x'_n \text{ for } i = 1, \cdots, n-1.$$

Putting $x'_n - x_n = c$, one has $x'_i = x_i + c$ for $i = 1, \cdots, n$.
Therefore,

$$x' = gx, \quad g \in G.$$

Therefore, the statistic y is maximal invariant.

It can be seen that the each of the statistic $z = (x_1 - x_2, x_2 - x_3, \cdots, x_n - x_{n-}$
$h = (x_1 - \bar{x}, \cdots, x_n - \bar{x})$ is maximal invariant under G.

▶ **Example 18.12.3** *Let $x = (x_1, \cdots, x_n)$ and the G be the group of transformations*

$$gx = (x_1 c, \cdots, x_n c), \quad c \neq 0.$$

Consider the statistic $y = T(x) = (x_1/x_n, \cdots, x_{n-1}/x_n)$. Clearly, $T(x)$ is invaria under G.

Consider another point $x' = (x'_1, \cdots, x'_n)$.

$$gx' = (cx'_1, \cdots, cx'_n).$$

$$y' = (x'_1/x'_n, \cdots, x'_{n-1}/x'_n).$$

Suppose $y' = y$ i.e., $x_i/x_n = x'_i/x'_n, i = 1, \cdots, n-1$.

Putting $x'_n/x_n = c$, $x_i/x'_i = 1/c$ or $x'_i = cx_i$ i.e., $x' = gx, g \in G$. Thus, the statisti is maximal invariant.

▶ **Example 18.12.4** *Let $x = (x_1, \cdots, x_n)$ where x_i's $(i = 1, \cdots, n)$ are all distin Consider $G = \{g\}$ where*

$$gx = (x'_1, \cdots, x'_n), x'_i = f(x_i)$$

f being continuous and a strictly increasing function of x_i. Consider the statistic

$$y = T(x) = (r_1, r_2, \cdots, r_n) \text{ when } r_i \text{ is rank of } x_i (i = 1, \cdots, n).$$

learly y is invariant under G. Consider another point $Z = (Z_1, \cdots, Z_n)$.

$$gZ = (Z'_1, \cdots, Z'_n) \text{ when } Z'_i = f(Z_i).$$

$$y' = T(Z) = (r_1^0, r_2^0, \cdots, r_n^0), \text{ when } r_i^0 \text{ is rank of } Z_i (i = 1, \cdots, n).$$

Suppose $r_i = r_i^0$ i.e., rank of x_i = rank of $Z_i, i = 1, \cdots, n$. Clearly then Z_i can be)tained from x_i by a transformation of the form $Z_i = x_i + (Z_i - x_i) = f(x_i)$ so that $= gx, g \in G$.

The class of all invariant functions can be obtained as the totality of all functions a maximal invariant statistic $T(x)$. Therefore, the class of all invariant tests is the tality of all tests that depend only on the maximal invariant statistic T.

Example 18.12.5 *Let* $\mathbf{X} = (X_1, \cdots, X_n)$ *and suppose the density of* $\mathbf{X}$ *is* $f_i\ (x_1 - \theta,$ $\cdot, x_n - \theta)$ *under* $H_i\ (i = 0, 1)$, $-\infty < \theta < \infty$.

Consider the group G of transformations

$$gx = (x_1 + c, \cdots, x_n + c), -\infty < c < \infty.$$

The induced transformation on the parameter space is

$$\bar{g}\theta = \theta + c.$$

Under this group of transformations clearly both H_o and H_1 remain invariant (as remains the same, $i = 0, 1$). By example 18.12.2, a maximal invariant statistic is $= (X_1 - X_n, \cdots, X_{n-1} - X_n)$. The distribution of Y is independent of θ and under $_i$ has the density

$$\int_{-\infty}^{\infty} f_i(y_1 + Z, \cdots, y_{n-1} + Z, Z)\, dZ$$

ıere $Y_i = X_i - X_n$ takes the value $y_i (i = 1, \cdots, n-1)$ and $Z = X_n - \theta$ takes the lue Z.

When, considered in terms of the statistic Y, the problem of testing H_o against $_1$ becomes one of testing a simple hypothesis against a simple alternative. The rresponding MP test is independent of θ, and therefore, UMP (with respect to variations θ) among all invariant tests. Its rejection region by Neyman-Pearson lemma is

$$\frac{\int_{-\infty}^{\infty} f_i(y_1 + Z, \cdots, y_{n-1} + Z, Z)\, dZ}{\int_{-\infty}^{\infty} f_0(y_1 + Z, \cdots, y_{n-1} + Z, Z)\, dZ} = \frac{\int_{-\infty}^{\infty} f_i(x_1 + u, \cdots, x_n + u)\, du}{\int_{-\infty}^{\infty} f_0(x_1 + u, \cdots, x_n + u)\, du} > C.$$

Example 18.12.6 *Let* $X_1, \cdots, X_n$ *be a random sample from* $N(\mu, \sigma^2)$. *The hypothesis* $: \sigma^2 \geq \sigma_0^2$ *is invariant under the transformation* $X'_i = X_i + c, -\infty < c < \infty$. *ıe sufficient statistic* $(\bar{X}, S^2)$ *reduce to* $(Y = \bar{X} + c, S'^2 = S^2)$ *under this group of ınsformations.* S^2 *is a maximal invariant statistic. It follows that there exists a* UMP *st with the region of rejection* $S^2 \leq c$. *Since* S^2 *is a maximal invariant statistic the test* UMP *among the class of all invariant tests.*

▶ **Example 18.12.7** *Let $(X_1, \cdots, X_m)$, $(Y_1, \cdots, Y_n)$ be samples from $N(\mu, \sigma_1)$, $N(\mu_1, \sigma$ respectively. A set of sufficient statistic is $T_1 = \bar{X}, T_2 = \bar{Y}$, $T_3 = \sqrt{\Sigma(X_i - \bar{X}}$ $T_4 = \sqrt{\Sigma(Y_i - \bar{Y})^2}$. The problem of testing $H : \sigma_2^2/\sigma_1^2 \leq \lambda_0$ remains invariant unc the group of transformations G_1 such that*

$$T_1' = \bar{X}_1 + c_1, \quad T_2' = \bar{Y} + c_2, \quad T_3' = T_3, \quad T_4' = T_4, \quad -\infty < c_1, \quad c_2 < \infty$$

and also under the group of transformations G_2 which involve a common change of sc for all the four statistics. A maximal invariant with respect to G_1 is (T_3, T_4). In space of this maximal invariant, the group of scale changes induces the transformatic $T_3'' = cT_3$, $T_4'' = cT_4$, $c > 0$ and for this the maximal invariant is the ratio T_4/T_3. T statistic $Z = \sigma_1^2\{T_4^2/(n-1)\}/\{T_3^2/(m-1)\}\sigma_2^2$ has a F distribution. The rejection reg for the test is $Z > C$. This is a UMP *test. This is therefore,* UMP *among all invaric tests. Also this is a* UMP *unbiased test.*

Sufficient statistic provide a simplification of the problem by reducing the sam space. This process involves no change in the parameter space. The principle invariance reduces the data to a maximal invariant statistic whose distribution m depend only on a function of the parameter and thus generally reduces the parame space. This is stated in the following theorem.

Theorem 18.12.2 *If $T(x)$ is invariant under G and if $v(\theta)$ is maximal invariant un the induced group $\bar{G}$, then the distribution of $T(X)$ depends only on $v(\theta)$.*

Proof: Let $v(\theta_1) = v(\theta_2)$. Then $\theta_2 = g\theta_1$ and hence

$$P_{\theta_1}\{T(X) \in B\} = P_{\theta_1}\{T(gX) \in B\} = P_{\bar{g}\theta_1}\{T(X) \in B\} = P_{\theta_2}\{T(X) \in B\}.$$

Thus, in example 18.12.6, the maximal invariants $T(X)$ and $\sigma = v(\theta)$ under G a $\bar{G}$ are real-valued and the family of p.d.f.'s $p_\sigma(t)$ of T has monotone likelihood ra For testing $H_o : \sigma \leq \sigma_0$, there exists a UMP test among those depending only on T, a hence a UMP invariant test. Its rejection region is $t \geq c$ when $P(T|\sigma_0) = \alpha$.

18.13 Monotonicity of Power Functions

When a null hypothesis (simple or composite) is tested it is a natural requirement t the power of the test should increase with increase in the divergence of the true value o from the value θ_0 under H_o. Thus, for simplicity, if H_o is $(\theta = \theta_0)$ against $H(\theta \neq \theta_0)$ tl one would require that the power $P_\theta(w)$ should be a monotonically increasing funct of $|\theta - \theta_0|$. It is easy to check that all UMP or UMPU tests derived so far have this prope (vide, for example, examples 17.8.1, 17.8.3).

Theorem 18.13.1 *A consistent estimate provides a consistent test but the convers not necessarily true.*

If the statistic t_n based on the sample size n be a consistent estimate of θ then the following critical region w_n supplies a consistent test for $H_\circ : \theta = \theta_0$ against $H : \theta = \theta_1$.

$w(n)$:

(i) $\{x : t_n \geq c_n\}$ if $\theta_1 > \theta_0$ when $P\{t_n \geq c_n|\theta_0\} = \alpha$;

(ii) $\{t_n \leq c'_n\}$ if $\theta_1 < \theta_0$ when $P\{t_n \leq c'_n|\theta_0\} = \alpha$.

Consider the case when $\theta_1 > \theta_0$. Assume that $H_\circ$ is true.

Since t_n is a consistent estimate of θ, $P_{\theta_0}\{|t_n - \theta_0| > \varepsilon\} < \eta$ for $n > n_0(\varepsilon, \eta)$, when $\varepsilon, \eta \to 0$ as $n \to \infty$.

Now, as $\alpha = P_{\theta_0}\{t_n \geq c_n\} = P_{\theta_0}\{|t_n - \theta_0| \geq |c_n - \theta_0|\}$ is a fixed quantity, we must have $|c_n - \theta_0| < \varepsilon$ where $\varepsilon \to 0$ as $n \to \infty$. Hence c_n converges to θ_0 as $n \to \infty$. Hence, given an $\varepsilon(> 0)$, however small, there exists a value n_1 of n such that

$$c_n < \theta_0 + \epsilon \text{ for } n \geq n_1(\varepsilon). \tag{18.13.1}$$

Again, when θ_1 is the true value of θ, t_n must also converge to θ_1 i.e.,

$$P\{|t_n - \theta_1| < \varepsilon|\theta_1\} > 1 - \eta \text{ for } n \geq n_2(\varepsilon, \eta)$$
$$\text{i.e., } P\{t_n > \theta_1 - \varepsilon|\theta_1\} > 1 - \eta \text{ for } n \geq n_2(\varepsilon, \eta). \tag{18.13.2}$$

Let $n_0 = \max(n_1, n_2)$. Hence (18.13.1) and (18.13.2) hold simultaneously for $n \geq n_0$. Let us now choose ε such that $\theta_0 + \varepsilon < \theta_1 - \epsilon$. Then from (18.13.1) and (18.13.2),

$$P\{t_n \geq c_n|\theta_1\} > P\{t_n > \theta_0 + \varepsilon|\theta_1\} > P\{t_n > \theta_1 - \varepsilon|\theta_1\} > 1 - \eta \text{ for } n \geq n_0.$$

Therefore,

$$\lim_{n\to\infty} P\{t_n \geq c_n|\theta_1\} = 1.$$

Hence, the critical region w_n provides a consistent test for $H : \theta_1 > \theta_0$. Therefore, a consistent estimate provides a consistent test.

To prove that the converse is not necessarily true we take an example.

Let $x_1, \cdots, x_n$ is be i.i.d $N(0, 1)$. It is required to test $H_\circ : (\mu = 0)$ against $H(\mu \neq 0)$. We use the statistic

$$T = \sqrt{\frac{1}{n}\sum_1^n x_i^2} \text{ for } H_\circ.$$

Now, $E(x^2) = 1 + \mu^2$. Hence, by Khintchen's w.l.l.n.,

$$\frac{1}{n}\sum_1^n x_i^2 \xrightarrow{P} 1 + \mu^2;$$

therefore, $T \xrightarrow{P} \sqrt{1 + \mu^2} = \theta$. Now $H_\circ : \mu = 0$ against $H : \mu = 1$ is equivalent to $H_\circ : \theta = 1$ against $H : \theta \neq 1$ and since T is a consistent estimate of θ it provides a consistent test for $H : \theta = 1$. However, T is not a consistent estimate of μ. Hence, we have a consistent test for μ which does not provide a consistent estimate for the same.

Exercise

1. Let X, Y have independent exponential distributions with parameters $\mu, \theta\mu$ respectively. The null hypothesis is $H_\circ : \theta = 1$, the alternative hypothesis is $H : \theta = \theta_1 > 1$. Show that the statistic $T = X + Y$ is sufficient for μ when $\theta = 1$. Considering that the conditional distribution of Y given $T = t$ is uniform over $(0, t)$ show that the UMP test of size α for $H_\circ$ against H is to reject $H_\circ$ if $y/t \leq c$ (whatever be the value of μ).

2. Let $X_1, \cdots, X_n$ be a random sample of size n from a $N(\mu, \sigma^2)$ distribution. The null hypothesis is $H_\circ(\sigma^2 = \sigma_0^2)$ where σ_0^2 is a known constant, the alternative hypothesis is $H(\sigma^2 \neq \sigma_0^2)$ where μ is unspecified. Show that the likelihood ratio statistic is

$$\lambda = (\tfrac{\hat{\sigma}^2}{\sigma_0^2})^{n/2} \exp\left[-\tfrac{n}{2}(\tfrac{\hat{\sigma}^2}{\sigma_0^2} - 1)\right]$$

where $\hat{\sigma}^2 = \Sigma_{i=1}^n (x_i - \bar{x})^2/n$. Verify that λ is not a monotonic function of $\hat{\sigma}^2$ but has a single maximum. Hence, show that an equivalent test may be based on $n\hat{\sigma}^2/\sigma_0^2$, which when $\sigma^2 = \sigma_0^2$ has a χ^2_{n-1} distribution and explain how to construct a size α-test.

3. A random sample of 3 observations is drawn from the distribution with p.d.f.

$$f(x|\theta_1, \theta_2) = \tfrac{1}{\theta_2} \exp\{-[x - \theta_1]/\theta_2\}, \quad \theta_2 > 0, x > \theta_1.$$

If Y_1, Y_2, Y_3 are the order statistics, show that the p.d.f. of $U_1 = Y_1$, $U_2 = Y_3 - Y_1$, $U_3 = Y_2 + Y_3 - 2Y_1$ is

$$h(u_1, u_2, u_3) = \tfrac{6}{\theta_2^3} \exp\{-[u_3 + 3u_1 - 3\theta_1]/\theta_2\}, \quad \theta_1 < u_1, \quad \tfrac{u_3}{2} < u_2 < u_3.$$

Hence, or otherwise, deduce that U_3, U_1 are independent and that U_1 is sufficient for θ_1. Find the MP similar test of $H_\circ(\theta_2 = \theta_2')$ against $H(\theta_2 = \theta_2'' > \theta_2')$ and discuss whether it is UMP similar against $\theta_2 > \theta_2'$. Discuss briefly how this test is to be constructed.

4. Let $(X_1, \cdots, X_4), (Y_1, \cdots, Y_4)$ be two independent random samples of size 4 from uniform distributions $U[0, \theta], U[0, \phi]$ respectively. Show that the likelihood ratio test of $H_\circ(\theta = \phi)$ against $H(\theta \neq \phi)$ may be reduced to the form: reject $H_\circ$ unless $K^{-1}X_{(4)} < Y_{(4)} < KX_{(4)}$, where $X_{(1)} \leq \cdots \leq X_{(4)}$ and $Y_{(1)} \leq \cdots \leq Y_{(4)}$. Write down the joint density of $X_{(4)}$ and $Y_{(4)}$. Determine the value of K that gives a test of size 5%, and show that if $\theta = 3, \phi = 2$, this test has power approximately equal to 1/8.

5. Let $(X_1, \cdots, X_n)$ be a random sample from $N(\mu_1, \sigma_1^2)$ distribution and $(Y_1, \cdots, Y_m)$ be an independent random sample from $N(\mu_2, \sigma_2^2)$ distribution, when $-\infty < \mu_i < \infty$, $\sigma_i^2 > 0$ $(i = 1, 2)$. The null hypothesis is that $\sigma_1^2 = \sigma_2^2$, the alternative is that $\sigma_1^2 \neq \sigma_2^2, \mu_1, \mu_2$ being unspecified under both hypotheses. Show that the likelihood

ratio test can be based on the statistic

$$F = \frac{\sum_1^n (X_i - \bar{X})^2/(n-1)}{\sum_1^m (Y_i - \bar{Y})^2/(m-1)}.$$

Show that under H_o, F has a Snedecor's F distribution with $\{(n-1),(m-1)\}$ d.f.

6. Let $X_1, \cdots, X_n$ be a random sample from $N(\mu_1, \sigma^2)$ and $Y_1, \cdots, Y_m$ be an independent random sample from $N(\mu_2, \sigma^2)$, where $-\infty < \mu_i < \infty$, $(i = 1, 2), \sigma^2 > 0$. The null hypothesis is that $\mu_1 = \mu_2$, the alternative is that $\mu_1 \neq \mu_2$, σ^2 being unspecified. Show that the unrestricted likelihood is maximised when

$$\mu_1 = \hat{\mu}_1 = \bar{x}, \quad \mu_2 = \hat{\mu}_2 = \bar{y},$$
$$\sigma^2 = \hat{\sigma}^2 = \left[\sum_{i=1}^n (x_i - \bar{x})^2 + \sum_{i=1}^m (y_i - \bar{y})^2\right] / (m+n),$$

and that the likelihood under null hypothesis is maximised when

$$\mu_1 = \mu_2 = \tilde{\mu} = (n\bar{x} + m\bar{y})/(m+n),$$
$$\sigma^2 = \tilde{\sigma}^2 = \left[\sum_{i=1}^n (x_i - \tilde{\mu})^2 + \sum_{i=1}^m (y_i - \tilde{\mu})^2\right] / (m+n).$$

Hence show that the likelihood ratio statistic can be written as

$$\left[\frac{n+m-2}{(n+m-2)+t^2}\right]^{(n+m)/2}$$

where

$$T = (\bar{X} - \bar{Y}) / \left[\left(\frac{1}{n} + \frac{1}{m}\right)\left\{\frac{\Sigma(X_i - \bar{X})^2 + \Sigma(Y_i - \bar{Y})^2}{n+m-2}\right\}\right]^{1/2}.$$

Show that T has Student's t distribution with $(n+m-2)$ d.f. when $\mu_1 = \mu_2$.

7. Define a similar test. Let X, Y be independent r.v.'s having the Poisson distributions with parameters λ and μ, respectively. Obtain the UMP size α similar test for testing $\lambda = \mu$ against $\lambda > \mu$.

8. Suppose x_1, x_2 are independent random variables having Binomial distributions $B(n_1, p_1)$, $B(n_2, p_2)$ respectively $(0 \leq p_i \leq 1,\ i = 1, 2)$. Consider $H_o(p_1 = p_2)$. Show that the likelihood ratio test is given by

$$\lambda(\boldsymbol{x}) = \left(\frac{x}{n}\right)^x \left(1 - \frac{x}{n}\right)^{n-x} \Big/ \prod_{i=1}^2 \left(\frac{x_i}{n_i}\right)^{x_i} \left(1 - \frac{x_i}{n_i}\right)^{n_i - x_i}$$

where $x = x_1 + x_2$, $n = n_1 + n_2$. Show also that the limiting distribution of $-2\log_e \lambda$ as $n_1, n_2 \to \infty$ is a $\chi^2_{(1)}$ distribution if H_o is true.

9. Generalise the above problem and obtain its solution where $x_1, \cdots, x_k$ are independent r.v.'s from binomial distributions $B(n_1, p_1), \cdots, B(n_k, p_k)$.

10. Let X_1, X_2, X_3, X_4 be independent normal variables with a common variance σ^2 and mean given by

$$E(X_1) = \theta_1 + \theta_2,$$
$$E(X_2) = \theta_1 - \theta_2$$
$$E(X_3) = \theta_1$$
$$E(X_4) = \theta_2$$

where θ_1, θ_2 and σ^2 are all unknown, $-\infty < \theta_1, \theta_2 < \infty$, $\sigma^2 > 0$. Derive the LR test statistic for testing $H_o : \theta_1 = \theta_2$ versus $H_1 : \theta_1 \neq \theta_2$.

11. Let $(X_1, \cdots, X_m)$ and $(Y_1, \cdots, Y_n)$ be the random samples from the densities $\theta_1 e^{-\theta_1 x}$ and $\theta_2 e^{-\theta_2 y}$ respectively. Assume that the samples are independent.

(a) Derive the likelihood ratio test for $H_o : \theta_1 = \theta_2$ against $H : \theta_1 \neq \theta_2$.

(b) Show that the test can be expressed in terms of the statistic $T = \Sigma X_i / \Sigma Y_i$ and argue that the distribution of T does not involve any unknown parameter where H_o is true.

12. *Tests for independence in a $r \times s$ contingency table.* Suppose $(n_{ij}, i = 1, \cdots, r;\ j = 1, \cdots, s)$ is a $(rs - 1)$-dimensional random vector having the multinomial distribution

$$\frac{n!}{\prod_{i=1}^{r}\prod_{j=1}^{s} n_{ij}!} \prod_{i=1}^{r}\prod_{j=1}^{s} p_{ij}^{n_{ij}}$$

where $\sum_i \sum_j n_{ij} = n$, $p_{ij} > 0 \ \forall\ i, j$ and $\sum_i \sum_j p_{ij} = 1$.

Let H_o be the hypothesis $(p_{ij} = p_i q_j)$ when $\sum_j p_{ij} = p_i, \sum_i p_{ij} = q_j$ and $\sum_i p_i = \sum_j q_j = 1$.

Show that the LR test is given by

$$\lambda = \frac{\prod_{i=1}^{r}\left(\frac{n_{i0}}{n}\right)^{n_{i0}} \prod_{j=1}^{s}\left(\frac{n_{oj}}{n}\right)^{n_{oj}}}{\prod_{i=1}^{r}\prod_{j=1}^{s}\left(\frac{n_{ij}}{n}\right)^{n_{ij}}}$$

when $n_{i0} = \sum_j n_{ij}, n_{0j} = \sum_i n_{ij}$. Show also that the limiting distribution of $-2 \log_e \lambda$ is $\chi^2_{(r-1)(s-1)}$ if H_o is true.

13. Suppose $(n_1, \cdots, n_k)$ is a $(k-1)$-dimensional random vector having multinomial distribution

$$\frac{n!}{\prod_{i=1}^{k} n_i!} \prod_{i=1}^{k} p_i^{n_i}$$

when $\Sigma n_i = n$, $p_i > 0 \; \forall \; i$ and $\Sigma p_i = 1$. Consider $H_o(p_1 = p_{10}, \cdots, p_k = p_{ko})$. Show that the LR test is

$$\lambda = \left[\prod_{i=1}^{k} \left(\frac{p_{i0}}{\hat{p}_i}\right)^{\hat{p}_i}\right]^n$$

when $\hat{p}_i = n_i/n (i = 1, \cdots, k)$. Show that the limiting distribution of $-2 \log_e \lambda$ as $n \to \infty$ is $\chi^2_{(k-1)}$ if H_o is true.

14. Let $(X_1, \cdots, X_n)$ be a random sample from $N(\mu, \sigma^2)$ distribution. Let $\Omega = \{(\mu, \sigma^2), -\infty < \mu < \infty, \sigma^2 \leq \sigma_0^2\}$, $\Omega_0 = \{(\mu, \sigma^2), -\infty < \mu < \infty, \sigma^2 = \sigma_0^2\}$. Show that using the LR test for $H_o(\theta \in \Omega_0)$ against $H(\theta \in \Omega - \Omega_0\}$ when $\theta = (\mu, \sigma^2)$ is equivalent to using the ratio $(n-1)s^2/\sigma_0^2$ which has the $\chi^2_{(n-1)}$ distribution if H_o is true. Hence the critical region is the set of values of s^2 for which $(n-1)s^2/\sigma_0^2 > \chi^2_{(n-1),\alpha}$.

15. Let $\zeta = \{\boldsymbol{x} = (x_1, \cdots, x_n) : x_i > 0 \; \forall \; i\}$. Let G be the group of transformations $g\boldsymbol{x} = (cx_1, \cdots, cx_n), c > 0$. Show that a maximal invariant statistic under G is (sgn $x_n, x_1/x_n, \cdots, x_{n-1}/x_n$), where sgn of Z is $+1$ or -1 according as Z is positive or negative. **[Lehmann, (1959)]**

16. Let $\boldsymbol{X} = (X_1, \cdots, X_n)$ have a p.d.f. $\frac{1}{\theta^n} f\left[\left(\frac{x_1-\mu}{\theta}\right), \cdots, \left(\frac{x_n-\mu}{\theta}\right)\right]$ where $\mu (\in R_1)$, $\theta (> 0)$ are unknown and f is a even function. Show that the problem of testing $f = f_0$ against $f = f_1$ remains invariant under the transformation $x'_i = ax_i + b (i = 1, \cdots, n)$, $a \neq 0, -\infty < b < \infty$ and the MP invariant test is given by the rejection region

$$\frac{\int_{-\infty}^{\infty} \int_0^{\infty} v^{n-1} f_1(vx_1 + u, \cdots, vx_n + u)\, du\, dv}{\int_{-\infty}^{\infty} \int_0^{\infty} v^{n-2} f_0(vx_1 + u, \cdots, vx_n + u)\, du\, dz} > c$$

[Lehmann, (1959)]

17. Let $(X_1, \cdots, X_m)$, $(Y_1, \cdots, Y_n)$ be respectively samples from exponential populations

$$f(x|\mu, \sigma) = \tfrac{1}{\sigma} e^{-(x-\mu)/\sigma}, \; x \geq \mu$$
$$f(y|\nu, \theta) = \tfrac{1}{\theta} e^{-(y-\nu)/\theta}, \; y \geq \nu.$$

Show that for testing $H_o(\theta/\sigma \leq \Delta)$ against $H(\theta/\sigma > \Delta)$, there exists a UMP invariant test with respect to the group $G : X'_i = aX_i + b, Y'_j = aY_j + c, a > 0,$

$-\infty < b,\ c < \infty$ and its rejection region is

$$\sum_j [y_j - y_{(1)}] \Big/ \sum_i [x_i - x_{(1)}] > c$$

where $y_{(1)} = \min_i y_i,\ x_{(1)} = \min_i x_i$.

Show that this test is also UMP unbiased. **[Lehmann, (1959)]**

18. Let s_i^2 be the sample variance of independent samples of sizes n_i, from the distributions $N(\mu_i, \sigma_i^2)$, $-\infty < \mu_i < \infty$, $\sigma_i^2 > 0$ $(i = 1, \cdots, k)$. Show that the LR test of the hypothesis $H_\circ(\sigma_1^2 = \cdots = \sigma_k^2)$ is given by

$$\lambda = \prod_{i=1}^{k} \left[\frac{(n_i - 1)s_i^2}{n_i s_0^2} \right]^{n_i/2}$$

where

$$s_0^2 = \frac{\sum_{i=1}^{k} (n_i - 1)s_i^2}{\sum_{i=1}^{k} n_i}.$$

19. Consider a random sample of size 1 from the p.d.f. $f(x|\theta) = \frac{2}{\theta^2}(\theta - x),\ 0 < x < \theta$. Derive the LR test for $H_\circ(\theta = \theta_0)$ against $H(\theta \neq \theta_0)$.

20. Suppose $(x_1, \cdots, x_{n_1})$, $(y_1, \cdots, y_{n_2})$ are independent samples from Poisson distributions $P(\lambda_1), P(\lambda_2)$ respectively, $\lambda_1 > 0, \lambda_2 > 0$. Consider the null hypothesis $H_\circ(\lambda_1 = \lambda_2)$. Show that the LR test is given by $\lambda = \frac{\bar{Z}^{n\bar{Z}}}{\bar{x}^{n_1\bar{x}} \cdot \bar{y}^{n_2\bar{y}}}$ where $\bar{x}, \bar{y}$ are the means of the two samples and $\bar{Z} = \frac{n_1\bar{x} + n_2\bar{y}}{n_1 + n_2}$, $n = n_1 + n_2$. Show that the limiting distribution of $-2\log_e \lambda$ as $n_1, n_2 \to \infty$ is the $\chi^2_{(1)}$ distribution if $H_\circ$ is true. Generalise the result to k samples.

21. Show that a necessary and sufficient condition for densities $p_\theta(x)$ to have monotone likelihood ratio in x, is that $\frac{\partial^2 \log p_\theta(x)}{\partial\theta\partial x}$ if they exist are greater than or equal to zero for all θ and x.

22. The unbiased estimates of the mean vector $\boldsymbol{\mu}$ and variance-covariance matrix $\boldsymbol{\Sigma}$ of a trivariate normal distribution are given by

$$\bar{\boldsymbol{x}} = (3, 41, 2.23, 1.49)'$$

$$\boldsymbol{s} = \begin{bmatrix} 16.90 & 22.01 & 12.35 \\ & 51.72 & 19.46 \\ & & 28.73 \end{bmatrix}.$$

Test the following hypotheses at 5% level:

(i) $\mu_1 = \mu_2 = \mu_3$; (ii) $\mu_1 - \mu_2 = \mu_2 - \mu_3$; (iii) $\mu_1 = \mu_2 = \mu_3 = 3$.

23. Let $(X_1, \cdots, X_m), (Y_1, \cdots, Y_n)$ be independently normally distributed with means μ_1 and μ_2 and variances σ_1^2, σ_2^2 respectively, and consider the hypotheses $H : \sigma_2 \leq \sigma_1$ against $K : \sigma_1 < \sigma_2$. Show that if μ_1, μ_2 are known there exists a UMP test given by the rejection region $\Sigma(Y_i - \mu_2)^2/\Sigma(X_i - \mu_1)^2 \geq C$. Show also that no UMP test exists when μ_1, μ_2 are unknown.

24. Let $(X_1, \cdots, X_n)$ be a sample from an exponential density $f(x; a, b) = ae^{-a(x-b)}, x \geq b$.

    (a) Determine the UMP test for testing $H_\circ(b = b_0)$ agianst $H(b \neq b_0)$ when a is assumed known.

    (b) Determine the UMP test for testing $H_\circ(a = a_0, b = b_0)$ against the alternative $H(a > a_0,\ b < b_0)$. Explain the (very unusual) existence in this case of a UMP test in two-parameter problem.

[*Hints:* (i) The variable $Y_i = e^{-aX_i}$ are a sample from the uniform distribution on $(0, e^{-ab})$].

[Lehmann, (1959)]

25. Let X, Y be independently distributed according to one-parameter exponential family so that their joint distribution is given by

$$f(x_1, x_2|\theta_1, \theta_2) = \left\{C(\theta_1)e^{\theta_1 T(x)}\right\}\left\{C(\theta_2)e^{\theta_2 U(y)}\right\}.$$

Show that a UMPU test does not exist for $H_\circ : (\theta_1 = a, \theta_2 = b)$ against the alternative $H(\theta_1 \neq a \text{ or } \theta_2 \neq b)$.

26. Let $\boldsymbol{X}_1, \cdots, \boldsymbol{X}_N (N > p)$ be a random sample from $N_p(\boldsymbol{\mu}, \boldsymbol{\Sigma})$ when $\boldsymbol{\mu}$ and $\boldsymbol{\Sigma}$ are unknown. Show that $T^2 = N\left[\bar{\boldsymbol{X}}'\boldsymbol{S}^{-1}\bar{\boldsymbol{X}} - \dfrac{(\bar{\boldsymbol{X}}'\boldsymbol{S}^{-1}\boldsymbol{1})^2}{\boldsymbol{1}'\boldsymbol{S}^{-1}\boldsymbol{1}}\right]$ can be used for testing $H_\circ(\mu_1 = \cdots = \mu_p)$ against H[not all μ's are equal], when $\bar{\boldsymbol{X}}$ and $\boldsymbol{S}$ are the sample mean vector and the unbiased estimator of $\boldsymbol{\Sigma}$ respectively and $\boldsymbol{1} = (1, \cdots, 1)'$. Derive the non-null distribution of the test statistic.

27. Let $\bar{\boldsymbol{X}}$ and $\boldsymbol{A}$ represent the mean vector and the matrix of corrected sum of squares (S.S.) and sum of products (S.P.) of observations in a sample of size n from $N_p(\boldsymbol{\mu}, \boldsymbol{\Sigma})$. Show that Hotelling's T^2 for testing $H_\circ : \boldsymbol{\mu} = \boldsymbol{\mu}_0$, given by

$$\frac{T^2}{n-1} = n(\bar{\boldsymbol{X}} - \boldsymbol{\mu}_0)'\boldsymbol{A}^{-1}(\bar{\boldsymbol{X}} - \boldsymbol{\mu}_0)$$

is obtained as

$$\max_{\boldsymbol{a}} n(n-1)\frac{[\boldsymbol{a}'(\bar{\boldsymbol{X}} - \boldsymbol{\mu}_0)]^2}{\boldsymbol{a}'\boldsymbol{A}\boldsymbol{a}}$$

where '$\boldsymbol{a}$' is a non-null $p \times 1$ vector.

A statistician, by mistake, uses the statistic

$$\frac{T'^2}{n-1} = n(\bar{\boldsymbol{X}} - \boldsymbol{\mu}_0)' \boldsymbol{A}_0^{-1} (\bar{\boldsymbol{X}} - \boldsymbol{\mu}_0)$$

where $\boldsymbol{A}_0$ is the matrix of S.S. and S.P. of the observations measured about the true means μ_0.

Find the relation between T^2 and T'^2 and derive the distribution of T'^2, assuming the distribution of T^2.

28. Let $\bar{\boldsymbol{x}} = (\bar{x}_1, \cdots, \bar{x}_p)'$ and $\boldsymbol{s} = ((s_{ij}))$ $(i, j = 1, \cdots, p)$ be based on n random observations from $N_p(\boldsymbol{\mu}, \boldsymbol{\Sigma})$. Let $\boldsymbol{x} = (x_1, \cdots, x_p)'$ be an additional observation from the same distribution. Show that $(\boldsymbol{x} - \bar{\boldsymbol{x}})$ is distributed according to $N_p(\boldsymbol{0}, \frac{n+1}{n}\boldsymbol{\Sigma})$. Verify that $\frac{n}{n+1}(\boldsymbol{x} - \bar{\boldsymbol{x}})'\boldsymbol{s}^{-1}(\boldsymbol{x} - \bar{\boldsymbol{x}})$ has T^2 distribution with $(n-1)$ d.f.

29. The following table gives the estimated dispersion matrices of two anthropometric characters: maximum head length x_1 and maximum head breadth x_2 for Muslims in two districts of Bangla Desh:

District : Dacca *Sample size* : 357 Estimated Dispersion Matrix			*District* : Mymensingh *Sample size* : 299 Estimated Dispersion Matrix		
	x_1	x_2		x_1	x_2
x_1	41.09	7.05	x_1	40.37	4.16
x_2		27.72	x_2		25.37

Are the two dispersion matrices significantly different?

Chapter 19

Confidence Interval Estimation

19.1 Introduction

In the theory of point estimation, we are interested in estimating the value of a parametric function $\gamma(\theta)$ by a single value t based on the observations $x_1, \cdots, x_n$ when the samples are drawn from a density $f(x;\theta)$, $\theta \in \Theta$. In practice, one is not generally interested in finding a point estimate of $\gamma(\theta)$, but a set of values, say, $H(\theta)$, such that $H(\theta)$ contains the true value of the parameter $\gamma(\theta)$ with a high probability. Problems of this type are called problems of confidence interval (set) estimation. When the set $H(\theta)$ is an interval, it is called a confidence interval.

In section 19.2 we introduce the basic concepts relating to estimation of a confidence interval. Section 19.3 considers the method of pivotal quantity for finding a confidence interval. The method of finding a confidence belt for θ based on a suitable statistic is considered in Section 19.4. Section 19.5 considers approximate confidence intervals in large samples. Theory of confidence sets is considered in sections 19.6, 19.7 and 19.8. The concluding section considers tolerance intervals.

19.2 Basic Concepts

Definition 19.2.1 (***Confidence Interval***). *Let $\boldsymbol{X} = (X_1, \cdots, X_n)$ be a random sample from a population with density function $f(x;\theta)$, $\theta \in \Theta$. Let $T_1 = t_1(X_1, \cdots, X_n)$, $T_2 = t_2(X_1, \cdots, X_n)$ be two statistics satisfying $T_1 \leq T_2$ such that*

$$P_\theta\left[T_1 \leq \gamma(\theta) \leq T_2\right] = 1 - \alpha \;\; \forall\, \theta \in \Theta \tag{19.2.1}$$

where $(1-\alpha)$ does not depend on θ. Then the random interval (T_1, T_2) is called the $100(1-\alpha)\%$-confidence interval for $\gamma(\theta)$. The quantity $(1-\alpha)$ is called the confidence coefficient of this interval. The statistic T_1, T_2 are respectively called lower and upper confidence limits for $\gamma(\theta)$. For a given sample observation $\boldsymbol{x} = (x_1, \cdots, x_n)$, the values of the statistic $T_1(\boldsymbol{x})$, $T_2(\boldsymbol{x})$ are the confidence limits for $\gamma(\theta)$.

Usually α is taken to be a very small quantity, $0.05, 0.01$ (say) so that $1-\alpha$ is $0.95, 0.99$, etc. In some cases, any of the two statistic T_1, T_2 may be a constant; however T_1, T_2 cannot both be constants.

Definition 19.2.2 (*One-sided confidence interval*). *Let* $\boldsymbol{X}$ *be a random sample from a population with p.d.f.* $f(x;\theta), \theta \in \Theta$. *Let* $T_1 = t_1(\boldsymbol{X})$ *be a statistic such that*

$$P_\theta[T_1 \leq \gamma(\theta)] = 1-\alpha \ \forall\, \theta \in \Theta \tag{19.2.2}$$

where α *does not depend on* θ. *Then* T_1 *is called the one-sided lower confidence limit for* $\gamma(\theta)$. *Thus, the interval* (T_1, ∞) *covers* $\gamma(\theta)$ *with probability* $1-\alpha$. *Similarly, let* $T_2 = t_2(\boldsymbol{X})$ *be a statistic such that*

$$P_\theta[T_2 \geq \nu(\theta)] = 1-\alpha \ \forall\, \theta \in \Theta \tag{19.2.3}$$

where α *does not depend on* θ. *Then* T_2 *is called the one-sided upper confidence limit for* $\gamma(\theta)$. *Here, the interval* $(-\infty, T_2)$ *covers* $\gamma(\theta)$ *with probability* $(1-\alpha)$.

Note that θ may be a vector of parameters.

In making probability statements (19.2.1), (19.2.2), (19.2.3) we do not mean that $\gamma(\theta)$ is a random variable. The statement (19.2.1) means that the probability is $(1-\alpha)$ that the random interval (T_1, T_2) will cover $\gamma(\theta)$, whatever the true value of the parameter θ may be. Similarly (19.2.2) [(19.2.3)] means that the statistic $T_1(\boldsymbol{X})$ will be less than [greater than] or equal to $\gamma(\theta)$ with probability $(1-\alpha)$ whatever the true value of θ may be.

► **Example 19.2.1** *Let* $(X_1, \cdots, X_n)$ *be a random sample from* $N(\mu, \sigma^2)$ *population when* σ^2 *is known. We want to find a* $100(1-\alpha)\%$ *confidence interval for* μ. *It is known that*

$$P_\mu\left[\bar{X} - \tau_{\alpha/2}\frac{\sigma}{\sqrt{n}} < \mu < \bar{X} + \tau_{\alpha/2}\frac{\sigma}{\sqrt{n}}\right] = 1-\alpha \ \forall\, \mu.$$

when $\tau_{\alpha/2}$ *is the upper* $100\left(\frac{\alpha}{2}\right)\%$ *per cent probability point a standard normal distribution. Hence the interval*

$$\left(\bar{X} - \frac{\sigma}{\sqrt{n}}\tau_{\alpha/2},\ \bar{X} + \frac{\sigma}{\sqrt{n}}\tau_{\alpha/2}\right)$$

is $100(1-\alpha)\%$ *confidence interval for* μ.

Clearly, any random interval

$$\left(\bar{X} - \frac{\sigma}{\sqrt{n}}\tau_{\alpha_1},\ \bar{X} + \frac{\sigma}{\sqrt{n}}\tau_{\alpha_2}\right)$$

where $\alpha_1 + \alpha_2 = \alpha$ is $100(1-\alpha)\%$ confidence interval for μ. Again,

$$P_\mu\left[\bar{X} - \frac{\sigma}{\sqrt{n}}\tau_\alpha \leq \mu\right] = 1-\alpha \ \forall\, \mu.$$

Also

$$P_\mu\left[\bar{X} + \frac{\sigma}{\sqrt{n}}\tau_\alpha \geq \mu\right] = 1 - \alpha \ \forall\, \mu.$$

Therefore,

$$T_1 = \bar{X} - \frac{\sigma}{\sqrt{n}}\tau_\alpha, \ \ T_2 = \bar{X} + \frac{\sigma}{\sqrt{n}}\tau_\alpha$$

are respectively the lower and upper confidence limits for μ.

▶ **Example 19.2.2** *Let $(X_1, \cdots, X_n)$ be a random sample from $N(\mu, \sigma^2)$ where μ is known. We want to find a $100(1-\alpha)\%$ confidence interval for σ^2.*

It is known that $\sum_{i=1}^{n}(X_i - \mu)^2/\sigma^2$ has a $\chi^2_{(n)}$ distribution. Hence,

$$P\left\{\chi^2_{n,1-\frac{\alpha}{2}} \leq \sum_{i=1}^{n}(X_i - \mu)^2/\sigma^2 \leq \chi^2_{n,\alpha/2}\right\} = 1 - \alpha$$

$$\text{or } P\left\{\sum_{i=1}^{n}(X_i - \mu)^2/\chi^2_{n,\frac{\alpha}{2}} \leq \sigma^2 \leq \sum_{i=1}^{n}(X_i - \mu)^2/\chi^2_{n,1-\frac{\alpha}{2}}\right\} = 1 - \alpha$$

when $\chi^2_{n,\alpha}$ is $100(1-\alpha)$-percent point of $\chi^2_{(n)}$ distribution (i.e., $\chi^2_{n,\alpha}$ is the upper $100\alpha\%$ point of $\chi^2_{(n)}$ distribution).

In fact,

$$P\left\{\sum_{i=1}^{n}(X_i - \mu)^2/\chi^2_{n,\alpha_1} \leq \sigma^2 \leq \sum_{i=1}^{n}(X_i - \mu)^2/\chi^2_{n,1-\alpha_2}\right\} = 1 - \alpha$$

where $\alpha_1 + \alpha_2 = \alpha$.

In case of discrete random variables, it is evident, that it is not possible to construct confidence intervals of exact confidence coefficient $(1-\alpha)$ for each $0 < \alpha < 1$. In this case, one may construct confidence intervals of confidence coefficient measuring at least $(1-\alpha)$. The statistics (T_1, T_2) will, therefore, provide confidence limits to a parameter $\gamma(\theta)$ if

$$P_\theta[T_1 \leq \gamma(\theta) \leq T_2] \geq 1 - \alpha \ \ \forall\, \theta \in \Theta. \tag{19.2.4}$$

Similarly, $T_1(T_2)$ will be a lower (upper) confidence limit with confidence coefficient $(1-\alpha)$ if

$$P_\theta[T_1 \leq \gamma(\theta)] \geq 1 - \alpha \quad (P_\theta[T_2 \geq \gamma(\theta)] \geq 1 - \alpha) \ \forall\, \theta \in \Theta. \tag{19.2.5}$$

The case of discrete distributions have been considered in details in example 19.4.2.

It may be noted that if (T_1, T_2) is a $100(1-\alpha)\%$-point confidence interval for θ, then $(g(T_1), g(T_2))$ is a $100(1-\alpha)\%$-point confidence interval for $g(\theta)$, provided $g(\theta)$ is a monotonically increasing function of θ.

19.3 Pivotal Quantity Method of Finding Confidence Interval

Suppose there exists a statistic T and a function $\psi(T,\theta)$ of T and θ, which is measurable for each $\theta \in \Theta$ and whose distribution is independent of θ. The statistic $\psi(T,\theta)$ is called a pivotal quantity.

One can then find two quantities k_1, k_2 depending on α, but independent of θ, such that

$$P_\theta[\psi(T,\theta) \le k_1] = \alpha_1 \;\; \forall\, \theta \tag{19.3.1}$$

and

$$P_\theta[\psi(T,\theta) \ge k_2] = \alpha_2 \;\; \forall\, \theta \tag{19.3.2}$$

so that

$$P_\theta\left[k_1 \le \psi(T,\theta) \le k_2\right] = 1-\alpha \;\; \forall\, \theta \tag{19.3.3}$$

where $\alpha = \alpha_1 + \alpha_2$. Suppose that the pivot $\psi(T,\theta)$ can be inverted or pivoted so that

$$\begin{aligned} &k_1 \le \psi(T,\theta) \le k_2 \\ &\Leftrightarrow c_1(T) \le \gamma(\theta) \le c_2(T) \end{aligned} \tag{19.3.4}$$

where $c_1(T), c_2(T)$ are independent of θ. Then

$$P_\theta[c_1(T) \le \gamma(\theta) \le c_2(T)] = P_\theta[k_1 \le \psi(T,\theta) \le k_2] = 1-\alpha \;\; \forall\, \theta. \tag{19.3.5}$$

Hence, the random interval $(c_1(T), c_2(T))$ may be taken as $100(1-\alpha)\%$ confidence interval for $\gamma(\theta)$.

It is evident from (19.3.1), (19.3.2) that one can find many confidence intervals with the confidence coefficient $(1-\alpha)$. One obvious criterion for choosing among different rival confidence intervals is to choose the one with the smallest length or the smallest expected length.

► **Example 19.3.1** *Let $(X_1, \cdots, X_n)$ be a random sample from a $N(\mu, \sigma^2)$ population where σ^2 is unknown. It is required to find a confidence interval for μ.*

Here $\boldsymbol{\theta} = (\mu, \sigma^2)$ and $\gamma(\boldsymbol{\theta}) = \mu$.
It is known that $T = (\bar{X} - \mu)\sqrt{n}/S$ when

$$S^2 = \frac{1}{n-1}\sum_{i=1}^{n}(X_i - \bar{X})^2$$

follows Student's t distribution with $(n-1)$ d.f. The distribution is independent of μ and σ^2. Hence, the statistic T may be taken as a pivotal quantity. Now

$$P_\theta\left\{t_{(n-1),\alpha_1} \le \frac{(\bar{X}-\mu)\sqrt{n}}{S} \le t_{(n-1),1-\alpha_2}\right\} = 1-\alpha \;\; \forall\, \boldsymbol{\theta} \tag{19.3.6}$$

provided $\alpha_1 + \alpha_2 = \alpha$ where $t_{n-1,\alpha}$ is the $100\alpha\%$-percent point of $t_{(n-1)}$ distribution. The statement (19.3.6) implies

$$P_\theta\left\{\bar{X} - t_1\frac{S}{\sqrt{n}} \le \mu \le \bar{X} + t_2\frac{S}{\sqrt{n}}\right\} = 1 - \alpha \ \forall\, \boldsymbol{\theta}.$$

where $t_1 = t_{(n-1),\alpha_1}$, $t_2 = t_{(n-1),(1-\alpha_2)}$.
The length of the confidence interval is

$$L = \frac{S}{\sqrt{n}}(t_2 - t_1). \tag{19.3.7}$$

To find the shortest $100(1-\alpha)\%$ confidence interval we are to minimise L with respect to t_1, t_2 subject to the condition

$$\int_{t_1}^{t_2} f_T(t)dt = 1 - \alpha \tag{19.3.8}$$

where $f_T(t)$ is distribution of T. It is evident that for minimising L, t_1, t_2 should be symmetrically placed around the origin. However, the following argument may be useful.

Equation (19.3.8) gives t_2 as a function of t_1. Differentiating (19.3.8) with respect to t_1 gives

$$f_T(t_2)\frac{dt_2}{dt_1} - f_T(t_1) = 0. \tag{19.3.9}$$

To minimise L with respect to t_1, the minimising equation is

$$\frac{\partial L}{\partial t_1} = \frac{S}{\sqrt{n}}\left(\frac{\partial t_2}{\partial t_1} - 1\right) = 0 \tag{19.3.10}$$

$$\text{or, } \frac{S}{\sqrt{n}}\left(\frac{f_T(t_1)}{f_T(t_2)} - 1\right) = 0 \text{ [by (19.3.9)]}$$

which holds iff $f_T(t_2) = f_T(t_1)$ i.e., iff $t_1 = -t_2$ since distribution of T is symmetric about origin. Again, t_1, t_2 have to satisfy (19.3.8); thus, $t_1 = t_{(n-1),\alpha/2}$, $t_2 = t_{(n-1),1-\frac{\alpha}{2}}$.

▶ **Example 19.3.2** *Let $(X_1, \cdots, X_n)$ be a random sample from $N(\mu, \sigma^2)$, when both (μ, σ^2) are unknown. It is desired to find $100(1-\alpha)\%$ confidence interval for σ^2.*

Here

$$\boldsymbol{\theta} = (\mu, \sigma^2),\ \ \Theta = \{(\mu, \sigma^2)| -\infty < \mu < \infty,\ \ 0 \le \sigma^2 < \infty\},\ \ \gamma(\theta) = \sigma^2.$$

Recall that

$$T = \sum_{i=1}^{n}(X_i - \bar{X})^2/\sigma^2 = \frac{(n-1)S^2}{\sigma^2} \tag{19.3.11}$$

has a $\chi^2_{(n-1)}$ distribution and this distribution is independent of $\boldsymbol{\theta}$. Hence T can be taken as a pivotal quantity. Also

$$\left\{k_1 < \frac{(n-1)S^2}{\sigma^2} < k_2\right\} \Leftrightarrow \left\{\frac{(n-1)S^2}{k_2} < \sigma^2 < \frac{(n-1)S^2}{k_1}\right\}$$

so that

$$\left(\frac{(n-1)S^2}{k_2}, \frac{(n-1)S^2}{k_1}\right) \tag{19.3.12}$$

is a $100(1-\alpha)\%$ confidence interval for σ^2, where k_1, k_2 are such that

$$P[k_1 < T < k_2] = 1 - \alpha. \tag{19.3.13}$$

The quantities k_1, k_2 are often selected so that

$$P(k_1 < T) = \frac{\alpha}{2} = P(T > k_2).$$

Such a confidence interval is referred to as the equal-tailed confidence interval or central confidence interval for σ^2. To obtain the shortest-length confidence interval note that length of the interval (19.3.12) is

$$L = (n-1)s^2\left(\frac{1}{k_1} - \frac{1}{k_2}\right) \tag{19.3.14}$$

which we are to minimise subject to the condition

$$\int_{k_1}^{k_2} f_T(t)dt = 1 - \alpha \tag{19.3.15}$$

when $f_T(t)$ is the p.d.f. of T given in (19.3.11).
The equation (19.3.15) gives k_2 in terms of k_1. Differentiating (19.3.15) with respect to k_1,

$$\frac{dk_2}{dk_1} f_T(k_2) - f_T(k_1) = 0. \tag{19.3.16}$$

The minimising equation is

$$\begin{aligned}\frac{\partial L}{\partial k_1} &= (n-1)s^2\left(-\frac{1}{k_1^2} + \frac{1}{k_2^2}\frac{\partial k_2}{\partial k_1}\right)\\ &= (n-1)s^2\left(-\frac{1}{k_1^2} + \frac{1}{k_2^2}\frac{f_T(k_1)}{f_T(k_2)}\right) \quad \text{[by (19.3.16)]}\\ &= 0\end{aligned} \tag{19.3.17}$$

which implies that

$$k_1^2 f_T(k_1) = k_2^2 f_T(k_2). \tag{19.3.18}$$

The length of the confidence interval (19.3.12) will be minimum if k_1, k_2 are chosen satisfying (19.3.18) subject to (19.3.13). Note that these do not give a equal-tailed confidence interval.

The equations (19.3.18) (with (19.3.13)) can be solved numerically for each n. The solution for large n is close to the symmetric solution, where

$$k_1 = \chi^2_{(n-1),\alpha/2}$$
$$k_2 = \chi^2_{(n-1),1-\alpha/2}.$$

Table A.21 gives the optimal left cut off probability α_1, and the corresponding confidence limits $k_1 = \chi^2_{\alpha_1,\nu}$ and $k_2 = \chi^2_{\alpha_2,\nu}$ where $\alpha_2 = (1-\alpha) - \alpha_1$, for $\alpha = 0.1$ and $\alpha = 0.05$ and various degrees of freedom (ν). The table also gives the relative gain resulting from using the shortest confidence interval (with length proportional to $(\frac{1}{k_1} - \frac{1}{k_2})$), as compared to confidence interval with equal cut off probabilities.

► **Example 19.3.3** *Let $(X_1, \cdots, X_n)$ be a random sample from a $N(\mu, \sigma^2)$ population, where both μ and σ^2 are unknown. Find a joint confidence interval for (μ, σ^2).*

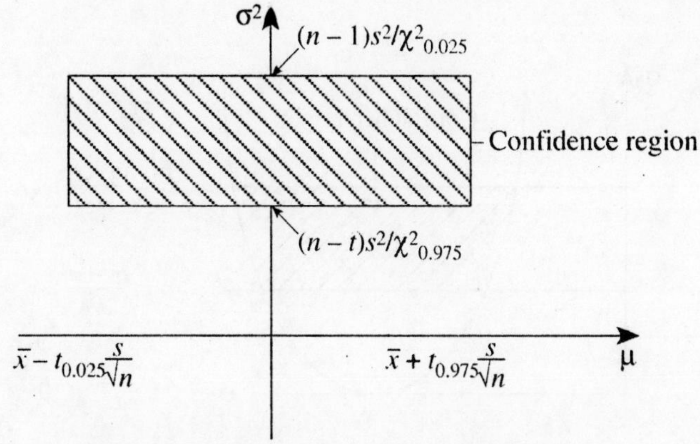

Fig. 19.1 Showing a joint confidence region for μ and σ²

One may use the results obtained in examples 19.3.1 and 19.3.2. Thus, one may obtain a confidence region given in figure 19.1 by using the two relations.

$$P\left[\bar{X} - t_{0.025}\sqrt{\frac{s^2}{n}} < \mu < \bar{X} + t_{0.975}\sqrt{\frac{s^2}{n}}\right] = 0.95$$

$$\text{and } P\left[\frac{(n-1)s^2}{\chi^2_{0.975}} < \sigma^2 < \frac{(n-1)s^2}{\chi^2_{0.025}}\right] = 0.95.$$

However, since t and χ^2 are not independent, the confidence coefficient of this region will not be $(0.95)^2$.

However, using the property of independence of

$$\tau = \frac{(\bar{X}-\mu)\sqrt{n}}{\sigma} \quad \text{and} \quad \chi^2 = \frac{(n-1)s^2}{\sigma^2}$$

one may set up another confidence region for (μ, σ^2) whose confidence-coefficient can be readily evaluated.

$$P\left\{-k_1 < \frac{(\bar{X}-\mu)\sqrt{n}}{\sigma} < k_1\right\} = (1-\alpha_1)$$

$$\text{and } P\left\{k_2 < \frac{(n-1)s^2}{\sigma^2} < k_3\right\} = (1-\alpha_2)$$

where k_1, k_2, k_3 are suitably determined. Hence,

$$P\left[-k_1 < \frac{(\bar{X}-\mu)\sqrt{n}}{\sigma} < k_1;\ k_2 < \frac{(n-1)s^2}{\sigma^2} < k_3\right] = (1-\alpha_1)(1-\alpha_2). \quad (19.3.19)$$

The four inequalities in (19.3.19) determine the confidence region as shown in figure 19.2, below.

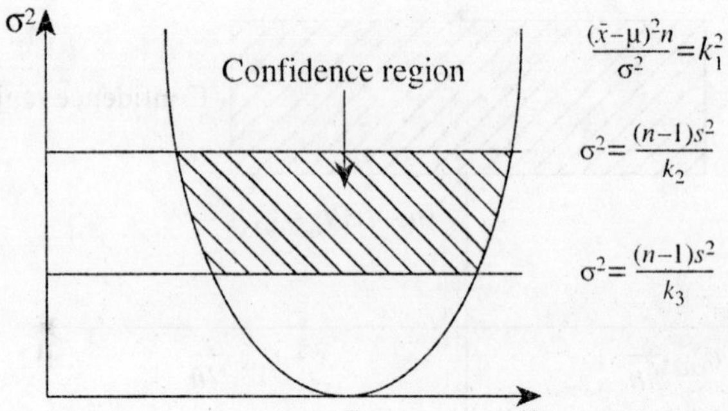

Fig. 19.2 Showing a simultaneous confidence region for (μ, σ^2)

The region so constructed does not, however, have the minimum area among all confidence regions with confidence coefficient $(1-\alpha_1)(1-\alpha_2)$.

A confidence region for (μ, σ) may be obtained in exactly the same way. Here, the graphs have to be plotted in terms of σ rather than in terms of σ^2.

► **Example 19.3.4** *Let $(X_1, \cdots, X_m)$, $(Y_1, \cdots, Y_n)$ be two independent samples from $N(\mu_1, \sigma^2)$, $N(\mu_2, \sigma^2)$ respectively where μ_1, μ_2, σ^2 are all unknown. We want to find a confidence interval for $\mu_1 - \mu_2$.*

It is known that

$$(\bar{X}-\bar{Y}) \cap N\left(\mu_1-\mu_2,\ \sigma^2\left(\frac{1}{m}+\frac{1}{n}\right)\right).$$

Also

$$\sum_{i=1}^{m}(X_i - \bar{X})^2/\sigma^2 \cap \chi^2_{(m-1)}, \quad \sum_{i=1}^{n}(Y_i - \bar{Y})^2/\sigma^2 \cap \chi^2_{(n-1)}.$$

Therefore,

$$\left[\sum_{i=1}^{m}(X_i - \bar{X})^2 + \sum_{i=1}^{n}(Y_i - \bar{Y})^2\right]/\sigma^2 = \frac{S_0^2}{\sigma^2} \text{ (say) } \cap \chi^2_{m+n-2}.$$

Hence

$$T = \frac{[(\bar{X} - \bar{Y}) - (\mu_1 - \mu_2)]/\sqrt{\sigma^2\left(\frac{1}{m} + \frac{1}{n}\right)}}{\sqrt{\frac{\chi^2}{m+n-2}}}$$

follows Student's $t_{(m+n-2)}$. The distribution is independent of μ_1, μ_2, σ^2 and hence T can be taken as the pivotal quantity. Hence

$$P\left\{-t_{\alpha/2} < T < t_{(1-\alpha/2)}\right\} = 1 - \alpha$$

from which it follows then

$$\left(\bar{X} - \bar{Y} - t_{\alpha/2}\sqrt{\left(\frac{1}{m} + \frac{1}{n}\right)S_0^2},\ \bar{X} - \bar{Y} + t_{(1-\frac{\alpha}{2})}\sqrt{\left(\frac{1}{m} + \frac{1}{n}\right)S_0^2}\right)$$

is a $100(1-\alpha)\%$ confidence interval for $\mu_1 - \mu_2$.

▶ **Example 19.3.5** *Let $(X_i, Y_i), i = 1, \cdots, n$ be a random sample of size n from a bivariate normal population $(\mu_1, \mu_2, \sigma_1^2, \sigma_2^2, \rho)$. Find a confidence interval for $\mu_1 - \mu_2$.*

Let $D_i = X_i - Y_i$. Then $D_1, \cdots, D_n$ are independent samples from $N(\mu_1 - \mu_2, \sigma_1^2 + \sigma_2^2 - 2\rho\sigma_1\sigma_2)$. Here σ_1, σ_2, ρ are unknown. Proceeding as in example 19.3.1 (also vide section 11.5.2), a $100(1-\alpha)\%$ confidence interval for $\mu_1 - \mu_2$ is

$$\left(\bar{D} - t_{(n-1),\alpha/2}\sqrt{\frac{s_D^2}{n}},\ D + t_{(n-1),(1-\alpha/2)}\sqrt{\frac{s_D^2}{n}}\right)$$

where

$$s_D^2 = \frac{1}{n-1}\sum_{i=1}^{n}(D_i - \bar{D})^2.$$

19.3.1 A General Method of finding a Pivotal Statistic

Let X be a random variable with continuous distribution function $F_X(x;\theta)$, $0 \le F_X(x;\theta) \le 1$. Then the random variable $Y = F_X(x;\theta)$ has a rectangular distribution over $(0,1)$ [vide section 8.3]. The variable Y obtained in this way is called the probability integral transformation of X.

Now, for $u > 0$,

$$\begin{aligned} P[-\log F_X(\cdot\,;\theta) \geq u] &= P[\log F_X(\cdot\,;\theta) \leq -u] \\ &= P[Y \leq e^{-u}] = e^{-u}. \end{aligned}$$

Hence, $-\log F_X(x;\theta)$ has an exponential p.d.f. $e^{-u}, 0 < u < \infty$.
Therefore, if $(X_1, \cdots, X_n)$ is a random sample from a distribution $F_X(x;\theta)$ then

$$T = -\sum_{i=1}^{n} \log F_{X_i}(x_i;\theta) \tag{19.3.20}$$

has a gamma distribution with parameters $(1, n)$. Distribution of T is independent of θ. Hence, T or $\prod_{i=1}^{n} F_{X_i}(x_i;\theta)$ may be taken as a pivotal quantity. Similar result holds for the random variable $W = -\sum_{i=1}^{n} \log G_X(x_i;\theta)$ where $G = 1 - F_X(x, \theta)$.

□ **Remark 19.3.1** *The above result shows that if a sample is taken from a population having a continuous cumulative distribution, then a pivotal quantity T given in* (19.3.20) *always exists. However, this does not guarantee that T can be inverted to give a confidence interval for θ. However, if $T = T(x_1, \cdots, x_n)$ is a monotonic function in θ for a given $\boldsymbol{x}$ then T (in* (19.3.20)*) can always be inverted to give a confidence interval for θ.*

► **Example 19.3.6** *Let $(X_1, \cdots, X_n)$ be a random sample from the p.d.f. $f(x;\theta) = \theta x^{\theta-1}$, $0 < x < 1$. Then $F(x;\theta) = x^{\theta}$. Since $T = -\sum_{i=1}^{n} \log F_{X_i}(x_i, \theta)$ follows a Gamma distribution with parameters $(1, n)$ we can determine k_1, k_2 such that*

$$\begin{aligned} 1 - \alpha &= P\left[k_1 < \sum_{i=1}^{n} F_{X_i}(x_i;\theta) < k_2\right] = P\left[k_1 < \prod_{i=1}^{n} X_i^{\theta} < k_2\right] \\ &= P\left[\log k_1 < \theta \log \prod_{i=1}^{n} X_i < \log k_2\right] \\ &= P\left[-\log k_2 < -\theta \log \prod_{i=1}^{n} X_i < -\log k_1\right] \\ &= P\left[\frac{\log k_2}{\log \prod_{i=1}^{n} X_i} < \theta < \frac{\log k_1}{\log \prod_{i=1}^{n} X_i}\right]. \end{aligned}$$

Hence,

$$\left(\frac{\log k_2}{\log \prod_{i=1}^{n} X_i}, \frac{\log k_1}{\log \prod_{i=1}^{n} X_i}\right)$$

may be taken as a $100(1-\alpha)\%$-confidence interval for θ.

We now consider if pivotal random variables always exist. For this we consider a few definitions. θ is called a location parameter of a p.d.f. $f(x, \theta)$ if

$$f(x, \theta) = h(x - \theta)$$

for some function h. Similarly θ is called a *scale* parameter if

$$f(x, \theta) = \frac{1}{\theta} h\left(\frac{x}{\theta}\right)$$

for some function h.

The mean of a distribution may or may not be its location parameter. For instance, for normal distribution mean is the location parameter. However, in case of exponential distribution mean is a scale parameter, while for gamma distribution mean is neither a scale parameter nor a location parameter.

Theorem 19.3.1

1. *If θ is a location parameter and $\hat{\theta}$ is its m.l.e., then $\hat{\theta} - \theta$ is pivotal.*
2. *If θ is a scale parameter and $\hat{\theta}$ is its m.l.e., then $\hat{\theta}/\theta$ is pivotal.*
3. *Let θ_1 be a location parameter and θ_2 be a scale parameter, so that $f(x; \theta_1, \theta_2) = \frac{1}{\theta_2} h\left(\frac{x-\theta_1}{\theta_2}\right)$ for some function h. If $\hat{\theta}_1$, $\hat{\theta}_2$ are m.l.e.'s of θ_1, θ_2 respectively then $(\hat{\theta}_1 - \hat{\theta}_2)/\hat{\theta}_2$ is pivotal.*

However, confidence intervals can be constructed even if pivotal random variables do not exist.

▶ **Example 19.3.7** *Let $X_1, \cdots, X_n$ be a random sample from a distribution with cdf $F(t)$ and let $X_{(1)} \leq X_{(2)} \leq \cdots \leq X_{(n)}$ be the order statistics of the sample. Assume that F is continuous and strictly increasing and let $\xi_{1/2}$ be the median of the distribution F. Thus $\xi_{1/2}$ is the unique solution of the equation $F(t) = \frac{1}{2}$. We shall construct confidence interval for $\xi_{1/2}$.*

The event $X_{(a)} \leq \xi_{1/2}$ occurs iff at least a elements in the sample are below the median. Similarly, the event $\xi_{1/2} \leq X_{(b)}$ occurs iff at least $(n - b)$ elements in the sample are above the median. It we call an observation below the median a success(S) and an observation above the median a failure (F) then $X_{(a)} \leq \xi_{1/2} \leq X_{(b)}$ occurs iff $a \leq S \leq b$. Now S has a binomial distribution $\left(n, \frac{1}{2}\right)$.

Therefore,

$$P\left(X_{(a)} \leq \xi_{1/2} \leq X_{(b)}\right) = \sum_{k=a}^{b} \binom{n}{k} \frac{1}{2^n}.$$

We can find a and b suitably for a given value of confidence coefficient.

One sided confidence bounds for $\xi_{1/2}$ are obtained in the same way. For example, $X_{(a)}$ is a lower bound for $\xi_{1/2}$ with the confidence level $\sum_{k=a}^{n} \binom{n}{k} \frac{1}{2^n}$.

If we want to find confidence interval for the pth quantile ξ_p, we have

$$P\left\{X_{(a)} \le \xi_p \le X_{(b)}\right\} = \sum_{k=a}^{b} \binom{n}{k} p^k (1-p)^{n-k}.$$

For $n = 10$, $p = \frac{1}{4}$,

$$\sum_{k=3}^{7} \binom{10}{k} \left(\frac{1}{4}\right)^k \left(\frac{3}{4}\right)^{10-k} = 0.526,$$

so that (3, 7) is a confidence interval with confidence coefficient 52.6%. Clearly, in this case it is better to take confidence intervals based on order statistics that are not equidistant from the extremes.

19.4 Method of Finding Confidence Belt

Let $(X_1, \cdots, X_n)$ be a random sample from a p.d.f. $f(x;\theta)$, $\theta \in \Theta \in R_1$. We seek an interval estimate of θ. Let $T = t(X_1, \cdots, X_n)$ be some statistic. The statistic T may be a sufficient statistic or may be the maximum likelihood estimate (m.l.e.) of θ if no sufficient statistic exists.

Let $f_T(t;\theta)$ be the p.d.f. of T. Assume that T is a continuous r.v. The case of discrete r.v. will be illustrated subsequently.

Define two functions $h_1(\theta), h_2(\theta)$ satisfying $h_1(\theta) < h_2(\theta)$ and

$$\int_{-\infty}^{h_1(\theta)} f_T(t;\theta)dt = \alpha_1, \quad \int_{h_2(\theta)}^{\infty} f_T(t;\theta)dt = \alpha_2 \tag{19.4.1}$$

where α_1, α_2 are two fixed numbers satisfying $0 < \alpha_1, \alpha_2 < 1$ and $\alpha_1 + \alpha_2 < 1$.

The quantities $h_1(\theta), h_2(\theta)$ can be plotted as functions of θ. We assume that both $h_1(\theta)$, $h_2(\theta)$ are strictly monotonic increasing functions of θ (Fig. 19.3).

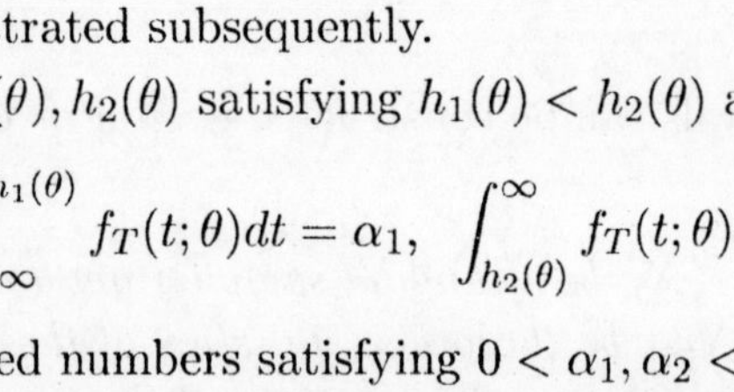

Fig. 19.3 A confidence belt for θ

For the observed $\boldsymbol{x} = (x_1, \cdots, x_n)$, calculate the value of the statistic $T(\boldsymbol{x}) = t$ (say). Find the value of the abscissa of the curves $h_1(\theta)$, $h_2(\theta)$ corresponding to ordinate t. Suppose that the required points on the curves $h_2(\theta), h_1(\theta)$ are (u_1, t), (u_2, t) respectively. Thus, $t = h_2(u_1) = h_1(u_2)$. Then (U_1, U_2) is the $100(1-\alpha_1-\alpha_2)\%$ confidence interval for θ i.e., $P\{U_1 < \theta < U_2\} = 1 - \alpha_1 - \alpha_2$.

To see this consider a point $\theta_0 \in (u_1, u_2)$. The ordinates corresponding to abscissa $\theta = \theta_0$ on the curves $h_1(\theta)$, $h_2(\theta)$ are t_1, t_2 respectively. Hence

$$\begin{aligned} P\{U_1 < \theta < U_2\} &= P\{h_1(\theta) < T < h_2(\theta)\} \\ &= 1 - \alpha_1 - \alpha_2 \end{aligned}$$

by (19.4.1).

Note that the above procedure would work even if $h_1(\cdot)$, $h_2(\cdot)$ are not monotonic functions. In that case the confidence region may come in the form of union of a number of intervals instead of a single interval.

Example 19.4.1 *Let $(X_1, \cdots, X_n)$ be a random sample from the rectangular population $f(x;\theta) = \frac{1}{\theta}$, $0 < x < \theta$. It is required to find a confidence interval for θ.*

Consider $T = X_{(n)} = \max\limits_i X_i$. This statistic is sufficient for θ. Also T is maximum likelihood estimator of θ.

$$f_T(t;\theta) = n\left(\frac{t}{\theta}\right)^{n-1} \cdot \frac{1}{\theta}, \quad 0 < t < \theta.$$

$h_1(\theta), h_2(\theta)$ are found by solving the equations (19.4.1).

$$\alpha_1 = \int_0^{h_1(\theta)} nt^{n-1}\theta^{-n} dt \Rightarrow h_1(\theta) = \theta\alpha_1^{1/n}.$$

Similarly,

$$\alpha_2 = \int_{h_2(\theta)}^{\theta} nt^{n-1}\theta^{-n}\, dt \Rightarrow h_2(\theta) = \theta(1-\alpha_2)^{1/n}.$$

For observed value of $T(x_1, \cdots, x_n) = t$ (say),

$$h_1(u_2) = t \Rightarrow u_2 = t\alpha_1^{-1/n}.$$

Similarly,

$$h_2(u_1) = t \Rightarrow u_1 = t(1-\alpha_2)^{-1/n}.$$

Therefore, $100(1-\alpha_1-\alpha_2)\%$ confidence interval for θ is

$$\left(X_{(n)}(1-\alpha_2)^{-1/n},\ X_{(n)}\alpha_1^{-1/n}\right).$$

The length of the confidence interval is

$$L = X_{(n)}\left[\alpha_1^{-1/n} - (1-\alpha_2)^{-1/n}\right].$$

The length will be shortest if α_1, α_2 are chosen so as to minimise $\left[\alpha_1^{-1/n} - (1-\alpha_2)^{-1/n}\right]$ subject to the condition $\alpha_1 + \alpha_2 = \alpha$. This is achieved if $\alpha_2 = 0$, $\alpha_1 = \alpha$ (since $\frac{dL}{d\alpha_1} < 0$).

We note that for finding confidence intervals one need not find $h_1(\theta), h_2(\theta)$. For a given observed value of the statistic $T = t$ we need only to find u_1, u_2. These are obtained by solving the equations

$$\int_{-\infty}^{t_0} f_T(t;\theta)dt = \alpha_1 \tag{19.4.2}$$

$$\int_{t_0}^{\infty} f_T(t;\theta)dt = \alpha_2 \tag{19.4.3}$$

in θ, where t_0 is the observed value of T. The solution of the equation (19.4.2) gives u_2 and that of (19.4.3) gives u_1.

▶ **Example 19.4.2** *Let $X_1, \cdots, X_{10}$ be a random sample from a point Binomial distribution $P[X=1] = \theta = 1 - P[X=0]$, $0 < \theta < 1$. We know that $T = \sum_{i=1}^{10} X_i$ is a sufficient statistic for θ and $f_T(t;\theta) = \binom{10}{t}\theta^t(1-\theta)^{10-t}, t = 0, 1, \cdots, 10$. We want a confidence interval for θ.*

We first determine for a suitable set of values of θ, the quantities

$$\sum_{0}^{h_1(\theta)} \binom{10}{t}\theta^t(1-\theta)^{10-t} \leq 0.05 \tag{19.4.4}$$

$$\sum_{h_2(\theta)}^{10} \binom{10}{t}\theta^t(1-\theta)^{10-t} \leq 0.05 \tag{19.4.5}$$

where α_1 and α_2 have both been taken as 0.05; however, since exact probabilities α_1, α_2 cannot be achieved, we have taken these as the upper bounds of the probabilities. In calculating $h_1(\theta), h_2(\theta)$ those values of $h_1(\theta), h_2(\theta)$ should be taken for a given value of θ which satisfy the inequalities (19.4.4), (19.4.5) as closely as possible. The resulting confidence region will have the probability,

$$P\{U_1 < \theta < U_2\} \geq 0.90.$$

The plot of $h_1(\theta)$, $h_2(\theta)$ against θ will give the 90% confidence belt for θ.

The following table gives values of $h_1(\theta)$, $h_2(\theta)$ for different values of θ. These values have been obtained from the table A.2 of binomial probabilities given in the appendix.

Table 19.1 *Showing Values of $h_1(\theta), h_2(\theta)$*

θ	$h_1(\theta)$	$h_2(\theta)$
0.1	0	3
0.2	0	4
0.3	1	5
0.4	2	7
0.5	2	8
0.6	3	8
0.7	5	9
0.8	6	10
0.9	7	10

Suppose $T = t_0 = 5$ is observed. Therefore, $h_1(\theta) = 5$ gives $u_2 = 0.7$ and $h_2(\theta) = 5$ gives $u_1 = 0.3$. Hence,

$$P\{0.3 < \theta < 0.7\} \geq 0.90.$$

In practice one need not calculate the values of the functions $h_1(\theta)$ and $h_2(\theta)$ for various values of θ. By substituting $h_1(\theta) = h_2(\theta) = t_0$ (observed value of T), one needs only find solutions for θ which satisfy the relations (19.4.4) and (19.4.5) as closely as possible.

19.5 Approximate Confidence Intervals in Large Samples

It has been seen that sometimes it is possible to find a sequence of estimators $\{t_n\}$ that are asymptotically normally distributed about the true value of the parameter with an asymptotic variance $\sigma_n^2(\theta)$, which depends on n and θ. In particular, we noted that under certain regularity conditions, maximum likelihood estimators are asymptotically normally distributed about θ and have asymptotic variances

$$\sigma_n^2(\theta) = \frac{1}{nE_\theta\left[\frac{\partial \log f(x;\theta)}{\partial\theta}\right]^2} = -\frac{1}{nE_\theta\left[\frac{\partial^2 \log f(x;\theta)}{\partial\theta^2}\right]}. \tag{19.5.1}$$

When such asymptotically normal estimators exist, $(t_n - \theta)/\sigma_n(\theta)$ may be taken as a pivotal quantity and a $100(1-\alpha)\%$ confidence coefficient of θ may be taken as approximately

$$\left\{T_n - \tau_{\alpha/2}\sigma_n(\theta),\ \ T_n + \tau_{(1-\alpha/2)}\sigma_n(\theta)\right\}.$$

The above method provides a large-sample confidence interval so long as $(t_n - \theta)/\sigma_n(\theta)$ can be inverted.

▶ **Example 19.5.1** *Let $X_1, \cdots, X_n$ be a random sample from $N(0, \sigma^2)$. We are interested in finding a central confidence interval for σ^2 (i.e., a confidence interval with equal tail probabilities) with an approximate confidence coefficient $1-\alpha$.*

Let

$$L = \prod_{i=1}^{n} f(x_i;\theta); \quad \frac{\partial \log L}{\partial \theta} = -\frac{n}{2\theta} + \frac{\Sigma x_i^2}{2\theta^2}$$

when $\theta = \sigma^2$. Therefore, m.l.e. of θ is

$$\hat{\theta} = \sum x_i^2/n; \; E\left(\frac{\partial^2 \log L}{\partial \theta^2}\right) = -\frac{n}{2\theta^2}.$$

Hence,

$$P\left\{-\tau_{\alpha/2} < \left(\Sigma\frac{X_i^2}{n} - \theta\right) \Big/ \left(\frac{2\theta^2}{n}\right)^{1/2} < \tau_{\alpha/2}\right\} = 1 - \alpha$$

$$\text{i.e., } P\left\{\theta\left(1 - \sqrt{\frac{2}{n}}\tau_{\alpha/2}\right) < \frac{\Sigma X_i^2}{n} < \theta\left(1 + \sqrt{\frac{2}{n}}\tau_{\alpha/2}\right)\right\} = 1 - \alpha.$$

The required confidence interval is

$$\left\{\frac{\Sigma X_i^2/n}{1 + \sqrt{\frac{2}{n}}\tau_{\alpha/2}}, \frac{\Sigma X_i^2/n}{1 - \sqrt{\frac{2}{n}}\tau_{\alpha/2}}\right\}. \quad (19.5.2)$$

It may be noted that the confidence interval obtained by this procedure is not invariant under transformation of parameters. Thus, the confidence interval (19.5.2) is not the same as the confidence interval obtained by squaring up the end points of the interval for σ in a $N(0, \sigma^2)$ population [vide exercise 19]

We now consider construction of large-sample confidence intervals which are invariant under transformation of parameters.

Suppose that the p.d.f. $f(x;\theta)$ is such that

$$E\left(\frac{\partial \log f(X;\theta)}{\partial \theta}\right) = 0$$

and

$$V\left(\frac{\partial \log f(X;\theta)}{\partial \theta}\right) = -E\left(\frac{\partial^2 \log f(X;\theta)}{\partial \theta^2}\right) = K^2 \text{ (say) } < \infty.$$

Let $X_1, \cdots, X_n$ be a random sample from $f(x;\theta)$ and $L = \prod_{i=1}^{n} f(x_i;\theta)$. Clearly then each of the random variables $\partial \log(X_i;\theta)/\partial\theta$ $(i = 1, \cdots, n)$ has mean zero and variance K^2. Therefore, by the central limit theorem, their sample mean

$$\frac{1}{n} \cdot \frac{\partial \log L}{\partial \theta} \sim N\left(0, \frac{K^2}{n}\right) \text{ i.e., } \frac{\partial \log L/\partial\theta}{\sqrt{\left\{-E\left(\frac{\partial^2 \log 2}{\partial\theta^2}\right)\right\}}} \sim N(0,1). \quad (19.5.3)$$

Using this property one can get a large sample confidence interval for θ. Note that the maximum likelihood estimate of θ has not been used here.

Let $\phi(\theta)$ be a strictly increasing function of θ. Now

$$\frac{\partial \log L}{\partial \theta} = \frac{\partial \log L}{\partial \phi} \cdot \frac{\partial \phi}{\partial \theta}.$$

$$\frac{\partial^2 \log L}{\partial \theta^2} = \frac{\partial^2 \log L}{\partial \phi^2}\left(\frac{\partial \phi}{\partial \theta}\right)^2 + \frac{\partial \log L}{\partial \phi}\left(\frac{\partial^2 \phi}{\partial \theta^2}\right);$$

hence

$$E\left(\frac{\partial^2 \log L}{\partial \theta^2}\right) = \left(\frac{\partial \phi}{\partial \theta}\right)^2 E\left(\frac{\partial^2 \log L}{\partial \phi^2}\right).$$

Therefore, if (θ_1, θ_2) is a confidence interval for θ as is obtained from (19.5.3), then $(\phi(\theta_1), \phi(\theta_2))$ remains a confidence interval for $\phi(\theta)$.

Example 19.5.2 *Let $X_1, \cdots, X_n$ be a random sample from a $N(0, \sigma^2)$ population. Here*

$$\frac{\partial \log L}{\partial \sigma} = -\frac{n}{\sigma} + \frac{\Sigma x_i^2}{\sigma^3};\ E\left(\frac{\partial^2 \log L}{\partial \sigma^2}\right) = -\frac{2n}{\sigma^2}.$$

Hence, by (19.5.3),

$$\frac{(\Sigma X_i^2/\sigma^2) - n}{\sqrt{2n}} \sim N(0, 1)$$

A central $100(1-\alpha)\%$ confidence interval for σ is, therefore,

$$\left\{\left(\frac{\Sigma X_i^2}{n + \sqrt{(2n)}\tau_{\alpha/2}}\right)^{1/2}, \left(\frac{\Sigma X_i^2}{n - \sqrt{(2n)}\tau_{\alpha/2}}\right)^{1/2}\right\}. \tag{19.5.4}$$

Note that this is not the same confidence interval as given in exercise 19, although for large sample, the differences between the two are negligibly small.

If the variance σ^2 is treated as parameter, then the method yields $100(1-\alpha)\%$ confidence interval for σ^2 as

$$\left\{\frac{\Sigma X_i^2/n}{1 + \left(\sqrt{\frac{2}{n}}\right)\tau_{\alpha/2}}, \frac{\Sigma X_i^2/n}{1 - \sqrt{\left(\frac{2}{n}\right)}\tau_{\alpha/2}}\right\}. \tag{19.5.5}$$

Note that the end-points are squares of those in (19.5.4).

It may be noted that the large-sample confidence intervals based on maximum likelihood estimators will be shorter on an average than the large-sample confidence intervals based on any other estimator.

19.6 Theory of Confidence Sets

It is clear by now that we are really interested in a set of values of parameters o a subspace of the parameter space Θ, determined in the light of the observation $\boldsymbol{x}$ that may be supposed to cover the true value(s) of the parameter(s). Thus, for eac observation $\boldsymbol{x}$, we determine a subset of the parameter space $S(\boldsymbol{x}) \in \Theta$ (say), which ma be considered as covering the true value of the parameter with a high probability. Suc a set is called a confidence set. When the set is an interval, it is called a confidenc interval. The theory of confidence sets was formulated mainly in the work of Neyma (1939) and was explicitly stated in the work of Lehman (1977).

We shall write $S \ni \theta$ to denote that the set S, a subset of Θ covers or includes θ Assume without any loss of generality that $\theta = (\theta_1, \cdots, \theta_k)$ and that $\Theta \subseteq R^k$.

Definition 19.6.1 *A family of subsets $S(\boldsymbol{x})$, for varying $\boldsymbol{x}$ over $\mathcal{X}$, of $\Theta \subseteq R^k$ is sai to constitute a family of confidence sets at the confidence level $1 - \alpha$ if*

$$P_\theta\{S(\boldsymbol{x}) \ni \theta\} \geq 1 - \alpha \ \forall \, \theta \in \Theta \tag{19.6.1}$$

i.e., the random set $S(\boldsymbol{x})$ covers the true parameter value θ with probability $\geq 1 - \alpha$.

If $k = 1$ and $S(\boldsymbol{x})$ is of the form $S(\boldsymbol{x}) = (\underline{\theta}(\boldsymbol{x}), \bar{\theta}(\boldsymbol{x}))$ then we call it a confidenc interval at confidence level $1 - \alpha$, provided that

$$P_\theta\{\underline{\theta}(\boldsymbol{x}) < \theta < \bar{\theta}(\boldsymbol{x})\} \geq 1 - \alpha \ \forall \, \theta \tag{19.6.2}$$

and the quantity

$$\inf_\theta P_\theta\{\underline{\theta}(\boldsymbol{x}) < \theta < \bar{\theta}(\boldsymbol{x})\} \tag{19.6.3}$$

is referred to as the confidence coefficient associated with the random interval.

To make a choice among the families of confidence sets in (19.6.1) consider

Definition 19.6.2 *A family of sets $S_0(\boldsymbol{x})$, for varying $\boldsymbol{x}$ over $\mathcal{X}$, of the parameter spa Θ is said to constitute a family of uniformly most accurate*(UMA) *or uniformly mo selective confidence sets at level $(1 - \alpha)$ if*

$$P_\theta[S_0(\boldsymbol{X}) \ni \theta] \geq 1 - \alpha \ \forall \, \theta \in \Theta \tag{19.6.4}$$

and

$$P_{\theta'}[S_0(\boldsymbol{X}) \ni \theta] \leq P_{\theta'}[S(\boldsymbol{X}) \ni \theta] \ \forall \, \theta, \theta' \in \Theta (\theta \neq \theta'), \tag{19.6.5}$$

whatever be the other family of sets $S(\boldsymbol{x})$, satisfying (19.6.4). *The implication of* (19.6.5) *is that $S_0(\boldsymbol{X})$ has a smaller probability of containing a wrong value of θ, when son other value θ' is true, than any other set $S(\boldsymbol{X})$ of the same confidence coefficient.*

It will be seen that in most cases, a family of UMA confidence sets cannot be obtaine Hence, we introduce the concept of unbiased confidence sets.

Definition 19.6.3 *A family of sets $S(\boldsymbol{x})$, for varying $\boldsymbol{x}$ over $\mathcal{X}$, of Θ is said to constitute a family of unbiased confidence sets of level $(1-\alpha)$, if*

$$P_{\theta'}[S(\boldsymbol{X}) \ni \theta] \leq 1-\alpha \;\; \forall\, \theta, \theta' \in \Theta, \;\; \theta \neq \theta'. \tag{19.6.6}$$

Hence, $S(\boldsymbol{X})$ is a family of unbiased confidence sets iff the probability for $S(\boldsymbol{X})$ to cover θ when some alternative value θ' is true does not exceed the same probability when θ itself is true. This is, obviously, a desirable property of a family of confidence sets.

In the absence of a family of UMA confidence sets, one looks for the family of UMA-unbiased confidence sets.

Definition 19.6.4 *A family of sets $S_0(\boldsymbol{x})$, for varying $\boldsymbol{x}$ over $\mathcal{X}$, of Θ is said to constitute a family of uniformly most accurate unbiased* (UMAU) *confidence sets of level $(1-\alpha)$ if*

$$P_\theta[S_0(\boldsymbol{X}) \ni \theta] \geq 1-\alpha \;\; \forall\, \theta \in \Theta \tag{19.6.7}$$

$$P_{\theta'}[S_0(\boldsymbol{X}) \ni \theta] \leq 1-\alpha \;\; \forall\, \theta, \theta' \in \Theta \;\; (\theta \neq \theta') \tag{19.6.8}$$

and

$$P_{\theta'}[S_0(\boldsymbol{X}) \ni \theta] \leq P_{\theta'}[S(\boldsymbol{X}) \ni \theta] \;\; \forall\, \theta, \theta' \in \Theta \;\; (\theta \neq \theta') \tag{19.6.9}$$

whatever be the other family of sets $S(\boldsymbol{x})$, satisfying (19.6.7) *and* (19.6.8).

▶ **Example 19.6.1** *Let X_1, X_2 be a random sample from $f(x;\theta) = \frac{1}{\theta}\, e^{-x/\theta}$, $x>0$, $\theta>0$. It can be shown that distributions of both the statistic U/θ and V/θ where $U = X_1 + X_2$ and $V = \min(X_1, X_2)$ does not depend on θ. Also*

$$P_\theta[U/k > \theta] = e^{-k}(1+k)$$

and

$$P_\theta[lV > \theta] = e^{-2/l}.$$

Hence both the intervals $(0, U/k)$ and $(0, lV)$ are confidence intervals for θ and have the same confidence coefficient if $e^{-k}(1+k) = e^{-2/l}$ (exercise 22).
Now

$$\begin{aligned} P_{\theta'}\left[\frac{U}{k} > \theta\right] &= P_{\theta'}\left[(X_1+X_2) > k\theta\right] \\ &= \int_{k\theta}^{\infty} \frac{u}{\theta'^2} e^{-u/\theta'}\, du = \left(1 + \frac{k\theta}{\theta'}\right) \exp\left(-\frac{k\theta}{\theta'}\right) \end{aligned} \tag{19.6.10}$$

because when $\theta = \theta'$, $U = X_1 + X_2$ has the density

$$\frac{u}{\theta'^2} \exp(-u/\theta');$$

also

$$P_{\theta'}(lV > \theta) = \exp\left(\frac{-2\theta}{l\theta'}\right). \tag{19.6.11}$$

Of course, we require that (19.6.10) and (19.6.11) should be equal when $\theta' = \theta$. Thus

$$\exp\left(\frac{-2\theta}{l\theta'}\right) = \left[\exp\left(\frac{-2}{l}\right)\right]^{\theta/\theta'} = \left[(1+k)e^{-k}\right]^{\theta/\theta'} = (1+k)^{\theta/\theta'} \exp\left[-\frac{k\theta}{\theta'}\right].$$

Since $k > 0$, when $\theta > \theta'$,

$$(1+k)^{\theta/\theta'} > 1 + \frac{k\theta}{\theta'}.$$

Therefore, the interval $\left(0, \frac{U}{k}\right)$ is superior in the sense that it is less likely to trap wrong value of θ when another value θ' is true, so long as $\theta > \theta'$. Since this result hol for all $\theta > \theta'$, the upper limit U/k gives a uniformly more accurate confidence interv than lV.

19.7 Relationship between Confidence Estimation and Hypothesis Testing

We investigate in this section the relationship between confidence estimation an hypothesis testing. More precisely, given a level α of test $H_\circ(\theta = \theta_0)$ is it possib to construct a $(1-\alpha)$ level confidence interval for θ_0 and conversely? This is considere in the following theorems.

Theorem 19.7.1 *For each $\theta_0 \in \Theta$, let $A(\theta_0)$ be the acceptance region of an α-level te for testing $H_\circ(\theta = \theta_0)$. Again, for each $\boldsymbol{x} \in \mathcal{X}$, let $S(\boldsymbol{x})$ denote the set of paramet values defined by*

$$S(\boldsymbol{x}) = \{\theta | \boldsymbol{x} \in A(\theta), \theta \in \Theta\}. \tag{19.7.1}$$

Then $S(\boldsymbol{x})$ is a family of confidence sets for θ at confidence level $(1-\alpha)$. If, moreove $A(\theta_0)$ is UMP for $H_\circ : \theta = \theta_0$ against $H_1 : \theta \neq \theta_0$, then $S(\boldsymbol{x})$ defined in (19.7.1) is, f varying $\boldsymbol{x}$ over $\mathcal{X}$, a family of UMA confidence sets of level $(1-\alpha)$.

Proof: We have

$$S(\boldsymbol{x}) \ni \theta \text{ iff } \boldsymbol{x} \in A(\theta).$$

Hence,

$$P_\theta\{S(\boldsymbol{x}) \ni \theta\} = P_\theta\{\boldsymbol{x} \in A(\theta)\} \geq 1 - \alpha.$$

Hence, $S(\boldsymbol{x})$, for varying $\boldsymbol{x} \in \mathcal{X}$, is a family of confidence sets for θ at level $(1-\alpha$

Let $S^*(\boldsymbol{x})$ be any other family of $(1-\alpha)$-level confidence sets. Let $A^*(\theta)$ $\{\boldsymbol{x} : S^*(\boldsymbol{x}) \ni \theta\}$. Then

$$P_\theta\{\mathbf{X} \in A^*(\theta)\} = P_\theta\{S^*(\boldsymbol{x}) \ni \theta\} \geq 1 - \alpha.$$

Again, since $A(\theta_0)$ is UMP for $H_\circ(\theta = \theta_0)$ against $H(\theta \neq \theta_0)$, it follows that

$$P_\theta\{\mathbf{X} \in A^*(\theta_0)\} \geq P_\theta\{\mathbf{X} \in A(\theta_0)\} \;\; \forall\, \theta \neq \theta_0.$$

Hence,

$$P_\theta\{S^*(\boldsymbol{X}) \ni \theta_0\} \geq P_\theta\{\boldsymbol{X} \in A(\theta_0)\} = P_\theta\{S(\boldsymbol{X}) \ni \theta_0\} \ \forall\, \theta \neq \theta_0.$$

Hence the proof.

The above theorem is not of much use, since a UMP test against two-sided hypothesis generally does not exist. The following theorem indicates how a family of most accurate unbiased confidence sets can be obtained from a family of UMPU tests.

Theorem 19.7.2 *Let $A_0(\theta_0)$ be the acceptence region for the level-α UMPU tests for $H_\circ(\theta = \theta_0)$ against $H(\theta \neq \theta_0)$ and let $S(\boldsymbol{x}) = \{\theta | \boldsymbol{x} \in A_0(\theta_0)\}$. Then the sets $S_0(\boldsymbol{x})$, for varying $\boldsymbol{x} \in \mathcal{X}$, are a family of UMAU confidence sets at the confidence level $(1-\alpha)$.*

Proof: Since $A(\theta)$ is the acceptance region of a unbiased test,

$$P_{\theta'}\{\boldsymbol{X} \in A(\theta)\} \leq 1 - \alpha, \ \forall\, \theta' \neq \theta.$$

Hence

$$P_{\theta'}\{S(\boldsymbol{X}) \ni \theta\} = P_{\theta'}\{\boldsymbol{X} \in A(\theta)\} \leq 1 - \alpha \ \forall\, \theta' \neq \theta.$$

Hence $S(\boldsymbol{X})$ is UMA. Let $S^*(\boldsymbol{x})$ be any other unbiased $(1-\alpha)$ level family of confidence sets. Let $A^*(\theta) = \{\boldsymbol{x} : S^*(\boldsymbol{x}) \ni \theta\}$. The $P_{\theta'}\{\boldsymbol{X} \in A^*(\theta)\} = P_{\theta'}\{S^*(\boldsymbol{X}) \ni \theta\} \leq 1 - \alpha$ and therefore, $A^*(\theta)$ is the acceptance region of an unbiased size α test.
Hence,

$$\begin{aligned} P_{\theta'}\{S(\boldsymbol{X}) \ni \theta\} &= P_{\theta'}\{\boldsymbol{X} \in A^*(\theta)\} \geq P_{\theta'}\{\boldsymbol{X} \in A(\theta)\} \\ &= P_{\theta'}\{S(\boldsymbol{X}) \ni \theta\}. \end{aligned}$$

The inequality holds since $A(\theta)$ is the acceptance region of a UMP unbiased test. Hence the proof.

▶ **Example 19.7.1** *Let $(X_1, \cdots, X_n)$ be a random sample from a $N(\mu, \sigma^2)$ population, where σ^2 is known. It has been noted that for testing $H_\circ(\mu = \mu_0)$ against $H(\mu \neq \mu_0)$, a UMPU level-α test has the acceptance region*

$$A(\mu_0) = \left\{\boldsymbol{x} : \frac{\sqrt{n}|\bar{x} - \mu_0|}{\sigma} \leq \tau_{\alpha/2}\right\}.$$

Now, for given μ, $\boldsymbol{x} \in A(\theta_0)$ iff

$$\bar{x} - \tau_{\alpha/2}\frac{\sigma}{\sqrt{n}} \leq \mu \leq \bar{x} + \tau_{\alpha/2}\frac{\sigma}{\sqrt{n}}.$$

Hence, for gives $\boldsymbol{x}$, the set

$$\delta(\boldsymbol{x}) = \left\{\mu : \bar{x} - \tau_{\alpha/2}\frac{\sigma}{\sqrt{n}} \leq \mu \leq \bar{x} + \tau_{\alpha/2}\frac{\sigma}{\sqrt{n}}\right\}$$

is a UMAU confidence set of level $(1-\alpha)$.

▶ **Example 19.7.2** *Let $(X_1, \cdots, X_n)$ be a random sample from a $N(\mu, \sigma^2)$ population when μ is known. It has been noted that for testing $H_o(\sigma^2 = \sigma_0^2)$ against $H(\sigma^2 \neq \sigma_0^2)$, a* UMPU *level α-test has the acceptence region*

$$W(\sigma_0^2) = \left\{ \boldsymbol{x} : \sum_{i=1}^{n} (x_i - \mu)^2 / \sigma_0^2 < \chi_1^2 \text{ or } \chi_2^2 \right\}$$

where χ_1^2, χ_2^2 are such that

$$\int_{\chi_1^2}^{\chi_2^2} f(\chi^2) d\chi^2 = 1 - \alpha \tag{19.7.2}$$

and

$$\chi_1^2 f(\chi_1^2) = \chi_2^2 f(\chi_2^2), \tag{19.7.3}$$

$f(\chi^2)$ being the p.d.f. of χ^2 with $(n-1)$ d.f.

Therefore, the acceptance region is

$$A(\sigma_0^2) = \left\{ \boldsymbol{x} : \chi_1^2 \leq \sum_{i=1}^{n} (x_i - \mu)^2 / \sigma_0^2 \leq \chi_2^2 \right\}.$$

Hence, for given σ^2, $\boldsymbol{x} \in A(\sigma^2)$ iff

$$\sum_i (x_i - \mu)^2 / \chi_2^2 \leq \sigma^2 \leq \sum_i (x_i - \mu)^2 / \chi_1^2.$$

Thus, for given $\boldsymbol{x}$, the set

$$S(\boldsymbol{x}) = \left\{ \sigma^2 : \sum_i (x_i - \mu)^2 / \chi_2^2 \leq \sigma^2 \leq \sum_i (x_i - \mu)^2 / \chi_1^2 \right\} \tag{19.7.4}$$

is a UMAU confidence set of level $(1 - \alpha)$.

Note that (19.7.4) is not the shortest length confidence interval given in (19.3.18) where χ_1^2, χ_2^2 are to obey the relation (19.7.2) and

$$\chi_1^4 f(\chi_1^2) = \chi_2^4 f(\chi_2^2). \tag{19.7.5}$$

19.8 Confidence Sets in the Presence of Nuisance Parameters

Assume that the parameter vector $\theta = (\xi, \eta)$ when ξ is the set of parameters that are to be estimated and η is the set of nuisance parameters. We have now the following definitions.

Definition 19.8.1 *A family of sets $S(\boldsymbol{x})$, for varying $\boldsymbol{x}$ over $\mathcal{X}$, of the space of ξ is said to be a family of confidence sets for ξ at the confidence level $(1 - \alpha)$ if*

$$P_{(\xi,\eta)}[S(\boldsymbol{X}) \ni \xi] \geq 1 - \alpha \;\; \forall \, \xi, \eta. \tag{19.8.1}$$

Definition 19.8.2 *A family of sets $S_0(\boldsymbol{x})$, for varying $\boldsymbol{x}$ over $\mathcal{X}$, of the space of ξ is said be a family of* UMA *confidence sets if*

$$P_{(\xi,\eta)}[S_0(\boldsymbol{X}) \ni \xi] \geq 1-\alpha \ \forall\, \xi \text{ and } \forall\, \eta \tag{19.8.2}$$

and

$$P_{(\xi',\eta)}[S_0(\boldsymbol{X}) \ni \xi] \leq P_{(\xi',\eta)}[S(\boldsymbol{X}) \ni \xi] \ \forall\, \xi, \xi'(\xi \neq \xi') \text{ and all } \eta \tag{19.8.3}$$

whatever be the other family of sets $S(\boldsymbol{x})$ satisfying (19.8.2).

Definition 19.8.3 *A family of sets $S(\boldsymbol{x})$, for varying $\boldsymbol{x}$ over $\mathcal{X}$, of the space of ξ is said to be a family of unbiased confidence sets of level* $(1-\alpha)$ *if*

$$P_{(\xi',\eta)}[S_0(\boldsymbol{X}) \ni \xi] \leq 1-\alpha \ \forall\, \xi, \xi'(\xi \neq \xi') \text{ and all } \eta. \tag{19.8.4}$$

Definition 19.8.4 *A family of sets $S_0(\boldsymbol{X})$, for varying $\boldsymbol{x}$ over $\mathcal{X}$, of the space of ξ is said to be a family of* UMAU *confidence sets of ξ with confidence coefficient* $(1-\alpha)$ *if*

$$P_{(\xi,\eta)}[S_0(\boldsymbol{X}) \ni \xi] \geq 1-\alpha \ \forall\, \xi, \eta \tag{19.8.5}$$

$$P_{(\xi',\eta)}[S_0(\boldsymbol{X}) \ni \xi] \leq 1-\alpha \ \forall\, \xi, \xi'(\xi \neq \xi') \text{ and all } \eta \tag{19.8.6}$$

and

$$P_{(\xi',\eta)}[S_0(\boldsymbol{X}) \ni \xi] \leq P_{(\xi',\eta)}[S(\boldsymbol{X}) \ni \xi] \ \forall\, \xi, \xi'(\xi \neq \xi') \text{ and all } \eta \tag{19.8.7}$$

whatever be the family of sets $S(\boldsymbol{x})$, satisfying (19.8.5) *and* (19.8.6).

The following theorems show connection between the confidence sets in the presence of nuisance parameters and similar tests in the presence of such parameters.

Theorem 19.8.1 *For each ξ_0, let $A(\xi_0)$ be the acceptance region of a level-α similar test for the hypothesis $H_o(\xi = \xi_0)$ and for each $\boldsymbol{x} \in \mathcal{X}$, let $S(\boldsymbol{x})$ denote the set of values of ξ defined by*

$$S(\boldsymbol{x}) = \{\xi : \boldsymbol{x} \in A(\xi)\}.$$

Then $S(\boldsymbol{x})$ is a family of confidence sets for ξ at level $(1-\alpha)$.

Theorem 19.8.2 *For each ξ_0, let $A_0(\xi_0)$ be the acceptance region of the level-α* UMP *similar test for $H_o(\xi = \xi_0)$ against alternative $H(\xi \neq \xi_0)$. Then $S_0(\boldsymbol{x})$ defined by*

$$S_0(\boldsymbol{x}) = \{\xi : \boldsymbol{x} \in A_0(\xi)\},$$

is, for varying $\boldsymbol{x} \in \mathcal{X}$, a family of UMA *confidence sets for ξ at the level* $(1-\alpha)$.

Theorem 19.8.3 *For each ξ_0, let $A_0(\xi_0)$ be the acceptance region of the level-α* UMPU *similar tests for $H_o(\xi = \xi_0)$ against $H(\xi \neq \xi_0)$. Then $S_0(\boldsymbol{x})$ defined by*

$$S_0(\boldsymbol{x}) = \{\xi : \boldsymbol{x} \in A_0(\xi)\},$$

is, for varying $\boldsymbol{x} \in \mathcal{X}$, a family of UMAU *confidence sets for ξ with confidence coefficient* $(1-\alpha)$.

► **Example 19.8.1** *Let $(X_1, \cdots, X_n)$ be a random sample from $N(\mu, \sigma^2)$ when both μ, σ^2 are unknown. It is known that the* UMPU *similar test of level α for testing $H_\circ(\mu = \mu_0)$ against $H(\mu \neq \mu_0)$ has the acceptance region*

$$A(\mu_0) = \{\boldsymbol{x} : \sqrt{n}|\bar{x} - \mu_0|/s \leq t_{n-1,\alpha/2}\}.$$

Now, for given $\mu, \boldsymbol{x} \in A(\mu_0)$ iff

$$\bar{x} - (t_{n-1,\alpha/2})s/\sqrt{n} \leq \mu \leq \bar{x} + (t_{n-1,\alpha/2})s/\sqrt{n}.$$

Hence, for given $\boldsymbol{x}$,

$$S_0(\boldsymbol{x}) = \left\{\mu : \bar{x} - (t_{n-1,\alpha/2}) \cdot \frac{s}{\sqrt{n}} \leq \mu \leq \bar{x} + (t_{n-1,\alpha/2}) \cdot \frac{s}{\sqrt{n}}\right\}$$

is a UMAU confidence set for μ with confidence coefficient $(1 - \alpha)$.

Note. In some cases while

$$P[T_1 < \theta < T_2] = 1 - \alpha,$$

the sample $\boldsymbol{X}$ may provide additional information allowing us to claim higher coverage probability than $(1 - \alpha)$.

This is illustrated by the following example from De Groot (1986).

► **Example 19.8.2** *Suppose two independent observations are taken from a uniform distribution $\left[\theta - \frac{1}{2}, \theta + \frac{1}{2}\right]$. Here $[X_{(1)}, X_{(2)}]$ is a 50% confidence interval for θ. In fact, if $X_{(2)} - X_{(1)} > \frac{1}{2}$, then the interval $[X_{(1)}, X_{(2)}]$ covers θ with probability 1. However, definition of confidence interval requires us to report the coverage probability of such an interval as 50% rather than 100%.*

19.9 Tolerance Interval

Definition 19.9.1 *Let the random variable X have a continuous df $F_\theta(x)$, $\theta \in \Theta$. A $100\beta\%$ tolerance interval for F with confidence coefficient γ is a random interval $(L_1, L_2), L_1 < L_2$, such that the probability is γ that this interval covers at least 100β percentage of the distribution i.e.,*

$$P_\theta\left[\int_{L_1}^{L_2} dF(x) \geq \beta\right] = \gamma \qquad \forall\, \theta \in \Theta$$

$$\text{i.e.,} \quad P_\theta\left[F_\theta(L_2) - F_\theta(L_1) \geq \beta\right] = \gamma \quad \forall\, \theta \in \Theta. \tag{19.9.1}$$

It is clear that even for a given (β, γ), there would be an infinite number of intervals (L_1, L_2) for which (19.9.1) is satisfied. Among all such intervals one should choose one for which the length $L_2 - L_1$ is minimum.

Example 19.9.1 *Let $X_1, \cdots, X_n$ be a random sample from a $N(\mu, \sigma^2)$-population with nown μ, but unknown σ^2. Find a (β, γ)-tolerance interval for F.*

onsider tolerance limits of the form

$$L_1(x) = \mu - \lambda s_0$$
$$L_2(x) = \mu + \lambda s_0$$

here $s_0^2 = \frac{1}{n}\sum_{i=1}^{n}(x_i - \mu)^2$.

is to be so chosen that

$$P_{\sigma^2}\left[\frac{1}{\sqrt{2\pi}\sigma}\int_{\mu-\lambda s_0}^{\mu+\lambda s_0} e^{-\frac{1}{2\sigma^2}(x-\mu)^2}dx \geq \beta\right] = \gamma.$$

ow,

$$\frac{1}{\sqrt{2\pi}\sigma}\int_{\mu-\lambda s_0}^{\mu+\lambda s_0} e^{-\frac{1}{2\sigma^2}(x-\mu)^2}dx = \frac{1}{\sqrt{2\pi}}\int_{-\lambda s_0/\sigma}^{\lambda s_0/\sigma} e^{-y^2/2\sigma^2}dy$$
$$= \Phi\left(\frac{\lambda s_0}{\sigma}\right) - \Phi\left(-\frac{\lambda s_0}{\sigma}\right)$$
$$= 2\Phi\left(\frac{\lambda s_0}{\sigma}\right) - 1$$

a monotonically increasing function of $\lambda s_0/\sigma$. Hence

$$\frac{1}{\sqrt{2\pi}}\int_{-\lambda s_0/\sigma}^{\lambda s_0/\sigma} e^{-y^2/2\sigma^2}dy \geq \beta$$

f

$$\frac{\lambda s_0}{\sigma} \geq \tau_{(1-\beta)/2}$$

e.,

$$\frac{\lambda^2 s_0^2}{\sigma^2} \geq \tau^2_{(1-\beta)/2}.$$

herefore, λ is to be such that

$$P\left\{\frac{\lambda^2 s_0^2}{\sigma^2} \geq \tau^2_{(1-\beta)/2}\right\} = \gamma$$

$$\text{i.e., } P\left\{\frac{ns_0^2}{\sigma^2} \geq \frac{n}{\lambda^2}\tau^2_{(1-\beta)/2}\right\} = \gamma$$

$$\text{i.e., } P\left\{\chi^2_{(n)} \geq n\tau^2_{(1-\beta)/2}/\lambda^2\right\} = \gamma$$

since ns_0^2/σ^2 has a $\chi^2_{(n)}$ distribution. Hence $n\tau^2_{(1-\beta)/2}/\lambda^2$ must be the upper γ-point of a $\chi^2_{(n)}$ distribution i.e.,

$$\frac{n\tau^2_{(1-\beta)/2}}{\lambda^2} = \chi^2_{(n);1-\gamma}$$

$$\text{i.e.,}\quad \lambda^2 = \frac{n\tau^2_{(1-\beta)/2}}{\chi^2_{(n);1-\gamma}}.$$

Hence the (β, γ)-tolerance limits for the distribution obtained from any set of n observations is

$$\mu \pm \frac{\sqrt{n}\tau_{(1-\beta)/2}}{\chi_{(n);1-\gamma}} s_0.$$

Wald and Wolfwitz (1946) considered the problem of finding tolerance limits for normal distribution where both μ and σ^2 are unknown. Tables connected with computing these limits have been given by Bowker (1947).

We now consider use of order statistics for finding tolerance limits. Let $X_{(1)}, \cdots, X_{(n)}$ be the order statistics corresponding to observations $X_1, \cdots, X_n$ drawn from a population with continuous d.f. F. If $(X_{(r)}, X_{(s)})$, $r < s$ is a (β, γ)-tolerance interval,

$$P\left[F(X_{(s)}) - F(X_{(r)}) \geq \beta\right] = \gamma \quad \forall\, \theta \in \Theta. \tag{19.9.2}$$

Since F is continuous, $F(x)$ has a uniform distribution over $(0, 1)$ [section 8.3]. Therefore,

$$F\left(X_{(s)}) - F(X_{(r)}\right) = U_{(s)} - U_{(r)},$$

where we write $F(X_{(t)}) = U_{(t)}$ and hence $U_{(1)} < U_{(2)} < \cdots < U_{(n)}$ are the order statistics from a $U(0, 1)$-distribution. It is required that

$$P\left[U_{(s)} - U_{(r)} \geq \beta\right] = \gamma. \tag{19.9.3}$$

From (11.6.16), the p.d.f. of $W = U_{(s)} - U_{(r)}$ is given by

$$dF(w) = \frac{n!}{(s-r-1)!(n-s+r)!} w^{s-r-1}(1-w)^{n-s+r} \quad 0 < w < 1. \tag{19.9.4}$$

Therefore, by (19.9.3), we require

$$\gamma = \int_\beta^1 n \binom{n-1}{s-r-1} w^{s-r-1}(1-w)^{n-s+r} dw \tag{19.9.5}$$

$$= \sum_{i=s-r-1}^{n} \binom{n}{i} (1-\beta)^i \beta^{n-i} = \sum_{i=0}^{s-r-1} \binom{n}{i} \beta^i (1-\beta)^{n-i}$$

by (5.4.5). However, given n, β and γ it may not be always possible to find r, s such that the exact tolerance coefficient γ is realised.

▶ **Example 19.9.2** *Let $X_1, \cdots, X_7$ be a sample from a continuous d.f. F. Find $r < s$ such that $X_{(s)} - X_{(r)}$ covers at least 40% of the population values with a probability 0.9.*

We require to find r, s such that

$$P\left[U_{(s)} - U_{(r)} \geq 0.4\right] = 0.9$$

$$\text{i.e.,} \quad \sum_{i=0}^{s-r-1} \binom{7}{i} (0.4)^i (0.6)^{7-i} = 0.9$$

It follows from table A.2 that for $s - r = 5$, $\gamma = 0.9037$. Hence each of intervals $(X_{(6)} - X_{(1)})$, $(X_{(7)} - X_{(2)})$ will cover at least 40% of the population values with probability 0.9037.

Exercise

1. Let $(X_1, \cdots, X_n)$ be a random sample from the exponential distribution $f(x;\theta) = \theta e^{-\theta x}$, $0 \leq x < \infty$. Using the m.l.e. of θ as $1/\bar{X}_n$, show that a $100\gamma\%$ large-sample confidence interval of θ is

$$\left(\frac{1/\bar{X}_n}{1 + \frac{Z}{\sqrt{n}}}, \frac{1/\bar{X}_n}{1 - \frac{Z}{\sqrt{n}}} \right)$$

where Z is such that $\Phi(Z) - \Phi(-Z) = \gamma$.

2. Let X be an observation from $f(x;\theta) = \theta x^{\theta-1}, 0 < x < 1, \theta > 0$.

   (a) Find a confidence interval for θ by the pivotal quantity method.

   (b) Show that $\left(-\frac{1}{2\log X}, -\log X\right)$ is a confidence interval for θ. Find its confidence coefficient.

3. Let $X_1 . X_2$ be a random sample from a rectangular distribution with p.d.f.

$$f(x;\theta) = \begin{cases} 1/\theta, \text{ if } 0 \leq x \leq \theta \\ & 0 < \theta < \infty. \\ 0, \text{ otherwise.} \end{cases}$$

Show that $(x + x_2)/[1 \mp (1 - \sqrt{\alpha})]$ are confidence limits to θ with confidence coefficient $(1 - \alpha)$.

Show also that if $x_{(2)}$ is the largest of the two observations, then $(x_{(2)}, x_{(2)}/\sqrt{\alpha})$ is also confidence interval to θ with the same confidence coefficient.

4. Obtain the shortest expected length confidence interval for θ of confidence level p based on a random sample of size n from $U(0, \theta)$ distribution.

5. X is an observation from $f(x;\theta) = \theta e^{-\theta x}, 0 < x < \infty, \theta > 0$. Show that $(X, 2X)$ is a confidence interval for $1/\theta$. Find its confidence coefficient. Find another confidence interval for $1/\theta$ that has the same confidence coefficient but smaller expected length.

6. Let $X_1, \cdots, X_n$ have exponential distribution with density $\theta e^{-\theta x}$. Let $G(x,\theta) = 1 - F(x,\theta)$. Considering the distribution of the pivotal statistic $W = -2\sum_{i=1}^{n} \log G(X_i,\theta)$ show that $(1-\alpha)$-level confidence interval for θ is

$$\left(\chi^2_{2n,1-\frac{\alpha}{2}} \Big/ 2\sum_{i=1}^{n} X_i, \quad \chi^2_{2n,\frac{\alpha}{2}} \Big/ 2\sum_{i=1}^{n} X_i\right).$$

Note that the use of $T = -2\sum_{i=1}^{n} \log F(x_i,\theta)$ does not give an explicit expression for confidence interval for θ in this case.

7. Let $(X_1, \cdots, X_n)$ be a random sample from a Bernoulli distribution, $P(X = 1) = \theta = 1 - P(X = 0)$. Using $\bar{X}_n$, the maximum likelihood estimate of θ, show that a $100\gamma\%$ confidence-interval of θ, obtained by inverting the inequality in $P\left\{-Z < \frac{\hat{\theta}_n - \theta}{\sqrt{\frac{\theta(1-\theta)}{n}}} < Z\right\} \simeq \gamma$ is

$$\frac{2n\hat{\theta}_n + Z^2 \mp Z\sqrt{4n\hat{\theta}_n + Z^2 - 4n\hat{\theta}_n^2}}{2(n+Z^2)}.$$

If the terms of order $1/\sqrt{n}$ and lower are neglected, show that the above limits reduce to

$$\hat{\theta}_n \mp Z\sqrt{\frac{\hat{\theta}_n(1-\hat{\theta}_n)}{n}}.$$

8. Let $(X_1, \cdots, X_m)$, $(Y_1, \cdots, Y_n)$ be independent random samples from $N(\mu_1, \sigma_1^2)$, $N(\mu_2, \sigma_2^2)$, when both σ_1^2, σ_2^2 are known. Find a $100(1-\alpha)\%$-confidence interval for $(\mu_1 - \mu_2)$. Consider also the case when both σ_1^2, σ_2^2 are unknown and are unequal.

9. Let $(X_1, \cdots, X_n)$ be a sample from a gamma population $G(1,\theta)$. Find the shortest confidence interval for θ at level $(1-\alpha)$, based on sufficient statistic for θ.

10. Let μ be the unknown median of a random variable X whose distribution is continuous but unknown. Describe, after deriving necessary results, how you may set confidence limits for μ on the basis of a random sample of size five from the distribution.

11. In problem 8 find a confidence interval for the ratio σ_2^2/σ_1^2, both when μ_1, μ_2 are known and when μ_1, μ_2 are unknown. Consider also the case when either μ_1 or μ_2 is unknown but the other is known.

12. Find a $100(1-\alpha)\%$ confidence interval for θ when sampling from $f(x;\theta) = 1$, $\theta - \frac{1}{2} < x < \theta + \frac{1}{2}$.

13. Let X have a p.d.f. $f(x;\theta) = \frac{1}{\theta}e^{-x/\theta}, x > 0,\ \theta > 0$. Show that

$$P\{K_1X < \theta < K_2X\} = e^{-1/K_2} - e^{-1/K_1}\ \forall\,\theta \in \Theta.$$

14. Let $X_1, \cdots, X_n$ be a sample from a discrete uniform distribution $f(x;N) = \frac{1}{N}$, $x = 1, 2, \cdots, N,\ N = 1, 2, \cdots$. Find an upper $100(1-\alpha)\%$ confidence limit for N, based on max $(X_1, \cdots, X_n)$.

15. Let $X_1, \cdots, X_{10}$ be a random sample from a Poisson distribution with parameter λ. Find a confidence belt for λ. Hence find a confidence interval for λ when $\sum_{i=1}^{10} X_i = 7$.

16. Let $X_1, \cdots, X_n$ be a random sample from the exponential p.d.f. $f(x;\theta) = \theta e^{-\theta x}$, $x > 0,\ \theta > 0$. Show that a central confidence interval for θ with an approximate confidence coefficient $(1-\alpha)$ is $\left(\frac{\sqrt{n}}{\bar{X}(\sqrt{n}+\tau_{\alpha/2})}, \frac{\sqrt{n}}{\bar{X}(\sqrt{n}-\tau_{\alpha/2})}\right)$.

17. Let $X_1, \cdots, X_n$ be a random sample from

$$f(x;\theta) = \begin{cases} e^{-(x-\theta)}, & x > \theta \\ 0, & \text{otherwise.} \end{cases}$$

Find the shortest-length confidence interval for θ at level $(1-\alpha)$, based on a sufficient statistic for θ.

18. Let $X_1, \cdots, X_n$ be a random sample from a Poisson distribution with parameter λ. Show that a central confidence interval for λ with an approximate confidence coefficient $(1-\alpha)$ is

$$\left\{\bar{X} + \left(\tau^2_{\alpha/2}/2n\right) \pm \sqrt{\frac{\bar{X}\tau^2_{\alpha/2}}{n} + \frac{\tau^4_{\alpha/2}}{4n^2}}\right\}.$$

Replacing λ by $\hat{\lambda}$, the m.l.e. of λ, in the variance of the asymptotic distribution of $\hat{\lambda}$, show that a central confidence interval for λ with an approximate confidence coefficient $(1-\alpha)$ is

$$\left(\bar{X} - \tau_{\alpha/2}\sqrt{\frac{\bar{X}}{n}}, \bar{X} + \tau_{\alpha/2}\sqrt{\frac{\bar{X}}{n}}\right).$$

19. Let $X_1, \cdots, X_n$ be a random sample from a $N(0, \sigma^2)$ population. Show that a central confidence interval for σ with an approximate confidence coefficient $1-\alpha$ is

$$\left(\frac{(\Sigma X_i^2/n)^{1/2}}{1+\tau_{\alpha/2}/\sqrt{2n}}, \frac{(\Sigma X_i^2/n)^{1/2}}{1-\tau_{\alpha/2}/\sqrt{2n}}\right).$$

Hence, considering example 19.5.1 show that this confidence interval is not invariant under transformation.

20. Let $X_1, \cdots, X_n$ be a random sample from a Bernoulli distribution with parameter θ Show that m.l.e. of θ is $\bar{X}$ and $E\left(\frac{\partial^2 \log L}{\partial\theta^2}\right) = -\frac{n}{\theta(1-\theta)}$. Hence find a centra confidence interval for θ with an approximate confidence coefficient $(1-\alpha)$.

21. In nine independent Bernoulli trials with constant probability of success θ, fou successes are observed. Show that an approximate 95% confidence interval for θ i (0.14, 0.79).

22. Let X_1, X_2 be two independent r.v.'s with p.d.f. $f(x;\theta) = \frac{1}{\theta}e^{-x/\theta}$, $x > 0$, $\theta > 0$ Show that the distribution of the statistic U/θ, as well as that of the statistic V/θ where $U = X_1 + X_2$ and $V = \min(X_1, X_2)$, are independent of θ. Hence, for all θ

(i) $P(U/k > \theta) = e^{-k}(1+k)$.
(ii) $P(lV > \theta) = e^{-2/l}$.

[*Hints:* If

$$e^{-K}(1+K) = e^{-2/l} \qquad \text{(a}$$

both the confidence intervals $\left(0, \frac{U}{K}\right)$ and $(0, lV)$ have the same confidence coefficient.

For $K = 2$, $l \simeq 2.2$. In this case if $X_1 \simeq X_2$, then $U/2 < 2.2V$ and the confidence interval give by (i) is shorter than the one in (ii). On the other hand if $X_1 > 4X_2$ then $U/2 > 2.5V > 2.2V$ and the situation is reversed.

Moreover, $E\left(\frac{U}{K}\right) = \frac{2\theta}{K}$; $E(l.V) = \frac{l\theta}{2}$.

Again, from the relation (a),

$$\exp\left(\frac{K}{2} - \frac{2}{l}\right) = (K+1)\exp\left(-\frac{K}{2}\right) \qquad \text{(b}$$

The expression on the right-hand side of (b) first increases with K and then decreases. Therefor $\frac{2}{K} \nleq \frac{l}{2}$ for all values of K. Therefore, none of the confidence intervals in (a) and (b) is uniforml shorter than the other.]

23. Find the smallest value of n such that the interval $(X_{(1)}, X_{(n)})$ covers at least 80 of the population values with probability 0.9.

24. Find the relation between n and p such that $(X_{(1)}, X_{(n)})$ covers at least p per cen of the distribution with probability $\geq 1-p$.

25. Let $Y_1 < \cdots < Y_p$ be the order statistic of a sample from the strictly increasin continuous function $F(x)$.

(a) Show that the probability is $1 - n\beta^{n-1} + (n-1)\beta^n$ that $P(Y_n) - P(Y_1) > \beta$
(b) Show that $P|Y_r < Z_p| = I_p(r, n-r+1)$, when $I_p(a,b)$ is the incomplete bet function and $F(Z_p) = p$.

26. Let $(X_1, \cdots, X_n)$ be a random sample of size n from a population with continuou distribution function $F(x)$ with the median μ_e, i.e., $F(\mu_e) = \frac{1}{2}$. Obtain $P[X_{(1)} < \mu_e < X_{(n)}]$ where $X_{(1)} < X_{(2)} < \cdots < X_{(n)}$ denotes the order statistics of th sample. Determine the minimum sample size n such that $(X_{(1)}, X_{(n)})$ is a non parametric confidence interval for the population median μ_e at level 0.99.

Chapter 20

Some Large Sample Procedures

20.1 Introduction

In this chapter we shall consider two major types of large sample procedures. One is testing of different hypotheses related to categorical data and the other is transformation of statistics in order to achieve some desirable properties in the large sample. Sections 20.2 to 20.4 deal with the χ^2-tests for testing hypotheses related to categorised data. Section 20.5 considers some other large sample tests based on χ^2 statistic. Measures of association of attributes in a contingency table are also considered in this section. Transformation of statistics and variables are considered in Sections 20.6 and 20.7.

20.2 Test for Goodness of Fit

We shall first consider different tests of significance based on chi-square distribution in the analysis of categorical data. By categorical or categorised data we mean the data which are presented in the form of frequencies falling in different categories or classes. The categorisation may be with respect to one or more characters, which may be attributes or variables or both. A two-way table of categorical data is called a *contingency* table. In this section we consider χ^2-test for goodness of fit.

Suppose that it is required to test the hypothesis that the given data has been obtained by random sampling from an infinite population with known values of parameters. The given data can usually be arranged in the form of a frequency distribution in various classes or cells. The corresponding theoretical frequencies are obtained from a knowledge of the p.d.f. of the population and our problem is to test the compatibility of observed and theoretical frequencies. To be more explicit we want to test whether the deviations between the observed frequencies and the expected frequencies may be ascribed due to sampling fluctuations or are large enough to suggest that perhaps the data could not have possibly come from the assumed theoretical population. This is the problem of testing goodness of fit. Note that we are assuming here that the forms of the theoretical population distributions as well as all of its parameters are complete known.

Let us suppose that the elements of an infinite population are divided into k mutua exclusive sets $S_i (i = 1, \cdots, k)$ such that the probability that an element belongs to is $\pi_i (> 0)$ where $\Sigma \pi_i = 1$. A random sample of size n is drawn from this populati and it is found that f_i elements in the sample belongs to S_i. Therefore, the observ frequency in the cell S_i is f_i and the expected frequency is $n\pi_i$ $(i = 1, \cdots, k)$. As measure of goodness of fit between the observed frequency and the expected frequen Karl Pearson suggested the following statistic

$$\chi^2 = \sum_{i=1}^{k} \frac{(O_i - e_i)^2}{e_i} = \sum_{i=1}^{k} \frac{(f_i - n\pi_i)^2}{n\pi_i} \qquad (20.2.$$

where O_i and e_i denote respectively the observed and the expected frequency in S_i.

20.2.1 Derivation of Distribution of χ^2

The probability of obtaining f_i observations in $S_i (i = 1, \cdots, k)$ in a random sample size n is given by the multinomial distribution as

$$P = \frac{n!}{f_1! \cdots f_k!} \pi_1^{f_1} \cdots \pi_k^{f_k} \qquad (20.2$$

where $\Sigma f_i = n$, $\Sigma \pi_i = 1$. If the sample is sufficiently large so that $f_1, \cdots, f_k$ are n small, then using Stirling's approximation for factorials, we have

$$P \sim \frac{\sqrt{2\pi} e^{-n} n^{n+\frac{1}{2}}}{\prod_{i=1}^{k} \left[\sqrt{2\pi} e^{-f_i} f_i^{f_i + \frac{1}{2}} \right]} \pi_1^{f_1} \cdots \pi_k^{f_k}$$

$$= K \left(\frac{n\pi_1}{f_1} \right)^{f_1 + \frac{1}{2}} \cdots \left(\frac{n\pi_k}{f_k} \right)^{f_k + \frac{1}{2}} \qquad (20.2$$

where K is a constant independent of $f_1, \cdots, f_k$. Define

$$X_i = \frac{f_i - n\pi_i}{\sqrt{n\pi_i}} = \frac{f_i - e_i}{\sqrt{e_i}} \text{ where } e_i = n\pi_i$$

so that

$$\chi^2 = \sum_{i=1}^{k} X_i^2.$$

Taking logarithm of both sides of (20.2.3),

$$\log_e \left(\frac{P}{K} \right) \sim \sum_i \left(f_i + \frac{1}{2} \right) \log_e \left(\frac{e_i}{f_i} \right)$$

$$= - \sum_i \left[\left(e_i + \frac{1}{2} + X_i \sqrt{e_i} \right) \log_e \left(1 + \frac{X_i}{\sqrt{e_i}} \right) \right].$$

If e_i is large, X_i will be small compared to $\sqrt{e_i}$ and on expanding the logarithms, we have, to first order of approximation,

$$\log_e\left(\frac{P}{K}\right) \sim -\sum_i\left[\left(e_i+\frac{1}{2}+X_i\sqrt{e_i}\right)\left(\frac{X_i}{\sqrt{e}}-\frac{X_i^2}{2e_i}+\cdots\right)\right]$$
$$= -\frac{1}{2}\sum_i X_i^2 - \sum_i X_i\sqrt{e_i} - 0\left(\frac{1}{\sqrt{e_i}}\right).$$

Now,

$$\sum_i X_i\sqrt{e_i} = \sum_i (f_i - e_i) = 0. \tag{20.2.4}$$

Hence to order $1/\sqrt{e_i}$, we have

$$P \sim K\exp\left[-\frac{1}{2}\sum_{i=1}^{k} X_i^2\right]. \tag{20.2.5}$$

This show that $X_1, \cdots, X_k$ are distributed as independent $N(0,1)$ variables in the large sample. The variables X_i are, however, subject to the linear constraint (20.2.4). Hence

$$\chi^2 = \sum_{i=1}^{k} X_i^2$$

follows in the large sample a χ^2 distribution with $(k-1)$ d.f.

If the fit between the observed frequencies and expected frequencies is good, value of χ^2 would be expected to be small. Hence, one would reject the hypothesis of goodness of fit if the observed value of Pearsonian χ^2 exceeds the tabulated value $\chi^2_{(k-1),1-\alpha}$ at $100\alpha\%$ level of significance.

Note that χ^2 in (20.2.1) can also be written as

$$\chi^2 = \sum_{i=1}^{k}\frac{O_i^2}{e_i} - n = \sum_{i=1}^{k}\frac{f_i^2}{n\pi_i} - n. \tag{20.2.6}$$

In general, the cell probabilities $\pi_1(\boldsymbol{\theta}), \cdots, \pi_k(\boldsymbol{\theta})$ will involve some unknown parameters $\boldsymbol{\theta} = (\theta_1, \cdots, \theta_s)$ [We have assumed above that $\boldsymbol{\theta}$ is known.] When $\boldsymbol{\theta}$ is estimated from the frequencies $f_1, \cdots, f_k$, $\hat{\pi}_i$ becomes functions of the sample frequencies $f_1, \cdots, f_k$. In such a case the asymptotic distribution of

$$\sum_{i=1}^{k}(f_i - n\hat{\pi}_i(\hat{\theta}))^2/n\hat{\pi}_i(\hat{\theta})$$

will more or less depend on the method of estimation of $\boldsymbol{\theta}$. The problem of finding the limiting distribution in these situations was first considered by Fisher. It has been shown that for an important class of methods of estimation, the estimation of each parameter will impose a linear restriction on the cell frequencies $f_1, \cdots, f_k$. The asymptotic distribution of

$$\sum_{i=1}^{k}(f_i - n\pi_i(\hat{\boldsymbol{\theta}}))^2/n\pi_i(\hat{\boldsymbol{\theta}})$$

will be a χ^2 distribution with $(k-1-s)$ d.f., s d.f. being lost for the estimation of s parameters.

We recall the following theorem due to Crámer (1946).

Theorem 20.2.1 *Let the cell probabilities $\pi_1(\boldsymbol{\theta}), \cdots, \pi_k(\boldsymbol{\theta})$ be specified functions of $s(<k)$ unknown parameters $\boldsymbol{\theta} = (\theta_1, \cdots, \theta_s)$ such that for all points of a non-degenerate interval I in the s-dimensional parameter space of $\boldsymbol{\theta}$, π_i satisfy the following conditions:*

(i) $\pi_i(\boldsymbol{\theta}) > c^2 > 0, i = 1, \cdots, k$;

(ii) every π_i has continuous derivatives $\frac{\partial \pi_i}{\partial \theta_j}$ and $\frac{\partial^2 \pi_i}{\partial \theta_j \partial \theta_k}$;

(iii) the matrix $\left(\left(\frac{\partial \pi_i}{\partial \theta_j}\right)\right)$ has rank s.

Let f_i be the observed cell frequency for the ith category. Then the modified χ^2 method, which is equivalent to maximum likelihood method, gives a unique estimator $\hat{\boldsymbol{\theta}}$ such that $\hat{\boldsymbol{\theta}}$ converges in probability to $\boldsymbol{\theta}_0$, the true value of $\boldsymbol{\theta}$, an inner point of I, as $n \to \infty$.

Also

$$\sum_{i=1}^{k} \frac{(f_i - n\pi_i(\hat{\boldsymbol{\theta}}))^2}{n\pi_i(\hat{\boldsymbol{\theta}})} \tag{20.2.7}$$

has asymptotically a $\chi^2_{(k-s-1)}$ distribution.

■ NOTE 20.2.1 *In the above theorem θ's have been estimated by using the statistic $f_1, \cdots, f_k$ rather than using the actual values of the variable $X_1, \cdots, X_n$. If, however observations $X_1, \cdots, X_n$ were available, then one could estimate $\theta_i (i = 1, \cdots, s)$, more efficiently by using, say, maximum likelihood estimators based on $X_1, \cdots, X_n$. In such a case, the limiting distribution of the statistic* (20.2.7) *is no longer a χ^2 distribution with $(k-s-1)$ d.f.; instead the limiting distribution is bounded between a $\chi^2_{(k-s-1)}$ distribution and a $\chi^2_{(k-1)}$ distribution. For further discussions, the reader may refer to Kendall and Stuart, vol.* 2.

■ NOTE 20.2.2 (a) *In order that the χ^2 test for goodness of fit is applicable, the following conditions must be satisfied:*

(i) *The members of the sample should be independent.*

(ii) *Constraints on cell frequencies, if any, should be linear.*

(iii) *Since a discrete distribution is being approximated by a continuous χ^2 distribution, which is valid in large sample, n should be sufficiently large, say, ≥ 50. Again, the cell-frequencies should not be too small, should be, say ≥ 5. To avoid irregularities due to discontinuity the expected frequencies $e_i \geq 5$ for all i.*

It has been found empirically, that χ^2-approximation will be satisfactory if all $f_i \geq 5$, $e_i \geq 5$ and $k \geq 5$. If k is less than 5, it is desirable to have e_i somewhat larger than 5.

If some of the cell-frequencies do not exceed 5, they may be combined with other cell-frequencies so that this condition is satisfied.

(b) *Since the distribution of χ^2 defined in* (20.2.1) *does not depend on the form of the distribution of the population from which the sample has been drawn, the above large sample test may be regarded as a non-parametric procedure.*

▶ **Example 20.2.1** *The Mendelian theory states that the frequencies in the four phenotypic clases AB, Ab, aB and ab should be in the ratio* 9 : 3 : 3 : 1. *Test whether the observed frequencies,* 105, 25, 28 *and* 5 *in the respective clases, are in agreement with the theory.*

The observed value of

$$\chi^2_{(3)} = \sum_{i=1}^{4} \frac{(n_i - np_i)^2}{np_i} = \sum_{i=1}^{4} \frac{n_i^2}{np_i} - n = 5.8016$$

where $p_1 = \frac{9}{16}$, $p_2 = p_3 = \frac{3}{16}$, $p_4 = \frac{1}{16}$. Since observed $\chi^2_{(3)} < \chi^2_{(3),.95} = 7.815$ the hypothesis is accepted at 5% level of significance.

▶ **Example 20.2.2** *One thousand families were classified according to sex and colour-blindedness.*

	Male	Female
Normal	442	514
Colour-Blind	38	6

According to a genetic model, this distribution should have the following relative frequency distribution where $q = 1 - p$ is the proportion of colour-blind individuals in the population.

	Male	Female
Normal	$p/2$	$p^2/2 + pq$
Colour-Blind	$q/2$	$q^2/2$

Are the data consistent with the model?

The likelihood function is

$$L = \text{Const.}\left(\frac{p}{2}\right)^{n_1}\left(\frac{p^2}{2}+pq\right)^{n_2}\left(\frac{q}{2}\right)^{n_3}\left(\frac{q^2}{2}\right)^{n_4}$$

$$\frac{\partial \log L}{\partial p} = 0 \Rightarrow Ap^2 + Bp + C = 0$$

where

$$\begin{aligned} A &= n_1 + 2n_2 + n_3 + 2n_4 = 1514 \\ B &= -(3n_1 + 4n_2 + 2n_3 + 4n_4) = -3482 \\ C &= 2(n_1 + n_2) = 1912. \end{aligned}$$

This gives

$$\hat{p} = 0.90607.$$

The expected frequencies are : $n_1^0 = 453$, $n_2^0 = 496$, $n_3^0 = 47$, $n_4^0 = 4$.

Hence,

$$\chi^2_{(2)} = \sum_{i=1}^{4} \frac{n_i^2}{n_i^0} - n = 3.643 < 5.99 = \chi^2_{(2),0.95}.$$

The hypothesis is accepted at 5% level of significance.

20.2.2 Derivation of Pearsonian χ^2-distribution by Poisson Approximation

Consider P given in (20.2.2),

$$\begin{aligned} P\left[f_1,\cdots,f_k \middle| \sum f_j = n\right] &= \frac{n!}{f_1!\cdots f_k!}\pi_1^{f_1}\cdots\pi_k^{f_k} \\ &= \frac{n!}{f_1!\cdots f_k!}\left(\frac{m_1}{n}\right)^{f_1}\cdots\left(\frac{m_k}{n}\right)^{f_k} \\ &\qquad \text{where } m_j = n\pi_j (j = 1,\cdots,k) \\ &= \frac{n!}{n^n e^{-n}}\left(\frac{e^{-m_1}m_1^{f_1}}{f_1!}\right)\cdots\left(\frac{e^{-m_k}m_k^{f_k}}{f_k!}\right) \qquad (20.2.8) \end{aligned}$$

(20.2.8) represents the conditional probability distribution of k Poisson variates $f_i \sim P(m_i)$ subject to the condition that $f_1 + \cdots + f_k = n$ (vide also exercise 35, chapter 7). If m_i's are large, ≥ 5 $(\forall\, i)$, then each Poisson m.f. can be approximated by a normal density function. Therefore, (20.2.8)

$$\simeq \frac{n!}{n^n e^{-n}} \prod_{i=1}^{k} \left[\left\{ \frac{1}{\sqrt{2\pi m_i}} \exp\left(-\frac{(f_i - m_i)^2}{2m_i} \right) \right\} \left(1 + 0\left(\frac{1}{\sqrt{m_i}} \right) \right) \right]$$

$$\simeq \frac{\sqrt{2\pi} e^{-n} n^{n+\frac{1}{2}}}{e^{-n} n^n} \left(\frac{1}{2\pi} \right)^{k/2} \left\{ \frac{1}{\sqrt{m_1 \cdots m_k}} \exp\left(-\frac{1}{2} \sum_{i=1}^{k} \frac{(f_i - m_i)^2}{m_i} \right) \right\}$$

$$\left\{ 1 + 0\left(\frac{1}{\sqrt{m_1 \cdots m_k}} \right) \right\} \text{ (using Stirling's approximation for factorials)}$$

$$\simeq \frac{\sqrt{n}}{2\pi^{(k-1)/2}} \cdot \frac{1}{\sqrt{m_1 \cdots m_k}} \exp\left(-\frac{1}{2} \sum_{i=1}^{k} \frac{(f_i - m_i)^2}{m_i} \right). \tag{20.2.9}$$

Let us put

$$\chi^2 = \sum_{i=1}^{k} \frac{(f_i - m_i)^2}{m_i}.$$

We want to find the probability that

$$\left(\chi - \frac{1}{2} d\chi \right)^2 \leq \chi^2 \leq \left(\chi + \frac{1}{2} d\chi \right)^2. \tag{20.2.10}$$

This is given by

$$\sum P\left(f_1, \cdots, f_k | \sum f = n \right)$$

where the summation is taken over all f_j's satisfying (20.2.10). Now

$$\sum \frac{\sqrt{n}}{(2\pi)^{(k-1)/2} \sqrt{m_1 \cdots m_k}} \exp\left(-\frac{1}{2} \sum_{i=1}^{k} \frac{(f_i - m_i)^2}{m_i} \right)$$

$$= \sum \frac{\sqrt{n}}{(2\pi)^{(k-1)/2}} \exp\left(-\frac{1}{2} \sum_{i=1}^{k} x_i^2 \right) \Pi \Delta x_i \tag{20.2.11}$$

where

$$x_i = \frac{f_i - m_i}{\sqrt{m_i}}, \quad \Delta x_i = x_i + 1 - x_i = \frac{f_i + 1 - f_i}{\sqrt{m_i}} = \frac{1}{\sqrt{m_i}}.$$

In order to replace the summation involved in (20.2.11) by integration we require that k, the number of classes should be sufficiently large. Hence, for large k, (20.2.11) can be written as

$$\text{Const.} \int \cdots \int_R \exp\left(-\frac{1}{2} \sum_{i=1}^{k} x_i^2 \right) \Pi dx_i \tag{20.2.12}$$

where the region of integration R is defined by (20.2.10). Since within R, $\sum_{i=1}^{k} x_i^2$ is (nearly) a constant, χ^2, the expression (20.2.12) is, approximately,

$$\text{Const.} \exp\left(-\frac{1}{2} \chi^2 \right) \left[\int \cdots \int_R \Pi dx_i \right]. \tag{20.2.13}$$

The quantity [ ] is the volume of the annular space between two concentric hyperspheres of radii $\chi - \frac{1}{2}d\chi$ and $\chi + \frac{1}{2}d\chi$. Again, since

$$\sum_{i=1}^{k} f_i = \sum_{i=1}^{k} m_i (= n), \quad \sum_{i=1}^{k} \sqrt{m_i} x_i = 0.$$

Therefore, variables $x_1, \cdots, x_k$ are subject to a linear constraint and hence both the above-mentioned hyperspheres are of dimension $(k-1)$. By (12.3.7), the volume of this annular space is proportional to $\chi^{k-2} d\chi$. Hence, the probability in (20.2.13) is

$$\text{Const.} \exp\left(-\chi^2/2\right) \chi^{k-2} d\chi, \quad 0 < \chi < \infty.$$

This shows that

$$\sum_{1}^{k} (f_i - m_i)^2 / m_i \sim \chi^2_{(k-1)} \text{ in large sample.} \tag{20.2.14}$$

□ **Remark 20.2.1** *In deriving* (20.2.14) *we have made the assumptions that* (i) *expected frequency* m_i *in each class must be large* ≥ 5; (ii) *the total number of classes* k *must be large.*

20.3 A Test of Independence in Contingency Tables

Let the data be classified into classes $A_1, \cdots, A_r$ according to category A and into classes $B_1, \cdots, B_s$ according to category B. We have thus a frequency distribution in $(r \times s)$ classes, a typical class being represented by $A_i B_j$. A two-way table of this type is called a (two-way) *Contingency* table.

Class	B_1	B_2	$\cdots$	B_j	$\cdots$	B_s	Totals
A_1	n_{11}	n_{12}		n_{1j}		n_{1s}	n_{1o}
A_2	n_{21}	n_{22}		n_{2j}		n_{2s}	n_{2o}
$\vdots$							
A_i	n_{i1}	n_{i2}		n_{ij}		n_{is}	n_{io}
$\vdots$							
A_r	n_{r1}	n_{r2}		n_{rj}		n_{rs}	n_{ro}
Totals	n_{o1}	n_{o2}		n_{oj}		n_{os}	N

Let the observed frequency in the class $A_i B_j$ be n_{ij} and the probability of an element falling in this class be p_{ij}. Also let

$$\sum_j n_{ij} = n_{io}, \quad \sum_i n_{ij} = n_{oj}, \quad \sum_i n_{io} = \sum_j n_{oj} = N,$$

$$\sum_i p_{ij} = p_{oj}, \quad \sum_j p_{ij} = p_{io}, \quad \sum_j p_{oj} = \sum_i p_{io} = 1.$$

The probability of the observed frequency is given by the multinomial law as

$$P\{n_{ij}, i=1,\cdots,r; j=1,\cdots,s\} = \frac{N!}{\prod\limits_{ij}(n_{ij})!}\prod_{i=1}^{r}\prod_{j=1}^{s}p_{ij}^{n_{ij}}. \tag{20.3.1}$$

Consider the following three models.

Model I. The total number of observations N is a fixed quantity while the marginal totals n_{io}'s $(i = 1,\cdots,r)$ and n_{oj}'s $(j = 1,\cdots,s)$ are random variables, subject to the constraints

$$\sum_i n_{io} = \sum_j n_{oj} = N.$$

Model II. One set of marginal total say, $\{n_{io}$'s $(i = 1,\cdots,r)\}$, are given fixed quantities while the other set of marginals are random variables.

Model III. Both the sets of marginals are fixed quantities.

We want to find the probability

$$P\left\{n_{ij}|\sum_j n_{ij} = n_{io}, \sum_i n_{ij} = n_{oj}, i=1,\cdots,r; j=1,\cdots,s\right\} \tag{20.3.2}$$

in the first two cases under the hypothesis that the classificatory characteristics are independent. Under the hypothesis that A and B are independent,

$$P\{A_iB_j\} = P\{A_i\}P\{B_j\} \text{ i.e., } p_{ij} = p_{io}\times p_{oj} \text{ for all } i,j. \tag{20.3.3}$$

In model III we want to find the probability of obtaining the given configuration subject to fixed marginal totals.

Model I. Under (20.3.2),

$$\begin{aligned} P\{n_{ij}, i=1,\cdots,r; j=1,\cdots,s\} &= \frac{N!}{\prod\limits_i\prod\limits_j n_{ij}!}\prod_i\prod_j(p_{io}p_{oj})^{n_{ij}} \\ &= \frac{N!}{\prod\limits_i\prod\limits_j n_{ij}!}\prod_i p_{io}^{n_{io}}\prod_j p_{oj}^{n_{oj}}. \end{aligned} \tag{20.3.4}$$

Now, the joint probability of marginal totals is

$$P\{n_{io}, n_{oj}, i=1,\cdots,r, j=1,\cdots,s\} = \sum P\{n_{ij}\}$$

where the sum is over n_{ij}'s satisfying

$$\begin{aligned} &\sum_j n_{ij} = n_{io}, \\ &\sum_i n_{ij} = n_{oj}, i=1,\cdots,r; j=1,\cdots,s, \end{aligned} \tag{20.3.5}$$

$$\therefore \quad \sum P\{n_{ij}\} = N!\left(\prod_i p_{io}^{n_{io}}\right)\left(\prod_j p_{oj}^{n_{oj}}\right)\sum \frac{1}{\prod_i\prod_j n_{ij}!}. \tag{20.3.6}$$

Lemma 20.3.1

$$S_1 = \sum \frac{1}{\prod_i\prod_j n_{ij}!} = \frac{N!}{\left(\prod_i n_{io}!\right)\left(\prod_j n_{oj}!\right)}. \tag{20.3.7}$$

Proof: Consider

$$\left(\sum_{i=1}^{r}\sum_{j=1}^{s} a_i b_j\right)^N = \sum{}' \sum \frac{N!}{\prod_i\prod_j n_{ij}!}\prod_i\prod_j (a_i b_j)^{n_{ij}}$$

$$= \sum{}' \left\{\left(\sum \frac{N!}{\prod_i\prod_j n_{ij}!}\right)\prod_i a_i^{n_{io}}\prod_j b_j^{n_{oj}}\right\} \tag{20.3.8}$$

where $\sum$ is as defined in (20.3.5) and $\sum'$ is over all possible n_{io}'s, n_{oj}'s subject to the total N.
Equating coefficients of

$$\left(\prod_i a_i^{n_{io}}\prod_j b_j^{n_{oj}}\right)$$

in the expression in the extreme right and extreme left of (20.3.8), for a fixed set of values of n_{io}'s and n_{oj}'s,

$$\sum \frac{N!}{\prod_i\prod_j n_{ij}!} = N!S_1 = \text{coeff. of } \left(\prod_i a_i^{n_{io}}\prod_j b_j^{n_{oj}}\right) \text{ in } \left(\sum_i\sum_j a_i b_j\right)^N$$

$$\left(= \left(\sum a_i\right)^N \left(\sum b_j\right)^N\right)$$

$$= \left\{\text{coeff. of } \prod_i a_i^{n_{io}} \text{ in } \left(\sum a_i\right)^N\right\}\left\{\text{coeff. of } \prod_j b_j^{n_{oj}} \text{ in } \left(\sum b_{ij}\right)^N\right\}$$

$$= \frac{N!}{\prod_i n_{io}!} \times \frac{N!}{\prod_j n_{oj}!}.$$

Therefore, from (20.3.6)

$$P\{n_{io}, n_{oj}, i = 1, \cdots, r; j = 1, \cdots, s\} = \frac{(N!)^2}{\prod_i n_{io}!\prod_j n_{oj}!}\prod_i p_{io}^{n_{io}}\prod_j p_{oj}^{n_{oj}}. \tag{20.3.9}$$

The conditional probability (20.3.2) is, therefore, from (20.3.4) and (20.3.9),

$$\frac{\prod_i n_{io}! \prod_j n_{oj}!}{N! \prod_i \prod_j n_{ij}!} = p(n_{ij}) \text{ (say)}. \tag{20.3.10}$$

Interestingly, the above expression is independent of the hypothetical values of population proportions.

Model II. Here also we have to find (20.3.2). For a fixed i,

$$P\{n_{ij}, j = 1, \cdots, s\} = \frac{n_{io}!}{\prod_j n_{ij}!} \prod_j p_{oj}^{n_{ij}} \tag{20.3.11}$$

because, $p_{ij} = p_{io} \times p_{oj}$ and the probability that a random observation belongs to B_j given that it is in A_i is p_{oj}. Hence,

$$P\{n_{ij}, i = 1, \cdots, r, j = 1, \cdots, s\} = \prod_i n_{io}! \prod_j p_{oj}^{n_{oj}} \frac{1}{\prod_i \prod_j n_{ij}!}. \tag{20.3.12}$$

Again, the joint probability distribution of n_{oj}'s $(j = 1, \cdots, s)$ is

$$P\{n_{oj}, j = 1, \cdots, s\} = \prod_i n_{io}! \prod_j p_{oj}^{n_{oj}} \sum{}'' \frac{1}{\prod_i \prod_j n_{ij}!}$$

where $\sum''$ is over all n_{ij}'s such that $\sum_j n_{ij} = n_{io}$.

Lemma 20.3.2

$$S_2 = \sum{}'' \frac{1}{\prod_i \prod_j n_{ij}!} = \frac{N!}{\prod_i n_{io}! \prod_j n_{oj}!}. \tag{20.3.13}$$

Proof: Consider the expansion

$$\begin{aligned}
\left(\sum_j b_j\right)^N &= \left(\sum_j b_j\right)^{n_{1o}} \cdots \left(\sum_j b_j\right)^{n_{ro}} \\
&= \left(\sum_{\Sigma n_{1j} = n_{1o}} \frac{n_{1o}!}{\prod_j n_{ij}!} b_j^{n_{ij}}\right) \cdots \left(\sum_{\Sigma n_{rj} = n_{ro}} \frac{n_{ro}!}{\prod_j n_{rj}!} b_j^{n_{rj}}\right) \\
&= \sum{}'' \frac{\prod_i n_{io}!}{\prod_i \prod_j n_{ij}!} \prod_i \prod_j b_j^{n_{ij}} = \sum{}'' \frac{\prod_i n_{io}!}{\prod_i \prod_j n_{ij}!} \prod_j b_j^{n_{oj}}
\end{aligned} \tag{20.3.14}$$

Equating coefficient of $\prod_j b_j^{n_{oj}}$ on both sides of (20.3.14)

$$\frac{N!}{\prod_j n_{oj}!} = \sum'' \frac{\prod_i n_{io}!}{\prod_i \prod_j n_{ij}!} = \prod_i n_{io}! S_2.$$

Hence, the lemma.

From (20.3.12) and (20.3.13), the conditional probability (20.3.2) is, again, given by the expression (20.3.10).

Model III. Here the marginal totals n_{io}'s, n_{oj}'s are given fixed quantities. For a given set of n_{ij}'s satisfying

$$\sum_j n_{ij} = n_{io}, \quad \sum_i n_{ij} = n_{oj},$$

the total number of ways in which the cells can be filled up with these frequencies n_{ij}'s is

$$\frac{N!}{\prod_i \prod_j n_{ij}!}. \tag{20.3.15}$$

Again the total number of ways in which the cells can be filled up subject to the restrictions that marginal totals are n_{io}'s and n_{oj}'s is

$$\sum \frac{N!}{\prod_i \prod_j n_{ij}!} \tag{20.3.16}$$

where the sum is over all n_{ij}'s satisfying (20.3.5). By lemma 20.3.1, (20.3.16) is

$$\frac{(N!)^2}{\prod_i n_{io}! \prod_j n_{oj}!}.$$

Hence

$$P\left\{ n_{ij} | \sum_j n_{ij} = n_{io}, \sum_i n_{ij} = n_{oj} \right\} = \frac{\prod_i n_{io}! \prod_j n_{oj}!}{N! \prod_i \prod_j n_{ij}!} \tag{20.3.17}$$

which is the same expression as (20.3.10).

Under models I and II we are interested in testing the hypotheses that the two attributes are independent i.e., $H_\circ = p_{ij} = p_{io} \times p_{oj} \;\; \forall \; i, j$.

The hypothesis involves $(r+s)$ parameters p_{io}, p_{oj} $(i=1,\cdots,r; j=1,\cdots,s)$ of which $(r+s-2)$ are independent (because of the constraints $\Sigma p_{io} = \Sigma p_{oj} = 1$). The hypothesis H_o is a composite hypothesis of $(r+s-2)$ d.f. and we can test this hypothesis by a statistic based on sufficient statistic. The conditional probability $p(n_{ij})$ is independent of parameters p_{io}'s, p_{oj}'s and as such $(n_{io}, i=1,\cdots,r; n_{oj}, j=1,\cdots,s)$ is a set of sufficient statistic for p_{io}'s, p_{oj}'s.

Under the hypothesis of independence, expected frequency in the (i,j)th cell is $e_{ij} = n\hat{p}_{io}\hat{p}_{oj} = n_{io}n_{oj}/N$, since for the distribution (20.3.1), the maximum likelihood estimator (or least square estimator or modified Chi-square estimator) of p_{io}, p_{oj} are n_{io}/N, n_{oj}/N respectively. Hence, from (20.2.1),

$$\chi^2 = \sum_i \sum_j (O_{ij} - e_{ij})^2/e_{ij} = \sum_i \sum_j \left(n_{ij} - \frac{n_{io}n_{oj}}{N}\right)^2 \Big/ \left(\frac{n_{io}n_{oj}}{N}\right). \tag{20.3.18}$$

By replacing the distribution involved in equation (20.3.10) by its multivariate normal approximation it can be shown that χ^2 given in (20.3.18) follows approximately a χ^2 distribution with $(rs-1)-(r+s-2) = (r-1)(s-1)$ d.f.

■ **NOTE 20.3.1** *Consider the following probability model for the problem. The N observations may be regarded as a random sample $(X_1,\cdots,X_N)$ from the multinomial distribution $(p_{ij},\ i=1,\cdots,r;\ j=1,\cdots,s)$. The r.v. X_t takes value $x_{ij} = 1(0)$ if the same belongs to cell (i,j) (otherwise) with $\sum_i \sum_j x_{ij} = 1$. The p.d.f. of X is*

$$f_X(x_{ij}) = \prod_i \prod_j p_{ij}^{x_{ij}}.$$

For n observation $(X_1,\cdots,X_N)$, the likelihood of the sample is, therefore,

$$L = \prod_i \prod_j p_{ij}^{n_{ij}}.$$

The complete parameter space Θ is of dimension $rs-1$ $\left(\text{because } \sum_i \sum_j p_{ij} = 1\right)$. Also m.l.e. of p_{ij} under Θ is $\hat{p}_{ij} = n_{ij}/N$.

Under H_o, the parameter space is of dimension $r+s-2$ (because $\Sigma p_{io} = \Sigma p_{oj} = 1$). The likelihood is

$$L(H_o) = \left(\prod_i p_{io}^{n_{io}}\right)\left(\prod_i p_{oj}^{n_{oj}}\right).$$

The m.l.e. of p_{io} and p_{oj} are $\hat{p}_{io} = n_{io}/N$, $\hat{p}_{oj} = n_{oj}/N$. The generalised likelihood ratio is, therefore,

$$\lambda = \frac{\sup\limits_{\theta\in\Theta_0} L(\boldsymbol{x}|\theta\in\Theta)}{\sup\limits_{\theta\in\Theta} L(\boldsymbol{x}|\theta\in\Theta)} = \frac{\left(\prod_i n_{io}^{n_{io}}\right)\left(\prod_j n_{oj}^{n_{oj}}\right)}{\prod_i \prod_j n_{ij}^{n_{ij}}}.$$

However, the distribution of λ under $H_\circ$ is not unique, because the hypothesis is composite and its exact distribution involves the unknown parameters p_{io}, p_{oj}. It is, therefore, difficult to find λ_0 such that

$$\sup_{\theta \in \Theta_0} P[\lambda \leq \lambda_0] = \alpha.$$

However, for large samples, $-2 \log \lambda \sim \chi^2_{(r-1)(s-1)}$ and this property can be used in testing $H_\circ$. This test differs from the χ^2 test in (20.3.18) by terms of order $1/\sqrt{n}$.

► **Example 20.3.1** *A group of* 1650 *school children were classified according to their performance in school tests and family economic level. Test if there is any association between these two categories.*

Economic	Performance in Tests				Total
level	Very Good	Good	Average	Poor	
Very Rich	4 (4.95)	7 (19.82)	16 (14.75)	25 (12.48)	52
Rich	13 (9.22)	37 (77.00)	79 (57.29)	73 (48.48)	202
Average	105 (90.39)	372 (362.15)	298 (269.45)	175 (228)	950
Poor	35 (42.44)	213 (170.02)	75 (126.50)	123 (107.04)	446
Total	157	629	468	396	1650

The expected frequencies under the hypothesis of independence have been shown in the parenthesis

$$\chi^2 = \sum_{i=1}^{4} \sum_{j=1}^{4} \frac{o_{ij}^2}{e_{ij}} - N = 127.61 > 25.0 = \chi^2_{(15),0.95}.$$

Hence, the hypothesis of independence is rejected at 5% level of significance.

► **Example 20.3.2** *It is desired to test the hypothesis that an observed random sample $x_1, \cdots, x_n$ has been drawn from a normal population with unknown μ and σ. The sample observations have been grouped into k classes, $A_i = (Z_{i-1}, Z_i), i = 1, \cdots, k$ with $Z_0 = -\infty$, $Z_k = \infty$ and let n_i be the frequency in A_i. The probability that a random observation falls in A_i is, therefore, $p_i = \Phi(\tau_i) - \Phi(\tau_{i-1})$ where $\tau_i = (Z_i - \mu)/\sigma$.*

Let $\hat{\mu}, \hat{\sigma}$ be m.l.e.'s of μ and σ respectively based on $n_1, \cdots, n_k$. Then $\hat{p}_i = \Phi(\hat{\tau}_i) - \Phi(\hat{\tau}_{i-1})$, $\hat{\tau}_i = (Z_i - \hat{\mu})/\hat{\sigma}$. Hence

$$Q = \sum_{j=1}^{k} (n_j - \hat{p}_j)/n\hat{p}_j \cap \chi^2_{(k-3)}.$$

$H_\circ$ is rejected if observed $\chi^2 > \chi^2_{(k-3),(1-\alpha)}$. If, however, μ and σ are estimated on the basis of data $X_1, \cdots, X_n$, then the asymptotic distribution of Q falls between $\chi^2_{(k-1)}$ and $\chi^2_{(k-3)}$ and $H_\circ$ is rejected if $\chi^2 > C$ where C lies between $\chi^2_{(k-1),(1-\alpha)}$ and $\chi^2_{(k-3),(1-\alpha)}$. However, for large k, there is little difference between these two limits.

20.3.1 χ^2 test in a 2×2 Contingency Table

Consider the contingency table 20.1.

Table 2.1 A 2×2 table

a	b	$a+b$
c	d	$c+d$
$a+c$	$b+d$	$a+b+c+d=N$

$$n_{1o} = a+b \qquad n_{2o} = c+d$$
$$n_{o1} = a+c \qquad n_{o2} = b+d.$$

Hence,

$$\begin{aligned}
\chi^2 &= \frac{\left\{a - \frac{(a+b)(a+c)}{N}\right\}^2}{\frac{(a+b)(a+c)}{N}} + \text{sum of three other similar terms} \\
&= \frac{(ad-bc)^2}{N}\left[\frac{1}{(a+b)(a+c)} + \frac{1}{(a+b)(b+d)} + \frac{1}{(c+d)(a+c)} + \frac{1}{(c+d)(b+d)}\right] \\
&= \frac{N(ad-bc)^2}{(a+b)(a+c)(b+d)(c+d)}.
\end{aligned}$$

Yates Correction for continuity

The distribution of all frequencies in a contingency table is necessarily discontinuous whereas the χ^2 distribution is a continuous one. The approximation of this distribution to a χ^2 distribution is something like approximating a binomial distribution to a normal distribution where the calculated frequency between two values x_1 and x_2 (both values inclusive) is given by the area under the corresponding normal curve between values $x_1 - \frac{1}{2}$ and $x_2 + \frac{1}{2}$.

In a 2×2 table (Table 20.1) there is only one d.f. and the frequency of only one cell can be assigned arbitrarily for fixed marginals. For such a table Yates suggested that the approximation to χ^2 is improved by replacing one cell frequency, say, d by $d \pm \frac{1}{2}$ according as $ad <$ or $> bc$ and adjusting the other frequencies so that the marginal totals remain the same.

After correcting to continuity, one has the following contingency table, assuming $ad < bc$.

Table 2.2

(Table 20.1 after correction for continuity)

$a+\frac{1}{2}$	$b-\frac{1}{2}$	$a+b$
$c-\frac{1}{2}$	$d+\frac{1}{2}$	$c+d$
$a+c$	$b+d$	$a+b+c+d=N$

Hence,

$$\chi^2_{(\text{corrected})} = \frac{N\left[\left(a+\frac{1}{2}\right)\left(d+\frac{1}{2}\right)-\left(b-\frac{1}{2}\right)\left(c-\frac{1}{2}\right)\right]^2}{(a+b)(a+c)(b+d)(c+d)}$$

$$= \frac{N\left[(ad-bc)+\frac{N}{2}\right]^2}{(a+b)(a+c)(b+d)(c+d)}.$$

V.M. Dandekar suggested a slightly different correction for continuity in a 2×2 table. This involves the calculation of χ_0^2, χ_{-1}^2 and χ_1^2 for the observed frequencies, and those obtained by increasing and decreasing the smallest frequency by unity respectively. The corrected χ^2 is

$$\chi_c^2 = \chi_0^2 - \frac{\chi_0^2 - \chi_{-1}^2}{\chi_1^2 - \chi_{-1}^2}(\chi_1^2 - \chi_0^2) \tag{20.3.19}$$

In general, Dandekar's correction is slightly better than that of Yates, although the method due to Yates is a simpler one.

The Exact treatment for a 2×2 Contingency table

Consider the table 20.1. By (20.3.10), the conditional probability of observing the given frequencies for fixed marginals is

$$\frac{(a+c)!(b+d)!(a+b)!(c+d)!}{N!a!b!c!d!}. \tag{20.3.20}$$

If the probability of the observed and more extreme frequencies is very small, the hypothesis of independence is rejected.

Fisher's exact test consists in computing the total probability of the variable taking values from zero up to the smallest frequency in the table, say, a. This probability $P = p_0 + p_1 + \cdots + p_a$ corresponds to one tail of the distribution and this is comparable with half the probability calculated from χ^2 since the later corresponds to both tails. Fisher's test can be used to test (i) independence of two attributes (ii) homogeneity of two parallel samples i.e., if the two samples have come from the same binomial population (vide section 20.4).

► **Example 20.3.3** *In an experiment on the immunisation of goats from anthrax the following results were obtained. Derive your inference on the efficiency of vaccine.*

	Died	Survived	Total
Inoculated	2	10	12
Not inoculated	6	6	12
Total	8	16	24

$$\chi^2 = \frac{24(60-12)^2}{12\times 12\times 8\times 16} = 3.$$

After applying Yates correction the adjusted table is

	Died	Survived	Total
Inoculated	2.5	9.5	12
Not Inoculated	5.5	6.5	12
Total	8	16	24

The corrected χ^2 is

$$\chi^2_{(C)} = \frac{24(5.5 \times 9.5 - 2.5 \times 6.5)^2}{12 \times 12 \times 8 \times 16} = 1.69.$$

None of the uncorrected and corrected value of χ^2 exceeds $\chi^2_{(1),0.95} = 3.84$. Hence, at this level, the inocculation may be said to be ineffective against the disease.

We consider now the exact treatment for this 2×2 table. The probability for the number of dead-inoculated to be equal to or less than 2 is

$$\begin{aligned}
&\frac{12!12!8!16!}{24!}\left[\frac{1}{0!12!8!4!} + \frac{1}{1!11!7!5!} + \frac{1}{2!10!6!6!}\right] \\
&= \frac{12!}{24!}\left[\frac{16!}{4!} + \frac{12 \times 8 \times 16!}{5!} + \frac{(12 \times 11 \times 8 \times 7) \times 16!}{2! \times 6!}\right] \\
&= \frac{4.79 \times 10^8}{6.20 \times 10^{23}}\left[\frac{2.09 \times 10^{13}}{24} + \frac{12 \times 2.09 \times 10^{13}}{120} + \frac{7392 \times 2.09 \times 10^{13}}{2 \times 720}\right] \\
&= \frac{10.01 \times 10^{21}}{6.20 \times 10^{23}}[0.04167 + 0.1 + 5.1333] \\
&= \frac{52.802 \times 10^{21}}{6.20 \times 10^{23}} = 0.08516.
\end{aligned}$$

The value of standard normal deviate corresponding to uncorrected χ^2 is 1.73205. The probability of getting a higher value of standard normal deviate is 0.042. With correction this probability is 0.1. Thus Yates correction gives a value of probability close to the exact probability calculated by Fisher's test.

It may be noted here that since in the exact test we have found the probability of 'a' being less than or equal to 2, to compare it with the probability for a given χ^2 we use either one tail of the corresponding normal curve or half the probability obtained from the χ^2 table.

20.3.2 χ^2 in a $2 \times s$ Contingency Table

Classes	B_1	B_2	$\cdots$	B_s	Total
A_1	a_1	a_2		a_s	T_a
A_2	b_1	b_2		b_s	T_b
Total	T_1	T_2		T_s	T

$$\chi^2 = \sum_{i=1}^{s}\left(a_i - \frac{T_iT_a}{T}\right)^2 \Big/ \left(\frac{T_iT_a}{T}\right) + \sum_{i=1}^{s}\left(b_i - \frac{T_iT_b}{T}\right)^2 \Big/ \left(\frac{T_iT_b}{T}\right)$$

$$= \left\{\sum_{i=1}^{s}\frac{a_i^2T}{T_iT_a} + \sum_{i=1}^{s}\frac{T_iT_a}{T} - 2\sum_{i=1}^{s}a_i\right\} + \left\{\sum_{i=1}^{s}\frac{b_i^2T}{T_iT_b} + \sum_{i=1}^{s}\frac{T_iT_b}{T} - 2\sum_{i=1}^{s}b_i\right\}$$

$$= \left\{\frac{T}{T_a}\sum\frac{a_i^2}{T_i} - T_a\right\} + \left\{\frac{T}{T_b}\sum\frac{b_i^2}{T_i} - (T - T_a)\right\}$$

$$= \frac{T}{T_a}\sum\frac{a_i^2}{T_i} + \frac{T}{T_b}\left\{\sum_i\frac{(T_i - a_i)^2}{T_i}\right\} - T$$

$$= \sum\frac{a_i^2}{T_i}\left(\frac{T}{T_a} + \frac{T}{T_b}\right) + \frac{T^2}{T_b} - T - \frac{2T_aT}{T_b}$$

$$= \frac{T^2}{T_aT_b}\sum\frac{a_i^2}{T_i} - \frac{TT_a}{T_b} = \frac{T^2}{T_aT_b}\left\{\sum\frac{a_i^2}{T_i} - \frac{T_a^2}{T}\right\}, d.f.(s-1).$$

This is known as Brandt and Snedecor's formula.

► **Example 20.3.4** *Show that for a* $2 \times n$ *table*

$$\sum_j \frac{n_1n_2\left(\frac{a_{1j}}{n_1} - \frac{a_{2j}}{n_2}\right)^2}{a_{1j} + a_{2j}} = \chi^2_{(n-1)}.$$

Class	B_1	B_2	$\cdots$	B_j	$\cdots$	B_n	Total
A_1	a_{11}	a_{12}	$\cdots$	a_{ij}	$\cdots$	a_{1n}	n_1
A_2	a_{21}	a_{22}	$\cdots$	a_{2j}	$\cdots$	a_{2n}	n_2
Total	a_{o1}	a_{o2}	$\cdots$	a_{oj}	$\cdots$	a_{on}	N

Proof:

$$\chi^2 = \sum_{i=1}^{2}\sum_{j=1}^{n}\left[\frac{\left(a_{1j} - \frac{n_ia_{oj}}{N}\right)^2}{\frac{n_ia_{oj}}{N}}\right] = \sum_{j=1}^{n}\left\{\frac{\left(a_{1j} - \frac{n_1a_{oj}}{N}\right)^2}{\frac{n_1a_{oj}}{N}} + \frac{\left(a_{2j} - \frac{n_2a_{oj}}{N}\right)^2}{\frac{n_2a_{oj}}{N}}\right\}$$

$$= \sum_{j=1}^{n}\left\{\frac{\left[a_{1j} - \frac{n_1(a_{1j}+a_{2j})}{N}\right]^2}{\frac{n_1a_{oj}}{N}} + \frac{\left[a_{2j} - \frac{n_2(a_{1j}+a_{2j})}{N}\right]^2}{\frac{n_2a_{oj}}{N}}\right\}$$

$$= \sum_{j=1}^{n}\left\{\frac{(n_2a_{1j} - n_1a_{2j})^2}{n_1n_2a_{oj}}\right\}.$$

20.4 The Chi-square Test for Homogeneity

Suppose we have samples from $p(\geq 2)$ independent multinomial populations, the probability that an element in the jth population belongs to class i is p_{ij} $(i = 1, \cdots, k;$ $j = 1, \cdots, p)$. A random sample of size n_j is drawn from the jth population and let n_{ij} be the number of observations in class i. Our object is to test that all the p samples come from the same multinomial population i.e., $p_{ij} = p_i (j = 1, \cdots, p) \; \forall \; i = 1, \cdots, k$. This is the test for homogeneity of p multinomial populations. If the parameters $p_{ij} (i = 1, \cdots, k)$ are known, then provided n_j is large

Population	Classes			
	1	2	$\cdots$	k
j	p_{1j}	p_{2j}		p_{kj}

$$\sum_{i=1}^{k} (n_{ij} - n_j p_{ij})^2 / n_j p_{ij} \sim \chi^2_{(k-1)}$$

and hence

$$Q = \sum_{j=1}^{p} \sum_{i=1}^{k} (n_{ij} - n_j p_{ij})^2 / n_j p_{ij} \sim \chi^2_{p(k-1)}$$

provided n_j's $(j = 1, \cdots, p)$ are sufficiently large. In general, p_{ij}'s will not be known and they have to be estimated from the data. Under H_o, $p_{ij} = p_i (j = 1, \cdots, p) \; \forall \; i = 1, \cdots, k$, the maximum likelihood estimate of p_i is

$$\hat{p}_i = \frac{n_{io}}{N} \text{ when } n_{io} = \sum_{j=1}^{k} n_{ij}, N = \sum_i \sum_j n_{ij}.$$

Replacing p_{ij} by $\hat{p}_i$ we obtain

$$Q' = \sum_{j=1}^{p} \sum_{i=1}^{k} (n_{ij} - n_j n_{io}/N)^2 / (n_j n_{io}/N).$$

It can be shown that Q' has a limiting chi-square distribution with $p(k-1) - (k-1) = (p-1)(k-1)$ distribution. (This result is not a direct corollary of theorem 20.2.1, but, however, is a corollary of a generalisation of theorem 20.2.1 from one to p populations).

The above is a test of homogeneity of p multinomial populations. This test can be generalised to test for homogeneity of p arbitrary populations, say, binomial, Poisson, Normal etc. (not necessarily multinomial). For example, if one wants to test the hypothesis that p samples come from the same Poisson (λ) population, then assuming the hypothesis to be true, one has to estimate λ on the basis of the given categorical data and find the expected frequency for each class and then use the statistic similar to Q', which follows asymptotically $\chi^2_{p(k-1)-1}$, only one parameter λ being estimated in this case.

Example 20.4.1 *The following data have been taken from Mood, Graybill and Boes. A serum supposed to have some effect in preventing colds was tested on* 500 *individuals* *and their records for* 1 *year were compared with the records of* 500 *untreated individuals as follows:*

	No Colds	One Cold	More than one Cold
Treated	252	145	103
Untreated	224	136	140

It is required to test whether the two parallel samples may be regarded as coming from the same population.

Here

$$\chi^2_{(2)} = \sum \frac{O_i^2}{e_i} - n$$
$$= 7.4 > 5.99 = \chi^2_{(2),0.95}; \quad \text{also } 7.4 < 9.21 = \chi^2_{(2),0.99}.$$

The hypothesis of homogeneity of the two samples is rejected at 5% level of significance; however, the hypothesis is accepted at 1% level of significance.

20.5 Other Tests of Significance Based on χ^2 Distribution

20.5.1 Combination of Probabilities from independent tests of significance

Sometimes we have a number of independent tests of significance for the same hypothesis, giving different probabilities for the statistic if the hypothesis is true. The χ^2 test is used to combine these various probabilities to obtain a single probability from which to decide if the hypothesis is true.

In testing a hypothesis we are generally interested in finding the probability that the value of the test statistic T does not exceed or exceeds the observed value of t. Let $f_T(t)$ the p.d.f. of T and let

$$y = \int_{-\infty}^{t} f_T(t)\,dt.$$

Let Y be the random variable whose one particular value is y (depending on t) and let $g_Y(y)$ be its p.d.f., $0 \le y \le 1$. We have

$$dy = f(t)\,dt. \tag{20.5.1}$$

Now

$$g(y)\,dy = f(t)\,dt.$$

Hence by (20.5.1)

$$g(y) = 1, \quad 0 \le y \le 1$$

i.e. Y has a rectangular distribution over $[0, 1]$.

Let us put

$$u = -2\log_e y \text{ i.e., } y = e^{-u/2},$$

$$\left|\frac{dy}{du}\right| = \frac{y}{2} = \frac{1}{2}e^{-u/2}.$$

The p.d.f. of U is

$$h_U(u) = \frac{1}{2}e^{-u/2}.$$

Hence U has a $\chi^2_{(2)}$ distribution. If we combine k independent U's, then

$$\sum_{i=1}^{k} U_i = -2\sum_{i=1}^{k} \log Y_i \sim \chi^2_{(2k)}.$$

Similarly, if

$$y' = \int_t^{\infty} f_T(t)\, dt$$

and

$$Q = \prod_{i=1}^{k} y_i' = \prod_{i=1}^{k}(1 - y_i)$$

then it can be shown that $-2\log_e Q \sim \chi^2_{(2k)}$.

The statistic $-2\log_e y_i$ is known as Pearson's P_λ. The P_λ test can, therefore, be applied to get a single test based on a number of independent tests.

20.5.2 Testing for the Mean Vector of a Multivariate Normal Distribution

Consider a p-variate normal distribution with mean $\boldsymbol{\mu}$ and known dispersion matrix $((\sigma_{ij}))$. For testing $H_o(\mu_1 = \mu_1^0, \cdots, \mu_p = \mu_p^0)$, the test statistic is

$$\chi^2 = n\sum_i \sum_j \sigma^{ij}(\bar{x}_i - \mu_i^0)(\bar{x}_j - \mu_j^0)$$

which follows $\chi^2_{(p)}$ where $((\sigma^{ij})) = ((\sigma_{ij}))^{-1}$ and $\bar{x}_i$ is the sample mean of X_i.

20.5.3 Testing homogeneity of several independent estimates of population variance : Bartlett test

Let k independent samples give unbiased estimates $S_1^2, \cdots, S_k^2$ of the population variance based on $\nu_1, \cdots, \nu_k$ d.f. respectively. It is required to test that all samples come from the same population and hence have the same value of the population variance σ^2. Under this null hypothesis, an unbiased estimate of σ^2 is

$$S^2 = \sum_{i=1}^{k} \nu_i S_i^2/\nu \text{ where } \nu = \sum_{i=1}^{k} \nu_i.$$

Bartlett has shown that under H_o,

$$\sum_i \left(\nu_i \log_e \frac{S^2}{S_i^2}\right) \Big/ \left[1 + \frac{1}{3(k-1)}\left\{\left(\sum_{i=1}^{n} \frac{1}{\nu_i}\right) - \frac{1}{\nu}\right\}\right]$$

is distributed approximately as a $\chi^2_{(k-1)}$.

20.5.4 Testing Equality of Several Binomial Proportions

Let $X_1, \cdots, X_k$ be independent r.v.'s with $X_i \sim (n_i, p_i), i = 1, \cdots, k$. It is required to test $H_o(p_1 = \cdots = p_k)$. If $n_1, \cdots, n_k$ are all large and if the common value of P_i's, p is known, then under H_o,

$$Y = \sum_1^k \frac{(X_i - np)^2}{n_i p(1-p)} \sim \chi^2_{(k)}.$$

If, therefore, $Y \geq \chi^2_{(k),\alpha}$, H_o is rejected. If p is unknown, the maximum likelihood estimate of p is

$$\hat{p} = \frac{\sum_1^k x_i}{\sum_1^k n_i}.$$

Under certain regularity conditions (Cramer, 1946, p. 426-27), it can be shown that under H_o,

$$Y = \sum_1^k \frac{(X_i - n\hat{p})^2}{n_i \hat{p}(1-\hat{p})} \sim \chi^2_{(k-1)}.$$

20.5.5 Measures of Association of Attributes in Contingency Tables

For measuring the degree of association of two attributes A and B in a 2×2 contingency table, the data may be arranged as follows:

	B	$\bar{B}$(= not B)
A	a	b
$\bar{A}$ (= not A)	c	d

The following measures of association between A and B have been proposed.

(i) Yule's coefficient of association, $Q = \dfrac{ad - bc}{ad + bc}$.

(ii) Yule's coefficient of colligation, $Y = \left(1 - \sqrt{\frac{bc}{ad}}\right) \Big/ \left(1 + \sqrt{\frac{bc}{ad}}\right)$.

(iii) $V = \frac{ad - bc}{+\sqrt{(a+b)(a+c)(b+d)(c+d)}}$.

If the attributes are independent, Q, Y and V all vanish. For a $r \times s$ contingency table, measures are χ^2 or mean-square contingency χ^2/N. Karl Pearson proposed,

$$C = \sqrt{\frac{\chi^2}{N + \chi^2}}$$

called Pearson's coefficient of contingency, $0 < C < 1$. Tschuprow proposed the measure,

$$T = \left[\frac{\chi^2}{N(s-1)(t-1)}\right]^{1/2}.$$

20.6 Transformation of Statistics

Consider k statistic $T_1, \cdots, T_k$. Let the joint distribution of $\boldsymbol{T} = (T_1, \cdots, T_k)$ tend to k-variate normal distribution with mean vector $\boldsymbol{\theta} = (\theta_1, \cdots, \theta_k)$ and dispersion matrix $((\sigma_{ij}/n))$ where n is the sample size on which the statistics $T_1, \cdots, T_k$ are based. This means that the distribution of the random vector

$$(\sqrt{n}(T_1 - \theta_1), \cdots, \sqrt{n}(T_k - \theta_k)) \tag{20.6.1}$$

tend to $N_k(\mathbf{0}, ((\sigma_{ij})))$.

Lemma 20.6.1 *If $f(T_1, \cdots, T_k)$ is a continuous function with continuous partial derivatives then the variable*

$$U = \sqrt{n}\,[f(T_1, \cdots, T_k) - f(\theta_1, \cdots, \theta_k)]$$

tends to be distributed normally about zero mean and variance

$$\sum_i \sum_j \left(\frac{\partial f \partial f}{\partial T_i \partial T_j}\right)_{\boldsymbol{\theta}} \sigma_{ij}.$$

Proof: Since $f(T_1, \cdots, T_k)$ has continuous partial derivatives in the neighbourhood of $\boldsymbol{\theta}$, by expanding f about $\boldsymbol{\theta}$, by the Mean Value Theorem

$$f(T_1, \cdots, T_k) = f(\theta_1, \cdots, \theta_k) + \sum (T_i - \theta_i)\left(\frac{\partial f}{\partial \theta_i} + \eta_i\right) \tag{20.6.2}$$

where $\frac{\partial f}{\partial \theta_i}$ means $\left.\frac{\partial f}{\partial T_i}\right]_{\boldsymbol{\theta}}$ and $\eta_i \to 0$ as $T_i \to \theta_i$.

$$U - \sqrt{n}\sum(T_i - \theta_i)\frac{\partial f}{\partial \theta_i} = \sqrt{n}\sum(T_i - \theta_i)\eta_i \tag{20.6.3}$$

so that U and

$$\sqrt{n}\sum(T_i - \theta_i)\frac{\partial f}{\partial \theta_i}$$

have the same limiting distribution if $\sqrt{n}\sum(T_i - \theta_i)\eta_i$ converges in probability to zero. For this it is sufficient to prove that

$$\sqrt{n}(T_i - \theta_i)\eta_i \xrightarrow{P} 0 \text{ for each } i.$$

Now,

$$P\left[|\sqrt{n}(T_i - \theta_i)\eta_i| < \varepsilon\right] > P\left[|\sqrt{n}(T_i - \theta_i)| < \frac{\varepsilon}{\varepsilon'}, |\eta_i| \leq \varepsilon'\right]$$

(where ε' is a small positive quantity)

$$\begin{aligned} &= P(AB) \text{ (say)} \\ &> P(A) - P(\bar{B}) \\ &= P\left[|\sqrt{n}(T_i - \theta_i)| < \frac{\varepsilon}{\varepsilon'}\right] - P\left[|\eta_i| > \varepsilon'\right]. \end{aligned} \tag{20.6.4}$$

Since $\sqrt{n}(T_i - \theta_i) \sim N(0, \sigma_{ii})$, in the limit $P[|\sqrt{n}(T_i - \theta_i)| < G]$ can be made greater than $1 - \delta$ by choosing G sufficiently large.
Also since $T_i \to \theta_i$ stochastically, $\eta_i \to 0$ stochastically and hence

$$P\{|\eta_i| > \varepsilon'\} < \delta' \quad \text{for large } n,$$

however small ε' may be. We now choose ε' such that $\varepsilon/\varepsilon' = G$. Hence the right hand side of (20.6.4) $> 1 - \delta - \delta' > 1 - \delta''$ where $\delta'' > \delta + \delta'$ for n sufficiently large.
Therefore, U and

$$\sqrt{n}\sum(T_i - \theta_i)\frac{\partial f}{\partial \theta_i}$$

have the same limiting distribution. By virtue of the assumption (20.6.1),

$$\sum(T_i - \theta_i)\frac{\partial f}{\partial \theta_i}$$

tends to be normally distributed with mean 0 and variance

$$\frac{1}{n}\sum_{ij=1}^{k}\sum \frac{\partial f}{\partial \theta_i}\frac{\partial f}{\partial \theta_j}\sigma_{ij}.$$

Corollary 20.6.1 *If T is asymptotically $N\left(\theta, \frac{\sigma^2}{n}\right)$ then any function $f(T)$ of T is asymptotically $N\left(f(\theta), [f'(\theta)]^2\frac{\sigma^2}{n}\right)$ where $f'(\theta) = \left.\frac{df(T)}{dT}\right]_{T=\theta}$, provided $f(T)$ is continuous with continuous first order derivative in the neighbourhood of θ. (A comparable result is given in corollary* 10.11.2.)

In transformation of variables one generally transforms from one variable X to another variable Y in such a way that the distribution of the transformed variable Y is much simpler than that of X. Therefore, one may use the transformed variable Y rather than X. In the case of transformation of statistics, one uses transformation mainly for the following reasons.

(i) Stability of variance. The transformation is so chosen that asymptotic variance of the transformed statistic $g(T_n)$ is independent of the population parameter. Thus, if a sequence $\{T_n\}$ for estimating θ is asymptotically normal $(\theta, \psi(\theta))$, then for large samples, the normal distribution can be used for testing hypothesis regarding $g(\theta)$ (and, hence, regarding θ) if asymptotic variance of $g(T_n)$ is independent of θ. Also, in the analysis of variance, where assumption of homoscedasticity is made, such transformation of statistics may be useful.

(ii) The limiting distribution of the transformed statistic $g(T_n)$ approaches normality more rapidly than the original statistic T_n. Thus large-sample tests using $g(T_n)$ can be made even for moderately large sample sizes.

Let $g(T)$ be a transformed statistic such that asymptotic variance of $g(T)$ is independent of θ. By lemma 20.6.1,

$$\begin{aligned} V[g(T)] &\simeq [g'(\theta)]^2 V(T) \\ &= c^2 \text{ (a constant) (say).} \end{aligned} \tag{20.6.5}$$

On solving the differential equation (20.6.5), one gets the function g. Thus,

$$\frac{dg}{d\theta} = \frac{c}{\sqrt{V(T)}} \Rightarrow g(\theta) = \int \frac{c d\theta}{\sqrt{V(T)}} + k \tag{20.6.6}$$

where k is a constant of integration. The constants c and k are so chosen that $g(T)$ is of a simple form. Clearly, asymptotic variance of $g(T)$ is c^2.

20.6.1 $\sin^{-1}\sqrt{p}$ transformation of Binomial Proportion

In sampling from a Binomial (n, π) distribution, the statistic $p = r/n$ where r is the number of success has mean π and variance $\pi(1-\pi)$. The transformation obtained from (20.6.6) is

$$g(\pi) = c \int \frac{\sqrt{n} d\pi}{\sqrt{\pi(1-\pi)}} = 2c\sqrt{n}\sin^{-1}\sqrt{\pi}.$$

Choosing $c = \frac{1}{2\sqrt{n}}$, one gets $g(\pi) = \sin^{-1}\sqrt{\pi}$ and hence $g(p) = \sin^{-1}\sqrt{p}$. Asymptotically,

$$\begin{aligned} E[\sin^{-1}\sqrt{p}] &\simeq \sin^{-1}\sqrt{\pi}, \\ V[\sin^{-1}\sqrt{p}] &\simeq \frac{1}{4n}. \end{aligned} \tag{20.6.7}$$

In case of testing $H_o : \pi = \pi_0$, one may use the statistic

$$\tau = (\sin^{-1}\sqrt{p} - \sin^{-1}\sqrt{\pi_0})\sqrt{4n}$$

which is asymptotically $N(0,1)$ under H_o.

Ansocombe (1948) has shown that a slightly better transformation is

$$\sin^{-1}\sqrt{\left(r+\frac{3}{8}\right)\Big/\left(n+\frac{3}{4}\right)}$$

which has asymptotic variance $(4n+2)^{-1}$.

20.6.2 Square-root Transformation of Poisson Variate

If X is a Poisson (λ) variate, then $E(X) = \lambda$, $V(X) = \lambda$ and $E(\bar{X}) = \lambda$, $V(\bar{X}) = \lambda/n$. Also $\sqrt{n}(\bar{X}-\lambda)/\sqrt{\lambda} \sim N(0,1)$ as $n \to \infty$. By (20.6.6), the transformation is

$$g(\lambda) = c\int \frac{\sqrt{n}d\lambda}{\sqrt{\lambda}} = 2c\sqrt{n\lambda}.$$

By choosing $c = \frac{1}{2\sqrt{n}}$, $g(\lambda) = \sqrt{\lambda}$ and hence $g(\bar{X}) = \sqrt{\bar{X}}$

$$E(\sqrt{\bar{X}}) = \sqrt{\lambda} + 0\left(\frac{1}{\sqrt{\lambda}}\right)$$

$$V(\sqrt{\bar{X}}) = \frac{1}{4n} + 0\left(\frac{1}{n\lambda}\right).$$

To test the hypothesis $H_o(\lambda = \lambda_0)$, one may use the statistic $(\sqrt{\bar{X}} - \sqrt{\lambda})\sqrt{4n}$ which follows $N(0,1)$ asymptotically under H_o.

Anscombe (1948) observed that the transformation $\sqrt{X+b}$ where b is a suitably chosen constant, has some advantages in achieving greater stability in variance. Writing $X - \lambda = U$, $\lambda + b = \lambda'$, and using Taylor's expansion,

$$\sqrt{X+b} = \sqrt{U+\lambda'} = \sqrt{\lambda'} + \frac{U}{2\sqrt{\lambda'}} - \frac{U^2}{2^2.2!\lambda'^{3/2}} + \frac{1.3.U^3}{2^3.3!\lambda'^{5/2}} - \cdots \quad (20.6.8)$$

Since

$$E(U) = 0,\ \ E(U^2) = \lambda,\ \ E(U^3) = \lambda,\ \ E(U^4) = 3\lambda^2 + \lambda,$$

we have

$$E(\sqrt{X+b}) = \sqrt{\lambda+b} - \frac{1}{8\sqrt{\lambda}} + \frac{24b-7}{128\lambda^{3/2}} + \cdots$$

$$V(\sqrt{X+b}) = \frac{1}{4}\left[1 + \frac{3-8b}{8\lambda} + \frac{32b^2-52b+17}{32\lambda^2} + \cdots\right].$$

If we choose $b = \frac{3}{8}$, the expression reduces to

$$E\left(\sqrt{X+\frac{3}{8}}\right) = \sqrt{\lambda+\frac{3}{8}} - \frac{1}{8\sqrt{\lambda}} + \frac{1}{64\lambda^{3/2}} + \cdots$$

$$V\left(\sqrt{X+\frac{3}{8}}\right) = \frac{1}{4}\left(1+\frac{1}{16\lambda^2}+\cdots\right) = \frac{1}{4} + 0\left(\frac{1}{\lambda^2}\right).$$

Thus the variance of $\sqrt{X+\frac{3}{8}}$ is more stable then that of $\sqrt{X}$, since $V(\sqrt{X}) = \frac{1}{4} + 0\left(\frac{1}{\lambda}\right)$.

20.6.3 Logarithmic Transformation

Let $X_1, \cdots, X_n$ be a random sample of size n from a $N(\mu, \sigma^2)$ population. It is known that the sampling variance $s^2 = \frac{1}{n-1}\sum(X_i - \bar{X})^2$ has $E(s^2) = \sigma^2$, $V(s^2) = 2\sigma^4/(n-1)$. (20.6.6) gives the transformation,

$$g(\sigma^2) = c\int \frac{d\sigma^2(\sqrt{n-1})}{\sqrt{2\sigma^4}} = c\sqrt{\frac{n-2}{2}}\log_e \sigma^2.$$

Choosing $c = \sqrt{\frac{2}{n-2}}$, one gets $g(\sigma^2) = \log_e \sigma^2$ and $g(s^2) = \log_e s^2$.
Clearly,

$$E[\log_e s^2] \simeq \log_e \sigma^2, \quad V(\log_e s^2) \simeq \frac{2}{(n-1)}.$$

In testing $H_o(\sigma = \sigma_0)$, one may use the statistic

$$\left(\log_e(s^2) - \log_e(\sigma_0^2)\right)\sqrt{(n-1)/2} \sim N(0,1) \text{ under } H_o.$$

An empirical transformation is $\log_e(1+s^2)$.

20.6.4 $\tanh^{-1}$ Transformation of the Correlation Coefficient

Let r be the sample correlation coefficient based on n pairs of observations (X_i, Y_i) $(i = 1, \cdots, N)$ taken from a bivariate normal population with correlation coefficient ρ. It is known that in large samples,

$$E(r) = \rho, \quad V(r) = \frac{(1-\rho^2)^2}{(n-1)}.$$

(20.6.6) gives the transformation

$$g(\rho) = c\int \frac{\sqrt{n-1}\,d\rho}{(1-\rho^2)} = c\sqrt{n-1}\tan h^{-1}\rho = c\sqrt{n-1}\cdot\frac{1}{2}\log_e\frac{1+\rho}{1-\rho}.$$

Choosing $c = \frac{1}{\sqrt{n-1}}$, one gets

$$g(\rho) = \frac{1}{2}\log_e \frac{1+\rho}{1-\rho} = \tan h^{-1}\rho = \zeta \text{ (say)}$$

$$\text{and } g(r) = \frac{1}{2}\log_e \frac{1+r}{1-r} = \tanh^{-1} r = Z \text{ (say)}.$$

This is the well-known Z-transformation of Fisher. Note that $-\infty \le Z \le \infty$ (while $-1 \le r \le 1$).

It has been found,

$$E(Z) = \zeta + \frac{\rho}{2(n-1)}\left\{1 + \frac{5+\rho^2}{4(n-1)} + \cdots\right\},$$

$$V(Z) = \frac{1}{n-1}\left\{1 + \frac{4-\rho^2}{2(n-1)} + \cdots\right\},$$

$$\mu_3(Z) = \frac{\rho^3}{(n-1)^3} + \cdots,$$

$$\mu_4(Z) = \frac{1}{(n-1)^2}\left\{3 + \frac{14-3\rho^2}{n-1} + \frac{184-48\rho^2-21\rho^4}{4(n-1)^2} + \cdots\right\},$$

so that,

$$\beta_1 = \frac{\rho^6}{(n-1)^3} + \cdots, \quad \beta_2 = 3 + \frac{2}{n-1} + \frac{4+2\rho^2-3\rho^4}{(n-1)^2} + \cdots.$$

Since β_1 and $\beta_2 - 3$ are small, even for moderate $n(n = 5, 6$, say) it follows that the distribution of $U = Z - \zeta$ is approximately normal with

$$E(U) = \frac{\rho}{2(n-1)}$$

$$V(U) = \frac{1}{n-1} + \frac{4-\rho^2}{2(n-1)^2} \simeq \frac{1}{n-3}. \tag{20.6.9}$$

Some uses of Z-transformation

(a) To test the hypothesis $H_\circ : \rho = \rho_0$ regarding a bivariate normal population, use

$$\tau = \left\{Z - \left(\tanh^{-1}\rho_0 + \frac{\rho_0}{2(n-1)}\right)\right\}\sqrt{n-1}$$

which is asymptotically $N(0,1)$ under $H_\circ$. The correction for bias $\rho_0/2(n-1)$ is important if n is not large.

(b) Two samples of sizes n_1, n_2 are taken from two bivariate normal populations with correlation coefficients ρ_1 and ρ_2 respectively, T_0 test $H_\circ(\rho_1 = \rho_2)$ calculate

$$Z_i = \frac{1}{2}\log_e \frac{1+r_i}{1-r_i}, \quad i = 1, 2,$$

where r_i is the sample correlation coefficient for the ith sample. If the samples are not small, the statistic

$$(Z_1 - Z_2)/\sqrt{\frac{1}{n_1-3}+\frac{1}{n_2-3}} \sim N(0,1) \text{ under } H_o.$$

(c) Let $r_1, \cdots, r_k$ be a set of sample correlation coefficients based on $n_1, \cdots, n_k$ observations from k bivariate normal populations with correlation coefficients $\rho_1, \cdots, \rho_k$ respectively. It is required to test $H_o(\rho_1 = \cdots = \rho_k)$. If n_i's are not small, the biases $E(Z_i - \zeta_i) = \rho_i/2(n_i - 1)$ can be neglected and the maximum likelihood estimate of $\tanh^{-1}\rho$ where ρ is the common correlation coefficient, assuming that H_o holds, is

$$\bar{Z} = \frac{\sum_i (n_i - 3)Z_i}{\sum_i (n_i - 3)}. \tag{20.6.10}$$

For testing H_o, one may use the statistic

$$\chi^2 = \sum_i (n_i - 3)(Z_i - \bar{Z})^2 \tag{20.6.11}$$

which follows approximately a $\chi^2_{(k-1)}$ under H_o. One can write

$$\chi^2 = \sum_i (n_i - 3)Z_i^2 - \frac{\left\{\sum_i (n_i - 3)Z_i\right\}^2}{\sum_i (n_i - 3)}.$$

□ **Remark 20.6.1** *For moderately small values of n (e.g., $n = 10, 12$ etc.), the distribution of Z can be satisfactorily approximated to a normal distribution, whereas that of r being extremely skew, can not be approximated to a normal distribution for that sample size. The bias of Z is of order $\frac{1}{n}$ which tends to zero as $n \to \infty$.*

20.7 Transformation of Variables

(i) $\sqrt{2\chi^2}$ transformation.
It has been found that for a $\chi^2_{(n)}$ variable

$$\mu_1' = n, \quad \mu_2 = 2n, \quad \mu_3 = 8n, \quad \mu_4 = 48n + 12n^2, \quad \beta_1 = \frac{8}{n}, \quad \beta_2 = 3 + \frac{12}{n}. \tag{20.7.1}$$

Hence, for large $n(> 30)$, the standardised variate $\tau = (\chi^2 - n)/\sqrt{2n}$ is asymptotically $N(0,1)$.

Alternatively, characteristic function of τ is

$$\phi(t) = E\exp\left[it(\chi^2 - n)/\sqrt{2n}\right]$$
$$= \exp\left(-\frac{nit}{\sqrt{2n}}\right)\left(1 - \frac{2it}{\sqrt{2n}}\right)^{-n/2} \quad ; \text{ [vide example 4.6.2]}$$

$$\log\phi(t) = -\frac{nit}{\sqrt{2n}} - \frac{n}{2}\left\{-\frac{2it}{\sqrt{2n}} - \frac{1}{2}\left(\frac{2it}{\sqrt{2n}}\right)^2 + \cdots\right\}$$
$$\to -\frac{t^2}{2} \text{ as } n \to \infty.$$

The approach of $(\chi^2 - n)/\sqrt{2n}$ to normality is slow and the approximation may not hold good for moderate sample size. Fisher has suggested that $\sqrt{2\chi^2} - \sqrt{2n-1}$ approaches normality more rapidly then τ.

Proof: The distribution of χ is

$$f(\chi)d\chi = \frac{1}{2^{(n-2)/2}\Gamma\left(\frac{n}{2}\right)}\exp\left(-\frac{\chi^2}{2}\right)\chi^{n-1}d\chi$$

$$\mu_r' = 2^{r/2}\frac{\Gamma\left(\frac{n+r}{2}\right)}{\Gamma\left(\frac{n}{2}\right)};$$

hence

$$\mu_1' = \sqrt{2}\frac{\Gamma\left(\frac{n+1}{2}\right)}{\Gamma\left(\frac{1}{2}\right)}.$$

Using Stirling's approximation

$$\log_e \Gamma(n+1) \simeq \frac{1}{2}\log_e 2\pi + \left(n + \frac{1}{2}\right)\log_e n - n + \frac{1}{12n} - \frac{1}{360n^3} + \cdots$$

we get, after some reduction,

$$\mu_1' = \sqrt{n}\left(1 - \frac{1}{4n} + \frac{1}{32n^2} + \frac{5}{128n^3} + \cdots\right)$$
$$\mu_1'^2 = n\left(1 - \frac{1}{2n} + \frac{1}{8n^2} + \frac{1}{16n^3} + \cdots\right).$$

Also,

$$\mu_2' = n, \;\; \mu_3' = (n+1)\mu_1', \;\; \mu_4' = (n+2)n,$$
$$\mu_2 = \frac{1}{2} - \frac{1}{8n} - \frac{1}{16n^2} + \cdots, \;\; \mu_3 = \frac{1}{4\sqrt{n}} + \cdots,$$
$$\mu_4 = \frac{3}{4} - \frac{3}{8n} + 0(n^{-2}).$$

Hence for the distribution of $\sqrt{2\chi^2} = \sqrt{2}\chi$, we have

$$\left.\begin{aligned} \mu_2 &= 1 - \tfrac{1}{4n} + 0\left(\tfrac{1}{n^{3/2}}\right), \\ \gamma_1 &= \sqrt{\beta_1} = \tfrac{1}{\sqrt{2n}} + 0\left(\tfrac{1}{n^2}\right), \\ \gamma_2 &= \beta_2 - 3 = 0(n^{-2}). \end{aligned}\right\} \tag{20.7.2}$$

Comparing (20.7.1) and (20.7.2) we see that $\sqrt{2\chi^2}$ tends to normality more rapidly than χ^2.

Again, taking μ_1' of χ as approximately

$$\sqrt{n\left(1 - \frac{1}{2n}\right)} = \sqrt{n - \frac{1}{2}}$$

and μ_2 as approximately $\frac{1}{2}$,

$$\tau = \frac{\chi - \sqrt{n - \frac{1}{2}}}{\sqrt{\frac{1}{2}}} = \sqrt{2\chi^2} - \sqrt{2n-1} \sim N(0, 1).$$

A third and fairly accurate approximation has been suggested by Wilson and Helferty (1941).

Let $\delta = \chi^2 - E(\chi^2) = \chi^2 - n$. Consider the distribution of $U = \left(\frac{\chi^2}{n}\right)^h$, where h is to be suitably determined.

$$U = \left(1 + \frac{\delta}{n}\right)^h = 1 + \frac{h}{n} \cdot \delta + \frac{h(h-1)}{2} \frac{\delta^2}{n^2} + \cdots.$$

Taking expectation,

$$\begin{aligned} E(U) &= 1 + \frac{h}{n} E(\delta) + \frac{h(h-1)}{2n^2} E(\delta^2) + \frac{h(h-1)(h-2)}{6n^3} E(\delta^3) + \cdots \\ &= 1 + \frac{h(h-1)}{n} + \frac{h(h-1)(h-2)(3h-1)}{6n^2} \\ &\quad + \frac{h^2(h-1)^2(h-2)(h-3)}{6n^3} + 0\left(\frac{1}{n^4}\right). \end{aligned}$$

We choose h such that the third term is zero. This gives $h = \frac{1}{3}$. On substituting $h = 1/3$, it is found that for distribution of U,

$$\begin{aligned} \mu_1' &= 1 - \frac{2}{9n} + \frac{80}{3^7 n^3} + 0\left(\frac{1}{n^4}\right), \\ \mu_2' &= 1 - \frac{2}{9n} + \frac{4}{3^4 n^2} + \frac{56}{3^7 n^3} + 0\left(\frac{1}{n^4}\right), \\ \mu_3' &= 1, \\ \mu_4' &= 1 + \frac{4}{9n} - \frac{4}{3^3 n^2} + \frac{80}{3^7 n^3} + 0\left(\frac{1}{n^4}\right). \end{aligned}$$

Hence the central moments of distribution of U are

$$\mu_2 = \frac{2}{9n} - \frac{104}{3^7 n^3} + 0\left(\frac{1}{n^4}\right),$$

$$\mu_3 = \frac{32}{3^6 n^3} + 0\left(\frac{1}{n^4}\right),$$

$$\mu_4 = \frac{4}{3^3 n^2} - \frac{16}{3^6 n^3} + 0\left(\frac{1}{n^4}\right).$$

For this distribution

$$\gamma_1 = \sqrt{\beta_1} = \frac{2^{7/2}}{3^3 n^{3/2}} + 0(n^{-2}),$$

$$\gamma_2 = \beta_2 - 3 = -\frac{4}{9n} + 0(n^{-3/2}).$$

Hence the convergence to normality is more rapid than that of $\sqrt{2\chi^2}$, because fo convergence the skewness is more important than Kurtosis. Here

$$\tau = \left\{\left(\frac{\chi^2}{n}\right)^{1/3} - \left(1 - \frac{2}{9n}\right)\right\}\left(\frac{9n}{2}\right)^{1/2} \sim N(0,1).$$

Exercise

1. A die is rolled 1200 times with the following results:

	1	2	3	4	5	6
Frequency	205	279	217	257	133	109

Test if the die is unbiased.

2. A sample survey revealed the following information about the distribution of midd class families by expenditure level for the years 1949 and 1953. Investigate if the is any change in the pattern of expenditure between the years.

Expenditure / Year	0-50	51-100	101-150	151-200	201-300	301-400	401-500	501-600	601-70
1949	152	227	202	181	119	94	93	73	63
1953	66	347	347	273	191	135	154	129	55

[$\chi^2 = 89.9672$

3. Columns (2) and (3) of the following table show the distribution by age group and habits of a sample of 580 from cancer patients of a particular type. Columns (4) and (5) show the same for a sample of size 299 from normal persons. Assuming the samples to be taken at random (a) test whether habit changes with age (b) assuming further that habit does not change with age, test whether some of the habits are peculiar to the class of cancer patients.

Type of	Cancer Patients		Normal Persons	
habit	Below 50 years	50 years are above	Below 50 years	50 years and above
A	11	18	10	6
B	5	11	8	5
C	96	189	58	25
D	10	19	20	10
E	42	126	42	25
F	10	43	61	29
	174	406	199	100

4. The following data give mortality rates among vaccinated and non-vaccinated patients. Analyse the data.

	Living	Dead	Total
Vaccinated	325	135	460
Non-vaccinated	113	210	343
Total	438	345	803

5. Four interviewers determined whether subjects had a certain characteristic (Y) with the results in the following table. Is there any evidence that the interviewers do not have random samples from the same population? Indicate other reasons for the proportions to be unequal.

	A	B	C	D
No. interviewed	305	312	250	278
No. with Y	50	72	120	90

6. Mendelian theory states that certain variety of pea ought to be of four types by shape and color—A (round and yellow), B (round and green), C (angular and yellow), D (angular and green) in the ratio 9 : 3 : 3 : 1.

For $n = 685$, the following frequencies were observed $A : 412$, $B : 116$, $C : 125$, $D : 32$.

Test if the observed frequencies are consistent with the Mendelian hypothesis.

7. The following table gives the results of a series of controlled experiments. Discuss whether the treatment may be considered to have any positive effect.

	Positive	No effect	Negative	Total
Treatment	9	2	1	12
Control	3	6	3	12
Total	12	8	4	24

8. Samples of sizes 75, 125, 100, 150 and 130 from 5 Poisson populations give the mean values as 115/75, 197/125, 141/100, 237/150 and 185/130 respectively. Do the populations have the same mean value?

[*Hints:* Under H_o, an estimate of population mean for the common Poisson population is $\bar{X} = \frac{\Sigma n_i \bar{X}_i}{\Sigma n_i} = \frac{\Sigma T_i}{n} = \hat{\lambda}$ where $T_1 = 115$, $n_1 = 75$, etc. Under H_o, $E(T_i) = E(X_{i1} + \cdots + X_{in_i}) = n_i\lambda$ where $X_{i1}, \cdots, X_{in_i}$ are observations in the ith sample.
Now conditional p.m.f. of $T_1, \cdots, T_5$ given that $T_1 + \cdots + T_5 = T$ is multinomial with parameters T, $n_i\lambda_i/\Sigma n_i\lambda_i$. Hence $\sum_{i=1}^{5}(O_i - e_i)^2/e_i \cap \chi^2_{(4)}$.]

9. The following table shows the distribution of 2610 married women in Denmark according to area of residence and type of occupation. Test the hypothesis that the distribution is the same for all the five living areas.

Living area	House-wife	Occupation part-time worker	Full-time worker	In husbands' business
Copenhagen	188	66	139	16
Subarban Copenhagen	145	32	98	14
Cities	401	114	239	65
Townships	81	24	53	25
Farms	364	118	164	164

[Anderson, 1980

10. The following table gives Batson's distribution of sweet pea plants obtained from an intercross. The marginal frequencies are expected to be in the ratio 3 : 1, and if the two characters, flower colour and the pollen shape, are independently inherited then the cell frequencies are expected to be in the ratio 9 : 3 : 3 : 1.

Pollen Shape	Flower Colour Purple	Red	Total
Long	296	27	323
Round	19	85	104
Total	315	112	427

Carry out a suitable statistical analysis of the data and write a short note on your findings.

11. The following table gives the result of 4 Danish political polls, published by Weekendavisen, Saturday, October 12, 1974.

Party	Gall up	Survey Observe	Company AIM	K. Vilstrup
Labour	444	403	282	182
Liberals	129	104	93	63
Conservatives	105	104	68	56
Socialists	68	65	60	42
Communists	75	65	52	56
Christian Democrats	80	52	47	35
Farmers	225	286	196	140
Progress Party	280	143	132	84
Others	94	78	70	42
Sample size	1500	1300	1000	700

Test if the results given by the four survey companies are homogeneous.

[Anderson, 1980]

12. When women were first employed in air plane factories in the United States, one official predicted that the number of accidents involving women would be larger than the number involving men. Use the data given in the table below to test the hypothesis that the number of women involved in accidents is as large as the number of men involved in accidents. Assume that the factories were of the same size and employed the same number of men and women.

Number of persons involved in accidents

Factories	Male	Female
A	15	22
B	12	19
C	18	28
D	20	32
E	18	26

13. Let $T_1, \cdots, T_k$ be totals based on samples of sizes $n_1, \cdots, n_k$ from k Poisson populations. How will you test whether the k populations have the same mean value?

14. Let $\pi_i (> 0)$ be the probability that a randomly selected individual belongs to the ith of k mutually exclusive and exhaustive classes, $\pi_1 + \pi_2 + \cdots + \pi_k = 1$. In a random sample of n individuals, let n_i be the number of individuals belonging to the ith class, $\sum n_i = n$.

Let $A = \sum a_i n_i$, $B = \sum b_j n_j$ where a_i, b_j are known constants. Show that

$$V(A) = n\left[\sum a_i^2 \pi_i - \left(\sum a_i \pi_i\right)^2\right]$$

$$\text{Cov}(A, B) = n\left[\sum a_i b_i \pi_i - \left(\sum a_i \pi_i\right)\left(\sum b_i \pi_i\right)\right].$$

Use the above result to express $T = \sum (n_i - n\pi_i)^2 / n\pi_i$ in the form

$$T = y_1^2 + y_2^2 + \cdots + y_{k-1}^2$$

where y_i's are non-homogeneous linear functions of $n_1, \cdots, n_k$ each with mean zero, variance unity and any two of the y_i's are mutually uncorrelated.

15. Prove that for a $s \times t$ table, $\chi^2 \le \min[N(t-1), N(s-1)]$, where N is the total frequency.

[*Hints:* $e_{ij} = \frac{A_i \times B_j}{N}$; $O_{ij} \le A_i$, $O_{ij} \le B_j$.

$$\chi^2 = \sum_i \sum_j \frac{O_{ij}^2}{e_{ij}} - N = N \sum_i \sum_j \frac{O_{ij}}{A_i} \cdot \frac{O_{ij}}{B_j} - N$$

$$\le N\left\{\sum_i \sum_j \frac{O_{ij}}{A_i} - 1\right\} = N\left\{\sum_{i=1}^{s} \frac{O_i}{A_1} - 1\right\} = N(s-1).$$

Similarly, $\chi^2 \le N(t-1)$].

16. Consider the following $2 \times r$ contingency table

				(Total)
a_1	a_2	$\cdots$	a_r	a
b_1	b_2	$\cdots$	b_r	b
n_1	n_2		n_r	n
(total)				

Show that the value of χ^2 for testing homogeneity is

$$\chi^2 = \sum_{i=1}^{r} w_i (p_i - \bar{p})^2$$

when $p_i = a_i/n_i$, $\bar{p} = a/n$, $w_i = n_i/p_i q_i$, $q_i = b_i/n_i$, $\bar{q} = b/n$.

17. Prove Brandt and Snedecor's formula for a $2 \times r$ contingency table

			(Total)
a_1	$\cdots$	a_r	n
a'_1	$\cdots$	a'_r	n'

$$\chi^2 = \frac{1}{\bar{p}\bar{q}}\left[\sum_{i=1}^{r}(a_i p_i) - n\bar{p}\right], \quad \text{where } \bar{p} = \frac{n}{n+n'}, \quad p_i = \frac{a_i}{a_i + a'_i}, \quad \bar{q} = 1 - \bar{p}.$$

18. The marginal probabilities of an $r \times s$ table, p_{io}, p_{oj}'s are known, and it is required to estimate the cell probabilities p_{ij}'s. A sample of n observations is taken from a correspondingly-classified population with probabilities p_{ij}. Show that the maximum likelihood estimate $\hat{p}_{ij}$ of p_{ij} are the solutions of $(r-1)(s-1)$ equations

$$\frac{n_{ij}}{p_{ij}} - \frac{n_{is}}{p_{is}} - \frac{n_{rj}}{p_{rj}} + \frac{n_{rs}}{p_{rs}} = 0, \quad i = 1, \cdots, r-1; j = 1, \cdots, s-1.$$

19. Suppose $n_{ij}(i = 1, \cdots, r;\ \ j = 1, \cdots, s)$ are a $(rs-1)$-dimensional random variable having the multinomial distribution

$$\frac{n!}{\prod_i \prod_j n_{ij}!} \prod_i \prod_j p_{ij}^{n_{ij}},$$

where $\sum_i \sum_j n_{ij} = n$, each $p_{ij} > 0$ with $\sum_i \sum_j p_{ij} = 1$. Show that the Likelihood Ratio criterion for testing $H_o : p_{ij} = p_{io}p_{oj}$ where $\sum_i p_{io} = \sum_j p_{oj} = 1$, is given by

$$\lambda = \frac{\prod_i \left(\frac{n_{io}}{n}\right)^{n_{io}} \prod_j \left(\frac{n_{oj}}{n}\right)^{n_{oj}}}{\prod_i \prod_j \left(\frac{n_{ij}}{n}\right)^{n_{ij}}}$$

and that, if H_o is true, the limiting distribution of $-2\log_e \lambda \cap \chi^2_{(r-1)(s-1)}$.

Show also that if H_o is true, $-2\log_e \lambda$ and χ^2 in (20.3.18) converge together in distribution.

20. Consider a χ^2 variate with n d.f. and the transformation $\alpha\sqrt{n}\left\{\left(\frac{\chi^2}{n}\right)^{\beta} + \left(\frac{\gamma}{n}\right) - 1\right\}$. By comparing the first three moments of the statistic with those of a $N(0,1)$-distribution obtain the values of α, β and γ. Show that $\alpha = \sqrt{\frac{9}{2}}$, $\beta = \frac{1}{3}$ and $\gamma = \frac{2}{9}$ is a good choice for rapid approach of the transformed statistic to normality.

21. Show that

$$\left[\left(\frac{13\chi^2-\nu}{12\nu}\right)^{5/13}+\frac{5}{18\nu}\left(1+\frac{7}{48\nu}\right)-1\right]\frac{6\nu\sqrt{2}}{5}\left(1-\frac{1}{18\nu}\right),$$

where ν is the d.f. of χ^2, is approximately a $N(0,1)$-variable.

22. If T is a statistic which in large samples tends to a population parameter θ, and variance of T is a function of θ say, $\frac{\psi(\theta)}{n}+0\left(\frac{1}{n^2}\right)$, show that the variable

$$Z=\int_{-\infty}^{T}\{\psi(\theta)\}^{-1/2}d\theta$$

has variance $\frac{1}{n}+0(n^{-2})$. Hence, show that the transformation $Z=\tan h^{-1}r$ has variance $\frac{1}{n}+0(n^{-2})$.

23. The correlations obtained from 6 samples of sizes 10, 14, 16, 25, 28, 20 were found to be 0.318, 0.106, 0.253, 0.116, 0.112 and 0.340 respectively. Can these be considered homogeneous? If so, obtain a combined estimate of ρ.

Chapter 21

Theory of Linear Estimation

21.1 Introduction

In many situations the variable of interest is expressible as a linear function of several unknown parameters with known coefficients apart from an error term. We shall investigate in this chapter point estimation of linear functions of these parameters by considering statistics which are linear functions of observations. In Section 21.2 we consider Gauss-Markoff linear model. Section 21.3 considers the theory of best linear unbiased estimation. Sections 21.4 through 21.6 deal with theory of least squares estimation. The sum of squares (SS) due to best linear unbiased estimates, errors and their properties have been dealt with in Section 21.7. The next section introduces the analysis of covariance model. The uses of orthogonal polynomials in fitting least squares curvilinear regressions have been considered in the last section.

21.2 Gauss-Markoff Linear Model

Consider a set of n independent random variables $Y_1, Y_2, \cdots, Y_n$ whose expectations are given as linear functions of p unknown parameters $\beta_1, \cdots, \beta_p (p \leq n)$ with known coefficients a_{ij}'s and whose variances are a constunt, σ^2. Thus

$$E(Y_i) = a_{i1}\beta_1 + a_{i2}\beta_2 + \cdots + a_{ip}\beta_p, i = 1, \cdots, n$$

$$V(Y_i) = \sigma^2, \text{ Cov}(Y_i, Y_{i'}) = 0 \ (i \neq i'). \tag{21.2.1}$$

Writing

$$\boldsymbol{Y} = (Y_1, \cdots, Y_n)', \ \boldsymbol{\beta} = (\beta_1, \cdots, \beta_p)'$$

$$\boldsymbol{\alpha}_j = (a_{1j} \ a_{2j} \ \cdots \ a_{nj})', \ j = 1, \cdots, p \ (\leq n)$$

$$\boldsymbol{A} = (\boldsymbol{\alpha}_1, \cdots, \boldsymbol{\alpha}_p),$$

we can write the model (21.2.1) as

$$\begin{aligned} E(\boldsymbol{Y}) &= \boldsymbol{\eta} \text{ (say)} = \boldsymbol{\alpha}_1\beta_1 + \cdots + \boldsymbol{\alpha}_p\beta_p = \boldsymbol{A\beta} \\ D(\boldsymbol{Y}) &= \sigma^2 I \end{aligned} \tag{21.2.2}$$

where $D(\boldsymbol{Y})$ denotes the dispersion matrix of $\boldsymbol{Y}$ and I is the identity matrix of order n. Obviously, the model (21.2.1) can also be written as

$$\begin{aligned} \boldsymbol{Y} &= \boldsymbol{A\beta} + \boldsymbol{e} \\ E(\boldsymbol{e}) &= \boldsymbol{0}, \\ D(\boldsymbol{e}) &= \sigma^2 I \end{aligned} \tag{21.2.3}$$

where $\boldsymbol{e} = (e_1, \cdots, e_n)'$ is a column vector of n independently distributed errors $e_1, \cdots, e_n$. The model (21.2.1) (or (21.2.2) or (21.2.3)) is known as Gauss-Markoff (GM) linear model. If we assume that $\boldsymbol{\alpha}_1, \cdots, \boldsymbol{\alpha}_p$ are linearly independent ($p \leq n$), then the model (21.2.3) is a GM-model with full rank.

The unknown parameters β_j's are called effects (due to jth cause) or regression parameters, $j = 1, \cdots, p$. When all the parameters β's are fixed constants, the model (21.2.1) is called a *fixed effect model.* In such cases, if one set of the coefficients $\{a_{ij}, i = 1, \cdots, n\}$, say, $a_{i1} = 1 \ \forall\, i = 1, \cdots, n$ so that $E(Y_i) = \beta_1 + a_{i2}\beta_2 + \cdots + a_{ip}\beta_p$, then β_1 is an additive constant and is called a *general effect.*

If we assume that $\beta_1, \cdots, \beta_p$ themselves are random samples from some distributions then the model (21.2.1) is called a *random effect* model. If some of the parameters $\beta_1, \beta_2, \cdots, \beta_p$ are fixed effects while others are random samples from some distributions, model (21.2.1) is called a *mixed effect* model. We shall in this chapter consider fixed effect model only.

We now classify the models according to values of a_{ij}'s. Sometimes a_{ij}'s are indicator variables, $a_{ij} = 1(0)$ if the jth effect β_j is present (absent) in Y_i. Suppose the yield (per unit area) of a crop is considered as the result of application of three fertilisers n, p and k, all of which may or may not be present.

Assume that

$$\begin{aligned} E(Y_1) &= \beta_n + \beta_p, \\ E(Y_2) &= \beta_n + \beta_k, \end{aligned}$$

etc. In Y_1, the fertilisers n and p have been applied, in Y_2 the fertilisers n and k. Clearly, $a_{1n} = 1$, $a_{1p} = 1$, $a_{1k} = 0$, $a_{2n} = 1$, $a_{2p} = 0$, $a_{2k} = 1$. Such a fixed effect model where $a_{ij} = 1$ or $0 \ \forall\, i, j$, is called a *Analysis of Variance* model or *Experimental Design* model.

Sometimes, a_{ij}'s are observed values of some independent variables (also called 'concommittant' variables or 'regressor' variables) x_j ($j = 1, \cdots, p$). Thus, one may associate yield of a crop in a plot with the area under the crop (x_1), the amount of rainfull (x_2), the soil humidity (x_3), etc. Here

$$E(Y_i) = \beta_1 x_{i1} + \beta_2 x_{i2} + \beta_3 x_{i3} + \cdots$$

where x_{i1} is the measurement of the area under the crop, x_{i2} is the amount of rainfull, x_{i3}, a measure of soil humidity for the ith crop, etc. Such a model is called a *Regression* model.

Sometimes, some of the a_{ij}-values are values of indicator variables and others are values of regressor variables. This is called an *Analysis of Covariance* model.

In the theory of linear estimation we find unbiased linear estimators of estimable linear parametric functions, starting from a model of the form (21.2.1). Among all such unbiased linear estimators we find one having the minimum possible variance. Such an estimator is called the best linear unbiased estimator (BLUE) of the parametric function.

21.3 Best Linear Unbiased Estimation

Definition 21.3.1 *A* ***linear parametric function***

$$\boldsymbol{l}'\boldsymbol{\beta} = l_1\beta_1 + \cdots + l_p\beta_p \tag{21.3.1}$$

where $\boldsymbol{l} = (l_1, \cdots, l_p)'$ *is a linear function of parameters.*

Definition 21.3.2 ***Linear function of observations.*** *Let y be a linear function of observations $y_1, \cdots, y_n$ (where y_i denotes an observation corresponding to a random variable Y_i), $y = C_1y_1 + \cdots + C_ny_n = \boldsymbol{C}'\boldsymbol{y}$ where $\boldsymbol{C} = (C_1, \cdots, C_n)'$, $\boldsymbol{y} = (y_1, \cdots, y_n)'$. If we take Y as the random variable corresponding to y,*

$$\begin{aligned} E(Y) &= \sum_{i=1}^{n} C_iE(Y_i) = \boldsymbol{C}'E(\boldsymbol{Y}) = \boldsymbol{C}'\boldsymbol{A\beta} = \boldsymbol{C}'(\boldsymbol{\alpha}_1 \;\; \boldsymbol{\alpha}_2 \;\; \cdots \;\; \boldsymbol{\alpha}_p)\boldsymbol{\beta} \\ &= (\boldsymbol{C}'\boldsymbol{\alpha}_1 \;\; \boldsymbol{C}'\boldsymbol{\alpha}_2 \;\; \cdots \;\; \boldsymbol{C}'\boldsymbol{\alpha}_p)\boldsymbol{\beta} = \beta_1\boldsymbol{C}'\boldsymbol{\alpha}_1 + \beta_2\boldsymbol{C}'\boldsymbol{\alpha}_2 + \cdots + \beta_p\boldsymbol{C}'\boldsymbol{\alpha}_p \end{aligned} \tag{21.3.2}$$

$$V(Y) = V(\boldsymbol{C}'\boldsymbol{Y}) = \boldsymbol{C}'D(\boldsymbol{Y})\boldsymbol{C} = \sigma^2\boldsymbol{C}'\boldsymbol{I}\boldsymbol{C} = \sigma^2\sum_{i=1}^{n} C_i^2.$$

Definition 21.3.3 *As* ***estimable*** *(linear)* ***parametric function*** *is a parametric function $\boldsymbol{l}'\boldsymbol{\beta}$ for which an unbiased linear estimator $\boldsymbol{C}'\boldsymbol{Y}$ exists, i.e., if*

$$E_{\boldsymbol{\beta}}(\boldsymbol{C}'\boldsymbol{Y}) = \boldsymbol{l}'\boldsymbol{\beta} \tag{21.3.3}$$

independently of the parameters $\boldsymbol{\beta}$. The function $\boldsymbol{l}'\boldsymbol{\beta}$ is also called a ***linearly estimable function.***

□ **Remark 21.3.1** *We are here mainly interested in the linear unbiased estimation of estimable linear parametric functions.*

Theorem 21.3.1 *A necessary and sufficient condition for a parametric function $\boldsymbol{l}'\boldsymbol{\beta}$ to be estimable is that*

$$\textit{Rank}\ (\boldsymbol{A}') = \textit{Rank}\ (\boldsymbol{A}', \boldsymbol{l}). \tag{21.3.4}$$

Proof: The function $\boldsymbol{l}'\boldsymbol{\beta}$ is estimable if there exists a linear function $\boldsymbol{C}'\boldsymbol{Y}$ such that $E(\boldsymbol{C}'\boldsymbol{Y}) = \boldsymbol{l}'\boldsymbol{\beta}$ independently of $\boldsymbol{\beta}$.

This means,

$$C'E(Y) = C'A\beta = l'\beta \text{ identically in } \beta$$
$$\text{i.e., } C'A = l'$$
$$\text{i.e., } A'C = l. \tag{21.3.5}$$

This is a set of p equations in n unknowns $C_1, \cdots, C_n (p \leq n)$. For the function $l'\beta$ to be estimable, the equations (21.3.5) must be solvable in $C_1, \cdots, C_n$. A *n.s.* condition for this is (21.3.4).

□ **Remark 21.3.2** *We must have $n \geq p$.*

Corollary 21.3.1 *If $R(A) = p$, every linear function of the parameters is linearly estimable.*

Proof: If $R(A) = p$, then $R(A', l)$ is necessarily p for all l and hence the result.

□ **Remark 21.3.3** *If $R(A) = p = n$, the equations* (21.3.5) *have a unique solution. If $R(A) = r < n$, there may be an infinite number of solutions or no solution of the equation* (21.3.5) *depending on whether conditions* (21.3.4) *hold or not. If there is an infinite number of solutions, only $(n - r)$ solution vectors among them will be linearly independent. Thus, in this case, there will be $(n - r)$ independent unbiased estimators of the parametric function $l'\beta$. Clearly $r \leq \min(n, p)$.*

Corollary 21.3.2 *Every linearly estimable parametric function is of the form $l'\beta = b'A\beta$ where $b = (b_1, \cdots, b_n)'$, a set of constants.*

Proof: The function $l'\beta$ is estimable if l is linearly dependent on the columns of A' i.e., if l belongs to the column-space of A' (as can be seen from the condition (21.3.5)) i.e., if $l = A'b$ or if $l'\beta = b'A\beta$ for some set of constants $b = (b_1, \cdots, b_n)'$.

Definition 21.3.4 ***Estimation Space and Error Space.*** *The space generated by the column vectors $\alpha_1, \cdots, \alpha_p$ of A is called the estimation space of A and is denoted by $V(\alpha_1, \cdots, \alpha_p) = V_p$. Obviously, $V_p \subset \mathbb{R}^n$ (because, $p \leq n$).*

Any point $\tau = (\tau_1, \cdots, \tau_n)'$ belongs to V_p if it is a linear combination of α_j i.e., if

$$\tau = \sum_{j=1}^{p} \theta_j \alpha_j \text{ for some real } \theta_1, \cdots, \theta_p.$$

Thus,

$$V_p = \{\theta_1\alpha_1 + \cdots + \theta_p\alpha_p | \theta_1, \cdots, \theta_p \subset \mathbb{R}^1\}.$$

Clearly, $\eta \in V_p$. Rank of the estimation space is the number of independent vectors among $\alpha_1, \cdots, \alpha_p$.

The vector space E that is orthogonal to the vector space $\boldsymbol{V}_p$ is called the error space. Thus, if γ is any vector in E, $\gamma'\boldsymbol{\alpha}_j = 0 \ \forall\, j = 1, \cdots, p$. If r is the rank of V_p, then rank of error space E is $n - r$.

▶ **Example 21.3.1** $E(\boldsymbol{Y}) = \beta_0\mathbf{1} + \beta_1\boldsymbol{x}$ *where* $\boldsymbol{x} = (x_1, \cdots, x_n)'$. *Here* $\boldsymbol{A} = (\mathbf{1}\ \boldsymbol{x})$. *Estimation space* $V_2 = \{\tau = \theta_1\mathbf{1} + \theta_2\boldsymbol{x} | \theta_1, \theta_2 \in R^1\} \subset R^n$ *and contains the point* $\eta = \beta_0\mathbf{1} + \beta_1\boldsymbol{x}$. *The error space is the space which is orthogonal to* V_2 *i.e.,* $E = \{\boldsymbol{a} : \boldsymbol{a}'\mathbf{1} = 0, \boldsymbol{a}'\boldsymbol{x} = 0\}$ (Fig. 21.1).

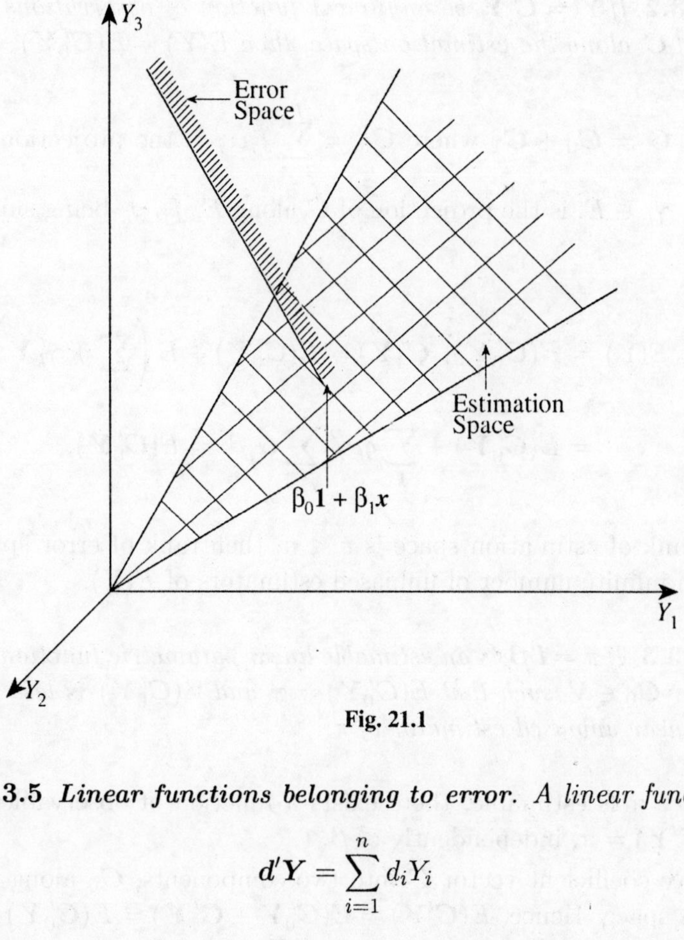

Fig. 21.1

Definition 21.3.5 ***Linear functions belonging to error.*** *A linear function*

$$\boldsymbol{d}'\boldsymbol{Y} = \sum_{i=1}^{n} d_i Y_i$$

is said to belong to error if its co-efficient vector $\mathbf{d}$ *belongs to error space* E, *i.e., if* $\boldsymbol{d}'\boldsymbol{\alpha}_j = 0,\ j = 1, \cdots, p.$

Hence,

$$E(\boldsymbol{d}'\boldsymbol{Y}) = \boldsymbol{d}'\boldsymbol{A}\boldsymbol{\beta} = \boldsymbol{d}'(\boldsymbol{\alpha}_1, \boldsymbol{\alpha}_2, \cdots, \boldsymbol{\alpha}_p)\boldsymbol{\beta} = \boldsymbol{d}'\boldsymbol{\alpha}_1\boldsymbol{\beta} + \cdots + \boldsymbol{d}'\boldsymbol{\alpha}_p\boldsymbol{\beta} = 0$$

independently of $\boldsymbol{\beta}$. Thus $\boldsymbol{d}'\boldsymbol{Y}$ is an unbiased estimator of zero.

Conversely, suppose $E_{\boldsymbol{\beta}}(\boldsymbol{d}'\boldsymbol{Y}) = 0$ for all values of $\boldsymbol{\beta}$.

Then,

$$\boldsymbol{d}'A\boldsymbol{\beta} = \sum_{j=1}^{p} \boldsymbol{d}'\boldsymbol{\alpha}_j\beta_j = 0$$

for all values of $\beta_j(j = 1, \cdots, p)$. Hence $\boldsymbol{d}'\boldsymbol{\alpha}_j = 0,\ j = 1, \cdots, p$.

Therefore, a function $\boldsymbol{d}'\boldsymbol{Y}$ belongs to error space E iff it is an unbiased estimator of zero.

Theorem 21.3.2 *If $Y = \boldsymbol{C}'\boldsymbol{Y}$ be any linear function of observations and if $\boldsymbol{C}_0$ be th resolved part of $\boldsymbol{C}$ along the estimation space, then $E(Y) = E(\boldsymbol{C}_0'\boldsymbol{Y})$.*

Proof: Let $\boldsymbol{C} = \boldsymbol{C}_0 + \boldsymbol{C}_1$ where $\boldsymbol{C}_0 = \sum_{j=1}^{p} f_j\boldsymbol{\alpha}_j$ is the projection of $\boldsymbol{C}$ on V_p an $\boldsymbol{C}_1 = \sum_k g_k\boldsymbol{\gamma}_k,\ \boldsymbol{\gamma}_k \in E$, is the projection of $\boldsymbol{C}$ along E, f_j, g_k being suitable coefficient

Hence,

$$E(Y) = E(\boldsymbol{C}_0'\boldsymbol{Y} + \boldsymbol{C}_1'\boldsymbol{Y}) = E(\boldsymbol{C}_0'\boldsymbol{Y}) + E\left(\sum_k g_k\boldsymbol{\gamma}_k'\boldsymbol{Y}\right)$$

$$= E(\boldsymbol{C}_0'\boldsymbol{Y}) + \sum_k g_k\boldsymbol{\gamma}_k' \sum_{j=1}^{p} \boldsymbol{\alpha}_j\beta_j = E(\boldsymbol{C}_0'\boldsymbol{Y}).$$

Hence, if rank of estimation space is $r < n$, then rank of error space is $n - r(>$ and there is an infinite number of unbiased estimators of $E(Y)$.

Theorem 21.3.3 *If $\pi = \boldsymbol{l}'\boldsymbol{\beta}$ is an estimable linear parametric function, then there exis a unique vector $\boldsymbol{C}_0 \in V$ such that $E(\boldsymbol{C}_0'\boldsymbol{Y}) = \pi$ and $V(\boldsymbol{C}_0'\boldsymbol{Y})$ is less than the varian of any other linear unbiased estimator of π.*

Proof: Since π is estimable, there exists a function of observables, say, $Y = \boldsymbol{C}'$ such that $E(\boldsymbol{C}'\boldsymbol{Y}) = \pi$, independently of $\boldsymbol{\beta}$.

Now, resolve coefficient vector $\boldsymbol{C}$ into two components, $\boldsymbol{C}_0$ along estimation spac $\boldsymbol{C}_1$ along error space. Hence, $E(\boldsymbol{C}'\boldsymbol{Y}) = E(\boldsymbol{C}_0'\boldsymbol{Y} + \boldsymbol{C}_1'\boldsymbol{Y}) = E(\boldsymbol{C}_0'\boldsymbol{Y}) = \pi$ for all $\boldsymbol{\beta}$.

Again, the component $\boldsymbol{C}_0$ along the estimation space is unique. If not, let $\boldsymbol{C}_2$ another component of $\boldsymbol{C}$ along the estimation space. Then $E(\boldsymbol{C}_2'\boldsymbol{Y}) = E(\boldsymbol{C}_0'\boldsymbol{Y})$ $\pi\ \forall\,\boldsymbol{\beta}$.

Hence, $E(\boldsymbol{C}_2 - \boldsymbol{C}_0)'\boldsymbol{Y} = 0\ \forall\,\boldsymbol{\beta}$ i.e., $(\boldsymbol{C}_2 - \boldsymbol{C}_0) \in E$.

Thus $(\boldsymbol{C}_2 - \boldsymbol{C}_0)$ belongs to V (since, both $\boldsymbol{C}_0, \boldsymbol{C}_2 \in V$) as well as E. But this impossible unless $\boldsymbol{C}_0 - \boldsymbol{C}_2$ is $\boldsymbol{0}$. Hence $\boldsymbol{C}_0$ is unique.

Again, $V(\boldsymbol{C}'\boldsymbol{Y}) = \boldsymbol{C}'D(\boldsymbol{Y})\boldsymbol{C} = \sigma^2\boldsymbol{C}'\boldsymbol{C}$.

Also, $V(\boldsymbol{C}_0'\boldsymbol{Y}) = \sigma^2\boldsymbol{C}_0'\boldsymbol{C}_0$.

Now

$$V(\boldsymbol{C}'\boldsymbol{Y}) = V\{(\boldsymbol{C}_0 + \boldsymbol{C}_1)'\boldsymbol{Y}\} = \sigma^2(\boldsymbol{C}_0 + \boldsymbol{C}_1)'(\boldsymbol{C}_0 + \boldsymbol{C}_1)$$
$$= \sigma^2(\boldsymbol{C}_0'\boldsymbol{C}_0 + \boldsymbol{C}_1'\boldsymbol{C}_1) \text{ (since } \boldsymbol{C}_0'C_1 = \boldsymbol{C}_1'\boldsymbol{C}_0 = 0)$$
$$\geq \sigma^2\boldsymbol{C}_0'\boldsymbol{C}_0.$$

Hence, $V(\boldsymbol{C}'\boldsymbol{Y}) \geq V(\boldsymbol{C}_0'\boldsymbol{Y})$, equality holds iff $\boldsymbol{C}_1'\boldsymbol{C}_1 = 0$ i.e., $\boldsymbol{C}_1$ is a null vector i.e., iff $\boldsymbol{C} = \boldsymbol{C}_0$.

Definition 21.3.6 *The function $Y = \boldsymbol{C}'\boldsymbol{Y}$ is the best linear unbiased estimator* (BLUE) *of π if $E(\boldsymbol{C}'\boldsymbol{Y}) = \pi$ and $V(\boldsymbol{C}'\boldsymbol{Y})$ is less than the variance of any other linear unbiased estimator of π.*

□ **Remark 21.3.4** *From theorem* 21.3.3 *it is seen that the statistic $Y = \boldsymbol{C}'\boldsymbol{Y}$ is the* BLUE *of π if its coefficient vector $\boldsymbol{C} \in V$. Theorem* 21.3.3 *can, therefore, be expressed in the following form.*

Theorem 21.3.4 *Of all the unbiased linear estimators of an estimable function π, the one which has its coefficient vector along the estimation space, is the* BLUE *of π.*

Corollary 21.3.3 *The* BLUE *of any estimable function $\pi = \boldsymbol{l}'\boldsymbol{\beta}$ must be of the form $\boldsymbol{q}'\boldsymbol{A}'\boldsymbol{Y}$ where $\boldsymbol{q} = (q_1, \cdots, q_p)'$ and satisfies*

$$\boldsymbol{q}'\boldsymbol{A}'\boldsymbol{A} = \boldsymbol{l}'. \tag{21.3.6}$$

Proof: Since the coefficient vector $\boldsymbol{C}$ of the BLUE lies in the column space of $\boldsymbol{A}$, $\boldsymbol{C} = \boldsymbol{A}\boldsymbol{q}$ for some vector $\boldsymbol{q}$ or $\boldsymbol{C}' = \boldsymbol{q}'\boldsymbol{A}'$.

Since,

$$E(\boldsymbol{C}'\boldsymbol{Y}) = E(\boldsymbol{q}'\boldsymbol{A}'\boldsymbol{Y}) = \boldsymbol{q}'\boldsymbol{A}'\boldsymbol{A}\beta = l'\beta \text{ (by unbiasedness)}$$

we must have $\boldsymbol{q}'\boldsymbol{A}'\boldsymbol{A} = \boldsymbol{l}'$.

Hence, the BLUE is

$$\hat{\pi} = \boldsymbol{l}'(A'A)^{-1}A'\boldsymbol{Y} \quad (\because \text{ by (21.3.6)}, q' = \boldsymbol{l}'(\boldsymbol{A}'\boldsymbol{A})^{-1}) \tag{21.3.7}$$

provided $\boldsymbol{A}$ is of full rank.

Corollary 21.3.4 *Let $\pi_i = \boldsymbol{l}_i'\beta$ $(i = 1, \cdots, k)$ be k estimable parametric functions and let $T_i = \boldsymbol{C}_i'\boldsymbol{Y}$ be the* BLUE *of $\pi_i (i = 1, \cdots, k)$. Then the* BLUE *of $\sum_{i=1}^{k} b_i\pi_i$ is $T = \sum_{i=1}^{k} b_iT_i$, where b_i's are arbitrary constants.*

Proof:

$$E(T) = \sum_i b_iE(T_i) = \sum_i b_i\pi_i.$$

Again,

$$T = \sum_i b_i(C_i'Y) = \sum_i (b_iC_i)'Y.$$

The coefficient vector of $T \in V$. Hence, by theorem 21.3.4, T must be BLUE of $\Sigma b_i \pi_i$.

■ NOTE 21.3.1 *Procedure for finding the* BLUE *of any estimable (linear) parametric function is, therefore, as follows. Find any vector* $\boldsymbol{C}$ *such that* $E(\boldsymbol{C}'\boldsymbol{Y}) = \pi$. *Resolve* $\boldsymbol{C}$ *along the estimation space. Let* $\boldsymbol{C}_0$ *be the resolved part. Then* $\boldsymbol{C}_0'\boldsymbol{Y}$ *is the* BLUE *of* π.

21.4 Solution by the Method of Least Squares

Consider the model (21.2.1) again, our problem being to estimate the unknown parameters $\beta_1, \cdots, \beta_p$. In the method of least squares (l.s.) $\beta_1, \cdots, \beta_p$ are so determined that the sum of squares

$$S = \sum_{i=1}^{n}(Y_i - E(Y_i))^2 = \sum_{i=1}^{n}\{Y_i - a_{i1}\beta_1 - \cdots - a_{ip}\beta_p\}^2 = (\boldsymbol{Y} - \boldsymbol{A\beta})'(\boldsymbol{Y} - \boldsymbol{A\beta})$$

is minimum among all the choices of $\boldsymbol{\beta}$. This is found by solving the equations

$$\frac{\partial S}{\partial \beta_j} = 0, \quad j = 1, \cdots, p$$

subject to the condition that the matrix

$$\left(\left(\frac{\partial^2 S}{\partial \beta_j \partial \beta_k}\right)\right)$$

is positive definite. It will be found that these lead to the same set of equations as (21.3.6). The l.s. estimate of any estimable parametric function will be found to be identical with its BLUE, as given by Theorem 21.3.4.

Theorem 21.4.1 *Consider the model* (21.2.3) *with rank* $(\boldsymbol{A}) = p$. *The best linear unbiased estimator of* $\boldsymbol{\beta}$ *is given by the least square estimate* $\hat{\boldsymbol{\beta}} = (\boldsymbol{A}'\boldsymbol{A})^{-1}\boldsymbol{A}'\boldsymbol{Y}$ *of* $\boldsymbol{\beta}$.

Proof: The l.s. estimate of $\boldsymbol{\beta}$ is obtained by minimising $S = (\boldsymbol{Y} - \boldsymbol{A\beta})'(\boldsymbol{Y} - \boldsymbol{A\beta})$ w.r.t. $\boldsymbol{\beta}$. The value of $\boldsymbol{\beta}$ that minimises S is given by the solution of $\frac{\partial S}{\partial \boldsymbol{\beta}} = 0$.

Hence,

$$\boldsymbol{A}'\boldsymbol{Y} - \boldsymbol{A}'\boldsymbol{A}\hat{\boldsymbol{\beta}} = 0. \tag{21.4.1}$$

The l.s. estimate is, therefore,

$$\hat{\boldsymbol{\beta}} = (\boldsymbol{A}'\boldsymbol{A})^{-1}\boldsymbol{A}'\boldsymbol{Y} = \boldsymbol{C}^{-1}\boldsymbol{A}'\boldsymbol{Y} \tag{21.4.2}$$

when $\boldsymbol{C} = \boldsymbol{A}'\boldsymbol{A}$.

The equations (21.4.1) are called the normal equations. The normal equations always admit a solution, since $\boldsymbol{A'Y}$ always lies in the column space of $\boldsymbol{A'A}$.

The solution of the normal equations namely, $\boldsymbol{\beta} = \hat{\boldsymbol{\beta}}$ gives an extreme value of S. To show that this extreme value is the minimum value of S consider as follows:

$$\begin{aligned}(\boldsymbol{Y} - \boldsymbol{A\beta})'(\boldsymbol{Y} - \boldsymbol{A\beta}) &= [\boldsymbol{Y} - \boldsymbol{A}\hat{\boldsymbol{\beta}} + \boldsymbol{A}(\hat{\boldsymbol{\beta}} - \boldsymbol{\beta})]'[\boldsymbol{Y} - \boldsymbol{A}\hat{\boldsymbol{\beta}} + \boldsymbol{A}(\hat{\boldsymbol{\beta}} - \boldsymbol{\beta})] \\ &= (\boldsymbol{Y} - \boldsymbol{A}\hat{\boldsymbol{\beta}})'(\boldsymbol{Y} - \boldsymbol{A}\hat{\boldsymbol{\beta}}) + (\hat{\boldsymbol{\beta}} - \boldsymbol{\beta})'\boldsymbol{A'A}(\hat{\boldsymbol{\beta}} - \boldsymbol{\beta}) \\ &\qquad \text{(the cross-product term vanishes)} \\ &= (\boldsymbol{Y} - \boldsymbol{A}\hat{\boldsymbol{\beta}})'(\boldsymbol{Y} - \boldsymbol{A}\hat{\boldsymbol{\beta}}) + [\boldsymbol{A}(\hat{\boldsymbol{\beta}} - \boldsymbol{\beta})]'[\boldsymbol{A}(\hat{\boldsymbol{\beta}} - \boldsymbol{\beta})] \\ &\geq (\boldsymbol{Y} - \boldsymbol{A}\hat{\boldsymbol{\beta}})'(\boldsymbol{Y} - \boldsymbol{A}\hat{\boldsymbol{\beta}}). \end{aligned} \tag{21.4.3}$$

Therefore, the solution $\boldsymbol{\beta} = \hat{\boldsymbol{\beta}}$ minimises S.

Proof that $\hat{\boldsymbol{\beta}}$ is BLUE**:** Let $\boldsymbol{B}$ be any $p \times n$ constant matrix and let $\tilde{\boldsymbol{\beta}} = \boldsymbol{BY}$ be an estimate of $\boldsymbol{\beta}$. Clearly $\tilde{\boldsymbol{\beta}}$ is a vector of linear functions of $\boldsymbol{Y}$. Let $\boldsymbol{B} = \boldsymbol{C}^{-1}\boldsymbol{A}' + \boldsymbol{D}$ where $\boldsymbol{D}$ has to be suitably determined. For unbiasedness we must have,

$$E(\tilde{\boldsymbol{\beta}}) = E(\boldsymbol{BY}) = E[(\boldsymbol{C}^{-1}\boldsymbol{A}' + \boldsymbol{D})\boldsymbol{Y}] = (\boldsymbol{C}^{-1}\boldsymbol{A}' + \boldsymbol{D})\boldsymbol{A\beta} = \boldsymbol{\beta} + \boldsymbol{DA\beta} = \boldsymbol{\beta}$$

i.e.,

$$\boldsymbol{DA\beta} = \boldsymbol{0} \text{ for all } \boldsymbol{\beta}$$

i.e.,

$$\boldsymbol{DA} = \boldsymbol{0}. \tag{21.4.4}$$

To investigate into the property of bestness we must find $\boldsymbol{D}$ such that $V(\tilde{\beta}_j)$ is minimum among all the linear unbiased estimators of $\beta_j (j = 1, \cdots, p)$, subject to the restriction (21.4.4). Now

$$D(\tilde{\boldsymbol{\beta}}) = E[(\tilde{\boldsymbol{\beta}} - \boldsymbol{\beta})(\tilde{\boldsymbol{\beta}} - \boldsymbol{\beta})'] = E\{[(\boldsymbol{C}^{-1}\boldsymbol{A}' + \boldsymbol{D})\boldsymbol{Y} - \boldsymbol{\beta}][(\boldsymbol{C}^{-1}\boldsymbol{A}' + \boldsymbol{D})\boldsymbol{Y} - \boldsymbol{\beta}]'\}.$$

Substituting $\boldsymbol{Y} = \boldsymbol{A\beta} + \boldsymbol{e}$ and using (21.4.4),

$$\begin{aligned} D(\tilde{\boldsymbol{\beta}}) &= E(\boldsymbol{C}^{-1}\boldsymbol{A'ee'AC}^{-1} + \boldsymbol{Dee'D'} + \boldsymbol{C}^{-1}\boldsymbol{A'ee'D'} + \boldsymbol{Dee'AC}^{-1}) \\ &= \sigma^2(\boldsymbol{C}^{-1} + \boldsymbol{DD'}) = \sigma^2(\boldsymbol{C}^{-1} + \boldsymbol{G}) \text{ (say)} \end{aligned}$$

where $\boldsymbol{G} = ((g_{jk})) = \boldsymbol{DD'}$. The diagonal elements of $D(\tilde{\boldsymbol{\beta}})$ are respective variances of $\tilde{\beta}_j$. To minimise each $V(\tilde{\beta}_j)$, we must, therefore, minimise each diagonal element of $\boldsymbol{D}(\tilde{\boldsymbol{\beta}})$. Since σ^2 and $\boldsymbol{C}^{-1}$ are constants, we must find a matrix $\boldsymbol{G}$ such that each diagonal element of $\boldsymbol{G}$ is a minimum. Now $\boldsymbol{G} = \boldsymbol{DD'}$ is positive definite, hence $g_{jj} \geq 0$. Therefore, the diagonal elements of $\boldsymbol{D}(\tilde{\boldsymbol{\beta}})$ will attain their minimum value when $g_{jj} = 0 \ \forall \ j = 1, \cdots, n$. Again, if $\boldsymbol{D} = ((d_{jk}))$, then

$$g_{jj} = \sum_{k=1}^{n} d_{jk}^2.$$

Therefore, if $g_{jj} = 0 \ \forall \ j$, then $d_{jk} = 0 \ \forall \ j, k$. This implies that $\boldsymbol{D} = \boldsymbol{0}$. Therefore, $\boldsymbol{B} = \boldsymbol{C}^{-1}\boldsymbol{A}'$ and $\tilde{\boldsymbol{\beta}} = \boldsymbol{C}^{-1}\boldsymbol{A'Y} = \hat{\boldsymbol{\beta}}$.

■ NOTE 21.4.1 *The* BLUE *of* $\boldsymbol{\beta}$ *is defined as the estimator* $\hat{\hat{\boldsymbol{\beta}}}$ *for which* $V(\hat{\hat{\beta}}_j)$ *minimum among all linear unbiased estimators of* β_j *for all* $j = 1, \cdots, p$. *Therefor using* (21.3.7), BLUE *of* β_j *is* $(0 \cdots 010 \cdots 0)\boldsymbol{C}^{-1}\boldsymbol{A}'\boldsymbol{Y} = \hat{\beta}_j$.

Hence the BLUE of $\boldsymbol{\beta}$ is $\hat{\boldsymbol{\beta}}$, the l.s. estimator of $\boldsymbol{\beta}$. This provides another pro of BLUE-ness of $\hat{\boldsymbol{\beta}}$ using the theory of best linear unbiased estimation in Section 21. However, the proof given in theorem 21.4.1 is a direct proof of this result.

Theorem 21.4.2 (***Gauss-Markoff theorem***). *Under the model given in theorem 21.4 the best linear unbiased estimator of any linear parametric function* $\boldsymbol{l}'\boldsymbol{\beta}$ *is* $\boldsymbol{l}'\hat{\boldsymbol{\beta}}$ *where is the l.s. estimator of* $\boldsymbol{\beta}$ *as given in* (21.4.2).

In other words, the LSE of $\boldsymbol{l}'\boldsymbol{\beta}$ is identical with the minimum variance linear unbias estimator of $\boldsymbol{l}'\boldsymbol{\beta}$.

Proof: From corollary 24.3.4, the BLUE of $\boldsymbol{l}'\boldsymbol{\beta}$ is

$$\begin{aligned} \boldsymbol{q}'\boldsymbol{A}'\boldsymbol{Y} &= \boldsymbol{q}'\boldsymbol{A}'\boldsymbol{A}\hat{\boldsymbol{\beta}} \quad \text{[by (21.3.8)]} \\ &= \boldsymbol{l}'\hat{\boldsymbol{\beta}} \quad \text{[by (21.3.6)]} \end{aligned}$$

when $\hat{\boldsymbol{\beta}}$ is the l.s. estimator of $\boldsymbol{\beta}$, given by (21.4.2).

Alternative Proof

We note that $\boldsymbol{l}'\boldsymbol{\beta} = \hat{\pi}$ is linear in $\boldsymbol{Y}$, since $\hat{\pi} = \boldsymbol{C}'\boldsymbol{Y}$ where

$$\boldsymbol{C}' = \boldsymbol{l}'(\boldsymbol{A}'\boldsymbol{A})^{-1}\boldsymbol{A}'; \quad V(\hat{\pi}) = \sigma^2\boldsymbol{l}'(\boldsymbol{A}'\boldsymbol{A})^{-1}\boldsymbol{l}.$$

Suppose we have another unbiased estimator $\tilde{\pi} = \boldsymbol{d}'\boldsymbol{Y}$ ($\boldsymbol{d} \neq \boldsymbol{C}$). Since $\tilde{\pi}$ is unbiase $\pi = \boldsymbol{l}'\boldsymbol{\beta} = E(\tilde{\pi}) = \boldsymbol{d}'\boldsymbol{A}\boldsymbol{\beta}$ independently of $\boldsymbol{\beta}$.

Hence,

$$\boldsymbol{l}' = \boldsymbol{d}'\boldsymbol{A}.$$

Again, covariance between $\hat{\pi}$ and $\tilde{\pi}$ is

$$\begin{aligned} \operatorname{Cov}(\boldsymbol{C}'\boldsymbol{Y}, \boldsymbol{d}'\boldsymbol{Y}) &= E\boldsymbol{C}'(\boldsymbol{Y} - \boldsymbol{\eta})(\boldsymbol{Y} - \boldsymbol{\eta})\boldsymbol{d}] = \sigma^2\boldsymbol{C}'\boldsymbol{d} = \sigma^2\boldsymbol{l}'(\boldsymbol{A}'\boldsymbol{A})^{-1}\boldsymbol{A}'\boldsymbol{d} \\ &= \sigma^2\boldsymbol{l}'(\boldsymbol{A}'\boldsymbol{A})^{-1}\boldsymbol{l} \quad \text{[by (i)]} \\ &= V(\hat{\pi}). \end{aligned}$$

Now,

$$0 \le V(\tilde{\pi} - \hat{\pi}) = V(\tilde{\pi}) + V(\hat{\pi}) - 2\operatorname{Cov}(\tilde{\pi}, \hat{\pi}) = V(\tilde{\pi}) - V(\hat{\pi})$$

or,

$$V(\tilde{\pi}) \ge V(\hat{\pi})$$

with equality iff $\hat{\pi} = \tilde{\pi}$ i.e., iff $\boldsymbol{C} = \boldsymbol{d}$. Note that $\hat{\pi}$ is unique BLUE of π.

Explict expression for BLUE *of* $l'\beta$.
he normal equations (21.4.1) are:

$$\begin{aligned} \alpha_1'\alpha_1\beta_1 + \cdots + \alpha_1'\alpha_p\beta_p &= \alpha_1'Y \\ \cdots\cdots\cdots\cdots\cdots\cdots\cdots & \quad \cdots \\ \alpha_p'\alpha_1\beta_1 + \cdots + \alpha_p'\alpha_p\beta_p &= \alpha_p'Y \end{aligned} \tag{21.4.5}$$

here

$$\alpha_j'\alpha_k = \sum_{i=1}^{n} a_{ij}a_{ik}, \quad \alpha_j'Y = \sum_{i=1}^{n} a_{ij}Y_i.$$

riting $C^{-1} = ((c^{ij}))$,

$$\hat{\beta} = \begin{bmatrix} c^{11} & \cdots & c^{1p} \\ \cdot & \cdots & \cdot \\ c^{p1} & \cdots & c^{pp} \end{bmatrix} \begin{bmatrix} \alpha_1'Y \\ \cdot \\ \alpha_p'Y \end{bmatrix}$$

e.,

$$\hat{\beta}_i = \sum_{j=1}^{p} c^{ij}(\alpha_j'Y). \tag{21.4.6}$$

ence, BLUE of π,

$$\hat{\pi} = l'\hat{\beta} = \sum_{i=1}^{p}\sum_{j=1}^{p} l_i c^{ij}(\alpha_j'Y). \tag{21.4.7}$$

lue in the Determinant Form

onsider the determinant

$$|D| = \begin{vmatrix} 0 & \alpha_1'Y & \cdots & \alpha_p'Y \\ l_1 & c_{11} & \cdots & c_{1p} \\ \cdot & \cdot & \cdots & \cdot \\ l_p & c_{p1} & \cdots & c_{pp} \end{vmatrix}$$

$$= -l_1 \begin{vmatrix} \alpha_1'Y & \cdots & \alpha_p'Y \\ c_{21} & \cdots & c_{2p} \\ \cdot & \cdots & \cdot \\ c_{p1} & \cdots & c_{pp} \end{vmatrix} + l_2 \begin{vmatrix} \alpha_1'Y & \cdots & \alpha_p'Y \\ c_{11} & \cdots & c_{1p} \\ c_{31} & \cdots & c_{3p} \\ \cdot & \cdots & \cdot \\ c_{p1} & \cdots & c_{pp} \end{vmatrix} + \cdots + (-1)^p \begin{vmatrix} \alpha_1'Y & \cdots & \alpha_p'Y \\ c_{11} & \cdots & c_{1p} \\ \cdot & \cdots & \cdot \\ c_{(p-1)1} & \cdots & c_{(p-1)p} \end{vmatrix} \tag{21.4.8}$$

he first term in (21.4.8)

$$= -l_1 \left[\alpha_1'YC^{(11)} - \alpha_2'YC^{(12)} + \cdots + (-1)^{p-1}C^{(1p)}\right];$$

the second term in (24.4.8)

$$= l_2[\boldsymbol{\alpha}_1'\boldsymbol{Y}C^{(21)} - \boldsymbol{\alpha}_2'\boldsymbol{Y}C^{(22)} + \cdots + (-1)^{p-1}C^{(2p)}]$$

where $C^{(ij)}$ denotes cofactor of c_{ij} in $|\boldsymbol{C}|$. Hence,

$$|\boldsymbol{D}| = -\sum_{i=1}^{p}\sum_{j=1}^{p} l_i\boldsymbol{\alpha}_j'\boldsymbol{Y}C^{(ij)}.$$

Therefore,

$$\hat{\pi} = \sum_{i=1}^{p}\sum_{j=1}^{p} l_i\boldsymbol{\alpha}_j'\boldsymbol{Y}c^{ij} = -\frac{|\boldsymbol{D}|}{|\boldsymbol{C}|}. \qquad (21.4.9$$

■ NOTE 21.4.2 *Note that no assumption about the distribution of* **e** *except that* *has zero mean and covariance matrix* $\sigma^2\boldsymbol{I}$ *has been made. If, further,* **e** *is normall distributed* $N(\boldsymbol{0}, \sigma^2\boldsymbol{I})$, *then the model* (21.2.3) *is said to be a normal theory model o general linear-hypothesis model. In this case, the maximum likelihood estimator (m.l.e. of* $\boldsymbol{\beta}$ *is the same as the l.s. estimator* $\hat{\boldsymbol{\beta}}$.

It is also the minimum variance unbiased estimator of $\boldsymbol{\beta}$. *More specificall estimators* $\hat{\boldsymbol{\beta}}$ *and* $\hat{\sigma}^2 = [\boldsymbol{Y}'(I - \boldsymbol{A}\boldsymbol{C}^{-1}\boldsymbol{A}')\boldsymbol{Y}]/(n-p)$ *have the following properties. The are* (i) *consistent* (ii) *efficient* (iii) *unbiased* (iv) *sufficient* (v) *complete* (vi) *minimur variance unbiased. Also* $\hat{\boldsymbol{\beta}}$ *and* $\hat{\sigma}^2$ *are independently distributed,* $\hat{\boldsymbol{\beta}} \sim N_p(\boldsymbol{\beta}, \sigma^2\boldsymbol{C}^{-1})$ *an* $(n-p)\hat{\sigma}^2/\sigma^2 \sim \chi^2_{(n-p)}$. *Note that* $\hat{\sigma}^2$ *is not the m.l.e. of* σ^2, *it is* $\hat{\hat{\sigma}}^2 = (n-p)\hat{\sigma}^2/n$. *Th estimator* $\hat{\sigma}^2$ *has been derived in Section* 21.7.

Again, because of the invariance property of the m.l.e.'s $\boldsymbol{l}'\hat{\boldsymbol{\beta}}$ *is also the m.l.e. of* $\boldsymbol{l}'\boldsymbol{\beta}$ *The situation for l.s.e.'s is not as simple. However, in case of linear function of* β_i*'s the invariance property holds for l.s.e.'s as well. This has been stated in theorem* 21.4.2

■ NOTE 21.4.3 *In theorem* 21.4.1 *we assumed rank* $(\boldsymbol{A}) = p$. *In this case* $\boldsymbol{C}^{-1}$ *exist and the solution of the normal equation* $\hat{\boldsymbol{\beta}} = \boldsymbol{C}^{-1}\boldsymbol{A}'\boldsymbol{Y}$ *is unique. In case* $\boldsymbol{C}$ *is singula one will have to find a generalised inverse* $\bar{\boldsymbol{C}}$ *of* $\boldsymbol{C}$ *(the generalised (g) inverse has bee discussed in chapter* 0). *It has been already observed that the normal equations* (21.4.1 *are consistent. Hence, if a g-inverse* $\bar{\boldsymbol{C}}$ *of* $\boldsymbol{C}$ *is found, a particular solution of the norm equation is given by* $\hat{\boldsymbol{\beta}}_z = \bar{\boldsymbol{C}}\boldsymbol{A}'\boldsymbol{Y}$. *Since a g-inverse is not unique, the solution of th normal equations is not unique in this case. However, if* $\hat{\boldsymbol{\beta}}_1$ *and* $\hat{\boldsymbol{\beta}}_2$ *are any two solution of normal equations, then*

$$(\boldsymbol{Y} - \boldsymbol{A}\hat{\boldsymbol{\beta}}_1)'(\boldsymbol{Y} - \boldsymbol{A}\hat{\boldsymbol{\beta}}_1) = (\boldsymbol{Y} - \boldsymbol{A}\hat{\boldsymbol{\beta}}_2)'(\boldsymbol{Y} - \boldsymbol{A}\hat{\boldsymbol{\beta}}_2).$$

This along with (21.4.3) shows that every solution of the normal equations is a set c least square estimators. Again, the theorem 21.4.2 holds for any l.s. estimator $\hat{\boldsymbol{\beta}}_z$ of $\boldsymbol{\beta}$

A general solution of the normal equations in this case is $\hat{\boldsymbol{\beta}}_z = \bar{\boldsymbol{C}}\boldsymbol{A}'\boldsymbol{Y} + [\bar{\boldsymbol{C}}\boldsymbol{C} - I]\boldsymbol{t}$ where $\boldsymbol{t}$ is any arbitrary $p \times 1$ vector.

In application it will not be necessary to obtain $\bar{C}$ in its general form. It will be found that the expression in which we are interested are invariant under a substitution of one g-inverse for the other. Hence, to obtain $\bar{C}$, it is only necessary to solve the equations $\boldsymbol{Cx} = \boldsymbol{z}$ keeping $\boldsymbol{z}$ general. We shall mainly consider non-singular $\boldsymbol{C}$ throughout. The case of a singular $\boldsymbol{C}$ has been considered in example 21.7.4.

21.4.1 Geometrical Interpretation

Consider as in example 21.3.1, $E(\boldsymbol{Y}) = \beta_0\mathbf{1} + \beta_1\boldsymbol{x}$. Here $V_2 = \{\theta_1\mathbf{1} + \theta_2\boldsymbol{x}|\theta_1, \theta_2 \in R^1\}$. Clearly $\boldsymbol{\eta} \in V_2$. We may wish to choose a point in the estimation space V_2 which is, in some sense, the best representation of $E(\boldsymbol{Y})$ i.e., which serves as a best estimate of $\boldsymbol{A\beta}$.

A natural choice is the orthogonal projection of $\boldsymbol{Y}$ on the estimation space. Since this projection lies in the estimation space we may write it as

$$\hat{\boldsymbol{\eta}} = \sum_{j=1}^{p} \hat{\beta}_j \boldsymbol{\alpha}_j = \boldsymbol{A}\hat{\boldsymbol{\beta}}.$$

Because of the above representation, $\boldsymbol{Y} - \hat{\boldsymbol{\eta}} = \boldsymbol{Y} - \boldsymbol{A}\hat{\boldsymbol{\beta}}$ is orthogonal to the estimation space. Therefore,

$$\boldsymbol{A}'(\boldsymbol{Y} - \hat{\boldsymbol{\eta}}) = \mathbf{0} \text{ or, } \boldsymbol{A}'\boldsymbol{Y} = \boldsymbol{A}'\boldsymbol{A}\hat{\boldsymbol{\beta}}. \tag{21.4.10}$$

These are the normal equations (21.4.1). Since we have assumed $\boldsymbol{A}$ is the full rank,

$$\hat{\boldsymbol{\beta}} = (\boldsymbol{A}'\boldsymbol{A})^{-1}\boldsymbol{A}'\boldsymbol{Y}.$$

Therefore, the projection of $\boldsymbol{Y}$ on the estimation space is

$$\hat{\boldsymbol{\eta}} = \boldsymbol{A}\hat{\boldsymbol{\beta}} = \boldsymbol{A}(\boldsymbol{A}'\boldsymbol{A})^{-1}\boldsymbol{A}'\boldsymbol{Y} = \boldsymbol{RY} \text{ (say)}$$

when $\boldsymbol{R} = \boldsymbol{A}(\boldsymbol{A}'\boldsymbol{A})^{-1}\boldsymbol{A}'$. It is readily seen that $\boldsymbol{R}$ is an idempotent and symmetric matrix (i.e., $\boldsymbol{R}'\boldsymbol{R} = \boldsymbol{R}$, $\boldsymbol{RR} = \boldsymbol{R}$). As noted, a symmetric and idempotent matrix defines an orthogonal projection matrix. The matrix $\boldsymbol{R}$ projects $\boldsymbol{Y}$ into V_p (since $\boldsymbol{RY} = \hat{\eta} = \boldsymbol{A}\hat{\boldsymbol{\beta}} \in V_p$).

Since $\boldsymbol{R}$ is symmetric idempotent, $\boldsymbol{I} - \boldsymbol{R} = \boldsymbol{Q}$ is also symmetric idempotent and hence a projection matrix. Since

$$\boldsymbol{A}'(\boldsymbol{I} - \boldsymbol{R}) = \mathbf{0}$$

$\boldsymbol{Q}$ projects on a space which is orthogonal to the estimation space i.e., on the error space. It follows from (21.4.10) that $(\boldsymbol{Y} - \hat{\boldsymbol{\eta}})$ belongs to the error space. Again $(I - R)\boldsymbol{Y} = \boldsymbol{Y} - \hat{\boldsymbol{\eta}}$.

Therefore, $\boldsymbol{Y} - \hat{\boldsymbol{\eta}} = \hat{\boldsymbol{e}}$ is the orthogonal projection of $\boldsymbol{Y}$ on the error space (see Fig. 21.2).

Properties of the matrices $\boldsymbol{R}$ *and* $\boldsymbol{Q}$:

$$\text{Rank } (\boldsymbol{R}) = tr(\boldsymbol{R}) = tr[\boldsymbol{A}'\boldsymbol{A}(\boldsymbol{A}'\boldsymbol{A})^{-1}] = tr(\boldsymbol{I}_p) = p$$
$$\text{Rank } (\boldsymbol{Q}) = tr(\boldsymbol{Q}) = tr(\boldsymbol{I} - \boldsymbol{R}) = tr(\boldsymbol{I}) - tr(\boldsymbol{R}) = n - p$$

since $\boldsymbol{R}$ and $\boldsymbol{Q}$ are idempotent. Also

$$\boldsymbol{RA} = \boldsymbol{A},\ \ \boldsymbol{A}'\boldsymbol{R} = \boldsymbol{A}'$$
$$\boldsymbol{QA} = \boldsymbol{O},\ \ \boldsymbol{A}'\boldsymbol{Q} = \boldsymbol{O}.$$

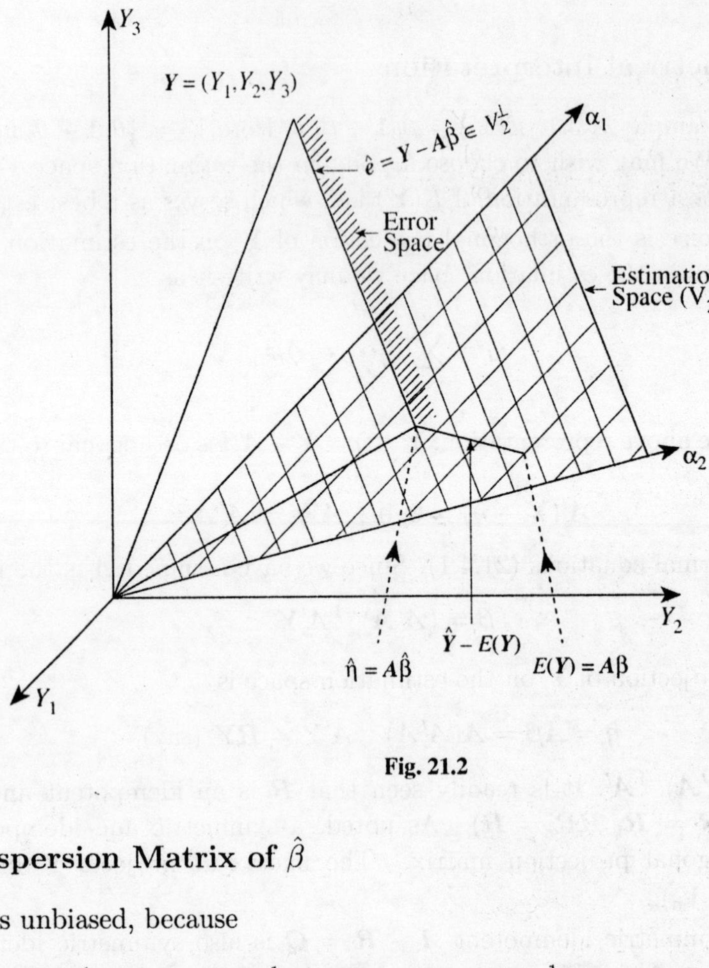

Fig. 21.2

21.4.2 Dispersion Matrix of $\hat{\beta}$

The l.s.e. $\hat{\boldsymbol{\beta}}$ is unbiased, because

$$E(\hat{\boldsymbol{\beta}}) = E(\boldsymbol{C}^{-1}\boldsymbol{A}'\boldsymbol{Y}) = E[\boldsymbol{C}^{-1}\boldsymbol{A}'(\boldsymbol{A\beta} + \boldsymbol{e})] = E(\boldsymbol{C}^{-1}\boldsymbol{C\beta}) + E(\boldsymbol{C}^{-1}\boldsymbol{A}'\boldsymbol{e}) = \boldsymbol{\beta}$$

$$D(\hat{\boldsymbol{\beta}}) = E(\hat{\boldsymbol{\beta}} - \boldsymbol{\beta})(\hat{\boldsymbol{\beta}} - \boldsymbol{\beta})' = E[\boldsymbol{C}^{-1}\boldsymbol{A}'\boldsymbol{ee}'\boldsymbol{AC}^{-1}] = \boldsymbol{C}^{-1}\boldsymbol{A}'E(\boldsymbol{ee}')\boldsymbol{AC}^{-1}$$
$$= \boldsymbol{C}^{-1}\boldsymbol{A}'\sigma^2\boldsymbol{I}\boldsymbol{AC}^{-1} = \sigma^2\boldsymbol{C}^{-1}. \tag{21.4.11}$$

Thus, $\hat{\beta}_j$ is a linear unbiased estimator of β_j with

$$V(\hat{\beta}_j) = \sigma^2 c^{jj};\ \ \text{Cov}(\hat{\beta}_j, \hat{\beta}_k) = \sigma^2 c^{jk}. \tag{21.4.12}$$

Again, if $\boldsymbol{l}_1'\hat{\boldsymbol{\beta}}$, $\boldsymbol{l}_2'\hat{\boldsymbol{\beta}}$ are the l.s.-estimators of two estimable functions $\boldsymbol{l}_1'\boldsymbol{\beta}$, $\boldsymbol{l}_2'\boldsymbol{\beta}$ respectively, then

$$\text{Cov}(\boldsymbol{l}_1'\hat{\boldsymbol{\beta}}, \boldsymbol{l}_2'\hat{\boldsymbol{\beta}}) = \boldsymbol{l}_1'D(\hat{\boldsymbol{\beta}})\boldsymbol{l}_2 = \sigma^2\boldsymbol{l}_1'\boldsymbol{C}^{-1}\boldsymbol{l}_2.$$

Again,

$$D(\hat{\eta}) = D(A\hat{\beta}) = \sigma^2 AC^{-1}A' = \sigma^2 R$$
$$D(\hat{e}) = D(Y - \hat{\eta}) = \sigma^2(I - R).$$

Note that $n \times n$ matrices $D(\hat{\eta})$ and $D(\hat{e})$ have ranks p and $(n-p)$ respectively and since $0 < p < n$, they are singular.

▶ **Example 21.4.1** *$Y_1, \cdots, Y_n$ are n independent r.v.'s with $E(Y_i) = \beta$, $V(Y_i) = \sigma^2$ $(i = 1, \cdots, n)$.*

Here $A = (1 \cdots 1)'$, normal equation is

$$A'A\beta = A'Y \quad \text{i.e.,} \quad n\beta = \sum_{i=1}^{n} Y_i;$$

hence l.s.e.

$$\hat{\beta} = \sum_{i=1}^{n} Y_i/n = \bar{Y}; \quad V(\hat{\beta}) = \sigma^2 C^{-1} = \sigma^2/n.$$

Hence rank $(V) = 1$, rank $(E) = n - 1$.

▶ **Example 21.4.2** *Samples of sizes n_1 and n_2 are drawn from two populations having means m_1 and m_2 respectively with a common variance σ^2. Find the* BLUE *of $l_1m_1 + l_2m_2$ along with its variance.*

Assume that samples are $(Y_1, \cdots, Y_{n_1})$ and $(X_1, \cdots, X_{n_2})$ respectively.

Then,

$$E(Y_i) = m_1, \ i = 1 \cdots n_1$$
$$E(X_j) = m_2, \ j = 1, \cdots, m_2$$
$$V(Y_i) = V(X_j) = \sigma^2.$$

Covariances are zero. Writing $X_j = Y_{n_1+j}$, $j = 1, \cdots, n_2$, $E(Y) = Am$ where

$$A = [\alpha_1 \ \alpha_2], \ \alpha_1 = (1, \cdots, 1, 0, \cdots, 0)', \ \alpha_2 = (0, \cdots, 0, 1, \cdots, 1)'$$
$$m = (m_1 \ m_2)'.$$

Normal equations: $A'Am = A'Y$

$$\text{i.e.,} \quad \begin{bmatrix} n_1 \ m_1 \\ n_2 \ m_2 \end{bmatrix} = \begin{bmatrix} \sum_{1}^{n_1} Y_i \\ \sum_{1}^{n_2} X_j \end{bmatrix}.$$

Hence, $\hat{m}_1 = \bar{Y}$, $\hat{m}_2 = \bar{X}$

BLUE of $\boldsymbol{l'm} = l_1\bar{Y} + l_2\bar{X}$

$$V(\boldsymbol{l'm}) = \sigma^2 \boldsymbol{l}\boldsymbol{C}^{-1}\boldsymbol{l} = \sigma^2(l_1\, l_2)\begin{bmatrix} 1/n_1 & 0 \\ 0 & 1/n_2 \end{bmatrix}\begin{bmatrix} l_1 \\ l_2 \end{bmatrix} = \sigma^2\left(\frac{l_1^2}{n_1} + \frac{l_2^2}{n_2}\right).$$

Hence, rank $(V) = 2$, rank $(E) = n_1 + n_2 - 2$.

21.5 Linear Estimation with Correlated Variables

We now consider the model $\boldsymbol{Y} = \boldsymbol{A\beta} + \boldsymbol{e}$ where,

$$\begin{aligned} E(\boldsymbol{Y}) &= \boldsymbol{A\beta} \\ D(\boldsymbol{Y}) &= \sigma^2\boldsymbol{V} \end{aligned} \tag{21.5.1}$$

where σ^2 is an unknown, positive constant, $\boldsymbol{V}$ is a known symmetric matrix. Assume that $\boldsymbol{V}$ is non-singular. This model can be reduced to the model (21.2.3) as follows.

Since $\boldsymbol{V}$ is a non-singular matrix of order n, there exists a non-singular matrix $\boldsymbol{H}$ of order n such that $\boldsymbol{H'VH} = I$. Consider the transformation $\boldsymbol{Z} = \boldsymbol{H'Y}$. Thus,

$$\begin{aligned} E(\boldsymbol{Z}) &= \boldsymbol{H'A\beta} = \boldsymbol{A^*\beta} \text{ (say)} \\ D(\boldsymbol{Z}) &= \boldsymbol{H'}D(\boldsymbol{Y})\boldsymbol{H} = \sigma^2\boldsymbol{H'VH} = \sigma^2\boldsymbol{I}. \end{aligned} \tag{21.5.2}$$

Again, rank $(\boldsymbol{A^*})$ = rank $(\boldsymbol{H'A})$ = rank $(\boldsymbol{A})$. Therefore, the transformed variables $\boldsymbol{Z}$ obey the model (21.2.3). All the results for the model (21.2.3), therefore, will hold for the model (21.5.2) if we replace $\boldsymbol{Y}$ by $\boldsymbol{Z}$ and $\boldsymbol{A}$ by $\boldsymbol{A^*}$. The l.s. estimators for $\boldsymbol{\beta}$ for the model (21.5.2) are found by minimising

$$\begin{aligned} (\boldsymbol{Z} - \boldsymbol{A^*\beta})'(\boldsymbol{Z} - \boldsymbol{A^*\beta}) &= (\boldsymbol{Y} - \boldsymbol{A\beta})'\boldsymbol{H'H}(\boldsymbol{Y} - \boldsymbol{A\beta}) = (\boldsymbol{Y} - \boldsymbol{A\beta})'\boldsymbol{V}^{-1}(\boldsymbol{Y} - \boldsymbol{A\beta}) \\ &= \sum_i\sum_j v^{ij}(Y_i - a_{i1}\beta_1 - \cdots - a_{ip}\beta_p) \\ &\qquad (Y_j - a_{j1}\beta_1 - \cdots - a_{jp}\beta_p) \end{aligned} \tag{21.5.3}$$

where $\boldsymbol{V}^{-1} = ((v^{ij}))$. The quantity (21.5.3) is called the *weighted sum of squares.*

If the variables are uncorrelated but do not have constant variance, then $\boldsymbol{V} =$ Diag$(v_1, \cdots, v_n)$ (say) and the weighted sum of the squares (21.5.3) becomes

$$\sum_i (Y_i - a_{i1}\beta_1 - \cdots - a_{ip}\beta_p)^2/v_i. \tag{21.5.4}$$

This is minimised to obtain the l.s. estimator (and also BLUE) of $\boldsymbol{\beta}$.
Under the model (21.5.1), the generalised BLUE of $\boldsymbol{\beta}$ is, therefore,

$$\hat{\beta}_G = (\boldsymbol{A'V^{-1}A})^{-1}(\boldsymbol{A'V^{-1}Y}) \tag{21.5.5}$$

when $\hat{\boldsymbol{\beta}}_G$ is Aitken's generalised least square estimator of $\boldsymbol{\beta}$.

$$D(\hat{\boldsymbol{\beta}}_G) = \sigma^2(\boldsymbol{A'V^{-1}A})^{-1}. \tag{21.5.6}$$

■ NOTE 21.5.1 *The ordinary or unweighted l.s.* (OLS) *estimator of* $\boldsymbol{\beta}$, $\hat{\boldsymbol{\beta}} = (\boldsymbol{A}'\boldsymbol{A})^{-1}\boldsymbol{A}'\boldsymbol{Y}$ *is also a linear unbiased estimator* $\boldsymbol{\beta}$ *under the model* (21.5.1) *but is not necessarily the* BLUE *under this model. It has been shown that if the column space of* $\boldsymbol{A}$ *is the the same as the column space of* $\boldsymbol{VA}$ *then* $\hat{\boldsymbol{\beta}}_G = \hat{\boldsymbol{\beta}}$ *(Zyskind, 1969). Rao (1967, 1968) showed that a necessary and sufficient condition that the* OLS *estimator* $\hat{\boldsymbol{\beta}}$ *is the same as the Aitken's generalised l.s.e.* $\hat{\boldsymbol{\beta}}_G$ *with probability unity is that*

$$\boldsymbol{A}'\boldsymbol{V}\boldsymbol{L} = \boldsymbol{0} \tag{21.5.7}$$

or equivalently $\sigma^2\boldsymbol{V}$ *be expressible in the form*

$$\boldsymbol{A}\boldsymbol{\Gamma}\boldsymbol{A}' + \boldsymbol{L}\boldsymbol{\theta}\boldsymbol{L}' + \omega^2\boldsymbol{I}_n \tag{21.5.8}$$

where $\boldsymbol{\Gamma}$ *and* $\boldsymbol{\theta}$ *are arbitrary symmetric matrices,* ω^2 *is an arbitrary scaler and* $\boldsymbol{L}$ *is a* $n \times (n-p)$ *matrix of rank* $(n-p)$ *such that* $\boldsymbol{A}'\boldsymbol{L} = \boldsymbol{0}$.

In particular, MeElroy (1967) showed that the OLS estimator $\hat{\boldsymbol{\beta}}$ is the same as the BLUE $\hat{\boldsymbol{\beta}}_G$ whenever $\boldsymbol{V} = [(1-\rho)\boldsymbol{I} + \boldsymbol{1}\boldsymbol{1}'\rho]$ for $-\frac{1}{n-1} < \rho < 1$, provided one column of $\boldsymbol{A}$ is proportional to $\boldsymbol{1}_n$ (vide exercises (16)–(18)).

■ NOTE 21.5.2 *Because of its simplicity and popularity, the* OLS *estimator* $\hat{\boldsymbol{\beta}}$ *may be preferred unless its relative efficiency compared to the generalised l.s.e.* $\hat{\boldsymbol{\beta}}_G$ *is substantially low. For any estimable function* $\boldsymbol{l}'\boldsymbol{\beta}$, *under model* (21.5.1) *its* OLSE *is* $\boldsymbol{l}'\hat{\boldsymbol{\beta}}$, *whereas its* BLUE *is* $\boldsymbol{l}'\hat{\beta}_G$.

The relative efficiency of $\boldsymbol{l}'\hat{\boldsymbol{\beta}}$ with respect to $\boldsymbol{l}'\hat{\beta}_G$ is

$$E = \frac{\text{var }(\boldsymbol{l}'\hat{\beta}_G)}{\text{var }(\boldsymbol{l}'\hat{\boldsymbol{\beta}})}.$$

Clearly $E \leq 1$ and under some conditions, $E = 1$. A lower bound for $E(\boldsymbol{l}'\hat{\beta})$, due to Magness and McGuire (1962) is

$$E \geq \frac{4\lambda_1\lambda_n}{(\lambda_1 + \lambda_n)^2} \tag{21.5.9}$$

where $\lambda_1 \geq \cdots \geq \lambda_n$ are the eigenvalues of V. The lower bound in (21.5.9) is obtained by using the Kantorovich inequality (vide subsection 0.4.6) and it holds for all $\boldsymbol{A}$ and all estimable functions $\boldsymbol{l}'\boldsymbol{\beta}$.

Further discussion on this model is continued in note 21.7.4.

21.5.1 Comparison between OLS Estimator $\hat{\beta}$ and the BLUE $\hat{\beta}_G$

For comparison between $\hat{\boldsymbol{\beta}}$ and $\hat{\boldsymbol{\beta}}_G$ consider the matrix mean square error (Mtx MSE) (introduced in section 15.8) of the estimators under the model (21.5.1).

$$\text{Mtx MSE }(\hat{\boldsymbol{\beta}}_G) = E(\hat{\boldsymbol{\beta}}_G - \boldsymbol{\beta})(\hat{\boldsymbol{\beta}}_G - \boldsymbol{\beta})' = \sigma^2(\boldsymbol{A}'\boldsymbol{V}^{-1}\boldsymbol{A})^{-1}$$

$$\text{Mtx MSE }(\hat{\boldsymbol{\beta}}) = E(\hat{\boldsymbol{\beta}} - \boldsymbol{\beta})(\hat{\boldsymbol{\beta}} - \boldsymbol{\beta})' = D(\hat{\boldsymbol{\beta}}) + [\text{Bias }(\hat{\boldsymbol{\beta}})][\text{Bias }(\hat{\boldsymbol{\beta}})]'.$$

Now,

$$D(\hat{\boldsymbol{\beta}}) = \sigma^2(\boldsymbol{A}'\boldsymbol{A})^{-1}\boldsymbol{A}'\boldsymbol{V}\boldsymbol{A}(\boldsymbol{A}'\boldsymbol{A})^{-1}$$
$$\text{Bias } (\hat{\boldsymbol{\beta}}) = \boldsymbol{0}.$$

Hence

$$\Delta = \text{Mtx MSE } (\hat{\boldsymbol{\beta}}) - \text{Mtx MSE } (\hat{\boldsymbol{\beta}}_G)$$
$$= \sigma^2\left[(\boldsymbol{A}'\boldsymbol{A})^{-1}\boldsymbol{A}'\boldsymbol{V}\boldsymbol{A}(\boldsymbol{A}'\boldsymbol{A})^{-1} - (\boldsymbol{A}'\boldsymbol{V}^{-1}\boldsymbol{A})^{-1}\right].$$

To show that Δ is positive definite we need to show that $\boldsymbol{\eta}'\Delta\boldsymbol{\eta} \geq 0$ where $\boldsymbol{\eta}$ is any $p \times 1$ non-zero constant vector.

$$\boldsymbol{\eta}'\Delta\boldsymbol{\eta} = \sigma^2\boldsymbol{\eta}'\left[(\boldsymbol{A}'\boldsymbol{A})^{-1}\boldsymbol{A}'\boldsymbol{V}\boldsymbol{A}(\boldsymbol{A}'\boldsymbol{A})^{-1} - (\boldsymbol{A}'\boldsymbol{V}^{-1}\boldsymbol{A})^{-1}\right]\boldsymbol{\eta} \geq 0$$
$$\Rightarrow \frac{\boldsymbol{\eta}'(\boldsymbol{A}'\boldsymbol{V}^{-1}\boldsymbol{A})^{-1}\boldsymbol{\eta}}{\boldsymbol{\eta}'(\boldsymbol{A}'\boldsymbol{A})^{-1}\boldsymbol{A}'\boldsymbol{V}\boldsymbol{A}(\boldsymbol{A}'\boldsymbol{A})^{-1}\boldsymbol{\eta}} \leq 1. \tag{21.5.10}$$

Assume without loss of generality $\boldsymbol{\eta}'\boldsymbol{\eta} = 1$. A sufficient condition for (21.5.10) to hold good is

$$\frac{\max\limits_{\boldsymbol{\eta}:\boldsymbol{\eta}'\boldsymbol{\eta}=1} \boldsymbol{\eta}'(\boldsymbol{A}'\boldsymbol{V}^{-1}\boldsymbol{A})^{-1}\boldsymbol{\eta}}{\min\limits_{\boldsymbol{\eta}:\boldsymbol{\eta}'\boldsymbol{\eta}=1} \boldsymbol{\eta}'(\boldsymbol{A}'\boldsymbol{A})^{-1}\boldsymbol{A}'\boldsymbol{V}\boldsymbol{A}(\boldsymbol{A}'\boldsymbol{A})^{-1}\boldsymbol{\eta}} \leq 1. \tag{21.5.11}$$

Numerator of (21.5.11) is

$$\lambda_1\left[(\boldsymbol{A}'\boldsymbol{V}^{-1}\boldsymbol{A})^{-1}\right] = \frac{1}{\lambda_p[\boldsymbol{A}'\boldsymbol{V}^{-1}\boldsymbol{A}]}.$$

Denominator of (21.5.11) is

$$\lambda_p\left[(\boldsymbol{A}'\boldsymbol{A})^{-1}\boldsymbol{A}'\boldsymbol{V}\boldsymbol{A}(\boldsymbol{A}'\boldsymbol{A})^{-1}\right] \tag{21.5.12}$$

where λ_1, λ_p denote respectively the largest and the smallest eigenvalues of the relevant matrices.

If $(\boldsymbol{A}'\boldsymbol{A})$ is orthogonal, (21.5.12) is

$$\lambda_p[\boldsymbol{A}'\boldsymbol{V}\boldsymbol{A}].$$

Hence the condition (21.5.11) reduces to

$$\lambda_p[\boldsymbol{A}'\boldsymbol{V}^{-1}\boldsymbol{A}]\lambda_p[\boldsymbol{A}'\boldsymbol{V}\boldsymbol{A}] \geq 1. \tag{21.5.13}$$

Comparison between $\hat{\boldsymbol{\beta}}, \hat{\boldsymbol{\beta}}_G$ have been studied by Baksalary and Kala (1980), Wang et al (1990), Mukhopadhyay and Schwabe (1997), among others.

21.6 Linear Estimation with Linear Restriction on Parameters

Consider the model (21.2.3) but suppose that the parameters in $\boldsymbol{\beta}$ are subject to m linear restrictions

$$\boldsymbol{H\beta} = \boldsymbol{h} \tag{21.6.1}$$

where $\boldsymbol{H}$ is a $m \times p$ matrix of rank m and $\boldsymbol{h}$ is a $m \times 1$ vector, both consisting of known non-stochastic numbers. The problem is to estimate the parameter vector $\boldsymbol{\beta}$ in (21.2.3) subject to the restriction (21.6.1). A rank restriction on $\boldsymbol{H}$ to make sure that constraints (21.6.1) are consistent is that rank $(\boldsymbol{H}) = m \le p$.

We minimise the squared distance between the observation vector and a point in the estimation space $(\boldsymbol{Y}-\boldsymbol{A\beta})'(\boldsymbol{Y}-\boldsymbol{A\beta})$ subject to the restriction (21.6.1) by the Lagrangian expression:

$$\begin{aligned} S^* &= (\boldsymbol{Y} - \boldsymbol{A\beta})'(\boldsymbol{Y} - \boldsymbol{A\beta}) + \boldsymbol{\mu}'(\boldsymbol{h} - \boldsymbol{H\beta}) \\ &= \boldsymbol{Y'Y} - 2\boldsymbol{\beta' A' Y} + \boldsymbol{\beta' A' A \beta} + \boldsymbol{\mu}'(\boldsymbol{h} - \boldsymbol{H\beta}) \end{aligned} \tag{21.6.2}$$

when $\boldsymbol{\mu}$ is a $m \times 1$ vector of m Lagrangian multipliers associated with m constraints along with m rows of $\boldsymbol{H\beta} = \boldsymbol{h}$.

We differentiate the Lagrangian w.r.t. $\boldsymbol{\beta}$ and $\boldsymbol{\mu}$ and equate it to zero. This gives the normal equations as:

$$\frac{\partial S^*}{\partial \boldsymbol{\beta}} = 0 \;\Rightarrow\; -2\boldsymbol{A'Y} + 2\boldsymbol{A'A}\hat{\boldsymbol{\beta}}_R - \boldsymbol{H'\mu} = \boldsymbol{0} \tag{21.6.3}$$

$$\frac{\partial S^*}{\partial \boldsymbol{\mu}} = 0 \;\Rightarrow\; \boldsymbol{h} - \boldsymbol{H}\hat{\boldsymbol{\beta}}_R = \boldsymbol{0} \tag{21.6.4}$$

where $\hat{\boldsymbol{\beta}}_R$ is the restricted l.s. estimate of $\boldsymbol{\beta}$. Pre-multiplying (24.6.3) by $(\boldsymbol{A'A})^{-1}$,

$$-2\hat{\boldsymbol{\beta}} + 2\hat{\boldsymbol{\beta}}_R - (\boldsymbol{A'A})^{-1}\boldsymbol{H'\mu} = \boldsymbol{0}. \tag{21.6.5}$$

Premultiplying (21.6.5) by $\boldsymbol{H}$,

$$\begin{aligned} &-2\boldsymbol{H}\hat{\boldsymbol{\beta}} + 2\boldsymbol{H}\hat{\boldsymbol{\beta}}_R - \boldsymbol{H}(\boldsymbol{A'A})^{-1}\boldsymbol{H'\mu} = 0 \\ \text{or,}\quad &-2\boldsymbol{H}\hat{\boldsymbol{\beta}} + 2\boldsymbol{h} - \boldsymbol{H}(\boldsymbol{A'A})^{-1}\boldsymbol{H'\mu} = 0 \;\text{ (using (21.6.4))} \\ \text{or,}\quad &2(\boldsymbol{h} - \boldsymbol{H}\hat{\boldsymbol{\beta}}) = \boldsymbol{H}(\boldsymbol{A'A})^{-1}\boldsymbol{H'\mu} \\ \text{or,}\quad &\boldsymbol{\mu} = 2[\boldsymbol{H}(\boldsymbol{A'A})^{-1}\boldsymbol{H}']^{-1}(\boldsymbol{h} - \boldsymbol{H}\hat{\boldsymbol{\beta}}) \end{aligned}$$

provided $(\boldsymbol{H}(\boldsymbol{A'A})^{-1}\boldsymbol{H}')^{-1}$ exists. A necessary (but not sufficient) condition for the existence of the inverse is that rank $(\boldsymbol{H}) = m \le p$, which has been assumed. Substituting this value of $\boldsymbol{\mu}$ in (21.6.5) and solving for $\hat{\boldsymbol{\beta}}_R$ we have

$$\hat{\boldsymbol{\beta}}_R = \hat{\boldsymbol{\beta}} - (\boldsymbol{A'A})^{-1}\boldsymbol{H}'[\boldsymbol{H}(\boldsymbol{A'A})^{-1}\boldsymbol{H}']^{-1}(\boldsymbol{H}\hat{\boldsymbol{\beta}} - \boldsymbol{h}). \tag{21.6.6}$$

The solution $\hat{\boldsymbol{\beta}}_R$ is called the restricted l.s. estimate of $\boldsymbol{\beta}$.

If we assume that Y_i's are normally distributed in addition to the assumption (21.2.3) and (21.6.1), then the restricted maximum likelihood estimator of σ^2 is

$$\hat{\sigma}_R^2 = \frac{1}{n}(\boldsymbol{Y} - \boldsymbol{A}\hat{\boldsymbol{\beta}}_R)'(\boldsymbol{Y} - \boldsymbol{A}\hat{\boldsymbol{\beta}}_R). \tag{21.6.7}$$

21.7 Sum of Squares due to BLUE's, Errors

Consider functions of observations, $y = \boldsymbol{C}'\boldsymbol{Y}$, $y_1 = \boldsymbol{C}_1'\boldsymbol{Y}, \cdots, y_k = \boldsymbol{C}_k'\boldsymbol{Y}$. A linear function y is said to depend on a number of linear functions $y_1, \cdots, y_k$ if

$$y = d_1 y_1 + \cdots + d_k y_k$$

$d_i \neq 0$ for at least one i.

A set of linear functions $y_1, \cdots, y_k$ is said to form a dependent set if

$$d_1 y_1 + \cdots + d_k y_k = 0,$$

$d_i \neq 0$ for at least one i.

A function $y_k = 0$ is said to be a dependent function.

Theorem 21.7.1 *The necessary and sufficient condition that the function y is dependent on the functions $y_1, \cdots, y_k$ is that the coefficient vector of y is dependent on the coefficient vectors of $y_1, \cdots, y_k$.*

Proof: Let

$$\begin{aligned} y &= C_1 Y_1 + \cdots + C_n Y_n \\ y_1 &= C_1^1 Y_1 + \cdots + C_n^1 Y_n \\ &\cdots\cdots\cdots\cdots\cdots\cdots \\ y_k &= C_1^k Y_1 + \cdots + C_n^k Y_n. \end{aligned}$$

If y is dependent on the set $y_1, \cdots, y_k$ we can write $y = d_1 y_1 + \cdots + d_k y_k$

$$\text{i.e., } \sum_{j=1}^{n} C_j Y_j = \sum_{i=1}^{k} d_i \sum_{j=1}^{n} C_j^i Y_j = \sum_{j=1}^{n} Y_j \sum_{i=1}^{k} d_i C_j^i.$$

This relation should hold for all values of $Y_1, \cdots, Y_n$. Hence we must have,

$$\begin{aligned} \text{i.e., } C_j &= \sum_{i=1}^{k} d_i C_j^i \\ \text{i.e., } \boldsymbol{C} &= d_1 \boldsymbol{C}_1 + d_2 \boldsymbol{C}_2 + \cdots + d_k \boldsymbol{C}_k \end{aligned}$$

where $\boldsymbol{C}_i = (C_1^i, C_2^i, \cdots, C_n^i)$, $i = 1, \cdots, k$.

Sufficiency can be proved in a similar way.

Theorem 21.7.2 *The necessary and sufficient condition that the linear functions (of observations) $y_1, \cdots, y_k$ form a dependent set is that the corresponding coefficient vectors form a dependent set.*

Proof: Similar.

Consider parametric functions $\pi = \boldsymbol{l}'\boldsymbol{\beta}$, $\pi_i = \boldsymbol{l}_i'\boldsymbol{\beta}$, $i = 1, \cdots, k$. The function π is said to be dependent on $\pi_1, \cdots, \pi_k$ if $\pi = d_1\pi_1 + \cdots + d_k\pi_k$ for some $d_1, \cdots, d_k$ ($d_i \neq 0$ for at least one i). The set of functions $\pi_1, \cdots, \pi_k$ form a dependent set if $d_1\pi_1 + \cdots + d_k\pi_k = 0$ for some $d_1, \cdots, d_k (d_i \neq 0$ for at least one i).

By corollary 21.3.4, if $y_1, \cdots, y_k$ are BLUE's for $\pi_1, \cdots, \pi_k$ respectively, then $\sum_{i=1}^{k} d_i y_i$ is BLUE for $\sum_{i=1}^{k} d_i\pi_i$. Hence, if $\pi_1, \cdots, \pi_k$ form a dependent set then for all values of $\boldsymbol{\beta}$ and given $\boldsymbol{A}$,

$$E\left(\sum_{i=1}^{k} d_i y_i\right) = \sum_{i=1}^{k} d_i\pi_i = 0 \text{ i.e., } E\left(\sum_{i=1}^{k} d_i \boldsymbol{C}_i'\right)\boldsymbol{Y} = \boldsymbol{0}$$

$$\text{i.e., } \left(\sum_{i=1}^{k} d_i \boldsymbol{C}_i'\right)\boldsymbol{A\beta} = \boldsymbol{0}$$

$$\text{i.e., } \sum_{i=1}^{k} d_i \boldsymbol{C}_i' = \boldsymbol{0}'. \tag{21.7.1}$$

The relation (21.7.1) means $\boldsymbol{C}_1, \cdots, \boldsymbol{C}_k$ form a dependent set. Hence the BLUE's $y_1, \cdots, y_k$ form a dependent set. Therefore, if $\pi_1, \cdots, \pi_k$ form a dependent set their BLUE's also form a dependent set. The converse is also true, i.e., if $y_1, \cdots, y_k$ form a dependent set of linear functions their expectations $\pi_1, \cdots, \pi_k$ ($\pi_i \neq 0, i = 1, \cdots, k$) also form a dependent set of parametric functions.

21.7.1 Linear Set and the Degrees of Freedom

A linear function $y = \boldsymbol{C}'\boldsymbol{Y}$ carries one degree of freedom (d.f.). Let us have k linear functions $y_1 = \boldsymbol{C}_1'\boldsymbol{Y}, \cdots, y_k = \boldsymbol{C}_k'\boldsymbol{Y}$. Of these, let r functions $y_1, \cdots, y_r$ be independent i.e.,

$$d_1 y_1 + \cdots + d_r y_r = 0 \;\Rightarrow\; d_1 = \cdots = d_r = 0.$$

The set of all linear functions dependent on the linear functions $y_1, \cdots, y_r$ is called the *linear set* (or *linear system*) generated by $y_1, \cdots, y_r$. The rank of a linear set is the number of independent linear functions in the generating set. This is called the d.f. carried by the linear set. It is also the rank of the vector space generated by $\boldsymbol{C}_1, \cdots, \boldsymbol{C}_k$.

Definition 21.7.1 ***The d.f. carried by estimators.*** *Let $y_1, \cdots, y_k$ be a set of unbiased linear estimators of estimable parametric functions $\pi_1, \cdots, \pi_k$ respectively. The d.f. carried by the set $y_1, \cdots, y_k$ is the d.f. carried by these set of estimators.*

Theorem 21.7.3 *The d.f. carried by all possible* BLUE*'s of estimable parametric functions (under the model 21.2.3) is the rank of the estimation space* $V(\boldsymbol{\alpha}_1, \cdots, \boldsymbol{\alpha}_p)$.

Proof: Let the estimation space $V(\boldsymbol{\alpha}_1, \cdots, \boldsymbol{\alpha}_p)$ be of rank n_e. Then there are n_e independent vectors $\gamma_1, \cdots, \gamma_{n_e}$ which form the basis of V. From the condition of estimability (21.3.5), we know that the vector $\boldsymbol{l}$ of an estimable parametric function $\boldsymbol{l}'\boldsymbol{\beta}$ must depend linearly on the columns of $\boldsymbol{A}'$ (i.e., $\boldsymbol{l}'$ must depend linearly on the rows of $\boldsymbol{A}$). Since rank $(\boldsymbol{A}) = n_e$, we can only find n_e independent $\boldsymbol{l}'$ vector i.e., there can be only n_e independent estimable parametric functions. Since there is a one-to-one correspondence between an estimable parametric function and its BLUE (corollary 21.3.4), we conclude there are n_e independent BLUE's and no more. Hence the d.f. carried by all possible BLUE's is equal to n_e.

■ NOTE 21.7.1 *The d.f. carried by all possible* BLUE*'s = d.f. carried by all estimable parametric functions = Rank (estimation space) =* n_e. *Hence if there are* $n_e + k$ *independent parametric functions, then k of these functions are not estimable.*

Definition 21.7.2 ***The d.f. due to error:*** *A linear function* $y = \boldsymbol{C}'\boldsymbol{Y}$ *such that* $E(y) = 0$ *independently of the values of the parameters carries one d.f. due to error. Similarly, the d.f. carried by a set of independent linear functions* $y_1, y_2, \cdots$, *all of which belong to error (i.e.,* $E(y_i) = 0$, $i = 1, 2, \cdots \forall \boldsymbol{\beta}$) *is called the d.f. due to error carried by this linear set.*

Theorem 21.7.4 *The d.f. carried by a set of all possible linear functions each belonging to error is equal to the rank of the error space i.e., d.f. due to error = rank of the error space.*

Proof: If $y = \boldsymbol{C}'\boldsymbol{Y} \in$ error space, $E(y) = \boldsymbol{C}'(\boldsymbol{\alpha}_1\beta_1 + \cdots + \boldsymbol{\alpha}_p\beta_p) = 0$ $(\forall\ \boldsymbol{\beta})$ i.e., $\boldsymbol{C}'\boldsymbol{\alpha}_i = 0$, $i = 1, \cdots, p$ i.e., $\boldsymbol{C}'$ is $\perp V(\boldsymbol{\alpha}_1, \cdots, \boldsymbol{\alpha}_p)$. Therefore, coefficient vectors of all linear functions belonging to error form a space V' which is orthogonal to $V(\boldsymbol{\alpha}_1, \cdots, \boldsymbol{\alpha}_p)$. Since rank $(V) = n_e$, rank $(V') = n - n_e$. Clearly $V' = E$. Therefore, d.f. due to all possible unbiased estimators of zero = d.f. due to error = rank of error space.

Two linear functions $\boldsymbol{C}_1'\boldsymbol{Y}$, $\boldsymbol{C}_2'\boldsymbol{Y}$ are said to be *orthogonal* if their coefficient vectors are orthogonal i.e., if $\boldsymbol{C}_1'\boldsymbol{C}_2 = 0$. If a linear function $\boldsymbol{b}'\boldsymbol{Y}$ is orthogonal to each of the linear functions $\boldsymbol{C}_1'\boldsymbol{Y}, \cdots, \boldsymbol{C}_k'\boldsymbol{Y}$ generating a linear set L, then we say that $\boldsymbol{b}'\boldsymbol{Y}$ is orthogonal to the set L. Clearly, $\boldsymbol{b}$ is orthogonal to the vector space generated by $(\boldsymbol{C}_1, \cdots, \boldsymbol{C}_k)$. Again, the set of linear functions $\boldsymbol{b}_1'\boldsymbol{Y}, \cdots, \boldsymbol{b}_m'\boldsymbol{Y}$ each orthogonal to a linear set L forms another linear set M, which is said to be the linear set orthogonal to the linear set L. Clearly, the space generated by $(\boldsymbol{b}_1, \cdots, \boldsymbol{b}_m)$ will be orthogonal to the space generated by $(\boldsymbol{C}_1, \cdots, \boldsymbol{C}_k)$. Hence,

Theorem 21.7.5 *If a set of linear functions* $\boldsymbol{C}_1'\boldsymbol{Y}, \cdots, \boldsymbol{C}_r'\boldsymbol{Y}$ *carries r d.f., then the set of linear functions orthogonal to the given set carries* $(n - r)$ *d.f.*

▶ **Example 21.7.1** *Find the d.f. carried by all the possible difference,* $Y_i - Y_j$ $(i \neq j = 1, \cdots, n)$.

Each of the differences (or contrasts) $y_{ij} = Y_i - Y_j$ is orthogonal to $Y_1 + \cdots + Y_n$, because the coefficient vector in $Y_i - Y_j$ is $(0, \cdots, 0, 1, 0, ; 0, -1, 0 \cdots 0)'$, that in $\sum_{i=1}^{n} Y_i$ is $(1 \cdots 1)'$ and these are orthogonal vectors. Similarly, coefficient vector in each y_{ji} is orthogonal to $\mathbf{1}_n$. Therefore, $R[V(y_{ij}, i \neq j = 1, \cdots, n)] = n - R(\mathbf{1}_n) = n - 1$.

Definition 21.7.3 ***Contrast.*** *A linear function* $\boldsymbol{C}'\boldsymbol{Y}$ *is said to be a contrast if* $\sum_{i=1}^{n} C_i = 0$ *i.e., if* $\boldsymbol{C}'\mathbf{1}_n = 0$.

Therefore, the coefficient vectors of all possible contrasts is each orthogonal to $\mathbf{1}_n$. Hence the vector space generated by the coefficient vectors of all possible contrasts is orthogonal to $V(\mathbf{1}_n)$ and is of rank $n - 1$. Therefore, there are $(n - 1)$ independent contrasts.

Theorem 21.7.6 *The coefficient vector of the* BLUE *of* β_i *is orthogonal to the space* $V(\boldsymbol{\alpha}_1, \cdots, \boldsymbol{\alpha}_{i-1}, \boldsymbol{\alpha}_{i+1}, \cdots, \boldsymbol{\alpha}_p)$.

Proof: From the normal equations (21.4.1),

$$\hat{\boldsymbol{\beta}} = \boldsymbol{C}^{-1}\boldsymbol{A}\boldsymbol{Y} \quad (\text{where } \boldsymbol{C} = \boldsymbol{A}'\boldsymbol{A})$$

$$\hat{\beta}_i = \sum_{j=1}^{p} c^{ij}\boldsymbol{\alpha}'_j\boldsymbol{Y} = \boldsymbol{C}'_i\boldsymbol{Y} \text{ (say) where } \boldsymbol{C}'_i = \sum_{j=1}^{p} c^{ij}\boldsymbol{\alpha}'_j. \tag{21.7.2}$$

Now

$$\boldsymbol{C}'_i\boldsymbol{\alpha}_k = \sum_{j=1}^{p} c^{ij}\boldsymbol{\alpha}'_j\boldsymbol{\alpha}_k = \sum_{j=1}^{p} c^{ij}C_{jk} = \delta_{ik} = \begin{cases} 0, \text{ if } i \neq k \\ 1, \text{ if } i = k. \end{cases}$$

Hence,

$$\boldsymbol{C}'_i\boldsymbol{\alpha}_i = 1, \quad \boldsymbol{C}'_i\boldsymbol{\alpha}_k = 0 \quad (k \neq i).$$

Theorem 21.7.7 *The* BLUE *of* $\pi = l_1\beta_1 + \cdots + l_k\beta_k$. $(k < p)$ *has its coefficient vector orthogonal to* $V(\boldsymbol{\alpha}_{k+1}, \cdots, \boldsymbol{\alpha}_p)$.

Proof: Similar to that of theorem 21.7.6.

21.7.2 Sum of Squares due to Linear Functions

Let $y = \boldsymbol{C}'\boldsymbol{Y}$ be a linear function of observations. The sum of squares (SS) due to y is

$$\frac{y^2}{\boldsymbol{C}'\boldsymbol{C}} = \frac{\left(\sum_{i=1}^{n} C_i Y_i\right)^2}{\sum_{i=1}^{n} C_i^2}.$$

Usually, this SS is referred to as the SS due to the single d.f. carried by $\boldsymbol{C}'\boldsymbol{Y}$.

Theorem 21.7.8 *The* SS *due to the d.f. carried by the linear function $y = \boldsymbol{C}'\boldsymbol{Y}$ is equal to the square of the length of projection of the observation vector $\boldsymbol{Y}$ upon the coefficient vector $\boldsymbol{C}$.*

Proof: The vector $\boldsymbol{Y}$ can always be resolved into two parts: one part along $\boldsymbol{C}$ and the other part along the space orthogonal to $\boldsymbol{C}$. Thus $\boldsymbol{Y} = k\boldsymbol{C} + \boldsymbol{C}_1$ where k is a scalar and $\boldsymbol{C}_1$ is orthogonal to $\boldsymbol{C}$ i.e., $\boldsymbol{C}'\boldsymbol{C}_1 = 0$. To evaluate k, multiply both sides by $\boldsymbol{C}$

$$\boldsymbol{Y}'\boldsymbol{C} = k\boldsymbol{C}'\boldsymbol{C} + \boldsymbol{C}_1'\boldsymbol{C} = k\boldsymbol{C}'\boldsymbol{C}$$

so that $k = \boldsymbol{Y}'\boldsymbol{C}/\boldsymbol{C}'\boldsymbol{C}$.
Hence,

$$k\boldsymbol{C} = \frac{(\boldsymbol{Y}'\boldsymbol{C})\boldsymbol{C}}{\boldsymbol{C}'\boldsymbol{C}}.$$

Therefore, square of the projection of $\boldsymbol{Y}$ on $\boldsymbol{C}$

$$= (k\boldsymbol{C}')(k\boldsymbol{C}) = \frac{(\boldsymbol{Y}'\boldsymbol{C})^2\boldsymbol{C}'\boldsymbol{C}}{(\boldsymbol{C}'\boldsymbol{C})^2} = \frac{(\boldsymbol{Y}'\boldsymbol{C})^2}{\boldsymbol{C}'\boldsymbol{C}} = \frac{y^2}{\boldsymbol{C}'\boldsymbol{C}} = \text{SS due to } \boldsymbol{y}.$$

Expectation of SS due to Linear Functions

$$\frac{E(\boldsymbol{C}'\boldsymbol{Y})^2}{\boldsymbol{C}'\boldsymbol{C}} = \frac{V(\boldsymbol{C}'\boldsymbol{Y}) + \{E(\boldsymbol{C}'\boldsymbol{Y})\}^2}{\boldsymbol{C}'\boldsymbol{C}} = \frac{\boldsymbol{C}'\boldsymbol{C}\sigma^2 + \{\boldsymbol{C}'E(\boldsymbol{Y})\}^2}{\boldsymbol{C}'\boldsymbol{C}}$$

$$= \sigma^2 + \{\boldsymbol{C}'E(\boldsymbol{Y})\}^2/\boldsymbol{C}'\boldsymbol{C} = \sigma^2 + \text{ SS due to } \boldsymbol{C}'E(\boldsymbol{Y}). \qquad (21.7.3)$$

In particular, if $E(y) = 0$, where $y = \boldsymbol{C}'\boldsymbol{Y}$, E (SS due to y) $= \sigma^2$.
Hence,

Theorem 21.7.9 *Expectation of* SS *due to a linear function carrying a d.f. belonging to error is σ^2, the common variance of the observations.*

ss due to a set of Linear Functions

Consider a linear set generated by the linear functions $y_1 = C_1'Y$, $y_2 = C_2'Y, \cdots$, $y_k = C_k'Y$ and carrying r d.f. We can always find $r(\leq k)$ mutually orthogonal linear functions $y_{(1)} = C_{(1)}'Y, \cdots, y_{(r)} = C_{(r)}'Y$ generating the linear set of which $y_1, \cdots, y_k$ are members. The SS due to the set $y_1, \cdots, y_k$ is the total of the SS due to r mutually orthogonal linear functions $y_{(1)}, \cdots, y_{(r)}$ and carries r d.f. This is given by $S^2 = S_{(1)}^2 + \cdots + S_{(r)}^2$ where

$$S_{(i)}^2 = \text{ SS due to } y_{(i)} = \frac{y_{(i)}^2}{C_{(i)}'C_{(i)}}. \tag{21.7.4}$$

The set of orthogonal functions $y_{(1)}, \cdots, y_{(r)}$ may be defined in an infinite number of ways but the SS obtained from all the sets is unique (as is proved below).
Let

$$S_r = \text{ vector space generated by } C_{(1)}, \cdots, C_{(r)}$$
$$Y_0 = \text{ projection of } Y \text{ on } S_r.$$

Thus,

$$Y = Y_0 + \boldsymbol{\lambda} = (d_1 C_{(1)} + \cdots + d_r C_{(r)}) + \boldsymbol{\lambda} \tag{21.7.5}$$

where $\boldsymbol{\lambda}$ is orthogonal projection of Y on the space which is orthogonal to S_r. To find the constant d_i, multiply both sides by $C_{(i)}'$.

$$C_{(i)}'Y = d_i C_{(i)}'C_{(i)} \text{ i.e., } d_i = \frac{C_{(i)}'Y}{C_{(i)}'C_{(i)}}.$$

Hence d_i is uniquely determined. Therefore,

$$Y_0 = \sum_{i=1}^{r} \frac{C_{(i)}'Y}{C_{(i)}'C_{(i)}} C_{(i)}$$

is uniquely determined. Hence the SS due to $y_{(1)}, \cdots, y_{(r)}$ is

$$\sum_{i=1}^{r} \frac{y_{(i)}^2}{C_{(i)}'C_{(i)}} = \sum_{i=1}^{r} \frac{(Y'C_{(i)})^2 (C_{(i)}'C_{(i)})}{(C_{(i)}'C_{(i)})^2} = \sum_{i=1}^{r} d_i^2 C_{(i)}'C_{(i)}$$
$$= \left(\sum_{i=1}^{r} d_i C_{(i)}\right)' \left(\sum_{i=1}^{r} d_i C_{(i)}\right) = Y_0'Y_0.$$

Hence the SS due to the set $y_{(1)}, \cdots, y_{(r)}$ is the square of the length of the projection of Y on $S_r(C_{(1)}, \cdots, C_{(r)})$ and is, therefore, uniquely determined.

Expectation of SS due to a set of linear functions carrying r d.f.

$$E\left(\sum_{i=1}^{r} S_{(i)}^2\right) = r\sigma^2 + \sum_{i=1}^{r} \left[C_{(i)}'E(Y)\right]^2 / C_{(i)}'C_{(i)} \quad \text{(from (21.7.3) and (21.7.4))}$$
$$= r\sigma^2 + \text{ SS due to } (C_{(1)}'E(Y), \cdots, C_{(r)}'E(r)).$$

In particular, if $E(C'_{(i)}Y) = 0$ $(i = 1, \cdots, r)$, i.e., if all the linear functions of the set belong to error, then $E(S^2) = r\sigma^2$.

■ NOTE 21.7.2 *Following theorem 21.7.8 the* SS *due to a set of linear functions may be defined as the square of the length of the projection of observation vector* $\boldsymbol{Y}$ *on the space generated by the coefficient vector of the linear functions.*

Suppose there are two linear sets, with r_1 d.f. and r_2 d.f. respectively, such that every linear function belonging to a set is orthogonal to every linear function belonging to the other set. In this case, it is said that r_1 d.f. for the first set are orthogonal to the r_2 d.f. for the second set.

Theorem 21.7.10 *If* S_1^2 *and* S_2^2 *are the* SS*'s corresponding to the two mutually orthogonal sets of linear functions carrying* r_1 *and* r_2 *d.f. respectively, then the* SS *corresponding to the* $(r_1 + r_2)$ *d.f. taken together is* $S^2 = S_1^2 + S_2^2$.

Proof: Let $(y_1, \cdots, y_{r_1})$ be the first set of orthogonal linear functions with r_1 d.f. Then

$$S_1^2 = S_{(1)}^2 + \cdots + S_{(r_1)}^2$$

where $S_{(i)}^2$ is SS due to y_i $(i = 1, \cdots, r_1)$. Similarly let $(z_1, \cdots, z_{r_2})$ be the second set of orthogonal linear functions. Then SS due to this set is

$$S_2^2 = S_{(1)}^{'2} + \cdots + S_{(r_2)}^{'2}$$

where $S_{(i)}^{'2}$ is SS due to z_i $(i = 1, \cdots, r_2)$. The combined set contains $(r_1 + r_2)$ mutually orthogonal linear functions. Hence SS due to this set is

$$S^2 = S_{(1)}^2 + \cdots + S_{(r_1)}^2 + S_{(1)}^{'2} + \cdots + S_{(r_2)}^{'2} = S_1^2 + S_2^2$$

and carries $(r_1 + r_2)$ d.f.

Conversely, if the combined set of $(r_1 + r_2)$ mutually orthogonal linear functions with SS equal to S^2 be split up into two sets of mutually orthogonal functions with r_1 and r_2 d.f. and SS equal to S_1^2, S_2^2 respectively, then $S^2 = S_1^2 + S_2^2$.

Obviously, the theorem can be generalized to partitioning into $k (> 2)$ sets of mutually orthogonal linear functions.

SS due to the totality of all BLUE's

Let us consider the set of all BLUE's of linear parametric functions under the model (21.2.3). Any BLUE of any estimable parametric function must be of the form (by theorem 21.3.2)

$$(q_1\boldsymbol{\alpha}_1 + \cdots + q_p\boldsymbol{\alpha}_p)'\boldsymbol{Y}.$$

Hence, by definition (note 21.7.2), the SS, S_0^2 due to all possible BLUES is the square of the length of the projection of the observation vector $\boldsymbol{Y}$ on the estimation space $V(\alpha_{(1)}, \cdots, \alpha_{(p)})$. Now, the projection of $\boldsymbol{Y}$ on V_p is $\hat{\eta} = A\hat{\beta}$ (vide section 21.4.1). Thus

$$S_0^2 = \hat{\eta}\eta = \boldsymbol{Y}'\hat{\eta} = \boldsymbol{Y}'A\hat{\beta} = \boldsymbol{Y}'A(A'A)^{-1}A'\boldsymbol{Y} = \boldsymbol{Y}'\boldsymbol{R}\boldsymbol{Y}.$$

The quantity S_R^2 is also called the SS due to regression (SS_R). The term derives from the fact that $\hat{\boldsymbol{\eta}}$ lies in the regression plane or estimation space V_p. The SS_R may be thought of as inner product of $\boldsymbol{Y}$ with $\boldsymbol{RY}$, the projection of $\boldsymbol{Y}$ into the estimation space.

$$E(SS_R) = \sigma^2 tr[\boldsymbol{A}(\boldsymbol{A}'\boldsymbol{A})^{-1}\boldsymbol{A}'] + \boldsymbol{\beta}'\boldsymbol{A}'\boldsymbol{A}\boldsymbol{\beta} = \sigma^2 p + \boldsymbol{\beta}'\boldsymbol{A}'\boldsymbol{A}\boldsymbol{\beta}.$$

The quantity SS_R/p is called the mean square due to regression MS_R. Hence $E(MS_R) = \sigma^2 + \frac{1}{p}\boldsymbol{\beta}'\boldsymbol{A}'\boldsymbol{A}\boldsymbol{\beta}$.

SS due to error

The SS due to all linear functions belonging to error is called the SS due to error. The linear functions belonging to error generate the error space of rank $n_0 = n - n_e$.
Now

$$\boldsymbol{Y} = \boldsymbol{A}\hat{\boldsymbol{\beta}} + \hat{\mathbf{e}} \text{ (where } \hat{\mathbf{e}} = \boldsymbol{Y} - \boldsymbol{A}\hat{\boldsymbol{\beta}} = \boldsymbol{Y} - \hat{\boldsymbol{\eta}}), \tag{21.7.6}$$

where $\boldsymbol{A}\hat{\boldsymbol{\beta}}$ is the projection of $\boldsymbol{Y}$ on the estimation space and $\hat{\mathbf{e}}$ is the projection of $\boldsymbol{Y}$ on the error space. Note that $\sum_{i=1}^{n} \hat{e}_i \neq 0$ in general, though $\boldsymbol{A}'\hat{\mathbf{e}} = \mathbf{0}$.
SS due to error,

$$\begin{aligned} SS_E &= \hat{\mathbf{e}}'\hat{\mathbf{e}} = (\boldsymbol{Y} - \boldsymbol{A}\hat{\beta})'(\boldsymbol{Y} - \boldsymbol{A}\hat{\beta}) = \boldsymbol{Y}'\boldsymbol{Y} - (\boldsymbol{A}\hat{\beta})'(\boldsymbol{A}\hat{\beta}) \\ &= \boldsymbol{Y}'\boldsymbol{Y} - \boldsymbol{Y}'\boldsymbol{A}\hat{\beta} = \boldsymbol{Y}'\boldsymbol{Y} - \hat{\eta}'\eta = \sum_{i=1}^{n} Y_i^2 - S_0^2 \\ &= \text{Total } SS(TSS) - SS \text{ due to all possible BLUE's.} \end{aligned}$$

It can be verified that

$$SS_E = \boldsymbol{Y}'(\boldsymbol{I} - \boldsymbol{R})\boldsymbol{Y} = \boldsymbol{Y}'\boldsymbol{Q}\boldsymbol{Y}.$$

Thus, the total SS has been broken into two parts: $\boldsymbol{Y}'\boldsymbol{R}\boldsymbol{Y}$ (or regression SS) and $\boldsymbol{Y}'\boldsymbol{Q}\boldsymbol{Y}$ (error SS). The quantity $\hat{\boldsymbol{Y}} = \boldsymbol{A}\hat{\boldsymbol{\beta}}$ contains information about the model $\boldsymbol{\eta} = \boldsymbol{A}\boldsymbol{\beta}$; $\hat{\mathbf{e}}$ reflects information about error only and hence $\hat{\mathbf{e}}'\boldsymbol{e}$ should give information about σ^2.

$$\begin{aligned} E(SS_E) &= E(\boldsymbol{Y}'(I - \boldsymbol{R})\boldsymbol{Y}) = \boldsymbol{\beta}'\boldsymbol{A}'(I - \boldsymbol{R})\boldsymbol{A}\boldsymbol{\beta} + tr\{(I - \boldsymbol{R})\sigma^2 I\} \\ &= \sigma^2[tr(I) - tr(\boldsymbol{R})] \text{ (since } (\boldsymbol{I} - \boldsymbol{R})\boldsymbol{A} = \mathbf{0}) \\ &= \sigma^2(n - p). \end{aligned}$$

Denoting the mean square due to error $SS_E/(n-p)$ as MSE,

$$E(MSE) = \sigma^2$$

so that MSE is unbiased for σ^2.

From (21.7.6),

$$\boldsymbol{Y}'\boldsymbol{Y} = \hat{\boldsymbol{\beta}}'\boldsymbol{A}'\boldsymbol{A}\hat{\boldsymbol{\beta}} + \hat{e}'\hat{e} \text{ i.e., } TSS = SS_R + SS_E.$$

The breakdown of the total sum of squares is usually shown in the analysis of variance table as follows.

Table 2.1 : Analysis of variance for the model $Y = A\beta + e$

Source	Degrees of freedom	Sum of Squares	Mean Square (M.S.)	Expected M.S.
Regression	p	$SS_R = \hat{\eta}'\eta = \hat{\boldsymbol{\beta}}'\boldsymbol{A}'\boldsymbol{A}\hat{\boldsymbol{\beta}} = \boldsymbol{Y}'\boldsymbol{R}\boldsymbol{Y}$	$MS_R = SS_R/p$	$\sigma^2 + \frac{1}{p}\boldsymbol{\beta}'\boldsymbol{A}'\boldsymbol{A}\boldsymbol{\beta}$
Error	$n-p$	$SS_E = \boldsymbol{Y}'\boldsymbol{Y} - \hat{\boldsymbol{\beta}}'\boldsymbol{A}'\boldsymbol{A}\hat{\boldsymbol{\beta}} = \boldsymbol{Y}'(\boldsymbol{I}-\boldsymbol{R})\boldsymbol{Y}$	$MS_E = SS_E/(n-p)$	σ^2
Total	n	$TSS = \boldsymbol{Y}'\boldsymbol{Y}$		

Suppose we wish to test the hypothesis $H_\circ(\boldsymbol{\beta} = \boldsymbol{0})$. If this hypothesis is true, both MS_R and MS_E would be unbiased estimates of σ^2. The likelihood ratio test criterion of this hypothesis uses the ratio MS_R/MS_E. If this ratio is unusually large, the null hypothesis is rejected.

If we assume further that $\boldsymbol{Y} \sim N_n(\boldsymbol{A\beta}, \sigma^2\boldsymbol{I})$, then under $H_\circ$, $SS_R \sim \sigma^2\chi^2_{(p)}$, $SS_E \sim \sigma^2\chi^2_{(n-p)}$ and these χ^2-variables are independent. Hence, under $H_\circ$, the ratio

$$F = \frac{MS_R}{MS_E} \sim F_{p,(n-p)}.$$

If the observed value of F exceeds the tabulated value $F_{p,(n-p);\alpha}$ we reject $H_\circ$ at $100(1-\alpha)\%$ level of significance. Otherwise $H_\circ$ is accepted.

■ NOTE 21.7.3 *In the case of a regression model, where a_{ij}'s are the observed values of some independent variable x_j $(j = 1, \cdots, p)$ (so that the model is $\boldsymbol{Y} = \boldsymbol{X\beta} + \boldsymbol{e}$, $E(\boldsymbol{e}) = \boldsymbol{0}$, $D(\boldsymbol{e}) = \sigma^2 I$), the square of the multiple correlation coefficient of Y on $x_1, \cdots, x_p$ is given by*

$$R^2 = \frac{SS_R}{TSS} = \frac{\hat{\boldsymbol{\beta}}'\boldsymbol{X}'\boldsymbol{X}\hat{\boldsymbol{\beta}}}{\boldsymbol{Y}'\boldsymbol{Y}} = 1 - \frac{\hat{e}\hat{e}}{\boldsymbol{Y}'\boldsymbol{Y}}.$$

If $\boldsymbol{X}$ contains $\mathbf{1}$, it is customary to define R^2 as

$$R^2 = 1 - \frac{\hat{\mathbf{e}}'\hat{\mathbf{e}}}{(\boldsymbol{Y} - \bar{Y}\mathbf{1})'(\boldsymbol{Y} - \bar{Y}\mathbf{1})} \quad \text{where } \bar{Y} = \frac{1}{n}\sum_{i=1}^{n} Y_i.$$

R^2 is close to unity, when $\hat{\eta}_j = \widehat{E(Y_j)}$ is a good predictor of Y_j (i.e., $\hat{e}_j$ is small) $(j = 1, \cdots, n)$ and is close to zero when it is a poor predictor.

■ NOTE 21.7.4 *In case of model* (21.5.1),

$$\begin{aligned} SS_R &= (\boldsymbol{A}\hat{\boldsymbol{\beta}}_G)'\boldsymbol{V}^{-1}(\boldsymbol{A}\hat{\boldsymbol{\beta}}_G) \\ SS_E &= (\boldsymbol{Y} - \boldsymbol{A}\hat{\boldsymbol{\beta}}_G)'\boldsymbol{V}^{-1}(\boldsymbol{Y} - \boldsymbol{A}\hat{\boldsymbol{\beta}}_G) \\ TSS &= \boldsymbol{Y}'\boldsymbol{V}^{-1}\boldsymbol{Y}. \end{aligned}$$

Also, an unbiased estimator of σ^2 is

$$\tilde{\sigma}^2 = \frac{1}{n-p}(\boldsymbol{Y} - \boldsymbol{A}\hat{\beta}_G)'\boldsymbol{V}^{-1}(\boldsymbol{Y} - \boldsymbol{A}\hat{\beta}_G). \tag{i}$$

Thus, if $\boldsymbol{V}$ is a known matrix, (21.5.5), (21.5.6) and the above results can be used in the standard way to make inference about $\boldsymbol{\beta}$ and σ^2. However, in many situations, $\boldsymbol{\Sigma}$ is unknown and must be computed from the data.

Suppose $\hat{\boldsymbol{V}}$ is a consistent estimator of $\boldsymbol{V}$. Define

$$\tilde{\beta}_G = (\boldsymbol{A}'\hat{\boldsymbol{V}}^{-1}\boldsymbol{A})^{-1}\boldsymbol{A}'\hat{\boldsymbol{V}}^{-1}\boldsymbol{Y} \tag{ii}$$

$$\hat{\tilde{\sigma}}^2 = \frac{1}{n-p}(\boldsymbol{Y} - \boldsymbol{A}\tilde{\beta}_G)'\hat{\boldsymbol{V}}^{-1}(\boldsymbol{Y} - \boldsymbol{A}\tilde{\beta}_G). \tag{iii}$$

In large samples, $\tilde{\beta}_G$ and $\hat{\tilde{\sigma}}^2$ preserve the Gauss-Markoff optimality property of $\hat{\boldsymbol{\beta}}_G$, $\tilde{\sigma}^2$. However, for small samples there is no guarantee that $\tilde{\boldsymbol{\beta}}_G$ and $\hat{\tilde{\sigma}}^2$ will yield more efficient estimates than $\hat{\boldsymbol{\beta}}_G$ and $\tilde{\sigma}^2$ respectively.

For estimation of $\boldsymbol{V}$, the estimator

$$\dot{\boldsymbol{V}} = \frac{1}{n-p}(\boldsymbol{Y} - \boldsymbol{A}\hat{\boldsymbol{\beta}})(\boldsymbol{Y} - \boldsymbol{A}\hat{\boldsymbol{\beta}})' \tag{iv}$$

can not be used for the following reasons. $\boldsymbol{V}$ is a $n \times n$ matrix and as $n \to \infty$, $\boldsymbol{V}$ becomes a matrix of infinite dimensions. Moreover, since $\boldsymbol{V}$ is a $n \times n$ symmetric matrix, it has $n(n+1)/2$ distinct parameters. As $n \to \infty$, the number of distinct parameters becomes unbounded. Again, for a given n a total number of $\left[\frac{n(n+1)}{2} + p\right]$ parameters are to be estimated by a total number of n observations, — which is an impossible task.

It can be verified that rank $\dot{\boldsymbol{V}} = 1$ so that $|\dot{\boldsymbol{V}}| = 0$ and hence in this case (ii) cannot be computed.

However, if $\boldsymbol{V}$ is a structured matrix [e.g., the intraclass correlation matrix (vide, Section 0.4)], number of parameters to be estimated becomes small, often independent of n.

▶ **Example 21.7.2** *Y_{ij}'s are uncorrelated with*

$$E(Y_{ij}) = \beta_j, \quad i = 1, \cdots, n_j; j = 1, \cdots, k,$$
$$V(Y_{ij}) = \sigma^2, \quad n = \sum_{j=1}^{k} n_j. \;.$$

Here

$$\boldsymbol{A} = \begin{bmatrix} \mathbf{1} & \mathbf{0} & \cdots & \mathbf{0} \\ \mathbf{0} & \mathbf{1} & \cdots & \mathbf{0} \\ \mathbf{0} & \mathbf{0} & \cdots & \mathbf{1} \end{bmatrix}$$

where **1** at the (j, j)th place is a unit vector of order n_j.

$$\boldsymbol{A}'\boldsymbol{A} = \text{Diag}\,(n_1, \cdots, n_k).$$

The LSE is

$$\hat{\boldsymbol{\beta}} = (\boldsymbol{A}'\boldsymbol{A})^{-1}\boldsymbol{A}'\boldsymbol{Y} = (\bar{Y}_1 \cdots \bar{Y}_k)'$$

where

$$\bar{Y}_j = \sum_{i=}^{n_j} Y_{ij}/n_j.$$

$$SS_R = \hat{\boldsymbol{\beta}}'\boldsymbol{A}'\boldsymbol{A}\hat{\boldsymbol{\beta}} = \sum_{j=1}^{k} n_j \bar{Y}_j^2, \quad df = k$$

$$SS_E = (\boldsymbol{Y} - \hat{\boldsymbol{\eta}})'(\boldsymbol{Y} - \hat{\boldsymbol{\eta}}) = \boldsymbol{Y}'\boldsymbol{Y} - \boldsymbol{Y}'\boldsymbol{R}\boldsymbol{Y}$$
$$= \sum_{j=1}^{k}\sum_{i=1}^{n_j} Y_{ij}^2 - \sum_{j=1}^{k} n_j \bar{Y}_j^2 = \sum_{j=1}^{k}\sum_{i=1}^{n_j} (Y_{ij} - \bar{Y}_j)^2, \quad df = n - k.$$

$$D(\hat{\boldsymbol{\beta}}) = \sigma^2 \boldsymbol{C}^{-1}; \quad \text{thus } V(\hat{\beta}_i) = \sigma^2/n_i, \quad \text{Cor}(\hat{\beta}_i, \hat{\beta}_{i'}) = 0 \;\; (i \neq i').$$

Table 21.2 : Analysis of Variance Table

Source	Degrees of freedom	Sum of squares	Expected Mean Squares
Regression	k	$\sum_{j=1}^{k} n_j \bar{Y}_j^2$	$\sigma^2 + \frac{1}{k}\sum_{j=1}^{k} n_j \beta_j^2$
Residual	$n - k$	$\sum_j \sum_i (Y_{ij} - \bar{Y}_j)^2$	σ^2

Under the hypothesis $H_\circ(\beta_1 = \cdots = \beta_k = 0)$, both SS_R and SS_E are unbiased estimates of σ^2. If we also assume that Y_{ij}'s are normally distributed $N(\beta_j, \sigma^2)$, then under $H_\circ$, $MS_R/MS_E \sim F_{k,n-k}$. Therefore, $H_\circ$ is rejected at $100(1-\alpha)\%$ level of significance if the observed F exceeds tabulated $F_{k,n-k;\alpha}$. Otherwise, $H_\circ$ is accepted.

▶ **Example 21.7.3** *Let* $\boldsymbol{Y} = (\boldsymbol{Y}'_1, \boldsymbol{Y}'_2, \cdots, \boldsymbol{Y}'_c)'$ *where*

$$\boldsymbol{Y}'_i = (Y_{i1}, \cdots, Y_{in_i}), \ i = 1, \cdots, c, \ n = \sum_{i=1}^{c} n_i$$

$$\boldsymbol{Y} = \boldsymbol{A\beta} + \boldsymbol{e}$$

where $\boldsymbol{Y}, \boldsymbol{e}$ *are* $n \times 1$ *column vectors,* $\boldsymbol{A}$ *is a known* $(n \times p)$ *design matrix of rank* p *and* β *is* $p \times 1$ *with* $E(\boldsymbol{e}) = \boldsymbol{0}$. *Assume that*

$$\begin{aligned} E(e_{ij}^2) &= \sigma^2 \ \forall \, i, j \\ E(e_{ij}e_{ik}) &= \rho\sigma^2 \ \ i = 1, \cdots, c; j \neq k = 1, \cdots, n_i \\ E(e_{ij}e_{i'k}) &= 0, \ \ i \neq i' = 1, \cdots, c; j = 1, \cdots, n_i, k = 1, \cdots, n_{i'}. \end{aligned}$$

Therefore,

$$D(\boldsymbol{Y}_i) = \sigma^2 \begin{bmatrix} 1 & \rho & \cdots & \rho \\ \rho & 1 & \cdots & \rho \\ \cdot & \cdot & \cdots & \cdot \\ \rho & \rho & \cdots & 1 \end{bmatrix} = \sigma^2 \Sigma_i(\rho) \text{ (say)}$$

is of order $n_i \times n_i$. Hence

$$D(\boldsymbol{Y}) = D(\boldsymbol{e}) = \text{diag}\left(\underset{1}{\boldsymbol{\Sigma}}(\rho), \cdots, \underset{c}{\boldsymbol{\Sigma}}(\rho)\right) = \sigma^2 \boldsymbol{\Sigma}(\rho)$$

is a block diagonal matrix. If ρ is known the BLUE of $\boldsymbol{\beta}$ is given by

$$\begin{aligned} \hat{\boldsymbol{\beta}}(\rho) &= (\boldsymbol{A}'\boldsymbol{\Sigma}^{-1}(\rho)\boldsymbol{A})^{-1}\boldsymbol{A}'\boldsymbol{\Sigma}^{-1}(\rho)\boldsymbol{Y} \\ D(\hat{\boldsymbol{\beta}}) &= \sigma^2(\boldsymbol{A}'\boldsymbol{\Sigma}^{-1}\boldsymbol{A})^{-1}\boldsymbol{A}'\boldsymbol{\Sigma}^{-1}\boldsymbol{\Sigma}\boldsymbol{\Sigma}^{-1}\boldsymbol{A}(\boldsymbol{A}'\boldsymbol{\Sigma}^{-1}\boldsymbol{A})^{-1} = \sigma^2(\boldsymbol{A}'\boldsymbol{\Sigma}^{-1}\boldsymbol{A})^{-1}. \end{aligned}$$

The SS due to BLUE's $= SS_R$

$$= (\boldsymbol{Y}'\boldsymbol{\Sigma}^{-1}\boldsymbol{A})(\boldsymbol{A}'\boldsymbol{\Sigma}^{-1}\boldsymbol{A})^{-1}\boldsymbol{A}'\boldsymbol{\Sigma}^{-1}\boldsymbol{Y}, \ \ df = p.$$

The SS due to error is $\boldsymbol{Y}'\boldsymbol{\Sigma}^{-1}\boldsymbol{Y} - SS_R$, d.f. $= n - p$.

▶ **Example 21.7.4** Y_1, Y_2, Y_3, Y_4 *are independent with*

$$\begin{aligned} E(Y_1) &= E(Y_2) = \theta_1 + \theta_2 \\ E(Y_3) &= E(Y_4) = \theta_1 + \theta_3 \\ V(Y_i) &= \sigma^2, \ i = 1, \cdots, 4. \end{aligned}$$

Determine the condition of estimability of parametric function $\boldsymbol{l}'\boldsymbol{\theta} = l_1\theta_1 + l_2\theta_2 + l_3\theta_3$. Obtain a solution of the normal equations and the SS due to error.

Here, $E(\boldsymbol{Y}) = \boldsymbol{A\theta}$ where

$$\boldsymbol{A} = \begin{bmatrix} 1 & 1 & 0 \\ 1 & 1 & 0 \\ 1 & 0 & 1 \\ 1 & 0 & 1 \end{bmatrix}, \ \boldsymbol{\theta} = \begin{bmatrix} \theta_1 \\ \theta_2 \\ \theta_3 \end{bmatrix}.$$

The condition of estimability of $\boldsymbol{l}'\boldsymbol{\theta}$ is rank $(\boldsymbol{A}')$ = rank $(\boldsymbol{A}' \, \boldsymbol{l})$.

Here rank $(\boldsymbol{A}') = 2$ with $[1\ 1\ 0]'$ and $[1\ 0\ 1]'$ as two independent column vectors of $\boldsymbol{A}'$. Hence $\boldsymbol{l}$ should depend linearly on them, i.e., $(l_1\ \ l_2\ \ l_3) = a(1\ \ 1\ \ 0) + b(1\ \ 0\ \ 1)$. Thus, $l_1 = a + b$, $l_2 = a$, $l_3 = b$. Hence, the condition of estimability is $l_1 = l_2 + l_3$.

$$\boldsymbol{A}'\boldsymbol{A} = \begin{bmatrix} 4 & 2 & 2 \\ 2 & 2 & 0 \\ 2 & 0 & 2 \end{bmatrix}$$

is singular. In order to solve the normal equations we are to find the conditional inverse of $\boldsymbol{C} = \boldsymbol{A}'\boldsymbol{A}$. It is required to solve the equations $\boldsymbol{Cx} = \boldsymbol{z}$, keeping $\boldsymbol{z}$ general. We solve the following simultaneous equations,

$$\begin{aligned} 4x_1 + 2x_2 + 2x_3 &= z_1 \\ 2x_1 + 2x_2 \qquad\quad &= z_2 \\ 2x_1 + 2x_3 \qquad\quad &= z_3. \end{aligned}$$

These give $0 = z_1 - z_2 - z_3$, $x_1 + x_2 = \frac{1}{2}z_2$ or, $x_2 = -x_1 + \frac{1}{2}z_2$; $x_1 + x_3 = \frac{1}{2}z_3$ or, $x_3 = -x_1 + \frac{1}{2}z_3$.

The general solution may be expressed as $x_1 = l$, $x_2 = -l + \frac{1}{2}z_2$, $x_3 = -l + \frac{1}{2}z_3$ where l is arbitrary. Taking l as zero, a particular solution is $x_1 = 0$, $x_2 = \frac{1}{2}z_2$, $x_3 = \frac{1}{2}z_3$ or as $\boldsymbol{x} = \bar{\boldsymbol{C}}\boldsymbol{z}$ when

$$\bar{\boldsymbol{C}} = \begin{bmatrix} 0 & 0 & 0 \\ 0 & \frac{1}{2} & 0 \\ 0 & 0 & \frac{1}{2} \end{bmatrix}.$$

It can be verified that $\boldsymbol{C}\bar{\boldsymbol{C}}\boldsymbol{C} = \boldsymbol{C}$. A set of solutions of normal equations is

$$\hat{\boldsymbol{\theta}} = \boldsymbol{C}^{-1}\boldsymbol{A}'\boldsymbol{Y} \text{ i.e., } \hat{\theta}_1 = 0,\ \hat{\theta}_2 = \frac{1}{2}(Y_2 + Y_3),\ \hat{\theta}_3 = \frac{1}{2}(Y_3 + Y_4)$$

and these are the BLUE's. The estimation space V is generated by columns of $\boldsymbol{A}$ and carries 2 d.f. We take $\boldsymbol{\alpha}_{(1)} = (1\ \ 1\ \ 0\ \ 0)'$, $\boldsymbol{\alpha}_{(2)} = (0\ \ 0\ \ 1\ \ 1)'$ as the two orthogonal vectors in V. The SS due to all possible BLUE's is

$$S_0^2 = \frac{(\alpha'_{(1)}\boldsymbol{Y})^2}{\boldsymbol{\alpha}'_{(1)}\boldsymbol{\alpha}_{(1)}} + \frac{(\boldsymbol{\alpha}'_{(2)}\boldsymbol{Y})^2}{\boldsymbol{\alpha}'_{(2)}\boldsymbol{\alpha}_{(2)}} = \frac{(Y_1 + Y_2)^2}{2} + \frac{(Y_3 + Y_4)^2}{2}$$

and carries 2 d.f. The SS due to error has 2 d.f. and is given by

$$\begin{aligned} SS_E = \boldsymbol{Y}'\boldsymbol{Y} - S_0^2 &= \sum Y_i^2 - \frac{1}{2}\left\{(Y_1 + Y_2)^2 + (Y_3 + Y_4)^2\right\} \\ &= \frac{1}{2}[(Y_1 - Y_2)^2 + (Y_3 - Y_4)^2]. \end{aligned}$$

This can also be determined by using the residual SS formula as

$$\sum_{i=1}^{4}(Y_i - \widehat{E(Y_i)})^2 = (Y_1 - \hat{\theta}_1 - \hat{\theta}_2)^2 + (Y_2 - \hat{\theta}_1 - \hat{\theta}_2)^2$$
$$+(Y_3 - \hat{\theta}_1 - \hat{\theta}_3)^2 + (Y_4 - \hat{\theta}_1 - \hat{\theta}_3)^2$$
$$= \frac{1}{2}[(Y_1 - Y_2)^2 + (Y_3 - Y_4)^2].$$

We can also determine the error SS by first considering two independent linear functions each belonging to error.

Note that each of $\alpha_{(3)} = (-1\ 1\ 0\ 0)'$, $\alpha_{(4)} = (0\ 0\ -1\ 1)'$ is orthogonal to each of $\alpha_{(1)}$, $\alpha_{(2)}$. Also $\alpha_{(3)}$, $\alpha_{(4)}$ are orthogonal among themselves. Hence,

$$SS_E = \frac{(\boldsymbol{\alpha}'_{(3)}\boldsymbol{Y})^2}{\boldsymbol{\alpha}'_{(3)}\boldsymbol{\alpha}_{(3)}} + \frac{(\boldsymbol{\alpha}'_{(4)}\boldsymbol{Y})^2}{\boldsymbol{\alpha}'_{(4)}\boldsymbol{\alpha}_{(4)}} = \frac{1}{2}[(Y_1 - Y_2)^2 + (Y_3 - Y_4)^2].$$

▶ **Example 21.7.5** (*Simple Linear Regression Model*). *$Y_1, \cdots, Y_n$ are independent variables with $E(Y_i) = \alpha + \beta x_i$; $V(Y_i) = \sigma^2$ where $x_1, \cdots, x_n$ are values of non-stochastic variable x on units $1, 2, \cdots, n$ respectively. Here $\boldsymbol{A} = [\mathbf{1}, x]$ when $\boldsymbol{x} = (x_1, \cdots, x_n)'$.*

Normal equations

$$\boldsymbol{A}'\boldsymbol{A}\boldsymbol{\beta} = \boldsymbol{A}'\boldsymbol{Y}, \text{ or, } \begin{bmatrix} n & \Sigma x_i \\ \Sigma x_i & \Sigma x_i^2 \end{bmatrix}\begin{bmatrix} \alpha \\ \beta \end{bmatrix} = \begin{bmatrix} \Sigma Y_i \\ \Sigma x_i Y_i \end{bmatrix}$$

$C = \boldsymbol{A}'\boldsymbol{A}$ is non-singular and

$$\boldsymbol{C}^{-1} = \left\{\sum(x_i - \bar{x})^2\right\}^{-1}\begin{bmatrix} \Sigma x_i^2/n & -\Sigma x_i/n \\ -\Sigma x_i/n & 1 \end{bmatrix}.$$

The unique l.s. estimates are

$$\hat{\alpha} = \left\{\bar{Y}\sum x_i^2 - \bar{x}\sum x_i Y_i\right\}/\sum(x_i - \bar{x})^2 = \bar{Y} - \hat{\beta}\bar{x}$$
$$\hat{\beta} = \sum(x_i - \bar{x})(Y_i - \bar{Y})/\sum(x_i - \bar{x})^2.$$

The dispersion matrix is

$$D(\hat{\boldsymbol{\beta}}) = \sigma^2\boldsymbol{C}^{-1}$$
$$V(\hat{\alpha}) = \sigma^2\left[\frac{1}{n} + \frac{\bar{x}^2}{\Sigma(x_i - \bar{x})^2}\right]$$
$$V(\hat{\beta}) = \sigma^2/\sum(x_i - \bar{x})^2, \quad \text{Cor}(\hat{\alpha}, \hat{\beta}) = -\sigma^2\bar{x}/\sum(x_i - \bar{x})^2.$$

The SS due to error,

$$SS_E = \sum(Y_i - \hat{\alpha} - \hat{\beta}x_i)^2 = \sum(Y_i - \bar{Y})^2 - \hat{\beta}^2(x_i - \bar{x})^2$$

and carries $(n-2)$ d.f. The MSE is an unbiased estimate of σ^2. The SS due to all possible BLUE's is

$$S_0^2 = SS_R = \sum Y_i^2 - SS_E = n\bar{Y}^2 + \hat{\beta}^2 \sum (x_i - \bar{x})^2$$

and has 2 d.f.

► **Example 21.7.6** (***Multiple Linear Regression Model***). *Suppose the random variable Y depends on k independent variables $x_1, \cdots, x_k$. The expected value of Y_i given $x_{i1}, \cdots, x_{ik}$ where x_{ij} is the value of x_j $(j = 1, \cdots, k)$ on unit i, is given by*

$$E(Y_i|\boldsymbol{x}_i) = \alpha + \beta_1(x_{i1} - \bar{x}_1) + \cdots + \beta_k(x_{ik} - \bar{x}_k).$$

Also

$$V(Y_i|\boldsymbol{x}_i) = \sigma^2; \ \text{Cov}(Y_i, Y_{i'}|\boldsymbol{x}_i, \boldsymbol{x}_{i'}) = 0 \ (i \neq i')$$

where

$$\boldsymbol{x}_i = (x_{i1}, \cdots, x_{ik}), \ \ \bar{x}_j = \sum_{i=1}^{n} x_{ij}/n, \ \ j = 1, \cdots, k.$$

Writing $x_{ij} - \bar{x}_j$ as x'_{ij}, the model is

$$E(\boldsymbol{Y}) = \boldsymbol{X}\gamma; \ \ D(\boldsymbol{Y}) = \sigma^2 \boldsymbol{I}$$

where

$$\boldsymbol{X} = \begin{bmatrix} 1 & x'_{11} & \cdots & x'_{1k} \\ \cdot & \cdot & \cdots & \cdot \\ 1 & x'_{n1} & \cdots & x'_{nk} \end{bmatrix}, \ \ \gamma = \begin{bmatrix} \alpha \\ \beta_1 \\ \vdots \\ \beta_k \end{bmatrix}.$$

Assume $n > k + 1$. Here

$$\boldsymbol{X}'\boldsymbol{X} = \begin{bmatrix} n & 0 & \cdots & 0 \\ 0 & s_{11} & \cdots & s_{1k} \\ \vdots & \cdots & \cdots & \cdots \\ 0 & s_{k1} & \cdots & s_{kk} \end{bmatrix} \ \text{when} \ s_{jj'} = \sum_{i=1}^{n} x'_{ij} x'_{ij'};$$

$$\boldsymbol{X}'\boldsymbol{Y} = [n\bar{Y}, \ \ s_{1y}, \ \ \cdots, \ \ s_{ky}] \ \text{when} \ s_{jy} = \sum_{i=1}^{n} x'_{iy} y_i.$$

The normal equations are

$$\boldsymbol{X}'\boldsymbol{X}\gamma = \boldsymbol{X}'\boldsymbol{Y} \ \text{i.e.,} \ n\hat{\alpha} = n\bar{Y}, \ \ \boldsymbol{s}\hat{\boldsymbol{\beta}} = \boldsymbol{s}_y$$

when

$$\boldsymbol{s} = ((s_{jj'})), \ \ \boldsymbol{s}_y = (s_{1y}, \cdots, s_{ky})', \ \ \boldsymbol{\beta} = (\beta_1, \cdots, \beta_k).$$

Solving, $\hat{\alpha} = \bar{Y}$, $\hat{\boldsymbol{\beta}} = \boldsymbol{C}\boldsymbol{s}_y$ when $\boldsymbol{C} = ((c_{ij})) = \boldsymbol{s}^{-1}$.

The BLUE of β_j is $\hat{\beta}_j = c_{j1}s_{y1} + \cdots + c_{jk}s_{yk}$, $j = 1, \cdots, k$.
The dispersion matrix of $\hat{\boldsymbol{\beta}}$ is $D(\hat{\boldsymbol{\beta}}) = \sigma^2 \boldsymbol{C}$.
The estimation space has rank $(k+1)$. The error SS is

$$SS_E = \sum (Y_i - \bar{Y} - \hat{\beta}_1 x'_{i1} - \cdots - \hat{\beta}_k x'_{ik})^2$$
$$= \sum_{i=1}^{n} (Y_i - \bar{Y})^2 - \sum_{j=1}^{k} \hat{\beta}_j s_{jy}$$

and has $(n-k-1)$ d.f. The MSE is an unbiased estimator of σ^2.
The SS due to all possible BLUE's is

$$S_0^2 = SS_R = \sum Y_i^2 - SS_E = n\bar{Y}^2 + \sum_{j=1}^{k} \hat{\beta}_j s_{jy}.$$

Since the rank of $\boldsymbol{X}$ is $(k+1)$, every linear parametric function is estimable.

The square of the multiple correlation coefficient of Y on $x_1, \cdots, x_k$, each variable being measured against the respective sample mean, is

$$R^2 = \frac{(\boldsymbol{X}^\circ \hat{\boldsymbol{\beta}})'(\boldsymbol{X}^\circ \hat{\boldsymbol{\beta}})}{(\boldsymbol{Y} - \bar{Y}\boldsymbol{1})'(\boldsymbol{Y} - \bar{Y}\boldsymbol{1})} = \frac{\boldsymbol{s}_y' \boldsymbol{C} \boldsymbol{s}_y}{\sum_{i=1}^{n} (Y_i - \bar{Y})^2} = \frac{\sum_i \sum_j c_{ij} s_{iy} s_{jy}}{\sum_{i=1}^{n} (Y_i - \bar{Y})^2}$$

when $\boldsymbol{X} = (\boldsymbol{1}, \boldsymbol{X}^\circ)$.

21.8 Analysis of Covariance Model

Suppose that apart from the main variable y we have k concommittant variables $x_j (j = 1, \cdots, k)$, x_{ij} denoting the value of x_j on unit i. Consider the model

$$E(\boldsymbol{Y}|\boldsymbol{X}) = \boldsymbol{A}\boldsymbol{\beta} + \boldsymbol{X}\gamma; \quad D(\boldsymbol{Y}|\boldsymbol{X}) = \sigma^2 I \tag{21.8.1}$$

where $\boldsymbol{X} = ((x_{ij}))$, a matrix of known values of $x_j (j = 1, \cdots, k)$, the ith row corresponding to the observed values of x_j's on the ith unit, $\boldsymbol{\gamma} = (\gamma_1, \cdots, \gamma_k)'$ a vector of regression coefficients, $\boldsymbol{\beta} = (\beta_1, \cdots, \beta_p)$, a set of parameters and $\boldsymbol{A} = ((a_{ij}))$ is the matrix of known coefficients (a_{ij}'s being generally values of indicator variables). The first equation of (21.8.1) may be expressed as

$$E(\boldsymbol{Y}|\boldsymbol{X}) = [\boldsymbol{A} \;\; \boldsymbol{X}] \begin{bmatrix} \boldsymbol{\beta} \\ \gamma \end{bmatrix}.$$

One may be interested in obtaining the BLUE of a parametric function $\boldsymbol{l}'\boldsymbol{\beta}$ and the BLUE of a linear function $\boldsymbol{l}'\boldsymbol{\gamma}$ of the regression coefficients. All these problems can be solved by using the general theory developed earlier. In the full rank case we assume rank $(\boldsymbol{A}) = p$ and rank $(\boldsymbol{X}) = k$, $n > p + k$.

The component $\boldsymbol{A\beta}$ in the right hand side of (21.8.1) arises from the classification of the data (vide e.g., problem 17 in the exercise). If the data are unclassified this part will be absent and the model will reduce to the regression model.

21.9 Orthogonal Polynomials

Consider again Y as the dependent variable, x as an explanatory variable and suppose we want to fit a polynomial in x, of an appropriate degree, to the values of Y. In this case, the model is

$$E(Y|x) = \beta_0 + \beta_1 x + \cdots + \beta_p x^p \tag{21.9.1}$$

where we assume that Y_i's are uncorrelated. For n observations (Y_i, x_i) $(i = 1, \cdots, n)$ $(n > p + 1)$, model is $E(\boldsymbol{Y}) = \boldsymbol{X\beta}$, $D(\boldsymbol{Y}) = \sigma^2 I$ where $\boldsymbol{X} = (\boldsymbol{1}, \boldsymbol{x}, \boldsymbol{x}^{(2)}, \cdots, \boldsymbol{x}^{(p)})$, $\boldsymbol{x}^{(j)} = (x_1^j, \cdots, x_n^j)'$, $j = 1, \cdots, p$. The BLUE of $\boldsymbol{\beta}$ is, therefore, $\hat{\beta}_{(p)} = (\boldsymbol{X}'\boldsymbol{X})^{-1}\boldsymbol{X}'\boldsymbol{Y}$.

In practice, the order of polynomial, p is unknown and is found by successively fitting polynomials of higher order and making appropriate tests for the significance of reduction in error SS. We stop at a point where further increase in the order of polynomial does not appreciably reduce the value of error SS. At each step we have to calculate values of $\hat{\boldsymbol{\beta}}$ afresh and hence calculate values of $(\boldsymbol{X}'\boldsymbol{X})^{-1}$ afresh. This is a serious practical disadvantage. If, instead, the matrix $\boldsymbol{X}$ is so chosen that $\boldsymbol{X}'\boldsymbol{X}$ is diagonal then for fitting a polynomial of higher degree only contributions from additional components of $\boldsymbol{X}$ have to be calculated afresh and the old terms can be retained. Thus, the disadvantage mentioned earlier can be removed.

Consider, therefore, polynomials $\psi_i(x)$ in x of degree $i (= 0, 1, \cdots, p)$ which are mutually orthogonal, i.e., such that

$$\sum_{k=1}^{n} \psi_i(x_k)\psi_j(x_k) = \begin{cases} 0, & \text{if } i \neq j \\ \sum_{k=1}^{n} \psi_i^2(x_k) = A_{ii} \text{ (say)}, & i = j. \end{cases} \tag{21.9.2}$$

We rewrite the model (21.9.1) as

$$Y_k = \alpha_0\psi_0(x_k) + \alpha_1\psi_1(x_k) + \cdots + \alpha_p\psi_p(x_k) + e_k \tag{21.9.3}$$

with $E(e_k) = 0$, $V(e_k) = \sigma^2$, $\text{Cov}(e_k, e_k') = 0$ $(k \neq k')$ when $\alpha_0, \alpha_1, \cdots, \alpha_p$ are new regression coefficients and have to be estimated. The model (21.9.3) is,

$$E(\boldsymbol{Y}) = \boldsymbol{\psi\alpha}; \; D(\boldsymbol{Y}) = \sigma^2 I \tag{21.9.4}$$

where $\boldsymbol{\alpha} = (\alpha_0, \cdots, \alpha_p)'$, $\boldsymbol{\psi} = ((\psi_j(x_i)))_{n \times (p+1)}$.

Clearly, α's are functions of β's. The BLUE's are $\hat{\boldsymbol{\alpha}} = (\boldsymbol{\psi}'\boldsymbol{\psi})^{-1}\boldsymbol{\psi}'\boldsymbol{Y}$.

Now,

$$\boldsymbol{\psi}'\boldsymbol{\psi} = \text{Diag}\left(\sum \psi_0^2(x_k), \sum \psi_1^2(x_k), \cdots, \sum \psi_p^2(x_k)\right).$$

Hence,

$$\hat{\alpha}_i = \sum_{k=1}^{n} \psi_i(x_k) Y_k / \sum_{k=1}^{n} \psi_i^2(x_k), \quad i = 0, \cdots, p \tag{21.9.5}$$

$$D(\hat{\boldsymbol{\alpha}}) = \sigma^2(\boldsymbol{\psi}'\boldsymbol{\psi})^{-1}.$$

Hence,

$$V(\hat{\alpha}_i) = \sigma^2 / \sum_{k=1}^{n} \psi_i^2(x_k)$$

$$\text{Cor}(\hat{\alpha}_i, \hat{\alpha}_j) = 0 \ (i \neq j). \tag{21.9.6}$$

Also

$$\begin{aligned} SS_E &= \boldsymbol{Y}'\boldsymbol{Y} - \hat{\boldsymbol{\alpha}}'\boldsymbol{\psi}'\boldsymbol{Y} \\ &= \sum_{k=1}^{n} Y_k^2 - \sum_{j=0}^{p} \left\{ \sum_{k=1}^{n} Y_k \psi_j(x_k) \right\}^2 \Big/ \sum_{k=1}^{n} \psi_j^2(x_k) \\ &= \sum_{k=1}^{n} Y_k^2 - \sum_{j=0}^{p} \hat{\alpha}_j^2 \left[\sum_{k=1}^{n} \psi_j^2(x_k) \right] \end{aligned} \tag{21.9.7}$$

with $(n-p-1)$ d.f.

It is found, therefore, that the values of $\hat{\alpha}_i$ and $V(\hat{\alpha}_i)$ depend only on the polynomial $\alpha_i(x_k)$. Therefore, if after fitting a polynomial of pth degree we want to fit a polynomial of $(p+1)$th degree, thereby adding a term $\alpha_{p+1}\psi_{p+1}(x)$ in (21.9.3), we are not to calculate $\hat{\alpha}_i$ and $V(\hat{\alpha}_i)$ $(i = 1, \cdots, p)$ afresh, but have only to calculate

$$\hat{\alpha}_{p+1} = \sum_{k=1}^{n} \psi_{p+1}(x_k) Y_k \Big/ \sum_{k=1}^{n} \psi_{p+1}^2(x_k)$$

$$V(\hat{\alpha}_{p+1}) = \sigma^2 \Big/ \sum_{k=1}^{n} \psi_{p+1}^2(x_k).$$

The new residual SS is

$$SS_E = \sum_{k=1}^{n} Y_k^2 - \sum_{j=0}^{p+1} \hat{\alpha}_j^2 \left[\sum_{k=1}^{n} \psi_j^2(x_k) \right] = (21.9.7) - \hat{\alpha}_{p+1}^2 \sum_{k=1}^{n} \psi_{p+1}^2(x_k) \tag{21.9.8}$$

with $(n-p-2)$ d.f.

Orthogonal polynomials can be fitted whether x-values are equally or unequally spaced. However, computations are greatly simplified when the x-values are equally spaced.

21.9.1 Polynomial regression in terms of orthogonal polynomials: Determination of orthogonal polynomials

Consider the bivariate distribution of (Y, X) where Y is considered a dependent variable and X the response variable.

If for the distribution of X, we have a sequence of polynomials $\{\psi_t(x)\}$ such that

(a) $\psi_r(x)$ is precisely of degree r,

(b) $E\{\psi_r(X)\psi_s(X)\} = 0$ for $r \neq s$,

then the sequence $\{\psi_t(x)\}$ is called a sequence of orthogonal polynomials associated with the distribution of X.

In fitting a parabolic regression equation of Y on X our object is basically to find a polynomial $g(X)$ of pth degree such that $E\{Y - g(X)\}^2$ is minimum, where expectation is taken w.r.t. the joint distribution of X and Y. Here we consider

$$g(x) = \alpha_0\psi_0(x) + \alpha_1\psi_1(x) + \cdots + \alpha_p\psi_p(x) \tag{21.9.9}$$

where α's are found such that

$$E\{Y - \alpha_0\psi_0(X) - \alpha_1\psi_1(X) - \cdots - \alpha_p\psi_p(X)\}^2 \tag{21.9.10}$$

is minimum. Differentiating (21.9.10) w.r.t. α_r and equating it to zero,

$$E\left[\psi_r(X)\left\{Y - \alpha_0\psi_0(X) - \cdots - \alpha_p\psi_p(X)\right\}\right] = 0$$

$$\text{i.e.,}\quad \alpha_r = \frac{E\{\psi_r(X)Y\}}{E\{\psi_r(X)\}^2},\quad r = 0, 1, \cdots, p. \tag{21.9.11}$$

Expression for α_r does not depend upon p. So if we increase the degree of the parabolic equation (21.9.9) from p to $p+1$, the coefficients $\alpha_0, \cdots, \alpha_p$ remain the same; we only require to calculate α_{p+1} afresh.

Determination of $\psi_r(x)$ in terms of moments of X-distribution

Condition (b) means

$$E\{\psi_0(X)\psi_r(X)\} = E\{\psi_1(X)\psi_r(X)\} = \cdots = E\{\psi_{r-1}(X)\psi_r(X)\} = 0 \tag{21.9.12}$$

Let us write

$$\psi_r(x) = C_{r0} + C_{r1}x + \cdots + C_{rr}x^r \tag{21.9.13}$$

where $C_{r0}, C_{r1}, \cdots, C_{rr}$ are not all zero. It follows that conditions (21.9.12) are equivalent to the conditions

$$E\{\psi_r(X)\} = E\{X\psi_r(X)\} = \cdots = E\{X^{r-1}\psi_r(X)\} = 0,\quad r = 1, 2, 3, \cdots \tag{21.9.14}$$

Therefore, using (21.9.14)

$$\left.\begin{aligned}
E\{\psi_r(X)\} &= C_{r0}+C_{r1}\mu_1'+C_{r2}\mu_2'+\cdots+C_{rr}\mu_r'=0\\
E\{X\psi_r(X)\} &= C_{r0}\mu_1'+C_{r1}\mu_2'+C_{r2}\mu_3'+\cdots+C_{rr}\mu_{r+1}'=0\\
&\cdots\cdots\cdots\cdots\cdots\cdots\cdots\cdots\cdots\cdots\\
E\{X^{r-1}\psi_r(X)\} &= C_{r0}\mu_{r-1}'+C_{r1}\mu_r'+C_{r2}\mu_{r+1}'+\cdots+C_{rr}\mu_{2r-1}'=0.
\end{aligned}\right\}\tag{21.9.15}$$

With this we consider the relation

$$C_{r0}+C_{r1}x+C_{r2}x^2+\cdots+C_{rr}\left(x^r-\frac{\psi_r(x)}{C_{rr}}\right)=0 \tag{21.9.16}$$

where $\mu_i'=\mu_{i0}'$. The determinant of the matrix of coefficient on the right-hand side of (21.9.15) and (21.9.16) is zero, because $(C_{r0},\cdots,C_{rr})\neq(0,\cdots,0)$. Therefore,

$$\begin{vmatrix} 1 & \mu_1' & \cdots & \mu_r' \\ \cdots & \cdots & \cdots & \cdots \\ \mu_{r-1}' & \mu_r' & \cdots & \mu_{2r-1}' \\ 1 & x & \cdots & x^r \end{vmatrix} - \frac{\psi_r(x)}{C_{rr}} \begin{vmatrix} 1 & \mu_1' & \cdots & \mu_{r-1}' \\ \mu_1' & \mu_2' & \cdots & \mu_r' \\ \cdots & \cdots & \cdots & \cdots \\ \mu_{r-1}' & \mu_r' & \cdots & \mu_{2r-2}' \end{vmatrix} = 0. \tag{21.9.17}$$

By lemma 7.5.1, the last determinant is non-zero in case the distribution of X is not confined to $(r-1)$ or less number of distinct points. In this case

$$\psi_r(x)=\frac{C_{rr}\left[\Delta_{r+1,1}^{(r)}+x\Delta_{r+1,2}^{(r)}+\cdots+x^r\Delta_{r+1,r+1}^{(r)}\right]}{\Delta^{(r-1)}} \tag{21.9.18}$$

where

$$\Delta^{(r)}=\begin{vmatrix} 1 & \mu_1' & \cdots & \mu_r' \\ \vdots & \vdots & \vdots & \vdots \\ \mu_{r-1}' & \mu_r' & \cdots & \mu_{2r-1}' \\ \mu_r' & \mu_{r+1}' & \cdots & \mu_{2r}' \end{vmatrix}, \tag{21.9.19}$$

$$\begin{aligned}
\Delta_{i,j}^{(r)} &= \text{cofactor of the } (i.j)\text{th element in } \Delta^{(r)},\\
\Delta^{(r-1)} &= \Delta_{r+1,r+1}^{(r)}.
\end{aligned}$$

Now,

$$\begin{aligned}
E\{\psi_r(X)\}^2 &= E\left[\{C_{r0}+C_{r1}X+\cdots+C_{rr}X^r\}\psi_r(X)\right]\\
&= C_{rr}E\{X^r\psi_r(X)\} \quad \text{(by (21.9.14))}\\
&= C_{rr}^2\Delta^{(r)}/\Delta^{(r-1)}.
\end{aligned}$$

If the distribution of X is confined to r distinct points, $\Delta^{(r)}$ will be zero so that $\psi_r(X)$ will be zero with probability one. We, therefore, assume that the distribution of X is confined to more than r distinct points. Now,

$$E\{\psi_r(X)Y\} = \frac{C_{rr}}{\Delta^{(r-1)}} \begin{vmatrix} 1 & \mu'_1 & \cdots & \mu'_{r-1} & \mu'_r \\ \cdots & \cdots & \cdots & \cdots & \cdots \\ \mu'_{r-1} & \mu'_r & \cdots & \mu'_{2r-2} & \mu'_{2r-1} \\ \mu'_{01} & \mu'_{11} & \cdots & \mu'_{r-11} & \mu'_{r1} \end{vmatrix}$$

$$= \frac{C_{rr}}{\Delta^{(r-1)}} \Delta^{(r)*} \quad \text{(say)}. \tag{21.9.20}$$

We choose C_{rr} such that

$$E\{\psi_r(X)\}^2 = 1. \tag{21.9.21}$$

This gives

$$C_{rr} = \sqrt{\frac{\Delta^{(r-1)}}{\Delta^{(r)}}}. \tag{21.9.22}$$

Hence,

$$\alpha_r = \frac{\Delta^{(r)*}}{\sqrt{\Delta^{(r-1)}\Delta^{(r)}}};$$

the condition (21.9.21) is, however, satisfied only in theory, not in practice. For practical purposes, C_{rr} is taken as unity.

For the standardised normal distribution $\mu'_1 = 0$, $\mu'_2 = \mu_2 = 1$, $\mu'_3 = \mu_3 = 0$ $\mu'_4 = \mu_4 = 3$, $\mu'_5 = \mu_5 = 0, \cdots$. Hence,

$$\psi_1(x) = \frac{\begin{vmatrix} 1 & 0 \\ 1 & x \end{vmatrix}}{1} = x,$$

$$\psi_2(x) = \frac{\begin{vmatrix} 1 & 0 & 1 \\ 0 & 1 & 0 \\ 1 & x & x^2 \end{vmatrix}}{\begin{vmatrix} 1 & 0 \\ 0 & 1 \end{vmatrix}} = x^2 - 1$$

$$\psi_3(x) = \frac{\begin{vmatrix} 1 & 0 & 1 & 0 \\ 0 & 1 & 0 & 3 \\ 1 & 0 & 3 & 0 \\ 1 & x & x^2 & x^3 \end{vmatrix}}{\begin{vmatrix} 1 & 0 & 1 \\ 0 & 1 & 0 \\ 1 & 0 & 3 \end{vmatrix}} = x^3 - 3x, \quad \text{etc..}$$

In practice, we are given a set of n values (y_i, x_i), $i = 1, \cdots, n$. In this case we determine the values of $\alpha_j (j = 0, \cdots, p)$ by the method of least squares without assuming any distribution of X i.e., by minimising

$$\sum_{i=1}^{n} \{y_i - \alpha_0\psi_0(x_i) - \alpha_1\psi_1(x_i) - \cdots - \alpha_p\psi_p(x_i)\}^2 . \tag{21.9.23}$$

Similar calculations taking $C_{rr} = 1$ will show that $\psi_r(x)$ is given from (21.9.17) with population moments μ'_j replaced by sample raw moment

$$m'_j = \frac{1}{n}\sum_{i=1}^{n} x_i^j \quad (j = 1, 2, \cdots).$$

We note that for practical purposes C_{rr} $(r = 1, 2, \cdots)$ is taken to be unity because of the following reason. The set of orthogonal polynomials up to degree p involve $\frac{(p+1)(p+2)}{2}$ constants (because $\psi_r(x)$ involves $r + 1$ constants $C_{r0}, C_{r1}, \cdots, C_{rr}$) while the equations similar to (21.9.15) (involving the sample moments m'_j) are $\frac{p(p+1)}{2}$ in number (corresponding to $\psi_r(x)$ there are r such equations). Hence the remaining $(p+1)$ constants are assigned arbitrary values. For convenience, each of C_{rr}, $(r = 0, \cdots, p)$ is given value unity.
Note that

$$\begin{aligned}
\psi_1(x) &= C_{10} + x, \quad (\because\ C_{11} = 1) \\
\psi_r(x) &= C_{r0} + C_{r1}x + C_{r2}x^2 + \cdots + C_{rr}x^r \quad (C_{rr} = 1) \\
&= d_{r0} + d_{r1}\psi_1(x) + d_{r2}\{\psi_1(x)\}^2 + \cdots + d_{rr}\{\psi_1(x)\}^r, \quad d_{rr} = 1
\end{aligned} \tag{21.9.24}$$

where d_{rj}'s are suitable constants.

Uniform distribution of X: Consider that X has a uniform distribution with $P(X = i) = \frac{1}{n}$, $i = 0, 1, \cdots, n-1$. The case will be similar to the case where we have n values (x_i, y_i), $i = 0, 1, \cdots, n-1$ with $x_i = i$ and we do not assume any probability distribution for X.

Such cases arise in time series where we have generally observations over equidistant time points or in grouped data where we have generally groups of equal length over an interval of x-values.
Here the condition (b) implies

$$\sum_{x=0}^{n-1} \psi_r(x)\psi_s(x) = 0, \quad s = 0, 1, \cdots, r-1. \tag{21.9.25}$$

Theorem 21.9.1 *In case X has a uniform distribution over $(0, 1, \cdots, n-1)$ the right hand side of* (21.9.24) *involves even (odd) powers of $\psi_1(x)$ where r is even (odd) i.e.,*

$$\psi_r(x) = \{\psi_1(x)\}^r + d_{rr-2}\{\psi_1(x)\}^{r-2} + \cdots .$$

Proof: First note that the conditions (21.9.25) are equivalent to the conditions

$$\sum_{x=0}^{n-1} \psi_r(x)\psi_1^s(x) = 0, \quad s < r. \tag{21.9.26}$$

By (21.9.24) this means

$$d_{r0}\sum_{x=0}^{n-1} \psi_1^s(x) + d_{r1}\sum_{x=0}^{n-1} \psi_1^{s+1}(x) + d_{r2}\sum_{x=0}^{n-1} \psi_1^{s+2}(x) + \cdots + d_{rr-1}\sum_{x=0}^{n-1} \psi_1^{s+r-1}(x) + \sum_{x=0}^{n-1} \psi_1^{s+r}(x) = 0, \quad s < r. \tag{21.9.27}$$

Now,

$$\sum_{x=0}^{n-1} \psi_1(x)\psi_0(x) = \sum_{x=0}^{n-1}(C_{10} + x) \quad (\because \ \psi_0(x) = 1).$$

$$\therefore \quad C_{10} = -\frac{n-1}{2}.$$

Hence,

$$\psi_1(x) = x - \frac{n-1}{2}.$$

This implies

$$\sum_{x=0}^{n-1} \psi_1^t(x) = 0 \quad \text{if } t \text{ is odd.}$$

Suppose r is even, say, $r = 2m$. By (21.9.27),

$$d_{2m,0}\sum \psi_1^s + d_{2m,1}\sum \psi_1^{s+1} + \cdots + d_{2m,2m-1}\sum \psi_1^{s+2m-1} + \sum \psi_1^{s+2m} = 0, \quad s < 2m \tag{21.9.28}$$

For $s = 1, 3, \cdots, 2m-1$ we have, respectively,

$$\left.\begin{aligned} d_{2m,1}\sum \psi_1^2 + d_{2m,3}\sum \psi_1^4 + \cdots + d_{2m,2m-1}\sum \psi_1^{2m} &= 0 \\ d_{2m,1}\sum \psi_1^4 + d_{2m,3}\sum \psi_1^6 + \cdots + d_{2m,2m-1}\sum \psi_1^{2m+2} &= 0 \\ \cdots\cdots\cdots\cdots\cdots\cdots\cdots\cdots\cdots\cdots\cdots\cdots & \\ d_{2m,1}\sum \psi_1^{2m} + d_{2m,3}\sum \psi_1^{2m+2} + \cdots + d_{2m,2m-1}\sum \psi_1^{4m-2} &= 0. \end{aligned}\right\} \tag{21.9.29}$$

The coefficient matrix of the equations (21.9.29),

$$\begin{vmatrix} \sum \psi_1^2 & \sum \psi_1^4 & \cdots & \sum \psi_1^{2m} \\ \sum \psi_1^4 & \sum \psi_1^6 & \cdots & \sum \psi_1^{2m+2} \\ \cdots & \cdots & \cdots & \cdots \\ \sum \psi_1^{2m} & \sum \psi_1^{2m+2} & \cdots & \sum \psi_1^{4m-2} \end{vmatrix} \tag{21.9.30}$$

s the matrix of the quadratic form

$$Q = \sum_x \left[u_1\psi_1 + u_3\psi_1^3 + \cdots + u_{2m-1}\psi_1^{2m-1}\right]^2$$

vhich is non-negative definite. The expression $Q = 0$ iff $u_1\psi_1(x) + \cdots + u_{2m-1}\psi_1^{2m-1}(x) =$) for $x = 0, 1, \cdots, n-1$ i.e., when $\psi_1(0), \psi_1(1), \cdots, \psi_1(n-1)$ are solutions of the equation

$$u_1 z + u_3 z^3 + \cdots + u_{2m-1} z^{2m-1} = 0. \qquad (21.9.31)$$

Now the maximum number of real distinct roots of the equation (21.9.31) is $2m-1 =$ $\cdot - 1 < n$. Hence the n distinct values $\psi_1(0), \cdots, \psi_1(n-1)$ can not be solutions of he equation (21.9.31). Therefore, Q is a positive definite form which means that the natrix (21.9.30) is non-singular. Hence the equations (21.9.29) have the unique solution $l_{2m,1} = d_{2m,3} = \cdots = d_{2m,2m-1} = 0$. The other coefficients in the expansion (21.9.28) or $r = 2m$, viz, $d_{2m,0}, d_{2m,2}, \cdots, d_{2m,2m}$ are all non-zero. To see this consider again the quation (21.9.28) for $s = 0, 2, 4, \cdots, 2m$.

$$s = 0 : d_{2m,0}n + d_{2m,2}\sum \psi_1^2 + \cdots + d_{2m,2m-2}\sum \psi_1^{2m-2} = \sum \psi_1^{2m}$$

$$s = 2 : d_{2m,0}\sum \psi_1^2 + d_{2m,2}\sum \psi_1^4 + \cdots + d_{2m,2m-2}\sum \psi_1^{2m} = \sum \psi_1^{2m+2}$$

$$\cdots\cdots \quad \cdots\cdots\cdots\cdots\cdots\cdots\cdots\cdots\cdots\cdots\cdots\cdots$$

$$s = 2m-2 : d_{2m,0}\sum \psi_1^{2m-2} + d_{2m,2}\sum \psi_1^{2m} + \cdots + d_{2m,2m-2}\sum \psi_1^{4m-4} = \sum \psi_1^{4m-2}.$$

These represent a set of m non-homogeneous equations in m unknowns, $d_{2m,0}$, $l_{2m,2}, \cdots, d_{2m,2m-2}$. The coefficient matrix is non-singular. As the vector in the right-hand side is non-null the solution must also be non-null. Therefore,

$$\psi_{2m}(x) = d_{2m,0} + d_{2m,2}\psi_1^2(x) + \cdots + d_{2m,2m-2}\psi_1^{2m-2}(x) + \psi_1^{2m}(x). \qquad (21.9.32)$$

A parallel result can be proved for $\psi_{2m+1}(x)$.

Thus

$$\psi_{2m+1}(x) = d_{2m,1}\psi_1 + d_{2m,3}\psi_1^3 + \cdots + \psi_1^{2m+1}.$$

Hence, the theorem holds.

■ NOTE 21.9.1 *This does not, however, mean that whenever we write $\psi_r(x)$ as a polynomial in x, the expression will only involve even or odd powers of x. Generally, all powers of x will be involved.*

If the x_i values are equally spaced, tables of orthogonal polynomials are available. The actual numerical values of $\psi_j(x_i)$ and A_{jj} as well as the general functional form of $\psi_j(x)$, for $j = 1, \cdots, 6$ and for $n \leq 52$ are given in Biometrika tables for Statisticians. Vol. 1, by *E.S. Pearson* and *H.O. Hartley*, Cambridge University Press, 1958. Similar values for $j \leq 5$ and $n \leq 75$ appear in Statistical Tables for Biological, Agricultural and

Medical Research, by *R.A. Fisher* and *F. Yates*, Hefner Publishing Co., N.Y., 1964 (6th Ed.). A short table of orthogonal polynomials for equally spaced data is given in the appendix (table A.22) for values of n up to 12.

The formulas for the orthogonal polynomials $\psi_j(x)$ up to order six for equally spaced x's and for any values of n are as follows:

$$\left.\begin{aligned}
\psi_0(x) &= 1\\
\psi_1(x) &= \lambda_1 x\\
\psi_2(x) &= \lambda_2\left\{x^2-\frac{1}{12}(n^2-1)\right\}\\
\psi_3(x) &= \lambda_3\left\{x^3-\frac{1}{20}(3n^2-7)x\right\}\\
\psi_4(x) &= \lambda_4\left\{x^4-\frac{1}{14}(3n^2-13)x^2+\frac{3}{360}(n^2-1)(n^2-9)\right\}\\
\psi_5(x) &= \lambda_5\left\{x^5-\frac{5}{18}(n^2-7)x^3+\frac{1}{1008}(15n^4-230n^2+407)x\right\}\\
\psi_6(x) &= \lambda_6\left\{x^6-\frac{5}{44}(3n^2-31)x^4+\frac{1}{176}(5n^4-110n^2+329)x^2\right.\\
&\qquad\left.-\frac{5}{14784}(n^2-1)(n^2-9)(n^2-25)\right\}
\end{aligned}\right\} \tag{21.9.33}$$

If $n \le 6$ we go only up to $j = n-1$. The appropriate λ-values are given at the bottom of each column and are chosen to ensure that the values in the tables are all integers. Above the λ-values are given A_{jj} values. X values are obtained from the relation $\psi_1(X) = \lambda_1 X$ when $\lambda_1 = 1(2)$ for n odd (even).

For further details on orthogonal polynomials the reader may refer to Applied Statistics Chapter 1 (Mukhopadhyay, 1999).

Exercise

1. For the model $E(\boldsymbol{Y}) = \boldsymbol{\eta} = \beta\boldsymbol{I}$, $D(\boldsymbol{Y}) = \sigma^2\boldsymbol{I}$, describe the estimation space and error space and find the l.s. estimator for β. Show that $\boldsymbol{A}$ and $\boldsymbol{Y}-\hat{\boldsymbol{\eta}}$ are orthogonal. Find $E(\hat{\beta})$, $E(SS_R)$, $E(SS_E)$.

2. Consider the model $\boldsymbol{Y} = \boldsymbol{A\beta} + \boldsymbol{e}$ when $\boldsymbol{A}(n\times p)$ is of full rank $p(<n)$ with $E(\boldsymbol{e}) = \boldsymbol{0}$, $D(\boldsymbol{e}) = \sigma^2 I$. If $\boldsymbol{H}$ is a non-singular matrix of order $p \times p$ and if $\boldsymbol{Y}$ is regressed on $\boldsymbol{W} = \boldsymbol{AH}$, then show that the regression and the error sum of squares remain invariant.

   [*Hints:* Since $\boldsymbol{H}$ is non-singular, $\boldsymbol{Y} = \boldsymbol{AHH}^{-1}\boldsymbol{\beta}+\boldsymbol{e} = \boldsymbol{W\phi}+\boldsymbol{e}$, $\boldsymbol{\phi} = \boldsymbol{H}^{-1}\boldsymbol{\beta}$. Now proceed to find the l.s.e. of $\boldsymbol{\phi}$.]

3. Let X_1, X_2, X_3 be three uncorrelated r.v.'s having common variance σ^2. If $E(X_1) = \theta_1 + \theta_2$, $E(X_2) = 2\theta_1 + \theta_2$, $E(X_3) = \theta_1 + 2\theta_2$, show that the l.s. estimates $\hat{\theta}_1$, $\hat{\theta}_2$ satisfy the equations

$$6\hat{\theta}_1 + 5\hat{\theta}_2 = x_1 + 2x_2 + x_3$$
$$5\hat{\theta}_1 + 6\hat{\theta}_2 = x_1 + x_2 + 2x_3.$$

Hence, show that $\hat{\theta}_1 = (x_1 + 7x_2 - 4x_3)/11$, $\hat{\theta}_2 = (x_1 - 4x_2 + 7x_3)/11$ are unbiased estimates. Verify that $V(\hat{\theta}_1) = V(\hat{\theta}_2) = 6\sigma^2/11$, $\text{Cov}(\hat{\theta}_1, \hat{\theta}_2) = -5\sigma^2/11$. Show that the minimum SS, $\boldsymbol{x}'\boldsymbol{x} - \boldsymbol{x}'\boldsymbol{A}\hat{\boldsymbol{\theta}}$ can be written as $(3x_1 - x_2 - x_3)^2/11$. Deduce that if X_1, X_2, X_3 are normally distributed, then $(3X_1 - X_2 - X_3)^2/11\sigma^2$ has a $\chi^2_{(1)}$ distribution.

4. Suppose $E(Y_1) = E(Y_2) = \theta$, but $V(Y_1) = 5\sigma^2$, $\text{Cov}(Y_1, Y_2) = \sigma^2$, $V(Y_2) = 2\sigma^2$. Show that the BLUE of θ is $\hat{\theta} = (Y_1 + 4Y_2)/5$.

5. Show that the linear unbiased estimator with the minimum variance has a fixed correlation with all linear unbiased estimators possessing the same variance.

6. Show that in the case of linear model $E(\boldsymbol{Y}) = \boldsymbol{A\beta}$, $D(\boldsymbol{Y}) = \sigma^2\boldsymbol{H}$ the BLUE of $\boldsymbol{l}'\boldsymbol{\beta}$ is $\boldsymbol{l}'(\boldsymbol{A}'\boldsymbol{H}^{-1}\boldsymbol{A})^{-}\boldsymbol{A}'\boldsymbol{H}^{-1}\boldsymbol{Y}$ with $V(\boldsymbol{l}\hat{\boldsymbol{\beta}}) = \sigma^2\boldsymbol{l}'(\boldsymbol{A}'\boldsymbol{H}^{-1}\boldsymbol{A})^{-}\boldsymbol{l}$ where $\boldsymbol{P}^{-}$ denotes g-inverse of $\boldsymbol{P}$.

7. Let $X_1, \cdots, X_n$ denote a random sample from a population with mean μ and variance $\sigma^2 (< \infty)$. Find the BLUE of μ.Using appropriate examples show that this estimator may attain Rao-Cramer lower bound for some distribution, while for some other distribution, there exists an unbiased estimator which is better than the BLUE.

8. When $\boldsymbol{y}$ has variance-covariance matrix $\boldsymbol{V}$, prove that the covariance of the BLUE of $\boldsymbol{p}'\boldsymbol{\beta}$ and $\boldsymbol{q}'\boldsymbol{\beta}$ is $\boldsymbol{p}'(\boldsymbol{A}'\boldsymbol{V}^{-1}\boldsymbol{A})^{-1}\boldsymbol{q}$.

9. Let $E(Y_i) = \sum_{k=1}^{p} a_{ik}\beta_k$, $i = 1, \cdots, n$, $\text{Cov}(Y_i, Y_j) = \sigma_{ij}$, $i, j = 1, \cdots, n$. Consider one-to-one transformation $X_i = \sum_{j=1}^{n} g_{ij}Y_j$. Show that the restriction $V(X_i) = 1$, $\text{Cov}(X_i, X_j) = 0$ $(i \neq j)$ imply $\sum_{s=1}^{n}\sum_{t=1}^{n} g_{is}g_{jt}\sigma_{st} = \delta_{ij}$ where $\delta_{ij} = 0(1)$ if $i \neq j$ (otherwise). Derive the normal equations corresponding to minimising $\sum_{i=1}^{n}[Y_i - E(Y_i)]^2$.

10. Let the normal equations be $\boldsymbol{A}'\boldsymbol{A\beta} = \boldsymbol{A}'\boldsymbol{Y}$ or $\boldsymbol{C\beta} = \boldsymbol{Q}$ where $\boldsymbol{A}'\boldsymbol{A} = \boldsymbol{C}$, $\boldsymbol{A}'\boldsymbol{Y} = \boldsymbol{Q}$. If $\boldsymbol{l}'\boldsymbol{\beta}$ is estimable, then there exists a vector $\boldsymbol{q}$ such that $\boldsymbol{l} = \boldsymbol{A}'\boldsymbol{A}\boldsymbol{q} = \boldsymbol{C}\boldsymbol{q}$. Show that $\boldsymbol{q}'\boldsymbol{Q}$ is the best unbiased estimator of $\boldsymbol{l}'\boldsymbol{\beta}$ and $\text{Var}(\boldsymbol{q}'\boldsymbol{Q}) = \sigma^2\boldsymbol{q}'\boldsymbol{l}$.

Show, further, that if $\boldsymbol{C}\boldsymbol{q}_1 = \boldsymbol{l}_1$ and $\boldsymbol{C}\boldsymbol{q}_2 = \boldsymbol{l}_2$, then $\boldsymbol{q}_1'\boldsymbol{Q}$ and $\boldsymbol{q}_2'\boldsymbol{Q}$ are, respectively, the BLUE's of $\boldsymbol{l}_1'\boldsymbol{\beta}$ and $\boldsymbol{l}_2'\boldsymbol{\beta}$ and $\text{Cov}(\boldsymbol{l}_1'\boldsymbol{\beta}, \boldsymbol{l}_2'\boldsymbol{\beta}) = \sigma^2\boldsymbol{q}_1'\boldsymbol{l}_2 = \sigma^2\boldsymbol{q}_2'\boldsymbol{l}_1$.

11. An experiment results in six independent observations $y_i (i = 1, \cdots, 6)$ such that

$$E(Y_i) = \alpha \cos\frac{2\pi_i}{6} + \beta \sin\frac{2\pi_i}{6}$$

$$V(Y_i) = \sigma^2.$$

Find the l.s. estimators of α and β and verify that each of these estimators has variance $\sigma^2/3$.

[$\hat{\alpha} = (y_1 - y_2 - 2y_3 - y_4 + y_5 + 2y_6)/6$; $\hat{\beta} = (y_1 + y_2 - y_4 - y_5)/2\sqrt{3}$]

12. Suppose an experiment is conducted with t treatments each replicated r times and a control (c) replicated s times. Let Y_{ij} be the jth observation on the ith treatment ($i = 1, \cdots, t$; $j = 1, \cdots, r$) and Y_{cj} be the jth observation on the control ($j = 1, \cdots, s$). Assume

$$Y_{ij} = \tau_i + e_{ij}$$

$$E(e_{ij}) = 0 \ \forall \ i, j$$

$$D(\boldsymbol{e}) = \sigma^2\boldsymbol{I}.$$

Show that the l.s. estimators of the difference $\theta_i = \tau_i - \tau_c$ is $\hat{\theta}_i = \bar{Y}_{io} - \bar{Y}_{co} (i = 1, \cdots, t)$. Prove also that for a given total number of experimental units, i.e., for constant $rt + s$, $\text{Var}(\hat{\theta}_i)$ is minimised if $s/r = \sqrt{t}$. **[Scheffé, 1961]**

13. Using a spring balance, a small container of unknown weight w_0 is weighed and the weight is recorded as x_1. An object of unknown weight w_1 is then put in the container and the two separate weighings are recorded as y_1, y_2. The object is then replaced with another of unknown weight and three separate weighings are recorded as z_1, z_2, z_3. If all readings are regarded as independent and subject to errors with zero mean and the same variance σ^2, find the l.s. estimates of w_0, w_1, w_2 and their variances and covariances.

[$\hat{w}_0 = X_1$; $\hat{w}_1 = (Y_1 + Y_2 - 2X_1)/2$; $\hat{w}_2 = (z_1 + z_2 + z_3 - 3X_1)/3$]

14. Let $y_i = \alpha + \beta x_i + e_{1i}$, $z_i = a + bx_i + e_{2i}$, $i = 1, \cdots, n$, denote standard regressions of y on x and y on z respectively; i.e., $E(e_{1i}) = 0$, $E(e_{2i}) = 0$, errors are independent and homoscedastic. Define the new variables as $y_i^* = y_i - \hat{\alpha} - \hat{\beta}x_i$, $z_i^* = z_i - \hat{a} - \hat{b}x_i$ where $\hat{\alpha}, \hat{\beta}, \hat{a}, \hat{b}$ are least square estimates of the respective parameters. The sample partial correlation coefficient of y and z holding x fixed is defined as the correlation coefficient between y^* and z^*, i.e.,

$$\rho_{yz.x} = \frac{\sum_{i=1}^{n} y_i^* z_i^*}{\sqrt{\sum_{i=1}^{n} {y_i^*}^2}\sqrt{\sum_{i=1}^{n} {z_i^*}^2}}.$$

Suppose (X, Y, Z) follow a trivariate normal distribution and x_i, y_i, z_i are their observed values. Find the conditional distribution of Y and Z given X and relate the correlation coefficient in this distribution to that found above.

15. Suppose you consider a true model $\boldsymbol{Y} = \boldsymbol{X\beta} + \boldsymbol{e}$ with no intercept and apply the l.s. method to obtain an estimate of $\boldsymbol{\beta}$. Another researcher uses this model, but before obtaining the l.s. estimates he takes deviations from the sample means for the variable y and those in the $\boldsymbol{X}$ matrix. What would be the effect on the efficiency of the estimates of $\boldsymbol{\beta}$ due to these approaches? Explain your model and estimators in the framework of restricted and unrestricted least square theory.

[Vinod and Ullah, 1981]

16. Consider the model $\boldsymbol{y} = \boldsymbol{A\beta} + \boldsymbol{e}$ with $E(\boldsymbol{e}) = \boldsymbol{0}, D(\boldsymbol{e}) = \sigma^2\boldsymbol{V}$.

Show that if the column-space of $\boldsymbol{A}$ is the same as the column space of $\boldsymbol{VA}$ then the generalised least square estimate of $\boldsymbol{\beta}, \hat{\boldsymbol{\beta}} = (\boldsymbol{X'V^{-1}X})^{-1}\boldsymbol{X'V^{-1}Y}$ becomes identical with the ordinary least square estimate $\hat{\boldsymbol{\beta}}_0 = (\boldsymbol{X'X})^{-1}\boldsymbol{X'Y}$.

17. The observations n in number are obtained from a two stage sample with c clusters, Y_{ij} denoting the jth observation on y in the ith cluster $(j = 1, \cdots, m_i; i = 1, \cdots, c)$. Consider the following model $\boldsymbol{Y} = \boldsymbol{X\beta} + \boldsymbol{e}$ when

$$\boldsymbol{Y} = [\boldsymbol{Y}_1' \ \boldsymbol{Y}_2' \ \cdots \ \boldsymbol{Y}_c']'$$
$$\boldsymbol{X} = [\boldsymbol{X}_1 \ \boldsymbol{X}_2], \quad \boldsymbol{\beta} = [\alpha_1, \cdots, \alpha_c, \beta_1, \cdots, \beta_p]'$$

$\boldsymbol{Y}_i$ is $m_i \times 1$, $\sum_{i=1}^{c} m_i = n$, $\boldsymbol{X}_1$ is $n \times c$, $\boldsymbol{X}_2$ is $n \times p$, $\boldsymbol{e}$ is $n \times 1$, the jth entry in ith column of $\boldsymbol{X}_1$ contains 1(0) if the observation Y_{ij} (jth entry in $\boldsymbol{Y}_i$) comes from the ith cluster (otherwise),

$$\boldsymbol{D}(\boldsymbol{e}) = \begin{bmatrix} \boldsymbol{V}_1 & \boldsymbol{O} & \cdots & \boldsymbol{O} \\ \boldsymbol{O} & \boldsymbol{V}_2 & \cdots & \boldsymbol{O} \\ \cdot & \cdot & \cdots & \cdot \\ \boldsymbol{O} & \boldsymbol{O} & \cdots & \boldsymbol{V}_c \end{bmatrix} = Blk \text{ diag } V_i = \bigoplus_{i=1}^{c} V_i$$

where

$$\boldsymbol{V}_i = \begin{bmatrix} 1 & \rho & \cdots & \rho \\ \rho & 1 & \cdots & \rho \\ \cdot & \cdot & \cdots & \cdot \\ \rho & \rho & \cdots & 1 \end{bmatrix}_{m_i \times m_i} = (1-\rho)I + \rho \boldsymbol{J}_i\boldsymbol{J}_i' \text{ where } J_i = (1 \cdots 1)'_{m_i \times 1}.$$

Here $\boldsymbol{X}$ denotes the matrix of values of auxiliary variables $x_1, \cdots, x_{p+c}$ on the units.

Show that in this case $C(\boldsymbol{VX}) \subset C(\boldsymbol{X})$ and hence the ordinary least square estimator $(\boldsymbol{X'X})^{-1}\boldsymbol{X'Y}$ and BLUE $(\boldsymbol{X'V^{-1}X})^{-1}\boldsymbol{X'V^{-1}Y}$ are identical.

18. Suppose that in model (21.5.1), the first column of $\boldsymbol{A}$ is $\mathbf{1}_n$ and $\boldsymbol{V} = (1-\rho)\boldsymbol{I}+\mathbf{1}\mathbf{1}'\rho$, $-\frac{1}{n-1} < \rho < 1$. Considering that $\boldsymbol{A}'\boldsymbol{L} = \mathbf{0}$ implies $\mathbf{1}_n'\boldsymbol{L} = \mathbf{0}$ where $\boldsymbol{L}$ is a $n \times (n-p)$ matrix of rank $(n-p)$, show that the condition (21.5.7) is always satisfied. Moreover, $\boldsymbol{V}$ can be written as $\boldsymbol{V} = \rho \boldsymbol{A}\boldsymbol{e}_1\boldsymbol{e}_1\boldsymbol{A}' + (1-\rho)\boldsymbol{I}_n$ where $e_1 = (1\ 0 \cdots 0)'_{n\times 1}$. Hence in (21.5.8), $\boldsymbol{\Gamma} = \rho \boldsymbol{e}_1\boldsymbol{e}_1'$, $\boldsymbol{\theta} = \mathbf{0}$, $w^2 = 1-\rho$. Hence, in this case the OLS $\hat{\boldsymbol{\beta}}$ coincides with Aitken's generalised l.s.e. $\hat{\boldsymbol{\beta}}_G$ with probability one.

19. Let y_j denote the change in the number of people residing in Calcutta from year j to year $j+1$ who would prefer to live in neighbouring subarban area, $j = 1, 2, 3$. Suppose

$$\underset{(3\times 1)}{\boldsymbol{y}} = \underset{(3\times 2)}{\boldsymbol{X}}\ \underset{(2\times 1)}{\boldsymbol{\beta}} + \underset{(3\times 1)}{\boldsymbol{e}}$$

where $\boldsymbol{y} = \begin{bmatrix} -2 \\ 3 \\ 1 \end{bmatrix}$ and $\boldsymbol{X} = \begin{bmatrix} 1 & 0 \\ 0 & 1 \\ 1 & 1 \end{bmatrix}$.

If $E(\boldsymbol{e}) = 0$, $D(\boldsymbol{e}) = \sigma^2\boldsymbol{\Sigma}$ where $\boldsymbol{\Sigma} = \begin{bmatrix} 2 & 1 & 2 \\ 1 & 3 & 1 \\ 2 & 1 & 4 \end{bmatrix}$ estimate (i) $\boldsymbol{\beta}$ (ii) σ^2. If instead of $\boldsymbol{\Sigma}$ being known, $\boldsymbol{\Sigma}$ is of the form $\boldsymbol{\Sigma} = \sigma^2 \begin{bmatrix} 1 & \rho & \rho \\ \rho & 1 & \rho \\ \rho & \rho & 1 \end{bmatrix}$, $-\frac{1}{2} < \rho < 1$ estimate $\boldsymbol{\beta}$.

20. Consider the model (21.5.1) with $\boldsymbol{V}$ as stated in problem (17), but $\boldsymbol{A}$ does not contain $\mathbf{1}_n$. Let $\boldsymbol{O}$ be an orthogonal matrix which diagonalised $\boldsymbol{V}$ i.e., $\boldsymbol{O}'\boldsymbol{V}\boldsymbol{O} = \wedge =$ Diag $(\lambda_1, \cdots, \lambda_n)$ [λ_i = latent root of $\boldsymbol{V}$]. Let $\boldsymbol{O} = [\mathbf{1}_n/\sqrt{n}, \boldsymbol{C}']'$. Premultiply (21.5.1) by $\boldsymbol{O}$ to obtain

$$\begin{bmatrix} \bar{Y} \\ \tilde{\boldsymbol{Y}} \end{bmatrix} = \begin{bmatrix} \bar{\boldsymbol{a}}' \\ \tilde{\boldsymbol{A}} \end{bmatrix} \boldsymbol{\beta} + \begin{bmatrix} \bar{e} \\ \tilde{\mathrm{e}} \end{bmatrix}$$

where

$$\bar{Y} = \mathbf{1}_n'\boldsymbol{Y}/\sqrt{n},\ \bar{\boldsymbol{a}}' = \mathbf{1}_n'\boldsymbol{A}/\sqrt{n},\ \bar{e} = \mathbf{1}_n'e/\sqrt{n},$$

$$\tilde{\boldsymbol{Y}} = \boldsymbol{C}\boldsymbol{Y}, \qquad \tilde{\boldsymbol{A}} = \boldsymbol{C}\boldsymbol{A}, \qquad \tilde{\mathrm{e}} = \boldsymbol{C}\boldsymbol{e}.$$

Show that

(i) $\bar{e}, \tilde{\mathbf{e}}$ are uncorrelated with $E(\bar{e}) = 0$, $E(\tilde{\mathbf{e}}) = \mathbf{0}$, $E(\tilde{\mathbf{e}}\tilde{\mathbf{e}}') = \sigma^2(1-\rho)I_m$, where $m = n-1$.

(ii) Minimising $\tilde{\mathbf{e}}'\tilde{\mathbf{e}}$, one obtains

$$\tilde{\boldsymbol{b}} = (\tilde{\boldsymbol{A}}'\tilde{\boldsymbol{A}})'\tilde{\boldsymbol{A}}'\tilde{\boldsymbol{Y}} = \left[\boldsymbol{A}'(I_n - \mathbf{1}_n\mathbf{1}_n'n^{-1})\boldsymbol{A}\right]^{-1} \boldsymbol{A}'(I_n - \mathbf{1}_n\mathbf{1}_n'n^{-1})\boldsymbol{Y} \tag{21.9.34}$$

as an unbiased estimator of $\boldsymbol{\beta}$.

(iii)

$$V(\tilde{b}) = \sigma^2(1-\rho)(A'A - A'1_n1_n'An^{-1})^{-1}$$
$$= \sigma^2(1-\rho)(A'A)^{-1}\left[I_p + (1_n'M1_n)^{-1}A'1_n1_n'A(A'A)^{-1}\right] \quad (21.9.35)$$

where $M = I_p - A(A'A)^{-1}A'$.

(iv) An unbiased estimator of $\sigma^2(1-\rho)$ is

$$(\tilde{Y} - \tilde{A}\tilde{b})'(\tilde{Y} - \tilde{A}\tilde{b})/(n-p-1) \quad (21.9.36)$$

(v)

$$E(\bar{Y} - \bar{a}'\tilde{b})^2 = E(\bar{e} - \bar{a}'(\tilde{A}'A)^{-1}\tilde{A}'\tilde{e})^2$$
$$= \sigma^2(1-\rho)[1 + \bar{a}'(\tilde{A}'\tilde{A})^{-1}\bar{a}'] + n\sigma^2\rho. \quad (21.9.37)$$

Consequently, an unbiased estimator of $\sigma^2\rho$ can be obtained using (21.10.3) and (21.10.4). Again, show that the OLSE $\hat{\beta} = (A'A)^{-1}A'Y$ has variance-covariance matrix

$$V(\hat{\beta}) = \sigma^2(A'A)^{-1}A'VA(A'A)^{-1}$$
$$= \sigma^2(1-\rho)(A'A)^{-1}\left[I_p + \rho(1-\rho)^{-1}A'1_n1_n'A(A'A)^{-1}\right]. \quad (21.9.38)$$

[Swamy, 1970]

Chapter 22

Nonparametric Inference

22.1 Introduction

In the previous chapters we considered the problem of drawing inference about population-characteristics (like mean, variance, skewness, kurtosis, etc.) on the basis of samples drawn from populations which follow some known probability distributions, some or all of whose model-parameters were possibly unknown and required estimation. In this chapter we do not assume any particular probability distribution for the population. We only assume that the sample is drawn from a general class of populations, e.g., class of all absolutely continuous distributions and/or discrete distributions. The problem is to draw inference about the population-characteristics on the basis of samples drawn from these general classes of populations. Such procedures are called *nonparametric* procedures. The term 'nonparametric' was first used by Wolfwitz (1942).

The theory of nonparametric estimation is considered in Section 22.2. The celebrated order statistic and U-statistic and their uses in estimation of parameters of distributions are considered in this section. The rest of the chapter deals with the testing of hypotheses. Section 22.3 is an introductory section. The next section considers some single-sample problems. Section 22.5 introduces two-sample problems and different types of related hypotheses. Tests useful mainly for location parameters are considered in subsections 22.5.1–22.5.11. Subsection 22.5.12 revisits the hypotheses related to scale parameters. The next four subsections describe some tests for dispersion. Section 22.6 introduces $C(\geq 2)$-sample problems. Subsections 22.6.1 and 22.6.2 consider two relevant tests.

22.2 Nonparametric Estimation

Let X be a random variable having a distribution function F belonging to a class of probability measures $\mathcal{F}$ with $\mathcal{X}$ as the sample space. A random sample of size $n, \boldsymbol{X} = (X_1, \cdots, X_n)$ is drawn from this distribution. The class of probability distribution of $\boldsymbol{X}$ may consist of absolutely continuous or discrete distributions or may be a subclass of these.

Of the different properties on which choice of estimators is based, unbiasedness is the most easily applied property in the nonparametric theory. We shall, therefore, first concentrate on unbiasedness.

Definition 22.2.1 *A real-valued parameter $g(F)$ is called estimable if it has an unbiased estimator i.e., if there exists a statistic $t(\boldsymbol{X})$ such that*

$$E_F\{t(\boldsymbol{X})\} = \int t(\boldsymbol{x}) \prod_{i=1}^{n} dF(x_i) = g(F) \quad \forall\, F \in \mathcal{F} \tag{22.2.1}$$

where $\boldsymbol{x} = (x_1, \cdots, x_n)$ and integration is taken over the product space $\mathcal{X}^n$. An estimable parameter is also called a ***regular parameter***.

The reader acquainted with parametric set up would like to see the parameter expressed as $g(\theta)$, rather than $g(F)$. Since θ's (parameters in F) are not explicitly involved here, a parameter function is characterised by F, the relevant distribution rather than θ.

Similarly, a vector-valued parameter $\boldsymbol{g}(F) = (g_1(F), \cdots, g_k(F))$ is called estimable if there exists a vector-valued statistic $\boldsymbol{t}(\boldsymbol{X}) = (t_1(\boldsymbol{X}), \cdots, t_k(\boldsymbol{X}))$ such that

$$E_F\{\boldsymbol{t}(\boldsymbol{X})\} = \left(\int t_1(\boldsymbol{x}) \prod_{i=1}^{n} dF(x_i), \cdots, \int t_k(\boldsymbol{x}) \prod_{i=1}^{n} dF(x_i)\right)$$
$$= \boldsymbol{g}(F) \quad \forall\, F \in \mathcal{F}.$$

► **Example 22.2.1** *If $\mathcal{F}$ be the class of all distributions for which the second moment exists, then it is known that population mean μ and population variance σ^2 are estimable because $E(\bar{X}) = \mu, E(S^2) = \sigma^2$ where $S^2 = \sum_{1}^{n}(X_i - \bar{X})^2/(n-1)$.*

Definition 22.2.2 *The degree $m(\geq 1)$ of an estimable parameter is defined to be the smallest sample size for which the parameter has an unbiased estimate, i.e., it is the smallest value of n such that there exists an unbiased estimator $t(X_1, \cdots, X_n)$ of $g(F)$.*

► **Example 22.2.2** *The population mean $\mu(F)$ has degree 1, because $E(X_1) = \mu(F)$ and just one observation is sufficient to produce an unbiased estimate of $\mu(F)$. Similarly $E(X_1X_2) = E(X_1)E(X_2) = \mu^2(F)$ and $\mu^2(F)$ has degree 2. Also $E(X^2) - (E(X))^2 = \mu_2(F)$ and therefore $E(X_1^2 - X_1X_2) = \mu_2(F)$ so that $\sigma^2(F)$ has degree 2. The parameter $g(F) = P_F(X \geq c)$ has degree 1, because $E(I_c) = g(F)$ where $I_c = 1(0)$ if $X \geq (<)C$.*

Definition 22.2.3 *An unbiased estimate of the parameter based on the minimum sample size m is called a* ***Kernel*** *of the estimate.*

► **Example 22.2.3** *In example 22.2.2, X_1 is a Kernel of the estimate of $\mu(F)$, $X_1^2 - X_1X_2$ is a Kernel of the estimate of $\mu_2(F)$.*

Lemma 22.2.1 *For any estimable parameter function $g(F)$, there always exists a symmetric Kernel.*

Proof: Suppose $t(X_1, \cdots, X_m)$ is a Kernel of the estimate of $g(F)$. Then it is always possible to find a symmetric Kernel

$$t_S(X_1, \cdots, X_m) = \frac{1}{m!} \sum_{\pi} t(X_{i_1}, \cdots, X_{i_m})$$

where the summation is over all $m!$ permutations $(i_1, \cdots, i_m)$ of $(1, \cdots, m)$. Clearly, $t_S(X_1, \cdots, X_m)$ is an unbiased estimator of $g(F)$ and being based on a sample of size m, is a symmetric Kernel.

▶ **Example 22.2.4** *A symmetric Kernel for $\mu_2(F)$ is*

$$\begin{aligned} t_S(X_i, X_j) &= \frac{1}{2!}\left[t(X_i, X_j) + t(X_j, X_i)\right] \\ &= \frac{1}{2}\left[X_i^2 - X_i X_j + X_j^2 - X_j X_i\right] \\ &= \frac{1}{2}(X_i - X_j)^2, (i \neq j). \end{aligned}$$

If $g_1(F), g_2(F)$ are estimable parameters of degrees m_1, m_2 respectively, then $h_1(F) = g_1(F)+g_2(F)$ has degree $m = \max(m_1, m_2)$ and $h_2(F) = g_1(F)g_2(F)$ has degree m_1+m_2. Clearly, both $h_1(F)$, $h_2(F)$ are estimable. This is, because, if $t_i(X_1, \cdots, X_{m_i})$ is a Kernel of $g_i(F)$, then

$$\int [t_1(x_1, \cdots, x_{m_1}) + t_2(x_1, \cdots, x_{m_2})] \prod_{i=1}^{n} dF(x_i) = g_1(F) + g_2(F)$$

and

$$\int [t_1(x_1, \cdots, x_{m_1}) t_2(x_{m_1+1}, \cdots, x_{m_1+m_2})] \prod_{i=1}^{m_1+m_2} dF(x_i) = g_1(F)g_2(F).$$

In general, it follows that any polynomial in estimable parameters is also an estimable parameter.

Definition 22.2.4 ***U-statistics.*** *Corresponding to any Kernel $t(X_1, \cdots, X_m)$ of an estimable function $g(F)$, we define an U-statistic for the sample $(X_1, \cdots, X_n)$ by*

$$U(X_1, \cdots, X_n) = \binom{n}{m}^{-1} \sum_{c} t_S(X_{i_1}, \cdots, X_{i_m})$$

where the summation c is over all $\binom{n}{m}$ combinations $(i_1, \cdots, i_m)$ of m integers chosen from $(1, \cdots, n)$ and t_S is the symmetric Kernel defined in lemma (22.2.1). Clearly, we *can also write*

$$U(X_1, \cdots, X_n) = \frac{(n-m)!}{n!} \sum_{\pi} t(X_{i_1}, \cdots, X_{i_m})$$

where the summation is over all permutations π of m integers $(i_1, \cdots, i_m)$ chosen from $(1, \cdots, n)$. Thus, $U(\boldsymbol{X})$ is a symmetrised form of $t(X_1, \cdots, X_m)$ considered as a function of $\boldsymbol{X}$. Clearly,

$$E_F\{U(\boldsymbol{X})\} = g(F) \quad \forall\, F \in \mathcal{F}$$

i.e., the corresponding U-statistic is an unbiased estimator of $g(F)$.

▶ **Example 22.2.5** *Consider estimation of $\mu(F), \mu^2(F)$. For $\mu(F)$, Kernel is $t(X_1) = X_1$ and U-statistic is*

$$\frac{1}{n}\sum_{i=1}^{n} X_i = \bar{X}.$$

For $\mu^2(F)$, Kernel is $t(X_1, X_2) = X_1 X_2$ and U-statistic is

$$\frac{1}{\binom{n}{2}} \sum\sum_{1<j} X_i X_j.$$

For $\mu_2(F)$, Symmetric Kernel is $t_S(X_1, X_2) = \frac{1}{2}(X_1 - X_2)^2$ and U-statistic is

$$\frac{1}{2\binom{n}{2}} \sum\sum_{i<j=1}^{n} (X_i - X_j)^2 = \frac{1}{n-1}\sum_{i=1}^{n}(X_i - \bar{X})^2 = S_x^2.$$

Similarly, for estimating the rth raw moment $\mu_r'(F) = E_F(X^r)$, the Kernel is $t(X_1) = X_1^r$ and U-statistic is

$$\frac{1}{n}\sum_{i=1}^{n} X_i^r = m_r'.$$

For $g(F) = F(t_0) = \int_{-\infty}^{t_0} dF(x) = P_F(X \le t_0)$ the Kernel $h(x) = I(x \le t_0)$ where $I(A)$ is the indicator function of the set A.

The corresponding U-statistic is

$$U(X_1, \cdots, X_n) = \frac{1}{n}\sum_{i=1}^{n} I(X_i \le t_0) = F_n(t_0),$$

where F_n denotes the sample distribution function.

We now consider the concept of sufficiency and complete sufficiency for the family of distributions $\mathcal{F}$.

Definition 22.2.5 *A statistic $T(\boldsymbol{X})$ is sufficient for the family of distributions $\mathcal{F}$ if the conditional distribution of $\boldsymbol{X}$ given $T(\boldsymbol{X}) = t$, is a constant, whatever $F \in \mathcal{F}$.*

Definition 22.2.6 *A family of distributions $\mathcal{F}$ is complete, if the only unbiased estimate of zero is the zero function itself, i.e., if*

$$\begin{aligned} E_F\{h(\boldsymbol{X})\} &= 0 \quad \forall\, F \in \mathcal{F} \\ \Rightarrow \quad h(\boldsymbol{x}) &= 0 \quad \textit{for all values } \boldsymbol{x} \textit{ of } \boldsymbol{X} \end{aligned}$$

except, possibly, for sets of probability measure zero for which $F \in \mathcal{F}$.

Definition 22.2.7 *A statistic $T(\boldsymbol{X})$ is said to be complete in relation to a class of distributions $\mathcal{F}$ if the class of induced distributions of T is complete.*

In Section 11.6 we defined the order statistics. Let $X_1, \cdots, X_n$ denote a random sample of size n from a distribution F in $\mathcal{F}$. The X-values arranged in order of magnitude from the least to the greatest are denoted as $X_{(1)}, \cdots, X_{(n)}$. The statistic $T(\boldsymbol{X}) = (X_{(1)}, \cdots, X_{(n)})$ is called the order statistic corresponding to the sample $\boldsymbol{X}$. We consider, here, some further properties of the order statistic.

Lemma 22.2.2 *Let $(X_1, \cdots, X_n)$ be a random sample from F belonging to the class of absolutely continuous distribution functions $\mathcal{F}$. If $h(x_1, \cdots, x_n)$ is a symmetric function of $(x_1, \cdots, x_n)$, then it can be written as a function of the order statistic $t(\boldsymbol{x}) = (x_{(1)}, \cdots, x_{(n)})$, i.e.,*

$$h(\boldsymbol{x}) = g\left[(x_{(1)}, \cdots, x_{(n)})\right];$$

conversely, if $h(\boldsymbol{x})$ can be written as a function of $(x_{(1)}, \cdots, x_{(n)})$ then it must be a symmetric function.

Proof: Follows easily by noting that $h(\boldsymbol{x})$ is a constant valued function for all the $n!$ permutations $(i_1, \cdots, i_n)$ of $(1, \cdots, n)$.

Theorem 22.2.1 *The order statistic $T = (X_{(1)}, \cdots, X_{(n)})$ is sufficient for the family of absolutely continuous distribution functions $\mathcal{F}$.*

Proof: Let $f(x)$ be the common p.d.f. of X. Since X_i's are i.i.d., the joint p.d.f. of $\boldsymbol{X}$ is

$$\prod_{i=1}^{n} f(x_i) = \prod_{i=1}^{n} f(x_{(i)}).$$

Again, the joint p.d.f. of $(X_{(1)}, \cdots, X_{(n)})$ is

$$n! \prod_{i=1}^{n} f(x_{(i)}), x_{(1)} < x_{(2)} < \cdots < x_{(n)} \quad \text{[vide (11.6.1)]}$$

Hence, the conditional p.d.f. of $\boldsymbol{X}$ given $T(\boldsymbol{X}) = (X_{(1)}, \cdots, X_{(n)})$ is

$$f(\boldsymbol{x}|T(\boldsymbol{X}) = t) = \frac{1}{n!},$$

which is a constant for all $F \in \mathcal{F}$. Therefore, the order statistic T is sufficient for the family of absolutely continuous d.f.'s $\mathcal{F}$.

Theorem 22.2.2 *The order statistic* $(X_{(1)}, \cdots, X_{(n)})$ *is complete for the family of distributions all of which are absolutely continuous or discrete.*

Proof: Omitted. The proof is available in Fraser (1957), p.27-30, 139-142.

▶ **Example 22.2.6** *Show that for a symmetrical distribution, the sample median, for an odd number of observations, is a minimum variance unbiased estimator of population median.*

Let $X_1, \cdots, X_{2m+1}$ be a random sample from $F_X(x)$ with

$$\int_{-\infty}^{\xi} dF(x) = \frac{1}{2}.$$

The sample median is $x_{(m+1)} = u$ (say). The distribution of u is

$$\begin{aligned} g(u)du &= \frac{(2n+1)!}{(n!)^2}\{F(u)\}^n\{1-F(u)\}^n dF(u) \\ &= \frac{(2n+1)!}{(n!)^2}\left[\int_{-\infty}^{u} f(x)\,dx\right]^n \left[\int_{u}^{\infty} f(x)\,dx\right]^n f(u)\,du. \quad \text{(i)} \end{aligned}$$

Let $u = \xi + Z$, $Z > 0$. Then

$$\int_{-\infty}^{u} f(x)\,dx = \frac{1}{2} + \int_{\xi}^{\xi+Z} f(x)\,dx$$

$$\text{and} \quad \int_{u}^{\infty} f(x)\,dx = \frac{1}{2} - \int_{\xi}^{\xi+Z} f(x)\,dx.$$

The expression (i) reduces to

$$\frac{(2n+1)!}{(n!)^2}\left[\frac{1}{4} - \left\{\int_{\xi}^{\xi+Z} f(x)\,dx\right\}^2\right]^n f(u)\,du$$

$$g(\xi+Z)\,du = \frac{(2n+1)!}{(n!)^2}\left[\frac{1}{4} - \left\{\int_{0}^{Z} f(\xi+x)\,dx\right\}^2\right]^n f(u)\,du. \quad \text{(i}$$

Similarly, if $u = \xi - Z$, $Z > 0$, then (i) reduces to

$$g(\xi-Z)\,du = \frac{(2n+1)!}{(n!)^2}\left[\frac{1}{4} - \left\{\int_{0}^{Z} f(\xi-x)\,dx\right\}^2\right]^n f(u)\,du. \quad \text{(ii}$$

Since $f(\xi+x) = f(\xi-x)$, $g(\xi+Z) = g(\xi-Z)$ so that distribution of sample media u is symmetric about ξ. Hence $E(u) = \xi$. The statistic $u = x_{(m+1)}$ being based on ord statistic is MVUE of its expectation, ξ.

□ **Remark 22.2.1** *In general, there does not exist an unbiased estimator for the median of an arbitrary continuous distribution on* R^1.

Theorem 22.2.3 *Let $\mathcal{F}$ be the class of all absolutely continuous or all purely discrete distribution functions on R. If $t(X_1, \cdots, X_n)$ is an unbiased estimator of a parameter $g(F)$, then the corresponding U-statistic is the unique uniformly minimum variance unbiased estimator among all unbiased estimators of $g(F)$.*

Proof: We have already proved that the order statistic is complete sufficient for $\mathcal{F}$. We are required only to prove that the U-statistic is the conditional expectation of $t(X_1, \cdots, X_n)$ given the order statistic $(x_{(1)}, \cdots, x_{(n)})$. Then the result will follow from Lehmann-Scheffé theorem (theorem 15.9.3).Now,

$$E\{t(X_1, \cdots, X_n)|(x_{(1)}, \cdots, x_{(n)})\} = \frac{1}{n!}\sum_{\pi} t(x_{i_1}, \cdots, x_{i_n}) \quad \text{(where } x_{i_j}\text{'s take one of the values } x_{(1)}, \cdots, x_{(n)}, \text{ each value being taken only once)}$$

$$= U(x_{(1)}, \cdots, x_{(n)})$$

(the summation being taken over all $n!$ permutations of $(1, 2, \cdots, n)$).

Alternative Proof

Let $t(x_{i_1}, \cdots, x_{i_m})$ be denoted as $t_i (= t)$, for the value of the statistic $t(x_1, \cdots, x_n)$ for the ith permutation $(i_1, \cdots, i_m)$ of $(1, 2, \cdots, n)$. Then

$$U(x_{(1)}, \cdots, x_{(n)}) = U = \frac{1}{n!}\sum_{i=1}^{n!} t_i.$$

Clearly, $E_F(U) = g(F)$ since $E(t_i) = g(F)$ $(i = 1, \cdots, n!)$.

$$\begin{aligned} V(U) &= E\left\{\left(\frac{1}{n!}\sum t_i\right)^2\right\} - (g(F))^2 \\ &\le E\left[\frac{1}{n!}\sum_{i=1}^{n!} t_i^2\right] - g(F)^2 \left(\text{since } \frac{1}{M}\sum_{i=1}^{M} Z_i^2 \ge \bar{Z}^2\right) \\ &= E(t_i^2) - g(F)^2 \\ &= V(t), \end{aligned}$$

the equality holds, iff

$$t_i = \text{ a constant } \quad \forall\, i = 1, \cdots, n!$$

i.e., if the statistic $t(x_{(1)}, \cdots, x_{(n)})$ is a symmetric function of x (for all points in the sample space except possibly for points on a null set). In this case, $t(x_{(1)}, \cdots, x_{(n)})$ itself is the U-statistic $U(x_{(1)}, \cdots, x_{(n)})$.

By theorem 22.2.2, there does not exist any other function of $(x_{(1)}, \cdots, x_{(n)})$ which is a unbiased estimator of $g(F)$. Hence the U-statistic is the unique UMVUE of $g(F)$ in the class of all unbiased estimators of $g(F)$.

Corollary 22.2.1 *Let $\mathcal{F}$ be the class of all absolutely continuous or all purely discrete distributions on R^1. Any estimable function $g(F)$, $F \in \mathcal{F}$ has a unique estimate that is unbiased and symmetric in the observations and has uniformly minimum variance among all unbiased estimators.*

Proof follows from lemma 22.2.2 and theorem 22.2.3.

□ **Remark 22.2.2** *In the theory of nonparametric estimation we should, therefore, only consider estimates that are symmetric in the observations and are unbiased. This will give an unbiased estimate with the least variance in the class of all unbiased estimators of the parameter.*

► **Example 22.2.7** *It follows from example 22.2.5 that $\bar{X}$, S_x^2 and m_r' are unique UMVUE of $\mu(F), \mu_2(F), \mu_r'(F)$ respectively. For the unbiased estimation of cumulants, Fisher (1928) proposed the k-statistics. They are unbiased estimators and because they are symmetric they are functions of the order statistics. It follows from theorem 22.2.3 that they are uniformly minimum variance unbiased estimators of the cumulants.*

Consider the parameter Δ_F which is a measure of spread or concentration of a distribution,

$$\Delta_F = E_F\{|X_1 - X_2|\} = \iint |X_1 - X_2| dF(x_1) dF(x_2), \quad F \in \mathcal{F}$$

where X_1, X_2 are two random samples from a distribution $F \in \mathcal{F}$. Clearly Δ_F is estimable and a Kernel of the estimator is $|X_1 - X_2|$. If we have a sample of size $n, X_1, \cdots, X_n$ from F, the corresponding U-statistic is

$$U = \frac{1}{\binom{n}{2}} \sum\sum_{i<j} |X_i - X_j|$$

which is Gini's mean difference. It has got minimum variance among all unbiased estimators of Δ_F.

Consider another parameter. For a distribution F, the probability measure of a set A is

$$P_F(A) = P\{X \in A\} = E\{I_A(X)\}$$

where $I_A(x) = 1(0)$ if $x \in A$ (otherwise). Clearly, a Kernel of the estimator is $I_A(X_1)$ and the corresponding U-statistic is

$$U = \frac{1}{n} \sum_{i=1}^{n} I_A(x_i) = \frac{m}{n}$$

where m is the number of observed x's belonging to set A. Therefore, $\frac{m}{n}$ is the UMVUE of $P_F(A)$.

Example 22.2.8 *Let $(X_1, \cdots, X_{n_1})$ be a random sample of size n_1 from F_1, $(Y_1, \cdots, Y_{n_2})$ be a random sample of size n_2 from F_2, F_1, $F_2 \in \mathcal{F}$, class of all absolutely continuous distributions. Assume that the variables X, Y are independent. For the X's it is known that the order statistic $t_1 = (x_{(1)}, \cdots, x_{(n_1)})$ is complete sufficient. Similarly, for the Y's, the order statistic $t_2 = (y_{(1)}, \cdots, y_{(n_2)})$ is complete sufficient. It follows that $t = \{(x_{(1)}, \cdots, x_{(n_1)}), (y_{(1)}, \cdots, y_{(n_2)})\}$ is complete sufficient for the combined experiment. It follows, therefore, that for any estimable parameter function, there is essentially only one unbiased estimate based on the statistic t and this estimate is* UMVUE. *It also follows that any function of this statistic is a symmetric function of x's and y's and conversely, any symmetric function of x's and y's is a function of this statistic. It follows, therefore, that any symmetric function of x's and y's is a* UMVUE *of its expectation.*

Consider estimation of $E(XY)$. A Kernel of the estimator is x_1y_1. By symmetrising this estimator with respect to y's, keeping x's fixed, we have

$$\frac{x_1}{n_2}\sum_{j=1}^{n_2} y_j$$

and by symmetrising this with respect to x's we have the symmetrised estimator

$$\left(\sum_{i=1}^{n_1} x_i \sum_{j=1}^{n_2} y_j\right) / n_1 n_2 = \bar{x}\bar{y}.$$

Obviously $\bar{x}\bar{y}$ is an unbiased estimator of $E(XY)$ and has minimum variance among all unbiased estimators.

Example 22.2.9 *Let $(X_1, Y_1), \cdots, (X_n, Y_n)$ be independent sample from the same bivariate distribution F over R^2, $F \in \mathcal{F}$, the class of all absolutely continuous bivariate distributions. The order statistic $\{(x_1, y_1), \cdots, (x_n, y_n)\}$ is a complete sufficient statistic. Any function of the order statistic is a symmetric function of the n entries $(x_1, y_1), \cdots, (x_n, y_n)$ and vice-versa. Consequently any symmetric function of (x_i, y_i)'s $(i = 1, \cdots, n)$ is* MVUE *of its expectation.*

Consider estimation of $E(XY)$. A Kernel of the estimator is x_1y_1 and the corresponding U-statistic is

$$\frac{1}{n}\sum_{i=1}^{n} X_iY_i$$

which is unique UMVUE of $E(XY)$. Similarly, for the estimation of σ^2_{X+Y}, a Kernel is $\frac{1}{2}[(x_1 + y_1) - (x_2 + y_2)]^2$ and the corresponding U-statistic is

$$\frac{1}{n-1}\sum_{i=1}^{n}(x_i + y_i - \bar{x}_i - \bar{y}_i)^2.$$

It has got MV in the class of all unbiased estimators.

The asymptotic distribution of the U-statistic has been derived by Hoeffding.

Theorem 22.2.4 (*Hoeffding*). *Let $X_1, \cdots, X_n$ be a random sample from the distribution F and $t_S(X_1, \cdots, X_m)$ be a real-valued symmetric statistic with expectation $g(F)$ and finite second moment. Then as $n \to \infty$, the distribution of U-statistics corresponding to t_S which is*

$$U(x_1, \cdots, x_n) = \frac{1}{\binom{n}{m}} \sum_c t_S(x_{i_1}, \cdots, x_{i_m})$$

(summation being taken over all combinations $(i_1, \cdots, i_m)$ from $(1, \cdots, n)$) is given by:

$$\sqrt{n}\,[U(X_1, \cdots, X_n) - g(F)]$$

is asymptotically normal with mean 0 and variance

$$m^2 V\left\{E(t_S(X_1, \cdots, X_m) | X_1 = x_1)\right\}.$$

For a proof see Fraser (1957).

The following theorem follows from theorem 22.2.4.

Theorem 22.2.5 *If $t(X_1, \cdots, X_m)$ is an unbiased estimator of $g(F)$ and if $E_F\{t^2(X_1, \cdots, X_m)\} < \infty$, then the corresponding U-statistic for a sample of size n is a consistent estimate of $g(F)$ as $n \to \infty$.*

We now consider random variables which would be U-statistics were it not for the fact that they contain an estimator. Let $X_1, \cdots, X_n$ denote a random sample from some population. Let $t_s(x_1, \cdots, x_m; \gamma)$ denote a symmetric Kernel of degree m with expected value

$$\theta(\gamma) = E_{\boldsymbol{\lambda}}[t_s(X_1, \cdots, X_m; \gamma)]$$

where $\boldsymbol{\lambda}$ denotes a p-parameter value and γ is a mathematical variable whose one particular value may be $\boldsymbol{\lambda}$. Both the Kernel and its expected value may depend on γ. The corresponding U-statistics is then

$$U_n(\gamma) = \frac{1}{\binom{n}{m}} \sum_c t_s(X_{i_1}, \cdots, X_{i_m}; \gamma).$$

Let $\hat{\boldsymbol{\lambda}}$ be a consistent estimator of $\boldsymbol{\lambda}$. It can be shown (Randles, 1982) that under some regularity conditions

$$\sqrt{n}\left[U_n(\hat{\boldsymbol{\lambda}}) - \theta(\boldsymbol{\lambda})\right] \xrightarrow{L} N(0, \tau^2)$$

provided $\tau^2 > 0$ where τ^2 is given by either

(a)

$$\tau^2 = \boldsymbol{D}'\boldsymbol{\Sigma}\boldsymbol{D} > 0,$$

$$\boldsymbol{D}' = \left(1, \frac{\partial\theta(\cdot)}{\partial\gamma_1}, \cdots, \frac{\partial\theta(\cdot)}{\partial\gamma_p}\right)\Big|_{\gamma=\boldsymbol{\lambda}}, \quad \gamma = (\gamma_1, \cdots, \gamma_p)',$$

Σ is the dispersion matrix of

$$\sqrt{n}\left(U_n(\boldsymbol{\lambda}) - \theta(\boldsymbol{\lambda}), \quad (\hat{\boldsymbol{\lambda}} - \boldsymbol{\lambda})'\right)_{1\times\overline{p+1}}$$

or, (b) $\tau^2 = \text{Var}\left[E\{h(X_1, \cdots, X_m; \lambda)|X_1\}\right]$.

22.3 Testing of Hypotheses

We now consider testing of hypotheses under non-parametric set-up. Section 22.4 considers single-sample problems, Section 22.5 two-sample problems. A few related definitions which will be used in the sequel are consider here.

Definition 22.3.1 (***Pitman*, 1948**) ***Power-efficiency.*** *Let $\{T_n\}$, $\{T_n^*\}$ be two sequences of test statistics of the same hypothesis $H_\circ$ at the same significance level α. Let the distributions be indexed by a real parameter δ such that $H_\circ$ corresponds to $\delta = 0$ and other δ_i's correspond to distributions under different alternative hypotheses. If for the same power with respect to the same alternative, the test T requires n_i observations and T^*, n_i^* observations, then the relative power efficiency of T with respect to T^* may be defined as*

$$e(T_n|T_n^*) = \frac{n_i^*}{n_i} = e(\delta_i, \alpha, n_i^*). \tag{22.3.1}$$

In general, e depends on δ_i, α, n_i^*. Some asymptotic values of e can be computed by keeping one of these three arguments fixed and letting the other two arguments tend to some limit.

Let the sequences of alternatives δ_i's be so indexed that $\delta_i \to 0$ as $i \to \infty$.

Definition 22.3.2 *Asymptotic relative efficiency* (ARE) *of a test T_n relative to the test T_n^* is the limiting value of $e(\delta_i, \alpha, n_i^*)$:*

$$\lim_{\delta_i \to 0, n_i^* \to \infty} (n_i^*/n_i) \tag{22.3.2}$$

provided the limit exists. In many cases (22.3.2) *does not depend on α. The* ARE *is also called the local asymptotic efficiency since it is the large sample power efficiency in the vicinity of $H_\circ$.*

22.4 Some Single Sample Problems

The term single sample problem is used customarily in estimation problem and tests of hypotheses involving the *c.d.f.*$F(x)$ for a given population. Thus, we may be interested in testing a hypothesis about the median of a distribution or that $F(x) = F_0(x) \quad \forall\, x$ where $F_0(x)$ is a known d.f. These hypotheses are tested on the basis of a random sample $X_1, \cdots, X_n$ from the same population.

22.4.1 A Test of Randomness

Suppose we want to test the null hypothesis that a given sequence of numbers is random against the alternative that the sequence is not random. The alternative hypotheses may also be H_+, the sequence exhibits tendency to mix or H_-, the sequence exhibits tendency to cluster.

Here the magnitude of each observation is compared with that of the one immediately preceding it in the sequence. If the preceding value is smaller, a run up is started, a '+' is noted; if the preceding value is larger, a run down is commenced, a '−' is noted. An arrangement of N different observations is recorded as a sequence of $(N-1)$, + and − signs. The test statistic is V, the total number of runs of + and −.

► **Example 22.4.1** 2 7 5 11 19 11 17

+ − + + − +

Total number of runs $= 5$.

Under the null hypothesis of randomness, every possible sequence of $(N-1)$ *symbols '+' and '−' is equally likely.*

If the sequence exhibits trend (either upword or downword), there will be fewer number of runs. This is the case of H_-. On the other hand, frequent fluctuations or cyclical movements (the case of H_+) would produce too many runs.

Under $H_o, E(V) = (2N-1)/3$, $V(V) = (16N-29)/90$. For large $N(\geq 25)$, the standardized variable with a correction for continuity, $\{V+0.5-(2N-1)/3\}/\sqrt{(16N-29)/90}$ $\sim N(0,1)$ and this can be used for testing the hypothesis.

Table A.8 in the appendix gives the null distribution of V for $N \leq 25$ as left hand probabilities for V less than or equal to the median and right tail probabilities for other values of V. For testing H_o, against $H_+(H_-)$, the appropriate P-values are obtained from the right (left) tail of the distribution.

► **Example 22.4.2** *Consider the sequence,* 3, 7, 11, 8, 9, 5, 2, 6, 9, 17, 21, 23.

We have + + − + − − + + + + +. Thus, $V = 5$. For testing H_o against H_- $P\{V \leq 5|N = 12\} = 0.0529$. The null hypothesis is just accepted at 5% level of significance.

22.4.2 Tests of Goodness of Fit

The problem of fit is to test the hypothesis that the sample observations $X_1, \cdots, X_n$ is from some specified distribution F_0 against the alternative that it is from some other distribution. If F_0 is completely specified, H_o is a simple hypothesis. If F_0 is not completely specified it may be required to estimate some unknown parameters from the sample in order to perform tests. We considered in section 20.2, χ^2-test of goodness of fit due to K. Pearson. Here follows another test.

Kolmogorov-Smirnov test

Let $X_1, \cdots, X_n$ be a random sample from $F(x)$. In Section 11.7 we defined the empirical d.f. of a random variable X as

$$F_n(x) = \frac{\text{no. of observations } X_j \leq x}{n}.$$

The Kolmogorov-Smirnov (KS) one sample test for goodness of fit is based on the absolute maximum deviation between $F_n(x)$ and $F(x)$.

Definition 22.4.1 *The statistic*

$$D_n = \sup_x |F_n(x) - F_0(x)| \tag{22.4.1}$$

is called the two-sided KS *statistics and is used for testing* $H_\circ : F(x) = F_0(x) \ \forall x$ *where* $F_0(x)$ *is the completely specified d.f. against the alternative* $H : F(x) \neq F_0(x)$ *for some* x.

For testing $H_\circ$ against $H_1 : F(x) \geq F_0(x) \ \forall x$, with strict inequality for some x, one-sided KS statistic is

$$D_n^+ = \sup_x [F_n(x) - F_0(x)]. \tag{22.4.2}$$

Similarly for testing $H_\circ$ against $H_2 : F(x) \leq F_0(x) \ \forall x$, with strict inequality for some x, one-sided KS statistic is

$$D_n^- = \sup_x [F_0(x) - F_n(x)]. \tag{22.4.3}$$

Theorem 22.4.1 *The* KS *statistics* D_n, D_n^+, D_n^- *are distribution-free, provided* F_0 *is any absolutely continuous d.f.*

Proof: Let $X_{(1)}, \cdots, X_{(n)}$ be the order statistics of $\boldsymbol{X} = (X_1, \cdots, X_n)$ and let $X_{(0)} = -\infty$, $X_{(n+1)} = \infty$. Then $F_n(x) = \frac{i}{n}$ for $X_{(i)} \leq x < X_{(i+1)}, i = 0, 1, \cdots, n$. Define

$$\begin{aligned} D_n^+ &= \max_{0 \leq i \leq n} \sup_{X_{(i)} \leq x < X_{(i+1)}} \left[\frac{i}{n} - F_0(x)\right] \\ &= \max_{0 \leq i \leq n} \left[\frac{i}{n} - \inf_{X_{(i)} \leq x < X_{(i+1)}} F_0(x)\right] \\ &= \max_{0 \leq i \leq n} \left[\frac{i}{n} - F_0(X_{(i)})\right] \\ &= \max\left\{\max_{1 \leq i \leq n} \left[\frac{i}{n} - F_0(X_{(i)})\right], 0\right\}. \end{aligned} \tag{22.4.4}$$

Similarly,

$$D_n^- = \max\left\{\max_{1 \leq i \leq n} \left[F_0(X_{(i)}) - \frac{i-1}{n}\right], 0\right\}. \tag{22.4.5}$$

Clearly,

$$D_n = \max(D_n^+, D_n^-).$$

Now, $0 < F_0(X_{(1)}) \le \cdots \le F_0(X_{(n)}) < 1$. Therefore, $F_0(X_{(i)})$ is the ith orde statistic of a sample from a uniform distribution (obtained by the probability integra transformation of X, vide Section 8.3), whatever F_0, so long as it is continuous Therefore, D_n^+ is independent of F_0. Similarly, if F_0 is a continuous c.d.f., each c D_n^- and D is distribution-free. We shall, henceforth, assume without loss of generalit F_0 to be a continuous uniform distribution over $(0, 1)$.

Theorem 22.4.2 (***Dvoretzky, Kiefer and Wolfowitz***). *Let F_0 be a continuous c.d. There exists a finite positive constant C (not depending on F_0) such that*

$$P(D_n > d) \le Ce^{-2nd^2}, \quad d > 0$$

for $n = 1, 2, \cdots$.

The derivation of exact distribution of D_n is tedius and is omitted. For a proof, th reader may see Gibbons (1971, p.77). The result is stated without proof.

Theorem 22.4.3 *If F_0 is a continuous c.d.f., then*

$$P\left[D_n < \frac{1}{2n} + c\right] = \begin{cases} 0 & \text{for } c \le 0 \\ \displaystyle\int_{(1/2n)-c}^{(1/2n)+c} \int_{(3/2n)-c}^{(3/2n)+c} \cdots \int_{(2n-1)/2n-c}^{(2n-1)/2n+c} f(u_1, \cdots, u_n) du_1, \cdots, du_n & \text{for } 0 < c < \frac{2n-1}{2n} \\ 1 & \text{for } c \ge \frac{2n-1}{2n} \end{cases} \quad (22.4.$$

where

$$f(u_1, \cdots, u_n) = \begin{cases} n!, & \text{for } 0 < u_1 < \cdots < u_n < 1 \\ 0, & \text{otherwise} \end{cases} \quad (22.4.$$

is the joint p.d.f. of the order statistic for a sample of size n from $U(0, 1)$.

Let $D_{n,\alpha}$ be the upper α%-point of the distribution of D_n, i.e., $P(D_n > D_n, \alpha) =$ Numerical values of $D_{n,\alpha}$ for $\alpha = 0.01$ and 0.05 for selected values of n have be tabulated by Birnbaum (1952). The large sample distribution of D_n was derived Kolmogorov (1933). Smirnov and Feller obtained the following approximation.

Theorem 22.4.4 *If F_0 is any continuous c.d.f., then for any $D \ge 0$*

$$\lim_{n \to \infty} P\left[\sqrt{n} D_n \le D\right] = 1 - 2\sum_{i=1}^{\infty} (-1)^{i-1} \exp\left[-2i^2 D^2\right]$$

$$= L(D) \ (say). \quad (22.4.$$

The approximation is close enough for $n \geq 35$. Tables have been constructed by mirnov.

Exact sampling distributions of D_n^+ and D_n^- are not difficult to derive. These are btained below.

heorem 22.4.5 *If F_0 is any continuous c.d.f.,*

$$P(D_n^+ < c) = \begin{cases} 0 & \text{if } c \leq 0 \\ \int_{1-c}^{1} \int_{(n-1)/n-c}^{u_n} \cdots \int_{2/n-c}^{u_3} \int_{1/n-c}^{u_2} f(u_1, \cdots, u_n) du_1, \cdots, du_n & \text{for } 0 < c < 1 \\ 1 & \text{for } c \geq 1 \end{cases} \tag{22.4.9}$$

here $f(u_1, \cdots, u_n)$ is given in (22.4.7).

Proof: Using (22.4.4) we can write

$$\begin{aligned} P(D_n^+ < c) &= P\left[\max_{1 \leq i \leq n}\left(\frac{i}{n} - F_0(X_{(i)})\right) < c\right] \\ &= P\left[\frac{i}{n} - F_0(X_{(i)}) < c \quad \forall\, i = 1, \cdots, n\right] \\ &= P\left[F_0(X_{(i)}) > \frac{i}{n} - c \quad \forall\, i = 1, \cdots, n\right] \end{aligned}$$

hich gives the result (22.4.9). In deriving (22.4.9) we have assumed without any loss generality that F_0 is a continuous uniform distribution over $(0, 1)$.

Birnbaum and Tingey tabulated values of $D_{n,\alpha}^+$ for $\alpha = 0.01, 0.05, 0.10$ for selected alues of n, where $P[D_n^+ > D_{n,\alpha}^+] = \alpha$. Smirnov gave the following asymptotic result.

heorem 22.4.6 *If F_0 is a continuous c.d.f., then for every $D \geq 0$*

$$\lim_{n\to\infty} P\left[\sqrt{n}D_n^+ < D\right] = 1 - \exp(-2D^2). \tag{22.4.10}$$

Because of symmetry, D_n^+ and D_n^- have identical distributions. For testing $H_\circ$ gainst $H^+(H^-)$, $H_\circ$ is rejected if $D_n^+ > D_{n,\alpha}^+(< D_{n,\alpha}^-)$. For both-sided alternative $: F(x) \neq F_0(x)$ for some x, $H_\circ$ is rejected if $D_n > D_{n,\alpha}$.

heorem 22.4.3 implies that for arbitrary $\varepsilon > 0$,

$$\lim_{n\to\infty} P[D_n < \varepsilon] = 1$$

e., $F_n(x)$ converges in probability to $F(x)$ $\forall\, x$, uniformly (Glivenko-Cantelli theorem, de theorem 11.7.2). This also implies the consistency of the KS-statistic D_n. Similarly, eorem (22.4.5) implies consistency of D_n^+ or D_n^-.

■ NOTE 22.4.1 *If $F(x)$ is assumed to be a discrete distribution, then the* KS*-statistic do not remain distribution-free. In this case, the sampling distribution of D_n (as also D_n^+ and D_n^-) depends on the original distribution. It follows that*

$$P\left[D_n^* \leq D\right] \geq P\left[D_n \leq D\right] \quad \forall\, D(>0)$$

where D_n^* is the KS*-statistic when $F(x)$ is a discrete distribution and D_n is the* KS *statistic when $F(x)$ is continuous. Therefore, $P\left[D_n^* > D_{n,\alpha}\right] \leq \alpha$, when $D_{n,\alpha}$ has bee defined before. For a discrete d.f. $F(x)$ the test becomes a conservative one.*

▸ **Example 22.4.3** *The following is a random sample of size* 20. *Test whether the samp can be considered as a sample from a $N(0,1)$-distribution.*

$$\begin{array}{rrrrr} 2.240, & -.400, & -1.152, & .980, & .361, \\ -.123, & -.625, & .682, & 2.323, & -1.053, \\ -.870, & -.164, & -.340, & -.041, & 1.405, \\ 1.187, & .323, & .270, & -.128, & .101. \end{array}$$

x	$F_0(x)$	$F_n(x)$	$\lvert F_0(x) - F_n(x)\rvert$
−1.152	.1247	1/20	.0747
−1.053	.1462	2/20	.0462
−.870	.1922	3/20	.0422
−.625	.2660	4/20	.0660
−.400	.3446	5/20	.0946
−.340	.3669	6/20	.0669
−.164	.4349	7/20	.0849
−.128	.4490	8/20	.0490
−.123	.4510	9/20	.0010
−.041	.5164	10/20	.0164
.101	.5402	11/20	.0098
.270	.6064	12/20	.0064
.323	.6266	13/20	.0234
.361	.6410	14/20	.0580
.682	.7523	15/20	.0023
.980	.8365	16/20	.0365
1.187	.8824	17/20	.0324
1.405	.9200	18/20	.0200
2.240	.9875	19/20	.0375
2.323	.9899	1	.0101

Here, $D_{20} = 0.0946 < 0.294 = D_{20,0.05}$ (vide table A.10). Hence $H_o(F(x)$ $F_0(x)\ \forall\, x)$ is accepted at 5% level of significance.

Comparison between χ^2-test and KS-test. The KS statistics examine the goodness of .t for each of the n-observations, whereas the χ^2-statistic examines this only when the .ata are classified into k classes and hence it (χ^2-statistic) sometimes loses information hrough grouping. The KS-statistic is applicable even in the case of very small samples, ut the χ^2-test is essentially for large samples. Moreover, the χ^2-statistic depends on he number of classes and is not unique if the categories are to be combined for having mall expected frequencies. The KS-statistics D_n^+ and D_n^- can test for deviations in a articular direction, whereas χ^2 always tests deviations in either directions.

With a continuous variable, if the sample is small and therefore adjacent categories ıust be combined for the χ^2-test, the χ^2-test is definitely less powerful than the KS-test. t appears that in all cases where it is applicable the KS-test is the most powerful test or goodness of fit.

Confidence-bounds for $F(x)$ from KS statistic

Suppose that we have a random sample of size n from a unknown distribution $F(x)$. t is intended to find a confidence bound for $F(x)$ with confidence coefficient α. Consider he Kolmogorov-Smirnov statistic,

$$D_n = \sup_x |F_n(x) - F(x)|$$

iven n and α we first find (from table A.10) $D_{n,\alpha}$ such that $P[D_n > D_{n,\alpha}] \leq \alpha$,

$$\text{i.e., } P\left[\sup_x |F_n(x) - F(x)| > D_{n,\alpha}\right] \leq \alpha$$

$$\text{i.e., } P\left[|F_n(x) - F(x)| > D_{n,\alpha} \ \forall\, x\right] \leq \alpha.$$

)efine

$$L_n(x) = \max\{F_n(x) - D_{n,\alpha}, 0\}$$

$$U_n(x) = \min\{F_n(x) + D_{n,\alpha}, 1\}.$$

Thus, the region between $L_n(x)$ and $U_n(x)$ can be taken as a confidence band for $F(x)$ with confidence coefficient $1-\alpha$.

Note. Another important class of tests for the hypothesis $H_o(F(x) = F_0(x) \ \forall\, x)$ is ased on the *Cramer-Von Mises test statistic*

$$C_n = n\int_{-\infty}^{\infty} [F_n(x) - F_0(x)]^2 dF_0(x).$$

ike D_n, C_n has a distribution not depending on F.

The following results hold:

(a) Finkelstein (1971): With probability 1,

$$\overline{\lim_{n\to\infty}} \frac{C_n}{2\log\log n} = \frac{1}{\pi^2}.$$

(b) Let ξ be a random variable represented as

$$\xi = \sum_{j=1}^{\infty} \frac{\chi_{1j}^2}{j^2\pi^2}$$

where $\chi_{11}^2, \chi_{12}^2, \cdots$ are independent $\chi_{(1)}^2$ variables. Then

$$\lim_{n\to\infty} P(C_n \le c) = P(\xi \le c), \quad c > 0.$$

Some excellent introductions to the theory underlying statistical tests based on F are the monograph of Durbin (1973), the survey paper of Gaenssler and Stute (197 and Bickel and Doksum (1977), among others.

22.4.3 The Problem of Location

Let $X_1, \cdots, X_n$ be a sample of size n from a population with continuous c.d.f. F. Let be a positive real number $0 < p < 1$ and let $\xi_p(F)$ be the quantile of order p for the d. F. The problem of location is to test $H_\circ : \xi_p(F) = \xi_0, \xi_0$ being a known number, again the alternatives $\xi_p(F) > \xi_0$ or $\xi_p(F) < \xi_0$ or $\xi_p(F) \ne \xi_0$. In non-parametric problems th median is often taken as the location parameter rather than the mean. The problem location and symmetry is to test $H_\circ : \xi_{.5}(F) = \xi_0$ and F is symmetric against $\xi_{.5}(F) \ne$ or F is not symmetric.

The Ordinary Sign Test

Let $\boldsymbol{X} = (X_1, \cdots, X_n)$ be a sample of size n from a population with an unknown c.d.f. and unknown median μ_e. Assume that F is absolutely continuous in the neighbourhoo of the population median. We are required to test the null hypothesis $H_\circ(\mu_e = \mu$ against one sided alternatives, $H_1(\mu_e > \mu_1^0)$, $H_2(\mu_e < \mu_e^0)$ or two-sided alternati $H(\mu_e \ne \mu_e^0)$.

Since we have assumed that X is a continuous r.v., $P(X = \mu_e^0) = 0$. If the samp comes from the population with $\mu_e = \mu_{e_0}$, then on the average half of the samp observations will be above μ_{e_0} and half below it. The test statistic $t(\boldsymbol{X})$ is the numb of positive signs S in the n differences $X_i - \mu_{e_0}$, $i = 1, \cdots, n$. If π is the probabilit $P[X > \mu_{e_0}]$ then S has a binomial distribution with parameters (n, π). Under $H_\circ, \pi =$ under $H_1, \pi > \frac{1}{2}$ etc. Therefore, critical region for $H_\circ$ against H_1 will be $S \ge S_\alpha$ whe S_α is the smallest integer for which

$$P\{S \ge S_\alpha | H_\circ\} = \sum_{S=S_\alpha}^{n} \binom{n}{S} \frac{1}{2^n} \le \alpha. \tag{22.4.1}$$

Because of the discreteness of the binomial distribution the test may not be exact size α. Also theoretically, there will be no zero difference $X - \mu_e$ because of the continui of the population; but in practice there may arise zero differences. The usual method is to ignore the zero differences and take $S + R$, R being the number of negative signs, as the sample size.

It follows similarly that for $H_\circ$ against H_2, the critical region is $S \leq \mathcal{S}'_\alpha$ where $\mathcal{S}'_\alpha$ is the largest integer for which

$$\sum_{S=0}^{\mathcal{S}'_\alpha} \binom{r+S}{S} \cdot \frac{1}{2^{r+S}} \leq \alpha. \tag{22.4.12}$$

For the two-sided alternative the equal-tailed size α critical region is

$$S \leq \mathcal{S}'_{\alpha/2} \cup S \geq \mathcal{S}_{\alpha/2}$$

where $\mathcal{S}'_{\alpha/2}$ and $\mathcal{S}_{\alpha/2}$ are defined as in (22.4.11) and (22.4.12).

For moderately large $n(\geq 12)$, $\tau = \left(S - \frac{n}{2}\right) / \sqrt{\frac{n}{4}} \sim N(0,1)$ and this property may be used for finding the critical regions.

Since S/n is a consistent estimator of $\pi(E(S/n) = \pi, V(S/n) = \pi(1-\pi)/n \to 0$ as $n \to \infty)$, the sign test which is based on S is a consistent test for hypotheses regarding π.

The test is uniformly most powerful (UMP) for the one-sided alternative and (UMP) unbiased against two-sided alternatives.

Clearly, the test can be extended to testing for quantile of any order. Suppose we want to test $H_\circ(\xi_p = \xi_p^0)$ against $H_1(\xi_p > \xi_p^0)$. Here, the test statistic is $t(\boldsymbol{X}) = S =$ number of positive signs in the differences $X_i - \xi_p^0 (i = 1, \cdots, n)$. Under $H_\circ$, $P(X > \xi_p^0) = 1 - p$, under H_1, this probability is $> 1 - p$. Hence critical region is $S \geq \mathcal{S}_{\alpha,p}$ where $\mathcal{S}_{\alpha,p}$ is the smallest integer for which

$$P\{S \geq \mathcal{S}_{\alpha,p} | H_\circ\} = \sum_{S=\mathcal{S}_{\alpha,p}}^{n} \binom{n}{S} (1-p)^S p^{n-S} \leq \alpha.$$

■ NOTE 22.4.2 *The test can be easily extended to a paired sample problem. Suppose that we have a random sample of size* $n, (X_1, Y_1), \cdots, (X_n, Y_n)$ *from a bivariate population. We want to test* $H_\circ$: *the median of the variable* $D = (X - Y)$, $\mu_e(X - Y)$ *is zero against the alternative* $\mu_e(X - Y) > 0$. *It is assumed that the population of the differences D is continuous so that* $P\{\mu_e(X - Y) = 0\} = 0$. *Note that median (D) is not necessarily equal to median (X) − median (Y), so that* $H_\circ$ *is not that med. (X) = med. (Y). All the procedures of sign-test is now applicable with* D_i *playing the role of* X_i *in the above situation. This test is comparable with the paired t-test (vide subsection* 11.5.2*) in the parametric set up where, of course, the normality is assumed and the problem is that of testing the difference between the means (and not the medians).*

Wilcoxon Signed-rank Test

The ordinary sign test does not make use of the magnitude of the difference between the observed value and the assumed value of the quantile. The Wilcoxon signed-rank test provides an alternative test of location (median of the distribution) and symmetry by taking into account the magnitude of the differences as well as their sign and as such is more efficient than the ordinary sign test.

Assume that $X_1, \cdots, X_n$ is a random sample from a continuous distribution which is symmetric about the median μ_e (Note that for sign test we did not assume the symmetry of the distribution about the median). The problem is to test $H_\circ(\mu_e = \mu_e^0)$ against the one-sided or two-sided alternatives. Under $H_\circ$, the difference $D = X - \mu_e^0$ is symmetrically distributed about mean zero. We first arrange $|D_1| < |D_2| < \cdots < |D_n|$ and rank those absolute values as $1, \cdots, n$. Let $T^+(T^-)$ be the sum of the ranks of the positive (negative) D_i values. Clearly,

$$0 \leq T^+ \leq \frac{n(n+1)}{2},$$

T^+ is 0 when all D_i's are negative, is $\frac{n(n+1)}{2}$ when all D_i's are positive. Similarly,

$$0 \leq T^- \leq \frac{n(n+1)}{2}.$$

Under $H_\circ$, T^+, T^- should be approximately equal. Clearly,

$$T^+ + T^- = \sum_{i=1}^{n} i = \frac{n(n+1)}{2}$$

always. Let $r(|D_i|) = r_i$ denote the rank of $|D_i|$. Let also $Z_i = 1(0)$ if $D_i > (<)0$. Then

$$T^+ = \sum_{i=1}^{n} r_i Z_i, \quad T^- = \sum_{i=1}^{n} r_i(1 - Z_i).$$

Thus

$$T^+ - T^- = 2\sum_{i=1}^{n} r_i Z_i - \frac{n(n+1)}{2}.$$

The statistic T^+ (or T^-) is known as Wilcoxon statistic. A large value of T^+ means that most of the deviations D_i are positive and this supports $H_1(\mu_e > \mu_0)$. The critical region is, therefore, $T^+ > c$ where $P(T^+ > c|H_\circ) = \alpha$. Similarly, critical regions for $H_2(\mu_e < \mu_{e_0})$ and $H(\mu_e \neq \mu_{e_0})$ can be found out.

Assume without loss of generality that $\mu_{e_0} = 0$.

To calculate the distribution of T^+ consider a random variable $Z_{(i)}$ defined on each ranked observation D_j as follows:

$$Z_{(i)} = \begin{cases} 1, \text{ if the observation } D_j(= X_j) \text{ whose rank } i \text{ is positive} \\ 0, \text{ otherwise.} \end{cases}$$

Then

$$T^{+} = \sum_{i=1}^{n} i Z_{(i)}$$

The $Z_{(i)}$'s are independent Bernoulli variables but not necessarily identically distributed. Clearly,

$$E(Z_{(i)}) = P(Z_{(i)} = 1) = P\{\text{the observation } D_j \text{ whose rank is } i \text{ is} > 0\} \quad (22.4.13)$$

$$= \frac{n!}{(i-1)!(n-1)!} \int_0^{\infty} [F_{|X|}(x)]^{i-1} [1 - F_{|X|}(x)]^{n-i} f(x)\, dx \quad (22.4.14)$$

where $f(x)$ is the p.d.f. of X and $F_{|X|}$ is the c.d.f. of $|X|$.
Now

$$F_{|X|}(x) = P\{|X| \leq x\} = F_X(x) - F_X(-x).$$

Hence, (22.4.14)

$$= \frac{n!}{(i-1)!(n-i)!} \int_0^{\infty} [F_X(x) - F_X(-x)]^{i-1} [1 - F_X(x) + F_X(-x)]^{n-i} f(x)\, dx. \quad (22.4.15)$$

Under $H_\circ$, X is symmetric about 0 so that $F(0) = \frac{1}{2}$.
Hence, $F(-x) = 1 - F(x) \;\; \forall \, x > 0$. Therefore,

$$E(Z_{(i)}|H_\circ) = \frac{n!}{(i-1)!(n-i)!} \int_0^{\infty} [2F(x) - 1]^{i-1} [2 - 2F(x)]^{n-i} f(x)\, dx. \quad (22.4.16)$$

Writing $v = 2F(x) - 1$, in (22.4.16)

$$E(Z_{(i)}|H_\circ) = \frac{1}{2}.$$

Also

$$V(Z_{(i)}) = E(Z_{(i)}) - \left(E(Z_{(i)})\right)^2.$$

Therefore,

$$V(Z_{(i)}|H_\circ) = \frac{1}{4}.$$

Hence

$$E(T^{+}|H_\circ) = \sum_{i=1}^{n} i E(Z_{(i)}|H_\circ) = \frac{n(n+1)}{4}$$

Similarly,

$$V(T^{+}|H_\circ) = \sum_{i=1}^{n} i^2/4 = \frac{n(n+1)(2n+1)}{24}.$$

For large n, $\{4T^{+} - n(n+1)\}/\sqrt{2n(n+1)(2n+1)/3} \sim N(0,1)$ and this is used for testing $H_\circ$. The distribution of T^{+} for small samples has been tabulated by Kraft and Van Eeden (1968).

Note that T^+ is completely determined by the values $(z_{(1)}, \cdots, z_{(n)})$ of the indicator random variables $(Z_{(1)}, \cdots, Z_{(n)})$. Under $H_\circ$, each of $Z_{(i)}$ takes value 1 with probability 1/2. Hence under $H_\circ$, each arrangement $(z_{(1)}, \cdots, z_{(n)})$ has the probability $\frac{1}{2^n}$. Therefore,

$$P\{T^+ = t|H_\circ\} = \frac{n(t)}{2^n}$$

where $n(t)$ is the number of ways in which we can assign $T^+ = t$.

Now if for a given assignment $(Z_{(1)}, \cdots, Z_{(n)})$ for which $T^+ = t$, we interchange the positive and negative signs we get value of T^+ as $\frac{n(n+1)}{2} - t$. Hence

$$P\left\{T^+ = \frac{n(n+1)}{2} - t|H_\circ\right\} = \frac{n(t)}{2^n}.$$

Therefore, distribution of T^+ is symmetric about its mean under $H_\circ$.

□ **Remark 22.4.1** *It has been assumed that under $H_\circ$, the distribution is symmetric about the median μ_{e_0}. Hence the statistic T^+ can also be used to test for the symmetry of the distribution about μ_{e_0}. Thus we may set the null hypothesis $H'_\circ$ as $H'_\circ : \mu_e = \mu_{e_0}$ and the distribution is symmetric about μ_{e_0} and the alternative hypotheses as*

$$H'_1 : \mu_e \neq \mu_{e_0} \text{ and } F \text{ is symmetric}$$
$$\text{or, } H''_1 : F \text{ is asymmetric.}$$

Since distribution of T^+ under $H'_\circ$ remains the same as that under $H_\circ$, the same statistic T^+ can be used for testing $H'_\circ$ against H'_1 or H''_1.

Let $T = \min(T^+, T^-)$. Table (A.14) gives various values of T and their associated levels of significance for $N = 6(1)25$. If observed $T \leq T_\alpha$, the null hypothesis may be rejected in favour of two-sided alternative at that level of significance.

Wilcoxon paired sample signed rank test: As in the case of sign test, this test can be extended to a paired sample. Suppose we have a sample of size n, $(X_1, Y_1), \cdots, (X_n, Y_n)$ from a bivariate population. We assume that the difference $Z_i = X_i - Y_i$ $(i = 1, \cdots, n)$ is a random sample from the population of $Z(= X - Y)$ with absolutely continuous d.f. which is symmetric about the medians $\mu_e(Z) = \mu_{e_0}$. We can then use the Wilcoxon statistic T^+ to test the null hypothesis $H_\circ(\mu_e(Z) = \mu_{e_0})$ against the usual hypotheses. Note, however, that $\mu_e(Z) \neq \mu_e(X) - \mu_e(Y)$ always.

► **Example 22.4.4** *The following data represent a random sample of size* 20: 0.436 0.278, 0.854, 0.142, 0.128, 0.875, 0.176, 0.598, 0.832, 0.653, 0.358, 0.572, 0.632, 0.192 0.482, 0.659, 0.791, 0.873, 0.859, 0.802. *Test $H_\circ(\mu_e = 0.5)$ against the alternative $H(\mu_e \neq 0.5)$.*

The number of positive signs S in $X_i - 0.5(i = 1, \cdots, 20)$ is 12. Hence $\tau = (12 - 10)/\sqrt{5} = 0.894$. Hence $H_\circ$ is accepted.

If we apply signed-rank test, the absolute values of $X_i - 0.5$, arranged in order of magnitude are: 0.018(–), 0.064(–), 0.072, 0.098, 0.132, 0.142(–), 0.153, 0.159, 0.222(–), 0.291, 0.302, 0.308(–), 0.324(–), 0.332, 0.354, 0.358(–), 0.359, 0.372(–), 0.373, 0.375.

Here $T^+ = 133$; under $H_\circ$, $E(T^+) = 105$, $V(T^+) = 717.5$.

Therefore, $\tau = (133 - 105)/\sqrt{717.5} = 1.045$. The $H_\circ$ is accepted.

▶ **Example 22.4.5** *During World War II bacterial polysaccherides were investigated as blood plasma extenders. Sixteen samples of hydrolized polysachharides supplied by various manufacturers in order to assess two chemical methods for determining the average molecular weight yielded the following result.*

Method A :	62,700	29,100	44,400	47,800	36,300	40,000
	43,400	35,800	33,900	44,200	34,300	31,300
	38,400	47,100	42,100	42,200		
Method B :	56,400	27,500	42,200	46,800	33,300	37,100
	37,300	36,200	35,200	38,000	32,200	27,300
	36,100	43,100	38,400	39,900		

Perform an appropriate test of the hypothesis that the two averages are the same against a one-sided alternative that the average of method A exceeds that of method B. (Natrella, M.G., 1963).

Let $C = A - B$. Let us test the hypothesis $H_\circ$: med. $(C) = 0$.

C :	6300	1600	2200	1000	3000	2900	6100	– 400
	–1300	6200	2100	4000	2300	4000	3700	2300

First consider the ordinary sign test. The number of C_i's > 0 is 14. We have $P\{$ Number of C_i's > 0 is $\geq 14|H_\circ\}$

$$= \sum_{k=14}^{16} \binom{16}{k} \left(\frac{1}{2}\right)^{16} = 0.00209.$$

Since this probability is less than 0.05, $H_\circ$ is rejected in favour of $H_1\{\text{med.}(C) > 0\}$.

Consider now the Wilcoxon signed-rank test. The absolute values of the differences in observations arranged in order of magnitude are:

400, 1000, 1300, 1600, 2100,
2200, 2300, 2900, 3000, 3700,
4000, 4000, 6100, 6200, 6300.

Here $T^+ = 122$; $E(T^+|H_\circ) = 68$, $V(T^+|H_\circ) = 374$. Therefore, $\tau = (122-68)/\sqrt{374} = 2.79$.

$H_\circ$ is, therefore, rejected at 5% level of significance. If paired t-test is considered, $t = \sqrt{n}\bar{C}/\mathcal{S}_e = 4.456 > 1.753 = t_{0.05}$. This test also rejects $H_\circ(\bar{A} = \bar{B})$ in favour of the alternative that $H(\bar{A} > \bar{B})$.

22.5 Two-Sample Problem

In the present section we shall investigate hypotheses relating to two independent samples from two populations. Let $X_1, \cdots, X_m$ be a random sample drawn from a population with continuous c.d.f. F_X and $Y_1, \cdots, Y_n$ be another random sample from the population with continuous c.d.f. G_Y. We are usually interested in the null hypothesis $H_o : F_X(x) = G_Y(x) \quad \forall\ x$ i.e., in the hypothesis that samples are drawn from the identical populations. We shall have three types of alternatives.

(I) The general type of alternative is that the two populations differ in any manner, say, in central tendency, in dispersion, in skewness, etc. The most general form of such a hypothesis is a two-sided alternative hypotesis and is given by

$$H : F_X(x) \neq G_Y(x) \quad \text{for some} \quad x.$$

We may also consider one-sided alternatives related to this two-sided alternative:

(i) $H_1 : F(x) \geq G(x) \quad \forall\ x$, with strict inequality for some x. The alternative H_1 states that the variable X is stochastically smaller than the variable $Y (X < Y)$. This is, because, $F(x) > G(x)$ means

$$P(X \leq x) > P(Y \leq x)$$

so that

$$P(X > x) < P(Y > x).$$

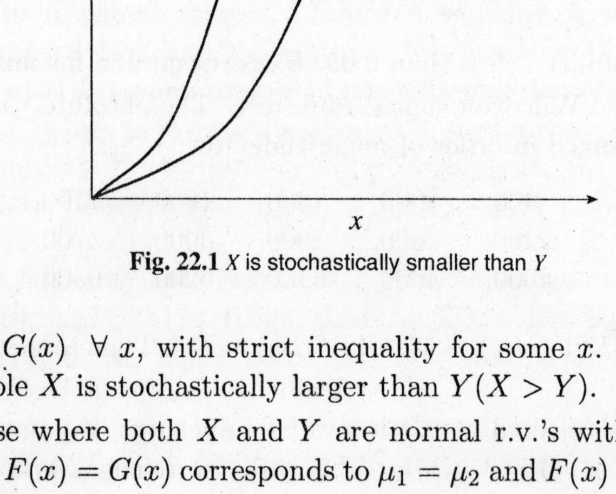

Fig. 22.1 *X* is stochastically smaller than *Y*

(ii) $H_2 : F(x) \leq G(x) \quad \forall\ x$, with strict inequality for some x. The alternative H_2 states that the variable X is stochastically larger than $Y (X > Y)$.

In the special case where both X and Y are normal r.v.'s with means μ_1, μ_2 and common variance σ^2, $F(x) = G(x)$ corresponds to $\mu_1 = \mu_2$ and $F(x) > G(x)$ corresponds to $\mu_1 < \mu_2$.

For these problems we shall discuss (i) the sign test (ii) the Wald-Wolfwitz run test (iii) the Mann-Whitney U-test (iv) the Kolmogorov-Smirnov two-sample test (v) Thompson's test (vi) Dixon's test.

(II) The second kind of alternative is called the location alternative, according to which the two distributions differ only in location. Thus, the alternatives are:

$$H_3 : F(x) = G(x + a), \ \ a > 0 \ \forall \, x$$
$$H_4 : F(x) = G(x - a), \ \ a > 0 \ \forall \, x.$$

Clearly, H_3 is a special case of H_1 and H_4 that of H_2. In this case, we consider the following tests: (i) the Wilcoxon two-sample (or rank sum test) (ii) the Mann-Whitney U-test (iii) the Median test (iv) the Terry-Hoeffding test. In $H_3(H_4)$, the c.d.f. G is shifted to left (right) of F. Such hypotheses are of great practical importance. Suppose we are interested in testing efficiency of a newly invented medical treatment. Thus, X may be measurements on control group (on which no treatment has been applied), while Y may be measurements on the group on which treatments have been applied. A shift $F(x)$ towords left implies that there has been improvement by applying the treatments.

(iii) The third type of alternative is called the scale alternative. According to this, the two populations differ only in their scale parameters and this alternative can be written.as

$$H_5 : F(x) = G(x/\theta) \text{ for some } \theta \neq 1.$$

The tests considered for this scale alternatives are: (i) the *Mood test* (ii) the *Siegel-Tukey* test (iii) the *Freund-Ansari test.*

A more general alternative than H_5 is

$$H_6 : G(x) = F\left(\frac{x - \theta_1}{\theta_2}\right),$$

the null hypothesis to be tested is $\theta_2 = 1$, whatever the (unknown) value of θ_1.

Clearly, the tests meant for the class of general alternatives H, H_1, H_2 can also be applied to test H_3, H_4, H_5, though they will be less efficient than the tests specially meant for testing location alternatives or scale alternatives (Moses, 1952).

In case the values of x and y are grouped into several classes, testing $H_\circ(F(x) = G(x))$ reduces to testing for homogeneity of two independent multinomial populations (obtained from the original c.d.f.'s, $F(x), G(x)$). This is done by the χ^2-test of homogeneity, considered in Section (20.4). For testing independence between X and Y we shall consider Kendall's τ-test and Spearman's rank correlation test.

22.5.1 Sign Test

It is the simplest possible test for the identity of two populations. Let X, Y be two random variables with continuous c.d.f. $F(x), G(y)$ respectively. We want to test the null hypothesis $H_\circ(F(x) = G(x) \ \ \forall \, x)$ against the alternative $H_1(F(x) \geq G(x))$ with strict inequality for some x i.e., X is stochastically smaller than Y or $H_2 : X$ is stochastically larger than Y.

Suppose we draw samples of size n from each population and pair the observations in an arbitrary way. Let the paired observations be $(x_1, y_1), \cdots, (x_n, y_n)$. For each pair, note the sign of $x_i - y_i$. We attach a (+)ve sign to the ith pair if $x_i - y_i > 0$, otherwise a negative sign. The case $x_i = y_i$ can be ignored since we assume that both $F(x)$ and $G(y)$ are continuous c.d.f.'s.

Now

$$\pi = P(X > Y) = \iint_{X>Y} dF(x)dG(y) = \int_0^1 \left[\int_{F(y)}^1 dF(x)\right] dG(y)$$

$$= \int_0^1 [1 - F(y)]\, dG(y) \tag{22.5.1}$$

Thus

$$\pi = \begin{cases} = \frac{1}{2} \text{ if } F(x) = G(x) \\ < \int_0^1 [1 - G(y)]dG(y) = \dfrac{1}{2} \text{ under } H_3 \\ > \int_0^1 [1 - G(y)]dG(y) = \dfrac{1}{2} \text{ under } H_4. \end{cases}$$

Under H_o, S, the number of positive signs follows a binomial distribution with $P(S = r|H_o) = \binom{n}{r}\left(\frac{1}{2}\right)^n$. Under $H_3(H_4)$, S will be small (large). Hence for testing H_o against H_3, critical region is $S \le S_0$ where $P\{S \le S_0|H_o\} = \alpha$ and for testing against H_4, critical region is $S \ge S_2$ where $P\{S \ge S_2|H_o\} = \alpha$. For large n, S follows under H_o, a normal distribution with mean $n/2$ and variance $n/4$ so that $\tau = (S - (n/2))/\sqrt{\frac{n}{4}} \sim N(0, 1)$. This statistic may therefore be used for testing H_o in large sample.

Consistency It is known that if t_n is a consistent estimator of $g(\theta)$, then the test based on t_n is consistent for testing any hypothesis regarding $g(\theta)$ [theorem (18.10.1)]. In the case of sign test statistic, $t_n = S/n =$ proportion of positive signs of $(x - y)$ in the sample. Hence $E(t_n) =$ probability of getting a positive sign of $(X - Y)$ in the population. Thus

$$\pi = E(t_n) = \begin{cases} p\{X > Y\} = \frac{1}{2} & \text{under } H_o \\ p_1 < \frac{1}{2} & \text{under } H_1 \\ p_2 > \frac{1}{2} & \text{under } H_2. \end{cases}$$

Again

$$V(t_n) = \frac{\pi(1 - \pi)}{n} < \frac{1}{4n} \to 0 \text{ as } n \to \infty.$$

Hence t_n is a consistent estimator of π. Therefore, test based on t_n is consistent for the hypothesis regarding π.

Sign test is uniformly most powerful for one-sided alternative.

22.5.2 Wald-Wolfwitz Run Test

Let X and Y be two independent random variables with absolutely continuous c.d.f.'s $F(x)$ and $G(y)$ respectively. We are interested in testing $H_\circ : (F(x) = G(x)\ \forall\, x)$ against the alternatives H, H_1 or H_2.

Let $(X_1, \cdots, X_m)$ be a random sample drawn from F and $(Y_1, \cdots, Y_n)$ be a random sample drawn from G. Consider the pooled ordered sequence of x's and y's and let it be $\boldsymbol{Z} = (Z_{(1)}, \cdots, Z_{(n+m)})$ where $Z_{(1)} < Z_{(2)} < \cdots < Z_{(n+m)}$ and each $Z_{(i)}$ is a either x or y. Replacing each x by a '0' and each y by a '1', one gets a sequence of m 0's and n 1's in $\boldsymbol{Z}$. Under $H_\circ$, all the possible $\frac{(m+n)!}{m!n!}$ sequences's of 0's and 1's have equal probability.

Consider U = total number of runs (of 0's and 1's) in $\boldsymbol{Z}$. A run is a sequence of identical letters preceded and followed by a different letter or by no letter at all.

Let us now consider the probability of having U runs, under $H_\circ$. We have,

$$P_{H_\circ}(U = 2k) = \frac{2\binom{m-1}{k-1}\binom{n-1}{k-1}}{\binom{m+n}{m}} \tag{22.5.2}$$

$$P_{H_\circ}(U = 2k+1) = \frac{\binom{m-1}{k-1}\binom{n-1}{k} + \binom{m-1}{k}\binom{n-1}{k-1}}{\binom{m+n}{m}} \tag{22.5.3}$$

(vide Feller, Vol. 1 (1968), p.62). This is so because the event $U = 2k$ may happen iff we have k runs of 0's and k runs of 1's. In this case if the first run is that of 0's the last run must be that of 1's. Suppose the runs start with a 0. Now to get k runs of 0's, we are to partition m 0's into k groups none of which is empty. This can be done by placing $(k-1)$ bars into the $m-1$ intermediate positions between m 0's. Therefore, the total number of distinguishable arrangements for k sequences of 0's is $\binom{m-1}{k-1}$. Now each bar must be occupied by at least one 1. Again, at the last place there must be at least one 1. Hence the possible number of distinguishable arrangements of k sequences of 1 is the number of ways in which n balls can be filled up in k cells, none of which is empty. This number is again $\binom{n-1}{k-1}$. Hence the total number of ways of k runs of 0's and k runs of 1's, which start with a 0 and end with a 1 is $\binom{m-1}{k-1}\binom{n-1}{k-1}$. Similar is the situation for runs starting with a 1 and ending with a 0. Hence the expression (22.5.2) is obtained. Similarly (22.5.3) is derived.

If the two distributions are identified, the number of runs will be generally large. If, however, the null hypothesis is false, there will be a tendency to have a smaller number of runs. So the critical region will be $U \leq u_\alpha$ where $P\{U \leq u_\alpha | H_\circ\} = \alpha$. Tables of critical values of U based on (22.5.2) and (22.5.3) have been derived by Swed and Eisenhart. This is reproduced in the appendix table A.15.

It is found

$$E(U) = \frac{2mn}{m+n} + 1, \quad V(U) = \frac{2mn(2mn - m - n)}{(m+n)^2(m+n-1)}. \tag{22.5.4}$$

A large sample approximation to the null distribution of U may be made by assuming both m, n are large (≥ 10) and $(m+n) \to \infty$ such that both $m/(m+n)$ and $n/(m+n)$ remain constant.

The asymptotic null distribution of U is normal with mean and variance given by (22.5.4). In large sample, therefore, $\tau = (U - E(U))/\sigma(U)$ may be used for testing of hypotheses.

Since the c.d.f.'s have been assumed to be continuous no tie should occur. But in practice, there may be ties. Ties within the same sample should pose no problem, because the number of runs remain unaffected whatever be the arrangement. If there are ties among observations from X and Y, then one does not get a unique value of U. In that case we should break ties in all the possible ways and compute U for each way of breaking ties. If all these U's give the same result, then there is no problem. Otherwise, one may use the largest value of U among all these values, for a conservative test.

The test also takes into account the differences in dispersion of the two distributions. If the dispersion of one distribution is smaller than the other, then also the number of runs will be small.

This test is developed for testing $H_\circ$ against general alternatives H or H_1 or H_2 and is thus less powerful for testing against alternatives of specific types, say, H_3 or H_4. Empirical studies show that the power efficiency of the Wald-Wolfwitz run test is about 75% for sample sizes near 20 (Smith, 1953).

► **Example 22.5.1** *Consider two samples as follows:*

$$\boldsymbol{X} = (1, 5, 7, 9, 15, 17, 21, 23),$$
$$\boldsymbol{Y} = (2, 6, 10, 12, 18, 20, 26, 28, 32).$$

The pooled ordered observation is

$$(1, 2, 5, 6, 7, 9, 10, 12, 15, 17, 18, 20, 21, 23, 26, 28, 32);$$

represented as a sequence of 0's (X's) and 1's (Y's) this is

$$(01010011001100111).$$

Here 0 and 1 have 5 runs each. So the total number of runs is 10. From the table A.15, 5% level of significance for $(m = 8, n = 9)$ is 5. Any value of U lower then 5 leads to the rejection of $H_\circ$ at 5% level of significance. Therefore, the null hypothesis is accepted.

22.5.3 The Mann-Whitney-Wilcoxon Test

Suppose we have two independent samples $\boldsymbol{X} = (X_1, \cdots, X_m)$ and $\boldsymbol{Y} = (Y_1, \cdots, Y_n)$ from two populations with continuous c.d.f.'s $F(x)$ and $G(y)$ respectively. We want to test the hypothesis $H_\circ(F(x) = G(x) \;\; \forall\, x)$ against the usual alternatives H or H_1 or H_2. However, this test is developed for testing $H_\circ$ against more specific alternatives, H_3 or H_4.

The $(m+n)$ observations are arranged according to the increasing order of magnitude and let $\boldsymbol{Z} = (Z_1, \cdots, Z_{m+n})$, $Z_1 < \cdots < Z_{m+n}$ be the combined ordered observations. To test H_o Wilcoxon proposed the statistic $T = \sum_{i=1}^{n} r_i$ where r_i is the rank of Y_i in $\boldsymbol{Z}$. Clearly if all the Y's are on the right side of X's i.e., if X is stochastically smaller than Y, than T will be large. On the other hand, if all the Y's are on the left side of X's i.e., if X is stochastically larger than Y, than T will be small. Thus, critical value of T for testing H_o against H_1 is $T \geq T_0$ where T_0 is such that $P\{T \geq T_0 | H_o\} = \alpha$. For testing H_o against $H_2 : (F(x) \leq G(x) \;\forall\, x$, with strict inequality for some $x)$ the critical region is $T \leq T_0'$ where $P\{T \leq T_0' | H_o\} = \alpha$.

We shall, however, consider Mann-Whitney modification of the test. This is based on U, the number of times Y precedes X. Then number of X's lying on the right hand side of Y_i is $m + n - r_i - (n - i) = m - r_i + i$. Hence

$$U = \sum_{i=1}^{n} (m - r_i + i) = mn + \frac{n(n+1)}{2} - T$$

where T is Wilcoxon's statistic.

Alternatively, for each pair (X_i, Y_j) define a indicator variable D_{ij} as follows:

$$D_{ij} = \begin{cases} 1, \text{ if } Y_j < X_i \\ & i = 1, \cdots, m; j = 1, \cdots, n. \\ 0, \text{ otherwise.} \end{cases} \tag{22.5.5}$$

Then the Mann-Whitney U-statistics is

$$U = \sum_{i=1}^{m} \sum_{j=1}^{n} D_{ij}. \tag{22.5.6}$$

Clearly, $U = 0$ if all the Y_j's are larger than X_i's and $U = mn$ if all the Y_j's are smaller than X_i's. Thus $0 \leq U \leq mn$. If U is large, the values of X tend to be larger than the values of Y (X is stochastically larger than Y) and this supports the alternative H_2. Similarly, if U is small, the X-values tend to be smaller than Y values and this supports the alternative H_1. These results may be summarised as follows:

$H_o : F = G$	$H_1 : F \geq G$	Critical region :	$U \leq C_2$
$H_o : F = G$	$H_2 : F \leq G$	"	$U \geq C_1$
$H_o : F = G$	$H \;: F \neq G$	"	$U \geq C_3$ or $\leq C_4$.

▶ **Example 22.5.2** *Let $m = 5$, $n = 6$ and the combined sample when ordered is as follows:*

$$x_3 < y_4 < x_1 < y_2 < x_4 < y_3 < y_5 < x_5 < y_6 < x_2 < y_1.$$

Then $T = 39$ and $U = 12$.

To compute the critical values we need the null distribution of U.

Distribution of U

Consider the ordered sequence of m x's and n y's. Since it is the ordering of x's and y's that matter, replace each x by a '0' and each y by a '1'. Thus, we get a sequence of $(m+n)$ elements with zeroes and one's. By definition, U counts the number of times 1's precede 0's.

Let $N_{m,n}(u)$ denote the number of sequences of m 0's and n 1's in which 1's precede 0's u times. Now, since the last term of the sequence may be a zero or one, we have the following recursion relation

$$N_{m,n}(u) = N_{m,n-1}(u) + N_{m-1,n}(u-n). \tag{22.5.7}$$

Under the null hypothesis, each of $\binom{m+n}{m}$ sequences of m 0's and n 1's are equally likely. Let $p_{m,n}(u)$ denote the probability that in a sequence of m 0's and n 1's, 1's will precede 0's u times under H_o. Thus, $p_{m,n}(u) = N_{m,n}(u)/\binom{m+n}{m}$. Hence, dividing (22.5.7) by $\binom{m+n}{m}$ one gets the recursion relation

$$p_{m,n}(u) = \frac{m}{m+n} p_{m-1,n}(u-n) + \frac{n}{m+n} p_{m,n-1}(u). \tag{22.5.8}$$

This relation together with the boundary conditions

$$\begin{aligned} & p_{i,j}(u) = 0, u < 0 \\ & p_{i,j}(o) = \frac{1}{2}, \qquad p_{j,u} = \frac{1}{2} \\ & p_{i,o}(u) = 0, u > 0; \quad p_{o,j}(u) = 1 \quad \text{for } u > 0 \end{aligned} \tag{22.5.9}$$

determine the distribution completely. We shall, however, consider only mean and variance of U.

$$E(U) = \sum_{i=1}^{m}\sum_{j=1}^{n} E(D_{ij}) = mnP\{X_i > Y_j\} = \frac{mn}{2} \text{ under } H_o, \tag{22.5.10}$$

$$\begin{aligned} E(U^2) = E\Bigg[& \sum\sum D_{ij}^2 + \sum_i \sum\sum_{j\neq j'} D_{ij}D_{ij'} + \sum\sum_{i\neq i'} \sum_j D_{ij}D_{i'j} \\ & + \sum\sum_{i\neq i'} \sum\sum_{j\neq j'} D_{ij}D_{i'j'} \Bigg]. \end{aligned} \tag{22.5.11}$$

Now,

$$E(D_{ij}^2) = P\{X_i > Y_j\} = \frac{1}{2} \text{ under } H_o.$$

$$\begin{aligned} E(D_{ij}D_{ij'}) &= P\{(X_i > Y_j) \cap (X_i > Y_{j'})\} = P\{Y_j < X_i) \cap (Y_{j'} < X_i)\} \\ &= \int_0^1 [G(x_i)]^2 dF(x_i) = \frac{1}{3} \text{ under } H_o. \end{aligned}$$

Similarly,

$$E(D_{ij}D_{i'j}) = \frac{1}{3}; \quad E(D_{ij}D_{i'j'}) = \frac{1}{4}.$$

Hence, from (22.5.11),

$$E(U^2) = \frac{mn}{2} + \frac{mn(n-1)}{3} + \frac{mn(m-1)}{3} + \frac{m(m-1)n(n-1)}{4}$$
$$V(U) = \frac{mn(m+n+1)}{12}.$$

The probability distribution of U is symmetric about $E(U) = mn/2$ in the null-case. Small-sample tables for U-statistic (for $m \le n$) have been given by Auble (1953), Mann and Whitney (1947). Mann found that β_1 and β_2 of the U-distribution tend respectively to 0 and 3 under H_o, even when $m = n = 8$. A generalisation of the Central Limit Theorem for dependent variables gives the asymptotic distribution of U (Note that D_{ij}'s are not independent variables). For large-samples

$$\tau = \{U - (mn/2)\}/\sqrt{\frac{mn(m+n-1)}{12}} \sim N(0,1).$$

Hence, for large samples, the test is performed by considering the distribution of τ.

Consistency of the test

Let us denote $P(D_{ij} = 1) = \pi$. Under H_o, $\pi = \frac{1}{2}$. Under H_1,

$$\pi = P(Y < X) = \int_0^1 G(x)dF(x) < \int_0^1 F(x)dF(x) = \frac{1}{2}.$$

Similarly, under H_2, $\pi > \frac{1}{2}$. The alternative hypothesis, in terms of π are : $H_1\left(\pi < \frac{1}{2}\right)$, $H_2\left(H > \frac{1}{2}\right)$.

Now

$$E(D_{ij}D_{ij'}) = \int_0^1 [G(x)]^2 dF(x) = \pi_1$$
$$E(D_{ij}D_{i'j}) = P\{(Y_j < X_i) \cap (Y_j < X_{i'})\}$$
$$= P\{(X_i > Y_j) \cap (X_{i'} > Y_j)\} = \int_0^1 \{1 - F(y)\}^2 dG(y) = \pi_2 \text{ (say)}.$$

Then, from (22.5.10)

$$E(U) = mn\pi;$$

from (22.5.11)

$$V(U) = mn[\pi - \pi^2(m+n-1) + (n-1)\pi_1 + (m-1)\pi_2].$$

Thus, $E(U/mn) = \pi$ and $V(U/mn) \to 0$ as $m, n \to \infty$. Thus, $U/(mn)$ is a consistent estimator of π and hence by theorem (18.13.1), U-test is consistent for alternatives H_1 and H_2.

Mood (1954) has shown that if U-test is applied to data for which t-test is the most powerful test, the power-efficiency of the test approaches $(3/\pi)\% = 95.5\%$ as $N(= m+n)$ increases and is close to 95% even for moderate N. It is, therefore, an excellent alternative to t-test.

► **Example 22.5.3** *Consider the data in example* 22.5.1.

Here $m = 8$, $n = 9$, $U = 5$. Under H_o, $E(U) = 36$, $V(U) = 108$. Hence $\tau = \frac{25-36}{\sqrt{108}} = -1.0585$. Since $|\tau| < 1.65$ we accept the null hypothesis that the two samples come from the some population at level $\alpha = 0.10$.

22.5.4 Median Test

Suppose we have two independent distributions $F(x)$and $G(y)$ and we want to test the hypothesis that the two distributions have the same median. Therefore, if the null hypothesis is formulated as $H_o : F(x) = G(x) \quad \forall\ x$, then the test will accept H_o if the medians of the distributions are equal even if they differ in shape. The alternative hypothesis may be that the two medians are unequal or one median is higher than the other.

The ordered sample $x_{(1)} < x_{(2)} < \cdots < x_{(n_1)}$ from $F(x)$ and $y_{(1)} < y_{(2)} < \cdots < y_{(n_2)}$ from $G(y)$ are combined together. Let $z_{(1)} < \cdots < z_{(n_1+n_2)}$ be the combined ordered sample. If $(n_1 + n_2)$ is odd, the median will be the $(n_1 + n_2 + 1)/2$th order statistic. If $n_1 + n_2$ is even, the median will be the mean of the two middle values. Let $z_{(m)}$ be the median of the z-series and n_{11}, n_{21} be the number of x's and y's respectively that exceed $z_{(m)}$. We shall find the distribution of n_{11} and n_{21}.

Observations	x	y
$> z_m$	n_{11}	n_{21}
$\leq z_m$	n_{12}	n_{22}
	n_1	n_2

Now z_m may belongs to series of x's or series of y's. When z_m belongs to the series of x's, the probability of getting n_{11} x's and n_{21} y's that exceed z_m is

$$\frac{n_1!}{(n_{12}-1)!n_{11}!}[F(z_m)]^{n_{12}-1}[1 - F(z_m)]^{n_{11}}dF(z_m)$$
$$\times \frac{n_2!}{n_{22}!n_{21}!}[F(z_m)]^{n_{22}}[1 - F(z_m)]^{n_{21}} = A \text{ (say)}. \tag{22.5.12}$$

Similarly, we can find the probability of getting n_{11} x's and n_{21} y's that exceed z_m when z_m belongs to series of y's.

Hence the joint distribution of Z_m, n_{11}, n_{21} is given by

$$dF(Z_m, n_{11}, n_{21}) = A + \frac{n_1!}{n_{12}!n_{11}!}[F(z_m)]^{n_{12}}[1-F(z_n)]^{n_{11}} \times \frac{n_2!}{(n_{22}-1)!n_{21}!}[F(z_m)]^{n_{22}-1} [1-F(z_m)]^{n_{21}}dF(z_m). \quad (22.5.13)$$

Integrating over $F(z_m)$ between 0 and 1 one gets the joint distribution of n_{11} and n_{21}. (Note that $F(z_m)$ may not be near 1/2, but may be any quantity in (0,1)). This comes out as

$$P(n_{11}, n_{21}|n_1, n_2) = \frac{\binom{n_1}{n_{11}}\binom{n_2}{n_{21}}}{\binom{n_1+n_2}{n_{11}+n_{21}}}, \quad \max(0, n_{11}+n_{21}-n_2) < n_{11} \le \min(N, n_{11}+n_{21}) \quad (22.5.14)$$

where $N = n_1 + n_2$.

In fact the result (22.5.14) is true if instead of Z_m we take $Z_{(r)}$, the rth order statistic in Z series ($r = 1, 2, \cdots$). Note that $n_{11}+n_{21}$ is a fixed quantity. For $Z_{(r)}, n_{11}+n_{21}$ is n_1+n_2-r. The expression (22.5.14) shows that the conditional distribution of n_{11} is hypergeometric with parameters n_1, n_2 and $n_{11}+n_{21}$. Hence

$$E(n_{11}) = \frac{n_1(n_{11}+n_{21})}{N}$$

$$V(n_{11}) = \frac{N-(n_{11}+n_{21})}{(N-1)} \times (n_{11}+n_{21})n_1\frac{n_2}{N^2}.$$

If $Z_{(r)}$ is Z_m, the sample median,

$$E(n_{11}) = \begin{cases} \frac{n_1}{2} - \frac{n_1}{2N}, & \text{if } N \text{ is odd} \\ \frac{n_1}{2}, & \text{if } N \text{ is even} \end{cases}$$

$$V(n_{11}) = \begin{cases} \frac{N+1}{4N^2}n_1n_2, & \text{if } N \text{ is odd} \\ \frac{n_1n_2}{4N}, & \text{if } N \text{ is even.} \end{cases}$$

Therefore, under H_o, for large sample size N, the quantity

$$\tau = \left(n_{11} - \frac{n_1}{2}\right)\Bigg/\sqrt{\frac{n_1n_2}{4N}} \sim N(0,1).$$

Hence to test H_o against the two-sided alternative H' (the medians are unequal) when N is large, one would reject H_o in favour of H' at α% level of significance if $|\tau| > \tau_{\alpha/2}$.

If again n_{11} is very small, it means that the actual median of X is smaller than that of Y. This means then X is stochastically smaller than Y. One, therefore, rejects H_o in favour of $H_1' : \mu_e(x) < \mu_e(y)$ if $\tau < -\tau_\alpha$. Similarly, for the other hypothesis.

Under H_o, one expects $\frac{n_1}{2}$-observations below the median and $\frac{n_1}{2}$ observations above the median. One can, therefore, use the χ^2 test of significance with 1 d.f. for testing H_o against the both-sided alternative. The test can be easily generalised to test for H_o'' : pth order percentile of the two distributions are equal.

If the total sample size, $N(= n_1 + n_2)$ is small, one may find the exact probability (22.5.14) for testing H_o.

Mood has shown that for large sample size $(n_1 + n_2)$, the asymptotic power efficiency of the test is $(2/\pi)\% = 63\%$.

► **Example 22.5.4** *For the data of example 22.5.5 the combined sample in the ordered form is as follows:*

70, 80, 80, 90, 100, 100, 110, 115, 120, 120, 125, 130, 130, 140, 160

The median is the 8th order value, namely, 115. There are two values of x which exceed the median. Here $n_{11} + n_{21} = 7$.

$$P\{n_{11} \le 2|H_o\} = \frac{\binom{7}{0}\binom{8}{7}}{\binom{15}{7}} + \frac{\binom{7}{1}\binom{8}{6}}{\binom{15}{7}} + \frac{\binom{7}{2}\binom{8}{5}}{\binom{15}{7}} = 0.397$$

Since $P_{H_o}\{n_{11} \le 2\} > 0.10$, we accept H_o that two samples come from the same population.

If we consider large sample test, $\tau = \left(2 - \frac{7}{2}\right) / \sqrt{\frac{7\times 8}{60}} = -1.55$; $|\tau| < 1.645$. The null hypothesis is accepted at 10% level of significance.

Considering the χ^2-test of independence between X and Y, which also covers the case that X and Y are identically distributed, the observed and the expected frequencies (in parentheses) are given in table 22.1.

Table 22.1

	no. of x's	no. of y's	Total
$> Z_m$	2 (3.3)	5 (3.7)	7
$\le Z_m$	5 (3.7)	3 (4.3)	8
Total	7	8	15

The observed value of $\chi^2_{(1)} = 1.82$. H_o is accepted at 10% level.

Under H_o, one would expect $\frac{n_1}{2}$ values of x below Z_m and $\frac{n_1}{2}$ values of x above Z_m. Similar is the case for y. (In general, if Z_p be the pth percentile for the Z series, then under H_o, one would expect $[n_1 p]$ observations of x below Z_p and $n_1 - [n_1 p]$ observations of x above Z_p and similarly for y). Calculating the expected frequencies in this way, one gets $\chi^2_{(1)} = 1.28$. H_o is again accepted.

22.5.5 Thompson's Test

This is a test for location of two distributions. If the alternative hypothesis H_1 stands for shift of location towards right (i.e., X is stochastically smaller than Y) then any percentile of X would be lower than the corresponding percentile of Y. Similarly, if H_2 stands for shift of location towards left (X is stochastically larger than Y) then the reverse situation would arise. Thus for alternative hypotheses concerning location of distributions a test based on percentiles can be made.

Here we take rth order statistic of $x, X_{(r)}$ and Sth order statistic of $y, Y_{(S)}$ and compare them. The critical region for the alternative H_1 is given by $x_{(r)} < y_{(S)}$ when $P\{X_{(r)} < Y_{(S)}|H_\circ\} = \alpha$ and r, S are to be suitably chosen.

As usual, let $(X_1, \cdots, X_{n_1})$ be a random sample from $F(x)$ and $(Y_1, \cdots, Y_{n_2})$ a random sample from $G(y)$. The null hypothesis is $F(x) = G(x)\ \forall\ x$. Now, under $H_\circ$

$$P_r\{Y_{(S)} = y_{(S)}\} = \frac{n_2!}{(S-1)!(n_2-S)!}\left[F(y_{(S)})\right]^{S-1}\left[1-F(y_{(S)})\right]^{n_2-S} dF(y_{(S)}).$$

Hence,

$$\begin{aligned} P\{X_{(r)} < Y_{(S)}|H_\circ\} &= \int_0^1 \sum_{i=r}^{n_1} \frac{n_1!}{i!(n_1-i)!}[F(y_{(S)})]^i[1-F(y_{(S)})]^{n_1-i} \\ &\quad \times \frac{n_2!}{(S-1)!(n_2-S)!}[F(y_{(S)})]^{S-1}[1-F(y_{(S)})]^{n_2-S}dF(y_{(S)}) \\ &= \sum_{i=r}^{n_1} \frac{n_1!n_2!}{(S-1)!(n_2-S)!i!(n_1-i)!} \\ &\quad \int_0^1 [F(y_{(S)})]^{i+S-1}[1-F(y_{(S)})]^{n_1+n_2-i-S}dF(y_{(S)}) \\ &= \sum_{i=r}^{n_1} \frac{n_1!n_2!}{(S-1)!(n_2-S)!i!(n_1-i)!}B(S+i, n_1+n_2-i-S+1) \\ &= \psi(r, S, n_1, n_2). \end{aligned}$$

Thus for given n_1 and n_2, r and S can be chosen suitably to find size α critical region for H_1.

For H_2, critical region is $x_{(r)} < y_{(S)}$ such that

$$P\{X_{(r)} > Y_{(S)}|H_\circ\} = \alpha.$$

For shift in both sides we generally divide the critical region in two equal parts by choosing the pairs (r_1, S_1) and (r_2, S_2) suitably. The above test is, however, not consistent for other types of changes, e.g., changes in dispersion, etc.

22.5.6 Dixon's Test

It is required to test the hypothesis that the two samples have come from the same continuous population. Let $\boldsymbol{X} = (X_1, \cdots, X_m)$ be a random sample from $F(x)$ and

$\boldsymbol{Y} = (Y_1, \cdots, Y_n)$ a random sample from $G(y)$. Without any loss of generality assume that $n \leq m$. Let $y_{(1)} < \cdots < y_{(n)}$. Let m_1 be the number of x-values less then $y_{(1)}$, m_{i+1} the number of x-values between $y_{(i)}$ and $y_{(i+1)}$, $(i = 1, 2, \cdots, n-1)$ and m_{n+1} be the number of values greater than $y_{(n)}$. The observed proportion of x-values between any two successive observations $y_{(i)}$ and $y_{(i+1)}$ is m_{i+1}/m.

From a fundamental property of order statistics (theorem (11.6.1)), it follows that under $H_\circ$, $(F(x) = G(x)\ \forall\ x)$, the expected proportion of sample observations between any two successive order statistics $y_{(i)}$ and $y_{(i+1)}$ $(i = 1, 2, \cdots)$ is $\frac{1}{n+1}$. Hence, the test can be carried out by considering the deviation between the observed frequencies and the expected frequencies. A test statistic is $C^2 = \sum_{i=1}^{n+1} \left(\frac{1}{n+1} - \frac{m_i}{m}\right)^2$ which is distributed as $\chi^2_{(n)}$ in large sample. The critical region is $C^2 \geq C^2_\alpha$ where $P\{C^2 \geq C^2_\alpha | H_\circ\} \leq \alpha$. Tables showing values of C^2_α for $\alpha = 0.05, 0.10$ and $n, m = 2, 3, 4, \cdots, 10$ have been prepared by Dixon by approximating the distribution of nkC^2 by $\chi^2_{(\nu)}$ where k, ν involve m, n.
It is found that

$$E(C^2) = \frac{n(n+m+1)}{m(n+1)(n+2)}$$

$$V(C^2) = \frac{4n(m-1)(m+n+1)(m+n+2)}{m^3(n+2)^2(n+3)(n+4)}.$$

22.5.7 Kolmogorov-Smirnov Two-sample Test

In one-sample case, the KS-statistic was used to test the goodness of fit. In the two-sample case, one may test the hypothesis that the two distributions are equal by considering the two empirical d.f.'s.

Let $X_1, \cdots, X_m$ and $Y_1, \cdots, Y_n$ be independent random samples from continuous c.d.f.'s F and G, respectively. Let the order statistics be $X_{(1)}, \cdots, X_{(m)}$ and $Y_{(1)}, \cdots, Y_{(n)}$ respectively. The empirical d.f. of X is

$$F_n(x) = \begin{cases} 0, & x < X_{(1)} \\ \frac{i}{m}, & X_{(i)} \leq x < X_{(i+1)} \\ 1, & x \geq X_{(m)} \end{cases}$$

Similarly, the empirical d.f. of Y is

$$G_n(x) = \begin{cases} 0, & x < Y_{(1)} \\ \frac{i}{n}, & Y_{(i)} \leq x < Y_{(i+1)} \\ 1, & x \geq Y_{(n)}. \end{cases}$$

These are, respectively, estimates of $F(x)$ and $G(x)$. Under $H_\circ : F(x) = G(x)\ \forall\ x$, one expects that the two sample d.f.'s will reasonably agree. The statistic

$$D_{m,n} = \sup_x |F_m(x) - G_n(x)|$$

is called the two-sided two-sample KS-statistic and is used for testing H_o against $H : F(x) \neq G(x)$ for some x. The test rejects H_o in favour of H at level α, if $D_{m,n} \geq D_{m,n;\alpha}$ where $P\{D_{m,n} \geq D_{m,n;\alpha}|H_o\} \leq \alpha$.

Similarly, for testing H_o against one-sided alternative $H^+ : F(x) \geq G(x) \ \forall \ x$, with strict inequality holding for some x, one-sided KS-statistics is

$$D^+_{m,n} = \sup_x [F_m(x) - G_n(x)].$$

For testing H_o against $H^- : F(x) \leq G(x) \ \forall \ x$, with strict inequality for some x, one-sided KS-statistics is

$$D^-_{m,n} = \sup_x [G_n(x) - F_m(x)].$$

H_o is rejected in favour of $H^+(H^-)$ if $D^+_{m,n}$ $(D^-_{m,n}) \geq D^+_{m,n;\alpha}(D^-_{m,n;\alpha})$. As in the case of single sample, the statistics are distribution-free if the c.d.f.'s are continuous.

Tables have been given by Massey (1952) for small samples. Values of $D_{m,n;\alpha}$ and $D^+_{m,n;\alpha}$ for selected values for m, n and α and for some values of $m = n$ and α have been given at tables A.11 and A.12.

For the asymptotic null distribution we have the following results similar to Theorems 21.4.3 and 21.4.5.

$$\text{(i)} \lim_{m,n\to\infty} P\left[\sqrt{\frac{mn}{m+n}} D_{m,n} \leq Z\right] = \begin{cases} 1 - 2\sum_{i=1}^{\infty}(-1)^{i-1}\exp[-2i^2Z^2], & Z > 0 \\ 0, & Z \leq 0. \end{cases}$$

$$\text{(ii)} \lim_{m,n\to\infty} P\left[\sqrt{\frac{mn}{m+n}} D^+_{m,n} \leq Z\right] = \lim_{m,n\to\infty} P\left[\sqrt{\frac{mn}{m+n}} D^-_{m,n} \leq Z\right]$$

$$= \begin{cases} 1 - \exp[-2Z^2], & Z > 0 \\ 0, & Z \leq 0. \end{cases}$$

▶ **Exampie 22.5.5** *The following data give the life-time of bulbs of two different brands.*

Brand 1 : 80, 100, 90, 110, 125, 130, 70;

Brand 2 : 100, 120, 80, 140, 130, 160, 115, 120.

Test whether the brands differ with respect to average life.

We apply KS-test to test H_o the two samples come from the same population.

x	$F_7(x)$	$G_8(x)$	$\lvert F_7(x) - G_8(x)\rvert$
70	1/7	0	1/7
80	2/7	1/8	9/56
90	3/7	1/8	17/56
100	4/7	2/8	9/28
110	5/7	2/8	13/28
115	5/7	3/8	19/56
120	5/7	5/8	5/56
125	6/7	5/8	13/56
130	1	6/8	1/4
140	1	7/8	1/8
160	1	1	0

Here

$$\sup_x |F_7(x) - G_8(x)| = \frac{13}{28}.$$

For two-sided test,

$$D_{m,n,0.10} = \frac{33}{56} > \frac{13}{28}.$$

Hence H_o is accepted at 10% level of significance.

If we consider, paramatric t test, $\bar{x} = 100.7, \bar{y} = 120.6$, $S_x^2 = 503.6, S_y^2 = 588.8$, $(m-1)S_x^2 + (n-1)S_y^2/(m+n-2) = 549.5$. Thus $|t_{14}| = 1.640 < 1.761$. Hence H_o is accepted at 10% level of significance.

22.5.8 Kendall's tau Statistic

Let $(X_1, Y_1), \cdots, (X_n, Y_n)$ be a random sample from a continuous bivariate population $F_{X,Y}(x, y)$. We want to test the null hypothesis $H_o(X, Y$ are independent) against the alternative hypothesis $H(X, Y$ are not independent). For this Kendall proposed the following test.

Definition 22.5.1 *The pairs (X_i, Y_i) and (X_j, Y_j) are said to be in concordance (agreement) if whenever $X_i > (<) X_j$, $Y_i > (<) Y_j$. The pairs (X_i, Y_i) and (X_j, Y_j) are said to be in discordance (disagreement) if whenever $X_i < (>) X_j$, $Y_i > (<) Y_j$.*

Let π_c be the probability that a randomly chosen pair (X_i, Y_i) and (X_j, Y_j) be in concordance. Similarly, let π_d be the probability that a random pair be in discordance. Then

$$\pi_c = P\{(X_j - X_i)(Y_j - Y_i) > 0\},$$
$$\pi_d = P\{(X_j - X_i)(Y_j - Y_i) < 0\}.$$

If the marginal distributions of X and Y are absolutely continuous so that no two x's y's) are equal,

$$\begin{aligned}\pi_c &= [P(X_j > X_i) - P\{(X_j > X_i) \cap (Y_j < Y_i)\}] \\ &\quad + [P(X_j < X_i) - P\{(X_j < X_i) \cap (Y_j > Y_i)\}] \\ &= 1 - \pi_d.\end{aligned}$$

Definition 22.5.2 *The measure of association between X and Y, defined by $\pi_c - \pi_d =$ is called Kendall's τ.*

the marginal distributions of X and Y are continuous, we can write

$$\tau = 1 - 2\pi_d = 2\pi_c - 1.$$

nder H_o,

$$\begin{aligned}\pi_c &= P\{X_i < X_j\}P\{Y_i < Y_j\} + P\{X_i > X_j\}P\{Y_i < Y_j\} \\ &= P\{X_i > X_j\}P\{Y_i < Y_j\} + P\{X_i < X_j\}P\{Y_i < Y_j\} \\ &= \pi_d,\end{aligned}$$

nce, $P(X_i < X_j) = P(X_i > X_j) = \frac{1}{2}$ (because of continuity). Therefore, under $H_o, \tau = 0$.

Note, however, that $\tau = 0$ does not necessary mean independence between X and Y. owever, if (X, Y) follows bivariate normal distribution then $\tau = 0$ iff $\rho(X, Y) = 0$ and this case $\rho(X, Y) = 0$ iff X, Y are independent.

In order to test for τ, we need some statistic which is an estimate of τ. For each pair $, j)$ $(i, j = 1, \cdots, n)$, define a statistic

$$t_{ij} = \begin{cases} 1, & \text{if } (X_i, Y_i), (X_j, Y_j) \text{ are in concordance} \\ -1, & \text{if } (X_i, Y_i), (X_j, Y_j) \text{ are in discordance} \\ 0, & \text{otherwise.} \end{cases}$$

Then $P\{t_{ij} = 1\} = \pi_c$, $P\{t_{ij} = -1\} = \pi_d$, $P\{t_{ij} = 0\} = 1 - \pi_c - \pi_d = 0$ if X and Y re absolutely continuous random variables.

Hence $E(t_{ij}) = \pi_c - \pi_d = \tau$. Clearly $t_{ij} = t_{ji}$ and $t_{ii} = 0$. Therefore, the statistic

$$T = \frac{1}{\binom{n}{2}} \sum\sum_{i<j} t_{ij} = \frac{S}{\binom{n}{2}}$$

a unbiased estimate of τ. Clearly, $S = P - N$ when $P(N)$ is the number of positive egative) values of t_{ij} $(i < j = 1, \cdots, n)$. If there are no ties, $P + N = \binom{n}{2}$ and

$$T = \frac{2P}{\binom{n}{2}} - 1 = 1 - \frac{2N}{\binom{n}{2}}$$

Clearly, $-1 \leq T \leq 1$. To test $H_o(X, Y$ are independent) we reject H_o if $|T|$ is larg Again, under $H_o, E(T) = 0$ and T is symmetric about 0. Therefore, H_o is rejected $\alpha\%$ level of significance if $|T| > T_{\alpha/2}$ where $T_{\alpha/2}$ is such that $P_{H_o}\{|T| > T_{\alpha/2}\} = \alpha$. F small values of n, exact probability distribution of T can be easily calculated. Tabl have been given by Kendall (1962) for $4 \leq n \leq 10$. Table A.20 gives the values of S where $P_{H_o}\{S > S_\alpha\} = \alpha$ for different values of n and α.

For large $n(\geq 10)$, it can be shown that T is asymptotically normally distribut about mean 0 and variance $2(2n+5)/[9n(n-1)]$ so that

$$3T\sqrt{n(n-1)}/\sqrt{2(2n+5)} \sim N(0,1)$$

and this may be used for testing the hypotheses. It is obvious that the statistic T invariant under any order-preserving transformations of values of X and values of Y One may, therefore, arrange the X-values in ascending order of magnitude and place th corresponding Y-values. Then P is the number of pairs (i, j) $(j > i)$ for which $Y_{(j)} > Y$ where $X_{(1)} > \cdots > X_{(n)}$.

► **Example 22.5.6** *Scores of students in mathematics and statistics are given below. Te whether the hypothesis that one's ability in mathematics is independent of one's abili in statistics is tenable.*

Mathematics	:	46	52	60	61	67	70	81	88
Statistics	:	49	45	67	72	82	65	78	92

Here, $P = 23$; $\tau = \frac{46}{28} - 1 = 0.64285$; at 5% level of significance, $S_\alpha = 16$ for bo sided alternative. Since, observed value of $S = 18$, the null hypothesis is rejected.

► **Example 22.5.7** *For $n = 3$, the distribution of T under H_o is as follows:*

P	T	*Number of Permutations*	*Probability* $P\{T = t\|H_o\}$
0	-1.00	1:(3, 2, 1)	1/6
1	-0.33	2:(2, 3, 1), (3, 1, 2)	2/6
2	0.33	2:(1, 3, 2), (2, 1, 3)	2/6
3	1.00	1:(1, 2, 3)	1/6

Note that since the distribution of T is symmetric we could do away with the la two rows.

22.5.9 Rank Correlation Test

Suppose we have n individuals whose ranks according to a character A are $X_1, \cdots, X$ and whose ranks according to another character B are $Y_1, \cdots, Y_n$. Thus, if we have tudents, $X_1, \cdots, X_n (Y_1, \cdots, Y_n)$ may be the rankings given by the examiner $A(B)$. The wo sets of rankings X, Y are independent.

'he values of X's range from $1, \cdots, n$ and similarly the values of Y. Thus,

$$\sum X_i = \sum i = \frac{n(n+1)}{2}, \quad \bar{X} = \frac{n+1}{2} = \bar{Y}.$$

Let $x_i = X_i - \bar{X}$, $y_i = Y_i - \bar{Y}$. Then $\Sigma x_i = \Sigma y_i = 0$. The correlation coefficient between two sets of ranks X's and Y's is called the Spearman's rank correlation oefficient.

Iow,

$$\sum x_i^2 = \sum \left(X_i - \frac{n+1}{2} \right)^2 = \frac{n(n^2-1)}{12} = \sum y_i^2. \tag{22.5.15}$$

et

$$d_i = x_i - y_i \tag{22.5.16}$$

$$\sum d_i^2 = \sum (x_i - y_i)^2 = \sum x_i^2 + \sum y_i^2 - 2 \sum x_i y_i.$$

Hence,

$$\sum x_i y_i = \frac{1}{2} \left[\sum x_i^2 + \sum y_i^2 - \sum d_i^2 \right] = \frac{1}{2} \left[\frac{n(n^2-1)}{6} - \sum d_i^2 \right].$$

'hus,

$$r = 1 - \frac{6 \sum d_i^2}{n(n^2-1)}. \tag{22.5.17}$$

We shall show that r is symmetrically distributed about zero. We consider the istribution of r with $n!$ permutation sets of rankings (fixing X's and permuting Y's). Ve shall show that if we get for a certain permutation the value of r as r', then there xists an inverse permutation which gives the value $-r'$.

We take rank's of X's as $1, \cdots, n$ and of Y's as $Y_1, Y_2, \cdots, Y_n$ (which are also the alues $1, 2, \cdots, n$ in some order). Consider the inverse permutation $Y_n, Y_{n-1}, \cdots, Y_1$. uppose the rank correlation coefficient for $[(1, Y_1), \cdots, (n, Y_n)]$ and $[(1, Y_n), \cdots, (n, Y_1)]$ re r' and r_0 respectively. Then

$$r' + r_0 = 1 - \frac{6 \sum d_i'^2}{n(n^2-1)} + 1 - \frac{6 \sum d_{i0}^2}{n(n^2-1)}.$$

Iow,

$$\begin{aligned} \sum d_i'^2 + \sum d_{i0}^2 &= \sum_1^n (Y_i - i)^2 + \sum_1^n \{Y_i - (n-i+1)\}^2 \\ &= 2\left(\sum Y_i^2 + \sum i^2 \right) + (n+1)^2 n - 4(n+1) \sum i \\ &= \frac{n(n^2-1)}{3}. \end{aligned}$$

Hence,

$$r' + r_0 = 0.$$

Thus, $r_0 = -r'$.

Therefore, the distribution of r is symmetrical about zero and hence all odd orde moments vanish. Again,

$$V(r) = E(r^2) = E\left\{\frac{(\sum xy)^2}{(\sum x^2)^2}\right\}.$$

Now,

$$E\left(\sum_{i=1}^{n} x_i^2 y_i^2\right) = (n-1)!\left(\sum_{i=1}^{n} x_i^2 \sum_j x_{ji}^2\right) \Big/ n! = \left(\sum_{i=1}^{n} x_i^2\right)^2 \Big/ n.$$

Alternatively,

$$E\left(\sum_{i=1}^{n} x_i^2 y_i^2\right) = \sum_{i=1}^{n} E(x_i^2 y_i^2) = \sum_{i=1}^{n} E(x_i^2)E(y_i^2) \quad \text{(by independence)}$$

$$= n[E(x_i^2)]^2 \quad \text{(since } x_i\text{'s and } y\text{'s are identically distributed.)}$$

$$= n \cdot \left[\frac{1}{n}\sum x_i^2\right]^2 = \frac{1}{n}\left(\sum_{i=1}^{n} x_i^2\right)^2. \tag{22.5.18}$$

Similarly,

$$E\left(\sum\sum_{i\neq j=1}^{n} (x_i y_i)(x_j y_j)\right) = \sum\sum_{i\neq j=1}^{n} E(x_i x_j)E(y_i y_j) \quad \text{(since } x_i x_j \text{ is independent of } y_i y_j)$$

$$= n(n-1)E(x_i x_j)E(y_i y_j)$$

$$= n(n-1)\left[\frac{1}{n(n-1)}\sum\sum_{i\neq j} x_i x_j\right]^2$$

$$= \frac{1}{n(n-1)}\left[\sum\sum_{i\neq j} x_i x_j\right]^2$$

$$= \frac{1}{n(n-1)}\left(\sum x_i^2\right)^2$$

$$\left(\because \sum\sum_{i\neq j} x_i x_j = \left(\sum x_i\right)^2 - \sum x_i^2 = -\sum x_i^2\right).$$

Hence,

$$V(r) = \frac{1}{(\sum x^2)^2}\left[\frac{(\sum x_i^2)^2}{n} + \frac{\left(\sum\sum_{i \neq j} x_i x_j\right)^2}{n(n-1)}\right] \quad \left(\because \sum x^2 \text{ is a constant}\right)$$

$$= \frac{1}{(\sum x^2)^2}\left[\frac{(\sum x_i^2)^2}{n} + \frac{(\sum x^2)^2}{n(n-1)}\right] = \frac{1}{n-1}.$$

It can be shown that the standardised variate $\{r - E(r)\}/\sqrt{V(r)}$ is approximately distributed as a standard normal variate whatever be the form of the parent distributions of X, Y.

We now consider some asymptotic results.

Some Asymptotic Results

We are often concerned with the limiting distribution of a statistic

$$L_n = C_{1n}X_1 + C_{2n}X_2 + \cdots + C_{nn}X_n$$

where $\mathcal{C}_n = \{C_{1n}, \cdots, C_{nn}\}$ is a set of constants and $\boldsymbol{X} = (X_1, \cdots, X_n)$ is a random vector which takes on values which are permutations of a set of quantities $\mathcal{A}_n = \{a_{1n}, \cdots, a_{nn}\}$ each with probability $1/n!$. Clearly,

$$E(L_n) = \frac{1}{n}\sum_{i=1}^{n} C_{in} \sum_{j=1}^{n} a_{jn}$$

$$V(L_n) = \sigma^2(L_n) = \frac{1}{n-1}\sum_{i=1}^{n}(C_{in} - \bar{C}_n)^2 \sum_{j=1}^{n}(a_{jn} - \bar{a}_n)^2$$

where

$$\bar{C}_n = \Sigma C_{in}/n, \quad \bar{a}_n = \frac{1}{n}\Sigma a_{in}.$$

The following theorem gives conditions for the asymptotic normality of the standardised variable $L_n^0 = \{L_n - E(L_n)\}/\sigma(L_n)$.

Consider the following conditions proposed by Wald and Wolfwitz.

A set of real numbers $\mathcal{W}_n = (w_{1n}, \cdots, w_{nn})$ is said to satisfy Wald-Wolfwitz conditions W if for all $r = 3, 4, \cdots$

$$W: \frac{\frac{1}{n}\sum_{i=1}^{n}(w_{in} - \bar{w}_n)^r}{\left\{\frac{1}{n}\sum_{i=1}^{n}(w_{in} - \bar{w}_n)^2\right\}^{r/2}} = 0(1) \tag{22.5.19}$$

where

$$\bar{w}_n = \frac{1}{n}\sum_{i=1}^{n} w_{in}.$$

The condition (22.5.19) means that the rth order central moment of w standardised with respect to its variance is bounded as $n \to \infty$.

Another condition suggested by Noether is:

$$N : \frac{\frac{1}{n}\sum_{i=1}^{n}(w_{in} - \bar{w}_n)^r}{\left\{\sum_{i=1}^{n}(w_{in} - \bar{w}_n)^2\right\}^{r/2}} = 0(1). \tag{22.5.20}$$

Hoeffding has shown that the condition N for $\mathcal{W}_n$ is equivalent to either of the following conditions:

$$(N.1) : \lim_{n\to\infty} \frac{\sum_{1}^{n}|w_{in} - \bar{w}_n|^r}{\left[\sum_{1}^{n}(w_{in} - \bar{w}_n)^2\right]^{r/2}} = 0 \quad \text{for } r > 2, \tag{22.5.21}$$

$$(N.2) : \max_{1\le i\le n} \frac{(w_{in} - \bar{w}_n)^2}{\sum_{1}^{n}(w_{in} - \bar{w}_n)^2} \to 0. \tag{22.5.22}$$

Theorem 22.5.1 (***Wald-Wolfwitz***). *If $\mathcal{C}_n$ satisfies conditions W and $\mathcal{A}_n$ satisfies any of the conditions N or $(N.1)$ or $(N.2)$ then $L_n^0 \sim N(0,1)$ asymptotically.*

There are extensions of theorem 22.5.1 which give the joint limiting distribution of a number of L_n-statistics. We consider one such theorem.

Theorem 22.5.2 *If $\mathcal{A}_n$ satisfies condition N or $(N.1)$ or $(N.2)$ and $\mathcal{C}_n$, $\mathcal{D}_n = \{d_{1n}, \cdots, d_{nn}\}$ satisfy conditions W and if the correlation between $\mathcal{C}_n$, $\mathcal{D}_n$,*

$$\rho_n = \frac{\sum_{1}^{n}(C_{in} - \bar{C}_n)(d_{in} - \bar{d}_n)}{\left[\sum_{1}^{n}(C_{in} - \bar{C}_n)^2 \sum_{1}^{n}(d_{in} - \bar{d}_n)^2\right]^{1/2}}$$

has a limit ρ, then the limiting distribution of L_n^0 and $L_n'^0 = \{L_n' - E(L_n')\}/\sigma(L_n')$ where $L_n' = \sum_{j=1}^{n} d_{jn}X_{jn}$, is bivariate normal $(0, 0, 1, 1, \rho)$.

Rank Correlation Test of independence

We have seen that if X, Y are independent, $E(r|H_\circ) = 0$. Thus, we should reject $H_\circ$ if $|r|$ is large, that is, if $|r| > r_{0,\alpha}$ where $P\{|r| > r_{0,\alpha}\} = \alpha$. To compute $r_{0,\alpha}$ one needs the null distribution of r. Without loss of generality one may assume $X_i = i$. Under $H_\circ$, X and Y are independent and all the $n!$ rankings (i, Y_i) are equally likely. Therefore, $P\{|r| = r_0\} = n_{r_0}/n!$ where n_{r_0} is the number of permutations for which $|r| = r_0$. Again, since the distribution of r is symmetric, one need only compute for positive (negative) ranks.

The rank correlation coefficient can also be written as

$$r = \frac{12}{n^3 - n}\sum_{i=1}^{n}\left(X_i - \frac{n+1}{2}\right)\left(Y_i - \frac{n+1}{2}\right) \quad \left(\text{since } \sum_{1}^{n}(X_i - \bar{X})^2 = \frac{n(n^2-1)}{12}\right)$$

$$= \frac{12}{n^3 - n}\sum_{i=1}^{n} X_i Y_i - \frac{3(n+1)}{n-1}. \tag{22.5.23}$$

An equivalent statistic that may be used is, therefore,

$$\sum_{i=1}^{n} X_i Y_i \quad \text{or} \quad \sum_{i=1}^{n} iY_i$$

where Y_i is the rank of Y corresponding to rank i of X.

Applying theorem 21.5.1 we note that (here $\mathcal{C}_n = \{1, \cdots, n\} = \mathcal{A}_n$)

$$\frac{1}{n}\sum\left(i - \frac{n+1}{2}\right)^2 = \frac{n^2-1}{12}, \quad \frac{1}{n}\sum i^r = 0(n^r),$$

so that conditions of theorem 21.5.1 are satisfied. Hence the statistic

$$Z = \sum_i iY_i$$

has a limiting normal distribution with mean

$$E(Z) = \frac{n(n+1)^2}{4} \sim \frac{n^3}{4},$$

and variance,

$$V(Z) = \frac{n^2(n-1)^2}{144(n-1)} \sim \frac{n^5}{144}$$

and the test statistic $\frac{Z - E(Z)}{\sqrt{V(Z)}}$ can be used for testing $H_\circ$. The approximation is good even for $n = 10$. For small values of $n (\leq 10)$, exact distribution of r has been tabulated by Kendall. Table A.19 gives values of r_α for selected values of n and α.

▶ **Example 22.5.8** *We compute the null distribution of r for $n = 3$. As usual, $\boldsymbol{X}$ is taken as* $(1, 2, 3)$

$\boldsymbol{Y}$	ΣiY_i	r
$(1, 2, 3)$	14	1.0
$(1, 3, 2)$	13	0.5
$(2, 1, 3)$	13	0.5

Thus, under $H_\circ$,

$$P\{r = 1\} = \frac{1}{6}, \quad P\{r = 0.5\} = \frac{2}{6}, \quad P\{r = -1\} = \frac{1}{6}, \quad P\{r = -0.5\} = \frac{2}{6}.$$

For $n = 4$, the distribution of r is obtained as follows:

$\boldsymbol{Y}$	ΣiY_i	r	$P\{R = r\}$
$(1, 2, 3, 4)$	30	1.0	1/24
$(1, 3, 2, 4), (2, 1, 3, 4), (1, 2, 4, 3)$	29	0.8	3/24
$(2, 1, 4, 3)$	28	0.6	1/24
$(1, 3, 4, 2), (1, 4, 2, 3), (2, 3, 1, 4)$	27	0.4	4/24
$(1, 4, 3, 2), (3, 2, 1, 4)$	26	2.0	2/24
$(3, 1, 4, 2), (2, 1, 4, 3)$	25	0.0	2/24

▶ **Example 22.5.9** *The ranking as given by two examiners to a set of* 5 *students are as follows. Test if the two populations of ranks are independent.*

$$\boldsymbol{X} : 1\ 2\ 3\ 4\ 5$$
$$\boldsymbol{Y} : 2\ 5\ 1\ 3\ 4$$

Here $r = -0.8$.

From table A.19, $P_{H_\circ}\{|r| > 0.9\} \leq 0.05$, $P_{H_\circ}\{|r| > 0.8\} \leq 0.10$; the null hypothesis is rejected at 10% level of significance.

22.5.10 Rank Statistic

Many non-parametric methods are based on ranks rather than on actual measurements. We have already encountered two such methods—signed rank test and Mann-Whitnay Wilcoxon test. This is because one feels that the scale underlying one's measurement is relatively arbitrary and one should come to the same conclusion even if a different scale of measurement is employed. This means that we want our statistical procedures to be invariant under order-preserving transformation of the observations. In addition, when quantitative measurements are not available, rank methods may still be applicable, if the subject matter of investigation allows for quantitative comparisons.

If there are N objects in a ranking and no two objects are tied for the same position, then rank $r(x_k)$ among observations $x_1, \cdots, x_N$ can be expressed as

$$r(x_k) = \sum_{h=1}^{N} t(x_k - x_h) = 1 + \sum_{h(\neq k)=1}^{N} t(x_k - x_h) \tag{22.5.24}$$

when $t(u)$ is the jump function defined by

$$t(u) = \begin{cases} 1, & u > 0 \\ 0, & u \le 0. \end{cases} \tag{22.5.25}$$

The above definition of rank is not satisfactory if there are ties among Z_k. Consider, therefore, the function

$$C(u) = \begin{cases} 0, & u < 0 \\ \frac{1}{2}, & u = 0 \\ 1, & u > 0. \end{cases} \tag{22.5.26}$$

In terms of $C(u)$, in the absence of ties, $r(x_k)$ becomes

$$\begin{aligned} r(x_k) &= 1 + \sum_{h(\neq k)} C(x_k - x_h) \\ &= \frac{1}{2} + \sum_{h=1}^{N} C(x_k - x_h). \end{aligned} \tag{22.5.27}$$

In the presence of ties, we shall use (22.5.27) as our definition of the rank of x_k.

The quantity $r(x_k)$ is defined as the midrank i.e., the average of the ranks that would be allocated to the tied observations had they been distinguishable.

The above definition of rank can be easily extended to the case of two samples. In the two-sample case when we have m-observations on X and n-observations on Y the rank of x_k in the combined sample is

$$r(x_k) = \sum_{l=1}^{m} t(x_k - x_l) + \sum_{l=1}^{m} t(x_k - y_l), \quad k = 1, \cdots, m \tag{22.5.28}$$

and similarly that of y_k in the combined sample is

$$r(y_k) = \sum_{l=1}^{m} t(y_k - y_l) + \sum_{l=1}^{m} t(y_k - x_l), \quad k = 1, \cdots, m. \tag{22.5.29}$$

▶ **Example 22.5.10** $\boldsymbol{x} = (3\ 7\ 11\ 11\ 11\ 11\ 15)$

$$r(x_1) = 1,\ r(x_2) = 2,\ r(x_3) = r(x_4) = r(x_5) = r(x_6) = 4\frac{1}{2},\ r(x_7) = 7.$$

Treatment for a two-sample problem will be similar.

In addition to r-ranks as above sometimes we will consider w-ranks as follows:

$$\begin{aligned} w(x_k) &= 2\left[r(x_k) - \frac{m+1}{2}\right] = \sum_{i=1}^{m} \text{sgn}\,(x_k - x_i) \\ &= \sum_{i(\neq k)=1}^{m} \text{sgn}\,(x_k - x_i) \end{aligned} \tag{22.5.30}$$

where the sign function sgn (u) is defined as

$$\text{sgn}\,(u) = \begin{cases} 1, & u > 0 \\ 0, & u = 0 \\ -1, & u < 0. \end{cases}$$

The w-ranks do not take fractional values in ties and their sum is zero.

▶ **Example 22.5.11** *For example* 22.5.10, $w(x_1) = -6$, $w(x_2) = -4$, $w(x_3) = 1$, $w(x_4) = 1$, $w(x_5) = 1$, $w(x_6) = 1$, $w(x_7) = 6$.

For a two-sample problem, for comparing every measurement of the first set (set of $\boldsymbol{x}$) with every measurement of the second set (set of $\boldsymbol{y}$), one may consider the w-rank sum

$$W = \sum_{k=1}^{m}\sum_{l=1}^{n} \text{sgn}\,(x_k - y_l) = W(\boldsymbol{x}, \boldsymbol{y}).$$

Obviously,

$$W = \sum_{k}\sum_{l}\{(\text{number of times } x_k > y_l) - (\text{numbers of times } x_k < y_l)\}.$$

Clearly,

$$W(\boldsymbol{x}, \boldsymbol{y}) = -W(\boldsymbol{y}, \boldsymbol{x})$$
$$W(\boldsymbol{x}, \boldsymbol{x}) = 0.$$

If the two set $\boldsymbol{x}$ and $\boldsymbol{y}$ are pooled together to form the set $\boldsymbol{z} = (z_1, \cdots, z_{m+n})$, we define w-rank of z_k as

$$w(z_k) = \sum_{l=1}^{N} \text{sgn}\,(z_k - z_l) = W((z_k),\ \boldsymbol{z} - (z_k))$$

where $\boldsymbol{z} = (z_1, \cdots, z_N)$ and $N = m + n$. Thus in the combined sample

$$w(x_k) = \sum_{l(\neq k)=1}^{m} \text{sgn}\,(x_k - x_l) + \sum_{l=1}^{n} \text{sgn}\,(x_k - y_l) = W((x_k), \boldsymbol{z}).$$

Similarly, for $w(y_k)$ in the combined sample.
Again,

$$\begin{aligned} W &= W(\boldsymbol{x}, \boldsymbol{y}) = W(\boldsymbol{x}, \boldsymbol{z}) \quad [\text{because } \boldsymbol{z} = \boldsymbol{x} + \boldsymbol{y} \text{ and } W(\boldsymbol{x}, \boldsymbol{x}) = 0] \\ &= W((x_1), \boldsymbol{z}) + \cdots + W((x_m), \boldsymbol{z}) \\ &= \sum_{k=1}^{m} w(x_k). \end{aligned}$$

Similarly,

$$w(y_1) + \cdots + w(y_n) = w(\boldsymbol{y}, \boldsymbol{x}) = -W.$$

▶ **Example 22.5.12**

$$\boldsymbol{x} = (3\ 5\ 7\ 9\ 11)$$
$$\boldsymbol{y} = (2\ 6\ 6\ 8\ 12\ 14)$$
$$\begin{matrix} \boldsymbol{z} = (2 & 3 & 5 & 6 & 6 & 7 & 8 & 9 & 11 & 12 & 14) \\ y & x & x & y & y & x & y & x & x & y & y \end{matrix}$$
$$W(\boldsymbol{x}, \boldsymbol{y}) = (-4) + (-4) + 0 + 2 + 2 = -4$$

$$\begin{array}{l|l} w(x_1) = -8 & w(y_1) = -10 \\ w(x_2) = -6 & w(y_2) = -3 \\ w(x_3) = 0 & w(y_3) = -3 \\ w(x_4) = 4 & w(y_4) = 2 \\ w(x_5) = 6 & w(y_5) = 8 \\ & w(y_6) = 10. \end{array}$$

We now consider relation between w-rank sum W and r-rank sums

$$R_x = \sum_{i=1}^{m} r(x_i) \quad \text{and} \quad R_y = \sum_{j=1}^{n} r(y_j)$$

in the case of two samples, $\boldsymbol{x}$ and $\boldsymbol{y}$.
By virtue of relation (22.5.30),

$$R_x = \sum_{i=1}^{m} \left[\frac{N+1}{2} + \frac{1}{2} w(x_k)\right] = \frac{m(N+1)}{2} + \frac{W}{2}, \qquad (22.5.31)$$

$$R_y = \frac{m(N+1)}{2} - \frac{W}{2}.$$

Defining

$$\bar{w}_x = \frac{1}{m}\Sigma w(x_i), \quad \bar{w}_y = \frac{1}{n}\Sigma w(y_i), \quad \bar{r}_x = R_x/m, \quad \bar{r}_y = \frac{R_y}{n}$$

we find

$$\bar{w}_x - \bar{w}_y = \frac{N}{nm} W.$$

For measuring association between $\boldsymbol{x} = (x_1, \cdots, x_m)$ and $\boldsymbol{y} = (y_1, \cdots, y_m)$ values we may sometimes consider the statistic,

$$S = \sum_{k<l=1}^{m}\sum \operatorname{sgn}(x_k - x_l) \operatorname{sgn}(y_k - y_l).$$

Linear Rank Statistic

Consider a two-sample problem. Let the values $(X_1, \cdots, X_m)$ and $(Y_1, \cdots, Y_n)$ be pooled together and arranged according to the increasing values of magnitudes and ranked. Let $z = (z_1, \cdots, z_{m+n})$ be the pooled ordered set of observations. Let ξ_i be a random variable which takes the values 1(0) if the ith observation in the combined ordered sample be on X (otherwise). Let a_i be the score of z_i in $\boldsymbol{z}$. The statistics

$$T_N(\xi) = \sum_{i=1}^{N} a_i \xi_i \tag{22.5.32}$$

where $N = m + n$ is called a linear rank statistic. Often a_i is i, i.e., rank of Z_i in $\boldsymbol{Z}$; but other types of scores are also used.

Clearly, under $H_\circ(F(x) = G(x))$, each ξ_i is a binomial variable, with

$$E(\xi_i) = P(\xi_i = 1) = \frac{m}{n},$$

$$V(\xi_i) = \frac{mn}{N^2}, \quad \text{Cov}(\xi_i, \xi_j) = -\frac{mn}{N^2(N-1)}.$$

Therefore, under $H_\circ$,

$$E(T_N) = \frac{m}{N}\sum_{i=1}^{N} a_i, \tag{22.5.33}$$

$$V(T_N) = \sum_{i=1}^{N} a_i^2 \left(\frac{mn}{N^2}\right) - \sum_{i \neq j = 1}^{N}\sum a_i a_j \frac{mn}{N^2(N-1)} \tag{22.5.34}$$

$$= \frac{mn}{N^2(N-1)}\left[N\sum_{i=1}^{n} a_i^2 - \left(\sum_{i=1}^{N} a_i\right)^2\right].$$

Under $H_\circ$, each possible value of $\boldsymbol{\xi} = (\xi_1, \cdots, \xi_N)$ has probability $1/\binom{N}{m}$. Therefore $P(T_n = t) = \nu_t/\binom{N}{m}$, when ν_t is the number of possible values of $\boldsymbol{\xi}$ for which $T_N = t$ The exact sampling distribution of T_N can thus be determined by direct enumeration.

T_N is a linear function of identically distributed (but not independent) random variables ξ_i. Then subject to certain regularity conditions $\{T_N - E(T_N)\}/\sqrt{V(T_n)} \sim N(0, 1)$ as $m, n \to \infty$ in such a way that m/n remains constant. Chernoff and Savage (1958) proved that the above result holds even in the non-null case, subject to certain regularity conditions.

22.5.11 Terry-Hoeffding Test

This is a test for location parameter. The null hypothesis is $H_\circ$, while alternative is H_3 or H_4. This is based on the statistic

$$C = \sum_{i=1}^{N} E(\zeta_{(i)})\xi_i$$

where $\zeta_{(i)}$ is the ith order statistic from a standard normal distribution. The test, therefore, uses the expected normal scores as weights a_i. The exact null distribution of C can be obtained by direct enumeration, using the tables of $E(\zeta_{(i)})$. The statistic C is symmetric about origin with

$$V(C) = mn \sum_{i=1}^{N} \left[E(\zeta_{(i)})\right]^2 /N(N-1).$$

For large sample, normal approximation may be used for the null distribution. The asymptotic relative efficiency of this test with respect to t-test under normality assumptions is 1.

22.5.12 Two-Sample Test for Dispersion

As noted in the beginning of Section 22.5, a general alternative for testing $H_\circ$ is

$$H_\circ : G(x) = F\left(\frac{x-\theta_1}{\theta_2}\right)$$

where $F(x)$ is an arbitrary distribution function and θ_1, θ_2 are unknown parameters. The two populations with d.f.'s F and G have the same dispersion if $\theta_2 = 1$, whatever θ_1. However, it is clear that even if $\theta_2 = 1$, H_6 does not reduce to $H_\circ$ unless $\theta_1 = 0$. Note also that θ_1 is the difference between two location parameters of the two distributions.

We shall first consider three tests for $\theta_2 = 1$ assuming $\theta_1 = 0$. After this we shall refer to tests for $\theta_2 = 1$ if θ_1 is unknown. (In the tests for location parameters so far considered we have assumed that scale-parameters have the some values). The classical parametric test for the scale parameters is the (Snedecor's) F-test which is not robust with respect to non-normality, though it requires no assumption about the location parameters. (The classical parametric test for the location parameter is the (Student's) t-test).

In the linear rank statistics, the ranks depend on the location parameters as well as the scale parameters of X and Y. We assume, therefore, that both X and Y have the same location parameters and they only differ in their scale values. The hypothesis is $H_\circ$ against $H_5 : (F(x) = G(x/\theta), \text{ for } \theta \neq 1)$. If $\theta > 1$, under alternative hypothesis $V(X)$ will be greater than $V(Y)$. In this case, one would expect some of the X-values to lie on both ends of the Y-values. If $\theta < 1$, the situation will be reverse.

It is, therefore, more appropriate to form a linear rank statistic when extreme values get high scores and middle values get low scores. The test statistic T in this case is the sum of the x-scores or $\sum_{i=1}^{m} a(x_i)$ where $a(x_i)$ is the weight associated with x_i. If by such an arrangement the value of T is high, $H_\circ$ should be rejected in favour of greater variability of x-population than y-population (i.e., $\theta > 1$), while a low rank sum indicates the reverse ($\theta < 1$).

Mood Test

In this and the next two tests we assume $\theta_1 = 0$. Consider a_i(in 22.5.32) $= \left(i - \frac{N+1}{2}\right)^2$ where i is the rank of Z_i. Clearly a_i is the least for the middle-most term and increases towards the two extremes. The test statistic is

$$M_N = \sum_{i=1}^{N} \xi_i \left(i - \frac{N+1}{2}\right)^2.$$

A large (small) value of M_N indicates $\theta > (<) 1$. Under H_o,

$$E(M_N) = m(N^2 - 1)/12$$
$$V(M_N) = mn(N+1)(N^2 - 4)/180.$$

For large samples, normal approximations may be used. For small samples, exact test may be made by direct computation. The test has a ARE of 76% relative to parametric F-test under usual assumptions.

Siegal-Tukey Tests

Consider the following two rank assignments, where $(Z_1, \cdots, Z_{m+n})$ is the set of pooled and ordered observations. Assume that there is no tie.

$$\begin{array}{cccccccc} Z_1 & Z_2 & Z_3 & Z_4 \cdots & Z_{N-3} & Z_{N-2} & Z_{N-1} & Z_N \\ \hline r_1 & r_4 & r_5 & r_8 \cdots & r_7 & r_6 & r_3 & r_2 \end{array} \qquad (22.5.35)$$

$$\begin{array}{cccccccc} r_2 & r_3 & r_6 & r_7 \cdots & r_8 & r_5 & r_4 & r_1 \end{array} \qquad (22.5.36)$$

Let

$$T_{ST} = \sum_{i=1}^{N} \xi_i r(Z_i)$$

be the rank-sum based on (22.5.35) and T'_{ST} be the rank-sum based on (22.5.36) where $r(Z_i)$ is the rank of Z_i. Since under H_o, all the possible assignments of ranks are equally likely, the distributions of T_{ST} and T'_{ST} are identical under H_o and each is same as the distribution of T under H_o in Wilcoxon's rank-sum test (obtained in subsection 22.5.3), provided there is no tie among the observations. Hence, under H_o,

$$E(T_{ST}) = E(T'_{ST}) = \frac{m(N+1)}{2}$$

$$V(T_{ST}) = V(T'_{ST}) = \frac{mn(N+1)}{12}.$$

It follows that the correlation between T_{ST} and T'_{ST} converges very rapidly to 1 as $N(= m+n)$ increases. Therefore, even for small sample sizes, the two tests based on T_{ST} and T'_{ST} are essentially equivalent.

The Freund-Ansari-Bradley Test

This test assigns average of ranks in (22.5.35) and (22.5.36) above to z-values. The ranks assigned are, therefore,

$$\frac{Z_1}{(r_1+r_2)/2} \; \frac{Z_2}{(r_3+r_4)/2} \; \frac{\cdots}{\cdots} \; \frac{Z_{N-1}}{(r_3+r_4)/2} \; \frac{Z_N}{(r_1+r_2)/2}. \qquad (22.5.37)$$

Let

$$T_{FAB} = \frac{1}{2}(T_{ST} + T'_{ST})$$

be the rank-sum based on (22.5.37). It follows that

$$E(T_{FAB}) = \frac{m(N+1)}{2}$$

$$V(T_{FAB}) = \frac{mn(N^2-4)}{12(N-1)}.$$

Usual normal approximations of the standardised variables can be used to test H_o, unless m, n are extremely small, in which case exact distributions of the test statistic have to be calculated. Both Siegel-Tukey test and this test are consistent against alternatives of the type H_5 when $\theta \neq 1$.

If ties do occur we can assign ranks as in (22.5.35) to each possible permutation of the tied observations and take their averages to obtain ranks in (22.5.35). The ranks in (22.5.36) and (22.5.37) can similarly be calculated.

A test for dispersion (when difference in location is unknown)

A test is as follows. Each of the two samples is divided into small subgroups of equal size and a suitable measure of dispersion is calculated for each subgroup. The two sets of dispersion measurements are then analysed by a two-sample test in order to find whether they come from identical populations or come from populations differing in dispersions.

The simplest such procedure uses subgroups of size 2, computes the absolute differences,

$$x'_1 = |x_1 - x_2|, \quad x'_2 = |x_3 - x_4|, \cdots$$
$$y'_1 = |y_1 - y_2|, \quad y'_2 = |y_3 - y_4|, \cdots$$

and apply Wilcoxon's rank-sum test to samples $x'_1, x'_2, \cdots, y'_1, y'_2, \cdots$.

22.6 C-sample Problems ($C \geq 2$)

We are given C random variables $X_1, X_2, \cdots, X_C$ with distribution functions $F_1(x)$, $F_2(x)$, $\cdots$, $F_C(x)$ respectively. We want to test the hypothesis.

$$F_1(x) = F_2(x) = \cdots = F_C(x), \; \forall\, x, \quad -\infty < x < \infty. \qquad (22.6.1)$$

If $C = 2$, this becomes a two-sample problem.

When specifying alternatives to (22.6.1), one may consider completely general alternative, stating simply that (22.6.1) is not true. However, one may consider more specific alternative concerning location parameters as follows. There exist C location parameters $\alpha_1, \cdots, \alpha_C$ such that $F_1(x) = F(x-\alpha_1), \cdots, F_C(x) = F(x-\alpha_C)$ where $F(x)$ is some arbitrary d.f. In this case, the null hypothesis becomes $\alpha_1 = \cdots = \alpha_C$.

22.6.1 Kruskal-Wallis test

Let $\boldsymbol{X}_i = (X_{i1}, \cdots, X_{in_i})$ be a random sample of size n_i from the population $F_i(x)$, $i = 1, \cdots, C$. We want to test the null hypothesis $H_\circ(F_1(x) = F_2(x) = \cdots = F_C(x) \ \forall\ x)$ against the general alternative that all the populations are not identical.
The w-rank sum for two sets $\boldsymbol{X}_i$ and $\boldsymbol{X}_j$ is

$$W_{(ij)} = \sum_{k=1}^{n_i}\sum_{l=1}^{n_j} \operatorname{sgn}(x_{ik} - x_{jl}) = W(\boldsymbol{X}_i, \boldsymbol{X}_j).$$

The w-rank sum for the set $\boldsymbol{X}_i$ with respect to all other observations is

$$\begin{aligned} W_i &= \sum_{j(\neq i)} \sum_{k=1}^{n_i}\sum_{l=1}^{n_j} \operatorname{sgn}(x_{ik} - x_{jl}) \\ &= \sum_{j(\neq i)=1}^{c} W_{(ij)} = W(\boldsymbol{X}_i, \boldsymbol{X} - \boldsymbol{X}_i) \end{aligned}$$

when $\boldsymbol{X} = (\boldsymbol{X}_1, \cdots, \boldsymbol{X}_c)$. Consider the test statistic,

$$T_C = \sum_{i=1}^{C} W_i^2/n_i = 4\sum_{i=1}^{C}\left(R_i - n_i\frac{N+1}{2}\right)^2$$

when

$$R_i = \sum_{j=1}^{n_i} r(x_{ij}),$$

$r(x_{ij})$ being the rank of x_{ij} in the combined sample $\boldsymbol{X}$ and $N = \sum_{i=1}^{C} n_i$.

$$\begin{aligned} T_C &= 4\sum_{i=1}^{C} n_i\left(\bar{r}_i - \frac{N+1}{2}\right)^2 \quad \text{where } \bar{r}_i = R_i/n_i \\ &= 4\sum_{i=1}^{C}\frac{R_i^2}{n_i} - N(N+1)^2. \end{aligned}$$

Consider the regular F-test based on the statistic

$$\sum_i n_i(\bar{x}_i - \bar{\bar{x}})^2 \Big/ \sum_i\sum_j (x_{ij} - \bar{\bar{x}})^2$$

but now applied to the rank-values $r(x_{ij})$. Here

$$\bar{\bar{r}} = \frac{N+1}{2}, \quad \sum_i \sum_j (r(x_{ij}) - \bar{\bar{r}})^2 = \frac{N^3 - N}{12}.$$

Therefore, apart from a constant, T_C is the numerator of a F test (now applied to ranks) whose denominator is a constant. As in the case of a F-test, H_o is rejected if the value of T_C is large. The test based on the statistic

$$H = \frac{3}{N(N+1)} T_C$$

is known as Kruskal-Wallis H-test (we have assumed no ties).

Kruskal has shown that if no n_i is small, then $H \sim \chi^2_{(C-1)}$. When C is small, say $C = 3$ and the n_i's are also small, the χ^2-approximation is not tenable. For such cases exact probability has to be computed. The test is consistent.

22.6.2 Friedman's Two-way Analysis of Variance by Ranks

For this test, the data are cast into a two-way table having n rows and C columns. Each row represents various subjects and each column various conditions.

Suppose we wish to study the scores of $3(= n)$ groups of students under $4(= C)$ different conditions, say; the scores of students of three different colleges who are pursuing their studies under four different conditions (say, economic conditions of their guardians).

condition group	1 2 3 4 $(= C)$
I	ranked scores $\{1, \cdots, C\}$
II	"
III $= (n)$	"

For each college (group), one student is sampled at random from each stratum (representing the specific condition) and his score is recorded. However, the scores are converted into ranks, and the data in each row is therefore the set $\{1, \cdots, C\}$ in some order.

If the null hypothesis that the different conditions are equivalent (i.e., they do not affect the scores and hence the ranks differently) is true, then the distribution of ranks in each column would be identical. The set of ranks in each column would then be a random sample of size n from the discrete uniform distributions at the points $\{1, 2, \cdots, C\}$. Under H_o, one expects that the mean rank $\bar{r}_j$ of each column would be equal $(j = 1, \cdots, C)$.

As an example from the design of experiments, each group represents a complete block and in each block a plot is allotted randomly to one of C treatments and its yield measured. The data on yield for each block are then ranked. The null hypothesis is that the treatments have no differential effect on yields. Under $H_\circ$, each sample (as represented by each column) will be a random sample from the same population. The null hypothesis is, therefore, $H_\circ(F_1(x) = \cdots = F_C(x) \ \forall\, x)$.

Friedman used the statistic

$$T_F = \frac{12}{nC(C+1)} \sum_{j=1}^{C} R_j - 3n(C+1)$$

where R_j is the total of ranks in the jth column. When n and C are not too small, T_F is distributed as $\chi^2_{(C-1)}$. If the observed value of $\chi^2 > \chi^2_{(C-1),\alpha}$ then $H_\circ$ is rejected at $100\alpha\%$ level of significance. When n or C is small, exact probability tables are available and these are reproduced at table A.18.

Non-parametric confidence intervals for population quantiles and tolerance intervals for population values based on order statistics have been considered in chapter 19.

Exercise

1. Let $(X_1, \cdots, X_{n_1}), (Y_1, \cdots, Y_{n_2})$ be random samples of sizes n_1, n_2 from F_1, F_2 respectively, $F_1, F_2 \in \mathcal{F}$ (class of all absolutely continuous distributions). Show that the minimum variance unbiased estimator of σ^2_{X+Y} is

$$\frac{\sum_1^{n_1}(x_i - \bar{x})^2}{n_1 - 1} + \frac{\sum_1^{n_2}(y_i - \bar{y})^2}{n_2 - 1}.$$

2. For the class of absolutely continuous distributions

   (a) on R^1, find the degrees of the parameters: $E(X^4)$, $E^4(X)$,

   $$E\{[X - E(X)]^4\},$$

   (b) on R^2, find the degrees of the parameters: $E^2(X_1)$, $E(X_1X_2)$,

   $$\sigma^2_{X_1},\ \sigma^2_{X_2},\ E^2(X_1)E^2(X_2).$$

   Find the minimum variance unbiased estimators of these parameters.

3. A computer program is written to generate X uniformly in the interval $0 \le X < 10$. From 250 consecutive values the following data are obtained:

X-values	0–1.99	2–3.99	4–5.99	6–7.99	8–9.99
Frequency	38	55	54	41	62

Do these data offer any evidence that the program is not written properly?

[Rohatgi, 1975]

4. For the data in example 22.4.3 construct a 0.90 level confidence band for $F(x)$.

5. For the data in problem 3 above test the hypothesis that the distribution is $U[0, 1]$. Take $\alpha = 0.05$.

6. A random sample gives the values (.126, −.302, .702, .982, −1.212, .759, 2.312, −.982). Using sign test, test the hypothesis that $\mu = 1.00$ against the both sided alternative $\mu \neq 1.00$.

7. The following data refer to the percentage of recovery of dextroamphetramine excreted after seven hours by a sample of children having organically related disorders and a sample of children having non-organic disorders.

Organic	:	17.53	20.60	17.62	20.93	27.10
Non-organic	:	15.59	14.76	13.32	12.45	12.79

   (a) Under the usual assumption of normality and equality of variances use the two sample t-test to test $H_o : \mu_1$ (organic) $= \mu_2$ (non-organic) against the alternative $H_1 : \mu_1 > \mu_2$. Use 5% level of significance [μ_i is the population mean of the ith group].

   (b) Notice the large difference between the two sums of squares and test at 5% level the hypotheses that the variances are equal.

   (c) Use Mann-Whitney-Wilcoxon test to test at $\alpha = 0.05$ the hypothesis that the two distributions are the same against the alternative that $\mu_1 > \mu_2$.

8. For the data in example 22.4.3 test the hypothesis that the quantile of order $p =$ .40 is .40.

9. The following are the data on weights of a batch of 10 guineapigs before and after giving them a certain diet.

Before:	5.8	6.4	4.9	4.2	7.2	8.4	7.5	6.8	6.5	6.9
After:	5.9	7.1	5.5	4.6	6.8	7.6	7.3	6.9	6.0	7.2

   By means of the sign test, test the hypothesis that the two populations have the same mean; compare the result with the one obtained from a paired t-test.

10. For the data in problem 6 use signed-rank test to test $H_o(\mu = 1.00)$ against $H(\mu \neq 1.00)$.

11. For the data in problem 9 use paired sample signed rank test to test that the population mean of the differences is zero.

12. The life in (000) kilometers of samples of two brands of tyres were found as follows:

Brand A:	8.1	7.6	9.2	6.8	7.5	8.1	7.9	6.8	7.2		
Brand B:	8.9	8.2	7.8	6.5	7.9	7.1	6.9	8.8	7.8	8.1	7.9

Apply (i) median test (ii) run test (iii) the Mann-Whitney-Wilcoxon test to test the null hypothesis that the samples come from the same population.

13. Use the data of problem 12 to apply the Kolmogorov-Smirnov test.

14. The following are the rankings given to 6 competitors by two judges:

Judge A:	6	5	3	1	4	2
Judge B:	5	6	2	3	1	4

Use both Kendall's tau and Spearman's rank correlation tests for testing that rankings of judges A and B were independent.

Chapter 23

Elements of Decision Theory

23.1 Introduction

All the problems of statistical inference can be looked upon as problems of theory of decision. The theory of decision was first introduced by A. Wald (1950). Section 23.2 introduces the basic concepts of decision rules, Section 23.3 considers different classes of decision rules. Bayes and Minimax decision procedures are covered in Section 23.4. Different types of loss functions are introduced in the next section. Section 23.6 describes the point estimation of parameters from the decision-theoretic point of view. Interval estimation of parametric functions is covered in Section 23.7. Lastly different decision rules for testing of hypotheses are considered.

23.2 Some Basic Concepts

Let $\{f_\theta, \theta \in \Theta\}$ be a family of p.d.f.'s and $X_1, \cdots, X_n$ be a random sample of size n from this distribution. In decision theory, the parameter space Θ is called the *space of all possible states of nature.* Once the sample point $\boldsymbol{x} = (x_1, \cdots, x_n)$ is observed the statistician takes an action or decision d on the basis of these data. Clearly, the decision $d = d(\boldsymbol{x})$ is a function of $\boldsymbol{x}$. Let $\mathcal{D} = \{d(\boldsymbol{x}), \boldsymbol{x} \in R^n\}$ be the set of all possible decisions under a given situation. We shall denote by $\delta(\boldsymbol{x})$ a function defined on R^n, called the decision function or decision rule (d.r.) which takes value $d = d(\boldsymbol{x})$ when $\boldsymbol{x} = \boldsymbol{x}$. Thus, range space of $\delta(\boldsymbol{x})$ is $\mathcal{D}$. Clearly, $\delta(\boldsymbol{x})$ is generally, a many to one function.

Definition 23.2.1 *A decision function $\delta(\boldsymbol{x})$ is a statistic that takes values in $\mathcal{D}$, that is δ is a Borel-measurable function that maps R^n into $\mathcal{D}$.*

▶ **Example 23.2.1** *In the problem of testing of hypothesis, $\mathcal{D} = \{d_1, d_2\}$ where d_1 means accepting H_o and d_2 rejecting H_o. If the sample point $\boldsymbol{x} \in$ critical region w, H_o is rejected; otherwise it is accepted. Thus, $\delta(\boldsymbol{x}) = d_2$ if $\boldsymbol{x} \in w$ and $\delta(\boldsymbol{x}) = d_1$ if $\boldsymbol{x} \in w^c$.*

➤ **Example 23.2.2** *In the problem of theory of estimation $\mathcal{D} = \Theta$.*

In taking the decision d when θ is the true state of nature, we may or may no incur an error. If there is an error we incur a loss. Let $L(d, \theta)$ be the loss incurred i making decision d when θ is the true state of nature. For a correct decision $L(d, \theta) = 0$ Otherwise its value is greater than zero. Clearly, $L(d, \theta)$ is defined on $\mathcal{D} \times \Theta$ and take values in R^1.

Definition 23.2.2 *Let $\mathcal{D}$ be an arbitrary space of decisions or actions. A non-negativ function L that maps $\mathcal{D} \times \Theta$ into R^1 is called a loss function.*

We shall consider different types of loss functions in Section 23.5.

Clearly, the loss $L(d(\boldsymbol{x}), \theta)$ is a random variable as the decision $d(\boldsymbol{x})$ depends on $\boldsymbol{x}$ For different observed points $\boldsymbol{x}$, the statistician would take different decisions $d(\boldsymbol{x})$ a dictated by the d.r. $\delta(\boldsymbol{x})$. Therefore, the average loss or risk associated with the d.r. when θ is the true value of the parameter is

$$r(\delta|\theta) = E_{\boldsymbol{x}|\theta}\left[L(\delta(\boldsymbol{x}), \theta)\right] = \int_{R^n} L(d(\boldsymbol{x}), \theta) f(\boldsymbol{x}|\theta) d\boldsymbol{x} \tag{23.2.1}$$

where $E_{\boldsymbol{x}|\theta}$ is the expectation operator w.r.t. the joint distribution of $\boldsymbol{X}$ when θ is th true value of the parameter.

Definition 23.2.3 *A d.r. δ is said to be uniformly better than a d.r. δ' if*

$$r(\delta|\theta) \leq r(\delta'|\theta) \ \forall \, \theta \in \Theta$$

with strict inequality holding for some θ.

Definition 23.2.4 *A d.r. δ^* is said to be uniformly best in a class of d.r.'s Δ, if δ^* uniformly better than any other d.r. $\delta \in \Delta$.*

A uniformly best d.r. does not exist in general.

Definition 23.2.5 *A d.r. δ is said to be admissible in a class Δ if there exists no othe d.r. in Δ which is uniformly better than δ.*

23.3 Classes of Decision Rules

A class C_1 of d.r.'s is said to be *complete* if given any d.r. δ' outside this class the exists at least one d.r. $\delta \in C_1$ which is uniformly better than δ'.

A class C_1' of d.r.'s is said to be *essentially complete* if given any d.r. δ' outside th class there exists at least one d.r. $\delta \in C_1'$ which is uniformly better than or equivale to δ'.

A class C_2 of d.r.'s is said to be minimal complete if it is complete and contains no subclass which is complete.

A class C_2' of d.r.'s is said to be essentially minimal complete if it is complete and contains no subclass which is essentially complete.

Theorem 23.3.1 *If a class of admissible d.r.'s is complete it must be minimal complete and conversely.*

Proof: Let C be the class of admissible d.r.'s which is complete. If it is not minimal complete, there exists a subclass C_1 of it which is complete. Let δ be a d.r. belonging to C but not to C_1. Then since C_1 is complete, these exists a d.r. $\delta' \in C_1$ which is uniformly better than δ. Thus, δ and δ' are d.r.'s belonging to the same admissible class of d.r.'s and yet δ' is uniformly better than δ. This leads to a contradiction. Hence, the first part of the theorem.

Conversely, if we consider a minimal complete class C of d.r.'s which is not admissible it must be possible to find two d.r.'s δ and $\delta' \in C$ such that one, say δ' is uniformly better than the other δ. Then if we consider a subclass C_1 of C containing δ' (and such other estimators) then, it follows that C_1 is a complete class of d.r.'s. Therefore, C_1 which is a subclass of a minimal complete class, is also complete. This is a contradiction.

23.4 Bayes and Minimax Decision Rules

We now consider θ as a random variable over the range space Θ.

If θ has a prior probability distribution $\xi(\theta)$, the average risk associated with d.r. δ is

$$\bar{r}_\xi(\delta) = \int_\Theta r(\delta|\theta)\xi(\theta)d\theta. \tag{23.4.1}$$

Definition 23.4.1 *A d.r.* $\delta_\xi \in \Delta$ *is said to be a Bayes d.r. w.r.t. prior probability distribution* ξ *if*

$$\bar{r}_\xi(\delta_\xi) = \min_{\delta \in \Delta} \bar{r}_\xi(\delta).$$

(We shall use the notation $\bar{r}_\eta(\delta_\lambda)$ to denote the average risk of the d.r. δ_λ with respect to prior distribution η of θ).

If no prior information regarding the distribution of θ is available, one might consider the maximum value of risks associated with d.r. δ for all possible values of θ. The desirability of a d.r. δ is then measured by $\sup_{\theta \in \Theta} r(\delta|\theta)$. Clearly, of two decision functions, one with the smaller value of the supremum will be preferable.

Definition 23.4.2 *A d.r.* $d_m \in \Delta$ *is said to be a minimax d.r. in the class* Δ *if*

$$\sup_{\theta \in \Theta} r(\delta_m|\theta) = \inf_{\delta \in \Delta} \sup_{\theta \in \Theta} r(\delta|\theta).$$

Since the maximum risk measures the worst possible average (w.r.t. distribution $f(x|\theta)$) loss or risk that can result from the use of a d.r., a minimax d.r. is the one which gives the greatest protection against the maximum risk. Fig 23.1 shows that such a d.r. may sometimes be very unreasonable. Here under most circumstances one would prefer δ_1 to δ_2 although its risk function has the largest maximum.

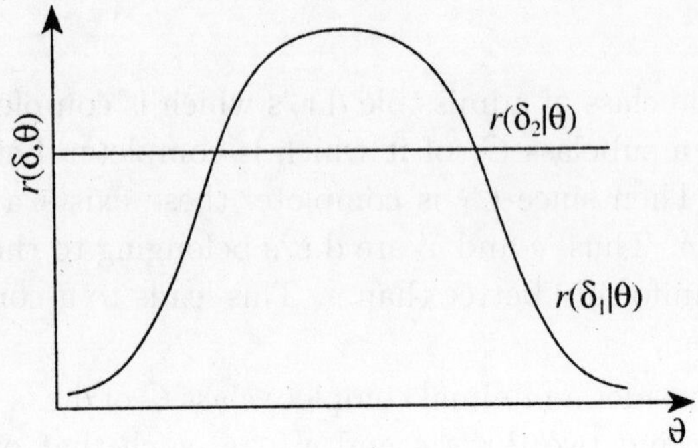

Fig. 23.1

The following theorems connect Bayes rules and minimax rules. First we consider:

Definition 23.4.3 ***Least Favourable Distribution.*** *Let δ_ξ and $\delta_{\xi'}$ be Bayes d.r.'s for a decision problem w.r.t. prior distribution ξ and ξ' respectively. Then ξ is called the least favourable distribution if $\bar{r}_\xi(\delta_\xi) \geq \bar{r}_{\xi'}(\delta_{\xi'})$ whatever the other prior distribution ξ' may be.*

Theorem 23.4.1 *If there exists a prior distribution ξ for θ and δ_ξ be a Bayes rule corresponding to ξ and if*

$$r(\delta_\xi|\theta) = constant \quad \forall\, \theta \in \Theta$$

then (a) δ_ξ is a minimax rule (b) ξ is a least favourable distribution.

Proof: (a) Since δ_ξ is a Bayes rule,

$$\bar{r}(\delta_\xi) \leq \bar{r}(\delta) \;\; \forall\, \delta.$$

Hence,

$$\bar{r}(\delta_\xi) = \sup_\theta r(\delta_\xi|\theta).$$

Hence,

$$\sup_\theta r(\delta_\xi|\theta) \leq \bar{r}(\delta) \leq \sup_\theta r(\delta|\theta) \;\; \forall\, \delta.$$

Therefore δ_ξ is minimax.

(b) For any prior ξ' with Bayes estimate $\delta_{\xi'}$ we have

$$\bar{r}_{\xi'}(\delta_{\xi'}) \leq \bar{r}_{\xi'}(\delta_\xi) \leq \sup_\theta r(\delta_\xi|\theta) = \bar{r}_\xi(\delta_\xi).$$

Hence ξ is a least favourable prior distribution.

Corollary 23.4.1 *If δ_ξ be a Bayes d.r. whose risk $r(\delta_\xi|\theta) =$ constant $\forall\ \theta \in \Theta$, then δ_ξ is a minimax d.r.*

Theorem 23.4.2 *For a given sequence of prior distributions $\xi_1, \xi_2, \cdots$ of θ over the space Θ let $\delta_n = \delta_{\xi_n}$ be the Bayes d.r. and $r_n = \bar{r}_{\xi_n}(\delta_{\xi_n})$ be the corresponding Bayes risk $(n = 1, 2, \cdots)$. If $\lim_{n\to\infty} r_n = r$ and if δ be any d.r. having maximum risk, $\sup_{\theta\in\Theta} r(\delta|\theta) \leq r$, then δ is a minimax d.r.*

Proof: We have

$$\sup_\theta r(\delta_0|\theta) \geq \int_\Theta r(\delta_0|\theta)\xi_n(\theta)d\theta \geq \int_\Theta r(\delta_n|\theta)\xi_n(\theta)d\theta = r_n$$

where δ_0 is any d.r. Therefore, taking limit of both sides as $n \to \infty$

$$\sup_\theta r(\delta_0|\theta) \geq r \geq \sup_\theta r(\delta|\theta).$$

Hence δ is a minimax d.r.

Corollary 23.4.2 *If δ is a minimax d.r. for θ varying over the set $\Theta_0 \subset \Theta$ and*

$$\sup_{\theta\in\Theta_0} r(\delta|\theta) = \sup_{\theta\in\Theta} r(\delta|\theta)$$

then δ is a minimax d.r. for θ varying over Θ.

Corollary 23.4.3 *If δ is a minimax d.r. for $\theta \in \Theta$ and $\sup_{\theta\in\Theta} r(\delta'|\theta) = \sup_{\theta\in\Theta_0} r(\delta'|\theta)$ for all δ', where $\Theta_0 \subset \Theta$, then δ is a minimax d.r. for θ varying over Θ_0.*

The proof of these corollaries are left as exercises.

Restricted Bayes Solution

This is possibly an intermediate procedure between Bayes solution and minimax solution. Past experience often provides an indication of what θ to expect. This information is not sufficiently precise and reliable to warrant the assumptions that Bayes solution requires. It seems desirable to make use of the available information without trusting it to such an extent that catastrophically high risk may result if it is inaccurate or misleading. To achieve this one can place a bound on the risk and restrict considerations to decision procedures δ for which $r(\delta|\theta) < C\ \ \forall\ \theta$ where C is a constant larger than the minimax risk for this problem. The experimenter can then use this knowledge to guess at a distribution $\xi(\theta)$ of θ. This leads to the solution of a procedure called 'Restricted Bayes Solution' which minimises the average risk for the a priori distribution subject to above conditions.

23.5 Different Types of Loss Functions

We have already noted that $L(d, \theta)$ should be a non-negative function. Since the loss should be higher, higher is the deviation $|d - \theta|$, a plausible form of loss function is:

$$L(d, \theta) = C(\theta) w(|d - \theta|) \tag{23.5.1}$$

where $C(\theta)$ is positive and $w(\cdot)$ is a strictly increasing function of $|d - \theta|$.

Special cases of (23.5.1) are

(i) Squared error loss function

$$L(d, \theta) = C(\theta)(d - \theta)^2.$$

In this case the risk function is

$$\begin{aligned} E_{\boldsymbol{X}|\theta} L\{\delta(\boldsymbol{X}), \theta)\} &= C(\theta) E\left[\delta(\boldsymbol{X}) - E(\delta(\boldsymbol{X})) + E(\delta(\boldsymbol{X})) - \theta\right]^2 \\ &= C(\theta)\left[V_{\boldsymbol{X}|\theta}\{\delta(\boldsymbol{X})\} + \{\text{Bias } (\delta(\boldsymbol{X}))\}^2\right]. \end{aligned}$$

(ii) $L(d, \theta) = C(\theta)|\theta - d|^p$, $p \geq 1$.

Some other loss functions are

$$\frac{|\theta - d|^2}{|\theta|}, \quad \left(\frac{|\theta - d|}{|\theta| + 1}\right)^{1/2}.$$

A simple loss function is

$$L(d, \theta) = \begin{cases} 0, \text{ if } d = \theta \text{ (or a correct decision is taken)} \\ 1, \text{ otherwise.} \end{cases}$$

Often a convex loss function $L(d, \theta)$ which is a convex function of d and θ is taken. For such functions it is known by Jensen's inequality (Section 3.2) that

$$E_{\boldsymbol{X}|\theta}\left[L(w(\boldsymbol{X}), \theta)\right] \geq L\left[E_{\boldsymbol{X}|\theta}(w(\boldsymbol{X}), \theta)\right].$$

It is easy to see that the following functions are convex over the indicated intervals:

Functions ϕ	Interval (a, b)
(i) $\|x\|$	$-\infty < x < \infty$
(ii) x^2	$-\infty < x < \infty$
(iii) $x^p, p \geq 1$	$0 < x$
(iv) $1/x^p, p > 0$	$0 < x$
(v) e^x	$-\infty < x < \infty$
(vi) $-\log x$	$0 < x < \infty$.

In all these cases, ϕ is strictly convex, except in (i) and (iii) with $p = 1$.

23.6 Estimation of Parameters

23.6.1 Preliminaries

Suppose that the density function $f(x|\theta)$ of a r.v. X under consideration involves an unknown parameter $\theta[\in \Theta]$ whose value we want to estimate. Corresponding to each point $\theta_i \in \Theta$ we can form a hypothesis H_i, thus obtaining an infinite number of hypotheses. Acceptance of any one of these hypotheses means the estimation of θ by the corresponding value θ_i.

Let $\delta(\boldsymbol{X}) = \theta(\boldsymbol{X})$ be a d.r. for estimating θ. In a non-randomised d.r. δ, whenever $\boldsymbol{X}$ is observed, an estimate $\theta(\boldsymbol{x})$ of θ is obtained with probability one. In a randomised d.r. $\delta_r(\boldsymbol{x})$, an estimate $\theta_r(\boldsymbol{x})$ of θ is obtained with probability $d\zeta(\theta_r(\boldsymbol{x}))$ for a given $\boldsymbol{x}$. Therefore, the expected value of the estimate of θ obtained by randomised d.r. $\delta_r(\boldsymbol{x})$ is

$$E_{\boldsymbol{X}|\theta}(\delta_r(\boldsymbol{X})) = \int_{R^n}\left[\int_{\Theta(\boldsymbol{x})} \theta_r(\boldsymbol{x})d\zeta(\theta_r(\boldsymbol{x}))\right] f(\boldsymbol{x}|\theta)d\boldsymbol{x} \tag{23.6.1}$$

if δ_r is an unbiased d.r. for θ. Here $\Theta(\boldsymbol{x})$ denotes the space of all estimates $\theta_r(\boldsymbol{x})$ for a given $\boldsymbol{x}$ (In general, $\Theta(\boldsymbol{x}) = \Theta$).

Let $L(\theta_r(\boldsymbol{X}), \theta)$ be the loss involved in estimating θ by $\theta_r(\boldsymbol{X})$ when the true value of the parameter is θ. Therefore, risk function of $\delta_r(\boldsymbol{X})$ i.e., the risk associated with $\delta_r(\boldsymbol{X})$ when true value of the parameter is θ is

$$r(\delta_r|\theta) = \int_{R^n}\left[\int_{\theta} L(\theta_r(\boldsymbol{x}), \theta)d\zeta(\theta_r(\boldsymbol{x}))\right] f(\boldsymbol{x}|\theta)d\boldsymbol{x}. \tag{23.6.2}$$

We want to find a d.r. $\delta \in \Delta$ for which the risk is minimum possible for all θ. If a d.r. $\delta \in \Delta$ exists for which

$$r(\delta|\theta) \le r(\delta'|\theta) \;\; \forall\, \theta \in \Theta \quad \text{and} \quad \forall\, \delta' \in \Delta$$

then δ is the uniformly minimum risk (UMR) estimate of θ and is the most desirable one, if it exists. However, no such UMR-estimate of θ exists in general.

23.6.2 Bayes Estimates and Minimax Estimates

Let $\xi(\theta)$ be a prior p.d.f. of $\theta, \theta \in \Theta$. The average risk of the d.r. δ for estimating θ is

$$\bar{r}_\xi(\delta) = \int r(\delta|\theta)\xi(\theta)d\theta. \tag{23.6.3}$$

Bayes solution is to choose that δ^0 for which $\bar{r}_\xi(\delta^0) = \min_\delta \bar{r}_\xi(\delta)$ for all d.r.'s δ. We shall henceforth consider non-randomised decision rules δ only. Now

$$\bar{r}_\xi(\delta) = \int_{\Theta}\left[\int_{R^n} L(\theta(\boldsymbol{x}), \theta)f(\boldsymbol{x}|\theta)d\boldsymbol{x}\right] \xi(\theta)d\theta \tag{23.6.4}$$

$$= \int_{R^n}\left[\int_{\Theta} L(\theta(\boldsymbol{x}), \theta)\xi(\theta)f(\boldsymbol{x}|\theta)\right] d\boldsymbol{x} \tag{23.6.5}$$

provided the change of order of integration is permissible. Now, posterior probability density function of θ (i.e., the conditional p.d.f. of θ given $\boldsymbol{x}$) is

$$\xi^*(\theta|\boldsymbol{x}) = \xi^*_{\boldsymbol{x}}(\theta) = \xi^*(\theta) = \frac{\xi(\theta)f(\boldsymbol{x}|\theta)}{f_\xi(\boldsymbol{x})} \tag{23.6.6}$$

where

$$f_\xi(\boldsymbol{x}) = \int_\Theta \xi(\theta)f(\boldsymbol{x}|\theta)d\theta. \tag{23.6.7}$$

[Note that posterior p.d.f. of θ is proportional to the joint p.d.f. of $\boldsymbol{x}$ and θ].

Clearly $f_\xi(\boldsymbol{x})$ is the unconditional joint p.d.f. of $\boldsymbol{X}$. It is also called the predictive distribution of $\boldsymbol{X}$. The posteriori risk in accepting $\theta(\boldsymbol{x})$ as an estimate of θ is

$$r^*(\theta(\boldsymbol{x})) = E_{\theta|\boldsymbol{x}}[L(\theta(\boldsymbol{x}), \theta)] = \int_\Theta L(\theta(\boldsymbol{x}), \theta)\xi^*(\theta)d\theta \tag{23.6.8}$$

when $E_{\theta|\boldsymbol{x}}$ denotes expectation of θ for given $\boldsymbol{x}$.

Now,

$$\begin{aligned}
\bar{r}_\xi(\delta) &= E_\theta E_{\boldsymbol{X}|\theta}[L(\theta(\boldsymbol{X}), \theta)] \quad \text{[from (23.6.4)]} \\
&= \int_{R^n}\left[\int_\Theta L(\theta(\boldsymbol{x}), \theta)\xi^*(\theta)d\theta\right] f_\xi(\boldsymbol{x})d\boldsymbol{x} \quad \text{[from (23.6.5)]} \\
&= E_{\boldsymbol{X}}\left[E_{\theta|\boldsymbol{X}}[L(\theta(\boldsymbol{X}), \theta)]\right] \quad \text{[from (23.6.8)]} \\
&= E_{\boldsymbol{X}}\left[r^*(\theta(\boldsymbol{X}))\right] = \int_{R^n} r^*(\theta(\boldsymbol{x}))f_\xi(\boldsymbol{x})d\boldsymbol{x}.
\end{aligned} \tag{23.6.9}$$

Here $E_{\boldsymbol{x}}$ denotes expectation w.r.t. unconditional joint distribution (23.6.7) of $\boldsymbol{X}$. For minimising $\bar{r}_\xi(\delta)$ we, therefore, minimise $r^*(\theta(\boldsymbol{x}))$ (with respect to choice of $\theta(\boldsymbol{x})$) for each given $\boldsymbol{x}$, $f_\xi(\boldsymbol{x})$ being a constant. Bayes a posterior solution (i.e., solution based on posterior risk) is, therefore, to choose that $\theta(\boldsymbol{x})$ as an estimate of θ for which the posterior risk $r^*(\theta(x))$ is minimum, i.e.,

$$r^*(\theta(\boldsymbol{x})) = \min_{\theta'(\boldsymbol{x})}[r^*(\theta'(\boldsymbol{x}))]. \tag{23.6.10}$$

Clearly for such a d.r. $\delta_\xi(\boldsymbol{x})[= \theta(\boldsymbol{x})$ satisfying (23.6.10) when $\boldsymbol{X} = \boldsymbol{x}]$, the average risk $\bar{r}_\xi(\delta)$ is minimum. The Bayes risk or the risk of $\delta_\xi(\boldsymbol{X})$ is given by (23.6.9) (which $\bar{r}_\xi(\delta_\xi)$).

Bayes estimates in case of squared error loss function.

Suppose $L(\theta(\boldsymbol{x}), \theta) = (\theta(\boldsymbol{x}) - \theta)^2$. In this case Bayes estimate $\delta_\xi(\boldsymbol{x})$ should be such that

$$\int_\Theta (\theta(\boldsymbol{x}) - \theta)^2\xi^*(\theta)d\theta$$

is minimum. It is known that the mean square of Z about A, $E(Z - A)^2$ is minimum when $A = E(Z)$. Hence,

$$\delta_\xi(\boldsymbol{x}) = \int_\Theta \theta\xi^*(\theta)d\theta = \frac{1}{f_\xi(\boldsymbol{x})}\int_\Theta \theta f(\boldsymbol{x}|\theta)\xi(\theta)d\theta \tag{23.6.11}$$

i.e., Bayes estimate is mean of the posterior distribution of θ given $\boldsymbol{x}$. Also Bayes risk is

$$E_{\boldsymbol{X}}\left[E_{\theta|\boldsymbol{X}}(\theta(\boldsymbol{X})-\theta)^2\right] = E_{\boldsymbol{X}}V_{\theta|\boldsymbol{X}}(\theta) \tag{23.6.12}$$

i.e., expected value of the variance of θ given $\boldsymbol{X}$ with respect to the conditional distribution of $\boldsymbol{X}$.

The theorems 23.6.1 and 23.6.2 follow directly from theorems 23.4.1 and 23.4.2 respectively.

Theorem 23.6.1 *If there exists a prior distribution ξ of θ and δ_ξ be Bayes estimator for θ corresponding to ξ and if*

$$r(\delta_\xi|\theta) = constant \;\; \forall\, \theta \in \Theta$$

then (a) δ_ξ is a minimax estimator of θ (b) ξ is a least favourable distribution.

Theorem 23.6.2 *For a given sequence of prior distribution $\{\xi_n\}$ of θ over Θ let $\delta_n = \delta_{\xi_n}$ be Bayes estimator of θ and $r_n = \bar{r}_{\xi_n}(\delta_{\xi_n})$ be the corresponding Bayes risk. If $\lim_{n\to\infty} r_n = r$ and if δ be any estimator with $r(\delta|\theta) \le r$, then δ is a minimax estimator of θ.*

Theorem 23.6.3 *Any Bayes estimator $\delta^0 (= \delta_\xi)$ with continuous risk function $r(\delta^0|\theta)$ and the prior $\xi(\theta) > 0 \;\forall\, \theta \in \Theta$ is admissible.*

Proof: If possible, let there be another d.r. δ which is uniformly better than δ^0 i.e.,

$$r(\delta|\theta) \le r(\delta^0|\theta) \;\; \forall\, \theta$$

with strict inequality for some θ. Suppose $r(\delta|\theta) < r(\delta^0|\theta)$ for $\theta = \theta_0$. Then because of continuity of risk function, $r(\delta|\theta) < r(\delta^0|\theta) \;\forall\, \theta \in (\theta_0 - \varepsilon, \theta_0 + \varepsilon)$, where ε is sufficiently small.

Hence,

$$\int_{\theta_0-\varepsilon}^{\theta_0+\varepsilon} r(\delta|\theta)\xi(\theta)d\theta < \int_{\theta_0-\varepsilon}^{\theta_0+\varepsilon} r(\delta^0|\theta)\xi(\theta)d\theta.$$

Therefore

$$\bar{r}_\xi(\delta) < \bar{r}_\xi(\delta^0) \quad (\text{ because } r(\delta|\theta) = r(\delta^0|\theta) \text{ for } \theta \notin (\theta_0 - \varepsilon, \theta_0 + \varepsilon))$$

which is a contradiction, since δ^0 is a Bayes estimator w.r.t. ξ.

Theorem 23.6.4 *A Bayes estimator $\delta^0 (= \delta^0_\xi)$ if it is unique is admissible.*

Proof: Suppose there exists a d.r. δ which is uniformly better than δ^0 i.e.,

$$r(\delta|\theta) \le r(\delta^0|\theta) \;\; \forall\, \theta$$

with strict inequality holding for some θ. Hence

$$\int r(\delta|\theta)\xi(\theta)d\theta \leq \int r(\delta^0|\theta)\xi(\theta)d\theta$$
$$\text{i.e.,}\quad \bar{r}(\delta) \leq \bar{r}(\delta^0)$$

which is a contradiction. Hence the proof.

■ NOTE 23.6.1 *Clearly theorems* 23.6.3, 23.6.4 *are quite general and are applicable for any Bayes d.r. (not only Bayes estimators).*

23.6.3 Conjugate Prior Distribution

Raiffa and Schlaifer (1961) noted the following desiderate for choosing a class G of priors.

(a) G should be analytically tractable in three respects: (i) posterior distribution should be easy to determine (ii) it should be possible to find expectation of some 'utility functions' w.r.t. members in G (iii) G should be closed in the sense that if a prior belongs to G, the posterior should also belongs to G.

(b) G should be rich, i.e., there should exist a member of G capable of expressing decision maker's belief and prior information.

(c) G should be parametrisable in a manner which can be readily interpreted, so that it will be easy to verify that the chosen member of the family is in close agreement with the decision-maker's prior judgements about Θ.

Suppose that the likelihood function $f(\boldsymbol{x}|\theta)$ can be written as $f(\boldsymbol{x}|\theta) = k(\boldsymbol{x}|\theta)\rho(\boldsymbol{x})$ when $\rho(\boldsymbol{x})$ is independent of θ. The function $k(\boldsymbol{x}|\theta)$ is called the Kernal of the likelihood and $\rho(\boldsymbol{x})$ is the residual of the likelihood. If $Y = \phi(\boldsymbol{x})$ is a sufficient statistic for θ, $f(\boldsymbol{x}|\theta) = k(y|\theta)\rho(\boldsymbol{x})$.

Instead of considering the Kernal $k(y|\theta)$ as a function of y with parameter θ on the sample space of Y, we can consider it as a function $k(\theta|y)$ of θ on Θ with parameter y. The density function

$$g(\theta|y) = k(\theta|y)N(y), \quad \theta \in \Theta$$

where $N(y)$ is a normalising constant,

$$N(y) = \left[\int_{\Theta} k(\theta|y)d\theta\right]^{-1}$$

and y is treated as a parameter is called the 'Natural conjugate of likelihood' $f(y|\theta)$ of y on Y.

► **Example 23.6.1** *Suppose that $X_1, \cdots, X_n$ are i.i.d. $N(\mu, \sigma)$ variables.*

$$f(\boldsymbol{x}|\mu, \sigma) = \left[\frac{\sqrt{n}}{\sqrt{2\pi}\sigma}e^{-\frac{n}{2}(\bar{x}-\mu)^2}\right]\rho(\boldsymbol{x}) = k(n, \bar{x})\rho(\boldsymbol{x})$$

so that the statistic $(n, \bar{x})$ is sufficient for μ.

Considering $k(n, \bar{x}|\mu, \sigma)$ as a function of μ, we let density function of μ over $(-\infty, \infty)$ or given $y[=(n, \bar{x})]$ looked upon as parameter as $g(\mu|y) \propto k(y|\mu, \sigma^2)$ i.e.,

$$g(\mu|y) = \frac{\sqrt{n}}{\sqrt{2\pi}\sigma} \exp\left[-\frac{n}{2\sigma^2}(\mu - \bar{x})^2\right]$$

s a density function of μ. This density function is a natural conjugate of likelihood $(\boldsymbol{x}|\mu)$.

The families of natural conjugate priors with parameter y (value of the sufficient tatistic) in the range of values of Y has the properties of tractability and closedness. Richness of families of natural conjugate densities can be greatly enhanced by increasing he domain of the parameter to include all values for which (i) $k(y|\theta)$ is non-negative ii) $\int_\Theta k(\theta|y)$ converges. One has not necessarily to restrict oneself to the domain of Y s arises from the given experiment from which the likelihood $f(\boldsymbol{x}|\theta)$ follows, but may lso restrict to the larger (conceivable) range of values of Y for which the properties (i) nd (ii) hold.

The table 23.1 gives some examples of natural conjugate priors along with the orresponding likelihood functions.

Table 13.1

Data Generating Process	Sufficient Statistics	Likelihood of the Sample $\propto$	Conjugate prior density
Bernoulli $p^x(1-p)^{n-x}$, $x = 0, 1$ $0 < p < 1$	$\left(n, r = \sum_{i=1}^n x_i\right)$	$p^r(1-p)^{n-r}$	$\xi_\beta(p, r', n')$ $\propto p^{r'-1}(1-p)^{n'-r'-1}$ (Beta 1 density), $0 < p < 1, n' > r' > 0$
Poisson $e^{-\lambda}\lambda^x/x!$, $x = 0, 1, 2, \cdots$ $\lambda > 0$	$\left(n, t = \sum_{i=1}^n x_i\right)$	$e^{-n\lambda}\lambda^t$	$\xi_r(\lambda \mid n', t')$ $\propto e^{-\lambda t'}(\lambda t')^{n'-1}$ (Gamma density), $\lambda > 0, n', t' > 0.$
	(a) μ unknown, σ known (n, m) when $m = \bar{x}$	$\exp\left[-\frac{n}{2\sigma^2}(\bar{x} - m)^2\right]$	$\xi_N(\mu \mid m', n'\sigma)$ $\propto \exp\left[-\frac{n'}{2\sigma^2}(\mu - m')^2\right]$, $-\infty < \mu < \infty, \sigma > 0$ $-\infty < m' < \infty, n' > 0$
Independent Normal $\frac{1}{\sqrt{2\pi}\sigma}\exp[-(x-\mu)^2/2\sigma^2]$ $-\infty < x < \infty,$ $-\infty < \mu < \infty, \sigma > 0$	(b) μ known σ unknown (n, v) where $v = \frac{1}{n}\sum_{i=1}^n (x_i - \mu)^2$	$e^{-nv/2\sigma^2}\sigma^{-n/2}$	$\xi_r(\sigma^2 \mid v', n')$ $\propto e^{-v'n'/2\sigma^2}(\sigma^2)^{-(n'/2-1)}$, $\sigma^2 > 0$ $v' > 0, n' > 0$
	(c) both μ, σ unknown $(n, m, \nu = n-1, w)$ where $w = \frac{1}{\nu}\sum_{i=1}^n (x_i - m)^2$	$\left\{\frac{1}{\sigma}e^{n(m-\mu)^2/2\sigma^2}\right\}$ $\left\{e^{-\nu w/2\sigma^2}\sigma^{-\nu/2}\right\}$	$\xi_{N\gamma}(\mu, \sigma^2 \mid m', w', n', \nu')$ $\propto \left\{\frac{1}{\sigma}e^{-n'(\mu-m')^2/2\sigma^2}\right\}$ $\left\{e^{-\nu' w'/2\sigma^2}e^{-(\nu'/2-1)}\right\}$ (Normal-gamma density) $-\infty < \mu < \infty, \sigma^2 > 0$ $-\infty < m' < \infty, w', n', \nu' > 0$

Bayesian Sufficiency

A statistic $Y = \phi(\boldsymbol{x})$ is said to be Bayesian sufficient for θ if the posterior distribution c θ depends on $\boldsymbol{X}$ only through $\phi(\boldsymbol{X})$. Clearly, if Y is sufficient for θ (in ordinary sense it is Bayesian sufficient for θ and vice-versa. Hence the definition of ordinary sufficienc and Bayesian sufficiency are equivalent.

► **Example 23.6.2** *Let $X_1, X_2, \cdots, X_n$ be n independent $N(\mu, \sigma^2)$ variables where μ unknown but σ^2 is known. Let prior distribution of μ be $N(\theta, \sigma_0^2)$. Find Bayes estimat of μ.*

The likelihood function of $\boldsymbol{x}$ is

$$f(\boldsymbol{x}|\mu) \propto \exp\left[-\frac{1}{2\sigma^2}\sum_{i=1}^{n}(x_i-\mu)^2\right] = \exp\left[-\frac{n}{2\sigma^2}(\bar{x}-\mu)^2 - \frac{1}{2\sigma^2}\sum_{i=1}^{n}(x_i-\bar{x})^2\right]$$

$$\propto \exp\left[-\frac{n}{2\sigma^2}(\bar{x}-\mu)^2\right].$$

The posterior distribution of μ given $\boldsymbol{x}$ is

$$f^*(\mu|\boldsymbol{x}) \propto f(\boldsymbol{x}|\mu)f(\mu|\theta)$$

$$\propto \exp\left[-\frac{n}{2\sigma^2}(\bar{x}-\mu)^2 - \frac{1}{\sigma_0^2}(\mu-\theta)^2\right]$$

$$\propto \exp -\frac{1}{2}\left(-\frac{n\sigma_0^2+\sigma^2}{\sigma_0^2\sigma^2}\right)\left(\mu - \frac{n\bar{x}\sigma_0^2+\theta\sigma^2}{n\sigma_0^2+\sigma^2}\right)^2.$$

which shows that posterior distribution of μ is Normal with mean $\frac{n\bar{x}\sigma_0^2+\theta\sigma^2}{n\sigma_0^2+\sigma^2}$ and posteri variance

$$V(\mu/\boldsymbol{x}) = \frac{\sigma_0^2\sigma^2}{n\sigma_0^2+\sigma^2} = \sigma^{*^2} \text{ (say)}.$$

Hence, Bayes estimator of μ is

$$\hat{\mu} = \frac{n\bar{x}\sigma_0^2+\theta\sigma^2}{n\sigma_0^2+\sigma^2} = \bar{x}\frac{(n/\sigma^2)}{n/\sigma^2+1/\sigma_0^2} + \theta\frac{1/\sigma_0^2}{n/\sigma^2+\frac{1}{\sigma_0^2}}.$$

Note that the maximum likelihood estimator of μ is $\bar{x}$.

Thus, Bayes estimator is a convex combination of $\bar{x}$ and θ, the weights bei proportional to inverse of the variances.

Again, since $V(\mu|\boldsymbol{x})$ is independent of $\boldsymbol{x}$, the Bayes risk of $\hat{\mu}$ is σ^{*^2}.

► **Example 23.6.3** *Let $X_1, \cdots, X_n$ be i.i.d. $N(\mu, \sigma^2)$ r.v.'s. Let a priori distribution o be $N(\theta, \tau^2)$. Find Bayes estimate of μ taking the loss function as $L(d(\boldsymbol{x}), \mu) = |d(\boldsymbol{x}) -$*

It follows that the conditional distribution of μ given $\boldsymbol{x} = (x_1, \cdots, x_n)$ is normal with mean

$$E(\mu|\boldsymbol{x}) = \frac{\theta(\sigma^2/n) + \bar{x}\tau^2}{\tau^2 + \sigma^2/n} = \mu^*_{\boldsymbol{x}}$$

and variance

$$V(\mu|\boldsymbol{x}) = \frac{\tau^2\sigma^2/n}{\tau^2 + \sigma^2/n}.$$

It follows that Bayes estimate is one which minimises the posterior risk $E_{\mu|\boldsymbol{x}}$ $[|d(\boldsymbol{x}) - \mu|]$ and it must be the median of the posterior distribution (since the mean deviation is minimum when taken about the median). Since the normal distribution is symmetric about the mean, the Bayes estimate is given by $\mu^*_{\boldsymbol{x}}$.

▶ **Example 23.6.4** *A lot contains N elements of which D are defectives. A random sample of size n produces X defectives. Find a minimax estimate of D, taking a squared error loss function.*

We have

$$P(X = k|D) = P_D(X = k) = \binom{D}{k}\binom{N-D}{n-k} \Big/ \binom{N}{n}.$$

It is known that

$$E_D(X) = n \cdot \frac{D}{N}, \quad V_D(X) = \frac{nD(N-n)(N-D)}{N^2(N-1)}.$$

We find a linear function of X with constant risk. It is found that

$$E_D(\alpha X + \beta - D)^2 = \beta^2$$

when

$$\alpha = \frac{N}{n + \sqrt{n(N-n)(N-1)}}, \quad \beta = \frac{N}{2}\left(1 - \frac{\alpha n}{N}\right). \tag{i}$$

It will now be shown that $\alpha x + \beta$ is the Bayes estimate corresponding to the a priori p.m.f.

$$P(D = d) = \frac{1}{B(a,b)} \int_0^1 \binom{N}{d} p^d (1-p)^{N-d} p^{a-1} (1-p)^{b-1} dp,$$

where $a, b > 0$. Clearly this a priori p.m.f. of D is obtained by considering $P(D = d|p)$ as Bin. (N, p) and prior of p as $B(a, b)$ and then calculating the unconditional distribution of D.

Since

$$\sum_{d=0}^{N} P(D = d) = 1, \quad \sum_d \binom{N}{d} \frac{B(a+d, N+b-a)}{B(a,b)} = 1.$$

The Bayes estimate is

$$D^* = E(D|X=k) = \frac{\sum_{d=k}^{N-n+k} d.\binom{d}{k}\binom{N-d}{n-k}\binom{N}{d}B(a+d, N+b-d)}{\sum_{d=k}^{N-n+k} \binom{d}{k}\binom{N-d}{n-k}\binom{N}{d}B(a+d, N+b-d)}.$$

Writing $d = (d-a)+a$ and using

$$\binom{d}{k}\binom{N-d}{n-k}\binom{N}{d} = \binom{n}{k}\binom{N-n}{d-k}\binom{N}{n}$$

gives

$$D^* = \frac{\sum_{i=0}^{N-n}\binom{N-n}{i}\overline{|(d+a+1)}\,\overline{|(N+b-d)}}{\sum_{i=0}^{N-n}\binom{N-n}{i}\overline{|(a+d)}\,\overline{|(N+b-d)}} - a = k\cdot\frac{a+b+N}{a+b+n} + \frac{a(N-n)}{a+b+n}.$$

Putting $\alpha = \frac{a+b+N}{a+b+n}$ and $\beta = \frac{a(N-n)}{a+b+n}$ and solving for a and b, we get

$$a = \frac{\beta}{\alpha - 1}, \quad b = \frac{N - \alpha n - \beta}{\alpha - 1}.$$

Therefore, $\alpha X + \beta$ is a Bayes estimator. Hence the estimator $\alpha X + \beta$ where α, β are given in (i) is a minimax estimator of D.

► **Example 23.6.5** *If $X_1, \cdots, X_n$ are n independent $N(\mu, \sigma^2)$ variables where μ is unknown but σ^2 is known find minimax estimator of μ.*

Consider a sequence of prior distributions $\{\xi_\lambda\}$ where ξ_λ is $N(0, \lambda^2)$.
The Bayes estimate of μ is

$$\delta_\lambda(\boldsymbol{x}) = E(\mu|\boldsymbol{x}) = \frac{n\bar{x}\lambda^2}{n\lambda^2 + \sigma^2}$$

with Bayes risk

$$\bar{r}(\delta_\lambda) = \frac{\lambda^2\sigma^2}{n\lambda^2 + \sigma^2}.$$

Now

$$\lim_{\lambda\to\infty} \bar{r}(\delta_\lambda) = \lim_{\lambda\to\infty}\left[\frac{\sigma^2}{n + \frac{\sigma^2}{\lambda^2}}\right] = \frac{\sigma^2}{n}.$$

Also

$$r(\bar{x}|\mu) = \frac{\sigma^2}{n} \quad \forall\, \mu.$$

Hence by theorem 23.6.2, $\bar{x}$ is a minimax estimator of μ.

► **Example 23.6.6** *If $\theta(X)$ is a minimax estimate of θ show that $f(X) = a\theta(x) + b$ is a minimax estimator of $a\theta + b$, $a(\neq 0), b$ being constants and the loss function being assumed to be of the squared error type.*

If possible, let there be another estimator $g(X)(\not\equiv f(X))$ which is a minimax estimator of θ. Let $r(g|\theta)$, $r(f|\theta)$ denote respectively the risks of g and f when θ is the true value of the parameter. Then

$$\sup_\theta r(g|\theta) \leq \sup_\theta r(f|\theta)$$

$$\text{or,} \quad \sup_\theta E[g(X) - a\theta - b]^2 \leq \sup_\theta E[a\theta(X) + b - a\theta - b]^2$$

$$= a^2 \sup_\theta E(\theta(X) - \theta)^2$$

$$\text{or,} \quad a^2 \sup_\theta E\left[\frac{g(X) - b}{a} - \theta\right]^2 \leq a^2 \sup_\theta E(\theta(X) - \theta)^2$$

$$\text{or,} \quad \sup_\theta E[\phi(X) - \theta]^2 \leq \sup_\theta E(\theta(X) - \theta)^2 \tag{i}$$

where $\phi(X) = \{g(X) - b\}/a$. Now $\phi(X)$ is also an estimate of θ. The expression (i) means $\sup_\theta r(\phi|\theta) \leq \sup_\theta r(\theta(X)|\theta)$ which is a contradiction.

► **Example 23.6.7** *Suppose r is a binomial variate with p.m.f. $\binom{n}{r}p^r(1-p)^{n-r}$, $r = 0, 1, \cdots, n$. Show that the minimax estimate of p is $p^*(r) = \frac{1}{1+\sqrt{n}}\left(\frac{r}{\sqrt{n}} + \frac{1}{2}\right)$; also show that the least favourable distribution of p is $dF(p) = \text{Const.}.p^{(\sqrt{n}/2)-1}(1-p)^{(\sqrt{n}/2)-1}dp$, $0 < p < 1$.*

How does the m.l.e. of p fare in this case?

Let us consider an estimate of the form $\hat{p}(r) = ar + b$. Risk function is

$$R(\bar{p}) = E(ar + b - p)^2$$

$$= p^2[(na-1)^2 - na^2] + p[2b(an-1) + a^2n] + b^2. \tag{i}$$

It is known that a constant risk Bayes estimator is a minimax estimator. To make (i) a constant (independent of p) we must have

$$(na-1)^2 = na^2 \tag{ii}$$

$$\text{and } 2b(an-1) + a^2n = 0 \tag{iii}$$

(ii) and (iii) give

$$a = \frac{1}{\sqrt{n}(\sqrt{n} \pm 1)}, \quad b = \pm\frac{1}{2(\sqrt{n}+1)} \tag{iv}$$

when

$$r(\hat{p}) = \frac{1}{4(\sqrt{n}+1)^2} \quad \text{and} \quad \hat{p}(r) = p^*(r).$$

We are to show that $p^*(r)$ is also a Bayes estimate. Now it is known that Bayes estimate of p,

$$\hat{p}_b = E(p|r) = \frac{\int_0^1 p\binom{n}{r}p^r(1-p)^{n-r}dF(p)}{\int_0^1 \binom{n}{r}p^r(1-p)^{n-r}dF(p)}$$

where $F(p)$ is the c.d.f. of the prior distribution of p. Taking

$$F(p) = \text{Const.}\{p(1-p)\}^{\alpha-1}, 0 < p < 1 \qquad \text{(v)}$$

where α is a suitable constant,

$$\hat{p}_b = \frac{\Gamma(r+\alpha+1)}{\Gamma(n+2\alpha+1)}\cdot\frac{\Gamma(n+2\alpha)}{\Gamma(r+\alpha)} = \frac{r+\alpha}{n+2\alpha} = \frac{1}{1+\sqrt{n}}\left(\frac{r}{\sqrt{n}}+\frac{1}{2}\right)$$

if

$$\left.\begin{aligned} \frac{1}{n+2\alpha} &= \frac{1}{\sqrt{n}(1+\sqrt{n})} \\ \text{and } \frac{\alpha}{n+2\alpha} &= \frac{1}{2(1+\sqrt{n})} \end{aligned}\right\}. \qquad \text{(vi)}$$

The equation (vi) give $\alpha = \frac{\sqrt{n}}{2}$. Hence for the prior distribution (v) with $\alpha = \frac{n}{2}$, $p^*(r)$ is a Bayes estimate with constant risk and hence is a minimax estimate.

The m.l.e. of $p, \hat{p} = \frac{r}{n}$. $V(\hat{p}) = \frac{p(1-p)}{n} = R(\hat{p}|p)$.

Now $R(p^*|p) < R(\hat{p}|p)$ if $\frac{1}{4(\sqrt{n}+1)^2} \leq \frac{p(1-p)}{n}$ i.e., if $\psi(p) = \frac{p^2}{n} - \frac{p}{n} + \frac{1}{4(\sqrt{n}+1)^2} \leq 0$.

The roots of the equation $\psi(p) = 0$ are $p = \frac{1}{2} \pm \frac{\sqrt{2\sqrt{n}+1}}{2(\sqrt{n}+1)}$.

Therefore, p^* will be a better estimate than the the m.l.e. $\hat{p}$ if

$$\frac{1}{2} - \frac{\sqrt{2\sqrt{n}+1}}{2(\sqrt{n}+1)} < p < \frac{1}{2} + \frac{\sqrt{2\sqrt{n}+1}}{2(\sqrt{n}+1)}.$$

Note that the m.l.e. $\hat{p}$ is also UMVUE of p. Moreover,

$$\frac{\sup_p R(\hat{p}|p)}{\sup_p R(p^*|p)} = \frac{1/4n}{1/[4(1+\sqrt{n})^2]} = \frac{n+2\sqrt{n}+1}{n} \quad \to \quad 1 \text{ as } n \to \infty.$$

Clearly, one should prefer the minimax estimator p^* if n is small, and UMVUE $\hat{p}$ for its simplicity, if n is large.

► **Example 23.6.8** *Let X, Y be independent binomial variates with $P(X = k) = \binom{n}{k} p^k (1-p)^{n-k}$, $P(Y = r) = \binom{n}{r} p^r (1-p)^{n-r}$. Show that the minimax estimate of $p_1 - p_2$ is*

$$p^{**} = \frac{\sqrt{2n}}{1+\sqrt{2n}}\left(\frac{X}{n} - \frac{Y}{n}\right).$$

We shall find minimax estimate of $p_1 - p_2$ when $p_1 = 1 - p_2 = p$ (say). In this case, $p_1 - p_2 = p - (1-p) = 2p - 1$.

Here $X \sim$ Bin (n, p) and $Y \sim$ Bin $(n, 1-p)$. Therefore, $n - Y \sim$ Bin (n, p). Hence $U = X + n - Y \sim$ Bin (N, p) when $N = 2n$ (theorem 5.4.1). By example (23.6.7), minimax estimate of p is

$$p^* = \frac{1}{\sqrt{N}+1}\left(\frac{u}{\sqrt{N}} + \frac{1}{2}\right).$$

By example (23.6.6), minimax estimate of $2p - 1$ is

$$2p^* - 1 = \frac{\sqrt{2n}}{1+\sqrt{2n}}\left(\frac{X}{n} - \frac{Y}{n}\right) = p^{**}.$$

We shall now show that

(a) $\sup\limits_{p_1,p_2} R(p^{**})$ is attained when $p_1 = 1 - p_2$,

(b) $\sup\limits_{p_1,p_2} R(p^{**}) = \inf\limits_{t} \sup\limits_{p_1,p_2} R(t)$

when $R(\cdot)$ denotes the risk of $(\cdot)$ in estimating $p_1 - p_2$.

Now

$$\begin{aligned}
R(p^{**}) &= E\left[\frac{\sqrt{2n}}{1+\sqrt{2n}}\left(\frac{X}{n} - \frac{Y}{n}\right) - (p_1 - p_2)\right]^2 \\
&= E\left[\frac{\sqrt{2n}}{1+\sqrt{2n}}\left\{\left(\frac{X}{n} - \frac{Y}{n}\right) - (p_1 - p_2)\right\} - \frac{1}{1+\sqrt{2n}}(p_1 - p_2)\right]^2 \\
&= \frac{2n}{(1+\sqrt{2n})^2} E\left[\left(\frac{X}{n} - \frac{Y}{n}\right) - (p_1 - p_2)\right]^2 + \frac{1}{(1+\sqrt{2n})^2}(p_1 - p_2)^2 \\
&= \frac{1}{(1+\sqrt{2n})^2}\left[(p_1 - p_2)^2 + 2p_1(1-p_1) + 2p_2(1-p_2)\right] \\
&= f(p_1, p_2) \text{ (say)}.
\end{aligned}$$

Again,

$$\frac{\partial f}{\partial p_1} = \frac{2}{(1+\sqrt{2n})^2}(1 - p_1 - p_2) = \frac{\partial f}{\partial p_2}.$$

Equating $\frac{\partial f}{\partial p_i} = 0$ $(i = 1, 2)$ it is found that $f(p_1, p_2)$ is maximum if $p_1 + p_2 = 1$.

Again, if $t = t(X,Y)$ is any estimate of $p_1 - p_2$ then

$$\sup_{p_1,p_2} R(t) \geq \sup_{\substack{p_1,p_2: \\ p_1+p_2=1}} R(t) \geq \sup_{\substack{p_1,p_2: \\ p_1+p_2=1}} R(p^{**}) = \sup_{p_1,p_2} R(p^{**}).$$

Hence,

$$\sup_{p_1,p_2} R(t) \geq \sup_{p_1,p_2} R(p^{**}) \;\; \forall\, t.$$

Hence the result.

▶ **Example 23.6.9** *Suppose we have samples of sizes $n_1, \cdots, n_k$ from k independen[t] populations $N(m_1, \sigma_1^2), \cdots, N(m_k, \sigma_k^2)$ and we want to find the minimax estimate [o]f $\boldsymbol{m} = (m_1, \cdots, m_k)$ when $\sigma_1^2, \cdots, \sigma_k^2$ are known.*

Let $\boldsymbol{t} = (t_1, \cdots, t_k)$ be an estimate and the corresponding loss function be

$$L(\boldsymbol{m}, \boldsymbol{t}) = \sum_{i=1}^{k} (t_i - m_i)^2.$$

The risk of the estimate is

$$r(\boldsymbol{t}|\boldsymbol{m}) = \sum_{1}^{k} E(t_i - m_i)^2$$

$$= \sum_{1}^{k} r_i(\boldsymbol{t}|\boldsymbol{m})$$

when $r_i(\boldsymbol{t}|\boldsymbol{m}) = E(t_i - m_i)^2$.

Now

$$r_i(\boldsymbol{t}|\boldsymbol{m}) \leq \sup_{\boldsymbol{m}} r_i(\boldsymbol{t}|\boldsymbol{m}).$$

Hence,

$$r(\boldsymbol{t}|\boldsymbol{m}) = \sum_{1}^{k} r_i(\boldsymbol{t}|\boldsymbol{m}) \leq \sum_{1}^{k} \sup_{\boldsymbol{m}} r_i(\boldsymbol{t}|\boldsymbol{m}).$$

Therefore,

$$\sup_{\boldsymbol{m}} r(\boldsymbol{t}|\boldsymbol{m}) \leq \sum_{1}^{k} \sup_{\boldsymbol{m}} r_i(\boldsymbol{t}|\boldsymbol{m}).$$

Now $\sup_{\boldsymbol{m}} E(t_i - m_i)^2$ is minimised for $t_i = \bar{x}_i$ (example, 23.6.5) and hence

$$\sum_{1}^{k} \sup_{\boldsymbol{m}} r_i(\boldsymbol{t}|\boldsymbol{m})$$

is minimised for $\boldsymbol{t} = (\bar{x}_1, \cdots, \bar{x}_k) = \bar{\boldsymbol{x}}$.

For this choice of $\boldsymbol{t}$,

$$\sum \sup_{\boldsymbol{m}} r_i(\boldsymbol{t}|\boldsymbol{m}) = \sum \frac{\sigma_i^2}{n_i}.$$

Again

$$r(\bar{\boldsymbol{x}}|\boldsymbol{m}) = \sum \frac{\sigma_i^2}{n_i}.$$

Therefore, $\bar{\boldsymbol{x}} = (\bar{x}_1, \cdots, \bar{x}_k)$ is minimax estimate of $\boldsymbol{\mu}$.

▶ **Example 23.6.10** *Minimax character of least square estimates:*

Consider n independent variables $Y_1, \cdots, Y_n$ with

$$E(Y_i) = \sum_{j=1}^{p} a_{ij}\beta_j = m_i \quad \text{and} \quad V(Y_i) \leq \sigma^2.$$

Distribution of Y_i may be anything within a class of distribution functions Γ including the normal distribution Γ_0 where $Y_i \sim N(m_i, \sigma^2)$. Assume rank of $\boldsymbol{A} = ((a_{ij})) = p$. Assume that the loss function is $L(\boldsymbol{t}, \boldsymbol{\beta}) = \Sigma(t_i - \beta_i)^2$. We first consider $\boldsymbol{Y} = (Y_1, \cdots, Y_n)'$ have a distribution in Γ_0.

From $\boldsymbol{Y}$ we pass on to new variables $(Z_1, \cdots, Z_n)'$ by an orthogonal transformation such that

$$E(Z_i) = \begin{cases} C_i\beta_i, & i = 1, \cdots, p \\ 0, & i = p+1, \cdots, n \end{cases}; \quad V(Z_i) = \sigma^2$$

and Z_i's are independently normally distributed. Hence, least square estimate of β_i is obtained by the solution of

$$\frac{\partial}{\partial \beta_i} \sum_{i=1}^{p} (Z_i - C_i\beta_i)^2 = 0$$

when $\hat{\beta}_i = Z_i/C_i, \quad i = 1, \cdots, p$.

Let $\hat{\boldsymbol{\beta}} = (\hat{\beta}_1, \cdots, \hat{\beta}_p)$. Risk of the estimate

$$\begin{aligned} \hat{\boldsymbol{\beta}} = r(\hat{\boldsymbol{\beta}}|\boldsymbol{\beta}) &= \sum_{i=1}^{p} E(\hat{\beta}_i - \beta_i)^2 = \sum_{i=1}^{p} V(\hat{\beta}_i) \\ &= \sigma^2 \sum_{i=1}^{p} \frac{1}{C_i^2} = \mu \text{ (say) (a constant).} \end{aligned}$$

Therefore,

$$\sup_{\boldsymbol{\beta}} r(\hat{\boldsymbol{\beta}}|\boldsymbol{\beta}) = \mu.$$

Let us now find the Bayes estimate $\hat{\boldsymbol{\beta}}^* = (\hat{\beta}_1^*, \cdots, \hat{\beta}_p^*)$, corresponding to a priori distribution

$$d\xi(\boldsymbol{\beta}) = \prod_{i=1}^{p} d\xi_i(\beta_i)$$

where $d\xi_i(\beta_i) = \text{Const.}.e^{-\beta_i^2/2\sigma^2}\, d\beta_i$.

$$\hat{\beta}_i^* = \frac{\int \beta_i \exp\left[-\frac{1}{2}\left\{\frac{(Z_i - C_i\beta_i)^2}{\sigma^2} + \frac{\beta_i^2}{\sigma_0^2}\right\}\right] d\beta_i}{\int \exp\left[-\frac{1}{2}\left\{\frac{(Z_i - C_i\beta_i)^2}{\sigma^2} + \frac{\beta_i^2}{\sigma_0^2}\right\}\right] d\beta_i}$$

$$= \frac{\int_{-\infty}^{\infty} \beta_i \exp\left[-\frac{C_i^2\sigma_0^2 + \sigma^2}{2\sigma_0^2\sigma^2}\left\{\beta_i - \frac{Z_iC_i\sigma_0^2}{C_i^2\sigma_0^2 + \sigma^2}\right\}\right]^2 d\beta_i}{\int_{-\infty}^{\infty} \exp\left[-\frac{C_i^2\sigma_0^2 + \sigma^2}{2\sigma_0^2\sigma^2}\left\{\beta_i - \frac{Z_iC_i\sigma_0^2}{C_i^2\sigma_0^2 + \sigma^2}\right\}\right]^2 d\beta_i}.$$

Hence, $\hat{\beta}_i^* = \lambda_i Z_i$ where

$$\lambda_i = \frac{C_i\sigma_0^2}{C_i^2\sigma_0^2 + \sigma^2}.$$

Let $\hat{\boldsymbol{\beta}}^{**}$ be any estimate. Then

$$\sup_{\boldsymbol{\beta}} r(\hat{\boldsymbol{\beta}}^{**}|\boldsymbol{\beta}) \geq \bar{r}(\hat{\boldsymbol{\beta}}^*|\boldsymbol{\beta}) = E_\xi E_Z \sum_{i=1}^{p}(Z_i\lambda_i - \beta_i)^2$$

$$= E_Z E_{\xi|Z} \sum_{i=1}^{p}(Z_i\lambda_i - \beta_i)^2 = \sum_{i=1}^{p} \frac{\sigma_0^2\sigma^2}{C_i^2\sigma_0^2 + \sigma^2} \quad \text{for all } \sigma_0^2. \quad \text{(i)}$$

Taking limit of both sides of (i) as $\sigma_0^2 \to \infty$.

$$\sup_{\boldsymbol{\beta}} r(\hat{\boldsymbol{\beta}}^{**}|\boldsymbol{\beta}) \geq \sigma^2 \sum \frac{1}{C_i^2} = \mu = \sup_{\boldsymbol{\beta}} r(\hat{\boldsymbol{\beta}}|\boldsymbol{\beta}).$$

Therefore, the least square estimator of $\boldsymbol{\beta}$ is minimax for the class Γ_0 of normal distributions.

For other types of populations in Γ the maximum risk of $\hat{\boldsymbol{\beta}}$ is less than μ (because $V(y_i) < \sigma^2$). Hence $\hat{\boldsymbol{\beta}}$ is the minimax estimate of $\boldsymbol{\beta}$ for all populations of $\boldsymbol{Y} \in \Gamma$.

We now consider some more theorems on d.r.'s for estimating θ.

23.6.4 Some Properties of Estimates

Theorem 23.6.5 *If the loss function $L(\theta(\boldsymbol{X}), \theta)$ is strictly convex, then the class of non-randomised d.r.'s for estimating θ is complete.*

Let $\delta(\boldsymbol{X})$ be a randomised d.r. for estimating θ which gives the value of estimate $\theta(\boldsymbol{X})$ with probability $d\xi\{\theta(X)\}$ when $\boldsymbol{X}$ is the sample point.

$$E_{\boldsymbol{X}}\{\delta(\boldsymbol{X})\} = \int_{\Theta(\boldsymbol{X})} \theta(\boldsymbol{X}) d\zeta\{\theta(\boldsymbol{X})\} = T(\boldsymbol{X}) \text{ (say)} \tag{23.6.13}$$

where $\Theta(\boldsymbol{X})$ is the set of possible values of $\theta(\boldsymbol{X})$ and $E_X(\cdot)$ denotes expectation of $(\cdot)$ over for fixed $\boldsymbol{X}$.

The risk function associated with $\delta(\boldsymbol{X})$ when θ is the true value of the parameter is

$$r(\delta|\theta) = \int_{\mathcal{X}} \left\{ \int_{\Theta(\boldsymbol{x})} L(\theta, \theta(\boldsymbol{x})) d\zeta\{\theta(\boldsymbol{x})\} \right\} f(\boldsymbol{x}|\theta) d\boldsymbol{x}. \tag{23.6.14}$$

where $\mathcal{X}$ is the sample space of $\boldsymbol{x}$. Clearly, $T(\boldsymbol{X})$ is a non-randomised decision rule. The risk function associated with $T(\boldsymbol{X})$ when θ is the true value of the parameter is

$$r(T|\Theta) = \int_{\mathcal{X}} L(\theta, T(\boldsymbol{x})) f(\boldsymbol{x}|\theta) d\boldsymbol{x}. \tag{23.6.15}$$

Again the loss function being a convex one

$$E_{\boldsymbol{X}}[L(\theta, \theta(\boldsymbol{X}))] \geq L(\theta, E_{\boldsymbol{X}}(\theta(\boldsymbol{X})))$$

i.e.,

$$\int_{\Theta(\boldsymbol{X})} L(\theta, \theta(X)) d\zeta\{\theta(\boldsymbol{X})\} \geq L(\theta, T(\boldsymbol{X})) \tag{23.6.16}$$

i.e.,

$$r(\delta|\theta) \geq r(T|\theta) \tag{23.6.17}$$

(follows from (23.6.14) and (23.6.15)), equality in (23.6.17) holds iff $\theta(\boldsymbol{X}) = T(\boldsymbol{X})$ with probability one for all $\boldsymbol{X}$ i.e., whenever $\boldsymbol{X}$ is observed, estimate is $\theta(\boldsymbol{X})(= T(\boldsymbol{X}))$ with probability one, when the d.r. becomes a non-randomised one. Hence, the proof.

■ NOTE 23.6.2 *If the loss function is convex, one should, therefore, consider only the class of non-randomised d.r.'s for estimating θ.*

Theorem 23.6.6 (***Rao-Blackwellisation***). *Given the loss function to be convex, if a sufficient statistic T for estimating θ exists, then the class of estimates based on T is complete.*

Proof: By virtue of theorem 23.6.5, we shall consider a non-randomised unbiased estimate $g(\boldsymbol{X})$ for θ. Let $h(T) = E\{g(X)|T\}$, the conditional expected value of $g(X)$ given T. Then

$$E\{h(T)\} = E_T E\{g(X)|T\} = E\{g(X)\} = \theta.$$

where E_T denotes expectation w.r.t. T.

Let $L(\theta, \theta(X))$ denote the loss involved in estimating θ by $\theta(X)$ when the true value of the parameter is θ. Let $r\{g|\theta\}$, $r\{h|\theta\}$ denote respectively the risks associated with g and h when θ is the value of the parameter.

$$\begin{aligned} r\{g|\theta\} &= E[L(g(X), \theta)] = E_T\{E[L(g(X), \theta)|T]\} \\ &\geq E_T\{L(\theta, E\{g(X)|T\})\} \quad \text{(since the loss function is convex)} \\ &= E_T\{L(\theta, h(T))\} = r\{h|\theta\}. \end{aligned} \tag{23.6.18}$$

Equality in (23.6.18) holds iff $P_\theta\{g(X) = h(T)\} = 1$, a.e. (θ).
Hence, the theorem.

Theorem 23.6.7 *If the parameter θ varies over a non-degenerate interval, then a uniformly minimum risk* (UMR) *estimator does not exist.*

Let θ vary over the interval (θ_1, θ_2) and the let θ_0 be a point in (θ_1, θ_2). Let $g(\boldsymbol{X})$ denote an estimate of θ. Let us take another estimate $h(\boldsymbol{X}) = \theta_0 \ \forall \boldsymbol{X}$. Then the risks

$$\begin{aligned} r(g|\theta) &= E\{L(g(\boldsymbol{X}), \theta)\}; \\ r(h|\theta) &= E\{L(\theta_0, \theta)\}. \end{aligned}$$

Clearly $r(h|\theta_0) = 0$.

Hence, if $g(\boldsymbol{X})$ is (UMR)-estimator, we must have $r(g|\theta_0) = 0$. Now since θ_0 is arbitrary, a UMR-estimator g must have $r(g|\theta) = 0 \ \ \forall \ \theta \in (\theta_1, \theta_2)$. This is possible iff $g(\boldsymbol{X}) = \theta$ when θ is the true value of the parameter. This is impossible since θ is not a degenerate point but varies over a non-degenerate interval. Hence no UMR-estimator of θ exists.

■ NOTE 23.6.3 *Note that we have considered here* UMR*-estimators, not* UMR*-unbiased estimators which may exist.*

Theorem 23.6.8 *If $g_1(\boldsymbol{X}), g_2(\boldsymbol{X})$ be two* UMR*-estimators of θ with convex loss function then $g_1(\boldsymbol{X}) = g_2(\boldsymbol{X})$ with porobability one.*

Proof: Consider another estimator of θ:

$$h(\boldsymbol{X}) = \alpha g_1(\boldsymbol{X}) + (1-\alpha) g_2(\boldsymbol{X}) \text{ where } 0 < \alpha < 1.$$

Since $g_1(\boldsymbol{X})$, $g_2(\boldsymbol{X})$ are both unbiased, $h(\boldsymbol{X})$ is also unbiased for θ. Let $r(g_1|\theta)$ $r(g_2|\theta)$, $r(h|\theta)$ denote the risk functions and $L(\theta(\boldsymbol{X}), \theta)$ the loss function when $\theta(\boldsymbol{X})$ is an estimate of θ. Now because of the convexity of the loss function

$$\alpha L(\theta, g_1(X)) + (1-\alpha) L(\theta, g_2(X)) \geq L(\theta, \alpha g_1(X) + (1-\alpha) g_2(X)) \tag{23.6.19}$$

for any two arguments $g_1(X)$, $g_2(X)$ of $L(\theta, \cdot)$ (vide section 3.2).

Taking expectation of both sides, $\alpha r(g_1|\theta) + (1-\alpha)r(g_2|\theta) \geq r(\alpha g_1 + (1-\alpha)g_2|\theta)$ i.e., $r(g_1|\theta) \geq r(h|\theta)$ since $r(g_1|\theta) = r(g_2|\theta)$; hence $h(\boldsymbol{X})$ is uniformly better than or equivalent to $g_1(X)$. However, since $g_1(\boldsymbol{X})$ is UMRU-estimator, $h(X)$ must be equivalent to $g_1(\boldsymbol{X})$ i.e., $P\{g_1(\boldsymbol{X}) = \alpha g_1(\boldsymbol{X}) + (1-\alpha)g_2(\boldsymbol{X})\} = 1$ i.e., $P\{g_1(\boldsymbol{X}) = g_2(\boldsymbol{X})\} = 1$ (since $\alpha \neq 1$).

Hence the theorem.

Theorem 23.6.9 (***Lehmann-Scheffé***). *If there is a complete sufficient statistic T for θ, then every estimable real parameter $g(\theta)$ has a unique unbiased estimate with minimum variance and minimum risk; for strictly convex loss function this estimator is the only unbiased estimator that is a function of T.*

Proof: Let $g(\theta)$ be estimable i.e., assume that there is at least one unbiased estimator $f(\boldsymbol{X})$ of $g(\theta)$. By Rao-Blackwell theorem (theorem 23.6.6) we known that if the statistic $h(T) = E\{f(\boldsymbol{X})|T\}$ is unbiased for $g(\theta)$, it has the property $r(h|g(\theta)) \leq r(f|g(\theta))$, if the loss function is convex, with strict inequality holding for at least one θ unless $f(\boldsymbol{X})$ can be written as a function of $h(T)$, a.e. for each θ. So in looking for minimum risk estimators we can confine our attention to the unbiased estimators based on sufficient statistics.

If there are two such estimators $h_1(T)$ and $h_2(T)$ then $E\{h_1(T)\} = E\{h_2(T)\} = g(\theta)$ or $E\{h_1(T) - h_2(T)\} = 0$. But from completeness of T it follows that $h_1(T) - h_2(T) = 0$, a.e. (θ). Thus, the unbiased estimators based on T are really the same and such an estimator has smaller variance or risk than any other unbiased estimator.

$$\begin{aligned} V[f(X)] &= E[f(X) - g(\theta)]^2 = E_T E\left[\{(f(X) - h_1(T)) + (h_1(T) - g(\theta))\}^2 |T\right] \\ &= E(f(X) - h_1(T))^2 + E(h_1(T) - g(\theta))^2 \geq V(h_1(T)). \end{aligned}$$

Also

$$r(f|g(\theta)) = E_T E\{L(f(\boldsymbol{X}), g(\theta))|T\} \geq E_T\{L(g(\theta), h(T))\} = r(h|g(\theta)).$$

Hence the proof.

Theorem 23.6.10 *If the convex loss function $L(\theta, T)$ is increasing for $T > \theta$ and increasing for $T < \theta$ (i.e., $L(\theta, T)$ is increasing on both sides of θ) when θ is the true value of the parameter, and if we have a prior information that $\alpha \leq \theta \leq \beta$ then the class of estimates T^* where $\alpha \leq T^* \leq \beta$ is essentially complete.*

Proof: Given an estimate T of θ we construct another estimates T^* such that

$$T^* = \begin{cases} T, \text{ if } \alpha < T < \beta \\ \alpha, \text{ if } T \leq \alpha \\ \beta, \text{ if } T \geq \beta. \end{cases}$$

If $r(\theta|T)$ be the risk of T, then

$$\begin{aligned} r(\theta|T) &= \int_{\alpha<T<\beta} L(\theta,T)f(x|\theta)dx + \int_{T\leq\alpha} L(\theta,T)f(x|\theta)dx + \int_{T\geq\beta} L(\theta,T)f(x|\theta)dx \\ &\geq \int_{\alpha<T<\beta} L(\theta,T)f(x|\theta)dx + \int_{T\leq\alpha} L(\theta,\alpha)f(x|\theta)dx + \int_{T\geq\beta} L(\theta,\beta)f(x|\theta)dx \\ &\qquad (\because\ L(\theta,\alpha) \leq L(\theta,T) \text{ for } T\leq\alpha \\ &\qquad \text{and } L(\theta,\beta) \leq L(\theta,\beta) \text{ for } T\geq\beta) \\ &= r(\theta|T^*). \end{aligned}$$

Therefore, for any given T there is an estimator T^*, where $\alpha \leq T^* \leq \beta$, whose risk is no more greater than that of T. Hence the theorem.

► **Example 23.6.11** *For the normal population $N(0,\sigma^2)$ if we want to estimate σ^2 when it is known that $0<\sigma^2<3$, the m.l.e. s^2 is inadmissible, as $0<s^2<\infty$. In this case, if we consider an estimator ${s^*}^2$ where*

$$\begin{aligned} {s^*}^2 &= s^2 \text{ if } 0<s^2<3 \\ &\quad 3 \text{ if } s^2\geq 3 \end{aligned}$$

then $r({s^*}^2|\sigma^2) \leq r(s^2|\sigma^2)$. Hence ${s^*}^2$ will be preferable.

23.6.5 Non-parametric Estimation by Decision Theoretic Approach

We consider here estimation problems in which the functional form of distribution of X is not known. Restrictions are generally imposed on the variables only to remove the existence of estimates with unbounded risk. The problems are treated under two different restrictions.

(1) the variables are bounded with known bounds;

(2) the variables have bounded variances.

In the first class we assume without loss of generality that the variables are distributed over the interval $[0,1]$. We consider here two examples.

► **Example 23.6.12** *Let $X_1,\cdots,X_n$ be independently distributed over $[0,1]$ according to a joint distribution belonging to a family $\mathcal{F}$. Suppose that $\mathcal{F}$ contains a subfamily $\mathcal{F}_0$ according to which $X_1,\cdots,X_n$ are i.i.d. with*

$$p(X_i=1)=p, \quad p(X_i=0)=1-p, \quad 0\leq p\leq 1.$$

Let

$$E(X_i)=\mu_i \quad \text{and} \quad \bar{\mu}=\frac{1}{n}\sum_{i=1}^{n}\mu_i.$$

Show that the minimax estimate of $\bar{\mu}$ *is*

$$\hat{\bar{\mu}} = \frac{1}{1+\sqrt{n}}\left(\sqrt{n}\bar{X} + \frac{1}{2}\right).$$

Proof: For the case $\mathcal{F}_0$, $\mu_i = p$, $\bar{\mu} = p$. In this case, we have seen that $\hat{\bar{\mu}}$ is a Bayes estimate of p with constant risk $\frac{1}{4(\sqrt{n}+1)^2}$ and hence is a Bayes estimate of p. Now, the unconditional risk corresponding to $\hat{\bar{\mu}}$ is

$$\begin{aligned} r(\hat{\bar{\mu}}|\bar{\mu}) &= E\left[\frac{\sqrt{n}\bar{X}}{1+\sqrt{n}} + \frac{1}{2(\sqrt{n}+1)} - \bar{\mu}\right]^2 \\ &= E\left[\frac{\sqrt{n}(\bar{X}-\bar{\mu})}{1+\sqrt{n}} + \frac{1}{2(1+\sqrt{n})} - \frac{1}{1+\sqrt{n}}\right]^2 \\ &= \frac{n}{(\sqrt{n}+1)^2}V(\bar{X}) + \left(\bar{\mu} - \frac{1}{2}\right)^2 \cdot \frac{1}{(1+\sqrt{n})^2}. \end{aligned}$$

Now,

$$V(\bar{X}) = \frac{1}{n^2}\sum V(X_i);$$

since $X_i \le 1, E(X_i^2) \le E(X_i) = \mu_i$; hence $V(X_i) \le \mu_i - \mu_i^2$. Therefore,

$$V(\bar{X}) \le \frac{n\bar{\mu} - \sum \mu_i^2}{n^2} = \frac{n\bar{\mu} - \sum(\mu_i - \bar{\mu})^2 - n\bar{\mu}^2}{n^2} \le \frac{\bar{\mu}(1-\bar{\mu})}{n}.$$

Hence

$$r(\hat{\bar{\mu}}|\bar{\mu}) \le \frac{\bar{\mu}(1-\bar{\mu})}{(\sqrt{n}+1)^2} + \frac{\left(\bar{\mu}-\frac{1}{2}\right)^2}{(\sqrt{n}+1)^2} = \frac{1}{4(\sqrt{n}+1)^2}.$$

Therefore, the risk function corresponding to $\hat{\bar{\mu}}$ is maximised under $\mathcal{F}_0$. Therefore, if $t(\bar{X})$ be any other estimator of $\bar{\mu}$, then

$$\sup_{F\in\mathcal{F}} r_F(t|\bar{\mu}) \ge \sup_{F\in\mathcal{F}_0} r_F(t|\bar{\mu}) \ge \sup_{F\in\mathcal{F}_0} r_F(\hat{\bar{\mu}}|\bar{\mu}) = \sup_{F\in\mathcal{F}} r_F(\hat{\bar{\mu}}|\bar{\mu}) \qquad \text{(i)}$$

when $r_F(\cdot|\bar{\mu})$ denotes the risk of $(\cdot)$ in estimating $\bar{\mu}$ when the joint distribution of $\boldsymbol{X}$ is F. The inequality in the middle of (i) has been proved in exercise (23.6.7). The last relation in (i) has been proved above. Hence the result.

► **Example 23.6.13** *Let* $X_1, \cdots, X_n$, $Y_1, \cdots, Y_n$ *be independently distributed over* $[0,1]$ *according to a joint distribution belonging to a family* $\mathcal{F}$ *of distributions. Suppose that* $\mathcal{F}$ *contains a subfamily* $\mathcal{F}_0$ *according to which* $(X_1, \cdots, X_n)$, $(Y_1, \cdots, Y_n)$ *are two samples from two distributions with* $P(X_i = 1) = p_1$, $P(X_i = 0) = 1 - p_1$, $P(Y_i = 1) = p_2$, $P(Y_i = 0) = 1 - p_2$, $0 \le p_1, p_2 \le 1$.

If for the general family $\mathcal{F}$,

$$E(X_i) = \mu_i, \quad E(Y_i) = \nu_i, \quad \frac{1}{n}\sum_{i=1}^{n} \mu_i = \bar{\mu}, \quad \frac{1}{n}\sum_{i=1}^{n} \nu_i = \bar{\nu},$$

then show that minimax estimate of $\bar{\mu} - \bar{\nu}$ is

$$\frac{\sqrt{2n}}{1+\sqrt{n}}(\bar{X} - \bar{Y}).$$

The reader may prove this result following examples 23.6.8 and 23.6.12.

23.7 Interval Estimation as a Decision Problem

Suppose that we want to find an interval estimate $[d_1, d_2]$ of a parameter θ where the sample observations $\boldsymbol{X} = (X_1, \cdots, X_n)$ have a joint density function $f(x_1, \cdots, x_n; \theta)$, $\theta \in \Theta$. Clearly, we must have $[d_1, d_2] \subset \Gamma$, the class of all conceivable intervals containing the possible values of θ.

The general form of a decision function is now $\delta(\boldsymbol{X}) = [\delta_1(\boldsymbol{X}), \delta_2(\boldsymbol{X})]$ where

$$P_\theta[\delta_1(\boldsymbol{X}) \le \theta \le \delta_2(\boldsymbol{X})] \ge 1 - \alpha,$$

$1 - \alpha$ being the size of the confidence interval. Consider the simple loss function

$$L(\delta(\boldsymbol{X}), \gamma(\theta)) = \begin{cases} 0, \text{ if } \delta_1(\boldsymbol{X}) \le \gamma(\theta) \le \delta_2(\boldsymbol{X}) \\ 1, \text{ otherwise.} \end{cases} \tag{23.7.1}$$

Then the risk of $\delta(\boldsymbol{X})$ for a given value of θ is

$$r(\delta(\boldsymbol{X})|\theta) = E_{\boldsymbol{X}|\theta} L(\delta(\boldsymbol{X}), \theta) = 1 - P_\theta[\delta_1(\boldsymbol{X}) \le \theta \le \delta_2(\boldsymbol{X})]. \tag{23.7.2}$$

Bayes confidence interval :

Let $\xi(\theta)$ be a priori probability distribution of θ. Then the posterior risk associated with d.r. $\delta(\boldsymbol{X})$ for given $\boldsymbol{X}$ is

$$r^*(d(\boldsymbol{x})) = 1 - P_{\theta|\boldsymbol{x}}[\delta_1(\boldsymbol{x}) \le \theta \le \delta_2(\boldsymbol{x})] = 1 - \int_{\delta_1(\boldsymbol{x})}^{\delta_2(\boldsymbol{x})} \xi^*_{\boldsymbol{x}}(\theta) d\theta \le \alpha \tag{23.7.3}$$

where $\xi^*_{\boldsymbol{x}}(\theta)$ is the posterior distribution of θ given $\boldsymbol{x}$ as shown in (23.6.6). Hence $(1-\alpha)$-level. Bayesian confidence interval of θ is $[\delta_1(\boldsymbol{x}), \delta_2(\boldsymbol{x})]$ where

$$P_{\theta|\boldsymbol{x}}[\delta_1(\boldsymbol{x}) \le \theta \le \delta_2(\boldsymbol{x})] \ge 1 - \alpha. \tag{23.7.4}$$

$\delta_1(\boldsymbol{x}), \delta_2(\boldsymbol{x})$ are called the upper and lower limits of the interval.

An optimum d.r. should be one for which the length of the confidence interval $[\delta_2(\boldsymbol{x}) - \delta_1(\boldsymbol{x})]$ is minimum subject to the constraint (23.7.4).

On the other hand suppose we are interested in confidence intervals of a specified length l (say). Here a d.r. is $\delta(\boldsymbol{X}) = [\delta_1(\boldsymbol{X}), \delta_2(\boldsymbol{X})]$ where

$$\delta_2(\boldsymbol{X}) - \delta_1(\boldsymbol{X}) = l = 2a \text{ (say) } \forall \boldsymbol{X}. \tag{23.7.5}$$

Considering the loss function in (23.7.1), Bayes posterior risk is given by (23.7.3). Bayes confidence interval is therefore, given by $[\delta_1(\boldsymbol{x}), \delta_2(\boldsymbol{x})]$ where $\delta_1(\boldsymbol{x})$, $\delta_2(\boldsymbol{x})$ are such that

$$\int_{\delta_1(\boldsymbol{x})}^{\delta_2(\boldsymbol{x})} \xi_x^*(\theta)d\theta \geq \int_{\delta_1'(\boldsymbol{x})}^{\delta_2'(\boldsymbol{x})} \xi_x^*(\theta)d\theta \tag{23.7.6}$$

for all other confidence intervals $[\delta_1'(\boldsymbol{x}), \delta_2'(\boldsymbol{x})]$ of length l.

▶ **Example 23.7.1** *Let $X_1, \cdots, X_n$ be i.i.d. $N(\theta, 1), \theta \in R$ and let the prior distribution of θ be $N(0,1)$. It is known that the posterior distribution of θ gives $\boldsymbol{x}$ is*

$$N\left(\frac{\sum_1^n x_i}{n+1}, \frac{1}{n+1}\right).$$

Hence a $(1-\alpha)$-level Bayesian confidence interval of θ is

$$\left(\frac{n\bar{x}}{n+1} - \frac{\tau_{\alpha/2}}{\sqrt{n+1}}, \frac{n\bar{x}}{n+1} + \frac{\tau_{\alpha/2}}{\sqrt{n+1}}\right).$$

If θ is fixed, the customary confidence interval is $(\bar{x} - \tau_{\alpha/2}/\sqrt{n}, \bar{x} + \tau_{\alpha/2}/\sqrt{n})$. Thus, the Bayesian interval is somewhat shorter in length.

▶ **Example 23.7.2** *Let $X_1, \cdots, X_n$ be i.i.d. Bernoulli $B(p)$ variables and p have a Rectangular $(0,1)$ prior distribution. The Posterior distribution of p given $\sum_{i=1}^n x_i = s$ is a Beta distribution*

$$\xi^*\left(p \middle| \sum x_i = s\right) = \frac{1}{B(s+1, n-s+1)} p^s (1-p)^{n-s}, \quad 0 < p < 1$$

from which $(1-\alpha)$-level confidence interval of p can be found out.

▶ **Example 23.7.3** *Let $X_1, \cdots, X_n$ be a random sample from a $N(\mu, \sigma^2)$ population where μ is unknown but σ^2 is known.*

Again, suppose that μ has a prior distribution $N(\theta, \sigma_0^2)$. Find a Bayes confidence interval for μ of fixed length $l = 2a$.

It is known that the posterior distribution of μ is

$$\xi^*(\mu) = N\left(\frac{n\bar{x}\sigma_0^2 + \sigma^2\theta}{n\sigma_0^2 + \sigma^2}, \frac{\sigma^2\sigma_0^2}{n\sigma_0^2 + \sigma^2}\right).$$

A Bayes d.r. for estimating μ is a confidence interval $(\theta_1(\boldsymbol{x}), \theta_2(\boldsymbol{x}))$ such that

$$\int_{\theta_{1\xi}(\boldsymbol{x})}^{\theta_{2\xi}(\boldsymbol{x})} \xi^*_{\boldsymbol{x}}(\theta)d\theta \geq \int_{\theta'_{1\xi}(\boldsymbol{x})}^{\theta'_{2\xi}(\boldsymbol{x})} \xi^*_{\boldsymbol{x}}(\theta)d\theta$$

for any other confidence interval $(\theta'_{1\xi}, \theta'_{2\xi})$ when both the intervals are of length $l = 2a$.

Since $\xi^*_{\boldsymbol{x}}(\mu)$ is normal, it is clear that the area under the curve (probability) over an interval is maximised when the centre of the interval coincides with the mean of the posterior distribution. Hence Bayes confidence interval is

$$I_\mu(\boldsymbol{x}) = \left\{\mu : \frac{n\bar{x}\sigma_0^2 + \theta\sigma^2}{n\sigma_0^2 + \sigma^2} - a \leq \mu \leq \frac{n\bar{x}\sigma_0^2 + \theta\sigma^2}{n\sigma_0^2 + \sigma^2} + a\right\}.$$

If $\sigma_0^2 \to \infty$,

$$I_\mu(\boldsymbol{x}) \to \{\mu : \bar{x} - a \leq \mu \leq \bar{x} + a\},$$

which is a confidence interval of the customary type.

23.8 Testing of Hypotheses

Let $H_1, \cdots, H_k$ be an exhaustive set of k mutually exclusive hypotheses completely specifying the density function of the r.v. X. The acceptance of one of these hypotheses is made on the basis of a random sample $\boldsymbol{X} = (X_1, \cdots, X_n)$. In a non-randomised decision procedure, the sample space χ is divided into k mutually exclusive and exhaustive subspaces $\chi_1, \cdots, \chi_k$ such that H_i is accepted if $\boldsymbol{X} \in \chi_i$. In the case of a randomised d.r., a set of k functions $\delta_i(\boldsymbol{X})$ $(i = 1, \cdots, k)$ is defined over each sample point $\boldsymbol{X}$ such that $\delta_i(\boldsymbol{X})$ is the probability that H_i is accepted. Clearly

$$\sum_{i=1}^{k} \delta_i(\boldsymbol{X}) = 1 \;\forall\, \boldsymbol{X}.$$

For a non-randomised d.r. $\delta_i(\boldsymbol{X}) = 1(0)$ if $\boldsymbol{X} \in \chi_i$ (otherwise). Our problem is to find $\delta_i(\boldsymbol{X})$ for different i and $\boldsymbol{X}$.

Clearly the probability of accepting H_i when H_j is true is

$$\int f(\boldsymbol{x}|H_j)\delta_i(\boldsymbol{x})dx = E_{H_j}(\delta_i(\boldsymbol{X})). \tag{23.8.1}$$

In case, there are only two hypotheses $H_\circ$ (null) and H_1 (alternative), $E_{H_\circ}(\delta_1)$, $E_{H_1}(\delta_0)$ denote respectively the probabilities of type I error and type II error.

Let L_{ji} denote the loss involved in accepting H_i when H_j is true. Then the average loss or risk involved in the decision rule δ when H_j is true is

$$\int \sum_{i=1}^{k} L_{ji}\delta_i(\boldsymbol{x})f(\boldsymbol{x}|H_j)d\boldsymbol{x} = r(\delta|H_j). \tag{23.8.2}$$

A d.r. δ is uniformly better than a d.r. δ' if

$$r(\delta|H_j) \le r(\delta'|H_j) \ \forall\, j = 1, \cdots, k,$$

with strict inequality holding for some j. If δ^* is uniformly better than any other d.r. $\delta \in \Delta$, then δ^* is uniformly best d.r. in Δ and is the most desirable one. A d.r. δ is said to be admissible in Δ, if there exists no other rule in Δ which is uniformly better than δ. Generally, a uniformly best d.r. does not exist.

Let π_j be the a priori probability that the hypothesis H_j is true, $\sum \pi_j = 1$. The average risk of d.r. δ corresponding to a priori p.d. $\{\pi_j\}$ is

$$\bar{r}_\pi(\delta) = \sum_{j=1}^{k} r(\delta|H_j)\pi_j. \tag{23.8.3}$$

A Bayes d.r. δ_π corresponding $\{\pi\}$ is one for which the average risk

$$\bar{r}_\pi(\delta_\pi) = \min_{\delta} \bar{r}_\pi(\delta).$$

Now

$$\begin{aligned}\bar{r}_\pi(\delta) &= \sum_{j=1}^{k} \pi_j \int \sum_{i=1}^{k} L_{ji}\delta_i(\boldsymbol{x}) f(\boldsymbol{x}|H_j) d\boldsymbol{x} \\ &= \sum_{i=1}^{k} \int \sum_{j=1}^{k} \pi_j L_{ji} f(\boldsymbol{x}|H_j)\delta_i(\boldsymbol{x}) d\boldsymbol{x} \\ &= \sum_{i=1}^{k} \int \phi_i(\boldsymbol{x})\delta_i(\boldsymbol{x}) d\boldsymbol{x}\end{aligned} \tag{23.8.4}$$

where

$$\phi_i(\boldsymbol{x}) = \sum_{j=1}^{k} \pi_j L_{ji} f(\boldsymbol{x}|H_j). \tag{23.8.5}$$

Bayes solution δ_π is one for which

$$\delta_{i(\pi)}(\boldsymbol{x}) = 1(0) \text{ if } \boldsymbol{x} \in w_i \text{ (otherwise)} \tag{23.8.6}$$

where

$$w_i = \left\{ \boldsymbol{x} : \phi_i(\boldsymbol{x}) = \min_{j=1,\cdots,k} \phi_j(\boldsymbol{x}) \right\}. \tag{23.8.7}$$

Clearly for any other rule δ,

$$\begin{aligned}\bar{r}_\pi(\delta) - \bar{r}_\pi(\delta_\pi) &= \sum_{i=1}^{k} \int \phi_i(\boldsymbol{x}) \left[\delta_i(\boldsymbol{x}) - \delta_{\pi(i)}(\boldsymbol{x})\right] d\boldsymbol{x} \\ &= \sum_{i=1}^{k}\sum_{j=1}^{k} \int_{w_j} \phi_i(\boldsymbol{x}) \left[\delta_i(\boldsymbol{x}) - \delta_{\pi(i)}(\boldsymbol{x})\right] d\boldsymbol{x} \\ &= \sum_{j=1}^{k} \int_{w_j} \sum_{i=1}^{k} \phi_i(\boldsymbol{x}) \left[\delta_i(\boldsymbol{x}) - \delta_{\pi(i)}(\boldsymbol{x})\right] d\boldsymbol{x}.\end{aligned}$$

Now, for $\boldsymbol{x} \in w_j$

$$\begin{aligned}\sum_{i=1}^{k} \phi_i(\boldsymbol{x})[\delta_i(\boldsymbol{x}) - \delta_{\pi(i)}(\boldsymbol{x})] &= \sum_{l(\neq j)=1}^{k} \phi_l(\boldsymbol{x})\delta_l(\boldsymbol{x}) + \phi_j(\boldsymbol{x})[\delta_j(\boldsymbol{x}) - 1] \\ &\geq \phi_j(\boldsymbol{x}) \sum_{l(\neq j)=1}^{k} \delta_l(\boldsymbol{x}) + \phi_j(\boldsymbol{x})[\delta_j(\boldsymbol{x}) - 1] \quad \text{[by (23.8.6)]} \\ &= \phi_j(\boldsymbol{x}) \left[\sum_l \delta_l(\boldsymbol{x}) - 1\right] = 0.\end{aligned}$$

Hence,

$$\bar{r}_\pi(\delta) \geq \bar{r}_\pi(\delta_\pi).$$

Therefore, δ_π is a Bayes d.r. This is the Bayes prior solution. We note that Bayes solution is a non-randomised d.r.

Bayes a Posteriori Solution

When $\boldsymbol{x}$ has been observed, a posterior probability that H_j is true is

$$\pi_j^* = \frac{\pi_j f(\boldsymbol{x}|H_j)}{\sum_{j=1}^{k} \pi_j f(\boldsymbol{x}|H_j)}. \tag{23.8.8}$$

The a posterior risk associated with d.r. δ in accepting H_i is

$$r_{\boldsymbol{x}}^*(H_i|\delta) = \sum_{j=1}^{k} \pi_j^* L_{ji} = \frac{\sum_j L_{ji}\pi_j f(\boldsymbol{x}|H_j)}{\sum_j \pi_j^* f(\boldsymbol{x}|H_j)}. \tag{23.8.9}$$

Bayes solution δ^* in terms of a posterior probability is that for the given sample point $\boldsymbol{x}$ accept that H_i for which $r^*(H_i|\delta^*)$ is minimum. Now $r_{\boldsymbol{x}}^*(H_i|\delta) = K\phi_i(\boldsymbol{x})$ where

$$K = \left[\sum_j \pi_j f(\boldsymbol{x}|H_j)\right]^{-1},$$

a constant. Therefore, when $\phi_i(\boldsymbol{x})$ is minimum $r^*(H_i|\delta)$ is minimum and vice-versa Hence it follows from (23.8.6) and (23.8.7) that Bayes solution δ_π in terms of 'prior risk and Bayes solution δ^* in terms of 'a posterior risk' are identical.

Theorem 23.8.1 *If $\pi_i > 0 \ \forall\, i$, Bayes d.r. δ_π is admissible in the class of all d.r.'s.*

Proof: If possible let δ_π be inadmissible. Then there exists a d.r. δ which is uniformly better than this, i.e.,

$$r(H_j|\delta) \leq r(H_j|\delta_\pi) \ \forall\, j = 1, \cdots, k$$

with strict inequality holding for some j. Hence for the set of prior probabilities $\{\pi\}$,

$$\sum \pi_j r(H_j|\delta) < \sum \pi_j r(H_j|\delta_\pi) \quad \text{i.e.,} \quad \bar{r}_\pi(\delta) < \bar{r}_\pi(\delta_\pi)$$

which is a contradiction.

Theorem 23.8.2 *If any Bayes solution is unique it must be admissible.*

Proof: Left as an exercise.

Minimax Solution

In the absence of the knowledge of a priori probabilities one way of getting around the difficulty of Bayes solution is given by the minimax rule.

If a uniformly best decision rule does not exist attention is focussed on the maximum risk involved in any d.r. The name minimax is based on the principle of minimising the maximum risk. For each d.r. we calculate the maximum risk. The d.r. $\delta^{(m)}$ for which the maximum risk is the smallest is called the minimax rule.

$$\max_i r(\delta^{(m)}|H_i) = \inf_\delta \max_i r(\delta|H_i). \tag{23.8.10}$$

The underlying assumption is that if the maximum is minimum, the other risks are also likely to be low. This may not, however, be completly true.

Theorem 23.8.3 *If the Bayes d.r. δ^0 is such that $r(\delta^0|H_i) = c$, a constant then δ^0 is a minimax d.r.*

Proof: Since δ^0 is a Bayes d.r.

$$\bar{r}(\delta^0) = \sum_{j=1}^{k} \pi_j r(\delta^0|H_j) \leq \bar{r}(\delta) \ \forall\, \delta$$

i.e., $c \leq \max_i r(H_i|\delta) \leq \bar{r}(\delta) \ \forall\, \delta$ i.e., $\max_i r(\delta^0|H_i) \leq \inf_\delta \max_i r(H_i|\delta)$.

Hence, δ^0 is minimax.

The case of two simple hypotheses

Suppose we have two hypotheses $H_1(\theta = \theta_1)$ (null), $H_2(\theta = \theta_2)$ (alternative). Let d_i be the decision that H_i is accepted with $L(d_i, \theta)$ as the loss involved in d_i when θ is the true value of the parameter. Clearly, $L(d_i, \theta_i) = 0$. The risk function associated with d.r. δ is, where $\delta(\boldsymbol{x})$ is the probability of accepting H_2 when $\boldsymbol{X} = \boldsymbol{x}$,

$$r(\delta|\theta) = \int \{L(d_2, \theta_1)\delta(\boldsymbol{x}) + L(d_1, \theta_2)(1 - \delta(\boldsymbol{x}))\} f(\boldsymbol{x}|\theta) d\boldsymbol{x}.$$

Therefore,

$$r(\delta|\theta_1) = \int L(d_2, \theta_1)\delta(\boldsymbol{x}) f(\boldsymbol{x}|\theta_1) d\boldsymbol{x} = L(d_2, \theta_1)\alpha,$$

$$r(\delta|\theta_2) = \int L(d_1, \theta_2)(1 - \delta(\boldsymbol{x})) f(\boldsymbol{x}|\theta_2) d\boldsymbol{x} = L(d_1, \theta_2)\beta$$

where α is the size of the test and β is the probability of type II error. It is desirable that one should choose that critical function for which both the risks are minimum possible. We cannot, however, simultaneously minimise both the risks.

At best we can only fix the risk of one kind and subject to this constraint minimise the risk of the other kind. If π_1, π_2 are the a priori probabilities that H_1, H_2 are true then average risk in δ is

$$\bar{r}(\delta) = \pi_1 L(d_2, \theta_1)\alpha + \pi_2 L(d_1, \theta_2)\beta.$$

After $\boldsymbol{X}$ has been observed the a posterior probability that π_i is true is

$$\pi_i^* = \frac{\pi_i f(\boldsymbol{x}|\theta_i)}{\pi_1 f(\boldsymbol{x}|\theta_1) + \pi_2 f(\boldsymbol{x}|\theta_2)}.$$

Therefore, a posterior risk in accepting H_1 is

$$r^*(\delta|H_1) = \pi_2^* L(d_1, \theta_2)$$

and in accepting H_2 is

$$r^*(\delta|H_2) = r_1^* L(d_2, \theta_1).$$

Accordingly we should accept H_1 if

$$r^*(\delta|H_1) < r^*(\delta|H_2)$$

i.e., if

$$\frac{f(\boldsymbol{x}|\theta_2)}{f(\boldsymbol{x}|\theta_1)} < \frac{\pi_1 L(d_2, \theta_1)}{\pi_2 L(d_1, \theta_2)}.$$

Note that the usual MP test for H_1 against H_2 is to reject H_1 if. $\frac{f(\boldsymbol{x}|\theta_2)}{f(\boldsymbol{x}|\theta_1)} > k$ where k is so chosen that the size of the test is α. Thus, Bayes solution belongs to the class of MP regions provided the constant $\pi_1 L(d_2, \theta_1)/\pi_2 L(d_1, \theta_2)$ is so chosen that the size of the critical region is α.

If the loss functions and α, β are so chosen that $L(d_2, \theta_1)\alpha = L(d_1, \theta_2)\beta$ then the Bayes solution δ has constant risk for $\theta = \theta_1$ and $\theta = \theta_2$ and is therefore, minimax.

▶ **Example 23.8.1** *Consider the problem of testing $H_\circ : \theta = \theta_0$ against $H_1 : \theta = \theta_1$. Let d_0, d_1 be the decisions denoting acceptance of $H_\circ$ and H_1 respectively. Consider $L(d_0, \theta_1) = a$, $L(d_1, \theta_0) = b$ and $L(d_0, \theta_0) = L(d_1, \theta_1) = 0$. Show that a minimax decision rule δ is to reject $H_\circ$ if*

$$\frac{f_{\theta_1}(\boldsymbol{x})}{f_{\theta_0}(\boldsymbol{x})} \geq k$$

provided the constant k is so chosen that

$$aP_{\theta_1}\left\{\frac{f_{\theta_1}(\boldsymbol{x})}{f_{\theta_0}(\boldsymbol{x})} < k\right\} = bP_{\theta_0}\left\{\frac{f_{\theta_1}(\boldsymbol{x})}{f_{\theta_0}(\boldsymbol{x})} \geq k\right\}. \tag{i}$$

For an arbitrary d.r. δ' the risk function is

$$\begin{aligned} R(\delta', \theta_0) &= bP_{\theta_0}\{\delta'(\boldsymbol{x}) = a_1\}, \\ R(\delta', \theta_1) &= aP_{\theta_1}\{\delta'(\boldsymbol{x}) = a_0\}. \end{aligned} \tag{ii}$$

A Bayes rule will be to accept $H_\circ$ if $f_{\theta_1}(\boldsymbol{x})/f_{\theta_0}(\boldsymbol{x}) < \frac{b\pi_0}{a\pi_1}$ where π_0, π_1 are a priori probabilities, $P\{\theta = \theta_0\}$, $P\{\theta = \theta_1\}$ respectively. In Neyman-Pearson theory, a MP critical region of size α is $\{\boldsymbol{x} : f_{\theta_1}(\boldsymbol{x})/f_{\theta_0}(\boldsymbol{x}) > K\}$ where K is so chosen that the desired size (α) is attained. A Bayes d.r. is of the same form as the MP critical function, where however, the bound of $f_{\theta_1}(\boldsymbol{x})/f_\theta(\boldsymbol{x})$ is different. A minimax d.r. δ given in (i) has, therefore, the property

$$R(\delta, \theta_1) = R(\delta, \theta_0) \quad \text{(by (ii))}.$$

We shall now show that δ given by (i) is minimax. Let δ^* be any other rule. If $R(\delta^*, \theta_0) > R(\delta, \theta_0)$ then

$$R(\delta, \theta_0) = R(\delta, \theta_1) < \max[R(\delta^*, \theta_0), R(\delta^*, \theta_1)]$$

and δ^* cannot be minimax. Thus $R(\delta, \theta_0) \geq R(\delta^*, \theta_0)$, which means that

$$P_{\theta_0}\{\delta^* = a_1\} \leq P_{\theta_0}\{\delta = a_1\} = P\{\text{reject } H_\circ | H_\circ \text{ is true}\}.$$

By the NP-lemma, rule δ is the MP of its size so that its power must be at least that of δ^*, i.e.,

$$P_{\theta_1}\{\delta = a_1\} \geq P_{\theta_1}\{\delta^* = a_1\}$$

so that

$$P_{\theta_1}\{\delta = a_0\} \leq P_{\theta_1}\{\delta^* = a_0\}.$$

It follows that

$$aP_{\theta_1}\{\delta = a_0\} \leq aP_{\theta_1}\{\delta^* = a_0\}$$

and hence that

$$R(\delta, \theta_1) \leq R(\delta^*, \theta_1).$$

This means that

$$\max[R(\delta,\theta_0), R(\delta,\theta_1)] = R(\delta,\theta_1) \le R(\delta^*,\theta_1)$$

and thus

$$\max[R(\delta,\theta_0), R(\delta,\theta_1)] \le \max[R(\delta^*,\theta_0), R(\delta^*,\theta_1)].$$

▸ **Example 23.8.2** *Let $\boldsymbol{X} = (X_1, \cdots, X_n)$ be a random sample from $N(\theta, 1)$.*

Consider testing of hypotheses $H_1(\theta = -1)$, $H_2(\theta = 0)$, $H_3(\theta = 1)$. Assume the loss function $L_{ij} = (\neq)0$ according $i = (\neq)0$ where L_{ij} is the loss in accepting H_j when H_i is true.

Here

$$f(\boldsymbol{x}|H_1) = k\exp\left[-\frac{1}{2}\sum(x_i+1)^2\right]$$

$$f(\boldsymbol{x}|H_2) = k\exp\left[-\frac{1}{2}\sum x_i^2\right]$$

$$f(\boldsymbol{x}|H_3) = k\exp\left[-\frac{1}{2}\sum(x_i-1)^2\right].$$

Let $\pi_i = a$ prior probability $\theta = \theta_i$ where $\theta_1 = -1$, $\theta_2 = 0$, $\theta_3 = 1$. The posterior probability $\pi_i^* = \pi_i f(\boldsymbol{x}|H_i)/\sum \pi_k f(\boldsymbol{x}|H_k)$.

Hence

$$\pi_1^*/\pi_2^* = \lambda_1 e^{-n\bar{x}} \quad \text{where } \lambda_1 = \pi_1 e^{-n/2}/n_2$$

$$\pi_1^*/\pi_3^* = \lambda_2 e^{-2n\bar{x}} \quad \text{where } \lambda_2 = \pi_1/\pi_3$$

$$\pi_2^*/\pi_3^* = \lambda_3 e^{-n\bar{x}} \quad \text{where } \lambda_3 = \pi_2 e^{n/2}/\pi_3.$$

The ratios are decreasing functions of $\bar{x}$. Let us divide the whole range of values of $\bar{x}$ by two points $\bar{x}_1$, $\bar{x}_2$ such that for $\bar{x}_1$, $\pi_1^* = \pi_2^*$, for $\bar{x}_2$, $\pi_2^* = \pi_3^*$. Hence, for $\bar{x} < \bar{x}_1$, $\pi_1^* > \pi_2^*$ and for $\bar{x} > \bar{x}_2$, $\pi_2^* < \pi_3^*$.
The a posteriori risks in accepting H_i,

$$r^*(H_i) \propto (1-\pi_i^*), \quad i = 1, 2, 3.$$

Hence for $\bar{x} \le \bar{x}_1$, π_1^* is maximum of $(\pi_1^*, \pi_2^*, \pi_3^*)$ and as such $r^*(H_1)$ is minimum. Hence the region of acceptance of H_1 is $\bar{x} \le \bar{x}_1$. Similarly, for $\bar{x} \ge \bar{x}_2$, π_3^* is maximum and so $r^*(H_3)$ is minimum. Therefore, H_3 is accepted if $\bar{x} \ge x_2$. For $\bar{x}_1 < \bar{x} < \bar{x}_3$, π_2^* is maximum and as such it provides the acceptance region of H_2. It may so happen that for some π_1, π_2, π_3, $\bar{x}_1$ might exceed $\bar{x}_2$. Then the region of acceptance of H_1 and H_3 would vanish and the test rule is to accept H_2.

Thus, given π_1^* and π_2^* we can always choose $\bar{x}_1$ and $\bar{x}_2$ such that the intervals define the acceptance regions for H_1, H_2 and H_3 for Bayes solution and conversely. Thus, Bayes solution is expressible in terms of a number of intervals.
The minimax solution is given by that Bayes solution δ_π for which

$$r(\delta_\pi|H_i) = \text{ constant } (i = 1, 2, 3)$$

$$\text{where} \quad r(\delta_\pi|H_1) = \int_{\bar{x}_1}^{\infty} \sqrt{\frac{n}{2\pi}} e^{-\frac{n}{2}(\bar{x}+1)^2} d\bar{x}$$

$$r(\delta_\pi|H_2) = \int_{\bar{x}_1}^{\bar{x}_2} \sqrt{\frac{n}{2\pi}} e^{-\frac{n}{2}\bar{x}^2} d\bar{x}$$

$$r(\delta_\pi|H_3) = \int_{\bar{x}_2}^{\infty} \sqrt{\frac{n}{2\pi}} e^{-\frac{n}{2}(\bar{x}-1)^2} d\bar{x}.$$

Case of Two Composite Hypotheses

Suppose Θ_0 and Θ_1 are two mutually exclusive sets of parameter space Θ and we want to test $H_o : \theta \in \Theta_0$ against $H_1 : \theta \in \Theta_1$. Let $d_0(d_1)$ denotes acceptance of $H_o(H_1)$. Consider the loss function

$$L(d_0, \theta) = \begin{cases} 0, & \text{if } \theta \in \Theta_0 \\ a(\theta), & \text{if } \theta \in \Theta_1 \end{cases}$$

$$L(d_1, \theta) = \begin{cases} b(\theta), & \text{if } \theta \in \Theta_0 \\ 0, & \text{if } \theta \in \Theta_1 \end{cases}$$

where $a(\theta)$, $b(\theta)$ are positive-valued functions of θ. The risk function associated with d.r. $\delta(\boldsymbol{x})$ is

$$r(\delta|\theta) = \int L(\delta(\boldsymbol{x}), \theta) f(\boldsymbol{x}|\theta) d\boldsymbol{x}$$

$$= \begin{cases} b(\theta) \int_W f(\boldsymbol{x}|\theta) d\boldsymbol{x}, & \text{if } \theta \in \Theta_0 \\ a(\theta) \int_{W^c} f(\boldsymbol{x}|\theta) d\boldsymbol{x}, & \text{if } \theta \in \Theta_1 \end{cases}$$

where W is the critical region of the test (i.e., W is the subset of $\boldsymbol{x}$ such that H_o is rejected if $\boldsymbol{x} \in W$) i.e., $\delta(\boldsymbol{x}) = d_1$ if $\boldsymbol{x} \in W$ and $\delta(\boldsymbol{x}) = d_0$ if $\boldsymbol{x} \in W^C$.

Clearly a uniformly minimum risk d.r. δ will be one for which $r(\delta|\theta) \leq r(\delta'|\theta)$ for all δ' with strict inequality holding for some θ. Such a rule is absent in general. Let $\xi(\theta)$ be a probability distribution of θ. Then average risk of δ is

$$\bar{r}_\xi(\delta) = \int_{\Theta_0} b(\theta) \left[\int_W f(\boldsymbol{x}|\theta) d\boldsymbol{x}\right] \xi(\theta) d\theta + \int_{\Theta_1} a(\theta) \left[\int_{W^C} f(\boldsymbol{x}|\theta) d\boldsymbol{x}\right] \xi(\theta) d\theta$$

$$= \int_W f_\xi(\boldsymbol{x}) \int_{\Theta_0} b(\theta) \xi_{\boldsymbol{x}}(\theta) d\theta + \int_{W^C} f_\xi(\boldsymbol{x}) \int_{\Theta_1} a(\theta) \xi_{\boldsymbol{x}}(\theta) d\theta$$

where $\xi_{\boldsymbol{x}}(\theta)$ is the posterior distribution of θ given $\boldsymbol{x}$ and $f_\xi(\boldsymbol{x})$ is the unconditional distribution of $\boldsymbol{x}$, as given in (23.6.6) and (23.6.7) respectively.

To obtain a Bayes d.r. we have to minimise $\bar{r}_\xi(\delta)$. This can be achieved by taking a d.r. $\delta_\xi(\boldsymbol{x})$ such that $H_\circ$ is accepted whenever

$$\int_{\Theta_0} b(\theta)\xi_{\boldsymbol{x}}(\theta)d\theta > \int_{\Theta_1} a(\theta)\xi_{\boldsymbol{x}}(\theta)d\theta \tag{23.8.11}$$

(i.e., such a point $\boldsymbol{x}$ is included in W^C) and $H_\circ$ is rejected if the inequality is reversed (i.e., such a $\boldsymbol{x}$ is included in W).
In case, equality holds we may take either of the decisions.

In case, the loss function has the simple form $a(\theta) = a \ \forall\ \theta \in \Theta_1$, $b(\theta) = b \ \forall\ \theta \in \Theta_0$, Bayes critical region W_ξ is defined by

$$\int_{\Theta_1} \xi_{\boldsymbol{x}}(\theta)d\theta > \frac{b}{a}\int_{\Theta_0} \xi_{\boldsymbol{x}}(\theta)d\theta.$$

If in addition, $H_\circ, H_1$ are simple hypotheses, the condition reduces to

$$\frac{f(\boldsymbol{x};\theta_1)}{f(\boldsymbol{x};\theta_0)} > \frac{b\xi(\theta_0)}{a\xi(\theta_1)}, \tag{23.8.12}$$

where $\xi(\theta_0)$, $\xi(\theta_1)$ are prior probabilities attached to θ_0, θ_1 respectively. The condition (23.8.12) is exactly similar to the condition of a MP-test obtained by NP-theorem. In that case the constant k of the inequality $f(\boldsymbol{x};\theta_1)/f(\boldsymbol{x};\theta_0) > k$ is so determined that the size of the test is of the desired value.

► **Example 23.8.3** *$X_1, \cdots, X_n$ be a random sample from $N(\mu, \sigma^2)$ where μ is unknown but σ^2 is known. Let the prior distribution of μ be $N(\theta, \sigma^2/n')$ where $n' > 0$ and $-\infty < \theta < \infty$ are known quantities.*

The posterior distribution of μ, $\xi_{\boldsymbol{x}}(\mu)$ is $N\left(\frac{n\bar{x}+n'\theta}{n+n'} = \theta^*_{\boldsymbol{x}},\ \frac{\sigma^2}{n+n'} = {\sigma^*}^2\right)$. Suppose we are to test $H_\circ : \mu \le \mu_0$ against $H_1 : \mu > \mu_0$. Consider the loss function is

$$L(\mu, d_0) = \begin{cases} a(\mu), \text{ for } \mu > \mu_0 \\ 0, \quad \text{otherwise,} \end{cases}$$

$$L(\mu, d_1) = \begin{cases} b(\mu), \text{ for } \mu \le \mu_0 \\ 0, \quad \text{otherwise,} \end{cases}$$

where $d_0(d_1)$ denotes the acceptance (rejection) of $H_\circ$. Assume for simplicity, $a(\mu) = a$, $b(\mu) = b$. The critical function $\delta_\xi(\boldsymbol{x})$ of a Bayes test is given by

$$a\int_{\mu_\theta}^{\infty} \xi_{\boldsymbol{x}}(\mu)d\mu > b\int_{-\infty}^{\mu_0} \xi_{\boldsymbol{x}}(\mu)d\mu$$

$$\text{i.e., } a\left[1 - \Phi\left(\frac{\mu_0 - \theta^*_{\boldsymbol{x}}}{\sigma^*}\right)\right] > b\Phi\left(\frac{\mu_0 - \theta^*_{\boldsymbol{x}}}{\sigma^*}\right)$$

$$\text{i.e., } \Phi\left(\frac{\mu_0 - \theta^*_{\boldsymbol{x}}}{\sigma^*}\right) < \frac{a}{a+b}$$

$$\text{or, } \Phi\left[\sqrt{n+n'}\left(\mu_0 - \frac{n\bar{x}+n'\theta}{n+n'}\right)\Big/\sigma\right] < \frac{a}{a+b}$$

$$\text{or, } \mu_0 - \frac{n\bar{x}+n'\theta}{n+n'} < \tau_{1-\alpha}\frac{\sigma}{\sqrt{n+n'}} \quad \left(\text{where } 1-\alpha = \frac{a}{a+b}\right)$$

Therefore, critical region is

$$W_\xi = \left\{\boldsymbol{x} : \frac{n\bar{x}+n'\theta}{n+n'} > \mu_0 + \tau_\alpha \frac{\sigma}{\sqrt{n+n'}}\right\}$$

$$= \left\{\boldsymbol{x} : \frac{n(\bar{x}-\mu_0)+n'(\theta-\mu_0)}{\sigma/\sqrt{n+n'}} > \tau_\alpha\right\}.$$

▶ **Example 23.8.4** *Consider $H_\circ(\theta = \theta_0)$ against $H(\theta \neq \theta_0)$, where θ varies continuously about θ_0. Let $L_1(L_2)$ be the loss in rejecting (accepting) $H_\circ$ when θ_0 is true (false). Show that any unbiased test of size $\alpha = L_2/(L_1+L_2)$ is a minimax procedure among the class of all decision procedures with continuous risk functions.*

Let δ_0 be a unbiased test of size α. Then the risk when $\theta = \theta_0$ is $r(\delta_0|\theta_0) = L_1\alpha$. Similarly, $r(\delta_0|\theta') = L_2\beta(\theta')$, where $\beta(\theta')$ is the probability of type II error when $\theta = \theta'(\neq \theta_0)$. Since δ_0 is unbiased, $\beta(\theta') < 1 - \alpha = \frac{L_1}{L_1+L_2}$, $\theta' \neq \theta_0$. Hence

$$r(\delta_0|\theta') < \frac{L_1L_2}{(L_1+L_2)} = r(\delta_0|\theta_0).$$

Therefore,

$$\max_\theta r(\delta_0|\theta_0) = L_1L_2/(L_1+L_2) = L_0 \text{ (say)}.$$

Consider any other test δ_0', biased or unbiased, of size $\alpha'(\theta_0) < \alpha$. If the test is biased, $\beta_{\delta_0'}(\theta') > 1-$ size $(\delta_0') > 1 - \alpha = \frac{L_1}{L_1+L_2}$. Hence,

$$r(\delta_0'|\theta') = L_2\beta_{\delta_0'}(\theta') > L_\theta.$$

Therefore,

$$\max_\theta r(\delta_0'|\theta') = \max(L_1\alpha, L_2\beta(\theta')) > L_0. \tag{i}$$

If δ_0' is unbiased and of size $\alpha'(\theta_0) < d$ we can find an $\varepsilon(> 0)$ arbitrarily small such that $\alpha'(\theta) < \alpha \;\forall\, \theta\varepsilon[\theta_0 - \varepsilon, \theta + \varepsilon]$. Therefore, in this neighbourhood, $\beta'(\theta) = 1 - \alpha'(\theta) > 1 - \alpha = L_1/(L_1+L_2)$ and hence, $\max(L_1\alpha', L_2\beta) > L_0$.

Therefore, $\max\limits_{\text{all}\theta} r(\delta_0'|\theta) > L_0$. Therefore, for any test δ_0' of size $(< \alpha)$, (i) holds.

Again, if δ'' is a test of size $\alpha''(\theta_0) > \alpha$, then $\max(\delta''|\theta) > L_0$. Therefore, the test δ_0 of size α provides a minimax test for $H_\circ(\theta = \theta_0)$ against $H(\theta \neq \theta_0)$ under the conditions stated.

Exercise

1. Let $X \sim$ Binomial (n, p) and $L(d(\boldsymbol{x}), p) = (d(\boldsymbol{x}) - p)^2$. Let the prior of p be $\pi(p) = 1$ for $0 < p < 1$. Show that Bayes estimator of p is $\frac{X+1}{n+2}$ and Bayes risk is $\frac{1}{6(n+2)}$.

2. Let $X \sim N(\mu, 1)$ and $\mu \sim N(0, 1)$. Show that Bayes estimate of μ is $n\bar{x}/(n+1)$ and Bayes risk is $\frac{1}{(n+1)}$.

3. Let $X_1, \cdots, X_n$ be a random sample from Poisson (λ) distribution. Consider a prior distribution of λ as $g(\lambda) = e^{-\lambda}, \lambda > 0$.

   Find Bayes estimate of λ. Find Bayes estimate of $\varphi(\lambda) = e^{-\lambda}$ with the same loss function and same priori p.d.f. of λ.

4. Let X follow a Poisson distribution with parameter λ and suppose prior distribution of λ is $dG(\lambda) = \dfrac{\alpha^\beta}{\lceil(\beta)} e^{-\alpha\lambda} \lambda^{\beta-1} d\lambda, \alpha, \beta > 0$.

   For a squared error loss function, show that Bayes estimate of λ is $\hat{\lambda}_b = \frac{\beta+x}{\alpha+1}$ with Bayes risk $r(\hat{\lambda}_b|\lambda) = \frac{\beta}{\alpha(\alpha+1)}$.

   Show that the $dG(\lambda)$ is the conjugate prior distribution of λ.

5. Let $X_1, \cdots, X_n$ be a random sample from a Poisson distribution with unknown parameter λ. Suppose that λ has an exponential (θ) prior, $\xi(\lambda) = \frac{1}{\theta} \exp(-\lambda/\theta)$, $\theta \leq \lambda < \infty$.

   (a) Suppose prior probability $P_\xi(\lambda \geq \lambda_0) = \frac{1}{2}$. Find the posterior probability $P_{\xi^*}(\lambda \geq \lambda_0)$ given that $\sum_{i=1}^{n} X_i = s$.

   (b) Obtain Bayes estimate δ_ξ of λ, for a simple loss function $L(d, \lambda)$. Is δ_ξ unbiased? Consistent?

   Does the risk $r(\delta_\xi, \lambda) \to 0$ as $n \to \infty$ for all λ?

6. Let $X_1, \cdots, X_n$ be a random sample from a Bernoulli (p)-distribution, $f_x(p) = p^x(1-p)^{1-x}$, $x = 0, 1$. Assume that p has a Beta (α, β) prior and the loss function is $L(d, p) = (d-p)^2 / \, p^r(1-p)^s$ when r, s are known integers. Find a Bayes estimator of p relative to this prior. (When $s = 0, r = 2$, the loss function is squared relative error).

   Find a minimax estimator of p when $r = s = 1$.

7. Show that for the above prior and the loss function $L(d, p) = 0(1)$ if $\frac{|\alpha - p|}{p} < (>)\varepsilon$, the Bayes estimator $\delta_\pi = \frac{S+\alpha}{n+\alpha+\beta-1}$ where $S = \sum_{i=1}^{n} X_i$. Is δ_π a consistent estimator?

8. Let X be a random variable with hypergeometric density function

$$P(X=k)=\frac{\binom{M}{k}\binom{N-M}{n-k}}{\binom{N}{n}}$$

where $\max(0, n-N+M) \leq k$ (integer) $\leq \min(M, n)$ and let $(\delta - M)^2$ be the loss function. Prove that $\frac{N}{n+\sqrt{n(N-n)/(N-1)}}X + \frac{N}{2}\left[1 - \frac{n}{n+\sqrt{(N-n)n/(N-1)}}\right]$ is the minimax estimate of M.

[*Hints:* Use

$$P(M=m)=C\int_0^1 \binom{N}{m} p^{m+a-1} q^{N-m+b-1}\,dp$$

as the prior distribution of M, where $a, b > 0$ and $C = \overline{|(a+b)}/\{\overline{|a}\,\overline{|b}\}$.

9. Let X be a random variable with density function

$$P(X=k)=qp^{k-1},\quad k=1,2\cdots,\infty; 0\leq p<1,\quad q=1-p.$$

For a squared-error loss function $(\delta - p)^2$, find the minimax estimate of p.

10. Let $X \sim$ Bin (n, θ) and suppose that a priori p.d.f. of θ is Rectangular $(0, 1)$. Find the Bayes estimate of θ, using loss function $L(\theta, d) = (\theta - d)^2/[\theta(1-\theta)]$. Find a minimax estimate of θ.

11. Let $X_1, \cdots, X_n$ be a random sample from Gamma $G(1, 1/\lambda)$. Let a priori p.d.f. of λ be $g(\lambda) = e^{-\lambda}, \lambda > 0$ and the loss function be squared error. Find Bayes estimate of λ.

12. Let X be a random variable with geometric density function

$$P(X=k)=pq^{k-1},\quad k=1,2,\cdots,\infty;\quad 0<p\leq 1,\quad q=1-p$$

and let $(\delta - p)^2$ be the loss function. Then prove that

$$p=\begin{cases} 3/4, \text{ when } k=1 \\ 1/4, \text{ when } k\geq 2\end{cases}$$

is a minimax estimate of p.

[*Hints:* Use the following a priori distribution of p:

$$p=\begin{cases} 1/4, \text{ with probability } 2/3 \\ 1, \quad \text{with probability } 1/3.]\end{cases}$$

13. Let X be a random variable with density function

$$P(X = 1) = p, 0 \leq p \leq 1,$$
$$P(X) = 0 = 1 - p = q.$$

Let $\delta(X)$ be an estimate of p and let the loss function be

$$L[p, \delta(X)] = \begin{cases} 1, \text{ if } |\delta(X) - p| \geq \frac{1}{4} \\ 0, \text{ if } |\delta(X) - p| < \frac{1}{4}. \end{cases}$$

Find a minimax estimate of p. Show that for every non-randomized estimate $\delta(X)$, $\sup_p E\ [L(p, \delta)] = 1$. Construct a randomized minimax estimate δ^* and compare $\sup_p E[L(p, \delta(X)]$ and $\sup_p E[L(p, \delta^*)]$.

14. Let X be a random variable with the following density functions. Compare the risk of the minimax estimate, the uniformly minimum-variance estimate and the maximum-likelihood estimate of the parameter involved taking the squared error as the loss function.

(a) $P(X = k) = \binom{n}{k} p^k q^{n-k}, k = 0, 1, \cdots, n, 0 \leq p \leq 1, q = 1 - p;$

(b) $P(X = k) = pq^{k-1}, k = 1, 2, \cdots, \infty, 0 < p \leq 1;$

(c) $P_\theta(x) = \frac{1}{\sqrt{2\pi}} e^{-(x-\theta)^2/2}, -\infty < x < \infty, -\infty < \theta < \infty.$

15. Prove that if a minimax estimator of a parametric function is unique, then it must be admissible.

16. Suppose X_1, X_2, X_3 have a trinomial distribution with parameters n, p_1, p_2, p_3. Find a Bayes estimator of $p_1 - p_2$ under squared-error loss, assuming a uniform prior distribution over the whole parameter space.

17. Two contestants, statistician (A) and nature (B) simultaneously put up either one or two fingures. The statistician wins if the sum of the digits showing is odd and the nature wins if the sum of the digits showing is even. The loss table is given by

A \ B	1	2
1	−2	3
3	3	−4

The statistician is allowed to ask the nature how many fingures he intends to put up and that nature must answer truthfully with probability 3/4. The statistician therefore, observes a random variable X (the answer nature gives) taking values 1 or 2.

$$P(X = 1|\theta = 1) = \tfrac{3}{4}, \quad P(X = 2|\theta = 1) = \tfrac{1}{4},$$
$$P(X = 1|\theta = 2) = \tfrac{1}{4}, \quad P(X = 2|\theta = 2) = \tfrac{3}{4}.$$

Characterise all non-randomised rules for estimating θ. Compute the risk table for all the non-randomised rules. Draw the risk set. Find the minimax rule.

18. Let X have p.d.f. f_θ, with θ taking one of the values $\theta_1, \cdots, \theta_m$ and consider the problem of determining the correct value of θ. Consider the gain function

$$G(d_i, \theta_j) = \begin{cases} g_i, & \text{if } i = j \\ 0, & \text{otherwise} \end{cases}$$

where d_i denotes the decision that $\theta = \theta_i$. Show that the Bayes estimator (which maximises the average gain) with respect to prior of θ, $P(\theta = \theta_i) = \frac{1}{m}, i = 1, \cdots, m$, coincides with the m.l.e. of θ.

19. Let $X_1, \cdots, X_n$ be i.i.d. observations with a common $N(\theta. 1)$ distribution. Assume that the parameter θ has a prior distribution that is $N(0, \tau^2 \text{ (known)})$. Compute the shortest length Bayesian credible interval for θ with credibility (i.e., posterior probability) 95%. What happens to this Bayesian credible interval as the prior s.d. τ tends to ∞? Interpret the result.

20. Let $X_1, \cdots, X_n$ be a sample from a Poisson distribution with parameter λ. Assume that λ has a Gamma distribution $G(\alpha, \beta)$. Find a Bayesian confidence interval for λ.

21. Let $X_1, \cdots, X_n$ be a random sample from a $N(\mu, 1)$ population and suppose that μ has a prior distribution $R(-1, 1)$. Find a Bayesian confidence interval for μ.

22. Suppose that we observe $n = 3$ Bernoulli trials with unknown probability of success θ, where θ has the prior distribution $B(2, 2)$. Assume that we recorded $x = 2$ successes. Show that Bayes estimate of θ is 0.57 and that the true value of θ lies between 0.2 and 0.8 with probability 0.884.

23. Let $X_1, \cdots, X_n$ be a random sample from a Geometric distribution with parameter θ. Assuming that θ has a prior p.d.f. $B(\alpha, \beta)$, find a Bayesian confidence interval for θ.

24. Let $X_1, \cdots, X_n$ be i.i.d. $N(\theta, 1)$. We want to estimate θ under the loss function $L(a, \theta) = |a - \theta|$. Show that $\bar{X}_n$ is an admissible estimator of θ.

25. A sample of six observations is taken on a Bernoulli $B(1, \theta)$ r.v. to test $H_\circ : \theta = \frac{1}{2}$ against $H_1 : \theta = \frac{3}{4}$,

    (a) Find the MP test of size $\alpha = 0.05$.

(b) If $L\left(\frac{1}{2},\frac{1}{2}\right) = L\left(\frac{3}{4},\frac{3}{4}\right) = 0,\ L\left(\frac{1}{2},\frac{3}{4}\right) = 1,\ L\left(\frac{3}{4},\frac{1}{2}\right) = 2$, find the minimax rule.

(c) If the prior probabilities of $\theta = \frac{1}{2}$ and $\theta = \frac{3}{4}$ are $\pi_0 = \frac{1}{3}$ and $\pi_1 = \frac{2}{3}$, respectively, find the Bayes rule.

26. Let $X_1, \cdots, X_n$ be a random sample from a population $f(x;\theta) = \theta e^{-\theta x}$, $x > 0$. Consider $H_o : \theta = 1, H_1 = \theta = 2$. Let the *a* prior p.m.f. of θ be $\pi_0 = \frac{2}{3}, \pi_1 = \frac{1}{3}$. Find Bayes solution if $a = b$. Find the power of the test at $\theta = 1$ and $\theta = 2$.

27. Suppose we have k independent samples $X_i \sim N(\theta_i, 1), i = 1, \cdots, k\ (\geq 3)$. For estimating $\boldsymbol{\theta} = (\theta_1, \cdots, \theta_k)$ by $\boldsymbol{\delta}(\boldsymbol{x}) = (\delta_1(\boldsymbol{x}), \cdots, \delta_k(\boldsymbol{x}))$, $\boldsymbol{x} = (x_1, \cdots, x_k)$ consider the loss function

$$L(\boldsymbol{\theta}, \boldsymbol{\delta}(\boldsymbol{x})) = \frac{1}{k}\sum_{1}^{k}(\delta_i - \theta_i)^2.$$

Show that the maximum likelihood estimator $\boldsymbol{\delta}^0(\boldsymbol{x}) = \boldsymbol{x}$ has the risk function $r(\boldsymbol{\delta}^0|\theta) = 1\ \forall\ \boldsymbol{\theta}$.

Assume now that $\boldsymbol{\theta}$ has a prior distribution such that $\theta_i \sim$ indep. $N(0, A)$. Then show that the posterior distribution of θ_i given $x_i \sim N((1-B)x_i, 1-B)$ where $B = 1/(1+A)$ and hence, Bayes estimator of θ_i is $\delta_i^* = (1-B)x_i$ with Bayes risk $1-B$. Therefore, $r(\boldsymbol{\delta}^*|\boldsymbol{\theta}) = 1-B$.

Suppose now B is unknown. Consider the unconditional distribution of x_i given A, which is indep. $N(0, A+1)$. Hence, show that $SB \sim \chi^2_{(k)}$ where $S = \sum_{i=1}^{k} x_i^2$ and therefore, $E\left(\frac{k-2}{S}\right) = B$. A suggested estimator of $\boldsymbol{\theta}$ is therefore, $\boldsymbol{\delta}^{(1)}$ where $\delta_i^{(1)} = \left(1 - \frac{k-2}{S}\right)x_i$. The estimator $\boldsymbol{\delta}^{(1)}$ is known as James-Stein (1969) estimator of $\boldsymbol{\theta}$. Show that $r(\boldsymbol{\delta}^{(1)}|\theta) = 1 - \frac{k-2}{k}B$. Therefore, JS-estimator has a lower average risk than $\boldsymbol{\delta}^{(0)}$.

28. Suppose that $\boldsymbol{y}(n \times 1)$ given $\boldsymbol{\theta}_1(p_1 \times 1)$ is a n-variate Normal $\boldsymbol{y}|\boldsymbol{\theta}_1 \sim N(\boldsymbol{A}_1\boldsymbol{\theta}_1, \boldsymbol{C}_1)$. Also $\boldsymbol{\theta}_1|\boldsymbol{\theta}_2 \sim N(\boldsymbol{A}_2\boldsymbol{\theta}_2, \boldsymbol{C}_2)$ and $\boldsymbol{\theta}_2|\boldsymbol{\theta}_3 \sim N(\boldsymbol{A}_3\boldsymbol{\theta}_3, \boldsymbol{C}_3)$.

Show that the posterior distribution of $\boldsymbol{\theta}_1$, given $\{A_i\}, \{C_i\}, \boldsymbol{\theta}_3$ and $\boldsymbol{y}$ is $N(\boldsymbol{Dd}, \boldsymbol{D})$ with $\boldsymbol{D}^{-1} = \boldsymbol{A}_1'\boldsymbol{C}_1^{-1}\boldsymbol{A}_1 + \{\boldsymbol{C}_2 + \boldsymbol{A}_2\boldsymbol{C}_3\boldsymbol{A}_2'\}^{-1}$ and $\boldsymbol{d} = \boldsymbol{A}_1'\boldsymbol{C}_1^{-1}\boldsymbol{y} + \{\boldsymbol{C}_2 + \boldsymbol{A}_2\boldsymbol{C}_3\boldsymbol{A}_2'\}^{-1}\boldsymbol{A}_2\boldsymbol{A}_3\boldsymbol{\theta}_3$.

Show also that an alternative expression for $\boldsymbol{D}^{-1}$ is

$$\boldsymbol{D}^{-1} = \boldsymbol{A}_1'\boldsymbol{C}^{-1}\boldsymbol{A}_1 + \boldsymbol{C}_2^{-1} - \boldsymbol{C}_2^{-1}\boldsymbol{A}_2(\boldsymbol{A}_2'\boldsymbol{C}_2^{-1}\boldsymbol{A}_2 + \boldsymbol{C}_3^{-1})^{-1}\boldsymbol{A}_2'\boldsymbol{C}_2^{-1}.$$

Chapter 24

Sequential Procedures

24.1 Introduction

In the problems of statistical inference considered so far we have assumed that a statistical decision (e.g., point estimation, confidence interval for a parameter, test of hypotheses) has to be made on the basis of a *fixed number* n of observed values $x_1, \cdots, x_n$. It seems obvious that advantageous possibilities are given away by not making use of the information gathered during the course of investigation (this is illustrated by a well-known example from statistical quality control, vide example 24.2.1). Thus, it may be disadvantageous in some problems to fix the sample size in advance. An alternative procedure is to sample sequentially, that is to take one observation after another and make a decision to stop after the mth observation by using some procedure that is based on the observations $x_1, x_2, \cdots, x_m$. This is known as sequential statistical inference.

The first formalised sequential procedure is the "double sample" inspection plan by Dodge and Roming (1929), which was subsequently generalised to a multistage procedure by Bartkey (1943). Although there were other suggestions of multi-stage procedures (e.g., Mahalanobis (1940), Hotelling (1941)), a systematic development of sequential methods started only in the last two years of World War II independently in USA and Great Britain. Abraham Wald developed a general theory of sequential analysis and in particular proposed the "*Sequential Probability Ratio Test*" (SPRT) taking cue from a problem posed by the Statistical Research Group at Columbia University in 1943. Works of Wald (1945, 1947, 1950), Wald and Wolfwitz (1948), Arrow, Blackwell and Girshick (1949) and their co-workers laid the foundation of sequential decision procedures.

In Section 24.2 we consider some preliminary notions. Section 24.3 deals with elements of theory of sequential estimation. Wald's SPRT is considered in the next section.

24.2 Some Fundamental Ideas

We first consider a few examples.

► **Example 24.2.1** *In an inspection plan a batch of items is rejected if a sample of 25 items contains more than 3 defectives; otherwise the batch is accepted. This is a* $(25, 3)$ *sampling plan.*

Consider the sample

$$E\,E\,D\,E\,E\,D\,E\,E\,D\,E\,E\,D\,E\,E\,E\,E\,E\,E\,D\,E\,E\,E\,E\,E\,E$$

where $E(D)$ denotes effective (defective). Here the twelfth value already determines the final decision (the batch is rejected). Thus, it is possible to save the time and cost of inspecting the additional 13 items. An obvious idea is, therefore, to curtail the (25,3) sampling plan by stopping the inspection as soon as either 4 defective items are found (and rejecting the batch) or 22 effective items are observed (and accepting the batch).

More generally a (n, a)-sampling plan may be modified as follows: inspect items successively until either the number of defective items is $a + 1$ or the number of effective items is $n - a$. Obviously, the sample size of a curtailed inspection plan N is a random variable with the property $N \leq n$ and an average sample number (ASN).

$$E(N) < n$$

for all possible distributions of the sample. Thus, the sequential procedure leads to more or less considerable saving of observations without any loss of statistical decision.

This example shows an essential aspect of sequential procedures. The sample size is not fixed in advance, but depends on the information contained at each stage.

► **Example 24.2.2** *Let* $X_1, \cdots, X_n$ *be a sample from a* $N(\mu, \sigma^2)$ *population, where both* μ *and* σ^2 *are unknown. The shortest confidence interval of level* $(1 - \alpha)$ *based on* $\bar{X}$ *for the parameter* μ *for a given* n *is*

$$\left(\bar{X} - \frac{S}{\sqrt{n}} t_{n-1,\alpha/2},\ \bar{X} + \frac{S}{\sqrt{n}} t_{n-1,\alpha/2}\right)$$

where S^2 *is the sample variance. The length of this interval is* $\left(\frac{2S}{\sqrt{n}}\right) t_{n-1,\alpha/2} = Z$ *(say) which is a random variable.*

If σ is known, the $(1 - \alpha)$-level confidence interval for μ based on $\bar{X}$ is

$$\left(\bar{X} - \frac{\sigma}{\sqrt{n}} \tau_{\alpha/2},\ \bar{X} + \frac{\sigma}{\sqrt{n}} \tau_{\alpha/2}\right)$$

which is of length $\left(\frac{2\sigma}{\sqrt{n}}\right)\tau_{\alpha/2}$. If in this case one requires a confidence interval of maximum length $2d$, then

$$\frac{\sigma}{\sqrt{n}}\tau_{\alpha/2} \leq d$$

$$\text{i.e., } n \geq \frac{\sigma^2}{d^2}\tau^2_{\alpha/2} = n_0 \text{ (say).}$$

Therefore taking $n \geq n_0$ one can always attain in this case a confidence interval of maximum length $2d$.

However, if σ is unknown, the length of the confidence interval Z is a random variable and the minimum sample size required to attain the interval of length $2d$ can not be worked by the non-sequential procedures (Dantzig, 1940).

However, there exist sequential confidence intervals, e.g., Stein's two-stage procedure (vide subsection 24.3.1) with the desired properties.

▶ **Example 24.2.3** *Suppose we want to test $H_\circ(\mu = \mu_0)$ against $H_1(\mu = \mu_1)$, $\mu_1 > \mu_0$, in sampling from a $N(\mu, \sigma^2)$ population with known variance σ^2. It is known that the* MP *test for $H_\circ$ is to reject $H_\circ$ if $\bar{X} > C$ where C is so chosen that the size of the test is α. Suppose it is also desired that the probability of type II error β is a fixed quantity.*

Thus

$$P\{\bar{X} > C|H_\circ\} = \alpha \text{ i.e., } P\left\{\frac{\sqrt{n}(\bar{X}-\mu_0)}{\sigma} > \frac{\sqrt{n}(C-\mu_0)}{\sigma}\right\} = \alpha$$

so that

$$\frac{\sqrt{n}(C-\mu_0)}{\sigma} = \tau_\alpha. \tag{i}$$

Similarly,

$$P\{\bar{X} \leq C|H_1\} = \beta \text{ i.e., } P\left\{\frac{\sqrt{n}(\bar{X}-\mu_1)}{\sigma} \leq \frac{\sqrt{n}(C-\mu_1)}{\sigma}\right\} = \beta$$

so that

$$\frac{\sqrt{n}(C-\mu_1)}{\sigma} = -\tau_\beta. \tag{ii}$$

One can solve out n from (i) and (ii) which gives

$$n = \frac{\sigma^2(\tau_\alpha + \tau_\beta)^2}{(\mu_1 - \mu_0)^2} = n_0 \text{ (say).}$$

Therefore, the above test based on a sample of size n_0 has the desired properties. However, in practice σ is not known and one cannot compute n_0.

In examples 24.2.2 and 24.2.3 we have seen that if we fix a sample of size n in advance it is not always possible to achieve the desired result. An alternative procedure is to proceed sequentially, i.e., take observations one at a time and use the information provided by the observations taken so far to determine whether a further observation is to be taken or to stop sampling and come to a decision.

Let $\boldsymbol{X} = (X_1, X_2, \cdots)$ be an infinite sequence of random variables, X_i denoting the r.v. corresponding to time i, $i = 1, 2, \cdots$. We assume that $X_1, X_2, \cdots$, are i.i.d. with a common p.d.f. (p.m.f.) $f(x; \theta), \theta \in \Theta$.

Let $\mathcal{D}$ be a set of terminal decisions taken on the basis of observation $(x_1, \cdots, x_n)$ after the sampling has stopped at the nth step.

We shall assume that a statistician takes at least one observation before taking a terminal decision.

A sequential decision procedures has two components.

(a) A stopping rule that specifies for every set of observed values $(x_1, \cdots, x_n), n \geq 1$, whether to stop sampling and choose a terminal decision $d_n(x_1, \cdots, x_n) \in \mathcal{D}$ or to take another observation.

(b) A (terminal) decision rule which specifies for each set of observations $(x_1, \cdots, x_n)$ the terminal decision $d_n(x_1, \cdots, x_n) \in \mathcal{D}$ to be taken after the sampling has been stopped at the nth observation $(n \geq 1)$.

We shall denote by N the stopping random variable, $N = 1, 2, \cdots$. Clearly the distribution of N will depend on the stopping rule applied. The event $\{N = n\}$ means the sampling is stopped at the nth step, i.e., after observing n values $x_1, \cdots, x_n$ and not before. We also note that the event $\{N = n\}$ and the event $\{N \leq n\} = \bigcup_{k=1}^{n} \{N = k\}$ depend only on the observations $X_1, \cdots, X_n$ and not on $X_{n+1}, X_{n+2}, \cdots$.

A sequential procedure is said to be *closed* if the samping eventually terminates with probability 1 i.e., if

$$P\{N < \infty\} = 1.$$

We shall consider throughout closed sequential procedures only.

Definition 24.2.1 *Let $\mathcal{R}_n \subset \mathcal{X}_n$, where $\mathcal{X}_n = \{(x_1, \cdots, x_n)\}$ is the sample space up to the nth stage of sampling (nth observation), $(n = 1, 2, \cdots)$ be a sequence of Borel-measurable sets such that the sampling is terminated if the observation $\boldsymbol{x}_n = (x_1, \cdots, x_n) \in R_n$ and a terminal decision is taken. If $\boldsymbol{x}_n \notin R_n$, another observation x_{n+1} is taken. The sets $\mathcal{R}_n (n = 1, 2, \cdots)$ are called stopping regions.*

Thus,

$$\begin{aligned}
&\{N = 1\} \Leftrightarrow x_1 \in \mathcal{R}_1 \\
&\{N = 2\} \Leftrightarrow x_1 \notin \mathcal{R}_1, x_2 \in \mathcal{R}_2 \\
&\{N = n\} \Leftrightarrow x_1 \notin \mathcal{R}_1, \boldsymbol{x}_2 \notin \mathcal{R}_2, \cdots, \boldsymbol{x}_{n-1} \notin \mathcal{R}_{n-1}, \boldsymbol{x}_n \in \mathcal{R}_n.
\end{aligned}$$

24.3 Sequential Estimation

The theory of sequential point estimation differs very little in principle from fixed sample size (FSS) estimation, though the mathematical details are more complicated. One takes observations sequentially and stops according to some well-defined stopping rule. The parameter of interest is then estimated on the basis of the sample thus far obtained by using some standard estimation procedure like the method of unbiased minimum variance estimation, maximum likelihood. The problem becomes complicated as the sample space changes with each observation and further one has to consider the distribution of the stopping r.v. N.

Let $X_1, X_2, \cdots$ be a sequence of independent r.v.'s with common p.d.f. (p.m.f) $f(X;\theta), \theta \in \Theta$. Let $\psi(\theta)$ be an estimable parametric function which we want to estimate. Confining our attention to unbiased estimators of $\psi(\theta)$ we would like to have a Fréchet-Cramer-Rao type lower bound to variance of unbiased estimators as in the case of a FSS procedure. For this we first consider some preliminary results.

Theorem 24.3.1 (***Wald's Equation***). *Let $X_1, X_2, \cdots$ be i.i.d. r.v.'s with $E|X_1| < \infty$. Let N be a stopping random variable and $S_N = \sum_{i=1}^{N} X_i$. If $E(N) < \infty$, then*

$$E(S_N) = E(X_1)E(N). \tag{24.3.1}$$

Proof: Define a r.v. Y_i such that

$$Y_i = \begin{cases} 1, \text{ if no decision is reached up to } (i-1)\text{th stage, i.e., if } N > (i-1) \\ 0, \text{ otherwise.} \end{cases} \tag{24.3.2}$$

Clearly, Y_i depends only on $X_1, \cdots, X_{i-1}$ and does not depend on X_i. Also

$$S_N = \sum_{n=1}^{\infty} X_n Y_n.$$

Hence,

$$E(S_n) = E\left(\sum_{n=1}^{\infty} X_n Y_n\right). \tag{24.3.3}$$

Now,

$$\begin{aligned}
\sum_{n=1}^{\infty} E|X_n Y_n| &= \sum_{n=1}^{\infty} E|X_n| E|Y_n| \qquad \text{(because } X_n \text{ and } Y_n \text{ are independent)} \\
&= E|X_1| \sum_{n=1}^{\infty} E|Y_n| \\
&= E|X_1| \sum_{n=1}^{\infty} P\{N \geq n\} \text{ (because } E|Y_n| = P(Y_n = 1) = P(N \geq n))
\end{aligned}$$

$$= E|X_1|\sum_{n=1}^{\infty}\sum_{k=n}^{\infty}P\{N=k\} = E|X_1|\sum_{n=1}^{\infty}nP\{N=n\}$$
$$= E|X_1|E(N) < \infty.$$

Therefore, $E(S_n)$ exists and we may change the order of operation of expectation and summation sign in (24.3.3). Hence,

$$E(S_n) = \sum_{n=1}^{\infty}E(X_nY_n) = E(X_1)\sum_{n=1}^{\infty}E(Y_n)$$
$$= E(X_1)\sum_{n=1}^{\infty}P(N \geq n) = E(X_1)E(N).$$

■ NOTE 24.3.1 *It follows that theorem* 24.3.1 *holds if only we assume* $E(X_n) = \mu, E|X_n| < \infty$ *and* $E(N) < \infty$ *and the assumption that* X_i*'s are i.i.d. is not necessary.*

Theorem 24.3.2 *Let* $X_1, X_2, \cdots,$ *be a sequence of i.i.d. r.v.'s with* $E(X_1) = 0, E(X_1^2) = \sigma^2 < \infty$. *For any stopping r.v.* N *with* $E(N) < \infty$, *if* $E\left\{\left(\sum_{k=1}^{N}|X_k|\right)^2\right\} < \infty$, *then*

$$E(S_N^2) = \sigma^2E(N). \tag{24.3.4}$$

Proof: Let Y_i be defined as in (24.3.1). Then

$$E(S_N^2) = E\left\{\left(\sum_{i=1}^{\infty}X_iY_i\right)\left(\sum_{j=1}^{\infty}X_jY_j\right)\right\}$$
$$= E\left(\sum_{i=1}^{\infty}X_i^2Y_i^2 + \sum\sum_{i\neq j}X_iY_iX_jY_j\right). \tag{24.3.5}$$

Now,

$$E|S_N^2| = E\left(\sum_{i=1}^{\infty}X_i^2Y_i^2 + \sum\sum_{i\neq j}|X_iX_j|Y_iY_j\right)$$
$$= E\left(\sum_{k=1}^{N}|X_k|\right)^2 < \infty \text{ (by assumption)}$$

Hence the order of operation of summation and expectation in (24.3.5) can be interchanged. Now,

$$E\left(\sum_{i=1}^{\infty}X_i^2Y_i^2\right) = E(X_1^2)E\left(\sum_{i=1}^{\infty}Y_i^2\right)$$
$$= \sigma^2E\left(\sum_{i=1}^{\infty}Y_i\right)$$
$$= \sigma^2E(N) \text{ (by theorem 24.3.1).}$$

Again,

$$E\left(\sum\sum_{i\neq j} X_iY_iX_jY_j\right) = 2E\left(\sum\sum_{i>j}^{\infty} X_iY_iX_jY_j\right) = 2\sum_{i=2}^{\infty}\sum_{j=1}^{i-1} E(X_iX_jY_i)$$

$$= 2\sum_{i=2}^{\infty}\sum_{j=1}^{i-1} E\{Y_iE\{X_iX_j|Y_i\}\} = 2\sum_{i=2}^{\infty}\sum_{j=1}^{i-1} E\{Y_iE(X_i)E(X_j|Y_i)\}$$

$$= 0. \qquad (\because \ X_j \text{ and } Y_i \text{ are independent of } X_i)$$

Hence the proof.

Theorem 24.3.3 (***Wolfwitz, 1947***). *Let $X_1, X_2, \cdots$ be an infinite sequence of i.i.d. r.v.'s with common p.d.f. (p.m.f.) $f(x;\theta), \theta \in \Theta$. Let N be a stopping r.v. associated with the sequential procedure with $E_\theta(N) < \infty$. Let also $h(X_1, X_2, \cdots)$ be an unbiased estimate of $\psi(\theta)$. Then under the regularity conditions stated below,*

$$\operatorname*{Var}_{\theta}\{h(X_1,\cdots,X_N)\} \geq \frac{\{\psi'(\theta)\}^2}{E_\theta(N)E_\theta[\partial \log f(X;\theta)/\partial\theta]^2} \ \forall\ \theta \in \Theta. \tag{24.3.6}$$

The regularity conditions are:

(i) Θ is an open interval of the real line which may consist of the entire line or of an entire half-line.

(ii) $f(x;\theta)$ is differentiable with respect to $\theta\ \forall\ \theta \in \Theta$. Whenever, $f(x;\theta) = 0$, define $\partial \log f(x;\theta)/\ \partial\theta = 0$ so that $\partial \log f(x;\theta)/\partial\theta$ is defined for all θ in Θ and all x.

(iii) The integral (the sum) $\int f(x;\theta)\,dx \left(\sum_x f(x;\theta)\right)$ can be differentiated under the integral (summation) sign.

(iv) $0 < E_\theta\left[\frac{\partial \log f(x;\,\theta)}{\partial\theta}\right]^2 < \infty\ \forall\ \theta$.

(v) For each $n = 1, 2, \cdots$ and all θ,

$$E_\theta\left[\sum_{i=1}^{n}\left|\frac{\partial \log f(X_i;\ \theta)}{\partial\theta}\right|\right]^2 < \infty.$$

(vi) The equation

$$E_\theta[h(X_1,\cdots,X_n)] = \psi(\theta)$$

can be differentiated under the expectation sign with respect to all θ in Θ and for each $n = 1, 2, \cdots$.

(vii) The series $\sum_{n=1}^{\infty} d\varphi_n(\theta)/d\theta$ converges uniformly on Θ, where

$$\varphi_n(\theta) = \int_{N=n} h(x_1, \cdots, x_n) \prod_{i=1}^{n} f(x_i; \theta)\, dx_i$$

$$\left[\text{or } \sum_{N=n} h(x_1, \cdots, x_n) \prod_{i=1}^{n} f(x_i; \theta)\right]$$

Proof: The r.v.'s $\partial \log f(X_i; \theta)/\partial\theta, i = 1, 2, \cdots$ are i.i.d. with mean 0 and finite variance. Since $E_\theta(N) < \infty$, by Wald's equation (24.3.1),

$$E_\theta\left[\sum_{i=1}^{N} \frac{\partial}{\partial\theta} \log f(X_i; \theta)\right] = E_\theta(N) E_\theta\left[\frac{\partial}{\partial\theta} \log f(X;\ \theta)\right] = 0 \ \forall\ \theta.$$

By condition (v) and theorem (24.3.2),

$$E_\theta\left[\sum_{i=1}^{N} \frac{\partial \log f(X_i;\ \theta)}{\partial\theta}\right]^2 = E_\theta(N) E_\theta\left[\frac{\partial}{\partial\theta} \log f(X_i; \theta)\right]^2. \tag{24.3.7}$$

Now, by Cauchy-Schwartz inequality,

$$\begin{aligned}
&\operatorname*{Cov}_{\theta}\left[h(X_1, \cdots, X_N),\ \sum_{i=1}^{N} \frac{\partial \log f(X_i; \theta)}{\partial\theta}\right] \\
&\le \{V_\theta\left[h(X_1, \cdots, X_N)\right]\}^{1/2} \left[V_\theta \sum_{i=1}^{N} \frac{\partial \log f(X_i; \theta)}{\partial\theta}\right]^{1/2} \\
&= \{V_\theta\left[h(X_1, \cdots, X_N)\right]\}^{1/2} \left\{E_\theta(N) E_\theta\left[\frac{\partial \log f(X_1; \theta)}{\partial\theta}\right]^2\right\}^{1/2} \quad \text{(by 24.3.7)}
\end{aligned} \tag{24.3.8}$$

for all θ. Now, by conditions (vi) and (vii),

$$\begin{aligned}
\varphi'(\theta) &= \frac{\partial}{\partial\theta} E_\theta[h(X_1, \cdots, X_N)] = \int h(x_1, \cdots, x_N) \frac{\partial}{\partial\theta} \prod_{i=1}^{n} f(x_i; \theta)\, dx_i \\
&= \int h(x_1, \cdots, x_N) \sum_{i=1}^{N} \frac{\partial \log f(x_i; \theta)}{\partial\theta} \prod_{i=1}^{n} f(x_i; \theta)\, dx_i \\
&= E_\theta\left[h(X_1, \cdots, X_N) \sum_{i=1}^{N} \frac{\partial \log f(X_i; \theta)}{\partial\theta}\right].
\end{aligned} \tag{24.3.9}$$

Hence, the result follows by (24.3.8) and (24.3.9).

■ NOTE 24.3.2 *The conditions* (i)–(iv) *are same as those assumed in theorem* 15.11.1. *The conditions* (v)–(vii) *are needed because of the sequential nature of the population.*

For fixed sample size problem the result (24.3.6) coincides with the inequality (15.11.1).

■ NOTE 24.3.3 *In the case of fixed sample size* (FSS) (n_0) *procedure,* $\underset{\theta}{\text{Var}}\{h(\boldsymbol{X})\} \geq \frac{\delta}{n_0}$ *whereas for a sequential procedure* $\underset{\theta}{\text{Var}}\{h(\boldsymbol{X})\} \geq \frac{\delta}{E(N)}$ *where*

$$\delta = \{\psi'(\theta)\}^2 / E_\theta[\partial \log f(X;\theta)/\partial\theta]^2.$$

Therefore, if $n_0 = E(N)$ (for a given sequential procedure) and if there exists a FSS (n_0) procedure which attains Rao-Cramer lower bound, nothing is gained by using that sequential procedure on an average. For a FSS (n)-procedure with $n \geq E_\theta(N)$, for every unbiased estimator $h(\boldsymbol{X})$ of θ,

$$\underset{\theta}{\text{Var}}\{h(\boldsymbol{X})\} \geq \frac{1}{E_\theta(N) E_\theta\left[\frac{\partial \log f(X,\theta)}{\partial\theta}\right]^2} \geq \frac{1}{n E_\theta\left[\frac{\partial \log f(X,\theta)}{\partial\theta}\right]^2}.$$

24.3.1 Sequential Estimation of the Mean of a Normal Population

Let $X_1, X_2, \cdots$ be i.i.d. $N(\mu, \sigma^2)$-variables where both μ and σ^2 are unknown.

(a) **Point estimation of** μ **:**

We consider the procedure suggested by Robbins (1959). Choose $\hat{\mu} = \bar{X}_n$ where n is to be so chosen that risk (average loss) is minimum. Consider loss function as

$$L(n) = A|\bar{X}_n - \mu| + n$$

where $A(> 0)$ is a constant and n is the cost of sampling assuming each observation costs one unit. The risk is

$$R(n) = EL(n) = E[A|\bar{X}_n - \mu| + n] = A\sqrt{\frac{2}{\pi}} \cdot \frac{\sigma}{\sqrt{n}} + n \text{ (vide section 6.7).}$$

Regarding n as a continuous variable,

$$\frac{\partial R(n)}{\partial n} = 0 \Rightarrow n = \left(\frac{A\sigma}{\sqrt{2\pi}}\right)^{2/3} = n_0 \text{ (say)}$$

which makes the average risk minimum.
For this value of n,

$$R(n_0) = 3n_0.$$

However, if σ is unknown, n_0 can not be used. For the case where σ is unknown, Robbins suggested the following sequential sampling procedure: sample sequentially, observing $x_1, x_2, \cdots$, and computing $\bar{x}_n$, $s_n = \frac{1}{n-1}\sum_{i=1}^{n}(\bar{x}_i - \bar{x}_n)^2, n = 2, 3, \cdots$; stop sampling the first time for $n \geq 2$ when

$$n \geq \left(\frac{As_n}{\sqrt{2\pi}}\right)^{2/3}.$$

It can be proved that this sequential procedure terminates with probability unity.

(b) **Interval estimation:**

Suppose that we require a confidence interval for μ of width at most $2d(d > 0)$ at confidence level $1-\alpha(0 < \alpha < 1)$. It has been shown in example 24.2.2 that the confidence interval $(\bar{X}_n - d, \bar{X}_n + d)$ assures this provided

$$n \geq \frac{\sigma^2}{d^2}\tau^2_{\alpha/2}.$$

If σ is unknown Stein (1945) suggested the following procedure. Take a preliminary sample of size m (fixed) and compute

$$\bar{X}_m = \frac{1}{m}\sum_{i=1}^{m} X_i, \ S_m^2 = \frac{1}{(m-1)}\sum_{i=1}^{m}(X_i - \bar{X}_m)^2.$$

Then take a further sample of size $N - m$, where N is the smallest integer $\geq \frac{S_m^2}{d^2}t^2_{m-1,\alpha/2}$ and use as confidence interval

$$\left(\bar{X}_N - \frac{S_m}{\sqrt{N}}t_{m-1,\alpha/2}, \ \bar{X}_N + \frac{S_m}{\sqrt{N}}t_{m-1,\alpha/2}\right), \tag{24.3.10}$$

where $\bar{X}_N = \frac{1}{N}\sum_{i=1}^{N} X_i$. It can be checked that if we choose N to be the smallest integer satisfying

$$N \geq \frac{t^2_{m-1,\alpha/2}S_m^2}{d^2}$$

the confidence interval will have length $\leq 2d$. If $N < m$, the confidence interval based on the first sample itself will have width $\leq 2d$.

To prove the result we consider the following lemmas.

Lemma 24.3.1 *As $n \to \infty$, $P_\sigma[N > n] \to 0$ for all σ.*

Proof: Since n is allowed to tend to ∞ we consider those values of N which are greater than n. Now

$$N > n \Leftrightarrow \frac{t^2 S_m^2}{d^2} > n \quad (\text{writing } t_{m-1,\alpha/2} \text{ as } t).$$

Hence,

$$P_\sigma(N > n) = P_\sigma\left(\frac{t^2 S_m^2}{d^2} > n\right)$$

$$= P_\sigma\left(\chi^2_{(m-1)} > \frac{n(m-1)d^2}{t^2\sigma^2}\right) \quad (24.3.11)$$

where $\frac{(m-1)S_m^2}{\sigma^2}$ follows a $\chi^2_{(m-1)}$ d.f.
Now, (24.3.11) $\to 0$ as $n \to \infty$.
Hence the result.

□ **Remark 24.3.1** *The lemma* 24.3.1 *implies that the procedure terminates with probability one.*

Lemma 24.3.2 *The random variable* $T_N = \sqrt{N}(\bar{X}_N - \mu)/S_m$ *has Student's t distribution with* $(m-1)$ *d.f.*

Proof: Let ν be a fixed integer $\geq m$. The random variable $\sqrt{\nu}(\bar{X}_\nu - \mu)/\sigma$ has a $N(0,1)$-distribution.
Now,

$$\nu\bar{X}_\nu = \sum_{i=1}^{m} X_i + \sum_{i=m+1}^{\nu} X_i$$

$$= m\bar{X}_m + \sum_{i=m+1}^{\nu} X_i.$$

Now, $X_{m+1}, \cdots, X_\nu$ are independent of $X_1, \cdots, X_m$ and $\bar{X}_m$ is independent of S_m^2.
Hence

$$T_\nu = \frac{\sqrt{\nu}(\bar{X}_\nu - \mu)}{S_m}$$

follows a Student's t distribution with $(m-1)$ d.f. This is the conditional distribution of T_N for $N = \nu$. Since this distribution does not depend on ν, this is also the unconditional distribution of T_N.

Theorem 24.3.4 $\bar{\mathbf{X}}_N$ *is an unbiased estimator of* μ *with variance* $\leq \frac{d^2(m-1)}{t^2(m-3)}$.

Proof: Since $\sqrt{N}(\bar{X}_N - \mu)/\sigma$ follows a $N(0,1)$-distribution,

$$E_\sigma\left[\frac{\sqrt{N}(\bar{X}_N - \mu)}{\sigma}\right] = 0.$$

Now,

$$E_\sigma\left[(\bar{X}_N-\mu)\right]=E_\sigma\left[\frac{\sqrt{N}(\bar{X}_N-\mu)}{\sigma}\cdot\frac{\sigma}{\sqrt{N}}\right]$$

$$=E_\sigma\left[\frac{\sqrt{N}(\bar{X}_N-\mu)}{\sigma}\right]E_\sigma\left[\frac{\sigma}{\sqrt{N}}\right]\quad\text{(because, distribution of }\tfrac{\sqrt{N}(\bar{X}_N-\mu)}{\sigma}\text{ is independent of that of }\tfrac{\sigma}{\sqrt{N}}\text{)}$$

$$=0 \text{ for all } \sigma.$$

Again,

$$E_\sigma(\bar{X}_N-\mu)^2=E_\sigma\left[\frac{N(\bar{X}_N-\mu)^2}{S_m^2}\cdot\frac{S_m^2}{N}\right]$$

$$\leq E_\sigma\left[\frac{N(\bar{X}_N-\mu)^2}{S_m^2}\cdot\frac{d^2}{t^2}\right]\quad\left(\because N\geq\frac{t^2S_m^2}{d^2}\right)$$

$$=\frac{d^2}{t^2}\cdot\frac{m-1}{m-3},$$

since $\frac{\sqrt{N}(\bar{X}_N-\mu)}{S_m}$ follows $t_{(m-1)}$ distribution.
By our choice of N, the length of the confidence interval (24.3.10) is $\leq d$.

Theorem 24.3.5 $P_\sigma\left[|\bar{X}_N-\mu|\leq d\right]\geq 1-\alpha$ *for all* σ.

Proof:

$$P_\sigma\left[|\bar{X}_N-\mu|\leq d\right]=P_\sigma\left[\frac{|\sqrt{N}(\bar{X}_N-\mu)|}{S_m}\leq\frac{\sqrt{N}d}{S_m}\right]$$

$$\geq P_\sigma\left[\frac{|\sqrt{N}(\bar{X}_N-\mu)|}{S_m}\leq t\right]=1-\alpha,\quad\text{for all }\sigma.$$

24.4 The Sequential Probability Ratio Test

The sequential test of testing a hypothesis H may be described as follows. A rule is made for making one of the following three decisions at the mth trial ($m=1,2,\cdots$) : (1) to accept H (2) to reject H (3) to continue the experiment by making an additional observation. The sample space $\mathcal{X}_m=\{(x_1,\cdots,x_m)\}, (m=1,2,\cdots)$ is divided into three mutually exclusive and exhaustive subspaces $\mathcal{R}_m^0, \mathcal{R}_m^1$ and $\mathcal{R}_m$. If $\boldsymbol{x}_m=(x_1,\cdots,x_m)$ ∈ $\mathcal{R}_m^0[\mathcal{R}_m^1]$ the hypothesis H is accepted [rejected]; if $\boldsymbol{x}_m\in\mathcal{R}_m$, a new observation is made. Clearly, the process is stopped when and only when either the first or second decision is made. A sequential test is, therefore, completely defined by defining the sets $\mathcal{R}_m^0$, $\mathcal{R}_m^1, \mathcal{R}_m (m=1,2,\cdots)$.

Definition 24.4.1 *Let X be a r.v. with p.d.f. $f(x;\theta), \theta \in \Theta$. Suppose we are interested in testing $H_o(\theta = \theta_0)$. Clearly, the probability of accepting H_o will be a function of the true value of the parameter θ. This function is called the Operative Characteristic (OC) function of a test and is generally denoted as $L(\theta)$. Since we shall consider only closed sequential procedures which terminate with probability one, the probability that the hypothesis is rejected when θ is the true value of the parameter is $1 - L(\theta) = \gamma(\theta)$, power of the test.*

For any parameter point θ the probability of making a correct decision can be obtained immediately from the OC function. If the parameter point θ is consistent with the hypothesis H_o, then the probability of making a correct decision is $L(\theta)$. If the true parameter point θ is not consistent with H_o, then the probability of making a correct decision is $1 - L(\theta)$. Clearly a favourable OC function gives high probability $L(\theta)$ for θ consistent with H_o and low value of $L(\theta)$ for θ inconsistent with H_o (fig. 24.1).

Fig. 24.1 An ideal OC function for $H_o : (0 \leq \theta \leq \theta_0)$

Definition 24.4.2 *The Average Sample Number (ASN) Function.*

The number of observations N required for a sequential test is a random variable. The expected value of $N, E_\theta(N)$ depends on the test procedure and distribution of X and is called the ASN function of the test.

We now consider the sequential probability ratio test (SPRT) for testing a simple hypothesis H_o against a simple alternative H_1 as developed by Wald (1945).

Let $X_1, X_2, \cdots$ be a sequence of i.i.d. r.v.'s with common p.d.f. (p.m.f) $f(x;\theta)$. We are interested in testing $H_o(\theta = \theta_0)$ against $H_1(\theta = \theta_1)$. For any positive integral n define the joint p.d.f. of $\boldsymbol{X} = (X_1, \cdots, X_n)$ under H_o and H_1 as

$$f_{jn}(\boldsymbol{x}) = \prod_{i=1}^{n} f(x_i;\theta_j), \quad j = 0, 1 \tag{24.4.1}$$

respectively, where $\boldsymbol{x} = (x_1, \cdots, x_n)$. Define

$$\lambda_n(\boldsymbol{x}) = \frac{f_{1n}(\boldsymbol{x})}{f_{0n}(\boldsymbol{x})}. \tag{24.4.2}$$

The SPRT for H_o against H_1 is defined as follows. Two positive constants A and $B (B < A)$ are suitably chosen. At nth trial λ_n is computed ($n = 1, 2, \cdots$). If

$$B < \lambda_n < A \tag{24.4.3}$$

the experiment is continued by taking an additional observation. If

$$\lambda_n \leq B \tag{24.4.4}$$

the process is terminated with acceptance of H_o. If

$$\lambda_n \geq A \tag{24.4.5}$$

the process is terminated with rejection of H_o (acceptance of H_1). [If for a particular sample $f_{1n} = f_{0n} = 0$, define $\lambda_n = 1$; if for some sample $f_{1n} > 0$ but $f_{0n} = 0$ assume (24.4.5) is fulfilled.]

The constants A and B are so determined that the test have strength (α, β) where α, β are the probabilities of type I error and type II error respectively. If N is the stopping r.v.

$$\alpha = P_{\theta_0}\{\lambda_N \geq A\}, \quad \beta = P_{\theta_1}\{\lambda_N \leq B\}. \tag{24.4.6}$$

Since for purposes of computations logarithms are convenient to deal with, one considers

$$Z_i = \log \frac{f(\theta_1; X_i)}{f(\theta_0; X_i)}. \tag{24.4.7}$$

Then the r.v.'s $Z_1, Z_2, \cdots$ are i.i.d. and

$$\log \lambda_n(\boldsymbol{X}) = \sum_{i=1}^{n} Z_i = S_n \text{ (say)}. \tag{24.4.8}$$

The SPRT can now be stated as follows. At each stage n compute $s_n = \sum_{i=1}^{n} z_i$. If

$$b = \log B < s_n < \log A = a \tag{24.4.9}$$

continue by taking observation Z_{n+1}. If $s_n \leq b$ accept H_o; if $s_n \geq a$ reject H_o.

■ NOTE 24.4.1 *Neyman-Pearson's* MP*-test procedure may be looked upon as a particular case of* SPRT *and as such the* SPRT *is suggested by* NP*-lemma. The assumption of independence and identical distribution of $X_1, X_2, \cdots$ made above is not necessary to define the test, but is only a simplifying assumption.*

■ NOTE 24.4.2 *The trivial case* $P\{Z_i = 0\} = 1$ *will be ignored. In this case* $f(x;\theta_1) = f(x;\theta_0)$ *and* H_o *will be accepted even without making any observation.*

■ NOTE 24.4.3 *If* λ_n *has a discrete distribution randomisation at the boundaries may be needed to achieve a pre-assigned strength* (α, β).

▶ **Example 24.4.1** *Let* $X_1, X_2, \cdots$ *be i.i.d. Bernoulli (p) variables, where* $p \in \Theta = \{p_0, p_1\}$. *We want to test* $H_o(p = p_0)$ *against* $H_1(p = p_1)$. *Here*

$$Z_i = \log\frac{P\{X_i = x; p_1\}}{P\{X_i = x; p_0\}} = \begin{cases} \log\frac{p_1}{p_0}, & \text{if } x = 1 \\ \log\frac{1-p_1}{1-p_0}, & \text{if } x = 0. \end{cases}$$

Let T_n be the number of 1's in the sequence of first n observations. Then

$$S_n = \sum_{i=1}^{n} Z_i = T_n \log\frac{p_1}{p_0} + (n - T_n)\log\frac{1-p_1}{1-p_0}.$$

Thus, at the nth stage H_o is accepted if

$$s_n = t_n \log\frac{p_1}{p_0} + (n - t_n)\log\frac{1-p_1}{1-p_0} \le b$$

and H_1 is accepted if $s_n \ge a$; otherwise another observation is taken.
On simplification, we obtain the following result. H_o is accepted if

$$t_n \le \frac{b + n\log\frac{1-p_0}{1-p_1}}{\log\frac{p_1}{p_0} - \log\frac{1-p_1}{1-p_0}}$$

H_1 is accepted if

$$t_n \ge \frac{a + n\log\frac{1-p_0}{1-p_1}}{\log\frac{p_1}{p_0} - \log\frac{1-p_1}{1-p_0}};$$

otherwise another observation is taken.

This sampling plan has been suggested in a number of sampling inspection tables (vide Hill, 1962).

▶ **Example 24.4.2** *Let* $X_1, X_2, \cdots,$ *be i.i.d.* $N(\theta, 1)$, *where* $\theta \in \Theta = \{\theta_0, \theta_1\}$. *To test* $H_o(\theta = \theta_0)$ *against* $H_1(\theta = \theta_1)$, *we have*

$$Z_i = \log\frac{f(X_i;\theta_1)}{f(X_i;\theta_0)} = (\theta_1 - \theta_0)X_i + \frac{1}{2}(\theta_0^2 - \theta_1^2).$$

$$\sum_{i=1}^{n} Z_i = (\theta_1 - \theta_0)\sum_{1}^{n} X_i + \frac{n}{2}(\theta_0^2 - \theta_1^2)$$

$$= \sum_{i=1}^{n-1} Z_i + (\theta_1 - \theta_0)X_n + \frac{1}{2}(\theta_0^2 - \theta_1^2).$$

Therefore, the quantity $\log\lambda_n(\boldsymbol{x})$ can be computed cumulatively by adding to the preceding value $\log\lambda_{n-1}(\boldsymbol{x})$ the quantity $(\theta_1 - \theta_0)x_n + \frac{1}{2}(\theta_0^2 - \theta_1^2)$.

▶ **Example 24.4.3** *A lot consisting of N units of manufactured products of which D (a unknown number) are defective is submitted for inspection. On the ith inspected unit we assign value 1(0) if it is a defective item (otherwise). Let d_i be the number of defective items observed in the set of first i observations. The joint distribution of $(x_1, \cdots, x_m)$ is given by*

$$f_m(\boldsymbol{x}) = f\left(x_1, \frac{D}{N}\right) f\left(x_2, \frac{D-d_1}{N}\right) f\left(x_3, \frac{D-d_2}{N}\right) \cdots f\left(x_m, \frac{D-d_{m-1}}{N}\right) \quad \text{(i)}$$

where

$$f(x,p) = \begin{cases} p, & \text{if } x = 1 \\ 1-p, & \text{if } x = 0. \end{cases}$$

Suppose we want to test $H_\circ(D = D_1)$ against $H_1(D = D_2)$. Using (i) one can calculate $f_{jm}(\boldsymbol{x})$ for $j = 0, 1$. The SPRT is based on the ratio f_{1m}/f_{0m}. In this case $X_1, X_2, \cdots$ are not i.i.d. variables.

Theorem 24.4.1 *For the* SPRT *with stopping bounds A and B(< A) and strength* (α, β),

$$A \leq \frac{1-\beta}{\alpha}, \quad B \geq \frac{\beta}{1-\alpha}, \quad 0 < \alpha < 1, \quad 0 < \beta < 1. \tag{24.4.10}$$

Proof: Let

$$\mathcal{R}_n^1 = \{(x_1, \cdots, x_n) : \quad \lambda_n(\boldsymbol{x}) \geq A\}$$
$$R_n^0 = \{(x_1, \cdots, x_n) : \quad \lambda_n(\boldsymbol{x}) \leq B\}.$$

If an observed $\boldsymbol{x} \in R_n^1$, it means after taking the nth observation $H_\circ$ is rejected. The process terminates with rejection of $H_\circ$ if observed value falls in either $\mathcal{R}_1^1, \mathcal{R}_2^1, \cdots$. Hence

$$\alpha = P\left\{\boldsymbol{x} \in \cup_{n=1}^{\infty} \mathcal{R}_n^1 | H_\circ\right\} = \sum_{n=1}^{\infty} P\{\mathcal{R}_n^1 | H_\circ\} \quad \text{(since the sets } \mathcal{R}_n^1, n = 1, 2, \cdots \text{ are mutually disjoint)}$$

$$= \sum_{n=1}^{\infty} \int_{\mathcal{R}_n^1} f_{0n}(\boldsymbol{x}) d\boldsymbol{x} \leq \frac{1}{A} \sum_{n=1}^{\infty} \int_{\mathcal{R}_n^1} f_{1n}(\boldsymbol{x}) d\boldsymbol{x} \qquad \left[\because \frac{f_{1n}(\boldsymbol{x})}{f_{0n}(\boldsymbol{x})} \geq A\right]$$

$$= \frac{1}{A} P\left\{\boldsymbol{x} \in \bigcup_{n=1}^{\infty} \mathcal{R}_n^1 | H_1\right\} = \frac{1}{A} P\{H_\circ \text{ is rejected } | H_1\}$$

$$= \frac{1-\beta}{A}.$$

In the above proof we have assumed that the SPRT terminates with probability 1 (this is proved in theorem 24.4.2) i.e., we have assumed

$$\sum_{n=1}^{\infty} P\{N = n\} = \sum_{n=1}^{\infty} \int_{\mathcal{R}_n^0 \cup \mathcal{R}_n^1} f_{jn}(\boldsymbol{x}) d\boldsymbol{x} = 1, \quad i = 0, 1.$$

Similarly,

$$\beta = P\left\{\bigcup_{n=1}^{\infty} \mathcal{R}_n^0 | H_1\right\} = \sum_{n=1}^{\infty}\int_{\mathcal{R}_n^0} f_{1n}(\boldsymbol{x})d\boldsymbol{x} \le B\sum_{n=1}^{\infty}\int_{\mathcal{R}_n^0} f_{on}(\boldsymbol{x})d\boldsymbol{x}$$
$$= BP\{H_\circ \text{ is accepted } |H_\circ\} = B(1-\alpha).$$

□ **Remark 24.4.1** *From* (24.4.10) *we have* $\alpha \le \frac{1-\beta}{A} \le \frac{1}{A}$ *and* $\beta \le B(1-\alpha)$. *Therefore, in the* (α, β)*-plane the shaded portion gives the values of* (α, β) *which satisfy the inequalities in* (24.4.10) *for a given value of* (A, B) *(fig. 24.2).*

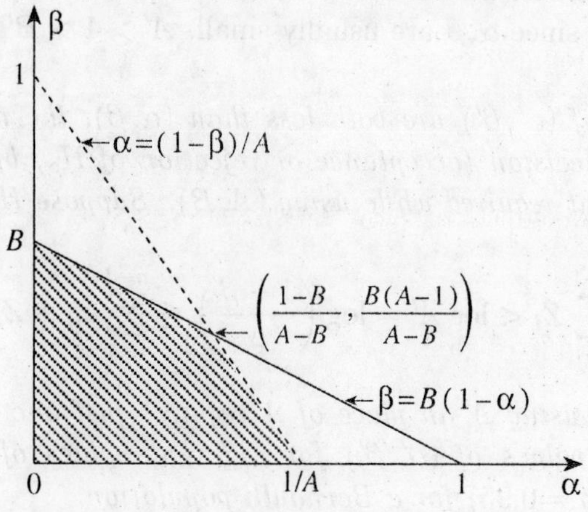

Fig. 24.2

■ NOTE 24.4.4 *It follows from* (24.4.10) *that the boundaries* A, B *can be approximated respectively by*

$$A' = \frac{1-\beta}{\alpha},\ B' = \frac{\beta}{1-\alpha}. \tag{24.4.11}$$

Let (α', β') be the true strength of a SPRT with boundaries (A', B').

Now,

$$\frac{\alpha'}{1-\beta'} \le \frac{1}{A'} = \frac{\alpha}{1-\beta}\ \left(\because \frac{1}{A} \ge \frac{\alpha}{1-\beta} \text{ by } (24.4.10)\right). \tag{24.4.12}$$

Also

$$\frac{\beta'}{1-\alpha'} \le B' = \frac{\beta}{1-\alpha}\ \left(\because B \ge \frac{\beta}{1-\alpha}\right). \tag{24.4.13}$$

Hence,

$$\alpha' \le \frac{\alpha}{1-\beta} \tag{24.4.14}$$

$$\beta' \le \frac{\beta}{1-\alpha}. \tag{24.4.15}$$

Multiplying (24.4.14) by $(1-\beta)(1-\beta')$ and (24.4.15) by $(1-\alpha)(1-\alpha')$ and adding

$$\alpha' + \beta' \leq \alpha + \beta. \tag{24.4.16}$$

The values of α and β are usually small in practice (ranging between 0.01 and 0.05) and thus α' and β' will be very nearly equal to α and β respectively. The equations (24.4.14) and (24.4.15) show that the amount by which $\alpha'(\beta')$ may exceed $\alpha(\beta)$ is negligibly small. The expression (24.4.16) shows that by using A', B' in place of A, B respectively, at most one of the probabilities of error (of two types) may increase. Thus, the use of A' and B' in place of A and B respectively as boundary values of SPRT does not result in any appreciable increase in the value of probability of error of either type I or type II. (Note that since α, β are usually small, $A' > 1 > B' > 0$).

■ NOTE 24.4.5 *Even if (α', β') are both less than (α, β), the number of observations required to come at a decision (acceptance or rejection of $H_\circ$) by using (A', B') cannot be much larger than that required while using (A, B). Suppose $H_\circ$ is rejected at the nth step. Then*

$$\sum_{i=1}^{n-1} Z_i < \log A' = \log\left(\frac{1-\beta}{\alpha}\right) \leq \sum_{i=1}^{n-1} Z_i + Z_n.$$

Therefore, error in using A' in place of A requires just one additional observation. Table 24.1 shows the values of (α', β') for different values of (α, β) for testing $H_\circ$ $(p = 0.90)$ against $H_1(p = 0.95)$ for a Bernoulli population.

□ **Remark 24.4.2** *In Neyman-Pearson theory the critical values of the test statistic depend on $H_\circ$ and α. In the present set up the boundary values A' and B' depend only on α and β and do not depend on form of the p.d.f. under $H_\circ$ or H_1.*

□ **Remark 24.4.3** *Results of theorem 24.4.1 and the results in (24.4.11)–(24.4.16) hold even for dependent observations provided we assume $P\{N < \infty | H_i\} = 1$, $i = 1, 2, \cdots$.*

Table 24.1 *The values of (α', β') for different values of (α, β) for testing $H_\circ(p = 0.90)$ against $H_1(p = 0.95)$ for a Bernoulli population (Eger, 1982).*

		B'	A'	α'	β'
$\alpha = 0, 10$	$\beta = 0.10$	1/9	9	0.0998	0.791
	$\beta = 0.05$	1/18	19/2	0.0987	0.396
	$\beta = 0.01$	1/90	9.9	0.0973	0.0079
$\alpha = 0.05$	$\beta = 0.10$	2/19	18	0.0499	0.0791
	$\beta = 0.05$	1/19	19	0.0493	0.0396
	$\beta = 0.01$	1/95	99/5	0.0487	0.0079

Some Properties of the SPRT

(1) SPRT terminates with probability one.

Theorem 24.4.2 *(Stein 1946). Let $X_1, X_2, \cdots,$ be i.i.d. r.v.'s with common p.d.f. (p.m.f) $f(x;\theta), \theta \in \Theta$. Suppose we are interested in testing $H_o(f = f(x;\theta_0) = f_0)$ against $H_1(f = f(x;\theta_1) = f_1)$. Let N be the number of observations required for a decision using a* SPRT. *Suppose $P\{|Z| > 0|\theta\} > 0 \ \forall \ \theta \in \Theta$ where $Z = \log[f_1(X)/f_0(X)]$. Then*

$$P_\theta\{N < \infty\} = 1 \tag{24.4.17}$$

and

$$E_\theta\{e^{tN}\} < \infty \text{ for } -\infty < t < t_0, t_0 > 0. \tag{24.4.18}$$

Proof: **First Part.** The r.v.'s $Z_i = \log[f_1(X_i)/f_0(X_i)]$ are i.i.d.

Let us denote $r(m,k) = [\frac{m}{k}]$ where $[h]$ denotes the largest integer contained in h and m, k are fixed positive integers. Consider the partial sums

$$S_k, S_{2k} - S_k, \cdots, S_{rk} - S_{(r-1)k}$$

where $S_n = \sum_{i=1}^{n} Z_i, n = 1, 2, \cdots$.

If $N > m$ then $b(= \log B) < S_i < a(= \log A), i = 1, 2, \cdots, m$ and in particular for $i = k, 2k, \cdots, rk$ for a fixed k. Hence

$$|T_j| = |S_{jk} - S_{(j-1)k}| < |b| + |a| = C \text{ (say)}, \quad (j = 1, 2, \cdots, r).$$

Now,

$$P\{N > m\} \le P\{|T_j| < C, \ j = 1, 2, \cdots, r\} = [P\{|T_1| < C\}]^r \tag{24.4.19}$$

since the T_j's are i.i.d. Therefore,

$$\lim_{m\to\infty} P\{N > m\} \le \lim_{r\to\infty} [P\{|T_1| < C\}]^r$$

for a fixed k (to be suitably chosen).

Now, since $P\{|Z| > 0\} > 0$, we can find an $\varepsilon(> 0)$ such that either $P\{Z > \varepsilon\} > 0$ or $P\{Z < -\varepsilon\} > 0$ (or both).

Hence

$$\begin{aligned} P\{|T_1| > C\} &= P\{|Z_1 + \cdots + Z_k| > C\} \\ &= P\left\{\sum_{i=1}^{k} Zi > C\right\} + P\left\{-\sum_{i=1}^{k} Z_i > C\right\} \\ &\ge P\left\{Z_i > \frac{C}{k}, i = 1, \cdots, k\right\} + P\left\{Z_i < -\frac{C}{k}, i = 1, \cdots, k\right\}. \end{aligned}$$

Now choose k an integer such that $C/k < \varepsilon$. Then

$$P\{|T_1| > C\} \geq [P\{Z_1 > \varepsilon\}]^k + [P\{Z_1 < -\varepsilon\}]^k > 0 \text{ (by assumption).}$$

Thus,

$$P\{|T_1| < C\} = \delta < 1$$

when δ is independent of m. Hence by (24.4.19)

$$\lim_{m\to\infty} P\{N > m\} \leq \lim_{m\to\infty} \delta^{r(m)} = 0,$$

where $r(m) = r(m,k)$. As k is fixed and as $m \to \infty$, $r \to \infty$. Thus

$$P\{N < \infty\} = 1 - P\{N = \infty\} = 1 - \lim_{m\to\infty} P\{N > m\} = 1.$$

Second Part. We have

$$P\{N > m\} \leq \delta^{r(m)} \quad \text{for every } m.$$

Hence,

$$\begin{aligned} E_\theta(e^{tN}) &= \sum_{m=1}^{\infty} e^{tm} P\{N = m\} \leq \sum_{m=1}^{\infty} e^{tm} P\{N > m-1\} \\ &= e^t \sum_{m=0}^{\infty} e^{tm} P\{N > m\} \leq e^t \sum_{m=0}^{\infty} e^{tm} \delta^{r(m)} \\ &= e^t \sum_{m=0}^{\infty} e^{tm} \delta^{m/k} \delta^{r-m/k} \leq e^t \delta^{-1/k} \sum_{m=0}^{\infty} (\delta^{1/k} e^t)^m < \infty \end{aligned}$$

provided $e^t\delta^{1/k} < 1$ or $t < -(\log \delta)/k = t_0$. Since $0 < \delta < 1$, $t_0 > 0$. Hence the proof.

■ NOTE 24.4.6 *Clearly, the probability generating function $E_\theta(t^N)$ of the r.v. N is finite for $0 < t < \delta^{-1/k}$. Hence, moments of all order exist for the stopping r.v. N.*

Corollary 24.4.1 *The SPRT terminates with probability 1 under both H_o and H_1.*

We have only to show that

$$P\{|Z| > 0|H_i\} > 0, \quad i = 0, 1.$$

For this we consider the following lemma, proof of which can be easily verified.

Lemma 24.4.1 *Let f and g be p.d.f.'s (p.m.f's) and let $C = \{x : f > g\}$. Then*

$$\int_C f(x) \log \frac{f(x)}{g(x)}\, dx \geq 0 \tag{24.4.20}$$

if f and g are p.d.f's and

$$\sum_{x\in C} f(x) \log \frac{f(x)}{g(x)} \geq 0 \tag{24.4.21}$$

if f and g are p.m.f's with equality iff $f(x) = g(x)$ almost everywhere.

It follows from lemma 24.4.1 that unless $P\{f_0(X)=f_1(X)\}=1$, we have $E\{Z|H_\circ\}<0$. This implies that $P\{Z<0|H_\circ\}>0$ or $P\{|Z|>0|H_\circ\}>0$. Similarly, putting $f=f_1$ and $g=f_0$ in lemma 24.4.1 we obtain $P\{|Z|>0|H_1\}>0$.

(2) **OC function of SPRT**

We shall consider the performance $(L(\theta))$ of SPRT for testing $H_\circ(\theta=\theta_0)$ against H_1 $(\theta=\theta_1)$ when θ is the true value of the parameter $\theta\in\Theta$.

We shall consider derivation of an approximate formula for $L(\theta)$, neglecting the excess of f_{1m}/f_{0m} over the boundaries A and B at the termination of the process. For this we first consider the following lemma.

Lemma 24.4.2 (***Wald* 1947**). *Let Z be a r.v. such that*

(i) *$E(Z)$ exists and $E(Z)\neq 0$;*

(ii) *there exists a positive δ such that $P[e^Z<1-\delta]>0$ and $P[e^Z>1+\delta]>0$;*

(iii) *for any real value h, $E(e^{hZ})=g(h)$ exists.*

Then there exists one and only one real value $H_\circ\neq 0$ such that

$$E(e^{H_\circ Z})=1. \tag{24.4.22}$$

Proof: For any positive h,

$$\begin{aligned} g(h)=E(e^{hZ}) &= \int e^{hZ}f(z)\,dz \\ &> \int_{e^Z>(1+\delta)} e^{hZ}f(z)\,dz>(1+\delta)^hP(e^Z>1+\delta). \end{aligned} \tag{24.4.23}$$

Since $P(e^Z>1+\delta)>0$,

$$\lim_{h\to\infty} g(h)=\infty.$$

Similarly, it can be seen that for any negative h,

$$g(h)>P(e^Z<1-\delta)(1-\delta)^h.$$

Since $P(e^Z<1-\delta)>0$,

$$\lim_{h\to-\infty} g(h)=\infty.$$

Now condition (iii) implies that moments of all order of Z exist and may be obtained by differentiating $g(h)$ under the expectation sign as many times as needed. Thus

$$g''(h)=E\{Z^2e^{hZ}\}>0 \ \forall \ \text{real } h.$$

Hence $g(h)$ is a convex function of h. Since $g(h)\to\infty$ as $h\to\pm\infty$, g has a minimum at $h=h^*$ where $g'(h^*)=0$. Since $g'(0)=E(Z)\neq 0$, we see that $h^*\neq 0$ and since $g(h^*)$ is minimum $g(h^*)<g(0)=1$.

Thus, g is strictly decreasing over $(-\infty, h^*)$ and strictly increasing over (h^*, ∞). Since $g(0) = 1$ and $g(h^*) < 1$, there exists exactly one real value $h_0 \neq 0$ such that $g(h_0) = 1$. Hence the lemma (fig. 24.3a and 24.3b).

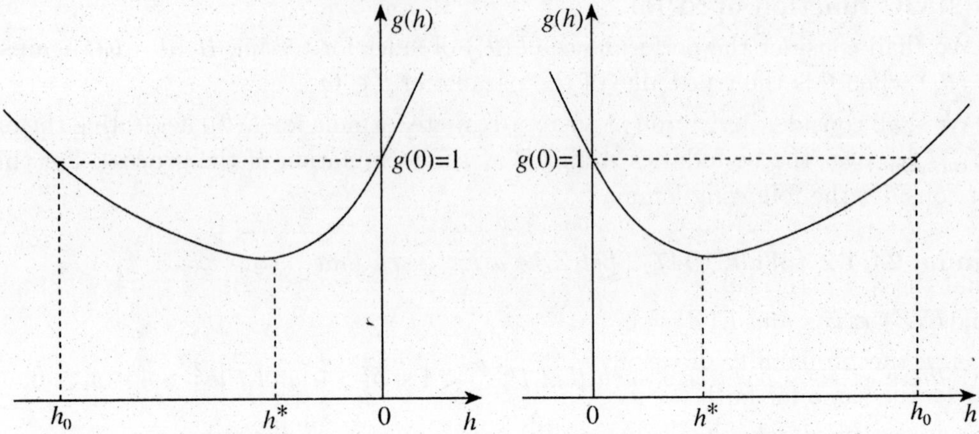

Fig. 24.3(a) **Fig. 24.3(b)**

■ NOTE 24.4.7 *It follows that if $h^* > 0$, then $h_0 > 0$ and if $h^* < 0$, then $h_0 < 0$. Furthermore, if $h^* > 0$, then $E(Z) = g'(0) < 0$ and if $h^* < 0$ then $E(Z) = g'(0) > 0$. Hence h_0 and $E(Z)$ are of opposite signs.*

We now apply lemma 24.4.2 to approximate $\beta(\theta) = 1 - L(\theta)$. Consider the equation

$$E_\theta\left[\left\{\frac{f_1(X)}{f_0(X)}\right\}^h\right] = \int_{-\infty}^{\infty}\left[\frac{f(x;\,\theta_1)}{f(x;\,\theta_0)}\right]^{h(\theta)} f(x;\,\theta)\,dx = E_\theta(e^{hZ}) = 1. \tag{24.4.24}$$

Let $h_0 = h_0(\theta)$ be a non-zero solution of (24.4.24). Then

$$\left[\frac{f_1(x)}{f_0(x)}\right]^{h_0} f_\theta(x) = f_{\theta^*}(x) \quad . \tag{24.4.25}$$

is a p.d.f (p.m.f). Suppose that $h_0 > 0$ (the other case can be treated similarly). Consider the SPRT with stopping bounds A^{h_0}, B^{h_0} for testing $H_0^*(f = f_\theta)$ against $H_1^*(f = f_{\theta^*})$. The test is to continue so long as

$$B^{h_0} < \frac{\prod_{i=1}^{n} f_{\theta^*}(x_i)}{\prod_{i=1}^{n} f_\theta(x_i)} = \lambda^*(\boldsymbol{x}) \text{ (say) } < A^{h_0};$$

to reject f_θ if

$$\lambda^*(\boldsymbol{x}) \geq A^{h_0}$$

and to reject f_{θ^*} if

$$\lambda^*(\boldsymbol{x}) \leq B^{h_0}.$$

Let α^* and β^* be the probabilities of type I error and type II error respectively for testing H_o^* against H_1^*. Then

$$B^{h_0} \simeq \frac{\beta^*}{1-\alpha^*} \quad \text{and} \quad A^{h_0} \simeq \frac{1-\beta^*}{\alpha^*}. \tag{24.4.26}$$

But

$$\lambda^*(\boldsymbol{x}) = \left[\frac{\prod_{i=1}^{n} f_1(x_i)}{\prod_{i=1}^{n} f_0(x_i)}\right]^{h_0} \tag{24.4.27}$$

so that the test is the same as the SPRT for $H_o(f = f_0)$ against $H_1(f = f_1)$ with stopping bounds A and B.

Since the probability of rejection of the hypothesis $(f = f_\theta)$ when in fact it is true must be the same for both the SPRT's

$$\beta(\theta) = \alpha^*.$$

However, on solving from (24.4.26),

$$\alpha^* \simeq \frac{1 - B^{h_0}}{A^{h_0} - B^{h_0}}. \tag{24.4.28}$$

Hence,

$$L(\theta) = 1 - \beta(\theta) \simeq \frac{A^{h_0} - 1}{A^{h_0} - B^{h_0}}. \tag{24.4.29}$$

Note that $h(\theta_0) = 1$ is a solution of

$$E_{\theta_0}\left[e^{h(\theta_0)Z}\right] = 1$$

and $h(\theta_1) = -1$ is a solution of

$$E_{\theta_1}\left[e^{h(\theta_1)Z}\right] = 1 \text{ [vide (24.4.24)]}.$$

Also $E_\theta[e^{hZ}] = 1$ when $h = 0$. However, $\beta(\theta)(= \alpha^*)$ is not defined for $h = 0$. In this case

$$\beta(\theta)]_{h=0} = -\frac{\log B}{\log A - \log B} \quad \text{(by L' Hospital rule).} \tag{24.4.30}$$

Also

$$\beta(\theta_0) = \alpha, \;\; \beta(\theta_1) = P\{\text{rejecting } H_1 | \theta = \theta_1\} = 1 - \beta.$$

Finally,

$$\beta(\theta) \simeq \left[\frac{1}{A^{h_0}} - \left(\frac{B}{A}\right)^{h_0}\right]\left[1 - \left(\frac{B}{A}\right)^{h_0}\right]^{-1} \to 0$$

as $h^0 \to \infty$. Similarly $\beta(\theta) \to 1$ as $h^0 \to -\infty$.

► **Example 24.4.4** *Let $X_1, X_2, \cdots$ be i.i.d. Bernoulli (θ)-variables. To test $H_o(\theta = \theta_0)$ against $H_1(\theta = \theta_1), \theta_1 > \theta_0$ we have*

$$E_\theta\left[\left\{\frac{f_{\theta_1}(X)}{f_0(X)}\right\}^{h(\theta)}\right] = E_\theta\left[\frac{\theta_1^X(1-\theta_1)^{1-X}}{\theta_1^X(1-\theta_0)^{1-X}}\right]^{h(\theta)}$$

$$= \theta\left(\frac{\theta_1}{\theta_0}\right)^h + (1-\theta)\left(\frac{1-\theta_1}{1-\theta_0}\right)^h = \psi(\theta) \text{ (say)}.$$

The equation (24.4.24) can now be written as

$$\psi(\theta) = 1. \tag{i}$$

To plot the OC function it is not necessary to solve the equation (i) with respect to $h(\theta)$. We may consider $h = h(\theta)$ as a parameter and solve (i) with respect to θ. This gives

$$\theta = \frac{1 - \left(\frac{1-\theta_1}{1-\theta_0}\right)^h}{\left(\frac{\theta_1}{\theta_0}\right)^h - \left(\frac{1-\theta_1}{1-\theta_0}\right)^h}. \tag{ii}$$

If $h \to 0$,

$$\theta = -\frac{\log[(1-\theta_1)/(1-\theta_0)]}{\log(\theta_1/\theta_0) - \log[(1-\theta_1)/(1-\theta_0)]}.$$

Letting $A = \frac{1-\beta}{\alpha}$, $B = \frac{\beta}{1-\alpha}$, (24.4.29) can be written as

$$L(\theta) \simeq \frac{\left(\frac{1-\beta}{\alpha}\right)^h - 1}{\left(\frac{1-\beta}{\alpha}\right)^h - \left(\frac{\beta}{1-\alpha}\right)^h}. \tag{iii}$$

For any arbitrarily chosen value of h, the point $(\theta, L(\theta))$ computed from (ii), (iii) will be a point in the OC function. The OC function can be drawn by considering sufficiently large number of points $(\theta, L(\theta))$ for various values of h.

► **Example 24.4.5** *Let $X_1, X_2, \cdots$ be i.i.d. $N(\mu, 1)$ r.v.'s. It is required to test $H_o(\mu = \mu_0)$ against $H_1(\mu = \mu_1), \mu_1 > \mu_0$. Here*

$$Z = \frac{\mu_0^2 - \mu_1^2}{2} + (\mu_1 - \mu_0)X$$

$$E_\mu(e^{hZ}) = \exp\left(\frac{\mu_0^2 - \mu_1^2}{2}h\right)\exp\left\{(\mu_1-\mu_0)h\mu + (\mu_1-\mu_0)^2\frac{h^2}{2}\right\}$$

$$= \exp\left\{(\mu_1-\mu_0)h\left(\mu - \frac{\mu_0+\mu_1}{2}\right) + \frac{(\mu_1-\mu_0)^2}{2}h^2\right\}.$$

Setting this equal to 1 and solving for h one gets

$$h = h(\mu) = \frac{\mu_0 + \mu_1 - 2\mu}{\mu_1 - \mu_0}.$$

The approximate power function is

$$\beta(\mu) \simeq \frac{1 - [\beta/(1-\alpha)]^{h(\mu)}}{[(1-\beta)/\alpha]^{h(\mu)} - [\beta/(1-\alpha)]^{h(\mu)}}$$

whenever $h(\mu) \neq 0$. If $h(\mu) = 0$,

$$\mu = \frac{\mu_0 + \mu_1}{2} = \mu^* \text{ (say)}$$

$$\beta(\mu^*) \simeq -\frac{\log B}{\log(A/B)} \quad \text{[by (24.4.30)]}$$

$$= \frac{\log[(1-\alpha)/\beta]}{\log[(1-\alpha)(1-\beta)/\alpha\beta]}.$$

(3) ASN for SPRT

We now consider evaluation of approximate sampling number for SPRT.

Approximating $S_N = \sum_{i=1}^{N} Z_i$ by $\log A$ if $H_\circ$ is rejected and by $\log B$ if $H_\circ$ is accepted,

$$E_\theta S_N \simeq \beta(\theta) \log A + [1 - \beta(\theta)] \log B. \tag{24.4.31}$$

Now, $E_\theta S_N = E_\theta(N) E_\theta(Z)$ (by theorem 24.3.1) if $E_\theta(Z) \neq 0$. Thus,

$$E_\theta(N) = \frac{E_\theta(S_N)}{E_\theta(Z)} \simeq \frac{\beta(\theta) \log A + [1 - \beta(\theta)] \log B}{E_\theta(Z)}. \tag{24.4.32}$$

Hence,

$$E_{\theta_0}(N) \simeq \frac{\alpha \log A + (1-\alpha) \log B}{E_{\theta_0}(Z)} \quad (\because\ \beta(\theta_0) = \alpha) \tag{24.4.33}$$

$$E_{\theta_1}(N) \simeq \frac{(1-\beta) \log A + \beta \log B}{E_{\theta_1}(Z)} \quad (\because\ \beta(\theta_1) \doteq 1 - \beta). \tag{24.4.34}$$

If $E_\theta(Z) = 0$, it follows that

$$E_\theta(N) \simeq -\frac{\log A \log B}{E_\theta(Z^2)}. \tag{24.4.35}$$

► **Example 24.4.6** *Consider example* 24.4.4. *Here*

$$E_\theta(Z) = \theta \log \frac{\theta_1}{\theta_0} + (1-\theta)\log \frac{1-\theta_1}{1-\theta_0}$$
$$= \theta \log \left[\frac{\theta_1(1-\theta_0)}{\theta_0(1-\theta_0)}\right] + \log \frac{1-\theta_1}{1-\theta_0}.$$
$$E_\theta(Z^2) = \theta\left(\log \frac{\theta_1}{\theta_0}\right)^2 + (1-\theta)\left(\log \frac{1-\theta_1}{1-\theta_0}\right)^2.$$

Note that calculation of $\beta(\theta)$ depends on the value h_0 of the equation (24.4.24). For this problem $h_0 = h$ is given in (ii) of example 24.4.4. Suppose $\theta_0 = 0.5$, $\theta_1 = 0.8$, $\alpha = 0.05, \beta = 0.05$

h	$-\infty$	-1	0	1	∞
θ	1.0	0.8	.25	.5	0
$\beta(\theta)$	1	0.95	.50	.05	0
$E_\theta(Z)$	.20412	.08371	−.24742	−.09691	−.39794
$E_\theta(N)$	6.26	13.75	0	11.88	3.21

► **Example 24.4.7** *Let* $X \sim N(\theta,1), \theta_0 = 0,\ \theta_1 = 2,\ \alpha = 0.05,\ \beta = 0.10.$

Then $A \simeq \frac{0.9}{0.05} = 18$, $B \simeq \frac{0.10}{0.95} = 0.10526$. Also

$$Z = \log \frac{f_{\theta_1}(X)}{f_0(X)} = X - \frac{1}{2}$$
$$E_\theta(Z) = \theta - \frac{1}{2}, E_{\theta_0}(Z) = -\frac{1}{2}, E_{\theta_1}(Z) = \frac{3}{2}.$$
$$E_\theta(Z^2) = E_\theta\left(X - \frac{1}{2}\right)^2 = \theta^2 - \theta + \frac{5}{4}$$
$$E_{\theta_0}(S_N) \simeq \alpha \log A + (1-\alpha)\log B = -0.86609$$
$$E_{\theta_0}(N) \simeq -\frac{0.86609}{(-1/2)} = 1.73218 = 1.732$$
$$E_{\theta_1}(S_N) \simeq (1-\beta)\log A + \beta \log B = 1.03198$$
$$E_{\theta_1}(N) \simeq 0.688.$$

We now consider fixed sample size procedure for which critical region is $\bar{X}_n > k$. Then

$$\alpha = P_{\theta_0}\{\bar{X}_n > k\} = P\left\{\frac{\bar{X}_n - 0}{(1/\sqrt{n})} > \frac{k}{(1/\sqrt{n})}\right\} = 0.05 \text{ i.e., } \sqrt{n}k = 1.65$$

and

$$\beta = P_{\theta_1}\{\bar{X}_n < k\} = P\left\{\frac{\bar{X}_n - 2}{1/\sqrt{n}} < \frac{k-2}{1/\sqrt{n}}\right\} = 0.10.$$

Solving $n = 10.9$.

In general, let $X_1, \cdots, X_n$ be a random sample from $N(\theta, 1)$. We are interested in testing $H_o(\theta = \theta_0)$ against $H_1(\theta = \theta_1)$, $\theta_0 < \theta_1$. We shall compare the expected number of observations required by the SPRT of strength (α, β) with the MP test of fixed size and of the same strength for testing H_o against H_1. MP critical region is

$$\{\boldsymbol{x} : \bar{x} \geq d\}$$

where d is such that

$$P\{\bar{X} \geq d | H_o\} = \alpha$$

$$\text{and} \quad P\{\bar{X} \geq d | H_1\} = 1 - \beta.$$

These give respectively

$$\Phi(\sqrt{n}(d - \theta_0)) = 1 - \alpha$$

$$\Phi(\sqrt{n}(d - \theta_1)) = \beta.$$

Let $\Phi(\lambda_0) = 1 - \alpha$, $\Phi(\lambda_1) = \beta$. Hence

$$\sqrt{n}(d - \theta_0) = \lambda_0$$

$$\sqrt{n}(d - \theta_1) = \lambda_1.$$

Solving sample size required for MP-test is

$$n = n(\alpha, \beta) = \frac{(\lambda_1 - \lambda_0)^2}{(\theta_0 - \theta_1)^2}. \tag{24.4.36}$$

If this expression is not an integer, $n(\alpha, \beta)$ is $[n(\alpha, \beta)] + 1$.

In considering the corresponding SPRT we shall use the approximations $A \simeq \frac{1-\beta}{\alpha}$, $B \simeq \frac{\beta}{1-\alpha}$. Since $\frac{1-\beta}{\alpha} \geq A$ and $\frac{\beta}{1-\alpha} \leq B$ by taking the approximate values of A, B we can only increase the number of observations required by the SPRT. Consequently, the saving affected by the use of these approximations cannot be lower than that resulting from the exact values of A and B.

Assume $|\theta_1 - \theta_0|$ is small so that the approximate formula (24.4.32) holds. It can be easily verified that

$$E_{\theta_0}(Z) = \frac{1}{2}(\theta_0 - \theta_1)^2$$

$$E_{\theta_1}(Z) = -\frac{1}{2}(\theta_0 - \theta_1)^2.$$

Hence, from (24.3.33), (24.3.34),

$$\frac{E_{\theta_0}(N)}{n(\alpha,\beta)} = \frac{2}{(\lambda_1-\lambda_0)^2}[\alpha\log A + (1-\alpha)\log B] \tag{24.4.37}$$

$$\frac{E_{\theta_1}(N)}{n(\alpha,\beta)} = \frac{-2}{(\lambda_1-\lambda_0)^2}[(1-\beta)\log A + \beta\log B]. \tag{24.4.38}$$

It is interesting to note that the ratios (24.4.37), (24.4.38) are independent of values of θ_0 and θ_1. The average savings of the SPRT as compared to the fixed-size test procedure is given by

$$100\left[1-\frac{E_{\theta_0}(N)}{n(\alpha,\beta)}\right] \text{ under } H_\circ \text{ and } 100\left[1-\frac{E_{\theta_1}(N)}{n(\alpha,\beta)}\right] \text{ under } H_1.$$

We state below an optimal probability of SPRT Theorem (Wald and Wolfwitz, 1948). Among all tests (sequential or non-sequential) of a simple hypothesis $H_\circ$ against a simple alternative H_1 for which

$$P[\text{Type I error}] \le \alpha, \quad P[\text{Type II error}] \le \beta$$

and $E_{H_\circ}(N) < \infty$, $E_{H_1}(N) < \infty$, the SPRT with strength (α,β) minimises both $E_{H_\circ}(N)$ and $E_{H_1}(N)$.

Exercise

1. Explain how the sequential test procedure differs from the Neyman-Pearson test procedure. Develop the SPRT for testing $H_\circ(\theta=\theta_0)$ against $H_1(\theta=\theta_1)(\theta_1>\theta_0)$ based on a random sample of size n from a population with p.d.f

$$f(x;\theta) = \frac{1}{\theta}e^{-x/\theta}, \quad x>0, \theta>0.$$

Also obtain its ASN and OC functions.

2. Let X be a Bernoulli variate with p.d.f. $f(x;\theta)=\theta^x(1-\theta)^{1-x}$, $x=0,1;\ 0\le\theta\le1$. Employ SPRT for testing $H_\circ(\theta=\theta_0)$ against $H_1(\theta=\theta_1)$ and obtain its ASN and OC functions.

3. Describe the SPRT for testing $H_\circ(\pi=\pi_0)$ against $H_1(\pi=\pi_1)$ based on a random sample from a binomial population with parameters (n,π), n being known. Obtain its OC and ASN functions.

4. Obtain SPRT for testing $H_\circ(\lambda=\lambda_0)$ against $H_1(\lambda=\lambda_1)(\lambda_1>\lambda_0)$ when λ is the parameter of a Poisson distribution. Find approximate expressions for OC function and ASN function of the test.

5. Consider the SPR test for testing $H_\circ:\theta=\theta_0$ against $H_1:\theta=\theta_1$ where θ is the mean of a normal distribution with known variance. If $\alpha=\beta$, prove that the ASN under $H_\circ, H_1$ are equal.

6. Construct SPRT for testing $H_\circ(\theta = \theta_0)$ against $H_1(\theta = \theta_1)$ $(0 < \theta_0 < \theta_1)$, on the basis of a random sample drawn from the Pareto distribution with density function

$$f(x;\theta) = \frac{\theta a^\theta}{x^\theta + 1}, \quad x \geq a.$$

Also obtain its OC and ASN functions.

7. Deduce the SPRT for $H_\circ : \rho = \rho_0$ against $H_1 : \rho = \rho_1$ (> 0) for a bivariate normal population with both means 0, variances 1 and unknown correlation coefficient ρ. How can you use the test graphically?

8. For testing the hypothesis $H_\circ(p = p_0)$ against $H_1(p = p_1)$, $p_0 < p_1$ where p $(0 < p < 1)$ is the proportion defective of a lot the following sequential test is considered. A sample of size n_0 (a fixed number) is drawn successively. If the mth element $(m \leq n_0)$ is defective, $H_\circ$ is rejected, if all the n_0 elements are good $H_\circ$ is accepted. Find the OC function of this test. If n is the number of observations which lead to either of these decisions, find $E(n)$. Compute α, β and $E(n)$ for $p_0 = 0.05$, $p_1 = 0.1$, $n_0 = 20$. Compare the power of this test with the power of the SPRT in example 24.4.1. **[Fisz, 1963]**

9. Modify the test of exercise (8) as follows. After drawing the mth $(m \leq n_0 - 1)$ element, continue the investigation if at most one of the drawn elements is defective and reject $H_\circ$ if more than one is defective. After drawing n_0 elements, accept $H_\circ$ if at most one element is defective and reject $H_\circ$ if more than one element is defective. Answer all the questions of exercise (8). **[Fisz, 1963]**

10. Suppose independent random variables X_i can be observed, which are normally distributed with $\sigma^2 = 1$ and mean θ unknown. You are required to test $H_\circ : \theta = -d/2$ against $H_1 : \theta = +d/2$ with probabilities of error $\alpha = \beta$, where θ is the unknown mean. Obtain (a) the fixed sample size test and (b) the SPRT, and its OC and ASN-functions. Show that at $\theta = 0$, there is a range of α for which the ASN of the SPRT is greater than the fixed sample size, independently of d. Find the value of α for which the ASN of the SPRT is equal to the fixed sample size. Comment on the practical importance of the result. **[Wetherill, 1966]**

11. Find the SPRT with stopping bounds $B < 1 < A$ of the simple hypothesis $H_\circ(\theta = \theta_0)$ against alternative hypothesis $H_1(\theta = \theta_1)$ in each of the following cases:

   (a) $X \sim N(\mu, \theta)$, μ known. (b) $X \sim$ Geometric distribution (θ).
   (c) $X \sim$ Rectangular $(0, \theta)$. (d) X has p.d.f.

$$f_\theta(x) = \begin{cases} e^{-(x-\theta)}, & x \geq \theta \\ 0, & \text{otherwise.} \end{cases}$$

12. Suppose that X_i's are i.i.d. Bernoulli (θ). Let N be the first n such that $\left|\sum_{i=1}^{n} X_i - \frac{n}{2}\right| \geq 2$. Prove that $E_\theta(N) = 2/[\theta^2 + (1-\theta)^2]$.

13. Let X_i be i.i.d. according to some $f(x, \theta)$, $\underline{\theta} \le \theta \le \bar{\theta}$ and consider $H_\circ : \theta = \theta_0 - \gamma_1$ against $H_1 : \theta = \theta_0 + \gamma_2$ $(\gamma_1 \ge 0, \gamma_2 \ge 0)$. Describe the regions $\mathcal{R}_n^0, \mathcal{R}_n^1$ and $\mathcal{R}_n$ of the sequential test S which at stage $n \ge 1$ accepts $H_\circ$ if $x_n \le \theta_0 - \gamma_1$ or if $\theta_0 - \gamma_1 < x_{n-1} \le \theta_0$ and $\theta_0 - \gamma_1 < x_n \le \theta_0$, rejects $H_\circ$ if $x_n > \theta_0 + \gamma_2$ or if $\theta_0 < x_{n-1} \le \theta_0 + \gamma_2$ and $\theta_0 < x_n \le \theta_0 + \gamma_2$. Then show that for all $\underline{\theta} \le \theta \le \bar{\theta}$,

$$
\begin{aligned}
L(\theta) &= [1 + \zeta_3(\theta)] \frac{\zeta_1(\theta) + \zeta_1(\theta)\zeta_2(\theta) + \zeta_2^2(\theta)}{1 - \zeta_2(\theta)\zeta_3(\theta)} \\
p(n; \theta) &= 1 - \zeta_2(\theta) - \zeta_3(\theta), \quad \text{for } n = 1 \\
&= [\zeta_2(\theta)\zeta_3(\theta)]^{(n/2)-1}[\zeta_2(\theta) - 2\zeta_2(\theta)\zeta_3(\theta) + \zeta_3(\theta)], \quad \text{for } n = 2, 4, \cdots \\
&= [\zeta_2(\theta)\zeta_3(\theta)]^{(n-1)/2}[2 - \zeta_2(\theta) - \zeta_3(\theta)] \quad \text{for } n = 3, 5, \cdots \\
E_\theta(N) &= [1 + \zeta_2(\theta)] \frac{1 + \zeta_3(\theta)}{1 - \zeta_2(\theta)\zeta_3(\theta)} \le 3
\end{aligned}
$$

where $p(n; \theta) = E_\theta(N = n)$,

$$
\begin{aligned}
\zeta_1(\theta) &= P(X \le \theta_0 - \gamma_1 | \theta), \\
\zeta_2(\theta) &= P(\theta_0 - \gamma_1 < X \le \theta_0 | \theta), \\
\zeta_3(\theta) &= P(\theta_0 < X \le \theta_0 + \gamma_2 | \theta).
\end{aligned}
$$

[Ghosh, 1970]

Chapter 25

Stochastic Process

25.1 Introduction

A stochastic process means a process whose results depend on some chance elements. The models of this process can, therefore, be used to describe, study and investigate many phenomena in real life such as size and composition of population, taking into account birth, death and migration, spread of epidemics, size of a queue in a railway booking counter, growth of a bacterial population in a laboratory culture, Brownian motion of a particle in a fluid, thermal noise in a electric circuit, etc. Here we study the probability distribution of a family of vector of random variables $\{X_t, Y_t, \cdots, t \in T\}$. indexed by a parameter t like time, space, etc.

In Section 25.2 we introduce definition and classification of stochastic processes. Section 25.3 considers Markov chains, Section 25.4 the simple random walk, Section 25.5 discrete branching process. Continuous time Markov processes are dealt with in Section 25.6.

25.2 Preliminary Notions

A stochastic process may be defined as a family of random variables $\{X_t, t \in T\}$, where T is an index set. For each t, let $\mathcal{X}_t$ be the sample space of $X_t, \mathcal{X}_t = \{x_t\}$, the set of all possible values x_t of X_t, which may be finite or infinite. The random variables $X_t, X_{t+r} (r > 0)$ may be dependent or independent.

The index set T may be finite or infinite. If T is countably infinite, e.g., $T = \{0, 1, 2, \cdots,\}$ or $T = \{0, \mp 1, \mp 2, \cdots\}$, then the stochastic process $\{X_t, t \in T\}$ is a discrete-parameter stochastic process. If T is any finite or infinite interval, e.g., $T = \{t : a < t < b\}$, $T = \{t, 0 \leq t < \infty\}$, the process is said to be a continuous parameter stochastic process. Often t is considered as a time parameter. The particular value x_t of X_t is often called state and the sample space $\mathcal{X}_t$ the state space of X_t.

For example, for a birth and death process if we assume that births (deaths) occur only at time point, $1, 2, \cdots$, then this is a discrete stochastic process. A Poisson process (e.g., number of calls entering a telephone exchange) where we assume a change may occur at any time, is a continuous time stochastic process.

Clearly, X_t may also be a vector and in this case $\{\boldsymbol{X}_t, t \in T\}$ is a multivariate stochastic process.

25.3 Markov Chains

Consider a stochastic process $\{X_n, n = 0, 1, 2, \cdots\}$ i.e., a sequence of random variables each defined on the sample space $\mathcal{X}$. The space $\mathcal{X}$ is called state space and the values x as different states. We shall consider here $x = 0, 1, 2, \cdots$. Hence the process is discrete both with respect to state variable x and time variable n.
Consider now the conditional probability

$$P[X_{n+1} = x_{n+1} | X_n = x_n, X_{n-1} = x_{n-1}, \cdots, X_0 = x_0]. \tag{25.3.1}$$

If the stochastic process is such that the conditional probability (25.3.1) depends only on the value of X_n and is independent of all previous values, we say that the process has *Markov property* and call it a *Markov chain.* In this case the condition probability (25.3.1) is

$$P[X_{n+1} = x_{n+1} | X_n = x_n]. \tag{25.3.2}$$

More generally, a stochastic process $\{X_n, n = 0, 1, \cdots,\}$ is a Markov chain of order r, if

$$\begin{aligned} P[X_{n+1} &= x_{n+1} | X_n = x_n, \cdots, X_0 = x_0] \\ &= P[X_{n+1} = x_{n+1} | X_n = x_n, \cdots, X_{n-r+1} = x_{n-r+1}] \end{aligned}$$

for all $n \geq r$ for all possible values of the random variables $X_0, \cdots, X_{n-r+2}$. A stochastic process $\{X_n\}$ with property (25.3.2) may be called a Markov chain of the first order. We shall here consider Markov chains of first order only.
Consider again special class of conditional probabilities as follows:

$$p_{ij} = P[X_{n+1} = j | X_n = i], \ i, \ j = 0, 1, \cdots \tag{25.3.3}$$

where

$$p_{ij} \geq 0; \ \sum_{j=0}^{\infty} p_{ij} = 1 \ \forall \, i = 0, 1, \cdots.$$

Note that the conditional probability p_{ij} that at time $(n+1)$ the system is at state j given that at time n it was at state i, depends only on i and j and not on n. The quantity p_{ij} is also called transition probability that the system passes from state i at time n to state j at time $(n+1)$ for any n. Such conditional probabilities are called constant or stationary conditional probabilities. Markov chains with this property are called homogeneous Markov chains. We shall confine our attention to homogeneous Markov chains of first order.

We shall denote the initial probabilities

$$P\{X_0 = j\} = q_j, \quad j = 0, 1, \cdots,$$
$$q_j \geq 0, \ \sum q_j = 1. \tag{25.3.4}$$

The unconditional or absolute probabilities at step n will be denoted as

$$P\{X_n = j\} = q^{(n)}(j), \quad j = 0, 1, \cdots,$$
$$q^{(n)}(j) \geq 0, \ \sum_{j=0}^{\infty} q^{(n)}(j) = 1. \tag{25.3.5}$$

We shall write

$$\boldsymbol{q} = (q_0, q_1, \cdots)'$$
$$\boldsymbol{q}^{(n)} = (q_0^{(n)}, q_1^{(n)}, \cdots)', \ n \geq 1$$

as vector of initial probabilities and n-step absolute probabilities respectively.

Transition Probability Matrix

Consider the matrix of transition probabilities p_{ij},

$$\boldsymbol{P} = \begin{bmatrix} p_{00} & p_{01} & p_{02} & \cdots \\ p_{10} & p_{11} & p_{12} & \cdots \\ p_{20} & p_{21} & p_{22} & \cdots \\ \cdots & \cdots & \cdots & \cdots \end{bmatrix}. \tag{25.3.6}$$

$\boldsymbol{P}$ is a square matrix of infinite order with non-negative elements with row sum equal to unity

$$\left(\sum_{j=0}^{\infty} p_{ij} = 1 \ \forall \, i \right).$$

A matrix satisfying these conditions is called a *stochastic* matrix or a *Markov* matrix. A Markov chain is completely specified by a matrix of transition porbabilities $\boldsymbol{P}$ and vector $\boldsymbol{q} = (q_0, q_1, \cdots)'$ of initial probabilities.

Higher order transition probabilities can be defined in a similar way. Let us denote $p_{ij}^{(n)}$ = probability that the system passes from state i to state j in n steps.

Clearly,

$$p_{ij}^{(1)} = p_{ij},$$
$$p_{ij}^{(n+1)} = \sum_{\nu=0}^{\infty} p_{i\nu}^{(n)} p_{\nu j}, \ n = 1, 2, \cdots \tag{25.3.7}$$

where $p_{kj} = 1(0)$ if $k = j$ (otherwise).

It follows by the assumption of stationarity of p_{ij} above,

$$p_{ij}^{(k)} = P\{X_{m+k} = j | X_m = i\} \ \forall\, k = 1, 2, \cdots, \tag{25.3.8}$$

whatever m. For $k = 1$, (25.3.8) is true by definition. Assume that (25.3.8) is true for $k = n$. Now

$$\begin{aligned} P\{X_{m+n+1} = j | X_m = i\} &= \sum_{\nu=0}^{\infty} P\{X_{m+n} = \nu | X_m = i\} P\{X_{m+n+1} = j | X_{m+n} = \nu\} \\ &= \sum_{\nu=0}^{\infty} p_{i\nu}^{(n)} p_{\nu j} = p_{ij}^{(n+1)}. \end{aligned}$$

Hence (25.3.8) follows by induction. In general, we have the following relation:

$$p_{ij}^{(m+n)} = \sum_{\nu=0}^{\infty} p_{i\nu}^{(m)} p_{\nu j}^{(n)}. \tag{25.3.9}$$

That (25.3.9) is true for all m can also be shown by induction. Denoting by $\boldsymbol{P}^n = \left(\left(p_{ij}^{(n)}\right)\right)$ the matrix of n-step transition probabilities, the equation (25.3.9) can be written as

$$\boldsymbol{P}^{m+n} = \boldsymbol{P}^m \boldsymbol{P}^n. \tag{25.3.10}$$

Equations (25.3.10) are called *Chapman-Kolmogorov* functional equations and these characterise Markov chains.

Clearly,

$$\boldsymbol{P}^n = \boldsymbol{P}^{n-1} \boldsymbol{P} \tag{25.3.11}$$

$$\boldsymbol{P}^n = \boldsymbol{P} \boldsymbol{P}^{n-1}. \tag{25.3.12}$$

The equation (25.3.11) and (25.3.12) are respectively called forward and backward equations.

The unconditional probabilities $q^{(n)}(j)$ are given by

$$q^{(n)}(j) = \sum_{i=0}^{\infty} q(i) p_{ij}^{(n)}. \tag{25.3.13}$$

Asymptotic values of $p_{ij}^{(n)}$ and $q^{(n)}(j)$ as $n \to \infty$ will be considered in subsection 25.3.2.

► **Example 25.3.1** *The simplest non-trivial Markov chain is one of two states, the transition matrix being necessarily of the form*

$$\boldsymbol{P} = \begin{bmatrix} 1-\alpha & \alpha \\ \beta & 1-\beta \end{bmatrix}, \ 0 < \alpha, \ \beta < 1. \tag{25.3.14}$$

The two states may correspond to two energy levels say 1, 2 of a molecule.

▶ **Example 25.3.2** *Random walk with absorbing barriers. Let the possible states be* $0, 1, \cdots, m$ *and consider*

$$P = \begin{bmatrix} 1 & 0 & 0 & 0 & \cdots & 0 & 0 & 0 \\ q & 0 & p & 0 & \cdots & 0 & 0 & 0 \\ 0 & q & 0 & p & \cdots & 0 & 0 & 0 \\ \cdots & \cdots & \cdots & \cdots & \cdots & \cdots & \cdots & \cdots \\ 0 & 0 & 0 & 0 & \cdots & q & 0 & p \\ 0 & 0 & 0 & 0 & \cdots & 0 & 0 & 1 \end{bmatrix}.$$

From each of the interior states $1, 2, \cdots, m-1$ transitions are possible to the right and left neighbours ($p_{i,i+1} = p$, $p_{i,i-1} = q$). However, no transition is possible from either states 0 or m to any other state.

▶ **Example 25.3.3** *In genetics Mendel's law states that inherited characters depend on genes (which always occur in pairs). In the simplest case every gene may be of two forms A or a. The possible genotypes are* (AA, Aa, aa). *An offspring receives one gene from each parent under the conditions of Bernoulli scheme. Suppose that the population consists of N elements in each generation. Thus, we have 2N genes in each generation. If in some generation there are i genes of the form A* $(0 \leq i \leq 2N)$, *the generation is said to be in state i. Thus, we have here a homogeneous Markov chain with* $(2N+1)$ *possible states:* $0, 1, \cdots, 2N$. *The probability of passing from state i in some generation to state j in the next generation is*

$$p_{ij} = \binom{2N}{j} \left(\frac{i}{2N}\right)^j \left(1 - \frac{i}{2N}\right)^{2N-j}.$$

Note that states 0 and 2N are here absorbing barriers. If in some generation the population is in one of these states, it will remain there forever.

25.3.1 Classification of States

Classification of states of a Markov chain is of great importance in the study of the asymptotic properties of $p_{ij}^{(n)}$.

Definition 25.3.1 *Two states i and j of a Markov chain are said to communicate* $(i \to j)$ *if* $p_{ij}^{(n)} > 0$ *and* $p_{ji}^{(m)} > 0$ *for some* $n(\geq 1)$ *and* $m(\geq 1)$. *The following relations hold for communicative states.*

(a) $i \to j \Rightarrow j \to i$;

(b) $i \to j$ *and* $j \to k \Rightarrow i \to k$;

(c) *if* $i \to j$ *for some* j *then* $i \to i$.

Definition 25.3.2 *A set of states $S \subset \mathcal{X}$ is called closed if $p_{ij} = 0$ for all $i \in S$ and $j \in \bar{S} = \mathcal{X} - S$. If the closed set S contains only one state i, then this state i is called an absorbing state. Clearly, a necessary and sufficient condition for state i to be an absorbing state is $p_{ii} = 1$. If the state space of a Markov chain contains two or more closed sets, the chain is called decomposible or reducible.*

The transition probability matrix associated with a decomposible chain can be written in the form of a partitioned matrix. For example, suppose

$$\boldsymbol{P} = \begin{bmatrix} \boldsymbol{P}_1 & \boldsymbol{O} \\ \boldsymbol{O} & \boldsymbol{P}_2 \end{bmatrix}.$$

Here $\boldsymbol{P}_1$, $\boldsymbol{P}_2$ represent Markov matrices which describe the transition within the two closed sets of states.

A chain or matrix which is not decomposible is called *indecomposible* or *irreducible.* A chain is irreducible iff every state can be reached from every other state.

For an arbitrary set $\mathcal{G}$ of states the smallest closed set containing $\mathcal{G}$ is called closure of $\mathcal{G}$. For a state i the closure of i will be denoted as $C(i)$. It is the set of all states that are accessible from i.

Definition 25.3.3 *If $i \to i$, the greatest common divisor of the set of integers n such that $p_{ii}^{(n)} > 0$ is called the period of the state i. We denote by $w(i)$ the period of the state i and say that the state i is periodic with period $w(i)$. If $i \not\to i$, we say $w(i) = 0$ and the state i is aperiodic.*

► **Example 25.3.4** *The simplest example of a chain with period 3 is one in which only the transitions $1 \to 2 \to 3 \to 1$ are possible. Here* $\boldsymbol{P} = \begin{bmatrix} 0 & 1 & 0 \\ 0 & 0 & 1 \\ 1 & 0 & 0 \end{bmatrix}$.

► **Example 25.3.5** *Consider the following transition matrix where a $*$ denotes a positive element and states are $1, \cdots, 9$.*

$$\boldsymbol{P} = \begin{bmatrix} 0 & 0 & 0 & * & 0 & 0 & 0 & 0 & * \\ 0 & * & * & 0 & * & 0 & 0 & 0 & * \\ 0 & 0 & 0 & 0 & 0 & 0 & 0 & * & 0 \\ * & 0 & 0 & 0 & 0 & 0 & 0 & 0 & 0 \\ 0 & 0 & 0 & 0 & * & 0 & 0 & 0 & 0 \\ 0 & * & 0 & 0 & 0 & 0 & 0 & 0 & 0 \\ 0 & * & 0 & 0 & 0 & * & * & 0 & 0 \\ 0 & 0 & * & 0 & 0 & 0 & 0 & 0 & 0 \\ 0 & 0 & 0 & * & 0 & 0 & 0 & 0 & * \end{bmatrix}.$$

It can be easily checked that 5 is an absorbing state ($p_{55} = 1$). The sets $(3, 8)$, $(1, 4, 9)$ are two closed sets.

The closure of state 2 is the set $(2, 3, 5, 8)$.

If a Markov chain is reducible and has k irreducible sets, then by a suitable relabelling of the states the transition probability matrix P may be written in the form

$$\boldsymbol{P} = \begin{bmatrix} \boldsymbol{P}_1 & 0 & 0 & \cdots & 0 & \cdots & 0 \\ 0 & \boldsymbol{P}_2 & 0 & \cdots & 0 & \cdots & 0 \\ \cdots & \cdots & \cdots & \cdots & \cdots & \cdots & \cdots \\ 0 & 0 & 0 & \cdots & \boldsymbol{P}_k & \cdots & 0 \\ \boldsymbol{A} & \boldsymbol{B} & \boldsymbol{C} & \cdots & \boldsymbol{G} & \cdots & \boldsymbol{H} \end{bmatrix}$$

when $\boldsymbol{P}_1, \cdots, \boldsymbol{P}_k$ are transition probability matrices for these k sets. The matrix $\boldsymbol{P}^n$ is of the same type with $\boldsymbol{P}_1, \boldsymbol{P}_2, \cdots, \boldsymbol{P}_k$ and $\boldsymbol{H}$ replaced by $\boldsymbol{P}_1^n, \boldsymbol{P}_2^n, \cdots, \boldsymbol{P}_k^n$, $\boldsymbol{H}^n$ respectively whereas $\boldsymbol{A}, \boldsymbol{B}, \cdots, \boldsymbol{G}$ have more complicated forms.

Theorem 25.3.1 *If in the matrix $\boldsymbol{P}^n$ of a reducible chain all rows and all columns corresponding to the states outside the closed set C are omitted, the elements of the resulting stochastic matrix also satisfy fundamental relations* (25.3.10)–(25.3.12).

This means we have a Markov chain defined on C and this subchain can be studied independently of the remaining states.

We now define several other probabilities. Let

$$f_{ij}^{(n)} = P\{X_n = j | X_0 = i, X_h \neq j, 0 < h < n\}. \tag{25.3.15}$$

Thus $f_{ij}^{(n)}$ is the conditional probability that the system reaches the state j for the first time at time n given that it started from state i at time zero. Let

$$f_{ij} = P\{X_h = j \text{ for at least one } h | X_0 = i\}. \tag{25.3.16}$$

f_{ij} is the probability that the system reaches the state j given that it was at state i at time zero. Clearly

$$f_{ij} = \sum_{n=1}^{\infty} f_{ij}^{(n)}. \tag{25.3.17}$$

Note that

$$p_{ij}^{(n)} = \sum_{\nu=1}^{n} f_{ij}^{(\nu)} p_{jj}^{(n-\nu)}. \tag{25.3.18}$$

We have $f_{ij} \leq 1$. When $f_{ij} = 1, \{f_{ij}^{(n)}\}$ is a proper probability distribution and can be referred to as first passage distribution for state j given that the system started from state i at time zero.

Let also

$$\begin{aligned} g_{ij} &= P\{X_h = j \text{ infinitely often } | X_0 = i\} \\ &= P\{X_h = j \text{ for an infinite number of values of } h | X_0 = i\}. \end{aligned} \tag{25.3.19}$$

Definition 25.3.4 *A state i is called recurrent (persistent) if $f_{ii} = 1$ or $g_{ii} = 1$ (this means a return to state i is certain); a state i is called non-recurrent (transient) if $f_{ii} < 1$ or $g_{ii} = 0$ (i.e., a return to state i is uncertain).*

The equivalence of conditions on f_{ii} and g_{ii} can be seen as follows.
Let

$$F_n = P\{X_h = i \text{ for at least } n \text{ values of } h | X_0 = i\}.$$

Then

$$F_1 = f_{ii}, \;\; F_2 = f_{ii}F_1, \cdots, \;\; F_n = f_{ii}F_{n-1} = (f_{ii})^n.$$

Hence,

$$g_{ii} = \lim_{n\to\infty} F_n = \lim_{n\to\infty} (f_{ii})^n = \begin{cases} 0, \text{ if } f_{ii} < 1 \\ 1, \text{ if } f_{ii} = 1. \end{cases}$$

Suppose a state i is recurrent. Given that $X_n = i$, let T_i be the waiting time (recurrence time) till the next occurrence of state i. Thus, $T_i = m$ if $X_{n+k} \neq i$ for $1 \leq k < m$ and $X_{n+m} = i$. Then $P\{T_i = m\} = p_{ii}^{(m)}$.
The mean recurrence for time i is

$$E(T_i) = \mu_i = \sum_{m=1}^{\infty} m f_{ii}^{(m)} \leq \infty. \tag{25.3.20}$$

Definition 25.3.5 *A recurrent state i is null if $\mu_i = \infty$. A recurrent state which is neither null nor periodic is called ergodic. Thus, an aperiodic persistent state i with $\mu_i < \infty$ is called ergodic.*

If $f_{ij} = \sum_{n=1}^{\infty} f_{ij}^{(n)} = 1$, we can define the mean first passage time from state i to state j as

$$f_{ij} = \sum_{n=1}^{\infty} n f_{ij}^{(n)}. \tag{25.3.21}$$

Theorem 25.3.2 *If the series*

$$\sum_{n=1}^{\infty} p_{ij}^{(n)} \tag{25.3.22}$$

converges then $g_{ij} = 0$ (this always holds if the state j is nonrecurrent). If (25.3.22) diverges, $g_{ij} > 0$ and $g_{jj} = 1$. If $i \to j$, $g_{ij} = g_{ji} = 1$.

Proof: Let E_n denote the event $(X_{m+n} = j, \;\; X_m = i)$.
Let $P(X_m = i) > 0$. Hence

$$P(E_n) = P(X_m = i) p_{ij}^{(n)}. \tag{25.3.23}$$

Therefore, if (25.3.22) converges, by Borel Cantelli lemma (vide section 9.7),

$$P\{E_n \text{ occurs infinitely often}\} = 0.$$

Hence $g_{ij} = 0$. Again, from (25.3.18), summing both sides over $n = 1, \cdots, N$,

$$\sum_{n=1}^{N} p_{ij}^{(n)} = \sum_{n=1}^{N}\sum_{\nu=1}^{n} f_{ij}^{(\nu)} p_{jj}^{(n-\nu)}$$

$$= \sum_{\nu=1}^{N} f_{ij}^{(\nu)} \sum_{n=\nu}^{N} p_{jj}^{(n-\nu)} \leq \sum_{\nu=1}^{N} f_{ij}^{(\nu)} \left(1 + \sum_{n=1}^{N} p_{jj}^{(n)}\right). \tag{25.3.24}$$

Allowing $N \to \infty$ and using (25.3.17)

$$\sum_{n=1}^{\infty} p_{ij}^{(n)} \leq f_{ij} \left(1 + \sum_{n=1}^{\infty} p_{jj}^{(n)}\right). \tag{25.3.25}$$

Putting $j = i$ in (25.3.24) and dividing both sides by $1 + \sum_{n=1}^{N} p_{ii}^{(n)}$,

$$\frac{\sum_{n=1}^{N} p_{ii}^{(n)}}{1 + \sum_{n=1}^{N} p_{ii}^{(n)}} \leq \sum_{\nu=1}^{N} f_{ii}^{(\nu)}.$$

Hence, if (25.3.22) diverges, it follows from (25.3.25) that $\sum_{n=1}^{\infty} p_{jj}^{(n)}$ diverges and hence

$$\sum_{\nu=1}^{\infty} f_{ii}^{(\nu)} = f_{ii} \geq 1.$$

Hence $f_{ii} = 1$ and $g_{ii} = 1$.

In general, if (25.3.22) diverges, $\sum_{n=1}^{\infty} p_{jj}^{(n)}$ diverges. Hence $g_{jj} = 1$. This implies the existence of infinitely many $p_{ij}^{(n)} > 0$ and hence $f_{ij} > 0$ and therefore, $g_{ij} = f_{ij}g_{jj} > 0$.

If $i \to j$, $f_{ij} > 0$, $f_{ji} > 0$. It follows that if $g_{ii} = 1$ and $f_{ij} > 0$, then $g_{ij} = 1$. Hence $g_{jj} = 1$ implies $g_{ji} = 1$. If the state j is nonrecurrent, $g_{ji} = 0$. Hence (25.3.22) must converge.

25.3.2 Limits of Higher Transition Probabilities

It is of interest in many practical applications to investigate the behaviour of $p_{ij}^{(n)}$ as $n \to \infty$. This is considered in this section.

We shall assume throughout that the process was at state i at time zero. We shall henceforth denote $f_{ii}^{(n)}$ as $f_i^{(n)}$, f_{ii} as f_i and $p_{ii}^{(n)}$ as $u_i^{(n)}$. From (25.3.18),

$$u_i^{(n)} = u_i^{(0)} f_i^{(n)} + u_i^{(1)} f_i^{(n-1)} + u_i^{(2)} fi^{(n-2)} + \cdots + u_i^{(n-1)} f_i^{(1)}, \quad n \geq 1 \tag{25.3.26}$$

where $u_i^{(0)} = 1$, $f_i^{(0)} = 0$. Consider the probability generating functions

$$U(S) = \sum_{n=0}^{\infty} u_i^{(n)} s^n \tag{25.3.27}$$

$$F(S) = \sum_{n=0}^{\infty} f_i^{(n)} s^n. \tag{25.3.28}$$

Hence,

$$U(S) = 1 + \sum_{n=1}^{\infty} u_i^{(n)} s^n = 1 + \sum_{n=1}^{\infty} \left\{ \sum_{j=1}^{n} u_i^{(j)} f_i^{(n-j)} \right\} s^n = 1 + U(S)F(S)$$

i.e.,

$$U(S) = \frac{1}{1 - F(S)}. \tag{25.3.29}$$

Theorem 25.3.3 *A state i is transient iff $\sum_{n=1}^{\infty} u_i^{(n)}$ converges. In this case*

$$f_i = \frac{u_i - 1}{u_i} \tag{25.3.30}$$

where

$$u_i = \sum_{n=0}^{\infty} u_i^{(n)}. \tag{25.3.31}$$

Proof: Suppose i is transient. Then $f_i < 1$. The series $F(S)$ converges and $F(1) = f_i < 1$. The series $U(S)$ has only non-negative coefficients, and since $U(s) \to \frac{1}{1-f_i}$ as $s \to 1$ it follows from Abel's theorem that the series converges for $s = 1$ and also $U(1) = u_i = \frac{1}{1-f_i}$.

If i is persistent, $f_i = 1$. As $s \to 1$, $F(s) \to 1$ and $U(s) \to \infty$ so that in this case $\sum_{n=0}^{\infty} u_i^{(n)}$ diverges.

□ **Remark 25.3.1** $\sum_{k=1}^{n} u_i^{(k)} = \sum_{k=1}^{n} p_{ii}^{(k)}$ *may be interpreted as the expected number of returns to step i during the course of n steps. Hence $\sum_{n=0}^{\infty} u_i^{(n)} - 1 = u_i - 1$ may be interpreted as the expected number of returns to step i (excluding the initial position) in an infinitely many steps.*

Theorem 25.3.4 (a) *If i is a persistent null state, then $u_i^{(n)} \to 0$ as $n \to \infty$.*

(b) *If i is ergodic, then $u_i^{(n)} \to 1/\mu_i$.*

Theorem 25.3.5 *If the state i is persistent and has period w, then as $n \to \infty$*

$$p_{ii}^{(nw)} = u_i^{(nw)} \to \frac{w}{\mu_i}$$

whereas $u_i^{(k)} = 0$ for every k which is not divisible by w.

Proof: Since i has period w, $F(s)$ contains only powers of s^w and therefore $F(s^{1/w}) = F_1(s)$ (say) is also a power series. Now $F_1(1) = 1$ so that

$$F_1(s) = \sum_{n=1}^{\infty} f_{i(1)}^{(n)} s^n,$$

($f_{i(1)}^{(n)}$ being suitably defined) is regarded as a probability generating function corresponding to a persistent state. Hence applying theorem 25.3.4, coefficients of

$$U_1(s) = \frac{1}{1 - F_1(s)}$$

tends to $1/\mu_{(1)}$ where

$$\mu_{(1)} = \frac{d}{ds} F_1(s)\bigg]_{s=1} = \frac{1}{w}\frac{d}{ds} F(s)\bigg]_{s=1} = \frac{\mu_i}{w}.$$

But the coefficient of s^n in $U_1(s)$ is the coefficient of s^{nw} in $U(s)$ i.e., $u_i^{(nw)}$. Hence the first part. The second part follows from the definition of a persistent state with period w.

Theorems 25.3.3–25.3.5 indicate that except when i is periodic, $u_i^{(n)}$ has a unique limit which is zero if i is null or is transient and is otherwise $1/\mu_i$. In the periodic case, a limit exists for the subsequence $n = w, 2w, 3w, \cdots$.

Theorem 25.3.6 *In an ergodic chain*

$$\lim_{n \to \infty} p_{ij}^{(n)} = \pi_j \tag{25.3.32}$$

exists and are independent of initial state i. Moreover

$$\pi_j = \frac{1}{\mu_j} \tag{25.3.33}$$

where μ_j is the mean recurrence time of j. Moreover, $\{\pi_j\}$ is a probability distribution and satisfy

$$\pi_j = \sum_{i=0}^{\infty} \pi_i p_{ij}. \tag{25.3.34}$$

For a proof the reader may refer to Feller (1968) vol. I, pp 393–394 or Bharucha-Reid (1960), pp 28–30.

25.3.3 Irreducible Chains

We consider in this section some theorems on the closure of sets and irreducible sets of states.

Theorem 25.3.7 *If C is an irreducible closed set of states and if $i \in C$, then $C = C(i)$.*

Proof: Suppose $j \in C(i)$. Then $j \in C$, for otherwise the root from j to i passes through one state outside C. This is impossible since C is irreducible and closed. Hence $C(i) \subset C$.

Suppose now $j \in C$. If $j \notin C(i)$, this means $C(i)$ is a closed proper subset of C. This is not possible since C is irreducible. Hence $C \subset C(i)$.

Therefore, $C = C(i)$.

Theorem 25.3.8 *If i is a persistent state then the closure of i, $C(i)$ is an irreducible set of states.*

Proof: Suppose the result is not true and there is a closed proper subset C of $C(i)$.

(1) Suppose $i \subset C$ and let $j \in (C(i) - C)$. Then since $j \in C(i)$, there is a finite sequence of transitions, each with positive probability, from $i \to j$. Thus, $p_{ij} > 0$, which is a contradiction.

(2) Suppose $i \in (C(i) - C)$ and $j \in C$.

Then there is a positive probability that the system will move from state i to state j. Since $j \in C$, transition from j to i is not possible. Then i cannot be a persistent state because $f_{ii} < 1$. This is a contradiction.

Theorem 25.3.9 *If C is a closed irreducible set of states, then all states in C have the same period.*

Proof: Suppose i and j are both in C. Then there exist positive integers M and N such that

$$p_{ij}^{(M)} > 0, \; p_{ji}^{(N)} > 0.$$

Obviously

$$p_{ii}^{(n+M+N)} \geq p_{ij}^{(M)} p_{jj}^{(n)} p_{ji}^{(N)} > 0. \tag{25.3.35}$$

and

$$p_{jj}^{(n+M+N)} \geq p_{ji}^{(N)} p_{ii}^{(n)} p_{ij}^{(M)} > 0. \tag{25.3.36}$$

Here, i, j, M, N are fixed while n is arbitrary. Let Ω_i be the set of integers n for which $p_{ii}^{(n)} > 0$. Similarly let Ω_j be the set of integers n for which $p_{jj}^{(n)} > 0$.

Let $\lambda_i = HCF(\Omega_i)$, $\lambda_j = HCF(\Omega_j)$ be the periodicities of state i and j respectively.

From (25.3.35), $(M+N) \in \Omega_i$; from (25.3.36) $(M+N) \in \Omega_j$. Hence $(M+N) \in \Omega_i \cap \Omega_j$. Again, if $n \in \Omega_j$, then $(n+M+N) \in \Omega_i$; hence n is divisible by λ_i. Thus λ_i divides every member in Ω_j and hence divides λ_j. By the same logic, λ_j divides λ_i. Hence $\lambda_i = \lambda_j$.

We shall define that two states are of the same type if they agree in all characteristics defined in subsection 25.3.1. Thus, two states of the same type have the same period or are aperiodic; both are transient or persistent. If they are persistent, both have mean recurrence time finite or infinite.

Theorem 25.3.10 *In an irreducible Markov chain all states are of the same type.*

Proof: Let i and j be two arbitrary states.

Since every state can be reached from every other state, there exist integers M and N such that $p_{ij}^{(M)} = \alpha > 0$ and $p_{ji}^{(N)} = \beta > 0$. Obviously

$$p_{ii}^{(n+M+N)} \geq p_{ij}^{(M)} p_{jj}^{(n)} p_{ji}^{(N)} = \alpha\beta p_{jj}^{(n)} . \tag{25.3.37}$$

If j is transient, $\sum_{r=1}^{\infty} p_{jj}^{(r)} < \infty$ and hence $\sum_{r=1}^{\infty} p_{ii}^{(r)} < \infty$. Further, if $p_{ii}^{(n)} \to 0$ then also $p_{jj}^{(n)} \to 0$. The same results hold if i and j are interchanged and hence either both are transient or persistent. If one is a null state, the other is also so.

Suppose i has period w. For $n = 0$, the right hand side of (25.3.37) is positive and hence $(M + N)$ is a multiple of w. Again, left hand side vanishes unless n is a multiple of w, i.e., j has a period w. By interchanging i and j we see that both the states have the same period.

□ **Remark 25.3.2** *For all practical purposes it is, therefore, always possible to restrict attention to states of one particular type only.*

Theorem 25.3.11 *In any Markov chain the persistent states can be divided in a unique manner into non-overlapping irreducible sets $C_1, C_2, \cdots$, such that any state in a given set can be reached from any other state in that set and such that all states in a given set are of the same type. Apart from the persistent states, there will, in general, remain transient states from which the persistent states can be reached but not vice versa.*

Proof: Let i be the state with the lowest suffix which is persistent. Consider closure of i, $C(i) = C_1$ (say). Now $C(i)$ is closed and since i is persistent, $C(i)$ is an irreducible set (theorem 25.3.8). Again all states of this set are of the same type (theorem 25.3.10).

Next consider state $i'(\notin C_1)$ with the lowest suffix which is persistent. Consider closure of i', $C(i') = C_2$ (say). C_2 is an irreducible set of states of the same type. Proceeding in this way one can divide the persistent states into sets $C_1, C_2, \cdots$ having the above-mentioned properties. Again such a division is unique.

▶ **Example 25.3.6** *Consider example* 25.3.1. *It can be shown by induction that*

$$P^n = \frac{1}{\alpha+\beta}\begin{bmatrix} \beta & \alpha \\ \beta & \alpha \end{bmatrix} + \frac{(1-\alpha-\beta)^n}{\alpha+\beta}\begin{bmatrix} \alpha & -\alpha \\ -\beta & \beta \end{bmatrix}, \quad 0 < \alpha,\ \beta < 1. \tag{25.3.38}$$

If $|1-\alpha-\beta|<1$, then

$$\boldsymbol{P}^n \to \frac{1}{\alpha+\beta}\begin{bmatrix} \beta & \alpha \\ \beta & \alpha \end{bmatrix} \text{ as } n \to \infty \tag{25.3.39}$$

so that the process is ergodic. This means whatever the initial distribution, the process tends to an equilibrium distribution between states with

$$\pi_1 = \frac{\beta}{\alpha+\beta}, \quad \pi_2 = \frac{\alpha}{\alpha+\beta}, \quad 0<\alpha,\ \beta<1.$$

The distribution satisfies the equilibrium equation (25.3.34).

In cases $\alpha=\beta=0$ or $\alpha=\beta=1$, the process is not ergodic. In the first case, all transitions are impossible and the molecule freezes in its initial position, whatever that is. The chain here is reducible. In the second case transition is certain and the molecule alternates regularly between two positions. In this case this chain is periodic.

► **Example 25.3.7** *Consider the following modification of the model of example* 25.3.1. *Suppose there are N molecules, each moving independently between two positions with transition matrix given in that example. The expanded process is still Markovian with* 2^N *states.*

If we neglect the distinction between molecules and simply consider that the process is at state m at time n (i.e., $X_n=m$) when there are m molecules at energy level 1 at that time ($m=0,1,\cdots,N$), then it can be shown that this is again a Markov process. The probability generating function of X_n is

$$P_{X_n}(Z) = [(1-\alpha_n)Z+\alpha_n]^{X_0}\,[\beta_n Z+(1-\beta_n)]^{N-X_0}, \quad |Z|\le 1$$

where

$$\boldsymbol{P}^n = \begin{bmatrix} 1-\alpha_n & \alpha_n \\ \beta_n & 1-\beta_n \end{bmatrix}$$

is given by (25.3.38) of previous example. In the ergotic case of (25.3.39)

$$P_{X_n}(z) \to \left(\frac{\alpha+\beta z}{\alpha+\beta}\right)^{X_0}$$

so that the equilibrium distribution of X (the number of molecules at energy level 1) is binomial. The model is known as Ehrenfest model.

► **Example 25.3.8** *Consider the case of pure inbreeding, in which an individual is mated with itself, as in plants. An offspring gets one allel from each parent at random. Thus, an individual with AA mated with itself* ($AA \times AA$) *produces only AA; similarly an aa produces only aa; an Aa (mated with itself) produces AA, Aa, aa in the proportion*

$\frac{1}{4}, \frac{1}{2}, \frac{1}{4}$. *If we consider only a single individual in each generation, then this constitutes a Markov chain with three states with transition matrix*

$$P = \begin{bmatrix} 1 & 0 & 0 \\ \frac{1}{4} & \frac{1}{2} & \frac{1}{4} \\ 0 & 0 & 1 \end{bmatrix}.$$

It can be seen that

$$P^n = \begin{bmatrix} 1 & 0 & 0 \\ \frac{1-2^{-n}}{2} & 2^{-n} & \frac{1-2^{-n}}{2} \\ 0 & 0 & 1 \end{bmatrix} \to \begin{bmatrix} 1 & 0 & 0 \\ \frac{1}{2} & 0 & \frac{1}{2} \\ 0 & 0 & 1 \end{bmatrix}.$$

Thus ultimately an AA or aa line remains unchanged while an Aa line becomes either AA or aa each with probability $\frac{1}{2}$. The effect of imbreeding is, therefore, to produce a pure (homozygous) line.

Consider now inbreeding such that all aa individuals are discarded and one continues breeding in a given generation until a non-aa individual is produced. Here

$$P = \begin{bmatrix} AA & Aa \\ 1 & 0 \\ \frac{1}{3} & \frac{2}{3} \end{bmatrix}$$

and the process is still Markovian.

$$P^n = \begin{bmatrix} 1 & 0 \\ 1-\left(\frac{2}{3}\right)^n & \left(\frac{2}{3}\right)^n \end{bmatrix} \to \begin{bmatrix} 1 & 0 \\ 1 & 0 \end{bmatrix}.$$

Thus, one ends with a pure AA line.

25.3.4 Martingales

Definition 25.3.6 ***Martingales.*** *A Markov chain is called a Martingale if for every j, the expectation of the conditional probability distribution $\{p_{jk}\}$ equals j, i.e., if*

$$E(X_{n+1}|X_n = j) = \sum_k k p_{jk} = j. \tag{25.3.40}$$

For simplicity consider a martingale with a finite number of states $0, 1, \cdots, a$. Putting $j = 0$ and $j = a$ in (25.3.40), we have

$$\sum_{k=0}^{a} k p_{0k} = 0 \;\Rightarrow\; p_{0k} = 0 \;\forall\, k \;\Rightarrow\; p_{00} = 1$$

$$\sum_{k=0}^{a} k p_{ak} = a \;\Rightarrow\; p_{aa} = 1.$$

Hence, a and 0 must be absorbing states.

Assume that the chain contains no further closed set. It follows that the interior states are transient states. Hence the process will terminate either at 0 or at a. From (25.3.40) it follows by induction that for all n,

$$\sum_{k=0}^{a} p_{jk}^{(n)} k = j. \tag{25.3.41}$$

But $p_{jk}^{(n)} \to 0$ for every transient state k. Hence (25.3.41) implies that for all $j > 0$,

$$p_{ja}^{(n)} \to \frac{j}{a}.$$

This means that if the process starts with j, the probabilities of ultimate absorption at 0 and a are $1 - \frac{j}{a}, \frac{j}{a}$ respectively.

25.4 Discrete Branching Process

The Markov process $\{X_n, n = 0, 1, 2, \cdots\}$ is said to represent a simple discrete branching process if the following conditions are satisfied:

(a) $X_0 = x_0 = 1$ with probability one;

(b) $p(x) = P\{X_1 = x\}, \;\; p(x) \geq 0, \;\; \sum_{x=0}^{\infty} p(x) = 1, x = 0, \;\; 1, 2, \cdots;$

(c) the conditional distribution of X_{n+1} given $X_n = j$ is the sum of j independent random variables, each having the same p.m.f as X_1.

Let us denote

$$p_m(r) = P\{X_m = r\}.$$

By assumption (c),

$$\begin{aligned} p_{n+1}(k) &= \sum_{j=0}^{\infty} p_n(j) P\{X_{n+1} = k | X_n = j\} \\ &= \sum_{j=0}^{\infty} p_n(j) P[Y_1 + \cdots + Y_j = k] \end{aligned}$$

where $Y_1, \cdots, Y_j$ are i.i.d. having p.m.f $p(x)$, $x = 0, 1, 2, \cdots$

$$= \sum_{j=0}^{\infty} p_n(j) \sum_{\substack{k_i \geq 0 \\ \Sigma k_i = k}} \{P(Y_1 = k_1) \cdots P(Y_j = k_j)\}$$

$$= \sum_{j=0}^{\infty} p_n(j) [p(k)]^{j*} \tag{25.4.1}$$

where $[p(k)]^{j*}$ denotes the j-fold convolution (defined below) of $p(k)$ with itself. (25.4.1) gives a recursion relation connecting p.m.f of X_{n+1} and X_n.

Here X_n denotes the number of persons at time (generation) n. For human population we consider sons (fathers) only.

Definition 25.4.1 ***Convolution.*** *Let $\{a_k\}$, $\{b_k\}$ be two numerical sequences. Define the sequence $\{c_k\}$ where*

$$c_r = a_0b_r + a_1b_{r-1} + \cdots + a_{r-1}b_1 + a_rb_0.$$

Then the sequence $\{c_r\}$ is called the convolution of $\{a_k\}$ and $\{b_k\}$ and is denoted as

$$\{c_k\} = \{a_k\} * \{b_k\}.$$

If we have sequences $\{a_k\}$, $\{b_k\}$, $\{c_k\}$, $\{d_k\}, \cdots$ we can form convolution $\{a_k\} * \{b_k\}$ and then convolution of this sequence with $\{c_k\}$, etc. If we have n such identical sequences $\{a_k\}$, then n-fold convolution of $\{a_k\}$ with itself is

$$\{a_k\}^{n*} = \{a_k\} * \cdots * \{a_k\}.$$

In particular if $Z_1, \cdots, Z_n$ are i.i.d. each with p.m.f $\{q_k\}$ $[P(Z_1 = k) = q_k]$ then the distribution of $\xi_n = Z_1 + \cdots + Z_n$ is $\{q_k\}^{n*}$.
Let $F(s)$ denote the probability generating function of X_1.

$$F(s) = \sum_{x=0}^{\infty} p(x)s^x, \quad |s| \leq 1.$$

Theorem 25.4.1 *The probability generating function $F_n(s)$ (p.g.f) of X_n, given that $X_0 = 1$ is*

$$F_n(s) = F[F_{n-1}(s)]. \tag{25.4.2}$$

(If $X_0 = x_0$, the p.g.f of X_n is $[F_n(s)]^{x_0}$.)

Proof: Consider X_2 which is determined by the number of offsprings of X_1. Let S_i be the number of offsprings of the ith individual in generation (time) 1. Hence

$$X_2 = \sum_{i=1}^{X_1} S_i.$$

By the result in example 7.4.4, p.g.f. of X_2 is

$$F_2(s) = F[F_1(s)].$$

Similarly,

$$\begin{aligned} F_3(s) &= F[F_2(s)] \\ \cdots &\quad \cdots\cdots\cdots \\ \cdots &\quad \cdots\cdots\cdots \\ F_n(s) &= F[F_{n-1}(s)]. \end{aligned}$$

Clearly, we can write

$$F_n(s) = F_{n-1}[F(s)]. \tag{25.4.3}$$

Moments of X_n

We have

$$\mu = E(X_1) = \left.\frac{dF_1(s)}{ds}\right]_{s=1} = F_1'(1) = F'(1).$$

Differentiating (25.4.2) with respect to s,

$$F_n'(s) = F'[F_{n-1}(s)]F_{n-1}'(s).$$

Hence,

$$\begin{aligned} F_n'(1) &= F'(1)F_{n-1}'(1) \quad [\because \ F_{n-1}(1) = 1] \\ &= \mu F_{n-1}'(1) = \mu^2 F_{n-2}'(1) = \cdots = \mu^{n-1}F_1'(1) = \mu^n. \end{aligned}$$

Therefore,

$$E(X_n) = \mu^n. \tag{25.4.4}$$

Again,

$$\sigma^2 = V(X_1) = F''(1) + F'(1) - [F'(1)]^2. \tag{25.4.5}$$

Differentiating $F_{n+1}(s) = F_n[F(s)]$ with respect to s and putting $s = 1$,

$$F_{n+1}'(1) = F_n'[F(1)]F'(1) = \mu^{n+1}.$$

Differentiating again,

$$F_{n+1}''(1) = F_n''(1)[F_n'(1)]^2 + F'(1)F_n''(1). \tag{25.4.6}$$

Similarly differentiating $F_{n+1}(s) = F[F_n(s)]$ twice,

$$F_{n+1}''(1) = F''(1)F_n'(1) + [F'(1)]^2 F_n(1). \tag{25.4.7}$$

If $\mu \neq 1$, one can equate (25.4.6) and (25.4.7) and solve for $F_n''(1)$. This gives

$$F_n''(1) = \sigma^2\mu^n\left(\frac{\mu^n - 1}{\mu^2 - \mu}\right) + \mu^n(\mu^n - 1). \tag{25.4.8}$$

Using this in the relation similar to (25.4.5) for $V(X_n)$,

$$V(X_n) = \sigma^2\mu^n\left\{\frac{\mu^2 - 1}{\mu(\mu - 1)}\right\}, \mu \neq 1. \tag{25.4.9}$$

If $\mu = 1$, L' Hôspital rule yields

$$V(X_n) = \sigma^2 n. \tag{25.4.10}$$

Let ρ be the probability that the population will ultimately be extinct, i.e.,

$$\rho = \lim_{n\to\infty} P\{X_n = 0|X_0 = 1].$$

Now,

$$\mu^n = E(X_n) = \sum_{i=1}^{\infty} iP[X_n = i] \geq \sum_{i=1}^{\infty} P[X_n = i] = P[X_n \geq 1].$$

Hence if $\mu < 1$,

$$P[X_n = 0] \geq 1 - \mu^n \to 1 \tag{25.4.11}$$

as $n \to \infty$. Thus if $\mu < 1$, the population will ultimately die out. In case $\mu > 1$, ρ may be found as follows.

$$\begin{aligned}\rho &= P[\text{Population ultimately dies out}]\\ &= \sum_{i=0}^{\infty} P[\text{Population ultimately dies out} \mid X_1 = i]p(i).\end{aligned} \tag{25.4.12}$$

Now when $X_1 = i$, the population dies out if and only if each of the i families started by i individuals in the first generation dies out. Since the individuals act independently this probability is ρ^i. Therefore, from (25.4.12)

$$\rho = \sum_{i=0}^{\infty} \rho^i p(i). \tag{25.4.13}$$

The solution ρ is the smallest positive number satisfying (25.4.13).

The extinction probability can be calculated more accurately as follows. Assume $X_0 = 1$. Let

$$\rho_n = P[X_n = 0] = p_n(0).$$

ρ_n is the probability of extinction by the nth generation and is not to be confused with the event 'extinction at the nth generation', which is $\rho_n - \rho_{n-1}$. Hence $\{\rho_n\}$ is a non-decreasing sequence and since it is bounded it has a limit

$$\rho = \lim_{n\to\infty} \rho_n.$$

Now,

$$\rho_{n+1} = F_{n+1}(0) = F(\rho_n) \ \text{(by (25.4.2))} \tag{25.4.14}$$

and this relation together with initial condition $\rho_0 = 0$ determines the sequence $\{\rho_n\}$.

By letting $n \to \infty$ in (25.4.14) we see that the limiting probability must be a root of the equation

$$\rho = F(\rho) \tag{25.4.15}$$

(25.4.15) directly relates the probability of extinction to the p.g.f. of the progeny.

Theorem 25.4.2 *If $F(s)$ is not identically equal to s (i.e., if a man does not have exactly one son with probability one) then the equation* (25.4.15) *has just two real positive roots (of which $\rho = 1$ is always one). The extinction probability ρ is the smaller of the two roots; also $\rho < 1$ if $\mu > 1$ and $\rho = 1$ if $\mu \leq 1$.*

Proof: Omitted. The reader may refer to Feller (1968), vol. I, pp 206–207 or Bharucha-Reid (1960), pp 25–26.

The problem of branching process has its origin in 1873-74 when Galton and de Candolle became interested in many instances of family names that had become extinct. The model for this situation was first proposed and partially analysed by H. W. Watson (a Clergyman) in 1874. The analysis was completed by J. F. Steffenson in 1930.

25.5 The Simple Random Walk

Consider a Markov process $\{X_n, n = \cdots, -1, 0, 1, \cdots\}$ for which the transition probabilities are

$$\begin{aligned} p_{jj+1} &= p_j, \\ p_{jj} &= r_j, \\ p_{jj-1} &= q_j, \\ p_{jk} &= 0 \ (k \neq j-1, j, j+1) \end{aligned} \tag{25.5.1}$$

where $p_j + r_j + q_j = 1$. If j represents the co-ordinate of a point on the x-axis, then the process describes that a point moves one step right with probability p_j, one step left with probability q_j and stays where it is with probability r_j.

The forward and backward equations (25.3.11) and (25.3.12) respectively, take the form

$$p_{jk}^{(n)} = p_{j,k-1}^{(n-1)} p_{k-1} + p_{jk}^{(n-1)} r_k + p_{j,k+1}^{(n-1)} q_{k+1} \tag{25.5.2}$$

$$p_{jk}^{(n)} = p_j p_{j+1,k}^{(n-1)} + r_j p_{jk}^{(n-1)} + q_j p_{j-1,k}^{(n-1)}. \tag{25.5.3}$$

The system (25.5.2) is more useful if we consider $p_{jk}^{(n)}$ for a fixed initial state j and a variable current state k. The system (25.5.3) is more useful in the reverse situation.

The model (25.5.1) has been used to represent Brownian motion, that is, the motion of a large colloidal particle subject to random forces. However, the model originated from the theory of games of chances.

Suppose two players A, B play a series of games of chance. At each game the winner takes unit stake from the loser. Then p_j, q_j, r_j respectively can be taken as the probabilities that A wins, loses, makes a draw when he holds a capital of j units. If the rule of the game is that the game must cease if $j = 0$ (because A is then ruined) then $p_0 = q_0 = 0$, $r_0 = 1$. Similarly, if a is the total capital of the two players, then play must also stop at $j = a$ (because then B is ruined) and hence $p_a = q_a = 0$, $r_a = 1$.

Generally, the probabilities p_j, q_j, r_j will be independent of j so long the game must continue.

The Markov chain here has three distinct set of states, $\{0\}, \{1, 2, \cdots, a-1\}, \{a\}$. The states 0 and a are persistent whereas the states $1, 2, \cdots, a-1$ are each transient. The states 0, a are absorbing barriers.

We shall calculate the probability that

$$A_j = \lim_{n \to \infty} p_{j,0}^{(n)} \tag{25.5.4}$$

i.e., the probability that the player is ultimately ruined, given that he started with an initial capital j. We shall assume $p_j = p$, $q_j = q$, $r_j = 0$, $j = 1, 2, \cdots, a-1$. We have

$$A_0 = 1, \quad A_a = 0. \tag{25.5.5}$$

We note that in view of independence of trials,

$$A_j = pA_{j+1} + qA_{j-1}$$

or equivalently

$$A_j - A_{j+1} = \frac{q}{p}(A_{j-1} - A_j), \quad j = 1, \cdots, a-1. \tag{25.5.6}$$

Using the boundary conditions (25.5.5),

$$\begin{aligned} A_1 - A_2 &= \frac{q}{p}(A_0 - A_1) = \frac{q}{p}(1 - q_1), \\ A_2 - A_3 &= \frac{q}{p}(A_1 - A_2) = \left(\frac{q}{p}\right)^2 (1 - q_1), \\ \cdots\cdots & \cdots\cdots\cdots\cdots\cdots\cdots\cdots\cdots \\ A_{a-1} - A_a &= \frac{q}{p}(A_{a-2} - A_{a-1}) = \left(\frac{q}{p}\right)^{a-1} (1 - q_1). \end{aligned} \tag{25.5.7}$$

Adding the first $(j-1)$ equation,

$$A_1 - A_j = (1 - A_1)\sum_{r=1}^{j-1}\left(\frac{q}{p}\right)^r \tag{25.5.8}$$

$$= \begin{cases} \frac{q}{p} \cdot \frac{(q/p)^{j-1}-1}{(q/p)-1}(1 - A_1), & \text{if } \frac{q}{p} \neq 1 \\ (j-1)(1 - A_1), & \text{if } \frac{q}{p} = 1. \end{cases}$$

Adding $(1 - A_1)$ to each side of (25.5.8)

$$1 - A_j = \begin{cases} \frac{(q/p)^j-1}{(q/p)-1}(1 - A_1), & \text{if } p \neq \frac{1}{2} \\ j(1 - A_1), & \text{if } p = \frac{1}{2}. \end{cases} \tag{25.5.9}$$

Putting $j = a$ in above equations,

$$1 - A_1 = \begin{cases} \frac{(q/p)-1}{(q/p)^a-1}, & \text{if } p \neq \frac{1}{2} \\ 1/a, & \text{if } p = \frac{1}{2}. \end{cases}$$

Hence, from (25.5.8),

$$A_j = \begin{cases} \frac{(q/p)^a-(q/p)^j}{(q/p)^a-1}, & \text{if } p \neq \frac{1}{2} \\ 1 - \frac{j}{a}, & \text{if } p = \frac{1}{2}. \end{cases} \tag{25.5.10}$$

A_j gives the probability of A's ruin. The probability A'_j of A's fortune eventually reaching a is equal to the probability of B's ruin (when B starts with an initial capital of $a - j$) and is obtained by replacing p, q and j by $q, p, a - j$ respectively in (25.5.10). This gives

$$A'_j = 1 - A_j = \begin{cases} \frac{(q/p)^j-1}{(q/p)^a-1}, & \text{if } p \neq \frac{1}{2} \\ j/a, & \text{if } p = \frac{1}{2}. \end{cases} \tag{25.5.11}$$

The gambler's (A's) ultimate gain is a random variable Z. Hence if A starts with initial capital j,

$$\begin{aligned} E(Z) &= (a - j)(1 - A_j) + A_j(-j) \\ &= a(1 - A_j) - j. \end{aligned}$$

Hence $E(Z) = 0$ iff $p = \frac{1}{2}$ so that under this system of gambling a fair game ($p = q$) remains fair ($E(Z) = 0$) and no unfair game ($p \neq q$) can be reduced to a fair one.

One special case of interest is that in which $a \to \infty$ sò that A is playing against an infinitely rich opponent. We find from (25.5.10) that in this case,

$$\lim_{a\to\infty} A_j = \bar{A}_j = \begin{cases} (q/p)^j, & (p > q) \\ 1, & (p \leq q). \end{cases} \tag{25.5.12}$$

Thus A loses with probability one even in case $p = q$, when the game is a fair one. If the game is advantageous to him ($p > q$), his chance of ruin decreases exponentially with the size of his original capital.

25.6 The Continuous Time Markov Process

Consider now t as a continuous parameter; X_t would, however, be assumed to be discrete (i.e., $\mathcal{X}_t$ is countable) taking values x_t. We shall assume that the process is time-homogeneous (stationary). Thus

$$P(X_{r+t} = x_k | X_r, r \leq t) = P(X_{s+t} = x_k | X_s = x_j) = p_{jk}(t), \quad t \geq 0. \tag{25.6.1}$$

The Markov property can be expressed as

$$E(Y_t|X_s; s \le t) = E(Y_t|X_s), \quad t \ge s$$

where Y_t is any random variable defined in terms of $X_s(t \ge s)$. This gives the Chapman-Kolmogorov equation

$$p_{jk}(s+t) = \sum_m p_{jm}(s)p_{mk}(t), \quad (s, t \ge 0). \tag{25.6.2}$$

We can rewrite this equation in the matrix form as

$$\boldsymbol{P}(s+t) = \boldsymbol{P}(s)\boldsymbol{P}(t) \tag{25.6.3}$$

where $\boldsymbol{P}(t)$ is the matrix of t-step transition probabilities.

Previously we considered one-step transition probabilities as given and used (25.3.11) to obtain n step transition probabilities. Here we shall allow both s and t tend to zero and obtain continuous time analogue of forward and backward equations (25.3.11) and (25.3.12).

We must have $p_{jk}(0) = \delta_{jk}$ (Kronceker's delta) i.e., $\boldsymbol{P}(0) = I$. Assume, therefore,

$$\boldsymbol{P}(t) = ((p_{jk}(t))) \to I \text{ as } t \to 0 \tag{25.6.4}$$

because one may reasonably expect that the probability that the process has left the initial state k in time t would tend to zero as $t \to 0$. We shall further assume that $\boldsymbol{P}(t)$ posseses a matrix of right derivatives, $\boldsymbol{Q} = ((q_{jk}))$ at time $t = 0$. Thus,

$$p_{jk}(t) = \delta_{jk} + q_{jk}t + 0(t), t \ge 0. \tag{25.6.5}$$

Since for $j \ne k$, $p_{jk}(0) = 0$, $p_{jk}(t) \ge 0$, we must have $q_{jk} \ge 0$. Hence for $j \ne k$, q_{jk} may be interpreted as the probability intensity of transition from j to k in the sense that $q_{jk}t + 0(t)$ is the probability that the process changes from j to k in time $(0, t)$, given that it was at state j at $t = 0$. The probability of no transition from j to k will, therefore, be

$$1 - t \sum_{\substack{k \\ (\ne j)}} q_{jk} + 0(t).$$

Hence we must have

$$q_{jj} = -\sum_{\substack{k \\ (\ne j)}} q_{jk} = -q_j \text{ (say)}$$

where q_j is the total intensity of transition out of state j i.e., the rate at which the state j is left for some other state.

By letting t and s tend to zero respectively in (25.6.3) one obtains the forward and backward equations

$$\dot{\boldsymbol{P}}(t) = \boldsymbol{P}(t)\boldsymbol{Q} \tag{25.6.6}$$

$$\dot{\boldsymbol{P}}(t) = \boldsymbol{Q}\boldsymbol{P}(t) \tag{25.6.7}$$

where $\dot{P}(t) = \frac{d}{dt}P(t)$. Written in full, this lead to

$$\dot{p}_{jk}(t) = \sum_{\substack{m \\ (\neq k)}} p_{jm}(t)q_{mk} - q_k p_{jk}(t) \tag{25.6.8}$$

$$\dot{p}_{jk}(t) = \sum_{\substack{m \\ (\neq j)}} q_{jm}p_{mk}(t) - q_j p_{jk}(t). \tag{25.6.9}$$

Note that the condition (25.6.5) is not necessary for a continuous time Markov process. However, almost all processes of physical interest have this property.

We shall subsequently consider X_t as the number of occurrences of an event of a given type up to time $t(t \geq 0)$. Thus, the process is at state j, if the number of events up to time t, X_t is j. Clearly j is integer valued and $X_t \leq X_{t+s}, s \geq 0$. Such a process is called a counting process. We shall consider here continuous time stochastic counting process. Such a process will be said to have independent increments if the number of events occurring in disjoint time intervals are mutually independent.

25.6.1 Poisson Process

The counting process $\{X_t, t \geq 0\}$ is said to be a Poisson process with rate (intensity) λ if

(i) $X(0) = 0$;

(ii) the process is stationary and has independent increments;

(iii) $P(X_t = 1) = \lambda t + 0(t)$;

(iv) $P(X_t \geq 2) = 0(t)$.

Here

$$q_{jk} = \begin{cases} \lambda \ (k = j+1) \\ -\lambda \ (k = j) \\ 0 \ (\text{otherwise}). \end{cases} \tag{25.6.10}$$

For example X_t may be the number of registrations on a Geiger counter, the number of calls entering a telephone exchange. Assumptions tacitly made in the model are (i) the occurrences of events after time t is not affected by number of events occurring up to time t (e.g., assumption that one telephone call cannot block the other) (ii) the events occur with constant intensity λ (iii) the multiple events (two telephone calls at the same time) occur with zero intensity. The forward equations (25.6.8) become

$$\frac{dP_0(t)}{dt} = -\lambda P_0 \tag{25.6.11}$$

$$\frac{dP_k(t)}{dt} = \lambda P_{k-1} - \lambda P_k$$

where $P_k(t) = p_{0k}(t) = P\{X_t = k\}$. In this notation we shall mean $P_x(t)$ as the probability that $X_t = x$ when it sturted with $X_0 = 0$. The initial condition is $P_0(0) = 1$, $P_k(0) = 0, k \neq 0$. The equations (25.6.11) have already been solved in section 5.6. The solution gives $P_x(t) = e^{-\lambda t}(\lambda t)^x/x!$ Some interesting properties of Poisson distribution have been given in section 5.6 and example 6.4.2.

25.6.2 Birth Process

Here the only possible transitions are of the form $j \to j+1$. The probability intensities are

$$q_{jk} = \begin{cases} \lambda_j \ (k = j+1) \\ -\lambda_j \ (k = j) \\ 0 \ (\text{otherwise}). \end{cases} \tag{25.6.12}$$

The forward equations (25.6.8) are

$$\dot{p}_{jk}(t) = \lambda_{k-1}p_{j,k-1}(t) - \lambda_k p_{jk}(t). \tag{25.6.13}$$

Note that the rate λ_j depends on j, not on t. The Poisson process is a particular case $(\lambda_j = \lambda)$ of this process. Another example is that of radioactive disintegration where a radioactive atom can pass through a number of states by successive disintegration. Here, however, (unlike the pure birth process) there is an absorbing state, after which further disintegration is not possible.

Suppose that the atom is at state 0 for $t = 0$ i.e., $X_0 = 0$ and m is the final state i.e., $k = 0, 1, \cdots, m$. Let as before, $P_k(t) = P\{X_t = k\}$. Then

$$P_k(t+\Delta t) = (1 - \lambda_k \Delta t)P_k(t) + \lambda_{k-1}P_{k-1}(t)\Delta t + 0(\Delta t) \tag{25.6.14}$$

By transposing $P_x(t)$, dividing by Δt and taking limit we obtain the system of differential equations

$$\begin{aligned} \dot{P}_0(t) &= -\lambda_0 P_0(t) \\ \dot{P}_k(t) &= \lambda_{k-1}P_{k-1}(t) - \lambda_k P_k(t), \quad t = 1, 2, \cdots, m-1 \\ \dot{P}_m(t) &= \lambda_{m-1}P_{m-1}(t) \\ P_k(0) &= 0 \text{ for } k \neq 0. \end{aligned} \tag{25.6.15}$$

The system can be solved recursively. For example,

$$\begin{aligned} P_0(t) &= e^{-\lambda_0 t} \\ P_1(t) &= \frac{\lambda_0}{\lambda_1 - \lambda_0}(e^{-\lambda_0 t} - e^{-\lambda_1 t}) \end{aligned} \tag{25.6.16}$$

An important special case is the simple (linear) birth process, where λ_j is proportional to j, say $\lambda_j = j\lambda$. Consider a culture of j bacteria each of which reproduces independently at the rate λ. Then the probability of one division at time t is

$$\binom{j}{1}[\lambda t + 0(t)][1 - \lambda(t) + 0(t)]^{j-1} = \lambda jt + 0(t)$$

so that the transistion $j \to j+1$ does not actually have intensity λj in this case. The probability of more than one birth is at least of order $(\lambda t)^2$, so that transition $j \to j+r$ has zero intensity, $r \geq 2$.

However, in the simple birth process we assume $\lambda_j = j\lambda\,(j > 1, \lambda > 0)$. Assume that $X_0 = 1$. Let $P_x(t)$ denote the probability that $X_t = x$. Then by solving the equations (25.6.14) one gets

$$P_x(t) = e^{-\lambda xt}\left[\lambda(x-1)\int_0^t P_{x-1}(t)e^{x\tau}\,d\tau + C_x\right], \quad x \geq 1. \tag{25.6.17}$$

From the initial condition $P_1(0) = 1$, $C_1 = 1$.
By induction one gets the Yule-Furry distribution

$$P_k(t) = \begin{cases} e^{-\lambda t}(1 - e^{-\lambda t})^{k-1}, & x = 1, 2, \cdots \\ 0, & \text{otherwise.} \end{cases} \tag{25.6.18}$$

It is easily seen that $P_k(t)$ is a geometric distribution with parameter $e^{-\lambda t}$.
Hence

$$\begin{aligned} E[X_t] &= e^{-\lambda t} \\ V[X_t] &= e^{\lambda t}(e^{\lambda t} - 1). \end{aligned} \tag{25.6.19}$$

Consider now $X(0) = i$. In this case Kolmogorov equations (25.6.13) has solution

$$\begin{aligned} P_{ij}(t) &= \binom{j-1}{i-1} e^{-i\lambda t}(1 - e^{-\lambda t})^{j-1}, \quad j \geq 1 \\ &= 0 \text{ elsewhere.} \end{aligned} \tag{25.6.20}$$

In this case

$$\begin{aligned} E(X(t)) &= ie^{\lambda t} \\ V(X(t)) &= ie^{\lambda t}(e^{\lambda t} - 1). \end{aligned}$$

We note the following condition for a pure birth process to be a proper probability distribution i.e.,

$$\sum_{x=0}^{\infty} P_x(t) = 1.$$

This is given by the following theorem.

Theorem 25.6.1 (***Feller-Lundberg***). *For a pure birth process with parameter* λ_x,

$$\sum_{x=0}^{\infty} P_x(t) = 1 \ \forall\, t$$

iff

$$\sum_{x=0}^{\infty} \frac{1}{\lambda_x(t)} = \infty.$$

For a proof the reader may refer to Bharucha-Reid (1960), pp 81–82.

25.6.3 Birth and Death Process

Here only possible transitions are:

$j \to j \pm 1$

$$q_{jk} = \begin{cases} \lambda_j & (k = j+1) \\ \mu_j & (k = j-1) \\ 1 - \lambda_j - \mu_j & (k = j) \\ 0 & (\text{otherwise}). \end{cases} \tag{25.6.21}$$

The state $x = 0$ is an absorbing state.

First we obtain the solution for $P_x(t)$, the probability for $X_t = x$. We have

$$P_x(t+\Delta t) = \lambda_{x-1}P_{x-1}(t)\Delta t + [1-(\lambda_x+\mu_x)\Delta t]P_x(t) + \mu_{x+1}P_{x+1}(t)\Delta t + 0(\Delta t). \tag{25.6.22}$$

This gives

$$P'_x(t) = \lambda_{x-1}P_{x-1}(t) - (\lambda_x + \mu_x)P_x(t) + \mu_{x+1}P_{x+1}(t), \ \ x = 1, 2, \cdots. \tag{25.6.23}$$

$$P'_0(t) = \mu_1 P_1(t) \tag{25.6.24}$$

since $\lambda_{-1} = \lambda_0 = \mu_0 = 0$. If $X_0 = i$, $P_i(0) = 0$,

$$P_i(0) = 1, \ \ P_x(0) = 0, \ \ x \neq i \tag{25.6.25}$$

When $\lambda_x = \lambda x$ and $\mu_x = \mu x, (\lambda, \mu > 0)$, it can be shown that

$$P_x(t) = [1 - \alpha(t)][1 - \beta(t)][\beta(t)]^{x-1}, x = 1, 2, \cdots \tag{25.6.26}$$

$$P_0(t) = \alpha(t) \tag{25.6.27}$$

where

$$\alpha(t) = \left[\mu(e^{(\lambda-\mu)t} - 1)\right]\left[\lambda e^{(\lambda-\mu)t} - \mu\right]^{-1} \qquad (25.6.28)$$

$$\beta(t) = \left[\lambda(e^{(\lambda-\mu)t} - 1)\right]\left[\lambda e^{(\lambda-\mu)t} - \mu\right]^{-1} \qquad (25.6.29)$$

$$E(X_t) = \psi_t = e^{(\lambda-\mu)t} \qquad (25.6.30)$$

$$V(X_t) = \frac{\lambda+\mu}{\lambda-\mu} e^{(\lambda-\mu)t}\left[e^{(\lambda-\mu)t} - 1\right]. \qquad (25.6.31)$$

It can be seen that

$$\lim_{t\to\infty} \psi_t = \begin{cases} 0, & \lambda < \mu \\ 1, & \lambda = \mu \\ \infty, & \lambda > \mu. \end{cases} \qquad (25.6.32)$$

Thus, for $\lambda = \mu$, the mean population size ψ_t is stationary.
The probability that the process will die out by time t is $P_0(t)$.

$$\lim_{t\to\infty} P_0(t) = \begin{cases} 1, & \lambda < \mu \\ \mu/\lambda, & \lambda > \mu. \end{cases} \qquad (25.6.33)$$

It can be shown that the limits

$$\lim_{t\to\infty} P_x(t) = \pi_x \qquad (25.6.34)$$

exist and are independent of the initial conditions (25.6.25). They satisfy the system of equations (25.6.23), (25.6.24) on replacing the derivatives on the left by zero.
For the pure process the forward equations are

$$\dot{p}_{jk} = \lambda_{k-1}p_{j,k-1} + \mu_{k+1}p_{j,k+1} - (\lambda_k + \mu_k)p_{jk}. \qquad (25.6.35)$$

The equations for the equilibrium probabilities π_k are

$$\lambda_{k-1}\pi_{k-1} + \mu_{k+1}\pi_{k+1} = (\lambda_k + \mu_k)\pi_k. \qquad (25.6.36)$$

Assume that $\mu_0 = 0$ (so that transition to $k < 0$ is impossible). Considering the equation (25.6.36) for $k = 0, 1, 2, \cdots$ one gets the simpler equations

$$\lambda_{k-1}\pi_{k-1} = \mu_k\pi_k \; (k = 1, 2, \cdots). \qquad (25.6.37)$$

Solving

$$\pi_k = \frac{\lambda_0\lambda_1\cdots\lambda_{k-1}}{\mu_1\mu_2\cdots\mu_k}\pi_0 \; (k = 1, 2, \cdots). \qquad (25.6.38)$$

π_0 is determined from the condition $\Sigma\pi_k = 1$. Relation (25.3.36) states that the probability flux from $k-1$ to k must balance that from k to $k-1$.

▶ **Example 25.6.1** (*A simple truncking problem*). *Calls are coming at a telephone exchange as a Poisson process with rate λ. The probability of a conversation ending between t and $t+h$ is $\mu h + 0(h)$. The durations of conversations are assumed to be independent. The system is at state n if n lines are busy. Assume that infinitely many truncks or channels are available.*

If n lines are busy, the probability that one of them will be freed within time h is $n\mu h + 0(h)$. The probability that within this time two or more lines will be freed is $0(h^2)$ and can be neglected. The probability of a new call arriving is $\lambda h + 0(h)$. The probability of a combination of several calls arriving or a conversation ending is $0(h)$. Hence

$$\lambda_n = \lambda, \quad q_{n,n+1} = \lambda, \quad q_{n,n-1} = n\mu, \quad q_{n,n} = 1 - \lambda - n\mu.$$

The limiting probability

$$\pi_n = e^{-\lambda/\mu} \frac{(\lambda/\mu)^n}{n!}. \tag{25.6.39}$$

▶ **Example 25.6.2** (*A waiting time problem*). *Consider the same model as in example 25.6.1 with modification that the number of channels is 'a' ($< \infty$). If all 'a' channels are busy each new call joins a waiting line and waits until a channel is freed.*

The system is in state $n(> a)$ if there are 'a' persons being served and $(n - a)$ persons in the waiting line. If at least one line is free the situation is the same as in previous example. If the system is at state $n(> a)$, $\mu_n = a\mu(n \geq a)$. The basic system of differential equations is given by (25.6.23), (25.6.24) for $n < a$ (defining the system similarly for $n < a$) and is

$$P_n'(t) = -(\lambda + a\mu)P_n(t) + \lambda P_{n-1}(t) + a\mu P_{n+1}(t) \tag{25.6.40}$$

for $n \geq a$. The limiting probabilities π_n satisfies (25.6.39) for $n < a$. Thus

$$\pi_n = \pi_0 \frac{(\lambda/\mu)^n}{n!}, n < a \tag{25.6.41}$$

$$\pi_n = \frac{(\lambda/\mu)^n}{a! a^{n-a}} \pi_0 \text{ (by using (25.6.39))}, \, n \geq a. \tag{25.6.42}$$

The series $\Sigma \frac{\pi_n}{\pi_0} < \infty$ only if $\frac{\lambda}{\mu} < a$. Hence a limiting distribution $\{\pi_x\}$ can not exist if $\lambda \geq a\mu$. In this case the distribution becomes concentrated on to large values of n which means the waiting line increases to infinity. If $\lambda < a\mu$, one can determine π_0 so that $\sum_0^\infty \pi_n = 1$.

▶ **Example 25.6.3** (*A single server problem*). *Consider servicing of automatic machines. At an epoch t a machine in the working state calls for service with probability $\lambda h + 0(h)$ during the time $(t, t+h)$. Again, when the machine is being serviced, the* *probability that the servicing time terminates during* $(t, t+h)$ *is* $\mu h + 0(h)$. *For an efficient machine the servicing factor* λ/μ *should be small. Assume that* m *machines are working independently and are serviced by a single repairman (when required). The system is said to be in state* n *if* n *machines are not working. When* $1 \leq n \leq m$, *this means one machine is being repaired and* $(n-1)$ *machines are in the waiting line for servicing. Thus, transition* $n \to n+1$ *occurs by a breakdown of one of the working* $(m-n)$ *machines;* $n \to n-1$ *if the machine being serviced reverts to working state.*

Thus,

$$\lambda_n = (m-n)\lambda, \quad \mu_0 = 0, \quad \mu_1 = \cdots = \mu_m = \mu.$$

▶ **Example** 25.6.4 (***A multiple server problem***). *Consider the same situation as in example* 25.6.3 *but now there are* r *repairman* $(r < m)$. *It can be checked that*

$$\begin{aligned}\lambda_0 &= m\lambda, \quad \mu_0 = 0 \\ \lambda_n &= (m-n)\lambda, \quad \mu_n = n\mu \ \ (1 \leq n \leq r) \\ \lambda_n &= (m-n)\lambda, \quad \mu_n = r\mu \ \ (r \leq n \leq m).\end{aligned}$$

In the discontinuous Markov processes with continuous time parameters considered above we have assumed that for a small time interval Δt the probability of no change far exceeds the probability of a change of state. However, if a change takes place it may be rather striking. In the Markov processes which are both continuous in time and space (also called diffusion processes) it is assumed that some changes will take place in any interval Δt and if Δt is small the change of state will also be small with probability one. We shall consider only one such process, called the normal process.

25.6.4 Normal Process

A stochastic process $\{X_t, t \in T\}$ is said to be a normal process if for any integer n and any subset $\{t_1, t_2, \cdots, t_n\}$ of T the n-random variables $X_{t_1}, X_{t_2}, \cdots, X_{t_n}$ are jointly normally distributed. As a result the probability law of the process is completely specified by specifying the mean value function $E(X_t)$ and the covariance kernel $\text{Cov}(X_S, X_t)$.

An example of a normal process is Weiner process which provides a model for Brownian motion and thermal noise in electric circuits, among other applications. A tiny particle (of say, diameter 10^{-4} cm) when immersed in a liquid or fluid undergoes ceaseless irregular motion due to the continuous bombardment of the molecules of the surrounding medium. This is known as Brownian motion.

A process $\{X_t; t \geq 0\}$ is said to be a Wiener process if

(i) it has stationary independent increments;

(ii) for $s < t, X_t - X_s$ is normally distributed;

(iii) $E(X_t) = 0 \ \forall\, t > 0$;

(iv) $X_0 = 0$.

The first condition means that for every choice of indices $t_0 < t_1 < \cdots < t_n$ and for every n, the n random variables $X_{t_0} - X_{t_1}, X_{t_2} - X_{t_1}, \cdots, X_{t_n} - X_{t_{n-1}}$ are independently distributed and also for every choice of $t_1 < t_2, X_{t_2+h} - X_{t_1+h}$ has the same distribution as $X_{t_2} - X_{t_1}$.

Given n points $t_1 < t_2 < \cdots < t_n$, let $Z_1 = X_{t_1} = X_{t_1} - X_0$, $Z_2 = X_{t_2} - X_{t_1}, \cdots$, $Z_n = X_{t_n} - X_{t_{n-1}}$. By hypothesis $Z_1, Z_2, \cdots, Z_n$ are independently normally distributed. Now $X_{t_1}, X_{t_2}, \cdots, X_{t_n}$ are linear combinations of $Z_1, \cdots, Z_n$. Hence $X_{t_1}, \cdots, X_{t_n}$ are jointly normally distributed. Therefore, $\{X_t; t \geq 0\}$ is a normal process. It follows that

$$E[X_t - X_s] = 0.$$

The fact that X_t has independent increments shows that there is some positive constant σ^2 such that for $t \geq s \geq 0$,

$$\mathrm{Var}[X_t - X_s] = \sigma^2 |t - s|.$$

This follows because $\mathrm{Var}[X_t]$ satisfies the functional form $f(t_1 + t_2) = f(t_1) + f(t_2)$. Again,

$$\begin{aligned}\mathrm{Cov}[X_s, X_t] &= \mathrm{Cov}[X_s, X_t - X_s + X_s] \\ &= \mathrm{Cov}[X_s, X_t - X_s] + \mathrm{Var}(X_s) \\ &= \mathrm{Var}(X_s).\end{aligned}$$

Therefore, the covariance kernel is

$$K(s, t) = \sigma^2 \min(s, t) \ \forall\ s, t \geq 0.$$

We now show that the Wiener process can be used as a model for Brownian motion of a particle on a line arising from the numerous random impacts with other particles. Assume that times between two successive impacts are independent random variables, exponentially distributed with mean $1/\mu$. It then follows that the number $N(t)$ of impacts during $(0, t)$ is a Poisson process with intensity μ. Assume that the effect of an impact on the particle is to change its position by either $+a$ or $-a$, each with probability $\frac{1}{2}$. The position X_t of the particle at time t may be represented as

$$X_t = \sum_{n=1}^{N(t)} Y_n,$$

where Y_n is the particle's change of position as a result of the nth impact. The random variables Y_n are assumed to be independently distributed with

$$P[Y_n = +a] = P[Y_n = -a] = \frac{1}{2}.$$

It follows that $\{X_t; t \geq 0\}$ is a stochastic process with stationary independent increments and characteristic function $\varphi_{X_t}(u)$ given by

$$\ln \varphi_{X_t}(u) = \mu t E\left[e^{iuY} - 1\right].$$

Now,

$$\begin{aligned} E\left[e^{iuY} - 1\right] &= iuE[Y] - \frac{u^2}{2}E[Y^2] + \theta|u|^3 E[|Y|^3] \\ &= -\frac{1}{2}u^2a^2 + \theta|u|^3a^3, \quad |\theta| \leq 1. \end{aligned}$$

Defining $\sigma^2 = a^2\mu$,

$$\ln \varphi_{X_t}(u) = -\frac{1}{2}u^2\sigma^2 t + a\theta|u|^3\sigma^2 t.$$

If we now let $\mu \to \infty$ and $a \to 0$ such that the product μa^2 (total mean square displacement of the particle per unit time) is constant, then

$$\ln \varphi_{X_t}(u) \to -\frac{1}{2}u^2\sigma^2 t.$$

Hence, $\{X_t; t \geq 0\}$ is approximately a Wiener process, since it is approximately normally distributed and has stationary independent increments. The above results are due to Parzen (1960).

Exercise

1. (a) Let $\{X_n\}$ be a Markov chain with

$$P\{X_0 = 0\} = \frac{1}{2} = P\{X_0 = 1\}$$

and state space $(0, 1)$. If the transition probability matrix is

$$\begin{bmatrix} 0 & 1 \\ 1-p & p \end{bmatrix}$$

obtain $P[X_n = 0]$ and $P[X_n = 1]$ and hence their limits as $n \to \infty$.

(b) Three children $(1, 2, 3)$ arranged in a circle play a game of throwing a ball to one another. At each stage the child having the ball is equally likely to throw it to any one of the other two children. Suppose X_0 denotes the child who had the ball initially and $X_n (n \geq 1)$ denotes the child who had the ball after n throws. Show that $\{X_n\}$ forms a Markov chain. Find the transition probability matrix. Calculate the probability that the child who had originally the ball will have it after two throws.

2. For the following stochastic matrices classify all the states. Check whether the ergodic theorem holds. If so, find ergodic probabilities.

$$\text{(a) } \boldsymbol{P} = \begin{bmatrix} \frac{1}{4} & \frac{1}{2} & 0 & \frac{1}{4} \\ \frac{1}{5} & 0 & \frac{1}{3} & \frac{7}{15} \\ 0 & \frac{2}{3} & \frac{1}{3} & 0 \\ \frac{1}{4} & \frac{1}{4} & \frac{1}{4} & \frac{1}{4} \end{bmatrix}; \quad \text{(b) } \boldsymbol{P} = \begin{bmatrix} \frac{1}{3} & \frac{2}{3} & 0 & 0 \\ \frac{3}{4} & \frac{1}{8} & \frac{1}{8} & 0 \\ 0 & 0 & \frac{1}{2} & \frac{1}{2} \\ 0 & 0 & \frac{1}{3} & \frac{2}{3} \end{bmatrix}; \quad \text{(c) } \boldsymbol{P} = \begin{bmatrix} \frac{2}{5} & 0 & \frac{2}{5} & 0 & 0 \\ \frac{1}{5} & \frac{3}{5} & \frac{1}{5} & 0 & 0 \\ \frac{1}{5} & 0 & \frac{4}{5} & 0 & 0 \\ 0 & 0 & 0 & \frac{3}{5} & \frac{2}{5} \\ 0 & 0 & 0 & \frac{2}{5} & \frac{3}{5} \end{bmatrix}.$$

3. Consider a homogeneous Markov chain with a countable number of states with transition matrix

$$\boldsymbol{P} = \begin{bmatrix} \frac{1}{2} & \frac{1}{2} & 0 & 0 & 0 & \cdots & 0 \\ \frac{1}{2} & 0 & \frac{1}{2} & 0 & 0 & \cdots & 0 \\ \frac{1}{3} & 0 & 0 & \frac{2}{3} & 0 & \cdots & 0 \\ \frac{1}{4} & 0 & 0 & 0 & \frac{3}{4} & \cdots & 0 \\ \cdots & \cdots & \cdots & \cdots & \cdots & \cdots & \cdots \end{bmatrix}.$$

Show that all states are recurrent null states and hence $\lim_{n\to\infty} p_{ij}(n) = 0$ $(i, j = 1, 2, \cdots)$. **[Fisz, 1963]**

4. A matrix is said to be doubly stochastic if in addition to be property $\sum_j p_{ij} = 1$, we have $\sum_i p_{ij} = 1$. Show that a finite Markov chain with a doubly stochastic matrix has no transient states. Show also that if the chain is irreducible and aperiodic then $p_{ij}^{(n)} \to N^{-1}$ where N is the number of states. On the other hand, if an irreducible infinite chain has a doubly stochastic matrix, then its states are either null or transient. **[Prabu, 1965]**

5. Consider example 25.3.8. Suppose that the initial individual is chosen randomly from a population in which three genotypes AA, Aa, aa occur in the proportion p_0, q_0, r_0 respectively. Show that the probability that the line ultimately tends to the AA-strain is $p_0 + \frac{1}{2}q_0$ which is also the A-gene frequency in the original population.

6. Suppose that X_n is a numerical valued Markov process with transition matrix (25.3.14). Show that for the ergodic case the correlation coefficient between X_n and X_{n+m} is $(1-\alpha-\beta)^m$, $(m > 0)$ when n is so large that the equilibrium has been attained.

7. (a) Consider a finite Markov chain with one-step transition probability matrix $\boldsymbol{P} = ((p_{ij}))$ and state space $S = \{1, 2, \cdots, N\}$. Let the states $\{1, 2, \cdots, t\}$ be all the transient states. Show that $(\boldsymbol{I}_t, -\boldsymbol{Q})$ is a non-singular matrix where $\boldsymbol{I}_t$ is a $t \times t$ identity matrix and $\boldsymbol{Q} = ((p_{ij}))$, $i, j = 1, \cdots, t$.

(b) Consider the Markov chain with state space $\{0, 1, \cdots, d\}$ and stationary transition probabilities given by

$$p_{oo} = 1, \quad p_{oj} = 0, \quad 1 \le j \le d$$
$$p_{jk} = \frac{1}{j}, \quad 0 \le k \le j-1, \quad 1 \le j \le d.$$

Find $E(T|X_0 = i)$, $i = 1, \cdots, d$ where the random variable T is defined by

$$T = n \text{ iff } X_n = 0 \text{ and } X_j \neq 0, \text{ for } j = 1, \cdots, n-1.$$

8. (A nonhomogeneous random walk). Let $p_{00} = q_0$, $p_{01} = p_0$, $p_{i,i-1} = q_i$, $p_{i,i+1} = p_i$ $(i = 1, 2, \cdots)$, where $p_i = 1 - q_i (i \ge 1)$.

   Further let $L_i = (q_1 q_2 \cdots q_i)(p_1 p_2 \cdots p_i)^{-1} (i \ge 1)$. Show that the states are (i) transient if the series $\Sigma L_i < \infty$; (ii) persistent null if $\Sigma L_i = \infty$, $\Sigma (P_i L_i)^{-1} = \infty$ and (iii) persistent nonnull if $\Sigma L_i = \infty$, $\Sigma (P_i L_i)^{-1} < \infty$.

   Find the stationary distribution in the last case.

9. Let Z denote a random variable whose possible values are the integers $0, \pm 1, \pm 2, \cdots$. Assume that $E[|Z|] < \infty$.

   Let $Z_1, Z_2, \cdots$ be a sequence of independent random variables distributed identically as Z. Define $X_n = Z_1 + Z_2 + \cdots + Z_n$.

   Describe the transition probability matrix of the Markov chain $\{X_n, n = 1, 2, \cdots, \}$. Show that the chain is recurrent iff $E(Z) = 0$.

10. For a certain Markov chain,

$$\boldsymbol{P} = \begin{bmatrix} p_0 & p_1 & p_2 & \cdots & p_{m-1} \\ p_{m-1} & p_0 & p_1 & \cdots & p_{m-2} \\ \cdots & \cdots & \cdots & \cdots & \cdots \\ p_1 & p_2 & p_3 & \cdots & p_0 \end{bmatrix}.$$

   Determine $\boldsymbol{P}^n$ and prove that $\lim_{n\to\infty} P[X_n = j] = \frac{1}{m}, \quad (j = 1, \cdots, m)$.

11. Suppose that X_n can take only discrete values $\{x_k\} (k = 1, 2, \cdots)$ for all n. Show that for a Markov process

$$P(X_j = x_{kj},\ j = 1, \cdots, n | X_0 = x_0) = \prod_{j=1}^{n} P(X_j = x_{kj} | X_{j-1} = x_{kj-1})$$

12. Show that for the discrete branching process considered in section 25.4, the relation (25.4.2) can be generalised to

$$F_{m+n}(s) = F_m[F_n(s)] \quad (m, n = 0, 1, 2, \cdots).$$

13. Suppose that

$$F(s) = \frac{ps+q}{1+c-cs}.$$

Show that this is the probability generating function of a non-negative r.v., if $p+q=1$; $p, q, c \geq 0$. Show that the nth iterate of $F(s)$,

$$F^{(n)}(s) = F[F[F\cdots[F(s)]\cdots]\cdots] = \frac{A_n s + B_n}{C_n s + D_n}$$

where

$$\begin{bmatrix} A_n & B_n \\ C_n & D_n \end{bmatrix} = \begin{bmatrix} p & -c \\ q & 1+c \end{bmatrix}^n.$$

14. Show that

$$F(s) = 1 - c(1-s)^{\delta}$$

is the probability generating function of a non-negative r.v. if $0 \leq c, \delta \leq 1$ and evaluate its nth iterate $F^{(n)}(s)$ (defined in exercise 13).

15. Suppose there are two types of individuals A and B, whose numbers in the nth generation have joint p.g.f. $F_n(s,t)$. If individuals of type A, B have progeny p.g.f.'s $G_A(s,t)$, $G_B(s,t)$ respectively, show that

$$F_{n+1}(s,t) = F_n[G_A(s,t), G_B(s,t)].$$

Make the same types of assumption as in section 25.4.

16. Using equation (25.5.3) show that if M_j is the expected duration of game given that A started with an initial capital of j units, show that

$$M_j = 1 + pM_{j+1} + rM_j + qM_{j-1}, \quad (j = 1, 2, \cdots, a-1)$$

with $M_0 = M_a = 0$. Solve for M_j.

17. Consider the model (25.5.1) to represent the Brownian motion of a particle in a one-dimensional container with walls at $j=0$ and a so that $q_0 = p_a = 0$. Show that if π_j is the equilibrium probability that the particle is at position j, then

$$\pi_j = \frac{p_0 p_1 \cdots p_{n-1}}{q_1 q_2 \cdots q_j} \pi_o, \quad (j = 1, 2, \cdots, a).$$

18. (a) For a Poisson process $X(t), t \geq 0$, let W_r be the random time of the occurrence of the rth Poisson event. Let $0 \leq s \leq t$ and $1 \leq r \leq n$. Show that

$$P\{W_r \leq s | X(t) = n\} = \sum_{k=r}^{n} \binom{n}{k} \left(\frac{s}{t}\right)^k \left(1 - \frac{s}{t}\right)^{n-k}.$$

(b) Consider a birth and death process $X(t)$ with 3 states 0,1 and 2 with birth rates $\lambda_0, \lambda_1, 0$ and death rates $0, \mu_1, \lambda_0$ respectively. Write down the forward equation. Using these find

$$p_{oj}(t) = P\{X_{(t)} = j | X_{(0)} = 0\} \text{ for } j = 0, 1, 2.$$

19. Suppose that the Poisson process is modified to allow multiple events, so that

$$q_{jk} = \begin{cases} \lambda_{k-j} & (k > j) \\ 0 & (k < j). \end{cases}$$

Show that the probability generating function of the number of events in $(0, t)$ is

$$P(s,t) = \exp\left[t \sum_r \lambda_r (s^r - 1)\right], \quad |s| \le 1.$$

20. For the simple death process:

$$q_{jk} = \begin{cases} \mu j & (k = j-1) \\ -\mu j & (k = j) \\ 0 & \text{(otherwise)}. \end{cases}$$

Show that the probability generating function is

$$\pi_j(s,t) = E\left(s^{X_t} | X_0 = j\right) = \sum_k p_{jk}(t) s^k = \left[1 + e^{\mu t}(s-1)\right]^j.$$

21. Show that if $q_{jj+1} = \lambda(N - j)$, $q_{jj-1} = j\mu (0 \le j \le N)$, and $X_0 = 0$, then th distribution of X_t is binomial

$$\binom{N}{k} p^k q^{N-k} \quad (0 \le k \le N)$$

when $p = \lambda\left[1 - e^{-(\lambda+\mu)t}\right] / (\lambda + \mu)$ and $q = 1 - p$.

22. Let $X(t) = (-1)^{N(t)}$, where $N(t)$ is a Poisson process with parameter λ. Le $Y(t) = \alpha X(t)$, where α is a random variable taking the values ± 1 with probabilit $\frac{1}{2}$ each, independent of $X(t)$. Show that

(a) $E(Y(t)) = 0$,

(b) obtain $E(X(t))$, $\text{Var}(X(t))$,

(c) $\text{Cov}(Y(s), Y(t)) = \exp(-2\lambda |t - s|)$.

23. Consider a pure birth process with $\lambda_j = \lambda + j\alpha (j \geq 0)$. Show that the transition probabilities of X_t are given by

$$p_{jk}(t) = e^{-(\lambda+jk)t} \binom{-j-(\lambda/\alpha)}{k-j} (e^{-\alpha t} - 1)^{k-j} \ (k \geq j).$$

24. Show that in a pure death process with $\mu_j = j\mu$, the transition probabilities are given by

$$p_{jk}(t) = \begin{cases} \binom{j}{k} e^{-k\mu t}(1-e^{-\mu t})^{j-k}, & (k \leq j) \\ 0, & (k > j). \end{cases}$$

Show that the time for extinction has the distribution

$$e^{\mu t}(1-e^{-\mu t})^{j-1} j\mu dt \ (0 < t < \infty).$$

25. Show that the times spent in states $j, j+1, \cdots$, of a birth process are independent exponential variables with parameters $\lambda_j^{-1}, \lambda_{j+1}^{-1}, \cdots$.

26. (a) A blood culture is started with one red cell initially. At the end of unit time the red cell dies and is replaced by

either (i) 2 red cells;

or (ii) one red cell and one white cell;

or (iii) 2 white cells

with probabilities $\frac{1}{4}, \frac{2}{3}, \frac{1}{12}$, respectively. Subsequently, each red cell reproduces as above and each white cell dies at the end of unit time without reproduction. Show that the probability of the extinction of the entire culture is $\frac{1}{3}$.

(b) Let $N(t)$ be a renewal process with inter-arrival time $\tau_1, \tau_2, \cdots$ where each τ_i has distribution function $F(x)$ and density function $f(x)$. Let $E(\tau_i) = \mu < \infty$ and let $S_n = \tau_1 + \cdots + \tau_n$, $n \geq 1$.

Prove the following:

i. $E(N(t)) = \sum_1^\infty P(S_n \leq t)$.

ii. $\frac{N(t)}{t} \to \frac{1}{\mu}$ almost surely.

27. Let $k(>0)$ denote the number of customers in a queue and suppose that customers leave the queue with intensity μ and join it with intensity λ (for $k < a$) or zero ($k \geq a$). Obtain the differential equations for this system. Show that the limiting probability

$$\pi_k = \frac{(1-\rho)\rho^k}{1-\rho^{a+1}}, \quad k = 0, 1, \cdots, a$$

where $\rho = \lambda/\mu$, the traffic intensity. Show that when $a \to \infty$ and $\rho < 1$ the limiting distribution is a geometric distribution. However, if $a \to \infty$ and $\rho \geq 1$ the distribution becomes concentrated over indefinitely large values of k.

28. For the Gambler's ruin problem, let μ_i denote the expected number of games to be played by A until the gambler either goes broke or reaches a fortune a, given that he started with an initial capital i. Show that μ_i satisfies

$$\mu_0 = \mu_a = 0$$
$$\mu_i = 1 + p\mu_{i+1} + q\mu_{i-1}, \quad (i = 1, 2, \cdots, a-1).$$

Hence, show that

$$\mu_i = \begin{cases} i(a-i), & \text{if } p = \frac{1}{2} \\ \frac{i}{q-p} - \frac{a}{q-p} \times \frac{1-\left(\frac{q}{p}\right)^i}{1-\left(\frac{q}{p}\right)^a}, & \text{if } p \neq \frac{1}{2}. \end{cases}$$

29. Let the smaller root of equation (25.4.15) be ξ. Then $F'(\xi)$ exists and $0 < F(\xi) \leq 1$.

Show that $\xi \geq p(X_1 = 0) = p_0$. Show also that in the range $0 \leq s \leq \xi$,

$$\xi + F'(\xi)(s-\xi) \leq F(s) \leq \xi + \frac{\xi - p_0}{\xi}(s-\xi). \tag{i}$$

Using this inequality in (25.4.14), show that

$$\left(\frac{\xi - p_0}{\xi}\right)(\xi - \rho_n) \leq (\xi - \rho_{n+1}) \leq F'(s)(s - \rho_n). \tag{ii}$$

The first inequality in (ii) indicates that ξ is the only possible limit for the sequence $\{\rho_n\}$ and $\rho = \xi$ (ρ is defined in 25.4.12).

Appendix

Table A.1 Random Digits and Digit Permutations

row number	10-digit permutations	random digits									
1	7513462980	9787	3792	5241	0556	7070	0786	7431	7157	8539	4118
2	4310765982	4479	1397	8435	3542	8435	6169	7996	3314	1299	1935
3	2731469508	0191	2800	1056	2735	4816	1979	0042	5824	6636	2332
4	6014738925	8710	6903	1347	9332	6962	6786	9875	7565	8683	6490
5	8467523019	4656	5960	0812	5144	5355	3335	4784	7573	3841	4255
6	0689351274	9974	9239	8049	4971	7555	3935	9405	8545	4329	5358
7	0978346251	8493	7128	3654	8976	1901	5496	3453	7539	3255	6742
8	2637098154	6135	6954	3436	3841	9009	3768	9256	3631	9066	7153
9	8761492305	1217	2748	3864	4752	7407	9975	6372	3308	0000	4734
10	6321850749	2623	1282	4389	8889	0764	2328	2140	8843	4986	4413
11	9123075684	1144	5336	4426	9003	6956	9406	8464	8827	3143	4754
12	3047921685	5854	9981	9079	2908	4755	4620	6455	6793	7539	4031
13	9714850236	0615	8188	2812	0270	5733	5339	1175	2919	7343	0477
14	1684205973	3624	0853	3128	7952	2678	3011	7710	9734	6386	8400
15	6132758904	1185	6832	4918	9236	3026	5795	0352	7533	4435	0306
16	0674951832	7391	3210	9540	4085	9234	4892	3962	3883	4538	8286
17	1694508372	7195	1986	6146	0946	5421	8430	2128	7602	5609	7064
18	3168752094	6137	7286	5283	0609	0941	4935	2521	7937	2153	2629
19	4750823961	7401	8099	7482	2210	3662	8253	7507	7809	0094	4101
20	2604381975	0192	9452	7189	9552	7498	0105	8295	9762	7434	3518
21	9708245361	3621	3037	2274	3803	0946	9874	4911	6797	1227	8494
22	3859761402	2661	0047	6628	6199	2526	5631	8334	7668	3994	7439
23	8245139076	8072	5085	3576	4939	0352	7386	7690	7108	6668	8246
24	2409873165	0839	5224	9768	3839	8495	1668	6957	7031	2032	1468
25	2864935170	2354	9266	8034	3813	3648	7825	6156	3605	7796	1645
26	9164078352	9050	6800	0490	3261	7748	3609	1050	0591	3799	2827
27	6053894271	7174	7703	1540	8001	6230	0387	9553	7447	0240	2511
28	2674159083	3465	7017	2278	0357	5800	1048	8382	8800	7608	4325
29	8703615942	8805	1265	5202	6872	3282	5331	5398	1426	2805	2110
30	7410239586	0250	4100	5263	8506	9848	2451	2031	2026	8661	4163
31	8219476035	6088	8366	7751	1577	9534	2458	1886	1522	4161	8726
32	7140532968	8833	3449	3499	4223	2854	6855	4042	1294	1728	5494
33	2709538146	4675	2535	1915	9783	9754	2790	6856	0352	9628	8342
34	5704196823	8990	4993	2922	8842	9904	8442	0105	3308	3320	6361
35	7163482059	1790	8590	5792	0983	3494	0945	4966	2194	9823	2599
36	5017249683	9276	3967	2486	6242	3276	1884	1847	8922	7356	1528
37	3541076298	2965	7991	3777	9303	0536	1517	0570	7212	7593	0566
38	0187326495	6620	4234	8407	6890	6904	8599	5876	2608	7329	6117
39	4791635802	4706	8319	6252	3177	9108	3069	0910	8241	9842	0895
40	6701439825	8395	3882	0259	2092	4885	3434	0879	0000	0790	0735
41	0251784693	3991	3406	0151	2594	9137	9924	2393	7699	6116	9655
42	9452783601	9644	6763	3512	0139	4119	2722	3219	0070	3830	7997
43	8360527149	3658	7813	0207	0357	8225	4497	2435	5121	4776	3611
44	3592076148	5728	1882	9120	7893	3503	8579	9070	1952	8390	5517
45	8456091273	6221	6366	8192	8429	4387	5484	7553	4053	9458	2292
46	7831506249	7635	5248	1750	0868	0173	4989	2300	3916	6732	8284
47	2754319608	4368	3113	5887	8439	0026	1902	4114	3127	5140	6684
48	5207934861	8635	9723	2550	8216	7531	7732	3963	4014	2099	3030
49	6274095183	3304	3254	3936	9361	9771	8255	4592	8808	3803	4010
50	3486157209	9336	5666	1349	1932	7326	2151	1573	3045	8746	8059
Column no.:		1–4	5–8	9–12	13–16	17–20	21–24	25–28	29–32	33–36	37–40

Table A.2 Cumulative Binomial Probabilities, $\Sigma_{x=0}^{r}\binom{n}{x}p^x(1-p)^{n-x}$, $r = 0, 1, 2, \cdots, n-1$

		p								
n	r	.01	.05	.10	.20	.25	.30	.333	.40	.50
2	0	.9801	.9025	.8100	.6400	.5625	.4900	.4444	.3600	.2500
	1	.9999	.9975	.9900	.9600	.9375	.9100	.8888	.8400	.5000
3	0	.9703	.8574	.7290	.5120	.4219	.3430	.2963	.2160	.1250
	1	.9997	.9928	.9720	.8960	.8438	.7840	.7407	.6480	.5000
	2	1.0000	.9999	.9990	.9920	.9844	.9730	.9629	.9360	.8750
4	0	.9606	.8145	.6561	.4096	.3164	.2401	.1975	.1296	.0625
	1	.9994	.9860	.9477	.8192	.7383	.6517	.5926	.4742	.3125
	2	1.0000	.9995	.9963	.9728	.9492	.9163	.8889	.8198	.6875
	3		1.0000	.9999	.9984	.9961	.9919	.9877	.9734	.9375
5	0	.9510	.7738	.5905	.3277	.2373	.1681	.1317	.0778	.0312
	1	.9990	.9774	.9185	.7373	.6328	.5283	.4609	.3370	.1874
	2	1.0000	.9988	.9914	.9421	.8965	.8370	.7901	.6826	.4999
	3		.9999	.9995	.9933	.9844	.9693	.9547	.9130	.8124
	4		1.0000	1.0000	.9997	.9990	.9977	.9959	.9898	.9686
6	0	.9415	.7351	.5314	.2621	.1780	.1176	.0878	.0467	.0156
	1	.9986	.9672	.8857	.6553	.5340	.4201	.3512	.2333	.1094
	2	1.0000	.9977	.9841	.9011	.8306	.7442	.6804	.5443	.3438
	3		.9998	.9987	.9830	.9624	.9294	.8999	.8208	.6563
	4		.9999	.9999	.9984	.9954	.9889	.9822	.9590	.8907
	5		1.0000	1.0000	.9999	.9998	.9991	.9987	.9959	.9845
7	0	.9321	.6983	.4783	.2097	.1335	.0824	.0585	.0280	.0078
	1	.9980	.9556	.6554	.5767	.4450	.3294	.2633	.1586	.0625
	2	1.0000	.9962	.8503	.8520	.7565	.6471	.5706	.4199	.2266
	3		.9998	.9743	.9667	.9295	.8740	.8267	.7102	.5000
	4		1.0000	.9973	.9953	.9872	.9712	.9547	.9037	.7734
	5			.9998	.9996	.9987	.9962	.9931	.9812	.9375
	6			1.0000	1.0000	.9999	.9998	.9995	.9984	.9922
8	0	.9227	.6634	.4305	.1678	.1001	.0576	.0390	.0168	.0039
	1	.9973	.9427	.8131	.5033	.3671	.2553	.1951	.1064	.0352
	2	.9999	.9942	.9619	.7969	.6786	.5518	.4682	.3154	.1445
	3	1.0000	.9996	.9950	.9437	.8862	.8059	.7413	.5941	.3633
	4		1.0000	.9996	.9896	.9727	.9420	.9120	.8263	.6367
	5			1.0000	.9988	.9958	.9887	.9803	.9502	.8555
	6				1.0000	.9996	.9987	.9974	.9915	.9648
	7					1.0000	.9999	.9998	.9993	.9961
9	0	.9135	.6302	.3874	.1342	.0751	.0404	.0260	.0101	.0020
	1	.9965	.9287	.7748	.4362	.3004	.1960	.1431	.0706	.0196
	2	.9999	.9916	.9470	.7382	.6007	.4628	.3772	.2318	.0899
	3	1.0000	.9993	.9916	.9144	.8343	.7296	.6503	.4826	.2540
	4		.9999	.9990	.9805	.9511	.9011	.8551	.7334	.5001
	5		1.0000	.9998	.9970	.9900	.9746	.9575	.9006	.7462
	6			.9999	.9998	.9987	.9956	.9916	.9749	.9102
	7			1.0000	1.0000	.9999	.9995	.9989	.9961	.9806
	8					1.0000	.9999	.9998	.9996	.9982

Contd.

Contd.

		p								
n	*r*	.01	.05	.10	.20	.25	.30	.333	.40	.50
10	0	.9044	.5987	.3487	.1074	.0563	.0282	.0173	.0060	.0010
	1	.9958	.9138	.7361	.3758	.2440	.1493	.1040	.0463	.0108
	2	1.0000	.9884	.9298	.6778	.5256	.3828	.2991	.1672	.0547
	3		.9989	.9872	.8791	.7759	.6496	.5592	.3812	.1719
	4		.9999	.9984	.9672	.9219	.8497	.7868	.6320	.3770
	5		1.0000	.9999	.9936	.9803	.9526	.9234	.8327	.6231
	6			1.0000	.9991	.9965	.9894	.9803	.9442	.8282
	7				.9999	.9996	.9984	.9966	.9867	.9454
	8				1.0000	1.0000	.9998	.9996	.9973	.9893
	9						1.0000	.9999	.9999	.9991
11	0	.8954	.5688	.3138	.0859	.0422	.0198	.0116	.0036	.0005
	1	.9948	.8981	.6974	.3221	.1971	.1130	.0752	.0320	.0059
	2	.9998	.9848	.9104	.6174	.4552	.3128	.2341	.1189	.0327
	3	1.0000	.9984	.9815	.8389	.7133	.5696	.4726	.2963	.1133
	4		.9999	.9972	.9496	.8854	.7897	.7110	.5328	.2744
	5		1.0000	.9997	.9884	.9657	.9218	.8779	.7535	.5000
	6			1.0000	.9981	.9924	.9784	.9614	.9007	.7256
	7				.9998	.9988	.9947	.9912	.9707	.8867
	8				1.0000	.9999	.9994	.9986	.9941	.9673
	9					1.0000	.9999	.9999	.9993	.9941
	10						1.0000	1.0000	1.0000	.9995
12	0	.8864	.5404	.2824	.0687	.0317	.0139	.0077	.0022	.0002
	1	.9938	.8816	.6590	.2749	.1584	.0850	.0540	.0196	.0032
	2	.9998	.9804	.8892	.5584	.3907	.2528	.1811	.0835	.0193
	3	1.0000	.9978	.9744	.7946	.6488	.4925	.3931	.2254	.0730
	4	1.0000	.9998	.9957	.9806	.8424	.7237	.6315	.4382	.1939
	5	1.0000	1.0000	.9995	.9961	.9456	.8822	.8223	.6652	.3872
	6			1.0000	.9994	.9858	.9614	.9336	.8418	.6128
	7				.9999	.9972	.9905	.9812	.9427	.8062
	8				1.0000	.9996	.9983	.9962	.9848	.9270
	9					1.0000	.9998	.9995	.9972	.9807
	10						1.0000	.9999	.9997	.9968
	11							1.0000	1.0000	.9998
13	0	.8775	.5134	.2542	.0550	.0238	.0097	.0052	.0013	.0000
	1	.9928	.8746	.6214	.2337	.1267	.0637	.0386	.0126	.0017
	2	.9997	.9755	.8661	.5017	.3326	.2025	.1388	.0579	.0112
	3	1.0000	.9969	.9659	.7473	.5843	.4206	.3224	.1686	.0462
	4		.9997	.9936	.9009	.7940	.6543	.5521	.3531	.1334
	5		1.0000	.9991	.9700	.9198	.8346	.7587	.5744	.2905
	6			.9999	.9930	.9757	.9376	.8965	.7712	.5000
	7			1.0000	.9988	.9944	.9818	.9654	.9024	.7095
	8				.9998	.9990	.9960	.9912	.9679	.8666
	9				1.0000	.9999	.9994	.9984	.9922	.9539
	10					1.0000	.9999	.9998	.9987	.9888
	11						1.0000	1.0000	.9999	.9983
	12								1.0000	.9999

Contd.

Contd.

n	r	p .01	.05	.10	.20	.25	.30	.333	.40	.50
14	0	.8687	.4877	.2288	.0440	.0178	.0068	.0034	.0008	.0000
	1	.9916	.8470	.5847	.1979	.1010	.0475	.0274	.0081	.0009
	2	.9997	.9700	.8416	.4480	.2812	.1608	.1054	.0398	.0065
	3	1.0000	.9958	.9559	.6982	.5214	.3552	.2612	.1243	.0287
	4		.9996	.9908	.8702	.7416	.5842	.4755	.2793	.0898
	5		1.0000	.9986	.9562	.8884	.7805	.6898	.4859	.2120
	6			.9998	.9884	.9618	.9067	.8506	.6925	.3953
	7			1.0000	.9976	.9897	.9686	.9424	.8499	.6048
	8				.9996	.9979	.9917	.9826	.9417	.7880
	9				1.0000	.9997	.9984	.9960	.9825	.9102
	10					1.0000	.9998	.9993	.9961	.9713
	11						1.0000	.9999	.9994	.9936
	12							1.0000	.9999	.9991
	13									.9999
15	0	.8601	.4633	.2059	.0352	.0134	.0048	.0023	.0005	.0000
	1	.9904	.8291	.5491	.1672	.0802	.0353	.0194	.0052	.0005
	2	.9996	.9638	.8160	.3980	.2361	.1268	.0794	.0271	.0037
	3	1.0000	.9946	.9444	.6482	.4613	.2969	.2092	.0905	.0176
	4		.9994	.9873	.8358	.6865	.5255	.4041	.2173	.0592
	5		1.0000	.9978	.9390	.8516	.7216	.6184	.4032	.1509
	6			.9997	.9820	.9434	.8689	.7970	.6098	.3036
	7			1.0000	.9958	.9827	.9500	.9118	.7869	.5000
	8				.9992	.9958	.9848	.9692	.9050	.6964
	9				.9999	.9992	.9964	.9915	.9662	.8491
	10				1.0000	.9999	.9993	.9982	.9907	.9408
	11					1.0000	.9999	.9997	.9981	.9824
	12						1.0000	1.0000	.9997	.9963
	13								1.0000	.9995
	14									1.0000

Source: For $n = 2$ through 10, E. Parzen, *Modern Probability Theory and Its Applications,* John Wiley, New York, 1962. For $n = 11$ through 15, *Tables of Cumulative Binomial Probability Distribution,* Harvard University Press, Cambridge, Mass., 1955.

Table A.3 Ordinates of the Normal Density Function

$$\phi(x) = \frac{1}{\sqrt{2\pi}} e^{-x^2/2}$$

X	.00	.01	.02	.03	.04	.05	.06	.07	.08	.09
.0	.3989	.3989	.3989	.3988	.3986	.3984	.3982	.3980	.3977	.3973
.1	.3970	.3965	.3961	.3956	.3951	.3945	.3939	.3932	.3925	.3918
.2	.3910	.3902	.3894	.3885	.3876	.3867	.3857	.3847	.3836	.3825
.3	.3814	.3802	.3790	.3778	.3765	.3752	.3739	.3725	.3712	.3697
.4	.3683	.3668	.3653	.3637	.3621	.3605	.3589	.3572	.3555	.3538
.5	.3521	.3503	.3485	.3467	.3448	.3429	.3410	.3391	.3372	.3352
.6	.3332	.3312	.3292	.3271	.3251	.3230	.3209	.3187	.3166	.3144
.7	.3123	.3101	.3079	.3056	.3034	.3011	.2989	.2966	.2943	.2920
.8	.2897	.2874	.2850	.2827	.2803	.2780	.2756	.2732	.2709	.2685
.9	.2661	.2637	.2613	.2589	.2565	.2541	.2516	.2492	.2468	.2444
1.0	.2420	.2396	.2371	.2347	.2323	.2299	.2275	.2251	.2227	.2203
1.1	.2179	.2155	.2131	.2107	.2083	.2059	.2036	.2012	.1989	.1965
1.2	.1942	.1919	.1895	.1872	.1849	.1826	.1804	.1781	.1758	.1736
1.3	.1714	.1691	.1669	.1647	.1626	.1604	.1582	.1561	.1539	.1518
1.4	.1497	.1476	.1456	.1435	.1415	.1394	.1374	.1354	.1334	.1315
1.5	.1295	.1276	.1257	.1238	.1219	.1200	.1182	.1163	.1145	.1127
1.6	.1109	.1092	.1074	.1057	.1040	.1023	.1006	.0989	.0973	.0957
1.7	.0940	.0925	.0909	.0893	.0878	.0863	.0848	.0833	.0818	.0804
1.8	.0790	.0775	.0761	.0748	.0734	.0721	.0707	.0694	.0681	.0669
1.9	.0656	.0644	.0632	.0620	.0608	.0596	.0584	.0573	.0562	.0551
2.0	.0540	.0529	.0519	.0508	.0498	.0488	.0478	.0468	.0459	.0449
2.1	.0440	.0431	.0422	.0413	.0404	.0396	.0387	.0379	.0371	.0363
2.2	.0355	.0347	.0339	.0332	.0325	.0317	.0310	.0303	.0297	.0290
2.3	.0283	.0277	.0270	.0264	.0258	.0252	.0246	.0241	.0235	.0229
2.4	.0224	.0219	.0213	.0208	.0203	.0198	.0194	.0189	.0184	.0180
2.5	.0175	.0171	.0167	.0163	.0158	.0154	.0151	.0147	.0143	.0139
2.6	.0136	.0132	.0129	.0126	.0122	.0119	.0116	.0113	.0110	.0107
2.7	.0104	.0101	.0099	.0096	.0093	.0091	.0088	.0086	.0084	.0081
2.8	.0079	.0077	.0075	.0073	.0071	.0069	.0067	.0065	.0063	.0061
2.9	.0060	.0058	.0056	.0055	.0053	.0051	.0050	.0048	.0047	.0046
3.0	.0044	.0043	.0042	.0040	.0039	.0038	.0037	.0036	.0035	.0034
3.1	.0033	.0032	.0031	.0030	.0029	.0028	.0027	.0026	.0025	.0025
3.2	.0024	.0023	.0022	.0022	.0021	.0020	.0020	.0019	.0018	.0018
3.3	.0017	.0017	.0016	.0016	.0015	.0015	.0014	.0014	.0013	.0013
3.4	.0012	.0012	.0012	.0011	.0011	.0010	.0010	.0010	.0009	.0009
3.5	.0009	.0008	.0008	.0008	.0008	.0007	.0007	.0007	.0007	.0006
3.6	.0006	.0006	.0006	.0005	.0005	.0005	.0005	.0005	.0005	.0004
3.7	.0004	.0004	.0004	.0004	.0004	.0004	.0003	.0003	.0003	.0003
3.8	.0003	.0003	.0003	.0003	.0003	.0002	.0002	.0002	.0002	.0002
3.9	.0002	.0002	.0002	.0002	.0002	.0002	.0002	.0002	.0001	.0001

Source: Fisher, R. A. (1953): Statistical Tables for Biological, Agricultural and Medical Research, Oliver an
Boyd, London.

Table A.4 Cumulative Normal Distribution

$$\Phi(x) = \int_{-\infty}^{x} \frac{1}{\sqrt{2\pi}} e^{-t^2/2}\, dt$$

X	.00	.01	.02	.03	.04	.05	.06	.07	.08	.09
.0	.5000	.5040	.5080	.5120	.5160	.5199	.5239	.5279	.5319	.5359
.1	.5398	.5438	.5478	.5517	.5557	.5596	.5636	.5675	.5714	.5753
.2	.5793	.5832	.5871	.5910	.5948	.5987	.6026	.6064	.6103	.6141
.3	.6179	.6217	.6255	.6293	.6331	.6368	.6406	.6443	.6480	.6517
.4	.6554	.6591	.6628	.6664	.6700	.6736	.6772	.6808	.6844	.6879
.5	.6915	.6950	.6985	.7019	.7054	.7088	.7123	.7157	.7190	.7224
.6	.7257	.7291	.7324	.7357	.7389	.7422	.7454	.7486	.7517	.7549
.7	.7580	.7611	.7642	.7673	.7704	.7734	.7764	.7794	.7823	.7852
.8	.7881	.7910	.7939	.7967	.7995	.8023	.8051	.8078	.8106	.8133
.9	.8159	.8186	.8212	.8238	.8264	.8289	.8315	.8340	.8365	.8389
1.0	.8413	.8438	.8461	.8485	.8508	.8531	.8554	.8577	.8599	.8621
1.1	.8643	.8665	.8686	.8708	.8729	.8749	.8770	.8790	.8810	.8830
1.2	.8849	.8869	.8888	.8907	.8925	.8944	.8962	.8980	.8997	.9015
1.3	.9032	.9049	.9066	.9082	.9099	.9115	.9131	.9147	.9162	.9177
1.4	.9192	.9207	.9222	.9236	.9251	.9265	.9279	.9292	.9306	.9319
1.5	.9332	.9345	.9357	.9370	.9382	.9394	.9406	.9418	.9429	.9441
1.6	.9452	.9463	.9474	.9484	.9495	.9505	.9515	.9525	.9535	.9545
1.7	.9554	.9564	.9573	.9582	.9591	.9599	.9608	.9616	.9625	.9633
1.8	.9641	.9649	.9656	.9664	.9671	.9678	.9686	.9693	.9699	.9706
1.9	.9713	.9719	.9726	.9732	.9738	.9744	.9750	.9756	.9761	.9767
2.0	.9772	.9778	.9783	.9788	.9793	.9798	.9803	.9808	.9812	.9817
2.1	.9821	.9826	.9830	.9834	.9838	.9842	.9846	.9850	.9854	.9857
2.2	.9861	.9864	.9868	.9871	.9875	.9878	.9881	.9884	.9887	.9890
2.3	.9893	.9896	.9898	.9901	.9904	.9906	.9909	.9911	.9913	.9916
2.4	.9918	.9920	.9922	.9925	.9927	.9929	.9931	.9932	.9934	.9936
2.5	.9938	.9940	.9941	.9943	.9945	.9946	.9948	.9949	.9951	.9952
2.6	.9953	.9955	.9956	.9957	.9959	.9960	.9961	.9962	.9963	.9964
2.7	.9965	.9966	.9967	.9968	.9969	.9970	.9971	.9972	.9973	.9974
2.8	.9974	.9975	.9976	.9977	.9977	.9978	.9979	.9979	.9980	.9981
2.9	.9981	.9982	.9982	.9983	.9984	.9984	.9985	.9985	.9986	.9986
3.0	.9987	.9987	.9987	.9988	.9988	.9989	.9989	.9989	.9990	.9990
3.1	.9990	.9991	.9991	.9991	.9992	.9992	.9992	.9992	.9993	.9993
3.2	.9993	.9993	.9994	.9994	.9994	.9994	.9994	.9995	.9995	.9995
3.3	.9995	.9995	.9995	.9996	.9996	.9996	.9996	.9996	.9996	.9997
3.4	.9997	.9997	.9997	.9997	.9997	.9997	.9997	.9997	.9997	.9998

x	1.282	1.645	1.960	2.326	2.576	3.090	3.291	3.891	4.417
$\phi(x)$	.90	.95	.975	.99	.995	.999	.9995	.99995	.999995
$2[1-\phi(x)]$	.20	.10	.05	.02	.01	.002	.001	.0001	.000001

Source: Fisher, R. A. (1953): Statistical Tables for Biological, Agricultural and Medical Research, Oliver and Boyd, London.

Table A.5 Critical Values under Chi-Square Distribution

For degrees of freedom greater than 30, the expression $\sqrt{2}\chi^2 - \sqrt{(2n-1)}$ may be used as a normal deviate with unit variance, where n is the number of degrees of freedom.

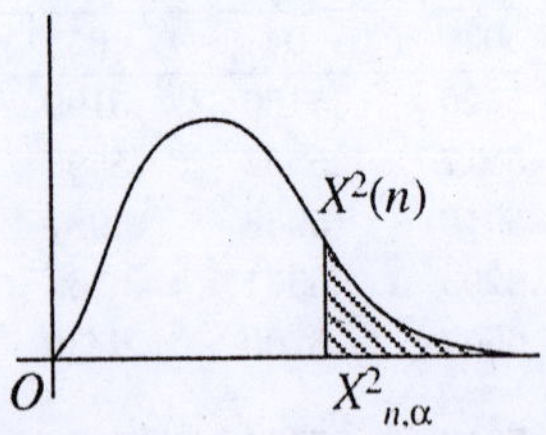

Degrees of freedom	$\alpha = 0.99$	0.98	0.95	0.90	0.80	0.70	0.50	0.30	0.20	0.10	0.05	0.02	0.01
1	0.000157	0.000628	0.00393	0.0158	0.0642	0.148	0.455	1.074	1.642	2.706	3.841	5.412	6.635
2	0.0201	0.0404	0.103	0.211	0.446	0.713	1.386	2.408	3.219	4.605	5.991	7.824	9.210
3	0.115	0.185	0.352	0.584	1.005	1.424	2.366	3.665	4.642	6.251	7.815	9.837	11.341
4	0.297	0.429	0.711	1.064	1.649	2.195	3.357	4.878	5.989	7.779	9.488	11.668	13.277
5	0.554	0.752	1.145	1.610	2.343	3.000	4.351	6.064	7.289	9.236	11.070	13.388	15.086
6	0.872	1.134	1.635	2.204	3.070	3.828	5.348	7.231	8.558	10.645	12.592	15.033	16.812
7	1.239	1.564	2.167	2.833	3.822	4.671	6.346	8.383	9.803	12.017	14.067	16.622	18.475
8	1.646	2.032	2.733	3.490	4.594	5.527	7.344	9.524	11.030	13.362	15.507	18.168	20.090
9	2.088	2.532	3.325	4.168	5.380	6.393	8.343	10.656	12.242	14.684	16.919	19.679	21.666
10	2.558	3.059	3.940	4.865	6.179	7.267	9.342	11.781	13.442	15.987	18.307	21.161	23.209
11	3.053	3.609	4.575	5.578	6.989	8.148	10.341	12.899	14.631	17.275	19.675	22.618	24.725
12	3.571	4.178	5.226	6.304	7.807	9.034	11.340	14.011	15.812	18.549	21.026	24.054	26.217
13	4.107	4.765	5.892	7.042	8.634	9.926	12.340	15.119	16.985	19.812	22.362	25.472	27.688
14	4.660	5.368	6.571	7.790	9.467	10.821	13.339	16.222	18.151	21.064	23.685	26.873	29.141
15	5.229	5.985	7.261	8.547	10.307	11.721	14.339	17.322	19.311	22.307	24.996	28.259	30.578
16	5.812	6.614	7.962	9.312	11.152	12.624	15.338	18.418	20.465	23.542	26.296	29.633	32.000
17	6.408	7.255	8.672	10.085	12.002	13.531	16.338	19.511	21.615	24.669	27.587	30.995	33.409
18	7.015	7.906	9.390	10.865	12.857	14.440	17.338	20.601	22.760	25.989	28.869	32.346	34.805
19	7.633	8.567	10.117	11.651	13.716	15.352	18.338	21.689	23.900	27.204	30.144	33.687	36.191
20	8.260	9.237	10.851	12.443	14.578	16.266	19.337	22.775	25.038	28.412	31.410	35.020	37.566
21	8.897	9.915	11.591	13.240	15.445	17.182	20.337	23.858	27.171	29.615	32.671	36.343	38.932
22	9.542	10.600	12.338	14.041	16.314	18.101	21.337	24.939	27.301	30.813	33.924	37.659	40.289
23	10.196	11.293	13.091	14.848	17.187	19.021	22.337	26.018	28.429	32.007	35.172	38.968	41.638
24	10.856	11.992	13.848	15.659	18.062	19.943	23.337	27.096	29.553	33.196	36.415	40.270	42.980
25	11.524	12.697	14.611	16.473	18.940	20.867	24.337	28.172	30.675	34.382	37.652	41.566	44.314
26	12.198	13.409	15.379	17.292	19.820	21.792	25.336	29.246	31.795	35.563	38.885	42.856	45.642
27	12.879	14.125	16.151	18.114	20.703	22.719	26.336	30.319	32.912	36.741	40.113	44.140	46.963
28	13.565	14.847	16.928	18.939	21.588	23.647	27.336	31.391	34.027	37.916	41.337	45.419	48.278
29	14.256	15.574	17.708	19.768	22.475	24.577	28.336	32.461	35.139	39.087	42.557	46.693	49.588
30	14.953	16.306	18.493	20.599	23.364	25.508	29.336	33.530	36.250	40.256	43.773	47.962	50.892

Source: R. A. Fisher, *Statistical Methods for Research Workers,* 14th Ed., Hafner Press, 1972.

Table A.6 Student's t-Distribution

The first column lists the number of degrees of freedom (n). The headings of the other columns give probabilities (α) for t to exceed the entry value, use symmetry for negative t values.

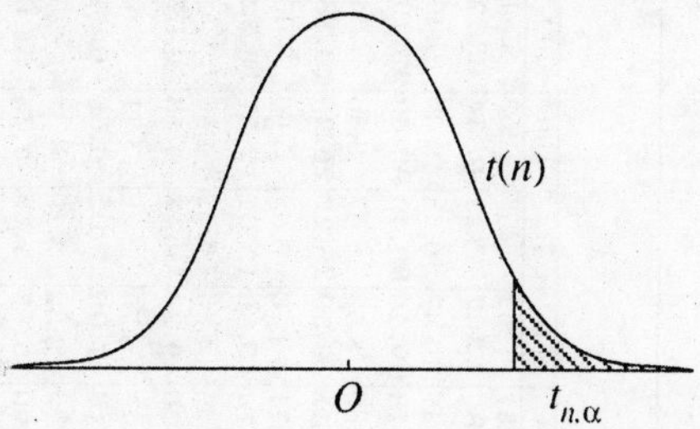

n \ α	.10	.05	0.025	.01	.005
1	3.078	6.314	12.706	31.821	63.657
2	1.886	2.920	4.303	6.965	9.925
3	1.638	2.353	3.182	4.541	5.841
4	1.533	2.132	2.776	3.747	4.604
5	1.476	2.015	2.571	3.365	4.032
6	1.440	1.943	2.447	3.143	3.707
7	1.415	1.895	2.365	2.998	3.499
8	1.397	1.860	2.306	2.896	3.355
9	1.383	1.833	2.262	2.821	3.250
10	1.372	1.812	2.228	2.764	3.169
11	1.363	1.796	2.201	2.718	3.106
12	1.356	1.782	2.179	2.681	3.055
13	1.350	1.771	2.160	2.650	3.012
14	1.345	1.761	2.145	2.624	2.977
15	1.341	1.753	2.131	2.602	2.947
16	1.337	1.746	2.120	2.583	2.921
17	1.333	1.740	2.110	2.567	2.898
18	1.330	1.734	2.101	2.552	2.878
19	1.328	1.729	2.093	2.539	2.861
20	1.325	1.725	2.086	2.528	2.845
21	1.323	1.721	2.080	2.518	2.831
22	1.321	1.717	2.074	2.508	2.819
23	1.319	1.714	2.069	2.500	2.807
24	1.318	1.711	2.064	2.492	2.797
25	1.316	1.708	2.060	2.485	2.787
26	1.315	1.706	2.056	2.479	2.779
27	1.314	1.703	2.052	2.473	2.771
28	1,313	1.701	2.048	2.467	2.763
29	1.311	1.699	2.045	2.462	2.756
30	1.310	1.697	2.042	2.457	2.750
40	1.303	1.684	2.021	2.423	2.704
60	1.296	1.671	2.000	2.390	2.660
120	1.289	1.658	1.980	2.358	2.617
∞	1.282	1.645	1.960	2.326	2.576

Source: P. G. Hoel. *Introduction to Mathematical Statistics,* 4th Ed., John Wiley, New York, 1971, page 393.

Table A.7 *F*-Distribution : 5% (Lightface Type) and 1% (Boldface Type) Points for the Distribution of *F*

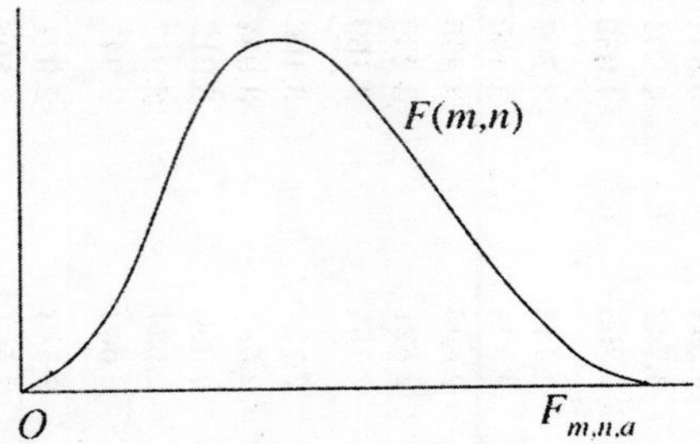

Degree of freedom for denominator (n)	Degrees of freedom for numerator (m): 1	2	3	4	5	6	7	8	9	10	11	12	14	16	20	24	30	40	50	75	100	200	500	∞
1	161	200	216	225	230	234	237	239	241	242	243	244	245	246	248	249	250	251	252	253	253	254	254	254
	4052	**4999**	**5403**	**5625**	**5764**	**5859**	**5928**	**5981**	**6022**	**6056**	**6082**	**6106**	**6142**	**6169**	**6208**	**6234**	**6258**	**6286**	**6302**	**6323**	**6334**	**6352**	**6361**	**6366**
2	18.51	19.00	19.16	19.25	19.30	19.33	19.36	19.37	19.38	19.39	19.40	19.41	19.42	19.43	19.44	19.45	19.46	19.47	19.47	19.48	19.49	19.49	19.50	19.50
	98.49	**99.01**	**99.17**	**99.25**	**99.30**	**99.33**	**99.34**	**99.36**	**99.38**	**99.40**	**99.41**	**99.42**	**99.43**	**99.44**	**99.45**	**99.46**	**99.47**	**99.48**	**99.48**	**99.49**	**99.49**	**99.49**	**99.50**	**99.50**
3	10.13	9.55	9.28	9.12	9.01	8.94	8.88	8.84	8.81	8.78	8.76	8.74	8.71	8.69	8.66	8.64	8.62	8.60	8.58	8.57	8.56	8.54	8.54	8.53
	34.12	**30.81**	**29.46**	**28.71**	**28.24**	**27.91**	**27.67**	**27.49**	**27.34**	**27.23**	**27.13**	**27.05**	**26.92**	**26.83**	**26.69**	**26.60**	**26.50**	**26.41**	**26.30**	**26.27**	**26.23**	**26.18**	**26.14**	**26.12**
4	7.71	6.94	6.59	6.39	6.26	6.16	6.09	6.04	6.00	5.96	5.93	5.91	5.87	5.84	5.80	5.77	5.74	5.71	5.70	5.68	5.66	5.65	5.64	5.63
	21.20	**18.00**	**16.69**	**15.98**	**15.52**	**15.21**	**14.98**	**14.80**	**14.66**	**14.54**	**14.45**	**14.37**	**14.24**	**14.15**	**14.02**	**13.93**	**13.83**	**13.74**	**13.69**	**13.61**	**13.57**	**13.52**	**13.48**	**13.46**
5	6.61	5.79	5.41	5.19	5.05	4.95	4.88	4.82	4.78	4.74	4.70	4.68	4.64	4.60	4.56	4.53	4.50	4.46	4.44	4.42	4.40	4.38	4.37	4.36
	16.26	**13.27**	**12.06**	**11.39**	**10.97**	**10.67**	**10.45**	**10.27**	**10.15**	**10.05**	**9.96**	**9.89**	**9.77**	**9.68**	**9.55**	**9.47**	**9.38**	**9.29**	**9.24**	**9.17**	**9.13**	**9.07**	**9.04**	**9.02**
6	5.99	5.14	4.76	4.53	4.39	4.28	4.21	4.15	4.10	4.06	4.03	4.00	3.96	3.92	3.87	3.84	3.81	3.77	3.75	3.72	3.71	3.69	3.68	3.67
	13.74	**10.92**	**9.78**	**9.15**	**8.75**	**8.47**	**8.26**	**8.10**	**7.98**	**7.87**	**7.79**	**7.72**	**7.60**	**7.52**	**7.39**	**7.31**	**7.23**	**7.14**	**7.09**	**7.02**	**6.99**	**6.94**	**6.90**	**6.88**
7	5.59	4.74	4.35	4.12	3.97	3.87	3.79	3.73	3.68	3.63	3.60	3.57	3.52	3.49	3.44	3.41	3.38	3.34	3.32	3.29	3.28	3.25	3.24	3.23
	12.25	**9.55**	**8.45**	**7.85**	**7.46**	**7.19**	**7.00**	**6.84**	**6.71**	**6.62**	**6.54**	**6.47**	**6.35**	**6.27**	**6.15**	**6.07**	**5.98**	**5.90**	**5.85**	**5.78**	**5.75**	**5.70**	**5.67**	**5.65**
8	5.32	4.46	4.07	3.84	3.69	3.58	3.50	3.44	3.39	3.34	3.31	3.28	3.23	3.20	3.15	3.12	3.08	3.05	3.03	3.00	2.98	2.96	2.94	2.93
	11.26	**8.65**	**7.59**	**7.01**	**6.63**	**6.37**	**6.19**	**6.03**	**5.91**	**5.82**	**5.74**	**5.67**	**5.56**	**5.48**	**5.36**	**5.28**	**5.20**	**5.11**	**5.06**	**5.00**	**4.96**	**4.91**	**4.88**	**4.86**
9	5.12	4.26	3.86	3.63	3.48	3.37	3.29	3.23	3.18	3.13	3.10	3.07	3.02	2.98	2.93	2.90	2.86	2.82	2.80	2.77	2.76	2.73	2.72	2.71
	10.56	**8.02**	**6.99**	**6.42**	**6.06**	**5.80**	**5.62**	**5.47**	**5.35**	**5.26**	**5.18**	**5.11**	**5.00**	**4.92**	**4.80**	**4.73**	**4.64**	**4.56**	**4.51**	**4.45**	**4.41**	**4.36**	**4.33**	**4.31**
10	4.96	4.10	3.71	3.48	3.33	3.22	3.14	3.07	3.02	2.97	2.94	2.91	2.86	2.82	2.77	2.74	2.70	2.67	2.64	2.61	2.59	2.56	2.55	2.54
	10.04	**7.56**	**6.55**	**5.99**	**5.64**	**5.39**	**5.21**	**5.06**	**4.95**	**4.85**	**4.78**	**4.71**	**4.60**	**4.52**	**4.41**	**4.33**	**4.25**	**4.17**	**4.12**	**4.05**	**4.01**	**3.96**	**3.93**	**3.91**
11	4.84	3.98	3.59	3.36	3.20	3.09	3.01	2.95	2.90	2.86	2.82	2.79	2.74	2.70	2.65	2.61	2.57	2.53	2.50	2.47	2.45	2.42	2.41	2.40
	9.65	**7.20**	**6.22**	**5.67**	**5.32**	**5.07**	**4.88**	**4.74**	**4.63**	**4.54**	**4.46**	**4.40**	**4.29**	**4.21**	**4.10**	**4.02**	**3.94**	**3.86**	**3.80**	**3.74**	**3.70**	**3.66**	**3.62**	**3.60**
12	4.75	3.88	3.49	3.26	3.11	3.00	2.92	2.85	2.80	2.76	2.72	2.69	2.64	2.60	2.54	2.50	2.46	2.42	2.40	2.36	2.35	2.32	2.31	2.30
	9.33	**6.93**	**5.95**	**5.41**	**5.06**	**4.82**	**4.65**	**4.50**	**4.39**	**4.30**	**4.22**	**4.16**	**4.05**	**3.98**	**3.86**	**3.78**	**3.70**	**3.61**	**3.56**	**3.49**	**3.46**	**3.41**	**3.38**	**3.36**
13	4.67	3.80	3.41	3.18	3.02	2.92	2.84	2.77	2.72	2.67	2.63	2.60	2.55	2.51	2.46	2.42	2.38	2.34	2.32	2.28	2.26	2.24	2.22	2.21
	9.07	**6.70**	**5.74**	**5.20**	**4.86**	**4.62**	**4.44**	**4.30**	**4.19**	**4.10**	**4.02**	**3.96**	**3.85**	**3.78**	**3.67**	**3.59**	**3.51**	**3.42**	**3.37**	**3.30**	**3.27**	**3.21**	**3.18**	**3.16**
14	4.60	3.74	3.34	3.11	2.96	2.85	2.77	2.70	2.65	2.60	2.56	2.53	2.48	2.44	2.39	2.35	2.31	2.27	2.24	2.21	2.19	2.16	2.14	2.13
	8.86	**6.51**	**5.56**	**5.03**	**4.69**	**4.46**	**4.28**	**4.14**	**4.03**	**3.94**	**3.86**	**3.80**	**3.70**	**3.62**	**3.51**	**3.43**	**3.34**	**3.26**	**3.21**	**3.14**	**3.11**	**3.06**	**3.02**	**3.00**

Contd.

Degree of freedom for denominator (n)	Degrees of freedom for numerator (m)																							
	1	2	3	4	5	6	7	8	9	10	11	12	14	16	20	24	30	40	50	75	100	200	500	∞
15	4.54	3.68	3.29	3.06	2.90	2.79	2.70	2.64	2.59	2.55	2.51	2.48	2.43	2.39	2.33	2.29	2.25	2.21	2.18	2.15	2.12	2.10	2.08	2.07
	8.68	**6.36**	**5.42**	**4.89**	**4.56**	**4.32**	**4.14**	**4.00**	**3.89**	**3.80**	**3.73**	**3.67**	**3.56**	**3.48**	**3.36**	**3.29**	**3.20**	**3.12**	**3.07**	**3.00**	**2.97**	**2.92**	**2.89**	**2.87**
16	4.49	3.63	3.24	3.01	2.85	2.74	2.66	2.59	2.54	2.49	2.45	2.42	2.37	2.33	2.28	2.24	2.20	2.16	2.13	2.09	2.07	2.04	2.02	2.01
	8.53	**6.23**	**5.29**	**4.77**	**4.44**	**4.20**	**4.03**	**3.89**	**3.78**	**3.69**	**3.61**	**3.55**	**3.45**	**3.37**	**3.25**	**3.18**	**3.10**	**3.01**	**2.96**	**2.89**	**2.86**	**2.80**	**2.77**	**2.75**
17	4.45	3.59	3.20	2.96	2.81	2.70	2.62	2.55	2.50	2.45	2.41	2.38	2.33	2.29	2.23	2.19	2.15	2.11	2.08	2.04	2.02	1.99	1.97	1.96
	8.40	**6.11**	**5.18**	**4.67**	**4.34**	**4.10**	**3.93**	**3.79**	**3.68**	**3.59**	**3.52**	**3.45**	**3.35**	**3.27**	**3.16**	**3.08**	**3.00**	2.92	**2.86**	**2.79**	**2.76**	**2.70**	**2.67**	**2.65**
18	4.41	3.55	3.16	2.93	2.77	2.66	2.58	2.51	2.46	2.41	2.37	2.34	2.29	2.25	2.19	2.15	2.11	2.07	2.04	2.00	1.98	1.95	1.93	1.92
	8.28	**6.01**	**5.09**	**4.58**	**4.25**	**4.01**	**3.85**	**3.71**	**3.60**	**3.51**	**3.44**	**3.37**	**3.27**	**3.19**	**3.07**	**3.00**	**2.91**	**2.83**	**2.78**	**2.71**	**2.68**	**2.62**	**2.59**	**2.57**
19	4.38	3.52	3.13	2.90	2.74	2.63	2.55	2.48	2.43	2.38	2.34	2.31	2.26	2.21	2.15	2.11	2.07	2.02	2.00	1.96	1.94	1.91	1.90	1.88
	8.18	**5.93**	**5.01**	**4.50**	**4.17**	**3.94**	**3.77**	**3.63**	**3.52**	**3.43**	**3.36**	**3.30**	**3.19**	**3.12**	**3.00**	**2.92**	**2.84**	**2.76**	**2.70**	**2.63**	**2.60**	**2.54**	**2.51**	**2.49**
20	4.35	3.49	3.10	2.87	2.71	2.60	2.52	2.45	2.40	2.35	2.31	2.28	2.23	2.18	2.12	2.08	2.04	1.99	1.96	1.92	1.90	1.87	1.85	1.84
	8.10	**5.85**	**4.94**	**4.43**	**4.10**	**3.87**	**3.71**	**3.56**	**3.45**	**3.37**	**3.30**	**3.23**	**3.13**	**3.05**	**2.94**	**2.86**	**2.77**	**2.69**	**2.63**	**2.56**	**2.53**	**2.47**	**2.44**	**2.42**
21	4.32	3.47	3.07	2.84	2.68	2.57	2.49	2.42	2.37	2.32	2.28	2.25	2.20	2.15	2.09	2.05	2.00	1.96	1.93	1.89	1.87	1.84	1.82	1.81
	8.02	**5.78**	**4.87**	**4.37**	**4.04**	**3.81**	**3.65**	**3.51**	**3.40**	**3.31**	**3.24**	**3.17**	**3.07**	**2.99**	**2.88**	**2.80**	**2.72**	**2.63**	**2.58**	**2.51**	**2.47**	**2.42**	**2.38**	**2.36**
22	4.30	3.44	3.05	2.82	2.66	2.55	2.47	2.40	2.33	2.30	2.26	2.23	2.18	2.13	2.07	2.03	1.98	1.93	1.91	1.87	1.84	1.81	1.80	1.78
	7.94	**5.72**	**4.82**	**4.31**	**3.99**	**3.76**	**3.59**	**3.45**	**3.35**	**3.26**	**3.18**	**3.12**	**3.02**	**2.94**	**2.83**	**2.75**	**2.67**	**2.58**	**2.53**	**2.46**	**2.42**	**2.37**	**2.33**	**2.31**
23	4.28	3.42	3.03	2.80	2.64	2.53	2.45	2.38	2.32	2.28	2.24	2.20	2.14	2.10	2.04	2.00	1.96	1.91	1.88	1.84	1.82	1.79	1.77	1.76
	7.88	**5.66**	**4.76**	**4.26**	**3.94**	**3.71**	**3.54**	**3.41**	**3.30**	**3.21**	**3.14**	**3.07**	**2.97**	**2.89**	**2.78**	**2.70**	**2.62**	**2.53**	**2.48**	**2.41**	**2.37**	**2.32**	**2.28**	**2.26**
24	4.26	3.40	3.01	2.78	2.62	2.51	2.43	2.36	2.30	2.26	2.22	2.18	2.13	2.09	2.02	1.98	1.94	1.89	1.86	1.82	1.80	1.76	1.74	1.73
	7.82	**5.61**	**4.72**	**4.22**	**3.90**	**3.67**	**3.50**	**3.36**	**3.25**	**3.17**	**3.09**	**3.03**	**2.93**	**2.85**	**2.74**	**2.66**	**2.58**	**2.49**	**2.44**	**2.36**	**2.33**	**2.27**	**2.23**	**2.21**
25	4.24	3.38	2.99	2.76	2.60	2.49	2.41	2.34	2.28	2.24	2.20	2.16	2.11	2.06	2.00	1.96	1.92	1.87	1.84	1.80	1.77	1.74	1.72	1.71
	7.77	**5.57**	**4.68**	**4.18**	**3.86**	**3.63**	**3.46**	**3.32**	**3.21**	**3.13**	**3.05**	**2.99**	**2.89**	**2.81**	**2.70**	**2.62**	**2.54**	**2.45**	**2.40**	**2.32**	**2.29**	**2.23**	**2.19**	**2.17**
26	4.22	3.37	2.89	2.74	2.59	2.47	2.39	2.32	2.27	2.22	2.18	2.15	2.10	2.05	1.99	1.95	1.90	1.85	1.82	1.78	1.76	1.72	1.70	1.69
	7.72	**5.53**	**4.64**	**4.14**	**3.82**	**3.59**	**3.42**	**3.29**	**3.17**	**3.09**	**3.02**	**2.96**	**2.86**	**2.77**	**2.66**	**2.58**	**2.50**	**2.41**	**2.36**	**2.28**	**2.25**	**2.19**	**2.15**	**2.13**
27	4.21	3.35	2.96	2.73	2.57	2.46	2.37	2.30	2.25	2.20	2.16	2.13	2.08	2.03	1.97	1.93	1.88	1.84	1.80	1.76	1.74	1.71	1.68	1.67
	7.68	**5.49**	**4.60**	**4.11**	**3.79**	**3.56**	**3.39**	**3.26**	**3.14**	**3.06**	**2.98**	**2.93**	**2.83**	**2.74**	**2.63**	**2.55**	**2.47**	**2.38**	**2.33**	**2.25**	**2.21**	**2.16**	**2.12**	**2.10**
28	4.20	3.34	2.95	2.71	2.56	2.44	2.36	2.29	3.24	2.19	2.15	2.12	2.06	2.02	1.96	1.91	1.87	1.81	1.78	1.75	1.72	1.69	1.67	1.65
	7.64	**5.45**	**4.57**	**4.07**	**3.76**	**3.53**	**3.36**	**3.23**	**3.11**	**3.03**	**2.95**	**2.90**	**2.80**	**2.71**	**2.60**	**2.52**	**2.44**	**2.35**	**2.30**	**2.22**	**2.18**	**2.13**	**2.09**	**2.06**
29	4.18	3.33	2.93	2.70	2.54	3.43	2.35	2.28	2.22	2.18	2.14	2.10	2.05	2.00	1.94	1.90	1.85	1.80	1.77	1.73	1.71	1.68	1.65	1.64
	7.60	**5.52**	**4.54**	**4.04**	**3.73**	**3.50**	**3.33**	**3.20**	**3.08**	**3.00**	**2.92**	**2.87**	**2.77**	**2.68**	**2.57**	**2.49**	**2.41**	**2.32**	**2.27**	**2.19**	**2.15**	**2.10**	**2.06**	**2.03**
30	4.17	3.32	2.92	2.69	2.53	2.42	2.34	2.27	2.21	2.16	2.12	2.09	2.04	1.99	1.93	1.89	1.84	1.79	1.76	1.72	1.69	1.66	1.64	1.62
	7.56	**5.39**	**4.51**	**4.02**	**3.70**	**3.47**	**3.30**	**3.17**	**3.06**	**2.98**	**2.90**	**2.84**	**2.74**	**2.66**	**2.55**	**2.47**	**2.38**	**2.29**	**2.24**	**2.16**	**2.13**	**2.07**	**2.03**	**2.01**
32	4.15	3.30	2.90	2.67	2.51	2.40	2.32	2.25	2.19	2.14	2.10	2.07	2.02	1.97	1.91	1.86	1.82	1.76	1.74	1.69	1.67	1.64	1.61	1.59
	7.50	**5.34**	**4.46**	**3.97**	**3.66**	**3.42**	**3.25**	**3.12**	**3.01**	**2.94**	**2.86**	**2.80**	**2.70**	**2.62**	**2.51**	**2.42**	**2.34**	**2.25**	**2.20**	**2.12**	**2.08**	**2.02**	**1.98**	**1.96**
34	4.13	3.28	2.88	2.65	2.49	2.38	2.30	2.23	2.17	2.12	2.08	2.05	2.00	1.95	1.89	1.84	1.80	1.74	1.71	1.67	1.64	1.61	1.59	1.57
	7.44	**5.29**	**4.42**	**3.93**	**3.61**	**3.38**	**3.21**	**3.08**	**2.97**	**2.89**	**2.82**	**2.76**	**2.66**	**2.58**	**2.47**	**2.38**	**2.30**	**2.21**	**2.15**	**2.08**	**2.04**	**1.98**	**1.94**	**1.91**
36	4.11	3.26	2.86	2.63	2.48	2.36	2.28	2.21	2.15	2.10	2.06	2.03	1.89	1.93	1.87	1.82	1.78	1.72	1.69	1.65	1.62	1.59	1.56	1.55
	7.39	**5.25**	**4.38**	**3.89**	**3.58**	**3.35**	**3.18**	**3.04**	**2.94**	**2.86**	**2.78**	**2.72**	**2.62**	**2.54**	**2.43**	**2.35**	**2.26**	**2.17**	**2.12**	**2.04**	**2.00**	**1.94**	**1.90**	**1.87**
38	4.10	3.25	2.85	2.62	2.46	2.35	2.26	2.19	2.14	2.09	2.05	2.02	1.96	1.92	1.85	1.80	1.76	1.71	1.67	1.63	1.60	1.57	1.54	1.53
	7.35	**5.21**	**4.34**	**3.86**	**3.54**	**3.32**	**3.15**	**3.02**	**2.91**	**2.82**	**2.75**	**2.69**	**2.59**	**2.51**	**2.30**	**2.32**	**2.22**	**2.14**	**2.08**	**2.00**	**1.97**	**1.90**	**1.86**	**1.84**
40	4.08	3.23	2.84	2.61	2.45	2.34	2.25	2.18	2.12	2.07	2.04	2.00	1.95	1.90	1.84	1.79	1.74	1.69	1.66	1.61	1.59	1.55	1.53	1.51
	7.31	**5.18**	**4.31**	**3.83**	**3.51**	**3.29**	**3.12**	**2.99**	**2.88**	**2.80**	**2.73**	**2.66**	**2.56**	**2.49**	**2.37**	**2.29**	**2.20**	**2.11**	**2.05**	**1.97**	**1.94**	**1.88**	**1.84**	**1.81**

Contd.

Degree of freedom for denominator (n)	Degrees of freedom for numerator (m)																							
	1	2	3	4	5	6	7	8	9	10	11	12	14	16	20	24	30	40	50	75	100	200	500	∞
42	4.07	3.22	2.83	2.59	2.44	2.32	2.24	2.17	2.11	2.06	2.02	1.99	1.94	1.89	1.82	1.78	1.73	1.68	1.64	1.60	1.57	1.54	1.51	1.49
	7.27	**5.15**	**4.29**	**3.80**	**3.49**	**3.26**	**3.10**	**2.96**	**2.86**	**2.77**	**2.70**	**2.64**	**2.54**	**2.46**	**2.35**	**2.26**	**2.17**	**2.08**	**2.02**	**1.94**	**1.91**	**1.85**	**1.80**	**1.78**
44	4.06	3.21	2.82	2.58	2.43	2.31	2.23	2.16	2.10	2.05	2.01	1.98	1.92	1.88	1.81	1.76	1.72	1.66	1.63	1.58	1.56	1.52	1.50	1.48
	7.24	**5.12**	**4.26**	**3.78**	**3.46**	**3.24**	**3.07**	**2.94**	**2.84**	**2.75**	**2.68**	**2.62**	**2.52**	**2.44**	**2.32**	**2.24**	**2.15**	**2.06**	**2.00**	**1.92**	**1.88**	**1.82**	**1.78**	**1.75**
46	4.05	3.20	2.81	2.57	2.42	2.30	2.22	2.14	2.09	2.04	2.00	1.97	1.91	1.87	1.80	1.75	1.71	1.65	1.62	1.57	1.54	1.51	1.48	1.46
	7.21	**5.10**	**4.24**	**3.76**	**3.44**	**3.22**	**3.05**	**2.92**	**2.82**	**2.73**	**2.66**	**2.60**	**2.50**	**2.42**	**2.40**	**2.22**	**2.13**	**2.04**	**1.98**	**1.90**	**1.86**	**1.80**	**1.76**	**1.72**
48	4.04	3.19	2.80	2.56	2.41	2.30	2.21	2.14	2.08	2.03	1.99	1.96	1.90	1.86	1.79	1.74	1.70	1.64	1.61	1.56	1.53	1.50	1.47	1.45
	7.19	**5.08**	**4.22**	**3.74**	**3.42**	**3.20**	**3.04**	**2.90**	**2.80**	**2.71**	**2.64**	**2.58**	**2.48**	**2.40**	**2.28**	**2.20**	**2.11**	**2.02**	**1.96**	**1.88**	**1.84**	**1.78**	**1.73**	**1.70**
50	4.03	3.18	2.79	2.56	2.40	2.29	2.20	2.13	2.07	2.02	1.98	1.95	1.90	1.85	1.78	1.74	1.69	1.63	1.60	1.55	1.52	1.48	1.46	1.44
	7.17	**5.06**	**4.20**	**3.72**	**3.41**	**3.18**	**3.02**	**2.88**	**2.78**	**2.70**	**2.62**	**2.56**	**2.46**	**2.39**	**2.26**	**2.18**	**2.10**	**2.00**	**1.94**	**1.86**	**1.82**	**1.76**	**1.71**	**1.68**
55	4.02	3.17	2.78	2.54	2.38	2.27	2.18	2.11	2.05	2.00	1.97	1.93	1.88	1.83	1.76	1.72	1.67	1.61	1.58	1.52	1.50	1.46	1.43	1.41
	7.12	**5.01**	**4.16**	**3.68**	**3.37**	**3.15**	**2.98**	**2.85**	**2.75**	**2.66**	**2.59**	**2.53**	**2.43**	**2.35**	**2.23**	**2.15**	**2.06**	**1.96**	**1.90**	**1.82**	**1.78**	**1.71**	**1.66**	**1.64**
60	4.00	3.15	2.76	2.52	2.37	2.25	2.17	2.10	2.04	1.99	1.95	1.92	1.86	1.81	1.75	1.70	1.65	1.59	1.56	1.50	1.48	1.44	1.41	1.39
	7.08	**4.98**	**4.13**	**3.65**	**3.34**	**3.12**	**2.95**	**2.82**	**2.72**	**2.63**	**2.56**	**2.50**	**2.40**	**2.32**	**2.20**	**2.12**	**2.03**	**1.93**	**1.87**	**1.79**	**1.74**	**1.68**	**1.63**	**1.60**
65	3.99	3.14	2.75	2.51	2.36	2.24	2.15	2.08	2.02	1.98	1.94	1.90	1.85	1.80	1.73	1.68	1.63	1.57	1.54	1.49	1.46	1.42	1.39	1.37
	7.04	**4.95**	**4.10**	**3.62**	**3.31**	**3.09**	**2.93**	**2.79**	**2.70**	**2.61**	**2.54**	**2.47**	**2.30**	**2.37**	**2.18**	**2.09**	**2.00**	**1.90**	**1.84**	**1.76**	**1.71**	**1.64**	**1.60**	**1.56**
70	3.98	3.13	2.74	2.50	2.35	2.32	2.14	2.07	2.01	1.97	1.93	1.89	1.84	1.79	1.72	1.67	1.62	1.56	1.53	1.47	1.45	1.40	1.37	1.35
	7.01	**4.92**	**4.08**	**3.60**	**3.29**	**3.07**	**2.91**	**2.77**	**2.67**	**2.59**	**2.51**	**2.45**	**2.33**	**2.28**	**2.15**	**2.07**	**1.98**	**1.88**	**1.82**	**1.74**	**1.69**	**1.63**	**1.56**	**1.53**
80	3.96	3.11	2.72	2.48	2.33	2.21	2.12	2.05	1.99	1.95	1.91	1.88	1.82	1.77	1.70	1.65	1.60	1.54	1.51	1.45	1.42	1.38	1.35	1.32
	6.96	**4.88**	**4.04**	**3.56**	**3.25**	**3.04**	**2.87**	**2.74**	**2.64**	**2.55**	**2.48**	**2.41**	**2.32**	**2.24**	**2.11**	**2.03**	**1.94**	**1.84**	**1.78**	**1.70**	**1.65**	**1.57**	**1.52**	**1.49**
100	3.94	3.09	2.70	2.46	2.30	2.19	2.10	2.03	1.97	1.92	1.88	1.85	1.79	1.75	1.68	1.63	1.57	1.51	1.48	1.42	1.39	1.34	1.30	1.28
	6.90	**4.82**	**3.98**	**3.51**	**3.20**	**2.99**	**2.82**	**2.69**	**2.59**	**2.51**	**2.43**	**2.36**	**2.26**	**2.19**	**2.06**	**1.98**	**1.89**	**1.79**	**1.73**	**1.64**	**1.59**	**1.51**	**1.46**	**1.43**
125	3.92	3.07	2.68	2.44	2.29	2.17	2.08	2.01	1.95	1.90	1.86	1.83	1.77	1.72	1.65	1.60	1.55	1.49	1.45	1.39	1.36	1.31	1.27	1.25
	6.84	**4.78**	**3.94**	**3.47**	**3.17**	**2.95**	**2.79**	**2.65**	**2.56**	**2.47**	**2.40**	**2.33**	**2.23**	**2.15**	**2.03**	**1.94**	**1.85**	**1.75**	**1.68**	**1.59**	**1.54**	**1.46**	**1.40**	**1.37**
150	3.91	3.06	2.67	2.43	2.27	2.16	2.07	2.00	1.94	1.89	1.85	1.82	1.76	1.71	1.64	1.59	1.54	1.47	1.44	1.37	1.34	1.29	1.25	1.22
	6.81	**4.75**	**3.91**	**3.44**	**3.13**	**2.92**	**2.76**	**2.62**	**2.53**	**2.44**	**2.37**	**2.30**	**2.20**	**2.12**	**2.00**	**1.91**	**1.83**	**1.72**	**1.66**	**1.56**	**1.51**	**1.43**	**1.37**	**1.33**
200	3.89	3.04	2.65	2.41	2.26	2.14	2.05	1.98	1.92	1.87	1.83	1.80	1.74	1.69	1.62	1.57	1.52	1.45	1.42	1.35	1.32	1.26	1.22	1.19
	6.76	**4.71**	**3.88**	**3.41**	**3.11**	**2.90**	**2.73**	**2.60**	**2.50**	**2.41**	**2.34**	**2.28**	**2.17**	**2.09**	**1.97**	**1.88**	**1.79**	**1.69**	**1.62**	**1.53**	**1.48**	**1.39**	**1.33**	**1.28**
400	3.86	3.02	2.62	2.39	2.23	2.12	2.03	1.96	1.90	1.85	1.81	1.78	1.72	1.67	1.60	1.54	1.49	1.42	1.38	1.32	1.28	1.22	1.16	1.13
	6.70	**4.66**	**3.83**	**3.36**	**3.06**	**2.85**	**2.69**	**2.55**	**2.46**	**2.37**	**2.29**	**2.23**	**2.12**	**2.04**	**1.92**	**1.84**	**1.74**	**1.64**	**1.57**	**1.47**	**1.42**	**1.32**	**1.24**	**1.19**
1000	3.85	3.00	2.61	2.38	2.22	2.10	2.02	1.95	1.89	1.84	1.80	1.76	1.70	1.65	1.58	1.53	1.47	1.41	1.36	1.30	1.26	1.19	1.13	1.08
	6.66	**4.62**	**3.80**	**3.34**	**3.04**	**2.82**	**2.66**	**2.53**	**2.43**	**2.34**	**2.26**	**2.20**	**2.09**	**2.01**	**1.89**	**1.81**	**1.71**	**1.61**	**1.54**	**1.44**	**1.38**	**1.28**	**1.19**	**1.11**
	3.84	2.99	2.60	2.37	2.21	2.09	2.01	1.94	1.88	1.83	1.79	1.75	1.69	1.64	1.57	1.52	1.46	1.40	1.35	1.28	1.24	1.17	1.11	1.00
	6.64	**4.60**	**3.78**	**3.32**	**3.02**	**2.80**	**2.64**	**2.51**	**2.41**	**2.32**	**2.24**	**2.18**	**2.07**	**1.99**	**1.87**	**1.79**	**1.69**	**1.59**	**1.52**	**1.41**	**1.36**	**1.25**	**1.15**	**1.00**

Source: George W. Snedecor and William G. Cochran, *Statistical Methods*, 6th Ed., (c) 1967, Iowa State University Press, Ames, Iowa.

Table A.8 Number of Runs Up and Down Distribution

Entries in the table labelled P are the cumulative probability from each extreme to the value of V. Left-tail probabilities are given only for V less than or equal to its median; right-tail probabilities are given for all other values of V. N is the number of observations for which successive differences are non-zero, and hence one more than the number of symbols of signs of differences.

N	V	Left-tail P	V	Right-tail P	N	V	Left-tail P	V	Right-tail P	N	V	Left-tail P	V	Right-tail P
3	1	.3333	2	.6667	14	1	.0000				10	.2443	13	.3152
4			3	.4167		2	.0000				11	.4568	12	.5432
	1	.0833	2	.9167		3	.0000			19	1	.0000		
5	1	.0167	4	.2667		4	.0007	13	.0046		2	.0000		
	2	.2500	3	.7500		5	.0079	12	.0391		3	.0000		
6	1	.0028				6	.0441	11	.1536		4	.0000		
	2	.0861	5	.1694		7	.1534	10	.3722		5	.0000	18	.0005
	3	.4139	4	.5861		8	.3633	9	.6367		6	.0003	17	.0056
7	1	.0004	6	.1079	15	1	.0000				7	.0025	16	.0308
	2	.0250	5	.4417		2	.0000				8	.0137	15	.1055
	3	.1909	4	.8091		3	.0000				9	.0523	14	.2546
8	1	.0000				4	.0002				10	.1467	13	.4663
	2	.0063	7	.0687		5	.0027	14	.0029		11	.3144	12	.6856
	3	.0749	6	.3250		6	.0186	13	.0267	20	1	.0000		
	4	.3124	5	.6876		7	.0782	12	.1134		2	.0000		
9	1	.0000				8	.2216	11	.2970		3	.0000		
	2	.0014				9	.4520	10	.5480		4	.0000		
	3	.0257	8	.0437	16	1	.0000				5	.0000		
	4	.1500	7	.2347		2	.0000				6	.0001	19	.0003
	5	.4347	6	.5653		3	.0000				7	.0009	18	.0038
10	1	.0000				4	.0001	15	.0019		8	.0058	17	.0218
	2	.0003	9	.0278		5	.0009	14	.0182		9	.0255	16	.0793
	3	.0079	8	.1671		6	.0072	13	.0828		10	.0821	15	.2031
	4	.0633	7	.4524		7	.0367	12	.2335		11	.2012	14	.3945
	5	.2427	6	.7573		8	.1238	11	.4631		12	.3873	13	.6127
11	1	.0000				9	.2975	10	.7025	21	1	.0000		
	2	.0001			17	1	.0000				2	.0000		
	3	.0022	10	.0177		2	.0000				3	.0000		
	4	.0239	9	.1177		3	.0000				4	.0000		
	5	.1196	8	.3540		4	.0000				5	.0000		
	6	.3438	7	.6562		5	.0003	16	.0012		6	.0000		
12	1	.0000				6	.0026	15	.0123		7	.0003	20	.0002
	2	.0000				7	.0160	14	.0600		8	.0023	19	.0025
	3	.0005				8	.0638	13	.1812		9	.0117	18	.0154
	4	.0082	11	.0113		9	.1799	12	.3850		10	.0431	17	.0591
	5	.0529	10	.0821		10	.3770	11	.6230		11	.1202	16	.1602
	6	.1918	9	.2720	18	1	.0000				12	.2622	15	.3293
	7	.4453	8	.5547		2	.0000				13	.4603	14	.5397
13	1	.0000				3	.0000			22	1	.0000		
	2	.0000				4	.0000				2	.0000		
	3	.0001	12	.0072		5	.0001				3	.0000		
	4	.0026	11	.0568		6	.0009	17	.0008		4	.0000		
	5	.0213	10	.2058		7	.0065	16	.0083		5	.0000		
	6	.0964	9	.4587		8	.0306	15	.0431		6	.0000	21	.0001
	7	.2749	8	.7251		9	.1006	14	.1389		7	.0001	20	.0017

N	V	Left-tail P	V	Right-tail P
	8	.0009	19	.0108
	9	.0050	18	.0437
	10	.0213	17	.1251
	11	.0674	16	.2714
	12	.1661	15	.4688
	13	.3276	14	.6724
23	1	.0000		
	2	.0000		
	3	.0000		
	4	.0000		
	5	.0000		
	6	.0000		
	7	.0000	22	.0001
	8	.0003	21	.0011
	9	.0021	20	.0076
	10	.0099	19	.0321
	11	.0356	18	.0968
	12	.0988	17	.2211
	13	.2188	16	.4020
	14	.3953	15	.6047
24	1	.0000		
	2	.0000		
	3	.0000		
	4	.0000		
	5	.0000		
	6	.0000		
	7	.0000		
	8	.0001	23	.0000
	9	.0008	22	.0007
	10	.0044	21	.0053
	11	.0177	20	.0235
	12	.0554	19	.0742
	13	.1374	18	.1783
	14	.2768	17	.3405
	15	.4631	16	.5369
25	1	.0000		
	2	.0000		
	3	.0000		
	4	.0000		
	5	.0000		
	6	.0000		
	7	.0000	24	.0000
	8	.0000	23	.0005
	9	.0003	22	.0037
	10	.0018	21	.0170
	11	.0084	20	.0564
	12	.0294	19	.1423
	13	.0815	18	.2852
	14	.1827	17	.4708
	15	.3384	16	.6616

For $N > 25$, the probabilities are found from the table as follows:

$$z_L = \frac{V + 0.5 - (2N - 1)/3}{\sqrt{(16N - 29)/90}}; \quad z_R = \frac{V - 0.5 - (2N - 1)/3}{\sqrt{(16N - 29)/90}}.$$

Desired	**Approximated by**
Left-tail probability for V	Left-tail probability for z_L
Right-tail probability for V	Right-tail probability for z_R

Source: E. S. Edgington (1961), Probability table for number of runs of signs of first differences, *Journal of the American Statistical Association,* 56, 156–159.

Table A.9 Table of Probabilities associated with Values as small as observed values of x in the Binomial Test

Given in the body of this table are one-tailed probabilities under H_0 for the binomial test when $P = Q = \frac{1}{2}$. To save space, decimal points are omitted in the p's.

N \ x	0	1	2	3	4	5	6	7	8	9	10	11	12	13	14	15
5	031	188	500	812	969	†										
6	016	109	344	656	891	984	†									
7	008	062	227	500	773	938	992	†								
8	004	035	145	363	637	855	965	996	†							
9	002	020	090	251	500	746	910	980	998	†						
10	001	011	055	172	377	623	828	945	989	999	†					
11		006	033	113	274	500	726	887	967	994	†	†				
12		003	019	073	194	387	613	806	927	981	997	†	†			
13		002	011	046	133	291	500	709	867	954	989	998	†	†		
14		001	006	029	090	212	395	605	788	910	971	994	999	†	†	
15			004	018	059	151	304	500	696	849	941	982	996	†	†	†
16			002	011	038	105	227	402	598	773	895	962	989	998	†	†
17			001	006	025	072	166	315	500	685	834	928	975	994	999	†
18			001	004	015	048	119	240	407	593	760	881	952	985	996	999
19				002	010	032	084	180	324	500	676	820	916	968	990	998
20				001	006	021	058	132	252	412	588	748	868	942	979	994
21				001	004	013	039	095	192	332	500	668	808	905	961	987
22					002	008	026	067	143	262	416	584	738	857	933	974
23					001	005	017	047	105	202	339	500	661	798	895	953
24					001	003	011	032	076	154	271	419	581	729	846	924
25						002	007	022	054	115	212	345	500	655	788	885

Source: Table IV B, of Walker, Helen, and Lev, J. 1953. *Statistical Inference.* New York: Holt, p. 458.

† 1.0 or approximately 1.0.

Table A.10 Critical values of the Kolmogorov-Smirnov One Sample Test Statistics

This table gives the values of $D^+_{n,\alpha}$ and $D_{n,\alpha}$ for which $\alpha \geq P\{D^+_n > D^+_{n,\alpha}\}$ and $\alpha \geq P\{D_n > D_{n,\alpha}\}$ for some selected values of n and α.

One-Sided Test:											
$\alpha =$	.10	.05	.025	.01	.005	$\alpha =$	.10	.05	.025	.01	.005
Two-Sided Test:											
$\alpha =$	.20	.10	.05	.02	.01	$\alpha =$	.20	.10	.05	.02	.01
$n = 1$	.900	.950	.975	.990	.995	$n = 21$	.226	.259	.287	.321	.344
2	.684	.776	.842	.900	.929	22	.221	.253	.281	.314	.337
3	.565	.636	.708	.785	.829	23	.216	.247	.275	.307	.330
4	.493	.565	.624	.689	.734	24	.212	.242	.269	.301	.323
5	.447	.509	.563	.627	.669	25	.208	.238	.264	.295	.317
6	.410	.468	.519	.577	.617	26	.204	.233	.259	.290	.311
7	.381	.436	.483	.538	.576	27	.200	.229	.254	.284	.305
8	.358	.410	.454	.507	.542	28	.197	.225	.250	.279	.300
9	.339	.387	.430	.480	.513	29	.193	.221	.246	.275	.295
10	.323	.369	.409	.457	.489	30	.190	.218	.242	.270	.290
11	.308	.352	.391	.437	.468	31	.187	.214	.238	.266	.285
12	.296	.338	.375	.419	.449	32	.184	.211	.234	.262	.281
13	.285	.325	.361	.404	.432	33	.182	.208	.231	.258	.277
14	.275	.314	.349	.390	.418	34	.179	.205	.227	.254	.273
15	.266	.304	.338	.377	.404	35	.177	.202	.224	.251	.269
16	.258	.295	.327	.366	.392	36	.174	.199	.221	.247	.265
17	.250	.286	.318	.355	.381	37	.172	.196	.218	.244	.262
18	.244	.279	.309	.346	.371	38	.170	.194	.215	.241	.258
19	.237	.271	.301	.337	.361	39	.168	.191	.213	.238	.255
20	.232	.265	.294	.329	.352	40	.165	.189	.210	.235	.252
				Approximation for $n > 40$:			$\frac{1.07}{\sqrt{n}}$	$\frac{1.22}{\sqrt{n}}$	$\frac{1.36}{\sqrt{n}}$	$\frac{1.52}{\sqrt{n}}$	$\frac{1.63}{\sqrt{n}}$

Source: Table 1 of Leslie H. Miller, Table of Percentage points of Kolmogorov statistics, *J. Am. Stat. Assoc.* 5 (1956), 111–121.

Table A.11

Critical values of the Kolmogorov-Smirnov One Sample Test Statistics for Two Samples of Equal Size

This table gives the values of $D^+_{n,n,\alpha}$ and $D_{n,n,\alpha}$ for which $\alpha \geq P\{D^+_{n,n} > D^+_{n,n,\alpha}\}$ and $\alpha \geq P\{D_{n,n} > D_{n,n,\alpha}\}$ for some selected values of n and α.

One-Sided Test:											
$\alpha =$	.10	.05	.025	.01	.005	$\alpha =$	.10	.05	.025	.01	.005
Two-Sided Test:											
$\alpha =$	.20	.10	.05	.02	.01	$\alpha =$	.20	.10	.05	.02	.01
$n = 3$	2/3	2/3				$n = 20$	6/20	7/20	8/20	9/20	10/20
4	3/4	3/4	3/4			21	6/21	7/21	8/21	9/21	10/21
5	3/5	3/5	4/5	4/5	4/5	22	7/22	8/22	8/22	10/22	10/22
6	3/6	4/6	4/6	5/6	5/6	23	7/23	8/23	9/23	10/23	10/23
7	4/7	4/7	5/7	5/7	5/7	24	7/24	8/24	9/24	10/24	11/24
8	4/8	4/8	5/8	5/8	6/8	25	7/25	8/25	9/25	10/25	11/25
9	4/9	5/9	5/9	6/9	6/9	26	7/26	8/26	9/26	10/26	11/26
10	4/10	5/10	6/10	6/10	7/10	27	7/27	8/27	9/27	11/27	11/27
11	5/11	5/11	6/11	7/11	7/11	28	8/28	9/28	10/28	11/28	12/28
12	5/12	5/12	6/12	7/12	7/12	29	8/29	9/29	10/29	11/29	12/29
13	5/13	6/13	6/13	7/13	8/13	30	8/30	9/30	10/30	11/30	12/30
14	5/14	6/14	7/14	7/14	8/14	31	8/31	9/31	10/31	11/31	12/31
15	5/15	6/15	7/15	8/15	8/15	32	8/32	9/32	10/32	12/32	12/32
16	6/16	6/16	7/16	8/16	9/16	34	8/34	10/34	11/34	12/34	13/34
17	6/17	7/17	7/17	8/17	9/17	36	9/36	10/36	11/36	12/36	13/36
18	6/18	7/18	8/18	9/18	9/18	38	9/38	10/38	11/38	13/38	14/38
19	6/19	7/19	8/19	9/19	9/19	40	9/40	10/40	12/40	13/40	14/40
						Approximation for $n > 40$:	$\frac{1.52}{\sqrt{n}}$	$\frac{1.73}{\sqrt{n}}$	$\frac{1.92}{\sqrt{n}}$	$\frac{2.15}{\sqrt{n}}$	$\frac{2.30}{\sqrt{n}}$

Source: Table 2 and 3 of Z. W. Birnbaum and R. A. Hall, Small sample distributions for multisample statistics of the Smirnov type, *Ann. Math. Stat.* 31 (1960), 710–720.

Table A.12 Critical Values of the Kolmogorov-Smirnov Test Statistic for Two Samples of Unequal Size

This table gives the values of $D^+_{m,n,\alpha}$ and $D_{m,n,\alpha}$ for which $\alpha \geq P\{D^+_{m,n} > D^+_{m,n,\alpha}\}$ and $\alpha \geq P\{D_{m,n} > D_{m,n,\alpha}\}$ for some selected values of N_1 = smaller sample size, N_2 = larger sample size, and α.

One-Sided Test:						
	α =	.10	.05	.025	.01	.005
Two-Sided Test:						
	α =	.20	.10	.05	.02	.01
$N_1 = 1$	$N_2 = 9$	17/18				
	10	9/10				
$N_1 = 2$	$N_2 = 3$	5/6				
	4	3/4				
	5	4/5	4/5			
	6	5/6	5/6			
	7	5/7	6/7			
	8	3/4	7/8	7/8		
	9	7/9	8/9	8/9		
	10	7/10	4/5	9/10		
$N_1 = 3$	$N_2 = 4$	3/4	3/4			
	5	2/3	4/5	4/5		
	6	2/3	2/3	5/6		
	7	2/3	5/7	6/7	6/7	
	8	5/8	3/4	3/4	7/8	
	9	2/3	2/3	7/9	8/9	8/9
	10	3/5	7/10	4/5	9/10	9/10
	12	7/12	2/3	3/4	5/6	11/12
$N_1 = 4$	$N_2 = 5$	3/5	3/4	4/5	4/5	
	6	7/12	2/3	3/4	5/6	5/6
	7	17/28	5/7	3/4	6/7	6/7
	8	5/8	5/8	3/4	7/8	7/8
	9	5/9	2/3	3/4	7/9	8/9
	10	11/20	13/20	7/10	4/5	4/5
	12	7/12	2/3	2/3	3/4	5/6
	16	9/16	5/8	11/16	3/4	13/16
$N_1 = 5$	$N_2 = 6$	3/5	2/3	2/3	5/6	5/6
	7	4/7	23/35	5/7	29/35	6/7
	8	11/20	5/8	27/40	4/5	4/5
	9	5/9	3/5	31/45	7/9	4/5
	10	1/2	3/5	7/10	7/10	4/5
	15	8/15	3/5	2/3	11/15	11/15
	20	1/2	11/20	3/5	7/10	3/4

Contd.

Contd.

One-Sided Test: α =		.10	.05	.025	.01	.005
Two-Sided Test: α =		.20	.10	.05	.02	.01
$N_1 = 6$	$N_2 = 7$	23/42	4/7	29/42	5/7	5/6
	8	1/2	7/12	2/3	3/4	3/4
	9	1/2	5/9	2/3	13/18	7/9
	10	1/2	17/30	19/30	7/10	11/15
	12	1/2	7/12	7/12	2/3	3/4
	18	4/9	5/9	11/18	2/3	13/18
	24	11/24	1/2	7/12	5/8	2/3
$N_1 = 7$	$N_2 = 8$	27/56	33/56	5/8	41/56	3/4
	9	31/63	5/9	40/63	5/7	47/63
	10	33/70	39/70	43/70	7/10	5/7
	14	3/7	1/2	4/7	9/14	5/7
	28	3/7	13/28	15/28	17/28	9/14
$N_1 = 8$	$N_2 = 9$	4/9	13/24	5/8	2/3	3/4
	10	19/40	21/40	23/40	27/40	7/10
	12	11/24	1/2	7/12	5/8	2/3
	16	7/16	1/2	9/16	5/8	5/8
	32	13/32	7/16	1/2	9/16	19/32
$N_1 = 9$	$N_2 = 10$	7/15	1/2	26/45	2/3	31/45
	12	4/9	1/2	5/9	11/18	2/3
	15	19/45	22/45	8/15	3/5	29/45
	18	7/18	4/9	1/2	5/9	11/18
	36	13/36	5/12	17/36	19/36	5/9
$N_1 = 10$	$N_2 = 15$	2/5	7/15	1/2	17/30	19/30
	20	2/5	9/20	1/2	11/20	3/5
	40	7/20	2/5	9/20	1/2	
$N_1 = 12$	$N_2 = 15$	23/60	9/20	1/2	11/20	7/12
	16	3/8	7/16	23/48	13/24	7/12
	18	13/36	5/12	17/36	19/36	5/9
	20	11/30	5/12	7/15	31/60	17/30
$N_1 = 15$	$N_2 = 20$	7/20	2/5	13/30	29/60	31/60
$N_1 = 16$	$N_2 = 20$	27/80	31/80	17/40	19/40	41/80
Large-sample approximation:		$1.07\sqrt{\frac{m+n}{mn}}$	$1.22\sqrt{\frac{m+n}{mn}}$	$1.36\sqrt{\frac{m+n}{mn}}$	$1.52\sqrt{\frac{m+n}{mn}}$	$1.63\sqrt{\frac{m+n}{mn}}$

Source: F. J. Massey, Distribution table for the deviation between two sample cumulatives, *Ann. Math. Stat.* **23** (1952), 435–441.

Table A.13 Critical Values of the Wilcoxon Signed Ranks Test Statistic

This table gives values of t_α for which $P\{T^+ > t_\alpha\} \leq \alpha$ for selected values of n and α. Critical values in the lower tail may be obtained by symmetry from the equation $t_{1-\alpha} = n(n+1)/2 - t_\alpha$.

	α			
n	.01	.025	.05	.10
3	6	6	6	6
4	10	10	10	9
5	15	15	14	12
6	21	20	18	17
7	27	25	24	22
8	34	32	30	27
9	41	39	36	34
10	49	46	44	40
11	58	55	52	48
12	67	64	60	56
13	78	73	69	64
14	89	84	79	73
15	100	94	89	83
16	112	106	100	93
17	125	118	111	104
18	138	130	123	115
19	152	143	136	127
20	166	157	149	140

Source: Table 1 of R. L. McCornack, Extended tables of the Wilcoxon matched pairs signed-rank statistics, *J. Am. Stat. Assoc.* 60 (1965), 864–871.

Table A.14 Table of Critical Values of T in the Wilcoxon Matched-Pairs Signed-Ranks Test

N	Level of significance for one-tailed test		
	.025	.01	.005
	Level of significance for two-tailed test		
	.05	.02	.01
6	0	—	—
7	2	0	—
8	4	2	0
9	6	3	2
10	8	5	3
11	11	7	5
12	14	10	7
13	17	13	10
14	21	16	13
15	25	20	16
16	30	24	20
17	35	28	23
18	40	33	28
19	46	38	32
20	52	43	38
21	59	49	43
22	66	56	49
23	73	62	55
24	81	69	61
25	89	77	68

Source: Table I of Wilcoxon, F. 1949. *Some Rapid Approximate Statistical Procedures.* New York : American Cyanamid Company, p. 13.

Table A.15 Table of Critical Values of r in the Wald-Wolfowitz Two-Sample Runs Test

Any value of r which is equal to or smaller than that shown in the Table is significant at the .05 level.

n_1 \ n_2	2	3	4	5	6	7	8	9	10	11	12	13	14	15	16	17	18	19	20
2											2	2	2	2	2	2	2	2	2
3					2	2	2	2	2	2	2	2	2	3	3	3	3	3	3
4				2	2	2	3	3	3	3	3	3	3	3	4	4	4	4	4
5			2	2	3	3	3	3	3	4	4	4	4	4	4	4	5	5	5
6		2	2	3	3	3	3	4	4	4	4	5	5	5	5	5	5	6	6
7		2	2	3	3	3	4	4	5	5	5	5	5	6	6	6	6	6	6
8		2	3	3	3	4	4	5	5	5	6	6	6	6	6	7	7	7	7
9		2	3	3	4	4	5	5	5	6	6	6	7	7	7	7	8	8	8
10		2	3	3	4	5	5	5	6	6	7	7	7	7	8	8	8	8	9
11		2	3	4	4	5	5	6	6	7	7	7	8	8	8	9	9	9	9
12	2	2	3	4	4	5	6	6	7	7	7	8	8	8	9	9	9	10	10
13	2	2	3	4	5	5	6	6	7	7	8	8	9	9	9	10	10	10	10
14	2	2	3	4	5	5	6	7	7	8	8	9	9	9	10	10	10	11	11
15	2	3	3	4	5	6	6	7	7	8	8	9	9	10	10	11	11	11	12
16	2	3	4	4	5	6	6	7	8	8	9	9	10	10	11	11	11	12	12
17	2	3	4	4	5	6	7	7	8	9	9	10	10	11	11	11	12	12	13
18	2	3	4	5	5	6	7	8	8	9	9	10	10	11	11	12	12	13	13
19	2	3	4	5	6	6	7	8	8	9	10	10	11	11	12	12	13	13	13
20	2	3	4	5	6	6	7	8	9	9	10	10	11	12	12	13	13	13	14

Source: Swed, Frieda S., and Eisenhart, C. 1943. *Tables for testing randomness of grouping in a sequence of alternatives. Ann. Math. Statist.*, **14**, 83–86.

Table A.16 Table of Probabilities associated with values as small as observed values of U in the Mann-Whitney Test

$n_2 = 3$

U \ n_1	1	2	3
0	.250	.100	.050
1	.500	.200	.100
2	.750	.400	.200
3		.600	.350
4			.500
5			.650

$n_2 = 4$

U \ n_1	1	2	3	4
0	.200	.067	.028	.014
1	.400	.133	.057	.029
2	.600	.267	.114	.057
3		.400	.200	.100
4		.600	.314	.171
5			.429	.243
6			.571	.343
7				.443
8				.557

$n_2 = 5$

U \ n_1	1	2	3	4	5
0	.167	.047	.018	.008	.004
1	.333	.095	.036	.016	.008
2	.500	.190	.071	.032	.016
3	.667	.286	.125	.056	.028
4		.429	.196	.095	.048
5		.571	.286	.143	.075
6			.393	.206	.111
7			.500	.278	.155
8			.607	.365	.210
9				.452	.274
10				.548	.345
11					.421
12					.500
13					.579

$n_2 = 6$

U \ n_1	1	2	3	4	5	6
0	.143	.036	.012	.005	.002	.001
1	.286	.071	.024	.010	.004	.002
2	.428	.143	.048	.019	.009	.004
3	.571	.214	.083	.033	.015	.008
4		.321	.131	.057	.026	.013
5		.429	.190	.086	.041	.021
6		.571	.274	.129	.063	.032
7			.357	.176	.089	.047
8			.452	.238	.123	.066
9			.548	.305	.165	.090
10				.381	.214	.120
11				.457	.268	.155
12				.545	.331	.197
13					.396	.242
14					.465	.294
15					.535	.350
16						.409
17						.469
18						.531

$n_2 = 7$

U \ n_1	1	2	3	4	5	6	7
0	.125	.028	.008	.003	.001	.001	.001
1	.250	.056	.017	.006	.003	.001	.001
2	.375	.111	.033	.012	.005	.002	.001
3	.500	.167	.058	.021	.009	.004	.002
4	.625	.250	.092	.036	.015	.007	.003
5		.333	.133	.055	.024	.011	.006
6		.444	.192	.082	.037	.017	.009
7		.556	.258	.115	.053	.026	.013
8			.333	.158	.074	.037	.019
9			.417	.206	.101	.051	.027
10			.500	.264	.134	.069	.036
11			.583	.324	.172	.090	.049
12				.394	.216	.117	.064
13				.464	.265	.147	.082
14				.538	.319	.183	.104
15					.378	.223	.130
16					.438	.267	.159
17					.500	.314	.191
18					.562	.365	.228
19						.418	.267
20						.473	.310
21						.527	.355
22							.402
23							.451
24							.500
25							.549

	$n_2 = 8$									
U \ n_1	1	2	3	4	5	6	7	8	t	Normal
0	.111	.022	.006	.002	.001	.000	.000	.000	3.308	.001
1	.222	.044	.012	.004	.002	.001	.000	.000	3.203	.001
2	.333	.089	.024	.008	.003	.001	.001	.000	3.098	.001
3	.444	.133	.042	.014	.005	.002	.001	.001	2.993	.001
4	.556	.200	.067	.024	.009	.004	.002	.001	2.888	.002
5		.267	.097	.036	.015	.006	.003	.001	2.783	.003
6		.356	.139	.055	.023	.010	.005	.002	2.678	.004
7		.444	.188	.077	.033	.015	.007	.003	2.573	.005
8		.556	.248	.107	.047	.021	.010	.005	2.468	.007
9			.315	.141	.064	.030	.014	.007	2.363	.009
10			.387	.184	.085	.041	.020	.010	2.258	.012
11			.461	.230	.111	.054	.027	.014	2.153	.016
12			.539	.285	.142	.071	.036	.019	2.048	.020
13				.341	.177	.091	.047	.025	1.943	.026
14				.404	.217	.114	.060	.032	1.838	.033
15				.467	.262	.141	.076	.041	1.733	.041
16				.533	.311	.172	.095	.052	1.628	.052
17					.362	.207	.116	.065	1.523	.064
18					.416	.245	.140	.080	1.418	.078
19					.472	.286	.168	.097	1.313	.094
20					.528	.331	.198	.117	1.208	.113
21						.377	.232	.139	1.102	.135
22						.426	.268	.164	.998	.159
23						.475	.306	.191	.893	.185
24						.525	.347	.221	.788	.215
25							.389	.253	.683	.247
26							.433	.287	.578	.282
27							.478	.323	.473	.318
28							.522	.360	.368	.356
29								.399	.263	.396
30								.439	.158	.437
31								.480	.052	.481
32								.520		

Source: Mann, H. B., and Whitney, D. R. 1947. *On a test of whether one of two random variables is stochastically larger than the other. Ann. Math. Statist.*, **18**, 52–54.

Table A.17 Table of Critical Values of U in the Mann-Whitney Test

Table A.17a Critical values of U for a One-Tailed Test at $\alpha = .001$ or for a Two-Tailed Test at $\alpha = .002$

n_1 \ n_2	9	10	11	12	13	14	15	16	17	18	19	20
1												
2												
3									0	0	0	0
4		0	0	0	1	1	1	2	2	3	3	3
5	1	1	2	2	3	3	4	5	5	6	7	7
6	2	3	4	4	5	6	7	8	9	10	11	12
7	3	5	6	7	8	9	10	11	13	14	15	16
8	5	6	8	9	11	12	14	15	17	18	20	21
9	7	8	10	12	14	15	17	19	21	23	25	26
10	8	10	12	14	17	19	21	23	25	27	29	32
11	10	12	15	17	20	22	24	27	29	32	34	37
12	12	14	17	20	23	25	28	31	34	37	40	42
13	14	17	20	23	26	29	32	35	38	42	45	48
14	15	19	22	25	29	32	36	39	43	46	50	54
15	17	21	24	28	32	36	40	43	47	51	55	59
16	19	23	27	31	35	39	43	48	52	56	60	65
17	21	25	29	34	38	43	47	52	57	61	66	70
18	23	27	32	37	42	46	51	56	61	66	71	76
19	25	29	34	40	45	50	55	60	66	71	77	82
20	26	32	37	42	48	54	59	65	70	76	82	88

Table A.17b Critical Values of U for a One-tailed Test at $\alpha = .01$ or for a Two-tailed Test at $\alpha = .02$

n_1 \ n_2	9	10	11	12	13	14	15	16	17	18	19	20
1												
2					0	0	0	0	0	0	1	1
3	1	1	1	2	2	2	3	3	4	4	4	5
4	3	3	4	5	5	6	7	7	8	9	9	10
5	5	6	7	8	9	10	11	12	13	14	15	16
6	7	8	9	11	12	13	15	16	18	19	20	22
7	9	11	12	14	16	17	19	21	23	24	26	28
8	11	13	15	17	20	22	24	26	28	30	32	34
9	14	16	18	21	23	26	28	31	33	36	38	40
10	16	19	22	24	27	30	33	36	38	41	44	47
11	18	22	25	28	31	34	37	41	44	47	50	53
12	21	24	28	31	35	38	42	46	49	53	56	60
13	23	27	31	35	39	43	47	51	55	59	63	67
14	26	30	34	38	43	47	51	56	60	65	69	73
15	28	33	37	42	47	51	56	61	66	70	75	80
16	31	36	41	46	51	56	61	66	71	76	82	87
17	33	38	44	49	55	60	66	71	77	82	88	93
18	36	41	47	53	59	65	70	76	82	88	94	100
19	38	44	50	56	63	69	75	82	88	94	101	107
20	40	47	53	60	67	73	80	87	93	100	107	114

Table A.17c Critical Values of U for a One-Tailed Test at $\alpha = .025$ or for a Two-Tailed Test at $\alpha = .05$

n_1 \ n_2	9	10	11	12	13	14	15	16	17	18	19	20
1												
2	0	0	0	1	1	1	1	1	2	2	2	2
3	2	3	3	4	4	5	5	6	6	7	7	8
4	4	5	6	7	8	9	10	11	11	12	13	13
5	7	8	9	11	12	13	14	15	17	18	19	20
6	10	11	13	14	16	17	19	21	22	24	25	27
7	12	14	16	18	20	22	24	26	28	30	32	34
8	15	17	19	22	24	26	29	31	34	36	38	41
9	17	20	23	26	28	31	34	37	39	42	45	48
10	20	23	26	29	33	36	39	42	45	48	52	55
11	23	26	30	33	37	40	44	47	51	55	58	62
12	26	29	33	37	41	45	49	53	57	61	65	69
13	28	33	37	41	45	50	54	59	63	67	72	76
14	31	36	40	45	50	55	59	64	67	74	78	83
15	34	39	44	49	54	59	64	70	75	80	85	90
16	37	42	47	53	59	64	70	75	81	86	92	98
17	39	45	51	57	63	67	75	81	87	93	99	105
18	42	48	55	61	67	74	80	86	93	99	106	112
19	45	52	58	65	72	78	85	92	99	106	113	119
20	48	55	62	69	76	83	90	98	105	112	119	127

Table A.17d Critical Values of U for a One-tailed Test at $\alpha = .05$ or for a Two-tailed Test at $\alpha = .10$

n_2 \ n_1	9	10	11	12	13	14	15	16	17	18	19	20
1											0	0
2	1	1	1	2	2	2	3	3	3	4	4	4
3	3	4	5	5	6	7	7	8	9	9	10	11
4	6	7	8	9	10	11	12	14	15	16	17	18
5	9	11	12	13	15	16	18	19	20	22	23	25
6	12	14	16	17	19	21	23	25	26	28	30	32
7	15	17	19	21	24	26	28	30	33	35	37	39
8	18	20	23	26	28	31	33	36	39	41	44	47
9	21	24	27	30	33	36	39	42	45	48	51	54
10	24	27	31	34	37	41	44	48	51	55	58	62
11	27	31	34	38	42	46	50	54	57	61	65	69
12	30	34	38	42	47	51	55	60	64	68	72	77
13	33	37	42	47	51	56	61	65	70	75	80	84
14	36	41	46	51	56	61	66	71	77	82	87	92
15	39	44	50	55	61	66	72	77	83	88	94	100
16	42	48	54	60	65	71	77	83	89	95	101	107
17	45	51	57	64	70	77	83	89	96	102	109	115
18	48	55	61	68	75	82	88	95	102	109	116	123
19	51	58	65	72	80	87	94	101	109	116	123	130
20	54	62	69	77	84	92	100	107	115	123	130	138

Source: Tables 1, 3, 5 and 7 of Auble, D. 1953. *Extended tables for the Mann-Whitney statistic. Bulletin of the Institute of Educational Research at Indiana University,* **1,** No. 2.

Table A.18

Table of Probabilities associated with values as large as observed values of χ_r^2 in the Friedman two-way analysis of variance by ranks

Table A.18.1 k = 3

$N=2$		$N=3$		$N=4$		$N=5$	
χ_r^2	p	χ_r^2	p	χ_r^2	p	χ_r^2	p
0	1.000	.000	1.000	.0	1.000	.0	1.000
1	.833	.667	.944	.5	.931	.4	.954
3	.500	2.000	.528	1.5	.653	1.2	.691
4	.167	2.667	.361	2.0	.431	1.6	.522
		4.667	.194	3.5	.273	2.8	.367
		6.000	.028	4.5	.125	3.6	.182
				6.0	.069	4.8	.124
				6.5	.042	5.2	.093
				8.0	.0046	6.4	.039
						7.6	.024
						8.4	.0085
						10.0	.00077

$N=6$		$N=7$		$N=8$		$N=9$	
χ_r^2	p	χ_r^2	p	χ_r^2	p	χ_r^2	p
.00	1.000	.000	1.000	.00	1.000	.000	1.000
.33	.956	.286	.964	.25	.967	.222	.971
1.00	.740	.857	.768	.75	.794	.667	.814
1.33	.570	1.143	.620	1.00	.654	.889	.865
2.33	.430	2.000	.486	1.75	.531	1.556	.569
3.00	.252	2.571	.305	2.25	.355	2.000	.398
4.00	.184	3.429	.237	3.00	.285	2.667	.328
4.33	.142	3.714	.192	3.25	.236	2.889	.278
5.33	.072	4.571	.112	4.00	.149	3.556	.187
6.33	.052	5.429	.085	4.75	.120	4.222	.154
7.00	.029	6.000	.052	5.25	.079	4.667	.107
8.33	.012	7.143	.027	6.25	.047	5.556	.069
9.00	.0081	7.714	.021	6.75	.038	6.000	.057
9.33	.0055	8.000	.016	7.00	.030	6.222	.048
10.33	.0017	8.857	.0084	7.75	.018	6.889	.031
12.00	.00013	10.286	.0036	9.00	.0099	8.000	.019
		10.571	.0027	9.25	.0080	8.222	.016
		11.143	.0012	9.75	.0048	8.667	.010
		12.286	.00032	10.75	.0024	9.556	.0060
		14.000	.000021	12.00	.0011	10.667	.0035
				12.25	.00086	10.889	.0029
				13.00	.00026	11.556	.0013
				14.25	.000061	12.667	.00066
				16.00	.0000036	13.556	.00035
						14.000	.00020
						14.222	.000097
						14.889	.000054
						16.222	.000011
						18.000	.0000006

Table A.18.2 k = 4

$N = 2$		$N = 3$		$N = 4$			
χ_r^2	p	χ_r^2	p	χ_r^2	p	χ_r^2	p
.0	1.000	.2	1.000	.0	1.000	5.7	.141
.6	.958	.6	.958	.3	.992	6.0	.105
1.2	.834	1.0	.910	.6	.928	6.3	.094
1.8	.792	1.8	.727	.9	.900	6.6	.077
2.4	.625	2.2	.608	1.2	.800	6.9	.068
3.0	.542	2.6	.524	1.5	.754	7.2	.054
3.6	.458	3.4	.446	1.8	.677	7.5	.052
4.2	.375	3.8	.342	2.1	.649	7.8	.036
4.8	.208	4.2	.300	2.4	.524	8.1	.033
5.4	.167	5.0	.207	2.7	.508	8.4	.019
6.0	.042	5.4	.175	3.0	.432	8.7	.014
		5.8	.148	3.3	.389	9.3	.012
		6.6	.075	3.6	.355	9.6	.0069
		7.0	.054	3.9	.324	9.9	.0062
		7.4	.033	4.5	.242	10.2	.0027
		8.2	.017	4.8	.200	10.8	.0016
		9.0	.0017	5.1	.190	11.1	.00094
				5.4	.158	12.0	.000072

Source: Friedman, M. 1937. *The use of ranks to avoid the assumption of normality implicit in the analysis variance. J. Amer. Statist. Ass.,* **32,** 688–689.

Table A.19 Table of Crirtical Values of r_s, the Spearman-Rank Correlation Coefficient

N	Significance level (one-tailed test)	
	.05	.01
4	1.000	
5	.900	1.000
6	.829	.943
7	.714	.893
8	.643	.833
9	.600	.783
10	.564	.746
12	.506	.712
14	.456	.645
16	.425	.601
18	.399	.564
20	.377	.534
22	.359	.508
24	.343	.485
26	.329	.465
28	.317	.448
30	.306	.432

Source: Olds, E. G. 1938. ***Distributions of sums of squares of rank differences for small numbers of individuals.*** *Ann. Math. Statist.*, **9,** 133–148, and Olds, E. G. 1949. The 5% significance levels for sums of squares of rank differences and a correction. *Ann. Math. Statist.*, **20,** 117–118.

Table A.20 Critical Points of Kendall's Tau Statistic

This table gives the values of S_α for which $P\{S > S_\alpha\} \leq \alpha$, where $S = \binom{n}{2}T$, for some selected values of α and n. Values in the lower tail may be obtained by symmetry, $S_{1-\alpha} = -S_\alpha$.

	α			
n	.100	.050	.025	.01
3	3	3	3	3
4	4	4	6	6
5	6	6	8	8
6	7	9	11	11
7	9	11	13	15
8	10	14	16	18
9	12	16	18	22
10	15	19	21	25

Source: Table 1, page 173, of M. G. Kendal, *Rank Correlation Methods,* 3rd Ed., Griffin, London, 1962. For values of $n \geq 11$, see W. J. Conover, *Practical Non-parametric Statistics,* John Wiley, New York, 1971, page 390.

A.21 Quantiles of Chi-square distribution for determining the shortest confidence interval for σ

ν	$\alpha = 0.10$				$\alpha = 0.05$			
	α_1	$\chi^2_{\alpha_1,\nu}$	$\chi^2_{\alpha_2,\nu}$	%	α_1	$\chi^2_{\alpha_1,\nu}$	$\chi^2_{\alpha_2,\nu}$	%
1	0.09998	0.016	18.189	24.92	0.04999	0.004	19.511	24.99
2	0.09988	0.211	18.056	49.02	0.04998	0.103	21.640	49.48
3	0.09948	0.582	17.647	61.21	0.04988	0.351	20.726	61.82
4	0.09882	1.056	18.100	68.50	0.04969	0.708	21.047	69.10
5	0.09800	1.594	18.907	73.41	0.04943	1.139	21.806	73.94
6	0.09708	2.175	19.871	76.95	0.04911	1.623	22.736	77.41
7	0.09612	2.788	20.927	79.65	0.04876	2.147	23.791	80.04
8	0.09516	3.426	22.041	81.77	0.04839	2.703	24.910	82.11
9	0.09420	4.084	23.182	83.48	0.04802	3.284	26.083	83.78
10	0.09328	4.758	24.352	84.90	0.04764	3.886	27.270	85.16
11	0.09238	5.447	25.530	86.09	0.04726	4.505	28.472	86.32
12	0.09152	6.147	26.719	87.11	0.04689	5.141	29.689	87.31
13	0.09070	6.858	27.915	87.98	0.04653	5.790	30.915	88.16
14	0.08990	7.579	29.109	88.75	0.04619	6.451	32.153	88.91
15	0.08916	8.308	30.313	89.42	0.04585	7.123	33.386	89.56
16	0.08844	9.045	31.514	90.02	0.04552	7.804	34.619	90.14
17	0.08774	9.788	32.711	90.55	0.04521	8.495	35.859	90.67
18	0.08710	10.539	33.916	91.03	0.04490	9.193	37.091	91.13
19	0.08648	11.295	35.117	91.46	0.04461	9.899	38.328	91.56
20	0.08588	12.056	36.315	91.85	0.04433	10.612	39.563	91.94
21	0.08530	12.823	37.510	92.21	0.04405	11.331	40.791	92.29
22	0.08476	13.595	38.708	92.54	0.04379	12.056	42.024	92.61
23	0.08424	14.371	39.903	92.84	0.04354	12.787	43.254	92.91
24	0.08374	15.151	41.095	93.12	0.04329	13.523	44.478	93.18
25	0.08326	15.935	42.286	93.37	0.04306	14.264	45.706	93.43
26	0.08278	16.723	43.471	93.61	0.04283	15.009	46.928	93.67
27	0.08234	17.514	44.659	93.83	0.04261	15.759	48.149	93.89
28	0.08192	18.310	45.846	94.04	0.04240	16.513	49.369	94.09
29	0.08150	19.108	47.027	94.23	0.04219	17.271	50.583	94.28
30	0.08110	19.909	48.208	94.41	0.04199	18.032	51.797	94.46
40	0.07780	28.063	59.927	95.75	0.04031	25.823	63.834	95.78
50	0.07536	36.418	71.499	96.57	0.03903	33.855	75.694	96.59
60	0.07346	44.918	82.949	97.13	0.03803	42.064	87.418	97.14
80	0.07066	62.232	105.571	97.83	0.03653	58.859	110.535	97.84
100	0.06866	79.847	127.915	98.25	0.03544	76.015	133.321	98.26

(For an explanation of various terms vide example 19.3.2)

Table A.22 Table of Orthogonal Polynomial Coefficients

$n=3$		$n=4$			$n=5$				$n=6$				
ψ_1	ψ_2	ψ_1	ψ_2	ψ_3	ψ_1	ψ_2	ψ_3	ψ_4	ψ_1	ψ_2	ψ_3	ψ_4	ψ_5
−1	1	−3	1	−1	−2	2	−1	1	−5	5	5	1	−1
0	−2	−1	−1	3	−1	−1	2	−4	−3	−1	7	−3	5
1	1	1	−1	−3	0	−2	0	6	−1	−4	4	2	−10
		3	1	1	1	−1	−2	−4	1	−4	−4	2	10
					2	2	1	1	3	−1	−7	−3	−5
									5	5	5	1	1
2	6	20	4	20	10	14	10	70	70	84	180	28	252
1	3	2	1	10/3	1	1	5/6	35/12	2	3/2	5/3	7/12	21/10

Table of Orthogonal Polynomial Coefficients

$n=7$						$n=8$					
ψ_1	ψ_2	ψ_3	ψ_4	ψ_5	ψ_6	ψ_1	ψ_2	ψ_3	ψ_4	ψ_5	ψ_6
−3	5	−1	3	1	1	−7	7	−7	7	−7	1
−2	0	1	−7	4	−6	−5	1	5	−13	23	−5
−1	−3	1	1	−5	15	−3	−3	7	−3	−17	9
0	−4	0	6	0	−20	−1	−5	3	9	−15	−5
1	−3	−1	1	5	15	1	−5	−3	9	15	−5
2	0	−1	−7	−4	−6	3	−3	−7	−3	17	9
3	5	1	3	1	1	5	1	−5	−13	−23	−5
						7	7	7	7	7	1
28	84	6	154	84	924	168	168	264	616	2184	264
1	1	1/6	7/12	7/20	77/60	2	1	2/3	7/12	7/10	11/60

Table of Orthogonal Polynomial Coefficients

$n=9$						$n=10$					
ψ_1	ψ_2	ψ_3	ψ_4	ψ_5	ψ_6	ψ_1	ψ_2	ψ_3	ψ_4	ψ_5	ψ_6
−4	28	−14	14	−4	4	−9	6	−42	18	−6	3
−3	7	7	−21	11	−17	−7	2	14	−22	14	−11
−2	−8	13	−11	−4	22	−5	−1	35	−17	−1	10
−1	−17	9	9	−9	1	−3	−3	31	3	−11	6
0	−20	0	18	0	−20	−1	−4	12	18	−6	−8
1	−17	−9	9	9	1	1	−4	−12	18	6	−8
2	−8	−13	−11	4	22	3	−3	−31	3	11	6
3	7	−7	−21	−11	−17	5	−1	−35	−17	1	10
4	28	14	14	4	4	7	2	−14	−22	−14	−11
						9	6	42	18	6	3
60	2772	990	2002	468	1980	330	132	8580	2860	780	600
1	3	5/6	7/12	3/20	11/60	2	1/2	5/3	5/12	1/10	11/240

Table of Orthogonal Polynomial Coefficients

$n = 11$						$n = 12$					
ψ_1	ψ_2	ψ_3	ψ_4	ψ_5	ψ_6	ψ_1	ψ_2	ψ_3	ψ_4	ψ_5	ψ_6
−5	15	−30	6	−3	15	−11	55	−33	33	−33	11
−4	6	6	−6	6	−48	−9	25	3	27	57	−31
−3	−1	22	−6	1	29	−7	1	21	33	21	11
−2	−6	23	−1	−4	36	−5	−17	25	−13	−29	25
−1	−9	14	4	−4	−12	−3	−29	19	12	−44	4
0	−10	0	6	0	−40	−1	−35	7	28	−20	−20
1	−9	−14	4	4	−12	1	−35	−7	28	20	−20
2	−6	−23	−1	4	36	3	−29	−19	12	44	4
3	−1	−22	−6	−1	29	5	−17	−25	−13	29	25
4	6	6	−6	−6	−48	7	1	−21	−33	−21	11
5	15	30	6	3	15	9	25	−3	−27	−57	−31
						11	55	33	33	33	11
110	858	4290	286	156	11220	572	12012	5148	8008	15912	4488
1	1	5/6	1/12	1/40	11/120	2	3	2/3	7/24	3/20	11/360

Bibliography

Books

1. Adke, S.R. and Manjunath, S.M. (1984): *Introduction to Finite Markov Processes,* Wiley Eastern, Calcutta.
2. Anderson, E.B. (1980): *Discrete statistical Models with Social Science Applications*, North Holland.
3. Anderson, T.W. (1974): *An Introduction to Multivariate Statistical Analysis,* Wiley Eastern, 2nd edition, New Delhi.
4. Apostol, T.M. (1974): *Mathematical Analysis,* 2nd edition, Addison-Wesley, Massachusetts.
5. Bailey, N.T.J. (1964): *The Elements of Stochastic Processes,* John Wiley, N.Y.
6. Bharucha-Reid, A.T. (1960): *Elements of the Theory of Markov Processes and their Applications,* McGraw-Hill, N.Y.
7. Bickel, P. and Doksum, K. (1977): *Mathematical Statistics*, Basic Ideas and Selected Topics. Holden-Day, Sanfrancisco.
8. Birnbaum, Z.W. (1962): *Introduction to Probabilities and Mathematical Statistics,* Harper and Bros.
9. Bodewig, E. (1959): *Matrix Calculus,* North-Holland.
10. Breiman, Leo (1968): *Probability,* Reading, Massachusetts, Addison-Wesley.
11. — (1969): *Probability and Stochastic Process: With a view to Applications,* Boston Houghton Mifflin.
12. Bremud, Pierre (1988): *Introduction to Probabilitic Modelling,* Springer Verlag.
13. Chung, K.L. (1960): *Markov Process with Stationery Transition Probabilities,* Springer-Verlag.
14. — (1974): *A Course in Probability Theory,* 2nd edition, Academic Press, N.Y.
15. Cox, D.R. and Hinkley, D.V. (1974): *Theoretical Statistics,* Chapman and Hall.
16. Cramer, H. (1950): *Mathematical Methods of Statistics,* Princeton University Press, 1946 and Asia Publishing House, 1950.

17. DeGroot, M.H. (1986): *Probability and Statistics*, Addison-Wesley.

18. Dudewicz, E.J. (1976): *Introduction to Statistics and Probability*, Holt, Rinehart and Winston, N.Y.

19. Dudewicz, E.J. and Misra, S.N. (1988): *Modern Mathematical Statistics*, John Wiley & Sons, N.Y.

20. Durbin, J. (1973): *Distribution Theory for Tests based on the Sample Distribution Functions*, SIAM, Philadelphia, P.A.

21. Dwass, M. (1970): *Probability and Statistics: An undergraduate course*, W.A. Benjamin, Inc.

22. Efron, B. (1982): *The jack knife, the bootstrap and other resampling plans.* SIAM, monograph no. 38, Philadelphia.

23. Feller, W. (1968): *An Introduction to Probability Theory and its Applications*, Vol. 1 (3rd edition), Wiley Eastern.

24. — (1971): *An Introduction to Probability Theory and its Applications*, Vol. 2, 2nd edition, John Wiley, New York.

25. Ferguson, T.S. (1967): *Mathematical Statistics*, Academic Press.

26. Fisher, R.A. (1925): *Statistical Methods for Research Workers*, Oliver and Boyd, Edinborough.

27. — (1959): *Statistical Methods and Scientific Inference*, Oliver and Boyd, Edinburgh and London.

28. Fisz, M. (1963): *Probability Theory and Mathematical Statistics*, John Wiley and S. Chand.

29. Fraser, D.A.S. (1957): *Nonparametric Methods in Statistics*, John Wiley & Sons, N.Y.

30. Freund, J.E. (1992): *Mathematical Statistics, Chiffs*, Prentice Hall 1962, Prentice Hall of India 1992.

31. Fry, T.C. (1928): *Probability and its Engineering uses*, Van Nostrand.

32. Gaybill, F.A. (1961): *An Introduction to Linear Statistical Models*, Vol. I, McGraw-Hill.

33. Ghosh, B.K. (1970): *Sequential Estimation of Statistical Hypothesis*, Addision-Wesley Reading, Mass.

34. Gibbons, J.D. (1971): *Nonparametric Inference*, McGraw-Hill, N.Y.

35. Giri, N.C. (1974): *Introduction to Probability and Statistics* Part 1: *Probability*, Marcel Dekker, Inc. New York.

36. Gnedenko, B.V. (1962): *Theory of Probability*, Chesla Publishing Co. and Mir Publishers

37. Goldberg, S. (1960): *Probability: An Introduction*, Prentice Hall, Inc. N.J.

38. Goon, A.M., Gupta, M.K. and Dasgupta, B. (1977): *An Outline of Statistical Theory* Vol. 1, 2, World Press.

39. Gupta, J. and Panchapakesan, S. (1979): *Multiple Decision Procedures*, John Wiley.

40. Guttman, I. (1982): *Linear Models: An Introduction,* John Wiley, N.Y.

41. Halmos, P.R. (1964): *Measure Theory,* D. Van Nostrand, 1950 and East-West Press, 1964.

42. Harris, B. (1966): *Theory of Probability,* Addison-Wesley.

43. Hastings, N.A.J. and Peacock, J.B. (1975): *Statistical Distributions,* Longom-Butterworths.

44. Hodges, J.L. (Jr.) and Lehmann, E.L. (1965): *Elements of Probability,* Holden Day, Inc.

45. Hoel, P.G., Port, S.C. and Stone, C.J. (1971): *Introduction to Probability Theory,* Houghton Mifflin Company.

46. Hogg, R.V. and Craig, A.T. (1965): *Introduction to Mathematical Statistics,* Macmillan and Amerind.

47. Johnson, N.L. and Kotz, S.M. (1970): *Discrete Distributions,* John Wiley.

48. — (1970): *Continuous Multivariate Distributions,* John Wiley.

49. — (1970): *Continuous Univariate Distributions,* 1, 2, John Wiley.

50. — (1977): *Urn Models and their Applications: An Approach to Modern Discrete Probability Theory,* John Wiley.

51. Johnson, R.A. and Wichern, D.W. (1988) : *Applied Multivarite Statistical Analysis*, 2nd edition, Prentice Hall International, New Jersey.

52. Kemeny, J.G. and Snell, J.L. (1960): *Finite Markov Chains,* Van Nostrand.

53. Kendall, M.G. (1962): *Rank Correlation Methods,* 3rd edition, Griffin, London.

54. Kendall, M.G. and Stuart, A. (1958): *Advanced Theory at Statistics,* Vol. 1, Charles Griffin & Co., London.

55. — (1961): *The Advanced Theory of Statistics,* Vol. 2, Charles Griffin & Co., London.

56. Keynes, J.M. (1921): *A Treatise of Probability,* Macmillan.

57. Kolmogorov, A.N. (1956): *Foundations of Theory of Probability,* Chesla Publishing Co.

58. Korovkin, P.P. (1975): *Inequalities,* Mir Publishing, Moscow.

59. Lamperti, J. (1966): *Probability,* W.A. Benjamin, Inc.

60. Larson, H.J. (1982): *Introduction to Probability Theory and Statistical Inference,* 3rd edition, John Wiley.

61. Lehmann, E.L. (1977): *Testing Statistical Hypothesis,* John Wiley (1959), Wiley Eastern (1977).

62. Lo'eve, M. (1963): *Probability Theory,* D Van Nostrand (1963) and East-West Press (1968).

63. Lukacs, E. (1960): *Characteristic Functions,* Charles Griffin & Co. Ltd.

64. Mann. N.R., Schafer, R.E. and Singpurwalla, N.D. (1974): *Methods for Statistical Analysis of reliability and life data,* John Wiley, N.Y.

65. Mardia, K.V., Kent, J.T. and Bibby, J.M. (1979): *Multivariate Analysis*, Academic Press.

66. McCord, J.R. (III) and Moroney, R.M. (Jr.) (1964): *Probability Theory,* The Macmillan Co. N.Y.

67. McShane, E.J. (1944): *Integration,* Princeton University Press.

68. Medhi, J. (1982): *Stochastic Processes,* Wiley Eastern.

69. Meyer, P.L. (1965): *Introductory Probability and Statistical Application,* Addison-Wesley.

70. Mood, A.M., Graybill, F.A. and Boes, D.C. (1974): *An Introduction to Theory of Statistics,* McGraw-Hill and Tata McGraw-Hill.

71. Moor, E.H. (1935): *General Analysis*, American Phil. Soc., Philadelphia.

72. Mosteller, F., Rourke, R.E.K. and Thomas, G.B. (Jr.) (1961): *Probability and Statistics,* Addison-Wesley.

73. Muirhead, R.J. (1982): *Aspects of Multivariate Statistical Theory,* John Wiley, N.Y.

74. Mukhopadhyay, P. (1991): *Theory of Probability,* New Central Book Agency, Kolkata.

75. Mukhopadhyay, P. (2005): *Applied Statistics*, 2nd edn., Books and Allied, Kolkata.

76. Munroe, M.E..(1953): *Introduction to Measure and Integration,* Addison-Wesley.

77. Natrella, M.G. (1963): *Experimental Statistics,* Natl. Bur. Stand. Handbook '91, Washington D.C.

78. Noether, G.E. (1967): *Elements of Nonparametric Statistics*, Wiley, N.Y.

79. Papoulis, A. (1965): *Probability, Random Variables and Stochastic Process*, McGraw-Hill.

80. Parzen, E. (1960): *Modern Probability Theory and its Applications,* John Wiley and Sons, Inc.

81. Parzen, E. (1962): *Stochastic Processes,* Holden-Day.

82. Pfeiffer, P.E. (1965): *Concepts of Probability Theory,* McGraw-Hill.

83. Piskunov, N. (1981): *Differential and Integral Calculus,* Vol. I and II, 4th printing, Mir Publishing, Moscow.

84. Pitman, E.J.G. (1948): *Non-parametric Statistical Inference*, University of North Carolina, Institute of Statistics (mimeographed lecture notes).

85. Pitman, E.J.G. (1979): *Some Basic Theory for Statistical Inference,* Chapman and Hall, London.

86. Pitt, H.R. (1963): *Integration, Measure and Probability Theory,* Oliver and Boyd.

87. Prabhu, N.U. (1963): *Stochastic Processes,* Macmillan.

88. Rahaman, N.A. (1967): *Exercise in Probability and Statistics,* Griffin.

89. Raiffa, H. and Schlaiffer, R. (1961): *Applied Statistical Decision Theory,* Havard University Press, Boston.

90. Rao, C.R. (1963): *Linear Statistical Inference and its Applications,* John Wiley & Sons.

91. Rao, M.M. (1984) : *Probability Theory with Applications*, Academic Press, Inc.

92. Rohatgi, V.K. (1976): *An Introduction to Probability, Theory and Mathematical Statistics,* Wiley Eastern.

93. Ross, S.M. (1972): *Introduction to Probability Models,* Academic Press.

94. Rozanov, Y.A. (1969): *Introductory Probability Theory,* (Translated and edited by R.A. Silverman) Prentice Hall.

95. Sankaranarayanan, G. (1966): *Elementary Theory of Mathematical Statistics and its Application,* Oxford IBH Publication.

96. Savage, L.J. (1954): *The Foundations of Statistics,* John Wiley.

97. Scheffé, H. (1961): *The Analysis of Variance,* John Wiley, N.Y.

98. Schmitz, N. (1993): *Optimal Sequentially Planned Decision Procedures,* Springer Verlag, Berlin.

99. Searle, S.R. (1971): *Linear Models,* John Wiley, N.Y.

100. Uspensky, J.V. (1937): *Introduction to Mathematical Probability,* McGraw-Hill and Tata McGraw-Hill.

101. Vinod, H.D. and Ullah, A. (1981): *Recent Advances in Regression Methods,* Marcell Dekker, N.Y.

102. Wald, A. (1947): *Sequential Analysis,* Wiley, N.Y.

103. — (1950): *Sequential Decision Function,* Wiley, N.Y.

104. Wasen, M.T. (1970): *Parametric Estimation,* McGraw-Hill Book Co., N.Y.

105. Weatherill, G.B. (1966): *Sequential Methods in Statistics,* Methuen, London.

106. Wedder, D.V. (1946): *The Laplace Transform,* Princeton University Press, Princeton, N.J.

107. Whittle, P. (1976): *Probability,* John Wiley and Sons.

108. Wilks, S.S. (1968): *Stochastic Convergence,* Health Mathematical Monograph.

109. — (1976): *Mathematical Statistics,* John Wiley and Toppan.

Papers

110. Anscombe, F.J. (1948): *The transformation of Poisson, Binomial and Negative Biomial Data,* Binometrika, **35,** 246–254.

111. Arrow, K.I., Blackwell, D. and Girshick, M.A. (1949): *Bayes and minimax solutions of sequential decision problems,* Econometricka, **17,** 213–244.

112. Auble, J.D. (1953): *Extended tables for the Mann-Whitney statistic,* Bull. Inst. Educ. Res. **1,** i–iii, 1–39.

113. Baksalary, J.K. and Kala, R. (1980): *A new bound for the Eucledian norm of the difference between the least squares and the best linear unbiased estimators,* Ann. Stat., **8**, 679–681.

114. Barlett, M.S. (1938): *Further aspects of the theory of multiple regression,* Proceedings of the Cambridge Philosophical Society, **34**, 33–40.

115. Bartkey, W. (1943): *Multiple sampling with constant probability,* Ann. Math. Stat., **14**, 363-377.

116. Bhattacharyya, A. (1946): *On some analogues of the amount of information and their use in statistical estimation,* Sankhya A, **8,** 1–14.

117. Birnbaum, Z.W. (1952): *Numerical tabulation of the distribution of Kolmogorov's statistic for finite sample size,* Jour. Amer. Stat. Assoc., **17,** 425–441.

118. Blackwell, D. (1947): *Conditional expectation and unbiased sequential estimation,* Ann. Math. Stat. **18,** 105–110.

119. Bloomfield, P. and Watson, G. S. (1975): *The efficiency of least squares,* Biometrika **62,** 121–128.

120. Bose, R.C. and Roy, S.N. (1938): *The exact distribution of the Studenlized D^2-Statistic,* Sankhya **4,** 10–38.

121. Bowker, A.H. (1960): *A representation of Hotelling's T^2 and Aderson's classification statistic W in terms of simple statistic in Contribution to Probability and Statistic,* Essays in honour of Harold Hotelling, ed. I, Olkin et al, Stanford, California: Stanford University Press, 142–149.

122. Chapman, D.G. and Robbins, H. (1951): *Minimum variance estimation without regularity assumptions,* Ann. Math. Stat., **22,** 581–586.

123. Chernoff, H. and Savage, I.R. (1958): *Asymptotic normality and efficiency of certain non-parametric test statistics,* Ann. Math. Stat., **29,** 972–994.

124. Cramer, H. and Wald, H. (1936): *Some Theorems on Distribution Functions* : J. London Math. Soc., **11**, 290–295.

125. Dantzig, G.B. (1940): *On the nonexistence of tests of student's hypothesis having power function independent of σ^2,* Ann. Math. Stat., **II,** 186–192.

126. Dodge, H.F. and Romig, H.G. (1929): *A method of sampling inspection,* Bull Syst. Techn. J., **8,** 613–631.

127. Dvoretzky, A.K. Kiefer, J. and Wolfwitz, J. (1956): *Asymptotic minimax character of the sample distribution function and of the classical multinomial estimator,* Ann. Math. Stat., **27**, 352–374.

128. Essen (1945): *Fourier analysis of the distribution functions,* Acta Math., **77**, 1–125.

129. Finkelstein, J.M. (1971): *The law of the iterated logarithms for empirical distributions,* Ann. Math. Stat., **42**, 607–715.

130. Fisher, R.A. (1912): *On an absolute criterion for fitting frequency curves,* Messanger of Math., **41,** 155.

131. — (1921): *On the mathematical foundation of theoretical statistics,* Phil. Trans. Royal Soc., London, **A,** 309.

132. — (1922): *On the mathematical foundation of theoretical statistics,* Phil. Trans. Royal Soc. London, Ser. A, Vol. **222,** pp 309–368.

133. — (1925): *Theory of statistical estimation,* Proc. Cambridge Phil. Soc., **22,** 700.

134. — (1928): *Moments and product moments of sampling distribution,* Proc. Lond. Math. Soc. (2), **30,** 199.

135. Fréchet, M. (1937): *Rechérches théoriques modernes Sur la théorie des probabilés,* Forms Vol. I, Part III, Paris.

136. — (1951): *Sur les tableaux de correlation dont les mages sont donnecs,* Annals de l'Universite' de Lyon, Section A. Ser. 3, **14**, 53–77.

137. Fuller, W.A. and Battese, G.E. (1973): *Transformation for estimation of linear models with nested error structure,* J. Amer. Stat. Assoc., **68**, 626–632.

138. Gaenssler and Slute (1979): *Empirical processes: a survey of results from independent and identically distributed random variables,* Ann. Prob. **7**, 193–243.

139. Gart, J. J. (1959): *An estension of the Rao Inequality,* Ann. Math. Stat., **30,** 367–380.

140. Ghosh, M. and Meeden, G. (1986): *Empirical Bayes estimation in finite population sampling,* J. Amer. Stat. Assoc., **81**, 1058–1062.

141. Hadamard, J. (1893): *Résolution dune Question Relative aux Determinants,* Bull. Scientific Mathematics, Vol. **2**, 240–248.

142. Hill, I.D. (1962): *Sampling inspection and defence specifications DEF*-131, J. Roy Stat. Soc. A, **125,** 31–87.

143. Hoeffding, W. (1950): *Optimum nonparametric tests.* Proc. 2nd Berkley Symp. Math. Stat., **19**, 293.

144. Hotelling, H. (1931): *The generalisation of Student's ratio,* Ann. Math. Stat, **2**, 360.

145. Hotelling, H. (1941): *Experimental determination of the maximum of a function,* Ann. Math. Stat., **12**, 20–45.

146. Hsu, P.L. (1938): *Notes on Hotelling's generalised* T^2, Ann. Math. Stat. **9,** 231–243.

147. Huzurbazar, V.S. (1948): *The likelihood equation, consistency and the maxima of the likelihood function,* Ann. Eugen., **14**, 185–200.

148. Kale, B.R. (1961): *On the solution of the likelihood equation by iteration processes,* Binometrika, **48**, 425–456.

149. Khatri, C.G. and Rao, C.R. (1981): *Some extension of the Kantorovich inequality and statistical applications,* J. Multivariate Analysis, **11**, 498–505.

150. Kiefer, J. (1952): *On minimum variance estimators,* Ann. Math. Stat., **23,** 627–629.

151. Kingman, A. and Graybill, F.A. (1970): *A non-linear characterisation of the normal distribution,* Ann. Math. Stat., **41,** 1889–1895.

152. Knott, M. (1975): *On minimum efficiency of least squares,* Biometrika **62,** 129–132.

153. Kolmogorov, A.N. (1933): *Sulla determinazone empirica di una di distribuzione, Giorn,* Inst. Ital, Attuari **4** (1933), 83–91.

154. — (1950): *Unbiased estimates (in Russian),* Izv. Akad. Nauk. USSR, **14,** 303.

155. Koopman, B.O. (1956): *On distributions admitting a sufficient statistic,* Trans. Amer. Math. Soc. **39,** 399.

156. Kulldorff, G. (1957): *Contributions to the theory of grouped and partially grouped samples,* Almqvist and Wiksell, Stockholm.

157. Laha, R.G. (1957): *On a characterisation of the normal distribution from properties of suitable linear statistics,* Ann. Math. Stat., **28**, 126-139.

158. Lehmann, E.L. and Scheffé, H. (1950): *Completeness, Similar regions and unbiased estimation,* Part I, Sankhya, **10,** 305–340.

159. Lukacs, E. (1956): *Characterisation of populations by properties of suitable statistics,* Proc. Third Berkley Symp. **2,** 195–214.

160. Magness, T.A. and McGuire, J.B. (1962): *Comparison of least squares and minimum variance estimates of regression parameters*, Ann. Math. Stat., **33**, 462–470.

161. Mahalanobis, P.C. (1940): *A sample survey of acreage under jute in Bengal with discussion on planning of experiments*, Proc. (second) Indian Stat. Conf., Calcutta.

162. Mann, H.B. and Whitney, D.R. (1947): *On a test whether one of two random variables is stochastically larger than the other,* Ann. Math. Stat., **18,** 50–60.

163. Massey, F.G. (1952): *Distribution table for the deviation between two sample cumulatives,* Ann. Math. Stat., **23,** 435–441.

164. Mood, A.M. (1954): *On the asymptotic efficiency of certain nonparametric two-sample tests,* Ann. Math. Stat., **25,** 514–522.

165. Moses, L.E. (1952): *Nonparametric statistics for psychological research,* Psychol. Bull., **49,** 122–143.

166. Mukhopadhyay, P. and Schwabe, R. (1998): *On the performance of the ordinary least squares method under an error component model*, Metrika, **47**, 215–226.

167. Neyman, J. (1935): *Su un teorema concernente le cosiddette statistiche sufficienti,* Inst, Ital. Atti Giorn., Vol. **6,** 320–324.

168. — (1949): *Contribution to the theory of the χ^2-test,* Proc. (First) Berkely Symp. on Math. Stat. Prob. 239–273.

169. Neyman, J. and Pearson, E.S. (1928): *On the use and interpretation of certain test criteria for purpose of statistical inference,* Biometrika, **20A,** 175–240 and 263–294.

170. — (1933): *On the problem of the most efficient tests of statistical hypothesis Phil,* Trans. Roy. Soc. **A231,** 289–337.

171. Neyman, J. and Scott. (1948): *Consistent estimates based on partially consistent observations*, Econometrika, **16**, 1–32.

172. Patil, G.P. and Boswell, M.T. (1970): *A characteristic property of the multivariate normal density funciton and some of its applications*, Ann. Math. Stat. **41,** 1970–1977.

173. Pearson, Karl (1900): *On the criterion that a given system of deviations from the probable in the case of a corrected system of variables is such that it can be reasonably supposed to have arisen from random sampling,* Phil. Mag. V. **50,** p. 157.

174. Penrose, R. (1955): *A generalised inverse for matrices*, Proc. Camb. Phil. Soc., **51,** 406–413.

175. Quenouille, M.H. (1956): *Notes on bias in estimation,* Biometrika **43,** 353–355.

176. Randles, R.H. (1982) : *On the asymptotic normality of statistics with estimated parameters,* Ann. Stat., **10**, 462–474.

177. Rao, C.R. (1945): *Information and the accuracy attainable in the estimation of statistical parameters,* Bull. Calcutta Math. Soc. **37,** 81–91.

178. — (1957): *Maximum likelihood estimation for multinomial distribution,* Sankhya **18,** 139–148.

179. — (1969): *Some chracterisation of the multivariate normal distribution in*, P. R. Krishnaiah (ed.) Multivariate Analysis, Vol. **II,** 321–328. Academic Press, New York.

180. — (1972): *Recent trends of resarch work in multivariate analysis,* Biometrics, **28,** 3–22.

181. — (1972a): *Estimation of variance and covariance components in linear models*, J. Amer. Stat. Assoc., **67**, 112–115.

182. — (1973): *Representation of best linear unbiased estimators in the Gauss-Markoff model with a singular dispersion matrix*, J. Multivariate Analysis, **3**, 276–292.

183. Robbins, H. (1959): *Sequential estimation of the mean of a normal population in Probability and Statistics*—the Harold-Cramer volume, Almquist and Wilsell, Uppsala, 233–245.

184. Roy, J. and Mitra, S.K. (1957): *Unbiased minimum variance estimation in a class of discrete distributions,* Sankhya A, **18,** 371–378.

185. Scheffe, H. (1943): *On solutions of Behrens-Fisher problem based on the t-distribution,* Ann. Math. Stat, **13,** 371–388.

186. Sedransk, J. and Meyer, J. (1978): *Confidence intervals for quantities of a finite population for simple random sampling and stratified random sampling,* J. Amer. Stat. Assoc., **73,** 239–272.

187. Smith, K. (1953): *Distribution-free statistical methods and the concept of power efficiency in,* L. Festinger and D. Katzs (eds). Research Methods in the Behavioural Sciences, New York: Dryden, 563–577.

188. Stein, C. (1945): *A two-sample test for a linear hypothesis whose power is independent of the variance,* Ann. Math. Stat., **16**, 243–258.

189. Stain, C. (1946): *A note on cumulative sums,* Ann. Math. Stat., **17,** 489–499.

190. Swed, F.S. and Eisenhart, C. (1943): *Tables for testing ramdomness of grouping in a sequence of alternatives,* Ann. Math. Stat., **14,** 66–87.

191. Wald, A. (1941a): *Asymptotically most powerful tests of statistical hypotheses,* Ann. Math. Stat., **12,** 1–19.

192. — (1941b): *Some examples of asymptotically most powerful tests,* Ann. Math. Stat., **12,** 396–408.

193. — (1945): *Sequential tests of statistical hypotheses,* Ann. Math. Stat., **16**, 117–186.

194. — (1949): *Note on the consistency of the maximum likelihood estimate,* Ann. Math. Stat., **20,** 595–601.

195. Wald, A. and Wolfwitz, J. (1940): *On a test whether two samples are from the same population,* Ann. Math. Stat., **11,** 147–162.

196. — (1946): *Tolerance limits for a normal distribution*, Ann. Math. Stat., **17**, 208.

197. — (1948): *Optimal character of the sequential probability ratio test,* Ann. Math. Stat., **19,** 326–339.

198. Wang, S.G., Chow, S.C. and Tse, S.K. (1994): *On ordinary least squares methods for sample surveys*, Stat. Probab. Letters., **20**, 173–182.

199. Wilks, M.B., Ganadesiken, R. and Huyett, M.J. (1962): *Estimation of parameters of the gamma distribution using order statistics,* Biometrika, **49,** 525–545.

200. Wilks, S.S. (1938): *The large sample distribution of the likelihood ratio for testing composite hypotheses,* Ann. Math. Stat. **9,** 60–62.

201. Wolfwitz, J. (1942): *Additive partition functions and a class of statistical hypotheses,* Ann. Math. Stat., **13,** 247–279.

202. — (1947): *The efficiency of sequential estimates and Wald's equation for sequentia processes,* Ann. Math. Stat., **18,** 215–230.

203. Zyskind, G. (1967): *On canonical forms, non-negative convariance matrices and best anc simple least square estimators in linear models*, Ann. Math. Stat., **38**, 1092–1109.

Subject Index